Advances in Intelligent Systems and Computing

Volume 794

D1799781

Series editor

Janusz Kacprzyk, Polish Academy of Sciences, Warsaw, Poland
e-mail: kacprzyk@ibspan.waw.pl

The series "Advances in Intelligent Systems and Computing" contains publications on theory, applications, and design methods of Intelligent Systems and Intelligent Computing. Virtually all disciplines such as engineering, natural sciences, computer and information science, ICT, economics, business, e-commerce, environment, healthcare, life science are covered. The list of topics spans all the areas of modern intelligent systems and computing such as: computational intelligence, soft computing including neural networks, fuzzy systems, evolutionary computing and the fusion of these paradigms, social intelligence, ambient intelligence, computational neuroscience, artificial life, virtual worlds and society, cognitive science and systems, Perception and Vision, DNA and immune based systems, self-organizing and adaptive systems, e-Learning and teaching, human-centered and human-centric computing, recommender systems, intelligent control, robotics and mechatronics including human-machine teaming, knowledge-based paradigms, learning paradigms, machine ethics, intelligent data analysis, knowledge management, intelligent agents, intelligent decision making and support, intelligent network security, trust management, interactive entertainment, Web intelligence and multimedia.

The publications within "Advances in Intelligent Systems and Computing" are primarily proceedings of important conferences, symposia and congresses. They cover significant recent developments in the field, both of a foundational and applicable character. An important characteristic feature of the series is the short publication time and world-wide distribution. This permits a rapid and broad dissemination of research results.

More information about this series at http://www.springer.com/series/11156

Tareq Z. Ahram · Christianne Falcão
Editors

Advances in Usability, User Experience and Assistive Technology

Proceedings of the AHFE 2018 International
Conferences on Usability & User Experience
and Human Factors and Assistive Technology,
Held on July 21–25, 2018,
in Loews Sapphire Falls Resort at Universal Studios,
Orlando, Florida, USA

 Springer

Editors
Tareq Z. Ahram
University of Central Florida
Orlando, FL, USA

Christianne Falcão
Catholic University of Pernambuco
Boa Viagem, Pernambuco, Brazil

ISSN 2194-5357   ISSN 2194-5365 (electronic)
Advances in Intelligent Systems and Computing
ISBN 978-3-319-94946-8   ISBN 978-3-319-94947-5 (eBook)
https://doi.org/10.1007/978-3-319-94947-5

Library of Congress Control Number: 2018947439

Printed on acid-free paper

This Springer imprint is published by the registered company Springer International Publishing AG part of Springer Nature
The registered company address is: Gewerbestrasse 11, 6330 Cham, Switzerland

Advances in Human Factors and Ergonomics 2018

AHFE 2018 Series Editors

Tareq Z. Ahram, Florida, USA
Waldemar Karwowski, Florida, USA

9th International Conference on Applied Human Factors and Ergonomics and the Affiliated Conferences

Proceedings of the AHFE 2018 International Conference on Usability & User Experience and Human Factors and Assistive Technology, Held on July 21–25, 2018, in Loews Sapphire Falls Resort at Universal Studios, Orlando, Florida, USA

Advances in Affective and Pleasurable Design	Shuichi Fukuda
Advances in Neuroergonomics and Cognitive Engineering	Hasan Ayaz and Lukasz Mazur
Advances in Design for Inclusion	Giuseppe Di Bucchianico
Advances in Ergonomics in Design	Francisco Rebelo and Marcelo M. Soares
Advances in Human Error, Reliability, Resilience, and Performance	Ronald L. Boring
Advances in Human Factors and Ergonomics in Healthcare and Medical Devices	Nancy J. Lightner
Advances in Human Factors in Simulation and Modeling	Daniel N. Cassenti
Advances in Human Factors and Systems Interaction	Isabel L. Nunes
Advances in Human Factors in Cybersecurity	Tareq Z. Ahram and Denise Nicholson
Advances in Human Factors, Business Management and Society	Jussi Ilari Kantola, Salman Nazir and Tibor Barath
Advances in Human Factors in Robots and Unmanned Systems	Jessie Chen
Advances in Human Factors in Training, Education, and Learning Sciences	Salman Nazir, Anna-Maria Teperi and Aleksandra Polak-Sopińska
Advances in Human Aspects of Transportation	Neville Stanton

(continued)

(continued)

Advances in Artificial Intelligence, Software and Systems Engineering	*Tareq Z. Ahram*
Advances in Human Factors, Sustainable Urban Planning and Infrastructure	*Jerzy Charytonowicz and Christianne Falcão*
Advances in Physical Ergonomics & Human Factors	*Ravindra S. Goonetilleke and Waldemar Karwowski*
Advances in Interdisciplinary Practice in Industrial Design	*WonJoon Chung and Cliff Sungsoo Shin*
Advances in Safety Management and Human Factors	*Pedro Miguel Ferreira Martins Arezes*
Advances in Social and Occupational Ergonomics	*Richard H. M. Goossens*
Advances in Manufacturing, Production Management and Process Control	*Waldemar Karwowski, Stefan Trzcielinski, Beata Mrugalska, Massimo Di Nicolantonio and Emilio Rossi*
Advances in Usability, User Experience and Assistive Technology	*Tareq Z. Ahram and Christianne Falcão*
Advances in Human Factors in Wearable Technologies and Game Design	*Tareq Z. Ahram*
Advances in Human Factors in Communication of Design	*Amic G. Ho*

Preface

Successful interaction with products, tools, and technologies depends on usable designs and accommodating the needs of potential users without requiring costly training. In this context, this book is concerned with emerging ergonomics in design concepts, theories, and applications of human factors' knowledge focusing on the discovery, design, and understanding of human interaction and usability issues with products and systems for their improvement.

The Human Factors and Assistive Technology promotes the exchange of ideas and techniques which enable humans to communicate and interact with each other in almost every aspect. The new relationship between humans and technology added convenience for many, and for those with impairments, modern-day technology has transformed their daily living into a journey toward capability instead of disability. Assistive technology assessment focuses on the examination of problems in designing and providing assistive devices and services to individuals with disabilities or impairment, to assist mobility, communication, positioning, environmental control, and daily living. The conference addresses a wide spectrum of theoretical and practical topics related to assistive technologies. It provides an excellent forum for combining real experience and academic research, while examining how we can adapt to machinery and increase the technology acceptance, effectiveness, and efficiency. The conference aims at investigating how psychological factors can affect the efficiency and acceptability of assistive technology.

This book will be of special value to a large variety of professionals, researchers, and students in the broad field of human modeling and performance, who are interested in feedback of devices' interfaces (visual and haptic), user-centered design, and design for special populations, particularly the elderly. We hope this book is informative, but even more that it is thought-provoking. We hope it inspires, leading the reader to contemplate other questions, applications, and potential solutions in creating good designs for all. The book is organized into nine sections that focus on the following subject matters:

Section 1: UX Evaluation and Design Thinking
Section 2: Human Machine Interfaces
Section 3: Usability Evaluation and User-Centered Design
Section 4: Virtual Reality and Interaction Design
Section 5: User Experience in Healthcare and Learning
Section 6: User Experience and Visualization in Automotive Industry
Section 7: Eye Tracking and Visualization
Section 8: Assistive Technology and Design Solutions
Section 9: Assistive Design Solutions and Prosthetic Environments

This book will be of special value to a large variety of professionals, researchers, and students in the broad field of human–computer interaction, usability engineering, and user experience research, who are interested in feedback of devices' interfaces (visual and haptic), user-centered design, and design for special populations, particularly the elderly.

Each section contains research papers that have been reviewed by members of the International Editorial Board. Our sincere thanks and appreciation to the board members as listed below:

Usability & User Experience

Hanan A. Alnizami, USA
Wolfgang Friesdorf, Germany
S. Fukuzumi, Japan
Sue Hignett, UK
Wonil Hwang, South Korea
Yong Gu Ji, South Korea
Bernard C. Jiang, Taiwan
Ger Joyce, UK
Chee Weng Khong, Malaysia
Zhizhong Li, PR China
Nelson Matias, Brazil
Abbas Moallem, USA
Beata Mrugalska, Poland
Francisco Rebelo, Portugal
Valerie Rice, USA
Emilio Rossi, Italy
Javed Anjum Sheikh, Pakistan
Alvin Yeo, Malaysia
Wei Zhang, PR China

Assistive Technology

Hanan A. Alnizami, USA
Wolfgang Friesdorf, Germany
S. Fukuzumi, Japan
Sue Hignett, UK
Matteo Zallio, Ireland

We hope this book is informative, but even more thought provoking to inspire the reader to contemplate other questions, applications, and potential solutions in creating good designs for all.

July 2018   Tareq Z. Ahram
                                                                        Christianne Soares Falcão

Contents

Virtual Reality and Interaction Design

User Experience in Healthcare and Learning

Assistive Design Solutions and Prosthetic Environments

UX Evaluation and Design Thinking

V.2 Evaluation and Design Tracking

Idealization Effects in UX Evaluation at Early Concept Stages: Challenges of Low-Fidelity Prototyping

Lara Christoforakos[(✉)] and Sarah Diefenbach

Department Psychology, Ludwig-Maximilians-University Munich,
Leopoldstr. 13, 80802 Munich, Germany
{Lara.Christoforakos,Sarah.Diefenbach}@psy.lmu.de

Abstract. Early stage prototyping gains ever more importance in product development and User Experience (UX) evaluation. Especially in innovative technology and service domains (e.g., VR, AR, IoT) with increasing relevance of experiential aspects, early prototyping and evaluation is crucial to assess a product idea's success potential. A central question is which prototyping approach best represents the product idea and allows its valid yet cost- and time-efficient evaluation. While low-fidelity prototyping supports low-cost adjustments, a potential biasing factor are idealization tendencies: UX evaluation subjects may idealize unspecified product aspects following their imagination, possibly reducing results' validity. This study ($N = 255$) examines effects of prototype fidelity within early product development comparing different product concept representations systematically. Results imply that the lower the fidelity, the more people idealize a product idea, having numerous implications for prototype use in UX design and research. Practices to counteract idealization tendencies and optimize low-fidelity prototyping are discussed.

Keywords: UX evaluation · Early stage prototyping · Fidelity
Idealization effects · Validity

1 Introduction

Along with the rapid growth of technological innovations and new emerging product concepts, prototyping in early development stages increasingly gains importance. Especially for novel technologies such as Augmented Reality (AR), Virtual Reality (VR) or in general the Internet of Things (IoT), design heuristics and previous experience are limited. Therefore, an early and valid evaluation of product feasibility and sustainability becomes ever more important [1]. Furthermore, the evaluation of such innovative products often emphasizes new experiential factors of product use [2], such as consequences of a product's use on its user's social contexts [3]. For example, whether new services such as AR functionality in shop windows will actually be successful on the market, does not only depend on technical feasibility and usefulness, but also very much on the experiential value that people will see in the service.

The assessment of such prospective user experiences through prototypes, representing the product concept, does not necessarily require prototypes of high fidelity.

© Springer International Publishing AG, part of Springer Nature 2019
T. Z. Ahram and C. Falcão (Eds.): AHFE 2018, AISC 794, pp. 3–14, 2019.
https://doi.org/10.1007/978-3-319-94947-5_1

With the term prototype fidelity, we refer to the prototype's similarity to the final product, as proposed in the definition of prototype fidelity by Hochreuter and colleagues [4]. Studies show that for a general evaluation of a product idea, products focusing on experiential components can be prototyped validly using low-fidelity methods, which in many ways have only little similarity to the product itself [3, 5]. Even without a tangible product representation, prototypes in the form of narrative representations through texts or storyboards can provide helpful insights, as during early development phases the envisioned experience is more essential than usability issues. Thus, many companies make use of low-fidelity prototypes as an approach of early and cost-efficient prototyping [6]. In general, prototypes may support different purposes within the product development process. This may include ideation and idea generation, demonstration and communication, thereby keeping stakeholders involved in the development process or acquiring new costumers, as well as prototypes as a basis for the evaluation of product concepts [7]. Regarding the latter, low-fidelity prototypes might come with crucial challenges, complicating the interpretation of User Experience (UX) evaluation results. For example, studies indicate that when using low-fidelity prototypes in evaluation studies, participants tend to idealize the product idea according to their personal imagination and aspiration, resulting in a more positive but biased evaluation [3, 8]. Typically, these studies compared product evaluation based on different prototyping methods of the same product idea and measured product evaluation based on first-impression or visual attractiveness. Though based on the contrast of single dimension's fidelity (e.g., visual refinement of the user interface), such findings must be taken seriously. If participants are evaluating the product concept based on a personal imaginary vision of the product, which may differ from person to person and, even more importantly, may also differ from the later developed product, it becomes difficult to interpret the evaluation results and asses the concept's actual potential. Hence, it is of essential importance for UX research to advance these first insights and take a closer look at the phenomenon of idealization tendencies in UX evaluation at early concept stages and asses the consequences for the same product's evaluation at a later development stage. Prior findings imply that product expectations play an important role in customer satisfaction [9–11]. Thus, product evaluations based on low-fidelity prototypes might affect product evaluation at a later development stage.

This study aims to explore how prototype fidelity influences users' reactions to a product concept represented by a prototype as well as the same product concept at a later development stage. Hereinafter, we summarize previous research on idealization tendencies in low-fidelity prototyping by Sauer and Sonderegger [8] as well as Diefenbach and colleagues [3] and derive our hypotheses. Afterwards, we present an experimental study conducted to advance these findings by systematically manipulating prototype fidelity. Finally, we discuss implications of our findings for UX research and product development.

2 Related Research

Mixed Fidelity. Prototype fidelity, namely the similarity between a product concept's prototype and its final version [1, 12] is often used for prototype categorization, and a broad differentiation between high and low fidelity prototypes (e.g., functional prototype versus paper prototype). While in general, low-fidelity prototypes are not very similar to the final product, high-fidelity prototypes have many similarities to the products they represent. If closely considered, however, the differences between prototypes are more complex than the simple *"high versus low"*- differentiation suggests. In fact, prototypes can have varying fidelities depending on the considered product aspect, which is expressed in the mixed fidelity concept by McCurdy et al. [1]. While a website prototype might look like the final product and have a high fidelity in its visualization, it might be a marketing-screen without functionalities, thus being low-fidelity in this aspect. To acknowledge this complexity, Hochreuter et al. [4] pick up the work of McCurdy et al. [1] and Lim et al. [12] and propose the filter-fidelity-model [4] that sees prototypes as filtering specific product dimensions. Based on this, every prototype can be characterized by its filter-fidelity-profile, specifying its fidelity regarding the dimensions appearance, data, functionality, interactivity and spatial structure. A textual prototype, namely a narrative representation of a product concept describing the product and its use through a specific scenario, has a low fidelity regarding many dimensions, as it only describes appearance, data, functionalities, interactivities and spatial structures but does not materially embody them.

Fidelity and Evaluation Insights. Many studies discuss the appropriate fidelity level in view of valid prototyping. Findings show that low prototype fidelity does not necessarily reduce evaluation insights, as for example the same usability issues were revealed using a paper-prototype as by using a finalized product [13]. Similarly, Walker et al. [5] compared different prototyping methods (paper vs. computer display) as well as the general prototype fidelity (low vs. high) in the development of an online-banking website and did not find a significant difference in the detected usability issues of the product. Some studies even show advantages of low-fidelity compared to high-fidelity prototyping. Whereas with high-fidelity prototypes subjects' attention might be caught by currently irrelevant prototype details and lead to confiding insights [14], low-fidelity prototypes can support the focus on single product aspects, thus leading to more valid results to be clearly affiliated to certain product aspects. Furthermore, prototyping with low-fidelity has the potential to motivate subjects in engaging with product details, as it might stir up curiosity by presenting only few product components [7]. Thus, low-fidelity prototyping has great potential regarding early evaluation of innovative products' success potential as it allows time- and cost-efficient but valid insights.

Idealization Tendencies at Early Stage Low-Fidelity Prototyping. Other studies highlight undeniable challenges of low-fidelity prototyping. Sauer and Sonderegger [8] and Diefenbach et al. [3] have found subjects' idealization tendencies when confronted with low-fidelity prototypes, meaning that subjects idealize product concepts, as they perceive only few product components. For example, Sauer and Sonderegger [8] evaluated product concepts of two different mobile phones which were first represented by two varying paper-prototypes and then two varying final products. Obvious differences in visual attractiveness in the final products were not detected in the evaluation based on paper-prototypes, suggesting that subjects may have idealized the low-fidelity representation according to their ideal imagination. Diefenbach et al. [3] used varying prototyping methods (e.g. textual prototype, comic-story, video) to collect evaluations of a product idea. Results showed that the lower the prototyping method's fidelity, the more positive the evaluation was, implying that not only might subjects use their imagination more with low- rather than high-fidelity prototypes, but that they probably also idealize product concepts more, the lower the prototype's fidelity is, respectively the less they know about the product.

Interpretation of Idealization Tendencies in Light of Psychological Theory. The found idealization tendencies in UX evaluation comply with psychological research such as the control theory [15], positing that people strive for explicability, predictability and suggestibility and implying that an evaluation task with more freedom of imagination will be preferred to a more restrictive one [15]. Thus, people might react more positively to low-fidelity than high-fidelity prototypes, since they feel freer in imagining the final design. On the other hand, goal setting theories such as the high-performance-cycle [16] state that the more specific tasks are set, the more motivated people are to solve them and the likelier they will be solved. Comparing low-fidelity prototyping to a more unspecific task than high-fidelity prototyping could imply that people will be less motivated when confronted with a product evaluation based on a low-fidelity prototype than a high-fidelity one, so that high-fidelity prototypes should be rated more positively. Furthermore, previous studies regarding the expectancy disconfirmation theory [10, 11] have shown that the more people's expectations of products are confirmed or positively disconfirmed, the more positively these products are evaluated. The more negatively such expectations are disconfirmed, the more negatively the relevant products are evaluated [9]. Accordingly, low-fidelity prototypes leading to idealizations tendencies regarding the product concept might not be very beneficial for stakeholders' satisfaction with the final product.

Altogether, considering the crucial consequences of possible idealization effects for prototyping validity, it is essential to examine the generalizability of these findings and explore possible countermeasures. In addition, extended research insights could offer guidelines regarding when to use which level of prototype fidelity and maximize the potential of low-fidelity prototyping as a valid method for early and cost-efficient evaluation, especially for currently trending innovational technologies.

3   Hypotheses

Based on the above described HCI studies on idealization tendencies [3, 8] and psychological approaches such as the control theory [15] and expectancy disconfirmation theory [10, 11] we hypothesize the following:

> *H1: Variations in prototype fidelity come along with variances in product evaluation.* This includes (H1a) the global product evaluation, i.e., a "bad-good" rating of the product concept as well as (H1b) product purchase intention, whereby we

assume (H1c) the relation between prototype-fidelity and product purchase intention to be mediated through global product evaluation.

H2: The higher the perceived similarity of product concepts based on two proto-types is, the more positive the product evaluation based on the second prototype will be. More specifically, we assume that (H2a) the higher a textual prototype's fidelity is, the higher the perceived similarity of product concepts after confrontation with a second photograph-prototype will be. Also, the higher the perceived similarity of product concepts based on the two prototypes (textual prototype, photograph-prototype) is, the more positive (H2b) the global product evaluation as well as the as (H2c) the product purchase intention based on the second prototype (photograph-prototype) will be, whereby we assume (H2d) the relation between the perceived similarity and the product purchase intention to be mediated through global product evaluation.

4   Study

Our study's aim was to advance the findings of idealization tendencies in low-fidelity prototyping and consider possible consequences of low-fidelity prototyping regarding global product attributes. Thus, our research question is how prototype fidelity affects users' reactions to a product concept represented by a prototype and respectively the same product concept at a later development stage regarding global product evaluation.

4.1   Methods

Participants. Two hundred fifty-five participants (175 female) aged between 17 and 69 ($M = 28.57$; $SD = 11.03$) were recruited through social media as well as an information-platform of the Ludwig-Maximilians-University. The study's aim was presented as an exploration of innovative product ideas and incentives of Amazon gift coupons between ten and fifty Euros were raffled among all participants.

Experimental Design. We conducted an experimental study with textual prototype fidelity as experimental factor and global product evaluation, product purchase intention and perceived product similarity as dependent measures to test our hypotheses. In addition, we surveyed visual attractiveness for exploratory analyses. Textual prototype fidelity was varied at three levels (low fidelity, medium fidelity and high fidelity). The product concept we used to assess the effect of prototyping fidelity was the *money-saving plant*. Its intention is to support its users in saving a specific amount of money within a defined time frame. One can water the plant by inserting coins in the plants crock and therefore causing a container to water the plant, creating and supporting feelings of competence and pride in its users. A basic motivation for choosing especially this product was focusing on a concept with many experiential aspects, such as the feeding of the plant and the achievement of the long awaited goal of money-saving. This allowed studying UX evaluation beyond solidly visual product components. Another reason was its missing existence on the market in order for subjects not to have a prior mental representation of it, possibly impairing results' validity. The textual prototypes described the plant and characteristic situations of use in various fidelity levels. They were created based on a systematic procedure of continuously increasing the fidelity regarding the data dimension according to Hochreuter et al. [4] describing product details regarding visual, functional and usage components. After evaluating the concept based on the textual prototype, participants were confronted with a second prototype of the product. This was the same for participants in all conditions, i.e., a photograph-prototype (see Fig. 1). It included a picture of the plant and a mock-up of the according smartphone-application, representing the product idea at a later stage of development. Thus, the time of UX measurement (t1: after textual prototype, t2: after photograph prototype) built a second experimental factor, realized within subjects.

Fig. 1. Photograph-prototype of the money-saving plant as presented to participants at t2.

Procedure. Participants were randomly assigned one of three textual prototypes of low, medium or high fidelity describing the money-saving plant. Global product evaluation, purchase intention and the product's visual attractiveness were measured. Afterwards, all subjects were shown the same photograph-prototype of the product. Global product evaluation, product purchase intention and visual attractiveness were measured, this time based on the impression caused by the photograph-prototype. Then participants were asked to state perceived concept similarity based on the prototypes (textual prototype, photograph-prototype). An analysis of variance (ANOVA) and two Chi-Square tests showed that the three conditions of textual prototype fidelity did not differ significantly regarding age ($F(2,249) = 0.04$, $p = .964$, $\eta^2 < .001$), sex ($X^2(2, N = 252) = 1.21$, $p = .545$) or student status ($X^2(2, N = 252) = 1.37$, $p = .503$).

Measures and instruments. *Global product evaluation* was measured with a "bad-good"-item on a 7-point scale (1 = "bad"; 7 = "good"). For *product purchase intention,* we used the item "I would like to own the money-saving plant", "I like the visual design of the money-saving plant" for *visual attractiveness* and "The image of the money-saving plant has high similarity with what I imagined it to be based on the text describing the product idea" for *perceived product similarity* based on the prototypes, all rated on a 5-point Likert scale (1 = "does not apply at all"; 5 = "applies fully").

4.2 Results

Table 1 lists the descriptive data of product evaluations for the three fidelity conditions.

Table 1. Descriptive data of product evaluation for low (n = 86), medium (n = 82) and high (n = 84) fidelity of textual prototype.

Measure and time of measurement	Experimental condition: fidelity of textual prototype		
	Low *M(SD)*	Medium *M(SD)*	High *M(SD)*
t1: After textual prototype			
Global product evaluation	5.56 (1.37)	5.07 (1.39)	4.92 (1.53)
Product purchase intention	3.24 (1.23)	2.61 (1.21)	2.73 (1.28)
Product visual attractiveness	3.37 (1.01)	3.05 (1.08)	3.10 (1.08)
t2: After photograph prototype			
Global product evaluation	5.59 (1.38)	5.18 (1.43)	5.26 (1.39)
Product purchase intention	3.33 (1.31)	2.71 (1.29)	2.80 (1.30)
Product visual attractiveness	2.97 (1.51)	2.51 (1.29)	2.80 (1.28)
Perceived similarity of product concepts	3.09 (1.01)	3.01 (1.06)	3.05 (1.14)

H1: Product idealization in low fidelity-prototyping. In line with H1a an ANOVA with textual prototype fidelity as independent variable showed that the fidelity of the textual prototypes had a significant negative effect on global product evaluation ($F(2,249) = 4.64$, $p = .011$, $\eta^2 = .036$) and, as assumed in H1b, also on product purchase intention after perceiving the textual prototype ($F(2, 249) = 6.26$, $p = 0.002$, $\eta^2 = .048$). Post-hoc Tukey tests showed that for global product evaluation only the low and high fidelity conditions differed significantly ($p = .011$) whereas for purchase intention, the low-fidelity condition differed significantly from the medium-fidelity one as well as the low-fidelity one from the high-fidelity one. In order to test the assumed mediation (H1c) we conducted a path analysis. It showed that the standardized regression weight $\beta = -.26$ ($p = .006$) of the relation between textual prototype fidelity and product purchase intention after perception of the textual prototype decreased to $\beta = -.10$ ($p = .221$) after consideration of global product evaluation as a mediator. The indirect connection between product fidelity and product purchase intention after perception of the textual prototype through global product evaluation became significant ($\beta = .16$, $p = .004$). Thus, as assumed in H1c, we found a total mediation, meaning that the effect of textual prototype fidelity on product purchase intention was caused by global product evaluation. Figure 2 illustrates this mediation.

H2: Product Expectation and Product Evaluation. Contrary to our hypothesis (H2a) the ANOVA with textual prototype as independent variable showed that textual prototype fidelity had no significant influence ($F(2,249) = 0.12$, $p = .997$, $\eta^2 < .001$) on perceived similarity of product concepts based on the two prototypes (textual prototype, photograph-prototype). Also against our expectations (H2b), the Pearson-Correlation showed no significant relation between perceived product similarity based

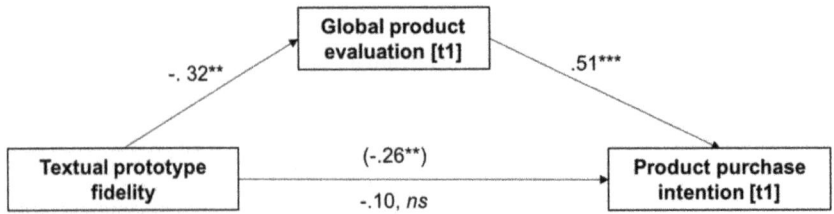

Fig. 2. Beta coefficients for mediation model of the relationship between prototype fidelity and product purchase intention at t1 through global product evaluation. Values in parentheses represent total effects. Note: *p < .05, **p < .01, ***p < .001.

on the two prototypes (textual prototype, photograph-prototype) and global product evaluation after perceiving the second prototype (photograph-prototype) ($r = .10$, $p = .107$), however, in line with H2c, a significant positive correlation between perceived similarity and product purchase intention ($r = .15$, $p = .016$). Given that the expected correlation between perceived similarity and global evaluation (H2b) was not significant, we did not test the mediation model assumed in H2d, where global evaluation was expected to be a mediator on purchase intention.

Exploratory Analyses. An ANOVA with textual prototype fidelity as independent variable showed no significant effect of textual prototype fidelity on the rating of visual attractiveness of the product idea after perceiving the textual prototype ($F(2,249) = 2.33$, $p = .010$, $\eta^2 = .018$) as well as the photograph-prototype ($F(2,249) = 2.27$, $p = .106$, $\eta^2 = .018$). We also conducted an ANOVA with textual prototype fidelity and time of measurement (t1, t2) as independent factors with dates of measurement being the perception of the textual prototype (t1) and the perception of the photograph-prototype (t2). Regarding both global product evaluation ($F(1, 498) = 1.66$, $p = .198$, $\eta^2 = .003$) and product purchase intention ($F(1,498) = 0.54$, $p = .462$; $\eta^2 = .007$) time of measurement did not have a significant effect.

5   Discussion

Main results showed that the lower the textual prototype fidelity was, the more positive was the global product evaluation and the higher was the product purchase intention. The latter relation was mediated through global product evaluation. Textual prototype fidelity did not have a significant effect on perceived similarity of product concepts based on the two prototypes at different times of measurement (textual prototype, photograph-prototype). Nor did we find a significant relation between perceived similarity of product concepts based on the two prototypes (textual prototype, photograph-prototype) and global product evaluation after perception of the second prototype (photograph-prototype), yet we did with product purchase intention.

The results of our study support the phenomenon of idealization tendencies in low-fidelity prototyping according to previous studies [3, 8] and are compatible with the theoretical approach of the control theory [15]. The fact that the found idealization tendencies are not only based on global product evaluation but also on product purchase intention, could be a sign of robustness of the effect as product purchase intention implies planned behaviour. The results showing no significant variance in visual attractiveness depending on the condition of textual prototype fidelity imply that the effect might reach beyond visual product components and therefore be generalizable to global product evaluation. Furthermore, the effects of prototype fidelity on global product evaluation and product purchase intention remain significant after the perception of both prototypes. Also, there is no significant within-subjects variance in global product evaluation or product purchase intention between the prototype perceptions. These findings further support the significance of the effect of prototype fidelity.

Moreover, results imply that perceived similarity between product concepts based on prototypes in various stages of product development might not depend on prototype fidelity. Based on our findings on idealization tendencies in low-fidelity prototyping, a possible explanation could be that certain product components first shown in the photograph-prototype were already imagined when confronted with the low-fidelity prototype. Such an effect could have erased differences in perceived similarity between the prototypes. According to the hindsight-bias [17], another possible explanation could be that participants viewed the product at its more developed stage as having been predictable after seeing the photograph-prototype. This would have erased between-subject differences in perceived similarity of the prototypes.

Following the expectancy disconfirmation theory [10, 11] the positive relation between perceived similarity of product concepts based on the prototypes and product purchase intention could be explained through confirmation or positive disconfirmation of expectations having a positive effect on product satisfaction [9]. This phenomenon might explain the positive effect of perceived product similarity based on the two prototypes on product purchase intention. This relation not being significant for global product evaluation might be an issue of personal involvement [18, 19] as a global product evaluation measured through the 7-point "bad-good"-item probably comes with less involvement than self-disclosure about wishing to possess a product, measured through product purchase intention.

In sum, we could advance findings regarding the phenomenon of idealizations tendencies in low-fidelity prototyping [3, 12]. It seems that choosing a low fidelity in early stage prototyping can positively affect purchase intentions of potential users. Yet, such a phenomenon might negatively influence prototyping validity. Furthermore, choosing a low fidelity in early stage prototyping might negatively affect peoples' reactions regarding the same product concept at later stages of development. Still, interpretations resulting from our results concern prototyping as means of evaluation and should be accepted with reservation for different prototyping purposes.

6   Limitations

Due to the online-questionnaire we could not ensure total controllability of subjects' conditions. We do not know how intensively participants engaged with the prototypes or under which circumstances they completed the questionnaire. However, random condition assignment controlled such potential biasing factors to a certain extent.

Furthermore, textual prototype fidelity was varied only between-subjects, impeding direct comparison of dependent variables depending on manipulation. Note, however, that this rather underlines the robustness of the effect of idealization tendencies, since effects are generally revealed more easily by direct compared to indirect contrast.

Another aspect which might have affected our results, is the restrictive character of the money-saving plant, additionally imposing goals on the user (save money, caring for a plant) that may not appear meaningful to everyone. This might have negatively influenced global product evaluation and restricted variances of this variable. Also, regarding the found idealization tendencies it should be kept in mind that there might generally be a narrower bandwidth to criticize with low-fidelity prototypes.

7 Implications for Research and Practice

Our results have numerous theoretical and practical implications. We advanced empirical findings regarding idealization tendencies in early stage low-fidelity prototyping. We demonstrated that the phenomenon already exists at a stage of textual concept representation and small differences in prototype fidelity were sufficient to induce such, revealed for several measures such as global product evaluation and product purchase intention. Thus, people do not only idealize the product on a visual level, but generally have a more positive impression. In sum, results underline the essential role of prototyping methods and fidelity for the validity of the whole prototyping process.

Our results have further implications for prototyping as a part of product development, as a communication artefact between company departments, as a trigger for inspiration and design, or as an object of co-creation. In line with control theory [15], results indicate that people prefer situations, with more creative freedom rather than less.

For practitioners, this means that it is of great necessity to choose prototyping methods and fidelity in consideration of prototyping motivation and context. Especially, as the choice of fidelity might affect internal organizational components such as an efficient process of product development but also external ones like client communication and satisfaction, which are crucial for an organization's success.

Regarding the many advantages of low-fidelity prototyping such as early, valid and cost- as well as time-efficient results, we recommend their usage. Yet due to the challenge of idealization tendencies it seems best to use them in purpose of ideation, internal communication of product ideas or acquisition of potential customers. Low-fidelity prototypes seem less ideal for product demonstrations in front of stakeholders, who might be confronted with the same product at a later development stage as their product expectations might not be confirmed due to possible idealization tendencies. Also, when using low-fidelity prototypes for UX evaluation it is of great importance to consider possible effects of idealization tendencies influencing the evaluation in a positive manner, leading to less valid insights as the evaluations may root in the imaginary, ideal product impression more than the product concept itself.

In case of further empirical results on idealization tendencies due to low-fidelity prototyping, such findings might be generalizable and therefore relevant for e.g. advertising or the film industry. Namely, it might be of great advantage for these industries to use advertisements or movie-trailers presenting very little of the product or awaited movie. According to idealization tendencies people would idealize the product or movie suiting their aspirations and imagination and thus desire the product more or intend to watch the movie more highly than following a more detailed trailer.

8 Conclusion and Future Research

Along with innovative technologies facilitating novel product concepts with only few existing design heuristics, early stage prototyping becomes increasingly important. Yet findings show that to collect valid results and thus ensure a cost- and time-efficient and respectively sustainable product development process, the choice of prototyping method and fidelity is essential [3, 8, 12].

In this paper, we showed that prototype fidelity can influence user's reactions regarding a represented product concept and respectively the same product concept at a later stage of development. Namely, low-fidelity prototypes can lead to people ideal-izing product concepts. This can have negative consequences regarding the validity of prototyping insights as well as stakeholders' later satisfaction with the final product.

Yet, low-fidelity prototypes have great potential in validly evaluating a product concept's success potential without even representing the final product in a tangible manner [3]. Therefore, further research and practical experience is needed to find countermeasures for the mentioned, possible negative consequences of low-fidelity prototyping and maximize its potential for an effective, efficient and innovative process of product development. Given that technological advances allow innovative ways to envision and design new kinds of positive experiences, the question of early UX evaluation and validity of prototyping will gain ever more importance.

Acknowledgments. Part of this research has been funded by the German Federal Ministry of Education and Research (BMBF), project ProFI (FKZ: 01IS16015).

References

1. McCurdy, M., Connors, C., Pyrzak, G., Kanefsky, B., Vera, A.: Breaking the fidelity barrier: an examination of our current characterization of prototypes and an example of a mixed-fidelity success. In: Proceedings of the SIGCHI Conference on Human Factors in Computing Systems, pp. 1233–1242 (2006)
2. Blomkvist, J., Holmlid, S.: Existing prototyping perspectives: considerations for service design. Nordes **4**, 1–10 (2011)
3. Diefenbach, S., Chien, W.C., Lenz, E., Hassenzahl, M.: Prototypen auf dem Prüfstand. Bedeutsamkeit der Repräsentationsform im Rahmen der Konzeptevaluation. i-com Zeitschrift für interaktive und kooperative Medien **12**(1), 53–63 (2013)
4. Hochreuter, T., Kohler, K., Maurer, M.: Prototypen im Kontext begreifbarer Interaktion besser verstehen. In: Mensch and Computer, pp. 169–180 (2013)
5. Walker, M., Takayama, L., Landay, J.A.: High-fidelity or low-fidelity, paper or computer? Choosing attributes when testing web prototypes. In: Proceedings of the Human Factors and Ergonomics Society Annual Meeting, vol. 46, no. 5, pp. 661–665 (2002)

6. Väänänen-Vainio-Mattila, K., Roto, V., Hassenzahl M.: Towards practical user experience evaluation methods. Meaningful measures: Valid useful user experience measurement (VUUM), pp. 19–22 (2008)
7. Diefenbach, S., Hassenzahl, M.: Psychologie in der nutzerzentrierten Produktgestaltung: Mensch-Technik-Interaktion-Erlebnis. Springer, Berlin (2017)
8. Sauer, J., Sonderegger, A.: The influence of prototype fidelity and aesthetics of design in usability tests: effects on user behaviour, subjective evaluation and emotion. Appl. Ergon. **40** (4), 670–677 (2009)
9. Cardozo, R.N.: An experimental study of customer effort, expectation, and satisfaction. J. Mark. Res. **2**, 244–249 (1965)
10. Oliver, R.L.: Effect of expectation and disconfirmation on post exposure product evaluations: an alternative interpretation. J. Appl. Psychol. **62**(4), 480–486 (1977)
11. Oliver, R.L.: A cognitive model of the antecedents and consequences of satisfaction decisions. J. Mark. Res. **17**(4), 460–469 (1980)
12. Lim, Y.K., Stolterman, E., Tenenberg, J.: The anatomy of prototypes: prototypes as filters, prototypes as manifestations of design ideas. ACM Trans. Comput.- Hum. Interact. (TOCHI) **15**(2), 7–27 (2008)
13. Virzi, R.A., Sokolov, J.L., Karis, D.: Usability problem identification using both low- and high-fidelity prototypes. In: Proceedings of the SIGCHI Conference on Human Factors in Computing Systems: Common Ground, pp. 236–243 (1996)
14. Struckmeier, A.: Warum "gutes Aussehen" nicht immer von Vorteil ist. Über den Einfluss der optischen Gestaltung von Prototypen auf das Nutzerverhalten im Usability-Test. In: Brau, H., Lehmann, A., Petrovic, K., Schroeder, M.C. (eds.) Usability Professionals 2011, pp. 52–57. German Chapter der Usability Professionals' Association e.V., Stuttgart (2001)
15. Frey, D., Jonas, E.: Die Theorie der kognizierten Kontrolle. In: Frey, D., Irle, M. (eds.) Theorien der Sozialpsychologie, pp. 13–50. Huber, Bern (2002)
16. Locke, E.A., Latham, G.P.: Building a practically useful theory of goal setting and task motivation: a 35-year odyssey. Am. Psychol. **57**(9), 705 (2002)
17. Fischhoff, B.: Hindsight is not equal to foresight: the effect of outcome knowledge on judgment under uncertainty. J. Exp. Psychol.: Hum. Percept. Perform. **1**(3), 288–299 (1975)
18. Oliver, R.L., Bearden, W.O.: The role of involvement in satisfaction processes. Adv. Consum. Res. **10**, 250–255 (1983)
19. Spreng, R.A., Sonmez, E.: The moderating effect of involvement on the consumer satisfaction formation process. In: American Marketing Association Conference Proceedings, vol. 11, pp. 168–174 (2000)

Heuristic Evaluation for Mobile Applications: Extending a Map of the Literature

Ger Joyce[✉], Mariana Lilley, Trevor Barker, and Amanda Jefferies

School of Computer Science, University of Hertfordshire, College Lane,
Hatfield, Hertfordshire AL10 9AB, UK
gerjoyce@outlook.com, {m.lilley, t.l.barker,
a.l.jefferies}@herts.ac.uk

Abstract. Ensuring that mobile applications are as usable as possible is an important area of Human-Computer Interaction research. Part of that research effort is to consider how traditional, tried-and-tested usability evaluation approaches might be applied to newer technologies, including mobile applications. The contribution of this work is to further the work of other researchers by discovering if heuristic evaluation is commonly applied to mobile applications by Human-Computer Interaction practitioners. Additionally, the authors empirically test the suggestion that Nielsen's heuristics may be generic enough for the usability evaluation of mobile applications.

Keywords: Heuristic evaluation · Mobile applications · Usability

1 Introduction

Given the proliferation of mobile devices in our everyday lives, exploring approaches that allow for an effective, holistic usability evaluation of mobile applications is an exciting area of research. To achieve this, one of the usability evaluation methods that might be used by the Human-Computer Interaction (HCI) community is heuristic evaluation [1]. During a heuristic evaluation, one or more usability experts evaluate an application using a set of guidelines [2], with the recommended number of evaluators being three to five [3]. In addition to being an inexpensive, relatively fast, and effective way to surface usability issues, heuristic evaluation is also holistic in nature, allowing the evaluation of a full application across the entire product design and development life cycle. This mitigates two of the weaknesses of usability testing, namely the tendency to be narrow in focus [4], and that the method can only be used relatively late in the software development life cycle [5].

These strengths lend heuristic evaluation to the usability evaluation of mobile applications. Yet, there are extensive dissimilarities between the desktop-based software and websites of the early 1990s, which heuristic evaluation was designed for, and the mobile applications of today. This presents a challenge for HCI researchers, educators, and practitioners [6, 7]. To that end, there has been considerable interest in defining heuristics for mobile technologies [8–12].

Two informative research papers from Salgado et al. [13, 14] have extensively mapped these mobile heuristic sets. Based on a mapping of the literature, Salgado

© Springer International Publishing AG, part of Springer Nature 2019
T. Z. Ahram and C. Falcão (Eds.): AHFE 2018, AISC 794, pp. 15–26, 2019.
https://doi.org/10.1007/978-3-319-94947-5_2

et al. [13] conclude that heuristic evaluation is commonly applied to mobile technologies, with traditional heuristics from Nielsen and Molich [1] being the most utilized. As this conclusion was based on a mapping of the literature, it is possible that the conclusion may only apply to the academic HCI community. Secondly, Salgado et al. [14] suggest that traditional heuristics [1] may be generic enough to be effective in surfacing usability issues within different contexts, including mobile technologies, yet this was not empirically tested. To that end, the authors wish to take the work of Salgado et al. [13, 14] further by addressing the following research questions:

- RQ1. To what extent is heuristic evaluation used by HCI practitioners when evaluating the usability of mobile applications?
- RQ2. Are Nielsen's heuristics as effective in surfacing usability issues within mobile applications as sets of heuristics designed specifically for mobile?

2 Theoretical Background

Usability guidelines, including heuristic evaluation, are essentially checklists for HCI experts. Therefore, the theoretical lens through which this work is viewed is that of Michael Scriven, one of the well-known theorists of program and product evaluation. Scriven [15] has argued that checklists allow evaluators of all experience levels to remember all elements of an evaluation. Without the list, the theorist suggests that novice and experienced evaluators will overlook key areas during an evaluation. Further, Scriven [15, p. 4] states that "checklists can contribute substantially…to the validity, reliability, and credibility of an evaluation".

Yet, Scriven [15] also urges caution—the theorist contends that checklists, and thus sets of heuristics, need to add value, but also be easy-to-use. Therefore, if a list of heuristics is too long, or any of the individual heuristics are too verbose, the entire list can be impractical to use. Indeed, unworkable lists of guidelines are not uncommon. During the 1980s, as early user interfaces were being developed, standard usability guidelines were needed [16]. However, initial sets of usability guidelines were unusable and costly to implement [17]. For instance, one set alone contained 944 guidelines [18]. In a bid to cut costs and allow sets of usability guidelines to be easier to work with, researchers began to define shorter sets of guidelines that addressed the most common issues, such as heuristic evaluation [1] and cognitive principles [19]. Today, expert-based usability inspection methods are well established. In particular, heuristic evaluation is popular as it is widely known for being fast, inexpensive, and easy-to-learn [20], and for the ability to find more usability problems compared to other usability evaluation methods [21].

As with any usability evaluation method, heuristic evaluation is not without detractors. Several HCI researchers have argued that results from heuristic evaluation are too subjective [22]. This claim has been counter-argued, however, whereby other researchers have suggested that this difference in perspective enables the discovery of more diverse usability issues [23]. Additionally, despite the suggestion that heuristic evaluation may not be as effective as it claims [24], the method remains popular for desktop-based web applications.

As mobile devices become prevalent, Nielsen's [25] set of heuristics might also be used to evaluate the usability of mobile applications. However, researchers have argued that the heuristics need to be modified before they can be used for this purpose [26–28]. To that end, several sets of heuristics have been defined to evaluate the usability of mobile technologies, each of which have focused on different areas of mobile usability. The research undertaken by Salgado et al. [13, 14] is important to the HCI community in that it extensively maps these sets of heuristics. This paper simply extends that work.

3 Understanding HCI Practitioners' Usage of Heuristic Evaluation to Evaluate the Usability of Mobile Applications

3.1 Approach

To address the first research question, namely, to what extent is heuristic evaluation used by HCI practitioners when evaluating the usability of mobile applications, 16 semi-structured interviews were conducted. This surpasses the number of interviews suggested by Kuzel [29], who recommends six to eight interviews should a sample be homogenous. The interviews were held over two rounds several months apart, with ten participants in round one (5 male, 5 female), and six participants in round two (2 male, 4 female).

To ensure the validity of the findings, the sampling techniques chosen were purposive, systematic, stratified, and snowball sampling [30]. The initial non-probability sampling technique, purposive sampling, helped to ensure that the sample chosen had experience within HCI. This was achieved by conducting the semi-structured interviews of HCI practitioners at the UXPA conference in Boston, Massachusetts on May 15, 2015 [31]. Subsequent probability sampling removed bias by interviewing every tenth attendee (systematic sampling) based on gender (stratified sampling). Following this, only qualitative data from participants with experience in mobile HCI were analyzed. Consequently, the interview transcript from one participant was excluded from the analysis as the participant had no experience of mobile HCI. In addition, even though the participant was attending a HCI/UX conference, the participant's role was primarily a software engineer. Further, to ensure the reliability of findings, three of the participants from the first round were invited back for the second round of interviews conducted on September 1 and 2, 2015. The answers given in round two were cross-checked with the answers given in round one to ensure consistency. Each of the three returning participants was also requested to invite an HCI colleague (snowball sampling) to participate in the second round of interviews. The level of experience of the 15 remaining interviewees, both in terms of general Human-Computer Interaction and mobile Human-Computer Interaction, is shown in Table 1.

Framework analysis [32] was employed to analyze the data, as it is commonly used to analyze interview transcripts [33]. This analysis method follows a standard thematic analysis [34], with the main difference being that the data is placed in a matrix, where patterns can clearly be recognized. During the implementation of framework analysis, thematic categories were defined within Nvivo 10 based on the interview questions and probes. Interview participants were labelled P1 to P13, with P3, P8, and P10 having

Table 1. Interviewees' HCI experience.

Experience	HCI (years)	Mobile HCI (Years)
Range	1–20	1–9
Mean	8.66	3.33
SD	5.97	2.42

been interviewed twice, and P6 being removed from the analysis. Case nodes were then created, based on individual participants. This allowed the analysis of interview data on a case-by-case and a thematic basis. As is common within framework analysis, matrix headings initially reflected interview and probing questions. As the analysis continued, the matrix was re-coded multiple times to reflect emerging themes.

3.2   Results and Discussion

From the analysis, it appears that heuristic evaluation is rarely used by HCI practitioners to evaluate the usability of mobile applications. Based on the themes surfaced from the qualitative data, the reasons why heuristic evaluation is rarely used were:

Heuristic Evaluation was not Normally Associated with Mobile Applications:
Heuristic evaluation was originally defined as a usability evaluation approach for desktop-based software [1]. This was later modified for website usability evaluations [25]. Consequently, it seems that many HCI practitioners continue to perceive heuristic evaluation as a method to evaluate the usability of desktop-based software and websites. In fact, some HCI practitioners hesitated before answering, and seemed quite surprised about being asked if they had used heuristic evaluation for mobile applications.

> P1: "I find that if you are going to do a heuristic evaluation, they are going to ask for it for the main site, the main application."
> P8: "I don't think I've ever used heuristics in order to evaluate mobile applications."
> P9: "I've done heuristic evaluations for websites. I can't say I recall doing one specifically and only for mobile devices."

Usability Testing with Representative Users was Often Considered Good Enough:
Heuristic evaluation was originally defined as a usability evaluation approach for desktop-based software [1]. Usability testing tends to be narrow in focus [4], whereby a small number of tasks are undertaken by a small number of participants over a short period. This results in a relatively small number of usability issues being surfaced [21]. Critical design decisions are subsequently based on this small data set. Heuristic evaluation can complement this and other evaluation methods by holistically finding more usability issues before usability testing is used later in the design and development life cycle. Yet, despite the limitations of usability testing, HCI practitioners view usability testing with representative users as the 'Gold standard'.

P1: "Nothing is better than user testing in front of the actual people it's intended for."
P8: "[I] use prototyping, then we show it to users" and "You know what I did on mobile applications were more like usability studies."
P9: "We were more interested in looking at our audience to see what our audience would think about an application than an expert review."

Good Mobile Design was Perceived as Enclosing Heuristics:

Several participants felt that good mobile design and the design process meant that heuristic evaluation was no longer required. Mobile application design has indeed matured in recent years. Nevertheless, usability issues continue to be surfaced, even within designs created by well-experienced teams that have multiple feedback loops built into their process.

Additionally, many participants felt that they knew what the usability of a mobile application entailed. Therefore, those participants felt that they did not need to consult a checklist. To argue against this, Wharton and Lewis [35] contend that "Designers think they know it all...they do not". This claim by participants also brings us back to Scriven's [15] theory, and the suggestion that both novice and experienced evaluators will overlook key areas during a usability evaluation without a checklist.

P2: "We don't have anything written down."
P4: "It's some of my own discovery."
P5: "Just based on my understanding of usability."
P9: "There's no call or drive to do a heuristic evaluation. Somebody is already doing the design, I guess some of those practices are being incorporated into it."
P10: "I feel like the practice of design has moved forward to the point where there's a shared understanding of what constitutes good design."
P12: "I use my gained knowledge over the years."

4 Discovering the Efficiency of Using Nielsen's Heuristics When Evaluating the Usability of Mobile Applications

4.1 Approach

To address the second research question, are Nielsen's heuristics as effective in surfacing usability issues within mobile applications as sets of heuristics designed for mobile, the usability of a mobile travel app from a well-established provider was evaluated. The heuristic evaluation was conducted using three sets of heuristics by six HCI practitioners in a within-subjects study (2 male, 4 female). This surpasses the recommended five HCI evaluators [3]. The experience level of the evaluators is shown in Table 2.

Table 2. Evaluators' HCI experience.

Experience	HCI (years)	Mobile HCI (Years)
Range	1–20	0–6
Mean	7.5	6.9
SD	2.91	2.2

The three sets of heuristics selected for comparison, were Nielsen [25] (slightly modified from Nielsen and Molich's traditional set to account for websites [1]), Bertini et al. [10], and Joyce et al. [36]. This allowed for the comparison of the number of usability issues found by Nielsen's [25] heuristics, which were not defined for mobile applications, against two sets of heuristics which were defined for mobile applications. The reasoning behind the choice of the other two sets of heuristics was that Bertini et al. [10] developed one of the first sets of mobile-specific heuristics, and Joyce et al. [36] defined a set of heuristics for mobile applications, which included heuristics for the latest mobile technologies. To avoid recognition bias, each set of heuristics was labelled with a letter: Nielsen-Set A; Bertini et al.-Set B; Joyce et al.-Set C. Furthermore, to avoid learning bias the heuristic sets were counterbalanced for every two evaluators (Table 3).

Table 3. Counterbalanced order of heuristics sets.

Participant	Order of heuristic sets
P1, P6	Set B, Set C, Set A
P2, P5	Set A, Set B, Set C
P3, P4	Set C, Set A, Set B

In addition to the heuristic evaluation, participants stated which heuristic set they perceived as the most applicable for mobile applications within a post-evaluation survey. Counterbalancing the heuristic sets, therefore, guarded against recently bias within the survey.

4.2   Results and Discussion

Using the three sets of heuristics, 145 usability issues (Mean = 48, SD = 9) were found in a well-known travel application on the Android platform (Fig. 1).

The greatest number of usability issues, including critical issues, was surfaced by a set of heuristics defined for mobile applications, namely Joyce et al. [36]. While not privy to the overall results, a one-tailed Friedman test ($\alpha = 0.05$) of the post-evaluation survey results indicated that participants perceived the heuristics from Joyce et al. [36] as being the most applicable for mobile applications, $X^2(2) = 12.000$, $p = 0.001$. Comments from participants, labelled P14 to P19 to avoid confusion with interview participants, supported this perception:

P15: *"Set C [Joyce et al.] covers essential evaluations for mobile applications."*
P17: *"Heuristic [set] A [Nielsen] is too broad to apply to the mobile experience."*

A one-tailed Friedman test ($\alpha = 0.05$), however, suggested that the actual difference between the number of usability issues found was only significant at $X^2(2) = 3.739$, $p = 0.077$. This could be due to the small number of evaluators as statistical significance is easier to achieve with larger sample sizes [37]. Indeed, G*Power [38] suggests that 28 evaluators might have been more applicable if a parametric test was used with the significance level was set at 5% ($\alpha = 0.05$), a medium effect size was sought

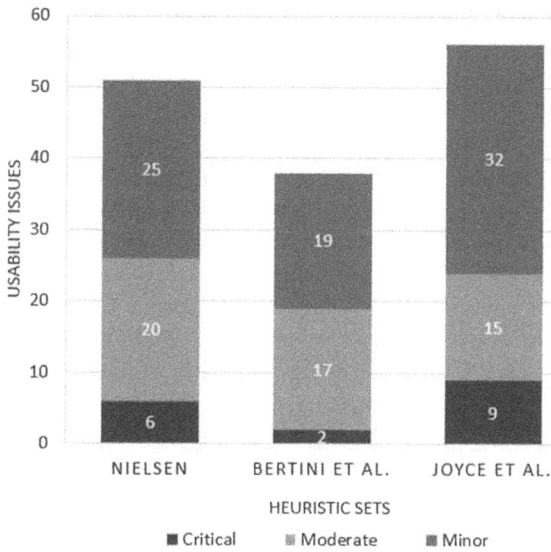

Fig. 1. Usability issues found during the heuristic evaluation.

(Cohen's $f = 0.25$), and statistical power was high $(1-\beta = 0.8)$. To that end, this study seems to have been underpowered. Arguably, while researchers have recruited this number of evaluators for a heuristic evaluation [12], it does not reflect a real-world scenario, given that the recommended number of evaluators for a heuristic evaluation is three to five [3].

Based on their mapping of the literature, Salgado et al. [14] suggest that the number of evaluators is generally three. During our interviews with HCI practitioners, the number of evaluators that participated in heuristic evaluations, regardless of the type of application being evaluated (e.g. mobile, software, website etc.), was often less:

P1: *"I have done it on my own. No one is going to do the test again"*
P3: *"One or two people"*
P8: *"Usually, it's a one-man show....evaluation"*
P9: *"We would have one person do it"*

Subsequently, it could be argued that Salgado et al. [14] are correct—It matter little which heuristic sets are used when evaluating the usability of mobile applications. Another viewpoint, however, might be that practical significance may be more important than statistical significance. Practical significance, in this case, consists of three parts. First and foremost, the perception that a set of heuristics is more applicable toward a specific domain could increase the sense that the heuristics contribute more substantially "to the validity, reliability, and credibility of an evaluation", as per Scriven's [15, p. 4] theory. The results of the post evaluation survey indicated that evaluators considered the heuristic set from Joyce et al. [36] as being most applicable for mobile applications.

Secondly, even if statistical significance is not achieved, any difference between the number of usability issues found would be detected by effect sizes. As illustrated by Vacha-Haase and Thompson [39, p. 473], effect sizes should be reported for every statistical study, even for results that are not statistically significant, as "statistical testing cannot evaluate result importance". This is supported by Durlak [40, p. 917] who contend that "There is no straightforward relationship between a p-value and the magnitude of effect". To that end, a post hoc analysis with Wilcoxon signed-rank tests was conducted. A Bonferroni adjustment was applied, which was set to $p = 0.025$, as we were only interested in the results of Bertini et al. [10] vs. Joyce et al. [36], as well as Nielsen's [25] vs Joyce et al. [36]. Based on Pearson's Correlation Coefficient r, a large effect size was evident between the number of usability issues found by Bertini et al. [10] vs. Joyce et al. [36], $Z = 1.802$, $p = 0.036$, $r = -0.520$. In contrast, the effect size between Nielsen's [25] vs. Joyce et al. [36] was small, $Z = 0.422$, $p = 0.337$, $r = -0.122$ While this study may have been underpowered, the effect size in regard to the difference in the number of usability issues between Bertini et al. [10] vs. Joyce et al. [36] may indicate that the choice of heuristic sets is important.

Finally, the types of issues surfaced are an essential consideration. Given the Universal nature of Nielsen's [25] heuristics, evaluators surfaced several issues that might be relevant to any type of user interface. However, evaluators also documented several issues that are associated more so with desktop-based software and websites than mobile applications. In addition, several issues applicable to mobile applications were missed, which were surfaced by the other two sets of heuristics. Moreover, in some cases, evaluators using Nielsen's [25] heuristics catalogued mobile-specific issues within the closest heuristic, even if that heuristic was a poor fit. The next section contains multiple examples of the points raised above.

Usability Issues Surfaced by Nielsen's Heuristics:
The heuristic "User control and freedom" was designed to capture issues that allowed users to "leave the unwanted state without having to go through an extended dialogue" as well as a call for the support of undo and redo. Yet, an evaluator used it to log a mobile-specific issue, as no other heuristics within that set were suitable:

> P15: *"The Done button is on the top-right corner which makes it hard to reach."*

Additionally, mobile application users tend to expect in-context help and tutorials over traditional online help and documentation [41]. This was reflected in issues raised for the heuristic "Help and documentation":

> P18: *"[I] suggest in-context help."*
> P19: *"I have no idea where the Help is, or where to find it."*

Usability Issues Surfaced by Bertini et al. Heuristics:
While the heuristics from Bertini et al. [10] surfaced the least amount of issues, arguably because the set consists of the least amount of heuristics, in many cases the issues were more relevant to mobile applications than Nielsen's [25] heuristics. For example, the heuristic "Ease of input, screen readability and glancability" surfaced issues related to the ability to read information while on the go:

P14: "You can only see two results at a time. It would be nice to see more results above the fold."
P19: "I would need to stop for a moment to read more carefully."

The heuristic "Aesthetic, privacy and social conventions" surfaced an issue about privacy that is becoming more important to mobile users depending on the type of data collected [42]:

P19: "I have connected via my Facebook account, so I am assuming my interactions are being tracked somewhere, so not private."

While the heuristic "Good ergonomics and minimalist design" could be clearer in terms of software, not hardware ergonomics, several evaluators brought up related issues:

P15: "I couldn't swipe to navigate even though it looks like it's swippable."
P17: "On the map, hotel markers were close together and difficult to target."

The heuristic "Flexibility, efficiency of use and personalization" is important for mobile users. For instance, one evaluator mentioned:

P19: "I don't see a list of most recently selected airports, or travel routes. It would be nice if the app recognized my regular travel and asked me if I wanted to start from that point."

Usability Issues Surfaced by Joyce et al. Heuristics:
The heuristics from Joyce et al. [36] were designed specifically for mobile applications, which was reflected in the issues surfaced. For instance, mobile devices nowadays are packed with technology that can reduce the burden on users. The heuristic from Joyce et al. [36] recognizes this with the heuristic "Use the camera, microphone and sensors when appropriate to lessen the user's workload". Several evaluators surfaced related issues:

P14: "The ability to use microphone to input the name of locations might be helpful."
P19: "GPS...never told me where I was located." and P19: "[If] I could take a photo of the location and add to the review - that would be neat."

As previously stated, traditional help and documentation are less suited to mobile application users. Related issues were raised around this with the heuristic "Display an overlay pointing out the main features when appropriate or requested":

P18: "It would be ideal to have some contextual help indicators for date range selector."
P19: "There was no tutorial on how the Reviews work."

Given the small size of today's mobile devices, designers should strive to "Facilitate easier input". Several issues surfaced remind us why:

P15: "The keyboard "feels" small, and there is no auto-correct."
P18: "Target for Room selector seems too small."

The heuristic "Cater for diverse mobile environments" recognizes that mobile applications are often used in changing contexts-of-use. While it is difficult to consider different mobile use scenarios when evaluating a mobile application in lab-like conditions, one evaluator mentioned the following issue:

P19: "I saw no indication of the app change to adapt to various scenarios."

5 Conclusion

To ensure that HCI community have the right tools to evaluate the usability of mobile applications, several sets of heuristics have been defined in recent years. Two papers from Salgado et al. [13, 14] contributed considerably to the literature by mapping these mobile heuristic sets. The contribution of this work was to further the work of Salgado et al. [13, 14]. Firstly, the authors of this paper set out to discover if the claim that heuristic evaluation is commonly applied to mobile technologies was also applicable to HCI practitioners, not just academic HCI researchers. Secondly, the authors set out to empirically test the suggestion that Nielsen's [25] heuristics may be generic enough for mobile technologies.

To address the claim that heuristic evaluation is commonly applied to mobile technologies, 15 semi-structured interviews were conducted with HCI practitioners. Few HCI practitioners of those interviewed used heuristic evaluation to evaluate the usability of mobile applications. The primary reasons given were that heuristic evaluation was not normally associated with mobile applications, usability testing with representative users was often considered good enough, and good design was perceived as enclosing heuristics. While this research does not attempt to generalize across the entire HCI population, it calls into question the claim that heuristic evaluation is commonly applied to mobile technologies.

To empirically test the suggestion that Nielsen's [25] heuristics may be generic enough for mobile technologies, six HCI practitioners participated in a within-subjects study. Based on the results, the authors demonstrate how the types of issues found and practical significance are importance considerations when deciding upon a set of heuristics when evaluating the usability of mobile applications.

References

1. Nielsen, J., Molich, R.: Heuristic evaluation of user interfaces. In: Proceedings of the SIGCHI Conference on Human Factors in Computing Systems (CHI 1990), Seattle, WA, USA, 1–5 April (1990)
2. Maguire, M.: Methods to support human-centred design. Int. J. Hum.-Comput. Stud. **55**(4), 587–634 (2001)
3. Nielsen, J.: Finding usability problems through heuristic evaluation. In: Proceedings of the SIGCHI Conference on Human Factors in Computing Systems (CHI 1992), Monterey, California, 3–7 June (1992)
4. Dicks, S.: Mis-usability: on the uses and misuses of usability testing. In: Proceedings of the 20th International Conference on Systems Documentations (SIGDOC 2002), Toronto, ON, Canada, 20–23 October (2002)

5. Jeffries, R., Miller, J., Wharton, C., Uyeda, K.: User interface evaluation in the real world: a comparison of four techniques. In: Proceedings of the SIGCHI Conference on Human Factors in Computing Systems (CHI 91), New Orleans, LA, USA, April 27–May 2 (1991)
6. Bernhaupt, R., Mihalic, K., Obrist, M.: Usability evaluation methods for mobile applications. In: Handbook of Research on User Interface Design and Evaluation for Mobile Technology, vol. 44, pp. 745–758 (2008)
7. Baharuddin, R., Singh, D., Razali, R.: Usability dimensions for mobile applications—a review. Res. J. Appl. Sci. Eng. Technol. 5(6), 2225–2231 (2013)
8. Weiss, S.: Handheld Usability. Wiley, Hoboken (2003)
9. Ji, Y.G., Park, J.H., Lee, C., Yun, M.H.: A usability checklist for the usability evaluation of mobile phone user interface. Int. J. Hum.-Comput. Interact. 20(3), 207–231 (2006)
10. Bertini, E., Gabrielli, S., Kimani, S.: Appropriating and assessing heuristics for mobile computing. In: Proceedings of the Working Conference on Advanced Visual Interfaces (AVI 2006), Venezia, Italy, 23–26 May (2006)
11. Joyce, G., Lilley, M.: Towards the development of usability heuristics for native smartphone mobile applications. In: Proceedings of the 16th International Conference on Human-Computer Interaction (HCII2014), Heraklion, Crete, Greece, 22–27 June (2014)
12. Inostroza, R., Rusu, C., Roncagliolo, S., Rusu, V., Collazos, C.A.: Developing SMASH: a set of SMArtphone's uSability Heuristics. Comput. Stand. Interfaces 43, 40–52 (2016)
13. Salgado, A., Freire, A.: Heuristic evaluation of mobile usability: a mapping study. In: Proceedings of the International Conference on Human-Computer Interaction, pp. 178–188, June 22 (2014)
14. Salgado, A., Rodrigues, S., Fortes, R.: Evolving heuristic evaluation for multiple contexts and audiences: perspectives from a mapping study. In: Proceedings of the 34th ACM International Conference on the Design of Communications, p. 19. ACM (2016)
15. Scriven, M.: The logic and methodology of checklists (2005). http://preval.org/documentos/2075.pdf. Accessed 20 Feb 2016
16. Gould, J., Lewis, C.: Designing for usability: key principles and what designers think. Commun. ACM 28(3), 300–311 (1985)
17. Quinn, C.N.: Pragmatic evaluation: lessons from usability. In: Proceedings of the 13th Annual Conference of the Australasian Society for Computers in Learning in Tertiary Education (ASCILITE 1996), Adelaide, Australia, 2–4 December (1996)
18. Smith, S.: Standards versus guidelines for designing user interface software. Behav. Inf. Technol. 5(1), 47–61 (1986)
19. Gerhardt-Powals, J.: Cognitive engineering principles for enhancing human-computer performance. Int. J. Hum.-Comput. Interact. 8(2), 189–211 (1996)
20. Hollingsed, T., Novick, D.G.: Usability inspection methods after 15 years of research and practice. In: Proceedings of the 25th annual ACM international conference on Design of communication, (SIGDOC 2007), El Paso, Texas, USA, 22–24 October (2007)
21. Jeffries, R., Desurvire, H.: Usability testing vs. heuristic evaluation: was there a contest? ACM SIGCHI Bull. 24(4), 39–41 (1992)
22. Kirmani, S., Rajasekaran, S.: Heuristic evaluation quality score (HEQS): a measure of heuristic evaluation skills. J. Usability Stud. 2(2), 61–75 (2007)
23. Wilson, C.: User Interface Inspection Methods: A User-Centered Design Method. Morgan Kaufmann, Burlington (2013)
24. Law, E.L.C., Hvannberg, E.T.: Analysis of strategies for improving and estimating the effectiveness of heuristic evaluation. In: Proceedings of the 3rd Nordic conference on Human-Computer Interaction (NordiCHI 2004), Tampere, Finland, 23–27 October (2004)
25. Nielsen, J.: 10 Usability Heuristics for User Interface Design (1995). http://www.nngroup.com/articles/ten-usability-heuristics/. Accessed 25 Sept 2015

26. Beck, E., Christiansen, M., Kjeldskov, J., Kolbe, N., Stage, J.: Experimental evaluation of techniques for usability testing of mobile systems in a laboratory setting. In: Proceedings of the New Directions in Interaction: Information Environments, Media and Technology Conference (Ozchi 2003), Brisbane, Australia (2003)
27. Ketola, P., Röykkee, M.: The three facets of usability in mobile handsets. In: Workshop at Mobile Communications: Understanding Users, Adoption & Design (CHI 2001), Seattle, Washington, USA, 31 March–5 April (2001)
28. Po, S., Howard, S., Vetere, F., Skov, M.B.: Heuristic evaluation and mobile usability: bridging the realism gap. In: Proceedings of the 6th International Symposium of Mobile Human-Computer Interaction (MobileHCI 2004), Glasgow, UK, 13–16 September (2004)
29. Kuzel, A.: Sampling in qualitative inquiry. In: Benjamin F. Crabtree, William L. Miller (eds.) Doing Qualitative Research (2nd edn.), pp. 33–45. Sage Publications (1992)
30. Creswell, J.: Research Design: Qualitative, Quantitative, and Mixed Methods Approaches. Sage Publications, Thousand Oaks (2013)
31. UXPA.: User Experience Professionals Association Conference, Boston, Massachusetts, 15 May 2015 (2015)
32. Ritchie, J., Lewis, J.: Qualitative Research Practice: A Guide for Social Science Students and Researchers. Sage Publications, Thousand Oaks (2013)
33. Gale, N., Heath, G., Cameron, E., Rashid, S., Redwood, S.: Using the framework method for the analysis of qualitative data in multi-disciplinary health research. BMC Med. Res. Methodol. 13(1), 117 (2013)
34. Blaxter, L.: How to Research. McGraw-Hill Education, Maidenhead (2010)
35. Wharton, C., Lewis, C.: The role of psychological theory in usability inspection methods. In: Usability Inspection Methods, pp. 341–350 (1994)
36. Joyce, G., Lilley, M., Barker, T., Jefferies, A.: Mobile application usability: heuristic evaluation and evaluation of heuristics. In: Advances in Human Factors, Software, and Systems Engineering, pp. 77–86 (2016)
37. Field, A.: Discovering Statistics using IBM SPSS Statistics, 4th edn. SAGE Publications, Thousand Oaks (2013)
38. Faul, F., Erdfelder, E., Buchner, A., Lang, A.G.: Statistical power analyses using G*Power 3.1: tests for correlation and regression analyses. Behav. Res. Methods 41, 1149–1160 (2009)
39. Vacha-Haase, T., Thompson, B.: How to estimate and interpret various effect sizes. J. Couns. Psychol. 51(4), 473 (2004)
40. Durlak, J.A.: How to select, calculate, and interpret effect sizes. J. Pediatr. Psychol. 34(9), 917–928 (2009)
41. Joyce, G., Lilley, M., Barker, T., Jefferies, A.: Mobile application tutorials: perception of usefulness from an HCI expert perspective. In: Proceedings of the International Conference on Human-Computer Interaction (HCII2016), pp. 302–308. Springer International Publishing (2016)
42. Martin, K., Shilton, K.: Putting mobile application privacy in context: an empirical study of user privacy expectations for mobile devices. Inf. Soc. 32(3), 200–216 (2016)

Empirical Study on the Documentation Phase in the Human-Centered Design Process

Gabriela Viana[1(✉)], Dory Azar[2], and Kristin Morin[2]

[1] Kronos Inc., 3535 Queen Mary Rd, Montreal, QC H3V 1H8, Canada
Gabriela.Viana@Kronos.com
[2] Kronos Inc., 900 Chelmsford St Lowell, Lowell, MA 01851, USA
{Dory.Azar,Kristin.Morin}@Kronos.com

Abstract. Documentation is an important artifact of the design process, especially when designing robust applications developed by companies with offices across the world. The documentation phase of the design process is how designers communicate their work to different teams, yet this phase is not described in the UCD process. While documenting, designers face several challenges, like identifying the right amount of information different stakeholders need. To overcome this and other challenges, the Interaction Design Team at Kronos analyzed the design process and interviewed different stakeholders to understand their expectations of the design documentation. As a result of these interviews, we changed our design process and the way we document. In this case study, we share on how we evolved our documentation, how it impacted our design process, and how it impacted our communication with stakeholders.

Keywords: User experience · Documentation · Process · Agile
Design thinking · Design specification

1 Introduction

Documentation is an important phase in the UX Process for the designers, especially when designing robust applications. In some cases, it is the primary point of reference for development teams to understand design intent, more likely so when the teams are spread in different geographies and time zones. The documentation describes the overall idea of the project, what the users need, the current state of the project, upcoming design activities, previous design decisions, how these designs relate to other designs, and major challenges the design project faced.

Our observations in the field exposed challenges with our documentation approach. Documenting designs can be time consuming, yet still fail to fulfill the needs of different stakeholders. To better comprehend stakeholders' preceptions of both benefits and challenges, we decided to run a focus group. But first, let's examine how the literature assesses the importance of documentation.

© Springer International Publishing AG, part of Springer Nature 2019
T. Z. Ahram and C. Falcão (Eds.): AHFE 2018, AISC 794, pp. 27–35, 2019.
https://doi.org/10.1007/978-3-319-94947-5_3

2 Related Work

Despite the potential benefits for documenting designs, one trendy opinion is that designers should focus efforts on activities that directly end up in products, rather than on documenting designs [1, 2, 4].

Benefits. Documented designs facilitate the process of sign-offs on key deliverables. Most of the time, stakeholders won't have time to participate in ideation sessions like brainstorms, and will only see the final deliverables. Therefore, they expect final deliverables to convey the rationale of how designs meet user needs [5]. Documentation conveys the key functions of the design to other stakeholders, even when designers are not available because of differences in location and time zone [8]. Key components like site maps and task flows help illustrate the big picture of the problem and identify the scenarios which will be further developed via interactive prototypes [12]. Documentation helps companies to visualize the design evolution and capture key decisions made throughout the process. This history helps teams avoid making the same mistakes again, and provides both continuity and a central reference point, even as the project evolves [3]. Documentation records the strategic and creative process to keep teams focused on the problem that needs to be solved, their recommended solutions, user goals, and guidelines for implementation. It also empowers collaboration and efficient team revisions [5]. Documentation helps set expectations and provide clarity because the fallibility of human memory means stakeholders often forget small decisions that took place across numerous meetings. Once we capture those decisions and review with stakeholders, the documented alignment creates a safety net against future changes to scope [5]. Documentation helps new team members understand the design and the decisions around it when they join the project.

Challenges. Despite a designer's best efforts, documentation may not have enough information for development teams to start implementation. In these situations, it is often more efficient for teams to work collaboratively in short, low-fidelity cycles and receive feedback from all stakeholders [4]. Even when designers try to keep documentation consistent, maintaining it can be more challenging and time consuming than creating and improving the design itself. All team members need to tell the same story, and contribute to the same vision. File-sharing and collaborating on a single source file always poses a challenge. In addition, designers must tailor deliverables specifically for audiences, but struggle to meet varied stakeholders' needs and expectations with a single document [3, 9, 10]. Because design deliverables cannot be "shipped" and documentation comments can be embedded in the product's code, some thought UX leaders argue that separate documentation is a waste of time.

Lean UX. In Agile, and more recently in Lean UX methodologies, some authors consider documentation an overhead to the design process. Agile teams iterate swiftly during product development and don't often refer to outdated design artifacts from previous sprints. The end user's experience is ultimately what matters, and design documentation—a literal description, rather than the experience itself—diverts effort from solving real design problems [4].

The authors also argue that once the prototype is built and shared with the team—along with the flows, user motivations, and design reasoning—any further explanation developers need can be achieved with collaboration. Designers don't need to spend six weeks creating massive requirements documents. Instead, a prototype explains interactions more clearly and succinctly, and it illustrates more accurately how users will experience the product. It drives more insightful discussions among teams and can be tested directly with users. Moreover, research teams can leverage the prototype for user testing [4, 9].

On the other hand, some authors argue that designers should not measure the benefits of documentation solely by the time spent to create it, but rather on its ability to expose design problems and articulate goals in writing [5]. Moreover, Lean UX focuses strictly on the design phase of the software development process [4]. But to be successful with Lean UX, the teams need strong collaboration. In an ideal world, the teams are always collocated or available via real-time communication tools. But the reality is, when stakeholders and design teams only have cumbersome communication methods like email available or infrequent access to each other, Lean UX cannot function properly. This is especially true for many companies that are distributed across multiple locations [4].

Considering both the benefits and the challenges of documenting designs, we rethought our process and considered ways to make our documentation more efficient, less time-consuming to create, and more useful to various stakeholders.

Identifying an opportunity to better understand their needs and expectations for our design specifications, we scheduled a focus group with some stakeholders that we describe in the next section.

3   Methodology

We ran a focus group with different audiences that consume our documentation often enough to give us meaningful feedback.

We conducted two focus groups on July 25, 2016. We uncovered results around two main points: how participants use the documentation and how we could improve it.

Seven senior practitioners in Development (3) and Quality Assurance (4), aged 30 to 55 years, female (2) and male (5), working at Kronos, participated in this study. We divided participants into two groups by their area of expertise because we wanted to understand the different needs of each team.

Each focus group lasted an hour and a half (1.5 h). We discussed the agenda with designers ahead of time to make sure we addressed their open questions about documentation during the focus group. We exposed these questions to both teams of participants via slides. For each session, we included three facilitators: a moderator, a note-taker, and a time-keeper. The sessions were recorded through the software GoToMeeting[1].

---

[1] GoToMeeting is a service created by Citrix Systems. It is online meeting software that allows the user to meet with other users via the Internet in real time.

4 Findings

Our research with Developers and Quality Assurance professionals revealed six key opportunities to improve our documentation.

Interactions are Not Clear. Even though the text was well-written, stakeholders often struggled to understand the interaction annotations. They explained that supplementing the document with an interactive prototype would illustrate complex interactions more clearly.

Time Consuming. Because updates were cumbersome, and thus our turn-around time for changes was often slow, the design deprecated quickly relative to what the development team was working on.

Lack of the Big Picture. Even though our documentation included detailed descriptions about the interactions, someone unfamiliar with the product struggled to quickly understand the product's core purpose and how the screen flows related to each other and to the overall system.

Edge Cases. They requested more elaborate explanations and better guidance around edge cases, as those are the situations where they need the most help.

Component Re-usage Not Clear. Designs lacked clear indications of whether the components illustrated new functionality vs. ones that other teams had already built and were eligible for re-use.

Track Changes. Change management proved particularly confusing, so they requested we find a way to keep them informed of changes in the design.

5 Results

Because of our study, we realized that we needed to make changes not only in our documentation, but also to our process. We established criteria for re-thinking our documentation: minimalistic (only the relevant information), effortless (not a lot of reading, easy to find content), accessible (available to the team anytime), and development-centered (make the developers' work easier) [11].

5.1 Changes in Our Process

While our process evolved substantially, what we learned directly inspired two key changes outside of our Documentation Phase.

Design Kick-Off Phase: Brief. We added the Design Brief as part of our formalized Design Kick-off Phase. Designers create a strategic brief to articulate the business and design objectives, differentiate intended audiences for the product or feature (the primary personas), acknowledge any assumptions we made about the user or the product, and consider any risks that could impact the design solution. The design brief also outlines the activities we plan to do, and exposes product dependencies. By aligning with stakeholders on these fundamental details before we even begin any design discovery, we minimize scope creep and miscommunication around design goals in later phases.

Design Phase: Storyboard. We formalized the Storyboard phase for designers create interactive prototypes to illustrate how personas perform tasks throughout the product. The interactive nature clarifies interactions, reduces the amount of text, leads to fewer misinterpretations, and accelerates our ability to test with users. Developers and QA now understand the interactions with less reliance on the design team, and devote less time to reading extensive annotations.

5.2   Changes in the Documentation Phase

However, while interactive prototypes do clarify complex interactions for major user scenarios, they don't explain every critical detail. They still leave gaps like a summary of the product's purpose, the product's relationship to other products in the suite, secondary screenflows, persona tasks and goals, edge cases, states of content, error and success messaging, product-specific actions, or use of common components. Documentation fills these critical gaps, and acts as a comprehensive reference for both developers and future designers (Fig. 1).

Table of Contents

Fig. 1. Table of contents of the new documentation. The documentation now includes a combination of the artifacts created in the previous phases, the Change History, Brief, Task Analysis, Wireflows, and the Design Specification, Customizations Options, Dependencies (list of pattern and re-use of components).

Change History. To track changes, we added a table which includes the version number, the date, and who created each version of the documentation. We also created a list of all design projects on a Confluence[2] page, enabling all stakeholders to "watch" the page and receive updates when documentation changes.

---

[2] Confluence is an online team collaboration software.

Brief. Documentation includes a summary of the design brief we created during the Design Kick-Off Phase to remind stakeholders of the personas and main use cases.

Task Analysis. Analyzing and deconstructing the tasks in a complex transactional product allows us to understand the mental model personas have accomplishing those tasks. Exposing the task analysis diagram in the documentation clarifies the important steps and focuses our design on right perspective and priority (Fig. 2).

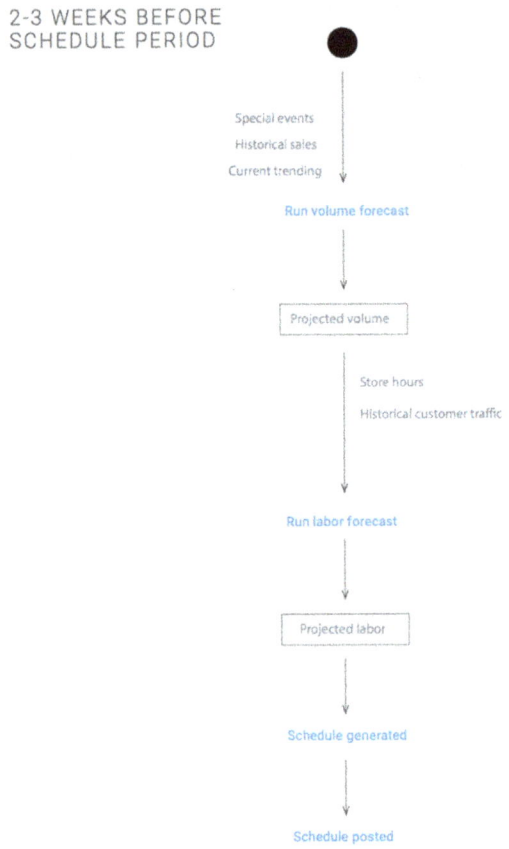

Fig. 2. Example of a task analysis from our current documentation

Wireflows. Similar to a site map for a website, an application wireflow gives stakeholders a high-level illustration of the relationship between screens. At each step of the flow, we include a thumbnail of the wireframe, and identify both paths between pages and dependencies on external content [7] (Fig. 3).

Design Specification. Design specifications comprise the bulk of our documentation. They describe key elements of each page, and the elements' interactions and relationships. They convey information hierarchy, page layout and content, relative

Wireflow

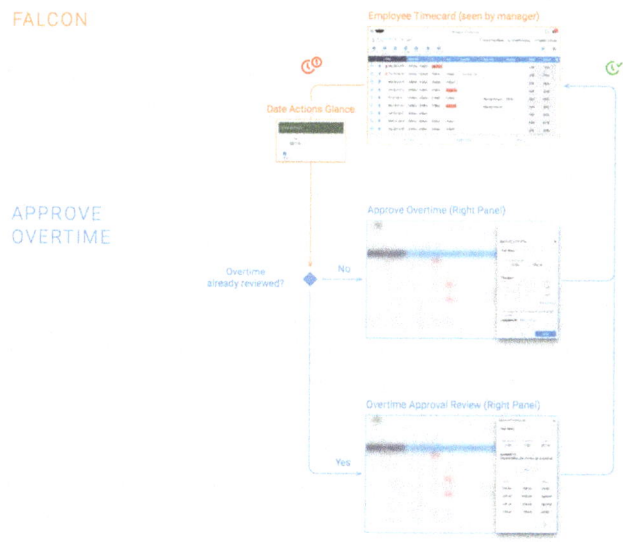

Fig. 3. Example of a wireflow from our current documentation

importance of elements, primary actions for users to perform priority tasks, and responsive behaviors for mobile and tablet versions of the UI. Design specifications supplement our interactive prototypes, rather than acting as the only source of reference, so we can eliminate repetitive pages and heavy annotations. We instead focus on components and reference screens [8].

Customization Options. Our customization options table contains edge cases (errors, empty states), and outlines configuration options that help Developers and QAs identify the flexibility the product will need to support.

Dependencies. The dependency reference table identifies the components and content used within the product, and shows how content from this product will be used across different products in the larger suite.

With these changes, we have already witnessed some improvements in the team workflow:

Interactive Screens. Product Owners and other stakeholders appreciate how interactive storyboards let them to simulate real product functionality during presentations to customers and prospective customers, or even conduct tests to validate workflow and usability.

More Confidence. Developers and QAs are asking designers fewer questions about straightforward screens that follow standard design patterns. They feel more confident about what they need to build and they understand interactions correctly right from the start.

Easier Maintenance. By shifting the focus of documentation, we have reduced the average number of pages from 150 pages to 50 pages—an improvement of 65%—and eliminated most of the redundancies. With that, we made the documentation not just more usable for stakeholders, but easier for our designers to maintain.

6   Conclusion

The Documentation Phase is important to our process as it allows designers to communicate and record design decisions, especially in cases when teams are distributed in different locations and time zones. During our focus group sessions, we learned the benefits and challenges developers and QAs faced when using our documentation. We uncovered several issues including too much text, unclear interactions, lack of the big picture, missing edge cases, and lack of visibility into when changes occurred.

Based on what we learned, we have five key recommendations for better documentation. First: Provide minimalist, accessible, and development-ready documentation. Heavy documentation that describes all interaction is ineffective and cumbersome to maintain. Second: Prototype. With the numerous prototyping tools at our disposal, it is far more efficient to give a feel of the experience than describe it in writing. This helps in design walkthroughs, presentations, and user research. Third: Supplement prototypes with streamlined design specifications. Even with prototypes, developers and QAs still need some guidance on the nuanced specifications of the design like primary actions, edge cases, customization options, and dependencies. Forth: Provide the overall picture. Developers and QA still need to understand the overall aspects of the project, such as the strategic brief, the task analysis, the wireflows. Fifth: Provide clear change history. Developers and QAs can't divert time from developing to search for documents, check for the latest version, or puzzle out what the changes are. Linking to our documentation on a Confluence page helps them easily find the documentation and get push notifications about updates.

This exercise taught us that prototypes provide a richer way to illustrate an experience but do not replace documentation. They work in concert to drive communication among teams but they fulfill different needs. We also learned that both the documentation and the prototype preserve a design legacy important for any company that continuously improves its products.

Given the lack of information on this subject and the lack of tangible examples on how organizations document their designs, we encourage that other UX teams share their best practices, challenges, and successes for documentation in their design process.

Acknowledgements. We thank Kronos for the support, and our colleagues who participated in the study and improvement of our documentation and UX Process.

References

1. Bank, C.: (2016). https://speckyboy.com/guide-ux-design-process-documentation-2/
2. Bank, C., Cao, J.: The guide to UX Design Process & Documentation (2014). https://www.uxpin.com/studio/ebooks/guide-to-ux-design-process-and-documentation/
3. Brown, D.: Communicating Design: Developing Web Site Documentation for Design and Planning, 2nd edn., New Riders, Berkeley, CA, pp. 210–230 (2001). ISBN 13: 978-0-321-71246-2, was 1
4. Gothelf, J.: Lean UX – Getting out of the Deliverables Business (2011). https://www.smashingmagazine.com/2011/03/lean-ux-getting-out-of-the-deliverables-business/
5. Halvorsen, T.: Why Documentation Is Important (2011). https://www.fastspot.com/publications/why-documentation-is-important/
6. Laubheimer, P.: Which UX Deliverables Are Most Commonly Created and Shared? https://www.nngroup.com/articles/common-ux-deliverables/ (2015). was 3
7. Laubheimer, P.: Wireflows: A UX Deliverable for Workflows and Apps (2016). https://www.nngroup.com/articles/wireflows/
8. Miller, E.: Improve your design process with these 4 deliverables (2016). https://www.shopify.ca/partners/blog/improve-your-design-process-with-these-4-deliverables
9. Schoen, I.: How Prototyping is Replacing Documentation (2015). https://uxmag.com/articles/how-prototyping-is-replacing-documentation
10. Spool, J.: Design's Fully-Baked Deliverables and Half-Baked Artifacts (2014). https://articles.uie.com/artifacts_and_deliverables/. was 2
11. Treder, M.: Agile UX & Documentation (2017). https://www.slideshare.net/uxpin/agile-ux-documentation-best-practices
12. Unger, R. Chandler, C.: A Project Guide to UX Design: For User Experience Designers in the Field or in the Making, New Riders, 2 edn. (2012). ISBN-10: 0321815386
13. Miller, E.: (2016). https://www.shopify.ca/partners/blog/improve-your-design-process-with-these-4-deliverables
14. Unger, R., Chandler, C.: A Project Guide to UX Design: For User Experience Designers in the Field or in the Making. 2nd edn. New Riders, 19 March 2012, ISBN-10: 0321815386

In Search of the User's Language: Natural Language Processing, Computational Ethnography, and Error-Tolerant Interface Design

Timothy Arnold[1,2（✉）] and Helen J. A. Fuller[1]

[1] National Center for Patient Safety, Ann Arbor, USA
{Timothy.Arnold4,Helen.Fuller}@va.gov
[2] College of Pharmacy, University of Michigan, Ann Arbor, USA

Abstract. Teams draw from many disciplines in working to design usable and error-tolerant systems. Anthropology is one such field, and ethnographic methods are often used and modified for this purpose. Linguistics expertise is commonly described as helpful; however, Natural Language Processing (NLP) and Computational Linguistics (CL) methods have rarely been described for aiding design efforts. Computational ethnography is described as using large bodies of data to provide insight into the routine of users for use in subsequent design efforts. Narrative text is often a subset of this data; NLP/CL methods are well-matched for analyzing bodies of existing user language. Sharing our previous and new thoughts on these methods in the facilitation of design team understanding can contribute to the discussion on computational ethnography. Both computational ethnography and linguistics can provide insight into error-tolerant system design.

Keywords: Human factors · Human computer interaction
User-centered design · Error-tolerant design · Natural Language Processing
Computational Linguistics · Computational ethnography · Language awareness

1 Introduction

Spoken and written language is ubiquitous in health care settings, just as in the larger world. Language is used to communicate between people and is found in tools and information sources such as user-facing health information technology and clinical decision support systems. Deciding on user-facing language in products and systems is critical and time consuming. In health care systems, designing for error tolerance should be a priority. Suboptimal language selection, organization, or orchestration has been reported to contribute to medical and medication errors. Current approaches in selecting language elements include design team wordsmithing activities, drawing from standard terminologies, and word burden reduction or decluttering. Language selection can be evaluated through human-computer interaction evaluations, for example heuristic and usability evaluations. These approaches are valuable indeed but often capture feedback, perceptions, and behaviors of only a fraction of potential users. Even with our best

© Springer International Publishing AG, part of Springer Nature (outside the USA) 2019
T. Z. Ahram and C. Falcão (Eds.): AHFE 2018, AISC 794, pp. 36–43, 2019.
https://doi.org/10.1007/978-3-319-94947-5_4

efforts, confusing user-facing terminology continues to be described as a contributing factor to medical adverse events. Nielsen provided guidance in suggesting "the system should speak the user's language," but how [1]? Previously, we provided examples in which a body of user language is analyzed using Natural Language Processing (NLP) methods to extract features that may facilitate language awareness amongst design team members [2]. Ethnographic study has been used to explore human-computer interaction (HCI) and design work, but the time necessary to conduct such work can be a limitation. Zheng et al. describe examples of computational methods as efficient approaches to sequester ethnography in HCI work and describe examples of common sources of and the analysis of computational ethnographical data [3].

Ethnography is the study and written observations of people and cultures, including language. Understanding and describing languages can be time consuming, especially when the language is ethnographically foreign or includes domain sublanguages such as medical languages and clinical sublanguages. Our previous discussion on NLP in design contributes to this discussion on computational ethnography by describing and evaluating user language through computational methods to inform design teams and provide recycled elements for user-facing terminology [2]. We provide examples of NLP methods for visualizing the users' language for ease of consumption with an emphasis on facilitating language awareness and possible integration into this computational ethnographic approach. We will share examples in which these methods can facilitate language awareness amongst design team members and guide selection of user-facing terminology with an emphasis on error prevention. Adding on to computational ethnographic approach, our goal is to share insights in the facilitation of error-tolerant and human-centered design through use of computational methods to view bodies of user language.

2 Error-Tolerant System Design in Health Care

Error-tolerant system design is crucial in health care. Wood and Keiras state that error-tolerant systems should facilitate error prevention, reduction, detection, identification, recovery, and mitigation [4]. User-facing language elements can facilitate error prevention or in contrast can contribute to error. For example, a seemingly simple medication instruction can be confusing to patients and contribute to a dosing error.

"Take one tablet daily with meals"

Should the patient take one tablet a day with a meal or one tablet with every meal? How many meals does the patient consume on a daily basis? What if the patient does not know what "daily" means or misses that word when reading the instructions? If the patient is not hungry, is a snack sufficient?

Language elements embedded within dialog boxes can also facilitate or impede the identification and correction of system failures or errors [5]. Because of the abundance of language elements within health care tools and environments, these elements are likely to play a greater role in system performance. User variability in terms of clinical domains, work environments, geographical regions, and native language increases the challenge of selecting standard user-facing terminologies. Just as it is important to understand the user group characteristics, language elements need to be evaluated within their context of use: the equipment, task, and environment. For example, electronic health record and order entry systems rely heavily on language elements. Clinical decision support systems and templates for documentation often contain a considerable number of language elements that are intended to support tasks for working through clinical diagnoses, clinical questions, and therapeutic options. Because of this language rich environment, selecting user-facing terminology becomes an even tougher task. This also presents challenges in the design of information architecture and deciding between tradeoffs in the volume and orchestration of terminology on each page and number of clickable layers while working through complex diagnostic and therapeutic pathways.

Tradeoffs exist between standardization and the necessity for local customization, considering geographical differences in language use and clinical sublanguages. User-facing terminology selection and influence on error-tolerance at the macro level are similar but also differ in the challenges at the local design and customization level. These factors should be taken into consideration. Will macro efforts and standardization, the autonomy to customize to local language use, or some hybrid provide the safer solution? Finding a balance between standardization and local customization and assessing system error-tolerance within this context and the influence of interspersion of language elements is a challenge.

3 Ethnography and Design

Our intention in this section is to enter and continue the conversation on ethnography and design and to describe characteristics, whatever the label, in an attempt to increase our understanding of this complex landscape. As we seek greater understanding, we look for connections to our work, but we claim no expertise in ethnography. Button discusses the entry of ethnography in design and the controversy and discussion surrounding the value of these methods [6]. Dourish provides thought-provoking conversation on the use of ethnography for design purposes [7]. This is important discussion and critically relevant to keep in mind as we proceed, but it is not the focus of this paper. Zheng et al. describe methodologies that they refer to as computational ethnography, harnessing the flood of big data and its potential for informing design [3]. These naming conventions of ethnography and computational ethnography are resonant of a discussion by Uszkoreit (2009) regarding linguistics and computational linguistics [8]. We are not suggesting an exact match but instead reflecting on the changes in meaning of words and phrases over time and across disciplines, semantic drift across disciplines. The methodologies described as computational ethnographic methods discuss the evaluation of computer logs, screen activities, eye tracking, and motion capture in the study of everyday activity as a means for informing design. Language is often an element explored and described in ethnographic work. How might analysis of user language fit into this computational framework? In a previous discussion, we described computational means for exploring user language, facilitating language awareness amongst design team members, and potentially recycling language features into subsequent design [2]. One goal of this paper is to provide a starting point for a conversation between computational ethnography and linguistics and their potential future in facilitating safer designs in health care.

4 Language and Design

The study and evaluation of HCI has long been a multidisciplinary endeavor. Human factors, sociology, and disciplines included now in the cognitive sciences such as computer science, anthropology, and linguistics have been instrumental [9]. Nielsen describes the distillation of usability heuristics from multiple sources [1]. In his analysis, match between the system and the real world was one of the most crucial factors, and he specified that the system should speak the user's language, contain familiar terms and natural language, and understand the user's language. General guidance from web usability is provided by Krug and Redish, such as simplifying, decluttering, and talking to the user using their words [10, 11]. Johnson simplifies user-centered design guidance, discussing how terminology should be familiar and task-oriented and how unfamiliar vocabulary can disrupt reading [12]. He provides guidance in typographical, formatting, and contrast components that influence readability.

The question remains: What guidance is available specific to the selection of user-facing language elements in health care and interface design? Similar recommendations as those provided above exist, such as the value of simplification, the provision of concise language, and formatting guidance for enhanced readability. HCI evaluations can provide the context in which the language is used and downstream detection of confusing language but may only assess a small number of users. Descriptive methods for informing decisions on initial user-facing language are sparingly provided in the literature and, when provided, are general in nature. Utilizing NLP methods to explore bodies of existing user language can be helpful early on during conceptual modeling and to compliment or evaluate questions brought up during heuristic and usability evaluations. NLP can complement HCI evaluations by providing an aggregate view of user language. We attempted to share a glimpse into our approach in exploring user language by using methods in NLP in a previous paper [2].

5 NLP in Facilitating Language Awareness

This approach can be evaluated for use wherever language is present and is not restricted to HCI design. We will discuss this approach for our example above that describes directions on patient-facing prescription labels.

"Take one tablet daily with meals"

This is an example that system users could be exposed to in an electronic format which may have different implications, but the final goal is for these instructions to be present on the prescription bottle and interpreted accurately by the patient or caregiver. The final product is a paper label with limited real estate wrapped around a cylindrical prescription bottle.

Analyzing a body of user language could be used to inform changes to these instructions. First, identifying an applicable body of language is necessary for further analysis. This presents a challenge in that language sources often contain a hybrid from different users with different tasks in mind. These language sources can also contain a mix of naturally occurring and standardized language. Standardization is inherent to language, but standardized terminologies in medical languages may not completely coincide with that of the users. For purposes of this discussion and considering utility, a database file that contains patient instructions is a practical place to start. This body of language could be analyzed to look for themes that relate to the word "meals." To start, we might search for synonyms to meals such as "food," or verb phrases such as "eat with" or "take with." Instructions containing negations, spelling errors, and words of interest in context that is not semantically related can be reviewed and processed iteratively to optimize the categorization of the training text.

"Take with food"
"with food"
"with meals"
"food/ drug interaction alert"
"give with food"
"before meals"
"do **not** take on an empty stomach"

Once the training text has been separated into groups and categorization tested, the remaining body of text can be automatically classified. This classification task can help identify language elements that we originally missed that refer to "taking with food." Language used in these messages can be quantified and their structure analyzed for themes by looking at different or a combination of language features. If we can identify which words and phrases are used to indicate that a medication should be taken with food, then this may help facilitate the selection of less error-prone user-facing language. We may also like to understand the prevalence of negation within instructions and avoid this because of reported higher rates of error in interpreting these sentences [14].

Text analysis may also assist in understanding what not to do, which can be just as important as understanding what to do. For example, when a safety concern is identified, system designers may wish to alert users of a potential interim solution in an effort to mitigate risk in the short-term until stronger solutions can be implemented. One proposed solution is modifications to existing warning messages. In this example, visualization of the word co-occurrence network (Fig. 1 shows an example) [15] or auto taxonomy construction can be a valuable tool for the design team in deciding if modifying existing messages is a viable and safe option. Quantification of and thematic analysis might show there are several large themes already present, and replacing existing warning messages might introduce risk. This concept coincides with Dourish's point that the most important feature of ethnographic study might suggest "what should not be built" [7].

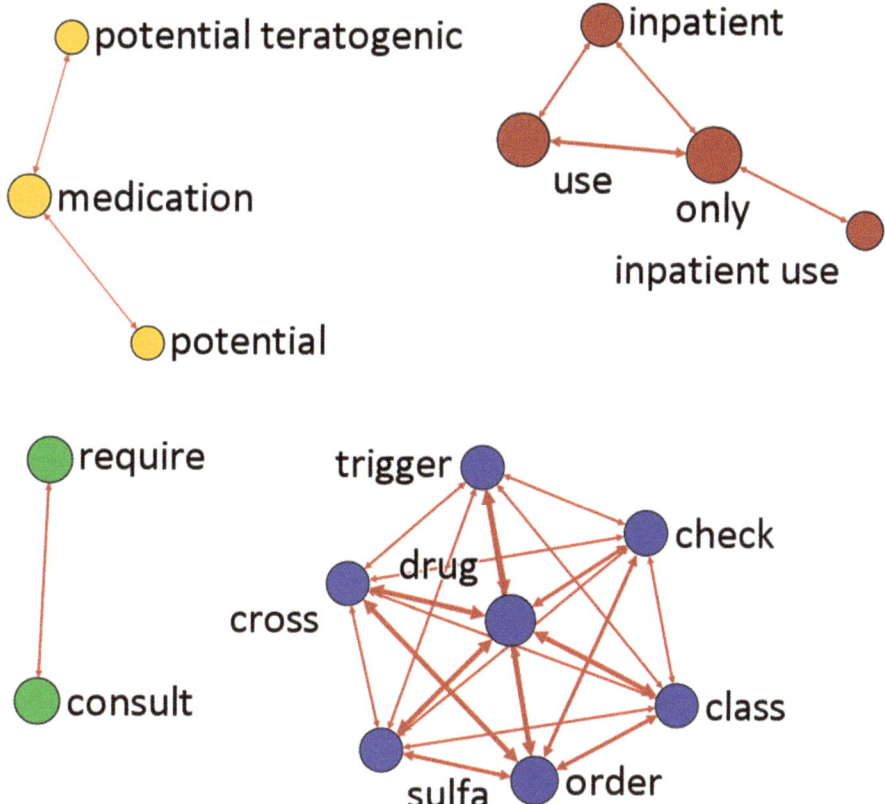

Fig. 1. Example word co-occurrence network for exploring drug comments. This network visualization was created using PolyAnalyst™ (Version 6.5.2030; Megaputer Intelligence)

6 Conclusion

We present a reflection on our last five years of attempting to interlace the analysis of user language with NLP methods for informing subsequent user-facing language selection. An unintentional and beneficial consequence of attempting to expand our understanding is we finally found a much earlier discussion on this interchange between HCI and NLP: Ozkan and Paris provide a thoughtful discussion on the pairing of these two disciplines, how each can benefit, and the limitations of each approach [16]. Abramson et al. discuss the promise of computational ethnography and the potential for these techniques to aid in text rich computational inquiry [17]. Equally important, this exercise catalyzed the gathering of these publications into one place. These publications describe limitations of the computational ethnographic and linguistic approaches outlined [2, 3, 7, 16] and [17]. For example, challenges exist to transferring methodological approaches across disciplines, and there are concerns over the potential to stifle creativity and innovation. Also, the human to human communication process may just be so different from our interaction with computers that other strategies should be considered [5, 16]. Other limitations or concerns include the potential of motivating linguistic changes or drift if questionable trust in the system arises. In addition, our work, although often tedious, may not provide any design insight. Although, unbeknown at the time, these processes and exploration may provide pieces of the puzzle for future operational work. Finally, how do we ensure these activities fall within an ethical imperative? How do we design our work to reduce bias in the processing of language and ensure we are aware of ethical principles surrounding these techniques [18]? With the growing use of speech recognition and natural language generation in health care, there continues to be a need for designing error-tolerant systems with language in mind. Balancing design within the complexity of these dynamic social-technical systems while attempting to maintain and improve safety within health care systems will continue to be a multidisciplinary challenge.

Acknowledgments. We would like to thank everybody at the National Center for Patient Safety for their commitment to patient safety. There were no relevant financial relationships or any source of support in the forms of grants, equipment, or drugs. The authors declare no conflicts of interest. The opinions expressed in this article are those of the authors and do not necessarily represent those of the Veterans Administration.

References

1. Nielsen, J.: Enhancing the explanatory power of usability heuristics. In: Proceedings of the SIGCHI Conference on Human Factors in Computing Systems, pp. 152–158. ACM, April 1994
2. Arnold, T., Fuller, H.J.: Local lexicon extraction and language processing in facilitating language awareness and informing user-centered design in the health care environment. In: Proceedings of the International Symposium on Human Factors and Ergonomics in Health Care, vol. 6, No. 1, pp. 97–103. Sage India, SAGE Publications, New Delhi, India, June 2017
3. Zheng, K., Hanauer, D.A., Weibel, N., Agha, Z.: Computational ethnography: automated and unobtrusive means for collecting data in situ for human-computer interaction evaluation studies. In: Patel, V., Kannampallil, T., Kaufman, D. (eds.) Cognitive Informatics for Biomedicine, pp. 111–140. Springer, Cham (2015)
4. Wood, S.D., Kieras, D.E.: Modeling human error for experimentation, training, and error-tolerant design. In: Proceedings of the Interservice/Industry Training, Simulation, and Education Conference, pp. 1075–1085, December 2002
5. Shneiderman, B., Plaisant, C., Cohen, M., Jacobs, S., Elmqvist, N., Diakopoulos, N.: Designing the User Interface: Strategies for Effective Human-Computer Interaction. Pearson, Boston (2016)
6. Button, G.: The ethnographic tradition and design. Des. Stud. 21(4), 319–332 (2000)
7. Dourish, P.: Implications for design. In: Proceedings of the SIGCHI Conference on Human Factors in Computing Systems, pp. 541–550. ACM, April 2006
8. Uszkoreit, H.: Linguistics in computational linguistics: observations and predictions. In: Proceedings of the EACL 2009 Workshop on the Interaction between Linguistics and Computational Linguistics: Virtuous, Vicious or Vacuous? pp. 22–25 (2009)
9. Miller, G.A.: The cognitive revolution: a historical perspective. Trends Cogn. Sci. 7(3), 141–144 (2003)

10. Krug, S.: Don't make me think, revisited: a common sense approach to Web usability. Pearson Education, San Francisco, California (2014)
11. Redish, J.G.: Letting Go of the Words: Writing Web Content that Works, 2nd edn. Morgan Kaufmann, Waltham (2012)
12. Johnson, J.: Designing with the Mind in Mind: Simple Guide to Understanding User Interface Design Guidelines. Elsevier, Amsterdam (2013)
13. Rosenbloom, S.T., Miller, R.A., Johnson, K.B., Elkin, P.L., Brown, S.H.: A model for evaluating interface terminologies. J. Am. Med. Inf. Assoc. 15(1), 65–76 (2008)
14. Tettamanti, M., Manenti, R., Della Rosa, P.A., Falini, A., Perani, D., Cappa, S.F., Moro, A.: Negation in the brain: modulating action representations. Neuroimage 43(2), 358–367 (2008)
15. Megaputer Intelligence. PolyAnalyst™ Professional: Version 6.5.2030 [Data & Text Analysis software]. Bloomington, Indiana (2018)
16. Ozkan, N., Paris, C.: Cross-fertilization between human computer interaction and natural language processing: Why and how. Int. J. Speech Technol. 5(2), 135–146 (2002)
17. Abramson, C.M., Joslyn, J., Rendle, K.A., Garrett, S.B., Dohan, D.: The promises of computational ethnography: improving transparency, replicability, and validity for realist approaches to ethnographic analysis. Ethnography 19, 254–284 (2017). https://doi.org/10.1177/1466138117725340
18. Leidner, J.L., Plachouras, V.: Ethical by design: ethics best practices for natural language processing. In: Proceedings of the First ACL Workshop on Ethics in Natural Language Processing, pp. 30–40 (2017)

Disrespectful Technologies: Social Norm Conflicts in Digital Worlds

Sarah Diefenbach[1(✉)] and Daniel Ullrich[2]

[1] Department of Psychology, Ludwig-Maximilians University,
Leopoldstr. 13, 80802 Munich, Germany
sarah.diefenbach@lmu.de
[2] Institute of Media Informatics, Ludwig-Maximilians University,
Frauenlobstr. 7a, 80337 Munich, Germany
daniel.ullrich@ifi.lmu.de

Abstract. Social norms are the informal understandings that govern the behavior within a society and are crucial for the feeling of togetherness and social cohesion. However, technology-mediated behavior often stands in conflict (e.g., parallel smartphone use undermining the norm of full attention for the conversation partner). Under the umbrella term 'disrespectful technologies', the present research discusses potential conflicts between technology use and social norms. We present four general types of frequent social norm conflicts, illustrate these by examples from research in different fields of HCI and a user survey, and discuss possibilities to address these in design. Altogether, we see the integration of social norm considerations in the user centered design approach as a long-term goal in HCI, with important consequences on an individual and societal level.

Keywords: Technology-mediated behavior · Social interaction
Social norms · Social conflict · Disrespectful technologies

1 Introduction

User experience (UX) research and design increasingly shifts its focus on emotional values and critical constituents of a fulfilling experience of technology use. If considered from a social dynamics perspective, namely effects of technology-mediated behavior on interaction between people, an important aspect are subtle side-effects of technology use and their potential conflicts with social norms. Social norms are the informal understandings that govern the behavior of society members, i.e., the shared beliefs regarding appropriate ways to feel, think and behave [1]. As such, they are crucial for the feeling of togetherness and social cohesion. While usually not being aware of the many unwritten laws underlying our behavior, one becomes typically aware of social norms in the case of conflict: if someone behaves in a way, that contradicts our informal understanding of what is appropriate (e.g., cutting in line, entering an office without knocking, starting the meal before everybody is at the table).

If closely considered, the gradual development of routines around technology use, particularly in the field of communication technology often breaks up such established

© Springer International Publishing AG, part of Springer Nature 2019
T. Z. Ahram and C. Falcão (Eds.): AHFE 2018, AISC 794, pp. 44–56, 2019.
https://doi.org/10.1007/978-3-319-94947-5_5

social norms. A typical example is the commitment to fixed appointments (old norm) versus the nowadays-convenient last minute cancelling or delay of a date via WhatsApp. Another example is the increasingly rare commitment of full attention for the conversation partner, often undermined by parallel smartphone use. Any impulse from the digital world – a phone call, a push notification, a text message – evokes instant reaction, independent of how the direct conversation partner may feel if the attention repeatedly shifts away from an emotional meaningful moment to the technology-mediated communication. Accordingly, studies already demonstrated that a silent phone lying on the table has negative impacts on the conversation atmosphere [2]. In contrast to "full attention for the conversation partner" (old norm), "digital communication comes first" seems to have become a new norm, especially among younger users.

The example of technology-mediated activities in parallel to conversation demonstrates the process of norm erosion and then norm fragmentation within a society. Some time ago, large parts of the society probably expressed commitment to the norm of full attention for the conversation partner, aiming to avoid interruptions through phone calls or messaging and feeling bad about it if it still happened. Over time, however, some people may have started enjoying this kind of norm-violation and interruptions of a conversation through the mobile phone. In fact, some may have discovered the benefits of this habit, e.g., using the smartphone as a welcome retreat or even scapegoat when the direct conversation partner's story gets boring. Over time, these people established a new, alternative norm – at least in their own subculture. It seems that the formerly norm-violating behavior has become the new norm and "appropriate". However, since not all members of a society undergo this process of norm erosion in the same speed – or even not at all, since they do not want to accept the new behavior described above and give up the established norm – norm fragmentation and diverging views on what is appropriate can result.

In addition to norm fragmentation resulting from heterogeneous norm erosion, technology also creates new spaces of interaction for which no norms exist to date. People may try to find parallels and transfer social norms from the digital world; however, there is not always an unambiguous parallel. For example, which norms should apply for chat conversations? Should one consider it a personal face-to-face conversation, thus having to say goodbye before leaving the (chat) room? Or is it a non-committal open channel, an occasional meeting zone such as a marketplace, where people are strolling around, may exchange a few words with one person or the other, without a felt obligation to say goodbye to everybody before leaving the zone.

The naturally resulting conflicts can be compared to intercultural conflicts stemming from varying social norms between cultures. For example, while it is appropriate to yell and make loud comments in the movie theaters in some African countries, it would be rude to do so in the United States [3]. However, while most people are aware of cultural differences and try to take these differences into account when interpreting others' behavior, there is less sensitivity (and probably less acceptance) for different norms of technology-mediated behavior. In general, one does not expect the people one is surrounded with to be very different – and feels totally puzzled if this turns out to be the case. One example of such a clash of social norms between social media users of different age is "fraping", i.e., the unauthorized alteration of content on a person's social media profile such as changing the profile picture, typically happening in a situation when the "victim" has left his or her phone unlocked. While many users generally consider fraping to be an unacceptable breach of trust, it is an acceptable social practice and considered a funny joke (even by the victims) among some younger users [4]. In sum, contrary to intercultural differences, social norm conflicts related to technology use could be even more unexpected, and thus easily result in an escalating process. If two people both feel that the other is wrong, each one might distance him- or herself, waiting for the other to apologize or somehow make up for the norm violation. The missing action of the other may be taken as another norm violation and thus cause further escalation.

2 Disrespectful Technologies

Under the umbrella term 'disrespectful technologies', the present research discusses potential conflicts between technology use and social norms. For example, such conflicts may be due to technology features that imply or enable undermining established norms (e.g., smartphone pop-up message interrupting a face-to-face conversation). In addition, unclear technology design, missing clear cues to transferrable concepts and norms from the non-digital world, could pave the way for alternative interpretations and related norm conflicts. For example, as discussed above, thinking about a chat room in parallel to a personal conversation implies norms of face-to-face discussions such as saying goodbye. In contrast, considering a chatroom a place for noncommittal occasional meetings such as a market place does not imply a personal good-bye after a chat.

Such social conflicts related to technology use are of broad relevance for private as well as professional settings. Given that the informal compliance to shared social norms is a basic fundament of feelings of togetherness and solidarity, the feeling that the other one does not respect such norms can have dramatic consequences. The felt offense creates distance between people and negative feelings – whereby, the true cause and norm conflict underlying the irritating behavior is probably not evident in most cases. From this perspective, social norm conflicts are an important societal phenomenon of relevance for studying conflicts in different contexts and scopes such as within families, intergenerational perspectives, or in business contexts, where perceived norm violation can destroy the team spirit. Thus, the discussion of technology-use related norms and a "digital etiquette" may be considered an important aspect of business culture design.

3 Outlook and Methodological Approach

In the following chapters we first present an overview and possible categorization of different types of social norm conflicts related to technology-use. For this purpose, the present analysis combines different methodological approaches. First, we performed a literature review, searching for indicators of social norms conflicts in various areas of HCI research (e.g., social media, technology-mediated communication). Second, we conducted a survey where we asked a convenience sample (N = 32) to report on recently experienced social conflicts related to technology use. Each participant provided 3–4 conflict reports, related to different (assumed) reasons of conflict mentioned in the instructions (i.e., divergent expectations, divergent perceptions of behavior, technological feature, general/other), resulting in a total of 110 reports. After a qualitative description of the conflicting situation, participants rated affect valence (1 = negative, 5 = positive) and feelings of closeness (1 = distanced, 5 = close) regarding their own and (expected) others' feelings on a 5-point scale. Besides exemplification of different social norms conflicts by cases from the literature and single narratives of our survey (Sect. 4), we present a summary analysis across all narratives and general tendencies in the experience of social norm conflicts (Sect. 5). In addition, we discuss possible starting points for more respectful technology design (Sect. 6), and future research directions (Sect. 7).

4 Categorization: Different Types of Social Norm Conflicts Related to Technology Use

4.1 Type 1: Technology-Initiated Norm Violation

For every social context there are particular norms implying how people behave and interact. As such, social norms may be seen as a corset that usually keeps behavior in a particular range and avoids conflict. However, even if people are generally eager to respect such norms, external impulses can undermine or lead people to deviate from such norms.

Examples. The ringing cellphone lying on the table interrupts an intense face-to-face conversation, without any respect for the emotional subject and the conversation partners' feelings.

Conflict interpretation: Violated norm: full attention for the conversation partner.

The WhatsApp Read Receipts feature (blue tick marks) creates a pressure for immediate answers, especially if the sender is an important person (e.g. close other). Thus, the recipient interrupts the current face-to-face conversation to write a (more or less thoughtful) message.

Conflict interpretation: Violated norm: full attention for the conversation partner, full thought for messaging important persons.

4.2 Type 2: Norm Erosion: Norm Violation Becomes a New Norm

If in a given social context technology-initiated norm violation happens over and over again, norm violation starts to become accepted. The norm violation becomes a new norm, where the formerly norm-violating behavior has become acceptable. In this process of norm erosion, different people may develop different norm attitudes. While some may still favor the old established norm (e.g., no smartphones at the dinner table), others may start preferring the technology-initiated norm erosion and its possibilities to override the rules. In consequence, the process of norm erosion is characterized by confusion and uncertainty how to behave or feel towards the behavior of others, including a wide variety of attitudes. Some may still fight for respecting the old norms, some may accept the new norm as an inevitable development, some may favor the tendency towards the new norm, and some may even demonstratively support the new norm and consider it a progressive behavior, standing for a modern society.

Examples. Instant messaging enabling last minute cancelation of appointments or information about delays. Being late becomes more and more normal, since you can always inform the other about being late in a convenient way.

Conflict interpretation: Violated norm: Punctuality, commitment.

> "I wanted to talk to my friend about an important issue and tell her how I feel. However, she was checking Facebook and Instagram in parallel. It really hurt me to see her attention drifting away from me all the time. I felt less important than the news feed" [P4].

Conflict interpretation: Violated norm: Full attention for the conversation partner.

4.3   Type 3: Norm Fragmentation: Old Norms Versus New Norms in Parallel

As the process of norm erosion proceeds, the formerly norm-violating behavior is very likely to ultimately replace the old norm, which will appear as outdated and not valid anymore – at least for those surrounded by heavy norm erosion or promoting such erosion themselves. The process of norm erosion however does not proceed in the same speed and manner for everyone. A survey on smartphone behavior in public places showed some diverging views on whether a behavior is appropriate or not. While about 50% see smartphone use in a restaurant as appropriate, the other 50% do not [5]. So what appears polite and acceptable for one person might be interpreted as a signal of rudeness by the other. The result is a society characterized by norm fragmentation: people living side by side but not sharing the same standards – and probably not even being aware of it. If the person next to us "breaks the norm" (according to our personal view), this is considered unacceptable and attributed to discourtesy, carelessness or impertinence – in short, bad character. It does not appear to us that the other one might just hold contrary norms, and sees nothing wrong or unusual in his or her behavior. The experience of confrontation between people from different norm segments may be comparable to a cultural shock. However, since not being attributable to "cultural differences" the confrontation may rather result in negative attributions regarding the other.

Examples. "My parents met my boyfriend for the first time. We were in a restaurant at the dinner table when his phone rang. He answered the phone started talking while we were eating. A very bad first impression" [P14].

Conflict interpretation: Old norm: Full attention for the others at the dinner table – no newspapers, no other distractions. New norm: phone use at the dinner table is normal. Presumably, the young man was eager to make a good impression when meeting his girlfriend's parents for the first time, for example by demonstrating his willingness to answer a work-call at any time of the day. However, he obviously was not aware that phone use at the dinner table could in general undermine the good impression.

"Me, my partner and my cousin met to watch a movie at home. My cousin used his smartphone all the time, busily texting with friends. It was an uncomfortable situation. We wanted to discuss the movie but he couldn't join the discussion since he didn't catch much of the content. Though sitting together, it was a feeling of separation, knowing that he had been somewhere else with his thoughts all the time" [P17].

Conflict interpretation: Old norm: Joint experience of joint events, no side activities. New norm: Physically close but mentally distant. Side activities are normal; meeting at a place does not imply doing things together and showing interest for each other (i.e., the usual picture in many Starbucks and coffee shops, where peers meet to drink coffee together but most of the time are absorbed by their smartphone).

"I was on holiday abroad around New Years Eve and turned off my mobile in order to really come down and get away from it all. Hence, I wrote my new year's greetings a few days later – which upset many people. Some of my friends were really angry and disappointed ("I thought I am not important to you")" [P2].

Conflict interpretation: Old norms: Vacation is meant to be relaxing – no duties, no obligatory contact. New year greetings are exchanged with the people directly around you. New norms: Staying in (digital) contact with your friends is obligatory, anytime, worldwide. New years greetings are exchanged and expected from all people that are important to you.

As Sabra [6] explored social norms of grief in the context of social media and revealed significant differences between "netiquettes" of mourning and memorialization among Facebook users. Several participants mentioned discomfort regarding intensity, timing and the proper placement of grief expressions, as for example: "If one day you are 'in grief' and express it in a ten line status update and the next day upload a picture of your delicious tuna sandwich, I, personally, find it difficult to take the first mentioned seriously" [6] (p. 28). On the other hand, those close to the bereaved often feel offended when being confronted with the emotional online expressions of grief by distant friends. Online mourning may even evoke the expression of exploiting death to get likes, a clear conflict with social norms of mourning and grief. One participant stated: "I hate when they do that! Tasteless. Why must other people be meddled in this? To get likes? Empty comments from people who say they are there for you, sweety" [6] (p. 31).

Conflict interpretation: Old norm: Expression of grief according to established social norms and rules of grief (also see [7]). For example, very emotional statements are usually "reserved" for close friends and family members. New norm: Emotional expressions online as a sign of solidarity, also by only distantly related "friends". No felt conflict between emotional grief statements and then going back to business as usual in the next minute. The latter pattern is similar to the concept of "virtue signaling" on social media, namely, taking "a conspicuous but essentially useless action ostensibly to support a good cause" such as changing one's profile picture to show support for refugees, without but any real action of support such as donating time or money [8].

4.4   Type 4: Norm Confusion: Alternative Interpretations of New Fields of Technology-Mediated Social Interaction

Besides the existence of old and new norms in parallel, another type of norm conflict can emerge if new fields of social interaction emerge, so that no previous experience and no clear norms exist yet, here called norm confusion. This could be situations emerging from new technologies, business models or any practice related to social interaction. Examples are the invention of e-mail, the raise of repair cafés where people assist each other in repairing things, or the internet platform for couch surfing (couchsurfing.com), building a global community of travelers and hosts offering complimentary overnight stays. If people have diverging mental models and interpretations of such newly emerging fields of social interaction, leading to alternative social norm assumptions, conflicts are likely. If one couchsurfer interprets couchsurfing as a pragmatic network of mutual support for inexpensive travelling, whereas another couchsurfer interprets it as a chance for social interaction and "visiting a friend", the experience will be unsatisfying for both of them and both will find each other's behavior inappropriate. While one might critique the lack of interest in conversation or joint activity, the other might critique the indirect implication of an inappropriate social burden.

Such alternative interpretations of new fields of social interaction can generally occur in the digital as well as in the non-digital world. However, there are several reasons why especially the digital world is prone to social norm conflicts of this form. First, the digital world is characterized by the fast development of technological innovation, leaving no time for norms to emerge and be established. All the time, new fields emerge where new norms have to be created and negotiated – and once they are established, the next innovation emerges and new norms have to be found. Second, conflicts due to norm confusion are more likely in the digital world, because the confusion becomes less obvious. In direct interaction between people, one usually recognizes the other's irritation due to a norm violation and there is a chance to resolve the conflict right in time. In the technology-mediated interaction, many of the non-verbal cues regulating conversation are missing [9] and conflicts can quickly escalate. As noted above, the risk for norm confusion may depend on whether there are evident corresponding fields for norm transfer, so that for the digital world, the risk for norm confusion may depend on how easy it is to identify evident real world parallels.

Examples. Lee et al. [10] studied interaction with a mobile remote presence (MRP) systems in the workplace, i.e., a physically embodied videoconferencing systems for remote coworkers. Through using the MRP, the remote coworkers (here called "pilots") can wander the hallways and spontaneously engage in interactions and communicate with the local coworkers. As their field observation and interviews showed, people seem to rely on different mental models when interacting with the MRP. While "the MRP was sometimes perceived as being the remote pilot (e.g., referring to the MRP by the pilot's name), which can encourage social norms of face-to-face interaction; at other times, the system was perceived as being an object—furniture, a device, or a robot (e.g., addressing the MRP as "Robot" or "it," resting feet on its base, or leaning on it)" [10] (p. 39).

Conflict interpretation: no clear non-digital world correspondence. Interpretation 1: MRP as a representation of a person, activating the same demands for courtesy as when directly interacting with a person. Interpretation 2: MRP as a technology, which creates a connection to another person, but does not represent the person itself, similarly to a telephone.

Already in 2004, where the technological opportunities and emerging fields for social interaction where way more restricted than today, Preece [11] discussed the conflicts and confusion around social norms in communication technologies. This for example refers to expectations in the context of email, where one of her interviewees stated that "… not addressing me by my name and ending without a farewell greeting and the sender's name – that's rude and unfriendly" [11] (p. 58). However, an online discussion around the question "How do you address someone in your reply to their emails?" [12] shows that the norm of addressing people by name in email is (no longer) shared by all users: One forum user states "It depends entirely upon the relationship and what kind of email", expressing that there are no general rules. Another one sees addressing by name as a sign of anachronism, stating that "No name whatsoever. I just reply without any kind of anachronistic salutation line. Death to the letter and every-thing it stood for." Finally, other users emphasize on how *not* addressing the other by name, is even a sign of valuation and intimacy: "In personal email that's a reply, I often don't even use the name. You've emailed me. You know me. I know you." [12]. Similarly, another makes up a parallel to face-to-face conversation: "I don't think I have ever put a formal greeting in an email, even back in olden times. Of course, I don't use someone's name when talking to them in person either, they know who they are, no need to remind them." Thus, what is experienced as rudeness by one person may be a sign of friendship for the other. Besides changes in norms from more formal to informal communication over time, again, also the unclear correspondence to non-digital world examples may contribute to the norm confusion.

Conflict interpretation: no clear non-digital world correspondence. Interpretation 1: Email as letter, implying the same norms as for handwritten or printed letters. Inter-pretation 2: Email as a new communication channel, open to interpretation by its users.

"My *message* was read, there were the blue tick marks, but I didn't get an answer for hours. I was waiting for the answer, in order to make plans for the day (together). I felt terribly ignored and neglected" [P1]. Obviously, the WhatsApp Read Receipts (blue tick marks) feature might no only create a pressure for immediate answers for the recipient, but also feelings of disregard for the sender. Consequently, some people may see it is a general obligation and question of appreciation to answer a message as soon as noticed and feel offended if the recipient does not answer immediately (which of course disregards the recipient's current situation and the potential reasons for a delayed answer). Again, the absence of clear correspondences in the non-digital world make up the room for different interpretations and norms to answer messages in WhatsApp and other instant messaging services. Instead of efficient communication, frequent misunderstandings require a lot of meta-communication.

Conflict interpretation: no clear non-digital world correspondence. Interpretation 1: Instant messaging as synchronous conversation, applying rules from face-to-face conversation: You heard my question and I expect an answer. Interpretation 2: Instant messaging as asynchronous conversation, leaving the speed up to the conversation partners. Messages file up like letters that one answers if the time is right.

> "When I woke up on my birthday morning a checked by smartphone for news, also Facebook. With surprise I realized that my father had posted a photo of my birthday breakfast table. And it got numerous likes. I was really angry that he shared this family moment with the whole world without my permission – and that all these people had seen *my* birthday breakfast even before myself. But when I confronted him, he just argued that it was his Facebook account where he could post whatever he likes" [P9].

Conflict interpretation: no clear non-digital world correspondence. Trying to transfer the father's behavior to the non-digital world does not lead to convincing parallels (e.g., taking polaroid pictures of the birthday table and showing them to the neighbors, inviting the neighbors to see the birthday table while the daughter is sleeping), and thus no basis to judge the behavior in light of existing norms.

> "I *created* a Facebook event to invite about 20 people to my birthday. Some friends replied the invitation with a positive or negative answer, others ignored the event and didn't reply at all. I personally expect people to reply such an invitation and get sad if this does not happen. Others might not care about it" [P24].

Conflict interpretation: unclear non-digital world correspondence, diverging interpretations of the concept Facebook event. Interpretation 1: Invitation, implies active reply. Interpretation 2: Event notice, implies passive acknowledgement.

5 Consequences: Individual Experience of Social Conflicts Related to Technology Use

As the many examples discussed above show, there are different types of social conflicts related to technology use, with different origins and starting points. We showed how many of these can be interpreted as social norm conflicts, ranging from technology-initiated norm-violation, to norm erosion, norm fragmentation, or norm confusion. While the qualitative reports from our survey participants provided a good overview of the wide range of such everyday conflicts, the analysis of their numerical ratings for own and others' expected feelings with regard to such situations provides deeper insight into the distancing power of such conflicts.

Unsurprisingly for conflict reports, the experience was rated as more negative than positive and characterized by feelings of distance rather than closeness. In general, participants estimated their own negative affect as more severe than that of others. Also, the perceived feeling of distance was rated as more severe for oneself (see Fig. 1). The own-other-gap was particularly pronounced for conflicts that participants saw as related to divergent perceptions of behavior, possibly indicating a feeling of being misunderstood and unbridgeable differences. While the sample of conflict reports covered a wide range of situations, also including complementary positions related to the same situation (e.g., conflicts related to the WhatsApp Read Receipts feature from sender and recipient perspective), the pattern of individual experience highlights a central characteristic of social norm conflicts. From the individual perspective, oneself is the one suffering more than the other, and oneself has the right to feel offended and keep distance, expecting the other to apologize for the disrespectful behavior. As already discussed in the introduction, this may end up in an escalating process, where feelings of distance and possibly also aggression grow over time.

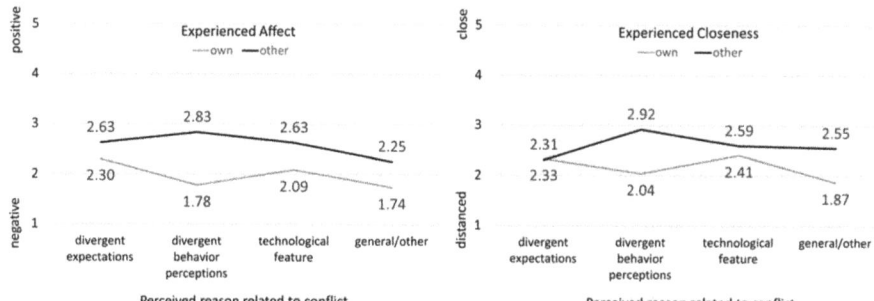

Fig. 1. Mean ratings for affect (left) and feelings of distance (right) for own and others' experience, categorized for different reasons of social conflict related to technology-use.

6 (Designerly) Ways Out of the Conflict?

Amongst others, disrespectful technologies as outlined above are a result of the currently established design process where social norms, potential norm conflicts or in general the emerging subjective experience are often not explicitly targeted. Instead, many companies seem to follow the logic of inventing ever-new features and filling the market with ever-new technological "solutions" – which might later turn out to be (social) problems. Messenger read receipts features resulting in unwanted side effects such as a pressure for immediate and consequently often overhasty and ambiguous answers are just one example. Such side effects could have been anticipated if designers felt responsible to reflect on the effects of technological features in full consequence. As already remarked by others [13], being aware of the normative powers of design is crucial. As Hassenzahl [13] (p. 63) states: "Each product is a proposition, and we cannot escape the fact that it has the power to change how people feel, think, and act. To do this consciously is important." Following this line of thought, it appears natural to also reflect on consequences for social norms. In analogy to Watzlawick's theorem that "one cannot not communicate" [14] designers need to acknowledge that "one cannot not influence", meaning that their products will automatically interfere and possibly change social interaction. One might argue that considering the normative power of design bears the risk of paternalism: Designers deciding at their own discretion how they would like to change society through their technologies, holding a great potential for political power. However, the power of design is not created by considering it – it exists either way, but might have unwanted effects if ignored. Hence, the most responsible thing for designers is to use this power in a conscious way. This includes deliberately forming social interaction and reflecting on which norms are desirable and which are not – which norms are worth protecting, and which we might want to get rid of.

Different design considerations are relevant depending on the context: While conservation of existing socials norms can be a goal in some cases, in others, technology design could address likely conflicts and help to prevent these. Given that the most severe social conflicts in our survey were often related to divergent behavior perceptions, this is one issue technology could address in its design. Obviously, users do not reflect their own behavior in the same degree that observers do (e.g., people answer a phone call in public and speak way too loud, unaware of their inappropriate behavior). A support of self-monitoring and feedback could help leveling such discrepancies, for example in online discussion forums. As Voggeser et al. [9] discuss, a failure to apply to social norms in online discussions can also be related to a lack in self-control, whereby the users' available self-control capacity may depend on the properties of the relevant communication platform (e.g., extraneous self-control demands through a confusing interface). In addition, we should probably think of approaches that support a shared understanding of social norms and contradict norm confusion. An example could be technology design supporting shared interpretations in analogy to non-digital world examples or making norms more unambiguous (e.g., through metaphor or usage rules). Altogether, the present design considerations also underline the need for interdisciplinary. For example, even if designers are motivated to design "respectful" technologies, they may lack the psychological knowledge to conceptualize it accordingly.

7 Conclusion and Future Research Directions

Digital technology has the power to connect people across continents. It facilitates contact between different nations and creates an advanced basis for intercultural influence and understanding. At the same time, the digital space opens opportunities for new subcultures emerging every day and existing in parallel, developing own norms and views on what is appropriate. Often, norm differences as a reason for conflict may not even become obvious, just leaving people with a feeling of anger, irritation and somehow not being treated right. We believe it to be important to recognize this harmful potential.

For researchers in HCI, psychology and media research, this implies the consideration of social norm differences as a potential explanation for (seemingly) aggressive behavior and conflicts in the digital world. The present categorization may provide a helpful starting point for further research. However, besides the here presented basic types of conflicts, many additional factors and psychological mechanisms could be relevant and related to the scheme. One could be the fundamental attribution error, i.e., the neglect to account for situational factors in explaining others' behavior, as also reflected in some of the present study's narratives. For example, several participants described conflicts due to the misinterpretation of (missing) emoticons in others' messages. The recipient typically searches for emotional reasons for a message without smileys (Is he angry with me? What have I done wrong?), while later it turned out that the missing smileys were just due to the situation (being in a hurry, just texting a quick answer). Moreover, "norms" of appropriate technology usage may also vary within one person from situation to situation, depending on their current role. For instance, in one of our previous studies on smartphone usage routines [5], only 11% rated their own smartphone usage in social settings as critical, realizing how their phone use takes a lot of their attention. In contrast, 87% stated that the phone use of others hurts the conversation "frequently" or at least "sometimes". Such double standards for own versus others' behavior can be an additional reason for conflict. DAlso individual attitudes of responsibility attribution (i.e., do people perceive the technology as responsible for its effects – or the one using the technology?) could be relevant. If one does not feel responsible for the effects of his or her technology use, one may not see any reason to align it with social norms. Finally, future studies could explore whether there are typical patterns or processual sequences between different types of conflicts (e.g., norm fragmentation following norm erosion).

Besides the research perspective, for all of us in our role as technology users the present work may help to interpret others' behavior in a more understanding manner, realizing that there may be larger differences in social norms than one generally assumes among people living close by and sharing the same cultural background. For designers, we hope the present work to be helpful to consider and foresee potential norm conflicts, and address those in the conceptualization of technological features. User centered design from a holistic perspective implies to also consider the broader consequences for users and how to use technology in line with their psychological needs. One of these needs is feeling related and understood by relevant others, being treated with respect and in line with social norms. If technology supports such behavior, it may be termed respectful technology.

Acknowledgments. Part of this research has been funded by the German Federal Ministry of Education and Research (BMBF), project Kommunikado (FKZ: 01IS15040D).

References

1. Turner, J.C.: Social Influence, vol. xvi. Thomson Brooks, Belmont (1991)
2. Przybylski, A.K., Weinstein, N.: Can you connect with me now? How the presence of mobile communication technology influences face-to-face conversation quality. J. Soc. Pers. Relatsh. **30**(3), 237–246 (2013)
3. Sparknotes. http://www.sparknotes.com/sociology/society-and-culture/section3/
4. Moncur, W., Orzech, K.M., Neville, F.G.: Fraping, social norms and online representations of self. Comput. Hum. Behav. **63**, 125–131 (2016)
5. Diefenbach, S., Christoforakos, L., Ullrich, D.: Digitale Disbalance - Herausforderungen der Smartphone-Ära. Wirtschaftspsychologie aktuell **17**(3), 36–42 (2017)
6. Sabra, J.B.: "I Hate When They Do That!" Netiquette in mourning and memorialization among Danish Facebook users. J. Broadcast. Electron. Media **61**(1), 24–40 (2017)
7. Lofland, L.H.: The social shaping of emotion: the case of grief. Symb. Interact. **8**, 171–190 (1985)
8. Urban Dictionary. https://www.urbandictionary.com/define.php?term=Virtue%20Signalling
9. Voggeser, B.J., Singh, R.K., Göritz, A.S.: Self-control in online discussions: disinhibited online behavior as a failure to recognize social cues. Front. Psychol. **8**, 2372 (2017)

10. Lee, M.K., Takayama, L.: Now, i have a body: uses and social norms for mobile remote presence in the workplace. In: Proceedings of the SIGCHI Conference on Human Factors in Computing Systems, pp. 33–42. ACM (2011)

11. Preece, J.: Etiquette online: from nice to necessary. Commun. ACM **47**(4), 56–61 (2004)

12. Ars Technica Open Forum: How do you address someone in your reply to their emails? https://arstechnica.com/civis/viewtopic.php?f=23&t=1244027

13. Hassenzahl, M.: Experience design: technology for all the right reasons. Synth. Lect. Hum. Cent. Inform. **3**(1), 1–95 (2010)

14. Watzlawick, P., Bavelas, J.B., Jackson, D.D.: Pragmatics of Human Communication: A Study of Interactional Patterns, Pathologies and Paradoxes. WW Norton & Company, New York (2011)

The Effects of Icon Characteristics
on Search Time

Mick Smythwood$^{(\boxtimes)}$ and Mirsad Hadzikadic

University of North Carolina at Charlotte, 9201 University City Blvd,
Charlotte, NC 28223, USA
{ksmythwo, mirsad}@uncc.edu

Abstract. The use of icons is ubiquitous in today's computing world. A handful of icon characteristics have been found to predict icon performance. Visual complexity, concreteness, familiarity, and aesthetic value have all been shown to affect performance in searching for and locating icons. We replicate previous studies that examined icon characteristics' effects on search time but use our own up-to-date icon stimulus set taken from existing mobile application icons. Our results verify most previous findings. The results on icon attractiveness, however, contradict findings that the advantage provided by attractiveness is only present when the icon is already difficult to find. We find the opposite—that simple icons alone are afforded an advantage by also being attractive. These confounding findings provide insight into the fundamental differences between mobile icons and icons traditionally used in search experiments.

Keywords: Aesthetic · Human-computer interaction
Human factors · Icon design · Visual complexity · Visual search

1 Introduction

Icons can be found on desktops, on mobile devices, and even on traffic signs. The use of the icon as a menu option on mobile devices is pervasive. Therefore, studying icon characteristics that contribute to icon usability benefits both designers of icons as well as users of icon interfaces. Numerous recommendations and guidelines exist concerning mobile application icon design. Both Microsoft and Apple recommend designers focus on simplicity in developing quality icons for use on personal computers, tablet and mobile phone devices. Microsoft recommends using a low level of complexity in icon design. Similarly, Apple's design guidelines recommend designers "embrace simplicity."

Visual complexity of an image correlates strongly with visual search—the more complex the image, the longer it will take to find that image in an array of other images [1]. These findings support existing guidelines for simplicity in icon design. Furthermore, icon complexity ratings correlate significantly with their respective search times [1].

Guidelines for application icon design often promote another icon characteristic, concreteness—how realistic an icon looks, or how similar pictorially the icons are to their counterparts in the real world [1]. Since a realistic icon is typically more complex than an abstract icon, should designers concentrate on making icons abstract instead of

© Springer International Publishing AG, part of Springer Nature 2019
T. Z. Ahram and C. Falcão (Eds.): AHFE 2018, AISC 794, pp. 57–67, 2019.
https://doi.org/10.1007/978-3-319-94947-5_6

concrete? This would make sense according to the original recommendations by Microsoft and Apple. How then can icon designers represent complex functions without increasing the level of complexity inadvertently?

The effects of complexity and concreteness combine to have an effect on icon usability. Balancing these two icon characteristics in icon design has been studied previously [1, 2]. Besides concreteness and complexity, aesthetic appeal also affects users' search of icons [3].

2 Purpose

Creating icons that are easily found promotes usability. Because the functions and applications behind mobile application icons are themselves increasingly more complex, the icons similarly have increasing levels of visual complexity. Mobile gaming applications are a perfect example of this trend. Additionally, as the number of mobile applications increases, the design of the icons representing them increases in complexity as well. As the design space narrows with every new icon, icon designers naturally create more complex icons than they did before. It follows then that an understanding of how an icon's visual complexity interacts with other icon characteristics stands to benefit the designer, especially in mobile computing.

3 Related Work

Visual perception research has primarily focused on three main icon characteristics that affect performance: visual complexity, concreteness, and familiarity [1, 2, 4]. Since the time to locate an icon in an interface is indicative of its ease of processing [5, 6], discovering icon characteristics that correspond strongly with search time becomes paramount.

Concreteness and complexity were once considered the same icon characteristic. Garcia's group actually went as far as to develop a concreteness metric [7]. It measured the complexity of icons by counting lines, arcs, letters, etc. Recent studies have demonstrated how complexity and concreteness are separate icon characteristics that affect performance in different ways [1, 8–10].

Visual complexity refers to the amount of detail or intricacy within an icon whereas concreteness/abstractness refers to how real it appears. Both of these icon characteristics have been used in explaining the perception of visual complexity [1]. The fact that icon concreteness allows users to access existing knowledge about real world objects infers that concreteness involves meaning for the users. If being able to ascribe meaning to an icon made it more usable, then why do many documented guidelines recommend simplicity in icon design when concrete icons are typically more complex? Byrne discovered that simpler icons can be identified more easily than complex icons and furthermore, that visually simple icons are easier to find in visual search [11]. Although visual complexity does not directly relate to icon identification, it is greatly important in visual search [1, 12]. Perceived icon complexity has more to do with search efficacy than concreteness does.

Previous work found that when the concreteness and complexity of icons have been properly controlled, they each behave in very different ways in their effects on usability [1, 12]. His team tested how icon concreteness and complexity affect user performance. The effects of concreteness were mainly associated with the initial grasp of meaning, whereas visual complexity effects persisted and were connected to search efficacy. They used a search-and-match task where participants are presented with an icon (or its function name) and are asked to select the corresponding icon from a search set.

Although icon complexity has been shown to affect search time to a greater extent than concreteness, we include the concreteness characteristic for examination since it has a long history and has been shown to have some effect on search time.

In addition, recent research has focused on icon appeal and its interplay with each of the three main icon characteristics relevant to search performance: visual complexity, concreteness, and familiarity [3, 13–16]. For example, visually complex icons were searched for and located better if the icon was also appealing. For simple icons, there was no advantage found by increasing aesthetic appeal. This study also found that aesthetic appeal had significant interactions with both complexity and concreteness [3].

4 Hypotheses

Given what we know from previous studies, we posited the following hypotheses:

1. Complex icons are found more slowly than simple icons.
2. Neither concreteness nor appeal alone contribute significantly to search time.
3. Concreteness and attractiveness have a significant interaction.
4. Complexity and attractiveness have a significant interaction.
5. Attractive, complex icons are found faster than unattractive complex icons.

5 Method

5.1 Materials

Researchers concerned with investigating icon search and identification typically choose an icon stimulus set already in wide use to provide results that can be easily compared with others' research [4, 17]. While this is a great way to communicate findings effectively, in this study, we selected real-world mobile application icons to provide a more up-to-date stimulus set that is more indicative of icons used in everyday life. Old stimulus sets such as Snodgrass and Vanderwart's set or McDougall's icon set consist of black and images [17]. Our stimulus set consisted of grey-scale, mobile application icons from Google's GooglePlay and Apple's AppStore.

Mechanical Turkers rated 200 icons on visual complexity, concreteness, familiarity, and aesthetic appeal. Instructions for rating icons on the four characteristics of interest were as follows:

(i) Visual Complexity: Rate the icon's visual complexity, its level of detail (-2 = very simple, 2 = very complex)

(ii) Aesthetic Appeal: Rate the aesthetic value, beauty, attractiveness of the icon (–2 = very unappealing, 2 = very appealing)

(iii) Familiarity: Rate how familiar you are with the icon, or how often you have seen it before (–2 = very unfamiliar, 2 = very familiar)

(iv) Concreteness: Rate the concreteness/abstractness of the icon, how realistic it looks (–2 = very abstract, 2 = very concrete).

Sixty-four out of the 200 icons were used in the actual experiment. By choosing less familiar icons from the larger sample, we were able to keep familiarity a relative constant in the stimulus set. Icons were then selected for each group according to whether they were complex/simple, attractive/unattractive, and concrete/abstract.

Each of the eight groups resulting from the different combination of complex/simple, concrete/abstract and attractive/unattractive consisted of eight icons each. The resulting groups were complex-attractive-concrete (CAC), complex-attractive-abstract (CAA), complex-unattractive-concrete (CUC), complex-unattractive-abstract (CUA), simple-attractive-concrete (SAC), simple-attractive-abstract (SAA), simple-unattractive-concrete (SUC), and simple-unattractive-abstract (SUA). An example from each group is presented in Fig. 1. Table 1 lists the acronyms for the groups for reference.

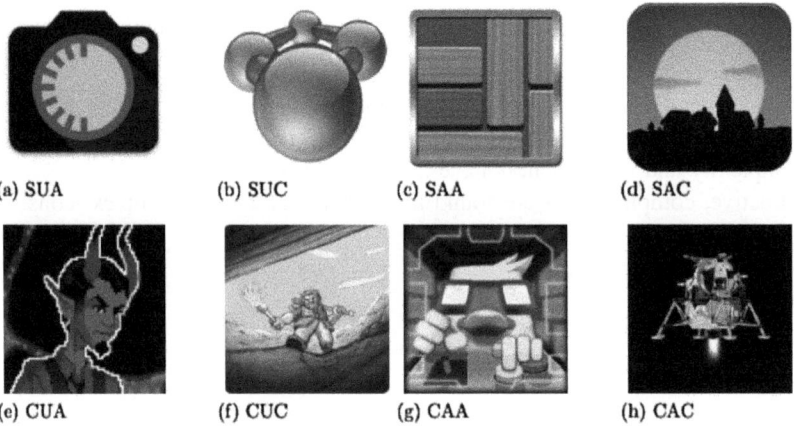

| (a) SUA | (b) SUC | (c) SAA | (d) SAC |
| (e) CUA | (f) CUC | (g) CAA | (h) CAC |

Fig. 1. Examples of icons used. One icon from each group.

The groups' descriptive statistics are listed in Table 2. A set of univariate ANOVAs revealed differences between the icon groups in terms of visual complexity, Concreteness, aesthetic attractiveness, and no differences between groups for familiarity. Since familiarity has an overriding effect on search time for icons [1], we kept this variable constant in our experiment.

Table 1. Acronyms for the eight groups across complexity, attractiveness, and concreteness.

Acronym	Complexity	Attractiveness	Concreteness
SUA	Simple	Unattractive	Abstract
SUC	Simple	Unattractive	Concrete
SAA	Simple	Attractive	Abstract
SAC	Simple	Attractive	Concrete
CUA	Complex	Unattractive	Abstract
CUC	Complex	Unattractive	Concrete
CAA	Complex	Attractive	Abstract
CAC	Complex	Attractive	Concrete

Table 2. Means for each of the four icon characteristics used to compose the complexity-attractiveness-concreteness groupings (on a scale of –2 to 2).

	Complexity	Attractiveness	Concreteness	Familiarity
SUA	–0.88	–0.35	–0.91	–0.79
SUC	–0.02	–0.40	0.26	–0.68
SAA	–0.61	0.34	–0.63	–0.60
SAC	–0.19	0.26	0.49	–0.24
CUA	0.29	–0.54	–0.64	–0.66
CUC	1.04	–0.24	0.78	–0.71
CAA	0.68	0.38	–0.24	–0.70
CAC	0.86	0.43	1.00	–0.38

5.2   Participants

Fifteen Mechanical Turkers participated in the experiment. Nine male and six female Americans completed the online experiment. Their ages ranged from 25 to 64 with the majority of participants falling in the 25–34 age range. All participants were experienced mobile phone users with normal or corrected-to-normal vision.

5.3   Design

This study replicated icon visual search experiments conducted by McDougall and others. These studies typically included blocks of trials in order to examine how performance changed over time. This sort of experiment conducted in a controlled environment (a lab) did not easily extend to online trials. Original experiments required participants to come back day after day to perform blocks of trials over time. This was not a realistic expectation from Mechanical Turkers. Each HIT or Human Intelligence Task began and ended in the same session.

Therefore, instead of $2 \times 2 \times 9$ group design for example (complex/simple, attractive/unattractive, nine blocks of trials over several days), we chose a $2 \times 2 \times 2$ experimental design. Each participant received the same, randomized treatments during a single session. Our design was within-subjects with a dependent variable of response time. The three primary icon characteristics of complexity, concreteness, and aesthetic served as independent variables.

5.4   Procedure

After the participant accepted the Mechanical Turk HIT, they were told they would be presented with an icon for 2 s before they would be expected to click the "Next" button to continue to a 3×3 matrix of icons (See Fig. 2 for an example trial). They were instructed to click on the target icon as quickly as possible once they clicked the "Next" button. Their first choice was the only icon selection they were allowed to make, after which they could continue to the next trial by clicking another "Next" button.

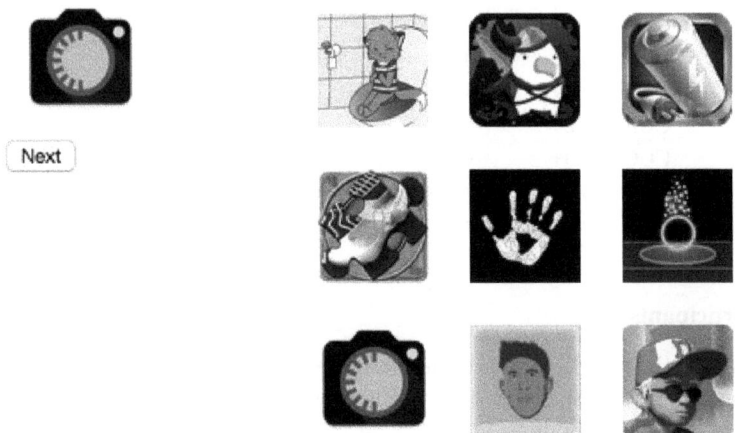

Fig. 2.   Example of an experiment trial.

Each participant was compensated $3.00 for completing the experiment. From the beginning, they were informed that if too many of their icon selections took too long, they would not receive compensation. Additionally they were informed that if they did not complete all trials, this too would prevent them from receiving payment. These efforts were made to encourage attentiveness and provide incentive for completing all trials.

Participants were encouraged to pause, if necessary, before continuing to the next trial. This instruction was included to encourage the participant to focus when presented with the new target icon and to allow them to take a break if needed. Since the online experiment was conducted in a single session, having the ability to pause served to prevent the participant from being distracted during the search task.

6   Results

Errors accounted for 2.10% of all trials. There were no differences in errors between any of the eight conditions ($p > 0.05$). Correct group means are shown in Table 3.

Table 3. Means and standard deviations for each complexity-attractiveness-concreteness group in milliseconds.

	Mean	Std. dev.
Concrete-Attractive-Concrete	1723	729
Concrete-Attractive-Abstract	1855	878
Concrete-Unattractive-Concrete	1778	734
Concrete-Unattractive-Abstract	1636	718
Simple-Attractive-Concrete	1504	666
Simple-Attractive-Abstract	1469	599
Simple-Unattractive-Concrete	1603	636
Simple-Unattractive-Abstract	1569	786

We used an alpha level of 0.05 for all statistical tests and partial eta-squared as a measure of effect size. Bonferroni corrections were used throughout.

A within-subjects ANOVA on correct RT (Table 4) showed no significant three-way interaction between icon characteristics, $F(1, 116) = 3.87$, $p > 0.05$, $\eta^2 = 0.03$. The main effect of Complexity was significant, $F(1,116) = 20.30$, $p < 0.01$, $\eta^2 = 0.15$. Neither concreteness nor aesthetic value's main effects were significant: concreteness, $F(1,116) = 0.34$, $p > 0.05$, $\eta^2 = 0.00$; aesthetic value, $F(1,116) = 0.07$, $p > 0.05$, $\eta^2 = 0.00$.

Table 4. Tests of within-subjects contrasts.

	F	Sig.	η^2
Complexity	20.30	0.00	0.15
Aesthetic	0.07	0.80	0.00
Concreteness	0.34	0.56	0.00
Complexity * Aesthetic	10.78	0.00	0.09
Complexity * Concreteness	0.16	0.70	0.00
Aesthetic * Concreteness	5.51	0.02	0.05
Complexity * Aesthetic * Concreteness	3.87	0.05	0.03

Interestingly, there was a significant interaction between Complexity and Aesthetic Value ($F = 10.78$, $p < 0.01$, $\eta^2 = 0.85$), with shorter RT for appealing than unappealing simple icons, $t(233) = -2.70$, $p < 0.01$. There was no such difference for complex icons.

7 Discussion

Not surprisingly, the first hypothesis proved true. As in most studies comparing visual complexity and search time, the difference in complex versus simple icons was significant. This test also served as a sanity check for this study. Ensuring that there was a difference for complexity, we can then examine possible interactions between other icon characteristics and complexity.

In keeping with previous findings, concreteness did not have a significant effect on search time. Others have found noticeable differences in the effect of concreteness on search time over blocks of trials. Since our approach focused on effects in a single trial, it made sense that concreteness did not reveal any real effect on search time.

Attractiveness alone did not have a significant effect on search time. So the second and third hypotheses that neither concreteness nor attractiveness would have a significant effect on search time proved true.

Unlike in previous studies, concreteness and attractiveness did not have a significant interaction. The third hypothesis was not proved correct.

Complexity and attractiveness had a significant interaction, proving the fourth hypothesis true.

The most interesting finding from the results was the significant difference found between simple-appealing and simple-unappealing icon groups. This trend was found in older studies but for complex icons. Complex-appealing icons were found faster than complex-unappealing icons.

Oddly enough, this trend among complex icons was not apparent in our study. There was no significant difference found between appealing and unappealing icons for the complex group. The last hypothesis was not proved true.

Another study, largely replicated here, comparing aesthetic value and concreteness' effects on search time, found a significant interaction between them [3]. Contrary to these findings, we discovered no significant interaction. Since familiarity did not vary significantly between icon groups, it makes sense that concreteness, which has traditionally been considered a correlate of familiarity, did not prove especially impactful in this study. Aesthetic appeal, traditionally an icon characteristic associated with familiarity, did not have a significant effect by itself either; however, aesthetic appeal did have a significant interaction with complexity.

In previous studies in controlled environments, accurate measurement of search time was much easier to accomplish than in an online survey. This may account for difference in results between our study and others.

8 Conclusions

According to the results from our study, we can give recommendations to icon designers that want to create easily findable icons. The number one recommendation would be to keep the icon simple, rather than complicate perception with complexity.

In addition, since we found attractive icons to be located more quickly than unappealing ones in the case of simple icons, we conclude that the aesthetic appeal of an icon does influence its performance. Although the former study revealed that aesthetic appeal affected task performance only when the task was difficult, such as when the icon was complex, our study's findings reflect that icons that are not difficult to find because of their simplicity can also benefit from the advantage afforded by attractiveness. Motivation to search for attractive icons may have been greater for simple icons.

The present study supports the proposition that aesthetic attractiveness can bias perceptual systems by giving priority to attractive stimuli. Emotion can give processing priority to positive or negative emotion in face processing [18, 19]. The effect of emotion on face processing, then, is apparent in our study for simple icons where there is little to no task difficulty. Visual processing of stimuli like faces with high evolutionary relevance seems to translate to the role of attractiveness on performance, even if it is only in icon search.

It is odd, however, that the same advantage was not found for complex icons. In reconciling the difference in findings between complex icons that perform better when also aesthetically appealing, and simple icons that perform better when attractive, it is worth noting that our stimulus set included icons that were more complex in general. Black and white icon images are marginally simpler than the graphic mobile application icons used today. Making one of these graphically complex icons aesthetically pleasing did not significantly assist in its performance.

Another source of difference between study results may originate from the wording used to collect ratings on aesthetic value. In the previous study aesthetic value was defined as "appealing" and "unappealing," whereas in our study we used the words "attractive" and "unattractive" [3]. Although appealing and attractive are in the same family of meaning, the connotation of "attractiveness" may differ from the connotation of "appealing" to participants.

An important distinction to make between studies is that the previous study used an old set of icons whose main characteristic ratings were collected several years ago. This study's appeal ratings were collected recently. In our study, individuals recently and during a single session, rated a corpus of 200 icons from which our stimulus set was chosen. Our stimulus set, as opposed to their old icon set, had greater cohesion among ratings as they were all made together.

Reducing the speed of processing in locating icons might seem a minimal advantage to overall usability. However, users notice performance costs as small as 150 ms [20]. Since icon search is a task users perform repeatedly often in the same session, time advantages add up over time. This study's findings reveal that simplicity and attractiveness improve performance in tasks common to real-world settings, where searching for and locating icons within interfaces is a constant requirement.

Implications for Icon Design. Icons designed with particular design characteristics in mind facilitate the visual processing involved in icon menu search. This research has shown that the visual search for an icon is likely to be affected by:

(i) the icon's complexity, with simple icons found faster than complex ones
(ii) the icon's appeal, which may not affect search times for *all* icons but may affect users' attitudes towards the display [3].

9 Future Plans

While the effects observed of icon complexity on interface search were very much in agreement with previous research, the effects of icon appeal on visual search appeared to be much more equivocal. One possible reason for the lack of an appeal effect for complex icons was that it is a challenge to create a stimulus set of icons with pronounced differences in appeal ratings. This is particularly true of well-designed icons currently used on mobile applications. Our next aim therefore is to examine the effect of icon appeal with a stimulus set consisting of icons differing widely on appeal.

The role of icon appeal in determining users' abilities to identify icons, their emotional responses, and their perceptions of icon interface usability, will be examined.

Previous studies have examined performance over blocks of trials as well. Usability studies conducted in a controlled environment will be necessary for a more detailed look at icon search over repeated exposure.

The original icons from our stimulus set were in color even though we used a greyscale version of each icon in the actual experiment. It would be interesting to study the difference that color makes in terms of complexity and aesthetic.

In the end, collecting data on how icon characteristics affect the search and use of icons can help in predicting overall icon performance.

Acknowledgements. The authors would like to thank Mark Faust for his assistance with the statistics involved in creating the icon stimulus set.

References

1. McDougall, S.J.P., de Bruijn, O., Curry, M.B.: Exploring the effects of icon characteristics on user performance: the role of icon concreteness, complexity and distinctiveness. J. Exp. Psychol. Appl. **6**, 291–306 (2000)
2. McDougall, S., Isherwood, S.: What's in a name? The role of graphics, functions, and their interrelationships in icon identification. Behav. Res. Methods **41**, 325–336 (2009)
3. Reppa, I., McDougall, S.: When the going gets tough the beautiful get going: aesthetic appeal facilitates task performance. Psychon. Bull. Rev. **22**, 1243–1254 (2015)
4. McDougall, S.J.P., Curry, M.B., Bruijn, O.: Measuring icon and icon characteristics: Norms for concreteness, complexity, meaningfulness, familiarity and semantic distance for 239 icons. Behav. Res. Methods Comput. Instrum. **31**, 487–519 (1999)
5. Hertwig, R., Herzog, S.M., Schooler, L.J., Reimer, T.: Fluency heuristic: a model of how the mind exploits a by-product of information retrieval. J. Exp. Psychol. Learn. Mem. Cognit. **34**, 1191–1206 (2008)
6. McDougall, S. Reppa, I.: Ease of processing can predict icon appeal. In: HCII 2013, Las Vegas, USA. Springer, 21–26 July 2013
7. Garcia, M., Badre, A.N., Stasko, J.T.: Development and validation of icons varying in their abstractness. Interact. Comput. **6**(2), 191–211 (1994)
8. Donderi, D.C.: Visual complexity: a review. Psychol. Bull. **132**, 73–96 (2006)
9. Forsythe, A.: Visual complexity: is that all there is? In: Engineering Psychology and Cognitive Ergonomics. LNCS, vol. 5639, pp. 158–166 (2009)

10. Forsythe, A., Mulhern, G., Sawey, M.: Confounds in pictorial sets: the role of complexity and familiarity in basic-level picture processing. Behav. Res. Methods **40**, 116–129 (2008)
11. Byrne, M.B.: Using icons to find documents: simplicity is critical. In: Interact '93 and CHI '93 Conference on Human Factors in Computing Systems, pp. 446–453, ACM, New York (1993)
12. Isherwood, S.J., McDougall, S.J., Curry, M.B.: Icons identification in context: the changing role of icon characteristics with user experience. J. Hum. Factors Ergon. Soc. **49**, 465–476 (2007)
13. Forsythe, A., Nadal, M., Sheehy, N., Cela-Conde, C.J., Sawey, M.: Predicting beauty: fractal dimension and visual complexity in art. Br. J. Psychol. **102**(1), 49–70 (2011)
14. Graham, D.J., Redies, C.: Statistical regularities in art: relations with visual coding and perception. Vis. Res. **50**(16), 1503–1509 (2010)
15. Hagerhall, C.M., Purcell, T., Taylor, R.: Fractal dimension of landscape silhouette outlines as a predictor of landscape preference. J. Environ. Psychol. **24**(2), 247–255 (2004)
16. Joosten, J.J., Soler-Toscano, F., Zenil, H.: Fractal dimension versus computational complexity. arXiv preprint arXiv:1309.1779 (2013)
17. Snodgrass, J.G., Vanderwart, M.: A standardized set of 260 pictures: norms for name agreement, image agreement, familiarity and visual complexity. J. Exp. Psychol. Hum. Learn. Mem. **6**, 174–215 (1980)
18. Gray, W.D., Boehm-Davis, D.A.: Milliseconds matter: an introduction to microstrategies and to their use in describing and predicting interactive behavior. J. Exp. Psychol. Appl. **6**(4), 322–335 (2000)
19. Becker, D.V., Anderson, U.S., Mortensen, C.R., Neufeld, S.L., Neel, R.: The face in the crowd effect unconfounded: happy faces, not angry faces, are more efficiently detected in single and multiple target visual search tasks. J. Exp. Psychol. Gen. **140**(4), 637 (2011)
20. LeDoux, J.E.: The Emotional Brain. Simon and Schuster, New York (1996)

The Limited Rationale in Decision Making, Impacts on the Evaluation of Artifacts in the Design Process

Walquir Fernandes[✉], Walter Correia, and Fabio Campos

Universidade Federal de Pernambuco, Pós Graduação em Design, Av.
Acadêmico Hélio Ramos, 50670-420 Recife, Brazil
fernandeswalquir@hotmail.com, wfmc10@gmail.com,
fc2005@gmail.com

Abstract. Would the user able to make decisions rationally? If not, how, then, are your decisions actually made? What happens if the decision is made on the basis of intuition? Situations involving judgment and decision-making are daily in the lives of human beings, including in matters of consumption. The judgment and decision-making are complex functions that imply in the analysis of the characteristics of each of the options for a particular decision-making task, as well as the estimation of the consequences of the choice to be made and the origin of these studies is Microeconomics. In cognitive psychology, decision-making studies have been devoted to investigating how human beings make decisions in reality, not following rules, but seeking rationality within limits. This article aims to bring to light some discussions that elucidates more clearly how these decision making processes can be efficient.

Keywords: Limited rationale · Limited rationale in decision making
Anchoring heuristics

1 Introduction

Would the user be able to make decisions rationally? If not, how their decisions are actually made? What happens if the decision is made by intuition?

Situations that involve judgment and decision making are present everyday in the lives of human beings, even in a matter of consumption. Decision making, can be a simple task in some circumstances, such as going out for lunch or ordering phone delivery, but it involves complex cognitive processing, as the existing options have to go through a judgment process, followed by decision-making between those alternatives. The judgment and decision-making are complex functions that imply the analysis of the characteristics of each options for a particular decision-making task, as well as the estimation of the consequences of the choice to be made [1, 2], and the origin of these studies is the Microeconomics.

In Design, Herbert Simon, a Nobel laureate in economics, an American economist known for his work in many fields such as cognitive psychology, economics and design, played an important role in understanding limited rationality in the field of

© Springer International Publishing AG, part of Springer Nature 2019
T. Z. Ahram and C. Falcão (Eds.): AHFE 2018, AISC 794, pp. 68–78, 2019.
https://doi.org/10.1007/978-3-319-94947-5_7

design. [3], describing what he named as "the sciences of the artificial", drew attention to the fact that Design as an area that it is the heart of professional training, with activity centered on modifying existing situations, leading to those that are preferred or desired by the market, using, for that, artifacts.

2 Theoretical Rationale and State of the Art of Limited Rationale in User Decision Making

2.1 Consumer Behavior

Currently in the field of marketing and design these areas provides the most academic research is the area that works with consumer behavior. The theory of consumer behavior originated in economic theory, based on the microeconomic principle that the consumer is endowed with limited income, which must be allocated between goods and services in order to seek a constant maximization of their "welfare" [4].

According to [4], the first studies in this area concentrated their efforts in the attempt to make consumers became loyal and thus acquire competitive differentials for the organizations. The authors further complement that, the understanding of their profile, their motivations and expectations about a particular product would be crucial. From the 1980s, research in the area began to include aspects such as: symbolism, past experiences and impacts of religions and emotions [5]. In addition, [6] point out that, consumer behavior can be understood as the result of a process of convergence in several factors, among which social, psychological, personal and cultural factors, that this last one, according to [7] exerts greater influence on individual behavior. Currently, this field of study has become broader, especially from contributions received from Psychology and Neuroscience, which study non-declarative aspects in decision making.

The Rational Versus Emotional Dichotomy in Decision Making. The processes of intuition, commonly referred in the literature as an opposition to reasoning processes, seem to be fundamental for the understanding of limited rationality [8]. According to the authors, the central point of the distinction between intuition and reasoning seems to lie in the interference of affection in decision making. In order to explore how individuals make decisions and what leads them to make logical reasoning errors, some researchers focused on the study of the operations of intuition and reasoning [9], a distinction previously made by [10] such as System 1 and System 2, respectively.

This division of a theory based on associative decision-making systems was first described by [11] who classified humans into two distinct groups of decision makers: those who are driven by intuition and emotional involvement, and those who take decision based on a rational analysis of the facts. To facilitate understanding, [10] labeled System 1 and System 2 these ways as people structure their thinking in the face of a decision-making situation. While System 1 is based on intuition, System 2 is more strongly based on reason.

Anchoring Heuristics. Behavioral economics is based on the science of heuristic judgment (or mental shortcuts, golden rules) that most people trust reflexively [12].

Heuristics are characterized as an "intuitive, fast and automatic system" [13], which "reduce the complex tasks of evaluating probabilities and prediction of values to simplify the operations of judgment' [14]. Although the use of golden rules reduces cognitive and time constraints, sometimes these golden rules can lead to serious and systematic errors, such as prejudices, biases and fallacies in decision making [14].

The idea of heuristics was originally created by [3], who proposed a behavioral model of rational choice, using arguments of a "limited" rationality, where decisions are derived from processes of dynamic adjustment, both external (environmental)and internal factors (human characteristics) factors. [15]. The anchorage heuristic, is a phenomenon very common in human judgments, with strong influence in these judgments.

Following the study [14] many studies (see Table 1) have illustrated the prevalence of the anchor effect in human decision processes.

Table 1. Literature available on the anchoring effect in the various areas

Area	Author(s)	Examples researched/objectives used
General knowledge/factual questions	Blankenship et al. (2008), Wegner et al. (2001)	The age he had when George Washington died The initial annual salary variation of graduates in the United States The age she had when Amelia Erhardt disappeared by riding around the Earth
	Tersky e Khaneman (1974)	What percentage of African nations make up the United Nations?
Probable estimates	Chapman e Johnson (1999 - experimento 2)	What is the likelihood next year of sending US troops to the areas of former Yugoslavia?
	Plous (1989)	What is the possibility of a nuclear war?
Decisões judiciais	Englich e Mussweiler (2001), Englich e Soder (2009), Englich et al. (2005, 2006)	Indication of a defense lawyer in the sentence issued (incarcerated or conditional sentence) Point out the duration of the sentence that a judge should convict (imprisonment or conditional penalty)
	Hastie et al. (1999), Marti e Wissler (2000)	Judge the punishment and liability for damages and assign a dollar value when damages are assessed
Evaluation and purchase decision	Ariely et al. (2003)	Willingness to pay for a diversity of products Make financial evaluations under loud noise
	Mussweiler et al. (2000)	Evaluation of the value of a used car manufactured 10 years ago
	Wasink et al. (1998)	Decision to purchase groceries

(continued)

Table 1. (*continued*)

Area	Author(s)	Examples researched/objectives used
Preditions	Critcher e Gilovitch (2008)	Estimates of an athlete's performance Forecasts of domestic sales of a given product Estimates of expenses in a restaurant
Negotiation	Galinsky e Mussweiler (2001)	The steps involved in negotiating the purchase of a drug factory
		Negotiation on payment of performance bonuses for employees
Self-efficacy	Cervone e Peake (1986)	How many items of an initial task (anagrams and graphs) did they think they are able to solve?

These have demonstrated the effect of anchoring in a variety of domains.

3 Synthesis of the Experiment (Anchorage and Adjustment Heuristics)

Impact of the Adjustment and Anchoring Effect on the responses of research questionnaires.

When organizing questionnaires items in attitude/opinion, do the questionnaire designers influence respondents' responses by simply choosing to organize their research in a specific way? We hypothesize that under specific conditions that occur frequently, respondents employ an anchoring and adjustment strategy in which their response to an initial research item provides a cognitive anchor from which they (insufficiently) adjust to respond to the subsequent item of the research questionnaire.

Three experiments suggest that respondents anchor and adjust insufficiently in certain situations, anchoring and adjusting lead to a greater inter-item correlation between adjacent items, and these inflated correlations may (spuriously) increase the reliability estimate of the scale they comprise and correlations with other measures. These effects are not consistently accounted for an "upper memory search" explanation. Organizing a research, researchers may want to combat this bias by mixing differently designed items, but with related constructs.

3.1   Context of the Experiment

The prevalence of opinion research within psychology is vast [16]. By studying the interviewees, psychologists-particularly in personality and social subfields-often attempt to measure attitudes, opinions, beliefs, and other "diffuse" constructs. To assess these types of constructs that are often unclear in the minds of respondents, difficult to access in their memories and/or unstable over time, scholars often try to mitigate the error by asking a series of questions that they later aggregate into one classification scale [17]. Despite the penetration of scales, researchers have little empirical guidance when deciding how to organize such sets of similar questions in a survey. Should they group all items in the same building sequentially, or should they mix items from different builds? What consequences come from one approach to the other?

To address these issues, it is important to understand how people respond to surveys. The prevailing theory is that respondents engage in four processes to develop responses to the research items. Specifically, respondents should understand the item that is requested, search their memories to retrieve relevant information, consolidate this information into a judgment, and select a response to present [18]. When fatigue or disinterest occurs during the response process, respondents may be tempted to take shortcuts or rely on heuristics to alleviate the cognitive load. In other words, respondents can get involved in the survey [19] called "satisficing" - failing to optimize the effort in conducting research.

This experiment investigates whether respondents satisfy (in a previously undocumented manner) relying on anchoring and adjusting when adjacent items in a survey are similar. We articulate how anchoring and tuning (as well as the concurrent explanation that focuses on memory retrieval) can apply to the research design context. Finally, we will present our research questions and the results of three studies designed to test these questions. As that the anchoring and adjustment phenomenon are generalized to the field of opinion research, this experiment contributes to the anchoring and adaptation of the literature. However, the main contribution of this experiment (and the focus of the discussion) is to better understand the impact that heuristic has on data fidelity and to provide practical guidance for Design research.

4   Research Questions

The present research uses a divided ballot project in which participants are randomly assigned to take one of two forms of a research. Form 1 places the pairs of similar items adjacent to each other, thereby facilitating anchoring and adjustment. Form 2 and Form 3 in the case of Experiment 3 separated the pairs of items of interest with interim items that were dissimilar in some way, assuming that this would attenuate the anchoring and adjustment. None of the scales examined will include items with reverse punctuation. We will use this research project to investigate to what extent the anchorage and adjusting heuristics occur in the questionnaire responses, what are the consequences for the data, and whether these consequences can be reasonably accounted for the explanation of the superior memory search.

Within each experiment, we will test three main research questions:

(1) To what extent do anchoring and (insufficient) adjusting occur between pairs of adjacent items that focus on similar topics and use similar response scales?
(2) Will anchoring and adjustment lead to greater correlations between items within the scale (and therefore to greater reliabilities assessed by the alpha coefficient)?
(3) Can the explanation of upper memory research adequately explain any differences in response patterns found between the two forms of survey?

5   Considerations

The analytical approach will be similar across all three experiments. We will describe the differences between the respondents and the differences between the pairs of relevant items in each search form throughout the results. As the distances between items in the two survey forms will be approximately distributed, we will report parametric tests of mean differences (t-tests and ANOVA) for these analyzes. For research questions 2 and 3 we will compare sets of correlation coefficients against each other. In these cases, we are reluctant to assume that these distributions are normal and therefore will conduct our statistical tests using non-parametric tests as a more conservative approach. Specifically, we will use Wilcoxon's rank test for search questions 2 and 4, where the correlations will be paired and the Wilcoxon (also known as Mann-Whitney-Wilcoxon or Mann-Whitney U) scoring test for the research question 3 where they do not paired.

6   Experiment 1

Initially we will investigate anchoring and adjusting within an assessment study of artifacts (SW Usability Assessment. We randomly assign professionals to two different forms of research.) We present the item pairs of interest to the respondents as being adjacent to each other in the Form 1 and non-adjacent in Form 2.

6.1   Method

Participants. The participants (N = 40) professionals of the area of diagnostic imaging that work with Magnetic Resonance Equipment in the states of Pernambuco, Alagoas and Paraiba.

6.2   Measures

First the author and a collaborator will develop 50 items to evaluate the usability of a Design artifact (usability in SW) to be applied in professionals who use SW in reference in five different areas with 10 items each. Based on the theory of artifact evaluation. The response anchors for the four constructions will be formed by adding each building marker to the following 5-point response anchors: not at all, slightly, moderately, completely, and extremely (e.g. "not at all aware", "slightly aware" ", etc.). The fifth scale, evaluating the frequency with which professionals take the measurements, will use "almost never", "from time to time", "sometimes", "often" and "almost all" response anchors.

Participants will be randomly assigned to complete Form 1 (n = 20) or Form 2 (n = 20). These forms will only vary in the order in which the items will be presented. Form 1, items will be grouped with questions that were intended to address the same construction being put together in a cohesive item block, while Form 2 mixed those

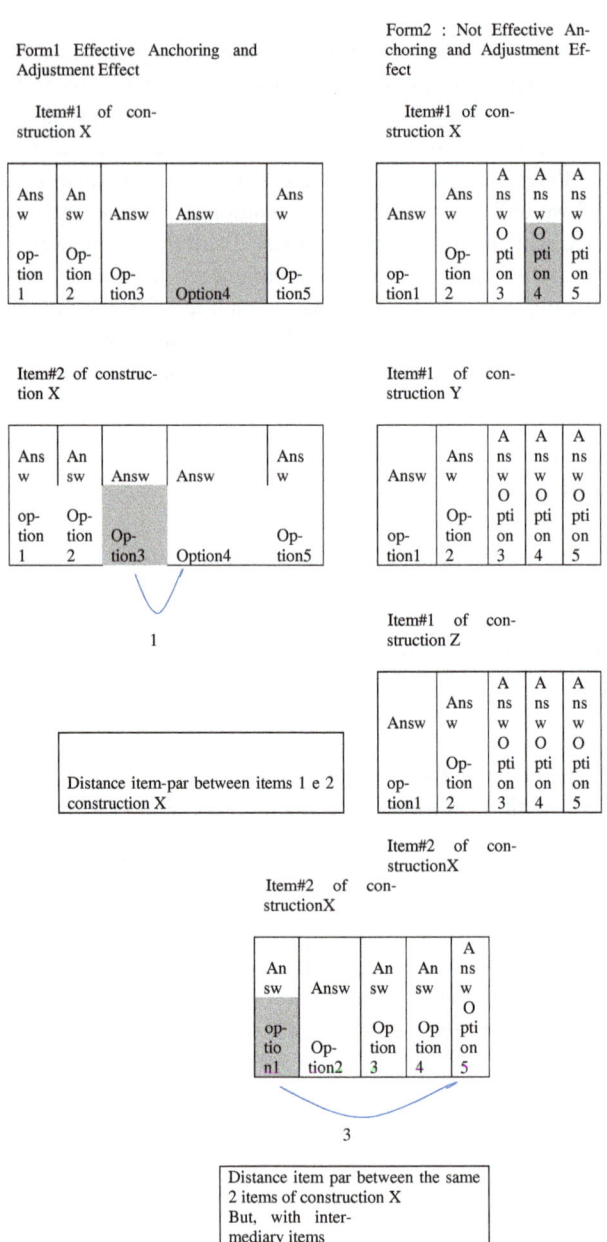

Fig. 1. Comparing how different organizations of identical research items can lead respondents to anchor and in some search conditions. (Gray cells indicate selected responses in the item pairs of interest.)

items so that items from the same construction are not adjacent. More specifically, Form 1 will present all items referring to the 1st area of artifact evaluation. Each subsequent construction proceeded in the same manner. This organization resulted in Form 1 with 20 pairs of items that were similar in the specific content of the question and in the formulation of response anchors. Thus, we examined each of these 20 pairs of items that were adjacent on Form 1 and not adjacent on Form 2.

6.3   Analysis of Results

Research Question 1. To investigate the extent to which anchoring and insufficient fit will occur, we will compare the difference in the distances between pairs of focal items between Form 1 (where items were adjacent) and 2 (where items were not adjacent). First, we will calculate the absolute value of the difference between the pairs of items of interest for each interviewee. In other words, if a respondent marked response option 4 for the initial item and the third response option for the subsequent item on the pair, the absolute value of the difference would be 1 (as shown on the left side of Fig. 1). Then the scores for each item - interest pair will be computed and aggregated so that each participant would receive a general "anchor and fit" score representing the average absolute difference between the pairs of items of interest to that person. These average scores will then be compared among respondents who completed Form 1 versus Form 2. If respondents anchored and then adjusted insufficiently on Form 1 as expected, then their mean anchorage and adjustment scores would be lower than for the respondents Form 2. Following this procedure, we will find solid evidence of anchorage and fit in the Formulary. By disaggregating these results to specific interest-item pairs, it will be verified that Form 1 will have minor differences between items.

7   Research Question 2

We will then investigate whether these insufficient adjustments will lead to stronger item correlations in Form 1 on Form 2 and whether these correlations will in turn affect the internal consistency of the scales. As expected, the correlations between pairs of adjacent items on Form 1 will be greater than when those items were not adjacent on Form 2. A Wilcoxon confirmation test would confirm that the mean correlation for all relevant item pairs on Form 1 will be significantly higher than for the corresponding item pairs on Form 2.

Using the Feldt test [20], we will compare the reliabilities of the five scales: area of evaluation artifact 1 (to be defined), area of evaluation artifact 2 (to be defined), area of evaluation artifact 3 (to be defined), area of evaluation artifact 4 (to be defined) and area of evaluation artifact 5 (to be defined), through the two forms. The reliability of Form 1 will be significantly higher than that of Form 2.

8 The Third Research Question

The third research question predicted that such differences in item pair distances, item pairs correlations, and reliability of scales could not be attributed to respondents who performed further memory surveys (although they could be explained by anchoring and adjustment). To test this possibility, we only analyzed the respondents who completed Form 1. If these individuals engaged in a superior memory search, then all inter-item correlations for items within that range should be similar. In other words, since all items belong to the same topic in Form 1, if a higher recovery process is responsible for those results, the average distance between items within a block should be approximately the same for pairs of adjacent items and not provided, where there are no intervention items on a different topic to stop the recovery process.

9 Conclusion and General Discussion and Implications for Researchers

Taken as a whole, these results illustrate that anchoring and tuning will occur in attitude/opinion questionnaires between adjacent items that use the same set of anchors responses and contain related content. Specifically, when respondents face items that are grouped according to the constructs they are intended to measure they invoke a heuristic in which they use their response to an initial item as an anchor. When responding to the subsequent item that is presented, they (insufficiently) conform to that anchor. These findings differ from the "straight answer," in which respondents mark the same response throughout a section or a full survey - we selected all respondents before beginning our analysis.

The concern with research researchers is that when anchoring and adjustment occur, data may be compromised, particularly for shorter scales (e.g. 3–7 items). Specifically, as respondents give artificially similar responses to adjacent items, they introduce error in their responses. This error will lead to spurious high correlations between items within the scale and may artificially inflate estimates of the internal consistency of the scale. In other words, researchers may be fooled into thinking that their scales are significantly more reliable than they really are.

For each experiment, we will also examine an alternative explanation that presenting conceptually similar items adjacent to each other could facilitate the cognitive capabilities of respondents [18, 21]. Potentially, this superior memory search process, not anchoring and tuning, could explain the similarity of respondents' responses. Although it remains plausible that this approach to organizing examinations facilitates searches of the respondents' memory, it will not explain the effects we find. In each experiment, when we will compare the adjacent and non-adjacent items within the focal lengths for Form 1 respondents, we will find that the adjacent distances between items will be smaller than their nonadjacent equivalents. As described above, these adjacent/nonadjacent differences should not result from differences in the way respondents search their memories - all issues within the item-block belong to the same construction - although these differences are expected as a consequence of anchoring and adjustment.

By relating the focal ranges of interest to other measures, we found no evidence that the (ostensibly) more reliable scales of Form 1 respondents produced correlations between stronger scales. On the contrary, we find substantial evidence of the exact opposite - in Experiments 2 and 3, correlations between focal scales and other measures will be stronger in the ways in which anchoring and adjustment will be mitigated.

We postulate that this potentially counterintuitive finding is due to the fact that the reliabilities were artificially inflated by anchoring and adjustment and that the true reliabilities of these measures are smaller than (or towards the lower limit) of the alpha coefficient estimate. The data [21], are also consistent with this conclusion. That the ordering of survey items can have such a substantial impact on the results of correlation findings demonstrates how important it is for researchers to address the order of items in the preparation of their surveys. In our final experiment, we will also get an insight into the behaviors associated with anchoring and tuning during a search administration. Specifically, since anchoring and tuning are mental shortcuts, we reasoned that those who employed anchoring heuristics and tuning would complete the relevant research sections faster than those who were not employing heuristics. Experiment 3 supported the notion that the anchorage and adjustment heuristic serves as a time-saving technique for respondents.

Data analysis indicates that questionnaire makers need to strategically think about the best way to organize survey instruments to alleviate anchorage and adjustment. Based on the evidence, the results suggest that search designers can avoid grouping items that evaluate the same concept together - data fidelity can be substantially degraded by doing so.

References

1. Plous, S.: The Psychology of Judgment and Decision Making. McGraw-Hill, New York (1993)
2. Tversky, A.D., Kahneman, D.: The framing of decisions and the psychology of choice. Science **211**(4481), 453–458 (1981)
3. Simon, H.A.: The Sciences of the Artificial, 3rd edn. MIT Press, Cambridge (1969)
4. Engel, J.F., Kollat, D.T., Blackwell, R.D.: Consumer behavior, 2nd edn. Holt, Rinehard & Winston, Oxford (1973)
5. Sheth, H., Gardner, D.M., Garret, D.E.: Marketing Theory: Evolution and Evaluation. Wiley, Canada (1998)
6. Gonçalves, M., Menezes, J., Marques, C.: Grocery consumer relational perceptions in green consumption context. Tourism & Management Studies, Portugal (2015)
7. Fernandes, P., Correia, L.: Consumer attitudes toward the marketing practices in Portugal. Tourism & Management Studies (2013)
8. Epstein, S., Lipson, A., Holstein, C., Huh, E.: Irrational reactions to negative out-comes evidence for two conceptual systems. J. Pers. Soc. Psychol. **62**, 328–339 (1992)
9. Kahneman, D.: A perspective on judgment and choice: mapping bounded rationality. Am. Psychol. **58**(9), 697–720 (2003)
10. Stanovich, K.E., West, R.F.: Individual differences in reasoning: implications for the rationality debate. Behav. Brain Sci. **23**, 645–726 (2000)

11. Sloman, S.A.: The empirical case for two systems of reasoning. Psychol. Bull. **119**(1), 3–22 (1996)
12. Belsky, G., Golivich, T.: Why smart people make big money mistakes – and how to correct them. Lessons from the New Science of Behavioural Economics (1999)
13. Shiloh, S., Salto, E., Sharabi, D.: Individual differences in rational and intuitive thinking styles as predictors of heuristic responses and framing effects. Pers. Individ. Differ. **32**, 415–429 (2002)
14. Tversky, A., Kahneman, D.: Judgment under uncertainty: heuristics and biases. Science **185** (4157), 1124–1131 (1974)
15. Todd, P., Gigerenzer, G.: Bounded rationality to the world. J. Econ. Psychol. **24**, 143–165 (2003)
16. Schwarz, N., Strack, F., Report of subjective Well Being Judgmental Process and Their Methodological Implications, (editora) (1999)
17. Spector, P., Jex, E., Steve, M.: Development of four self-report measures of job stressors and strain: interpersonal conflict at work scale, organizational constraints scale, quantitative workload inventory, and physical symptoms inventory. J. Occup. Health Psychol. **3**(4), 356–367 (1998)
18. Tourangeau, R., Rips, L.J., Rasinski, K.A.: The psychology of survey response. Cambridge University Press, New York (2002)
19. Krosnick, J.A.: Response strategies for coping with the cognitive demands of attitude measures in surveys. Appl. Cognit. Phychol. **5**, 213–236 (1991). Ohio State University
20. Hsu, T.-C., Feldt, L.S.: The effect of limitations on the number of criterion score values on the significance level of the F-Test. Am. Educ. Res. J. **6**(4), 515–527 (1969)
21. Harrison, D.A., McLaughlin, M.E.: Cognitive processes in self-report responses: tests of item context effects in work attitude measures. J. Appl. Psychol. **78**(1), 129–140 (1993)

Usability Heuristics for M-Commerce Apps

Samar I. Swaid[1(✉)] and Taima Z. Suid[2]

[1] Departmnet of Math and Computer Science, Philander Smith College,
Little Rock, AR, USA
sswaid@philander.edu
[2] Departmnet of Computer Science, University of Arkansas at Little Rock,
Little Rock, USA
txsuid@ualr.edu

Abstract. Usability studies found that mobile users suffer from poor user experience which might explain the low spending of mobile-based commerce (m-commerce). This study focuses on identifying the usability heuristics to apply when evaluating the usability of m-commerce mobile apps. An integrated approach is applied considering the general usability heuristics developed by Nielsen, the Google Android design guidelines and Apple human interface guidelines. The proposed 13 heuristics are: "Visibility", "Matching-Real-World", "User-Control", "Error-Prevention", "Recognition", "Flexibility-and-Efficient Use", "Minimal Design", "Diagnose-and-Recover", "Help", "Performance", "Information–and-Visual-Hierarchy", "Natural-Interaction", and "Dynamic-Engagement". Four usability experts test empirically the developed mobile usability heuristics against two mobile apps that are used for secondhand retailing where users can sell and buy used items. Usability violations are identified and severity is ranked. Mobile apps usability is an ever evolving and dynamic research playground. Many of research challenges posed by lack of a comprehensive tool to understand the usability in the context of m-commerce. This is the contribution of the present paper.

Keywords: Usability · Evaluation · Mobile apps · Heuristics
Secondhand retailing · M-Commerce

1 Introduction

> *"It's failure that gives you the proper perspective on success."*
>
> *Ellen DeGeneres.*

Mobile apps are software applications designed to run on smartphones, tablets and other mobile devices. It is expected by 2020, mobile apps to generate around 189 billion U.S. dollars in revenues via app stores and in-app advertising 0. The explosion of mobile apps is seen in just about every industry such as retail, media, travel, education, healthcare, finance and social. As of June 2016, there were 2.2 million available apps at Google Play store and two billion apps available in the Apple's App store, the two leading app stores in the world 0. The increasing number of mobile apps

© Springer International Publishing AG, part of Springer Nature 2019
T. Z. Ahram and C. Falcão (Eds.): AHFE 2018, AISC 794, pp. 79–88, 2019.
https://doi.org/10.1007/978-3-319-94947-5_8

indicates that many businesses have deployed mobile applications to gain competitive advantage and its share of mobile-based revenue.

In contrast to the exploded growth of mobile apps, mobile user experience research shows that users shied away from the increasing complex unusable apps. According to MobileCommerceDaily, about 45% of mobile app users dislike their mobile app experience 0 and average Android app loses about 77% of its daily active users within the first three days after the install 0. Unsurprisingly, now, mobile app usability testing is integrated in the apps development process and conducted iteratively, especially where agile software development is present 0 to ensure the quality of the software. Platform providers such as Google/Android and Apple developed a set of design guidelines for their mobile app developers00. Yet, a set of comprehensive guidelines that would support testing the usability of mobile apps across platforms to ensure pleasant user experience is absent. Due to limitation of space, this paper will not fully describe usability guidelines provided by Google/Android and Apple, but will focus on the development of usability heuristics for mobile apps integrating these guidelines.

The remaining of the paper is organized as follows. Second section introduces concepts of usability, usability heuristics, and industry design guidelines. The section also discusses the limitation of mobile apps that suggests more innovative approaches to usability evaluation. Next, the current study is presented describing the integration of Nielsen heuristics 0 with Google/Android 0 and Apple design guidelines 0. The empirical testing of the developed heuristics for m-commerce apps is presented as well. Finally, the paper ends with conclusion and suggestions for further research.

2   Background

The ISO 9241-11 defines usability as the "extent to which a product can be used by specified uses to achieve specified goals with effectiveness, efficiency and satisfaction in a specified context of use 0. Others look to usability as a "function of the context in which the product is used...It is a property of the system: it is the quality of use in context"0. Nielsen 0suggests five attributes of usability:

- Efficiency: Resources expanded in relation to the accuracy and completeness with which users achieve goals;
- Satisfaction: Freedom from discomfort, and positive attitudes towards the use of the product.
- Learnability: The system should be easy to learn so that the user can rapidly start getting work done with the system.
- Memorability: The system should be easy to remember so that the casual user is able to return to the system after some period of not having used it without having to learn everything all over again.
- Errors: The system should have a low error rate, so that users make errors they can easily recover from them. Further, catastrophic errors must not occur.

Unusable systems are probably the key variable that explains the failure of a system or software 0. Research shows that usability is not given a simple absolute definition, but it is found to be relative to the users, goals and contexts of use0 0, thereby it is contextualized 00 0.

Generally, usability evaluation can be categorized to: (i) expert-based techniques such as cognitive walkthrough and heuristics inspection; and (ii) user studies such as user testing, interviews and experimentation. Compared to user studies, heuristics evaluation is a cost-effective method allowing more than 60% usability deficiencies to be identified 0. Nielsen 0 developed via a multi-phase study a set of usability heuristics for user interfaces. The suggested practice when applying heuristics evaluation is the following: (i) engage a small set of evaluators as many as five to examine the interface or software and judge its compliance with usability heuristics; (ii) evaluators aggregate the usability evaluation results in a corresponding list of usability deficiencies and its corresponding violated heuristics; (iii) usability deficiencies are ranked in terms of severity from 0 to 4, where 0 is considered 'cosmetic' issue and 4 a 'usability catastrophe' that is imperative to fix. Usability heuristics has been found efficient to evaluate usability of video-games, medical devices, shared workplace groupware, children learning applications, large screen exhibits, visualization software and virtual reality systems 0 0 00 0 0.

2.1 The Rationale for an Integrated Approach to Usability Heuristics

Mobile computing devices are smart consumer products 0 that are usually used by heterogeneous group of users. This introduces three main reasons that signify the importance of an integrated approach to usability heuristics for mobile apps. First, mobile devices have inherited limitations due to the nature of mobile devises themselves such as the small screen size of the device, display resolution, limited input mechanisms, connectivity-based issues, security, and limited performance capabilities 00. Other constraints to take into account when discussing mobile devices are the huge variability among the different brands and variability within one brand. For example, when designing apps for iPhone, the user interface is standardized. However, when designing for Android or Blackberry phones, there are different screen sizes, and interaction models to consider. Also, it is challenging to conduct user studies in laboratories settings due to the variety of consumers who use mobile apps 0, difficulty to capture data while user in interaction mode with mobile apps 0 which suggests to apply an expert-based evaluation techniques, such as usability heuristics. Finally and most importantly, major platform providers offered design guidelines to follow when design android and iOS apps. However, these guidelines are inconsistent and not comprehensive. Thus, an integrated approach to generate usability heuristics should provide comprehensive set of usability rules 0, which is expected to help understand the user experience in mobile apps context.

Given the above-mentioned limitations, this discussion confirms the fact that there is a lack of usability guidelines for mobile applications. This study adapt and integrate the general rules proposed by Nielsen 0, the design guidelines suggested by Google/Android 0 and Human Factors Deign of Apple 0 to generate comprehensive usability heuristics for m-commerce apps. This work is expected to add to the theoretical understanding of usability measurement and guides the practical application to usability evaluation for m-commerce apps.

3   The Study

Below, the two-phase study is described. The first phase of the study demonstrates the integration of the 10-rule heuristics of Nielsen 0 with the Google/Androids' guidelines 0 and human interface guidelines of Apple 0. The second phase tests and validates the developed heuristics for the context of second-hand retailing mobile apps. Rationale and actions as they relate to the procedure applied are provided.

3.1   Phase I: Development of Mobile Apps Heuristics

One of the pioneers who attempted to develop usability heuristics to objectively evaluate interfaces is Nielsen 0 with his ten rules of usability heuristics. Although these rules are still hold up today, user additional needs and device characteristics, need to be factored into consideration. The procedure applied is presented below.

Procedure. The usability guidelines of Google and human factors guidelines of Apple, were linked to the ten-rule usability heuristics of Nielsen 0 where possible. The ones that are not linked to Nielsen rules suggest additional heuristics that are fit the context mobile-devices and mobile apps. Next, we operationalized each rule with the details provided by Google and Apple on how to apply the design rules. This provided a list of checklist to apply when testing the heuristics rules. Finally, a cleaning process is applied to remove any duplicates or ambiguous items (see Tables 1 and 2).

Table 1. Sample of mobile apps usability heuristics

Usability rule	Nielsen rule (1992)	Google usability guidelines	Apple usability guide lines
User Control	User control and freedom	Decide for me but let me have the final word	User control
Consistency	Consistency and standards	- App utilizes consistent design, typography, colour function, and content	Consistency
		- App uses familiar paradigm (navigation bars, buttons and other elements) follow iOS UI standards	
		If it looks the same, it should act the same	
Information and visual hierarchy	–	Information hierarchy and structure	–

3.2   Phase II: Empirical Validation

To validate the developed usability heuristics, four usability evaluators inspected the usability of two different mobile apps that are used for second hand retailing. All usability evaluators had experience in HCI theories, usability evaluation, and heuristics procedure through a senior course work and industry and research experience. Nevertheless, a training session on usability heuristics and Nielsen's heuristics was provided. Participant

Table 2. Selected usability heuristics with checklist measures

Usability Rule	Definition	Usability Check List
Visibility	The mobile app should always keep users informed about what is going on, through appropriate feedback within reasonable time	- The app provides the status of mobile app at any stage of usage
		- The app provides feedback to user based on user actions taken
		- The app provides feedback when device status is changes (e.g., charged, put in sleeping mode.. etc.)
		- The App provides user with feedback when app crashes or re-starts
Information and visual hierarchy	The mobile app should present information and visual objects in a hierarchy fashion based on users functions of searching, sorting, filtering, zooming and swiping and	- Text and content is easy to read
		- Only primary content on screen and other secondary content to be accessible when needed
		- App provides user with mechanisms to explore infinite stream content with no dead-ends
		- App segments processes to make it easy for user to complete (e.g., filing forms, displaying information.. etc.)
		- App provides mechanism to search filter sort, zoom, swipe s horizontally and vertically
		- App provides alternatives when there are no matching search results
		- App saves and displays to users recent searches
		- App displays sharp and hero images
		- App removes any distraction from users information search
Dynamic engagement	The mobile app should provide users with easy to use features to engage in two-way communications with widgets and other users to fulfil a complete engaged experience	- App is easy to use to connect and share via social media
		- App displays friendly toast notifications, alerts and notifications to users when appropriate
		- App integrates other widgets and apps to engage user
		- App is easy to use to create accounts using social-media existing accounts
		- App provides functions to engage the user in two-way communication with app, other apps and users

evaluators applied the following steps in usability inspection: (i) identify usability deficiencies; (ii) rate severity of usability issues applying Nielsen severity ranking system 0; and (iii) conduct a debriefing session where all evaluators meet and agree on usability issues identified, its category and severity rank.

Mobile App Selection: The Case of Secondhand Retailing Mobile Apps. Secondhand retailing is more than $18 billion industry that is projected to grow by 11% every year to reach at least $33 billion industry on 20210. The United States giant retailers are suffering while secondhand commerce is thriving. For example, the total secondhand apparel market is expected to grow from $18 billion in 2016 to $33 billion in 2021. The phenomena of secondhand retailing can be explained by factors of millennial penetration, desire for value and cost savings, and the entertainment factor of the treasure hunt for unique items. The emergence of websites such as ebay.com, Poshmark.com, Tradesy.com and Bonanza.com has streamlined the secondhand retailing, Similarly, mobile apps such as Letgo, Offerup, Gone, Mercari and Depop, among others, made it easy for non-professional sellers to sell their used items. Such peer-to-peer mobile apps offer innovative technological features and social media functionality for users that made it easy for unprofessional sellers to develop Mobile apps such as LetGo, which connects buyers and sellers, enables instant uploading for pictures, supports adding item description, powers users with instant messaging and provides automatic tagging for products using a proprietary software.

For this study mobile apps of LetGo0 and OfferUp 0 are selected to test the developed usability heuristics. These mobile apps enable users to buy and sell items locally using the GPS feature and/or zip code to locate items. Both LetGo and OfferUp allow users to make offers via text and negotiate prices in real-time text conversation. Users can apply filtering to the displayed items based on different criteria such as geographic distance, price and item type. Both apps allow sellers to share their items via social networking sites such as Facebook and Twitter. Users can share item found with others via email and can setup push notifications to be alerted when an offer is made on an item, or item status changes.

Evaluators download both apps on android and iOS mobile devises. Usability evaluators conducted a cognitive walkthrough to learn the apps and then conducted a usability inspection for each using the thirteen heuristics developed by the study. The evaluators conducted the usability inspection individually, and then, they met to debrief their findings and rate the severity of each of the identified usability violation.

Usability evaluation resulted in identifying a number of usability violations for LetGo and OfferUp apps. Usability deficiencies varied in terms of severity (see Tables 3 and 4). For example, when users browse the items saved on their favorites list, they are given the opportunity to click on any items to get more details. However, when exiting, the user is returned back to the beginning of the favorite items list, not to where the user was browsing (see Fig. 1a–d). In addition, when favorites item is sold, the user get notification, but user to browse the details of item sold is not possible. Instead, the user is given the option to browse other similar items. Both usability problems show the violation to usability rules of user control and user's ability to recognize rather than recall.

Table 3. Sample of usability violations

Usability rule	LetGo	OfferUp
User control	- Users when exit browsing an item from the favourite list, the app brings the list from the beginning. Then user needs to scroll down where s/he was originally	- Saved item to favourites was difficult to be removed from list of Board items (favourites)
	- When user adds an item to favourite and item is sold, a notification is sent. However, when user clicks on notification it does not show details of item, but other items that are recommended by the app	- Items listed for sale are difficult to be deleted; however, the app provides a feature to archive them. Searching Help did not produce any information about how to delete
Performance	- In some cases, users receive message notification, but when search for it, it is not found	- It allows users to like their own listed items
	- When in searching items in different zipcode and distant it takes long time to load pictures	
Error prevention	- Users can post items without selecting category of item	

Table 4. Number of usability violation and rank

Rank [0–4]	Number of usability violations in LetGo	Number of usability violations in OfferUp
4:Usability catastrophe 0	0	0
3: Major usability problem:	1	2
2: Minor usability problem	3	3
1:Cosmetic problem only	5	4
0: Not a usability problem	0	0

Deleting favorite item on OfferUp requires multi-steps from the user. First, user needs to open the board, and then selects the item to be deleted. Then, user to click on the heart icon to indicate deletion. If the item is on two boards, the user needs to replicate the steps per board. It will be more efficient, if item to delete is selected once, to be deleted from all boards (see Fig. 2a–d).

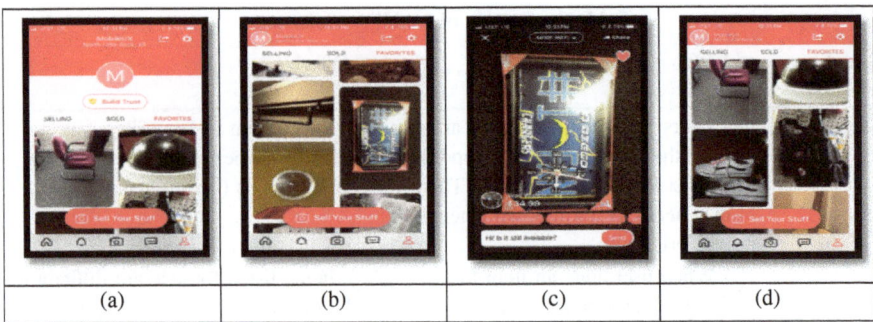

Fig. 1. Elected usability problems with LetGo App

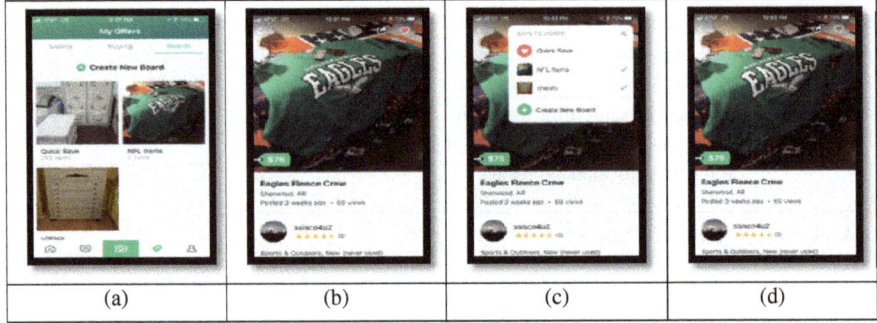

Fig. 2. Selected usability problems with OfferUp App

4 Conclusion

Usability is a key factor in the quality of mobile apps, especially for its users. However, research on mobile apps usability is still fragmented and inconsistent. In this study, a usability heuristics model is developed based on an integrative approach using Nielsen 10-rules heuristics, and the deign guideline put by Google and human-factors deign of Apple for their mobile apps developers. Using the developed 13 usability heuristics allows us to assess the usability of mobile apps and identify any deficiencies. The study supports the consistent and unambiguous compilation of the usability heuristics that consider the special characteristics of the handheld devices. The usefulness of this approach is demonstrated by the empirical testing of two mobile apps used for secondhand retailing. For future work, I plan to operationalize each of the heuristics based on the flow and stages of m-purchase transaction of: On-board, Use, Transact and Return 0. The developed heuristics and its checklist will be empirically tested using other types of m-commerce apps to validate our m-commerce usability principles and its checklist.

Mobile applications are growing in number and complexity, and becoming the standard for media consumption, retailing, virtual learning, and social networking, among others. The increasing spread of mobile app development by non-developer experts, make it essential to identify the usability principles to be considered in early stages of mobile apps software development to guarantee a positive user experience. As Steve Jobs advise, "You have to start with the customer experience and work your way back to technology".

References

1. Alsumait, A., Al-Osaimi, A.: Usability heuristics evaluation for child elearning applications. J. Softw. **5**(6), 654–661 (2010)
2. Android: Google Android Design Guidelines. https://developer.android.com/design/index.html. Accessed 28 Feb 2018
3. Apple: Human Interface Guidelines. https://developer.apple.com/design/. Accessed 28 Feb 2018
4. Baker, K., Greenberg, S., Gutwin, C.: Empirical development of a heuristics evaluation methodology for shared workspace groupware. In: Proceedings of the ACM CSCW 2002, New Orleans, Louisiana, USA, November, pp. 96–105 (2002)
5. Bertin, E., Gabrielli, S., Kimani, S.: Appropriating and assessing heuristics for mobile computing. In: Proceedings of the Working Conference on Advanced Visual Interfaces (AVI 2006), pp. 119–126. ACM, New York, May 2006
6. Bevan, N., Macleod, M.: Usability measurement in context. Behav. Inf. Technol. **13**, 132–145 (1994)
7. Bevan, N.: What is the difference between the purpose of usability and user experience evaluation methods. UXEM 2009 Workshop, INTERACT 2009, Uppsala (2009)
8. Desurvire, H., Caplan, D., Toth, J.: Using heuristics to improve the playability of games. In: CHI Conference, 2004. Vienna, Austria, April 2004. https://www.forbes.com/forbes/welcome/?toURL=https://www.forbes.com/sites/richardkestenbaum/2017/04/11/fashion-retailers-have-to-adapt-to-deal-with-secondhand-clothes-sold-online/&refURL=https://www.google.com/&referrer=https://www.google.com/. Accessed 28 Feb 2018
9. Forbes: Fashion Retailers Have To Adapt To Deal With Secondhand Clothes Sold Online (2017). www.forbes.com. Accessed 28 Feb 2018
10. Göransson, B., Gulliksen, J., Boivie, I.: The usability design process – integrating user-centered systems design in the software development process. Software Process Improvement and Practice **8**, 111–131 (2003)
11. Griffiths, S.: Mobile App UX Principles. 2015. https://storage.googleapis.com/think-emea/docs/article/Mobile_App_UX_Principles.pdf. Accessed 28 Feb 2018
12. Harrison, R., Flood, D., Duce, D.: Usability of mobile applications: literature review and rationale of a new usability model. Journal of Interaction Science **1**(1), 1 (2013)
13. ISO 9241-11: Guidelines for Specifying and Measuring usability (1998)
14. Letgo: 2018. www.letgo.com
15. Madrigal, D. McClain, B.: Usability of mobile devices: insights from research. https://www.uxmatters.com/mt/archives/2010/09/usability-for-mobile-devices.php. Accessed 28 Feb 2018
16. Mobile Commerce Daily: More Than Half of Consumers Dissatisfied with Mobile Retail Experiences: Adobe. https://www.retaildive.com/ex/mobilecommercedaily/more-than-half-of-shoppers-are-dissatisfied-with-mobile-retail-experiences-adobe. Accessed 28 Feb 2018
17. Nielsen, J.: Heuristic evaluation. In: Nielsen, J., Mack, R.L. (eds.) Usability Inspection Methods. Wiley, New York (1994)

18. Nielsen, J.: Severity Rankings for Usability Problems. http://www.useit.com/papers/heuristic/severityrating.html. Accessed 28 Feb 2018
19. Nielsen, J.: Usability Engineering. Morgan Kaufmann, San Diego (1993). Offerup.2018. www.OfferUp.com
20. Nielsen, J.: Finding usability problems through heuristics evaluation. In: CHI 1992, pp. 373–380. ACM Press, New York (1992)
21. Quettra: Losing 80% of mobile users is normal, and why the best apps do better (2015). https://www.linkedin.com/pulse/losing-80-mobile-users-normal-why-best-apps-do-better-andrew-chen
22. Po, S.: Mobile usability testing and evaluation. Master's thesis, University of Melbourne, Australia (2003)
23. Seffah, A., Donyaee, M., Kline, R., Padda, H.: Usability measurement and metrics: a consolidated model. Softw. Qual. J. **14**, 159–178 (2005)
24. Somervell, J., Wahid, S., McCrickard, D.: Usability heuristics for large screen information exhibits. In: Proceedings of Human-Computer Interaction (Interact 2003) Zurigo, Svizzera, pp. 904–907 (2003)
25. Statista: Mobile commerce in the United States - Statistics & Facts (2018). https://www.statista.com/topics/1185/mobile-commerce/
26. Sutcliff, A., Gault, B.: Heuristics evaluation of virtual reality applications. Inter. Comput. **16**, 381–849 (2004)
27. Swaid, S., Maat, M., Krishnan, H., Ghoshal, D., Ramakrishnan, L.: Usability heuristics of data analyses and visualization tools. In: 8th International Conference on Applied Human Factors and Ergonomics (AHFE 2017), Los Angeles, California, USA, 17–21 July 2017
28. Trivedi, M., Khanum, M.: Role of context in usability evaluations: a review. Adv. Comput. Int. J. (ACIJ) **3**, 2 (2012)

Usability and Design Guideline for Designing Single Handle Faucet's and Handle Shape

Riku Takagi[1]($\boxtimes$), Noriko Hashida[2], and Hiroyuki Takeuchi[3]

[1] Shibaura Institute of Technology Graduate School of Engineering and Science,
3-7-5, Toyosu, Koto-Ku, Tokyo, Japan
cyl4220@shibaura-it.ac.jp
[2] Shibaura Institute of Technology Engineering and Design, 3-7-5, Toyosu,
Koto-Ku, Tokyo, Japan
[3] TAKAGI CO., LTD, 2-4-51, Ishidaminami, Kitakyushu-Shi, Fukuoka, Japan

Abstract. The reason why there are many kinds of handle shapes for the single handle faucets is because there are no design guidelines for their usability. This study is on making a design guideline for the new single handle faucet which is now developing at TAKAGI CO., LTD. in Japan. First, a questionnaire survey on the usability of 10 basic handle shapes was developed. We made the test subjects evaluate those handle's usability in 5 levels, and compared them to find out which component improves their usability. Secondly, 4 advanced handle shapes were developed using the results of the basic shape questionnaire survey. We made the test subjects evaluate those handle's usability in 5 levels and by the image words we selected which may have a connection between the handle's appearance. Furthermore, we used the principle component analysis to find out which image word has a strong mutual relation between the appearance of the handle and their usability. Consequently, we were able to find out the component which will improve the handle's usability, and the image words which may have a strong mutual relationship between the appearance of the handle and their usability.

Keywords: Single handle faucet · Usability · Correlation

1 Introduction

This study is about finding the guidelines in designing the handle shapes for the user's usability. There are too many kinds of shapes in single handle faucets, and this is because there are no guidelines in designing the handle shapes for their usability. This is a serious matter to designers and to consumers, because it is hard in both developing and purchasing the best designs for usability. It also takes innumerable time to develop one handle which matches to the whole faucet image and fulfills their usability. But it will make easier and faster to develop the handles by creating a guideline. This study is a collaborative project with TAKAGI CO., LTD., an industry developing faucets in Japan. So it was decided to refer the cartridge and the size which is used in TAKAGI CO., KTD for our design condition, and make a guideline which matches best to the cartridge developed in TAKAGI CO., LTD.

© Springer International Publishing AG, part of Springer Nature 2019
T. Z. Ahram and C. Falcão (Eds.): AHFE 2018, AISC 794, pp. 89–98, 2019.
https://doi.org/10.1007/978-3-319-94947-5_9

2   Research on the Usability of Basic-Form Single Handle Faucets

2.1   Survey Content and Purpose

10 basic forms of handles, which are different in width, tip shape, and hole existence, were developed. The size and component used in the basic handle shapes were decided from the angles and the faucet size which TAKAGI CO., LTD. is developing. We made the test subjects evaluate those handle's usability in 5 levels, and compared them to find out which component will improve the usability. The contents of this research were as follows:

- Evaluation of each handle's usability in 5 levels
- Necessity on the existence of the ring and their comments
- Most favorite and least favorite handle shape from the 10 basic-forms and their comments

2.2   Survey Method

We made the test subject evaluate the usability in the handle shapes with 10 basic handle shapes (all handle has the same Length: 145 cm; Thickness: 8 mm; Angle: 10°). The specifications of each were as follows (Fig. 1):

Sample A (Width: 50 cm; Existence of ring: None; Shape of tip: Square)
Sample B (Width: 50 cm; Existence of ring: Have; Shape of tip: Square)
Sample C (Width: 50 cm; Existence of ring: None; Shape of tip: Round)
Sample D (Width: 50 cm; Existence of ring: Have; Shape of tip: Round)
Sample E (Width: 31 cm; Existence of ring: None; Shape of tip: Square)
Sample F (Width: 31 cm; Existence of ring: Have; Shape of tip: Square)
Sample G (Width: 31 cm; Existence of ring: None; Shape of tip: Round)
Sample H (Width: 31 cm; Existence of ring: Have; Shape of tip: Round)
Sample I (Width: 13 cm; Existence of ring: None; Shape of tip: Square)
Sample J (Width: 13 cm; Existence of ring: None; Shape of tip: Round)

Fig. 1.

2.3 Result

Figure 2 shows the average level of each handle's usability and the favorite, least favorite handle shape which was chosen by the test subjects. From this research, the most usable handle shapes were Sample D, G, H and their average levels were about 3.50. Moreover, the 3 handles which had the highest scores on the usability were most chosen as the most favorite handle shapes. In addition, they were least chosen as the least favorite handles. From this result, we found out that the basic handle shapes D, G, H are the most acceptable shapes to the users.

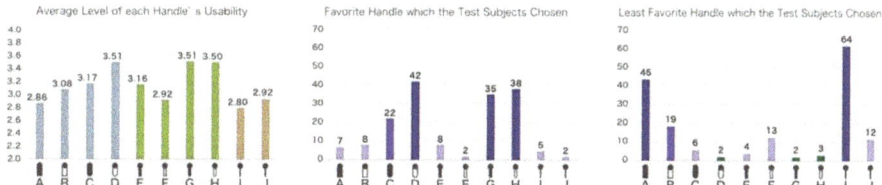

Fig. 2. Average level of each handle's usability, favorite and least favorite handle

On the opposite side, sample lever I had the lowest average level on the usability of the handle shape, and was most chosen as the least favorite handle.

Figure 3 shows the comparison of the handle shapes with each component. By grouping 2 handles by each component's difference and comparing the change in the usability, we were able to find out which component will improve the usability.

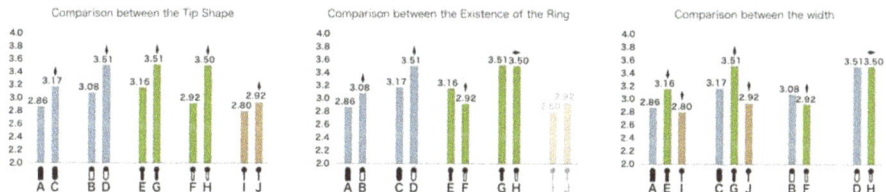

Fig. 3. Comparison between each handle shape's component

When we compared the handles by their tip shape, the usability increased when they were round. When we compared the handles by the existence of the ring, the usability increased on the wide-type handle when there was a ring. But, the usability decreased on the mid-width-type, when there was a ring. There were comments such as "The hole was too thin, the finger didn't go through", "It would be better to use when the finger went through". Referencing from these comments and the results, the usability will approve only when the width of the hole is wide enough to insert the finger. When we compared the handles by the width, the usability increased from the wide-type to the mid-width-type, and they decreased from the mid-width-type to the narrow-type handle.

Figure 4 shows the comparison of each age-groups rating. From this result, there was only a slight difference in all the age-group's rating. Therefore, we figured we can propose a universal guideline for handle shape design.

Fig. 4. Comparison of each age-groups rating

According to this survey, we found out the most acceptable handle shapes were samples D, G, H. On the opposite side, sample lever I was the least acceptable handle shape. The most user-friendly components for the handles are round tips, the width of the handle between 31 mm to 50 mm, and the width of the ring which people can easily put their fingers through. Therefore, on the next survey for the advanced-forms, they will be an improved design based on Sample D, G, H, I, which were chosen as the most favorite and the least favorite handle shapes. The reason for taking Sample I in the advanced-form research is, because there is a similar design in TAKAGI CO., LTD.

3   Research on the Usability and the Images of Advanced-Form Single Handle Faucets

3.1   Survey Content and Purpose

4 advanced forms of chrome plated handles are developed. The test subjects evaluate those handle's appearance and usability in 5 levels, and by the image words we selected which may have a connection between the handle's appearance. The evaluation on the appearance was based on whether the handle looked easy to use or not. The contents of this research were as follows:

- Evaluation of each handle's appearance and their usability in 5 levels
- Most favorite and least favorite handle shape from the 4 advanced-forms
- Evaluation of each handle's image words in 5 levels

Furthermore, we used the principle component analysis to find out which image word has a strong mutual relation between the appearance of the handle and their usability.

3.2 Survey Method

We made the test subject evaluate the appearance, usability and image words with 4 advanced handle shapes (all handle has the same length: 145 cm). The specifications of each were as follows (Fig. 5):

Sample A (Width: 56 cm; Existence of ring: Have; Shape of tip: Round)
Sample B (Width: 41 cm; Existence of ring: Have; Shape of tip: Round)
Sample C (Width: 31 cm; Existence of ring: None; Shape of tip: Round)
Sample D (Width: 15 cm; Existence of ring: None; Shape of tip: Square)

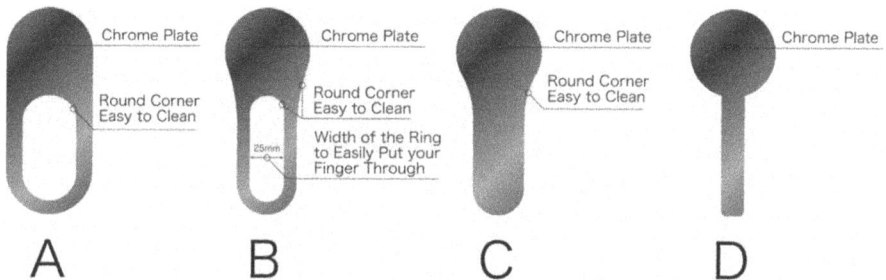

Fig. 5. 4 Advanced-forms of the single handle faucet's handle shape

3.3 Result

Figure 6 shows the average level of each handle's usability and their appearance, and the favorite, least favorite handle shape which was chosen by the test subjects. From this research, the most usable handle shape was Sample B and the average level for the appearance was about 3.50, and 4.1 for the usability. Moreover, the handle which had the highest score on the appearance and usability, was most chosen as the most favorite handle shapes. In addition, it was least chosen as the least favorite handle. From this result, we found out that the advanced handle shape B is the most acceptable shape to the users.

Table 1 shows the results for the correlation between the appearance, usability and the image words for advanced-form handle B. The numbers inside each cell stands for the correlation factors for each appearance, usability, and the image words. When the correlation factor gets bigger it able to know which image word has a strong mutual relation between the appearance and the usability [1, 2].

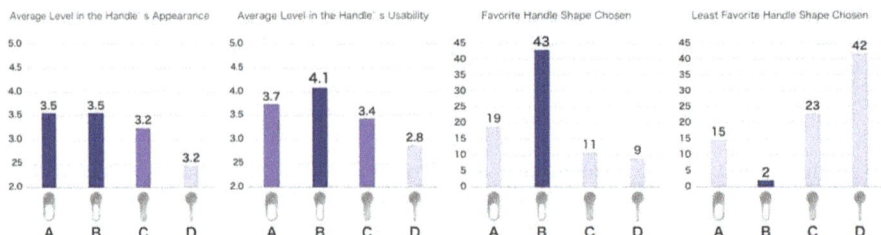

Fig. 6. Average level in the single handle faucet's appearance and their usability, most favorite and the least favorite handle shape

According to the analysis, we found out the image words which has a strong mutual relation between the appearance and the usability are, "fresh in shape", "high grade", and "maturity".

Table 1. The correlation between the appearance, usability and the image words for advanced-form handle B

	B Appearance	B Usability
B Appearance	1.000	0.471
B Usability	0.471	1.000
B Soft	0.096	0.068
B Fresh in Shape	0.532	0.376
B Neat Shape	0.323	0.082
B Natural	0.040	0.261
B Delicate	0.271	0.075
B High Grade	0.527	0.373
B Womanly	0.102	0.084
B Round	0.087	0.102
B Maturity	0.362	0.356
B Static	0.130	0.053

Therefore, those image words will help to develop designs of single handle lever shape and improve those appearance and usability.

Through this survey, we obtained the components and image words which will improve the usability as given in Table 2. In this survey, since we evaluated each handle by their appearance, usability and image words, we were able to find various factors for each item.

As a result, we were able to find out the design guideline on the components and the image words, which will improve the single handle faucet's usability. By developing a handle based on the components above (Table 2) and designing it in the directions of "fresh in shape", "high grade", and "maturity", we will be able to design a handle which is good in both appearance and usability.

Table 2. Design guideline on the components and image words

Design Guideline on the Component

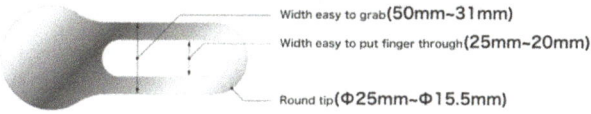

Width easy to grab(50mm~31mm)

Width easy to put finger through(25mm~20mm)

Round tip(Φ25mm~Φ15.5mm)

Design Guideline on the Image Words

	Appearance	Usability
Soft	➤	➤
Fresh in Shape	➤	➤
Neat Shape	➤	➤
Natural	➤	➤
Delicate	➤	➤
High Grade	➤	➤
Womanly	➤	➤
Round	➤	➤
Maturity	➤	➤
Static	➤	➤

4   Research on the Usability and Appearance of Final-Form Single Handle Faucets

4.1   Survey Content and Purpose

In the final research, 2 handles based on the guidelines in the shape and the image words of the directions, were developed. The difference of these to levers are whether it is straight or round from the side. We also used the advanced-form handle B and the existing single handle faucet which is developed in LIXIL Corporation, to check on the effects of the design guideline which was made. The test subjects evaluate those handle's appearance and usability in 5 levels, and by the image words we selected which may have a connection between the handle's appearance. The evaluation on the appearance was based on whether the handle looked easy to use or not. The contents of this research were as follows:

- Evaluation of each handle's appearance and their usability in 5 levels
- Most favorite and least favorite handle shape from the 4 advanced-forms
- Evaluation of each handle's image words in 5 levels

4.2   Survey Method

We made the test subject evaluate the appearance, usability and image words with 4 handle shapes. The specifications of each were as follows (Fig. 7):

Sample A (Advanced-form handle B; Width: 41 cm; Existence of ring: Have)
Sample B (Final-form straight-type; Width: 39 cm; Existence of ring: Have)
Sample C (Final-form round-type; Width: 39 cm; Existence of ring: Have)
Sample D (LIXIL single handle faucet; Width: 56 to 38 cm; Existence of ring: Have).

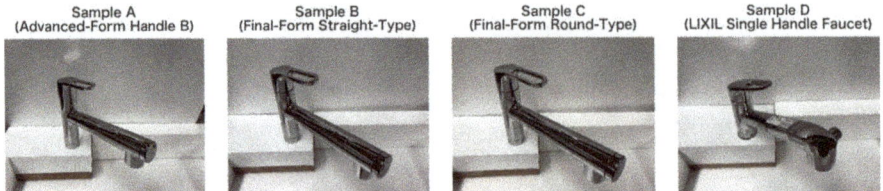

Fig. 7. Faucet handles which were used in the research

4.3   Result

Figure 8 shows the average level of each handle's appearance and usability, and the favorite, least favorite handle shape which was chosen by the test subjects. From this research, the most usable handle shape was Sample C and the average level for the appearance was about 4.00, and 4.33 for the usability. Comparing with Sample A (advanced-form handle B) and both final-form handles, the appearance and usability improved on both final-form handles. Moreover, a total of 80 percent of the test subjects chose the final-form handles as their most favorite handle shape and a total of 17 percent of the test subjects chose the final-form handles as their least favorite handle shape.

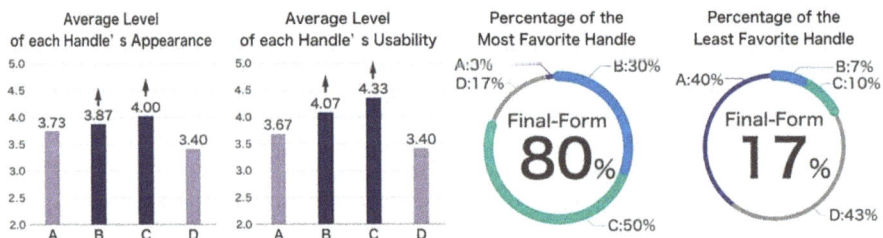

Fig. 8. Average level in the single handle faucet's appearance and their usability, Percentage of the most favorite and the least favorite handle shape.

Figure 9 shows the image word evaluation using the semantic differential method. When we compared the results between the advanced-form handle and the final-form handle, the "Fresh in Shape", "High Grade", "Maturity" image has improved on the final-form handle. These words were the image words which we abstracted that has a strong mutual relation between the image words.

From this result, we were able to make a handle shape which has a good appearance and usability, by using the design guideline.

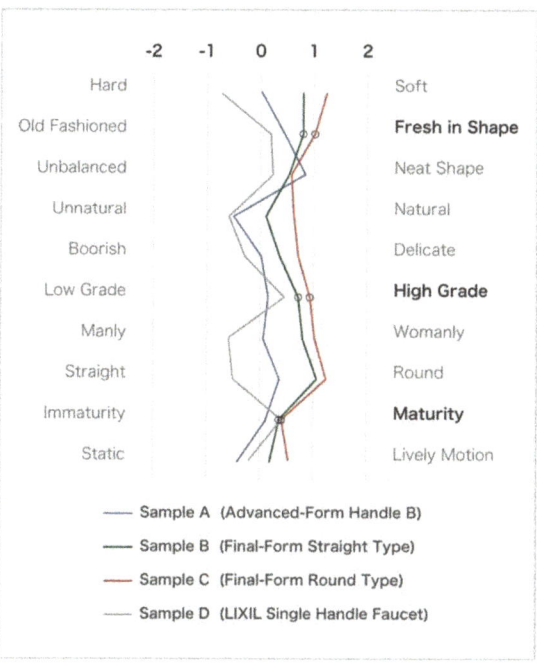

Fig. 9. The image word evaluation using the semantic differential method

5 Conclusion

In this study, we abstracted the components and image words which will improve the appearance and usability. We developed 10 basic-form handles and had a questionnaire survey on the usability, and found out the basic components which will improve the usability. In this survey, we found out the most user-friendly component for the handles are round tips, the width of the handle between 31 mm to 50 mm, and the width of the ring which people can easily put their fingers through. Moreover, by comparing each age-groups rating, there was only a slight difference in all the age-group's rating. Therefore, we figured we can propose a universal guideline for handle shape design. Next, we developed 4 advanced-forms of chrome plated handles referencing the results of the first questionnaire survey. From this survey, we found out that the handle with the ring which people can put one finger in was the most usable shape. Furthermore, we used the principle component analysis to find out which image word has a strong mutual relation between the appearance of the handle and their usability. According to the analysis, we found out the image words which has a strong mutual relation between the appearance and the usability are, "fresh in shape", "high grade", and "maturity". Therefore, those image words will help to develop designs of single handle lever shape and improve those usability. Finally, we made a design guideline for the handle shape's usability and developed 2 handle shapes using that guideline, and compared the usability with the advanced-shape handle which had the highest average level in the usability and the existing single handle faucet which was developed in LIXIL Corporation. As a result, the usability level of the designed handle was significantly improved thanks to the guidelines we developed. In conclusion, we believe our guideline will make the design work easier and adjustable for usability and design image at the same time.

References

1. Minrou, S.: A primer of multivariate analysis, studying from Exel
2. Meitetu, K.: Deta Science using R
3. Boger, J., Craig, T., Mihailidis, A.: Examining the impact of familiarity on faucet usability for older adults with dementia
4. https://bmcgeriatr.biomedcentral.com/articles/10.1186/1471-2318-13-63
5. Taati, B., Snoek, J., Mihailidis, A.: Video analysis for identifying human operation difficulties and faucet usability assessment
6. https://www.sciencedirect.com/science/article/pii/S0925231212003918DESIGN
7. Takumi, W., Masayuki, O., Ryouta, S.: An influence on Hot Water saving effect of operability saving hot water using single-Lever kitchen faucets, The society of Heating, Air-Conditioning and Sanitary Engineers of Japan
8. https://www.jstage.jst.go.jp/article/shasetaikai/2013.1/0/2013.1_13/_article/-char/ja/
9. Li, Z., Mihailidis, A., Boger, J.: The Usability Of Water Faucets For Older Adults With And Without Dementia: How Important Is Familiarity
10. http://citeseerx.ist.psu.edu/viewdoc/download?doi=10.1.1.514.8218&rep=rep1&type=pdf
11. Boger, J., Mihailidis, A.: Familiarity and Usability of Products by People with Dementia, Toronto Rehabilitation Institute, University of Toronto
12. Lo, S., Helander, M.G.: Developing a formal usability analysis method for consumer products
13. Chen, Y., Zhao, Z.M., Wang, X., Wu, G.X.: The production capacity analyses and balance improvement of faucet assembly line. Inst. Ind. Eng. Zhejiang University of Technology, Hangzhou 310032, China
14. http://en.cnki.com.cn/Article_en/CJFDTOTAL-GYGC200903020.htm
15. Soewardi, H., Pradana, V.: Developing features of water faucet by using user centered design approach, Department of Industrial Engineering, Industrial Technology Faculty, Islamic University of Indonesia, Yogyakarta, Indonesia. http://www.arpnjournals.org/jeas/research_papers/rp_2016/jeas_0416_4016.pdf

A Comparative Study of Product Usability and Ergonomic Assessment of Server Lifts

Dosun Shin[⊠]

The Design School, Arizona State University, Tempe 85287, USA
dosun.shin@asu.edu

Abstract. This paper provides an overview of the process of evaluating competing products from the perspectives of human factors and ergonomics, safety considerations, overall usability, and design characteristics. The project was conducted by faculty and students at a major university in U.S. and the product in question was a lift that is used for installation of servers in data centers. We conducted an in-depth study of three competing products on the basis of a series of criteria that were determined before the project was started. We developed a methodology for observation, testing and evaluation for the products, and in this paper, we will discuss the results of our study. In addition to the analysis, we also created a series of design recommendations for product development that built upon our findings. While this paper focuses on the specific case study of server lifts, the learnings from this study could have wide applications in terms of the methodology of how to evaluate the competitive products, how to develop test regimens, how to conduct observations to test usability, how to conduct analysis, and how to make recommendations. While there is a clear impact of this study in industry, it can also have value in education in helping students understand how to test competitive products, and how to learn from the testing to design their products for the future.

Keywords: Human factors · Usability test · Product comparison
Industrial design

1 Introduction

This paper explores the process of product evaluations comparing 3 different competing server lift products using human factors and ergonomics methods focusing on safety and usability testing. This independent study was conducted by faculty and student teams from Arizona State University's Industrial Design program and Engineering program collaborating to evaluate and analyze current designs based on the direct feedback from lift operators working in the field.

An executive comparison among all three competitors was conducted to provide data on the SL-350X's performance in comparison with its current lift market competitors. The Manual ServerLIFT SL-350X, manufactured by the sponsored company, RackLift's RL 600S lift and the Warehouse Lift constitute the most common alternatives, were tested and evaluated on their functions and product performance. All three lifts underwent

© Springer International Publishing AG, part of Springer Nature 2019
T. Z. Ahram and C. Falcão (Eds.): AHFE 2018, AISC 794, pp. 99–110, 2019.
https://doi.org/10.1007/978-3-319-94947-5_10

identical testing processes and were evaluated and documented based on the following categories.

- Functionality During Navigation
- Safety During Use
- Ergonomics
- Installation Functionality
- Crank Functionality
- Obstruction Avoidance Capability
- Feature Evaluation
- Server Installation
- Engineering Assessment

This paper will discuss the testing process and discuss the results of only the Navigation and Safety tests done on the three lift products, and an Ergonomics assessment of the ServerLIFT SL-350X product.

2 Competitor Overview

The following product images and descriptions are providing the overview of competitors that were tested and evaluated for this study (Table 1).

Table 1. Product competitor descriptions.

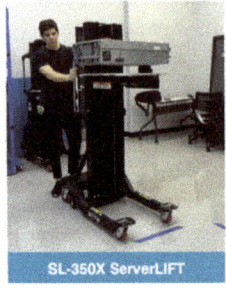  

Lift Capacity: 500lbs	Lift Capacity: 650lbs	Lift Capacity: 500lbs
Built speci cally for the data center, the SL-350X ServerLIFT has become the industry standard world-wide. This lift can lift blade chassis, switches, server hardware and any other heavy IT equipment.	The RL600S Manual Server Equipment Lift was created to the need to more safely, easily and productively lift up to 650 pounds in a data center or head end facility.	Great for shipping/receiving, heavy material on shelves, HVAC installation/ repair and more. Can be used as a hand truck, forklift or dolly. Non-marking rear wheels and dual-wheel, front-swivel casters.

3   Navigation Speed Test

The navigation speed test is a detailed assessment of the lift's efficiency in traveling between the narrow corridors of a data center. In this test, efficiency was evaluated using time. Variables impacting time such as change in the direction being traveled, made by the operator to avoid collision, were recorded to help evaluate where efficiency is lost. The result of this test compares the competitors using an average travel time derived from numerous runs with specified variables and constants.

Based on the overall aisle dimensions and layout of common server centers, the navigation course map was created in the testing room of the sponsor's company, and server racks were also installed. The team recorded the navigation time from the starting point and compared the speed with different server lift conditions. The testing procedure that was developed by industrial design team is shown below (Fig. 1).

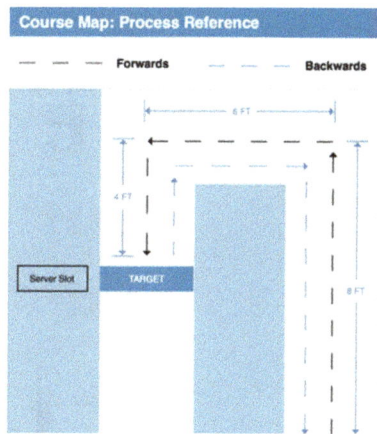

Fig. 1. Navigation test course map.

(1) Mark designated start point in adjacent aisle from targeted server slot. (18 ft Away)
(2) Place lift at designated starting point with the lift platform at the height of the intended server slot.
(3) Record the time lapsed as operator navigates through the course and up to the target Server Slot.
 4) During the test observe and record how many adjustment movements must be made by the operator to avoid any collisions.
(5) Simultaneously record the amount of time it takes for the operator to travel from START to TARGET.
(6) Repeat steps 1–5 moving in the opposite direction.
(7) Repeat steps 1–6 with the server loaded on the lift.
(8) Collect user feedback from lift operator about experience with each product being tested.

3.1   Navigation Time

Two professional server operators were recruited and participated in this test. The following data presents the speed comparisons of three products, and both forward and backward operation time were recorded with server and without server (Fig. 2).

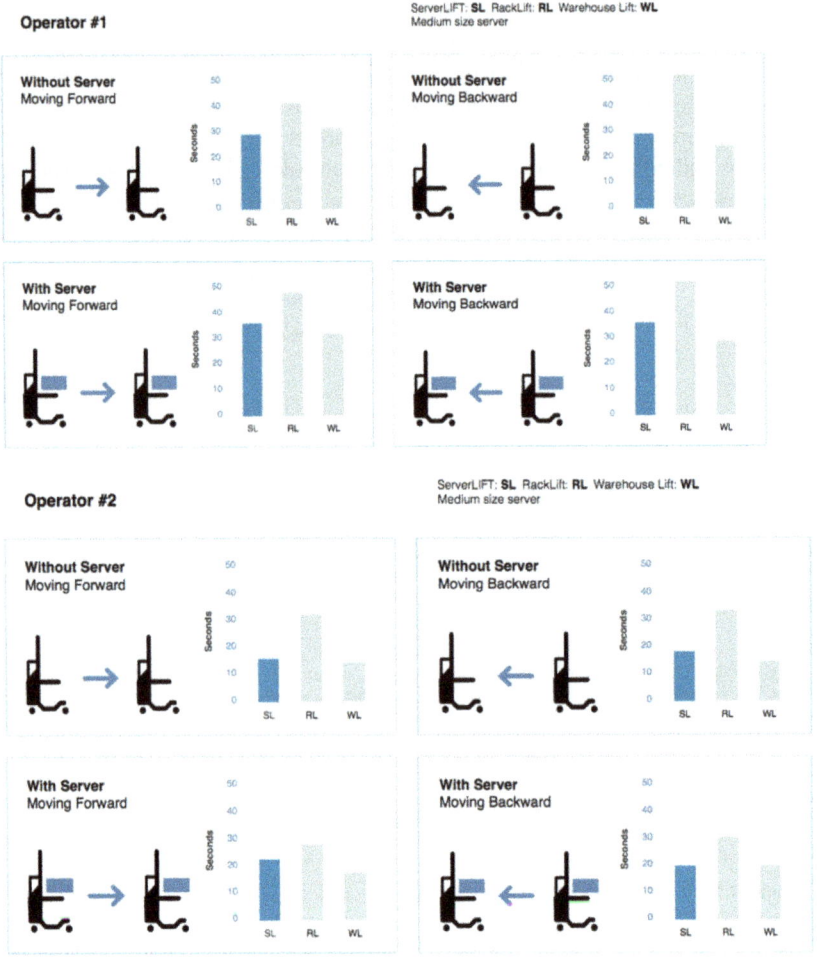

Fig. 2. Navigation time comparison.

This data presents Warehouse (WL) was the fastest, ServerLIFT (SL) was second and RackLift (RL) was the slowest. Both operators showed the same results regardless of lift operation directions and a server on the lifts.

3.2   Navigation Adjustments

In order to measure the difficulty in maneuverability of the products, the numbers of navigation adjustment were counted, and the majority of adjustments were observed in each corner of the aisle as shown in the Fig. 3 below.

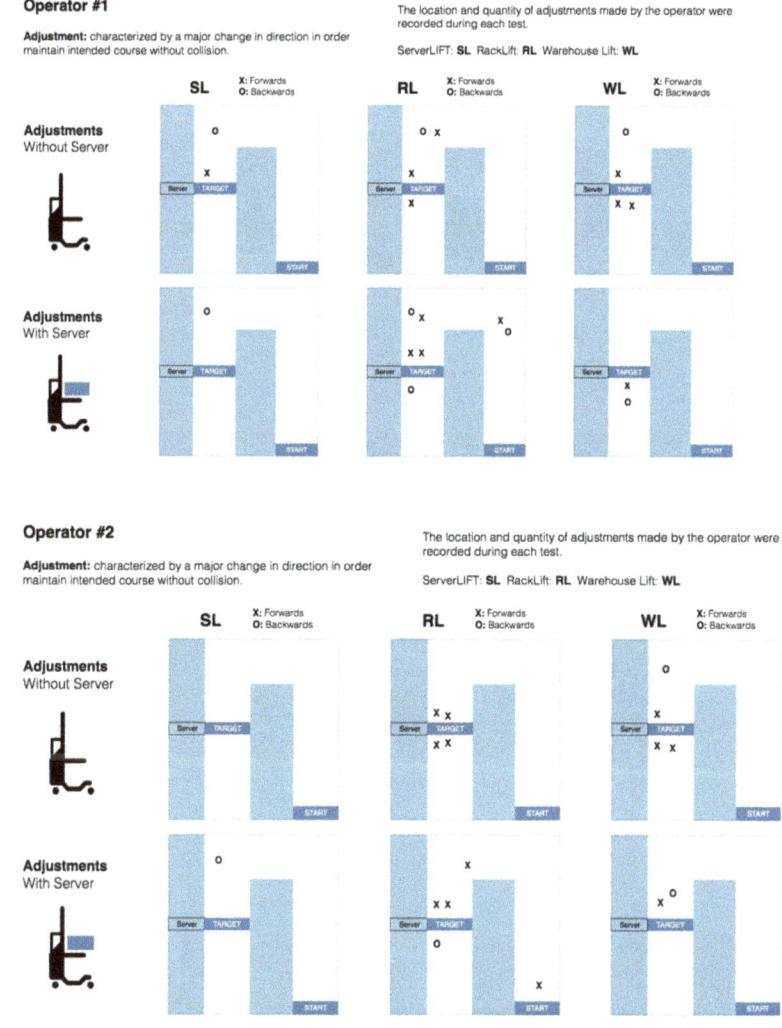

Fig. 3. Navigation adjustment numbers.

For operator 1, SL without a server had only one adjustment in forward and backward operation. SL with a server shows only one adjustment in backward.

For operator 2, SL without a server, there was no adjustment, and only one adjustment with a server in backward operation. The other two product competitors showed 4–7 adjustments in navigation test, and this data demonstrated SL was the easiest product among the three in regard to maneuverability.

4 Safety Test and Observations

In order to discuss the product safety, the test team conduced operational observations for all products, and looked for critical issues, such as pinch points, awkward positioning, poor ergonomics, etc. in evaluating safety characteristics (Figs. 4, 5 and 6).

Fig. 4. ServerLIFT safety test and observations

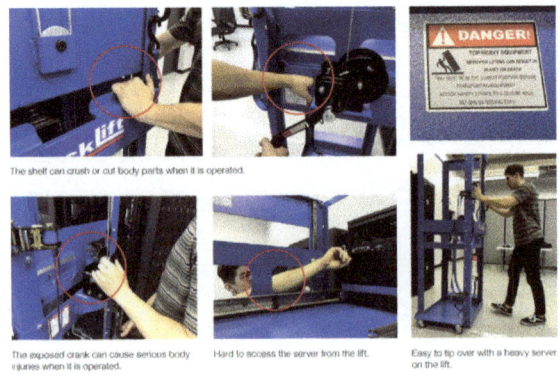

Fig. 5. RackLift safety test and observations

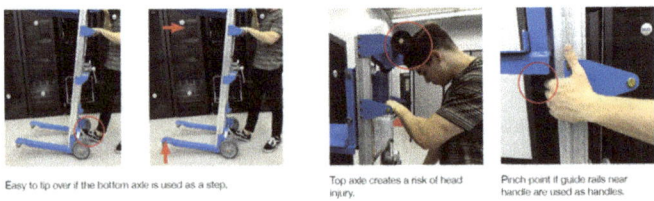

Fig. 6. Warehouse lift safety test and observations

5   Ergonomics Test

5.1   Testing Process

The objective of this test was to evaluate the ergonomics of each lift and compare their ease of use. This was done by observing the ability of lift operator to reach and use certain interaction points while driving, loading and unloading the lift at various heights. The testing procedure with the operators is listed below (Fig. 7).

(1)   Have the operator begin at start with the server shelf at the lowest point.
(2)   Observe as the operator drives the lift and aligns it with the server rack.
(3)   Observe as the operator applies the breaking mechanism
(4)   Observe as the operator adjusts the shelf height.
(5)   Observe as the operator loads the server onto the lift.
(6)   Starting from the beginning and complete steps 1–5 to reinstall the server into the housing.
(7)   Conduct steps 1–6 at the minimum, medium, and maximum heights with all three different servers.

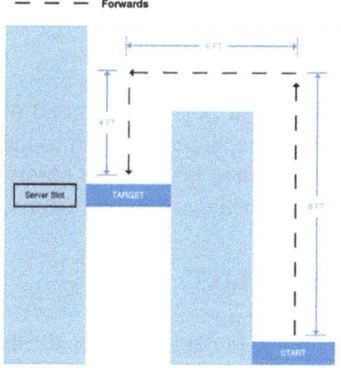

Fig. 7. Warehouse lift safety test and observations

5.2   Evaluation Process

During their operations following the process above, the following 5 areas that are directly related to the human interaction with the products were evaluated for ergonomics test, and the processes that were used for each part evaluation are listed below.

A. Handle Evaluation
   The testing team took the measurements of the steering handle and crank handle including length and diameter. Then, we compared the measurements to the average grip diameter of United States anthropometric data, and document the grip size and comfort while observing operator.

(1) Drive the lift from point A to B. (See the image in Handle Observation)
(2) Adjust the lifts position to load and unload lift
(3) Raise and lower the lift.

B. Crank Evaluation

The measurements to the average grip diameter of United States anthropometric data for the crank handles were compared and documented the grip size and comfort while observing operator:

(1) Adjust the lifts position to load and unload lift
(2) Raise and lower the lift.
    Evaluate the operators' ability to line up the server and operate the crank simultaneously.

C. Foot Operating Evaluation

The testing team evaluated foot clearance and comfort while operating lift brakes. We observed the foot space and contact while both driving the lift and installing the server.

D. Visibility Evaluation

The operators' visibility was observed while both driving the lift and installing the server at,

(1) Max Height
(2) Mid Height
(3) Low Height

E. Operation Evaluation

The testing team had the operator begin at start and observed any contact points made by natural tendencies while aligning the lift to the server housing. While the operator unloads the lift at all three heights, we observed any uncomfortable arm contact, angles, or obstacles experienced by the operator.

5.3 Handle Observation (ServerLIFT)

A. Current handles place additional pressure between the thumb and pointer finger and at the wrist.
B. Once the operator reached the housing, he would sometime push the lift from the side to get as close as possible to the server housing.

A.   B.

In general, operators felt comfortable with the amount of control provided by ServerLIFT's vertical handle orientation. They noted that the two vertical handles provided superior ability to pivot around corners, however, they would prefer if the bars did not spin freely. Adding handles or intended contact points to either side of the lift help take advantage of ServerLIFT's ability to move freely and quickly be nested closely with server housing, an area that other lifts struggled with.

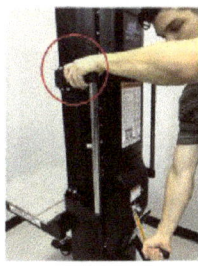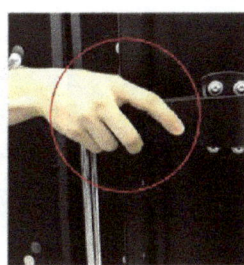

Operator naturally grabs onto the top handle brackets while operating the crank to lower or raise the platform, though this position is contrary to the published operating instructions. It was found that the top bracket of the handle encountered a high level of interaction with the operator's hand both during crank operation and navigation, although positioning one's hand there is contrary to the published operating instructions. This is a crucial point both used to support the weight of the operator and increase the leverage while making turns.

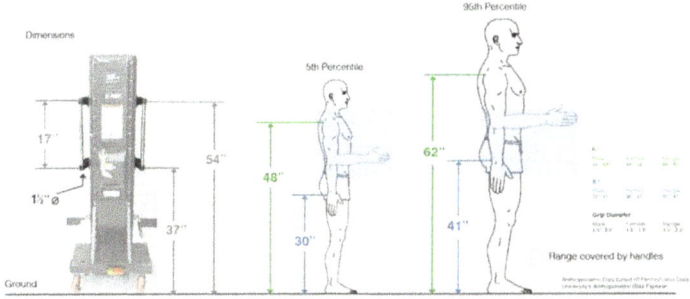

(1) ServerLIFT handles cover 53% of optimal arm height between the 5th and 95th percentile of users.
(2) ServerLIFT handles fall within the maximum and minimum grip width between the 5th and 95th percentile of users.

5.4 Crank Observations (ServerLIFT)

A. When operating the lift, the operator needs to be able to see the rails and lift the shelf to drop in the server simultaneously. When attempting to install or un-install the server off the left side of the lift, the operator has to reach further for the crank compromising level of visibility on the server rails.

B. While operating the crank the operator has a tendency to want to continue holding the handles as leverage and to support body weight bending over.

C. Operator can only line up one side of rails while operating crank simultaneously causing continuous trips back and forth to make small adjustments.

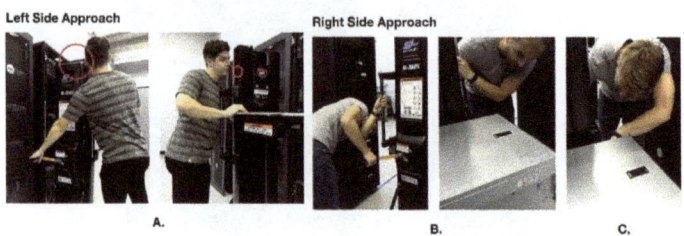

Small adjustments of the shelf height are crucial to lining up servers with the housing rails. Each small adjustment of the shelf height made by the crank is followed by re-evaluation and adjustments with the alignment of the rails on the side of the lift opposite the crank. The number of times traveled between the crank and the far side of the lift created a process that operators found tedious or annoying.

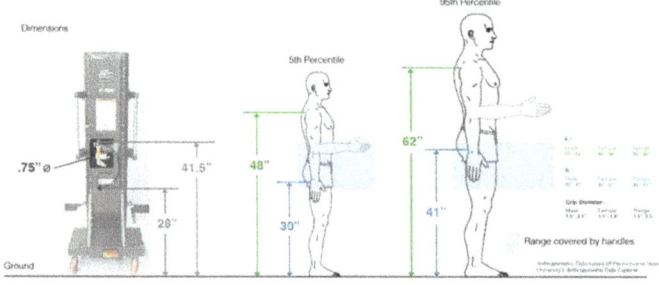

(1) While the crank is within the optimal height for lower percentile users, operators in the higher percentiles have to bend over to fully operate the crank.

(2) The ServerLIFT crank handle has an appropriately sized grip for lower percentile users (i.e. does not exceed the max. grip width), however, it may feel small for operators in the higher percentiles.

5.5   Foot Operating Observations (ServerLIFT)

A. Plenty of room for the operator's foot to engage and disengage the brake

B. While navigating the lift, there is enough clearance for the operator's foot to comfortably make steps underneath the frame.

A.   B.

According to the operators' feedback, they highly preferred ServerLIFT's easy to apply brakes, which were easy to access and quickly tell when they were and weren't apply to the lift.

Overall SeverLIFT offers plenty of foot space while operating the lift allowing for the operator to make navigational step within the boundaries of the lifts frame. The braking system and pedals on ServerLIFT receive superior reviews.

5.6   Accessory Observation (ServerLIFT)

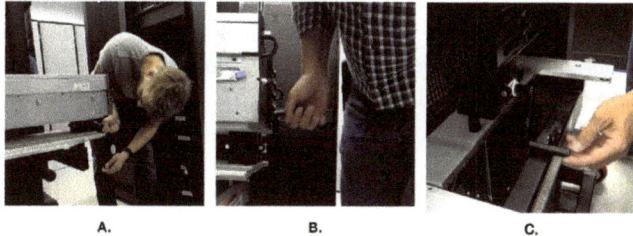

A.   B.   C.

A. Though, there is a decal on the unit that says "MUST READ & UNDERSTAND OPERATOR'S MANUAL BEFORE USING", operators were unclear as to the order in which the accessory should be used.
B. To use the accessory crank without obstruction the lift must be lined up far enough from the server housing that the server does not pass the end of the lift shelf when fully extended on the rails.
C. The accessory crank cannot be used simultaneously with the shelves sliding feature as it becomes obstructed by the frame.

Operators did not find the accessory intuitive, as it took some time for the operators to figure out the best way to make use of the accessory. Operators also tend to over-power the accessory when possible by lifting server themselves, but would find this accessory useful for the larger servers.

In general, both operators agreed that the use of the accessory could prove helpful after experience mastering the timing of its use. The largest obstacle with this accessory is knowing exactly where to the line up the lift, so that the server will not obstruct the path of the accessories crank. The crank of the accessory itself faced many obstacles.

5.7   Visibility Observation (ServerLIFT)

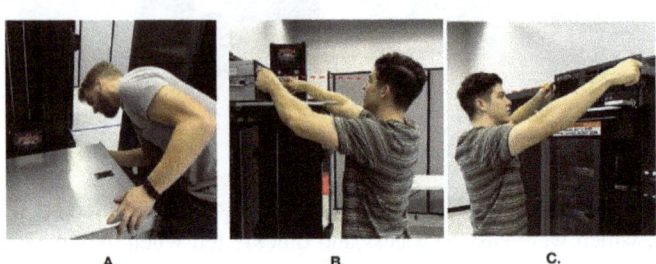

A.   B.   C.

A. Small amount of space between server and the lift frame on the right sight limits visibility.
B. The majority of operators have limited visibility when unloading server into the highest rack position.

The visibility of ServerLIFT suffers in three primary areas. Limited visibility server while operating the crank, limited visibility while installing and un-installing servers at maximum heights without a ladder or outside step, and visibility between the server and lift frame on the operator's side of the lift.

6   Conclusion

Product interaction, ergonomics, and usability are the most significant elements to be considered and evaluated in developing industrial products. The server lift products involve human interactions and must include a safety evaluation or their operation could cause serious body injuries. This study's thorough product evaluation and observation looking at product safety and ergonomics will provide continuing valuable information to the product developer, as well as, the potential product buyers and users.

In general, the ServerLIFT product shows better performance in regard to the navigation test using the measures of operation time and adjustments at the corners of aisles. A number of discussions and brainstorming activities were conducted among the team members to come up with testing procedures, data collection, analysis techniques, and the methods for conducting observations. Through this study, Industrial Design Students were able to understand the importance of product evaluation and gain valuable experience in developing testing methodologies.

User Evaluation of MyStudentScope: A Web Portal for Parental Management of Their Children's Educational Information

Theresa Matthews[1]([⊠]), Ying Zheng[2], Zhijiang Chen[1],
and Jinjuan Heidi Feng[1]

[1] Computer and Information Sciences Department, Towson University,
8000 York Rd, Towson, MD 21252, USA
tscott2@students.towson.edu, zhijiang@chen.me,
jfeng@towson.edu
[2] Computer Science and Information Technology, Frostburg State University,
101 Braddock Rd, Frostburg, MD 21532, USA
yzheng@frostburg.edu

Abstract. Parents and caregivers often struggle to successfully manage data from a variety of sources in a myriad of formats regarding their children's education. To address challenges that have been identified by previous research, a MyStudentScope (MSS) was designed with the integration of proposed solutions and recommendations from subject matter experts in education. In order to ensure that the system can fully meet users' needs, a user study was conducted investigating participants' perceptions of MSS. Because parents tend to use paper-based methods to archive and retrieve information regarding their children's education, the task performance through the use of the MSS web portal was compared to the paper-based methods. Situations parents/caregivers may encounter related to their children's education and extracurricular activities were simulated during the study. We present findings based on analysis of user responses and provide recommendations for improvement of the MSS design.

Keywords: Parents · Education · Personal information management
PIM · Information organization · Web portal

1 Introduction

Parents and caregivers often struggle to successfully manage data from a variety of sources in a myriad of formats regarding their children's education. Over the years, information can get lost or become extremely difficult to recall or retrieve for parents. To address challenges that have been identified by previous research (e.g., [3]), a web portal, named MyStudentScope (MSS), was designed with the integration of proposed solutions and recommendations from subject matter experts in education. Specifically, MSS has four primary functions: monitoring, retrieving, communication and decision-making. These functions are expected to help its users, namely parents, guardians or anyone else that is responsible for school-aged children, access and analyze collected data. In order to ensure that the design of MSS can fully meet users' needs, a user study

© Springer International Publishing AG, part of Springer Nature 2019
T. Z. Ahram and C. Falcão (Eds.): AHFE 2018, AISC 794, pp. 111–121, 2019.
https://doi.org/10.1007/978-3-319-94947-5_11

was conducted investigating participants' perceptions of MSS. Participants included both parents who have used a school-provided education management system and adults who may use a school-provided education management system in the future. We present findings based on our analysis of task performance as well as subjective questionnaires and comments. Participants' recommendations regarding functionality they expected but was missing form MSS is also presented. Our results indicate that improvements in parental management of information regarding a child's education were achieved using a technology-based educational information management solution tailored to parental needs.

2   Related Work

2.1   Challenges Managing Children's Education Information

Parental involvement in their children's learning results in better academic and behavioral performance of the children in school [5]. As parents acquire information regarding the education of their children, the ability to integrate the data is important to get the best use out of the information. Challenges related to dealing with the volume of information, recall and retrieval, communicating with educators and using the data to make decisions regarding children's education have been identified [3, 6].

A study was performed to determine perceived and realized challenges parents face when managing information regarding their children's education. In that study, 90% of parents surveyed indicated that they would be willing to dedicate some amount of time to organizing the educational information in effort to improve effectiveness in finding the information when needed. This indicates that there is room for improvement [3].

Although parents receive information regarding their children's education in a variety of forms including verbally, electronically and in hardcopy, per prior research results many parents use paper for data integration and archive. The parents reported that they rarely reviewed the files for updating or removal, they simply continued to add more paper to the archive [3]. As you would expect, this method of archival led to challenges with use of the information.

2.2   Why the Reliance on Paper? Support for Parents

Electronic student information systems available in most school districts are designed for the school and/or educator to provide information regarding the student's academics and attendance to parents on a frequent basis. They were not designed to optimize parental use of the provided information. Therefore, parents may not fully benefit from the wealth of information, or even worse, they may be overwhelmed by it. The electronic student information systems lack other information that student advocates and administrators recommend parents keep including documentation of teacher conferences/calls, copies of school work/assignments, official reports, major assessment results and benchmarks [1, 4, 7]. Currently available electronic student information systems allow parents to view their students' grades for all years the student is in the school system. However, if the student changes school districts, the parent is no longer able to view the students' academic history on one site.

A prior literature review revealed only one application that was specifically designed to assist parents in organizing and gathering information related to their child's education. That function of that application was to assist parents in preparing for Individualized Education Program (IEP) meetings. Recent inquiries indicate that the application is no longer available.

2.3 MyStudentScope Pilot

The MyStudentScope (MSS) web portal is intended to address challenges of (1) monitoring student's academic progress; (2) recalling and retrieving data; (3) communication; and (4) decision making with respect to the student's academic career. A prototype of the portal was developed. It was mocked up to appear to allow users to complete some of the functions that will be included in MSS. A preliminary study was conducted to gain preliminary understanding in task performance and user satisfaction when using the MSS web portal by simulating situations parents/caregivers may encounter related to their children's education and extracurricular activities.

3 MSS Design and Development

MSS is a web portal that was designed for parental management of information regarding their children's education. MSS is meant to be used with the current methods and systems via which parents receive information regarding their children's education like existing electronic student information systems. Instead of archiving information using paper, parents will archive the information in MSS by entering grades their student receives on assignments or in courses and uploading documents including samples of their student's schoolwork.

Based on feedback received from the review of the pilot system, a new user interface was designed to improve parents' interaction with the tool. The design premise for the functions remained largely the same, but the look and feel of the tool was modified to make it more user friendly and engaging.

3.1 MSS Functions

The portal's pages support the monitoring, retrieving, communication and decision-making functions. The mapping between the portal functions and pages is presented in Fig. 1.

Monitoring. Because the parent manages MSS, he/she uploads all of the students' grade information into the system. Regardless of the child's school or school system, the parent is able to view grades associated with the child's full academic career via the monitoring functions available in MSS. Parents are also able to view information related to their child's extracurricular activities, work samples and notes. All of the data the parent saves regarding his/her child's education is available for review.

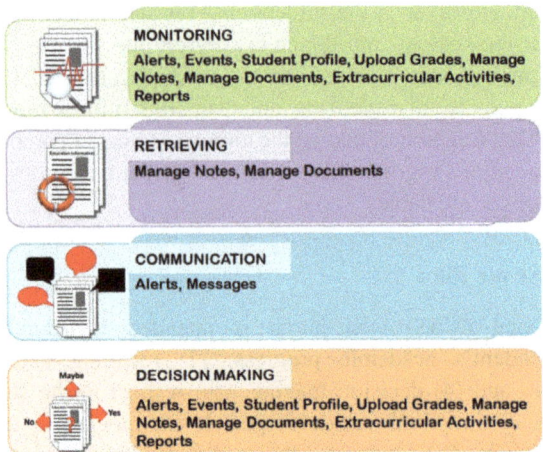

Fig. 1. Mapping between MSS functions and pages

Retrieving. Using search mechanisms, parents are able to retrieve previously saved information. In a previous study, the majority of parents surveyed said they would be willing to document the following information for each piece of educational information saved; date, source, category and description for items categorized as 'Other'. MSS was designed to allow parents to save and then later search and retrieve information based on these details.

Communication. The communication function allows parents to correspond with educators, coaches, and other providers from the tool. Because parents will ideally save important documents like work samples and assessments in the tool, the communication function provides a means for parents to attach these documents to messages with the goal of improving communication.

Decision Making. The decision making function of MSS is based on the extended data–information–knowledge–wisdom (DIKW) hierarchy as described by Mannion. The DIKW hierarchy is a method for describing how we move from data to information to knowledge to wisdom, but the extension includes decision-making, which reveals what direction to take in the future [2]. Via the decision making function, parents are able to observe trends and detect changes in their child's academic performance by viewing graphs and/or reports of the educational information stored in MSS. The graphical presentation of the data mitigates the need for parents to compare number values manually. For example, parents can view their child's average grades for all courses for all school years. However, if parents want to review numeric scores, they are able to search for them as needed.

3.2 MSS Interface Design

Many parents believe their children's education information is sensitive, therefore, each MSS account requires a username and password. All information saved in MSS is associated with a student or students. After account creation and initial login, the user is prompted to add a student to the account. A parent may add an unlimited number of students to his/her account. This flexibility allows parents of many children in different school systems to manage all of the information in one place. The general interface design of MSS is described in detail in a previous paper [4]. Some of the changes that were made to the interface as a result of preliminary user testing are described here.

Some existing education management systems allow users to export data; however, some do not. To accommodate both situations, MSS allows parents to enter assignment and course grades in bulk, file upload, or individually. The information requested and stored for both course and assignment grades includes Course, School, Teacher, School Year, Term and Grade. For assignments, the user also may enter an Assignment Name. Figure 2 shows how the grade input was improved from the pilot to the current version of the portal.

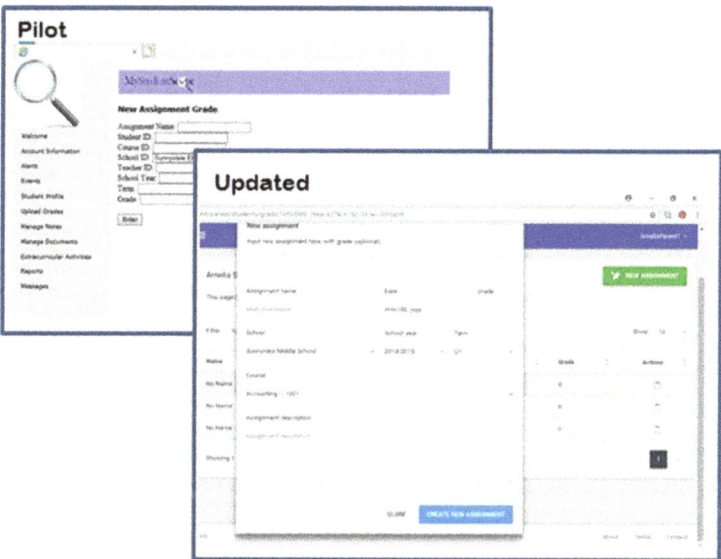

Fig. 2. Comparison between old and new assignment grade entry pages

For subsequent logins, the user is presented with the Dashboard page (Fig. 3). The Dashboard includes the course report as well as a graphical view of assignment grades that have been entered so that at a glance the user can see if anything is abnormal. The pivot tables at the bottom of each graph give the user the calculated average for the grade set they are currently viewing in the graph. The user has the option to filter each graph by school year.

Other improvements were made to the Events section of the portal. In the pilot, it was challenging for users to see details regarding events that had been added. The portal was therefore modified to give the user month, weekly and daily view options (see Fig. 4). By default, the user is still initially presented with the month view.

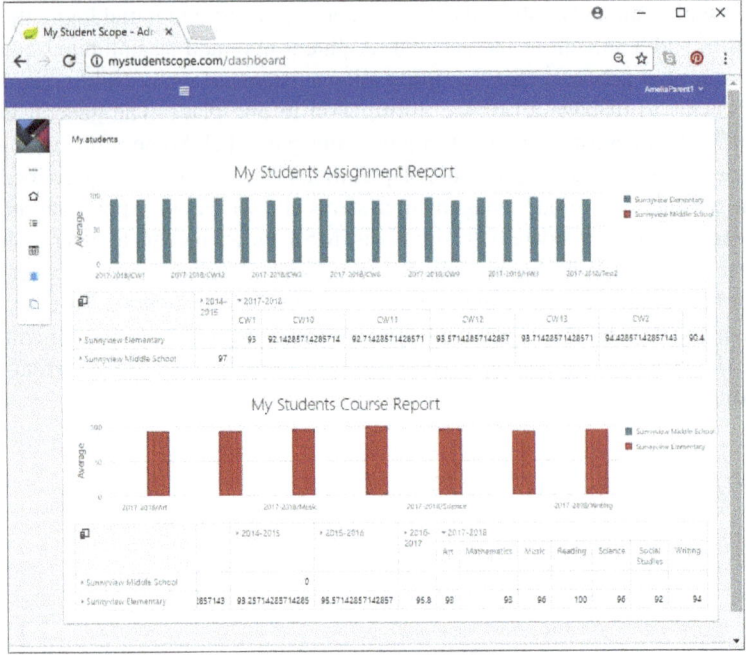

Fig. 3. MSS Dashboard page

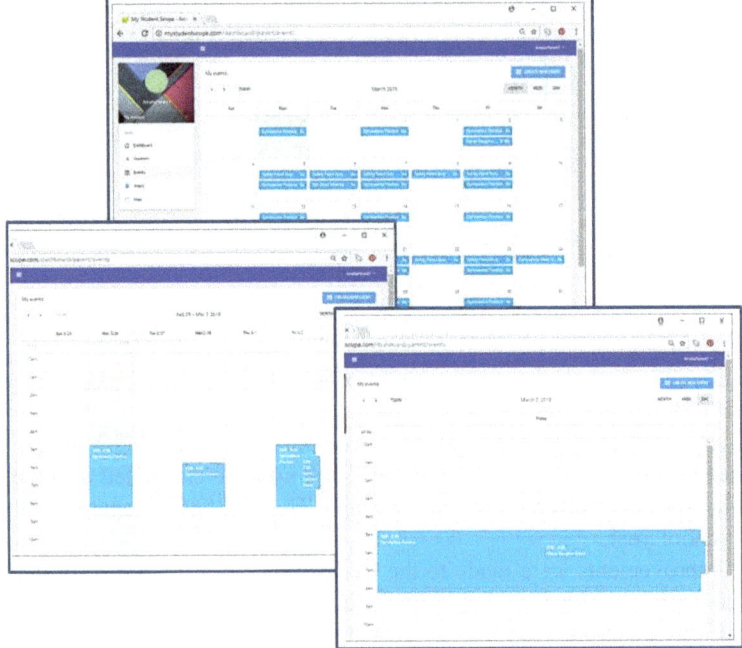

Fig. 4. MSS improved event viewing options

The user is permitted to schedule more than one event at the same time, but the conflict is visible to them.

4   Methods

We conducted a user study to evaluate the efficacy of the revised MSS portal as compared to traditional paper-based methods. We simulated situations parents/caregivers may encounter related to their children's education and extracurricular activities. The goals of the evaluation study are:

- To evaluate the overall functionality and interface design of MSS
- Collect user feedback on additional functions to implement in the portal
- Collect user feedback on communication functions currently under development

Regarding the third goal, we would like to collect information regarding how users currently and/or would like to record information regarding positive and negative events related to their children's education that they would like to or may need to recall later. This is particularly relevant to information that is not received in written or electronic form.

4.1   Participants

The typical user of MSS is a parent or guardian who is responsible for a school-aged child in grades Kindergarten through 12th grade. Eight parents (4 males) with at least one child in Kindergarten through 12th grade participated in the study. Six out of the eight participants were between the ages of 41–50 (average: 45, stdev: 5.41)

All participants have been using a computer, smart phone or tablet daily for more than ten years. Seven of the participants have an education management system available to them via their child's school. They all indicate that they accessed the system at least quarterly. Most access the system more frequently.

4.2   Tasks and Procedures

The user study consists of two conditions. In both conditions, the participants completed tasks requiring them to interact with information regarding their children's education. In one condition, they used paper resources to complete the tasks in the other condition they used data stored in MSS to complete the tasks.

The folder for the paper condition contained approximately 125 documents. The documents included report cards, interim reports, sample assignments, extracurricular schedules and sign-ups for the current school year and school newsletters for the current school year. For six of the eight participants, the documents were organized chronologically with the most recent documents on top. For two of the users the documents were further sorted by type. Data equivalent to the data in the paper folder was pre-loaded into MSS.

The order of conditions was balanced to control the learning effect. Four users completed the paper condition first and four users completed the MSS condition first.

Each user was given a brief demo of MSS prior to starting the MSS condition. For each condition, the participant was given a different sample student data set so the task results for both conditions would not be the same. The participants were not given any time constraints for task completion. If the participant asked for help or if we observed that the participant was not making progress toward task completion we would provide clarification on the task or guide the user to how they could solve the task.

At the end of each test condition, the participants completed a questionnaire to provide feedback on their experience. After completing both conditions, participants completed a survey comparing their experiences, reporting challenges and recommendations for changes or additional functions.

5   Results

5.1   Task Completion Time

The task listing and task completion times are reported in Table 1. A paired samples t test suggests that there is a significant difference between the MSS condition and the paper condition in the time it took to determine whether there are schedule conflicts $(t\ (7) = -3.45,\ p < 0.05)$ (Task 3). Participants took significantly shorter time to complete the task in the MSS condition than the paper condition. Paired samples t tests find no significant difference between the MSS condition and the paper condition in the time it took to complete the other tasks (Task 1: $t\ (7) = -1.53$, n. s.; Task 2: $t\ (7) = -0.91$, n. s.; Task 4: $t\ (7) = -0.14$, n. s.; Task 5: $t\ (7) = -0.94$, n. s.).

Table 1. Tasks with completion times (seconds) for each condition

ID	1. Determine the student's average grade in a specified subject area		2. Determine grade for specified grade level and marking period		3. Determine if there are schedule conflicts for specific date		4. Determine if recent grade is normal for student		5. State information used to determine if recent grades are normal	
	MSS	Paper	MSS	Paper	MSS	Paper	MSS	Paper	MSS	Paper
P1	159	46	47	79	30	65	56	125	41	44
P2	34	101	74	203	35	106	55	35	96	159
P3	171	352	160	56	42	375	56	140	68	69
P4	156	206	66	26	25	180	36	25	37	22
P5	136	357	170	224	21	236	253	90	67	48
P6	544	641	366	95	59	168	83	94	115	126
P7	120	300	60	60	60	180	60	120	60	180
P8	240	120	240	120	60	60	60	60	60	30

5.2   Participant Feedback

To understand the participants' preference for managing information and technology experience, each participant completed a questionnaire before the test. Responses to Likert scale questions from the pre-test questionnaire are summarized in Table 2. Similar to results seen in prior studies, most parents indicated that they use both paper and technology to manage information. Three participants disagreed or strongly disagreed that they have a tendency to use paper-based methods to organize information. All participants began the study with a positive opinion of the ease with which technology can be used to manage their children's educational information.

Table 2. Summary of answers to pre-test questionnaire Likert scale questions

ID	1. Tend to use paper to organize	2. Tend to use technology to organize	3. Manage education info like other info	4. Managing education info is important	5. Using technology to manage education info is easy
P1	2	5	4	5	5
P2	4	4	4	5	4
P3	1	5	5	5	4
P4	4	4	4	5	4
P5	*1*	4	4	5	5
P6	*4*	4	4	5	4
P7	*3*	4	4	4	4
P8	*3*	5	4	5	4

All participants answered a questionnaire after each test condition to evaluate their experience. The questionnaire after the MSS condition also asked participants to provide suggestions for improving the portal. The majority of the participant feedback was positive in favor of MSS. As shown in Table 3, all but one participant strongly agreed or agreed that using MSS to perform tasks was easier than using paper-based methods.

Participants also provided some recommendations for improving MSS. Some of the recommendations are already in development (e.g. communication function). Others were new. One participant recommended that parents be able to link to the school website from MSS. Another parent suggested that MSS have a designated place for IEP data.

Table 3. Summary of answers to MSS v. paper post-test comparison questionnaire Likert scale questions

ID	1. MSS was easier to use than paper	2. Completed task more quickly with paper	3. More productive with MSS than paper	4. Recovered from errors faster with paper	5. Easier to find information with MSS	6. More frustration using MSS than paper
P1	5	1	5	1	5	1
P2	2	2	2	4	2	4
P3	5	1	5	1	5	1
P4	5	2	5	2	5	1
P5	5	2	4	2	4	2
P6	4	2	4	2	4	2
P7	5	2	5	2	5	2
P8	5	4	5	2	5	2

6 Conclusions and Future Work

The user feedback indicates that MSS has great potential for improving how parents use the information they receive regarding their children's information. Most participants were able to complete tasks using MSS after only a brief demonstration of the tool. With more use and with more instructive on-screen documentation and prompts, we expect the benefits of using MSS to surpass the use of paper.

For future work, we will modify the portal based on feedback received during this study and challenges observed during the user test. We will complete a comprehensive study where users are asked to complete common tasks encountered by parents managing their children's educational information and extracurricular schedules using MSS. The communication and alerting capabilities that were not previously exercised will be included in that study.

References

1. Crabtree, R.K.: The Paper Chase: Managing Your Child's Documents (1998). http://www.wrightslaw.com/info/advo.paperchase.crabtree.htm. Accessed 14 Feb 2013
2. Mannion, P.: Optimal Analysis Algorithms are IoT's Big Opportunity, 12 January 2015. http://electronics360.globalspec.com/article/4890/optimal-analysis-algorithms-are-iot-s-big-opportunity
3. Matthews, T., Feng, J.H.: Understanding parental management of information regarding their children. human interface and the management of information: information, knowledge and interaction design. In: 19th International Conference, HCI International 2017, Proceedings, Part I, Vancouver, BC, Canada, 9–14 July, pp. 347–365. Springer International Publishing (2017)

4. Matthews, T., Feng, J.H., Zheng, Y., Chen, Z.: MyStudentScope: a web portal for parental management of their children's education information. human interface and the management of information: information, knowledge and interaction design. In: 20th International Conference, HCI International 2018, Proceedings, Las Vegas, Nevada, July 15–20 2018. Springer International Publishing (2018)

5. Patrikakou, E.N.: The Power of Parent Involvement: Evidence, Ideas, and Tools for Student Success. Lincoln: Center on Innovation and Improvement (2008)

6. Roshan, P.K., Jacobs, M., Dye, M., DiSalvo, B.: Exploring how parents in economically depressed communities access learning resources, pp. 131–141. ACM (2014)

7. Wright, P., Wright, P.: The Special Education Survival Guide: Organizing Your Child's Special Education File: Do It Right!, 21 July 2008. http://www.fetaweb.com/03/organize.file. htm. Accessed 14 Feb 2013

Usability Evaluation of a State University Grade Encoding System

John Paolo Isip[(⊠)] and Hazel Caparas

Bulacan State University, Guinhawa, Malolos City, Bulacan, Philippines
paoloisip7@gmail.com, hazelcescaparas@gmail.com

Abstract. Computerized encoding and posting of grades every end of an academic period is designed to provide convenience to teaching personnel with huge number of students like in the case of a state university. The initial assessment of usability experience showed that the rating was significantly below satisfactory level. Thus, the researchers conducted a usability test to evaluate the effectiveness, efficiency and satisfaction of the users. A three-phased usability test involved pre-task survey, scenario and tasks performance and post-task survey. The respondents evaluated the system in terms of different usability measures such as consistency, error tolerability, learnability, memorability, efficiency and likeability. The result primarily revealed that manual encoding and importing worksheet both occupied most of the performance time. In addition, learnability and memorability had the lowest satisfactory rating among respondents. As a result, this lead to the development of ergonomic interventions to enhance the usability experience of grade encoding system.

Keywords: Human-computer interaction · Usability · Interface design

1 Introduction

The way by which computer system communicates with human is through its user interfaces and this component highly affects the productivity of the users in performing their tasks. The growth of Human Computer Interaction (HCI) focuses on developing an effective, efficient and easy to use interactive computer system to avoid deterioration of user performance, confusion, panic, boredom, frustration, incomplete use of the system, system abandonment, modification of the task, compensatory actions and misuse of the system [1, 2]. The field of HCI intends to design a computer system that assists the users in performing their tasks safely and without wasting any resources such as time [3]. Increasing computer technology does not guarantee improvement in the productivity of the users. Thus, it is more essential to give consideration on the needs of the users rather the application of advancement in technology. The principles of HCI promotes designing user interface of computer systems that reflect on the appropriate application of available technologies [4]. Moreover, a well-developed interface design can provide substantial influence on learning duration, performance speed, minimization of errors, and level of user satisfaction [5].

Usability refers to the attribute of the product or a system of being easy to use. Based on ISO 9142, it is also parallel to the degree of effectiveness, efficiency and satisfaction

© Springer International Publishing AG, part of Springer Nature 2019
T. Z. Ahram and C. Falcão (Eds.): AHFE 2018, AISC 794, pp. 122–130, 2019.
https://doi.org/10.1007/978-3-319-94947-5_12

of the users to the product or system. Accordingly, a usability testing is a technique of collecting qualitative and quantitative data to gauge the usability experience of the users to identify problem areas and help the developers to modify and improve the product or system. In particular, the usability testing of user interfaces as components of machines and software is currently highly essential due to unceasing advancement in technology. Low usability interfaces may result to some inefficiency such as decrease in productivity, confusion, small number of user retentions, and wrong execution of task [2]. The fundamental goal of user interface design is to create a simple and efficient interaction between human and computer systems while performing a task [6].

The study focuses on measuring the usability of grade encoding system presently by a state university. One of the basic functions of the system is to aid the teaching personnel in periodic encoding and posting of final grades in the system. All faculty members have their own access to the system where they can either manually encode the grades or import the worksheet of student grades. An ergonomic audit on the usability of the system shows that more than half of the respondents prefer manual encoding rather than doing the faster way of importing the worksheets of grades. Most users respond that one reason is due to lack of familiarity to this feature since the user interface design itself cannot provide feedback on how to utilize importation of worksheet. Consequently, the users frequently use manual encoding which is an indication of alternative mode of performing the task. As a result, the current study aims to evaluate the usability of the present user interface design of grade encoding system in terms of consistency, error tolerability, efficiency, learnability, memorability, and likeability. Moreover, it intends to develop ergonomic interventions using various principles of human computer interactions in order to enhance the usability experience of the users.

2   Conceptual Model

In order to evaluate the usability experience of the users, identify possible error prone segments, and recommend abstract design principles in enhancing the overall usability of the system, the proponents developed a three-phased methodology usability test (Fig. 1).

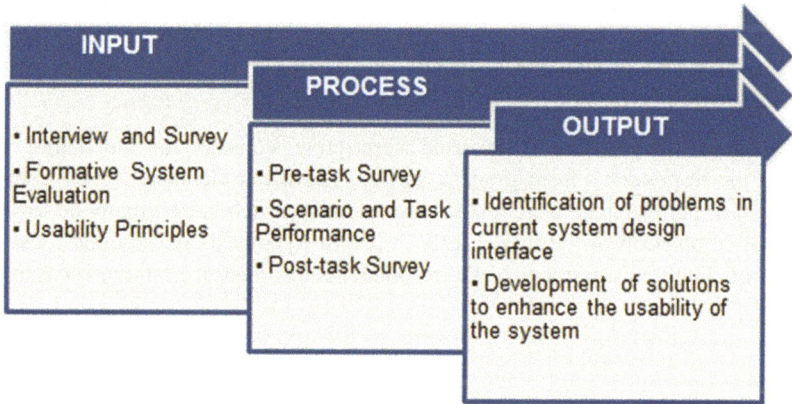

Fig. 1. Conceptual model of the grade encoding system usability test.

The methodology requires interview and survey questionnaire for pre-task and post-task survey. The scenario and tasks performance needs a formative system evaluation to assess the elemental activities of encoding and posting of grades. Lastly, the expected output of the study includes ergonomic intervention consisting of abstract design principles applicable in enhancing the current user interface of grade encoding system.

3 Methodology

Usability Test Three-Phase Procedure. The current study followed a three-phase procedure in evaluating the usability of the computer interface of grades encoding system. The methodology consisted of a pre-test survey, scenario and tasks performance and a post-task survey. The pre-task survey gathered the demographic profile of the respondents and initially evaluated their usage history of the system. Next, scenario provided the overall activity to be performed by the respondent. In this case, each respondent had to encode and post the final grades of a class, which consisted of 50 students either manual encoding of the final grades or importing the worksheet of final grades to be posted. The tasks involved the elements that each participant needed to perform during the usability test. Task performance involved actions that the evaluator asked the participants to execute on the interface under investigation [7]. The study developed prescribed tasks in which the proponents predetermined the tasks that the participants had to follow. Measuring the elemental time of each task identified which step consumed more time and helped in determining the problems in interface design. The last stage of the usability test involved post-task survey. This phase qualitatively evaluated the usability experience of the respondents while encoding and posting the final grades. The post-task survey instrument included usability measures such as consistency, error tolerability, efficiency, learnability, memorability, and likeability. The survey questionnaire used a 5-point Likert Scale with 5 being "Outstanding" and being "Needs Improvement".

Sampling. There was an estimated number of 803 faculty members during the time of the study. Using the Slovin's formula (Slovin, 1960) to calculate the sample size, it was approximated that 267 teaching personnel were required to obtain ±0.05 margin of error. However, the proponents gathered data from 135 respondents thus giving a margin of error of ±0.0785.

Data Analysis. The current study used descriptive statistics for the demographic profile of the respondents from pre-task survey. The mean elemental times from scenario and task performance were analyzed to identify which stage greatly added up to the overall completion of the task. Lastly, a test of hypothesis one-sample t-test validated the qualitative assessment of the respondents to different usability measures.

4 Result and Discussion

Descriptive Statistics for Pre-task Survey. The preliminary phase of the usability test involved pre-task survey, which described the characteristics of the respondents and displayed the overall assessment of the usability of the system design interface. The result showed that the highest number of the respondents was at the age of 51 and above which exhibited the most severe limitations on the system design interface. Moreover, 70.65% of the respondents have used the grade encoding system with utmost 10 times (Figs. 2 and 3).

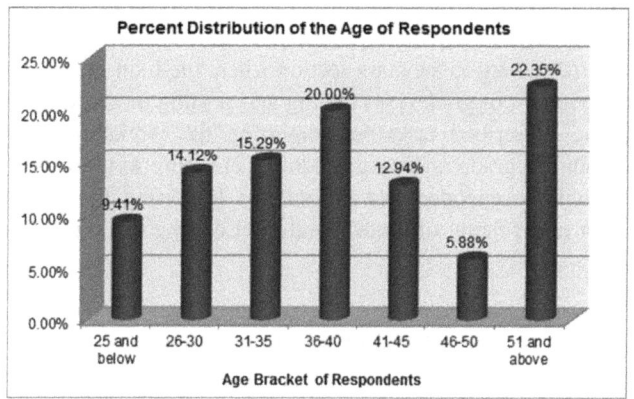

Fig. 2. Age of the respondents.

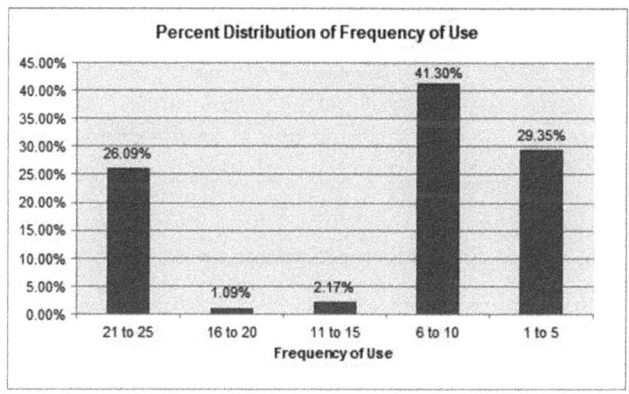

Fig. 3. Frequency of use of the system.

In addition, the pre-task survey initially evaluated the overall usability experience of the respondents to the grade encoding system. The result revealed that the mean score for satisfaction rating is 1.8112, which implied that the system design interface required improvement. A test of hypothesis one sample t-test verified that the mean score for satisfaction was statistically less than 3 or below satisfactory level (P-value = 0.000). Moreover, 67.32% of the respondents manually encoded grades and only 32.68% imported worksheet to the system. It indicated that there was a presence of workaround or the use of alternative method to complete the tasks in which manual encoding of grades represented the workaround option.

Scenario and Tasks Performance. The second stage of usability test consisted of scenario and tasks performance of encoding and posting of grades. The proponents established a scenario that comprised of encoding and posting of final grades of a class having an average of 50 students. Then, the scenario was divided into elemental parts called tasks. The proponents predetermined the tasks given to the participants. The prescribed tasks for manual encoding consist of the following: (1) Going to grade encoding module; (2) Going to the class section where the final grades will be encoded and posted; (3) Encoding of grades; (4) Saving and posting of grades; and (5) Logging out. Likewise, the prescribed tasks for importing the worksheet comprise of the following: (1) Going to grade encoding module; (2) Going to the class section where the final grades will be encoded and posted; (3) Importing the worksheet of final grades; (4) Saving and posting of grades; and (5) Logging out (Figs. 4 and 5).

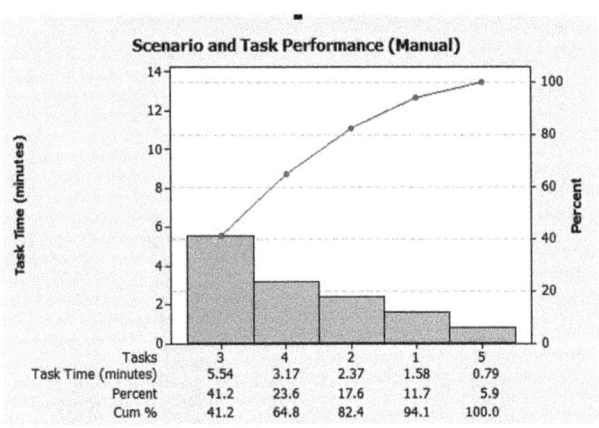

Fig. 4. Scenario and tasks performance for manual encoding

The result has shown that tasks of importing worksheet of final grades, saving and posting of grades and going to the class section consumed 80% of the performance time. Task 3 or importing the worksheet of final grades to the system occupied the greatest portion of performance time for the reason that the respondents encountered setback in verifying the imported worksheet. The respondents were required to input some character to proceed with the importation of worksheet in which this activity was not part of the normal process. Moreover, the respondents encountered slight difficulty in finding the location of icon for importing file. Similarly, scenario and task performance for manual encoding indicated that the tasks of manual encoding of grades,

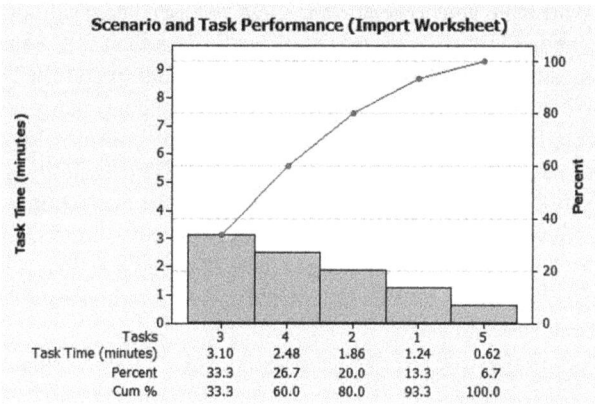

Fig. 5. Scenario and tasks performance for importing worksheet

saving and posting of grades and going to the class section occupied 82% of the performance time. Evidently, manual encoding of grades had the highest completion time, which was equivalent to 41% of the performance time. Manual encoding time for the given scenario was 78.71% higher than the importing the worksheet of final grades to the system.

Post-task Survey. The last phase was the post-task survey wherein respondents evaluated different usability measures such as consistency, error tolerability, efficiency, learnability, memorability, and likeability using a 5-point likert scale with 5 being the highest satisfactory rating. The first criterion was consistency of the user interface design of the grade encoding system, which referred to the similarity or the likeness of each components of the system. A test of hypothesis using one sample t-test was used and the result indicated that the consistency rating was statistically less than 3 or below satisfactory level (Consistency Rating = 1.89; P-value = 0.000). The main area to be enhanced in terms of consistency was the grouping and presentation of information in logical order. The second criterion was error tolerability of the grade encoding system wherein the system itself prevented the users from making errors and helped to avoid starting over from the beginning once an error occurred. The result revealed that the error tolerability rating was statistically less than 3 or below satisfactory level (Error Tolerability Rating = 1.99; P-value = 0.000). The main area to be improved was the presence of feedback when a nonconforming action was executed. For instance, there was no indication or feedback given by the system when the user failed to save the grades prior to posting. The third criterion was efficiency or the performance of the user while using the system without wasting any resources such as activity time. The outcome indicated that efficiency rating was statistically less than 3 or below satisfactory level (Efficiency Rating = 1.93; P-value = 0.000). Based on the qualitative data gathered from the respondents, the user interface design of the grade encoding system needed to be improved to cater both novice and elderly users efficiently and effectively. The fourth measures were the learnability and memorability of the grade encoding system design. These criteria focused on how fast an infrequent user could adapt to the design of the system to be able to perform the task at ease. The result identified that the learnability and memorability rating was statistically less than 3 or below satisfactory level (Learnability and Memorability Rating = 1.88; P-value = 0.000). The respondents reviewed that instructions about the use of the system should be readily retrievable when necessary particularly for limiting users such as novice and elderly users. The last measure was the likeability or the subjective rating of the usability experience and satisfaction of the users in using the system. The result of the evaluation revealed that the likeability or the overall satisfaction rating of the users was statistically less than 3 or below satisfactory level (Likeability Rating = 1.96; P-value = 0.000). Given the above discussion, the usability measures that obtained the lowest ratings were learnability and memorability of the system design.

5 Conclusion

After evaluating usability experience of the users following a three-phased usability testing including 135 participants out of 803 faculty members, the results showed that 62.33% of the respondents normally used manual encoding, which indicated presence of workaround in encoding and posting of grades. The general assessment of the respondents on the usability of grade encoding system indicated that the satisfaction rating was 1.8112 and statistically less than satisfactory level. Scenario and task performance phase revealed that elemental activity of manual encoding and importing worksheet both occupied most of the performance time. However, manual encoding time was 78.71% higher than importing worksheet of final grades into the system. The last phase of usability test showed that learnability and memorability (Learnability and Memorability Rating = 1.88; P-value = 0.000) had the lowest satisfactory rating among respondents. Accordingly, the findings of the study lead to the development of ergonomic interventions to enhance the usability experience of grade encoding system.

6 Recommendation

As from the results of the three-phased evaluation, the researchers found out that there are significant issues within the current interface that needs to be improved. Primary findings are low usability experience in terms of error tolerability, learnability and memorability. The used of feedbacks and dialog boxes are suggested so human error such as slip click or accidentally clicking of a specific command would be prevented. In addition, the users could recover easily once an error occurred [8] (Fig. 6).

Making the text and icon design scheme more visible and grouping these elements into logical order support in increasing the usability experience of the users in terms of consistency [9]. The interface design should assist novice and elderly users in executing the task. The system has to make all necessary steps visible [10]. This approach increases the efficiency of all types of users even the identified limiting users of the system.

One observation concerning the task of going to class sections showed that the execution of the activity was not straightforward thus added up to the performance time. The user had to search and select the current academic term to display the list of

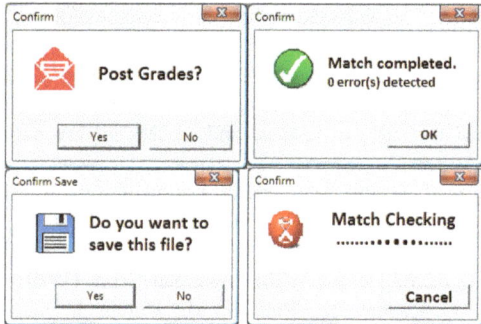

Fig. 6. Sample feedback and dialog boxes

class sections. Thus, the study recommended discrimination of the present period from previous academic periods. The system could provide direct selection or automatically displaying the list of class sections to avoid searching time of the users [11]. Important principle to be considered is the graying out of inactive sections to emphasize the area for the task.

The developer also needs to consider eliminating unnecessary visual elements of the user interface in order to easily find the needed functions and highlight the important information [12]. Other than text, graphical representations, sound, color and spatial position at the same time may increase the ease of use of the system [13]. The interface should also include large and vivid font and color for headings. Color helps us in memorizing certain information by increasing our attention levels. The more attention focused on a certain stimuli, the more chances of the stimuli to be transferred to permanent memory storage [14].

These abstract designs approach are seen to minimize the performance time and enhance the usability experience of the users in terms of consistency, error tolerability, efficiency, learnability, memorability and likeability. Nevertheless, area for future study could consider the technical feasibility of these recommendations to develop a high-fidelity usability test in order to evaluate the level of usability of the proposed system.

References

1. Nielsen, J.: Estimating the number of subjects needed for a thinking aloud test. Int. J. Hum. Comput. Stud. **41**, 385–397 (1994)
2. Galitz, W.: The Essential Guide to User Interface Design Second Edition, An Introduction to GUI Design Principles and Techniques. John Wiley & Sons, Inc., New York (2002)
3. Preece, J., Roger, Y., Benyon, D., Holland, S., Carey, T.: Human Computer Interaction. Addison-Wesley, Wokingham (1994)
4. Booth, P.: An Introduction to Human-Computer Interaction. Lawrence Erlbaum Associates Publishers, Hove/East Sussex (1989)
5. Shneiderman, B.: Designing the User Interface. Addison-Wesley, Reading (1992)
6. Kroemer, K., Kroemer, H., Kroemer-Elbert, K.: Ergonomics, How to Design for Ease and Efficiency, 2nd edn. Pearson Education, Asia Pte Ltd. (2001)

7. Nielsen, J., Norman, D.: Turn User Goals into Task Scenarios for Usability Testing. Nielsen Norman Group, Fremont (2014)
8. Kendall, K.E., Kendall, J.E.: Systems Analysis and Design, 8th edn. Pearson, London (2011)
9. Norman, D.: The Design of Everyday Things. The MIT Press, London (1998)
10. Gwizdka, J.: Implicit measures of lostness and success in web navigation. Interact. with Comput. **19**, 357–369 (2007)
11. Sandness, F.: Directional bias in scrolling tasks: a study of users' scrolling behaviour using a mobile text-entry strategy. Behav. Inf. Technol. **27**, 387–393 (2008)
12. Constantine, L., Lockwood, A.: Software for Use: A Practical Guide to the Models and Methods of Usage-Centered Design. ACM Press, New York (1999)
13. Smith, W., Dunn, J., Kirsner, K., Randell, M.: Colour in map displays: Issues for task/specific display design. Interact. Comput. **7**, 151–165 (1995)
14. Dzukifli, M., Muhammad Faiz, M.: The influence of colour on memory performance: a review. Malays. J. Med. Sci. **20**, 3 (2012)

How to Get to Know Your Customers Better? A Case Analysis of Smartphone Users with Chinese Input Method Based on Baidu Index

Wenchao Zuo[1,2(✉)], Yuhong Wang[1], and Yueqing Li[2]

[1] School of Business, Jiangnan University,
Wuxi 214122, Jiangsu, People's Republic of China
wenchaozuo@gmail.com, wyh2003@gmail.com
[2] Department of Industrial Engineering, Lamar University, Beaumont, TX, USA
yli6@lamar.edu

Abstract. The purpose of this paper aims to proposal a new idea to know the customers with using Chinese input method on smartphone better based on the big data analysis. Nowadays, China has the largest group of smartphone users. Everyday hundreds of millions of people get used to using their smartphones shopping, texting and searching for information and so on. A more efficient Chinese input method will definitely enhance their user experience and speed up the transactions. Furthermore, based on the massive user's behavior data analysis through Baidu Index, it can be identified the target-users attribute for the Chinese input method, and advance a more user-friendly Chinese input method of smartphones considering the social-cultural factors affecting users. So, it's meaningful to cluster the users and specify the user-based design features. Firstly, the most popular Chinese input method tools are introduced. Secondly, the research on the Chinese input method is analyzed to find the current research limitation. Thirdly, considering the limitation of the research on the Chinese input method, a big data analysis platform-Baidu Index is selected in order to research on the people's attributes and geographical distribution of using smartphone's Chinese input method. The conclusion and discussion are at the end of the paper.

Keywords: Chinese input method · Customers · User-friendly
Baidu index

1 Introduction

For 2017, the number of smartphone users in China reached 663.37 million based on the statistic by TABLEAU company [1]. Chinese input method, as a key tool for inputting information in a smartphone, is playing a more and more important role in people's daily life. For example, "WeChat", designed and operated by Tencent Company, is a messaging and calling App that allows you to easily connect with family and friends across the world. This App is the most famous smartphone Application in China and has 963 million monthly active users, among which over 50% are female

© Springer International Publishing AG, part of Springer Nature 2019
T. Z. Ahram and C. Falcão (Eds.): AHFE 2018, AISC 794, pp. 131–138, 2019.
https://doi.org/10.1007/978-3-319-94947-5_13

users [2]. When people want to text message with their friends using WeChat, they need to input the words using Chinese input method. Another interesting example is Alibaba Group's 11.11 Global Shopping Festival in China. Consumers including over 70% women spend RMB 168.2 billion (USD 25.87 billion) during the 24-h period and about 90% consumers used smartphones for their deals. So, obviously the target-users' attribute for different App is very different, and the social-cultural factors, such as geographical distribution, age, sex and so on will affect the usability. As a result, it's meaningful to identify the target-users attribute for the Chinese input method and advance a more user-friendly Chinese input method for smartphones.

2 Literature Review

The Chinese input method has many useful application scenarios on a smartphone, such as texting with friends, shopping online, searching for information, writing paper works and so on. When the user wants to edit a piece of Chinese words on mobile, Chinese input methods are always necessary. There are currently two kinds of Chinese input method: structure based input method and pronunciation based input method. The structure based on input method mainly includes Wubizixing (Chinese: 五笔字型), Wubihua (Chinese: 五笔划), Cangjie (Chinese: 仓颉), Sucheng (Chinese: 速成), Sanjiao (Chinese: 三角), Dayi (Chinese: 大易); The pronunciation based input method mainly includes Pinyin (Chinese: 拼音), Shuangpin (Chinese: 双拼), Jianpin (Chinese: 简拼), Bopomofo (Chinese: 注音). In iPhone's App store, the top-3 ranking in the Chinese input method are Sougou Pinyin Method (Chinese: 搜狗拼音输入法), Baidu Input Method (Chinese: 百度输入法) and IFLY Input Method (Chinese: 讯飞输入法); The ranking result of Chinese input method in Google Play is same as iPhone's App store (Fig. 1).

There are many studies in Chinese input method, and the outcome is significant. The research on Chinese input method mainly focuses on six aspects: (1) Spelling or typing correction in the Chinese words inputting process [3–5]; (2) Errors tolerant study, such as mistyped Pinyin [6], spelling errors [4, 7, 8]; (3) Input speed improving [9]; (4) Layout design [10]; (5) Pinyin-to-character conversion advance [5, 11, 12]; (6) Pinyin input method improving, such as neural network language model [13]. The previous work were summarized as follows: Chen and Lee set up a new Chinese language model based on the trigram language model and statistically segmentation, in order to solve the typing errors and spelling correction of Chinese input method. The result showed that both the total error rate and English error rate have reduced by 30% [3]. Zheng et al. proposed an error-tolerant Chinese pinyin input method named "CHIME" to solve the users' typing errors. The error rate in detecting mistyped pinyin and correcting mistyped pinyin methods are Significant lower than ever [6]. Wu et al. develop a Chinese input method Editor (IME), which can specify the precise Pinyin and promote advice on patterns of frequent spelling errors. The system advances and evolves interactive functions for Chinese input method by correcting the use of Pinyin and fixing spelling error automatically [7]. Based on an end-to-end method, Suzuki and Gao created a new online scenario referring to the spelling correction when typing, which could reduce the character error rate by 20% (from 8.9% to 7.12%) over the

Fig. 1. Texting with friend on WeChat's interface (the left pic) and shopping in Taobao's interface (the right pic) using Sougou Pinyin Method

previous noisy channel model [4]. Zhang, et al. developed a unified framework named "HANSpeller++" to help Chinese learners of non-native speakers check spell error. The experiments showed effective performance, such as error-detection: 19.92% improvement in precision and 14.4% improvement in recall rate; error-correction: 27.22% increase in precision and 16.67% increase in recall rate [14]. Considering that inefficient input system requires extra user's corrective effort and leads to a poor user experience, Jia et al. Applied the Markov hidden model and K-shortest path to the Pinyin-to-character conversion processing. Results outperformed both academic systems and existing commercial IME in reducing conversion errors. In contrast to the insufficient translation and recommendation sub-system leading to a wrong decoding, Yang et al. built a hybrid model (mini-path+LM) incorporating Minimized-Path Segmentation and Statistical Criteria. Results showed a common way to measure the spelling checking system's performance and achieved 12% improvement over the baseline.

That being said, there are still some areas that need further study: (1) No research has considered special attributes of different group of users. Since no product could satisfy all users, different design features should be considered for different group of users. For example, over 70% female clients spent 168.2 billion RMB (USD 25.87 billion) during the 24-h period on the Alibaba Group's 11.11 Global Shopping Festival in China [15], so the "TIANMAO" (Chinese: "天猫") App should pay more attention to women customers in the design feature to enhance the customer satisfaction. However, there are few papers talking about special group users. (2) Most of the research focused on the Chinese input method and advanced some model, such as Markov model and so on, but few evaluated the relationship between the usability and the design features of Chinese input method. Considering the limitation among the current research in the

Chinese input method design area, firstly, we will use the big data analysis platform-Baidu Index to research on the women's attributes and geographical distribution of using smartphone's Chinese input method; Secondly, we will analyze different design features of Chinese input method and find out how they affect the usability; Thirdly, a case study will be conducted to show the result of key design features for women user when they use the smartphone Chinese input method.

3 Big Data Analysis for Chinese Input Method

Baidu, set up on Jan, 18th, 2000, is the largest multinational technology company in China, also, is one of the largest internet companies all over the world. Baidu owns the most advanced search engines, which can get the massive users' behavior data for digging and analysis in different areas. The Baidu Index, based on the massive users' behavior data collected by people searching through Baidu search engines, can draw the trend and distribution of some special group users using algorithms. It's meaningful and optimized for us to find some rules through digging and analysis big data, and help us better know different group users' behavior habits and advance the Chinese input method's user-friendly on the mobile phone. Some interesting results can be found through the Baidu Index as follow.

Figure 2 shows the Top 3 Chinese input method Apps' using trend on the mobile side in the past 7 years. It can be found that the most popular App is Sougou Pinyin Method, and this App has the largest user group. The No. 2 popular App is Baidu Input Method. However, in 2016, the user number of IFLY Input Method is greater than Baidu Input Method in a period. In a word, the Sougou Pinyin Method is more popular than the others Apps in China.

Fig. 2. The trend of top 3 Apps with Chinese input method on the smartphone side in 2011–2017 (Bule line: Sougou Pinyin Method; Green line: Baidu Input Method; Orange line: IFLY Input Method)

Figure 3 shows the geographical distribution of Sougou Pinyin Method usage on the smartphone. The color from deep to light represents the number of users from high to low. The rank of the provinces based on the number of users from high to low is Guangdong, Henan, Shandong, Jiangsu, Hebei, Sichuan, Beijing, Zhejiang, Hunan and Anhui. Most users in Beijing and Shanghai use Sougou Pinyin Method.

Fig. 3. The geographical distribution of using Sougou Pinyin Method on the smartphone

Figure 4 shows the geographical distribution of Baidu Input Method usage on the smartphone. The color from deep to light represents the number of users from high to low. The rank of the provinces based on the number of users from high to low is Guangdong, Shandong, Henan, Jiangsu, Hebei. Beijing has the largest user group in all the cities in China.

Fig. 4. The geographical distribution of using Baidu Input Method on the smartphone

Figure 5 shows the geographical distribution of using IFLY Input Method on the smartphone. The color from deep to light represents the number of users from high to low. The rank of the provinces based on the number of users from high to low is Guangdong, Sichuan, Shandong, Henan, Hunan, Jiangsu, Hebei. Beijing and Chongqing have more people using IFLY Input Method.

Figure 6 shows the user age statistic of top 3 Apps with Chinese input method on the smartphones. It can be found that the younger people (age under 29 years old) preferred Sougou Pinyin Method. People between 30–39 years old is more likely using Baidu Input Method. IFLY Input Method is popular with the person who is above 40 years old. In a nutshell, people with different age will choose different Chinese input

Fig. 5. The geographical distribution of using IFLY Input Method on the smartphone

method to be their mainly Chinese words tools in the character editor processing. Generally speaking, 30–39 years old users are the largest group using Chinese input method. Together with the group of age between 40–49 years old, users between 30–49 years old are over 80% of the whole users. So how to improve the cell using group's usability becomes the key point for the designer.

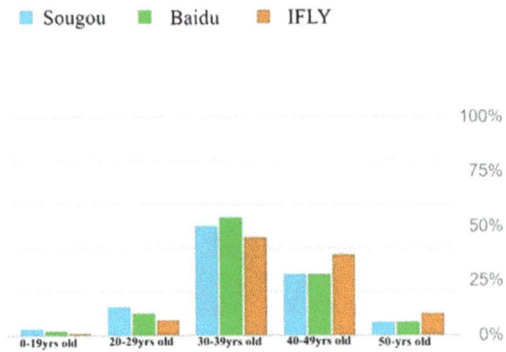

Fig. 6. The user bute statistic of top 3 Apps with Chinese input method on the smartphone side (Bule: Sougou Pinyin Method; Green: Baidu Input Method; Orange: IFLY Input Method)

4   Conclusion and Discussion

In this paper, a new design philosophy of Chinese input method is proposed.in order to improve the satisfaction of target-user and user experience. Based on the massive user behaviors data and big data analysis, we will know the potential users better, and design more user-friendly product in different application scenarios.

Based on the Baidu Index's big data analysis, the Chinese input method Apps could get to know better and serves better its customers. Some design feature suggestions are provided as follows in new application scenarios of shopping. (1) Build a new shopping-vocabulary of database systems with Chinese input method in order to reduce the input time when searching for shopping information online using smartphone; (2) Considering special attribute like age, sex, geographical distribution and so on, select property target-users group to investigate their preferred habit when they are shopping using smartphone. For example, the younger people (age under 29 years old) preferred Sougou Pinyin Method. If the background color of Sougou is attractive by them, they may willing to shopping in special application scenarios; (3) Crossing-area social culture need to pay more attention. The prefer of different areas in China emerge from using different Chinese input method tools. The crossing-area social culture influence on human's behavior and thinking, further human's using habit. How to affect the App's design feature using smartphone need to further study.

In the future, we can design more user group specific Chinese input method and build more rational model considering the human factors, such as age, sex, geographical distribution and so on.

Acknowledgments. This work is partially funded by the National Natural Science Foundation of China (71301061;71503103); Natural Science Foundation of Jiangsu Province (BK20150157); Social Science Foundation of Jiangsu Province (14GLC008); The research base of Chinese IOT development strategy (133930), The Fundamental Research Funds for the Central Universities (JUSRP11583; 2015JDZD004); Funding of Jiangsu Innovation Program for Graduate Education (KYZZ16_0305).

References

1. Zheng, Y., Li, C., Sun, M.: CHIME: An efficient error-tolerant Chinese pinyin input method. In: 22nd International Joint Conference on Artificial Intelligence, Barcelona, Catalonia, Spain (2011)
2. Suzuki, H., Gao, J.: A unified approach to transliteration-based text input with online spelling correction. In: Joint Conference on Empirical Methods in Natural Language Processing and Computational Natural Language Learning, Stroudsbu (2012)
3. Wang, Y., Xie, T., Yang, S.: Carbon emission and its decoupling research of transportation in Jiangsu Province. J. Cleaner Prod. **142**, 907–914 (2017)
4. Zhang, S., Xiong, J., Hou, J., Zhang, Q., Cheng, X.: HANSpeller++: a unified framework for Chinese spelling correction. In: The Eighth SIGHAN Workshop on Chinese Language Processing, Beijing (2015)
5. Jia, Z., Zhao, H.: A joint graph model for pinyin-to-Chinese conversion with typo correction. In: The 52nd Annual Meeting of the Association for Computational Linguistics (2014)
6. Wu, J., Kato, T., Yang, D.: Development of a smart classroom for chinese language learning using a smartphone & tablet. Educ. Tech. Res. **36**, 153–165 (2013)
7. Bi, X., Smith, B.A., Zhai, S.: Multilingual touchscreen keyboard design and optimization. Hum. Comput. Interact. **2**(4), 352–382 (2012)
8. Zhou, J., Rau, P.-L.P., Salvendy, G.: Older adults' text entry on smartphones and tablets: investigating effects of display size and input method on acceptance and performance. Int. J. Hum. Comput. Interact. **30**(9), 727–739 (2014)
9. Yang, S., Zhao, H., Lu, B.: A machine translation approach for chinese whole-sentence pinyin-to-character conversion. In: 26th Pacific Asia Conference on Language, Information, and Computation (2012)

10. Zhang, S.: Solving to the pinyin-to-Chinese-character conversion problem based on hybrid word lattice. Chin. J. Comput. **30**(7), 1145–1153 (2007)
11. Chen, S., Zhao, H., Wang, R.: Neural network language model for Chinese pinyin input method engine. In: 29th Pacific Asia Conference on Language, Information and Computation, Shanghai (2015)
12. Yang, S., Zhao, H., Wang, X., Lu, B.: Spell checking for Chinese. In: The International Conference on Language Resources and Evaluation, Istanbul (2012)
13. Statista: Number of smartphone users in China from 2013 to 2022 (in millions) (2018). https://www.statista.com/statistics/467160/forecast-of-smartphone-users-in-china/
14. Statistic: Number of monthly active WeChat users from 2nd quarter 2010 to 2nd quarter 2017 (in millions) (2018). https://www.statista.com/statistics/255778/number-of-active-wechat-messenger-accounts/
15. (2017). http://blog.csdn.net/clairliu/article/details/78582213

Human Machine Interfaces

Determining the Effect of Training on Uncertainty Visualization Evaluations

Stephen M. Fiore[1,2(✉)], Jihye Song[1,2], Olivia B. Newton[1,2],
Corey Pittman[1], Samantha F. Warta[1,2], and Joseph J. LaViola[1]

[1] University of Central Florida Orlando, Orlando, FL, USA
cpittman@knights.ucf.edu, jjl@cs.ucf.edu
[2] Institute for Simulation and Training Orlando, Orlando, FL, USA
{sfiore, csong, onewton, swarta}@ist.ucf.edu

Abstract. Traditional studies in uncertainty visualization often require naive participants to complete complex, domain-specific tasks in order to examine how effectively a visualization conveys uncertainty to support decision making. However, without assessing whether participants understand such tasks, it can be difficult to determine whether differences in performance are due to a given visualization or to varying degrees of comprehension. Although training is commonly administered to non-experts, to date, training has not been a focal point in uncertainty visualization research. In this paper, we evaluated how variations in training, coupled with assessments of knowledge acquisition and application, can inform uncertainty visualization research. Overall, we found significant performance differences based on training condition, illustrating how training influences task comprehension, which in turn influences decision making. This study serves to highlight training as a critical component of uncertainty visualization studies by quantifying performance variations due to training.

Keywords: Human factors · Visualization studies · Training and evaluation

1 Introduction

Decision making under uncertain conditions continues to be a pressing challenge for operators in complex environments. There has been significant interest in visualizing complex information [1–4] and examining uncertainty visualizations to support decision making [5–7]. However, studies in this area typically focus on *knowledge application*—that is, how a visualization alters performance, as opposed to how *knowledge acquisition* alters knowledge application. When assessing the efficacy of visualizations as decision aids, it is important to ensure performance differences can be attributed to the visualization of interest, and not to varying degrees of task understanding. This is particularly important for complex decision-making tasks requiring domain-specific knowledge that may be especially challenging for non-experts, necessitating greater care to ensure such participants understand the task at hand. In this paper, we seek to demonstrate how significant performance differences in uncertainty visualization studies may be attributed to training. We first review research related to this issue. We then outline the rationale and forms of training designed for this study,

© Springer International Publishing AG, part of Springer Nature 2019
T. Z. Ahram and C. Falcão (Eds.): AHFE 2018, AISC 794, pp. 141–152, 2019.
https://doi.org/10.1007/978-3-319-94947-5_14

as well as the assessments developed to evaluate knowledge acquisition and application. Additionally, we introduce a unique combinatory metric, *cognitive efficiency,* devised to more finely diagnose how training alters workload and performance.

1.1   Related Work

A number of challenges have been raised about how to adequately evaluate visualizations to improve their validity, with some calling for a critical examination of approaches with more rigorous experiments [8, 9]. We identify training as a key component of such evaluations, suggesting it can strengthen uncertainty visualization research. Prior work has examined the benefit of training for decision-making tasks using uncertainty visualizations [10]. Currently, however, while it is common practice to administer some form of introductory task or training prior to experimental tasks, there is significant variability across uncertainty visualization studies. First, some studies may provide relatively little training. Second, the format of training may be inconsistent across studies. Third, assessments may not be administered to determine participants' understanding of the training. Finally, because some publications only briefly mention training, it may be difficult to determine its form and how comprehensive it is [11]. Yet, training may have a significant effect on outcomes in visualization studies. For instance, the way participants are trained may alter how they use visualizations during tasks [12], suggesting that inconsistent training, even for the same task, may lead to performance differences. In a similar vein, lack of training has been explicitly listed as a potential cause of performance variation in tasks involving uncertainty visualizations [13]. One factor influencing performance is mental effort, or workload [14]. Effective uncertainty visualizations, combined with training, have the potential to improve decision making by externalizing information and reducing workload, and can lead to non-expert performance similar to that of experts [15]. However, without adequate training, introducing unfamiliar visualization systems to complex decision-making tasks may increase workload, even in experts [16]. Additionally, compared to experts, non-experts may be more susceptible to impaired performance and workload during decision-making tasks involving uncertainty, particularly when only limited training is provided [17].

In light of these issues, we propose training should be given closer attention as a factor influencing performance in uncertainty visualization studies. To investigate the effect of training on performance in this context, we developed a scenario designed for non-experts to assess performance, workload, and cognitive efficiency.

1.2   Scenario

One of our goals was to develop a realistic scenario that would lower the barrier of entry for a comparatively more complex scenario. To this end, we created a simulated civilian scenario to serve as an analogue to a Naval Intelligence Unit operational task—specifically, drug interdiction. Tasks of this nature require the integration of multiple components into a single decision outcome, sometimes referred to as Course of Action (COA) selection task [18]. Relevant decision processes in our scenario were meant to mimic those found in an environment where operators are faced with variable forms of uncertainty. Instead of intercepting drugs, our scenario requires participants to obtain supplies lost at sea in order to host a party on a tropical island. The COA selection task takes the form of a game that requires participants to retrieve a variety of supplies before their rivals steal them by sending vehicles and crew members to locations where they might be. Participants are instructed to maximize the amount and variety of assets (party supplies, alcohol, and food) while minimizing resource use (boats and crew members). They must consider weather conditions, travel time, location uncertainty, and rival boats when determining what decision to make. Our scenario was designed to be relatable to naive, civilian participants. Vignettes depicting scenes requiring complex decision making were created, and took the form of screenshots from the simulation testbed, which was developed in the Unity 3D game engine with an open source GIS plugin (see Fig. 1).

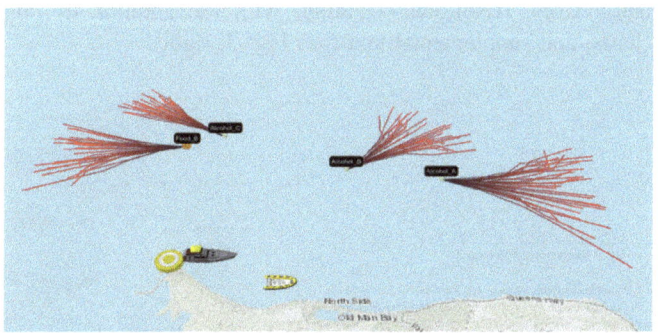

Fig. 1. Example a vignette extracted from the application illustrating the scenario.

2 Method

2.1 Experimental Design

We used a between-subjects design to evaluate the effects of training on decision making under uncertainty in a COA selection task.

Independent Variable. *Type of training* served as the independent variable and was manipulated at three levels: Control, Traditional, and Full.

Control Training. Control Training was the least informative condition. Participants read a brief introductory paragraph (see Fig. 2) before proceeding to the experimental tasks. We used this format based on previous studies with similar formats [11–13, 17].

Traditional Training. Traditional Training was designed to mimic what is sometimes found in the literature on visualization to support decision making [10]. Participants were presented with a text-only description of game components (see Fig. 3, left), along with rudimentary information about the game's rules and scoring system.

On the following pages will be a series of test questions related to a task having to do with decision making and uncertainty. These include questions about different kinds of boats, party supplies, their location in the ocean, and weather. We are testing how well people can interpret questions about uncertainty and these kinds of things when they have been given no background information.

Following the test questions, you will be presented with a series of scenarios called The Party Game. You will have the opportunity to earn points in the game scenarios.

In addition to the base pay of $2.00, if you score in the top 10%, you will receive a bonus of $1.00.

Fig. 2. Control training introduction page.

Full Training. Based on prior work on computer-based training for complex tasks [19], we designed a more comprehensive training to help participants make inferences about and comparisons across task elements. The aim was to more thoroughly ground participants in the task environment to aid learning [20]. Full Training included the same information from Traditional Training, with the addition of images, tables, comparison charts, and supplemental text (see Fig. 3, right).

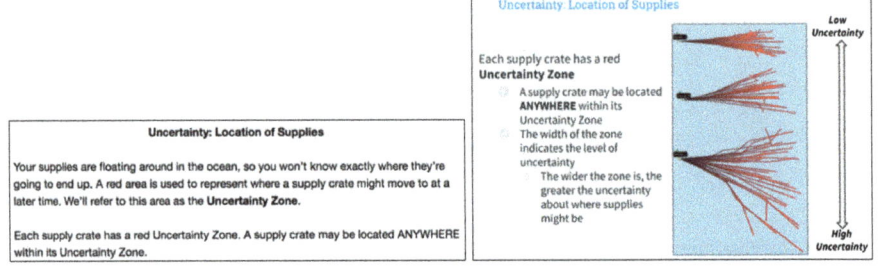

Fig. 3. Examples of equivalent information presented in Traditional Training (*left*) compared to Full Training (*right*).

In all conditions, training was self-paced with no time limit. Traditional and Full Training were split into multiple sections describing game elements, rules, and uncertainty visualization.

Dependent Variables. We investigated the effect of training on performance accuracy, workload, and cognitive efficiency. To obtain these measures, we developed assessments tapping both knowledge acquisition and knowledge application based on prior work on decision making in complex contexts [19, 20]. First, we had two measures of *knowledge acquisition*—the degree to which participants could recognize and understand basic concepts related to the decision-making task. First, the Recognition Knowledge Test assessed participants' recognition and identification of core training concepts (e.g., resources, supplies). Second, the Declarative Knowledge Test assessed learners' mastery of basic factual information associated with the training (e.g., value of assets). These tests each comprised ten multiple-choice questions with four response options, one of which was correct.

Next, we developed the Simulation Vignette Test to measure *knowledge application*, a complex form of assessment designed to tap a higher level of understanding by requiring integration of a variety of cues in a complex setting. Because uncertainty visualizations often support how to optimize COA selections, simulation vignettes were designed to mimic this while still being answerable by a naïve population. Participants were shown vignettes, consisting of images of a map with items (e.g., supplies, boats, etc.) with location uncertainty illustrated by spaghetti plots. The assessment consisted of ten multiple-choice questions. Participants were informed that for each vignette, there may be more than one correct response; however, they were instructed to select the *optimal* solution that maximized points. Each question had four response options: optimal, suboptimal (better), suboptimal (worse), and completely incorrect. For each vignette, all participants were provided with a "cheat sheet" with the information required to make a decision (e.g., capabilities of resources, value of supplies). We included this information so differences in performance could be credited to an understanding of how to apply knowledge acquired from training rather than memorization. Additionally, because the Control Training condition provided minimal information, this cheat sheet was necessary to allow all three conditions to complete the task. From these assessments, we obtained the following measures.

Performance Accuracy. We calculated the percent of correct responses for each test section (Recognition, Declarative, and Simulation Vignette). For the knowledge application (Simulation Vignette) assessment, performance accuracy was calculated as the percent of optimal uncertainty judgements.

Workload. Each question was accompanied by a subjective workload assessment asking participants to rate how easy or difficult it was to answer each item using a 7-point Likert-type scale. We calculated mean workload scores for each test section.

Cognitive Efficiency. We used a unique assessment of workload relative to performance. We drew from Cognitive Load Theory (CLT) and gains in understanding how human information processing influences decision making [21]. CLT attends to how a domain's complexity can alter workload; for example, dealing with multiple resources may overwhelm the human information processing system. We also leveraged the concept of *cognitive efficiency*, a metric evolving out of instructional efficiency measures that describes the relationship between workload and performance [22–24]. A cognitive efficiency score is obtained by computing the difference between standardized performance accuracy and workload scores. Positive scores indicate higher relative performance compared to relative workload and suggest more efficient cognitive processing. Negative scores indicate lower relative performance compared to relative workload.

2.2   Research Hypotheses

Hypothesis 1: Performance Accuracy. The Full Training condition will outperform Traditional Training and Control Training. Specifically:

1a. Training condition will have a significant effect on a naive population's overall knowledge assessment accuracy.

1b. Training condition will have a significant effect on knowledge acquisition for the Recognition Knowledge Test (REC) with Full Training outperforming Traditional Training and Control Training.

1c. Training condition will have a significant effect on knowledge acquisition for the Declarative Knowledge Test (DEC) with Full Training outperforming Traditional Training and Control Training.

1d. Training condition will have a significant effect on knowledge application in the Simulation Vignette Test (VIG) with Full Training outperforming Traditional Training and Control Training.

Hypothesis 2: Workload. The Full Training condition will lead to lower subjective workload during testing compared to Traditional Training and Control Training. Specifically:

2a. Training condition will have a significant effect on a naive population's overall workload experienced during knowledge assessment.

2b. Training condition will have a significant effect on workload during knowledge acquisition for the Recognition Knowledge Test (REC) with Full Training leading to lower workload compared to Traditional Training and Control Training.

2c. Training condition will have a significant effect on workload during knowledge acquisition for the Declarative Knowledge Test (DEC) with Full Training leading to lower workload compared to Traditional Training and Control Training.

2d. Training condition will have a significant effect on workload during knowledge application for the Simulation Vignette Test (VIG) with Full Training leading to lower workload compared to Traditional Training and Control Training.

Hypothesis 3: Cognitive Efficiency. The Full Training condition will lead to higher cognitive efficiency during testing compared to Traditional Training and Control Training. Specifically:

3a. Training condition will have a significant effect on a naive population's overall cognitive efficiency during knowledge assessment.

3b. Training condition will have a significant effect on cognitive efficiency during knowledge acquisition for the Recognition Knowledge Test (REC) with Full Training leading to greater cognitive efficiency compared to Traditional Training and Control Training.

3c. Training condition will have a significant effect on cognitive efficiency during knowledge acquisition for the Declarative Knowledge Test (DEC) with Full Training leading to greater cognitive efficiency compared to Traditional Training and Control Training.

3d. Training condition will have a significant effect on cognitive efficiency during knowledge application for the Simulation Vignette Test (VIG) with Full Training leading to greater cognitive efficiency compared to Traditional Training and Control Training.

2.3 Participants

We recruited 247 participants (132 female), ranging in age from 21 to 70 (mean age = 35.68 years) from Amazon's Mechanical Turk (AMT). 87.45% of participants were located in the United States. Participants were eligible if they identified English as their primary language. Also, because some task elements had alcohol related themes, participants had to be at least 21 years old. Participants were compensated $2.00 USD for completing the study. To increase effort and motivation, we awarded a $1.00 bonus to participants scoring in the top 10% on the COA task.

2.4 Procedure

First, participants were directed to our survey hosted on Qualtrics. Following confirmation of age and informed consent, participants were randomly assigned to Control, Traditional, or Full Training. Following training (or introductory paragraph), all participants completed the knowledge acquisition assessment, consisting of the Recognition Test and Declarative Knowledge Test. Prior to both tests, participants completed a practice question and were given feedback about the correct response. Following the knowledge acquisition assessment, participants completed a practice task to familiarize them with the knowledge application assessment, and were given feedback about the correct solution. Participants then completed the Simulation Vignette Test. For all assessment sections, question order was randomized, as was order of response options. Finally, participants completed a demographic questionnaire. To determine rankings for bonuses, partial points were awarded. For each vignette, optimal (best) solutions counted as 1 point, suboptimal solutions counted as 0.5 (better) and 0.25 (worse) points, and the completely incorrect option counted as 0 points. Participants who scored in the top 10% based on these criteria were awarded a $1.00 bonus.

3 Results

3.1 Performance Accuracy

To test Hypothesis 1, a MANOVA compared the effects of training type on accuracy for the three assessment types. There was a significant effect of training type on our three test types: Recognition Knowledge Test (REC), Declarative Knowledge Test (DEC), and Simulation Vignette Test (VIG). In support of Hypothesis 1a, training condition had a significant effect on overall performance accuracy (see Fig. 4), Wilk's Lambda $F(6, 484) = 8.81$, $p < .001$, observed power = 1.0. Accuracy was highest in Full Training ($M = .656$, $SD = .193$), next highest in Traditional Training ($M = .617$, $SD = .187$), and lowest in Control Training ($M = .514$, $SD = .170$).

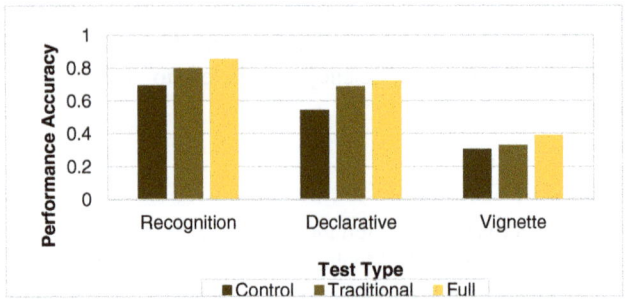

Fig. 4. Performance accuracy by condition and test type.

In support of Hypothesis 1b, training had a significant effect on REC performance accuracy, $F(2, 244) = 17.16$, $p < .001$, observed power = 1.00. REC accuracy was highest in Full Training ($M = .856$, $SD = .185$), next highest in Traditional Training ($M = .801$, $SD = .184$), and lowest in Control Training ($M = .695$, $SD = .162$). Post-hoc comparison showed accuracy for each condition was significantly different from one another.

In support of Hypothesis 1c, training had a significant effect on DEC performance accuracy, $F(2, 244) = 20.08$, $p < .001$, observed power = 1.00. DEC accuracy was highest in Full Training ($M = .723$, $SD = .213$), next highest in Traditional Training ($M = .686$, $SD = .191$ and lowest in Control Training ($M = .543$, $SD = .169$). Post-hoc comparison showed significant accuracy differences between Control Training and the others, but not between Traditional and Full Training.

In support of Hypothesis 1d, training had a significant effect on VIG performance accuracy, $F(2, 244) = 4.72$, $p < .01$, observed power = .786. VIG accuracy was highest in Full Training ($M = .389$, $SD = .181$), next highest in Traditional Training ($M = .328$, $SD = .177$), and lowest in Control Training ($M = .305$, $SD = .179$). Post-hoc comparison showed significant accuracy differences between Full Training and both other conditions, but not between Control and Traditional Training.

3.2 Workload

To test Hypothesis 2, a MANOVA compared the effects of training type on workload for the three assessment types. In support of Hypothesis 2a, training condition had a significant effect on overall workload (see Fig. 5), Wilk's Lambda $F(6, 484) = 12.49$, $p < .001$, observed power = 1.0. Workload was lowest in Full Training ($M = 3.05$, $SD = 1.35$), next lowest in Traditional Training ($M = 3.38$, $SD = 1.28$), and highest in Control Training ($M = 4.27$, $SD = 1.23$).

In support of Hypothesis 2b, training had a significant effect on REC workload, $F(2, 244) = 32.56$, $p < .001$, observed power = 1.00. REC workload was lowest in Full Training ($M = 2.23$, $SD = 1.32$), next lowest in Traditional Training ($M = 2.92$, $SD = 1.22$), and highest in Control Training ($M = 3.77$, $SD = 1.09$). Post-hoc comparison showed workload for each condition was significantly different from one another.

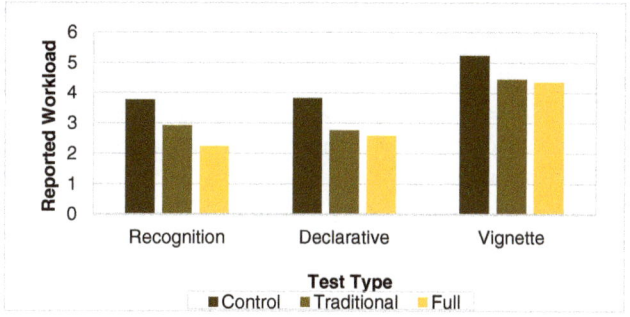

Fig. 5. Reported workload by condition and test type.

In support of Hypothesis 2c, training had a significant effect on DEC workload, F (2, 244) = 23.15, $p < .001$, observed power = 1.00. DEC workload was lowest in Full Training ($M = 2.58$, $SD = 1.25$), next lowest in Traditional Training ($M = 2.76$, $SD = 1.64$ and highest in Control Training ($M = 3.81$, $SD = 1.29$). Post-hoc comparison showed significant workload differences between Control Training and the other conditions, but not between Traditional and Full Training.

In support of Hypothesis 2d, training had a significant effect on VIG workload, $F(2, 244) = 9.53$, $p < .01$, observed power = .98. VIG workload was lowest in Full Training ($M = 4.35$, $SD = 1.48$), next lowest in Traditional Training ($M = 4.45$, $SD = 1.45$), and highest in Control Training ($M = 5.23$, $SD = 1.30$). Post-hoc comparison showed significant workload differences between Control Training and the other conditions, but not between Traditional and Full Training.

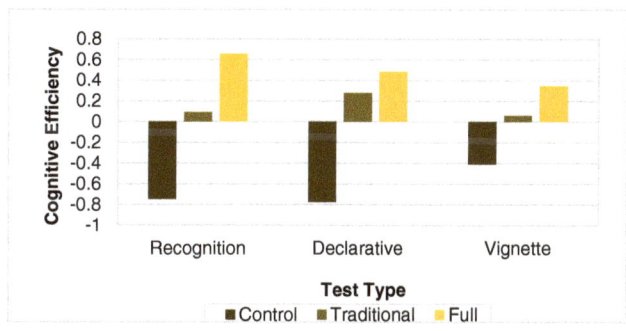

Fig. 6. Cognitive efficiency by condition and test type.

3.3 Cognitive Efficiency

In support of Hypothesis 3a, training condition had a significant effect on overall cognitive efficiency (see Fig. 6), Wilk's Lambda $F(6, 484) = 13.08$, $p < .001$, observed power = 1.0. Cognitive efficiency was highest in Full Training ($M = .497$, $SD = 1.2$), next highest in Traditional Training ($M = .142$, $SD = 1.05$), and lowest in Control Training ($M = -.642$, $SD = .97$).

In support of Hypothesis 3b, training had a significant effect on REC cognitive efficiency, $F(2, 244) = 33.74$, $p < .001$, observed power = 1.00. REC cognitive efficiency was highest in Full Training ($M = .658$, $SD = 1.25$), next highest in Traditional Training ($M = .091$, $SD = 1.11$), and lowest in Control Training ($M = -.746$, $SD = .890$). Post-hoc comparison showed cognitive efficiency for each condition was significantly different from one another.

In support of Hypothesis 3c, training had a significant effect on DEC cognitive efficiency, $F(2, 244) = 31.09$, $p < .001$, observed power = 1.00. DEC cognitive efficiency was highest in Full Training ($M = .485$, $SD = 1.20$), next highest in the Traditional Training ($M = .277$, $SD = 1.06$ and lowest in Control Training ($M = -.773$, $SD = 1.01$). Post-hoc comparison showed significant cognitive efficiency differences between Control Training and the others, but not between Traditional and Full Training.

In support of Hypothesis 3d, there was a significant effect of training on VIG cognitive efficiency, $F(2, 244) = 10.55$, $p < .001$, observed power = .99. Cognitive efficiency was highest on VIG in the Full Training condition ($M = .349$, $SD = 1.15$), next highest in Traditional Training ($M = .059$, $SD = .983$), and lowest in Control Training ($M = -.407$, $SD = 1.03$). Post-hoc comparison showed significant cognitive efficiency differences between Control Training and the other conditions, but not between Traditional and Full Training.

4 Discussion

Our overarching goal was to document the importance of training to support task comprehension and the ability to appropriately apply knowledge when making decisions under uncertainty. As predicted, the results document how appropriately designed training can facilitate both knowledge acquisition and knowledge application and thus contribute to improved evaluations of uncertainty visualization interventions.

Broadly, our results suggest that providing any training prior to a decision-making task involving uncertainty visualization improves overall performance. However, the effect of training on performance may vary for different tasks, and performance on certain tasks may benefit more from comprehensive training. Our results suggest traditional, text-only training may be sufficient to facilitate acquisition of factual knowledge when recognition of visual stimuli is not needed. However, when recognition and application of visual stimuli are required, traditional training may be insufficient. As such, to truly determine the effectiveness of visualizations, more comprehensive training may be needed for the complex tasks used in uncertainty visualization studies. Additionally, comprehensive training may aid in the reduction of cognitive effort during complex tasks compared to traditional training, and lack of adequate training can result in reduced information processing capacity, and thus, impaired performance. Importantly, our measure of cognitive efficiency provides an additional diagnostic for evaluating performance. As hypothesized, providing any training led to greater cognitive efficiency. Our results suggest basic training may sufficiently improve cognitive efficiency during tests of factual knowledge and decision-making tasks. However, comprehensive training was more beneficial for cognitive efficiency when identifying visual stimuli, and overall performance on decision-making tasks benefits more from comprehensive training.

With respect to potential limitations, because the study was run online, we could not monitor participants or control for external distractions. However, one of the advantages of an online study was that we were not limited to a narrow segment of the population and tested a wider demographic. Furthermore, the use of a scenario that is more relatable to a civilian and naive population, in lieu of a traditional military scenario, bolsters the ecological validity of this study. Essentially, this study serves as a jumping-off point for improving uncertainty visualization research by providing evidence of the impact of training in studies with non-expert participants. Despite a general understanding of the importance of training, there is not a standard, empirically-supported approach to training for experimental tasks in this area. This highlights the need for research that focuses on the impact of training on performance in decision-making tasks when evaluating uncertainty visualizations.

5   Conclusion

In sum, we submit that without adequately introducing participants to experimental tasks, it is unclear how uncertainty visualization interventions influence performance. The present research not only suggests that training has a significant effect on how participants perform on a task involving uncertainty visualization, as well as on task comprehension, but our research also provides evidence that training is a valuable tool for improving the assessment of uncertainty visualization.

Acknowledgments. This work was supported by the Office of Naval Research Grant N00014-15-1-2708, under the Command Decision Making program. The views and opinions contained in this article are the authors' and should not be construed as official or as reflecting the views of the University of Central Florida or the Office of Naval Research.

References

1. Andrienko, G., Andrienko, N., Demsar, U., Dransch, D., Dykes, J., Fabrikant, S.I., Jern, M., Kraak, M.J., Schumann, H., Tominski, C.: Space, time and visual analytics. Int. J. Geogr. Inf. Sci. **24**, 1577–1600 (2010)
2. Sedig, K., Parsons, P.: Interaction design for complex cognitive activities with visual representations: a pattern-based approach. AIS Trans. Hum. Comput. Interact. **2**, 84–133 (2013)
3. Kehrer, J., Hauser, H.: Visualization and visual analysis of multifaceted scientific data: a survey. IEEE Trans. Vis. Comput. Graph. **19**, 495–513 (2013)
4. Sanyal, J., Zhang, S., Bhattacharya, G., Amburn, P., Moorhead, R.J.: A user study to compare four uncertainty visualization methods for 1D and 2D datasets. IEEE Trans. Vis. Comput. Graph. **15**, 1209–1218 (2009)
5. Bisantz, A.M., Cao, D., Jenkins, M., Pennathur, P.R., Farry, M., Roth, E., Potter, S.S., Pfautz, J.: Comparing uncertainty visualizations for a dynamic decision-making task. J. Cogn. Eng. Decis. Mak. **5**, 277–293 (2011)

6. MacEachren, A.M., Robinson, A., Hopper, S., Gardner, S., Murray, R., Gahegan, M., Hetzler, E.: Visualizing geospatial information uncertainty: what we know and what we need to know. Cartogr. Geogr. Inf. Sci. **32**, 139–160 (2005)
7. Nadav-Greenberg, L., Joslyn, S.L.: Uncertainty forecasts improve decision making among nonexperts. J. Cogn. Eng. Decis. Mak. **3**, 209–227 (2009)
8. Isenberg, T., Isenberg, P., Chen, J., Sedlmair, M., Moller, T.: A systematic review on the practice of evaluating visualization. IEEE Trans. Vis. Comput. Graph. **19**, 2818–2827 (2013)
9. Andrews, K.: Evaluation comes in many guises. In: AVI Workshop on BEyond Time Errors, pp. 8–10 (2008)
10. Kwon, B.C., Lee, B.: A comparative evaluation on online learning approaches using parallel coordinate visualization. In: Proceedings of the 2016 CHI Conference on Human Factors in Computing Systems, pp. 993–997 (2016)
11. Netzel, R., Hlawatsch, M., Burch, M., Balakrishnan, S., Schmauder, H., Weiskopf, D.: An evaluation of visual search support in maps. IEEE Trans. Vis. Comput. Graph. **23**, 421–430 (2017)
12. Chang, C., Bach, B., Marriott, K., Dwyer, T.: Evaluating perceptually complementary views for network exploration tasks. In: Proceedings of the 2017 CHI Conference on Human Factors in Computing Systems, pp. 1397–1407 (2017)
13. Leitner, M., Buttenfield, B.P.: Guidelines for the display of attribute certainty. Cartogr. Geogr. Inf. Sci. **27**, 3–14 (2000)
14. Moray, N.: Subjective mental workload. Hum. Factors J. Hum. Factors Ergon. Soc. **24**, 25–40 (1982)
15. Kirschenbaum, S.S., Trafton, J.G., Schunn, C.D., Trickett, S.B.: Visualizing uncertainty: the impact on performance. Hum. Factors **56**, 509–520 (2014)
16. Karstens, C.D., Stumpf, G., Ling, C., Hua, L., Kingfield, D., Smith, T.M., Correia, J., Calhoun, K., Ortega, K., Melick, C., Rothfusz, L.P.: Evaluation of a probabilistic forecasting methodology for severe convective weather in the 2014 hazardous weather testbed. Weather Forecast. **30**, 1551–1570 (2015)
17. Loft, S., Morrell, D.B., Ponton, K., Braithwaite, J., Bowden, V., Huf, S.: The impact of uncertain contact location on situation awareness and performance in simulated submarine track management. Hum. Factors **58**, 1052–1068 (2016)
18. Beach, L.R., Mitchell, T.R.: A contingency model for the selection of decision strategies. Acad. Manag. Rev. **3**, 439–449 (1978)
19. Cuevas, H.M., Fiore, S.M., Bowers, C.A., Salas, E.: Fostering constructive cognitive and metacognitive activity in computer-based complex task training environments. Comput. Hum. Behav. **20**, 225–241 (2004)
20. Fiore, S.M., Cuevas, H.M., Scielzo, S., Salas, E.: Training individuals for distributed teams: problem solving assessment for distributed mission research. Comput. Hum. Behav. **18**, 729–744 (2002)
21. Paas, F., Renkl, A., Sweller, J.: Cognitive load theory and instructional design: recent developments. Educ. Psychol. **38**, 1–4 (2003)
22. Fiore, S.M., Scielzo, S., Jentsch, F., Howard, M.L.: Effects of discrimination task training on X-ray screening decisions. In: Proceedings of the 50th Annual Meeting of the Human Factors and Ergonomics Society, pp. 2610–2614 (2006)
23. Paas, F.G.W.C., Van Merriënboer, J.J.G.: The efficiency of instructional conditions: an approach to combine mental effort and performance measures. Hum. Factors J. Hum. Factors Ergon. Soc. **35**, 737–743 (1993)
24. Johnston, J.H., Fiore, S.M., Paris, C., Smith, C.A.P.: Application of cognitive load theory to develop a measure of team cognitive efficiency. Mil. Psychol. **25**, 252–265 (2013)

Adaptive Control Elements to Improve the HMI of an Agricultural Tractor

Timo Schempp[1]([✉]), Andreas Kaufmann[2], Ingmar Stoehr[3],
Markus Schmid[2], and Stefan Boettinger[1]

[1] Institute of Agricultural Engineering, University of Hohenheim,
Garbenstrasse 9, 70599 Stuttgart, Germany
{timo.schempp, stefan.boettinger,
info440a}@uni-hohenheim.de
[2] Institute for Engineering and Industrial Design, University of Stuttgart,
Pfaffenwaldring 9, 70569 Stuttgart, Germany
{andreas.kaufmann,
markus.schmid}@iktd.uni-stuttgart.de
[3] elobau GmbH & Co. KG, 88299 Leutkirch, Germany
i.stoehr@elobau.de

Abstract. Agricultural tractors are used for a multitude of work scenarios with appropriate implements. This leads to a large number of changing operating scenarios. However, the human-machine-interface (HMI) is mainly static and consequently a compromise solution for all possible operating scenarios. Therefore, it is often challenging drivers to understand the operating logic, the operating procedures, and the assignment of control elements (CE) to functions. This reduces efficiency and can lead to operating errors. An approach to avoid the previously mentioned ambiguities is an HMI comprising adaptive control elements that adapt to the operating scenario. This paper shows the automated and objective analysis of the operation of a state of the art tractor in 14 scenarios. Based on this, it is shown how the concept for an adaptive HMI can be derived from the analysis results.

Keywords: Human factors · Human-Machine-Interface
Cognitive ergonomics · Adaptive control elements · Operating analysis
Agricultural tractor

1 Introduction

The tractor plays a key role in the agricultural sector and is also used in the forestry sector, the municipal sector, and on construction sites. It is particularly designed to pull trailers and pull, push, carry, and operate implements [1]. According to this definition, the design of a tractor allows to couple it to suitable implements to successfully fulfill a given work task. For this sake, a standard tractor can offer the following standardized interfaces in the back and in the front: trailer hitch, three-point hitch (TPH), power take off shaft (PTO), and selective control valves (SCV) for driving hydraulic cylinders and motors. Besides the primary task of driving the tractor, these interfaces - with the exception of the trailer hitch - have to be operated by the driver as well. These four

© Springer International Publishing AG, part of Springer Nature 2019
T. Z. Ahram and C. Falcão (Eds.): AHFE 2018, AISC 794, pp. 153–165, 2019.
https://doi.org/10.1007/978-3-319-94947-5_15

main function groups driving, TPH, PTO, and SCVs are significant when operating a tractor to achieve added value. Figure 1 shows different contexts in which a driver has to operate these four main function groups. There are much more operating scenarios than the depicted ones but these set the frame of this research project.

Fig. 1. The 14 operating scenarios that were logged and analyzed in this research project. Beginning from top left: Tractor solo, manure tanker, universal spreader, 3x cultivator, 2x mower, loading wagon, large square baler, reversible plough, seeding combination, and 2x body tipper.

Considering all these operating scenarios, it is obvious that a static HMI is a compromise solution for the driver. Even within the same work scenario (i.e. cultivating 2nd picture line Fig. 1), there can be multiple operating scenarios due to different implements. Hence, a tractor is a perfect pilot machine when it comes to the development of haptic adaptive control elements.

2   The Idea of Adaptive Control Elements

The benefit of adaptive, haptic control elements is to provide an optimized usability to the operator in every context because the control elements become context sensitive. In particular, when the operator encounters changing operating scenarios, adaptive control elements will change in a way that the compatibilities of meaning, movement, and place always comply with the function that has to be controlled.

In a tractor, haptic control elements will never be completely replaced by a touch monitor because it is an off-highway vehicle.

Considering the technical design, at least one of the following features of a control element has to be changeable and then becomes an adaptivity feature: operating force, color, graphic, form, and position.

2.1   State of the Art of Adaptive Control Elements

Regarding haptic control elements that show adaptive behavior, it is mainly the research and development area where investigations, concepts, and prototypes can be found. Especially the features operating force [2–5] and form (Fig. 2) [6, 7] were investigated very thoroughly and their use for adaptive control elements was proven as advantageous for the operator. This is backed up by a basic research about the form of control elements in terms of their communicative impact [8]. The authors do not know publications with single adaptive control elements put together to an HMI as a whole.

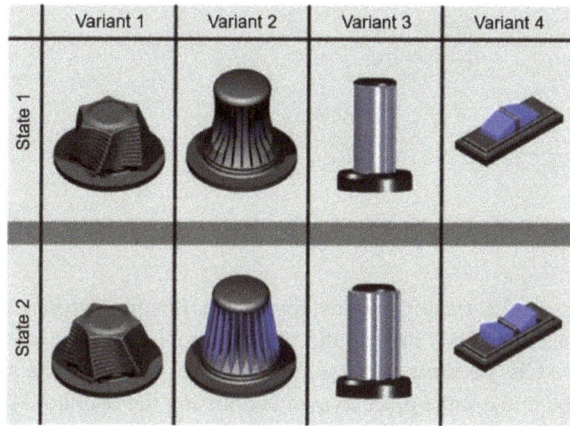

Fig. 2. Concept variants of adaptive control elements based on the adaptivity feature form [6].

Haptic standard control elements on one side and a software based terminal with touch display on the virtual side mark the frame of the range of how an operator can control a system. Whereas the haptic standard control element is permanently assigned to one function, the touch display can show an adaptive behavior. Coming from the touch display side, softkeys are the next step towards the haptic side of the range. Then there is a gap up to the standard control elements. Adaptive, haptic control elements can fill that gap. Both the interface designer and the operator will benefit from the ability to have the best of both worlds while designing or using an HMI with adaptive control elements.

2.2 Adaptive Control Elements in a Tractor's HMI

Figure 3 shows the overall idea of adaptive HMIs in agricultural tractors. For each operating scenario (plough, baler, manure tanker A, manure tanker B, etc.), the control elements adapt to the best settings concerning their adaptivity features. The result is an HMI with an optimized usability in every operating scenario.

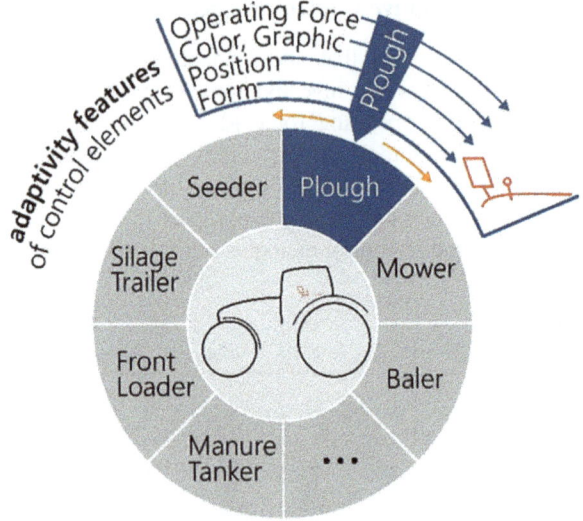

Fig. 3. The principle of an adaptive HMI in an agricultural tractor.

Coming back to the tractor and its four main function groups, a differentiated consideration is needed. The "driving the tractor" group does not offer a wide design scope for adaptive CEs because all these functions are needed in the same way all the time and should be in the same place to avoid confusing the operator. There is room for improvement in the operation logic but that does not belong to the adaptivity features as described above.

The same holds true for the "three-point hitch" group. Different implements can be coupled but it is always about lifting or lowering them and about the same settings on the tractor side.

For the PTO group, the state of the art is that the operator switches on/off the PTO. However, it is actually not about the PTO, it is about switching on/off the implement that is driven by the PTO. Hence, an adaptive graphic would help the operator to understand the actual user task namely to switch on/off the baler or the mower. This also helps when the PTO is not needed in an operating scenario: A prominent placed CE can be for the PTO driven function in one operating scenario and for another function in an operating scenario without the PTO.

The highest potential for adaptive control elements can be expected for the operation of the SCVs for hydraulic actuators. Because of their technical implementation, incompatibilities in the HMI arise when operating them. Three issues can be named when talking about operating the SCVs in a state of the art tractor - across all manufacturers:

- The HMI does not support the driver sufficiently in finding out which of the up to eight CEs for the SCVs belongs to which hydraulic actuator on the implement.
- The HMI does not support the driver sufficiently in finding out if the CE has to be pulled or pushed for example to lift or lower the support leg of a trailer. In addition, it can be that the CE has to be pulled to lower the support leg.
- The operation characteristics of the CEs are always the same whereas the characteristics of the controlled functions differ: reversing the plough is a black/white function, changing the working width of the plough is a greyscale function, and switching on/off the hydraulic oil flow for the manure tanker is yet another function characteristic.

3 Analysis of Operating a State of the Art Tractor

3.1 The Control Armrest in a State of the Art Tractor

For people not skilled in the use of agricultural tractors, a brief introduction to a control armrest of a Deutz-Fahr 9340 TTV Agrotron tractor is given. The control armrest, depicted in Fig. 4, can be seen as state of the art for tractors. The basic structure and functionalities also apply to control armrests of other tractor manufacturers. In general, the tractor and the implements are operated with the haptic CEs on the armrest. The touch monitor is used to adjust settings or for the use of a guidance system. If an implement is ISO-Bus compatible, a bus standard for implement control, it will load a graphical user-interface to the monitor to display information about the implement and to control a part of the functions or even all functions via softkeys on the monitor.

3.2 Data Acquisition and Analysis

In the first project phase, each operating step made with the 64 haptic CEs and the up to 12 ISO-Bus softkeys via the touch monitor on the control armrest (Fig. 4) was logged with a CAN-Bus data logger. For this, thirteen different implements were used in the field with 500 working hours, one tractor, and eight male operators (Table 1). An automated data processing tool was developed in Matlab to compute operating profiles for all operating scenarios in Table 1. Each profile comprises a GPS-Plot, a chart with the signal of each CE over time, a text file in which the absolute and relative frequency

Fig. 4. The control armrest of a Deutz-Fahr 9340 TTV Agrotron as a state of the art control armrest. The orange CEs belong to the main function group "driving", the green ones to "three-point hitch", the yellow ones to "power take-off", and the blue ones to the selective control valves for hydraulic purposes. The white dashed line marks the system boundary in this project.

for each CE is listed, a heatmap (Fig. 5), and a frequency distribution of the CEs regarding their number of operations (Fig. 6).

The logged data was separated for each operating scenario. An operating scenario starts shortly before an implement is coupled to the tractor and ends shortly after the tractor is decoupled from the implement. The remaining data is tractor solo. The time shortly before coupling and shortly after decoupling has to be taken into account because sometimes there are operations needed to bring the three-point hitch and the hydraulic top link in a suitable position for coupling an implement.

For example, the two topmost buttons on the joystick for the TPH (Fig. 5, #1) are just needed to couple the weight in the front but not during field use. The decoupling process is also the reason why seven CEs for the SCVs were operated at least one time although for this cultivator just three SCVs are needed: it is common practice to bring just all SCVs in float position when decoupling an implement and not just the ones where hydraulic actuators are coupled.

Based on the heatmap, one can see which control elements are actually used very often and which not. Moreover, by visualizing the frequency in different colors, one can see how the hotspots of the use of control elements are spread over the control armrest. Later in the project, it will be possible to make sure that the identified high frequency functions are always linked to adaptive CEs in prominent places of the new control armrest (see PTO example in Sect. 2.2). For a better distinction, the middle of the color scale is the mean of the number of operations of all CEs with at least one operation - like 206 in Fig. 5. Otherwise, outliers with a relative high number of operations lead to a very compressed area in the color scale for the other CEs without

Table 1. The available and analyzed dataset of operating and work scenarios with a tractor in the order of the pictures in Fig. 1.

Operating scenario	Work scenario	Logged working hours	Drivers
Deutz-Fahr 9340 TTV Agrotron	Tractor solo	57.57	8
Zunhammer MKE14PUL with ISO-Bus	Manure tanker	75.08	3
Bergmann TSW 5210 S with ISO-Bus	Universal spreader	21.50	2
Horsch Terrano 3 FX	Cultivator	14.78	1
Kerner Komet K420	Cultivator	18.00	2
Horsch Tiger 4 MT	Cultivator	36.92	2
Krone EasyCut 32 CV Float and EasyCut R 320 CV (Side mower)	Mower	11.78	1
Krone EasyCut 32 CV Float and EasyCut B 870 Cv Collect (Butterfly mower)	Mower	63.40	2
Krone MX 400 with ISO-Bus	Loading wagon	43.63	3
Krone BigPack 1270 XC with ISO-Bus	Large square baler	15.15	2
Lemken Juwel 8 with ISO-Bus	Plough	32.12	2
Lemken Solitair 9 with ISO-Bus	Seeding combination	32.73	3
Krampe Big Body 750	Body tipper	35.47	2
Wagner WK600	Body tipper	50.02	2
14	10	508.15	8

the chance to differentiate their coloring. Hence, if the number of operations of a CE is two times the mean or more, it belongs to the highest frequency class in dark red.

As the Horsch Tiger 4 MT is without ISO-Bus, this example is just about the 64 haptic CEs. The CE (Fig. 5, #2) for the second SCV was operated 6792 times up/down during these 36.92 working hours (equals 2215 min). According to the GPS-Plot the fields were pretty close to the farm so time for transport can be neglected. This yields to approximately 3 operations per minute on average – including higher and lower frequencies. The reason is that the depth adjustment of the cultivator was coupled to the second SCV. The depth of the cultivator needs to be controlled when there is heavy soil in the field to make sure the tractor does not get stuck. Usually, this is done by the electronic three-point hitch control when the cultivator is mounted to the TPH (i.e. Terrano 3 FX, Komet K420) but this can not be used for trailed cultivators.

Fig. 5. Computed heatmap of an operating profile for 36.92 working hours with a Horsch Tiger 4 MT cultivator (2nd line, 3rd column in Fig. 1).

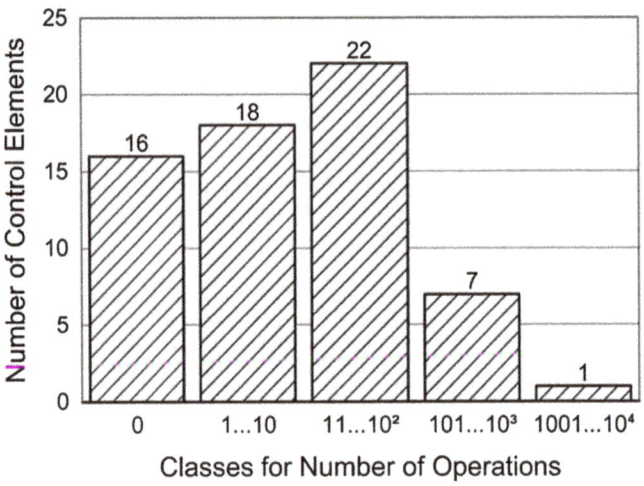

Fig. 6. Computed frequency distribution of the CEs regarding their number of operations in the operating profile for 36.92 working hours with a Horsch Tiger 4 MT cultivator (2nd line, 3rd column in Fig. 1).

Regarding errors in operation, the text file of the operating profile shows that this CE (Fig. 5, #2) for the depth control was brought to float position 87 times although float position is not needed except this one time for decoupling. Thus, it appears that it is too easy to get over the mechanical threshold for float position when pushing the CE (Fig. 5, #2) upwards.

The heatmap can also be used to quickly identify errors in operation. Although the cultivator was operated without headland management the button for it (Fig. 5, #3) was pressed a few times. An explanation could be that it was pressed accidentally when the driver did not have his eyes to support the placement of his thumb on the correct button when he actually wanted to hit forward, reverse, or cruise control.

Looking at the frequency distribution of the 64 CEs in Fig. 6, just 48 are used at all in this operating scenario: 18 CEs with a low frequency, 22 with a moderate frequency, seven with a high frequency, and one with a very high frequency.

The 16 not used CEs confirm the idea that good placed CEs - except the ones from the main function group "driving" - can be assigned to other functions with the adaptivity feature graphic and color depending on the operating scenario (see PTO example in Sect. 2.2). Regarding beneficial gripping areas for this operating scenario, there should be a spot in the most advantageous position for the hand-arm-system for the eight highly used CEs whereas the others can be grouped around.

4   Concepts for Adaptive Control Elements in a Tractor

4.1   The Method to Generate the Specification Sheet

First, all needed CEs in an operating scenario were placed on a grid of an empty armrest as if there were no other operating scenarios to take care of. For developing the adaptive HMI in this research project, the focus is on the CEs within the white dashed line in Fig. 4. Derived from the heatmaps the highly used CEs were placed in zones with a better reachability than the little used CEs. Within the zones, the CEs had to comply with the spatial compatibility regarding tractor and implement. The result was an armrest comprising 14 layers each with an optimized arrangement of the CEs.

Second, the characteristics of the functions controlled with the CEs were reduced to basic and universal operation characteristics as shown at the bottom in Fig. 7. Then, going through the layer of every operating scenario will bring up the needed operation characteristics for the adaptive control element at the considered place. The top disc in Fig. 7 can be seen as the specification sheet for the considered CE.

4.2   Technical Concepts for Adaptive Control Elements

Bringing the specification sheet together with the adaptivity features, the first technical design concepts can be developed. The adaptivity features graphic and sometimes color are needed for every adaptive control element because the CE has to tell the operator which function is linked to it when the operating scenario changes.

A technical concept for an adaptive control element comprising the operation characteristics back/forth (beige) and up/down (green) is depicted in Fig. 8. The blue

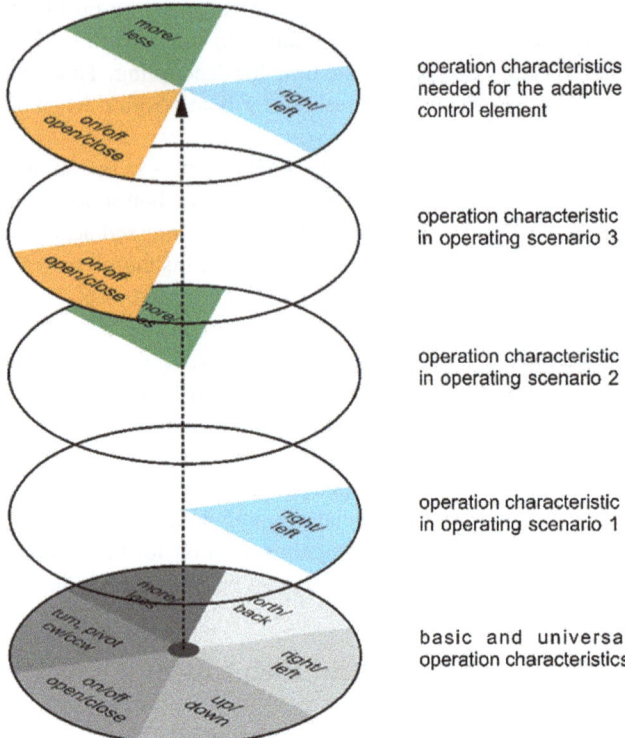

operation characteristics
needed for the adaptive
control element

operation characteristic
in operating scenario 3

operation characteristic
in operating scenario 2

operation characteristic
in operating scenario 1

basic and universal
operation characteristics

Fig. 7. Method to generate the specification sheet for an adaptive control element out of the basic and universal operation characteristics.

angle area can be used when the CE is linked to the function "tilting of body tipper". The change in the operation characteristics is realized with the adaptivity features operating force and position. Each angle area (position) has a characteristic curve that limits the angle of movement of the lever with a strongly increasing torque in opposite direction (operating force). Changing the operating scenario changes the characteristic curve of the lever depending on which operation characteristic is wanted.

Another technical concept comprising more operation characteristics and based on operating force is depicted in Fig. 9. The degrees of freedom that a rotary-push switch can have allow the integration of several operation characteristics without breaking with the compatibilities of meaning in ergonomics. The operator is informed with a graphical symbol which function currently lies on the rotary-push switch. According to that function, the characteristic curve of one operation characteristic is active. The characteristic curve can be modelled with snap-in points or with a continuous behavior. Coming back to the 3rd bullet point in Sect. 2.2, this would be a solution.

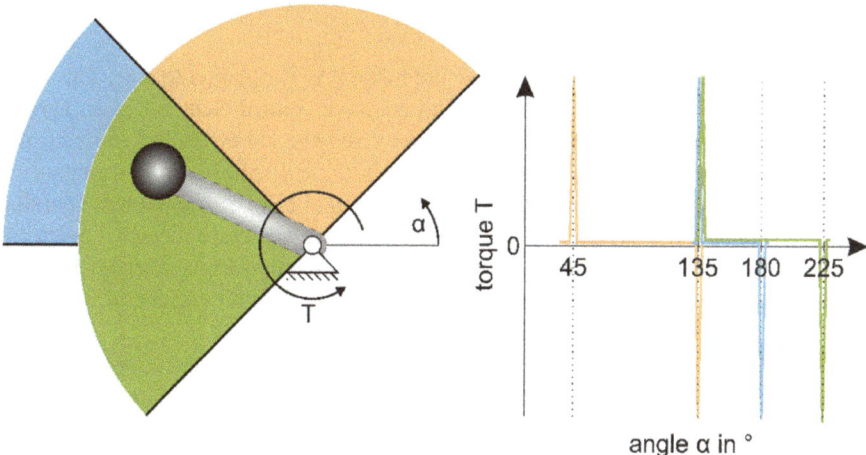

Fig. 8. Technical concept for an adaptive control element comprising the operation characteristics back/forth (beige) and up/down (green) realized with the adaptivity features operating force and position. The blue angle area can be used when the function "tilting of body tippers" is linked to the CE.

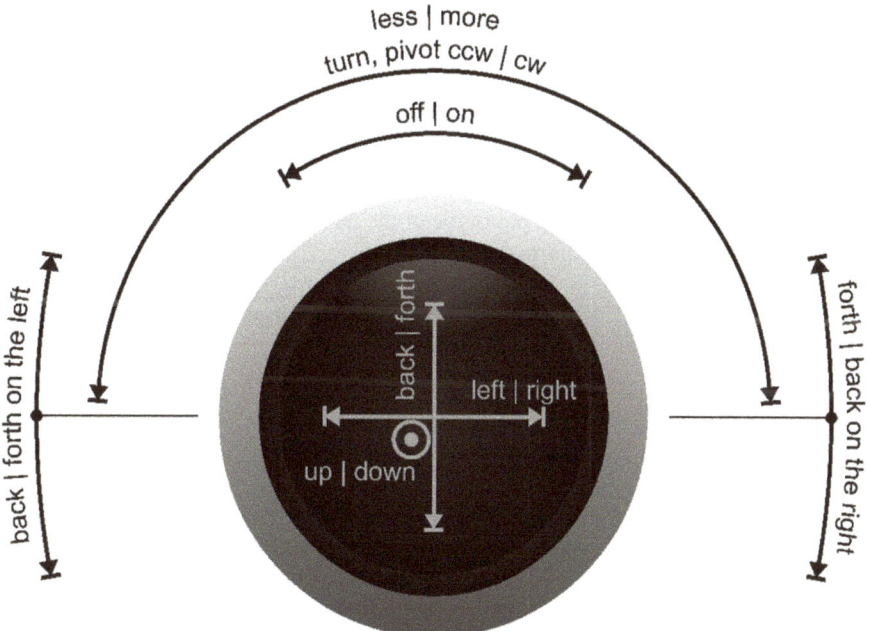

Fig. 9. Technical concept for an adaptive rotary-push switch (top view) offering the depicted operation characteristics.

5   Conclusion

Haptic and adaptive control elements have not made it to the market to date. This paper describes a research project about haptic and adaptive control elements in an agricultural tractor. Because the tractor is an off-highway vehicle, haptic control elements will never be completely replaced by a touch monitor. So haptic control elements are essential. Because the tractor is a universal machine that can be coupled to a countless number of implements, the operator encounters changing operating scenarios again and again. However, static HMIs are state of the art for tractors. Consequently, adaptive control elements are beneficial because they adapt to an operating scenario to provide the operator a better usability.

In the first phase of the project, the operation of a state of the art tractor was logged and analyzed: 14 operating scenarios in more than 500 working hours. For the analysis of the data, an automated tool was developed that computes profiles for each operating scenario. The data and results serve as a basis for the development of the HMI with adaptive control elements in the second project phase. For this, a new method had to be developed first as there is no publication known to the authors that describe the development of an HMI comprising adaptive control elements. Based on that method, it was possible to derive specification sheets for adaptive control elements. Bringing the specification sheets together with the adaptivity features, first technical concepts can be developed. When it comes to adaptive control elements the adaptivity feature graphic has to be used in any case to inform the operator about the changed functionality. In this paper, concepts based on the adaptivity features operating force and position are presented. It remains to be seen whether these concepts can be realized and implemented in a control armrest of a tractor. During the project, it will be worked on more concepts based on the adaptivity features form, operating force and position whereby graphic and sometimes color are needed for every adaptive control element as mentioned before. In the third phase, a prototype of the new control armrest will replace the current armrest for validation purposes in the field.

Acknowledgments. The project is supported by funds of the Landwirtschaftliche Rentenbank.

References

1. International Organization for Standardization: Tractors and machinery for agriculture and forestry - Basic types - Vocabulary. Beuth, Berlin 65.060.01 (2013)
2. Hampel, T., Maier, T.: Future contents of information of multi-functional control elements - the influence of adaptive operating forces and torques on control accuracy. In: Elektronik im Kraftfahrzeug, 14. internationaler Kongress, Baden-Baden, 7–8 Oktober 2009, Electronic systems for vehicles, 2075, pp. 721–732. VDI-Verl., Düsseldorf (2009)
3. Hampel, T., Maier, T.: Adaptive Stellmomente und deren Einflussgrößen auf den Komfort- und Qualitätseindruck bei multifunktionalen Drehstellern. In: Lichtenstein, A. (ed.) Der Mensch im Mittelpunkt technischer Systeme. 8th Berliner Werkstatt Mensch-Maschine-Systeme, 7–9. Oktober 2009, pp. 113–114. VDI-Verl., Düsseldorf (2009)

4. Hampel, T., Maier, T.: Einflussgrößen auf die Wahrnehmung von Drehmomentunterschieden bei Drehstellern. In: Der Fahrer im 21. Jahrhundert. Fahrer, Fahrerunterstützung und Bedienbarkeit; 5. VDI-Tagung, Braunschweig, 4–5 November 2009, pp. 269–276. VDI-Verl., Düsseldorf (2009)
5. Hampel, T.: Untersuchungen und Gestaltungshinweise für adaptive multifunktionale Stellteile mit aktiver haptischer Rückmeldung. IKTD, Stuttgart (2011)
6. Petrov, A., Maier, T.: Neue Stellteile - ein Blick in die Zukunft. In: Maier, T. (ed.) Human Machine Interaction Design. Von der Usability zur nutzergerechten Gestaltung, pp. 119–128. IKTD, Stuttgart (2009)
7. Sendler, J.: Entwicklung und Gestaltung variabler Bedienelemente für ein Bedien- und Anzeigesystem im Fahrzeug. dissertation.de, Berlin (2008)
8. Götz, M.: Die Gestaltung von Bedienelementen unter dem Aspekt ihrer kommunikativen Funktion. Techn. Univ., München (2007)

Research on Interface Design of Full Windshield Head-Up Display Based on User Experience

Ting Deng$^{(\boxtimes)}$, Wei Sun, Ruiqiu Zhang, and Yu Zhang

School of Design, South China University of Technology,
Guangzhou Higher Education Mega Centre, Panyu District, Guangzhou 510006,
People's Republic of China
dengtingcn@126.com, {sunwei,rqzhang}@scut.edu.cn,
867435436@qq.com

Abstract. HUD's technology has been more mature, the market is relatively stable, but penetration is still not high, while the more sophisticated WSD is still the concept stage. Based on the HUD/WSD technology, the research will build a simulation WSD driving simulation system, based on the OLED display, virtual traffic scene simulation software, ADAS and simulation cockpit, to test the visual guidance of distance, brightness, chromaticity and contrast of different images to the driver, to explore the information presentation mode to increase the driver's emotional cognition and driving performance. Finally, a development of a prototype of an all-window windshield display based on different target groups and the assessment of usability will provide. As a reference to other researchers, it will further improve the combination of driving assistance and the market, provide driver protection.

Keywords: Interface design · HUD/WSD · Car-assisted
Vehicle safety driving · Foveal/peripheral field of view

1 Introduction

Traffic accidents caused by vehicle driving become a serious problem all over the world. Research on road safety technology and vehicle safety driving have become a hot topic. In particular, research on vehicle-assisted driving technology has drawn great attention [1].

Head-Up Display (HUD) is a visual aid in car-assisted driving technology, can improve driving safety and user experience. It projects car-assisted information onto the car windshield, such as vehicle conditions, road conditions and external environment and so on. So that drivers do not need to bow down to see the dashboard. They can always maintain the posture of looking forward, avoid ignoring the environment changes and adjust the focal length delay and discomfort [2–4].

As technology advances, HUD can display more basic parameters of the car, navigation and complex traffic information. And its information display more in line with cognitive characteristics, for example, interact through the voice system, gesture recognition. Common HUD usually only 3–5 inches [5], it is small display areas that

© Springer International Publishing AG, part of Springer Nature 2019
T. Z. Ahram and C. Falcão (Eds.): AHFE 2018, AISC 794, pp. 166–173, 2019.
https://doi.org/10.1007/978-3-319-94947-5_16

impose displayed limitations on the amount of driving and safety-related information. At the same time, car-assisted information is spatially separated from the environment, and drivers need to distract themselves from looking for the location of the incident, for example, a pedestrian warning, the driver still needs time to search for the pedestrian's specific location.

Windshield Head-Up Display (WSD) finds a good solution to above problems. But the real-time conversion efficiency of real traffic and on-screen information is still a problem. For example, the combination of pattern color and real-world color can cause cognitive delay of drivers. Color mixing research has found four kinds of color shift phenomena: chromaticity erosion, brightness erosion, chroma and brightness erosion, and chromaticity linear conversion [6]. What's more, the impact of boundary vision on driving performance, most of the current WSD and HUD put information in the driver's vision center, the spatial resolution of human vision sharply reduced from the center to the edge [7], so in addition to the fovea Peripheral areas are not considered suitable for information display. However, peripheral vision plays an important role in the fovea visual field, and the non-fixed information of secondary and third tasks in driving is not displayed in the peripheral area [8].

2   The Current Research Situation of HUD and WSD

Head-Up Display (HUD) is a commonly used flight aid on aircraft. The information is projected onto the glass in front of the aircraft and reflected by the glass in front of the aircraft to reflect the projection information to the driver Eyes Can reduce the pilot's need to bow down to see the meter's frequency, avoid disrupting attention and losing awareness of Situation Awareness. Because of the convenience of the HUD and the ability to improve flight safety, the automotive industry and researchers are also researching and using HUDs one after another, with a growing consumer base.

Windshield Head-Up Display (WSD) is a product of further development of the HUD. There are two kinds of display modes at present, one is on the glass and the other is at a distance in front of the glass, relative to the HUD Larger display area [9, 10].

2.1   Research Status of Foreign Countries

HUD's first application to the car was Ford Motor Company's 1988 "Oldsmobile" model "Super Dagger" [11]. It has been nearly 30 years now. Some models of vehicle manufacturers such as BMW, Mercedes-Benz, GM, Honda, Toyota and Lexus are already equipped with HUDs. Most of the above models equipped with HUDs have also appeared in the domestic market [12].

BMW is not the forerunner, but it is a powerful promoter of car head-up system. BMW HUD covers almost all car models, and the practicality of BMW HUD system is greatly enhanced. BMW HUD head-up display system allows drivers to know important driving information from car windshield above steering wheel. Such as speed, navigation instructions, cruise control, driving diagnostic control and so on.

Drivers No longer need to distract themselves from front sight in driving when they worry about their speed. And for the first time, the information on the windshield is fully color-coded and clearly displayed under any lighting conditions. In addition, allowing the driver to read information faster while focusing on the road, BMW's head-up display system also has a height adjustable to ensure that information is projected to the driver's head-up range [13]. BMW night vision system is also very powerful, such as pedestrian identification warning information, or lane change warning function instruction information [14]. BMW Vision Next100, fully automatic driving concept car design, the screen is the windshield of a piece of the largest touch screen. The future touch BMW head-up display system uses the entire windshield to communicate with the driver. The system focuses on real driver-related information: ideal routes, turning points and speed. What's more, real-time connectivity, smart sensors and continuous data exchange capabilities allow the HUD to generate digital images of the vehicle's surroundings. For example, small obstacles such as stones appear in the road, and the driver can immediately feel it. Combined with a head-up display system, a subliminal communication with the driver takes place and an intuitive signal is used to anticipate an upcoming event [6].

Ford Motor Company is developing a laser-based WSD that shows the path of the pavement on the windshield in rainy days, snow, and even very hazy fog. Where visibility is unknown, driving safety is a major threat. The head-up display system jointly developed by GM and Southern California and Carnegie Mellon University. To assist driving effectively identify the poor front of the line of sight, it combines a number of sensors and cameras to detect the presence of the front windshield through the UV laser radiography.

2.2   Research Status of China

Many domestic enterprises and science and technology companies have been studying HUD. Since the second half of 2015, domestic HUD devices have come out one after another. Such as Carrobot, Hudim, CarPro, Carplus, HUDPLAY, Halo…… Known as the originator of domestic rear-mounted HUD equipment, Navdy landed on the crowd funding platform and won $2.4 million in August 2014. Same year, it got $6.5 million in angel round financing in November. In April 2015, another 20 million U.S. dollars of round A financing was obtained. It mainly due to his powerful HUD display capabilities, as well as dual-screen gesture control, voice recognition and other creative interactive entertainment. In the same period, some domestic teams such as Founder Technology, Roader Rover and Coagent, launched a reflection from on-board equipment to the windshield "HUD."

Such products are cheap, but along with poor experience and relatively simple interaction [15]. To sum up the domestic HUD industry, there are some well-developed companies, but are difficult to break the small screen limitations, research is also confined to the foveal field of HUD, there are few commercial, academic research on WSD.

3 HUD/WSD Development Related Technology and Theoretical Research Status Quo

Starting with the initial configuration of BMW's high-end cars, HUD's application in cars has been developed over a decade, many companies are beginning to step foot in this area. But from the current domestic market performance, HUD has slow development and low loading rate. The reason can be summarized as its user experience is not good enough, as a smart hardware to improve driving safety [16]. Specific factors that affect user experience include HUD display area, imaging quality of displayed images, system performance, information awareness, depth perception and parallax. The most current research status of these factors is showed blow.

3.1 Display Projection Technology

A new generation of car head-up display (HUD) technology can fundamentally change the HUD's user experience. With TI's DLP (Digital Light Procession) products, traditional HUDs have a significantly larger virtual display area and greatly improved image quality [17]. Based on Projection Emitter (EPD) technology, Sun Innovations has developed a full-screen head-up display to produce photographic quality images on a completely transparent RGB luminescent screen. With WSD you can display information without any viewing restrictions on any area of the windshield [18]. A New Laser Projection System Scanned-MEMS uses a laser diode tube driver to drive high-intensity red, green and blue (RGB) lasers that emit high-definition (HD) video to the windshield. These new real-time laser systems send messages without any delay. The AR-HUD maintains a fixed view of the real world. A transparent arrow points directly to the road ahead of the car, and color signals and navigation directions make it easier for the driver to perceive [19].

3.2 Advanced Driver Assistance System

Advanced Driver Assistance System (ADAS) senses the surrounding environment and collect data by a variety of sensors mounted on the vehicle, as the vehicle travels. Static and dynamic object identification, detection and tracking, combined with the navigator map data, the system operation and analysis. So that the driver can be detected potential danger in advance, increase the driving comfort and safety effectively. The ADAS market has grown rapidly in recent years, beginning with the limitation of such systems to the high-end market, which is now entering the mid-market. In the meantime, many low-tech applications are more common in the field of entry-level passenger cars. Improved new sensor-based transmission technologies are also creating new opportunities and strategies for system deployment [20]. Paul Weindorf proposed a mathematic framework for the automotive automatic brightness control system that automatically adjusts to normal information extraction at different ambient light levels [21].

The system proposed by Hye Sun Park et al. can enhance the identification of obstacles under harsh weather conditions and communicate the obstacle position information to the driver. The system consists of four modules: a ground obstacle detection module, an object decision module, an object recognition module and a display module. AR-HUD system proposed for driving safety information achieves a recognition rate of about 73%, and the system has a recognition speed of about 15 fps for vehicles and pedestrians [22]. Mainak Biswas uses a camera calibration algorithm that compensates for optical distortion with separate spatial geometry and perspective views, monitors the driver's eye position with the head tracker module, and detects lane markings on the road with the second module. This approach can reduce the haze, the impact of the rain and enhance the lighting effect, HUD projection of the composite image can be correctly presented by the vehicle coordinate perspective system [23]. The literature shows that driving safety monitoring and data processing modules are well established in a variety of low-visibility driving environments [24, 25], but drivers' ability to extract information and awareness of safety events still requires further work on the HUD interface the study.

3.3 HUD/WSD Interface Design and Research

The technical level of HUD is increased, the screen area is increased and the number of safety information is increased. Providing a suitable and effective interface design can reduce the traffic risk. Issues related to display transitions are also worth discussing [26]. Nowadays, The HUD system's virtual information on the market is displayed on the windshield's physical interface. Although in the driver's field of vision, the information displayed is still spatially separated. For example, the HUD has a warning that the driver still needs time to find the warning a real position after getting it. Windshield Display (WSD) allows the virtual warning to be displayed in a hazardous location. To this end, the authors conducted comparative tests and found no significantly faster response time [27]. Hyesun Park, Kyong-ho Kim tracked the driver's eye movements and used the RSME (Rating Scale Mental Effort) assessment [28] to understand the difference in driver's visual cognitive load at different locations, types and subjects of information.

AR-HUD defines new ergonomic display technologies and drives the current display technology. It sets four key areas of need, the image distance, brightness, chroma and contrast. The image distance requirement is derived from the relationship between the driver's reaction time and binocular parallax [29–31].

These factors affect the driver's situational awareness, there are many studies on this area. Most of the research on depth perception can be used directly in the field of car driving, but there are still some open-ended issues that require application and optimization design. There is a common misconception in the field of research that 3d stereoscopic display is superior to planar display [32]. Cutting and Vishton classify a large number of hints of eye-to-obstacle distance in stereoscopic and planar depth perceptions, dividing the visual space into three categories: personal space 0–1.5 m, behavioral space 1.5–30 m, long-range space > 30 m. When the spatial distance is more than 1.5 m, the information of the plane of single field of view is more efficient [33]. The focusing distance standard for early products is 2 m and the recommended distance given by the guild is 2–2.5 m. Although the HUD reduces the amount of time it takes to adjust the focal length, it still needs to adjust the line-of-sight between the real scene and the display. If the HUD shows that the image is two meters ahead of the car, then the driver's vision will be concentrated at a distance closer to the real obstacle, which is not appropriate [34]. HUD pattern colors and real-world color mixing can cause delays in driver perception, and color mixing studies have found four types of color shifting: chromaticity erosion, brightness erosion, chroma and brightness erosion, chromaticity linearity Conversion. At the same time, it also brings about the impact of HUD images and the real world on the contrast ratio, as well as contrast sensitivity (distinguishing objects with ambiguous boundaries) and visual acuity (ability to distinguish morphological details).

3.4   Interface Design Method Research

The aim was to find the correspondence between the simulator and the road driving performance, completing various driving tasks on two different platforms, including lane change, lane keeping, speed control, parking, turning. Conduct data and video censorship on a simulated driving platform using an electronic driver, make a false assessment, and have the professional car instructor make a false assessment on the real road driving platform. Finally, the degree of convergence of driving performance will be studied to prove that the simulator has a relative validity [35, 36]. In the study, three experiments were designed to quantify the expected return on the driver reaction time (RT) system. Finally, a quantitative model of predicted reaction time could be used to turn the HUD's ADAS off and on. Häuslschmid R proposed a method can help drivers to quickly recognize and understand, based on the large perspective of WSD, and he also put forward his views based on spatial theory, three-regional model, the information environment, but did not carry out empirical studies [37]. Zhou, et al. studied how to use advanced computer-based, physical and behavioral tools to determine a person's emotional/emotional state and real-time behavior, and in turn, how to adapt to human-computer interaction based on real- Cognitive knowledge needs. Their experiments collected both objective and subjective data, and the usability studies of vehicle system user interfaces proved to be of some value in the approach described [38].

4   Conclusion

Firstly, this paper reviews the research status of HUD and WSD systems at home and abroad. HUD technology has been relatively mature, its market is relatively stable, but the penetration rate is still not high. Most of the WSD is still concept stage, even though its function is more perfect. Secondly, based on the above situation, the research level of HUD related technologies is deeply analyzed. The data shows that the laser projection has a great improvement over the traditional optical reflection image quality. Advanced Assisted Driving System (ADAS) has developed rapidly, and driving safety monitoring has provided driver safety. The combination of information conversion with HUD/WSD is a challenge. Then, paper also summarizes the current situation of the interface design and design methods of HUD are preliminarily studied. The research flow, experimental setup and conclusions of these literatures have certain reference value for our further research.

References

1. Yu, H., Jun, K., Kongwu, J.: Headrest display system of vehicle research. Wuhan University of Technology (2012)
2. Jianmei, W., Weidong, L., Bo, C., et al.: Human interface design of vehicle HUD. Process Autom. Instrum. **36**(7), 85–87 (2015)
3. Alejandro Betancur, J., Villa-Espinal, J., Osorio-Gómez, G., Cuéllar, S., Suárez, D.: Effects of augmented-reality head-up display system use on risk perception and psychological changes of drivers (2016)
4. Nakamura, K., Ando, H., Kawahara, N.: Safe and comfortable picture image information presentation by "windshield display". Denso Technical Review, 10 (2005)
5. Häuslschmid, R., Osterwald, S., Lang, M., Butz, A.: Augmenting the driver's view with peripheral information on a windshield display. In: Proceedings of the 20th International Conference on Intelligent User Interfaces, pp. 311–321. ACM (2015)
6. Gabbard, J.L., Fitch, G.M., Kim, H.: Behind the glass: driver challenges and opportunities for ar automotive applications. Proc. IEEE **102**(2), 124–136 (2014)
7. Johnson, J., Yining, Z., Junfeng, W.: Cognition and Design. People's Posts and Telecommunications Press, Beijing, August 2014 (Translation)
8. Gibson, M., Lee, J., Venkatraman, V., et al.: Situation awareness, scenarios, and secondary tasks: measuring driver performance and safety margins in highly automated vehicles. Sae Technical Papers, 9(2016-01-0145) (2016)
9. Wintersberger, P., Frison, A.K., Riener, A., et al.: Towards a personalized trust model for highly automated driving. Mensch Und Computer (2016)
10. Tönnis, M.: Towards Automotive Augmented Reality (2008)
11. http://www.yicheshi.com/dealer/630/news_632299.html
12. Zengchen, C., Lei, D.: HUD technology automotive applications and development at home and abroad. China Build. Mater. Sci. Technol. **S2**, 16–18 (2010)
13. Betancur, J.A., Villa-Espinal, J., Osorio-G阅mez, G., et al.: Research topics and implementation trends on automotive head-up display systems. Int. J. Interact. Des. Manuf. **12**(1), 1–16 (2016)
14. Hosseini, A., Bacara, D., Lienkamp, M.: A system design for automotive augmented reality using stereo night vision. In: Proceedings of the Intelligent Vehicles Symposium. pp. 127–133. IEEE (2014)
15. http://www.leiphone.com/news/201601/SkQGrGefrLKIO1WE.html
16. http://wcm.cnautonews.com/pub/mobile/jdt/201607/t20160708_478068.html
17. Pettitt, G., Ferri, J., Thompson, J., et al.: Practical application of TI DLP technology in the next generation head-up display system. SID Int. Symp. Dig. Technol. Pap. **46**(2), 700–703 (2015)
18. Ted, X., Sun, A.: Novel full windshield heads-up display technology. SID Int. Symp. Dig. Technol. Pap. **46**(2), 712–715 (2015)
19. http://www.intersil.com/content/intersil/en/products/optoelectronics/implementing-laser-scanned-mems-in-automotive-head-up-displays-with-laser-diode-driver.html
20. http://baike.so.com/doc/23635166-24187828.html
21. Weindorf, P.: Forward looking light sensor utilization for automatic luminance control. SID Int. Symp. Dig. Technol. Pap. **46**(2), 846–849 (2015)
22. Park, H.S., Park, M.W., Won, K.H., et al.: In-Vehicle AR-HUD system to provide driving-safety information. ETRI J. **35**(6), 1038–1047 (2013)
23. Biswas, M., Xu, S.: World fixed augmented-reality HUD for smart notifications. SID Int. Symp. Dig. Technol. Pap. **46**(2), 708–711 (2015)

24. Charissis, V., Papanastasiou, S.: Human–machine collaboration through vehicle head up display interface. Cogn. Technol. Work **12**(1), 41–50 (2010)
25. Hosseini, A., Bacara, D., Lienkamp, M.: A system design for automotive augmented reality using stereo night vision. In: Intelligent Vehicles Symposium Proceedings, pp. 127–133. IEEE (2014)
26. Russo, P.M.: Megatrends driving automotive displays and associated mega issues. SID Int. Symp. Dig. Technol. Pap. **46**(2), 623–625 (2015)
27. Haeuslschmid, R., Schnurr, L., Wagner, J., et al.: Contact-analog warnings on windshield displays promote monitoring the road scene. In: Automotive User Interfaces, pp. 64–71 (2015)
28. Park, H., Kim, K.-h.: Efficient information representation method for driver-centered AR-HUD system. In: Design, User Experience, and Usability: User Experience in Novel Technological Environments, Part 3: Second International Conference On Design, User Experience And Usability (DUXU 2013), Held as part of 15th international conference on human-computer interaction (HCI International 2013), July 21–26, 2013, Las Vegas, pp. 393–400 (2013)
29. Pauzie, A.: Head Up display in automotive: a new reality for the driver. Design, User Experience, and Usability: Interactive Experience Design, pp. 505–516. Springer, Cham (2015)
30. Takaki, Y., Urano, Y., Kashiwada, S., et al.: Super multi-view windshield display for long-distance image information presentation. Opt. Express **19**(2), 704–716 (2011)
31. Broy, N., Schneegass, S., Alt F., et al.: FrameBox and MirrorBox: tools and guidelines to support designers in prototyping interfaces for 3D displays (2014)
32. Smith, M., Doutcheva, N., Gabbard, J.L., et al.: Optical see-through head up displays' effect on depth judgments of real world objects. In: IEEE Virtual Reality: 2015 IEEE Virtual Reality Conference (VR 2015), March 23–27 2015, Arles, France, pp. 401–405 (2015)
33. Cutting, J.E., Vishton, P.M.: Perceiving layout and knowing distances: the integration, relative potency and contextual use of different information about depth. In: Epstein, W., Rogers, S. (eds.) Perception of Space and Motion. Handbook of Perception and Cognition, vol. 5, pp. 69–117. Academic, San Diego (1995)
34. Swan, J.E., Jones, A., Kolstad, E., Livingston, M.A., Smallman, H.S.: Egocentric depth judgments in optical, see-through augmented reality. IEEE Trans. Vis. Comput. Graph. **13**(3), 429–442 (2007)
35. Aksan, N., Hacker, S., Sager, L., et al.: Correspondence between Simulator and On-Road drive performance: implications for assessment of driving safety. Geriatrics **1**(1), 8 (2016)
36. Halmaoui, H., Joulan, K., HautièRe, N., et al.: Quantitative model of the driver's reaction time during daytime fog – application to a head up display-based advanced driver assistance system. Intell. Transp. Syst. IET **9**(4), 375–381 (2014)
37. Häuslschmid, R.: Towards a Placement Strategy for Windshield Displays. In: Automotive User Interfaces (2015)
38. Zhou, F., Ji, Y., Jiao, R.J.: Augmented affective-cognition for usability study of in-vehicle system user interface. J. Comput. Inf. Sci. Eng. **14**(2), 617–628 (2014)

Study on Human-Computer Interaction in the Design of Public Self-service Equipment

Huaming Peng[✉], Shuxian Liu, and Tengfei Zhang

School of Design, Guangzhou Higher Education Mega Center,
South China University of Technology, Panyu District, Guangzhou 510006,
People's Republic of China
931889671@qq.com

Abstract. Public self-service equipment has a fixed use environment, complex product function, broad age group of users and other remarkable features compared to other products. This paper takes the increased amount of information and complicated operation of self-service equipment nowadays as the breakthrough point to conduct theory, case study and evaluation. This article studies the human-computer interaction factors in public self-service equipment from two aspects: appearance function design and interface interaction design. taking the bank self-service bank card machine as an example, and several simulation interface interactive systems are designed, then through the eye-tracking for testing, to analyze the data from the tests, according to the theoretical and experimental research results, the appearance and interactive interface of bank self-service card machine are designed. The results of this article greatly enhance the users' interactive experience, and achieve the optimization and upgrading of human-computer interaction, the theoretical results of this article are of reference, the experimental results are repeatable, which are conducive to different disciplines of reference and using for reference.

Keywords: Interaction design · Public self-service equipment

1 Introduction

1.1 Definition and Research Status of Public Self-help Facilities

Compared with manual service, self-service enjoys the advantages of low input cost, low labor pressure, high efficiency, low error and convenience. Public self-help facilities are a kind of new science and technology public facilities under the background of the continuous progress of science and technology and the rapid development of cities. These facilities have the characteristics of high technology, high efficiency, large amount of information, etc. Common public self-help facilities include: Automatic vending machine, information inquiry machine, self-service cash machine, bank self-collecting card machine, etc. The extensive use of public self-help facilities can not only improve the efficiency of work, but also effectively protect the privacy of users. The overall research framework of this paper is as follows (see Fig. 1):

© Springer International Publishing AG, part of Springer Nature 2019
T. Z. Ahram and C. Falcão (Eds.): AHFE 2018, AISC 794, pp. 174–183, 2019.
https://doi.org/10.1007/978-3-319-94947-5_17

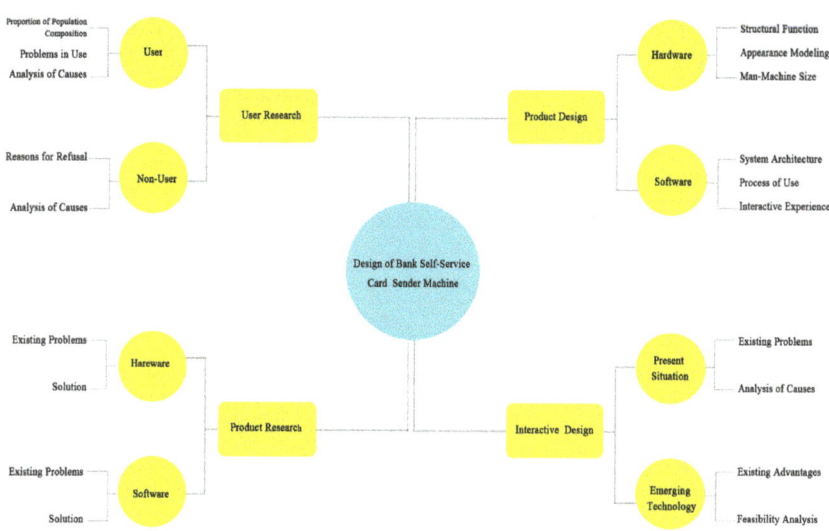

Fig. 1. Research framework of this study

2 Characteristics of Public Self-help Facilities

The convenience of payment means, some security risks also surfaced. At present, the consensus of the payment industry is that biometric identification will lead to a wave of mobile payments. The so-called biological characteristics, that is, each person's body can provide the unique characteristics, these characteristics through a series of algorithms to form a "unique" identification, the system through checking the identification to determine and operate.

2.1 Characteristics of Hardware and Software Interactions in Public Self-help Facilities

Self-help products are divided into two categories according to their functional properties (see Fig. 2). One is information facilities, such as library inquiry machine, scenic spot self-help electronic map, this kind of products mainly through the electronic screen to provide users with the information needed to assist the service work of related industries; One is consumer facilities, such as cash machines, ticket machines,, etc., this kind of products for the target users, to meet the specific consumer needs. Compared with general products, public self-help devices have the following characteristics: Numerous users, extensive user groups, high frequency of use, short operation interval, long product life cycle, fixed use environment, single function.

Unlike the one-to-one mode of use of other products, self-help facilities adopt a one-to-many mode of use, and the user community is complex, so the physical differences of

Device Type	Common Products
Commercial Equipment	ATM, Vending machine, Self-help ticket machine
Entertainment Equipment	Fitness facilities, Amusement facilities
Sanitary Equipment	Public toilet, Drinking water device
Management Equipment	Parking lot, Fire fighting equipment, Public transport
Information Equipment	Public sign, Opinion box, Message board

Fig. 2. Classification of public self-help facilities

different users should be fully taken into account in hardware design. Design a reasonable and comfortable operating area to avoid muscle fatigue; In the aspect of software interaction, the operation flow should be simple and smooth, easy to learn and operate, and the memory burden of users should be reduced. At the same time, the psychological demands of users' safety, privacy and trust should be satisfied.

2.2 Factors Affecting the Design of Public Self-help Facilities and Design Principles

The influencing factors of the design of public self-help facilities mainly include natural factors, spatial factors and behavioral factors

(1) Natural factors: The design of public facilities should take into account the natural environment factors, the selection of materials, structural planning and other aspects should be adapted to local conditions, reasonable planning.
(2) Spatial factors: In general, the public environment has only limited space and use area, so the design of public facilities should be planned from the whole to meet the needs of different activities.
(3) Behavioral factors: People's behavior is unpredictable, the same facility will often face a variety of different use behavior and treatment, so public facilities need to consider the characteristics of human behavior based on the characteristics of non-controllable and diverse.

The basic principles of public self-help facility design mainly include functional principle, security principle, aesthetic principle, sustainable principle and humanization principle.

3   Design of Bank Self-service Card Machine

Taking the subway gate system as an example to analyze the application of multi-channel biometric recognition technology in payment field. The following is through a comprehensive analysis of the fingerprint identification technology and the existing Metro ticket-buying and ticket-checking system, presenting a new type design plan of Metro ticket-buying and ticket-checking system based on fingerprint identification and palm recognition technology to solve some deficiencies in the current Metro ticket-buying system, so as to achieve the purpose of simplifying the subway ride process.

3.1   User Studies and Existing Problems of Bank Self-service Card Issuers

Bank self-service card issuers in China are a relatively new facility (see Fig. 3). Unlike ATM's simple deposit and withdrawal functions, self-service card issuers also have complex functions such as face recognition, fingerprinting and instant card issuance, of which the requirements for safety, stability and high efficiency are higher. The user group of bank self-service card issuers is huge, almost includes all kinds of people in society, the age, education level, behavior habit of users are uncertain, this design uses observation method and interview method to carry on the user research.

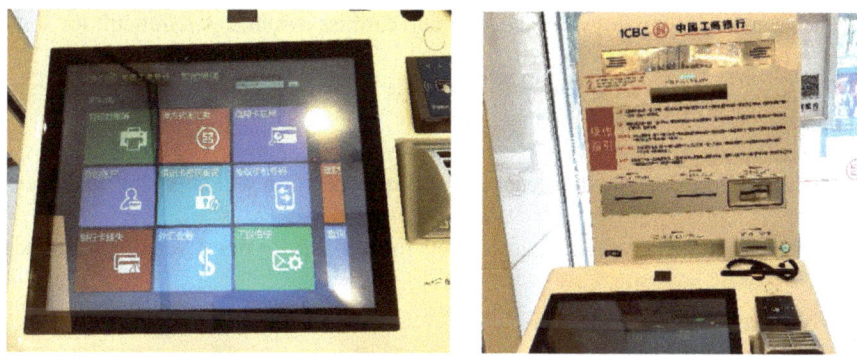

Fig. 3. Sample 1 (Self-service machine in ICBC in Guangzhou)

The author randomly selected 15 people of different ages from the users who had successfully picked up the card, interrupted the use of the card and handled the manual window, conducted interviews with them, modeled the users and established three user models as shown in the figure below (see Fig. 4):

Through the collation of the contents of the interview, it is found that the bank self-help card issuers have the following problems:

(1) Too many sockets and exits, disordered placement, and time consuming to find places;
(2) The operation process is complex and unclear;

Miss Tang
20 years old
Student

Mr Liang
34 years old
Businessman

Ms Lee
56 years old
Retired Worker

Miss Tang is a sophomore of design, she now has no income, monthly living expenses 1200 yuan given by her parents.

She went to the bank to the frequency of about a quarter, mainly for the loss of bank cards, mobile phone number to change .She also love self service machine, because she had plenty of time, and it is easy for her to understand how to operate the machine.

Mr. Liang is a sales businessman whose annual income is about 300 thousand yuan.

He went to the bank 3-4 times a month because of the business. He mainly deals with new cards and large amount of transfer accounts for customers. He seldom uses self-service machines, because he has a VIP card to enable him to enjoy special reception. He thinks that manual handling is faster and safer than machines.

Ms. Li is a retired company employee. Her job is the education industry. She is now earning about 100 thousand yuan a year.

At least once a month, she goes to the bank to remit her son to university. She prefers to line up for an hour to remit money or to operate with ATM, because she is worried that the machine is out of order. She has more trust in artificial services. But if there is the assistance of the bank staff, she is willing to try to operate the self-help machine.

Fig. 4. User modeling

(3) The "sense of technology" of the self-help card issuers makes it difficult for some elderly users to accept.

(4) The function is not complete, such as unable to deal with the business of changing the pre-stored telephone number, etc.;

(5) Screen tactile operation is not sensitive, need to force or repeat click;

(6) The interface of the previous user interrupting operation will affect the next user's use;

(7) Users have higher aesthetic requirements for self-service ticket vending machines.

3.2 External Design of Bank Self-service Card Issuers

This paper analyzes the exterior design of self-service card issuers from the aspects of shape, material and color. The vertical structure is adopted in shape, which is convenient to place and increase stability, and the bottom invisible wheel is easy to move and manage; The main materials are polymer engineering plastics and steel plate to enhance safety performance; Color is an important element of product appearance, which has a decisive impact on product modeling. The color selection of the card issuers design follows the following two elements: First, the use of contrast obvious colors, strong colors can make products stand out, in the environment to produce the focus of attention; The second is the use of neutral color, neutral color easy to blend into the environment, avoid causing part of the user group rejection. The design of the self-help card issuers is as follows (see Figs. 5, 6 and 7).

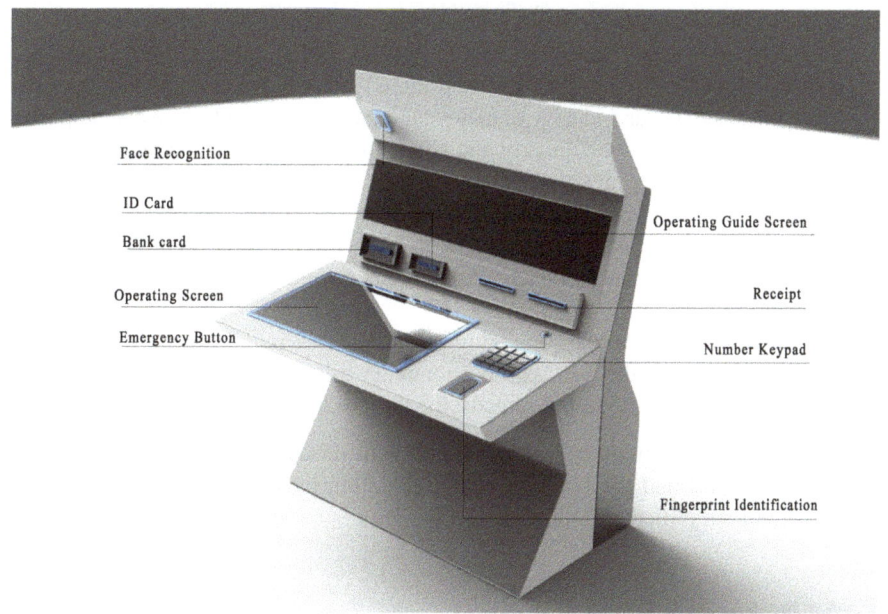

Face Recognition

ID Card

Bank card

Operating Guide Screen

Operating Screen

Receipt

Emergency Button

Number Keypad

Fingerprint Identification

Fig. 5. Functional explanation of bank self-service machine

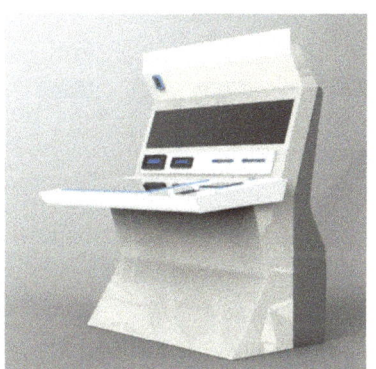

Fig. 6. Holistic diagram   **Fig. 7.** Side view

3.3 The Interface Design of Bank Self-service Card Issuers

Through the user investigation process and the research result analysis, the preliminary summary user to the self-help ticket vending machine function demand is as follows:

(1) Information guidance: Novice users and occasional users need to be alerted by the system to specific steps in the process of operation.

(2) Simplify the process: The main functions of the existing self-service card issuers are information query and instant card issuance. The simplification of the process includes the software system flow also includes the hardware operation process;

(3) Clear interface: Clear and easy to understand the hardware interface can help users to complete the operation quickly and efficiently, at the same time reduce the error rate, simple interface gives a friendly feeling;

(4) Timely feedback: The user should give timely feedback on every step of the operation behavior system, which is helpful for the user to make a correct judgment on the current state, make the operation smooth and reduce the interruption.

(5) Fault-tolerance: In the key operation steps, such as purchase and payment, the system should give the user a proper reminder to avoid the wrong operation, and at the same time provide the remedy for the possible misoperation.

After investigation and analysis, combined with this design, according to the user card handle process and the function of the operating parts, the software interface is rearranged and simplified as follows (see Fig. 8).

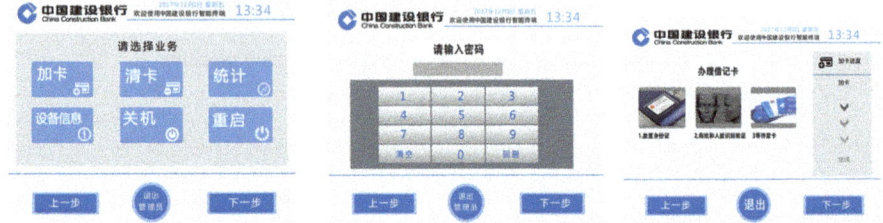

Fig. 8. Interface design of bank self-help machine

4 Testing

4.1 Test Facilities and Material

The experimental equipment and site conditions are shown in Fig. 3.1. Tobii-X120 eye tracker with a sampling frequency of 120 Hz was used to record the eye movement data of the subjects and the data was processed with the Tobii Studio 3.2 software. The high data accuracy of this eye tracker and the consistent tracking capability in a real-world test environment allowed it to capture high quality data (regardless of race or wearing glasses) across a wide range of subjects. The equipment was able to maintain the accuracy of the data and stable tracking ability regardless of subject's head movement or changes in light conditions.

4.2 Test Tasks and Indicators

Tasks: A total of 20 test users participated in the experiment, aged 20–55 years. All subjects had computer and ATM experience. The male and female ratio of both groups were close to 1: 1. The education level was high school and above. Based on the previous questionnaire design experiment task: Task 1: verify identity information; Task 2: check the account balance; Task 3: handle a debit card. The completion of all the three tasks was regarded as an accomplishment of the gross tasks. The subjects performed the above three task operations one by one, and the experiment contents were not allowed during the waiting process.

Indicators: The usability indicators used in the experiment are effectiveness, efficiency and satisfaction. The details are as follows (see Fig. 9).

Index	Index Content	Eye Movement Data
	Task completion rate	/
Validity	Task completion time	/
	Total access time	/
	Search efficiency of each interface target	First look at the start time (t1)
		Number of fixations (n)
Efficiency	Understanding efficiency of each interface area of interest	Total fixation time (t2)
		First fixation duration (t3)
	Understanding and operating efficiency of each interface target	The time from the first look to the next operation (t4)
Satisfaction	SUS : System-Usability-Scale	/

Fig. 9. Test index

4.3   Test Results and Conclusions

In the analysis of the results, we found that the correlation between the results and gender was low.

(1) Validity

The improved interface significantly improved the success rate in the task of verifying identity information and querying account balances (see Fig. 10). In the overall completion rate, the success rate of the improved interface was higher than that of the existing interface, which showed that the effectiveness of the improvement had a certain degree of improvement.

(2) Efficiency

The average time to verify the identity information in the improved interface was about 3 s less than that in the improved interface, but the statistical difference was not obvious. In the processing of debit card task, the time of the improved interface was less than that of the existing interface, and the difference was obvious at the level of

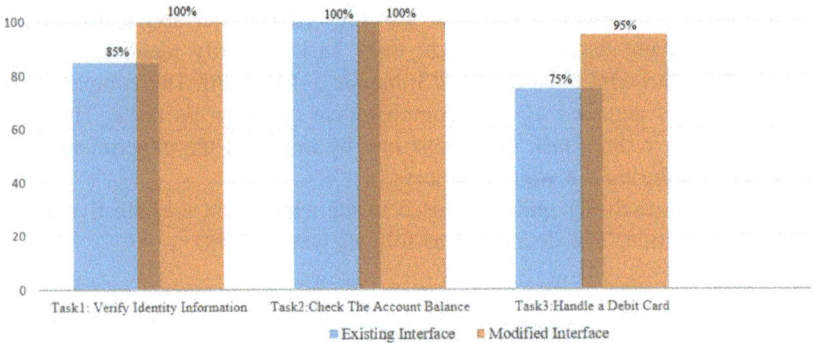

Fig. 10. Validity comparison

0.01, and the standard deviation of the task time was much smaller, which indicates that the task completion time of the improved interface was relatively stable. The main reason is that the task completed the obvious operation path standardization, so that different users to complete the same task according to the same operation path (see Fig. 11).

(3) Satisfaction

The statistical results of SUS scale showed that the two groups had similar satisfaction to the improved interface, and there was no significant difference (Fig. 12).

	Group	Mean Value (Unit : S)	Standard Deviation	Range	Significance (Bilateral)
Task 1	Existing Interface	9.95	0.94	3.03	0.614
	Modified Interface	7.22	0.40	1.53	
Task 1	Existing Interface	13.07	1.16	3.99	0.015*
	Modified Interface	11.41	1.11	3.82	
Task 1	Existing Interface	7.07	14.04	42.56	0.017*
	Modified Interface	60.57	8.36	20.10	
General task	Existing Interface	100.08	10.14	46.55	0.130
	Modified Interface	79.20	7.87	22.64	

*. Significant correlation on the 0.05 level (bilateral)

Fig. 11. Efficiency comparison

	Group	Mean Value (Unit : S)	Standard Deviation	Range	Significance (Bilateral)
System-Usability-Scale	Existing Interface	85.15	10.23	34.50	0.781
	Modified Interface	87.56	9.96	30.15	

Fig. 12. Satisfaction comparison

5   Conclusion

This paper mainly studies the design of public self-help facilities under the background of information technology. Based on the theory of human-factor engineering and interactive design, the user-centered self-service card issuers of the bank are improved. Based on the existing products and research, this paper analyzes the characteristics of the times and products of the public self-help facilities under the background of information technology, the paper studies the physiological characteristics of the user group and the needs of the users, and is guided by the theory of interactive design. In order to improve the frequency and efficiency of this kind of products and give users a more pleasant user experience, this paper optimizes the design of the product function, structure and usage flow of the self-service card issuing facilities of the bank. In the aspect of hardware design, a more reasonable structure is designed according to the man-machine dimension, so that the device can be applied to a wider group of users. Simplify hardware interface layout, reduce memory burden and improve product availability. In the aspect of software design, optimize the function structure and system operation flow, improve the system function and enhance the interactive experience according to the needs of users. Finally, the effectiveness, efficiency and satisfaction of the improved self-help card issuers are evaluated with eye movement tester. The research results of this paper greatly improve the user interaction experience and achieve the optimization and upgrading of human-computer interaction. The theoretical results of this paper are referential, and the experimental results are repeatable, which is conducive to the reference and reference of different disciplines.

References

1. Yarui, C., Jucheng, Y.: Multimodal biometrics recognition based on local fusion visual features and variational bayesian extreme learning machine. Expert Syst. Appl. **32**(5), 43–51 (2016)
2. Yaming, W., Fuqian, T.: Robust text-independent speaker identification in a time-varying noisy environment. J. Softw. **26**(9), 21–34 (2012)
3. Gaurav, G., Paritosh, M., Angshul, M.: Group sparse representation based classification for multi-feature multimodal biometrics. Inf. Fus. **32**, 3–12 (2016)
4. Ali, I.A.: From classical methods to animal biometrics: a review on cattle identification and tracking. Comput. Electron. Agric. **123**, 423–435 (2016)

Evaluating the Usability of a Head-Up Display While Driving a Vehicle

Guilherme Gattás Bara$^{(\boxtimes)}$, Patrícia Caetano Bara, José Castanõn,
and Maria Teresa Barbosa

Universidade Federal de Juiz de Fora, Juiz de Fora, Brazil
gbara@terra.com.br, jose.castanon@ufjf.edu.br,
teresa.barbosa@engenharia.ufjf.br

Abstract. This study highlights issues concerning the visual and tactile ergonomic comfort inside a vehicle, specifically in a human-technology system. Therefore, we evaluated a HUD in an imported vehicle sold in Juiz de Fora (Brazil), analyzing the pros and cons for users, who tend to show increased use of their spatial perception, which may alter their visual perception. Moreover, this study covers the symbolic as well as the technological evolution of motor vehicles in Brazil. Based on literature review, documentary analysis and a case study, this work reflected upon the technology available in Brazil in a vehicle with a factory-installed HUD.

Keywords: Usability · Vehicle · Head-Up Display

1 Introduction

The analytical study of a Head-Up Display (HUD) is the primary objective of this research, covering the advances in automotive technology, and the understanding of the relations between the HUD, its users, and the influences in Brazil in the end of the 20th century and the beginning of the 21st century.

During this time, Brazil was and still is under great external influence. This provides the theoretical basis necessary for the conscientious development of the process to construct this documentary narrative.

This study addresses the trajectory that led to the formation of the automobile user in Brazil, where cars became a national passion, exploring some details that drove this change.

The Second World War led to great advances in the studies of the relationships between human beings, machines, and designed environments, bringing about the so-called human engineering that, in the post-war period, promoted enormous technological growth [12].

Without the advent of algorithm, modern life would be different, as computers, internet, virtual reality, and artificial intelligence would not exist. Calculus is the "first greatest scientific idea of the West" and algorithm is the "second greatest scientific idea of the West", because calculus resulted in modern physics, but algorithm allowed computer and, later, software development. "An algorithm is an effective procedure. It is a way of doing something in a finite number of discreet steps" [7].

© Springer International Publishing AG, part of Springer Nature 2019
T. Z. Ahram and C. Falcão (Eds.): AHFE 2018, AISC 794, pp. 184–194, 2019.
https://doi.org/10.1007/978-3-319-94947-5_18

In Brazil, based on the study of the period mentioned previously, the government of President Getúlio Vargas and President Juscelino Kubitschek marked the years of 1953 and 1965, respectively. During this time, the European and American influence that, historically, was already part of our cultural references, became extremely important in the country under both of these governments.

Volkswagen introduced the 1953 models of the Beetle and the Kombi campervan to President Getúlio Vargas. These models were manufactured using imported auto parts at the factory in São Paulo. The German company was the first to accept the invitation to manufacture its vehicles in Brazil. During Juscelino Kubitschek's government (1965–60), the expansion project launched by the automobile industry was set off and several factories that had been set up in the country started manufacturing [8].

This was the beginning of the automobile history in Brazil. A history that developed from a new model of consumption that started with the trajectory of the Brazilian industry in the 1950's. There was a perception of great progress and people started migrating from the countryside to the cities, as the new developments greatly attracted the society, and people started concentrating in the main centers, accelerating industrialization and development. The Brazilian automobiles were easternized, as the East often influenced the design and the functionality of the automobiles, in addition to enriching the Brazilian culture for a whole period after the beginning of the 20th century.

However, this has been changing over the last years. Since the middle of the 20th century, there has been a revival of nationalism, a search for a new Brazilian design identity, which led to increased space in the national industrialization.

2 Objectives

Today, the number of people who buy an automobile has increased considerably, as there are many financing options and the ads are compelling. Media life is persuasive, and influences and encourages the user to buy leading-edge consumer goods, driving the Brazilian consumer to make use of all the resources offered by the technological growth. Among them, it is worth mentioning the inclusion of the HUD installed in a hybrid car that uses little fossil fuel.

This study highlights issues concerning the visual and tactile ergonomic comfort inside a vehicle, specifically in a human-technology system. Therefore, we evaluated a colorful HUD in an imported vehicle sold in Brazil, analyzing the pros and cons for users, who tend show increased use of their spatial perception, which may alter their visual perception.

Specifically, we propose questions and criteria to be taken into account as recommendations to use the HUD technology, its usability, and whether the user is prepared for the resources that the automobile industry updates every year.

3 Strategies/Methodology

Based on literature review, documentary analysis and a case study, this work reflected upon the technology available in Brazil in a vehicle with a factory-installed HUD. This exploratory and descriptive study investigates concepts and relevant cases, taking into account the historical context, delimiting the research phenomenon, and seeking to classify and interpret the data investigated. We also carried out some fieldwork to further our investigation in a car dealer in Juiz de Fora (Brazil).

Based on the analysis of the interpreted data, we described the qualitative diagnosis. We systematized the data based on the analysis of the research results, the confrontation between the theories present in the concepts to formulate questions, and the relevant criteria of the design applied to the HUD. The main option to develop an automotive HUD was determined during the documentary work and analyzed based on the considerations brought forth during the steps.

Together with the automotive technological development, artificial intelligence has been increasingly advancing. The advent of the HUD system made it possible to interact with a graphical user interface, using codes and symbols to help the driver, stimulating the senses and developing cognition [2].

The automotive electronic systems are part of the cars' usability, communicating through wires and wireless networks, using longitude and latitude to navigate, and interacting in three dimensions plus time. However, often, its main function of helping the driver to drive safely may fail, as this requires attention and human interaction [10].

4 Head-Up Display (HUD)

The HUD was first developed for use in the helmets of the pilots of military aircraft. The air space is virtually divided into corridors that are freer from human interference, making it necessary to have a direct view of the coordinates and other information under extreme circumstances that require reflexes and actions to prevent the unexpected. This is very different from driving a car, as the urban roads have more interference and objects around the vehicle that contribute to distracting the user [4–6].

Having artificial intelligence inside the vehicle may require more attention from the driver. This precaution starts with the capacity of the human brain to interact with new technologies that use leading-edge resources. Thus, more autonomous and safe vehicles are developed, with hybrid systems, on-board computers, state-of-the-art integrated GPS systems, access to Google Earth, cameras, front and lateral sensors, autopilot, GSM, among others [9].

Today, the ergonomics of car dashboards benefits the user, who can interact with the multi-function display, indicators, multimedia system, and buttons on the dashboard, the steering wheel, and the driver's door. With the advent of the Head-Up Display (HUD), we have an extra visual element that drives the human brain to work differently, devoting attention to more digital information, all of which are projected onto the windshield and have to be processed.

The user focuses on viewing the pedestrians and cyclists who travel on the urban roads, in addition to other vehicles, crossroads, traffic lights, signs, among others. Moreover, if the driver comes across an obstacle, an animal, an object or another human being in the road, this is going to distract the driver's attention and take his or her eyes off the road for a fraction of a second, even if the driver has good peripheral vision. As for driving in highways at night, excessive lighting may lead to drowsiness due to the several light options on the dashboard or the strange blindness caused by the headlights from other vehicles coming from the opposite direction. Alternatively, a lapse in concentration may reduce visibility depending on the present environmental conditions, as this becomes a natural reaction of the brain towards the optic nerve.

The ocular globe is a complex optical system composed of transparent media where light passes through and other parts that ensure protection, maintenance, mechanics etc. The ocular globe functions similarly to an analog camera. Thus, the objective combines a real object to a real image, but inverted in the back of the eye on the retina (sensitive membrane). The iris, whose central opening is the pupil, controls how much light enters the eye, working as a camera diaphragm. A person with normal vision is able to see objects between the least distance of distinct vision of 9.84 in. and the horizon line. To form the image on the retina, the focal length of the lens must be variable and this focalization system is called visual accommodation [11].

By looking at the projection surface, the observer sees the HUD through refraction in a certain region of the car windshield. This region is called visual field of the reflective mirror in relation to the observer. The visual field depends on the dimensions and the position of the observer (Fig. 1).

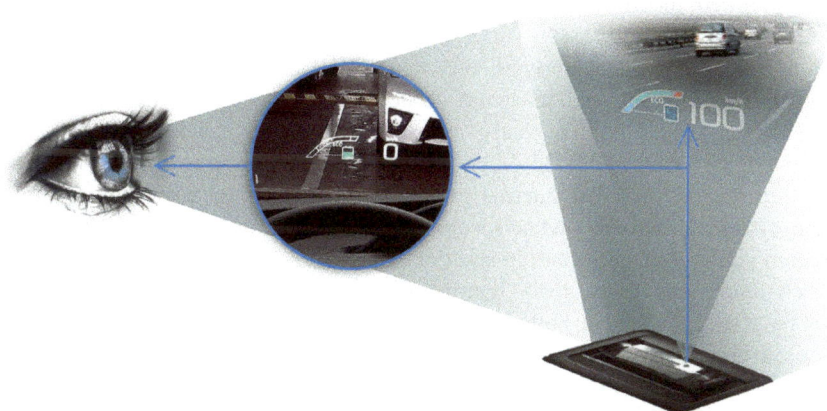

Fig. 1. Illustration of the sequence of the case-study HUD projection.

The eyesight is one of the most important organs to guide the driver towards the desired direction; in addition, the visualization of objects and everything around is necessary when driving a vehicle. However, the new technologies can be used to sharpen our other senses, such as our hearing, making these new technologies much more effective in the studies of more autonomous vehicles. Therefore, these superior technologies would be necessary for use in new environmental situations, physical or virtual, such as in augmented reality. Within the context of adaptive capacity, the knowledge practices and techniques employed by the visually and hearing impaired can be used, as they require much more from the other senses, to increase the command capacity of vehicle users. This requires further investigation and studies. A human being without any deficiency has a balanced body, but the absence of one of the senses makes the other cognitive capacities stand out.

Based on the current technologies available in the market, some generic classification criteria have been conveniently established. The displays are classified into two categories. The first includes the analog displays and the second the electro-optical displays. The automotive Head-Up Display is a device that is installed in the car's dashboard and reflects the information onto the windshield in the line of vision of the driver, in accordance with its ergonomic and functional proposal. The HUD, classified as an electro-optical display, offers better visualization from dawn. However, during this time, its projections conflict with the several night-lights. During the day, it has problems that arise from brightness, shadows, the reflection of the trees and buildings on the windshield, and natural lights [3].

Moreover, the HUD installed after the manufacturing of the vehicle is connected externally to the OBDI or OBDII port to read the speed of the automobile, use navigation data, infer the position of the car using GPS and also display the fuel consumption, among other information, according to each different model available in the market.

Thus, the HUD is a multimedia device positioned in front of the driver that makes the user glance quickly at the projection, reducing the need to keep on checking the activities required on the dashboard while driving and that are necessary for safe driving [13].

The new automotive technologies, such as the HUD, have been slowly adopted in Brazil. They have been used in sports models, hybrid cars, and medium sedans. However, the graphical interfaces vary according to each car brand and model.

For example, some factors are different from the subject of this case study. In this vehicle, the cover of the HUD internal compartment opens up and rises automatically once the system is started, and a special acrylic screen rises up, onto which the information you will be projected. In addition, the sensors to turn the HUD on and off, and the configurations for automatic lighting, day and night lighting, image positioning towards the left, right, top, or bottom are located in the multi-function display and are used by touching the touchscreen (Fig. 2).

The graphical representation of the HUD is cohesive and coherent with its representation of the virtual world, which, if otherwise ineffective, may lead to some visual disturbance that distracts the attention away. The HUD in passenger vehicles and military aircraft does its best to offer the information in a simple manner, trying to keep from distracting the user's attention away, unless in an emergency [1].

The HUD technology has to advance more and more. It can make the future easier by using technology to further this great progress, offering new value to automobile drivers. Still, the automotive resources have to comply with conventional traffic signs,

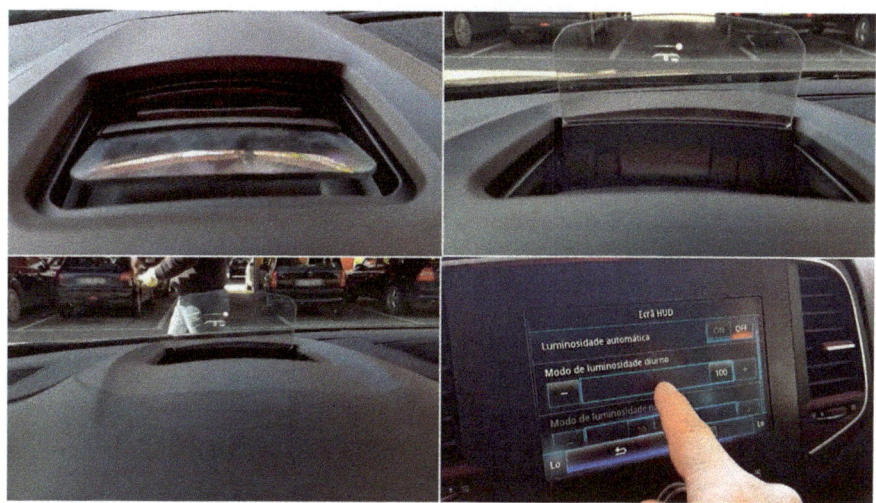

Fig. 2. Sequence of images of the HUD in another vehicle, a medium sedan.

reflective crosswalks, bright signs, and traffic controllers, among others, that help the drivers to keep organization, direction, and safety in traffic.

The introduction of new technologies is gradual; the human beings may improve their spatial ability to feel safe with everything that has been happening, as their media life tends to expand with all that.

At the same time, there is technological advance in the productive capacity of the population, as the user tends to stop using visible connection wires and simply starts preferring communication technologies that do not require wires, and the information is usually conveyed through wireless frequencies.

The technologies used in car dashboards have been increasingly growing over the last years. Our society is living a fast expansion in this sector; our cars communicate with increasingly more sophisticated codes and symbols, making it easier for the driver to arrive at his or her destination. However, due to the complexity of these highly sophisticated electronic devices, often, this does not happen as efficiently as planned. Human failure is part of the lack of knowledge to keep up with all the technological industrialization that is moving towards a more evolutionary growth.

All that may jeopardize safety, because, often, in some countries, the public roads are not as modern as this technology and have problems to keep up with it. Another major factor is the driver's attention, as sometimes the cognitive senses are not able to process too much information at the same time, and this may lead to distraction and potential accidents.

The user's ergonomics is favorable, offering proportions that help handle the multi-function display, the indicators, the multimedia system, and the buttons on the car dashboard. However, by using the HUD, which is an extra visual element for capacitation, the human brain has to work much harder, devoting attention to information that, perhaps, it is not able to process so quickly.

5   Case Study

The vehicle analyzed is classified as a medium sedan, has automatic transmission and 123 HP. It is manufactured in Japan and entered the Brazilian market in 2012 with the 2013 model that includes a factory-installed, monochromatic HUD.

Based on the fieldwork carried out in a car dealer in Juiz de Fora (Brazil) and a test-drive in a 2018 model manufactured in 2017, we were able to check how the colorful HUD worked, observing and analyzing its failures and qualities in relation to its usability for end consumers.

In the subject of this study, the HUD is located on the upper left side of the dashboard and projects the current speed and the variable information onto the vehicle windshield. According to the driver's choices, it shows several kinds of information that help driving the vehicle, such as the indicator of the hybrid system, the assistance to driving, and the navigation. Moreover, whenever the warning light turns on or flashes on the dashboard, its icon is also projected onto the windshield (Fig. 3).

Fig. 3. HUD projection onto the windshield displaying the speed and the hybrid system.

Only the driver is able to see the Head-Up Display over the steering wheel, reducing the need to look at the multi-function display, the indicators, or the multi-media system in the center of the dashboard. This projection window is due to the limiting dimensions (5.12 in. × 2.36 in.) of the device installed in the dashboard, preventing the passengers from seeing it (Figs. 4 and 5).

Fig. 4. HUD reflective internal device on the upper left side of the vehicle dashboard.

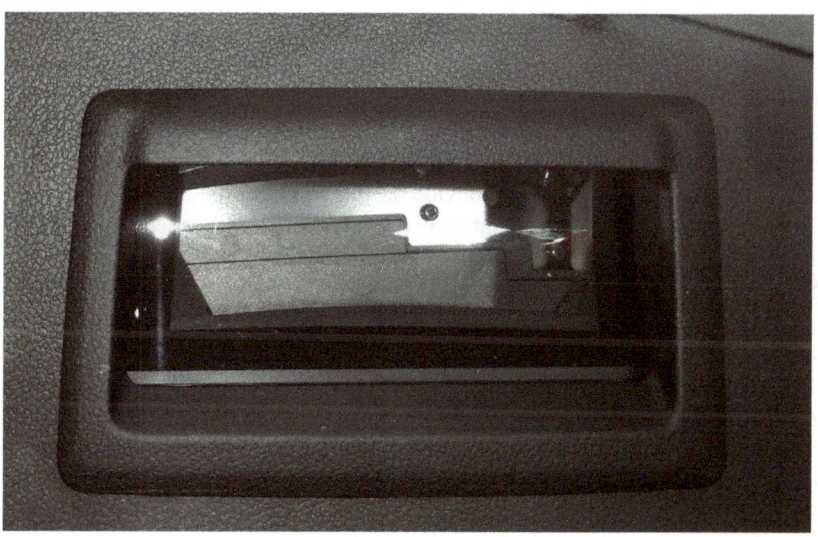

Fig. 5. Detail of the HUD internal device

The button to turn the HUD on/off is at the bottom left of the dashboard next to the steering wheel. When the device is turned off, the information is not projected onto the windshield until the button is used to turn the HUD on again, as this function is not linked to the engine start button (Fig. 6).

Fig. 6. HUD On/Off button on the left of the steering wheel.

6   Results

The machines have been addressing the human needs effectively, whether in driving vehicles and for other purposes. The HUD technology is one of them. However, to improve it, it is necessary to demand much more from the technological universe and think about the environment surrounding the cars. Based on that, it is important to emphasize that the Brazilian urban roads are not prepared for this kind of technology, making it unnecessary as one more device on the windshield.

These research results are qualitative and analytical and reflect upon the HUD technology available in the vehicle subject of this case study.

Some of the benefits of the Head-Up Display investigated include its graphical design and the brightness of the images projected that aims at reducing energy consumption.

The HUD adjusts the brightness on the windshield automatically based on the current use of the headlights (On/Off) and the lighting conditions in the surroundings of the vehicle. Even when the driver adjusts the brightness of the projection, it is auto-matically reduced when the vehicle is parked and, once it starts moving, from 5 km/h, it goes back to the brightness configured previously.

In addition, the graphical representation of the elements projected is cohesive and consistent, seeking to display the information in a simple manner. However, this efficacy may lead to some visual disturbance that distracts the driver's attention away, as follows.

Work, daily routine, and light and noise pollution may often make the human being tired and this may lead to lapse, as everybody is susceptible to human failure. By trying to use so many brain functions, there may be traffic accidents involving the user and other people.

Therefore, based on the analysis of this study, questions are raised regarding the drawbacks that may be harmful to the user and the human-technology system itself.

Staring at the projection on the windshield is not recommended, as this may lead to increased distraction in the users of passenger vehicles, reducing the driver's ability to see pedestrians, objects in the road, or other elements in front of the vehicle, and possibly leading to accidents with severe or deadly injuries. This is associated to the communication waves that travel the air as a huge net that covers all space-time, and everything can communicate without any difficulties; however, the human being is growing increasingly more isolated in a unique and digital world surrounded by superior technologies. When the human being immerses in these technologies, a flow state may be achieved and the space-time relationship becomes completely different, almost as a hypnotic effect. The HUD may be viewed in this context; so singular and imperceptible to the human mind.

The HUD projection may also be dark or blurry, making it difficult to see the information when wearing regular, or mainly polarized, sunglasses. This difficulty of visualization may also be present in extremely cold weather, as the presentation of the pre-selected information that is projected onto the windshield may experience visual distortions.

In addition, the GPS navigation is not displayed simultaneously on the HUD and the multi-function display, even if it is pre-selected on the multi-function display. This has a negative impact on the presentation of the navigation regarding spatial orientation, as the multi-function display is significantly larger than the HUD projection onto the windshield, in addition to showing more detailed and refined information.

Finally, another question was raised about cleaning the glass cover of the HUD, located on the upper left side of the car dashboard, which poses a handling problem, as the interior of the HUD projector should not be touched, otherwise leading to mechanical malfunction.

Self-sufficient technological tools are developed to help the human being, who should master them completely and adjust them to offer better usability and comfort to the user.

Based on that, the HUD technology has to improve, bringing more benefits to the driver, offering more safety to the user when driving the vehicle, and contributing to improved ergonomics and automotive interior design.

In future studies about the efficiency of the HUD, we recommend the use of other driver-related parameters, such as the length of time the driver has a driver permit; the driver's age; innate or acquired eyesight problems (myopia, photophobia, color-blindness, among others); any potential deficiencies; driver's reaction time; and social or cultural standards to check if this graphical language is going to be well absorbed and understood by the users.

References

1. de Araujo, R.P.: Imersão e Heads-Up Displays (HUDs) em videogames. Dissertação de Mestrado, Departamento de Design, Universidade de Brasília, DF, Brasília (2014)
2. Bertoldi, E.: Sistema anticolisão de alerta ao motorista com o uso de estímulo auditivo e háptico. Dissertação de Mestrado, Escola Politécnica, USP, São Paulo (2011)

3. Gutierrez, R.M.V., Monteiro Filha, D.C., Kauss, I.F., Oliveira, M.J.D.: Complexo eletrônico: displays e nanotecnologia. BNDES Setorial, Rio de Janeiro, no. 23, pp. 27–84, March 2006

4. IMOBILIS, Laboratório de Computação Móvel, UFOP. Introdução aos HUDs (Head Up Displays) e HMDs (Head Mounted Displays) – parte I. http://www.decom.ufop.br/imobilis/introducao-aos-huds-head-up-displays-e-hmds-head-mounted-displays-parte-i. Acesso em 10 May 2017 às 08:00

5. IMOBILIS, Laboratório de Computação Móvel, UFOP. Introdução aos HUDs (Head Up Displays) e HMDs (Head Mounted Displays) – parte II. http://www.decom.ufop.br/imobilis/introducao-aos-huds-head-up-displays-e-hmds-head-mounted-displays-parte-ii. Acesso em 10 May 2017 às 08:40

6. IMOBILIS, Laboratório de Computação Móvel, UFOP. Introdução aos HUDs (Head Up Displays) e HMDs (Head Mounted Displays) – parte III. http://www.decom.ufop.br/imobilis/introducao-aos-huds-head-up-displays-e-hmds-head-mounted-displays-parte-iii. Acesso em 10 May 2017 às 09:30

7. Leal, W.D.S.: O ensino de algoritmos no ensino médio: por que não? Dissertação de mestrado em Ensino de Ciências na Educação Básica – Universidade do Grande Rio, Duque de Caxias, RJ (2009)

8. de Mello, J.M.C., Novais, F.A.: Capitalismo Tardio e Sociabilidade Moderna in SCHWARCZ, Lilia Moritz (org.). História da Vida Privada no Brasil, vol. 4, Cia das Letras, São Paulo (2012)

9. de Moraes, A.M.; Quaresma, M.: A distração do motorista e o desenvolvimento de equipamentos eletrônicos em veículos. LEUI – Laboratório de Ergonomia e Usabilidade de Interfaces em Sistemas Humano-Tecnológicos da PUC, Rio de Janeiro (2006)

10. Quresma, M.M.R.: Avaliação da usabilidade de sistemas de informação disponíveis em automóveis: um estudo ergonômico de sistemas de navegação GPS. Tese de Doutorado, PUC, Rio de Janeiro (2010)

11. Ramalho, F.: IVAN, José. NICOLAU, Gilberto. TOLEDO, Paulo Antônio de. Os Fundamentos da Física 2. Termologia, Geometria da Luz e Ondas. Ed. Moderna, São Paulo (1979)

12. Santos, R.: Ergonomia e experiência do Usuário: Novas Fronteiras para o Design de Interface. 8º ErgoDesign & USIHC, São Luís, Maranhão (2015)

13. Souza, A.S.: TOSI, Sergio Rodrigues. Estudo sobre o uso de aplicativos em realidade aumentada para condução de veículos com ênfase na usabilidade. 15º ErgoDesign & USIHC, Recife, Pernambuco (2015)

Usability Evaluation and User-Centered Design

Getting the Complete Picture: Using Surveys as a Complementary Method for Assessing Usability

Courtney Titus[1(✉)], Mary Gordon[2], Krisanne Graves[2],
and Curt Braun[3]

[1] Baylor College of Medicine, 1102 Bates St., Suite, Houston, TX 940.35, USA
ctitus@bcm.edu
[2] Texas Children's Hospital, 6621 Fannin St., Houston, TX A2270, USA
{mdgordon, kxgraves}@texaschildrens.org
[3] Benchmark Research and Safety, Inc.,
1150 Alturas Dr., Suite 108, Moscow, ID, USA
cbraun@benchmarkrs.com

Abstract. Medical errors are an important area of study given their impact on patient safety. While new health technologies are created to reduce these errors, improperly designed solutions may unintentionally contribute to them. Current methods used by manufacturers to identify design issues, such as usability testing and heuristic evaluations, may not uncover issues that only arise when the product is exposed to a breadth of users, scenarios, and environments. Other methods, such as the System Usability Scale (SUS), have the potential to quickly assess a product's overall usability but provide little insight into specific aspects of the design that need improvement. Therefore, additional methods for gathering usability data, to complement current ones, need to be investigated. This study begins this investigation by evaluating the feasibility of an in-situ usability survey to provide insight into an infusion system's design issues. The findings and implications for health technology manufacturers are discussed.

Keywords: Usability · Survey · Medical errors · Medical device

1 Introduction

According to the 1999 Institute of Medicine report, approximately 98,000 patients die in hospitals each year as a result of preventable medical errors [1]. New health technologies are created to reduce these errors by streamlining workflow and automating processes. However, these same technologies have the potential to introduce new errors into hospital systems that can endanger patient safety.

One factor that contributes to these errors is poor design that prevents users from effectively using the technology. As demonstrated in a study by Kushniruk and colleagues, these usability problems can lead to actual medical errors [2]. Findings identified that the most common usability problem experienced by users of a handheld prescription writing program was display visibility (not seeing required information on

© Springer International Publishing AG, part of Springer Nature 2019
T. Z. Ahram and C. Falcão (Eds.): AHFE 2018, AISC 794, pp. 197–203, 2019.
https://doi.org/10.1007/978-3-319-94947-5_19

the screen) and this problem was closely associated with medication errors that went unnoticed and uncorrected by users [2].

The link between usability problems and medical errors highlights the importance of identifying and eliminating design issues before products are released. One way health technology manufacturers identify potential issues is through usability testing. The method typically involves a small sample of users performing predetermined tasks with the product in a controlled environment. Miller and colleagues showed this method to be effective when applied to evaluating infusion pump design [3]. In their study, 22 nurses from three hospital units performed 10 tasks designed to assess the function or operation of the pump and seven tasks designed to assess the nurses' ability to use the device to deliver care [3]. Sixty-eight percent of their participants experienced at least one error while performing functional tasks and 100% experienced an error while performing clinical tasks [3].

Although Miller and colleagues were able to identify design flaws with usability testing and, subsequently, develop nursing procedures and policies to mitigate the risks associated with these flaws, several limitations were noted [3]. The two primary limitations were the substantial time and financial resources required to implement the method [3]. Additionally, participants were asked to perform a number of tasks, during testing, that were not representative of those performed while on the job [3]. This discrepancy may have impacted the participants' performance during prescribed tasks more so than the pump's design.

In addition to the limitations noted by Miller et al., Schmettow and colleagues noted that sample sizes traditionally recommended for usability testing may not be sufficient for health technologies [4]. When usability testing is applied to commercial products, it is believed that only five users are needed to identify 85% of a product's usability problems. However, Schmettow and colleagues developed a statistical model that indicated larger sample sizes (25 or more users) are required for rigorous testing of complex, high-risk systems [4].

Other methods have addressed some of the limitations of usability testing. One such method is heuristic evaluation, where a small set of experts evaluate a product's compliance with recognized usability principles. According to usability practitioners, this approach can capture up to 60–75% of a product's usability problems [5]. Although this method has been commonly used to identify potential usability issues for websites and software applications, researchers have found it to be a successful approach for evaluating medical device design [5, 8]. Zhang and colleagues demonstrated that when using this approach to evaluate two 1-channel volumetric infusion pumps they could identify 89 usability problems for one pump and 52 problems for the other [5]. Identified issues were generally associated with the physical interface, primary screen, and pump personality screen for one pump and the options menu for the other [5].

Chan et al. successfully applied a similar approach to the evaluation of radiotherapy systems [6]. With this approach, two human factors experts were able to identify 37 low severity usability problems (problems that cause user frustration), 20 medium severity problems (problems that may contribute to user error but have no major impact on patient safety), and 18 high severity problems (problems that can have major impact on patient safety) [6].

Heuristic evaluation is often favored because it is low cost, easy to apply to a variety of products, and allows for the quick identification of usability problems. However, as noted by Chan et al., a limitation of this approach is that evaluators are often not the end users and may overlook problems that are only apparent to the intended population [6]. Also, heuristic evaluations often do not consider problems that may arise when the device is used in different environments [5].

The System Usability Scale, or SUS, is another method used for gathering usability data. This 10-item survey is generally administered to participants after a usability test. However, research conducted by Peres et al. suggests that since SUS scores are associated with participants' performance on tasks during usability testing it has the potential to be effectively used outside of the testing context [7].

Like heuristic evaluations, the SUS is favored because of its cost and minimal time commitment. Additionally, using it outside of a testing context, allows users to consider their actual experience using the device rather than relying on artificial experiences created for a usability test [7, 9, 10]. However, the primary limitation of the SUS is that it cannot provide manufacturers with insight into the specific issues users identified with their system or what specifically needs improvement.

Considering the importance of identifying usability problems for health technologies and the limitations of other methods, it is necessary for manufacturers to investigate additional approaches to gather usability data to complement current ones. The current study endeavors to begin this investigation by evaluating the feasibility of an in-situ usability survey to provide insight into the design issues associated with an infusion system.

2   Methods

2.1   Participants

Participants were identified for the study through an online scheduling system used by nurses at a Texas pediatric hospital. A stratified sample of 500 nurses from six hospital units (acute care, critical care, neonatal intensive care, ambulatory care, perioperative services, and women's services) were randomly selected to receive an email invitation to complete the survey through a web-based survey system.

Once respondents completed the survey, their name was entered into a drawing to win a $100 Amazon gift card. Three respondents were randomly selected to win a gift card.

2.2   Instrument

Survey items were derived from observations of current usage of the infusion system, a review of human factors literature, the SUS, and feedback from nurse leaders and human factors specialists. The survey underwent three rounds of review and revision by nursing staff to ensure items were understood and appropriately worded.

The final 59-item survey consisted of three types of questions – demographics, Likert scales, and open-ended. Example demographic questions were "years working at the hospital" and "did you attend a formal training [for the infusion system]." For the

Likert scales, respondents were asked to rate the extent to which they agreed or disagreed with statements regarding the infusion system's physical qualities, error prevention, alarms, emergency interventions, unexpected actions, compatibility with the environment, training requirements, and overall usability. Open-ended questions allowed respondents to indicate any changes they desired to see in the system's design to make it more user-friendly.

2.3 Device

The chosen system for the study was currently being used by nursing staff throughout the hospital. The system featured a programming unit, which allowed users to customize infusion delivery, and three modules – large volume pump module, syringe module, and patient-controlled analgesia (PCA) module. Each module could be connected to the programming unit, as needed, for patient care.

3   Results

3.1   Respondent Demographics

Table 1 provides survey respondents' demographic information. Respondents were on average 35.3 years old (range: 20–68 years old) and worked 9.7 years as a nurse. Most respondents were from the acute care unit and only 48.7% (n = 113) had attended formal classroom or superuser training for the infusion system.

Table 1. Demographics of survey respondents.

Demographics (N = 232)	Percent (n)	Mean ± SD
Age, years		35.3 ± 11.2
Years as nurse		9.7 ± 9.8
Unit type		
Acute care	37.1% (86)	
Critical care	28.0% (65)	
NICU	28.4% (66)	
Ambulatory care	1.7% (4)	
Perioperative services	3.9% (9)	
Attended formal training	42.4% (98)	
Attended superuser training	16.4% (38)	

3.2   System Utilization

Two hundred twenty respondents provided usable data for questions related to the number of infusions they administered with the system during a typical work week. As seen in Fig. 1, the majority of respondents reported using the system to administer 10 to 20 infusions per week. The average number of infusions, across respondents, was 16.8 (SD = 22.17).

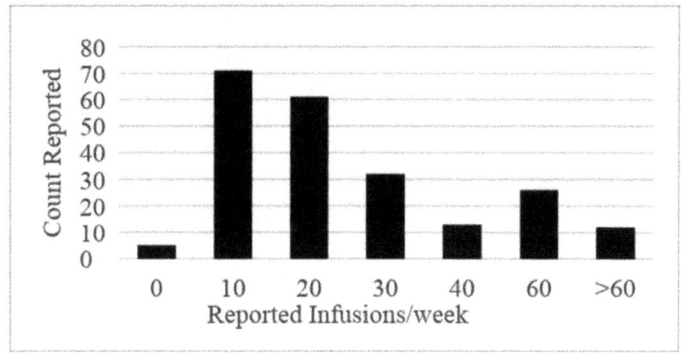

Fig. 1. Number of infusions administered per week by survey respondents.

3.3 Ease of Use

Overall, survey respondents perceived the infusion system's modules as easy to use. Specifically, 96.7% (n = 208) rated the large volume pump module as either "easy" or "extremely easy" to use. The syringe and PCA modules received similar ratings from 93.1% (n = 203) and 86.8% (n = 138) of respondents, respectively.

3.4 Alarm Management

Although the majority of respondents perceived the system as easy to use, specific questions about the system's alarms received the least consensus. The statement "I can identify the type of problem based solely on the alarm sound that the system makes" yielded 43.4% (n = 98) agreement and 38.4% (n = 89) disagreement.

When asked about the number of system alarms, 52.6% (n = 116) of respondents indicated the system produced either the right number of alarms or too few alarms. Forty-seven percent (n = 105), conversely, believed the system produced too many alarms.

The item "having different alarms for process conditions (approaching end of infusion) and for delivery problems (obstruction in the line) would helpful" elicited the most agreement among respondents. Seventy-four percent (n = 163) either agreed or strongly agreed with the statement while only 13.7% (n = 30) disagreed or strongly disagreed with it.

4 Discussion

4.1 Strengths

The aim of this research was to determine if an in-situ usability survey could help address some of the limitations identified for other usability methods. The data presented in this paper suggests that it can be.

Unlike usability testing, the survey required minimal cost and effort to perform. The researchers were able to administer the survey through a free online survey system provided by the hospital and, therefore, only needed to invest $300 to provide gift cards to drawing winners. Three email reminders, delivered over an eight week period, and one phone call were all that was needed to encourage participation in the survey and achieve a robust response rate of 46.4%.

The survey gathered data from a larger and more diverse sample than those typical of usability tests and heuristic evaluations. With over two hundred respondents from five different hospital unit types, the researchers gained better insight into how the infusion system worked within different contexts and with users of varying levels of experience with the device.

Similar to the SUS, the survey was able to provide insight into the overall usability of the infusion system. However, the survey had the advantage of revealing specific problematic areas of the infusion system, like the alarms, that require further investigation and may need refinement for future device iterations.

4.2   Limitations

As with other methods, the survey had limitations. One such limitation is the potential for nonresponse bias. Although 500 nurses were invited to complete the survey, 268 didn't respond to this invitation. It is possible that some non-respondents didn't use the infusion system as part of their duties. However, other factors may also be contributing to their lack of response.

The researchers called non-responders during work hours to encourage their participation. However, the end of the data collection period limited the number of calls that could be made. Additional effort is needed to ensure a true random sample is obtained and the full range of experiences with the infusion system are captured.

Another limitation is survey items related to general usability were less informative than more specific items related to system functionality. The responses to these non-specific usability items lacked variability and, therefore, suggest they may be less sensitive to respondents' perceptions. It is also possible that respondents interpreted these items as a measure of their individual competency rather than the system's usability.

5   Conclusions

Despite the limitations of this study, the findings suggest that an in-situ usability survey may be an inexpensive and efficient method for identifying usability issues for health technology manufacturers. The researchers caution that this method should not be used in place of more traditional techniques. Rather, it is an additional technique for manufacturers to add to their toolkits that can complement other methods and help provide a more complete picture of a technology's design issues.

References

1. Kohn, L.T., Corrigan, J.M., Donaldson, M.S.: To Err is Human: Building a Safer Health System. National Academy Press, Washington (1999)
2. Kushniruk, A., Triola, M., Stein, B., Borycki, E., Kannry, J.: The relationship of usability to medical error: an evaluation of errors associated with usability problems in the use of a handheld application for prescribing medication. Stud. Health Technol. Inform. **107**(2), 1073–1076 (2004)
3. Miller, K.E., Arnold, R., Campbell, M., Zern, S.C., Dressler, R., Duru, O.O., Ebbert, G., Jackson, E., Learish, J., Strauss, D., Wu, P., Bennett, D.A.: Improving infusion pump safety through usability testing. J. Nurse Care Qual. **32**(2), 141–149 (2017)
4. Schmettow, M., Vos, W., Schraagen, J.M.: With how many users should you test a medical infusion pump? sampling strategies for usability tests on high-risk systems. J. Biomed. Inform. **46**(4), 626–641 (2013)
5. Zhang, J., Johnson, T.R., Patel, V.L., Paige, D.L., Kubose, T.: Using usability heuristics to evaluate patient safety of medical devices. J. Biomed. Inform. **36**, 23–30 (2003)
6. Chan, A.J., Islam, M.K., Rosewall, T., Jaffray, D.A., Easty, A.C., Cafazzo, J.A.: Applying usability heuristics to radiotherapy systems. Radiother. Oncol. **102**(1), 142–147 (2012)
7. Peres, S.C., Pham, T., Phillips, R.: Validation of the System Usability Scale (SUS): SUS in the wild. In: Proceedings of the Human Factors and Ergonomics Society 57th Annual Meeting, pp. 192–196. SAGE, California (2013)
8. Graham, M.J., Kubose, T.K., Jordan, D., Zhang, J., Johnson, T.R., Patel, V.L.: Heuristic evaluation of infusion pumps: implications for patient safety in intensive care units. J. Med. Inform. **73**, 771–779 (2004)
9. Grier, R.A., Bangor, A., Kortum, P., Peres, C.: The system usability scale: beyond standard usability testing. In: Proceedings of the Human Factors and Ergonomics Society 57th Annual Meeting, pp. 187–191. SAGE, California (2013)
10. Predicting Task Completion with the System Usability Scale. https://measuringu.com/task-comp-sus/

The Effects of Grid- and List Design of E-Commerce Result Lists on Search Efficiency and Perceived Aesthetics

Friederice Schröder$^{(\boxtimes)}$, Anica Kleinjan, and Stefan Brandenburg

Technische Universität Berlin, Marchstraße 23, 10587 Berlin, Germany
friederice.schroeder@tu-berlin.de

Abstract. E-commerce became very important in recent years. About 40% of the customers state that they prefer online shopping over retail shopping. When searching for items in an online shop, the appearing result lists are an important part of the users' experience. The users' experience, in turn, is a crucial factor for a business' success. This online-study examines the search efficiency and perceived aesthetics of two different result list designs. A grid design, which shows more than one item in a row and a list design, which displays only one item per row. Participants completed a visual search task on a fictitious website using their own devices. Reaction times measured search efficiency and the VisAWI questionnaire assessed perceived aesthetics. Results showed that the grid design led to faster reaction times and higher values in perceived aesthetics than the list design. Implications of the results and future studies are discussed.

Keywords: Grid design · List design · Search efficiency · Perceived aesthetics
Online shopping · User experience

1 Introduction

High growth rates were recorded in E-commerce in recent years. In Germany for instance, online business sales increased about 50% from 2010 to 2016 [1]. Online shoppers rely heavily on information provided by the website. An online shops' interface design is, therefore, a highly important aspect of customers' choice in E-commerce. Liang and Lai [2] asked participants to buy books in different online shops. Afterwards the participants had to explain the reasons for their buying decision. Results showed, that –besides the price of a book– the websites' design was the most important reason for their decision. Similarly, Cooper-Martin [3] stated that customers' behavior in E-commerce is influenced by how specific product information is presented on a website. Result lists can present product information in two different designs: grid and list design. Grid designs show more than one result in a row. List designs show one result in a row. This distinction is important depending on the customers' behavioral intention. This means that it makes a difference whether customers have a clear shopping intention or not. With respect to the customers' intention, recent literature differentiates between browsing tasks and searching tasks [4]. The term browsing task means that a customer has a general but not explicit idea of what he or she is looking

© Springer International Publishing AG, part of Springer Nature 2019
T. Z. Ahram and C. Falcão (Eds.): AHFE 2018, AISC 794, pp. 204–211, 2019.
https://doi.org/10.1007/978-3-319-94947-5_20

for. In contrast a searching task is a task where the customer exactly knows what product or information he or she is looking for. In 2004 Hong, Thong and Tam examined the time needed to search for information and the possibility to recall product information among the two different designs and searching types [5]. The results of their study indicated that participants performed a browsing task faster when the results were presented in a list design. The grid design led to faster searching times for searching tasks.

Nowadays, it isn't enough to consider customers' efficiency when evaluating web-interfaces. Aesthetic aspects of interfaces are important for a customers' shopping experience as well. Consequently, these aspects have received more attention in research. Leder, Belke, Oeberst, and Augustin showed that aesthetic impressions form ad-hoc evaluations [6]. Moshagen and Thielsch state that aesthetic aspects influence a multitude of constructs [7] like perceived usability [8], satisfaction [9], the willingness to buy an offered product and the probability for recommendation [7]. Mahlke [10] found that the perceived aesthetics of a web-interface affects the intention to revisit the site. In addition, the results of Vogel [11] indicate that people tend to rate the aesthetics of an interface higher the more they become familiar with it. This effect shows that there might be differences in customers' perceived aesthetics depending on their online shopping experience.

The objectives of the present study are twofold. Firstly, we want to examine the effect of grid- and list design of E-commerce result lists on a customers' search efficiency. The present study complements Hong et al.'s work [5] regarding a couple of factors. It concentrates on searching task, but varies the length of the result lists and the presence or absence of the target. Corresponding to Hong et al. we expect a higher search efficiency with the grid design compared to the list design. Secondly, we want to examine the effects of grid- and list designs and the participants' E-commerce experience on perceived aesthetics. This objective is explorative because of missing literature on the topic.

2   Methods

Participants. A sample of 85 people participated in the online-experiment. The participants' age ranged from 19 to 61 years with a mean of 30.64 years ($SD = 10.33$). The sample was almost balanced with respect to gender (46 males and 37 females). Twenty-five participants stated to shop via E-commerce at least once a month. Sixty participants stated to shop less than once a month in E-commerce stores. Participants were recruited via Internet. They participated voluntarily and received no incentives.

Material. A searching task with a total of 48 result lists on a fictitious website was developed to test the efficiency and perceived aesthetics of the list and grid design. The items in the searching task were a collection of clothes as clothing is the most profitable product group in Germany [1]. The collection included shorts, skirts, shoes, sunglasses and bras of different price and size. Figure 1 visualizes an exemplary result list. The presented result lists consisted of three, six or nine items. Nine items were chosen to be the maximum number because participants should not have to scroll the page to see the whole list of items. In the list design, each row showed one item. In the grid design, each row consisted of three items. Figure 2a shows an exemplary list design and Fig. 2b visualizes the same result list as grid design.

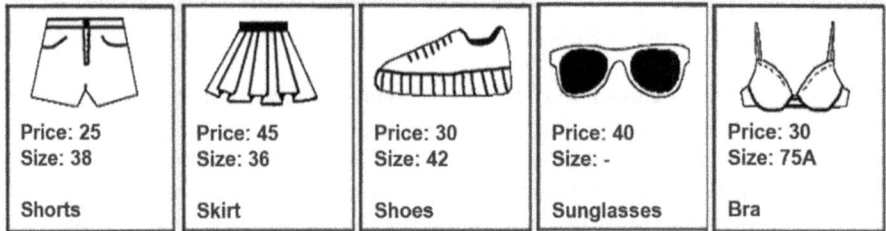

Price: 25	Price: 45	Price: 30	Price: 40	Price: 30
Size: 38	Size: 36	Size: 42	Size: -	Size: 75A
Shorts	Skirt	Shoes	Sunglasses	Bra

Fig. 1. An exemplary result list showing a collection of clothes.

The 48 result lists differed regarding their design, length, and the absence or presence of a target item, a skirt for €30. In both result list design conditions, the target was either present (18 result lists) or absent (6 result lists). If present, the target appeared in random position within but not between the participants. In addition, the result lists included two skirts with different price or two different items of €30 each as distractors. If the target was absent, the participants could only find a skirt with another price or another type of clothing of €30. Search efficiency was defined as the time a participant needs to find the target item, the skirt of €30, in a number of other items.

To assess perceived aesthetics of the result lists, the participants were asked to complete the Visual Aesthetics of Website Inventory (VisAWI) [8]. The VisAWI measures the subjective aesthetics of a web-interface. It defines aesthetics as a positive experience based on an object, a product in this case. The VisAWI has 18 items grouped into four facets: simplicity (5 items), diversity (5 items), craftsmanship (4 items) and colorfulness (4 items). Participants answer each of the items on a seven-point Likert scale ranging from 1 (strongly disagree) to 7 (strongly agree). Reliabilities of the sub-scales range from .85–.89 [8]. In the present study, the items were presented in different shades of grey. Therefore, the facets simplicity, diversity and craftsmanship were chosen for the evaluation only.

Procedure and Experimental Designs. Participants received a link to the online-study after they agreed to participate. They started the experiment by clicking on this link. Upon the start of the online-experiment, the participants were asked to read the written instructions on the screen of their own computer. These instructions included that they should only start the experiment if their monitor is sufficiently large and that they should not use the buttons of their browsers.

Then the participants were introduced to the procedure and the task. Their task was to search for a skirt of €30 among other items in a result list. Participants were instructed to decide whether the item of interest was in the list or not as fast and as correct as possible. They selected their answers by clicking on one of two buttons (yes/no).

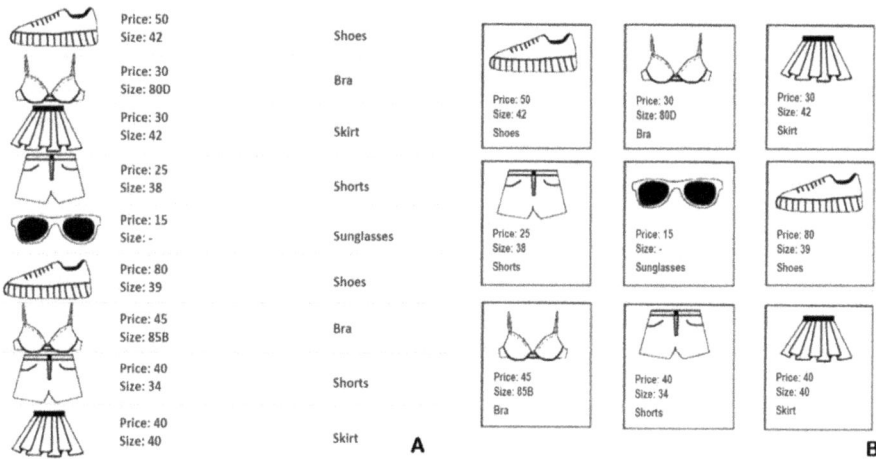

Fig. 2. Exemplary result lists in list - (A) and grid-design (B) with 9 items each and the item of interest at position 3.

After reading the instructions, the participants familiarized themselves with the task and practiced it by completing two searches. The main experiment consisted of two blocks of trials with 24 trials (searches) each. The design of the result list was the blocking factor. In each trial the participants saw the question whether a skirt of €30 was included in the result list on top of their monitor. Below the question one of the 48 result lists were shown. Finally, the two buttons yes (right) or no (left) were presented under the result list. A trial (and the assessment of the participants' reaction time) started with all three elements (question, result list, buttons) appearing on the participants' screen at once and ended after the participants pressed one of the two buttons.

After each block of trials the participants filled in the VisAWI and rated the perceived aesthetics of the previously presented result list. The blocks of trials were balanced across participants. This means that half of the participants searched the item of interest 24 times in a list design first (block 1), followed by searching the item of interest 24 times in the grid design (block 2). The other half of the participants saw the results presented in a grid design first (block 1), followed by a list design (block 2). At the end of the experiment the participants answered some demographic questions and reported how often they shop online. The whole experiment took about 10 to 15 min.

The experimental design for the effect of list design and list length on participants search efficiency was a 2 (result list design: grid or list) × 3 (list length: 3, 6 or 9 items) within-subjects design. The experimental design for the effects of grid- and list-design on perceived aesthetics was a 2 (list design: grid or list) × 2 (experience: less or more) mixed design with list design being a within-subject factor and experience being a between-subject factor.

3   Results

The analysis of the behavioral results was performed using the correct trials only. Correct means that participants did find the target item if it was present. We computed average reaction times across possible target positions for each list length.

Statistical data analysis included two steps. Firstly, behavioral and subjective data was screened for outliers on the participant and the group level. Outliers were defined as individual scores exceeding the group mean by more than ± 3 SD. Eight participants were excluded from further analysis of the behavioral data because more than 10% of their data exceeded the group mean by more than 3 SD. No participants were excluded from the analysis of the aesthetics data. Secondly, behavioral data was subject to analysis using ANOVAs. The effect size partial η^2 is reported for small ($0.01 < \eta^2_{part} < 0.06$), medium ($0.06 < \eta^2_{part} < 0.14$), and large ($\eta^2_{part} > 0.14$) effects [12].

3.1   The Effects of E-Commerce Result List Design and List Length on User Behavior

A 2 (design of the result list: grid or list) $\times$ 3 (list length: 3 or 6 or 9) ANOVA was calculated to examine the effects of the design of the result list and list length on the users' reaction time (RT). Results revealed main effects for the factors design of the result list ($F(1,76) = 4.30$, $p = .04$, $\eta^2_{part} = .05$), length of the result list ($F(2,152) = 111.49$, $p < .001$, $\eta^2_{part} = .59$), and an interaction effect of design and length of the result list on participants' search efficiency, $F(1,152) = 4.44, p = .01, \eta^2_{part} = .06$. Figure 3 visualizes the result of the analysis. It shows that participants' reaction time became significantly longer with increasing list length. In addition, grid design led to faster reaction times with increasing list length compared to list design.

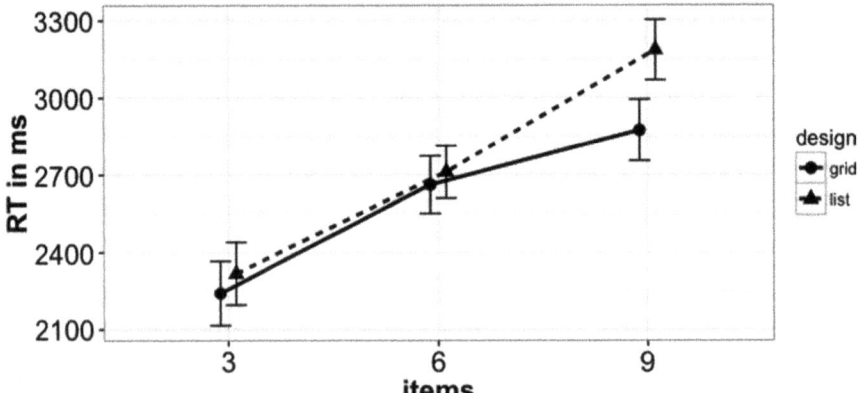

Fig. 3. Effects of the design of the result list and list length on participants' reaction time. Error bars indicate 95% confidence intervals.

3.2   The Effect of E-Commerce Result List Design on Perceived Aesthetics

The experience with online shopping was introduced as additional factor for the analysis of the aesthetic experience of grid and list design because online shopping experience might influence users' familiarity with the design of result lists. This might in turn affect their rating of perceived aesthetics. 25 of 85 (30%) reported to shop online less that once a month and 60 of 85 (70%) stated to shop online more than once a month. A 2 (list design: grid or list) × 2 (experience: less or more) mixed ANOVA on

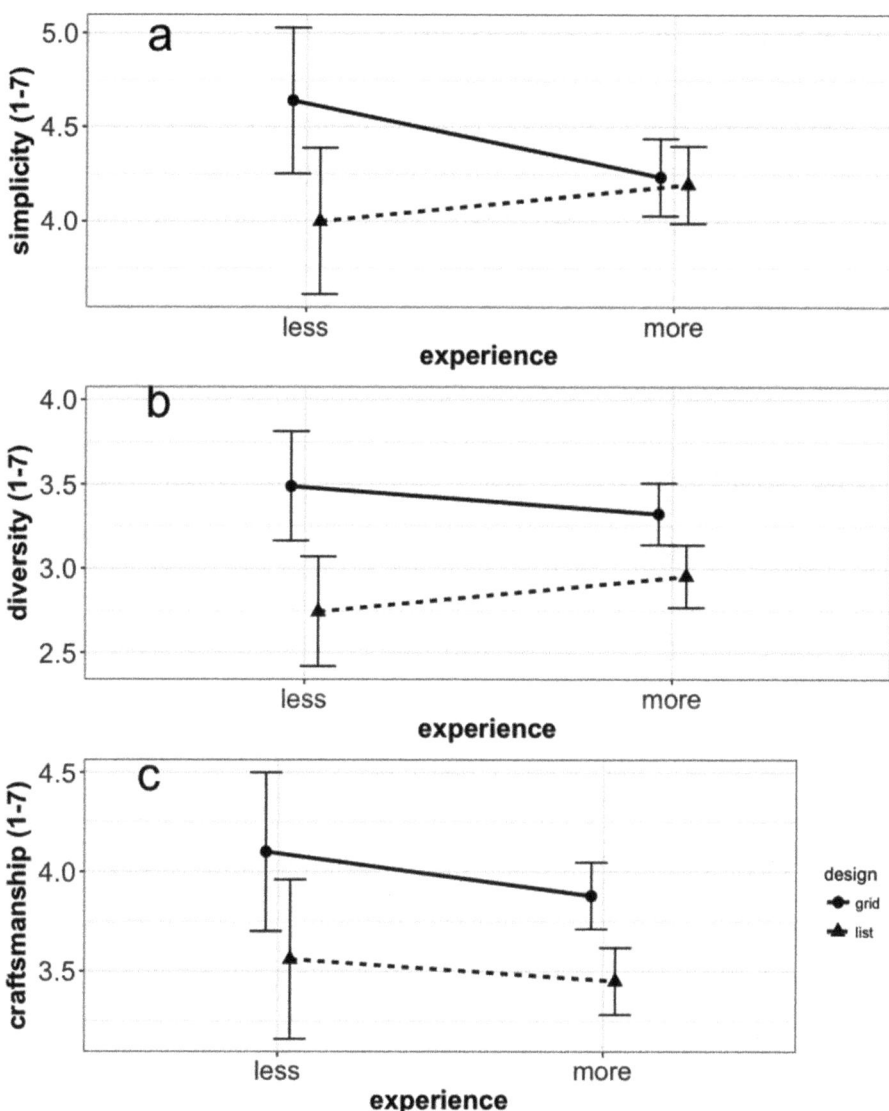

Fig. 4. The effects of result list design on (a) participants' simplicity rating, (b) their diversity ratings, and (c) their ratings of craftsmanship. Error bars show 95% confidence intervals.

the participants' simplicity ratings revealed a main effect for the design of the result list $(F(1,83) = 5.83, p = .02, \eta^2_{part} = .07)$ and an interaction effect of list design and online shopping experience, $F(1,83) = 4.54, p = .04, \eta^2_{part} = .05$. No main effect of online shopping experience on participants' simplicity rating was obtained, $p = .18$. Figure 4a visualizes the superiority of the grid design for participants with less shopping experience.

Another 2 (design) × 2 (experience) mixed ANOVA showed a main effect of list design on participants' diversity ratings, $F(1,83) = 20.24, p < .001, \eta^2_{part} = .20$. Again, no effect was obtained for their online shopping experience ($p = 0.92$) and no interaction was observed, $p = .13$. Figure 4b shows that the grid-design receives higher diversity ratings compared to the list design. Finally, the same analysis was repeated for participants' craftsmanship ratings. Here the ANOVA revealed another main effect for list design ($F(1,83) = 14.49, p = .003, \eta^2_{part} = .15$) but no effect of online shopping experience ($p = .51$) or an interaction of both factors, $p = .66$ (Fig. 4c).

4   Discussion

The present study had two objectives. Firstly, it examined the effects of grid- and list design of E-commerce result lists on peoples' search efficiency. Secondly, the study assessed the effects of result list design on perceived aesthetics considering the participants' online shopping frequency.

In accordance with the first objective the study replicates the results found by Hong et al. [5]. Participants had faster search times when a target item was presented in a grid design compared to a list design. This finding was affected by the number of items shown in the result list in both result list designs, complementing the findings of Hong et al. A higher number of items in the result list lead to longer reaction times in both designs. In addition, the benefit of the grid design increases with more items that are included in the result list. However, there were only three, six or nine items shown in the result lists to avoid effects caused by scrolling. Under natural circumstances E-commerce result lists are much longer. Therefore future studies should test the validity of our result for longer result lists. Moreover, in most cases the user interacts with the website and result list by scrolling, sorting and filtering. Regarding this issue, future studies should examine possible effects of user interactions on search efficiency for each result list design.

In addition to the search efficiency, the second objective of this study examined the perceived aesthetics of the two result list designs. Results revealed that the grid design was rated higher in each of the aesthetic dimensions of the VisAWI than the list design. However, the effect of the result lists' design on the aesthetic facet simplicity was affected by the participants' self-reported online shopping experience. In E-commerce, result lists of searches for clothes are often presented in grid design. It might have been that the grid design was more familiar to the participants and therefore might have been perceived as being simpler. Future work should examine whether the results of this study can be applied to online shops offering other product categories than clothes e.g. electronics or furniture.

Finally, most studies examined the effects of result list design on search efficiency on a computer desktop. In 2016, 33% of all E-commerce transactions in Germany included more than one device [13]. Consequently, further research should analyze the perception and efficiency of grid and list design on different devices with various screen sizes like smartphones, tablets or laptops and interaction modalities (i.e. tapping or clicking). In the long run a consistent cross-functional solution of result lists could be needed to guarantee a good customer experience on different devices.

References

1. HDE: Handel Digital - Online-Monitor 2017 (2017). https://de.statista.com/statistik/daten/studie/3979/umfrage/e-commerce-umsatz-in-deutschland-seit-1999/
2. Liang, T.-P., Lai, H.-J.: Effect of store design on consumer purchases: an empirical study of on-line bookstores. Inf. Manag. 39(6), 431–444 (2002)
3. Cooper-Martin, E.: Effects of information format and similarity among alternatives on consumer choice processes. J. Acad. Mark. Sci. 21(3), 239–246 (1993)
4. Carmel, E., Crawford, S., Chen, H.: Browsing in hypertext: a cognitive study. IEEE Trans. Syst. Man Cybern. 22(5), 865–884 (1992)
5. Hong, W., Thong, J.Y., Tam, K.Y.: The effects of information format and shopping task on consumers' online shopping behavior: a cognitive fit perspective. J. Manag. Inf. Syst. 21(3), 149–184 (2004)
6. Leder, H., Belke, B., Oeberst, A., Augustin, D.: A model of aesthetic appreciation and aesthetic judgments. Br. J. Psychol. 95(4), 489–508 (2004)
7. Moshagen, M., Thielsch, M.T.: Facets of visual aesthetics. Int. J. Hum.-Comput. Stud. 68(10), 689–709 (2010)
8. Moshagen, M., Musch, J., Göritz, A.S.: A blessing, not a curse: experimental evidence for beneficial effects of visual aesthetics on performance. Ergonomics 52, 1311–1320 (2009)
9. Cyr, D., Kindra, G.S., Dash, S.: Web site design, trust, satisfaction and e-loyalty: the Indian experience. Online Inf. Rev. 32, 773–790 (2008)
10. Mahlke, S.: Factors influencing the experience of website usage. In: CHI 2002 Extended Abstracts on Human Factors in Computing Systems, pp. 846–847. ACM, Minneapolis (2002)
11. Vogel, M.: Der "Mere-Exposure-Effekt" in der Mensch-Technik Interaktion und seine Auswirkungen auf das Nutzererleben. Dissertation, Technische Universität Berlin (2016)
12. Cohen, J.: Statistical Power for the Social Sciences. Lawrence Erlbaum (1989)
13. Criteo: The State of Cross-Device-Commerce (2017). http://www.criteo.com/media/6617/criteo-state-of-cross-device-commerce-2016-h2de.pdf

vis-UI-lise: Developing a Tool for Assessing User Interface Visibility

Ian Hosking$^{(\boxtimes)}$ and P. John Clarkson

Engineering Design Centre, Department of Engineering,
University of Cambridge, Trumpington Street, Cambridge CB2 1PZ, UK
{imh29, pjcl0}@cam.ac.uk

Abstract. Visibility in user interfaces (UI) is a critical element of making a product usable. However, the visibility of modern user interfaces can be compromised in a number of distinct ways. Firstly, some user interface controls have no visible attributes at all, from a visibility perspective they are effectively 'missing'. Secondly, if an element is present then it can be 'missed', thirdly, if it is seen it may be 'misunderstood' by the user. Previous work has seen the development of a model to represent this. This model is used to inform the development of an evaluation tool called vis-UI-lise that can be used to assess the visibility of user interfaces. It presents UI visibility as a series of 5 hurdles between the user and the interface that have to be overcome for a successful interaction. The output from this highlights to designers what the key issues are to help drive further development to improve usability.

Keywords: Usability · Visibility · Vision · Evaluation · Methods

1 Introduction

The importance of user interface (UI) visibility has been well established and the lack of it has been described as a crisis [1, 2]. With the development of modern UIs [3, 4], particularly smartphones, there is a worrying trend to interfaces with poor visibility. For example, analysis of the home screens of a popular smartphone showed that only 8% of the available functions were visible at the top level [5]. The problem of UI visibility can be broken down into three key aspects. Firstly, that some user interface elements are effectively 'missing' as in the example above. Secondly, they are 'missed' because they are not seen by the user, and thirdly they are seen but 'misunderstood' [6]. This categorization of visibility issues can be conveniently referred to as the 3 M's. These three categories of problems are addressed by previous work through a model of user interface visibility which is shown in Fig. 1.

'Missing' interface elements are addressed through understanding the functions the 'program' provides and the fact they lack visual 'presentation'. If a function is visible but 'missed' this can result from poor 'presentation', difficulties in the physical environment such as poor lighting (physics), the user's ability to resolve the image presented (physiology) and whether it grabs their attention (psychology). Finally, 'misunderstanding' comes primarily from the interpretation (psychology) of the 'presentation'.

© Springer International Publishing AG, part of Springer Nature 2019
T. Z. Ahram and C. Falcão (Eds.): AHFE 2018, AISC 794, pp. 212–221, 2019.
https://doi.org/10.1007/978-3-319-94947-5_21

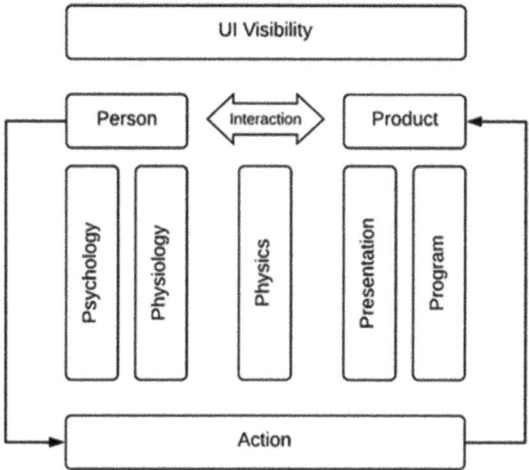

Fig. 1. The '7 P' model of UI visibility based on Hosking and Clarkson [6].

This conceptual model is helpful in bringing a more holistic view of visibility, in particular integrating the critical psychological layer. However, at a practical level, this needs to be translated into a tool to support user interface designers. This leads to the following research questions:

RQ1: Are there already existing tools that support this model?
RQ2: If not, what form should a new tool take?

2   Method

RQ1 is addressed by looking at different usability evaluation methods (UEMs) described in the literature. Specifically, it assesses whether a method directly addresses the range of visibility issues, previously identified, across a range of products and classified under the 3 M's categorisation [6].

RQ2 takes the form of an iterative process shown in Fig. 2. Drawing together the conceptual model described in Sect. 1 with the set of visibility issues. This was augmented with the underpinning literature from the fields of usability and vision. The outer arrows show the broad iterative loop that drives the input (inner arrows) to the development of the new tool. The key literature was mapped out using an Areas of Relative Contribution (ARC) Diagram [7]. The overarching question in the development process for the tool is, "will it identify the real-world issues identified from a range of products".

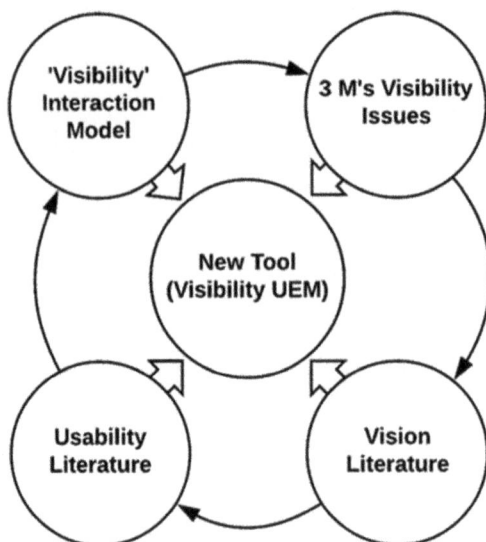

Fig. 2. Summary of research method showing the iterative process of drawing on multiple sources to develop the new tool (Visibility UEM).

3   Results and Analysis

With regards to RQ1 and assessing current UEMs, Fernandez et al. [8] provide a systematic mapping of UEMs for the web. The UEMs identified are not restricted to use on web interface evaluation so it provides a useful, generalizable categorisation and identification of tools. Table 1 is a summary based on this work. A review of these shows that in general they do not specifically address visibility. They are understandably more holistic, and outcome-oriented in nature. Visibility, although important, is only one of many contributory factors to overall usability. There is, however, scope for a specific UEM to augment these approaches and being more diagnostic in nature with regard to visibility. Having established a gap for a tool addressing visibility, with a focus on identifying underlying issues, helped set the direction for the work on RQ2.

The development of the new UEM followed the iterative approach outlined in Sect. 2 and shown diagrammatically by Fig. 2. There was a constant referencing back to the visibility issues identified across a range of products to see how the tool would identify these. Initial elements of the proposed UEM were derived from a basic understanding of human-computer interaction [9, 10] allied to task analysis [11, 12]. These are as follows:

- Establish a representative scenario for the use of the product and perform a task analysis to determine the steps of the task sequence

Then each task step is broken down into the following, based on the typical interactive loop:

- Can the user 'see' the function (interface control)

- Can the user 'see' how the function (interface control) operates
- Can the user 'see' feedback that the function (interface control) has been operated correctly

Table 1. A table summarizing the categorization of UEMs from Fernandez et al. [8].

Category	Usability evaluation method
Use Testing	
	Think-Aloud Protocol
	Question-Asking Protocol
	Performance Measurement
	Log Analysis
	Remote Testing
Inspection	
	Heuristic evaluation
	Cognitive Walkthrough
	Perspective-based inspection
	Guideline review
Inquiry methods	
	Questionnaire
	Interviews
	Focus group
Analytical Modeling	
	Cognitive Task Analysis
	Task environment analysis
	GOMS analysis
Simulation	
	Agents or algorithms whose intention is to simulate user behavior

The distinction between the visibility of the function and the operation of the function is important as the user may be able easily work out what a control does but not know how to operate it. This is particularly an issue with multi-finger, multi-gesture touch interfaces that have no or limited visibility with regard to the use of such gestures.

Drawing further on the human information processing model of Wickens [10] and the work of Green et al. [13] 'seeing' was broken down into 'sensation''perception' and 'attention'. This control level view was allied to higher order cognitive processes related to goals and prior knowledge. Review of this initial framework against the type of problems identified on various interfaces showed that it was too abstract for a designer to apply to the specifics of an interface design. A further iteration addressed this by considering it in terms of mapping the capability of the user in relation to the properties of the interface. This revised approach was influenced by the interface demand to user capability model outlined by Persad et al. [14] and the teaching of this in secondary school (11–14 years olds) which is discussed by Nicholl [15]. This educational work lead to the conceptualisation of interaction as being like a series of 'hurdles' that the user has to overcome. The greater the demand of the product (interface) the higher the hurdle the user has to overcome. Where the demand exceeds the user's capability then the user is excluded. For example, if the text is too small and low contrast this is high demand from a visibility perspective and may exclude the user. Increasing the text size and improving the contrast will reduce the demand

Looking at Fig. 3 from right to left, the interface shows five 'hurdles'. The implication is that the user has to get over each hurdle to reach effective comprehension. In reality this a simplification as cognitively there will be the process of perceptual exploration [13] i.e. it is not sequential but more of networked interaction between the elements. However, this is a compromise to produce something that designers can use.

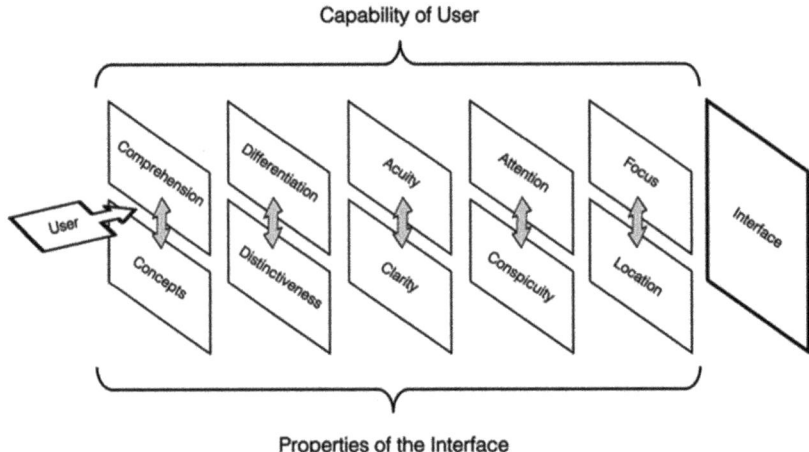

Fig. 3. UEM framework representing series of visibility 'hurdles' that the user has to overcome.

The five hurdles are described as a series of questions in Table 2 below. For each 'hurdle question', there are a series of sub-questions that aim to unpack the underlying contributory factors.

The process for answering these questions systematically is shown in Fig. 4.

The process starts by defining a scenario that represents typical use of the product. Multiple scenarios can be considered to explore a range of functionality including things such as potential error states that the user has to recover from. Based on the scenario a task analysis is performed breaking the scenario down into a series of task steps. For each task step, an image is captured (either via a screen grab or camera) of the state before and after the operation of the function related to the task step. The granularity of the tasks steps is based on their only being a single operation at the user interface level.

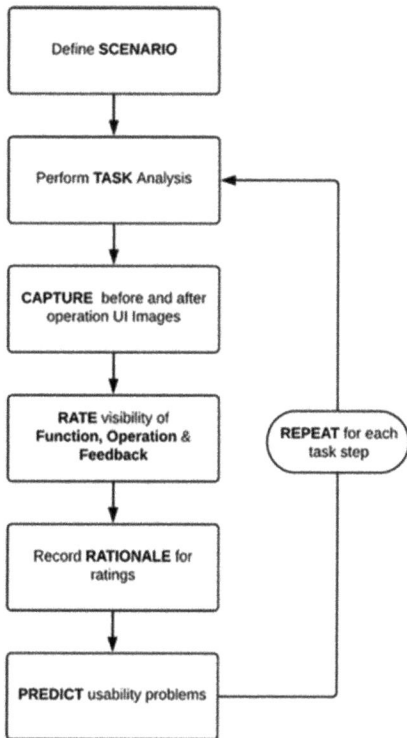

Fig. 4. vis-UI-lise process including the identification of global functions.

The pre-operation image is analysed using the questions outlined in Table 2. This is repeated for the visibility of the function and the visibility of the operation of the function. The feedback is analysed from the post operation image. In the case where feedback is more dynamic e.g. through the use of an animation, this is noted through a text-based description in addition to the image. In the current form of the tool, this is all recorded in a PowerPoint template. PowerPoint was chosen as the recording tool as it is widely used, flexible and is convenient for presenting back the results to relevant stakeholders. For each task step the following information is recorded in a series of slides as follows:

- Slide A: Pre and post operation images of the UI with summary ratings for each of the 5 hurdles
- Slide B: A table recording the answers and ratings to all the sub-questions for the visibility of the 'function'
- Slide C: A table recording the answers and ratings to all the sub-questions for the visibility of the of the 'operation' of the function
- Slide D: A table recording the answers and ratings to all the sub-questions for the visibility of the 'feedback'
- Slide E: A table of predictions of potential usability problems generated using insights from the analysis.

Table 2. The hurdle questions and their associated sub-questions

Hurdle Question and their associated sub-questions
1. Is the **location** of the user interface control such that the user can **focus** on it?
1.1 Does the user have to move to get line of sight?
1.2 Is the distance such that the user can focus on it?
2. Is the user interface control sufficiently **conspicuous** that it grabs the user's **attention** at the appropriate time?
2.1 Is it in the central visual field?
2.2 Is it where the user would expect it to be?
2.3 How many other related controls are there?
2.4 Does it stands out against other controls/background?
3. Are the key visual parts of the user interface control of sufficient **clarity** (size and contrast) that the user can **resolve** them? (i.e. within the range of the user's visual **acuity**)
3.1 What are the distinguishing graphical features of the UI control?
3.2 What size are they?
3.3 What is the level of contrast compared to their background?
4. Is the user interface control sufficiently distinctive from other controls that the user can correctly differentiate them?
4.1 How different is it from other controls visible at the same time?
4.2 How different is it from other controls visible at other times?
4.3 Could it be confused with commonly used graphics/symbols that indicate something different?
5. Are the **concepts** (metaphors) used to portray the function and operation of the control **comprehensible** to the user?
5.1 What are the concepts of the 'function' its 'operation' and the resultant 'feedback'?
5.2 How are these concepts conveyed visually?
5.3 Are they familiar concepts to the user? (check against real examples)
5.4 How well are these concepts represented and are there elements missing?
5.5 Are there general variations of this concept that could cause confusion?

Table 3. Visibility Rating Scale

Level	Colour	Description
0	Black	No visibility and is highly likely to cause usability issues
1	Red	An issue likely to contribute to some level of usability problem
2	Amber	An issue that may lead to usability problems
3	Green	Unlikely to directly cause usability problems

The rating scale is four levels and colour coded as shown in Table 3. The key thing is the rationale for the rating as much as the rating itself supporting the diagnostic purpose of the tool, i.e. to identify underlying visibility issues.

An example of type A slide is shown in Fig. 5

Fig. 5. An example task step analysis from the PowerPoint based tool (Type A slide) showing turning on a car heating and ventilation system (note: this has been converted into a grayscale format suitable for publishing in this journal)

In this particular example the scenario is to turn an automobile HVAC (Heating Ventilation Air Condition) system on and then turn the rear heated window on. The initial image at the top of Fig. 5 is of the HVAC in its off state. Of note is that the unit has a dedicated 'off' bottom (bottom right) but no dedicated 'on' button or 'on/off' button. Turning the unit on is done by pressing the circular 'auto' button in the centre of the right rotary control. For this reason, the on function is deemed to be invisible or 'missing' as there is no direct visual indication that this button turns the unit on. Therefore, the rating for function visibility in terms of concept, clarity, conspicuity and differentiation is rated at '0' or 'black'. In terms of focus it is give green because line of sight with the button is achievable and its distance from the driver makes focus achievable too. It is worth noting that in the case of an automobile line of sight can be problematic due to obscuration of controls by the steering wheel.

Having established the visibility of the function, the visibility of its operation is considered on the basis that the user knows what the control is. This at first may seem odd but it is important to distinguish the two as one or both may require work to improve them. In this case the operation is compromised by the fact that control in question is both a rotary control and push button (in the centre of the control). This duality of function results in a red rating for 'concept', 'conspicuity' and 'clarity'. However, once the button is correctly operated the unit displays the passenger and driver temperature, the fan speed and three LED indicators come on. This gives strong feedback compared to the pre-operation state and is rated at level 4 (green) for all five hurdles.

Taking into account the problem raised it is relatively easy to see a design alternative which would be to make the dedicated 'off' button an 'on/off' button which would immediately address the issues without any additional controls being required.

The later task step of turning on the rear heated window shows a problem of differentiation. This is because there is a button for front and rear window demisters. These are identified using a rectangular icon for the rear window and curved for the front window. This would require prior knowledge of what these icons mean to know which is which. The situation would be different if there was only a rear window demister option. This shows the importance of understanding the context of other controls that are present or used elsewhere that may hinder effective differentiation.

Having stepped through this example it is self-evident that the tool performs a very detailed analysis. This in part reflects the complexity of vision and the key role of the psychological layer outlined on the visibility model. However, this may be a barrier to use by designers and the implications of this are discussed in the next section.

4   Limitations and Further Work

Although the work to date has been informed by a set of visibility issues identified across a range of products, further work is required to determine the predictive reliability of the tool. This will be achieved by analysing a product using the vis-UI-lise tool and then testing with real users to see how many problems it predicts. Depending on that outcome, the tool will then be tested with user interface designers to see how repeatable the predictions are with different designers (inter-rater reliability), the usability of the tool and the likelihood of them using it. This work is all planned using the DRM framework [7].

It has already been stated that the aim of the tool is to augment existing UEMs to provide additional insights around visibility. However, further work is required to see how it would integrate in practice with other tools. The detailed nature of the tool may be a deterrent to designers and therefore evidence of how well it predicts problems may prove crucial in encouraging uptake. There is an interesting trade-off between the 'time' cost of using the tool versus the cost of testing with real users. The tool has the potential to spot problems early in the design cycle to reduce the amount of user testing and also to direct testing to areas that have been identified as potentially problematic.

5   Conclusion

It has been strongly argued that user interface visibility is critical for usability and that trends in user interface design represent a crisis with regard to visibility. Therefore, having tools that provide rigorous insight to visibility issues can help address this crisis providing designers with insights to where there may be problems in their design. vis-UI-lise is such a tool and has been developed to identify the range of visibility issues from missing to missed to misunderstood. It has been developed against a set of real-world examples. As such it represents an intermediate deliverable of on-going research. However, it is at a point where it can be critiqued, discussed and inform the thinking in this key area of user interface design.

References

1. Norman, D.A., Nielsen, J.: Gestural interfaces: a step backward in usability. Interactions **17**, 46–49 (2010)
2. Norman, D.A.: Natural user interfaces are not natural. Interactions **17**, 6–10 (2010)
3. van Dam, A.: Post-WIMP user interfaces. Commun. ACM **40**, 63–67 (1997)
4. van Dam, A.: User interfaces: disappearing, dissolving, and evolving. CACM **44**, 50–52 (2001)
5. Hosking, I.M., Clarkson, P.J.: Now you see it, now you don't: understanding user interface visibility. In: Antona, M., Stephanidis, C. (eds.) UAHCI 2017. LNCS, vol. 10279, pp. 436–445. Springer, Cham (2017). https://doi.org/10.1007/978-3-319-58700-4_35
6. Hosking, I.M., Clarkson, P.J.: Believing is seeing: rethinking the visibility of user interfaces. Presented at the Cambridge Workshop on Universal Access and Assistive Technology, Cambridge, 9 April (2018)
7. Blessing, L.T.M., Chakrabarti, A.: DRM, a Design Research Methodology. Springer Science & Business Media, New York (2009)
8. Fernandez, A., Insfran, E., Abrahão, S.: Usability evaluation methods for the web: a systematic mapping study. Inf. Softw. Technol. **53**, 789–817 (2011)
9. Card, S.K., Moran, T.P., Newell, A.: The human information-processor. In: The Psychology of Human-Computer Interaction, pp. 23–100. Lawrence Erlbaum Associates, Hillsdale (1983)
10. Wickens, C.D., Hollands, J.G., Banbury, S., Parasuraman, R.: Engineering Psychology and Human Performance. Pearson Education, Upper Saddle River (2013)
11. Annett, J., Duncan, K.D.: Task analysis and training design. Occup. Psychol. **41**, 211–221 (1967)
12. Shepherd, A.: HTA as a framework for task analysis. Ergonomics **41**, 1537–1552 (1998)
13. Green, M., Allen, J.M., Abrams, S.B., Weintraub, L.: Forensic Vision with Application to Highway Safety. Lawyers & Judges Publishing Company, Tucson (2008)
14. Persad, U., Langdon, P., Clarkson, J.: Characterising user capabilities to support inclusive design evaluation. Univ. Access Inf. Soc. **6**, 119–135 (2007)
15. Nicholl, B.: Empathy as an aspect of critical thought and action in design and technology. In: Williams, P.J., Stables, K. (eds.) Critique in Design and Technology Education, pp. 153–171. Springer, Singapore (2017). https://doi.org/10.1007/978-981-10-3106-9_9

An Investigation of Key Factors Influencing Aircraft Comfort Experience

Wenhua Li[✉], Jianjie Chu, Bingchen Gou, and Hui Wang

Shaanxi Engineering Laboratory for Industrial Design,
Northwestern Polytechnical University,
Youyi (West) Road 127, Xi'an 710072, People's Republic of China
{iliwenhua,cjj,Bingchengou,
wanghui0617}@mail.nwpu.edu.cn

Abstract. Comfort has been an increasingly important issue in air travel. Passenger comfort is clearly a key variable in research on user acceptance of transportation systems, and it is related to passenger's satisfaction and the willingness to fly again. After realizing its importance in capturing and retaining customers, a growing number of airlines are focusing on comfort. How to improve the experience of passenger comfort has become a major strategic goal for the airline management. This paper regards the passenger comfort in the aircraft cabin during the flight as an experience shaped by interaction between individual and the flight contextual situation. The aim of this study is prioritizing the factors affecting the passenger comfort experience, and sorts these factors into different categories. Firstly, the paper undertakes a systematic literature review to distinguish key critical factors affecting the aircraft passenger comfort, then classifies the factors into five categories by Kano model. At last, the managerial implication is given based on the result. The findings of this research provide important implications for the designers and airline managers and help them better understand the passenger comfort experience. The result can be used as a reference of symmetrically designing the aircraft interior and reasonably allocating the human and material resource.

Keywords: Comfort · Air travel · Passenger experience · Kano

1 Introduction

In recent years, de-regulation in the airline industry and the rapid expansion of low-cost carriers (LLC) have conspired to produce significant changes about the global airline market. Airlines are facing a fiercer competition in the international context. Comfort has been an increasingly important issue in air travel. Passenger comfort is clearly a key variable in research on user acceptance of transportation systems, and it is related to passenger's satisfaction and the willingness to fly again. After realizing its importance in capturing and retaining customers, a growing number of airlines are focusing on comfort. How to improve the experience of passenger comfort has become a major strategic goal for the airline management.

© Springer International Publishing AG, part of Springer Nature 2019
T. Z. Ahram and C. Falcão (Eds.): AHFE 2018, AISC 794, pp. 222–232, 2019.
https://doi.org/10.1007/978-3-319-94947-5_22

The solution of improving the aircraft comfort experience is to prevent the negative experience and to enhance the positive experience. The aim of this study is to present underlying factors influencing the passengers' experience during the flight.

This paper is organized as follows. First, the definition of comfort and aircraft comfort experience is addressed by prior literature. We formally state the problem that aircraft comfort experience is a core competitive power for airliner and it is a complex construct and involves multiple factors, then a variety of factors elicited from a review of literature in Sect. 2. The method of Kano and questionnaire was used in Sect. 3. Result and discussion were shown in Sect. 4. Finally, some conclusions and perspectives are given in Sect. 5.

2 Review of Comfort Experience Model and Factors Aircraft Comfort Experience

2.1 Comfort Experience Model

Comfort is a widely used term in daily life, and it means a desirable outcome or the optimum state for a given context. There are diverse definitions of comfort in previous study. Richards [1] describes that comfort is a state of a person involving a sense of subjective wellbeing, in reaction to an environment or situation. Slater [2] defines comfort as a pleasant state of physiological, psychological and physical harmony between a human being and its environment. A common agreed nature of comfort in the literature is that, comfort is a subjective experience, comfort is affected by factors of various nature (physical, physiological and psychological), and comfort is a reaction to the environment [3]. It must be clarified in advance that the term "comfort" in this paper refers to the all levels of comfort and discomfort. Another issue is the relationship between comfort and discomfort. Zhang [4] found that comfort and discomfort are two different entities. This paper is not taken this issue into consideration. Comfort is a complex and dynamic construct, and is associated with physical, psychosocial, physiological, cultural and social element, as well as environment and situational elements. From the time dimension, comfort can be studied in different durations, the first sight comfort, short-term comfort, and long-term comfort.

The term "comfort experience" was firstly proposed by Vinket et al. [5], comfort is described as a convenience experience that enhances product pleasure. Luttmann [6] regarded comfort as cherishment of human senses, freshness, satiation and tranquility concerned with the subjective aspects of bodily experience in product use. Comfort as an experience has been connected with different notions including customer satisfaction, the perception of the contextual factors, user usability and psychological concerns.

In this present study, aircraft comfort experience refers to passenger's experience during the flight between boarding and de-boarding.

Based on the models of De Looze [3] and Moes [7], Vink and Hallbeck [8] proposed a new comfort model (Fig. 1). The feeling of comfort, discomfort and no discomfort is defined as the interaction between "person", "product" and "usage/task". The interaction effects are the process of the responses of the human body through postures, movements and sense. These responses with the effect of expectation are then perceived subjectively and interpreted as three different kinds of comfort feelings. The interaction (I) with an environment is caused by the contact between the human and the product and its usage. This can result in internal human body effects (H), such as tactile sensations, body posture change and muscle activation. The perceived effects (P) are influenced by the human body effects, but also by expectations (E). These are interpreted as comfortable (C) or you feel nothing (N) or it can lead to feelings of discomfort (D). There is not one form of comfort or discomfort experience, but it can vary from almost uncomfortable to extremely comfortable and from no discomfort to extremely high discomfort. It could even be that both comfort and discomfort are experienced simultaneously. For instance, you may experience discomfort from your seat but have a feeling of comfort created by a nice flight attendant. The discomfort could result in musculoskeletal complaints (M). There is a circle around E-C as we believe expectations (E) are often linked to comfort (C). If discomfort is too high or the comfort not good enough there is a feedback loop to the person who could do something like shifting in the seat, adapt the product or to change the task/usage.

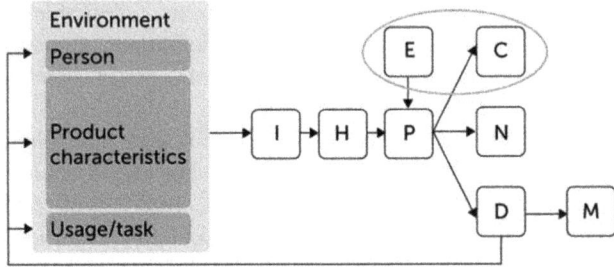

Fig. 1. The comfort model, Vink and Hallbeck [8] I = interaction, H = human body effects, P = perceived effects, E = expectations, C = comfort, N = nothing, D = discomfort, M = musculoskeletal complaints.

2.2 Factors Influencing Aircraft Comfort Experience

Aircraft is a kind of man-made environment; passenger comfort is influenced by a large number of inputs from the cabin interior environment and context. Several studies addressed the factors that affect passenger's comfort experience during flight. Table 1 provides some studies on passenger comfort experience. The results indicate that passenger's comfort experience is influenced by a variety of factors. Therefore, there is a need to prioritize the factors that affect passenger comfort experience. Via focus group, the extracted factors were shown in Table 2.

Table 1. Studies about factors influencing passenger comfort experience

Author	Transportation mode	Factors
Da Silva, Bortolotti, Campos and Merino [9]	Automobile	Physical facet (10), Psychological facet (8), Object facet (15), Environmental facet (3), Context facet (4)
Kolich [10]	Automobile	Vehicle/Package factors (4), Social factors (2), Individual factors (4), Seat factors (4)
Patel and D'Cruz [11]	Aircraft	Individual characteristics, Personal travel context, The pre-flight and in-flight environments, Interaction with others, Activities, current state/current needs and adaptive behaviours, Perceived control
Liu, Yu, Chu and Gou [12]	Aircraft	Environment factors, Products and service, Activities in cabin,
Ahmadpour and Robert [13]	Aircraft	Environment factors (3), Human factors (3), Expectation, Time/Activity
Li, Yu, Pei, Zhao and Tian [14]	Aircraft	Employees (5), Facilities (5), Flight schedule and information (4), Supporting service (1), Physical environment (3)
Vink, Bazley, Kampand Blok [15]	Aircraft	Legroom, hygiene, crew attention, seat/personal space

Table 2. Summary of factors influencing the aircraft comfort experience

Category	Factors
Physical Environment (6)	Temperature, Humidity, Odor, Air pressure, Noise, Lighting
Facilities and Seat (6)	Seat size, Legroom, comfort level of seat, adjustability of seat, usability of facility, hygiene
In-flight attendant (6)	Appearance, Attitude, Understanding of passengers' specific need, work skill, Broadcast, food and drink
Flight schedule (3)	Punctuality, Time of taking off and landing, flight seat occupied rate
Neighbor (2)	Noisy from talking or baby-crying, occupation of public armrest
De-boarding (1)	The waiting time of de-boarding

3   Method and Material

3.1   Kano Model

In this paper, it is believed good experience will produce satisfaction. Kano model was used to categorize the influential factors. In the early 1980's, Kano firstly proposed and developed the two-dimensional nonlinear relationship between qualities attributes of a product or service and the overall satisfaction of the customer. They discovered customer needs could be grouped into three different categories on different levels namely, must-be (M), one-dimensional (O) and attractive (A), and the discovery formed the initial basics, which was called Kano model. Lately they have developed it, and latest version of Kano model classified customer needs into five categories. The five-level Kano classification is Attractive quality (A), One-dimensional quality (O), Must-be quality (M), Indifferent quality (I) and Reverse quality (R). In practice, four types of product attributes are identified: (1) must-be attributes are expected by the customers and they lead to extreme customer dissatisfaction if they are absent or poorly satisfied, (2) one-dimensional attributes are those for which better fulfillment leads to linear increment of customer satisfaction, (3) attractive attributes are usually unexpected by the customers and can result in great satisfaction if they are available, and (4) indifferent attributes are those that the customer is not interested in the level of their performance. Depending on the product/service property being considered, different function-characteristics can be obtained. The Kano conceptual model employs inquiring techniques with pairs of functional and dysfunctional questions about each requirement; the functional situation considers the quality present or sufficient, while the dysfunctional situation supposes the quality to be absent or insufficient. The question of attribute is shown in Table 3. In a Kano questionnaire, both functional and dysfunctional questions are asked, customers choose one of the following responses to express their feelings: (a) I like it that way/I am satisfied; (b) it must be that way; (c) I am indifferent/neutral; (d) I can live with it; and (e) I dislike it/I am dissatisfied. For example, the functional form of the question is "If the food on flight is very delicious, how do you feel?" If the customer answers, for example, "I like it that way," and the dysfunctional form of the question is "If the food on flight don't taste well, how do you feel?" if the answers is "I am neutral," or "I can live with it that way," the combination of the responses in the Kano classification (Table 4) produces category A, indicating that delicious food on flight is an attractive customer requirement from the customer's viewpoint.

Table 3. Kano question and answer

Kano question	Answer
Functional form of the question (e.g., If the food on flight is delicious, how do you feel?)	(a) I like it that way (b) It must be that way (c) I am neutral (d) I can live with it that way (e) I dislike it that way
Dysfunctional form of the question (e.g., If the food on flight is not delicious, how do you feel?)	(a) I like it that way (b) It must be that way (c) I am neutral (d) I can live with it that way (e) I dislike it that way

Table 4. Five level Kano classification

Customer requirements		Dysfunctional (negative) question				
		Like	Must be	Neutral	Live with	Dislike
Functional (positive) question	Like	Q	A	A	A	O
	Must be	R	I	I	I	M
	Neutral	R	I	I	I	M
	Live with	R	I	I	I	M
	Dislike	R	R	R	R	Q

Note: A—attractive, O—one-dimensional, M—must be,
I—indifference, R—reversal, Q—questionable

3.2   Material and Procedure

The questionnaire consisted of 24 adjectives and items describing the passenger's experience, including both the functional and dysfunctional forms. The functional Kano questions is shown in Table 5. The items were derived from different research areas and were adapted to orient them to aircraft comfort experience context (Table 2). This study used a convenience sample technique to collect the data. The questionnaire was distributed to the passengers in the Xianyang International airport, this work was conducted by five students from Department of Industrial Design. Passengers who take airplane more than 4 times were selected to participate and complete the questionnaire. Each respondent was required to answer the Kano questions with respect to every item, and to give the perception of the importance of a factor using the self-stated importance (Table 6). A brief description about the research objectives and a definition of factors were explained by students. All passengers are short flight (<4 h) economic-class passengers.

Table 5. The functional questions used in the questionnaire.

Category	Code	Functional question
1. Physical environment	f1-1	The in-flight temperature is pleasant
	f1-2	The in-flight humidity is pleasant
	f1-3	The in-flight odor is pleasant
	f1-4	The in-flight air-pressure is normal
	f1-5	The in-flight noise is acceptable
	f1-6	The in-flight lighting is pleasant
2. Facilities and seat	f2-1	The seat is comfortable
	f2-2	The seat is easily adjustable
	f2-3	The legroom is appropriate
	f2-4	The seat size is right
	f2-5	The facilities are convenient
	f2-6	The facilities are clean

(continued)

Table 5. (*continued*)

Category	Code	Functional question
3. Attendant and service	f3-1	The appearance of attendant is charming
	f3-2	The attendant is thoughtful and caring
	f3-3	The work skill of attendant is professional
	f3-4	The attitude of attendant is enthusiastic and polite
	f3-5	The in-flight announcement is clear
	f3-6	The food and drink in-flight is delicious
4. Flight	f4-1	The flight is punctual
	f4-2	The time of flight is favorable
	f4-3	Some seats in-flight are empty
5. Neighbor	f5-1	The neighbor is quiet
	f5-2	The public armrest is empty
6. De-boarding	f6-1	The waiting time in line for de-boarding is short

Table 6. The level of importance

Not important	Somewhat important	Important	Very important	Extremely important
0.1	0.3	0.5	0.7	1

A total number of 150 responses were obtained. Six questionnaires were discarded due to being incomplete or containing unreliable answers. We report data from 144 participants. The data collected by questionnaire was shown in Table 7. Table 8 indicated the scores for the answers of the functional/dysfunctional features, it was used to calculate the total value of Kano classifier and Kano categories.

Table 7. Summary of the responses about the importance and level of Kano classification on every question. The value means the total number of every item.

No.	Code	a		b		c		d		e		Average of importance
		F	D	F	D	F	D	F	D	F	D	
1	f1-1	92	0	52	0	0	0	0	12	0	132	0.712
2	f1-2	67	0	55	0	22	12	0	34	0	98	0.486
3	f1-3	64	0	80	0	0	0	0	23	0	112	0.704
4	f1-4	56	0	88	0	0	0	0	5	0	139	0.72
5	f1-5	65	0	79	0	0	2	0	14	0	128	0.71
6	f1-6	56	0	65	0	23	21	0	45	0	78	0.35
7	f2-1	98	0	46	0	0	12	0	35	0	97	0.77
8	f2-2	35	0	78	0	31	10	0	56	0	78	0.74
9	f2-3	112	0	32	0	0	0	0	34	0	110	0.79
10	f2-4	20	0	92	0	32	1	0	23	0	120	0.634

(*continued*)

Table 7. (*continued*)

No.	Code	a		b		c		d		e		Average of importance
		F	D	F	D	F	D	F	D	F	D	
11	f2-5	67	0	74	0	3	0	0	11	0	135	0.68
12	f2-6	35	0	109	0	0	0	0	2	0	142	0.694
13	f3-1	27	0	30	0	87	34	0	78	0	32	0.448
14	f3-2	60	0	84	0	0	23	0	23	0	98	0.694
15	f3-3	65	0	79	0	0	0	0	21	0	123	0.75
16	f3-4	45	0	99	0	0	0	0	12	0	132	0.796
17	f3-5	35	0	78	0	31	34	0	47	0	63	0.62
18	f3-6	67	0	57	0	29	10	0	45	0	89	0.572
19	f4-1	90	0	54	0	0	0	0	5	0	139	0.878
20	f4-2	56	0	45	0	43	50	0	38	0	56	0.65
21	f4-3	46	0	0	34	98	102	0	42	0	0	0.19
22	f5-1	14	0	20	0	84	30	5	69	21	45	0.568
23	f5-2	56	0	21	0	67	79	0	34	0	31	0.316
24	f6-1	34	0	21	0	89	77	0	39	0	28	0.208

Note: a denotes I like it that way; b denotes It must be that way; c denotes I am neutral; d denotes I can live with it that way; e denotes I dislike it that way. F denotes functional question, D means dysfunctional question.

Table 8. Scores for functional/dysfunctional features

Kano question	Functional	Dysfunctional
I like it that way	1	−0.5
It must be that way	0.5	−0.25
I am neutral	0	0
I can live with it that way	-0.25	0.5
I dislike it that way	-0.5	1

4 Result and Discussion

A scatter dot plot on Kano classifier and Kano categories was showed in Fig. 1. The axis-X denoted the average score for dysfunctional feature of item, The axis-Y denoted the average score for functional feature of item. The four zones were divided by two lines and a curve, the curve was a quarter of a circle, the diameter was equal to the square side length. The angle formed by two blue lines and two sides was 30° The location of four zones was shown in Fig. 1.

There was no attractive factor among all these items (Fig. 2). Two factors were identified to belong to must-be zone, f_{5-1} (neighbor passenger) and f_{2-4} (seat size). The seat size had no change among aircrafts and airliners. Regarding to the neighbor passengers, if there were very noisey passengers, or baby crying, it would definitely decrease the satisfaction and comfort experience. It was shown that there were two factors fall into the Indifferent zone, they were f_{4-3} (flight seat occupied rate) and f_{6-1} (the waiting time for de-boarding). Passenger don't care whether there are seats are empty or not, the waiting time for de-boarding is another factor they don't consider. The rest of the items were one-dimensional. All physical environment factors (f_1) and facilities and seat factors (f_3) were categorized into the one-dimensional zone. It proved that these factors influence the passenger comfort experience, the little improvement for a factor will yield to increase the passenger satisfaction. There is still some room for improvement for facilities (seat) and physical environment to raise the status of air-liners [14].

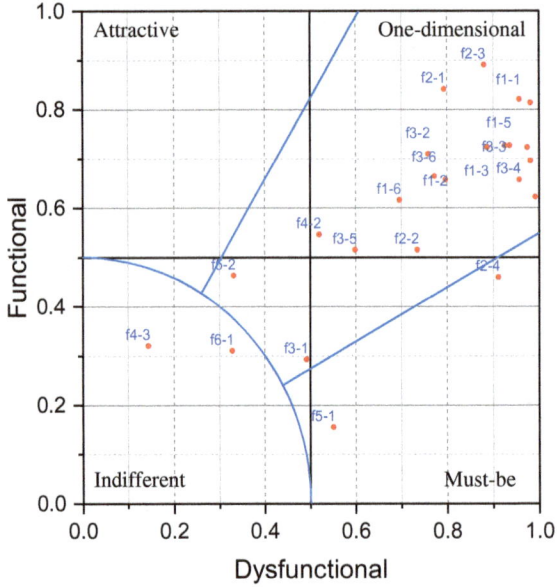

Fig. 2. Kano classifier and Kano categories

The importance of item is shown in Table 7. We can know the first three most importance factors are f_{4-1} (Punctuality of flight), f_{3-4} (Attitude of attendant) and f_{2-3} (Legroom). According to Vink and Van Mastrigt [16], this aligns with our finding that legroom is one of the most important factor influencing the comfort experience. Four least important factors (average importance <0.4) are f_{4-3} (The seat occupied rate), f_{1-6}(in-flight lighting), f_{5-2} (free public armrest), and f_{6-1} (waiting time in line for de-boarding). Among factor set of attendant and service, the appearance of attendant is not so important, compared the other factors.

Refers to the results of questionnaire and calculation, it can be concluded that the most influential factor in aircraft passenger experience are facility(seat), flight schedule, physical environment and attendant. In terms of the managerial implications, the findings of this study benefit practitioners in airline managerial position. Combine the average importance of item and Kano classification, it is found that the f_{4-3} (flight seat occupied rate) and $f_{6\text{-}1}$ (waiting time in line for de-boarding) are two factors which are less influential to passenger experience.

Further factors that influence the comfort experience are seat factor, vehicle factors, and the last is individual factor. The results also showed that people are still concerned and consider about the value of a prestigious when choosing a flight. These results are expected to be applied to the process of improvement and further development of the air travel comfort and may help to better focus on priority factors.

5 Conclusion

Airlines industry is customer-oriented and make efforts to improve passenger comfort experience, in order to attract more passengers. Therefore, identifying the factors affecting aircraft comfort experience is a core issue. This paper sorted the factors into four classifications based on Kano model and prioritized the factors influencing the aircraft comfort experience. It is believed that this result can help airliner managers achieve a clear understanding of their performance and help them reasonably allocate limited human and financial resource. It revealed that finding effective strategy to enhance factors which fell into the one-dimensional zone will improve the aircraft comfort experience.

Acknowledgments. The authors express special thanks to the participating passengers for their time and efforts. The authors would like to acknowledge support for this work from civil aircraft special scientific research program, China (Grant No. MJ-2015-F-018). The study is partly supported by the 111 Project, Grant No. B13044.

References

1. Richards, L.G.: On the psychology of passenger comfort. In: Oborne, D.J., Levis, J.A. (eds.) Human Factors in Transport Research, vol. 2, pp. 15–23. Academic Press, New York (1980)
2. Slater, K.: Human Comfort. CC Thomas, Springfield (1985)
3. De Looze, M.P., Kuijt-Evers, L.F.M., Van Dieën, J.: Sitting comfort and discomfort and the relationships with objective measures. Ergonomics **46**, 985–997 (2003)
4. Zhang, L., Helander, M.G., Drury, C.G.: Identifying factors of comfort and discomfort in sitting. Hum. Factors **38**, 377–389 (1996)
5. Vink, P., Overbeeke, C.J., Desmet, P.M.A.: Comfort Experience. In: Vink, P. (ed.) Comfort and Design - Principles and Good Practice, pp. 1–12. CRC Press, Boca Raton (2005)
6. Luttmann, A., Schmidt, K.-H., Jäger, M.: Working conditions, muscular activity and complaints of office workers. Int. J. Ind. Ergon. **40**, 549–559 (2010)
7. Moes, N.C.C.M.: Analysis of sitting discomfort, A review. In: Bust, P.D., McCabe, P.T. (eds.) Contemporary Ergonomics 2005, pp. 200–204. Taylor & Francis, London (2005)
8. Vink, P., Hallbeck, S.: Editorial: Comfort and discomfort studies demonstrate the need for a new model. Appl. Ergon. **43**, 271–276 (2012)
9. Da Silva, L., Bortolotti, S.L.V., Campos, I.C.M., Merino, E.A.D.: Comfort model for automobile seat. Work **41**, 295–302 (2012)
10. Kolich, M.: A conceptual framework proposed to formalize the scientific investigation of automobile seat comfort. Appl. Ergon. **39**, 15–27 (2008)

11. Patel, H., D'Cruz, M.: Passenger-centric factors influencing the experience of aircraft comfort. Transp. Rev. **38**, 252–269 (2018)
12. Liu, J., Yu, S., Chu, J., Gou, B.: Identifying and analyzing critical factors impacting on passenger comfort employing a hybrid model. Hum. Factors Ergon. Manuf. **27**, 289–305 (2017)
13. Ahmadpour, N., Robert, J.M., Pownall, B.: The dynamics of passenger comfort experience understanding the relationship between passenger and the aircraft cabin interior. In: The International Conference of Canadian Aeronautics and Space Institute, CASI Aero, Toronto, pp. 1–8 (2010)
14. Li, W., Yu, S., Pei, H., Zhao, C., Tian, B.: A hybrid approach based on fuzzy AHP and 2-tuple fuzzy linguistic method for evaluation in-flight service quality. J. Air Transp. Manag. **60**, 49–64 (2017)
15. Vink, P., Bazley, C., Kamp, I., Blok, M.: Possibilities to improve the aircraft interior comfort experience. Appl. Ergon. **43**, 354–359 (2012)
16. Vink, P., Van Mastrigt, S.: The aircraft interior comfort experience of 10,032 passengers. In: Proceedings of the Human Factors and Ergonomics Society 55th Annual Meeting, Las Vegas, pp. 579–583 (2011)

UX Evaluation of a New Rowing Ergometer: The Case Study of the Technogym "SkillRow"

Alessia Brischetto[1(✉)], Mattia Pistolesi[1], Giuseppe Fedele[2], and Francesca Tosi[1]

[1] Laboratory of Ergonomics and Design, Department of Architecture, University of Florence, Via Sandro Pertini 93, 50041 Florence, Calenzano, Italy
{alessia.brischetto,mattia.pistolesi, francesca.tosi}@unifi.it
[2] Scientific Research Department, Technogym S.P.A, Via Calcinaro, 2861, 47521 Cesena, Italy

Abstract. This paper demonstrates the results of workshop "UX Skillrow Evaluation" workshop, promoted by the Laboratory of Ergonomics and Design (LED) of the University of Florence in collaboration with Technogym, a leading-edge company that develops fitness equipment for any physical activity. The workshop aimed to define the current levels of usability and experience of use of rowing "Skillrow", through method of investigation and practice of Human-Centered Design and User Experience approaches. The predominant aim of work was to identify usability and user experience of rowing Skillrow and its user interfaces. Following this, to identify the potential, project proposals were conducted, brainstorming and focus group activities. During testing twenty-one users participated, aged between 22 and 30. The research goals were: measurements of current usability level and user experience of product-system interfaces, and definition of critical issues and implementation of the current user interfaces. Finally, the results from the evaluation phases allowed to get qualitative data on the levels of effective usability of the product, the components and its graphic interface. In the form of scenario-based design, solutions to improve the current high levels of usability of the user interface were also developed.

Keywords: Wellness · Human-Centered Design · User Experience User Observation · Focus group

1 Introduction

Human beings were not born for inactivity. Physical inactivity is nowadays identified as the fourth leading risk factor for global mortality and its levels are rising in many countries. This phenomenon has major implications for the prevalence of no communicable disease (NCDs) and the general health of the population worldwide [1]. On the contrary, movement and physical activity contribute to improving all aspects of quality of life, representing a strategic tool for healthy aging [2].

It is therefore necessary to educate as many people as possible to an active lifestyle even at an advanced age. In order to do so, it is important to develop high added-value

© Springer International Publishing AG, part of Springer Nature 2019
T. Z. Ahram and C. Falcão (Eds.): AHFE 2018, AISC 794, pp. 233–243, 2019.
https://doi.org/10.1007/978-3-319-94947-5_23

products for the wellbeing and health of people, and this goal can be achieved through an interdisciplinary approach between Design and Ergonomics [3].

Within the fitness industry one of the most complete UX markets in emerging, in the form of a gym or personal fitness environment. A new paradigm known as "smart gyms" are aimed at providing support for both the trainers and users, keeping track of all activity and later tailoring the experience to the direct specifications of the individual. The growth and development in digital technology has been expansive, assisting all individuals throughout a training session in addition to providing a more engaging and interactive experience. The advances of touch screen display not only offer information (requiring input) specific to the exercise however have additional features to allow a more connected experience with the digital ecosystem such a social networking and personalized multimedia content, making these tools particularly more interesting to the HF/E community. Unfortunately however, although there is a great increased interest in mobile technologies within fitness a cursory literature search identifies that the interest of technology within the fitness environment is directed towards the hardware components demonstrating a greater significance on biomechanics and ergonomics with reference to comfort and safety [4–6].

Additional contributions, commonly targeting the design of tools for a specific population, are rare (e.g. elderly, people with disabilities), in addition to the promotion and monitoring of physical activity.

This was the case with the "SkillRow UX evaluation" workshop, conducted and developed in collaboration with Technogym S.p.A. at the Laboratory of Ergonomics and Design (LED) of Florence University. Aim of the study was the evaluation of the overall user experience (UX) with the rowing ergometer SkillRow and the willingness to promote it.

2 Background: Usability and User Experience

To design an industrial product, it is fundamental knowing the specific needs of the addressed users.

The Italian standard UNI 11377-1:2010 [7] and international standard ISO 9241-210:2010 [8] define usability as: "extent to which a system, product or service can be used by specified users to achieve specified goals with effectiveness, efficiency and satisfaction in a specified context of use". Nielsen defines usability as the sum of 5 attributes [9]: learnability; efficiency; memorability; errors; satisfaction. Usability is not an absolute characteristic of the object, but it is always relative to the task, the user and the environment [10].

The process of identification and needs analysis on which the Human-centered Design is based, is carried out through the realization of usability tests that may be conducted by specialist and/or the direct involvement of a segment of users representing the targeted consumers [11]. The methods of usability verifications and safe check are based on the collection of information related to the modality with which the user interact with the product within a given context of use. By doing so it is possible to identify and analyze the behavior of users, their needs and finally the type and frequency of errors performed during the execution of the required tasks.

This information can be used to define the characteristics of the new product, to test prototypes, and/or to evaluate the existing products [11].

In regards to the User Experience, the international standard ISO 9241-210:2010 defines it as: "person's perceptions and responses resulting from the use and/or anticipated use of a product, system or service".

Whilst the usability focus on the degree to which a product can be used by specific users to achieve specific tasks with effectiveness, efficiency and satisfaction [8], the user-experience focuses on human factors such as emotional, affective and contextual aspects [12]. The evolution of this concept, as defined by ISO 9241-210:2010, takes into account that feeling positive or negative about a task can transform the user experience with the product/system [13].

3   Workshop SkillRow UX Evaluation

The "SkillRow UX Evaluation" workshop, organized by the University of Firenze's Ergonomic and Design Lab (LED) in partnership with Technogym SPA, had the purpose of evaluating the usability and User Experience (UX) of the "SkillRow" rowing machine. The test involved 21 subjects from various part of the World. The workshop was divided into 2 phases (phase 1 consisted of the measurement of the current usability and User Experience (UX) level of the product-interface system, while in phase 2 the criticality and margins of implementation of the current user interfaces were defined), organized in 4 days as followed:

- During the first day usability and thinking aloud tests were carried out. At the end of the session each subject submitted a user experience evaluation questionnaire;
- In regards to the second day, a brainstorming session was run, followed by a collective focus group one aimed at bringing out doubts, considerations, thoughts and difficulties encountered while interacting with the product and its components;
- During the third and fourth day, the Task Analysis, Personas and Scenario-based design methods were used in order to identify possible areas of intervention and future scenarios of use for the SkillRow.

For the evaluation of the usability level and the UX (step 1), referring to the norm ISO 9241-210:2010 [8], the following methods were selected: Questionnaire [14], Task Analysis – TA [15], User Observation [16] and Thinking aloud [17]. In regards to the to the user test sessions, a hybrid survey methodology was tested, which applied simultaneously the User Observation and the Thinking Aloud, following a heuristic approach. This approach allowed us to gather opinions, thoughts, expectations, critical points and intuitions useful for defining the requirements of the design concept. Step 2 focused on the definition of the possible areas of implementation in relation to the critical issues and needs that emerged in step 1. In this regard, the following methods were selected: Brainstorming [18] and Focus group [19], Personas [20] and Scenario-based design [21].

4 Methods

The workshop consisted of a preliminary phase which involved the analysis of the product and its components (phase 1). The results were useful for selecting the two most critical tasks that a user can face while using the rowing machine (phase 2). Once the tasks were defined, users run the test sessions. In order to increase the level of effectiveness of the analysis phase, using the tools of Thinking Aloud, each user was asked to talk loudly about the activities he was performing and the difficulties he was experiencing (phase 3). At the end of the test sessions a thematic questionnaire was submitted to each individual user (phase 4).

Once the test sessions were over, step 2 aimed to bring out needs and critical issues experienced during the tests with the rowing machine (phase 5) through brainstorming and focus groups. The last 3 phases of the workshop involved defining some possible areas of implementation and new usage scenarios in relation to the findings of phase 5; performing the Task analysis (phase 6); and using the Personas and Scenario-based design (phase 7).

4.1 Phase 1 and 2: Preliminary Analysis and Requirements Definition

A preliminary analysis of the product SkillRow (a) and its components SkillRow interface (b) and SkillRow app (c), was carried out by the Ergonomics and Design laboratory (LED) before proceeding to the test sessions (Fig. 1).

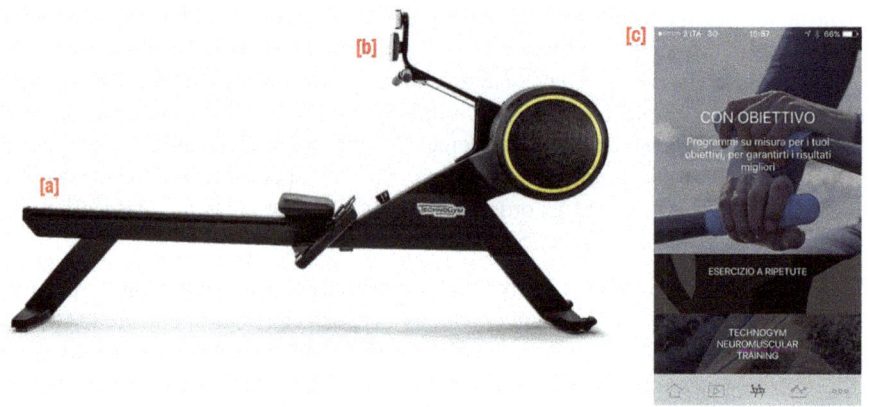

Fig. 1. The SkillRow Technogym machine and the related components

Afterwards, in collaborations with Technogym S.p.A. researchers, a recurrent task and a less recurrent one, representatives of the user experience, were identified:

- Task 1: distance training (700 m);
- Task 2: sending feedbacks to Technogym through the app.

4.2 Participants

The workshop involved 21 students from the University of Firenze (13 females and 8 males) aged between 22 and 30 years. 15 from Italy, 4 from Iran, 1 from Albany, and 1 from China. In order to avoid misunderstandings, non Italian subjects had to prove to be proficient in the Italian language.

4.3 Phase 3: Usability Test (User Observation and Thinking Aloud)

In this phase each subject had 15 min to complete the two task mentioned above. The test was run at the Ergonomics and Design Laboratory and was mediated by two researchers from the same lab. Each session had the following rules:

- set the target to 700 m (task 1);
- starting and ending the exercise (task 1);
- sending feedback to Technogym (task 2).

During the test sessions, a hybrid survey methodology was experimented. Following a heuristic approach, the User Observation and the Thinking aloud methods were simultaneously applied. This approach made it possible to gather opinions, thoughts, expectations, critical points and intuitions useful for defining the requirements of the design concept. The User Observation allowed us to observe how users were interacting with the rowing machine. [16–22] without interfering with the normal running of the test session. To increase the level of effectiveness of the analysis phase, the Thinking Aloud method was used simultaneously with the User Observation [17]. With this method the users were invited to express loudly their thoughts, feelings and frustrations, while interacting with the machine, the problems of usability of the product were identified, and we were able to observe at the same time these interactions. The role of the researchers was to stimulate each individual to express verbally his/her thoughts [22]. After the test session, a questionnaire was submitted to each user.

4.4 Phase 4: Questionnaire

Once the test sessions on the SkillRow ended, each user submitted a thematic questionnaire which included biographical data, user experience (UX) and comprehension questions. The questionnaire section related to personal data and understanding included both open and closed answers [22], while the UX section was developed following the NASA TLX evaluation method. It is a multidimensional assessment tool for the subjective workload [23, 24], which allows to evaluate the users' workload while interacting with the SkillRow integrated interface, the user-SkillRow app and with the SkillRow itself. The workload, defined as the effort sustained by the user to achieve a specific level of performance [23, 24], was calculated from the result of subjective responses weighted on the following five values: (1) mental request; (2) physical request; (3) global effort; (4) performance; (5) frustration.

4.5   Phase 5: Brainstorming e Focus Group

In order to highlight criticalities and needs emerged during the usability test, the participants, divided into 2 groups, were involved in the brainstorming and focus group activities, for a total duration of 4 h. One group dealt with the SkillRow and its app for IOS, while the other group dealt with the same rower and its application for Android.

4.6   Phase 6: Task Analysis

The next step consisted of performing 2 Task Analysis, one for the IOS group and one for the Android group. This method allowed each group to map the possible interactions of each user with the SkillRow app interface. The Task Analysis, additionally it was necessary to identify the critical issues and consequently the possible areas of implementation.

4.7   Phase 7: Personas and Scenario-Based Design

The Personas and Scenario-based design techniques are fundamental tools in the design process of a product/service. The personas represent fictitious profiles created to better represent needs, aspirations and behaviors of a particular segment of users, emerged in step 5. Four Personas were defined for the Android group and four for the IOS group, for a total of eight people and eight Scenario-based design. Thanks to this technique it is possible to describe in a realistic way the actions, or the sequence of actions, that a user makes while using a specific product/service (in this case SkillRow and SkillRow app), and therefore define how they should work in order to guarantee a satisfying user experience.

5   Results

5.1   Questionnaire

The questionnaire submitted to the users was of a thematic nature, divided into three parts, and aimed at incorporating personal data, user experience (UX), and comprehension information.

5.1.1   Personal Data

At the time the questionnaire was submitted 29% of the users trained between 2 and 6 h a week, while only 14% of them engaged into physical activity for more than 6 h a week. These activities took mainly place outdoor, followed by gyms and homes. 71% of the subjects were used to monitor their physical activity through apps or other devices, while the remaining 29% preferred not to use any monitoring devices and virtual trainers. Twenty users out of 21 were considered "novices" (as they had never used a rowing machine to perform physical activity), while one user was considered "competent".

5.1.2   User Experience

The 24% of users perceived a mental request of 5 (on a scale of 1 to 10 where 1 is the lowest level and 10 the highest level) while interacting with the integrated interface during training. Only 10% sensed a mental request of 10.

Data are similar in regards to physical request: 24% of users detected an request of 5 during the interaction with the machine and only a 5% has detected a level of 10.

Regarding the physical and mental requested to reach the personal level of performance during the interaction with the rower, 29% of the users recorded a level equal to 6 and only 5% a level equal to 10 (Fig. 2).

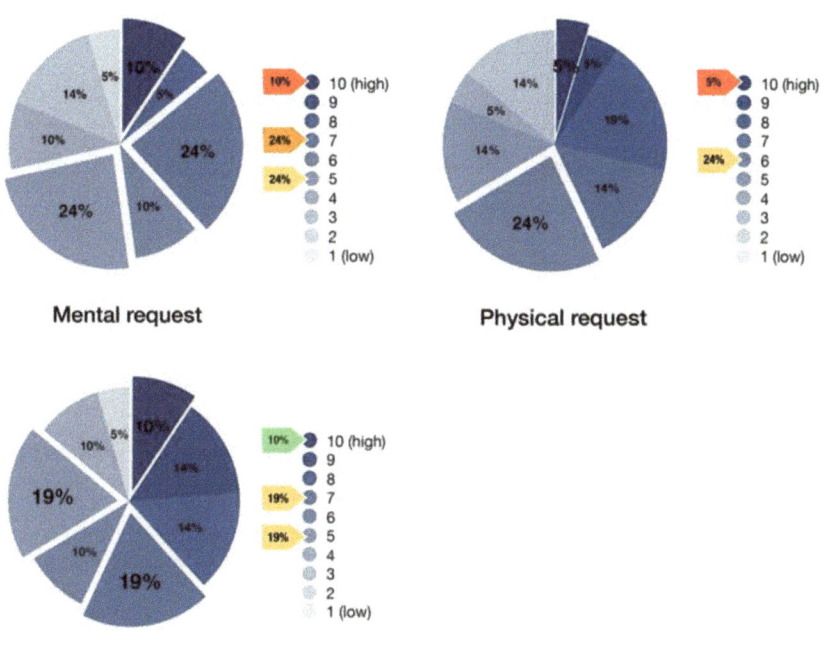

Fig. 2. Rower User Experience. Mental request, physical request and performance level charts.

Regarding the "performance level", which express how much each user believes to have been successful in completing the assigned task, 19% of users recorded a level of 5 and 7 while 10% had difficulty in bringing the task to completion. Most users have not experienced high levels of frustration while interacting with the product. With regard to the SkillRow app user experience, most users found the application easy to use, as the application itself guided users to achieve their goals. 5% of the users perceived a mental request of 1, 24% a level of 5 and 10% equal to 10. Only 19% recorded a level of physical exertion of 5 and 9. As shown in the chart, majority of the users did not register high levels of frustration (Fig. 3).

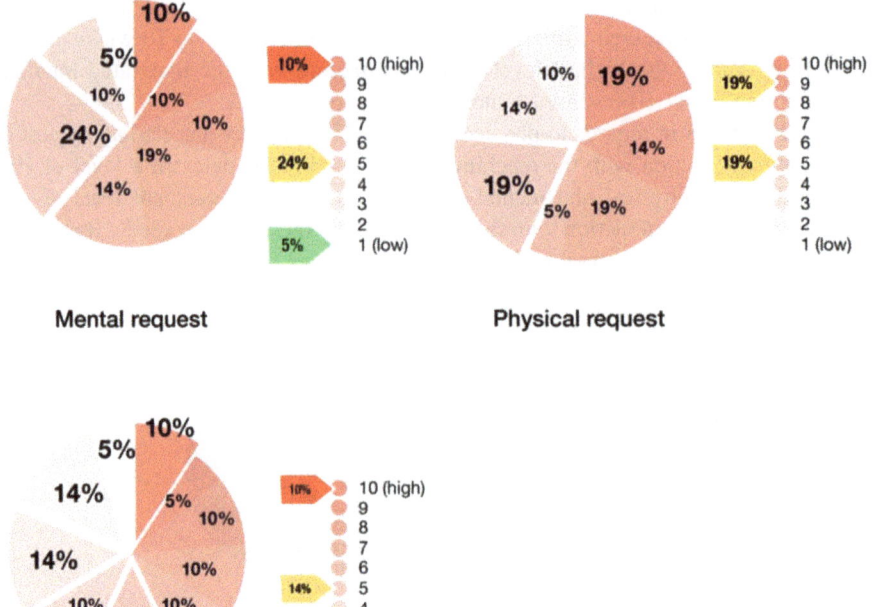

Mental request

Physical request

Performance level

Fig. 3. App *Skillrow* User Experience. Mental request, physical request and performance level charts

The physical request to interact with the application means aspects like visibility and reachability of the smartphone during the rowing exercise.

5.1.3 Comprehension

A significant portion of users did not experience difficulties in understanding how to use the rowing machine, how to turn it on and off, and how to regulate the physical activity intensity. As for the parameters displayed on the rowing machine's screen, almost all users recognize and understood the meaning of the icon and the numerical parameter of the time, distance and Kcal consumed. The majority of users also recognized the icon and understood the numeric parameter relative to both the 500 m split time and the W (power) generated.

On the other hand, 57% of users did not understand the numerical parameter, nor they recognized the icon, relative to the strokes per minute (spm). Majority of the users also had trouble understanding and/or recognizing the information relative to resistance level, AVG, DRAG, REPS.

The last section of the questionnaire covered the understanding of the Skillrow application. 76% of users recognized the icons and their meaning, but 62% of the interviewed users declared overall difficulties in using of the app. Finally, 38% of users consider the skillrow application to be fairly comprehensible.

5.1.4 Overall UX Evaluation and Net Promoter Score

As showed in *Overall User Experience* chart, 81% of users assessed the overall product user experience positively (5% a level of 10, 24% a level of 9, 38% a level of 8, 14% a level of 7), while 19% of users assessed it as sufficient (14% a level of 6 and 5% a level of 5). In the other chart, Net promote score, is clear how the 53% of users highly recommend the use of Skillrow to other people, while the rest of users suggest the use of the product with a varying level from 7 to 5 (Fig. 4).

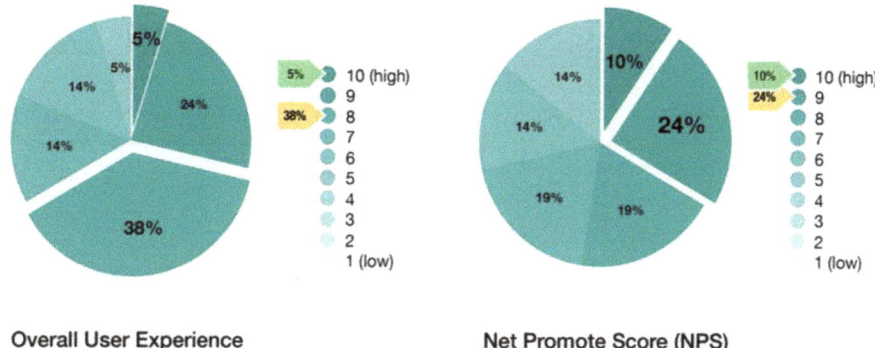

Overall User Experience   **Net Promote Score (NPS)**

Fig. 4. Overall User Experience and Net Promoter Score (NPS) charts

5.2 Brainstorming and Focus Group

This section highlighted that the problems experienced by the users were mainly due to the limited usability of the SkillRow application interface. In fact, most information could be understood only by advanced users. Some issues also arose with the SkillRow itself and with its interface. In fact, some users did not consider some of the information displayed on the screen or they misinterpreted them, some had trouble with the regulation of fundamentals components such as the pedals or the dumper settings (users tried to push or press the wheel instead of rotating it).

5.3 Task Analysis

What emerged from the two Task Analysis relative to the interaction with the SkillRow application interface, is that the application itself offers two types of interaction: press and scroll. The issues while using it were due to:

- Difficulty in visualizing the available exercises and in understanding them;
- Poor access intuitiveness to sub-menus;
- Difficulty in setting some parameters before physical activity;
- Difficulty in understanding certain specific acronyms and graphs relative to physical activity results.

5.4   Personas and Scenario-Based Design

In light of the critical issues emerged from the questionnaires, the Brainstorming and Focus Group, and the Task Analysis, the Personas and the Scenarios were designed. Thanks to this technique it was possible to define possible implementations for the rowing machine, the rower interface and the app so that they can be easy to use also for those with limited or little experience with this type of products and activities. Figure 5 represents only some of the Personas and the new Scenarios designed during the workshop.

Fig. 5. Personas (a) and Scenario-based design (b)

6   Conclusions

The paper proposes some methodological implications in the evaluation of usability and User Experience (UX) with the new SkillRow rowing machine. The empirical approach allowed us to determine criticalities and difficulties in addition to gather thoughts and suggestions from a segment of users representative of the target. These methodologies are also useful for improving the product usability, which enhances pleasure, satisfaction and user experience. In order to do that, it was necessary to adopt a Human-centered approach, aimed at indirectly or directly involving the user, as a partner of inestimable value, during all phases of the project. Furthermore, the Ergonomics for design and its methods offer many opportunities for intervention, thus allowing to outline new usage scenarios and new services-products.

References

1. World Health Organization: Global Recommendations on Physical Activity for Health. WHO, Geneva (2010)
2. World Health Organization: Health and Development Through Physical Activity and Sport. WHO, Geneva (2003)
3. Tosi, F., Rinaldi, A., Busciantella Ricci, D., Pistolesi M., Brischetto A.: Ergonomics evaluation and redesign of workstation to prototyping of luxury garments. In: XI Congresso nazionale SIE 2016, Napoli (2016)

4. Biscarini, A.: Measurement of power in selectorized strength-training equipment. J. Appl. Biomech. **28**(3), 229–241 (2012)
5. Carraro, A., Gobbi, E., Ferri, I., Benvenuti, P., Zanuso, S.: Enjoyment perception during exercise with aerobic machines. Percept. Mot. Skills **119**(1), 146–155 (2014)
6. Reilly, T., Lees, A.: Exercise and sports equipment: some ergonomics aspects. Appl. Ergon. **15**(4), 259–279 (1984)
7. UNI 11377-1:2010: Usabilità dei prodotti industriali, parte 1, Principi generali, termini e definizioni (2010)
8. ISO 9241-210:2010: Ergonomics of human-system interaction, part 210: human-centred design for interactive systems (2010)
9. Nielsen, J.: Usability Engineering. Morgan Kaufmann, Elsevier, Burlington (1994)
10. Polillo, R.: Facile da utilizzare. Una moderna introduzione generale all'ingegneria dell'usabilità, Apogeo Education, Milan (2010)
11. Tosi, F.: Ergonomia e progetto. Franco Angeli, Milan (2006)
12. Hollnagel, E., Woods, D.D.: Joint Cognitive Systems: Foundations of Cognitive System Engineering. CRC Press, Boca Raton (2005)
13. Triberti, S., Brivio, E.: User Experience, Psicologia degli oggetti, degli utenti e dei contesti d'uso, Apogeo Education, Maggioli Editore, Santarcangelo di Romagna (2016)
14. Stanton, N.A., Young, M.S., Harvey, C.: Guide to Methodology in Ergonomics. Taylor and Francis, Oxford (2014)
15. Hacksos, J.T., Redish, J.C.: User and Task Analysis for Interface Design. Wiley, New York (1998)
16. Rogers, Y., Sharp, H., Preece, J.: Interaction Design: Beyond Human Computer Interaction, 3rd edn. Wiley, Chichester (2011)
17. Ericsson, K.A., Simon, H.A.: Protocol Analysis: Verbal Reports as Data. MIT Press, Cambridge (1985)
18. Hartson, R., Pyla, P.S.: The UX Book: Process and Guidelines for Ensuring a Quality User Experience, pp. 280–284. Elsevier, Amsterdam (2012)
19. Cooper, L., Baber, C.: Focus Groups in Handbook of Human Factors and Ergonomics Methods. CRC Press, Boca Raton (2005)
20. Pruitt, J., Grudin, J.: Personas: practice and theory. In: Proceedings of the 2003 Conference on Designing for User Experiences, pp. 1–15. ACM, June 2003
21. Rosson, M.B., Carroll, J.M.: Usability Engineering: Scenario-Based Development of Human-Computer Interaction. Morgan Kaufmann, San Francisco (2002)
22. UNI 11377-2:2010: Usabilità dei prodotti industriali, parte 2, Metodi e strumenti di intervento (2010)
23. NASA: Task Load Index (TLX): Paper and Pencil Version, Moffett Field CA: NASA - Ames Research Center, Aerospace Human Factors Research Division (1986)
24. Hart, S.G., Stavelend, L.E.: Development of NASA-TLX (Task Load Index): results of empirical and theoretical research. In: Advances in Psychology, North-Holland, vol. 52, pp. 139–183 (1998)

Neck Flexion Angle and User Experience Compared on iPhone X and Samsung S8+

Saishyam Akurke$^{(\boxtimes)}$ and Yueqing Li

Department of Industrial Engineering, Lamar University, Beaumont, TX, USA
{sakurke,yueqing.li}@lamar.edu

Abstract. There have been a lot of research and explanation on Smartphone screen size and user experience of numerous Smartphones. This study considers the difference between neck flexion angle and its effects on human ergonomics while using two Smartphones with different screen sizes and display features (iPhone X and Samsung S8+). This study also collects user experience data through a questionnaire to generate the most ergonomic model among the Apple Inc.'s iPhone X and Samsung S8+. 8 participants were considered for this study. A digital goniometer was used to analyze participant's neck flexion angle while using both the Smartphones (iPhone X and Samsung S8+). A paired t-test of means was performed. Neck flexion angle was highly flexed for the participants while using iPhone X than compared to the neck flexion angle while using Samsung S8+. Participants felt more comfortable using both Smartphones. However, 'True-tone' technology in iPhone X eased the stress level of eyes according to the users. This study suggests a large screen smartphone with 'True-tone' or similar technology, which can be an iPhone X but with a larger screen size.

Keywords: Smartphone · Neck flexion angle · Digital goniometer

1 Introduction

In the contemporary world it is seen that there is greater penetration of the smartphones and the number of smartphone users have increased quite rapidly. Two of the most desirable and popular smartphones in the business are iPhone X and Samsung S8+. As the duration of the use of the smartphone is increasing, the number of issues with the musculoskeletal system has been increasing and that includes the damage to the neck fibers due to higher strain on the neck. It has been observed that while using a smartphone, the strain on the neck increases and that affects the neck fibers leading to increased neck flexion issues. It has been noticed that the degree of neck flexion angle depends on the concerned smartphone and the screen size of the smartphone. Keeping the following factors in consideration, there would be an attempted comparative examination of Samsung S8+ and iPhone X. The variables that would be selected for the study would be screen size of smartphone and tasks performed on smartphone to analyze which smartphone of the two would be more suitable for the users. Depending on the screen size of the two smartphones, the desirable model of the smartphone for avoiding the increased degree of neck flexion would also be determined in the study.

© Springer International Publishing AG, part of Springer Nature 2019
T. Z. Ahram and C. Falcão (Eds.): AHFE 2018, AISC 794, pp. 244–250, 2019.
https://doi.org/10.1007/978-3-319-94947-5_24

2   Literature Review

2.1   Screen Size

Screen size of a smartphone plays a vital role for users while choosing a smartphone. Also, there are wide variety of smartphones with different screen sizes available in market. But, do users consider all aspects and effects of a screen size while choosing a smartphone? Not much frequently. International Data Corporation (IDC) is a smartphone data forecaster company and has forecasted that large screen size of smartphone sale will be growing rapidly until 2021 and has started growing sales since 2015 [1]. IDC has also forecasted that small screen size devices will see dropping sales by 2021. Their analysis revealed that in 2017, 558.7 million of 5.5–6 inches smartphone were sold, and the number would rise to 749.3 million by 2021 (Figs. 1, 2 and 3).

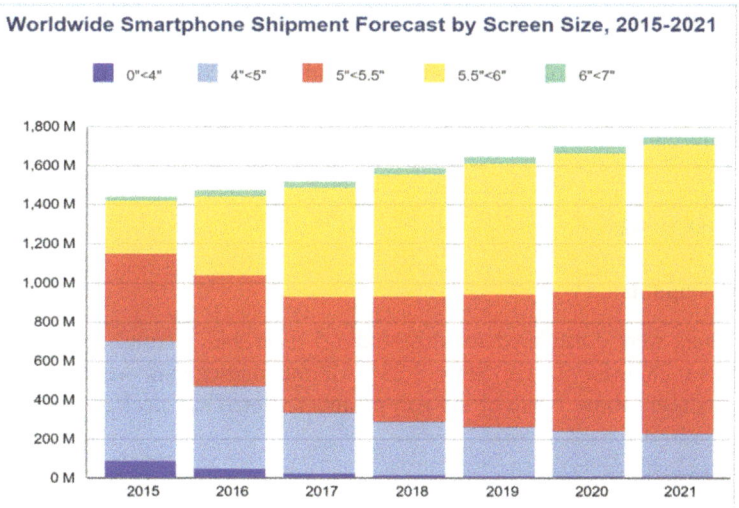

Fig. 1. Worldwide smartphone shipment forecast by screen size, 2015–2021 (source: International Data Corporation, www.idc.com).

One research studied the musculoskeletal effects on 292 participants. Their study found that small screen smartphone had more effect on back and users had to bend their head more to focus in small target points of small screen smartphone [2]. Another study found that large screen size of smartphone improves perceived control and affective quality of smartphone, but their study did not find any effects of screen size on the neck or upper extremity [3]. Also, one study has mentioned that connective tissues in neck can be damaged due to higher neck flexion angles while using smartphone [4]. However, not much has been studied regarding the neck flexion angle while using the large screen size smartphones. All above statistics and studies clearly show the higher adoption rate for larger screen size smartphone. But, none has considered the latest smartphone models with different display technology. Current study hence, considers ergonomic effects of these two latest smartphones in market- Apple's iPhone X (5.8 inches) which has made a competitive entry in market last year and Samsung's S8+ (6.2 inches) which is fulfilled with impressive features.

2.2   Tasks

There is not enough examination of tasks and its effects on neck flexion angle. One study found that texting task retains higher neck flexion angle. While video watching task retains least neck flexion angle [5]. However, studies are insufficient when active tasks such as gaming are considered, and no studies have been done on latest smartphone devices. Therefore, current study considers gaming and texting tasks to find the effect on neck flexion angle on iPhone X and Samsung S8+. One, study has already found higher muscle activities while texting compared to gaming, but their study did not consider neck flexion angle [6].

Regardless of the duration of the smartphone use and undertaking various tasks on a smartphone, it does not matter how long someone uses their smartphone. Studies did not find significant relation between duration and neck flexion for smartphone use [4]. Therefore, in-depth study with precise equipment and data collection techniques more exact relationship can be expected.

3   Methodology

3.1   Participants

8 participants were considered for this study, 5 male and 3 females. All participants aging 21 to 28 (M = 24.12; SD = 2.23) All the participants were provided with informed consent form. None of the participants reported any neck disorders anytime.

3.2   Equipment

A digital Goniometer (Biometrics ltd.) was used to measure the neck flexion angle. Smartphones used were- Apple's iPhone X (5.8 inches of screen size) and Samsung's S8+ (6.2 inches of screen size).

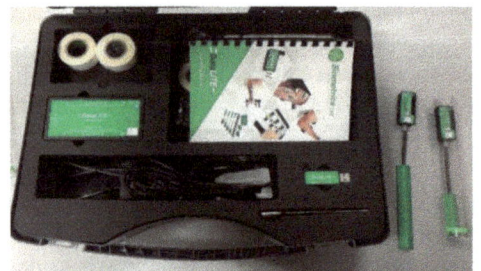

Fig. 2. Digital Goniometer (Biometrics ltd).

3.3   Independent Variables

Screen Size of smartphone: There are two levels of this independent variable- 5.8 inches of screen size (iPhone X) and 6.2 inches of screen size (Samsung S8+).

Tasks: There are two levels of this independent variable- Texting and Gaming.

3.4   Dependent Variable

Neck flexion angle: Neck flexion angle was calculated using the digital Goniometer. The device is wireless and attached at the back neck of the user. It transmits data from device to system and data can be collected and analyzed in the Biometrics Software or through other software. "Neck Flexion angle is the angle between global vertical and the vector pointing from C7 to OC1" [7].

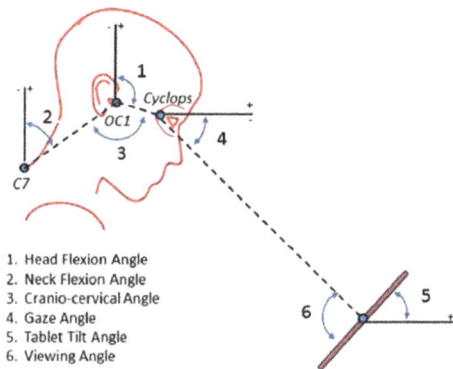

1. Head Flexion Angle
2. Neck Flexion Angle
3. Cranio-cervical Angle
4. Gaze Angle
5. Tablet Tilt Angle
6. Viewing Angle

Fig. 3. Neck flexion angle (source: young et al. 2012)

3.5   Tasks of Experiment

Participants were seated and performed texting and gaming tasks on iPhone X and Samsung S8+ separately. Participants texted for 60 s with the research volunteer. While, in gaming task participants played the 'Subway Surfers' game on both the smartphones for 60 s each. All the tasks were randomized for each participant.

3.6   Procedure

Participants were explained their tasks prior to experiment. Participants were seated on a chair with their back straight. The digital goniometer was then attached on their back of neck using the double-sided tape. Participants were seated straight and asked to maintain a 90-degree head flexion angle to ground and the goniometer reading was set to zero for each participant before beginning the experiment. Participants then performed tasks on both devices by randomized conditions (Fig. 4).

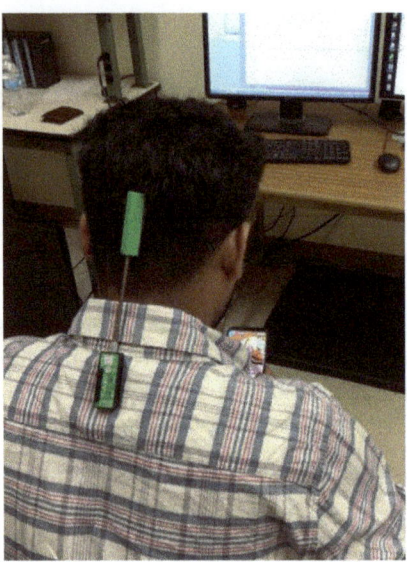

Fig. 4. Participant performing task while data was collected using digital goniometer

3.7   Data Acquisition and Analysis

Neck flexion angle data was recorded from the digital Goniometer. The sampling frequency was set to 1000samples/1 s. ASCII character encoding was used without filters to obtain engineering units from the Goniometer in the Biometrics ltd. software. While participants operated both the smartphones, the data for 60 s was recorded for each task. Sub-sampling setting was done which broke down the data to 100 samples for each task. Finally, the mean of the sub-sampled data was calculated. All the data recording and sub-sampling was done using the Biometrics ltd. data acquisition software. Paired t-test of means was performed on the mean of the sub-sampled data.

4   Results

Results found a significant effect of both the smartphones on neck flexion angle (p = 0.032) in texting task. In the texting task, the mean neck flexion angle for the iPhone was higher (M = 16.52) than that compared to mean neck flexion angle for the Samsung S8+ (M = 14.97). Also, in gaming task, mean neck flexion angle was higher for iPhone (M = 17.82) compared to mean neck flexion angle for Samsung S8+ (M = 16.03). The p-value for gaming task was p = 0.0052 which was also significant. Mean neck flexion for both the smartphones in gaming was higher compared to mean neck flexion in texting task.

From the questionnaire it was found that, more number of users would prefer large screen size smartphone preferably smartphone with larger screen size or equal to 5.5 inches.

5   Discussion

In the texting task, iPhone showed higher neck flexion angle among all the participants. The average neck flexion angle was also higher while texting on iPhone X compared to average neck flexion angle while texting on Samsung S8+. Possibly, the smaller screen size of iPhone X could have made users flex their neck more compared to lower neck flexion obtained on Samsung, which has a larger screen size. Larger screen size of smartphone is subjected to provide lower neck flexion angles. But, from the questionnaire survey its was found that users felt more comfortable while using iPhone X. Users also complained of higher eye stress on Samsung S8+ and convenient texting on iPhone X was found. Undoubtedly, this is because of the latest technology of iPhone X named 'True-tone' technology. 'True-tone' technology adjusts the display brightness and contrast automatically depending on the surrounding lighting and makes easier for eyes to read or watch content on the screen. On this technology, CNET said "The basic physics of the issues that Apple's highlighting are real: display contrast decreases as the light around you gets brighter, and whites look different under different light sources, regardless of whether you're viewing them on a reflective (paper) or light-emitting display" [8].

While, Samsung does not have any such feature users reported higher strain on eyes compared that to eye strain while using iPhone X. Similarly, in gaming task average neck flexion angle was higher while gaming on iPhone compared that to average neck flexion angle while gaming on Samsung S8+. Users also reported same kind of higher strain on eyes while gaming on Samsung S8+ compared that to gaming on iPhone X. iPhone X showed higher neck flexion angle but lower stress level. And Samsung showed lower neck flexion angles but higher stress levels. That clearly means a smartphone with a larger screen size but features an option like 'True-tone' technology would make ergonomically perfect smartphone model. In a study by Kenneth Hansraj, it was found that higher neck flexion angles create pressure on neck and spine. Dr. Hansraj also mentioned that after long time, this can also lead to chronic neurological with occipital neuralgia and can be treated only with surgery.

6   Conclusion

Neck flexion angle while texting on Apple's iPhone X and Samsung S8+ was examined in this study. The results found that users maintained higher neck flexion angle while texting on iPhone compared that to neck flexion angle while texting on Samsung S8+. Also, this study tested neck flexion angle while gaming on both the smartphones (iPhone X & Samsung S8+). Results showed higher neck flexion angle while gaming on iPhone compared to neck flexion angle while gaming on Samsung S8+. This study found that users preferred large screen size smartphone with display features like 'True-tone' technology which is available on iPhone X. This study suggests manufacturers to develop a large screen size device with display features which benefit users ergonomically.

Future studies considering large number of sample size and long task durations may help obtain more refined results and conclusions.

References

1. www.idc.com. (https://www.idc.com/getdoc.jsp?containerId=prUS42628117)
2. Kim, H., Kim, J.: The relationship between smartphone use and subjective musculoskeletal symptoms and university students. J. Phys. Ther. Sci. **27**(3), 575–579 (2015)
3. Kim, K.J., Sundar, S.: Does screen size matter for smartphones? Utilitarian and hedonic effects of screen size on smartphone adoption. Cyberpsychol. Behav. Soc. Netw. **17**, 466–473 (2014)
4. Lee, S., Shin, G.: Relationship between smartphone use and the severity of head flexion of college students. In: Proceedings of the Human Factors and Ergonomics Society, pp. 1788–1790 (2015)
5. Lee, S., Kang, H., Shin, G.: Head flexion angle while using a smartphone. Ergonomics **58**, 220–226 (2014)
6. Akurke, S., Li, Y., Craig, B.: Effect of smartphone use on upper extremity and neck. In: International Conference on Applied Human Factors and Ergonomics, pp. 241–249 (2017)
7. Young, J., Trudeau, M., Odell, D., Marinelli, K., Dennerlein, J.: Touch-screen tablet user configurations and case-supported tilt affect and neck flexion angles. Work **41**, 81–91 (2012)
8. www.cnet.com. (https://www.cnet.com/news/apples-true-tone-display-whats-the-deal/)

User Skill Characteristics Analysis by Mouse Operation Log Analysis Based on Algebraic Method

Takeshi Matsuda[1]($\boxtimes$), Michio Sonoda[2], Masasi Eto[2],
Hironobu Satoh[2], Tomohiro Hanada[2], Nobuhiro Kanahama[2],
and Hiroki Ishikawa[2]

[1] Department of Information Security, University of Nagasaki,
Manabino, Nagayo-cho, Nagasaki 851-2195, Japan
tmatsuda@sun.ac.jp
[2] National Cyber Training Center, National Institute of Information
and Communications Technology, Nukui-Kitamachi,
Koganei, Tokyo 184-8795, Japan

Abstract. This study proposed an investigation method that analyzes mouse locus data based on geometrical method. Moreover, we considered that the correspondence between the geometric features of the locus data of the mouse and the state of the user operation.

Keywords: GUI · Mouse locus · Homology · Gröbner basis
State of user operation

1 Introduction

The method of analyzing user's behavior focusing on the line of sight has been studied in fields such as usability evaluation and marketing research. Those studies investigate how users use the tools and which part of the content the user is looking at. In recent years, researches on utilization of such gaze data have been advanced in various fields such as medicine, art and educational support. However, the movement of the line of sight includes an ambiguous state whether the gaze state or not. Moreover, various costs are required for the collection of the line of sight data.

There are also studies focusing on mouse movement as a similar study concerning a line of sight. Fitts's law [1] is famous as the behavior model of the user concerning the graphical user interface. Fitts's law is used, for example, to design the position and size of buttons on a web page [2]. Therefore, this study had focus attention on the mouse operation log data of GUI tool user's, and investigated whether skill evaluation of users is possible. The log of a mouse operation is expressed by chain of coordinate on plane. When users uses GUI tools, the mouse log from a current operation to a next operation may include user's operation contents. This study regards the mouse log from a current operation to a next operation as a walk of graph theory, and investigates what kind of walk is generated by user's skill or work contents. The main contribute of this study is to propose an analysis method on a walk of mouse log based on geometrical method.

© Springer International Publishing AG, part of Springer Nature 2019
T. Z. Ahram and C. Falcão (Eds.): AHFE 2018, AISC 794, pp. 251–261, 2019.
https://doi.org/10.1007/978-3-319-94947-5_25

Moreover, this study considers a skill evaluation method of users by using our proposed method.

The reasons for focusing on mouse operation log data will be described below. First of all, since the operation of the mouse is indispensable in the PC operation designed by the GUI, the cost of data collection can be kept quite low. Mouse operation data also includes ambiguous data, but since mouse operation is necessary to accomplish the purpose of work, it is expected that data including some meaning will be observed. The contribution of this research mainly consists of the following two parts. One is to handle locus generated from mouse operation log data as geometric objects, and the other is to propose a method of quantitatively expressing data complexity by calculation algebraic method. Finally, we considered the user's skill evaluation method by examining the relationship between the analysis result and the user's behavior. The data acquired in this research is the mouse operation log of Wireshark used for analyzing network traffic data. Since Wireshark is equipped with many functions, it is expected that the presence or absence of user operation experience will be reflected in the acquired data. Mouse operation log data is composed of a plurality of coordinate data. By linking these coordinates with line segments, the locus of the mouse can be drawn. Since the coordinates are the vertices and the intersecting segments are the edges, the locus of the mouse is the graph itself. In the topology, there is a method to investigate the homology group and study the structure of the graph. The homology group of the graph can be obtained by solving simultaneous linear equations, but the calculation becomes more difficult as the number of data increases. Therefore, we proposed a method to express the state of the graph as a monomial and grasp the geometrical structure of the graph by using the property of homomorphism obtained from the relationship between vertices and edges. Although the homology group can extract the geometrical structure such as loops, our proposed method not only can express other geometric structures including loops but also can realize it by simple calculation. Moreover, we applied the proposed method to user's data without Wireshark operation experience, and confirmed that most data consists of complex geometric structure or very simple one.

2   Conventional Works

Firstly, we introduce Fitts's low that is very important index as usability. Fitts's low is expressed by the following equation

$$T = a + b \, \log\left(\frac{A}{W} + 1\right). \tag{1}$$

Fitts's low has 4 parameters. Parameters a, b indicate skill level of users. Parameter A and W indicate a moving distance and a target seize, respectively. Fitts's law is used to design UI that is easy to use for any user. The value of the parameter W becomes small, then the value of T becomes large, which means that it is difficult for the user to use. The value of parameter A becomes large, then the value of T also becomes larger. In order to improve the usability of arbitrary users, it is desirable to design such that the value of T is reduced. However, this research aims to consider about user evaluation method in GUI tools difficult to acquire skills. If the user's mouse operation log is smooth, it can be considered that the user can use the tool without hesitation. As described above, the purpose of this research is to investigate the operation log of the user and to establish how to evaluate the user skill by examining how the features of the locus of the mouse change.

3   Mathematical Preliminary

This study proposes the method that applies topological method to the analysis of mouse operation log data. Specifically, we propose a user's state estimation method using information of the homology group of the mouse locus.

3.1   Homology Group

Let us consider the n–simplex

$$\sigma = [x_0, x_1, \cdots, x_n] = \left\{ a_0 x_0 + a_1 x_1 + \cdots + a_n x_n \in R^N \mid a_0, \cdots, a_n \in R \right\}.$$

If τ is a simplex that composed of the subset of $\{x_0, x_1, \cdots, x_n\}$, τ is called the face of σ, and denoted by $\tau \leq \sigma$. Simplicial complex is very important to define homology group. Let $K \subset R^N$ be a simplex. K is a simplicial complex if and only if the following two conditions.

(1) τ is a face of $\sigma \in K$.
(2) $\sigma_1 \cap \sigma_2$ is a face of σ_1 and σ_2, $\sigma_1, \sigma_2 \in K$.

A free module

$$C_q(K) = \left\{ \sum_{i=1}^{n} a_i x_i \mid x_i \in K, a_i \in R \right\}$$

is called q-chain of K. The basis of $C_q(K)$ is q-simplex of K. Let

$$\partial_q : C_q(K) \to C_{q-1}(K)$$

be a homomorphism defined by

$$\partial_q([x_0, x_1, \cdots, x_n]) = \sum_{i=0}^{n} (-1)^i [x_0, x_1, \cdots, \hat{x}_i, \cdots, x_n].$$

∂_q is called a boundary operator. The symbol $\hat{x}_i$ means that removes x_i. Boundary operator has a following property

$$\partial_q \circ \partial_{q-1} = 0.$$

Therefore, we have $Im(\partial_q) \subset Ker(\partial_{q-1})$ in the sequence

$$\cdots \to C_q(K) \to C_{q-1}(K) \to C_{q-2}(K) \to \cdots.$$

Then the homology group of K is defined by

$$H_q(K) = Ker(\partial_q)/Im(\partial_{q+1}).$$

$H_q(K)$ is a finitely generated able group

$$H_q(K) = Z^{\oplus R_q} \oplus Z_{\theta_{q1}} \oplus \cdots \oplus Z_{\theta_{qj}}.$$

Here R_q and θ are called Betti number and torsion coefficient, respectively. This study treats the homology group of graph $G = (V, E)$. V and E are set of vertex and edge, respectively. Let $G_1 = (V_1, E_1)$ and $G_2 = (V_1, E_1)$ be graph. If $e_1 = (v_s, v_t) \in E_1, v_s, v_t \in V_1$ holds $v_s \neq v_t$, the operation that removes e_1 and redefines $v_s = v_t$ is called a retraction. Let G_2 be the graph obtained by the retraction of G_1. Then, we have

$$dimH_0(G_1) = dimH_0(G_2),$$
$$dimH_1(G_1) = dimH_1(G_2).$$

3.2 Toric Ideal

Let us consider the set

$$A = \{a_1, a_2, \cdots, a_n\} \subset Z^d.$$

If there exists $w \in R^d$ satisfies

$$\langle w, a_1 \rangle = \langle w, a_2 \rangle = \cdots = \langle w, a_n \rangle = 1,$$

A is called an arrangement of R^d. Here

$$\langle w, a_i \rangle = \sum_{j=1}^{d} w_j a_{ji}.$$

Let k be a field, and consider polynomial rings $k[E]$ and $k[V]$. Define the homomorphism

$$\phi : k[e_1, e_2, \cdots, e_s] \to k[v_1, v_2, \cdots, v_t, v_1^{-1}, v_2^{-1}, \cdots, v_t^{-1}]$$

in the following way.

$$\phi(e) = v_1^{a_{1i}} v_2^{a_{2i}} \cdots v_t^{a_{ti}}$$

for $a_i = (a_{1i}, \cdots, a_{ti})$. Then $Ker(\phi)$ is called a toric ideal of A. This paper considers the toric ideal of graph. Especially, we consider the walk of graph.

Let

$$v_{i_1}, v_{i_2}, \cdots, v_{i_{q+1}} \in V,$$
$$e_{i_1}, e_{i_2}, \cdots, e_{i_q} \in E.$$

If $e_{i_a} = \left(v_{i_a}, v_{i_{a+1}}\right)$, then
is called a walk. If $w' = \left(e_{j_1}, e_{j_2}, \cdots, e_{j_r}\right), r \leq q$ satisfies

$$e_{j_1}, e_{j_2}, \cdots, e_{j_r} \big| e_{i_1}, e_{i_2}, \cdots, e_{i_q},$$

then w' is called subway of w. If $v_{i_1} = v_{i_{q+1}}$, then w is called closed walk. The closed walk has special property. Let $w = \left(e_{i_1}, e_{i_2}, \cdots, e_{i_{2q}}\right)$ and

$$E^+ (w) = \prod_{k=1}^{q} e_{i_{2k-1}},$$
$$E^- (w) = \prod_{k=1}^{q} e_{i_{2k}}.$$

Then the binomial $B_w = E^+ (w) - E^- (w)$ belongs to toric ideal.

4   Geometry of Mouse Log

In GUI tool, the mouse log from a current position to a next click operation may include user's operation contents. For example, there are cases that user trace a character by a mouse cursor and scrolls the screen. In these cases the locus of mouse log will be distinctive.

Fig. 1. A locus including a state on tracing a character on the screen by using a mouse cursor.

When using the mouse scroll function, the locus of the mouse may be expressed as a point. We are interested in how the skill difference appears on the locus of the mouse. Figures 2 and 3 show the mouse locus of users with different skills. Although it cannot be said as always true, a locus of the user who has not experience operating a locus of the user who has not experience operating is including complicated shapes.

Fig. 2. Loci of the user who has experience operating the GUI tool

By considering Figs. 1, 2 and 3 as graphs, it is possible to calculate the homology group of each graph. However, since the graph contains many useless vertices and edges, the calculation cost is high. On the other hand, some information may be lost by retraction. However, retraction is convenient when it is only necessary to investigate whether or not the locus contains loop-like structures. However, there is a case that small loop is a noise data, so it is necessary to remove such data. Therefore, this paper proposes to convert the data by the following procedure.

Fig. 3. Loci of the user who has not experience operating the GUI tool

[Proposed algorithm]

(Step 1) Translate the mouse coordinate data to a range in the fourth quadrant
(Step 2) Consider the region

$$R_{i,j} = \{(x,y)|50(i-1) \leq x < 50i, 50(j-1) \leq x < 50j\},$$

If there is more than one mouse coordinate data in the region $R_{i,j}$, convert those coordinate data to (i,j). By using the converted coordinate data as a vertex, arrange it in chronological order and connect neighboring vertices with edges, generate graph.

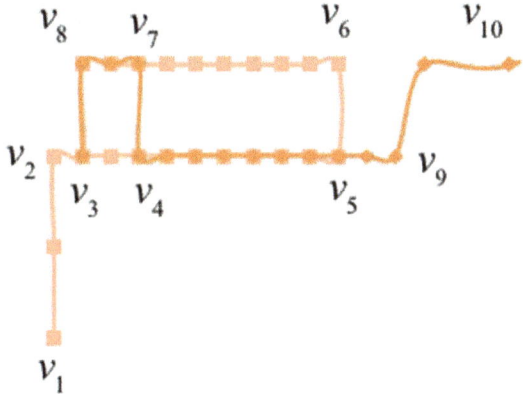

Fig. 4. The graph that applied the proposed algorithm to Fig. 1.

The graph $G = (V, E)$ of Fig. 4 is composed of

$$V = \{v_1, v_2, \cdots, v_{10}\},$$
$$E = \{e_1, e_2, \cdots, e_{12}\}.$$

From the walk $w = (e_1, e_2, \cdots, e_{12})$, we define

$$w = \sum_{j=1}^{12} k_j e_j,$$

and consider

$$0 \xrightarrow{\partial_2} C_1(G) \xrightarrow{\partial_1} C_1(G) \xrightarrow{\partial_0} O$$

Here, $e_1 = (v_1, v_2)$, $e_2 = (v_2, v_3)$, $e_3 = (v_3, v_4)$, $e_4 = (v_4, v_5)$, $e_5 = (v_5, v_6)$, $e_6 = (v_6, v_7)$, $e_7 = (v_7, v_8)$, $e_8 = (v_8, v_3)$, $e_9 = (v_3, v_8)$, $e_{10} = (v_8, v_7)$, $e_{11} = (v_7, v_4)$, $e_{12} = (v_4, v_5)$, $e_{13} = (v_5, v_9)$, $e_{14} = (v_9, v_{10})$, $e_{15} = (v_{10}, v_{11})$. Edges e_{13}, e_{14}, e_{15} are not used for convenience of calculation. The information of $Im(\partial_1)$ is obtained from the following matrix.

$$\begin{pmatrix}
-1 & 0 & 0 & 0 & 0 & 0 & 0 & 0 & 0 & 0 & 0 & 0 & 0 & 0 & 0 \\
1 & -1 & 0 & 0 & 0 & 0 & 0 & 0 & 0 & 0 & 0 & 0 & 0 & 0 & 0 \\
0 & 1 & -1 & 0 & 0 & 0 & 0 & 1 & -1 & 0 & 0 & 0 & 0 & 0 & 0 \\
0 & 0 & 1 & -1 & 0 & 0 & 0 & 0 & 0 & 0 & 1 & -1 & 0 & 0 & 0 \\
0 & 0 & 0 & 1 & -1 & 0 & 0 & 0 & 0 & 0 & 0 & 1 & -1 & 0 & 0 \\
0 & 0 & 0 & 0 & 1 & -1 & 0 & 0 & 0 & 0 & 0 & 0 & 0 & 0 & 0 \\
0 & 0 & 0 & 0 & 0 & 1 & -1 & 0 & 0 & 1 & -1 & 0 & 0 & 0 & 0 \\
0 & 0 & 0 & 0 & 0 & 0 & 1 & -1 & 1 & -1 & 0 & 0 & 0 & 0 & 0 \\
0 & 0 & 0 & 0 & 0 & 0 & 0 & 0 & 0 & 0 & 0 & 0 & 1 & -1 & 0 \\
0 & 0 & 0 & 0 & 0 & 0 & 0 & 0 & 0 & 0 & 0 & 0 & 0 & 1 & -1 \\
0 & 0 & 0 & 0 & 0 & 0 & 0 & 0 & 0 & 0 & 0 & 0 & 0 & 0 & 1
\end{pmatrix}$$

By elementary row operations, we have

$$\begin{pmatrix}
-1 & 0 & 0 & 0 & 0 & 0 & 0 & 0 & 0 & 0 & 0 & 0 & 0 & 0 & 0 \\
0 & -1 & 0 & 0 & 0 & 0 & 0 & 0 & 0 & 0 & 0 & 0 & 0 & 0 & 0 \\
0 & 0 & -1 & 0 & 0 & 0 & 0 & 1 & -1 & 0 & 0 & 0 & 0 & 0 & 0 \\
0 & 0 & 0 & -1 & 0 & 0 & 0 & 1 & -1 & 0 & 1 & -1 & 0 & 0 & 0 \\
0 & 0 & 0 & 0 & -1 & 0 & 0 & 1 & -1 & 0 & 1 & 0 & -1 & 0 & 0 \\
0 & 0 & 0 & 0 & 0 & -1 & 0 & 1 & -1 & 0 & 1 & 0 & -1 & 0 & 0 \\
0 & 0 & 0 & 0 & 0 & 0 & -1 & 1 & -1 & 1 & 0 & 0 & -1 & 0 & 0 \\
0 & 0 & 0 & 0 & 0 & 0 & 0 & 0 & 0 & 0 & 0 & 0 & -1 & 0 & 0 \\
0 & 0 & 0 & 0 & 0 & 0 & 0 & 0 & 0 & 0 & 0 & 0 & 0 & -1 & 0 \\
0 & 0 & 0 & 0 & 0 & 0 & 0 & 0 & 0 & 0 & 0 & 0 & 0 & 0 & -1 \\
0 & 0 & 0 & 0 & 0 & 0 & 0 & 0 & 0 & 0 & 0 & 0 & 0 & 0 & 0
\end{pmatrix}.$$

Therefore, we obtain

$$H_0(G) \simeq Z^{11}/Z^{10}.$$

Next, let us compute $Ker(\partial_1)$ by rewriting $w = \sum_{j=1}^{12} k_j e_j$ in the following way.

$$w = l_1 e_1 + \cdots + l_8 e_8 + l_9 e_{11} + l_{10} e_{13} + l_{11} e_{14} + l_{12} e_{15}.$$

Since it can be seen

$$l_1 = l_2 = l_{10} = l_{11} = l_{12} = 0,$$
$$l_3 = l_7 = l_8,$$
$$l_4 = l_5 = l_6,$$
$$l_9 = l_4 - l_3,$$

we get

$$Ker(\partial_1) = \{l_3(e_3 + e_7 + e_8 - e_{11}) + l_4(e_4 + e_5 + e_6 + e_{11}) | l_3, l_4 \in Z\}.$$

Therefore, we have

$$H_1(G) \simeq Z^2.$$

The graph of Fig. 5 is obtained by applying the proposed algorithm and complete retraction to Fig. 1.

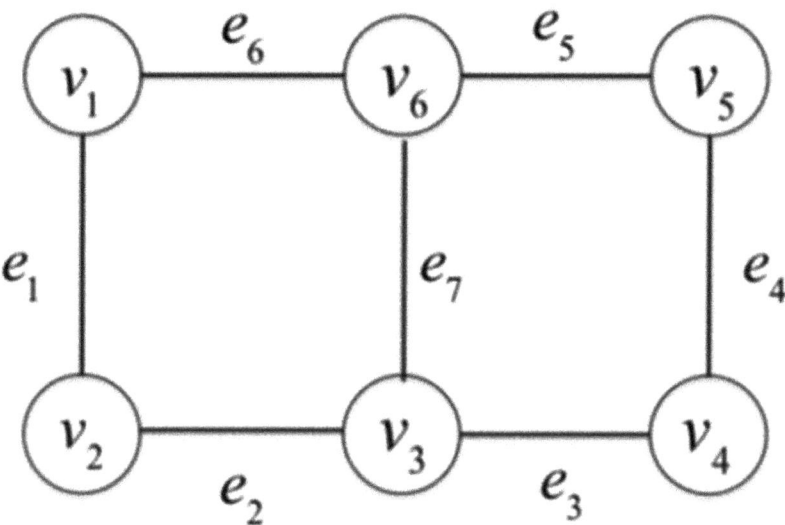

Fig. 5. Complete retraction of Fig. 4.

In order to calculate the toric ideal of Fig. 5, we define the following homomorphism.

$$\phi : k[e_1, e_2, \cdots, e_7] \rightarrow k[v_1, v_2, \cdots, v_6].$$

Here, if $e_k = (v_s, v_t)$, then $\phi(e_k) = v_s v_t$. Then, we have

$$\phi(e_1 e_7 - e_2 e_6) = 0,$$
$$\phi(e_3 e_5 - e_4 e_7) = 0.$$

Therefore, the toric ideal is

$$Ker(\phi) = \langle e_1 e_7 - e_2 e_6, e_3 e_5 - e_4 e_7 \rangle.$$

This paper introduces that the Gröbner basis of the toric ideal $Ker(\phi)$ has information on the structure of the loop of the graph $G = (V, E)$.

Firstly, let us define the term order in the following way.

$$e_1 > e_2 > \cdots > e_7.$$

Then, the Gröbner basis is given by

$$\{e_1 e_7 - e_2 e_6, e_3 e_5 - e_4 e_7\}. \tag{2}$$

Secondly, let us define the term order in the following way.

$$e_7 > e_6 > \cdots > e_1.$$

Then, the Gröbner basis is given by

$$\{e_1 e_7 - e_2 e_6, e_3 e_5 - e_4 e_7, e_2 e_4 e_6 - e_1 e_3 e_5\} \tag{3}$$

Equation (2) represents only two loop structures, but Eq. (3) also has information on the loop structure of the outer frame. Therefore, we can investigate the geometric structure of the data in more detail by changing the order of the terms. Finally, we propose a method to easily calculate topology information. Our goal is to find the structure of the loop of the mouse log data, it can be realized without calculating homology group.

Let us show a simple example.

We consider the graph $G_1 = (V_1, E_1)$, where

$$V = \{v_1, v_2, \cdots, v_6\},$$
$$E = \{e_1, e_2, \cdots, e_6\}.$$

We define the ring homomorphism $\phi : k[e_1, e_2, \cdots, e_6] \rightarrow k[v_1, v_2, \cdots, v_6]$ in the following way.

$$\phi(e_1) = v_1 v_2, \phi(e_2) = v_2 v_3, \phi(e_3) = v_3 v_4,$$
$$\phi(e_4) = v_4 v_5, \phi(e_5) = v_5 v_2, \phi(e_6) = v_2 v_6.$$

Then, we have

$$\phi\left(\prod_{i=1}^{6} e_i\right) = v_1 v_2^4 v_3^2 v_4^2 v_5^2 v_6.$$

Since $v_2^4 v_3^2 v_4^2 v_5^2 \,\Big|\, \phi\left(\prod_{i=1}^{6} e_i\right)$, we can see that the graph $G_1 = (V_1, E_1)$ has a loop. Moreover, $v_2^4 v_3^2 v_4^2 v_5^2 \,\Big|\, \phi\left(\prod_{i=1}^{6} e_i\right)$ means the walk $\{e_2, e_3, e_4, e_5\}$ is closed walk.

5 Discussion and Summary

This paper had proposed a method to calculate geometric features of mouse locus. The locus of a user without hesitation does not have a structure like a loop. The locus of users without skill may be observed as a point or have a loop structure. However, even with users with skills, it is not always the case that the mouse is moving. It is a future task to establish a method to evaluate skills by observing the time series change of the mouse locus and other information.

References

1. Paul, M.: Fitts: the information capacity of the human motor system in controlling the amplitude of movement. J. Exp. Psychol. **47**(6), 381–391 (1954)
2. Fu, J.: Parallax scrolling interface research based on Fitts' law. In: IEEE Advanced Information Management, Communicates, Electronic and Automation Control Conference (IMCEC), pp. 1370–1374 (2016)

Estimation of the Smartphone User' Satisfaction and Customer Intention on the Social Networking Service

Young-Hee Lee[1(✉)] and Ryang-Hee Kim[2]

[1] Department of Imaging Science and Arts,
Graduate School of Imaging Science, Multimedia and Film,
Chung-Ang University, 84 Heukseok-Ro, Dongjak-Gu, Seoul 156-756, Korea
yangheel003@naver.com
[2] Department of the Clothing and Textiles,
Smart and Sensibility Clothing and Textiles Lab., College of Human Ecology,
Yonsei University, # 262 Seongsanno, Seodaemun-Ku, Seoul 219-749, Korea
poolyh@hanmail.net

Abstract. Social-network service (SNS) is a service that allows user to strengthen personal relationships with friends, colleagues, and other people on the web, build new people, and build a broader human relationship. Also, SNS has been used for socializing and entertainment purposes, and information sharing all over the world. This study was to suggest basic data for user convenience/satisfaction of 'smartphones' and 'i-Phone' application. Findings showed that the users should be interested and satisfied in using SNS. Facebook showed twice the users with 'Very Satisfied' compared to those of 'KakaoTalk' in 'i-Phone' users. Overall results, 'KakaoTalk' seemed to have better usability than 'Facebook' in the Aged user. As a result, 'KakaoTalk' is superior to Facebook in terms of accessibility, application and use when using smartphone, and 'mobile phone model' and 'Age' could have a significant influence on 'usability' and 'convenience'.

Keywords: Social-Network Service
Estimation of satisfaction and customer intention · Usability evaluation

1 Introduction

In A social networking service (also social networking site, SNS or social media) is an online platform that people use to build social networks or social relations with other people who share similar personal or career interests, activities, backgrounds or real-life connections.

Social-network services, though in a broader sense, a social-network service usually provides an individual-centered service whereas online community services are group-centered. Social networking sites allow users to share ideas, digital photos and videos, posts, and to inform others about online or real-world activities and events with people in their network [1].

© Springer International Publishing AG, part of Springer Nature 2019
T. Z. Ahram and C. Falcão (Eds.): AHFE 2018, AISC 794, pp. 262–271, 2019.
https://doi.org/10.1007/978-3-319-94947-5_26

For an example, the main types of social networking services contain category places such as former school-year or classmates, means to connect with friends, and a recommendation system linked to trust. One can categorize social-network services into three types: socializing social network services used primarily for socializing with existing friends (e.g., Facebook).

Hence, the level of network sociability should determine by the actual performances of its users. According to the communication theory of uses and gratifications, an increasing number of individuals are looking to the Internet and social media to fulfill cognitive, affective, personal integrative, social integrative, and tension free needs [1, 2].

The purpose of this study was to provide a basic database for smartphone content applications that are easy to access and highly utilized. They were analyzed the current state of convenience and usability satisfaction of SNS users on the mobile phones that are currently used. From these results, we sought to improve the satisfaction and accessibility of users' SNS according to age and gender.

2 Literature Review

2.1 Social-Network Services in Ubiquitous

21st century is IT society, and demand of IT communication products is increasing, and need for service provider have been diversified and products with features corresponding to high-quality services are being released. Social networking services are Internet-based applications, and the potential for computer networking to facilitate newly improved forms of computer-mediated social interaction was suggested early on.

Early social networking on the World Wide Web began in the form of generalized online communities such as Theglobe.com (1995), Geocities (1994) and Tripod.com (1995). Many of these early communities focused on bringing people together to interact with each other through chat rooms, and encouraged users to share personal information and ideas via personal webpages by providing easy-to-use publishing tools and free or inexpensive web-space.

Web-based social networking services make it possible to connect people who share interests and activities across political, economic, and geographic borders. Through e-mail and instant messaging, online communities are created where a gift economy and reciprocal altruism are encouraged through cooperation. Information is suited to a gift economy, as information is a nonrival good and can be gifted at practically no cost. With Internet technology as a supplement to fulfill needs, it is in turn affecting everyday life, including relationships, school, church, entertainment, and family [3].

In the late 1990's, user profiles became a central feature of social networking sites, allowing users to compile lists of "friends" and search for other users with similar interests. With the rapid growth of social networking sites by 2005, Facebook launched in 2004 became the world's largest social networking site in early 2009. Facebook is a Harvard social networking site, and as soon as the term social media is first introduced, it has become a global social media communication medium.

Facebook and other social networking tools are increasingly the aim of scholarly research. Scholars in many fields have begun to investigate the impact of social networking sites, investigating how such sites may play into issues of identity, privacy, social capital, youth culture, and education. Research has also suggested that individuals add offline friends on Facebook to maintain contact and often this blurs the lines between work and home lives. According to a study in 2015, 63% of the users of Facebook or Twitter in the USA consider these networks to be their main source of news, with entertainment news being the most seen. In the times of breaking news, Twitter users are more likely to stay invested in the story. In some cases when the news story is more political, users may be more likely to voice their opinion on a linked Facebook story with a comment or like, while Twitter users will just follow the sites feed and/or retweet the article [1, 3].

KakaoTalk is a free mobile instant messaging application for smartphones with free text and free call features. It was launched on March 18, 2010 and is currently available on iOS, Android, Bada OS, BlackBerry, Windows Phone, Nokia Asha, Windows and macOS. As of May 2017, KakaoTalk had 220 million registered and 49 million monthly active users. It is available in 15 languages. The app is also used by 93% of smartphone owners in South Korea, where it is the number one messaging app. [4]

In addition to free calls and messages, KakaoTalk users can share diverse content and information including photos, videos, voice messages, location, URL links as well as contact information. Both one-on-one and group chats are available over WiFi, 3G or LTE and there are no limits to the number of friends who can join in on a group chat. KakaoTalk automatically synchronizes the user's contact list on their smartphones with the contact list on KakaoTalk to find friends who are on the service. Users can also search friends by KakaoTalk ID without having to know each other's phone number. The KakaoTalk service also allows its users to export their messages and save them for future reference.

KakaoTalk has targeted countries in Southeast Asia where no dominant mobile messenger service stands. KakaoTalk is forming strategic partnerships in Malaysia, Indonesia and the Philippines, as well. In 2013, KakaoTalk began airing TV commercials in Indonesia, the Philippines, and Vietnam featuring Big Bang. KakaoTalk has hit 13 million users and has potential to becoming KakaoTalk's largest market worldwide.

2.2   Usability and Satisfaction of Smartphone Applications

The number of smartphone users is forecast to grow from 2.1 billion in 2016 to around 2.5 billion in 2019, with smartphone penetration rates increasing as well. Just over 36% of the world's population is projected to use a smartphone by 2018, up from about 10 percent in 2011.

China, the most populous country in the world, leads the smartphone industry. The number of smartphone users in China is forecast to grow from around 563 million in 2016 to almost 675 million in 2019. The United States is also an important market for the smartphone industry, with around 247.5 million smartphone users by 2019 [5].

Google's Android and Apple's iOS are the two most popular smartphone operating systems in the industry. In 2016 alone, nearly 1.5 billion smartphones with either Android or iOS operating systems were sold to end users worldwide. Android, with 80% of all smartphones sales, leads the market. In contrast, about 15% of all smartphones sold to end customers have iOS as their operating system [7, 10].

Nearly half of Smartphone users responded that they use Smartphone to get in touch with their acquaintances. Trendmonitor made public that 45.9% of total respondents use text messages and SNS to form and maintain relationships, on survey of utilization of Smartphone. In other words, Smartphone has important role in expanding everyday social encounters into online webs. Smartphone were mostly used for their technical functions. 18.9% of respondents used Smartphone for daily life and time management, and 15.3% used Smartphone for their high tech functions. Users who use Smartphone for high tech, showed interests when having conversations with others about latest technology and trends [5].

In a 2015 study, 85% of people aged 18 to 34 use social networking sites to make purchasing decisions, and more than 65% of people over 55 years old are dependent on word of mouth. Some websites have begun to take advantage of the social networking model for charity. These models provide a means to connect small businesses with sellers who can reach more potential customers with users interested in buying things. In another study, social network systems provide another way for individuals to communicate digitally. These hypertext communities enable sharing of information and ideas that are old concepts that are placed in the digital environment.

However, along with remarkable use of Smartphone, the use of SNS mobile internet users is being amplified recently. Also, there is no heuristic interface standard evaluation of SNS communication, and there is even good understanding of usability of various SNS interface types in the fast growing ubiquitous environment of private Social Networking Service system. Therefore, the objectives of this study were to assess user satisfaction and usability evaluation of Social Networking Service [8, 9].

A study of how people use social networking to influence their feelings about loneliness has shown that the way people choose to use social networks can change the negative or negative feelings about loneliness. Some companies with mobile workers are encouraging employees to use social networking to feel workplace solidarity. Educators are using social networking to keep students connected, and individuals can take advantage of social networking Maintain close relationships through the applications you have already selected. In each social networking system, selected application users can create communities around personal identities that are created online [3, 4].

On the contrary, news reports stated that excessive usage of SNS sites may be associated with an increase in the rates of depression, to almost triple the rate for non-SNS users in 2016. Experts worldwide have said that people who use SNS more have higher levels of depression than those who use SNS less by 2030. At least one study went as far as to conclude that the negative effects of Facebook usage are equal to or greater than the positive effects of face-to-face interactions [4].

Recently, the use of SNS has been amplified by the portable Internet-users using the smart phone along with the smartphone hot air. In the early stage, it was mainly used for socializing and entertainment purposes, and information sharing. However, most of them use the latest information through SNS using smartphone. The purpose of this study is to construct and provide basic database for smartphone contents application which is convenient and accessible for the elderly in Asia, Europe and North America entering into aging society in 2020.

3   Research Method and Procedure

3.1   Survey Method

The subjects were selected as smartphone users (Android phone, iPhone) who have used SNS type (Facebook, KakaoTalk).

In evaluation Method, using the 5-point Likert scale, we will evaluate customer satisfaction and usability of Social Networking Service according to smartphone application [6–9].

Questionnaires for satisfaction of usability and convenience of smartphone application were consisted of 19 items: 1. Must be easy to learn., 2. Should be easy to use again., 3. When learning, there should be helps to learn easily., 4. System for on-screen menus and features the name, help should be clearly understood., 5. Structure and how it is used should be simple., 6. Should have elements that can be specified, 7. It should be interworking with many devices., 8. There should be no errors or bugs., 9. There should be a function to recover or cancel., 10. Compared with previously used SNS, functions should be felt not much different., 11. System needs to be quick in prosecuting its functions., 12. There should be no problem to use even without specific knowledge., 13. Need to feel affinity though it is first time using., 14. Should be interesting., 15. There should be additional explanation on detailed feature., 16. How satisfied are you for usage of this SNS?, 17. How satisfied are you with the visual aspect of the SNS of your smartphone, such as color?, 18. How satisfied are you for this SNS's visual aspect?, 19. Is this SNS easy to approach and convenient.

3.2   Analysis Tools

In this research, we have used district survey method to find for satisfaction of usability and convenience of smartphone application of gender and age. The survey was constructed with 5-scaled Likert scale, consisting of 6 demographic questions and 19 main questions.

The selected usability principles are classified systematically using statistical method SAS 9.1, t-test (two aging groups) selected by the results of Duncan's Multiple Range Test and Principle Components Analysis (PCA) using Factor Analysis.

240 subjects (Groups: Adolescence group (20–40's), Senior-age group (45 years and over)) were divided into the two groups: Adolescence group and Senior-age group, and carried out collaborative work academic institutes. And verify the hypotheses based on the results of Principle Components Analysis (PCA) using Factor Analysis.

4 Results and Discussion

4.1 Evaluation of Satisfaction of Usability and Convenience of 'Facebook'

In terms of gender, male respondents rated very satisfied, 6.22% satisfied, 26% satisfied, 44.67% dissatisfied, 20.44% dissatisfied and 2.67% dissatisfied. 8.66%, 30.67%, 38.67%, 19.11% and 2.89%, respectively (Fig. 1(A)). Both men and women were more likely to evaluate Facebook's ease of use and convenience.

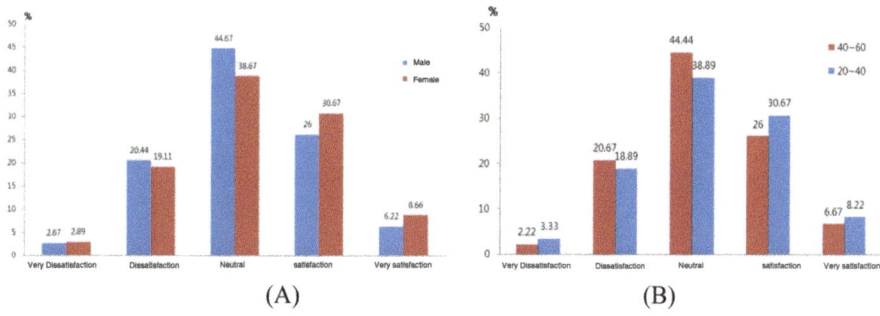

Fig. 1. Evaluation of satisfaction of usability and convenience of 'Facebook' according to sex (A) and the aged (B)

From the point of view of the age group of users, the satisfaction rate of the 20th to the 40th generation was 8.22%, the satisfaction was 30.67%, the average was 38.89%, the dissatisfaction was 18.89% and the very unsatisfied was 3.33%. In the 40 s and 60 s, the satisfaction rate was 6.67%, the satisfaction was 26%, the average was 44.44%, the dissatisfaction was 20.67%, and the dissatisfaction was 2.22% (Fig. 1(B)). It seems that people in their 20's and 60's tend to think that most of them are normal about usability and convenience.

According to the type of operating system of the smartphone used, Android users were very satisfied with 8.22%, satisfied with 29.11%, moderate with 36.67%, dissatisfied with 21.33% and very dissatisfied with 4.67%. In the case of iPhone, 19.34% satisfied, 46.44% satisfied, 26.44% normally, and 7.78% dissatisfied, and no one thought that it was very dissatisfied (Fig. 2). For Android users, there were more items evaluated as normal, and for iPhone users, more items were rated as satisfactory.

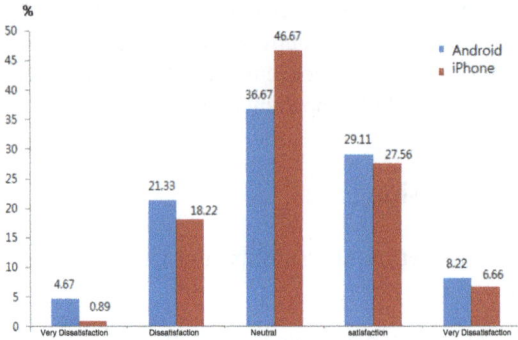

Fig. 2. Evaluation of satisfaction of usability and convenience of 'Facebook' according operating system of the smartphone used

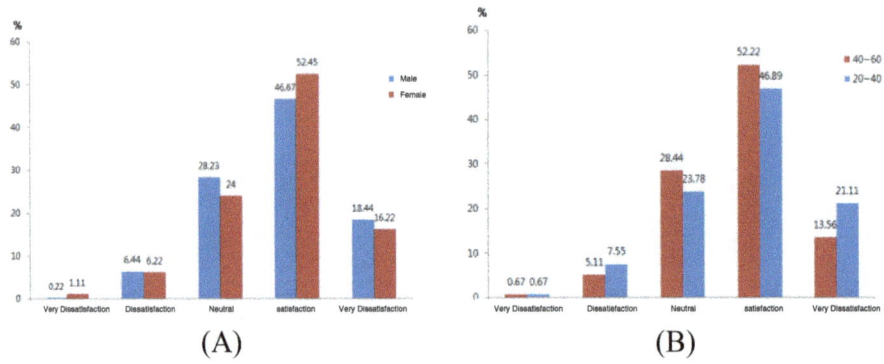

Fig. 3. Evaluation of user' preferences of colors according to sex (A) and the aged (B)

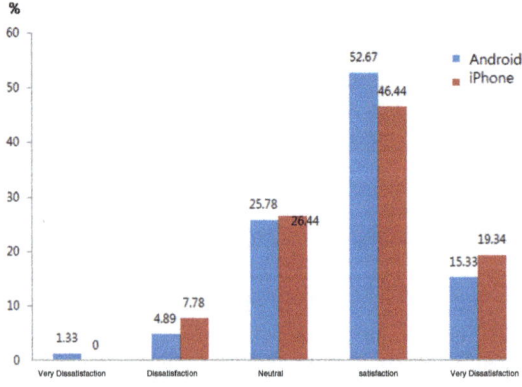

Fig. 4. Evaluation of satisfaction of usability and convenience of 'KakaoTalk' according operating system of the smartphone used

4.2   Evaluation of Satisfaction of Usability and Convenience of 'KakaoTalk

In terms of gender, men were very satisfied 18.44%, satisfied 46.67%, usually 28.23%, dissatisfied 6.44% and very dissatisfied 0.22%. The female respondents were 16.22% satisfied, 52.45% satisfied, 24.2% dissatisfied, 6.22% dissatisfied and 1.11% dissatisfied. There seems to be many items that both men and women are satisfied with the utility and convenience of KakaoTalk (Fig. 3(A)).

According to the age group, people in their 20s and 40s showed very satisfying 21.22%, satisfying 46.89%, usually 23.78%, dissatisfied 7.55% and very unsatisfied 4.17%. The users in the 40s and 60s were very satisfied, 13.56% satisfied, 52.22% satisfied, 28.44% satisfied, 5.11% dissatisfied, and 0.67% dissatisfied. From the point of view of age, users were more satisfied with KakaoTalk (Fig. 3(B)).

In terms of the operating system used, the Android phone users were found to be very satisfied 15.33%, satisfied 52.67%, average 25.78%, dissatisfied 4.89% and 1.33% very dissatisfied. For iPhone users, 19.34% were satisfied, 46.44% were satisfied, and 26.44% were dissatisfied and 7.78% were dissatisfied. In particular, it was found that users of Android phone rated the items of KakaoTalk as satisfied with more than half of items (Fig. 4).

5   Conclusion

Total usability items of 19 checklists for satisfaction of usability and convenience are enveloped with 5-point Likert scale.

PCA Factor analysis reduced the 19 usability measurements to four factors, and the Varimax rotation more clearly distinguished the usability components. Reliability within these factors were calculated by Cronbach's alpha. All of the reliability within these factors was very a high correlation respectively (Table 1).

Factor analysis results of subjective usability of scented sports-towel were named them respectively (1) "User-Performance", (2) "User-Cognition", (3) "User-Friendly", (4) "User–Satisfaction", The all 'eigen value's of the 4 factors were above 1.0, and showed as "User-Performance"(2.334), "User-Cognition" (2.047), "User-Friendly"(1.452), "User–Satisfaction"(2.1223).

Also, The results of comparison with the usability between two groups showed in There were significant differences between Youth and Elders Aged groups in the comparison and analysis of the usability of SNS ($p < .001***$).

Table 1. PCA factor analysis results of satisfaction of usability and convenience according operating system of the smartphone used

questionnaires	Factor 1 User-Performance	Factor 2 User-Cognition	Factor 3 User-Friendly	Factor 4 User–Satisfaction
10	.827			
9	.724			
12	.716			
13	.698			
5		.645		
6		.608		
4		.819		
7		.761		
14			.665	
17			.501	
16			.590	
18			.578	
3				.713
2				.663
1				.576
17				.526
ACCUMULATIVE VARIATION (%)	22.011	34.669	44.363	35.432
EIGEN VALUE	2.334	2.047	1.452	2.1223

1) * p< .05, *** p< .001

2) P. S. : Alphabet is the results of Duncan's Multiple Range Test

References

1. Kim, R.H., Kwon, Y.K., et al.: Assessment of customer satisfaction and usability evaluation of social networking service. In: Proceedings of the 4th International Conference on Applied Human Factors and Ergonomics ID:1193, San Francisco, USA (2012)
2. Hollerer, T., Feiner, S., Terauchi, T., Rashid, G., et al.: Exploring MARS: developing indoor and outdoor user interfaces to a mobile augmented reality system. Smartphone Secur. Graph. **23**(6), 779–785 (1999)
3. Choi, S.M.: Study on the effects of SNS on the social structure in the formation of social capital. J. Korean Acad. Assoc. Bus. Adm. **78**, 13–22 (2009)

4. Kim, W.K.: Physical disability and depression in older adults: predictability of structural and functional aspects of social support. Korean J. Clin. Psychol. **20**(1), 49–66 (2001)
5. Ji, Y.G., Jin, B.S., Ko, S.M.: Development of a AHP model for, telematic interface evaluation. In: Proceedings of HCII 2007 Conference on Human Computer Interaction. LNCS, vol. 4550. Springer, Heidelberg (2007)
6. Anderson, J.A., Wangner, J., Bessesen, M., Williams, L.C.: Usability testing in the hospital. J. Hum. Factors Ergon. Manuf. Serv. Ind. **22**(1), 1–12 (2011)
7. Kim, M.H., Ji, Y.G.: The Development of usability evaluation method for tangible user interface. Thesis, University of Yonsei (2007)
8. Kim, R.H., Kwon, Y.K., et al.: Users' awareness of computer security and usability evaluation. In: Proceedings of the 5th International Conference on Applied Human Factors and Ergonomics ID:1192, San Francisco, USA (2012)
9. Kim, R.H., Kwon, Y.K.: The heuristic evaluation methodology of the smartphone operating system on the user preferences and satisfaction of the security system. In: Proceedings of the 5th International Conference on Applied Human Factors and Ergonomics ID:1707, Kraków, Poland (2014)
10. Schmidt, A.D., Schmidt, H.G., et al.: Smartphone malware evolution revisited: Android next target? In: 4th International Conference on Malicious and Unwanted Software (MALWARE) (2009). ISBN:978-1-4244-5786-1. Accessed 30 Nov 2010

Adding Eye Tracking Data Collection to Smartphone Usability Evaluation: A Comparison Between Eye Tracking Processes and Traditional Techniques

Marcos Souza[✉] and Fracimar Maciel

Samsung SIDIA, Manaus, Brazil
{marcos.muniz, francimar.m}@samsung.com

Abstract. Due to the sensation of frustration caused by both technological changes and the lack of knowledge required to interact with new technologies, some confusing experiences with mobile computing systems tends to discourage the use of new interfaces. Therefore, it is necessary to offer products and services that can be fully controlled by users, transforming complex processes into simple and objective interactions. For this, initially a preparation of the case study to define the object to be evaluated, which concluded that an analysis of the use of the search tools of the Android system could meet the requirements of the proposed work. This work has as a premise to create a combination of usability assessment techniques which allowed adding eye tracking data from the users' experience to the experiments with traditional research methods. At the end of the study, it was possible to evaluate the contribution of each technique applied.

Keywords: Eye tracking · Usability · Usability test · HCI

1 Introduction

Figures on the consumption of services for mobile platforms emphasize the importance of understanding the visual aspects of interface design for smartphones.

According to an article published on the Global Web Index website [1], the time that Brazilians spend connected via smartphones tripled between 2012 and 2015 - about 3 h and 40 min per day, representing the third place in ranking of users with countries with more time online through mobile devices.

Fernando Meirelles, lecture at FGV-SP, believes that growth has occurred due to behavioral changes in consumption. People want, and some indeed need, to be more and more connected, so that access to information through computers no longer meets the needs and wants of an online generation.

The user-centered design process considers the people involved during the development process, their knowledge and their needs throughout the process, enabling the development of solutions to be guided less by technological possibilities and more by the needs of those who use [2].

© Springer International Publishing AG, part of Springer Nature 2019
T. Z. Ahram and C. Falcão (Eds.): AHFE 2018, AISC 794, pp. 272–282, 2019.
https://doi.org/10.1007/978-3-319-94947-5_27

The function of design, combined with art and science, can be understood as approximation of human beings to the conceptualization of goods that elevate the experience of their lives. Joining these areas of knowledge to achieve satisfactory results and monitoring the technological development demands a lot of research on the behavior of the users. Much of the day-to-day activities of people have been transported to computer interfaces, creating new challenges for designers. The smartphone became a popular item in all walks of life, prompting several people of varying degrees of familiarity with the technology to acquire a handset that would aid communication.

The democratization of access to mobile devices maximizes the challenge of building user-friendly interfaces. The design must ensure that the system has the correct representation, that is, it reveals the appropriate system image, allowing the user to create the correct model for interaction and understanding of the actions, remembering that all user knowledge starts from the perception of the image of the user. system. In this way, only the inherent possibilities of an interface must be visible, indicating how the user should interact with the device. Such possibilities of an interface can be explained through their mental processes, among which we highlight the perception [3].

This article refers to the evaluation of usability of interfaces as a necessary process to validate the expectations of the designer in front of the expectations of the users. This work will present a research process based on qualitative studies to better understand the perception of smartphone users, using the techniques of semi-structured interview, eye tracking and retrospective think alound.

2   Eye Tracking

The study of eye tracking initially was stimulated by clinical needs focused on physiological research applied to eye dysfunctions and reading disorders, and later was used in Psychology and Neurology, in the analysis of perceptual and cognitive processes.

The term eye tracking is a translation of the English term eye tracking, which can also be translated as eye tracking. It refers to a set of technologies that allow the measurement and recording of an individual's ocular movements before stimuli in a real or controlled environment, determining the areas of attention, time and flow of visual exploration [4].

Although not considered a novelty, the knowledge of eye tracking technology as a method of collecting and analyzing information, as well as its potential for implementation in different domains of the academic, scientific and commercial environment, are still in a nascent state.

The relevance of studying eye movements is based on the strong-minded hypothesis that what a person view is assumed as an indicator of current/prevalent thinking in cognitive processes [5]. Such a method means that the recording of eye movements provides a dynamic stroke where attention is directed to a particular field of vision. Measurement of other aspects associated with eye movements, such as fixations (times when eyes are relatively fixed, assimilating or "coding" the information), may also reveal the amount of processing applied to objects viewed.

In this work, the term eye tracking is applied to the technique of capturing the user's eye movements, which makes it possible to identify the observed points (fixations) by the user and the paths traveled between these points (balconies), in front of the interface that is being watched.

According to [6] the basic concept is based on video cameras to capture the reflexes generated by a light source purposely placed to illuminate the user's eye, causing highly reflective visible. The image taken by the camera is then used to identify the reflection of the light source in the cornea (brightness) and the pupil. It is possible to calculate a vector formed by the angle between the cornea and the pupil, combined with other geometrical characteristics of the reflections, to calculate the position and direction of the look.

3 Methodology

To prepare the experiment the research, it was necessary to define the study setting, as well as the methodology that would be applied to check for measurable evidence that the use of eye tracking combining other traditional methods research can improve test scores usability. In general terms, the case study was divided into the following stages:

- To define the object to be evaluated.
  - Online Questionnaire for prior validation of the author's personal hypothesis;
  - Definition of the operating system to be used;
  - Unstructured interview with users of the chosen system;
  - Definition of the version of the system to be used in the experiment;
  - Model and version of the smartphone to be used in the experiment;
- To define of the mix of techniques and usability evaluation method to be applied.
  - Semi-structured interview;
  - Usability testing using eye tracking;
  - Retrospective Think Aloud.
  - Definition of recruitment of participants and location of the case study.
  - Evaluation of results.
- Definition of recruitment of participants and location of the case study.
- Evaluation of results.

3.1 Case Study Preparation

To define the object of the case study, initially, it was necessary to define a scenario to be investigated within a context of specific use, which involved identifying models and versions of smartphone systems, as well as device models.

To do this, an online questionnaire was first shared among smartphone users aiming at partial understanding of the mental model created by users during the search for files (text, photos and videos), applications and system configuration functions of their device. The survey found that more than 60% of users do not use their smartphone's search engine when they need to conduct a survey and 23.17% of respondents say that their devices do not have a search tool available in their device settings.

It was also possible to identify the main operating systems and smartphone device models used by the respondents, which provided input for the direction the author needed to define his case study.

The details of this phase and of any preparation of the case study will be described below.

3.1.1   Online Survey

In all, 164 people between the ages of 18 and 55 answered the questionnaire that was accessible for 7 consecutive days, between March 30 and April 4, 2017.

When they come across a question about how they search for a text, image or video file on their smartphone, only 30% of the respondents said they use the search tool or use voice commands from their device's virtual assistant. The others replied that they search directly in the folders or by the gallery application, for case of images and videos.

The survey data allowed an individual analysis by file type, which showed that only 14.02% of the respondents used a search tool (search fields, Siri, S-Finder and etc.) to search for applications in their smartphones, 15.24% use search tools to find to search for a function in the settings menu and 23.17% state that there is no "Search".

Through the results obtained with the online questionnaire data, the opportunity to carry out a more in-depth study was identified, aiming to understand the navigation flow of smartphone users when they are searching for a local file. This study was mainly motivated by the number of users who said they did not have a search tool within the settings of their device, and based on the assumption that all models (Table 4) reported by the respondents have a search tool.

3.1.2   Unstructured Interviews with Android Smartphone User

To better understanding of the scenario presented by online research, which that users normally do not use or are unaware of the possibility of use the search tools of your devices to find applications, it was decided to explore through unstructured interview the form of navigation of smartphone users with Android system. This process involved 20 users.

During the interviews, people were invited to show the flow of navigation they followed to find some features and applications on your smartphone. It can be highlight some points:

- Although users have many applications, they can organize which use most frequently at one or at most two flaps of the screen of their devices to facilitate access;
- Most users conduct an exploratory survey within the when you need to change the status of system functionality;
- Few comment that they use the search engine when they are not exactly sure what to look for;
- Only 2 people commented that it would be nice to have a way to find the configuration functions in a quicker and simpler way.

This contact with Android smartphone users along with data collected in the online questionnaire were fundamental for formulating the hypothesis of this research, but not enough, it was still necessary to define which version of the system would be used during the experiment, having seen that both in the online questionnaire and unstructured interviews, many versions of systems were identified. This phase has been defined through a benchmarking will be described below.

3.1.3   Benchmarking

At the time of this research there were 10 versions of the Android system embedded in smartphones in the world (Android, 2017), without considering the customized by partner companies. In this way, the author of this research believe that benchmarking is necessary to better understand market share occupied by each version, so that you can choose the version of distribution.

With the result, the author of this research chose to select the most popular Android system raised, to be used during the in this case, version 6.0 was chosen (STATISTA, 2017).

3.1.4   Smartphone Device

Starting from the models identified in the first phase, an online questionnaire, was possible to evaluate that there were several models with the Android 6.0 system, but marks and visual configurations.

To meet these requirements the Motorola branded device and Moto model Maxx was selected for the case study. This source version was configured with version 4.0 of Android, but at the time of the experiment the device was upgraded to Google's native version 6.0.

4   Case of Study

After collecting the essential information in the preparation stages for the study of this work, an analysis of usability that would be necessary to contemplate the case study in order to validate the hypothesis of this research, which will be detailed below.

To validate that the addition of methods that add information cognitive evidence presents measurable evidence regarding the use of traditional methods of analysis in the process of evaluating the usability of interfaces, a descriptive research on the using a mix of techniques - unstructured interview, eye tracking data and think aloud.

It was necessary to define how the optimization of each technique would be detailed planning of each step (application and analysis). It was thought how would be used to allow application in the same session and with the same users, in order to optimize the study development time, but without disregard the recommendations of the literature for each of the methods.

4.1   Unstructured Interview

The interview was the first method of research to be applied, aiming to understand the participants' knowledge of the research object, context and model of use, besides the motivations and functions most used by the participants. At the time the interview was structured in 11 questions, being divided in 3 parts: analysis of the participants' profile (Table 1), search for functionality and application (Table 2) and issues arising from the search for features and applications (Table 2), all detailed below.

Table 1. Number of users by group and profile.

Group	Respondents	Profile	Main functions
A	7	Advanced	Bluetooth pairing, wifi router, security features, use of biosensors, stream and smart view
	3	Intermediate	Maps, alarm, calendar and calendar
B	8	Intermediate	Camera, agenda, e-mail, flashlight and calculator
	2	Beginners	Calls and messages

Table 2. Research mode for functions of system and applications.

Group	Respondents	Exploratory	By Tools	Comments
A	8	x		Usually exploits sessions within the device settings until you find what you want
	2	x	x	Usually explore area settings. Device, but when finds difficulty goes to YouTube
B	8	x		Usually exploits sessions within the device settings until you find what you want
	2		x	Enter in area of settings and uses the search tool

In the face of the analysis of the data, there is no doubt that the exploratory survey of the smartphone system configuration area is commonly used by most (90%) users. It should be noted that of all A and B participants, two people also use the search tool configuration system, and only one uses the search (Table 3).

Given this scenario, I was left wondering if users did not use search tool available in the configuration area of your device due to the to the adopted interaction model or to a system usability problem Android.

Table 3. Problems with search for features and applications.

Group	Respondents	Function	Application	Comments
A	2	x		They resolved by exploring the sessions in the settings
B	6	x		Usually exploits sessions within the device settings until you find what you want
	1		x	He had to call for your carrier's telephone operator to know if it was cell phone problem or operator

4.2   Usability Test by Eye Tracking

The usability test using eye tracking was selected with the objective of to verify the path taken by users to perform the tasks of the study of case, as well as validate the areas viewed by users during execution of the experiment to identify possible usability problems that do not stimulate the use of the search tools available on the Android system to smartphone. For the test were defined 4 tasks to be performed - disable keyboard tones, changing the device language, check updates for system and search by application. And for all questions users would have two possible flows to follow:

- Flow A, use the search tool.
- Flow B, navigate between sessions until you reach the desired option

4.2.1   Disable Keyboard Tones

Soon after the free navigation, the first task that the participants should perform to familiarize themselves with the smartphone, was presented as a second disable the keyboard tones of the device.

Considering groups A and B of participants, flow A was only followed immediately by one participant, other two participants resorted to flow A only after trying for more than 2 min to perform the task by flow B, three participants did not complete the task and the others completed following flow B.

Analyzing the heat map generated by the eye tracking program for each group of participants, it was possible to perceive the region of greater visual attention participants does not contemplate the area of the search tool.

4.2.2   Changing the Device Language

The second task requested from the participants was to find the changed the language of the device, and once more it was participants would have two possible streams to follow to complete the task.

Considering groups A and B of participants, flow A was only followed immediately by two participant, one of them being the one who turned the search engine after try for more than 2 min to solve task 1 by the navigation flow exploratory, and the others concluded following flow B.

Analyzing the heat map generated by the eye tracking program, both the participants in group A as the participants in group B presented areas of attention to the flow B of navigation. And the area where located the search tool again had little demarcation, just as the heat map generated for task 1.

4.2.3 Check Updates for System

The third and last task to be analyzed was to search for updates of the system. Participants were invited to check for updates available.

For the third and final task of the configuration system, 19 of 20 participants successfully completed using the exploratory navigation flow and only one participant, from group B, did not complete the task claiming they had no idea what a system upgrade would be.

Faced with the flow chosen by 95% of the participants, it was already expected that the region of the Android system search tool was not highlighted in the heat map generated.

4.2.4 Search by Application

Data on research for application were not interesting for further analysis during the usability tests with eye tracking, considering that the tasks do not present any difficulty between participants.

It is important to highlight, given the study carried out, that users daily searches between the main screens, organize applications, not needing search tools to find them.

4.3 Retrospective Think Aloud

A retrospective think aloud served mainly to understand with greater depth the path traveled by users during the usability test using eye tracking, for example, understand what the user thought when you focused on the search appliance of the configuration system, but not the used to finalize the task of the study.

This third and last phase of data collection of the experiment was divided into 3 moments: the first was to collect the first impressions of the participants related to the test, the second to know what they think about the ease of finding system features and installed applications, and finally, questions about the Android system search tool.

First impressions - In general, the participants did not experiment and stated that they had no difficulties understanding the tasks the driver and the author of the experiment. Participants who reported difficulties they claimed that they had never had the need to carry out that task and were surprised because they did not wait for the difficulties encountered to the location of the functions involved.

About the ease of finding system features and installed applications - It was verified that 80% of the participants expressed opinions about how they believed they could improve their experiences with research in their devices. Another point of view was also drawn to the participants in the who were asked about what they would keep from the current version of the their smartphones, considering the flow traveled when they need it functionality of the installed device or application. In this regard, 40% of participants said they would keep all the current system provides, 20% could not say what they would keep in a future version of the system and the others expressed various opinions on both interconnected functions related to the current system configurations regarding the organization and form of search for apps.

About the Android system search tool - the first question was asked whether the participants had the search tool in the device settings session. At that time, 60% of the participants in group A stated that they had seen and visually identify the tool. At the time they were asked to show on the device where this tool was located. Of these, only 33.3% during the study. For group B only 20% identified the tool, and only 10% used it.

Participants who did not identify the search tool in settings area were invited to review the video captured by the eye tracking during the execution of tasks, so they could see where they looked and be questioned what they understood by the magnifying glass icon.

Participants in group A said they had not realize the icon in the interface of the system, but that after presenting it understood that it was possible to search for device configuration functionality; 30% of group B participants did not they were able to inform the icon's functionality, and the others were surprised never noticed. The participants who realized through the software that at some point looked at the icon were surprised not to remember to have visa.

Participants in both group who was not have knowledge of the tool said they will probably use need to find some unusual functionality that needs to be configured on your device. Those who already knew recognized that using the tool can optimize to get the desired functionality, but which by custom prefer to search exploring through the sessions.

About using the search tool to find applications, 20% of the participants in group A and 80% of group B reported not having knowledge of a search tool to find applications device, and upon being presented to said search field by application, again reported not having noticed and confirmed that they organize their applications on the main screen of the device and have never had difficulty find one.

4.4   Comparison Between Methods

Each of the methods contributed data complementary to those collected by the previous method, which will be better presented through a table (Table 4) that synthesizes the main findings of the research and correlates with the applied technique which made it possible to collect this data.

The table enables the understanding that the only technique to explore profile of the users is the interview but that it cannot reveal the data relating to the user's visual attention during the use of a mobile application. Data those that only with the application of eye tracking were possible to be collected, and further, provided understanding of the path of the able to identify the areas of the interface that aroused greater interest visual. Finally, with the retrospective think aloud, the points of the evaluation proposed by the mix of methodologies are closed as the understanding of the fluxes and visual attention understood by this technique.

Table 4. Comparison between methods.

Group	Interviews	Eye tracking	Think Aloud
Profile of the participants	x		
Main functions used	x		
Organizational model	x		x
Use model	x	x	
Trouble finding features	x	x	
Paths traversed visually		x	
Areas of interest		x	
Impressions of the participants			x
Satisfaction of experience			x
Improvements points			x
Relationship between attention and decision making			x

5   Conclusion

Traditional research methods provide a broad and fundamental for usability studies, contributing with information indispensable for the knowledge of the profiles of the end users and development of the proposed interface design project. However, this study found that the addition of movement of the gaze information, such as those with the use of eye tracking, resulted in better usability of smartphone interfaces in comparison to the use of traditional, only.

The mix of techniques allowed not only to register that the participants the search tool during tasks related to system, as well as the non-use was due to the lack of function of knowledge and that the non-display of the search field on the application screen is user use model, which customize the main screens of your device by arranging the applications so that they see no need to support to find them.

We thank Samsung Electronics Amazonia Ltda., The part of the results presented in this study were obtained during the creation and definition of methods that can support the projects financed by the company under the Law 8387 (article 2)/91.

References

1. GLOBALWEBINDEX.: Fast-growth Nations Clock Up the Most Hours for Mobile Web Usage. https://blog.globalwebindex.net/chart-of-the-day/fastgrowth-nations-clock-up-the-most-hours-for-mobile-web-usage
2. Brennand, E., Lemos, G.: Televisão digital interativa: reflexões, sistemas e padrões. Horizonte, Vinhedo (2007)

3. Norman, D.: O design do dia-a-dia. Rocco, Rio de Janeiro (2006)
4. Jacob, R.: Eye tracking in advanced interface design. Virtual Environ. Adv. Interface Des. [S.L], 258–288 (1995)
5. Just, M.A., Caroenter, P.A.: A theory of reading: From eye fixations to comprehension (1980). http://repository.cmu.edu/cgi/viewcontent.cgi?article=1731&context=psychology
6. TOBII TECHNOLOGY.: How do Tobii eye tracking work? http://www.tobiipro.com/learn-and-support/learn/eye-tracking-essentials/how-do-tobiieye-trackers-work/

Usability Evaluation of Self-service Laundry System in Universities

Meiyu Zhou, Chao Li[✉], and Can Zhao[✉]

East China University of Science and Technology, Shanghai, China
zhoutc_2003@163.com, dreamingjiangnan@163.com,
1572030604@qq.com

Abstract. A set of usability evaluation index and model of Self-service laundry service system of colleges and universities are presented based on the Service system usability studies research summarized, combined with quantitative and qualitative methods in this paper. The usability evaluation model of the service system is used to further improve the practical operation of the usability evaluation of service system. Relevant theories of usability evaluation and service quality are referenced, through methods of user questionnaire survey and expert group discussion, the usability evaluation index of college students' self-service laundry service system was constructed from two dimensions of product and service, and multiple linear regression method is used to construct usability evaluation model, which is proved to be effective. The research shows that the availability evaluation model of University self-service laundry system obtained by this method has definite effectiveness.

Keywords: Self-service laundry system · Usability evaluation
Service quality · Regression analysis

1 Introduction

With the improvement of peple's life quality and the acceleration of their life pace, self-service mode which is independent and convenient is receiving more and more popularity by the market and is emerging rapidly as a service mode conforming to social development characteristics. At the same time, as the concept "internet+" is introduced, problems about the usability evaluation index system and evaluation model of the self-service system design that need to be further discussed have been exposed while there are increasingly more self-service modes. Thus, this paper attempts to explore usability evaluation methods of service system design by taking self-help laundry service as an example.

2 Usability Evaluation

Currently, widely used product usability evaluation standard models are proposed by Jakob Nielse and Brian Shackle (1993) who have evaluated the usability of products from five aspects, namely, usability, efficiency, memorability, error rate and subjective

© Springer International Publishing AG, part of Springer Nature 2019
T. Z. Ahram and C. Falcão (Eds.): AHFE 2018, AISC 794, pp. 283–292, 2019.
https://doi.org/10.1007/978-3-319-94947-5_28

satisfaction [1]. Professor Han et al. (2001) have come up with an index system of the usability evaluation level which includes user interface design elements and user usability dimension. By testing 36 products, an evaluation system framework of design elements and usability dimension has also been put forward [2]. Most domestic usability evaluation researches focus on webpage design and electronic products now, and main approaches taken by them are user survey and testing. Regarding the web-page design iteration process, via the comparative analysis before and after revision, the field usability test method can be taken to observe the user mode of users and find usability problems existing in the webpage information construction, so as to present design improvement suggestions [3]. With the method of user performance test, Ge Liezhong et al. (2006) have evaluated the usability degree of different electronic products by quantifying the user operation route and operation time, thus raising problems about product usability [4]. Cheng Shiwei and other scholars (2009) have evaluated and analyzed the usability of the mobile user interface by making use of the eye-gaze tracking technology. They have tried to establish the user interface usability evaluation model on the basis of interface visual search and information processing by combining the eye-gaze index, interaction task and visual cognitive theory. The sub-jective influence and error that may exist in the previous two usability evaluation ways can be prevented with this method, so that the inner cause of user interface cognition can be understood more effectively and the usability problem list can be obtained [5]. Miao Chongchong (2012) have studied the usability of the subway dispatching system by optimizing the established evaluation index system of the subway dispatching system via AHP evaluation index screening as well as improving the usability evalu-ation model of the subway dispatching system via the eye-gaze tracking test [6]. Throughout these usability evaluation researches, it can be found that most usability researches focus on traditional real objects. However, modern products are mostly the integration of products and services, so the "real object + service" product service system is taken as the research object in this paper.

3   Product Service System and Service Quality Evaluation

3.1   Product Service System

Product service system is a form that is produced in the process of socio economic transformation from commercial economy to service economy and integrates products with services. It is a result of the transformation from real products to service products, which provides new meanings for products, that is, products produced by enterprises are the integration of real objects and visional ones. Products convey not only the value of real objects but also the service value of products in the whole life cycle. Product design is actually an all-around solution meeting consumers' experience needs as well as a systematic design integrating products and services. It highlights the overall design and takes products and services under the new business mode as the purpose. In the design process, immaterial services and material products can be combined effectively, and systematic optimization for resources like the supply, production, marketing, users' experience, discard, recycling, government and public service organizations involved in the traditional product manufacturing mode can be realized [7].

3.2 Service Quality and Satisfaction Model

Service quality is a subjective quality based on customers' perception [8]. Service quality is not just an objective quality. It cannot be fixed on a material object, nor can it be defined by objective technological characteristics, instead, it is a perceptive quality based on customers' needs and expectation. Researches about service quality mainly focus on three aspects, namely, service quality connotation, service quality dimension and service quality evaluation. The evaluation on service quality is mostly from the perceptive feedback of users at home and abroad. Gronroos, a scholar from Finland, believes that service quality is a perceptive evaluation on services by customers and that it is not only closely associated with service results but also the service process [9].

Models studying the user need and user satisfaction mainly include KANO model [10], SERVQUAL evaluation model and SERVPERF model [11]. As a typical qualitative analysis model, KANO model reflects the nonlinear relationship between product performance and user satisfaction, and it is advantageous in helping service organizations to recognize customers' different levels of service needs and classify service quality attributes. Service quality is divided into five dimensions by SERVQUAL model, namely, tangible facilities, reliability, response, guarantee and empathy. Each dimension is refined to several problems to measure and evaluate service quality via the expectation value and actual feeling value of users for each problem. SERVPERF model also takes the above five dimensions as the basis, but it only stresses the perceptive quality of customers without taking expectation index into account. Although it can facilitate data collection and model application, its performance is not as comprehensive as SERVQUAL model I terms of the service quality evaluation, for it merely takes customers' perception into consideration.

4 Methods to Evaluate the Usability of Self-help Laundry Service in Colleges and Universities

4.1 Establishing the Usability Evaluation Index System of Self-help Laundry Service System in Colleges and Universities

4.1.1 Usability Evaluation Index Extraction Principles and Methods for Self-help Laundry Service System in Colleges and Universities

Via the survey on the mode, process characteristics, target users and use scenes of self-help laundry service as well as relevant researches on the product usability and service quality, the usability of self-help laundry service is mainly analyzed from the perspective of people - objects (or services) – environment. Through user interviews and field survey, the satisfaction is evaluated in accordance with the feedback of users in the self-help laundry service process. According to the people – machine – environment system, relevant theories are analyzed to explore factors influencing the usability of self-help laundry service in colleges and universities, including the product attribute (machine), user characteristics (people), environment characteristics (environment) and service quality. Thus, it is to set the usability evaluation index of self-help laundry service system in colleges and universities (Fig. 1).

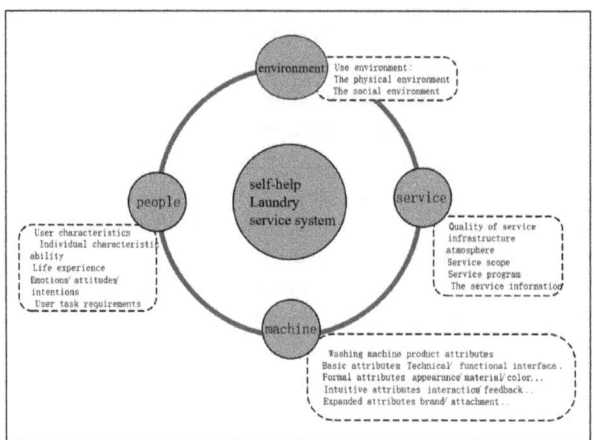

Fig. 1. Four factors influencing the usability of self-help laundry service system

According to basic requirements of the usability questionnaire scale, words and sentences for usability description should be extracted and screened. In this process, the researcher's original description connotation should be respected as far as possible along with the elimination of part with similar meanings and confusing definitions, thus getting the usability dimension in the initial state of self-help service washing machines. The expectation and need for services by users are usually their individual psychological experience which exists in their mind as implicit service standards. Therefore, it is hard to be described accurately by words or written materials, and it is also difficult to be perceived and imitated. KANO model is used in this research for statistical analysis and sorting of users' service expectation, which can help to analyze and identify quality characteristics of public self-help laundry service (e.g. essential quality, one-dimension quality and charm quality). Meanwhile, it is also conductive to turn these implicit service standards to explicit service usability evaluation indexes.

4.1.2 The Usability Evaluation Index Selection Process of Self-help Laundry Service System in Colleges and Universities

First, according to the KANO model, SEVQUAL model and usability evaluation model, 5 dimensions affecting the service quality of self-help laundry service system is put forward by combining the user need analysis and the evaluation on self-help laundry service quality, that is, reliability, response, usability, safety and empathy. On this basis, evaluation indexes of the service quality of self-help laundry service system are determined, as shown in Table 1.

Second, according to the analysis of service quality evaluation indexes of the product and service system, the usability of self-help laundry service is divided into

Table 1. Service quality indexes of self-help laundry machines

Main dimension	KK_reliability	KK_response	YY_usability	AQ_safety	YQ_empathy
Index	KK1_task achievements	XY1_complaint processing speed	YY1_clear indication	AQ1_environment and product sanitation	YQ1_attaching importance to customer needs
	KK2_task completion timeliness	XY2_transaction request processing speed	YY2_clear process	AQ2_system sterilization accomplishment	YQ2_personalized service function
	KK3_system stability	XY3_contact categories	YY3_system easy to be operated	AQ3_safety risks of washingmachines	YQ3_caring about customers

morphological cognition usability, washing performance usability and service perception usability by combining characteristics of self-help laundry service system itself. Considering that appearance is a key point of design, it is involved in perception usability after discussion by two interactive designers with design background and two user experiencers, thus initially determining 20 usability evaluation indexes of self-help laundry service.

Through the interview of managers, it is considered partial evaluation indexes have slight influence on service quality, so it is necessary to conduct dimension reduction processing for 20 indexes. The importance of the above usability indexes is surveyed by issuing 100 questionnaires. 67 have been received back, including 64 effective samples. The selection times of these usability indexes are sequenced with the abandoning of those with low attention degree by users. Finally, 12 indexes that are selected with most times are taken as usability indexes of the self-help laundry service system (see Table 2).

Table 2. Usability evaluation index system of self-help student laundry service system in colleges and universities

Evaluation index system		
x_1_no need to queue for laundry	x_5_clean washing machines	x_9_system service process meets users' thinking habits
x_2_positive reminding service for clothes fetching	x_6_simple operation of the service system	x_{10}_no safety risks in the use process
x_3_convenient payment	x_7_laundry task can be done in a short time	x_{11}_affordable
x_4_satisfactory washing quality	x_8_service system is easy to learn	x_{12}_easy to contact

4.2 Usability Evaluation Modeling Methods for Self-help Student Laundry Service System in Colleges and Universities

In mathematical statistics, regression analysis is a most basic and most important statistical method, through which problems with variable correlations can be processed by regression modeling and such relationship can be expressed by mathematical models. If x_1 ...... x_k is k independent variables that are correlative with a dependent variable y, then a linear regression model can be set:

$$y = \beta_0 + \beta_1 x_1 + \beta_2 x_2 + \ldots + \beta_k x_k + \varepsilon \tag{3.1}$$

Where, $x_1, x_2, \ldots\ldots, x_k$ represent influencing factors which can usually be controlled or preset, and they are known as explanatory variables or independent variables. y is the research object or the prediction target known as explained variable or dependent variable. ε is the sum of the influence on y by various random factors, known as the random error item that conforms to normal distribution, that is, $\varepsilon \sim N(0, \sigma^2)$; β_k is known as the regression coefficient of the multiple linear regression model; k is the number of explanatory variables in the multiple linear regression model. The estimated value $\hat{\beta_i}$ of the regression coefficient β_i can be obtained via the least square method, then the regression analysis model is:

$$y = \beta_1 x_1 + \beta_2 x_2 + \beta_3 x_3 + \ldots + \beta_i x_i \tag{3.2}$$

Where, y is the overall usability value (overall user satisfaction);
x_i is the quantized average value of the i-th index;
β_i is the coefficient of the i-th index quantized value, representing the importance of the index.

In this research, it is assumed the evaluation of each index is correlated with the overall satisfaction by users. 12 indexes are used for the evaluation on self-help laundry service usability in colleges and universities. By consulting relevant experts, 7-order Likert scale is used to design the questionnaire and measure the satisfaction of users. To ensure the reliability and malleability of the research result, questionnaire survey has been conducted for the self-help laundry service system of 15 colleges and universities in Shanghai, Nanjing and Changzhou, and 1 college is selected for verification. To neglect the possible influence on data by gender difference, 5 questionnaires are issued for boy users and girl users in each school, and 10 effective ones are finally obtained from each school. A total of 150 effective questionnaires are taken back in the survey, as shown in Table 3. As for the Cronbach's alpha coefficient α of the questionnaire data of the 15 colleges and universities, it is found the value α of 14 colleges and universities is larger than 0.8, and only that of the school G10 is smaller than 0.8 though it is still very close to 0.8, being 0.798. The data α of school G12, G3, G4 and G9 reaches 0.9 and higher, indicating the 15 groups of data specific to the 15 schools is authentic and valid and can be used for data analysis and calculation, as shown in Table 3.

4.3   Evaluation on Usability of the Self-help Laundry Service System in Colleges and Universities

In this research, it is assumed the overall usability quantitized value of self-help laundry service in colleges and universities y and the 12 usability evaluation indexes x meet the

Table 3. Average values of indexes

Colleges and universities	Index													
	x_1	x_2	x_3	x_4	x_5	x_6	x_7	x_8	x_9	x_{10}	x_{11}	x_{12}	y	α
G1	2.7	2.5	2.6	2.6	5.1	4.3	2.5	5.3	4.7	3.4	2.6	1.9	2.7	0.865
G2	3.9	3.1	2.3	4.1	2.5	4.6	2.5	4.2	4.3	5.1	2.6	4.0	2.9	0.805
G3	6.3	2.4	6.0	4.8	4.8	5.9	2.8	5.3	5.7	4.6	3.6	1.9	4.4	0.913
G4	3.5	2.5	5.1	2.7	5.2	5.9	2.2	6.2	5.7	5.0	2.7	1.7	2.8	0.938
G5	5.3	6.4	5.5	5.5	5.1	4.8	3.1	5.6	5.1	2.8	5.5	5.1	5.9	0.870
G6	5.1	2.6	5.5	4.4	5.2	4.9	4.2	4.1	4.1	5.5	3.0	2.9	4.1	0.814
G7	3.8	1.8	4.5	3.3	3.2	3.6	2.9	4.8	3.8	4.2	1.6	3.0	2.4	0.855
G8	4.5	2.1	1.9	3.0	3.3	5.4	2.5	5.9	4.6	3.7	3.2	1.6	2.9	0.876
G9	5.8	2.7	2.2	5.5	4.4	5.8	4.1	5.6	5.4	4.7	3.4	1.6	3.4	0.934
G10	5.2	3.1	5.7	3.9	4.5	4.2	3.1	3.4	4.6	2.8	3.8	3.2	3.8	0.798
G11	3.3	1.8	5.7	5.0	4.9	4.8	3.2	5.5	3.5	5.6	3.3	3.3	4.4	0.884
G12	5.8	6.5	5.2	5.3	5.9	4.6	3.1	5.9	4.9	3.0	6.4	5.5	6.2	0.878
G13	4.2	3.3	5.7	4.9	4.9	5.1	3.0	5.1	5.2	5.6	3.6	2.7	4.4	0.853
G14	4.2	1.6	2.4	3.7	3.9	3.8	2.7	4.1	3.9	4.5	1.9	3.0	2.4	0.930
G15	2.8	3.0	6.0	5.7	4.2	5.5	3.8	5.7	4.0	5.1	3.1	3.5	4.4	0.894

Note: G1–G15 is the name number of colleges and universities; α is the Cronbach's Alpha.

multiple linear regression relationship, then the value y and x will be in line with Formula 3.2.

By placing the self-help laundry service statistic date in colleges and universities of number ① ~ ⑭ to the above formula, the coefficient value of the indexes can be calculated by using the software SPSS and the least square method, results of which are shown Table 4:

Table 4. Usability index coefficients

Coefficient β					
Model	Nonstandardized coefficient		Standard coefficient	t	Sig.
	B	Standard, error	Trial version		
	5.554	3.068		1.810	.321
x_1	.268	.179	.235	1.494	.376
x_2	1.006	.475	1.270	2.120	.281
x_3	.209	.068	.277	3.073	.200
x_4	.611	.239	.513	2.559	.237
x_5	.278	.180	.215	1.547	.365
x_6	.256	.352	.155	.728	.599
x_7	−.917	.458	−.431	−2.001	.295
x_8	−.122	.161	−.083	−.755	.588

(*continued*)

Table 4. (*continued*)

Coefficient β					
Model	Nonstandardized coefficient		Standard coefficient	t	Sig.
	B	Standard, error	Trial version		
x_9	−1.456	.629	−.837	−2.315	.260
x_{10}	.106	.186	.089	.567	.671
x_{11}	.068	.359	.072	.190	.880
x_{12}	−.852	.462	−.866	−1.842	.317

a. Dependent variable: y

12 coefficients from index x_1 to index x_{12} are calculated. According to calculated regression coefficients, the usability evaluation model is obtained, and the standard regression coefficient formula is obtained:

$$y = 0.235x_1 + 1.270x_2 + 0.277x_3 + 0.513x_4 + 0.215x_5 + 0.155x_6$$
$$- 0.431x_7 - 0.083x8 - 0.837x_9 + 0.089x_{10} + 0.072x_{11} - 0.866x_{12} \tag{3.3}$$

By inputting data in the table to SPSS for analysis, it is obtained that the determination coefficient R square is 0.997 (as shown in Table 5), indicating the model can explain 99.7% data in the sample. It means the usability value calculated in Formula 3.3 and that evaluated by users are highly fitted, suggesting there's clear linear relationship between the single usability evaluation index value and the overall usability value (as shown in Fig. 2). Thus, the hypothesis is valid.

Table 5. Model accuracy test

Model	R	R square	Adjusting R square	Standard deviation
1	0.998[a]	0.997	0.979	0.17281

a. Predictive variable: (constant), x_{12}, x_8, x_7, x_1, x_3, x_{10}, x_5, x_9, x_6, x_4, x_{11}, x_2.

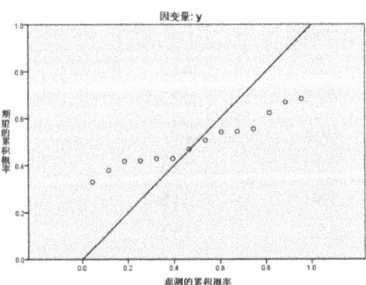

Fig. 2. Regression normalization residual PP diagram

4.4   Test of the Usability Evaluation Model of Self-help Student Laundry Service System in Colleges and Universities

The used data is placed to the model, and the user data that has not been used is used to check the accuracy of the usability evaluation model. The usability value calculated by the model and the value evaluated by users are listed in Table 6. Usability evaluation values of the two are highly similar, indicating e usability evaluation model constructed in this study is highly practical. As can be seen from the model accuracy test in Table 7, the R square value is 0.947, suggesting the fitness of the model is positive.

Table 6. Comparison of the two usability values

Sample	①	②	③	④	⑤	⑥	⑦	⑧	⑨	⑩	⑪	⑫	⑬	⑭	⑮
User score	2.70	2.90	4.40	2.80	5.90	4.10	2.40	2.90	3.40	3.80	4.40	6.20	4.40	2.40	4.40
Model calculation	2.68	2.87	4.38	3.11	5.99	3.73	2.19	3.09	3.49	3.09	4.45	6.40	4.58	2.35	4.11

Table 7. Model accuracy test

Model	R	R square	Adjusting R square	Standard deviation
1	0.973[a]	0.947	0.941	0.323

a. Predictive variable: (constant), V2.

5   Conclusions and Discussion

Based on the summarization of domestic and overseas product usability service quality evaluation theories and methods, the usability evaluation index system of self-help student laundry service system is constructed comprehensively from the perspective of service system quality in this paper. Through user interview and survey and discussion by expert teams, 12 system usability evaluation indexes are determined. Via questionnaires specific to students from 15 colleges in Shanghai, Nanjing and Changzhou, perception data of evaluation indexes is collected, and the multiple linear regression equation is used to build the usability evaluation model of self-help student laundry service system in colleges and universities. Based on the usability evaluation on the self-help laundry service system of three colleges and universities, the effectiveness of the usability evaluation model is verified, the result of which indicates the usability evaluation model of the self-help laundry service system constructed by this research is very effective and is able to be used to evaluate the existing self-help laundry system service in colleges and universities. As user characteristics of college student groups are very distinct and the difference in user characteristics is small, the need for self-help laundry service is with high polymerization degree. Therefore, usability study on other fields of product and service system can be carried out by subsequent researches by subdividing user groups, products and service places.

References

1. Nielsen, J.: Chapter 9–International user interfaces. Usability Eng. **C1**(1), 237–254 (1993)
2. Han, S.H., Yun, M.H., Kwahk, J., et al.: Usability of consumer electronic products. Int. J. Ind. Ergon. **28**(3–4), 143–151 (2001)
3. Yuxiang, Z., Yixin, X.: Study of usability on information architecture for academic library website:take fudan University library as an example. Libr. J. (1), 48–56 2009
4. Liezhong, G., Junkai, D., Zhe, W., et al.: Performance evaluation experiment of mobile phones usability. Ergonomics **12**(4), 8–10 (2006)
5. Shiwei, C., Yuanwu, S., Shouqian, S.: An approach to usability evaluation for mobile computing user interface based on eye-tracking. Chin. J. Electron. **37**(S1), 146–150 (2009)
6. Chongchong, M.: Research on the Availability of Subway Train Scheduling System. Beijing Jiaotong University (2012)
7. Xin, L., Jikun, L.: Possible opportunities: the concepts and practices of product service system design. Creativity Des.
8. Fuxiang, W.: Service Quality Evaluation and Management. People's Post and Telecommunications Press, pp. 76–89 (2005)
9. Raekallio, J., Gronroos, M., Turunen, S.: Treatment of menstrual problems and simultaneous sterilization of mentally disabled Women. In: Inter Noise, pp. 67–70 (1982)
10. Shuang, W., Guofu, Y., Zhongxiu, H.: Research on customer requirements' target system based on kano model. Packag. Eng. **27**(4), 209–210 (2006)
11. Xianfeng, G., Jincai, D., Zhizhe, H.: Composition of service quality and measurement methods. Indus. Technol. Econ. **26**(3), 111–113 (2007)

Agile Project Management: Better Deliveries to the End User in Software Projects with a Management Model by Scrum

Hugo Almeida[1(✉)] and Walter Correia[2(✉)]

[1] Centro de Estudos e Sistemas Avançados do Recife,
R. do Brum, 77, Recife, PE, Brazil
hugolnalmeida@gmail.com
[2] Universidade Federal de Pernambuco, Av. Prof. Moraes Rego,
1235, Recife, PE, Brazil
ergonomia@terra.com.br

Abstract. When managing software development projects, it is common to have some problems during the iterations of the project. This work presents an analysis and review of agile methodologies and some of the most well-known project management techniques, all these methodologies and techniques contextualized in the computational scope. Based on this information was possible to propose a new model more comprehensive that allows the mitigation of some recurring problems in the progress of software projects and give to final users good results. It was detailed a case study of an innovative project entitled "Calculadora CUG", where the combination of referenced techniques and concepts were introduced in the development process allowing more flow to some of the features of the project pointed out by the client (first final user) and to create, in a way, more efficient releases in involvement with the user.

Keywords: Project management · Agile methods · Software development
Scrum

1 Introduction

When the design is studied, there is a constant motivation to think initially about problems, problems in which it is possible to propose help to solve them, or at least to improve them. The work at the Institute SENAI of Innovation for Information and Communication Technologies (ISI-TICs), as a developer of innovation and technology, has allowed several experiences, including projects with deadlines, with high levels of complexity and with a series of difficulties inherent to them, thus justifying the study reported. One of the problems identified in the routine of the software development professional was the overworking with things already done before, that the development team ends up facing. Some measures were not enough to overcome this kind of problem, and while the most compelling solution proposal was being implemented and adapted to the project, other problems inherent to the developers were becoming more explicit.

© Springer International Publishing AG, part of Springer Nature 2019
T. Z. Ahram and C. Falcão (Eds.): AHFE 2018, AISC 794, pp. 293–305, 2019.
https://doi.org/10.1007/978-3-319-94947-5_29

The main purpose of this study is precisely to smooth out the problems that developers face during the product development cycle, generating better end-user deliverables.

Looking from the point of view of project managers, it is noted that some problems related to the function they play in the project are common. Since the part of the immersion in the technical part of the project until the effective management of the human resources, the manager has several challenges that must be overcome because their performance will affect, directly and indirectly, the performance of all the team and consequently the process of development of the managed project.

On the other side of the artifact development process, the client appears. Often, the customer is the end user of the product or service developed, but there are cases where the customer will continue what was delivered by the development team and will run the business, having customers themselves who will use their product/service. The need for more consistent and more anticipated products is one of the dilemmas faced by the project customer. Another very common issue that the customer experiences is the difficulty in following the steps, which generates consequences in the receipt of deliveries. When tracking is flawed throughout the process, the consequences are huge on the final delivery of the project and this situation creates a major trouble among all the stakeholders involved during the process.

From a practical point of view, the relevance of this work is related to the conclusions and discoveries of the case study carried out, as they demonstrate its usefulness to solve the day-to-day problems of the stakeholders involved in software development. From the theoretical point of view, the research project is justified by its ability to expand the intellectual horizons of project management and agile methodologies and how the good use of this knowledge interferes positively the user experience.

2   Theory Referencial

To substantiate the study well, it was necessary to conduct research in certain areas of knowledge. This is a case study of a project management model, therefore the present work makes use of the concepts and applications of the Project Management discipline.

Agile methodologies will also be approached conceptually, with a greater emphasis on Scrum, which served as the basis for the model to be presented.

2.1   Project Management

The set of tools that allows the executive sector to develop skills to deal with contingencies and with the new situations that the environment of continuous change imposes is called Project Management [1]. Project is a series of planned, executed and controlled actions that generate a unique result, involving a team and delivering or outcome possibly in the form of products or services [2]. The essence of project ideas is the opposite of repetitive routines [1].

Also called project administration, project management is considered the application of knowledge, skills and techniques in the formulation of related activities to achieve a set of predefined objectives. With the growth in demand for innovation and growing demand for a better competitive advantage, projects are very important in companies. The globalization of the market forces companies to meet local needs and make the same companies compete economically all over the world. In the same way, information technology and the Internet have generated a changes's revolution in today's companies. Projects are important because of their results: new products, new industries, new businesses and the improvement of existing products [3].

Nowadays, institutions are realizing that the use of project management has many advantages. Customers are increasingly demanding better products and services delivered more quickly. To follow the speed of the market, companies need greater efficiency in their production lines [4].

Faced with so many benefits from good project management, it is perceived that the efficiency of the management collaborates vehemently towards the successful conclusion of the project. Practices that can be included in project management tend to make project execution more assertive to estimated timeframes and costs, as well as satisfy customer.

2.2 Agile Methods

Software development needs to be analyzed as an unpredictable and complex process. Realizing that software is developed in different ways, with different teams, under various circumstances is a major shift from traditional thinking to software development. However, the most important thing is to recognize that the development process is an empirical process: it accepts unpredictability and has mechanisms of corrective action [5].

One characteristic of agile methodologies is that they are adaptive rather than predictive. Based on this concept, agile methodologies adapt to new factors during the development of the project, rather than obtaining prior analysis of all that may or may not happen during development. The prior analysis is always difficult and expensive, and it becomes a problem when it is necessary to make changes in the planning [5].

Agile methods are developed with the formulation of constant feedbacks, which allows the development team to quickly adapt to any changes in requirements. This type of change is often criticized in projects conducted using traditional methodologies that do not have the means to adapt quickly to change. Another positive point of the agile methodologies are the constant deliveries of operational parts of the software. With this, the client does not have to wait long to see parts of the software working and to conceive criticism that it was not quite what he expected of the product [6].

Agile methods share many features in common, although they have differences between their own practices. Iterative development, the focus on interactive communication and the reduction of effort employed in intermediary artifacts are some of these shared characteristics [7].

The application of agile methods in general can be examined from multiple perspectives. From a product perspective, agile methodologies are most appropriate when requirements are emerging and changing rapidly, although there is no complete consensus within this perspective [7].

2.3 Scrum

The term Scrum was first published in the article entitled "The New New Product Development Game" by Hirotaka Tekeuchi and Ikujiro Nonaka in the Harvard Business Review of 1986. Scrum was presented as an essential approach for companies looking to develop new products quickly and flexible.

Conceptualized as an agile framework, Scrum is still indicated for management of complex projects, although its structure is based on light and objective practices. Although widely used in software development projects, Scrum is a method related to several successful cases in the most varied areas of knowledge, helping to even write many books and scientific articles [8].

The practice of Scrum favors the organization of teams, provides greater transparency in the process and improves communication [9]. It maintains in its principles a solid consistency with the agile manifesto and guidelines for development activities. Within the development process itself, Scrum assists in the elaboration of requirements, in the analysis, design, evolution and delivery, and in each phase are delimited some tasks to be carried out within a standard period called Sprint [10].

Faced with a difficult scenario of prediction and future planning in which project management institutions are inserted, the Scrum framework imposes itself in traditional formats for managing software projects, representing a radical and modern approach, while providing the power of decision to the managers. Reducing defects, long-term maintenance costs and increased efficiency [11].

3 Study of Case of Specific Problem

SENAI (National Service of Industrial Learning) is a network of professional education famous in Brazil. Among the units that make up SENAI, the Institute SENAI of Innovation for Information and Communication Technologies (ISI-TICs), is the unit that aims to establish information and communication technology as the main competitiveness factor of the Brazilian Industry.

Many projects are in the portfolio of ISI-TICs, some already completed, others in the development phase. The case studied is part of this portfolio and presented several problems from the conception to the deliveries to the client. The challenges encountered by developers and management during most of the project execution, called the Calculadora CUG, were what motivated this research and generated benefits for the project itself implemented in the institute.

3.1 About CUG

The project itself aims to deliver a software that allows to provide accurate and agile construction cost estimates to support the feasibility study of real estate developments, providing benefits for all actors involved in the real estate development process or real estate base, from the stage of prospecting of the opportunity until the delivery of the work [12].

The tool adopts the Geometric Unit Cost Methodology (CUG), which gives its name to the implemented calculator, developed by the partners of CUG Consultoria (Proponent partner of this project) in a master's degree research at COPPE/UFRJ. Therefore, another objective that we can perceive in relation to the project is the insertion of the method in the market, concluding the innovation process triggered by the research that gave rise to it [12].

The most commonly used cost estimating practices condition the accuracy of the estimate to the volume of information about the project. CUG Modeling is a method that is innovative because it identifies the commitment of the final cost, delivering precision equivalent to the estimates based on all the projects developed. In the pilot model developed for the incorporator of one of the largest construction companies in the country, the maximum deviation was less than 4%, when the acceptable methodologies currently adopted in the preliminary study phase are up to 25% [12].

3.2 Technologies to CUG

The CUG was implemented with the use of some free access technologies available on the web. The definition of which technologies would be used in software development was based on the experience of the initial technical team. The CUG was developed largely in the Java language, which is known worldwide. Other languages have been used in the System such as JavaScript, SQL, HTML and CSS. The versioning system adopted was Git.

Jira software was used to manage the project. From this application it was possible to measure and illustrate the significant results of the research.

4 Solution Proposal: An Incremental Adapted Model

The proposed model is based on the agile Scrum methodology, its principles, techniques and recommendations. The template was forged based on the experience of its application in the CUG project. The idea was to increase in the model the improvements that were necessary. It is called the adapted model because it is a Scrum approach, and incremental because it receives improvements throughout the process. Scrum is very useful in project management but does not guarantee the team's hyperproductivity [13]. Due to this, by studying the CUG case, it was observed that some concepts and techniques aligned to management provide more support for the increase in technical productivity.

4.1 Techniques of Scrum

Based on the Scrum templates, some techniques were standardized to better collaborate with the progress of the project. The sprints were defined as one-week sizes and the post-it notes representing the activities were made according to Fig. 1, where #ID is the number issue identification registered in Jira, priority is a number that represents the priority of the issue (1-Lowest, 2-Low, 3-Medium, 4-High, 5-Highest), title of issue is self-explanatory, developer is the developer who developed the activity and tester is the

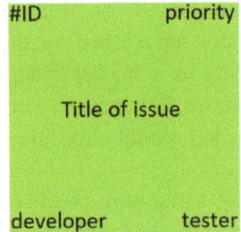

Fig. 1. Representation of a post-it template to the project

developer who tested the activity. The colors of the post-it notes varied according to their nature (User Stories: green, Tasks and Sub-Tasks: yellow, Bugs: red).

The Jira presents an activity flow signaling frame, called a Kanban, but it is important that in the team's work environment it has a version of the physical Kanban (Fig. 2) with the team committing to set up both frames. This facilitates meeting time, team motivation, and self-control of activities.

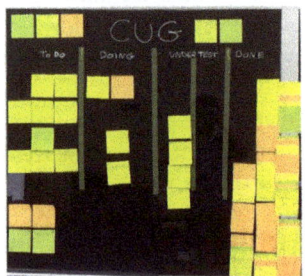

Fig. 2. Physical Kanban for the CUG Calculator

Other Scrum concepts should be preserved and adapted as much as possible, such as Daily Scrum, Product Backlog, Sprint Backlog, Sprint Planning Meeting, Sprint Retrospective, and Sprint Review Meeting.

4.2   Others Concepts

Some concepts have been studied and demonstrated to be effective, guaranteeing improvements in the working conditions of the developers and managers and consequently in the receptivity of the client with the deliveries. A series of 10 techniques and primordial approaches to the success of the case studied were selected and coupled to Scrum allowing a better management of the project:

1. Each developer has its tasks tested by another developer.
2. Record of the problems encountered during the tests.
3. Opening for pair programming.
4. Continuous integration.

5. Lean Thinking.
6. Feature-Driven Development (FDD).
7. Metrics Record.
8. Feedbacks.
9. Developer prioritization.
   Every developer should feel good about the activities they are running, the team needs to be motivated, and the ultimate success depends on the people who are engaged in it.
10. Long-term sprint planning.
    Although the proposal seeks to alleviate the problems reported in the introduction of this research, it is necessary to remember that several external factors strongly interfere in the project as the support of the management, communication and financial resources [14].

5 Sprint by Sprint

By initiating management with the proposed approach, the team quickly felt the effectiveness of testing practice. This then generated a change in Kanban where we could distinguish the phases Under Test, Testing and from the Testing phase pass the tasks to Done phase or return to To Do phase or Doing phase. The Jira's record of the problems encountered in the tests were also very well received by the team and collaborated with the task force idea that was sometimes performed to solve the bugs encountered during the project development. This task force in turn was strongly associated with peer programming that greatly aided in deliveries.

In the first few weeks the technical team found more points positives than negatives in the project's development, but then many negative feedbacks were reported throughout the project. Negative feedbacks appeared because of the change in scope that the technical team faced, coming from external factors, but which were very well circumvented. Many issues continued to be created in the project but the number of issues resolved was surpassed allowing management to see the progress of the project.

Since the sprints were weekly and there were weeks with only 4 or 3 business days due to holidays it was thought about changing the size of the sprints, but by keeping the sprints even weekly, the team was able to handle the activities, planning better for each week different and this did not disturb the extracted metrics, it maintained the schedules of meetings and of opening and closing of the sprints although with small differences.

Every day there was the daily meeting where participants were brief about what they did, what they were doing, what they were going to do, and whether they had any impediments in the development of activities. This brought the members closer and allowed some feedback to take place during the week, including personal issues. It was possible for one member to help the other on these issues and this positively influenced the progress of the project.

During the review, retrospective and sprint planning meetings, it was possible to analyze the metrics and then file them for further evaluation. During these meetings the developers were able to share the feedbacks that were annotated during the week and to devise solutions to feedbacks that were considered negative.

In the course of the sprints, positive feedbacks were noted as "the best sprint so far," and this motivated the team and the lead to try to beat the best sprint record the following week. The team was able to stay committed during the whole process, there were casualties in the team in some moments but in some moments were also added developers and this influenced directly in the planning of the sprints.

Over time the rework that the team always needed confront was diminished and until the end of the scrum management there was not more bug activities to be corrected.

6   General Results

In addition to the results found during the execution of sprints, interesting discoveries were made about the period in which the proposed model was applied. The Jira management system itself extracts detailed information about the various metrics being studied.

There are options for generating graphical illustrations by JIRA Reports session. In Figs. 3 and 4 are perceived two examples of Cumulative Flow Diagram that shows the status of issues over time. It is interesting to note that Fig. 3 represents the entire period in which the project was developed without the adoption of the adapted Scrum model, indicating the large increase of demand during the project that never seems to reach the complete resolution of the issues. In addition, it shows considerable ranges of issues that remain with the statuses marked Under Review.

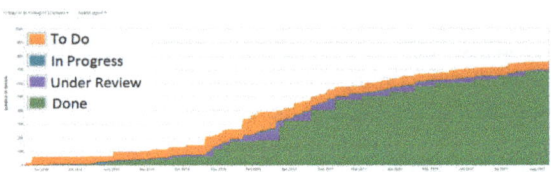

Fig. 3. Activity status progression from the first Sprint (May 2016) to the last day without using the Scrum-based management model (August 2017).

The Fig. 4 represents the entire period in which the project was developed with the adoption of the adapted Scrum model, indicating that there was an evident smoothness in the growth of demand during the project. It was also noted that issues were constantly revised since they did not retain the status of Under Review as in the past. Despite the explicit improvements, it is noticed that the project did not reach the totality of issues marked as Done.

When analyzing the control graph (Fig. 5) created by the Jira one can perceive a horizontal line that remains straight, it characterizes what the Jira calls the average time of rotation. When the other line (tortuous line of the image) is descending, it is a indicative of more efficiencies to the project, indicating improvements in the process.

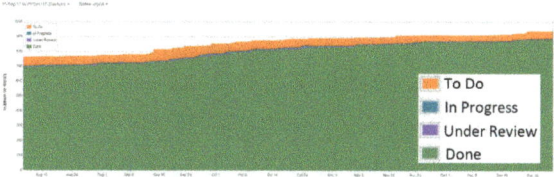

Fig. 4. Progression of activity statuses from the first Sprint using the Scrum-based management model (August 2017) to the last working day of 2017 (December 2017).

Points scattered around the image are clusters of issues and indicate the number of issues that have not been resolved at each sprint.

In Fig. 5, the moment in which the adapted Scrum model started to be used in the project is indicated by an arrow and a circle, indicating that after this there is an evident decrease in the average rotation line, indicating improvements in the process. In addition, the clusters became smaller, showing that a smaller set of activities remained unresolved by sprint with the adoption of the model.

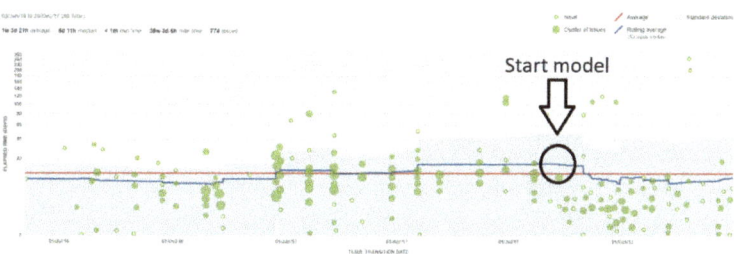

Fig. 5. Graphic illustration created by Jira that shows the development cycle indicating the generated clusters, the efficiency and predictability of the development.

It was still possible to generate a bar graph (Fig. 6), which clearly shows that the number of issues solved after the use of the adapted scrum incremental model increased.

The Fig. 6 depicts the results of exactly 5 months prior to using Scrum and the 5 months of Scrum usage. The number of issues did not stop growing as Fig. 4 demonstrated, but analyzing Fig. 7 it is evident that the number of issues solved overlapped the number of issues created after the management with Scrum.

In addition to the growth in the number of issues resolved, it is apparent from Fig. 8 that the number of development days in which issues were resolved and delivered increased. This explains how the management process severely interfered in the day-to-day of the project, reaping significant results daily. Before, the technical team took some days to complete a large group of tasks and spent many days with no completed tasks. This made it even easier the relationship to the customer who occasionally wanted to see something new.

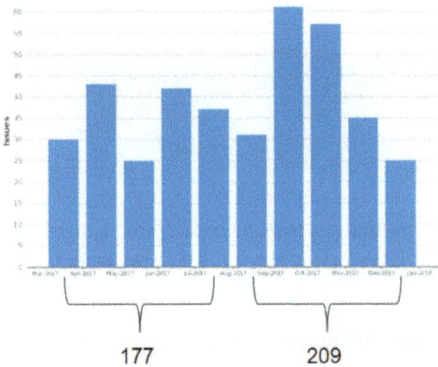

Fig. 6. Graphic illustration that shows the number of problems marked as solved per month daily in a time interval of 300 days during project development

Fig. 7. Number of issues created vs number of issues resolved daily in a time interval of 300 days during project development.

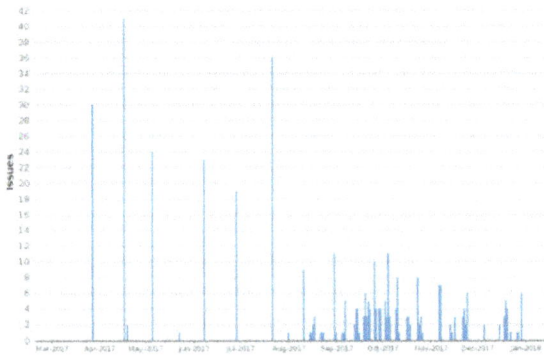

Fig. 8. Number of issues resolved daily in a time interval of 300 days during project development.

The more complex and time-consuming issues have been resolved with the implementation of the Scrum-adapted management model. The Fig. 9 shows that the activities took more days to complete, but in fact, shows that those activities that were underestimated for too long or that were maintained from sprint to sprint have finally been resolved. An example of this is that the largest bar of the graph in Fig. 9 indicates one of the first activities created in the project and that it was finally solved almost a year after its creation.

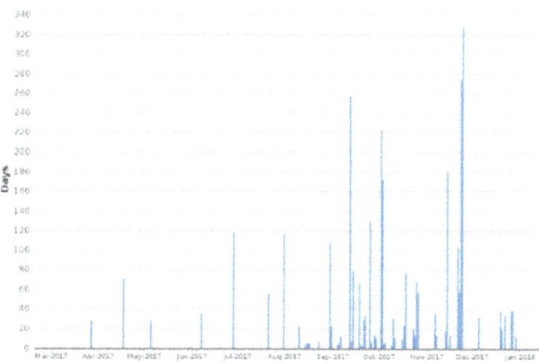

Fig. 9. Daily log of the average number of days the issues waited until they were resolved.

There were many results that greatly favored the view that management had of the project. It was possible to distinguish with numbers and evidence that the management with scrum brought innumerable advantages to the development of the project aiding even in its delivery to the client. With constant and consistent deliveries, the customer praised the team and returned to be more satisfied with the progress of the project.

The team of developers was able to get more motivated and better understand the requests of the project through this whole process, and the work of the developers was the biggest positive result that influenced the gains that the project management had and the final customer satisfaction, first end user of the System.

Usability Tests. In addition to the pilot model, the Calculadora CUG will be able to support plug-ins that have interaction with the system itself. A Sketchup model is an example that allows the extraction of geometric characteristics of the projects directly from the models drawn in the program [15]. The plugin, titled CUG BIM, is a Sketchup project that was started with the goal of interacting with the CUG if fed from its database.

The CUG BIM screens were developed in parallel to the CUG development and underwent usability tests with satisfactory results for ISI-TICs and for this research, since it was also managed through Scrum.

An innovation and technology developer and an industrial researcher participated in the usability test on a computer with support for running SketchUp, within a typical office environment. The participants interaction with the application was monitored by the facilitator sitting in the same office. A note taker and data logger monitored the sessions in the observation room.

The facilitator informed the participants that the intent of the tests was to evaluate the application, not evaluate the participant. The participants completed a pre-test questionnaire with 7 questions and an demographic questionnaire. The facilitator explained that the amount of time needed to complete the test task would be measured and that the exploratory behavior outside the task flow should not occur until the task is completed. The measure of time in the task began when the participant started the task.

The facilitator instructed the participant to "think out loud" so that there was a verbal record of their interaction with the application. The facilitator observed user behavior, recorded some comments and system actions on session records that contained task name information, scenario obtained, runtime, critical and non-critical errors encountered.

After each task, the note taker completed the task session document with the facilitator. After all task scenarios were attempted, the participant completed the post-test satisfaction questionnaire containing 23 questions. In general, the evaluations were positive in relation to the project and constructive criticism and praise was expressed by the participants.

Acknowledgments. To God for everything, my wife for the companionship and encouragement, to my family and friends for the cheering and the teachers of CESAR school for the knowledge and motivation.

References

1. Vargas, R.V.: Gerenciamento de Projetos com o MS Project 98, Estratégia, Planejamento e Controle, p. 302. Brasport, Rio de Janeiro (1998)
2. de Lima, R.M.: Avaliação do uso de Gestão Visual de Projetos no NTI-UFPE. Centro de Informática, Universidade Federal de Pernambuco, Recife (2017)
3. Shenhar, A.J., Dvir, D.: Reinventing Project Management: the Diamond Approach to Successful Growth and Innovation. Harvard Business School, Boston (2007)
4. Pmbok, G.: Um guia do conjunto de conhecimentos em gerenciamento de projetos. Project Management Institute (2004)
5. Libardi, P.L.O., Barbosa, V.: Métodos Ágeis. Campinas, São Paulo (2010)
6. Soares, M.S.: Comparação entre Metodologias Ágeis e Tradicionais para o Desenvolvimento de Software. Conselheiro Lafaiete (2004)
7. Cohen, D., Lindvall, M., Costa, P.: An introduction to agile methods. In: Advances in Computers, pp. 1–66. Elsevier Science, New York (2004)
8. Prikladnicki, R., Willi, R., Milani, F.: Métodos Ágeis Para o Desenvolvimento de Software. Bookman, Porto Alegre (2014)
9. Furuhjelm, J., Justice, J., Segertoft, J., Sutherland, J.J.: Owning the Sky with Agile: Building a Fighter Jet Faster, Cheaper. Better with Scrum, Global Scrum Gathering, San Diego, California (2017)
10. Pressman, R.S.: Engenharia de Software. 7. Pearson Makron Books, Porto Alegre (2011)
11. Ltda, B.L.M.: 40 + 16 Ferramentas e Técnicas de Gerenciamento. Brasport, Rio de Janeiro (2016)

12. Lima, F.S.A.: Custo Unitário Geométrico: Uma Proposta de Método de Estimativa de Custos na Fase Preliminar do Projeto de Edificações.bInstituto Alberto Luiz Coimbra de Pós-Graduação e Pesquisa em Engenharia, Universidade Federal do Rio de Janeiro, Rio de Janeiro (2013)

13. Downey, S., Sutherland, J.: Scrum Metrics for Hyperproductive Teams: How They Fly like Fighter Aircraft. Wailea, Maui, Hawaii (2013)

14. Ofori, D.F.: Project management practices and critical success factors – a developing country perspective. Int. J. Bus. Manag. (2013)

15. dos Santos, G.H.P.: Plataforma para estimativa de custo de empreendimentos imobiliários a partir do modelo de custo unitário geométrico. In 28th CONIC, Anais eletrônicos PIBIC: UFPE, Recife (2017)

The Contribution of Design in the Waiting Experience of Applicants to Parents in the Process of Adoption in Recife

Haidée Cristina Câmara Lima[1]([⊠]), Walquíria Castelo Branco Lins[1],
José Carlos Porto Arcoverde Jr.[1],
and Walter Franklin Marques Correia[2]

[1] CESAR School, Departamento de Design, Cais do Apolo, 77,
Recife, Pernambuco, Brazil
{haidee,walquiria.lins,mabuse}@cesar.school
[2] UFPE, Departamento de Design, Recife, Pernambuco, Brazil
ergonomia@terra.com.br

Abstract. The present study sought to contribute in improving the experience of the applicants in the adoption process, more specifically in the waiting period, between entering the National Register of Adoption (NRA) and the implementation of adoption itself, with the arrival of the expected child. This process has been investigated from the perspective of Design Anthropology which has the necessary means for a more in-depth research of such a complex problem.

Keywords: Family · Parenting · Adoption · Design · Anthropology
Design anthropology · Ethnography · User experience

1 Introduction

The engagement of this researcher in adoption process was the motivation for the present study. The very observation of the process and the own experience enabled the perception that there were aspects to be improved within the process, respecting the suitors' experience.

In the currently days, the focus of adoption is in the child. There are no searching children for suitors, but rather suitors for children sheltered in institutions and in need of a family. This shift in focus on adoption process in Brazil was very important and necessary. However, for this reason, the suitors have become somehow "abandoned", or so they feel like inside the process. Adoption is a system, where all parts are interconnected. Children, biological parents, suitors, and officials of Justice are part of a very complex process full of peculiarities.

Design, for its interdisciplinary characteristics, is able to build bridges, and dialogue with other fields of knowledge [1], like anthropology, which is what will be shown in this study. According to Cardoso [1], the complexity of the world needs answers generated from a teamwork that, precisely because of its interdisciplinarity, can find solutions more adequate to the current problems. For Megido [2] the design

© Springer International Publishing AG, part of Springer Nature 2019
T. Z. Ahram and C. Falcão (Eds.): AHFE 2018, AISC 794, pp. 306–317, 2019.
https://doi.org/10.1007/978-3-319-94947-5_30

must be the reflection of projects thought by people who seek to improve the lives of other people. And this is what the present study seeks to achieve.

2 Design Anthropology

Design Anthropology is the exercise of an interdisciplinary experimentation that comes from the dialogue between Design and Anthropology [3]. It's a new way of doing design and anthropology, and, not a submission of one discipline to another. From anthropology viewpoint, it is a way of being present in the world in a more inquisitive way, towards a transformation of the world, taking observation as starting point. For Ingold [4], quoted by Anastassakis [3], anthropology through design must be operationalized through design processes in experimental and improvisational ways. Hunt [5] comments that designers push the anthropologist into a more speculative mode of inquiry.

Designers materialize their research [6], although unlike the anthropologist, design research is not a quest for absolute truth, but rather for the insight that leads to possible futures [5]. Observation of the world is natural and fundamental so later these observations are materialized in artifacts, services, products etc.

Gunn [7], quoted by Anastassakis [3], considers that the objective of Design Anthropology is to reveal the "synergy" existing between disciplines, a synergy that is reflected not only in the products resulting from this partnership but also in the modes of knowledge production within a "context of ongoing interdisciplinary dialogue" [8]. The author comments that for Ingold, design, art, architecture, and anthropology observe, describe and propose, showing that all these disciplines work with forms of "exploratory engagement with our environment" [8].

Ventura and Bichard [9] also advocate the partnership between design and anthropology. This union makes the two disciplines stronger. Anthropology can take a more flexible and cooperative approach and design can gain a broader and deeper view of the world. Anthropology can contribute to the focus of observation on how people interact with each other and how they influence the environment through social relations and objects.

Otto and Smith [10] list three important points of the contribution of anthropology to design that deserves to be highlighted:

1. The theoretical and cultural interpretation role: design, unlike anthropology, has no tradition in the theorizing of contexts of use and interpretation of the cultural meaning of things. As you combine the two disciplines, contextualization and interpretation become part of the design tasks.
2. Research the past to understand the present and try to anticipate the future: expanding the temporal scope of research both forward and backward becomes a challenge of this new discipline.
3. Ethnography as a way to further enhance the human context in design.

One of the advantages of having diverse perspectives is that they can help to see situations through multiple glances and thus challenge conventional interpretations, revealing new possibilities that might otherwise be rejected [6]. Participants in interdisciplinary teams are usually encouraged to use their observation skills to discover problems collectively. For this, it is important to preserve flexibility and allow intuition as part of the research process, opening space and time for each one's personal look [6]. For Anastassakis [8] Design Anthropology is a proposal that emphasizes the practice of interdisciplinary as a diverse way of doing anthropology, which, working together with other disciplines, seeks to find shared spaces of understanding through an interdisciplinary practice that transcends the limits of the specific modes of work.

2.1   Ethnography Applied to Design

For Bichard and Gheerawo [11], the greatest asset of ethnography is its contextual nature, which can be applied anywhere, in any situation, involving anyone, and this aspect may be particularly interesting for designers, reinventing the research or finding new ways of applying it. One example is IDEO, who created an "ethnographic deck" to be used in research by design teams.

Anastassakis [3] comments that the concept created by Marcus [12] on the renewal of practice and research in anthropology, which he calls "design studio", argues that ethnography becomes a design practice so that it to become a "way to develop alternative ideas about research methods in a more shared and critical way."

2.2   What's the Contribution of Design Anthropology for This Study?

For Ventura and Bichard [9] Design Anthropology functions as a mediator between the parties involved within a design process. Anthropological designers should define their role as sociocultural mediators between the client and the user universe.

In the case of this study, it was concluded that the experience within an Adoption process required a deeper research to understand all the important elements that can contribute to experience of the applicants, such as waiting time, professionals involved, expectation of the process, etc. Ethnography, was the main element of the study, so that Design, as a catalyst item of changes, allowed the project vision from the research findings, so that suggestions for improvement could be found.

3   Methodological Procedures

The methodology was based on the ethnographic process, which included the following phases: selecting a problem, collecting cultural data, analyzing cultural data and make conclusions.

3.1   Select a Problem

The researcher realized that there was room for improvement in the experience of the suitors within the process, especially at the time of waiting, between entering the National Registry of Adoption (NRA) and adoption itself. This perception came both from a personal perspective, as well as from the observation of other suitors during moments as the course taught by professionals from the Children's and Youth Court of Recife, or the meetings organized by GEAD (adoption support group).

3.2 Collect Cultural Data

To make the best analises, the researcher decide to triangulate the data, through three different sources: specialized bibliography, semi-structured interview and digital ethnography in Whatsapp group. For the purpose of this article, we will detail only the interviews and digital ethnography.

3.3 Semi-structured Interview

For this research a semi-structured questionnaire was build with thirteen questions that were divided into the following subjects:

Motivation for Adoption. This theme refers to the issues on the motivation for adoption and its implications in the process and waiting period within the National Adoption Register, for each applicant.

Relationship with the Adoption Process. This topic deals with the relationship that suitors have with the process. We also sought to know how long they have been registered and how they are feeling or felt during the waiting period until the adoption.

Relationship with the Professionals of the Children's and Youth Court of Recife. This topic deals with the relationship that the suitors have with the professionals of the Children's and Youth Court of Recife with respect to the period of waiting. The researcher wants to perceive how the relationship between the suitors and the professionals influences the state of mind of the future parents in this period.

Relationship with Adoption Support Groups. The role of GEAD is fundamental throughout the adoption process, and with this theme, we try to find out if the suitors make use of this support or not and what it represents in terms of experience in the waiting period.

Research Participants. Nine people, six women, and three men participated in the interviews. Two of the respondents recently adopted, while others are waiting their turn to adopt. The time period of the applicants within the NRA varies from two months to six years. The interviews were conducted both in person and in the digital environment, for the convenience of the interviewees. The names presented in this study are all fictitious to preserve participants' privacy.

Digital Ethnography in Whatsapp Group. There is a WhatsApp group created by one of the Recife support groups for the exchange of information and knowledge about adoption. The group existed since 2014, but the research period is restricted to the months of December 2016 through February 2017.

4 Cultural Data Analysis

This section presents the results from ethnography, both from Whatsapp group observation and interviews.

4.1 Interviews

After the conclusion of the transcripts, a pre-analysis was conducted following the bibliographical orientation that consists of reviewing the records of the interviewees' speeches to search for cultural symbols and search for relationships between these symbols, and thus creates categories. For example, speeches dealing with motivation for adoption and infertility were grouped in the category "Desire", as both are related with the desire for a child. And so on. The next step was the treatment of the results and their interpretation. In this way, the collected data reflect the current situation of the applicants in this moment of waiting for the adopted child.

4.2 Categories

The following categories were found from the coding in the interviewees' speech:

Desire. Included the contents on the motivation for adoption, if dealing with infertility most of the time and the speech about the desired profile of the child.

Expectation. Were contemplated all statements related to the preparation for adoption, the perception of the process and waiting time, including fears and anxieties.

Frustration. The speeches about the difficulties, criticisms, and frustrations regarding the process and the professionals involved.

Hope. Included in this category are all suggestions for process improvement raised by the applicants and also the relationship with GAAs (adoption support groups).

4.3 Whatsapp Group

The observation period lasted three months, between December 2016 and February 2017, and the analysis of the results of the interviews was done using the method of ethnographic analysis previously with adaptations referring to the digital environment.

In addition to subjective observations made by the researcher during the observation period, the group's text file was used for coding the same.

4.4 Categories

From the coding the following categories were found in the interviewees' speech:

Expectations. All the statements related to the expectation of the process, such as anxiety, doubts, celebrations of achieved phases, satisfaction within the process and frustrations were contemplated here.

Hope. Included in this category are all statements regarding clarifications about the process, exchanges of experience and observations about the support group, as well as joint statements. Also discussed here are the experiences of parenting experiences, which, although not part of the waiting process, are reflected in the spirit of the applicants.

5 Conclusions

To understand the conclusions, it's important to point out that according to Strauss and Corbin [13] three aspects were part of the analysis:

(a) The data themselves, be they report on actual facts and actions, memories, texts, observations, videos, etc.
(b) The interpretations of the observers and the actors of these facts, objects, events, and actions.
(c) The interaction between the data and the researcher when collecting and analyzing this data.

5.1 Desire

As explained previously, the contents of the motivation for adoption are included here, being infertility most of the time, as well as the statements about the child's desired profile.

Adoption Motivation. The results of the qualitative research analysis corroborate with the research done by Weber [14] and Paiva [15] that showed that the main motivation for adoption is infertility. However, it was noticed that the choice for adoption was already part of the imagery of some suitors.

Child Profile. No specific question was asked about the profile of the child desired by the applicants, but this information came up spontaneously when asked about the waiting time. Sign that the two information are associated in the imaginary of the future parents.

It is possible to perceive that there is a reflection on the part of the suitors in relation to this information because, in almost all the speeches of the interviewees, there were changes regarding the initial profile in order to broaden the options. Both in age and in physical characteristics. Some of the changes occurred after attending GEAD meetings. Most suitors interviewed made an option for infants and children up to six years of age. However, even with the changes of openness to a greater age, almost all the interviewees started the profile with the preferences for babies.

This observation reflects the results of other researchers, such as that of Costa and Campos [16], in which the search of the pretenders for babies was still predominant. Hamad [17] says that the search for children at an early age can be due to the need for parents to shape the child in their image. A baby with no previous history and still no personality would be more conducive to this manipulation of parents. In addition, there is fear of children's previous history as Weber [18] comments. The greatest difficulty in adopting an older child is to face their history that precedes adoption and which is often constituted of rejection, pain, and loneliness.

By infertility being the main motivation for adoption, it also brings an expectation of parenting that fills this loss in some way. According to Schettini [19], most people base their family representations in the consanguineous affiliation. And as Levinzon [20] comments, there are conscious and unconscious feelings about the difference between the imagined child and the real child, and during adoption, a progressive accommodation is taking place regarding the reality that presents itself and the which had been previously imagined.

5.2 Expectation

Preparation for adoption, a perception of the process and waiting time, including fears and anxiety, were included here. And also the Whatsapp group talks about expectation with process, anxiety, doubts, celebrations of achieved phases, satisfaction within the process and frustrations.

Waiting Time. There is a difference in the expectation of the process, depending on the waiting time of each applicant within the National Register of Adoption. There are less anguish and suffering among those who have been recently enrolled or just entered.

In the same way, in the group, the applicants who just entered the NRA share their emotions by saying how they feel blessed and how they are calm because the child will arrive on time. It is easy to understand, after all, that the adopters who have just entered feel that they have passed through an important phase of the process and have been approved, so the spirit is of commemoration. They know that adoption takes time to happen so there is no point in raising the expectation that the child will arrive soon. In their speech, it's possible to see the belief in a greater force that will make everything happen in the right way. They have hope in the process and wait for their turn to be fathers and mothers. Speeches are common in which God is responsible for determining the time of arrival of the child.

Adoption Perception. In the suitors' speeches, there were moments in which they relate to what they think would be the relationship with this adopted son or daughter. There is an expectation that a similarity with the adopted parents will occur, even if it is not physical. Perhaps, the existence of a similarity as a need to confirm a bond. It has also been realized that there is an idealization of how the bond between the children and the parents happen. But biological parents also idealize the relations with the children when they are still in the belly. According to Gomes and Levy [21], parenting is the exercise of parents, both adoptive and biological, in which the child fits into a chain of desires, expectations, and fantasies.

Anxiety. As time goes by, anxiety begins to be part of the suitors' waiting process. Weber [18] comments that this waiting period generates anxiety because the suitors "do not have much to celebrate yet nor do they have many positive signs that they will really be the parents of a child." Unlike the biological pregnancy, the adopter awaits this child without signs of its physical presence and without the security of its arriving.

It is possible to notice in most of the lines that there is an understanding that the delay is part of the process. The suitors seem to understand that the profile they chose for the child does not allow an immediate adoption, since the majority of the children available for adoption in Recife are over the age of eight, and therefore not the profile of the majority of the interviewees. However, there is great dissatisfaction with the lack of transparency in the process.

5.3   Frustration

As explained earlier this category deals the difficulties, criticisms, and frustrations regarding the process and the professionals involved.

Invisibility of the Process. In, Brazil, during the waiting period until the arrival of the child, there is no visibility of what is happening in the process. The suitors can have access to their process, but they have to face the bureaucracy of justice and yet the available information does not diminishes their dissatisfaction since it does not bring any new information. Nothing related to the position of the applicant in the adoption queue. In a lecture about changes in the NRA that this researcher participated, the professionals of adoption made reference to a chronological order of entry in the register, so that this order would be responsible for the waiting time of each suitor. This queue is the only reference that the suitors have about the progress of their process that leads to the realization of the dream of parenthood. In Whatsapp group, questions about the participants' profiles are common, especially those who have received the long-awaited phone call or those who have just adopted. There is a need to make a comparison between profile and waiting time so that they can somehow perceive whether they are far away or close to their turn. This behavior shows that there is a great need to know what is happening, whether the "queue" is decreasing or not. This anxiety is understandable, after all the adopters are not waiting for the arrival of an object. It is a son, a cherished dream, perhaps, for a long time before the entry of the papers in the beginning of the process of registration in the NRA.

In addition, the applicant's registration is not available so that he can access without dependence on a professional of the court.

5.4   Hope

In this category are included all the suggestions of improvement of the process made by the applicants during the interviews and also the relationship with adoption support groups. The Whatsapp group includes all the statements regarding the clarification of the process, exchanges of experience and observations about the support group, as well as statements of solidarity. Also discussed here are the parenting experiences, which, although not part of the waiting process, have a positive impact on the spirit of the applicants. The relationship with adoption support groups is in the Hope category because the participation in these groups is one of the elements that enable the improvement in the waiting experience. In all the speeches of the interviewees who participate in the group meetings, both physical and virtual, they report that they are moments of renewing hope in the arrival of the expected child. As the meetings held by the groups deal with various issues related to adoption that are in the suitors' interest, such as fear of disclosure, distress generated by waiting time, revolt with brazilian justice, fear of losing the child to the original family [22], the participants of the groups have a space to share their feelings and thoughts and to ask questions about subjects that are directly related to them. The meetings also allow a maturation of the desire for adoption, the profile of the child and the emotional and mental preparation for the arrival of the child.

Suggestions for Process Improvement. From a specific question that concerned changes in the process, there came suggestions for improvement of the same. Most of the applicants' suggestions relate to the transparency of the information related to the process itself or the processes in general.

Another suggestion is the streamlining of the process, not only on the adopter's side but also on the child's side. Today the law obliges the professionals to exhaust all possibilities of replacement of the child back in its family of origin. This means that if the parents cannot stay with the child, it must be placed with the relatives. Then the justice needs to know if these relatives want to stay with this child. In this process of searching and evaluating, the time goes by and the child grows in the shelter losing the opportunity of a quicker adoption. And in order for processes to run faster, suitors suggest an increase in staff numbers. It means that they know that the work is too much for a small group to perform all the tasks related to the adoption process as a whole.

It is possible to notice that the applicants do not look for significant changes in the process. The main concern is to improve what already exists: giving visibility and increasing speed.

In this way, it is possible to realize that, although the adoption process has undergone several modifications over the years in order to organize and streamline legal procedures, the participants' emotional issues may have been neglected. The biggest annoyance of the suitors is not so much the delay of the process but the invisibility of the same.

6 Final Conclusions

The present study aimed to contribute to the experience of the applicants to parents during the waiting period in the Adoption process. From the triangulation of analysis of specialized literature, interviews and observations of the Whatsapp group, what has been learned is that, in fact, this is a delicate and anxious moment for suitors and their experience depends on several factors such as: What motivated the Adoption; The state of mind with which the suitor arrives at this moment; The time of enrollment in the NRA; Participation or not in support groups for Adoption; The relationship established with the psychosocial team during the evaluation period; And the experience with the process itself.

The reason for choosing adoption as an option for parenting brings a series of associated data, such as infertility, prejudice, attempts at biological pregnancy, loneliness, frustrations, etc. These emotional data and how the suitors work them internally influence the state of mind with which they arrive at the adoption process. The mourning for infertility needs to be worked out for adoption to succeed; otherwise, the real child will always live in the shadow of the idealized child.

Anxiety is present in almost all of the respondents' speeches and appears several times in the Whatsapp group conversations. From the research, it was possible to perceive that the degree of anxiety of the applicants in the waiting period varies, mainly, depending on the time of entry in the National Register of Adoption. People who have been in the registry for the longest time are more anxious than those who have recently entered. It is interesting to note that the time-out value for the onset of this anxiety is one year. In some of the statements it was commented, "up to a year, everything was quiet". This means that the applicants are aware that they will not find the child immediately after entering the register. They know there will be a waiting period, and perhaps even with the understanding that it will take more than a year, and as most of the time this desire for parenting started well before the Adoption initiative, a year in the register is already yours limit itself. As Maldonado [23] reminds us, a significant portion of the suitors goes through a long wait for the biological son who did not come. According to the author [23], "the time separating the abandonment of waiting for the 'child of the belly' and the decision to adopt a child can be short, long or very long."

Another important element in the waiting experience was participation in adoption support groups. Not all interviewees participate in groups, but all those who participate are grateful for their existence and feel welcomed in this space of knowledge exchange. Support groups are a privileged space when it comes to adoption. It is a place where you can hear speeches from experts on the subject, as well as adoptive parents telling their experiences so that the suitors can mature and strengthen the decision for adoption.

However, even with the participation in the groups, what was perceived is that the applicants interviewed feel a lack of contact with their process, of having visibility of the same, and certain autonomy. Only the support group is not enough to calm the anguish of waiting for the realization of the desire to be a parent. The suitors need to see the "pregnancy" happening in some way. And this monitoring can happen in a number of ways. The simplest would be the visibility of the process itself in the NRA, allowing the independent access without the need of the professionals of the court. Only this action would reduce the anguish of not knowing if their registration is "alive" or not. Another suggestion for visibility issue would be to send, from time to time, an email with the situation of the adopter in the NRA. It is a simple solution, although this researcher believes that the best option would be to give access to the applicant so that he or she feels more secure about their information and with some autonomy.

Another relatively simple action would be to allow applicants to change their own email and phone information. This is another reason for the parents' anguish, because with the phone outdated if a suitor is sought in case there is a child with his profile, he will not be found. One of the ways to solve this problem could be to empower the adopters themselves to change their contact information. It is true that the number of people is small for the amount of work existing in adoption offices, so why not reduce this responsibility of professionals sharing it with the suitors? They are most interested in keeping their contact information current, so nothing fairer than giving them the right and responsibility for them. The suitors, in suggesting improvements, were not asking for any significant changes that involved shifting the process itself. On the contrary, it was realized that they understood the need for the bureaucracy involved, but asked for only greater transparency and visibility of their lawsuit. These are small modifications that can greatly improve the waiting experience for these people and thus contribute to the improvement of the process as a whole.

References

1. Cardoso, R.: Design para um mundo complexo. Ubu Editora LTDA-ME (2011)
2. Megido, V.F.: A Revolução do Design: Conexões para o Século XXI. Editora Gente Live Edit Ltd. (2017)
3. Anastassakis, Z.: Laboratório de Design e Antropologia: preâmbulos teóricos e práticos. Arcos Design, Rio de Janeiro 7(1), 178–193 (2013)
4. Ingold, T.: Knowing From the Inside: Anthropology, Art, Architecture and Design. Aberdeen University, Scotland (2013). http://www.abdn.ac.uk/anthropology/postgrad/art-architecture-design.php
5. Hunt, J.: Prototyping the social: temporality and speculative futures at the intersection of design and culture. In: Clarke, A. (ed.) Design Anthropology. Object Culture in the 21st Century. Springer, Wien (2010). https://doi.org/10.1007/978-3-7091-0234-3_3
6. Suri, J.F.: Poetic observation: what designers make of what they see. In: Clarke, A. (ed.) Design Anthropology. Object Culture in the 21st Century. Springer, Wien (2010). https://doi.org/10.1007/978-3-7091-0234-3_3
7. Gunn, W. (ed.): Fieldnotes and sketchbooks: challenging the boundaries between descriptions and processes of describing. Peter Lang GmbH, Frankfurt am Main (2009)
8. Anastassakis, Z.: Design e Antropologia: Considerações Teóricas e Experimentações Práticas em Diálogo com a Perspectiva do Antropólogo Tim Ingold (2014)
9. Ventura, J., Bichard, J.-A.: Design anthropology or anthropological design? towards "Social Design." Int. J. Des. Creat. Innov., 1–13 (2016)
10. Otto, T., Smith, R.C.: Design anthropology: a distinct style of knowing. In: Gunn, W., Otto, T., Smith, R.C. (eds.) Design Anthropology: Theory and Practice. Bloomsbury Academic (2013)
11. Bichard, J.-A., Gheerawo, R.: The designer as ethnographer: practical projects from industry. In: Clarke, A. (ed.) Design anthropology. Object Culture in the 21st Century. Springer, Wien (2010)
12. Marcus, G.E., Rabinow, P.: Designs for an Anthropology of the Contemporary. Duke University Press, Durham and London (2008)
13. Strauss, A., Corbin, J.: Pesquisa Qualitativa: Técnicas e Procedimentos para o Desenvolvimento de Teoria Fundamentada. Artmed, Porto Alegre (2008)
14. Weber, L.N.D.: Aspectos Psicológicos da Adoção. Juruá Editora, Curitiba (2003)
15. de Paiva, L.D.: Adoção: significados e possibilidades. Casa do Psicólogo, São Paulo (2008)
16. Costa, L.F., Campos, N.M.V.: A Avaliação Psicossocial no Contexto da Adoção: Vivências das Famílias Adotantes. In: Psicologia: Teoria e Pesquisa, Brasília, vol. 19, pp. 221–230 (2003)
17. Hamad, N.A.: A Criança Adotiva e Suas Famílias. Companhia de Freud, Rio de Janeiro (2002)
18. Weber, L.N.D.: Adote com Carinho. Juruá Editora, Um Manual sobre Aspectos Essenciais da Adoção. Curitiba (2011)

19. Schettini, S.S.M.: Filhos Por Adoção: Um Estudo Sobre O Seu Processo Educativo. Universidade Católica de Pernambuco, Dissertação de Mestrado. Recife (2007)
20. Levinzon, G.K.: Adoção. Casa do Psicólogo, São Paulo (2009)
21. Gomes, I.C., Levy, L.: O mal-estar e a complexidade da parentalidade contemporânea. Cadernos de Psicanálise, Sociedade de Psicanálise da Cidade do Rio de Janeiro. v. 25, n. 28 (2009)
22. Scorsolini-Comin, F., Amato, L.M., Santos, M.A.d.: Grupo de apoio para casais pretendentes à adoção: a espera compartilhada do futuro. In: Revista da SPAGESP - Sociedade de Psicoterapias Analíticas Grupais do Estado de São Paulo, vol. 7, no. 2, pp. 40–50. July-December (2006)
23. Maldonado, M.T.: Comunicação entre pais e filhos: A linguagem do sentir. Saraiva, São Paulo (1998)

Difference of Sensitiveness Toward Information Based on User-Role

Yeongchae Choi[(⊠)] and Weonseok Yang

Shibaura Institute of Technology, Tokyo, Japan
yeongchae.choi@gmail.com, yang@shibaura-it.ac.jp

Abstract. User-role in recent online service had been diversified into provider and receiver, and significance of interaction between users had been grown. This is an explanatory study to understand both user-roles' difference of emotional response toward information on website based on online outsourcing platform. By using Brunswik's Lens Model, it compares each user-roles' assessment about information elements, depicted the similarity and difference of emotional response to provide insights for information design of platform which have various user-role.

Keywords: Customer-to-Customer Interaction · Online outsourcing
Information architecture · Brunswik's Lens Model

1 Introduction

1.1 Background

Business of trading information, product and service had been immigrated to online and increased globally since the invention of internet. Early form of e-commerce that company sold or provided products to customer had been evolved into platform business that companies operate online marketplaces on web, buyer and seller use websites to make exchange. In addition, goods of the online marketplace diversified into used products, contents, crowd-funding, sharing economy, furthermore outsourcing with micro job trend. In other words, people can trade service and skills via internet, too.

Most remarkable change in this situation would be differentiation of user-role. On e-commerce platform, Individual users can upload their products and services on platform to make a deal. So to say, individual users are not only buyer, but also they can be seller. To make successful trade, both individual users demand more information to have credibility and safety about partner or goods. In the same time, along with flooding information distributed in online marketplaces and online services' anonymity, individual users feel risks of miscommunication, fraud [1] or revealing personal information on public web.

There were many studies to find what information is valuable for buyer (end-user) of online marketplace, but we could not find study that revealed attitude of individual buyer and seller toward information. Both users had different role in online exchange, so their emotion and reaction toward information they used to make deals could be

© Springer International Publishing AG, part of Springer Nature 2019
T. Z. Ahram and C. Falcão (Eds.): AHFE 2018, AISC 794, pp. 318–325, 2019.
https://doi.org/10.1007/978-3-319-94947-5_31

different. In these reason, service operator should consider both user-role's response about information used in trading when designing online marketplaces. This study was an explanatory research to specify difference of both user-roles' emotion and sensitivity toward information on service touchpoint.

1.2 Customer-to-Customer Interaction (CCI) and Differences Between Provider and Receiver

Lusch and Vargo [2] suggested customer's role in servicescape as co-producer along with the Service-Dominant Logic. It means customer contributes the value creation of service, that is, other customer around specific customer could be considered as servicescape, too [3]. In recent online service, the new role of customer was more obvious. Companies, which operated services, do only provide and maintain platform and information system. Users in such service create, provide, and consume contents or products, that is, information. As main creator's role was moved into individual users, significance of Customer-to-Customer Interaction (CCI) was getting bigger. Many scholars performed studies that covered those agenda, but most of those studies were focused on interaction within receiver role [4], or between receiver and potential receiver [5]. However, few studies covered interactions within individual users in different user-role.

Leonard [6] told the differences of risk-taking attitude and credibility toward trade partner between seller and buyer. On the other hand, Huang and Liu (2010) found both user-role were effected by certain elements (i.e. interaction experience, reputation), [7] So to say, there might be common element that both user-role could be influenced, and element that could cause different response. Since e-commerce users communicate each other in same, fixed touchpoint, knowing what information element could influence on or bring out emotional response from each user-role.

2 Research Methodology

2.1 Research Framework

To understand the difference of emotion that provider and receiver perceived from website of online service, we applied Brunswik's Lens Model (Fig. 1). This model was introduced by Brunwik [8], and developed by multiple scholars [9] as a psychological research framework. This model could reveal how people perceive, think, predict about observation target.

This lens model provided the perspective that individuals who see the object could analyze only the cues they actually see, not the object's fundamental character. That is, this model suggests the frame that fundamental characters of the object and analysis of individuals could be different. Because of intuitive structure and wide implication for various situation or object, Brunswik's Lens Model was used in multiple researches about differences among people's recognition. For example, it accounts for how people perceive someone's sexual orientation based on hip, shoulder and walking motion [10], crafting organizational culture from design executive officer's leadership [11].

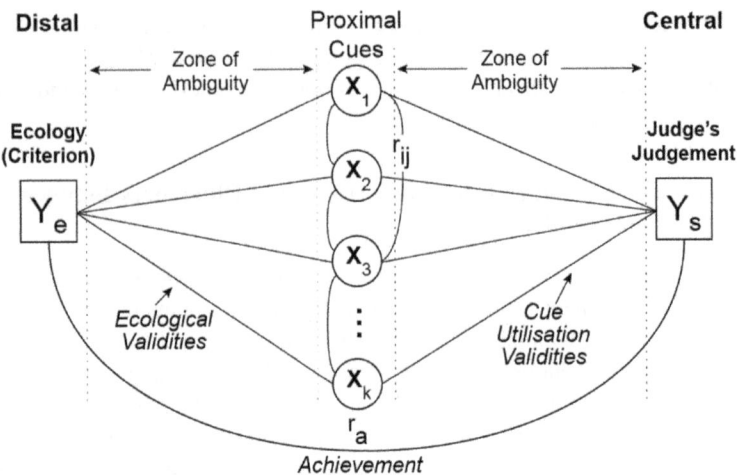

Fig. 1. Brunswik's Lens Model from Cooksey [9]

By implicating lens model into this study, provider (Y_e) and receiver (Y_s) communicate each other (Achievement) via website as a service touchpoint (Proximal Cues). This touchpoint consisted of series of information (X_k) such as platform, individual, goods, evaluation, and produces with intention of provider (Ecological Validity). On the other hand, this touchpoint is fixed after uploaded by provider, and does not give feedback about receiver's observation and analysis. Receiver embraces the cues on touchpoint along with its former experiences (Cue Utilization Validity), so there are some chances of difference between provider's intention and receiver's analysis. For instance, the sensitive information which provider does not want to reveal on web could be important clue for receiver to trust specific trade. We believed Brunswik's Lens Model could show us the importance, sensitivity of each cues on touchpoint, thus we would be able to provide insights of making more acceptable touchpoint for both provider and receiver.

2.2   Research Configuration

This study was to find differences of emotion by user-role about information in website. Online outsourcing platform was considered suitable to compare both user because main users of this service were individual; also, it contains both general information related to goods and sensitive personal information. Users in online service like shopping and reservation service were using external channel for searching reviews and recommendation. On the other hand, there was no remarkable external channel for both users in outsourcing service. In addition, the involvement of customer for outsourcing service (hereinafter referred as 'service') was high enough to see what information was more valuable than other information. Research process was consisted of three step (Fig. 2).

We gathered information elements from four online micro job marketplaces in Japan, and sorted into three categories such as platform, personal, work related, and evaluation based on purpose of information. This information were created by provider

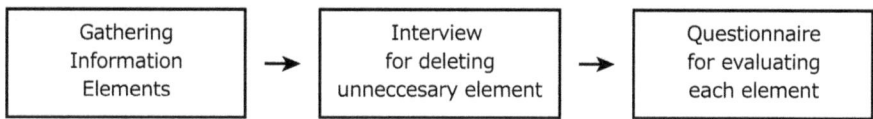

Fig. 2. Research configuration.

and publically distributed in web. Receiver's information was hidden and delivered via platform's system, so we focused on information on touchpoint. Details are provided below (Table 1).

Table 1. Information elements of provider in outsourcing service

Category	Element	Coconala	Crowdworks	Wowme	Bizseek
Platform related	Similar provider		O		
	Legal, Support	O			O
Personal	Age, Gender, Location		O		O
	Direct contact			O	O
	Profile, History		O	O	O
	Picture, character	O	O		O
Contract related	Price, Schedule	O	O	O	O
	Service contents	O	O	O	
	Output example	O		O	
	Portfolio		O		O
Evaluation related	Satisfaction, Bookmark	O	O	O	O
	Other user's review	O	O	O	

After gathering elements, each element was determined through interviews of how they use the service. Four interviews were performed with two designer (provider role) and two non-designer (receiver role) to know both user-roles' scenario. As a result, we decided not to use platform information in survey because both user-role did not show high attention to platform related information for making deal, even if platform information was important for platform operator to operate and defend legal issue. Based on these information elements, to understand what information was more critical and sensitive for each user-role, a questionnaire was performed.

24 people in their 20 s in Shibaura Institute of Tech participated in questionnaire, 12 people were design major student or designer as provider role, the other 12 people were engineering major student as receiver role. Questionnaire was made and delivered via Google Survey. We provided example and capture image of outsourcing service to participants, and then they chose whether they were designer or not (User-role). Participants read the user scenario given based on their user-role, after that, they answered series of questions. Questions were developed two by two matrix, sensitiveness toward information and effects on decision of provider and receiver. Participants received different questions based on their user-role (Fig. 3). Each section stood for information

categories, and participants evaluated every elements by 5-point Likert-scale, 1 for very low to 5 for very high. Data collection took 1 week.

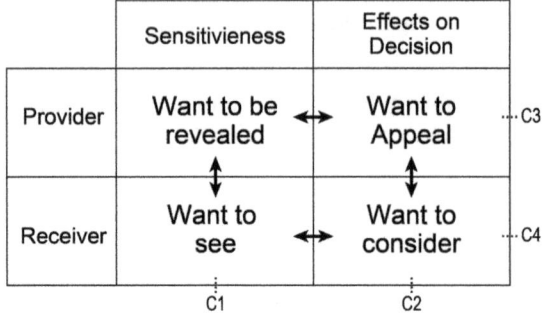

Fig. 3. Development of questions.

3   Result

3.1   Data Analysis

After collecting data, we compared each user-role's answers based on the matrix (Fig. 3).

The first comparison (C1, Fig. 4) was question 'Want to be revealed?' for provider, and 'Want to see?' for receiver. These questions were to find out and compare providers' sensitiveness toward revealing information and receivers' sensitiveness toward lack of information on public website.

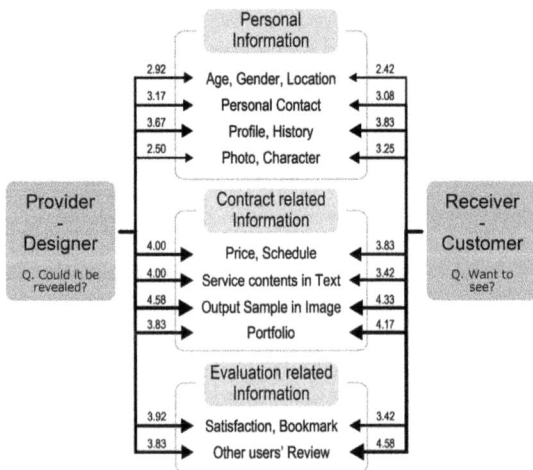

Fig. 4. Sensitiveness toward information elements.

We found out that the priority of each information categories were almost same for both user-role, but there were minor. Both user-role took seriously about contract and evaluation related information, and overall strength of elements were similar. Unlike 'Output Sample in Image' was most sensitive element for provider; 'Other Users' Review' was most sensitive element for receiver. In addition, provider did not want to show their facial photo, but receivers were tend to sense lack of credibility when there was no picture or logo of provider.

The second comparison (C2, Fig. 5) was question 'Want to appeal?' for provider, and 'Want to consider?' for receiver. This question was to find out and compare providers' prediction of what information would receivers consider and receivers' actual consideration while they make decision.

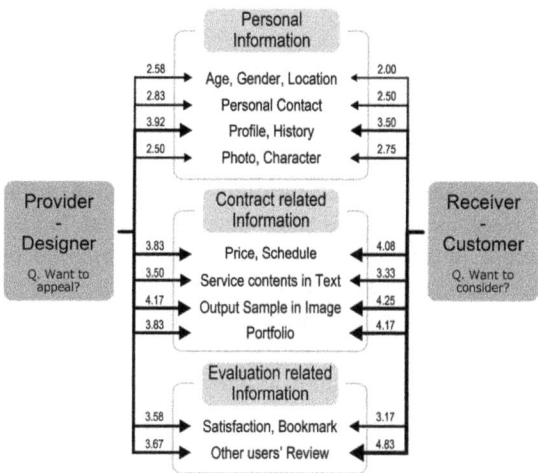

Fig. 5. Effects of information elements on decision.

In this comparison, it was notable that written information such as provider's profile, history, and service contents in text got lower assessment than graphical information. Receiver's actual consideration about personal information was lower than provider's expectation as opposed to contract related information was higher. In case of information elements, providers predicted 'Output Sample in Image' would make strongest effect on receivers' decision, but receivers assessed 'Other Users' Review' had highest effect on their decision.

Result of questions for each user-role to find out gap between sensitiveness and effects on decision showed similar shape (C3, C4). It can be seen as providers were appealing the information they wanted to reveal on web, and receivers were considering the information they wanted to see.

In summary, it is as follows.

- Both user-roles' priority of information category is similar, but priority of information elements was different.
- Receiver tend not to focus on written information.
- Receiver wanted to see provider's personal information, but it did not affect receiver while making decision.
- The biggest gap between both user-role was sensitiveness of facial photo of provider (personal information) and effects on decision of other user's review (evaluation related information).

4 Discussion

We collected emotional response from individual users in separated user-role, provider and receiver. By using Brunswik's Lens Model, we could have better understanding the difference of emotional response toward information on touchpoint between provider and receiver. Each comparison between both user-role provided chances to us.

This study was performed focused on online outsourcing platforms. These platforms were able to show us individual users' interaction well, but as they were newly developed servicescape, it was likely that users were influenced by the unexpected characteristics of outsourcing platforms only. That is, to understand general implication of individual user to user interaction may need additional case studies and factor analysis.

In the future, we will expand our study to other platform such as used goods trading and P2P banking, and widen the scope of provider role to corporate user. Through those additional studies, we expect that we could have insights for customer-to-customer interaction approach for web design and information architecture.

References

1. Yang, Z., Peterson, R.T., Cai, S.: Services quality dimensions of internet retailing: an exploratory analysis. J. Serv. Mark. **17**(7), 685–700 (2003)
2. Vargo, S.L., Lusch, R.F.: The Service-Dominant Logic of Marketing: Dialog, Debate, and Directions. Routledge, New York (2006)
3. Tombs, A., McColl-Kennedy, J.R.: Social-servicescape conceptual model. Mark. Theor. **3** (4), 447–475 (2003)
4. Nicholls, R.: Customer-to-customer interaction in the world of e-service. Serv. Manag. **3**, 97–104 (2008)
5. King, R.A., Racherla, P., Bush, V.D.: What we know and don't know about online word-of-mouth: a review and synthesis of the literature. J. Interact. Mark. **28**(3), 167–183 (2014)
6. Leonard, L.N.: Attitude influencers in C2C e-commerce: buying and selling. J. Comput. Inf. Syst. **52**(3), 11–17 (2012)
7. Huang, E., Liu, C.C.: A study on trust building and its derived value in C2C e-commerce. J. Glob. Bus. Manag. **6**(1), 1 (2010)

8. Brunswik, E.: Representative design and probabilistic theory in a functional psychology. Psychol. Rev. **62**, 193–217 (1955)
9. Cooksey, R.W.: Judgment Analysis: Theory, Methods, and Applications. Academic Press, San Diego (1996)
10. Johnson, K.L., Gill, S., Reichman, V., Tassinary, L.G.: Swagger, sway, and sexuality: judging sexual orientation from body motion and morphology. J. Pers. Soc. Psychol. **93**(3), 321 (2007)
11. Lee, Y., Joo, J.: How a design executive officer can craft an organizational culture. Des. Manag. J. **10**(1), 50–61 (2015)

Improvement Method for Business Operations Using User Experience Adaptive Information Sharing Terminals

Yuka Sugiyama[1](✉), Toshikazu Kato[2], and Takashi Sakamoto[3]

[1] Graduate School of Chuo University, Tokyo, Japan
y5g4@g.chuo-u.ac.jp
[2] Chuo University, Tokyo, Japan
t-kato@kc.chuo-u.ac.jp
[3] National Institute of Advanced Industrial Science and Technology,
Tokyo, Japan
takashi-sakamoto@aist.go.jp

Abstract. Cost reduction and operational efficiency improvement are important issues for corporate management. Hence, in this work, we focus on the time span of a momentary User Experience (UX). To our knowledge, there is no case study on incorporating the objective indicators based on individual experience and on improving the quality of teamwork. We propose a method to improve work by customizing the information presented to the users participating in a teamwork based on the experience and satisfaction of each user. As a method of evaluation, we use the user experience questionnaire. In addition, we perform a customer satisfaction analysis. We can extract the evaluation terms that largely improve the ease of viewing the screen and operability using this analysis. Moreover, from the evaluation terms, we can extract specific items that need to be improved preferentially as knowledge for the next design. Therefore, it is possible to improve both the work and user experience when sharing information.

Keywords: User centered design · Interface design

1 Introduction

Cost reduction and operational efficiency improvement are important issues for corporate management. However, if these are excessive, the workers will be exhausted and work will become stagnant. For example, a system with poor operability and discomfort would degrade the efficiency of the operations. Although the "workability" of the workers and managers working at the site should be incorporated into the business improvement plan as objective indicators, such methods have not been established yet.

Therefore, we focused on UX. UX is utilized in fields such as service products, interactive products, software, and business systems [1].

Business system development is progressing without reviewing a human-centered UX design perspective. Moreover, the process that a designer becomes involved in the late stage of development is common [2]. Thus, UX is not taken into consideration even in the structure construction and basic operation despite the expected natural and

© Springer International Publishing AG, part of Springer Nature 2019
T. Z. Ahram and C. Falcão (Eds.): AHFE 2018, AISC 794, pp. 326–330, 2019.
https://doi.org/10.1007/978-3-319-94947-5_32

intuitive movements for the user. For achieving easy to use experiences, it is necessary to reflect the perspective of the user in the screen design stage onward, such as for designing the screen transition and display elements in the screen.

UX can be categorized into "practical quality" including functionalities that can achieve the purpose, and "hedonic quality" that leads to stimulus and sensation [3]. It is reported that UX can, thus, refer to a specific variation in the experience during an interaction (momentary UX), an appraisal of a specific usage episode (episodic UX), or a view of the system as a whole after having used it for a period (cumulative UX). Anticipated UX may be related to the period before the first use or to any of the three other time spans of the UX [4]. In this research, we focus on momentary UX and measure the hedonic and emotional qualities of the users in a business system. We aim to construct a screen design that reduces the stress and weight experienced when using this business system and increases the user experience (Figs. 1 and 2).

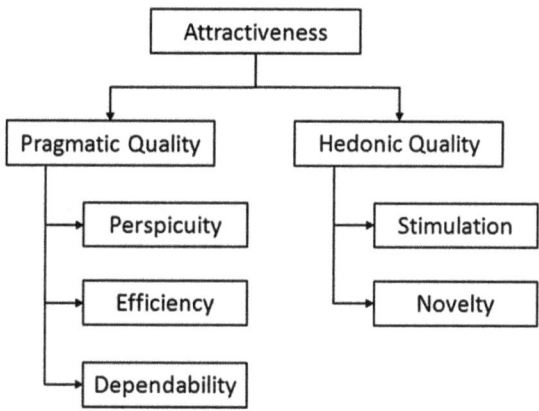

Fig. 1. Scale structure of the UEQ questionnaire [1].

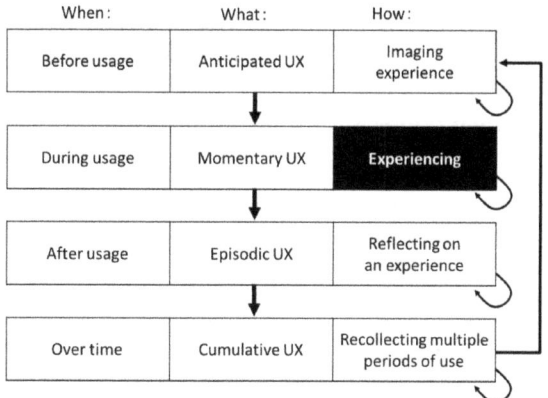

Fig. 2. Time spans of the user experience, adapted from [4].

2 Purpose of This Study

To our knowledge, there is no other case study on incorporating objective indicators based on individual experience (UX) and on improving the quality of teamwork (ease of working).

In this research, we propose a method to improve work by customizing the information presented to the users participating in a teamwork according to the experience and satisfaction of each user.

3 Experimental Method

Based on a human-centered design, hypotheses and verifications are performed from the point of view of a user in the following five stages [5]:

   ① Definition of user model
     The environment in which a user model shares and presents information in business is defined. In addition to the literature survey, interviews and observations are made, and the current business information of the user is collected.
   ② Definition of purpose and issue
     The issues identified from field the surveys and interviews clarified, and patterns of action scenarios are developed.
   ③ Preliminary survey
     A user is asked to use some screen samples, perform tasks, and fill a user experience questionnaire (UEQ).
   ④ Screen design · Prototype development
     A prototype is developed based on the user model, purpose, issue, and behavior scenario.
   ⑤ User evaluation
     For the hypothesis verification, a subject is asked to conduct a user test to evaluate the subjects in an environment similar to the actual user environment. The user is asked to use two tasks, an existing information sharing screen, and a screen improved for the task obtained in the interview.

3.1 Evaluation Method (Subjective)

As a method of evaluation, we use the international questionnaire, UEQ, available in several languages. The UEQ contains six scales, namely, attractiveness, efficiency, perspicuity, dependability, stimulation, and novelty, with 26 items in total. By using the UEQ, it is possible to evaluate the UX related to the information presentation. Moreover, it is possible to express the feelings, impressions, and attitudes of the user generated when information sharing is performed (Table 1).

Table 1. Japanese version of the UEQ [1].

	1		5	
annoying	O	...	O	enjoyable
not understandable	O	...	O	understandable
creative	O	...	O	dull
easy to learn	O	...	O	difficult to learn
valuable	O	...	O	inferior
boring	O	...	O	exciting
not interesting	O	...	O	interesting
unpredictable	O	...	O	predictable
fast	O	...	O	slow
inventive	O	...	O	conventional
obstructive	O	...	O	supportive
good	O	...	O	bad
complicated	O	...	O	easy
unlikable	O	...	O	pleasing
usual	O	...	O	leading edge
unpleasant	O	...	O	pleasant
secure	O	...	O	not secure
motivating	O	...	O	demotivating
meets expectations	O	...	O	does not meet expectations
inefficient	O	...	O	efficient
clear	O	...	O	confusing
impractical	O	...	O	practical
organized	O	...	O	cluttered
attractive	O	...	O	unattractive
friendly	O	...	O	unfriendly
conservative	O	...	O	innovative

3.2 Evaluation Method (Objective)

In addition to the questionnaire, to measure the operation and efficiency of the screen, the line of sight of the user performing the task is measured. It is possible to measure the task completion time by gaze measurement. Moreover, it is possible to estimate the gaze range and gaze time from the heat map. We use tobii X 3 - 120 made by Toby Technology for the measurement of the gaze.

4 Analysis Method

We use the customer satisfaction (CS) analysis method. Items that largely contribute to the ease of viewing the screen, operability, and smoothing information distribution can be extracted by performing the CS analysis. From these items, it is also possible to extract the concrete contents to be improved preferentially as knowledge for the next design. Using the UEQ and CS analysis, the display screen of the information sharing terminal is varied adaptively according to the result improving the work while improving the UX of the information sharing person.

To confirm the effectiveness of the proposed method, the above analysis is performed on the data obtained from the questionnaire survey.

5 Expected Case Study

In this paper, we explain the case studies of applying the business systems used in retail and distribution businesses. Therefore, we consider the issues of data input and data sharing. We define the users who work in the retail or distribution business, and the environment in which the data is to be entered. We collect and analyze information about the expected users by brainstorming. We perform observation and interview surveys and extract the issues at the work site.

We clarify the issue obtained there, create behavioral scenarios in the business, and develop prototypes based on the behavioral scenarios.

Subsequently, we ask the user to evaluate the existing and created prototypes, measure the line of sight during the evaluation, and perform the task AFTER completing the task, we ask the users to conduct the UEQ test.

Acknowledgments. I wish to thank Human Media Engineering Laboratory of the Faculty of Science and Technology, Chuo University, and the Kansei Robotics Research Center who supported the discussions and reviewed this study.

References

1. Rauschenberger, M., Schrepp, M., Perez-Cota, M., Olschner, S., Thomaschewski, J.: Efficient measurement of the user experience of interactive products. How to use the user experience questionnaire example. Spanish language version. Int. J. Interact. Multimed. Artif. Intell. 2(1), 39–45 (2013)
2. Shinku, K.: Evolution of business system by human-centered design process (2016). https://seleck.cc/710. Accessed 20 Feb 2018
3. Hassenzahl, M., Tractinsky, N.: User experience-a research agenda. Behav. Inf. Technol. **25** (2), 91–97 (2006)
4. Roto, V., Law, E., Vermeeren, A., Hoonhout, J.: User Experience White Paper – Bringing Clarity to the Concept of User Experience. Demarcating User eXperience, Schloss Dagstuhl (2010). Dagstuhl Seminar Abstracts Collection. Sections 2.1–2.5. http://www.allaboutux.org/uxwhitepaper
5. Kazutami, T.: How to make UI/UX contributing to business (2017)
6. https://www.nri.com/ ~ /mdia/PDF/jp/opinion/teiki/it_solution/2017/ITSF171005.pdf. Accessed 15 Feb 2018

Research on the Model Construction of Intelligent Home Product Service Based on User Value

Weiwei Wang$^{(\boxtimes)}$, Yunyan Zhang, and Ting Wei

College of Art and Design, Shaanxi University of Science and Technology,
Xi'an 710021, China
1095216825@qq.com, 649684039@qq.com,
suqier1102@foxmail.com

Abstract. With the rapid development of market economy, enterprises acquire competitive advantage through improving customers' loyalty. It is the key factor cannot be ignored to grasp the user value accurately. We extracted the central effective user value, established the product service system quickly to satisfy the users' needs and enhance the competitiveness of enterprises. Firstly, the user journey diagram was drawn to analyze the mental activities, by which we got the value requirements, and transformed it through the positive creative design thinking. Secondly, according to the positive value factors, the 'human-object' three-dimensional ecosphere was constructed, then we built the product service model. Finally, the intelligent air-housekeeper product service system verified the validity of the model and methods, and satisfied the value demands of users. Meanwhile, the model has provided some references for other products design, and it is benefit to improve the core competitiveness of product service.

Keywords: Product service model · User value
Positive creative design thinking · Intelligent Air-housekeeper product service

1 Introduction

Nowadays smart home market competition is intense, it has become the primary problem of enterprise concerned how to make product service attract consumer's purchase desire. Porter has referred to that the competitive advantage comes from the value enterprise created for consumers in the 'Competitive Advantage' [1]. In other word, the user intend to purchase and consume is not the product but the value of it. The central effective user value has become the focus of the theoretical realm and the business community. There are some studies on user value in the definition field. Woodruff presented that the customer value was user's perception and evaluation of the extent that achieved his purpose by the product, the service, and the application effect, which in a certain usage scenarios [2]. Dahai et al., put forward the customer value was the ratio of the effectiveness to the cost in the process of purchasing and using product service: customer value = effectiveness/cost [3]. In the field of user value acquisition and application, zeng li designed the small household electrical products as the object of study, through the methods of observation, empathy and interviews to obtain the

© Springer International Publishing AG, part of Springer Nature 2019
T. Z. Ahram and C. Falcão (Eds.): AHFE 2018, AISC 794, pp. 331–341, 2019.
https://doi.org/10.1007/978-3-319-94947-5_33

user value, and combined product attributes with user value to stimulate the user's active desire to buy [4]. Xiuli et al. [5], proposed a product service system modeling technology based on customer value, analyzed customer expectation to obtain the value factors, and established the value model according to the value hierarchy to realize the maximization of the customer value.

Smart home products service is a complete system, which including users, products, services, technology, support and other elements. Its purpose is to improve the competitiveness of enterprises, meet the needs of users, provide seamless service experience, reduce pollution and waste, etc. [6]. In the process of design, we need to define environment, target and characteristics from different fields and perspectives, and to give a definition of the functions and services that the system should have, and to characterize the interrelationships among different elements [7]. A large number of experts and scholars have studied on the smart home and the service system both here and abroad. Polaine et al. proposed an intelligent home environment service mechanism based on Service Oriented Architecture (SOA), which was used for the interaction among system components, and effectively solved the interaction difficulties between traditional structures [8]. Wenyan et al., researched on product service for user demand, put forward the method of user demand analysis and the Quality Function Deployment (QFD) model of service, and modularized the product and service. So as to reduce costs, improve design efficiency and meet the individual needs of users [9].

2 The Theories of Intelligent Home Product Service Model Construction

The smart home product service system integrates peoples, products and services into one entirety, forming a multi-factor ecosystem with user-centered. It is the key point of the system design research how to make the product service satisfy the user's value-experience demand. Firstly, based on the service design to draw the user journey and get mental activity. Secondly, the positive user value was abstracted based on positive design thinking and binary-flip method. And then we would excavate and integrate the value elements that satisfy the user's need, to create a service ecosystem, and define the product function service model. Therefore, this research model was established based on user value theory, combining service design theory, positive creative design thinking and service ecosphere theory.

2.1 Basic Concept of User Value

Woodruff connected together the product, the usage scenario, and the demand-oriented customer journey, in which he emphasized the customer's preference and evaluation [2]. In the field of smart home products, user value is all of the user's activities and feedback information, which produced in order to meet a certain demand. By studying the mental activities of users, we obtained the value proposition, covering four levels of value content of basic, expectation, demand, and unknown.

2.2   Basic Concept of User Journey

User journey (UJ) refers to a system description of a particular product service, consisting of all the value activities in chronological order [10], including the elements of time, value activity (contact, action), user mind, etc. With a storytelling method to present a complete service experience process, represented by symbols as:

$$\text{User Journey} = UJ(x) = \Sigma\, J_x\,, x = 1\ldots m = [J_1, J_2, \ldots J_m] \tag{1}$$

UJ(x) is a complete user journey, including m value activities, in which J1 is the starting point of product service, J2 to Jm is the process, and the Jm is the destination. The mental activity is divided into three dimensions: expectation, complaint, doubt, as shown in Table 1.

Table 1. Statements of user journey & mental journey

Item	User journey (UJ)			
Mental journey (MJ)	J_1	J_2	...	J_m
Expectation	$E(J_1)$	$E(J_2)$	...	$E(J_m)$
Complaint	$C(J_1)$	$C(J_2)$	...	$C(J_m)$
Doubt	$D(J_1)$	$D(J_2)$	...	$D(J_m)$

2.3   Basic Concept of Positive Creative Design Thinking

The positive creative design thinking proposed by professor Lu Dingbang of National Cheng Gong University, is a creative thinking that using the theories of humanistic logic and design rationality to draw forward the future desire and value proposition from the negative predicament [10]. It uses conversion thinking to translate user's expectation, complaint and doubt into user's satisfaction, desire and affirmation. In order to diverge thinking and open up creative service programs, the professor introduced binary proposition of logic, that is, 'negative-One' and 'not-One'. The 'negative-One' proposition means to find the solution completely eliminate user's complaints or problems. Yet the 'not-One' proposition refers to improve or alleviate the complaints and find an acceptable solution.

2.4   Basic Concept of Service Ecosphere

The service ecosphere is used to describe the relationship between the service participants and stakeholders, to study the relation among the various service elements, and reorganize the cooperation between the participants to produce innovative service concepts [8]. In the design of smart home product service model, there are three-layer relationship between people and objects, that is, object-object, people-object, people-people, forming a complete service system.

3   Construction of Smart Home Product Service Model

The framework of system modeling is divided into three parts: element domain, integration domain and scheme domain. The element domain is the extraction of user's core value proposition, and explores all aspects of user's value requirements. The integration domain associates each value element to establish the service ecosphere and construct the system framework. Yet the scheme domain defines the function model according to the system framework, and guides the product design, the technology application and the service realization [11] (Fig. 1 System modeling framework).

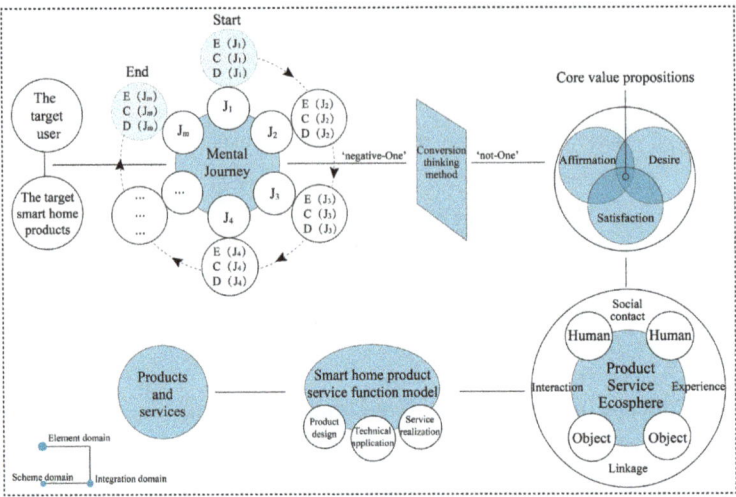

Fig. 1. System modeling framework

Element domain: In the section, we determine the target smart home product and its user group, draw the journey diagram, and record all the value activities from J1 to J2. That is, beginning with the product-service requirements and ending with user enjoyment. The users' mental activities are obtained by observation and interviews, and then apply the binary theory to convert the propositions from negative to positive, which choosing the appropriate binary proposition will reduce the workload and capture the user's value quickly. In order to ensure that the product service design does not deviate from the main line, we should simplify the value information.

Integration domain: In this section, the user value elements and product service design are combined to form a service ecosystem, which contains not only the three-dimensional relationship between human and object within the family, but also the outside. In the internal, the system should achieve intelligent linkage, software compatibility, friendly interaction, etc. In the external, it should cooperate and expand the users' social community, build the platforms of user and business, and it should be convenient to life payment, shopping, community services, and so on.

Scheme domain: The product function-service model is defined, and present the design scheme. According to the user value requirements to determine the product service functions, so that make it intelligent, easy to use and convenient.

4 Case Application

We took the intelligent Air-housekeeper product service system design as an example to show the model architecture and methods in detail (Fig. 2 Intelligent Air-housekeeper Service System Framework).

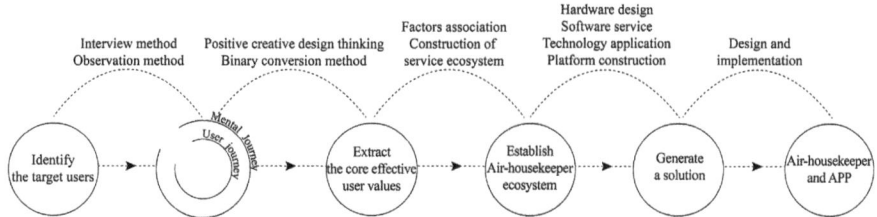

Fig. 2. Intelligent Air-housekeeper service system framework

4.1 The Target User

We have identified the target user by segmenting the market, which locked consumer groups between 18 to 45 years old. The main reasons were as followed: (1) The data showed that the group were more concerned about air quality and health; (2) 18 to 45 years old was the major age for young people to get married, buy houses and raising children, therefore, hey required high air quality in new houses; (3) They had a certain degree of education, the higher acceptance of smart home products, and the power of choice and purchasing.

To ensure that the reproduction of your illustrations is of a reasonable quality, we advise against the use of shading. The contrast should be as pronounced as possible. If screenshots are necessary, please make sure that you are happy with the print quality before you send the files.

4.2 Extracted the Core Effective User Values

We have drawn the user journey diagram that focuses on the user's healthy air demands, captured six major value activities: demand motivation, research and planning, purchase, use, new demand, and result. Mental activities were acquired through user interviews and observations (Fig. 3 User mental journey display).

(2) The users' mental journey in Fig. 3 was transformed by the binary theory, in which the users' expectation was directly converted into users' satisfaction and became the elements of value demand. As shown in Table 2.

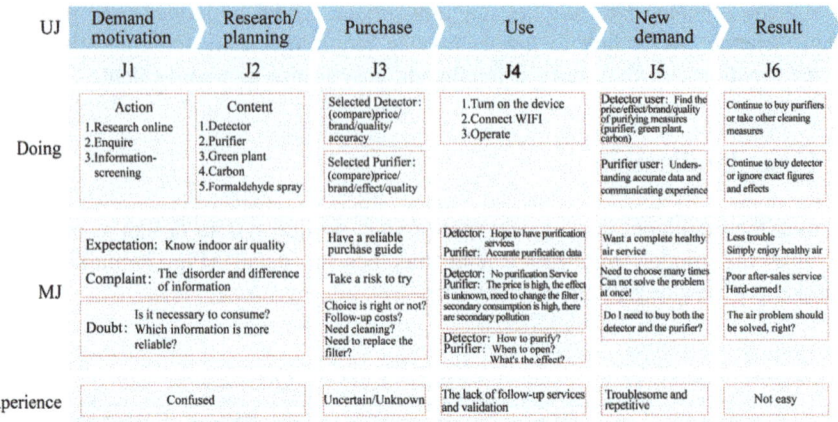

Fig. 3. User mental journey display

Table 2. Proposition flip

Item		Method	Negative proposition	Positive proposition
J₁ J₂	Expectation–Satisfaction	Demand	Know indoor air quality	Accurately detect air quality
	Complaint–Desire	'not-One'	The disorder and difference of information	Provide systematic information and guidance
	Doubt–Affirmation		Difficult to choose, is it necessary to buy?	
J₃	Expectation–Satisfaction	Demand	Have reliable purchasing instruction	Provide guidance
	Complaint–Desire	'not-One'	Take a risk to try	Experience first and buy later, communicate with the users
	Doubt–Affirmation	'negative-One'	Choice is right or not?/ Do i need to clean or replace the filter?/ What about the cost?	Green, safe and effective, without washing and changing
J₄	Expectation–Satisfaction	Demand	Detector: want a purification services Purifier: accurate cleaning data is required	Provide test data and purification services
	Complaint–Desire	'not-One'	Detector: no purification Purifier: expensive, there are secondary consumption/pollution, cleaning trouble	Green plant purification without pollution, price concessions

(*continued*)

Table 2. (*continued*)

Item		Method	Negative proposition	Positive proposition
	Doubt–Affirmation		Detector: how to purify Purifier: the use time and purification effect is unknown	Recommend purification packages, testing, intelligent linkage
J_5	Expectation–Satisfaction	Demand	Want a complete healthy air service	Provide complete service
	Complaint–Desire	'negative-One'	Need to choose many times	One-stop service
	Doubt–Affirmation	'not-One'	Do i need to buy both	Detector + green planting package
J_6	Expectation–Satisfaction	Demand	Less trouble, more enjoy	Provide complete services to enjoy
	Complaint–Desire	'not-One'	Poor after-sale service, hard-earned	Comprehensive and intimate service experience
	Doubt–Affirmation	'negative-One'	Has it solved the air problem	Yes

(3) Continued to integrate the positive propositions and extracted the core user value requirements through group discussion, comparison and selection. As shown in Table 3.

Table 3. Core value extraction

First streamlining	Second streamlining	Third streamlining
1. Accurately detect air quality 2. Provide air health knowledge and purchasing guidance 3. Recommend plants-purification packages 4. Experience first and buy later, communicate with the users 5. Intelligent linkage with the purifier 6. Provide complete healthy air service 7. One stop-service to solve the problem	1. Provide air detection 2.Provide air health guidance and purification recommendations 3. Build user platform to exchange experience 4. Intelligent linkage with home equipment 5. Provide a complete one-stop service	1. Complete air health service system 2. Provide detection, reminder, intelligent linkage 3. Offer green purification package 4. Build user platform and business platform

4.3   Construction of Service Ecosystem and Product Definition

We constructed ecosphere according to the core user values and the three-dimensional relationship between human and object, defined the product system and service ecosystem. We diverged the core user values and refined the content of the product service, in which there are many design elements. In order to realize the product service, we defined the support factors, which converged into a intelligent hardware and an APP (Fig. 4 Intelligent Air-housekeeper product service ecosystem).

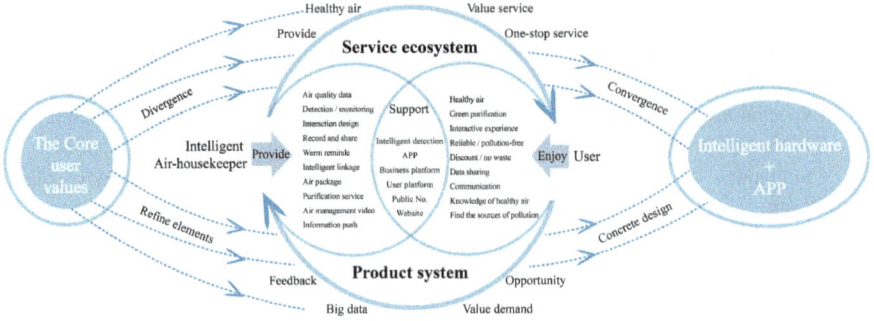

Fig. 4. Intelligent Air-housekeeper product service ecosystem

(1) Intelligent air-housekeeper provided users with air quality detection, data sharing, warm reminder, air packages, knowledge push, the experience of product interaction, home intelligent linkage and other services.

(2) Users through intelligent air-housekeeper to know the air quality data, and enjoy healthy air services, such as green purification, interactive experience.

(3) The supporting factors of service including business platform and the users', intelligent hardware, APP, Public No., website and so on, which links together users, air-housekeeper, and third-party service providers.

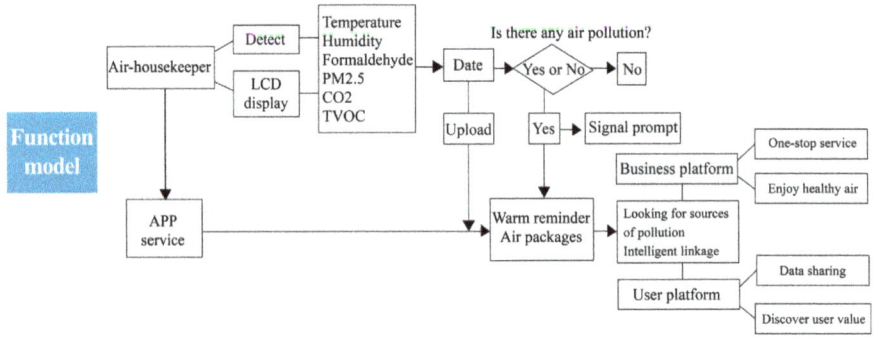

Fig. 5. The function model of Air-housekeeper and APP

4.4   Scheme Generation

We created a service function model for the Air-housekeeper and its APP (Fig. 5 The function model of Air-housekeeper and APP). The intelligent Air-housekeeper centered on user value, which formed a complete service system, including air detection, LCD screen display, data upload, warm reminder, intelligent linkage and air packages. Through APP design the business platform was built to achieve user online one-click purchase, and provided nearest delivery, which made the service more convenient and quick. Besides, we set up a user platform to support online communication, and regularly push air health information to achieve seamless after-sales service. We can mining more user value and promote product service update by mastering the user data.

Fig. 6. The sketches

We had determined the function model, then we investigated air health products in the market, and combined with the user's value needs to sketch the product and APP. The design of intelligent hardware followed the concept of green and people-oriented. So we chose the package design language with round shape for the hardware, which gave people a safe and healthy experience (Fig. 6 The sketches).

The detailed design process as shown in Fig. 7. The product was used the spraying of plastic material, there was a good interactive design that automatically saved the screen power through gesture sensor. Air-housekeeper designed with USB port charge, built-in battery, and small size can be portable. The air inlet was designed on the side to ensure the air intake, which can be measured accurately. The APP design was simple and fashionable, convenient for users to use. And it was the support of user platform and business platform, which combined the hardware to form a complete service system satisfied the users' value demands.

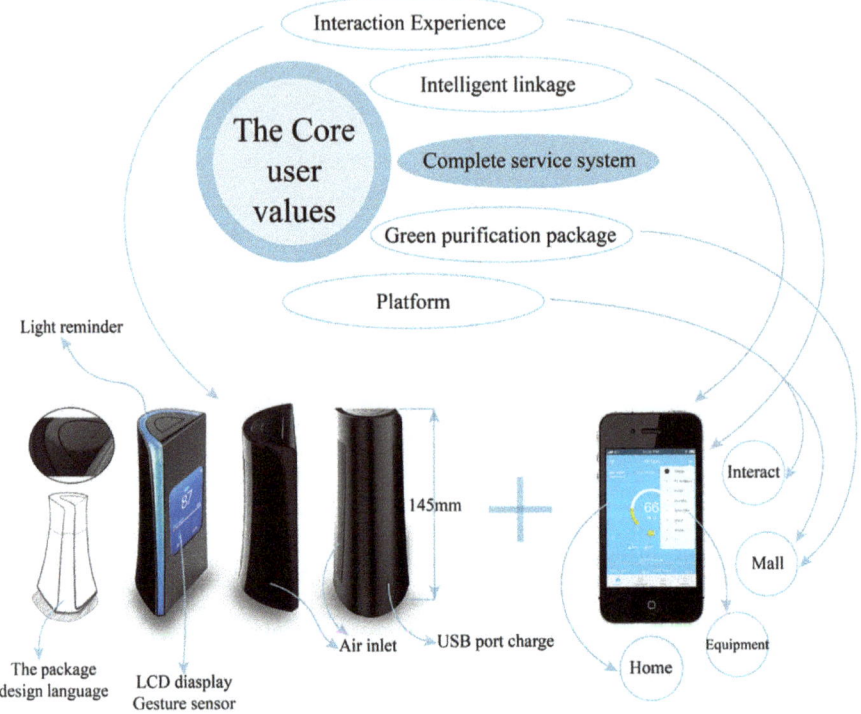

Fig. 7. Intelligent Air-housekeeper detailed design

5   Conclusion

Smart home products and services adhered to the 'people-oriented' concept, which based on user value and designed to enhance the competitiveness of enterprises. In this paper, the intelligent home product service model was created based on the theory of service design, positive creative design, service ecosystem and so on. The user journey was drawn, and the mental activity was captured. Therefore, we can extract the core value propositions to form a service ecosystem and finally created a product service model that satisfied the user's value. We took the intelligent Air-housekeeper as an example, around the core value of the user's healthy air demands to make detection, green purification, air health integrated into a complete service system. Finally, the service model was established quickly and effectively. This model can also be applied to other home product service system design, which can deal with the relationship between user value, product and service to provide users with better experience, and promote the development of smart home.

Acknowledgments. The author would like to thank the subjects for their participation in the experimental study. This research was supported by Ministry of education of Humanities and Social Science project (Semantic Analysis and Design Heritage of Han-Tang Culture 14YJC760008) and Doctoral research project of Shaanxi University of Science & Technology.

References

1. Porter, M.E.: Competitive Advantage. Free Press, London (1998). (in Chinese)
2. Woodruff, R.B., Gardial, S.F.: Know Your Customer. Wiley-Blackwell, New Jersey (1996). (in Chinese)
3. Dahai, D., Xiaoyan, Q., Xiaofei, Q.: Theory of customer value and its formation. J. Dali. Univ. Technol. (Soc. Sci.) **04**(004), 10–20 (1999). (in Chinese)
4. Zeng, L.: The Research on the Small Domestic Appliance Design Based on User Value. Hunan University, Hunan (2010). (in Chinese)
5. Xiuli, G.: Research on Modeling and Decision Making Technology of Product Service System Conceptual Design Driven by Customer Value. Shanghai Jiao Tong University, Shanghai (2012). (in Chinese)
6. Wu, C.L., Liao, C.F., Fu, L.C.: Service-oriented smart-home architecture based on OSGi and mobile-agent technology. IEEE Trans. Syst. Man Cybern. Part C **37**(2), 193–205 (2007)
7. Xinzhi, Z.: Research on the construction of demand model of industrial product service system. Manuf. Autom. **38**(6), 6–12 (2016). (in Chinese)
8. Polaine, A., Løvlie, L., Reason, B.: Service Design: From Insight to Implementation. Rosenfeld Media, Brooklyn (2013)
9. Wenyan, S.: Study on Methods and Technologies for Customer Requirement-Oriented Design of Product-Service Concept. Shanghai Jiao Tong University, Shanghai (2014). (in Chinese)
10. Dingbang, L.: Positive Creative/Mirror Theory, p. 34. Tsinghua University Press, Beijing (2015). (in Chinese)
11. Jinhua, Q., Suihuai, Y., Gang, L.: Construction of megacity sustainable pension service system model. Comput. Eng. Appl. (17), 1–10 (2016). (in Chinese)

The Effects of Response Time on User Perception in Smartphone Interaction

Zhengyu Tan[1], Jieru Zhu[2(✉)], Jun Chen[3], and Fusheng Li[3]

[1] State Key Laboratory of Advanced Design and Manufacturing for Vehicle Body, Hunan University, Changsha, China
Tanzy2004@126.com
[2] School of Design, Hunan University, Changsha, China
jieruzhu@hnu.edu.cn
[3] Huawei Device Co., Ltd., Pudong Shanghai, China
{aaron.chen,lifusheng}@huawei.com

Abstract. Responsiveness is an important factor to be considered in human-computer interfaces. With the advent of touch screen based smartphones and the evolve of more responsive technology, the user-acceptable levels of response time have changed considerably. However, previous studies have focused more on computers than on smartphones. The goal of this study was to explore the relationship between response time and user perception in the context of smartphone interactions. Different response times from 300 to 5400 ms were manipulated for four common task types in three applications, this simulation enables us to collect users' subjective evaluations and physiological measures including pupil dilation and galvanic skin response. The applicability of Web-Fechner Law is demonstrated when the response time is in milliseconds. Results also showed statistically significant effects from task type and loading animation. Users have lower tolerance when they swipe to switch pages and have greater tolerance on interfaces with a loading animation, which provide the developers with clearer design requirements guidance.

Keywords: Response time · Smartphone interaction · User perception

1 Introduction

Response time has been one of the most often-discussed subjects in human-computer interfaces since a long time. Android users often complain about the poor responsiveness of their phones, which has become an important defect of Android applications. Slow system response times may cause dissatisfaction among users [1]. Users' performance and behavioral intentions decreases as response time gets longer [2]. Lengthy waiting time interrupts the use of an application and forces the user to wait impatiently or leave the application, leading to significant poor user perception.

The main purpose of this paper is to explore the relationship between response time and user perception in the context of smartphone interaction, considering the factors of different tasks and the impact of loading animations. The result of the study will help the software and the hardware engineers to understand the different user-acceptable

T. Z. Ahram and C. Falcão (Eds.): AHFE 2018, AISC 794, pp. 342–353, 2019.
https://doi.org/10.1007/978-3-319-94947-5_34

levels of response time, so that the system hardware and software can be configured in accordance with specific requirements

2 Related Work

Response time (latency) is defined as the "delay between input and the output response" [3]. We begin by examining the different dimensions of user perception in smartphone interaction, in which perceived responsiveness is an important factor. Then, we examine prior researches on the acceptable response time in touch interfaces. Finally, we examine work on the factors affecting perceived responsiveness including different tasks and waiting time fillers.

2.1 Smartphone Interaction Perception

Interactivity is a state experienced by a user during his or her interaction with a mobile phone [4]. Wu proposed that the actual interactivity is dependent on the user, individuals can realize different levels of potential adequacy provided by actual interactivity [5]. However, few researchers have done research on the perceived interactivity in the context of smartphone interaction. Wu suggested that perceived interactivity is made up of three dimensions: perceived control, perceived responsiveness, and perceived personalization [5]. Johnson et al. found that responsiveness, nonverbal information, and speed of response have significant effects on perceived interactivity [6]. Song and Zinkhan measured interactivity on the three dimensions of communication, control, and responsiveness [7].

According to prior research, perceived responsiveness plays an important role in perceived interactivity. In addition, interactivity is one of the two key antecedents of smartphone usability [4]. Hence, perceived responsiveness has a great influence on the usability in the context of smartphone interaction.

2.2 Response Time in Touch Interfaces

Anderson et al. tested scrolling, panning and zooming gestures in touchscreen tasks such as web browsing and photo viewing on a multi-touch display and found that delay above 580 ms is unacceptable to the users [8]. Ng et al. studied the dragging task using a high-performance touch test system, finding that users highly appreciate lower latencies and they can perceive latency down to 6 ms [9]. However, they pointed out that users may actually notice the disparity between the finger and the object rather than the latency itself. Jota et al. also studied the dragging and tapping task using the same apparatus as Ng et al.'s and found that 85% of the participants could not notice the latency of 40 ms for tap-based input [10]. In their experiment, participants were asked to note the latency by the rectangle around their finger on the display after they touched the target, which is an initial input feedback of their gesture rather than a display of result to their command. Ritter et al. looked at the user-acceptable levels of latency of two tasks: tapping on button and simple dragging tasks performed on iPad Air and found the acceptable latency is 300 ms for tapping tasks and 170 ms for dragging task

[11]. Deber et al. focused on the comparison between direct input and indirect input for dragging and tapping tasks [12]. Form-factors (direct and indirect) and tasks (dragging and tapping) both had significant effects for latency perception and an improvement of a decrease of 33.3 ms in latency have an observable difference for users in tapping tasks.

2.3   User Perception of Response Time

Since the level of response time cannot reach the minimum perceivable latency as low as 6 ms proposed by Ng et al. under the condition of existing technology, the study of user's subjective perception in psychology has practical importance for improving the user perception of response time. It is acknowledged that subjective perception of a duration should never be assumed to be accurate and true to the actual duration. The applicability Web-Fechner Law has been demonstrated in many references. The work of [13] shows that the relationship between objective and perceived telephone waiting time for a commercial service could be described by a double-logarithmic function. In the context of interactive applications, Weber-Fechner Law can be applied to the relationship of waiting time and quality of experience for web-based services such as web browsing and online video services [14]. For using mobile applications, the relationship of response time and users' subjective evaluation fits Web-Fechner Law in response type, load type and download type of tasks [15].

Miller proposed that longer delays are more acceptable in a task after a closure than in the process of a task [16]. Shneiderman added that user-acceptable levels of latency are highly influenced by the tasks and he advised 1 s for frequent simple tasks and 2–4 s for common tasks [17]. The effects of waiting time fillers on users' perception have also been an interest of many scholars. Nah studied the users' waiting time for download of Web pages and found the presence of a feedback bar indicating that the system is carrying the request prolongs users' waiting time [18]. Lee et al. examined the distinctive effects of image, text and image motion in a filler interface on perceived waiting time [19]. Results demonstrated that users experience more focused immersion when they see a filler interface and moving image is a good design characteristic to manage perceived waiting time while wait online.

Such research has expanded the understanding of how to improve users' subjective perception. However, all of these studies adopted long time intervals such as 5 s and 15 s, and none of them demonstrated the applicability of Web-Fechner Law and how tasks and waiting time fillers affect time perception when the response time is in milliseconds. Besides, the majority of studies on touch interfaces focus on identifying the minimum perceivable "latency" for the gesture itself on high-performance touch devices, rather than the user experience in specific smartphone interactions such as selecting an icon from the menu. A developer would like to know how the users experience varies as the response time changes within the limitations of the existing technology. Our work aims to close this gap and to understand the relationship between user perception and response time in real life situation in the context of smartphone interaction.

3   Experiment

3.1   Participants

Twenty participants from college (9 male, 11 female), ranging in age from 18 to 26, took part in the study. The participants were selected through an online survey based on the questionnaire of the attention for the responsiveness and the satisfaction for their smartphone for the users. All of them had a great deal of experience using touch screen devices and owned one or more smartphones (such as IOS-based or Android-based phones), and they were familiar with the use of mobile applications. Each participant was paid $10 for their participation.

3.2   Apparatus

The experimental equipment includes mobile Android phones, a Tobii Pro Glasses 2, a MindWare Psychophysiology Lab System and a laptop. Participants executed the experimental tasks using Android phones of Huawei P9 with a 5.5-inch screen. The Tobii Glasses was used for tracking the eye position of participants, especially pupil dilation responses (changes in pupil sizes), and the MindWare System collected the measure of galvanic skin response (GSR) of participants. The flow of the experiment was controlled by a laptop, including the normal operation of the Tobii Glasses and the MindWare System.

3.3   Design

Tasks varied according to three independent variables: 3 mobile applications, 4 task types and 16 time intervals. The experiment adopted tasks in three applications: WeChat, Camera and Happy Elements. WeChat is a social media mobile application with platforms of instant messaging, commerce and payment services. Happy Elimination is a casual elimination game with simple rules. Both of the two applications have high popularity for all ages in China. We divided operation tasks into three types: enter type, exit type, switch type. Enter tasks and exit tasks are performed by tapping gestures and switch tasks are performed by swiping gestures. To compare the perception of response time on interfaces both with and without visual feedback, we set up experiments with the task type of entering a page with loading animation. Sixteen intervals of response times were designed for each task which is implemented by changing the setting of interface in Axure. Considering the presence of visual feedback, the sets of response time between the enter type with loading animation and the other three task types were different. Each set was used three times: one time in ascending order, the second time in descending order, and the third time in random order. All participants had to participate in the tasks of different time intervals in three orders. The setting of experiment tasks and response time is illustrated in Table 1.

The experimental design was fully repeated measures – all participants performed 3 repetitions of all combinations of levels of response times, task types, and applications in three days. The sets of response time were in ascending order for the first day, descending order for the second day and random order for the third day. The 12 operation tasks were randomly presented for each day. In summary, the overall design of the experiment was: 4 task types × 3 applications × 16 response time × 3 repetitions × 20 participants × = 11520 total trials.

The dependent variables were the participants subjective rating of each responsive time. After each trial, participants were asked to rate the acceptance level of responsiveness on five-point scale ('very annoying', 'annoying', 'slight annoying', 'perceptible but not annoying', 'imperceptible'). When experiencing the set of response time in ascending order, the trials ended when the participant rated a 1. When experiencing the set of response time in descending order, participants were presented with the third trial after the trial they rated a 1 on the first day as the starting trial, and the trials ended when they rated a 5. When experiencing the set of response time in random order, we arranged the tests according to the rating of the first two days.

Table 1. The setting of experiment tasks and response times.

Take type	Application	Operation task	Response time (ms)
Entering a page	WeChat Camera Happy Elimination	Tap on an icon to enter a page or an application	300, 330, 350, 365, 380, 390, 400, 420, 480, 590, 690, 800, 900, 1000, 1220, 1350
Exiting a page	WeChat Camera Happy Elimination	Tap on an icon to exit a page or an application	
Switch pages	WeChat Camera Happy Elimination	Swipe the screen to the left or right to enter a page	
Entering a page (with loading animation)	WeChat Camera Happy Elimination	Tap on an icon to enter a page or an application	380, 480, 590, 690, 800, 900, 1000, 1110, 1220, 1350, 1840, 2340, 2860, 3360, 4390, 5400

During the process of experience, we also adopted psychophysiological methods which are pupil dilation and GSR for the user perception evaluation. GSR is the electrical conductance of the skin directly influenced by the moist of skin, which is a linear correlate to arousal [20]. The differences in the users' GSR signal are correlated to the differences in their subjective responses for frustration [21]. Pupil dilation is known to quickly respond to arousal and mental workload. There is a significant linear negative correlation between the users' satisfaction and the pupil diameter fluctuation changes [22]. As a valuable complement to subjective reports, psychophysiological methods can be used to objectively measure user perception in the field of human computer interaction. In this experiment, the measurement of pupil dilation and GSR was mainly used to verify the reliability of subjective ratings.

3.4 Procedure

The experiment interfaces were made by Axure, a specialized prototyping software running on the computer. The program files were downloaded with a mobile application called Axure Share which can be installed from the Android market. Before the experiment, the participants had to take the pre-test of the Tobii Glasses due to the requirements of the participants' level of fatigue and physical characteristics for pupil dilation responses measurement. If the participants passed the pre-test, they were asked to complete a consent form and a questionnaire to collect demographic information. After the placement of surface electrodes of the MindWare System, they received instruction on how to operate the smartphone, and completed 10 training trials to practice the tapping and swiping gestures on different tasks to get familiar with the procedure. After that, the actual trial began.

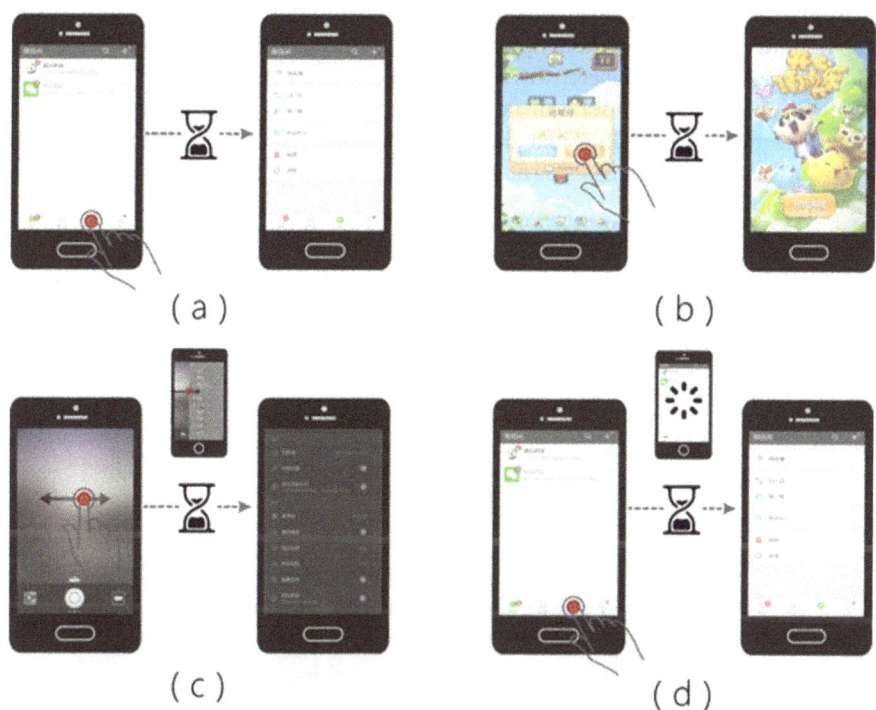

Fig. 1. The operation process for the experiment. (a) enter task. (b) exit task. (c) switch task. (d) enter task with loading animation.

The specific operation process of the experiment is shown in Fig. 1. The interfaces were designed to simulate the common operation processes of mobile applications to facilitate the effective collection of real data. Psychophysiology measurements of pupil dilation and GSR were collected along with the participant number, task type, application, response time and subjective ratings. Due to the fast manipulations and short response time of the task, each task was allowed to experienced repeatedly when the participant was unsure about the rating. The procedure lasted approximately 45 min and the entire session took less than 80 min including the pre-test of the Tobii Glasses and the training trials.

4   Results

To analyze the collection of physiological data, we processed the participants' signal in both GSR and pupil dilation to verify the reliability of subjective ratings. For example, one participant's physiological signal of the experiment when entering a page using WeChat application is shown in Fig. 2. When analyzing GSR signal, we selected the valid areas according to the points in time marked during the experiment. The participant exhibited lager response when the ratings changed as shown in areas labelled with "3", "7", "9" and "10". In contrast, the GSR curve became flat when the ratings remained the same. When analyzing the measurement of pupil dilation, we divided the data into three parts which were high ratings part, middle ratings part and low ratings part. The curve of pupil diameter was relatively flatter in the first part than the other two parts, indicating a better user satisfaction when they gave the responsiveness high ratings. Therefore, both GSR and dilation pupil had verified the reliability of the participant's subjective ratings. Through this analysis, we identified the reliable date to undertake further analysis.

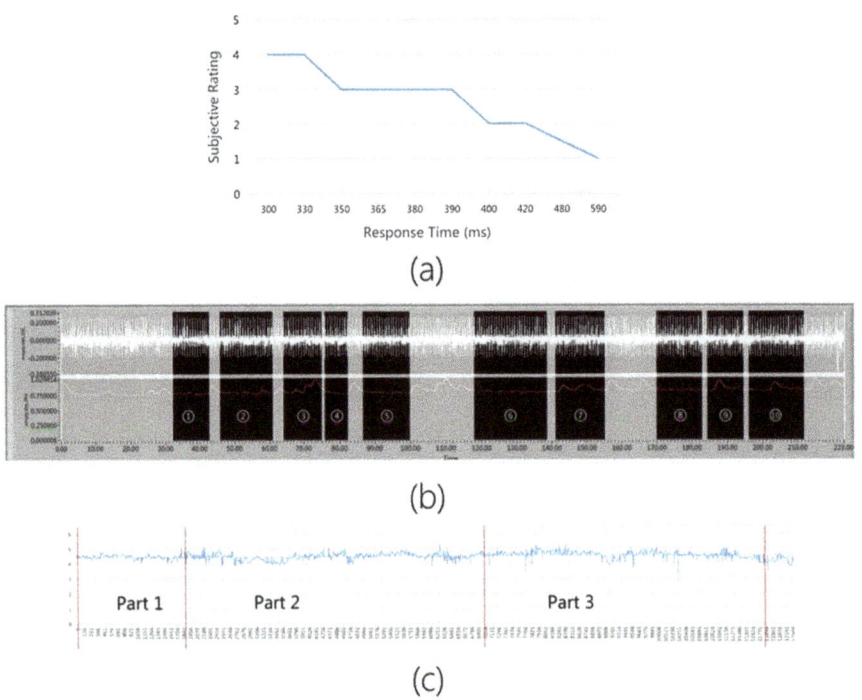

Fig. 2. Participant 11's subjective ratings and physiological measurement when entering a page in WeChat. (a) subjective ratings. (b) GSR signal. (c) dilation pupil signal.

We now examine the results of our experiment. First, we examine the effects of task type and visual feedback on responsiveness when using different applications. Secondly, we demonstrate the applicability of the logarithmic function relationship of subjective evaluation and response time by fitting the experiment results.

4.1   Analysis of Subjective Ratings

We performed three repeated-measures ANOVAs for three applications respectively to examine the effect from task type, using Task Type and Response Time as independent variables. For Wechat, we found a significant main effect from Task Type (F = 13.83, p < 0.01), with Exit Type being less annoying than the other two types. There was also a significant main effect from Response Time (F = 295.21, p < 0.01), with longer Response Time being associated with declining subjective ratings. A post-hoc pairwise comparison shows a significant difference between most pairs of Response Time when the time interval is at or below 1000 ms. And the interaction between Task Type and Response Time is significant (F = 9.25, p < 0.01). For Happy Elimination and Camera, we again found significance effects from each of Task Type, Response Time and the interaction between them.

Figure 3 shows the mean ratings for each Response Time by Task Type in using the applications. The ratings indicate the trend that exit type> enter type> switch type. The Enter/Switch Task was rated lower at 300 ms. With the response time getting longer, the mean ratings of Switch Task dropped lower than Enter Task. The response time above 1000 ms was very annoying for all the three tasks.

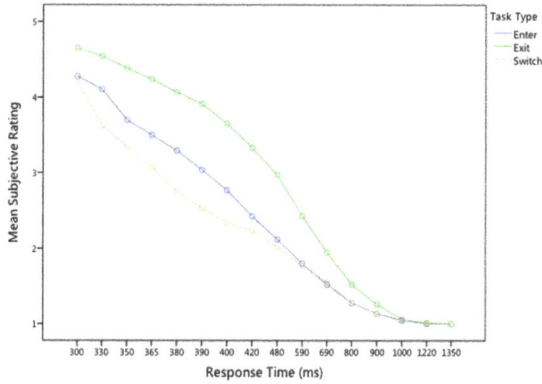

Fig. 3. Mean subjective Ratings for Enter Type, Exit Type and Switch Type, per response time level. Error bars show 95% confidence intervals.

To allow us to examine the effect of visual feedback on user perception of response time, we performed three more repeated-measures ANOVAs. The independent variables were Task Type consisting of enter task and enter task with loading animation and Response Time including 9 time intervals experienced in both two tasks. For Wechat, there were significant main effects from both Task Type (F = 46.19, p < 0.01)

and Response Time (F = 295.21, p < 0.01) on subjective ratings. For Happy Elimination and Camera, we again found significant effects from both loading animation and
Response Time on ratings.

Figure 4 shows the mean ratings for each Response Time with and without loading
animation. At each level of Response Time, the task with loading animation was rated
higher than the task without loading animation. But with the response time getting
longer, the difference of subjective ratings between the two task types gradually
decreased. At higher levels of Response Time, the responsiveness become annoying
even with the presence of loading animation.

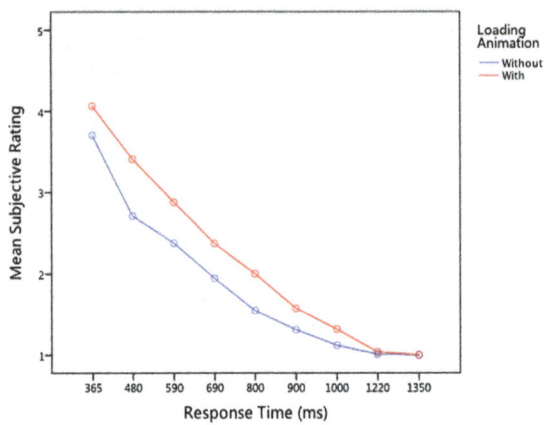

Fig. 4. Mean subjective Ratings for Enter Type with and without loading animation, per
response time level. Error bars show 95% confidence intervals.

4.2   The Relationship Between Subjective Evaluation and Response Time

Based on previous study on the relationship between response time and the user
experience, we demonstrated the applicability using the result of our experiment. The
relationship between subjective ratings and response time fits a logarithmic function:

$$MOS = b * \ln(RT) + a. \tag{1}$$

Where MOS is the average value of the subjective ratings of each task type, b is
coefficient and a is a constant. Table 2 presents the model summary and curve fitting of
the four task types respectively. The results imply that the logarithmic function fits well
when the response time is in milliseconds, which is consistent with Web-Fechner Law.
By using the logarithmic function, the response time can be obtained according to the
given subjective rating, and we can determine the subjective rating for a given response
time. For example, the levels of response time for evaluations of 'very annoying',
'annoying', 'slight annoying', 'perceptible but not annoying', and 'imperceptible'
when performing different tasks is shown in Table 3. Although we cannot manipulate
response time down to 300 ms, the imperceptible response time for users can be
determined by the function, which is 202 ms, 275 ms, 166 ms, and 278 ms.

Table 2. The model summary and curve fitting of different task types.

Task Type	R Square	F	Curve fitting
Enter	0.921	139.47	MOS = −2.748ln(RT) +19.592
Exit	0.987	946.70	MOS = −3.248ln(RT) +23.244
Switch	0.872	74.61	MOS = −2.511ln(RT) +17.842
Enter with loading animation	0.991	572.64	MOS = −3.063ln(RT) +22.236

Table 3. The levels of response time for different subjective ratings, separated by task type.

Task type	5	4	3	2	1
Enter	202 ms	291 ms	419 ms	603 ms	868 ms
Exit	275 ms	374 ms	509 ms	693 ms	942 ms
Switch	166 ms	248 ms	369 ms	550 ms	818 ms
Enter with loading animation	278 ms	385 ms	534 ms	740 ms	1026 ms

5   Discussion

In this paper, we have examined users' perception of response time when using mobile applications on smartphones for enter task, exit task and switch tasks, and under the effect of loading animation. A set of studies indicated that users have different levels of tolerance for the three tasks and the effect of loading animation is significant for touch screen user perception. Participants had higher tolerance for exit task and lower tolerance for switch task. The studies also demonstrated the applicability of Web-Fechner Law when the response time is in milliseconds. When setting design requirements for smartphone of various price ranges, developers can determine the suitable response time interval referring to Table 3. If the target users are high-end consumers, the response time rating 5 or 4 is the goal to be reached.

While the study is based on user experience, more improvements of both software and hardware can be made to meet the users' satisfaction. With regard to software, the front-end design, the back-end architecture and the design of database middleware can be optimized according to the user perception of different operation tasks to improve the responsiveness of an application. On the hardware side, the allocation of resource including CPU, RAM, IOPS and GPU can be more reasonable to meet the users' needs.

There are many deficiencies in this paper. Firstly, the manipulation of response time is restricted to the experimental equipment, and the minimum response time is 300 ms. Secondly, the experimental task is simple, participants may pay more attention to the response time in the test than in actual use. Participants may have a more appropriate evaluation when they perform a more complete task such as take a photo or play a game. Thirdly, although we considered tapping and swiping gestures in our research, but the gesture is treated as the operation form of different task types, not as a dependent variable. So the effect of gestures on the perception of response time has not been clarified.

Furthermore, there are numerous work needed to do on the user perception of response time when using smartphones. Additional research could explore several related areas such as other gestures, more complex tasks, and other forms of waiting time fillers. The commonly used gestures on touch screen include double tap, pinch, spread, press, rotate and so on. Smartphones are also used in shopping, phoning, web browsing and other categories. As the forms of waiting time fillers are becoming increasingly diverse, the difference of their benefits on the time perception to a user need to be understood.

Acknowledgments. The research was supported by National Natural Science Foundation of China (61402159, 51605154). We would like to thank all participants for their zeal during the experiments.

References

1. Hoxmeier, J.A., DiCesare, C.: System response time and user satisfaction: In: An experimental study of browser-based applications. AMCIS 2000 Proceedings, p. 347 (2000)
2. Galletta, D.F., et al.: Web site delays: how tolerant are users? J. Assoc. Inf. Syst. **5**(1), 1 (2004)
3. Rangardt, J., Czaja, M.: Empirical investigation of how user experience is affected by response time in a web application (2017)
4. Lee, D., et al.: Antecedents and consequences of mobile phone usability: linking simplicity and interactivity to satisfaction, trust, and brand loyalty. Inf. Manag. **52**(3), 295–304 (2015)
5. Wu, G., Wu, G.: Conceptualizing and measuring the perceived interactivity of websites. J. Curr. Issues Res. Advert. **28**(1), 87–104 (2006)
6. Johnson, G.J., Bruner II, G.C., Kumar, A.: Interactivity and its facets revisited: theory and empirical test. J. Advert. **35**(4), 35–52 (2006)
7. Song, J.H., Zinkhan, G.M.: Determinants of perceived web site interactivity. J. Mark. **72**(2), 99–113 (2008)
8. Anderson, G., Doherty, R., Ganapathy, S.: User perception of touch screen latency. In: International Conference of Design, User Experience, and Usability. Springer, Heidelberg (2011)
9. Ng, A., et al.: Designing for low-latency direct-touch input. In: Proceedings of the 25th annual ACM symposium on User interface software and technology. ACM (2012)
10. Jota, R., et al.: How fast is fast enough?: a study of the effects of latency in direct-touch pointing tasks. In: Proceedings of the SIGCHI Conference on Human Factors in Computing Systems. ACM (2013)
11. Ritter, W., Kempter, G., Werner, T.: User-acceptance of latency in touch interactions. In: International Conference on Universal Access in Human-Computer Interaction. Springer, Cham (2015)
12. Deber, J., et al.: How much faster is fast enough?: user perception of latency & latency improvements in direct and indirect touch. In: Proceedings of the 33rd Annual ACM Conference on Human Factors in Computing Systems. ACM (2015)
13. Antonides, G., Verhoef, P.C., Van Aalst, M.: Consumer perception and evaluation of waiting time: a field experiment. J. Consum. Psychol. **12**(3), 193–202 (2002)
14. Egger, S., et al.: Waiting times in quality of experience for web based services. In: Fourth International Workshop on Quality of Multimedia Experience (QoMEX). IEEE (2012)

15. Zhou, R., et al.: How to define the user's tolerance of response time in using mobile applications. In: IEEE International Conference on Industrial Engineering and Engineering Management (IEEM) 2016. IEEE (2016)
16. Miller, R.B.: Response time in man-computer conversational transactions. In: Proceedings of the December 9–11, 1968, fall joint computer conference, part I. ACM (1968)
17. Shneiderman, B., Designing the user interface: strategies for effective human-computer interaction. Pearson Education, India (2010)
18. Nah, F.F.-H.: A study on tolerable waiting time: how long are web users willing to wait? Behav. Inf. Technol. 23(3), 153–163 (2004)
19. Lee, Y., Chen, A.N., Ilie, V.: Can online wait be managed? the effect of filler interfaces and presentation modes on perceived waiting time online (2012)
20. Dawson, M., Schell, A., Filion, D.: The electrodermal system In: Cacioppo, J.T., Tassinary, L.G., Berntson, G.G. (eds.) Handbook of psychophysiology, pp. 159–181. Cambridge University Press, New York (2007)
21. Mandryk, R.L., Inkpen, K.M., Calvert, T.W.: Using psychophysiological techniques to measure user experience with entertainment technologies. Behav. Inf. Technol. 25(2), 141–158 (2006)
22. Wenjun, H., Xiaoyu, G., Tiemeng, L.: Customer satisfaction evaluation model based on pupil size changes. Space Med. Med. Eng. 5, 001 (2013)

User's Behavior Under Review: The Use of Instrument's to Evaluate Perception of Users

Marina Barros[1]([✉]), Walter Correia[2], and Fabio Campos[2]

[1] Physiotherapy Department, Universidade Católica de Pernambuco (UNICAP), Rua do Príncipe 526, Boa Vista, Recife, PE 50050-900, Brazil
marinalnbarros@gmail.com
[2] Design Department, Universidade Federal de Pernambuco (UFPE), Av. Prof. Moraes Rego 1235, Cidade Universitária, Recife, PE 50670-901, Brazil
ergonomia@terra.com.br, fc2005@gmail.com

Abstract. Considering that the process of perception is directly linked to the previous experiences of users, there is an entire process of selection, organization and interpretation of data that requires attention and dedication. This article, from a doctoral thesis, highlights that during the design process, it is necessary that there is respect for the perception of individuals who will use or evaluate the products or services offered. Thus, we used objective and subjective tools to evaluate the perception of users in a service field, in order to identify whether these results tend to the same or not, and from that, how does the identification of problems and solutions to facilitate the user's own understanding. A cross-sectional observation study with 1 evaluation and 2 reevaluations, with 31 users in the physiotherapy area, was carried out in a SUS (Unified Health System) outpatient clinic, in which they were initially interviewed and followed by a subjective self-assessment of strength and function and an objective measure of force in the three moments. With this it was possible to observe that, when the user does not know the service and/or the product, one cannot have the correct perception of these, demonstrating how important it is to evaluate the perception of the individuals in general.

Keywords: Artifact evaluation · Design · User perception
Design methodology

1 Introduction

Perception can be understood as the process by which sensations are selected, organized and interpreted, and because it is directly linked to the experiences lived in the past. So when an individual observes, buys, and uses a product, various feelings, sensations and emotions arising from the perception and information sent by the object are awakened, and perception is not influenced only by the tactile and visual elements, but also by the unconscious of it, through of the lived experiences [1, 2].

In this process of product development, perception evaluation criteria can be included, which can be divided into three components, subjective evaluation, this is not only a simple answer of yes or no, but an analysis of affections, symbols and semantic of people who evaluate, therefore, seeking different qualities such as functionality,

© Springer International Publishing AG, part of Springer Nature 2019
T. Z. Ahram and C. Falcão (Eds.): AHFE 2018, AISC 794, pp. 354–361, 2019.
https://doi.org/10.1007/978-3-319-94947-5_35

hedonism, reliability and among other things; the sensory environment through which the product is perceived includes touch, taste, smell, hearing, and vision; and the design element, depends on the complexity of the object and any detail can play a determining role in the evaluation of the product [3].

When thinking about product development or service evaluation, a concept that will generate great influence during the design process, is the User Experience (UX), which seeks to understand how the user will interpret/evaluate the use of the same, because each individual will have a different observance during this process.

And with this, the User Experience (UX) can be understood as a flow of feelings, thoughts and actions, and considered unconscious, but it is accessible to those who experience it. However, understanding this experience is a critical issue, especially for design, and in recent years there has been a growing interest in designing this experience, and some early efforts to create new theories can be observed. However, little has been done to expand this idea, requiring much more work in order to understand the human experience and the efforts to design for user thinking [4–6].

The success of a product on the market is determined not only by its technical and objective content, but also by aesthetic, emotional and other experiential factors. In the practice of designing, the development of new artifacts needs to take into account the balance between objective and subjective qualities, between the functionality of technology and emotional expressiveness, in an attempt to satisfy the demands and desires of potential users, already mentioned, "individual needs". It becomes imperative to capture relevant information and anticipate users' expectations. The various emotions triggered even by the appearance of the artifact to be acquired can increase the pleasure of buying, possessing and clearly, of using it.

According to the authors, one can have these emotional factors defined as very deep and intentional states, which involves a relationship between a human being and a given stimulus. There is a very singular relationship in dealing with concepts related to affectivity, as in the case of emotions, in which they are as intangible as they are at the same time quite attractive from the point of view of a deeper analysis. In fact, to have the ability to imbue emotional values with the design of artifacts, to understand the emotions of the users and to measure them become the main challenges, almost as important, if not more, as to the functionality, usability, quality, etc. [7].

When thinking about health perception, it is necessary to define and what is health, and the World Health Organization (WHO, 1946) defines it as a complete state of physical, mental and social well being and not merely the absence of disease. The behavior of a population through its health problems are built based on the health perception of this population, which rises from its sociocultural context [8, 9].

Previous knowledge of health perception can determine the thinking and action of the population in the face of the health-disease process, and is fundamental for the efficiency of health care and education actions. With this, a concern has grown not only with the frequency and severity of the diseases, but also with the evaluation of measures of disease impact and commitment of the activities, measures of health perception and functionality (2001).

With this, this research proposed to evaluate the perception of the users from the different instruments of perception capture.

2 Research Methods

The experiment is an observational, cross-sectional, descriptive-analytical study aimed at verifying whether the results of objective and subjective instruments of user perception analysis converge.

This experiment was approved by the ethics committee of the UFPE on the CAAE number: 45705715.0.0000.5208, Report No.: 1,144,880, project "Evaluation of the Behavior of Objective and Subjective Instruments in the Capture of Users Perception".

Five types of perception tests were chosen on the same object, three performed by the user, another by the expert, and another by an objective instrument based on a digital dynamometer specifically built for this experiment.

The experiment was carried out over a period of 7 months, from August 2015 to January 2016, at the Physiotherapy outpatient clinic specializing in the Single Health Service (SUS), at Getúlio Vargas Hospital, in which the three instruments were shown at different times. In each of these moments of application of the instruments of perception, they were applied in the following order: recovery evaluation scale, subjective evaluation by the user of force, subjective evaluation of function by the user, measurement of force by the digital dynamometer, subjective assessment of strength by the specialist.

This sample consisted of 31 individuals of both sexes, who had several hand pathologies, which had a decrease in the grip strength of digital pulp-to-pulp tweezers and were therefore performing physiotherapy in the hand physiotherapy outpatient clinic.

Before the participant entered the experiment, an explanatory lecture was held, explaining to the participant the points evaluated, how the information was collected and presenting potential benefits and risks of this experiment in the course of its treatment.

Afterwards, the users were asked if he would freely and willingly participate in the experiment. Those who agreed to participate were asked to sign the Informed Consent Form (TCLE).

Then the socio-clinical-demographic questionnaire was applied, in order to draw a profile of the sample. After the socio-clinical-demographic questionnaire, the first sequence of measurements was performed using all the instruments in the order shown in Fig. 1.

To avoid the user's response inducing the specialist or vice versa, the experiment predicted separate environments so that the user's and the expert's perception of strength were collected separately and in isolation.

Thus, first the participant was interviewed by the researcher in a room separated from the therapist, who was asked about his perception of strength and function and then had his strength measured by the digital dynamometer. Then the participant was taken to another environment to meet the therapist so that the patient, without knowing the results already obtained, made the evaluation of the patient's strength.

At the end the data were tabulated and stored in a spreadsheet, to feed the database that form the referred for statistical analysis. SPSS 13.0 Software for Windows and

Fig. 1. Measurement sequence

Excel 2010 were used for the analysis of the results and all the tests were applied with 95% confidence.

Numerical variables are represented by measures of central tendency and dispersion measures. The Kolmogorov-Smirnov Normality Test for quantitative variables, the Test between paired groups: Paired Student t Test (Normal Distribution) and Wilcoxon (Non-Normal), and the Correlation Coefficient of Pearson (Normal Distribution) and Spearman's (Non Normal). And for the Method for repeated measurements, the mixed linear regression model was used, which takes into account the possible correlation between the values of the response variable that constitute repeated measures.

3 Results

Of these 31 individuals, 19 were female and 12 were males, aged 22–78 years, mean age 48.8 and standard deviation of ±13.97, mostly: (83.9%) right-handed, and retirees/pensioners/home/unemployed (42%), who underwent surgery (61.3%). Of these individuals, 15 (48.4%) reported some improvement after physiotherapeutic treatment, with a mean of 12 (twelve) physiotherapy sessions (standard deviation of ±24.61).

When performing an analysis of the data of the hand treatment, it was possible to observe that there was a convergence between the different instruments in the three moments, they are: Strength in the hand evaluated by the user x Function of his hand user; Strength in Hand Evaluated Specialist x Arduino; Function of your user hand x Arduino; PSR x Function of its user hand, that is, in the three moments these presented a tendency to present similar results, or as close to the real as possible. Shown in Table 1.

The same was done with the contralateral hand, and it was possible to observe that there was a convergence between the instruments during the three moments, they are: Strength in the hand evaluated by the user x Arduino; Strength in Hand Evaluated Specialist x Arduino; Force in the hand evaluated by the user x Function of his user hand, that is, in the three moments these converged to similar results, or as close to the real as possible. Shown in Table 2.

Table 1. Correlation in the three moments – treatment hand

Variables	Treatment hand
Converge	• User-evaluated force on hand x Function of your user hand; • Hand in hand expert x Arduino; • Function of your user hand x Arduino; • PSR x Function of your user hand.
Diverge	• PSR x Strength in the hand evaluated by the user; • PSR x Strength in hand evaluated expert; • PSR x Arduino; • User-evaluated force on hand; • Specialist strength in hand evaluated; • User-evaluated force on hand x Arduino; • Strength in hand evaluated expert x Role of your hand user.

Table 2. Correlation in the three moments - Contralateral hand

Variables	Contralateral hand
Converge	• User-evaluated force on hand x Arduino; • Hand in hand expert x Arduino; • User-evaluated force on hand x Function of your user hand;
Diverge	• Arduino x Function of your user hand; • Strength at hand evaluated by the user x Strength at hand evaluated expert; • Strength in hand evaluated expert x Function of your hand user;

After analyzing the results of the convergence between the instruments of the hand treatment and contralateral, in the three moments, a comparative analysis was performed between these instruments that presented statistically significant correlation in both hands at the 3 different moments. And with that it was possible to detect, which instruments converged in both hands.

As can be seen in Table 3, the results that converged to the same when evaluated, were: Strength in Hand Evaluated Specialist x Arduino and Force in the hand evaluated by the user x Function of his hand user, pointing out that these results tend to be reliable, showing values closer to reality.

Table 3. Comparation between hands

Variables	Treatment hand	Contralateral hand
Converge	• Strength in hand evaluated specialist x Arduino;	• Strength in hand evaluated specialist x Arduino;
	• User-evaluated force on hand x Function of your user hand	• User-evaluated force on hand x Function of your user hand
Diverge	• User-evaluated force on hand x Arduino; • PSR x Function of your user hand; • Function x Arduino	

Already, force in the hand evaluated by the user x Arduino, PSR x Function of its user hand and Function x Arduino, although they are convergent, they are only in one hand, that is, this convergence could not be seen nor proven, in both hands (Table 3).

4 Discussion

When thinking about user surveys about perception, it is important to use tools that are easy to see but still require users to know about. That is, the user should not only perceive, but also understand the information presented. Perception and understanding depend on various factors intrinsic to people, but also the experience of each one in the execution of the required task or during the use of a given product [10].

For the visualization process or identification of the product it is necessary that there be information from specialists in the different areas of knowledge, such as psychologists, ethnologists, graphic designers and artists, among others [10].

Due to a growing relationship with the development of the function of the product, it is essential for the specialist to identify the feelings consumers have when they buy or use the products. However, there is a tendency for designers and consumers to diverge about the psychological perceptions of products they generate [11].

When analyzing several studies using similar instruments, it was observed that in the study by Shin et al. [12], it is observed that the number of individuals studied, resembles the group studied in this study, totaling a number of 30 individuals. This study differs from the works of Ferreira et al. [13], Gonçalves et al. [14], and Savian et al. [15] who studied with 199, 15 and 45 subjects, respectively.

When carried out in this work the comparative analyzes between the self-evaluation user force x expert assessment, self assessment user's force x self assessment user's function, expert evaluation x self assessment user's function, it can be seen that at all moments of the evaluation the results tended to diverge between them, that is, the expert's perception showed different from the user's perception. This type of analysis was not found in other authors, so that discussions on this topic were generated. However, when we think of the Kansei-engineering method, it is possible to observe that this lack of correlation between evaluator and evaluator is justified by the fact that each individual has different expectations and desires. And perception must be considered as a subjective variable, because it is intimately linked to individual psychological processes of each person, further justifying the differences of perception among the individuals of the research [11].

At the end, an analysis of the convergence of the results of all objective and subjective instruments was carried out, based on the correlation between the data evaluated in the three moments of the research and in both hands. And it could be observed that there was a convergence between Force in hand evaluated expert x Arduino (digital dynamometer) and Force in hand evaluated by the user x Function of his hand, user at all times of evaluation.

Therefore, with respect to the analysis to the finding that shows: Strength in the hand evaluated by the expert x Arduino (digital dynamometer), present a strong correlation, which can lead to understand that these results tend to be reliable, that is, very close to the truth. And this can be justified by the studies Savian et al. [15] and

Figueiredo et al. [16], which show that there is a tendency to similar results. But it can also be justified by the fact that the user experience counts a lot, because when we think of User Experience, we think that from previous experiences it can be understood that the individual manages to conceive better, in this case, the force for being a practice clinic, leading to a perception very close to reality [15, 16].

However, when one observes the correlation between the strength in the hand evaluated by the user and the function of his/her user hand, it can be justified by the fact that as he does not know how to evaluate his strength in the proper way, or suggested to evaluate his own strength, he may confuse force with function, as it is what interferes most in his daily activities, leading to an erroneous quantification of his own strength. As it can be observed, the results of the Force in the hand evaluated by the user were very different from those observed by the digital force dynamometer, as well as the results of the divergence function of the digital force dynamometer, which can prove this theory [4].

5 Conclusion

When we evaluated the user of the physiotherapy service about its recovery, its strength and its function, regarding the hand treatment, it could be observed that it cannot perceive how much its strength changed during the physiotherapeutic treatment, different recovery and function. So much so that it is plausible to think that the same cannot understand how to evaluate it, being able to be confusing with the function, because in performing the correlation between these 3 variables, the only one that converges is strength and function.

And with this, it was possible to observe that when correlating these two variables at the end of the three evaluations they tend to converge to the same result, in a statistically significant way. What cannot be seen with any other variable, when we correlate the expert's perception, and/or dynamometer, with the information obtained by the user.

Similar to what was validated on the user's perception, the idea of UX also justifies the fact that the specialist can understand the evolution more effectively, since the test used in the research is widely used in clinical practice by professionals, since the digital dynamometers are very expensive, it is out of the reality of several physiotherapy clinics, and consequently, in the public service would not be different. Therefore, the previous experience of the professional makes all the difference in the perception of what is happening with the individual.

And with this, it is suggested that it is possible to have an idea of how the perception of individuals who are unfamiliar with a given situation or product/artifact works, but never these opinions and information generated by it should be discarded, on the contrary, it is necessary to do more studies with the different types of individuals and their relations of perception that can be used as beacons/tools that assist as decision makers.

References

1. Solomon, M.R.: Consumer Behavior. Needham Heights, Allyn & Bacon (1994)
2. Lanutti, J., et al.: Intuitividade em produto de uso doméstico – um estudo de caso com ralador de queijo - 10° Congresso Brasileiro de Pesquisa e Desenvolvimento em Design, São Luís (MA) – P&D Design (2012)
3. Montignies, F., Nosulenko, V., Parizet, E.: Empirical identification of perceptual criteria for customer-centred design. Focus on the sound of tapping on the dashboard when exploring a car. Int. J. Ind. Ergon. **40**, 592–603 (2010)
4. Forlizzi, J., Battarbee, K.: Understanding experience in interactive systems. In: DIS 2004 Proceedings of the 5th Conference on Designing Interactive Systems: Processes, Practices, Methods, and Techniques, pp. 261–268. ACM, Estados Unidos (2004)
5. Hassenzahl, M., Diefenbach, S., Göritz, A.: Needs, affect, and interactive products – facets of user experience. Interact. Comput. **22**(5), 353–362 (2010). http://www.sciencedirect.com/science/journal/09535438/22/5
6. Forlizzi, J., Ford, S.: The building blocks of experience: an early framework for interaction designers. In: DIS 2000 Proceedings of the 3rd Conference on Designing Interactive Systems: Processes, Practices, Methods, and Techniques, pp. 419–423. ACM (2000)
7. Lu, W., Petiot, J.F.: Affective design of products using an audio-based protocol: application to eyeglass frame. Int. J. Ind. Ergon. **44**, 383–394 (2014)
8. Camara, A., et al.: Percepção do Processo Saúde-doença: Significados e Valores da Educação em Saúde. Revista Brasileira de Educação Médica **36**(1, Supl. 1), 40–50 (2012)
9. Fleck, M.: O instrumento de avaliação de qualidade de vida da Organização Mundial da Saúde (WHOQOL-100): características e perspectivas. Ciência & Saúde Coletiva. **5**(1), 33–38 (2000)
10. Fraunhofer, J., et al.: Research issues in perception and user interfaces. IEEE Comput. Graph. and Appl. **14**, 67–69 (1994)
11. Guo, F., Tian, T.: Consumer demand oriented study on mobile phones'form perception design method. IEEE (2010)
12. Shin, H., et al.: Reliability of the pinch strength with digitalized pinch dynamometer. Ann. Rehabil. Med. **36**, 394–399 (2012)
13. Ferreira, A., et al.: Força de preensão palmar e pinças em indivíduos sadios entre 6 e 19 anos. Acta Ortop Bras. **19**(2), 92–97 (2011)
14. Gonçalves, G., et al.: Força de preensão palmar e pinça digital em diferentes grupos de pilotos da Academia da Força Aérea brasileira. Fisioterapia e Pesquisa. **17**(2), 141–146 (2010)
15. Savian, N., et al.: A eficácia da dinamometria na avaliação da força muscular de diabéticos em relação ao teste de força manual. Colloquium vitae **4**(Especial), 79–83 (2012)
16. Figueiredo, I., Sampaio, R., Mancini, M., Nascimento, M.: Ganhos funcionais e sua relação com os componentes de função em trabalhadores com lesão de mão. Rev. Bras. Fisioter. **10** (4), 421–427 (2006)

A Study on Determining the Heuristics for Evaluating the Usability of Hot Drink Preparation Devices by Elderly Users

Aybegum Numanoglu[(✉)] and Cem Alppay

Faculty of Architecture, Industrial Product Design Department,
Istanbul Technical University,
Taskisla Campus, 34437 Beyoglu, Istanbul, Turkey
begumbiryol@gmail.com, calppay@itu.edu.tr

Abstract. This study focuses on determining usability problems resulting from the usage of coffee and tea preparation machines by elderly users. Both product types are widely available in Turkish market with international and local brands. However they are not widely used by elderly users due to various personal and social causes. The study consists from two data collection techniques. In the initial stage of the research, a face-to-face questionnaire study is conducted with 15 elderly users. All the participants were young-elderly users aged between 65–74 years. The second part of the study consist from an observation study. The data obtained from the questionnaire and the observation was used to document all the errors resulting from the usage of the products that are the focus of this study. Later these errors are classified according to universal design principles in order to obtain error categories to be a basis to formulate the heuristics to be used as the theoretical basis for the heuristic evaluation of hot drink preparation devices.

Keywords: Usability evaluation · Heuristics · Elderly users
Kitchen appliances

1 Introduction

In many countries the proportion of elderly population is increasing. For this reason, increasing research activity is being conducted to determine the problems that elderly users experience during the use of various products. As a result of various studies [5, 7], it was determined that elderly users experienced a great number of problems while using electronic products. Older users, for example, have difficulties when using electronic home appliances with small screens and buttons [8]. In order to create products that will enable the elderly to be able to live alone, designers need to understand the changes in psycho-motor, perceptual and cognitive abilities of older users. With the product designs made in consideration of the user's characteristics, physical and cognitive loss incompetent with age will be compensated. Because products can reduce sensory and perceptual losses; it can also compensate for the loss of physical strength and mobility. However, the design industry does not take into consideration elderly users' abilities [15]. Design methodologies that consider older users include user-centered design and inclusive design are widely accepted approaches.

© Springer International Publishing AG, part of Springer Nature 2019
T. Z. Ahram and C. Falcão (Eds.): AHFE 2018, AISC 794, pp. 362–373, 2019.
https://doi.org/10.1007/978-3-319-94947-5_36

Despite the high number of research focusing on elder users, the number of products that help elderly users to survive their own lives is inadequate. However, most of the elderly either live alone or remain alone during the day. For this reason, there is an important need to design products that will enable the elderly to live their lives alone [7]. Living at their own home alone requires that elderly people also be able to do some daily activities on their own. Preparing food and beverage is one of these activities. Providing preparing food and beverage for the elderly on their own will make their life easier without being dependent on anyone in their own homes. This paper proposes the heuristics to be used in evaluating the usability of household appliances have been revealed. Also, face-to-face questionnaires and observations are conducted for gathering the usability problems of hot beverage preparation devices.

2 Elderly Users

In many design projects, products are designed focusing on a hypothetical "typical user". These hypothetical user are presumed to be male or female, well-educated, capable of using computers and healthy in terms of mentally or physical capabilities However in reality this profile does not fit into elderly users. Elderly users are quite diffrerent in terms of physical and mental capabilities from young people. They also constitute different groups within themselves [6].

2.1 Grouping of Elderly Users

Çirput [3] has divided elderly people into three groups according to their age range: *young elderly aged* (aged between 65–74; this is the period when there is no loss of function); *advanced middle-aged* (aged between 75–84; loss of various functions is seen in elderly individuals in this period); *very old age* (aged 85 and above, the individuals are the 'very advanced age group').

On the other hand, Gregor and Newell [6] have also divided elderly people into three groups: *fit older people* (they do not show any signs of slowing down their cognitive and physical functions and do not think that their functions are restricted. However, cognitive and physical capacities are reduced compared to their youth); *frail older people* (It is thought to be functional loss due to illness or old age. All functions are reduced); *disabled people*, (they have long-term functional losses that affect older ages. Their function has been reduced compared to their previous possession).

World Health Organisation [18] divides elderly people into four groups: 60+ ages refer to elderly or older, 80+ ages refer to oldest old, 100+ ages refer to centenarian, 110+ ages refer to super-centenarian.

2.2 Differences Between Elderly Users

As the age progresses, the mental, physical and perceptual capacity variability of the individual increases. The rate of increase in functional losses also increase with age; rate differs from one people to another. The needs and desires of older people can be different from each other depending on the point they have reached in their lives [6].

Living and working environment is also one of the factors affecting demand and needs. For example; the needs of elderly people who live alone in their own homes are not the same as those of the elderly living in the nursing houses.

2.3 The Effect of Aging on Learning and Adaptation of New Technologies

Old age does not affect the cognitive abilities of individuals in the same way. The ability to learn new information and process existing information varies from one person to another. Memory plays a key role in processing and storing new and existing information. As the age progresses, memory's information processing speed and capacity decrease. For this reason, the ability to store and remember information weakens. Reduced memory capacity increases the time required to learn new information. The duration of completion of a task increases with age. It can be seen that as the difficulty of the task increases, the length of the task completion duration increases. Aging has an impact on the speed of information processing, but there is no direct impact on understanding and learning new information [9]. According to Lim [9], age increases semantic knowledge and aging has no major influence on recall capacity of previously learned information. The vast network of information that grows with age makes it easier to understand, analyze, integrate and remember new information. In recent years, the use of technology among generations is different because of the rapid development of technology. For this reason, the knowledge and abilities of elderly individuals cannot be directly transferred to the use of technological products. Learning occurs when new knowledge is built on past experiences. Experience is a key factor in determining whether a product is easy to use or difficult to use. It is easier for seniors with knowledge and experience in technology to use a new product with similar logic and interface [9].

Psychosocial and biophysical situations, competencies and the problems they face need to be taken into consideration in order to understand how the elderly interact successfully with technological products and systems [2].

3   Usability and Evaluating Usability

3.1   Usability

According to ISO 9241-11, ISO defines usability as *"The extent to which a product can be used by specified users to achieve specified goals with effectiveness, efficiency and satisfaction in a specified context of use."* In other words, the effectiveness, efficiency and satisfaction that users experience while performing their transactions determines the degree of usability of the product.

3.2   Heuristic Evaluation

Heuristic means the mental shortcuts used for decision-making or finding solutions. When a complex problem is encountered or there is not enough information, a solution is reached by using heuristics (mental shortcuts) obtained by evolutionary processes or gained by experience.

Heuristic evaluation is a method proposed by Nielsen and Molich [10] at the beginning of the 1990s for evaluating user interfaces. The purpose of the heuristic evaluation is to identify the usability problems of an interface design. So the problems can be solved [11]. Heuristic evaluation is made by usability experts. Experts evaluate interface designs with the accepted set of usability principles called heuristics [10]. The location and severity of each problem are noted, and experts also express ideas about how the interface can be improved. Nielsen and Molich claim that heuristic evaluations are less costly and more effective than empirical foral usability tests [10–12].

Usability heuristics are practical methods that describe the common characteristics of usable interfaces [10]. They also summarize the best practices of user interface designs [13]. However, heuristics help specialists to focus on the problem areas of user interfaces and to facilitate the perception of usability problems [1]. Contrary to heuristica, the design principles provide a set of rules that are derived from the acquired knowledge over time, and these principles are followed by the designers and developers [10]. Design principles often contain more rules than usual and often seem intimidating [10]. The primary users of the heuristics are usability practitioners and product designers. In 1989, Nielsen and Molich developed a list of 9 evaluation heuristics. These heuristics largely explain the problems found in the telephone and computer interfaces. and still used to explain the problems of today's advanced interfaces. Five years after the publication of the first heuristics, Nielsen [13] revised heuristics to increase their research power. Nielsen compared 7 sets of usability hysterics with a database containing 249 usability problems in 11 software projects. Nielsen classified the heuristics in 7 groups, explaining most of the usability problems after evaluating the usability problems in his study. Groups proposed by Nielsen are visibility of system status, match between system and the real world, user control and freedom, consistency and standards, error prevention, recognition rather than recall, flexibility and efficiency of use. Nielsen has argued that two important evaluation heuristics should be added as eighth and ninth heuristics [13]: aesthetic and minimalist design and help users recognize, diagnose, and recover from errors. In the Nielsen's Usability Inspection Methods book, the tenth usability heuristic has been proposed: help and documentation. Although Nielsen's heuristics are often used in interface design, they have some limitations. One of them is that the heuristics are determined based on only 11 software projects.

3.3 Adaptation of Heuristics

Nielsen's original heuristics have been adapted to many areas. Numerous successful heuristics adaptations are available. Common points to create a heuristic set are • Heuristics are practical methods that summarize the general properties of usable interfaces [10]. • Best interface design practices are compiled [13]. • Generally, the findings of usability tests are taken as a base [13]. • Evaluators are helped to be focused on the points that often cause problems at interfaces [1]. • Usability problems are perceived easier [1]. Heuristics are short, easy to understand and practical way of summarizing the general features of usable interfaces [10]. Heuristics summarize the best practices and principles in the field. Nielsen has based his original heuristics on his own personal experience and on widely known principles of usability [10]. Other usability professionals have based their heuristics onto the rules and common knowledge [1, 4].

4   Methodology

Data Collection
The study consisted from two data collection techniques. In the initial stage of the research, a face-to-face questionnaire was conducted with 15 young-elderly users. The second stage of the study, an observation was conducted. Users consisted of 14 female and 1 male users aged between 65–74 years. Tests were carried on at the participants' homes. Initially, researcher explained the purpose of the study and the process involved. Then, questions on the questionnaire were asked one by one. Answers were written on a paper. Questions are related to demographic characteristic of participants. After the questionnaire, participants were asked to use tea and filter coffee machine respectively. Researchers briefly informed participants about only the stages of filter coffee making. Because in Turkey usage of filter coffee machine is not common at all. Two user experience observation form including tasks was created for tea and filter coffee preparation separately (Table 1).

Table 1. Products that are used in the study

Household appliances	
Hot drink preparation devices	
Tea machine	Filter coffee machine

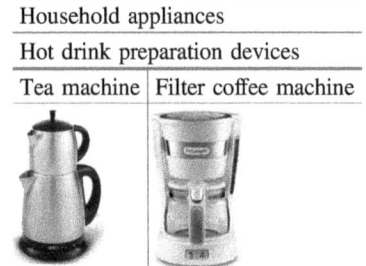

Twenty-nine tasks were grouped under two categories, based on the required activity steps for preparation of tea and filter coffee:

1. *Tasks for preparation of tea*: putting tea in tea pot, closing the tea pot lid, installing filter, filling with water, plug in, putting on the tea pot on water container, running on, waiting for boiling, brewing, adding water, placing water container on plate, waiting for boiling, filling the glass with tea as service, adding water into the glass as a service, cleaning the device.

2. *Tasks for preparation of filter coffee*: plug in, opening the filter receptacle, inserting the paper filter, batching and adding the coffee, batching and adding water, placing the water receptacle onto the hot plate, running on, waiting for coffee brewing,

service, throwing away the paper filter, taking out the filter receptacle, cleaning the filter receptacle, inserting the filter receptacle, cleaning.

Participants were monitored and recorded by video recorder while completing their tasks. Records were watched and evaluated for defining the non-achieved tasks and tasks' completion duration for each participant. All problems observed during the task performances were identified and listed. Additionally, tasks' completion duration for each participant were recorded. Data obtained from the tests were analyzed by different methods.

5 Findings

Usability tests were identified 90 usability problems. After classification, twenty-six distinct usability problems were discovered. Data obtained from the tests were basically analyzed in terms of two parameters: duration of task completion and task completion success grade. After the initial analysis, average and standard deviation of each task' completion duration and completion success grade for each separate task become visible. During the evaluation for calculating the completion success grade,

- not-achieved tasks were graded with '0',
- achieved tasks were graded with '1'.

Results were divided into three categories based on the total grade:

- 0–5 grade tasks have most-common usability problems,
- 6–9 grade tasks have common usability problems,
- 10–15 grade tasks have rare usability problems.

The completion success grade was used to identify tasks that have most-common usability problems. Task that have most-common usability problems, in other terms 0–5 graded tasks detected easily and they were all evaluated in terms of the universal design principles. During the study, universal design principles were thought as usability heuristics.

Most-common usability problems were evaluated one by one according to universal design problems by researchers. Since the usability problem conflicted with one of the universal design principles, problem was graded as '−1'; '0' grade means no conflict. Total score was calculated by multiplying the each problem's frequency by design principles conflict grade and all multiplications were collected. Conflict scores for tea preparation and filter coffee preparation tasks that have most-common usability problems were calculated The bigger negative score means more inconsistency in terms of heuristic the score belongs to. Total conflict scores sorted from biggest to smallest. Prioritization of heuristics has been determined by that sorting. Priority indicates the importance level of heuristics for young-elderly users.

Statistics of filter coffee and tea preparation tasks' completion durations were calculated. For filter coffee preparation tasks, last five tasks were included in cleaning task at all. The longest activity for filter coffee preparation is 'celaning'; the shortest activity is 'opening the filter receptacle'. Task with the highest and lowers standard deviation is 'inserting the paper filter' and 'opening the filter chamber' respectively. Since tea preparation task' are evaluated, it can be seen that the longest activity for tea preparation is 'cleaning the device'; the shortest activity is 'putting on the tea pot on water container'. Task with the highest and lowers standard deviation is 'cleaning the device' and 'putting on the tea pot on water container' respectively. Both analysis' results show that cleaning the products take long time. The reason is that the products consist more than one piece and individual piece needs to be cleaned.

Table 2 shows the total success grade for each filter coffee preparation task. During the evaluation, not-achieved tasks were graded with '0'; achieved tasks were graded with '1'. Results were divided into three categories based on the total grade: 1–5 grade tasks have most common usability problems, 6–9 grade tasks have were common usability problems, 10–15 grade tasks have rare usability problems. According to this categorization, only 'running on' task have most common usability problems.

Table 2. Filter coffee preparation tasks' completion success grade

Tasks	Users															
	K1	K2	K3	K4	K5	K6	K7	K8	K9	K10	K11	K12	K13	K14	K15	Total
Plug in	1	1	1	1	1	1	1	1	1	1	1	1	1	1	1	15
Opening the filter chamber	1	1	1	1	1	1	1	1	0	1	1	1	1	1	1	14
Inserting the paper filter	1	1	1	1	1	1	1	1	1	1	1	1	1	1	1	15
Batching and adding the coffee	1	1	1	1	1	1	1	0	1	1	1	1	1	1	1	14
Batching and adding water	1	1	1	1	1	0	0	0	0	0	0	0	0	0	0	5
Placing the water receptacle onto the hot plate	0	1	1	1	1	1	1	1	1	1	1	1	1	1	1	14
Running on	0	0	0	0	1	1	1	1	0	1	0	0	0	0	0	5
Service	1	1	1	1	1	1	1	1	0	1	1	1	1	1	1	14
Throwing away the paper filter	1	1	1	1	1	1	1	1	1	1	1	1	1	1	1	15
Taking out the filter receptacle	1	1	1	1	1	1	1	1	1	1	1	1	1	1	1	15
Cleaning the filter receptacle	1	1	1	1	1	1	1	1	1	1	1	1	1	1	1	15
Inserting the filter receptacle	1	1	1	0	1	0	1	1	1	1	1	1	1	0	0	11
Cleaning	1	1	1	1	1	1	1	1	1	1	1	1	1	1	1	15

10 out of 15 participants couldn't run the product correctly. 6 out of 10 push the handle mechanism instead of on/off button for running the product. They perceived the handle form as button. Figure 1 shows the failure.

Table 3 shows the total success grade for each tea preparation task. According to categorization, 'closing the pot lid', 'installing the filter', 'running on' and 'placing water container on plate' have most-common usability problems. Common usability problems were discovered during the 'brewing', 'filling the glass with tea as service' and 'adding water into the glass as a service'.

User-1   User-2

Fig. 1. Pushing the handle mechanism to run the product

Table 3. Tea preparation tasks' completion success grade

Tasks	Users															
	K1	K2	K3	K4	K5	K6	K7	K8	K9	K10	K11	K12	K13	K14	K15	Total
Putting tea in tea pot	1	1	1	1	1	1	1	1	1	1	1	1	1	1	1	15
Closing the tea pot lid	0	0	0	0	0	0	0	0	0	0	0	0	0	1	0	1
Installing filter	0	0	0	0	0	0	0	0	0	0	0	0	1	0	0	1
Filling with water	1	1	1	1	1	1	1	1	1	1	1	1	1	1	1	15
Plug in	1	1	1	1	1	1	1	1	1	1	1	1	1	1	1	15
Putting on the tea pot on water container	0	1	1	1	1	1	1	0	0	1	1	1	1	1	1	12
Running on	0	0	0	0	0	0	0	0	0	1	1	0	0	0	0	2
Brewing	0	0	1	1	1	0	0	0	1	0	1	1	0	0	0	6
Adding water	1	1	1	1	1	1	1	1	1	1	1	1	1	1	1	15
Placing water container on plate	0	0	0	0	1	1	0	0	0	0	0	0	0	0	0	2
Filling the glass with tea as service	1	0	1	1	1	1	0	0	0	0	1	1	1	0	1	9
Adding water into the glass as a service	1	1	0	0	0	1	0	0	1	1	1	0	1	1	1	9
Cleaning the device	0	1	1	0	1	1	1	0	0	0	1	1	1	1	1	10

It is seen that since the results are assessed, young-elderly participants were failing in activities requiring force and information about electrical devices. For both products, 'running on' the device couldn't be achieved by most of the participants. Because most of the participants don't know the on/off symbol. When there are more than one button, they get confused and they do not know which one must be pressed. On the other hand,

User-3   User-4

Fig. 2. Closing the pot lid and service

force-required activities such as closing tea pod lit and placing water-filled water container on electrical plate couldn't be achieved comfortably (Fig. 2).

Most common usability problems were evaluated one by one according to universal design problems. Since the usability problem conflict with one of the universal design principles, problem was graded as '−1'; '0' grade means no conflict. Table 4 shows the conflict scores for filter coffee preparation task that have most common usability problems. For 'running on' task, third and fourth principles's grade are −7. frequency of the pushing water receptacle instead of on/off button is 6.

Table 4. Conflict scores for filter coffee preparation task that has most common usability problems

Task	Observation of researcher	Frequency	1	2	3	4	5	6	7
Running on	Handle of the water receptable limits the users vision. So she couldn't see the on/off button	1	0	0	−1	−1	0	−1	0
	User pushes the aroma button instead off the on/off button.	3	0	0	−1	−1	0	−1	0
	At the first trial, she pushes the mechanism of the water receptacle	6	0	0	−1	−1	0	−1	0
	Total		0	0	−10	−10	0	−10	0

1 = Equitable use, 2 = Flexibility in use, 3 = Simple and intuitive use,
4 = Perceptible information, 5 = Tolerance for error, 6 = Low physical effort,
7 = Size and space for approach use

Conflict scores for tea preparation tasks that have most common usability problems were calculated. The highest negative score belongs to 'closing the pod lid' task. Closing the pot lid is a safety issue. The lid avoids spilling the boiled tea onto the user. Therefore, task must be required low physical effort and in accordance with equitable use, flexibility in use and tolerance for error principles.

Total conflict scores for hot drink preparation devices can be seen in Table 5. Conflict scores for tea preparation and filter coffee preparation tasks that have most-common usability problems were calculated.

Table 5. Total conflict scores for hot drink preparation devices

Hot drink preparation devices	1	2	3	4	5	6	7
Tea machine	−14	−14	0	0	−14	−14	0
	0	0	−11	−11	0	0	0
	0	0	−10	−10	0	0	0
	0	−9	0	0	−9	−9	0
Filter coffee machine	0	0	−10	−10	−9	−7	0
TOTAL	−14	−23	−31	−31	−32	−30	0
Prioritization	5.	4.	3.	3.	1.	2.	N/A

1 = Equitable use, 2 = Flexibility in use, 3 = Simple and intuitive use, 4 = Perceptible information, 5 = Tolerance for error, 6 = Low physical effort, 7 = Size and space for approach use

6 Results and Discussions

During the study, universal design principles have been thought as usability heuristics. Total conflict scores sorted from biggest to smallest. Prioritization has been determined by that sorting. Proposed heuristics can be seen in Table 6. The bigger total conflict score means more inconsistency in terms of heuristic the score belongs to.

Table 6. Original heuristics and proposed heuristics

	Original heuristics	Proposed heuristics
1	Equitable use	The design should minimize the hazard risk
2	Flexibility in use	The design should be used with minimum physical effort
3	Simple and intuitive use	The design should be easy and simple to understand
4	Perceptible information	The design should be used by the individuals that have distinct capabilities
5	Tolerance for error	
6	Low physical effort	
7	Size and space for approach use	

When the results are evaluated, it is seen that the principles that are most infringed are 3rd and 4th principles. It means that young-elderly participants have usability problems mostly in terms of 'simple and intuitive use' and 'perceptible information'. It means that they couldn't know easily what they will do next. In addition, most of them have no technology background. So, they couldn't guess how they use the product easily. On the other hand, physical loss makes them weaker and more open to the hazards.

References

1. Baker, K., Greenberg, S., Gutwin, C.: Empirical development of a heuristic evaluation methodology for shared workspace groupware. In: Proceedings of the 2002 ACM Conference on Computer Supported Cooperative Work, pp. 16–20. New Orleans, Louisiana (2002)
2. Chen, K., Chan, A.H.: A review of technology acceptance by older adults. Gerontechnology 10(1), 1–12 (2011)
3. Çiprut, H.: İstanbul kentinde farklı ekonomik-kültürel düzeyde yaşlılık ile ilgili sorunlar. (Yayınlanmamış Yüksek Lisans Tezi) İstanbul Üniversitesi Halk Sağlığı Anabilim Dalı, İstanbul (1996)
4. Desurvire, H., Caplan, M., Toth, J.: Using heuristics to improve the playability of games. In: CHI Conference (2004)
5. Freudenthal, A.: The design of home appliances for young and old consumers (1999)
6. Gregor, P., Newell, A.F.: Designing for dynamic diversity: making accessible interfaces for older people. In: Proceedings of the 2001 EC/NSF Workshop on Universal Accessibility of Ubiquitous Computing: Providing for the Elderly, pp. 90–92. ACM (2001)
7. Hotta, A.: The direction of design in aged society. Bull. JSSD 4(3), 35–42 (1997)
8. Lee, C.F., Kuo, C.C.: A pilot study of ergonomic design for elderly Taiwanese people. In: Proceedings of the 5th Asian Design Conference-International Symposium on Design Science, Seoul, Korea, TW-030 (2001)
9. Lim, C.S.C.: Designing inclusive ICT products for older users: taking into account the technology generation effect. J. Eng. Des. 21(2–3), 189–206 (2010)
10. Molich, R., Nielsen, J.: Improving a human-computer dialogue. Commun. ACM 33(3), 338–348 (1990)
11. Nielsen, J.: Finding usability problems through heuristic evaluation. In: Proceedings of the SIGCHI Conference on Human Factors in Computing Systems, pp. 373–380. ACM (1992)
12. Nielsen, J., Phillips, V.L.: Estimating the relative usability of two interfaces: heuristic, formal, and empirical methods compared. In: Proceedings of the INTERACT 1993 and CHI 1993 Conference on Human Factors in Computing Systems, pp. 214–221. ACM (1993)
13. Nielsen, J., Mack, R.L.: Usability inspection methods. John Wiley & Sons, New York (1994)
14. Nielsen, J.: Enhancing the explanatory power of usability heuristics. In: Proceedings of the SIGCHI Conference on Human Factors in Computing Systems: Celebrating Interdependence, pp. 152–158. ACM (1994)
15. Pinto, M.R., De Medici, S., Zlotnicki, A., Bianchi, A., Van Sant, C., Napou, C.: Reduced visual acuity in elderly people: the role of ergonomics and gerontechnology. Age Ageing 26 (5), 339–344 (1997)

16. Sayago, S., Sloan, D., Blat, J.: Everyday use of computer-mediated communication tools and its evolution over time: an ethnographical study with older people. Interact. Comput. **23**(5), 543–554 (2011)
17. Ten usability heuristics (2005). http://www.useit.com/papers/heuristic/heuristic_list.html
18. World Health Organisation. http://www.searo.who.int/entity/health_situation_trends/data/chi/elderly-population/en/

Comparative Study on Reading Performance of Different Electronic Ink Screens

Yuan Lyu[1], Yunhong Zhang[2(✉)], Wei Li[2], Zhongting Wang[2], and Lou Ding[3]

[1] Elementary Education College of Capital Normal University, Beijing, China
[2] AQSIQ Key Laboratory of Human Factor and Ergonomics,
China National Institute of Standardization, Beijing, China
zhangyh@cnis.gov.cn
[3] Tianjin Nomal University, Tianjin, China

Abstract. The study mainly discusses the difference between the comfort and visual fatigue in the process of reading with the two different mobile phone screens under the different conditions of the general mobile phone screen and the mobile electronic ink screen, in addition to test the performance difference between the electronic ink screen and the general mobile phone screen. This study simulated the ambient light conditions of the office, randomly selected 20 subjects from18 to 40 years old to explore the difference of reading task performance, flicker fusion frequency, visual fatigue subjective perception and electroencephalogram EEG, etc. under different conditions between the general mobile phone screen and mobile electronic ink screen, and also test the performance difference of them. The experiment adopted within-subjects design, the independent variables was different mobile phone screen (general screen, electronic ink screen), the dependent variable is the performance of the reading task. In summary, the reading experience of mobile phone with electronic ink screen is better when reading, and it is not easy to appear the phenomenon of visual fatigue, and the effect of user experience is better.

Keywords: Electronic ink screen · Mobile phone screen · Visual fatigue

1 Introduction

Since 1998, electronic paper book called Rocket EBook was launched in the United States. Then the e-book attracted people attention and developed rapidly with its unique way of reading. In 2004, SONY launched the first "electronic ink" electronic reader, can obtain the same paper book reading experience. In 2006, e-books began to enter people's daily lives [1]. Electronic paper book has many advantages, it is based on the technology of electronic ink display, and its reading experience is close to paper books, with the natures of environmental protection and comfortable, without injury to the eyes. The unique electronic ink screen is extremely thin, so it is very energy saving and environmental protection. The unique intelligent power management technology makes it only need to consume electricity when the text changes, and continuous standby time of up to ten days,in addition,the biggest advantage is to protect the eyes, and when

© Springer International Publishing AG, part of Springer Nature 2019
T. Z. Ahram and C. Falcão (Eds.): AHFE 2018, AISC 794, pp. 374–380, 2019.
https://doi.org/10.1007/978-3-319-94947-5_37

reading for a long time, there are almost no flicker, no radiation, no hurt the eyes and help you protect the eyesight [2]. Compared with the electronic reading books, mobile reading devices developed rapidly because of the advantages such as convenient carrying, variety of reading methods and flexible operation. The mobile phone reading has become the most popular reading method because of the portability, payment maturity and strong interaction [3]. The results showed that reading APP of different experience had different effects on the reading comprehension rate of literature. The reading comprehension rate of APP was higher than that of Kindle, but there was no difference on reading time [4]. There was no difference between the reading time of an electronic ink screen and that of an ordinary mobile phone screen, but the electronic ink screen may be better than the normal phone screen in terms of protecting eyesight and relieving fatigue. Therefore, some manufacturers try to combine the advantages of the two and design the electronic ink screen in mobile phones specifically for reading. Whether this new type of reading method can continue the advantages of protecting eyesight and prevent fatigue of the electronic ink screen and whether it can bring good reading experience, the effect remains need to be verified. This study used reading tasks to test the reading performance and visual fatigue of the electronic ink screen and that of the ordinary mobile phone screen, and it also tests whether there are advantages of the electronic ink screen reading in the mobile phone compared with that of the OLED screen reading of mobile phone.

2 Methods

2.1 Design

This experiment was to test the subjective and objective differences of reading task performance using mobile phone displays controlled by different displays (E-ink display and general OLED display) under indoor light uniform conditions. This experiment was a two-factor within-subject design. Two factors were different screen samples and time factors. Samples contain three different conditions: E-ink display and general OLED display, respectively. Time factor was before and after reading. Dependent variables were reading performance, visual fatigue, emotional response and comfort under different conditions of the task.

2.2 Participants

Seventeen ordinary right-handed adults from 18 to 35 years old (11 male and 9 female, mean age = 25.4, SD = 2.58) were recruited and paid to participate in the experiment. All of them had 4.8 normal or corrected-to-normal visual acuities and healthy physical conditions, without ophthalmic diseases.

2.3 Samples

The samples were a Normal OLED display phone, an E-ink display phone from a same company. Normal OLED display phone: screen is 5.5 in., resolution ratio is 720×1280

and electronic ink screen is 5.2 in., resolution ratio is 720×1280. The electronic ink screen of mobile phone selected the E-reader with the refreshing speed faster than Kindle, no screen flicker, while Kindle E-reader will appear the flashing screen full-screen refresh when reading. After testing under the condition of 1000 lx illumination, the performance index of mobile phone with electronic ink screen is superior to that of Kindle e-reader in general. The performance parameters are compared as follows (Table 1):

Table 1. Comparison of performance test parameters of electronic ink screen mobile phone and kindle electronic reader

	Splash screen	Remnant shadow	Multipoint touch control	Sliding operation	Video	Glare proof ability (Halo diameter)(Centimeter)	Brush screen speed (Millisecond)	Grey scale(Order)
Electronic ink screen of phone	None	Slight	Support	Support	Support	3.7	100	16
Kindle Voyage	Yes.	Obvious	Not supported	Not supported	Not supported	2.8	336	16

2.4   Apparatus

The research used Standard logarithmic visual acuity chart developed by the eye hospital of WMU [5], the BD-II-118 type critical fusion frequency, single electrode EEG equipment to record the EEG indicators during the experimental process. The visual fatigue scale developed by James E. Sheedy was used to test the visual fatigue after the task, including eye fatigue (such as eye burning, eye pain, eye strain, eye irritation, eye tearing, visual blur, double vision, eye dryness and headache, etc.) [6].

2.5   Procedures

Experiments were conducted in a quiet laboratory that experimental environment illumination value between 500 lx–700 lx. First, participants signed the informed consent and completed a general survey about their demographic information. The main task of using mobile phone were reading a popular science article "the Silk Road: an entirely new world history", and subjects were required to read randomly from a chapter of the reading material, read freely according to their usual reading habits. And, the location of the beginning and end of reading action were recorded, and the numbers of that the subjects had read was finally summarized. The reading time was about

40 min. The critical fusion frequency test was applied before and after the reading task. The subjects need wear EEG to record EEG changes in the entire mobile phone reading task process. The test sequence of samples in each subject was balanced by ABBA method. After completing the test of each sample, the subjective interviewed on the comfort and other aspects of all the samples. There was 20 to 30 min rest between each two tests to avoid visual fatigue. After completing the experiment, the subjects would got a certain reward.

2.6   Data Analysis

The changes value of visual search and visual fatigue data were analyzed by IBM SPSS 20 Statistics software (IBM-SPSS Inc. Chicago, IL). The repeated measured ANOVA analysis is applied to the experiment data to compare the differences of 2 factors from 3 quantitative indicators, namely: reading performance, visual fatigue, subjective report data and emotional response EEG and comfort EEG.

3   Results and Analysis

3.1   Reading Performance

The reading task test results showed that the reading performance of the subjects under the condition of normal OLED screen was slightly higher than that of the mobile phone sample with electronic ink screen. The repeated-measured ANOVA of attention index showed that there was no significant difference between the reading performance of the normal OLED screen condition (M = 381.93, SD = 122.177) and that of mobile phone with Electronic Ink screen (M = 381.37, SD = 175.706). The repeated-measured ANOVA results of emotion index showed that there was no significant difference in reading performance under different conditions (P = .985 > 0.05). The results displayed that there was no significant difference in reading performance between the mobile phone samples with ordinary OLED screen and electronic ink screen.

3.2   Degree of Objective Visual Fatigue

The repeated measurement ANOVA results of the critical fusion frequency after reading with normal OLED screen and the baseline of pre-task critical fusion frequency with repeated measurement showed that there was a significant difference between the two conditions (F = 14.610, p < 0.01). Critical fusion frequency of common OLED screen after reading task (M = 33.76, SD = 2.159) significantly lower than that before reading task (M = 34.56, SD = 2.552) (P < 0.01). The repeated measurement ANOVA of the critical fusion frequency of mobile phone samples with electronic ink screen after the reading task and before reading task showed that there was a significant difference between the two conditions (F = 7.814, p < 0.05). Critical fusion frequency after reading by using an electronic ink screen mobile phone sample (M = 33.83, SD = 2.178) significantly lower than that before reading task (M = 34.56, SD = 2.552) (P < 0.05).

The repeated measurement ANOVA of the critical fusion frequency after reading task by mobile phone sample with ordinary OLED screen and electronic ink screen with repeated measurement showed that there was no significant difference between them (F = 0.140, p > 0.05).

The repeated measurement ANOVA of the decreasing amplitude of critical fusion frequency under the conditions of two kinds of mobile phone screen showed that there was no significant difference (P = .713 > 0.05) between the decreasing amplitudes of critical fusion frequency under the condition of ordinary OLED screen (M = –0.800, SD = 0.936) and those of electronic ink screen (M = –0.723, SD = 1.156). The results showed that there was no significant difference in fatigue between ordinary OLED screen and electronic ink screen.

3.3   Reading Experience

According to the results of the reading experience of two mobile phone screens, the repeated measurement ANOVA of the reading experience under the two different mobile phone conditions showed that: (1) In terms of screen brightness, the screen brightness of ordinary OLED screen (M = 63.3, SD = 16.490) was significantly higher than that of electronic ink screen mobile phone (M = 40.85, SD = 12.708), F = 40.227, p < 0.001. (2) In terms of the screen without harsh, the electronic ink screen mobile phone sample (M = 87.7, SD = 14.165) was less than the ordinary OLED screen (M = 62.65, SD = 22.293), F = 29.413, p < 0.001. (3) In terms of the screen without flashing, the flicker frequency of the electronic ink screen mobile phone sample (M = 88.40, SD = 12.542) was significantly higher than that of the ordinary OLED screen (M = 79.65, SD = 19.597), F = 5.897, p < 0.05. (4) In terms of page refreshing speed, the ordinary OLED screen (M = 76.90 (SD = 14.542) was significantly faster than that of the electronic ink screen mobile phone sample (M = 30.55, SD = 22.049), F = 62.201, p < 0.001. (5) In the terms of page without blur flicker, electronic ink screen mobile phone samples (M = 60.15, SD = 25.938) was significantly higher than that of ordinary OLED screen image flicker (M = 82.00, SD = 22.946), F = 8.100, p < 0.05. (6) In the terms of touch sensitivity, the ordinary OLED screen M = 78.65 (SD = 10.703) was significantly higher than that of electronic ink screen, M = 41.00, SD = 20.787, F = 47.524, p < 0.001. (7) In terms of screen definition, the clarity of OLED M = 82.55 (SD = 12.878) was significantly higher than that of the electronic ink screen (M = 67.00, SD = 21.514), F = 10.291, p < 0.01. (8) In terms of color screen, electronic ink screen mobile phone sample (M = 46.15, SD = 14.702) was rather cool, while the ordinary OLED screen (M = 52.95, SD = 19.110) was rather warm, but the differences were not significant (F = 1.768, p > 0.05) (Fig. 1 and Table 2).

3.4   Reading Experience

The results of EEG data showed that under the condition of the mobile phone sample with electronic ink screen, the descend rang of attention index (M = –7.15, SD = 14.25) was larger than that of normal OLED screen (M = –5.75, SD = 11.65). The repeated measurement ANOVA of the indexes showed that there was no significant

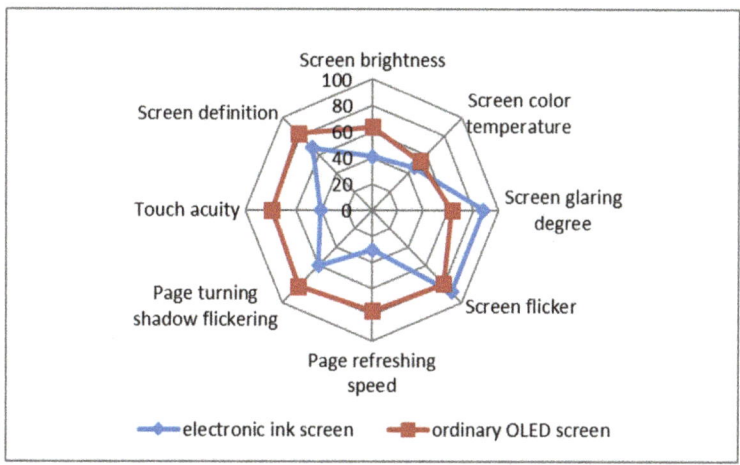

Fig. 1. Comparison of user reading experience results from different mobile phone conditions.

Table 2. Comparison of user reading experience results with different mobile phone conditions

	SS	df	MS	F value	Sig. (2-tailed)	η^2
Screen brightness	5040.025	1	5040.025	40.227	.000***	.679
Screen color temperature	462.4	1	462.4	1.768	.199	.085
Screen glaring degree	6275.025	1	6275.025	29.413	.000***	.608
Screen flicker	765.625	1	765.625	5.897	.025*	.237
Page refreshing speed	21483.225	1	21483.225	62.201	.000***	.766
Flip flashing blur	4774.225	1	4774.225	8.100	.010*	.299
Touch sensitive acuity	14175.225	1	14175.225	47.524	.000***	.714
Screen clarity	2418.025	1	2418.025	10.291	.005**	.351

Note1: "*" indicates that there was a significant difference at 95% confidence level.
"**" indicates that there was a significant difference at 99% confidence level.
"**" indicates that there was a significant difference at 99.9% confidence level.

difference between the two conditions (F = 0.141, p > 0.05). The mood index rose, the rising amplitude of the mood index under the condition of electronic ink screen (M = 6.45, SD = 14.277) was larger than that under the normal OLED screen condition (M = 2.85, SD = 14.110). The repeated measurement ANOVA results showed that there was no significant difference between the two conditions (F = 0.688, p > 0.05).

4 Test Conclusion and Discussion

This study compared two different types of mobile phone screen reading performance, EEG immediate response data, the critical fusion frequency data, etc. It was found that the reading task performance, the critical fusion frequency display of visual fatigue, attention and emotion index and color temperature perception of the ordinary OLED screen were superior to those of the electronic ink screen mobile phone samples. But in the subjective experience of dazzling and flashing, the performance of electronic ink screen mobile phone samples was better than that of ordinary OLED screen, indicating that user evaluation of electronic ink screen phone samples was more comfortable in dazzling and flashing. At the same time, it may be due to the characteristics of the electronic ink screen, the ordinary OLED screen was significantly superior to the mobile phone sample in terms of clarity and touch acuity. And the electronic ink screen flashed when reading, but it not appeared in the ordinary OLED screen. So the above three aspects should be improved by the electronic ink screen mobile phone samples. In terms of brightness, users also felt the brightness of the ordinary OLED screen was significantly higher than that of the electronic ink screen mobile phone. Therefore, although the electronic ink screen mobile phone sample looks more comfortable, but due to the influence of the residual scintillation, touch sensitive flip caused by low speed factors, the overall reading speed of electronic ink screen and visual fatigue advantages were not prominent, reading performance and critical fusion frequency difference display of visual fatigue did not reach the significant level.

Acknowledgments. The authors would like to gratefully acknowledge the support of the National Key R&D Program of China (2016YFB0401203), and China National Institute of Standardization through the "special funds for the basic R&D undertakings by welfare research institutions" (522018Y-5942, 712016Y-4940).

References

1. Qiao, T.: The research on interaction design of electric paper books. (Doctoral dissertation, HeFei University of Technology) (2012)
2. Meng, R.: Popular Research on the Development of Electronic Paper Books. (Doctoral dissertation, Henan University) (2011)
3. Shan, C., Jinjing, L.: The mobile reading era. Chin. New Commun. **13**(3), 52–57 (2011)
4. Dan, W., Liuxing, L.: A study on the influence of mobile reading tools on the reading efficiency of College students' academic literature (1), 64–72 (2017)
5. Standard Logarithmic Visual Acuity Chart Developed by the Eye Hospital of WMU. People's Medical Publishing House, July 2012
6. Sheedy, J.E., Hayes, J., Engle, J.: Is all asthenopia the same? Optom. Vis. Sci. **80**(11), 732–739 (2003)

Research on Interactive Innovation Design of Barrier-Free Products for Visually Impaired Groups

Zijie Xie$^{(\boxtimes)}$

The College of Fine Art, Hunan Normal University, Changsha 410012, China
695767314@qq.com

Abstract. According to the characteristics and needs of the visually impaired, an innovative design method for visually impaired groups with barrier-free products under the guidance of modern interactive design concepts is proposed to address barriers and problems when interacting with the products, and to create interactive products for visually impaired people with barrier-free products.

The study investigated and analyzed the characteristics and needs of typical visually-impaired people in daily life, studied current market conditions and problems of barrier-free products, developed a visually-impaired barrier-free product design under interactive design concepts. The barrier-free products for the visually impaired should be guided by the interactive design concept and interact innovative design from the perspectives of touch, hearing, taste and smell.

Keywords: Interactive innovation design · Barrier-free products
Visually impaired

1 Introduction

Barrier-free products refer to barrier-free urban architecture, transportation means and facilities in public environment especially designed for social-disadvantaged groups (including children, pregnant women, the handicapped and the elderly), such as sidewalks for the blind and toilets for the handicapped. Barrier-free emphasizes all planning and designing of public space environment, architectural facilities and devices, which are related to people's food, clothing, accommodation and traveling in the modern society with highly developed scientific technology, should give full considerations to people with different levels of physical disabilities and declining capacities. These facilities should be equipped with service functions and devices that are capable of meeting users' needs. More important, barrier-free designing aims at creating a modern, comfortable, convenient and living environment that is filled with love and care to keep humans safe [1].

Interactive innovation design is an increasingly important method for visually impaired groups to understand complex information from so called barrier-free products. Interactive design is an interdisciplinary field of industrial design, communication design, human-computer interaction, cognitive psychology, anthropology and sociology. It covers a wide range of specific application areas, from business to

© Springer International Publishing AG, part of Springer Nature 2019
T. Z. Ahram and C. Falcão (Eds.): AHFE 2018, AISC 794, pp. 381–389, 2019.
https://doi.org/10.1007/978-3-319-94947-5_38

telecommunications, entertainment and gaming to medicine. Successful interactive designs have simple, clearly defined goals, a strong purpose and intuitive screen interface.

Recently, more barrier-free products have been developed which could minimize the effects of a disability, improve quality of life, enhance social participation, and improve life skills, mobility and cognitive abilities, while providing a motivating and interesting experience for groups with disabilities. But, based on the market research on current products for the visually impaired, the author found out many significant problems in such products: (1) single functions; (2) outmoded styles and slow designing update; (3) poor user experience and complex operations; (4) threats to personal safety; (5) enhanced psychological burden; (6) polarized prices. These problems are caused by three reasons: a. The market pays inadequate attention to products for the visually impaired; b. Product development enterprises and designers lack enough knowledge of the visually impaired; c. Designers fail to designing innovative products for the visually impaired by combining interactive designing concepts [1, 2].

This chapter has the following organization. Section 2 Analyze the Characteristics and Needs of Visually Impaired Users. Section 3 Propose the principles and innovative ways of product interaction design for visually impaired groups. Finally, challenges are discussed in Sect. 4 and an extensive bibliography is provided.

2 Analyze the Characteristics and Needs of Visually Impaired Users

The thesis mainly researches a socially disadvantaged group, namely the visually impaired. The designing core is user-centered. The following designing research starts with the visually impaired.

2.1 Analysis of Visually Impaired Users

According to the statistics of survey results, there are over 500 million people with physical disabilities among six billion people in the world. Among 1.3 billion of Chinese, 80 million people have disabilities and 12.33 million people have visual impairment, which accounts for 14.86%. On the other hand, different levels of ophthalmocopia may also cause visual impairment to the elderly. Such shocking statistics shows the visually impaired group is a large-scale disadvantaged group, whose work and life urgently need humane and barrier-free service designing. It is necessary to follow client-centered designing principle in interactive designing methods to meet their basic needs in life. Regarding the regional distribution proportion of visually impaired users nationwide, Guangdong Province is home to the majority of visually impaired users in survey samples, which account for 21.09%. Beijing, Jiangsu and Shandong rank in a descending order.

Visually impaired users are scattered in the society. The questionnaire aims to study objects in several schools for the blind and shopping malls selling products for the visually impaired in Hunan, China. The author released 500 questionnaires in all and recycled 479 questionnaires. However, only 418 questionnaires were valid.

The questionnaire has four objectives: (1) survey users' basic information; (2) find out the difficulties faced by visually impaired users; (3) survey how visually impaired users use barrier-free products; (4) find out visually impaired users' opinions of and attitudes towards existing barrier-free products.

The visually impaired face many difficulties and dangers in daily activities, such as driving, reading, social contacts and walking. The author classified the typical problems faced by the visually impaired in daily life into three categories through survey, analysis and arrangement, including daily life problems, product application problems and psychological problems.

2.2   Analysis of Visually Impaired Users' Characteristics

(1) Analysis of Physiological and Behavioral Characteristics

Unlike common people, the visually impaired are unable to combine their visual sense, auditory sense, tactile sense, olfactory sense and sense of taste for sensory integration. The visually impaired perceive the environment by sensing the existence of objects, having stimulus reactions and receiving information [3]. Every step should be accurate in the perception process. Otherwise, it may affect the accuracy of information received by the visually impaired. On the other hand, the lack of good vision makes the visually impaired highly sensitive towards other senses, such as the auditory sense, tactile sense and memories. It is the so-called sensory compensation: the perception behaviors of the visually impaired replace and compensate for senses other than vision.

(2) Analysis of Psychological and Perceptual Characteristics

In general, the visually impaired have some psychological problems, including self-abasement, loneliness and psychological contradictions. Because of visual impairment, these people face many restrictions in life, which result in social-networking isolation. However, such interpersonal barriers have deepened the common people's stereotyped impressions of the visually impaired, who mistaken the visually impaired to be psychologically and physiologically unhealthy. In fact, the visually impaired only have visual problems. Apart from that, they are the same as common people, with dreams and objectives. In addition, the visually impaired give play to their unique capacities through training. Secondly, the visually impaired desire to use their favorite products the same as common people in daily life and find out information on products, which they care about. It is thus necessary to design products from the perspective of meeting users' psychological needs. Apart from considering existing visual sense, designers should also make full use of such users' normal and even better auditory sense, tactile sense and olfactory sense, so as to design barrier-free products suitable for the visually-impaired [5].

Users' needs for products include material needs (namely satisfaction from product functions) and spiritual needs (namely products should not be special and carry no obvious logo for disabilities). It fully represents Maslow's hierarchy of needs.

2.3   Analysis of Visually Impaired Users' Needs

Combining with visually impaired users' psychological and physical characteristics mentioned above, as well as the hierarchy of needs put forward by American psychologist Maslow, the author concluded four keywords for visually impaired users' needs: (1) psychological needs; (2) security needs; (3) emotional and social needs; (4) needs for respect. After a product achieves basic visualization, it should meet users' needs for self-realization and sense of achievement. It is thus necessary for products to meet users' needs for individuality and achieve good using experiences. Through interactive designing methods, designers should make products more interesting based on meeting users' basic needs. For one thing, it lowers the frequency for the visually impaired to use products in the wrong way. For another, it decreases users' psychological pressure, improves using efficiency and enhances using experiences Fig. 1.

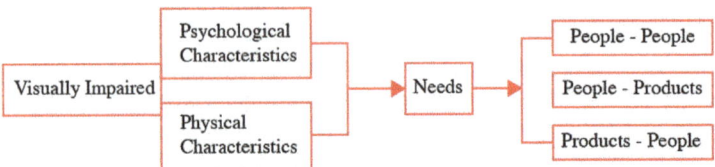

Fig. 1. Visually impaired user's needs

3   Interactive Design of Products for the Visually Impaired

Combining with analysis of the visually impaired and barrier-free products, namely the analysis of people-object, object-people and people-people, this chapter makes conclusions from the perspective of product interactive designing. In addition, it puts forward theoretical principles and innovative interactive modes to solve the problems mentioned above.

3.1   Interactive Design Principles on Products for the Visually Impaired

The style of human-machine interactions has shifted from order interface and graphical interface to multi-media interface. At present, it is developing towards virtual reality technology and multi-channel client interface [4]. In the current phase, the interactions between humans and products rely on the visual interface and users receive information through visual sense. As a result, an increasing number of users have poor vision and even suffer visual impairment. As we pay an excessive emphasis over interactions on visual interface, we also set a barrier for visually impaired users unconsciously. The multi-channel user interface helps more people to communicate with products more conveniently. In particular, gesture interaction, eye movement tracking and facial expression recognition are more worthy of promotion Fig. 2.

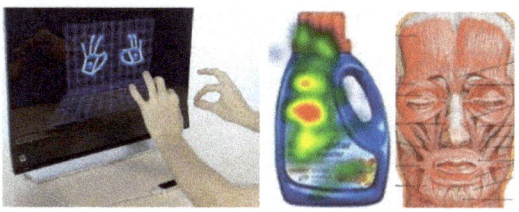

Fig. 2. Gesture interaction, eye movement tracking and facial expression recognition

(1) Safety:

Safety problems are elements considered by every user in the first place. Due to visual impairment, the visually impaired have poor capacity of identifying the external environment, which results in a weaker capacity of responding to external environment than common people. If a danger occurs, the visually impaired are unable to make responses in the first time. In addition, they may face the risk of wrong manual operations when operating a product. As a result, the visually impaired are unable to interact with products smoothly, have information communications or send out application orders normally. It is thus necessary for products to offer in-time feedbacks to operations and tell the visually impaired which step their operation is in. Conversely, the lack of in-time feedbacks for users may lead to operation difficulties and even wrong operations that cause potential safety hazards. For this reason, safety operation of products is the most fundamental principle of product designing [5].

(2) Usability:

Products should ensure users face no pressure and burden in operations and have no worry for difficulties and confusions in application. In this way, each user is able to experience products smoothly. For personal reasons, the visually impaired may have lower efficiency of inputting information than common users. In addition, there may be a higher rate of making errors. For this reason, decreasing users' input in application effectively helps to reduce the psychological pressure for visually impaired users in operations. How to help the visually impaired to decrease information recording in designing as much as possible will be worthy of considerations. The following details are worthy of noticing in the usability principle:

a. Replace text input orders with gesture interactions: Fast gesture operations help the visually impaired to reduce unnecessary trouble; b. Enough feedback reminders help and guide users to make operations; c. Automatically help users to save recorded information; d. Try to help users to turn the orders of recording information into selection orders; e. Try to use more sensory system modes and avoid using single-sensory channels.

(3) Perceptual Validity:

In information transmission, traditional interaction modes are unsuitable for visually impaired users. This is because users are unable to send out instruction orders and receive feedback information based on the information transmission via the interface.

For instance, the interaction modes of many mobile devices adopt "slide to unlock". In other words, users need to slide a password with sliding trajectory to unlock the device. However, such a designing lacks perceptual validity for visually impaired users. It is because such users have visual defects and can hardly participate in product interactions.

(4) Emotional Designing:

Among all users' interactive experiences, emotional designing accounts for a large proportion. The reason why a product is widely pursued by consumers not only lies in its functions that meet users' needs, but also in the care to visually impaired users' needs in user experiences. In fact, such care represents is a manifestation form of emotional designing. (a). Emotional product form: A form generally refers to an object's image, appearance and shape. It is understood as the emotional factor for product appearances. The thesis is more inclined towards understanding it as the combination of a product's internal qualities and other senses of the visually impaired; (b). Emotional product qualities: A product should be capable of conveying emotions, triggering memories and creating surprises. Designers should thus set up emotional connections between products, services and the visually impaired. In addition, inter-actions affect users' self-image, satisfaction and memories. As a result, users gain understandings of the brand and develop loyalty towards the brand. As time passes, brand becomes the representative or carrier of emotions; (c). Emotional product operations.

3.2 Innovation Ways of Interactive Design of Products for the Visually Impaired

Based on the above research, we propose five innovative ways of interaction design of products for the visually impaired: auditory interaction design, haptic interaction design, olfactory interaction design, taste interaction design, and multisensory experience interaction design Fig. 3.

(1) Auditory interactive design:

By utilizing the characteristics and advantages of the visually impaired's hearing, it is possible to better implement the functions of the products for the visually impaired and bring a good product interaction experience. In the real environment, about 70% of the information that humans obtain from the outside world comes from the visual sense with audition accounting for about 15% to 20%, which is the most important way of obtaining information in addition to visual accidents [7]. For the visually impaired, interaction is possible even without seeing or clearly seeing the interactive objects. Therefore, auditory interaction design is one of the important ways in the interactive design of accessibility products for the visually impaired. For the obstacle products, its role mainly includes: interactive feedback, information prompts, warnings, assistance, strengthen memory, share overload information and so on. Good auditory interaction design can not only improve the visually impaired work efficiency, reduce errors, but also enable them to enjoy hearing pleasure [8].

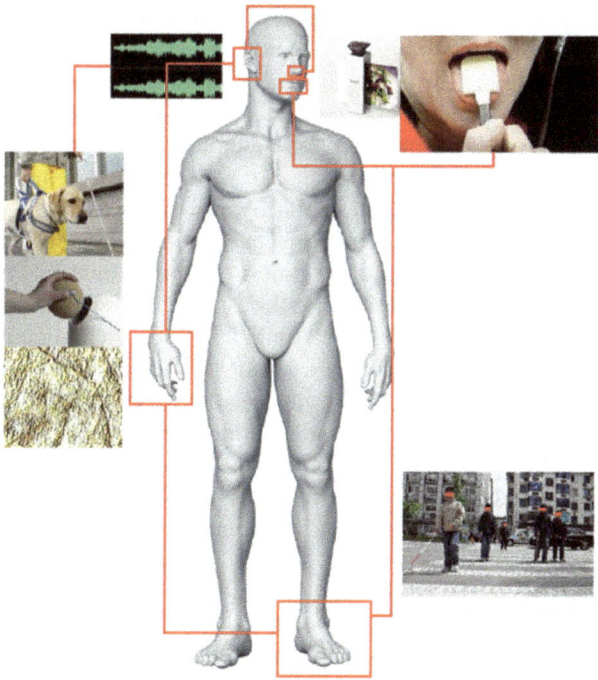

Fig. 3. Multisensory experience interaction design

Example 1: A visually impaired user with an acutely audible hearing ability can accurately identify the person who is around him based on the sound he or she hears, and can also identify the distance between the sound source and himself or herself by the distance of the sound. The sharpness of the auditory channel can help visually impaired users to better "observe" things and things around them; Example 2: Set up voice prompting and alarming systems in elevators, crossing streets, etc. so that people with visual impairments can know the environment they are in to quickly make necessary responses. In the process of using information products, visually impaired people also use auditory cognitive products to achieve human-computer interaction.

(2) Tactility interactive design:

"Unlike vision, haptic sensations are more sensitive and can directly measure and sense the various properties of things. It can accurately distinguish the shape, state, texture, texture, and material of a thing." The visually impaired user receives another major channel of information transmission, and the skin can transmit information to the brain by receiving external stimuli, which is another way for the visually impaired user to "observe". This is a very important reference for the interactive design of accessible products [10]. Example: Walking with the aid of a cane, they are usually used to walking along walls, railings or curbs. When walking, the cane swings from side to side in an arc and strikes the ground. The left and right scanning range is about 900–1500 mm. The cane can reach objects 150 mm away from the shoulders, so that you can feel the wall and obstacles whose height is less than 685 mm. In this process the tactility plays a very important role.

The product's tactility interactions should give a positive tactile sensation when the visually impaired is in contact with the product. This positive feeling is often reflected in the form, function, material, and safety of the product. In morphological treatment, because the visually-impaired people's limbs are very sensitive to the sense of touch, they can feel the product's appearance quickly and understand the details of the information conveyed through the touch. For the products designed for the visually impaired, the integrity, simplicity, and orderliness of product modeling are critical to a good using experience. In terms of functional design, emphasis should be placed on implementing product functions by enhancing the tactile sensation. For example, product haptic feedback methods are used to convey differentiated functional information by modeling the unevenness, temperature, softness, texture, and lightness of the product. In the selection of products' materials, soft materials that do not cause misuse of the product should be selected as much as possible. At the same time, dangerous behaviors can be avoided by restricting users' behavior. Combining the comprehensive tactile sensations of products' forms, functions, material, and security, a good using experience would be achieved.

(3) Olfactory interactive design:

"Smelling, as a receiver of human body's external odor information, can induce people to produce emotional reactions and bring a sensory experience." By strengthening the olfactory channel to replace the visual channel, it has an irreplaceable role in the design of barrier-free products giving people a more unique experience [8]. The olfactory experience design can emit different odors according to the product's own properties, identify and classify them to make up for visual deficiencies, and can also be used as operational feedback to remind users of the correctness of the operation.

(4) Taste interactive design:

Because people are more subjective to taste, it is difficult to apply it to the design of daily necessities. Therefore, there is little design using visual sense as visual compensation. With the development of science and technology, the concept of "seeing the world with the tongue" has been realized: The blind-aided tool BrainPortV200 utilizes driverless technology and applies it to a tongue chip that is connected to a mobile phone, using computer algorithms and image processing capabilities to present a virtual image for people to achieve "visual taste." Applying this new technology to the design of daily necessities of visually impaired groups will have tremendous potential for development [9].

(5) Multi-sensory experience interactive design:

Through the study of different sensory pairs, we explore more possibilities of multi-sensory fusion in the daily necessities' design of visually impaired people, and try to apply the designing strategy of multi-sensory combination to the daily necessities' design of visually impaired people, for example: Touch-and-sound combination, touch-and-smell combination, touch-and-taste combination and so on [8]. According to different using situations, the corresponding multi-sensory experience interactive mode is selected to meet the visually-impaired group's usable, easy-to-use, and enjoyable using experience, and barrier-free products that can compensate visual impairment by strengthening other sensory channels are designed [10].

4 Conclusion

Designers should be responsible for the visually impaired groups' attitude of responsibility and care, combined with modern interactive design concepts and methods. The perspective of hearing, smell, taste, touch is used to think about product interaction innovation. Designed products can meet the visually impaired characteristics and needs. At the same time, when designing products for the visually impaired, we should follow the principles of simplicity and ease of use and design products that meet their behavioral habits, so that visually-impaired groups can use products easily and happily, and coordinate the relationships between products, people, and the environment.

References

1. Welch, P., Palames, C.: A brief history of disability rights leg- islation in the United States. In: Welch, P. (ed.) Strategies for Teaching Universal Design. Adaptive Environments Center, Boston (1995)
2. Sodnik, J., Jakus, G., Tomazic, S.: Multiple spatial sounds in hierarchical menu navigation for visually impaired computer users. Int. J. Hum. Comput. Stud. **69**, 100–112 (2011)
3. Royal National Institute for the Blind, Communicating with blind and partially sighted people. Peterborough (2004)
4. Chisholm, W., Vanderheiden, G., Jacobs, I. (eds.): Web Content Accessibility Guidelines 1.0. http://www.w3.org. Accessed 15 May 2011
5. Wentz, B., Lazar, J.: Usability evaluation of email applications by blind users. J. Usability Stud. **6**(2), 75–89 (2011)
6. National Federation of the Blind, Assuring opportunities: A 21st Century strategy to increase employment of blind Americans. http://www.nfb.org. Accessed 5 May 2011
7. Gaudy, T., Natkin, S., Archambault, D.: Pyvox 2: an audio game accessible to visually impaired people playable without visual nor verbal instructions. Trans. Edutainment **2**, 176–186 (2009)
8. Gregor, P., Newell, A.F., Zajicek, M.: Designing for dynamic diversity: interfaces for older people. In: Proceedings of the Fifth International ACM Conference on Assistive Technologies, pp. 151–156. ACM (2002)
9. ISO: ISO/IEC Guide 71:2001. Guidelines for standards developers to address the needs of older persons and persons with disabilities (2001)
10. Lacey, G., Dawson-Howe, K.M.: The application of robotics to a mobility aid for the elderly blind. Robot. Auton. Syst. **23**(4), 245–252 (1998)

Harnessing Music to Enhance Speech Recognition

Vered Aharonson[1,2(✉)], Shany Mualem[2], and Eran Aharonson[3]

[1] School of Electrical and Information Engineering,
University of the Witwatersrand, Johannesburg, South Africa
Vered.aharonson@wits.ac.za
[2] Department of Biomedical Engineering, Afeka,
Tel Aviv Academic College of Engineering, Tel Aviv, Israel
shanimua@gmail.com
[3] Department of Software Engineering, Afeka,
Tel Aviv Academic College of Engineering, Tel Aviv, Israel
erana@afeka.ac

Abstract. The performance of automatic speech recognition highly depends upon the speaker's intelligibility and is affected by speech intensity and rate. Lombard reflex is an auditory feedback mechanism which is encountered when speakers spontaneously increase their voice in a noisy environment. We studied the feasibility of employing Lombard reflex to improve speech recognition without the speaker's conscious awareness of the process. Whereas previous studied employed noises to produce this reflex, which may be unpleasant to the speakers, we studied the effects of music-induced Lombard reflex. Twenty speakers were recorded when listening to two music types: a rhythmic dance music or a calm yoga music, as well as to white noise, metronome sound and silence, and the differences in the speakers' speech rate and intensity while listening to the different sounds were compared. Several cohort trends were observed: Speech intensity was particularly stronger in the rhythmic dance music condition for most subjects. This change was not observed for the metronome sound which had a similar rhythm. Speech rate was decreased for the yoga music condition for female speakers only. An examination of the changes in these prosodic variables for individual speakers yielded that most of them exhibited an increase in speech power and/or a decrease in speaking rate for at least one of the music types. This effect, when further explored, may be implemented in a personalized speech recognition engine, to enhance the usability of voice commands, dictation, and other speech based applications.

Keywords: Music effect on speech · Lombard reflex
Automatic speech recognition (ASR)

1 Introduction

There is a relentless effort to improve automatic speech recognition (ASR). A common cause for misrecognition or errors in ASR, as well as in human perception is when the speech is too soft [1] or too fast [2]. A solution to this problem may be implemented

© Springer International Publishing AG, part of Springer Nature 2019
T. Z. Ahram and C. Falcão (Eds.): AHFE 2018, AISC 794, pp. 390–396, 2019.
https://doi.org/10.1007/978-3-319-94947-5_39

within the user interface of ASR applications in the form of a correction mechanism prompting the speaker to speak more slowly or louder. This solution may be, however, irritating for the users of these applications, demanding them to repeat again and again their utterances. It will also considerably increase the recognition time thus further hampering the usability of the application.

In this research, we sought a technique which could induce speakers to change and improve the intelligibility of their speech in a more peasant manner. The technique chosen was based on employing the Lombard reflex [3, 4]. This reflex is frequently encountered in everyday life, when speakers spontaneously alter their voice in noisy environments such as loud parties or public spaces, in order to enhance their speech comprehensibility. The influence of the Lombard reflex on speech perception has been extensively studied [4–7]. A possible effect of the Lombard reflex on ASR has been proposed as well [8–10]. The studies reported various differences between Lombard-affected speech and normal speech. Among them were voice intensity, fundamental frequency, formant center frequencies and vowel duration. These differences, however, were observed to be highly speaker dependent [3–6, 11].

Different previous studies used different experimental settings and different auditory signals to produce the Lombard reflex. Therefore, they provided diverse results, especially when the contribution to ASR was discussed. Most results, however, support the hypothesis that the Lombard effect could be used in a controlled manner within an ASR system to improve the speaker's intelligibility.

In the current study we considered this hypothesis from the practical, application-wise point of view and sought for a virtual noisy environment which could be artificially generated and manipulated in a way that will induce users to alter their speech and consequently enhance their intangibility.

The ambient noises employed in former studies included noises, speech, tones, metronome sound. These noises would be rather unpleasant when used within a speech recognition application and may irritate users and hamper the usability of such system. In this study we chose to explore how would music induce Lombard reflex and affect speech in a manner that could enhance speech intelligibility. If successful, this music-induced Lombard effect could be implemented as an integral part of an ASR applications and would improve its recognition performance while still maintaining a pleasant user experience.

2 Methods

2.1 Experiment

Twenty users participated in the experiment: ten males and ten females. All participants were college students in the age range of 22 and 28 years old. The users were asked to read and recite aloud a text while wearing Logitech G430 headset with noise cancelling microphone. The prospective Lombard-inducing sounds were introduced into the closed-ear headphones and were therefore isolated from the recorded speech. The speech was captured by the noise cancelling microphone which reduced real environment noises from the recordings. The text recited consisted of four lines from the

"Hokey Pokey" nursery song: "You put your right hand in; You take your right hand out; You put your right hand in; And you shake it all about". This song was chosen since it was easy to recite without reading and had repeating words in each of the sentences that could later be processed and compared.

The experiment comprised alternating silence, music and noise that were introduced to the speakers via the headphones.

The music sounds were of two types: a rhythmic "dance" party music (Neelix - Sorry Hannes Mix ®) and a calming "yoga" music. These music types were chosen since they are purely instrumental, monotonous music, without verses or partitions. The difference between the two is in the rhythm: the dance music has a marked rhythm produced by electronic drum beats while the yoga music, consisting of long, calming melodic notes, has no rhythm. To examine the hypothesis that the differences between the two music types would be associated with the rhythm, a metronome signal of 60 beats per minutes, similar to the beats of the dance music, was induced as well. A random white noise sound was included to provide a comparison with previous studies. Five sound conditions were therefore introduced to the subjects: dance music, yoga music, metronome, white noise and silence. In each recording of a subject a different sound was introduced. A 3 min rest time was given to the subjects after each recording. The order in which the different sounds were introduced was randomly chosen among the subjects. The experiment yielded 5 recordings, for the 5 different headphones sound conditions, for each of the 20 subjects.

2.2 Speech Processing

The analysis aimed at quantifying acoustic differences between the speech produced in quiet conditions and the Lombard-affected speech. Since our goal was to study the effects that could improve speech intelligibility, we examined speech power and speaking rate as the measures that could contribute to this goal.

All processing stages were implemented using Matlab® software. The preprocessing included a segmentation of each recording into its 4 sentences. Only the 3 first sentences were included in our primary analysis, since they were almost identical, differing in two words only: "you X your right hand Y", where X was either "put" or "take" and Y was either "in" or "out". In addition, the identical parts of the 3 sentences, "put your right hand" were isolated to be processed separately. The signals for analysis thus consisted of 3 sentences, recorded in 5 different sound conditions (silence, dance music, yoga music, metronome and white noise); a total of 15 signals per subject. The features extracted from the signals were speech power, measured as the signals' norm divided by the signal length and speech rate, calculated in words per minute (wpm). These calculated speech features were normalized in order to handle the large inter-subject differences in speech intensity and rate values, as well as to highlight the effect of Lombard reflex: The power and rate values extracted from the Lombard induced speech were normalized by the values extracted from the speech in silence for each subject. These values, normalized by division to a reference, in this case – speech in silence, were considered adequate since prosodic features are perceived in log-scale by the human ear. The series of these relative values were then used to compare between the different sounds and between male and female speakers. Statistical significance for all comparisons were determined using a paired t-test, with a 0.05 confidence threshold].

3   Results

Figures 1 and 2 present the relative speech power and speaking rate, respectively, for the 4 sounds conditions, for female and male speakers. The figures demonstrate different patterns for speech power and speaking rate. The value ranges, conveying the Lombard speech features relative to silence are significantly smaller for speaking rate compared to speech power. Speaking rate changes in the presence of the various sounds reach 15–20% at max, and are much smaller than the range of the speech power values, which reach 50% and more.

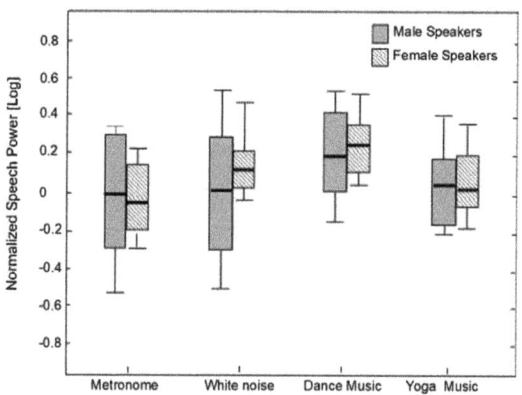

Fig. 1. Box-and-whisker plots of the changes in speech power for the 4 sound-induced Lombard speech recordings. Values are normalized by the speech intensity in silence conditions and presented as a log value. Values for male and female speakers are denoted by dots and line textured boxes, respectively.

The values themselves convey a mixed behavior where for some subjects, the speech power and/or rate were increased due to the presence of sounds whereas for other speakers it decreased. The averages of these values, as depicted in the "boxes" of the plots are close to zero. The exceptions to this observation are the dance music condition, where most values, as well as their average demonstrate an increase in power, the white noise, where the same trend can be observed, but only for female speakers and the yoga music, where speech rate was decreased for most female speakers (Figs. 3 and 4).

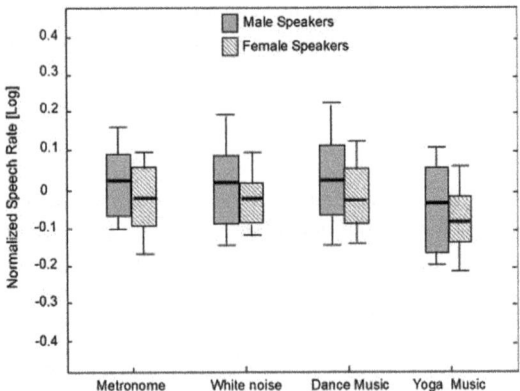

Fig. 2. Box-and-whisker plots of the changes in speech rate for the 4 sound-induced Lombard speech recordings. Values are normalized by the speech rates in silence conditions and presented as a log value. Values for male and female speakers are denoted by dots and line textured boxes, respectively

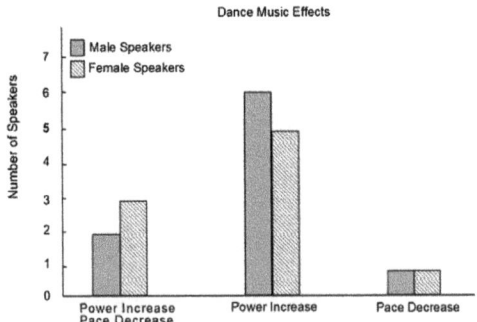

Fig. 3. A bar graph of the changes in speech power and pace of the male and female speakers, for the dance music.

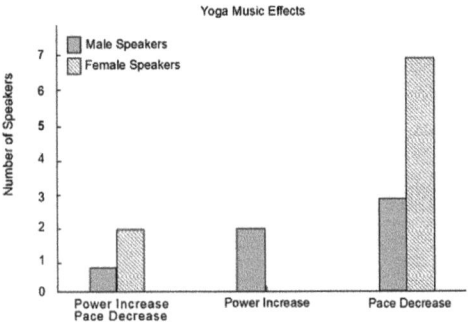

Fig. 4. A bar graph of the changes in speech power and rate of the male and female speakers, for the yoga music.

4   Discussion and Conclusions

This preliminary research explores the effects of music-induced Lombard's effect on features related to speech intelligibility: speech power and speech rate. Two different music types were studied, where the prominent difference between the two was in their rhythm: dance music and yoga music. The experiment consisted of a comparison of speech features in the speakers' recordings while they were listening to these two music types, and additionally, to metronome sound with similar rhythm as the dance music and to a white noise sound. The features were normalized using the silent, or no sound, condition, in order to quantify the Lombard reflex effect, if any, of the various sounds. The results could yield only a trend implying that the dance music produced an increase in speech intensity for most subjects, and white noise displayed a similar pattern but only for female speakers, whereas for other sounds: metronome and yoga, as well as white noise in the male speakers' case, the effect was mixed and could not portray a distinct trend in the cohort of subjects. The effect of the different sounds on speaking rate was considerably smaller than their effect on speech intensity and did not yield a consistent pattern, except for the case of yoga music, where speech rate was reduced in most female speakers' recordings.

Interestingly, our initial assumption that the difference between the two music types lies their rhythm, which was examined by comparing the effects induced by a metronome beat with a similar rhythm to the dance music, was not corroborated by the results: No significant trend of similar effects on speech for the dance music and the metronome sound was observed.

This preliminary study focused on "rhythm and blues", or on different types of music. More variables need to be assessed in further studies: particularly different intensities of the same music types, as well as other music types differing in their frequency ranges. The speakers' cohort in this study consisted of students of the same age group and demographics. Further experiments will include different age and cultural groups.

Previous studies have discussed the role of Lombard reflex as "enhancing speech communication" [5]. The authors argued that as such, the effect of this reflex would be more prominent in natural speech. In our experiment, the speakers recited a nursery rhyme, which made the speech more natural than in a read speech experiment. Still, an experiment in natural spontaneous speech is required to assess whether the effect would indeed be different from the results in the current study. Spontaneous speech will require a more robust processing, especially when speech rate is concerned [12].

The implementation of music-induced Lombard effect in a speech recognition experiment should also studied, in order to assess whether Lombard induced changes in speech could indeed improve recognition rate. Many potential users of speech based applications in computers and mobile devices work and play today while listening to music. Listening to music and speech commands could merge in a car driving scenario. The ASR in these circumstances may benefit from music-induced Lombard speech. The mixed behavior observed in the cohort of the current study may not be relevant in a practical ASR based application: the type and intensity of the music which may induce the desired effect may be personalized and tuned to a user. Music induced Lombard effect will therefore be further explored to assess its relevance and practical usage for speech recognition enhancement.

Acknowledgments. We thank Mr. Molefi Makuebu. University of the Witwatersrand, Johannesburg for his help in the signal preprocessing.

References

1. McCreery, R.W., Stelmachowicz, P.G.: Audibility-based predictions of speech recognition for children and adults with normal hearing. J. Acoust. Soc. Am. **130**(6), 4070–4081 (2011)
2. Bradlow, R., Torretta, G.M., Pisoni, D.B.: Intelligibility of normal speech I: Global and fine-grained acoustic-phonetic talker characteristics. Speech Commun. **20**(3–4), 255–272 (1996)
3. Egan, J.J.: The Lombard reflex: historical perspective. Arch. Otolaryngol. **94**(4), 310–312 (1971)
4. Brumm, H., Zollinger, S.A.: The evolution of the Lombard effect: 100 years of psychoacoustic research. Behaviour **148**(11–13), 1173–1198 (2011)
5. Junqua, J.C.: The Lombard reflex and its role on human listeners and automatic speech recognizers. J. Acoust. Soc. Am. **93**(1), 510–524 (1993)
6. Zhao, Y., Jurafsky, D.: The effect of lexical frequency and Lombard reflex on tone hyperarticulation. J. Phon. **37**(2), 231–247 (2009)
7. Junqua, J.-C., Fincke, S., Field, K.: Influence of the speaking style and the noise spectral tilt on the Lombard reflex and automatic speech recognition. In: Fifth International Conference on Spoken Language Processing, pp. 467–470 (1998)
8. Junqua, J.-C., Fincke, S., Field, K.: The Lombard effect: a reflex to better communicate with others in noise. In: Proceedings of the IEEE Acoustics, Speech and Signal Processing, pp. 2083–2086 (1999)
9. Junqua, J.-C.: Impact of the unknown communication channel on automatic speech recognition: a review. In: Proceedings of the European Conference on Speech Communication and Technology (Eurospeech), Rhodes, Greece, vol. 1, pp. KN29–KN32 (1997)
10. Vlaj, D., Kacic, Z.: The influence of Lombard effect on speech recognition. Speech Technologies: InTech, June 2011
11. Lane, H., Tranel, B.: The Lombard sign and the role of hearing in speech. J. Speech, Lang. Hearing Res. **14**(4), 677–709 (1971)
12. Aharonson, V., Aharonson, E., Raichlin-Levi, K., Sotzianu, A., Amir, O., Ovadia-Blechman, Z.: A real-time phoneme counting algorithm and application for speech rate monitoring. J. Fluen. Disord. **51**, 60–68 (2017)

Usability Study and Redesign of the Food Tray

Kimberly Anne Sheen, Yan Luximon$^{(\boxtimes)}$, Kar Hei Fung,
Shun Him Chak, Wai Yi Chiu, and Wing Sang Chan

School of Design, The Hong Kong Polytechnic University, Hung Hom,
Kowloon, Hong Kong SAR
yan.luximon@polyu.edu.hk

Abstract. Food trays can be found around the world in schools, canteens, and restaurants. While useful for carrying food to a table, the design has caused frustration in countless users. The aim of the research outlined in this paper was to identify design criteria for food trays and produce a redesign to improve the user experience with an emphasis on ergonomics. The investigation focused on four areas: loading, carrying, eating from, and cleaning and storing the food tray. Areas of concern and design criteria were found through materials research, observations, questionnaires, interviews, and prototype testing. Details of the findings and the ergonomically designed prototype are presented in this paper. By investigating food trays from the user and staff perspective, design criteria and a unique design were developed which users felt was more comfortable and fit cleaning and storage requirements.

Keywords: Usability study · Observations · Interviews · Product design

1 Introduction

Food trays are a common staple in schools, mall canteens, and restaurants. While used around the world, there tends to be very little variation in their design. While useful for carrying food to a table, the design has caused frustration to countless users while not only carrying the food but through all steps of use, yet there is very little research into this.

Research surrounding food trays tends to be very specific, such as prototyping and creating food trays to facilitate self-feeding in young children [1]. Even in the news, discussions surround food trays tend to be related to specialty trays such as airplanes and tied to their economic value [2]. The most recent news surrounding this common product was related to a trend in Singapore where they are starting to charge a deposit fee for the use of the trays. This has been so unpopular with the consumers, that 50% to 90% of patrons asked to pay the fee refused to take trays and instead moved their food in trips [3] or returned the trays yet left the plates and cutlery on the tables [4].

While trays are meant to ease the movement of food from a counter to a table, it is telling when people are willing to go without them when they feel they are inconvenienced with a returnable fee of 50 cents. And even though there have been patents for unique designs of food trays [5–7], they have not been adopted as the standard food tray. Taking all of this into account, the goal of the research project outlined in this paper was to identify specific ergonomic issues related to the design of food trays,

© Springer International Publishing AG, part of Springer Nature 2019
T. Z. Ahram and C. Falcão (Eds.): AHFE 2018, AISC 794, pp. 397–403, 2019.
https://doi.org/10.1007/978-3-319-94947-5_40

identify other areas of concern in the design of food trays, and produce a redesign to improve the user experience with an emphasis on ergonomics. This was done by ensuring the new design was fit for usage throughout the 4 main stages of use: loading food, carrying the tray, eating, and after-use management (to collect, clean, and store the tray).

2 Redesign Process

First, the identification of areas of concern and the criteria for the redesign were identified through several methods. These methods are as follows: materials research, observations, questionnaire, and interviews. Following this, usability testing with prototypes was conducted in two rounds. The investigation focused on four main areas: loading of the food tray, carrying a full tray of food, eating from the food tray, and the after-use management which includes cleaning and storage.

The investigation took place at one of the cafeterias located on a university campus. This study not only looked at the users of the food tray throughout their use experience but also at the staff who are tasked with cleaning and storing the food trays when they are not in use. Investigation began with observations of staff and users and materials research, which was then followed by an in-depth questionnaire. From this information, initial design criteria were developed. Following that, a prototype was developed and tested with users. Interviews were conducted during these user tests and data gathered was used to iterate the prototypes and create a final design. This final design and the criteria are presented in this paper.

2.1 Development of Design Criteria

Research into problem identification started with a review of existing products. This included the existing common food tray, patents for unique food tray designs, materials, and features in similar products. This research not only identified design criteria, but also looked into limitations applicable to the design. Following which, general observations of 13 people using the canteen was undertaken. Photos and notes were taken to identify the steps associated with using the food trays and issues that users encountered. Additionally, an in-depth online questionnaire was released to students and staff regarding usage of food trays. 41 respondents outlined their experience with food trays at the university. As diners are not the only stakeholders of the food tray, design criteria research was undertaken with canteen staff as well. Four members of canteen staff were interviewed about their experiences with the food trays and the cleaning and storage process. Additionally, researchers were allowed to observe the cleaning and storage process which the food trays go through (see Fig. 1).

Overall, the research for the design criteria found that the users report issues related to balance, drink and food spillage, drink and food slippage, cumbersome and uncomfortable grip area, and a lack of space for plates and personal belongings. The grip posture and location were identified along with the body posture while carrying the food tray and setting the food tray on tables. Based on the analysis of all of the results of the methods, four main criteria for the design were identified. First, the stability of

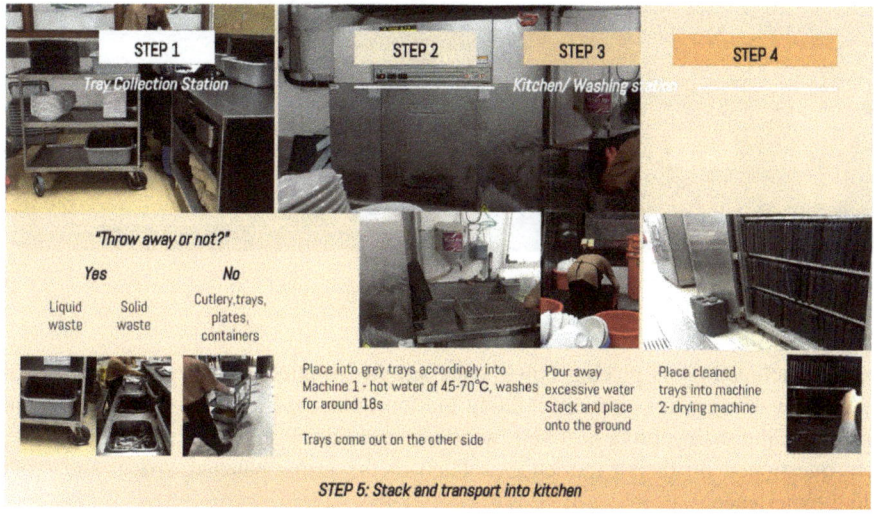

Fig. 1. Cleaning and storage process.

the tray and dishes set on the tray must be ensured with a slip-resistant design. Second, the design needs to allow for a flexible arrangement of food and other belongings on the tray. Third, the tray must have a hygienic surface for the safe placing of both food and belongings, including objects such as tableware and wallets. Finally, the new design must facilitate convenient after-use management and have a simple structure. After use management includes tray collection, cleaning stages, and storage. Due to the costs associated with the replacement of cleaning machinery, the design was restricted to a size which these machines could facilitate (44.5 cm × 34 cm). Additionally, a plastic based material was chosen for the design due to the manufacturing process, longevity of the product, and hygiene.

2.2   Initial Prototype Testing

Following development of the design criteria, three prototypes were created with a main focus on the different grip area types and a secondary focus on the eating experience (see Fig. 2). Prototype 1 used a common food tray and attached handles which aimed to correct the flexed angles of the wrist and place them in a more neutral position whilst keeping a larger food storage space. Prototype 2 was inspired by painting palettes for the grip area and included a cutout for drink placement so as to remove the likelihood of spillage. This design also included flat edges for comfort while eating. Finally, prototype 3 removed some of the space for food to create recessed butterfly-shaped grip area while removing some of the edges. The butterfly grip area was created to support the grip while correcting awkward wrist posture and the flattened edges were created to improve the eating experience.

After the prototypes were created, the three prototypes were tested with seven users to identify which grip area type and surface area were considered best. User testing

Fig. 2. Initial three prototypes.

sessions included users placing bowls with weights mimicking the weight of food on the trays, walking across a room holding the tray, placing the tray on the a table, and mimicking the eating process. Users were observed throughout the process and after were questioned on their experiences related to the grips, balance, space, and mock eating experience.

2.3   Prototype Iteration and Testing

Based on the findings from the initial prototype testing, prototype 3 was chosen to be expanded upon as the butterfly-shaped grip area provided a neutral position for the wrist and caused users to hold the tray closer to their body and in a more ideal position. Additionally, the butterfly-shape afforded comfort for both smaller hands and larger hands. Moreover, the space lost with the new grip design was considered acceptable to users. Finally, anti-slip materials and raised spaces for cutlery were added based on the original design criteria.

After the initial grip design was decided upon, rapid prototype testing and iteration was conducted (see Fig. 3). After testing, the prototypes were iterated to include a designated space for drinks, elevated cutlery rests to ensure the hygiene of the cutlery, and anti-slip material to assist in keeping the food and drink from spilling. Based on the rapid user testing, the grip area was developed further, and the bottom corners were recessed to avoid pain while eating but without lowering to the point where dishes may fall off the tray. Additionally, notches to hold the cutlery were added and a slip-resistant surface to hold dishes firmly was applied. Following the rapid prototyping, a 3D printed prototype was assembled for a final user testing session (see Fig. 4). This prototype was covered fully with slip-resistant silicone.

After the final user evaluation session with three participants, the thickness of the bezel was reduced, the recessed handles were smoothed, longer and steeper notches for cutlery were developed, the recessed handles were moved more towards the middle for better balance, and the handles were lengthened slightly for the comfort of larger hands.

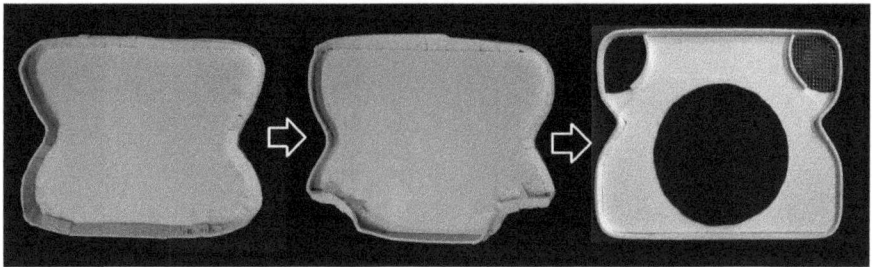

Fig. 3. Initial rapid iterations of prototypes used for testing.

Fig. 4. 3D printed prototype without the slip-resistant material applied.

3 Final Redesign

The final design is slip-resistant, has improved handles, and an elevated slot for cutlery (see Fig. 5). The slip-resistant silicone layer prevents cups, plates, and bowls from sliding on the tray and spilling their contents. Improved handling by a butterfly-shape recessed grip area allows users to hold the tray closer to their bodies and to the center of gravity of the tray for better balancing and a more neutral wrist position. The slot for cutlery has notches on both sides to allow users to place their cutlery on top of the elevated slots so that parts of the cutlery that are in contact with the food are raised so that they are not in contact with the tray surface. The trays remain the same width, length, and thickness so that current machinery will remain useful. The design is also stackable for easy storage.

Fig. 5. Final design of the food tray.

4 Conclusion

Overall, a new tray was developed using ergonomics principles to ease some of the frustrations of dealing with a cafeteria style food tray. The tray was designed with the users in mind from the very first step of loading the tray with food to the cleaning process conducted by the canteen staff. The unique butterfly-shape handle was designed for better balance and more comfort when carrying the tray to a table. The material of Polypropylene for the base was chosen to be light weight and durable while a silicone layer was added to be slip-resistant. Additionally, attention was paid to the size, stackability, and structure. This was done to facilitate convenient after-use management, storage, and to allow the trays to be used with current cleaning machinery. The ergonomically optimized design presented in this paper is an alternative to the common food tray found around the world.

Acknowledgments. The researchers would like to thank RISUD and the Hong Kong Polytechnic University for the support, and all participants for their assistance.

References

1. Gal-Oz, A., Weisberg, O., Keren-Capelovitch, T., Uziel, Y., Slyper, R., Weiss, P.L.T., Zuckerman, O.: ExciteTray: Developing an assistive technology to promote selffeeding among young children. In: Proceedings of the 2014 Conference on Interaction Design and Children, pp. 297–300. ACM (2014)
2. Estes, A.C.: How a Better Food Tray Is Saving Virgin Atlantic Millions. Gizmodo (2014). https://gizmodo.com/how-a-better-food-tray-is-saving-virgin-atlantic-millio-1588604293

3. Lee, G.: Pay a deposit for a tray? Then I won't take one, say patrons at 2 food centres. Singapore Times (2018). http://www.straitstimes.com/singapore/pay-a-deposit-for-a-tray-then-i-wont-take-one-say-patrons-at-two-food-centres
4. Ming, T.E.: Pay for your tray: Two hawker centres to start charging deposits for food trays. Times Online (2018). https://www.todayonline.com/singapore/pay-your-tray-two-hawker-centres-start-charging-deposits-food-trays
5. Bauman, C.E., Bauman, B.M.: Food and Beverage Tray: U.S. Patent No. 4,744,597. U.S. Patent and Trademark Office (1988)
6. Trivison, J.A.: Food Serving Tray or the Like: U.S. Patent Application No. 06/537,872. U.S. Patent and Trademark Office (1986)
7. Wilcox, C., Preusser, D.: Food Tray: U.S. Patent Application No. 29/238,379. U.S. Patent and Trademark Office (2006)

Evaluation of Aesthetic and Emotional Satisfaction of Mobile Phone Users

Young-Hee Lee[1] and Ryang-Hee Kim[2(✉)]

[1] Department of Imaging Science and Arts, Graduate School of Imaging
Science, Multimedia and Film, Chung-Ang University, 84 Heukseok-Ro,
Dongjak-Gu, Seoul 156-756, Korea
poolyh@hanmail.net
[2] Department of the Clothing and Textiles, Smart & Sensibility Clothing
and Textiles Lab, College of Human Ecology, Yonsei University, # 262
Seongsanno, Seodaemun-Ku, Seoul 219-749, Korea
yangheel003@naver.com

Abstract. Customer's satisfaction is necessary not only to satisfy physical
human characteristics, but also to satisfy human aesthetic and emotional characteristics. We surveyed the color, material and design of the mobile phone case,
which is the most important point in this paper, and the design satisfaction of the
mobile phone case. Findings were as follow that were appropriate for color and
material and appearance, depending on demographic variables. And the result of
SD method analysis, emotional vocabulary required especially in mobile phone
cases appeared 'modest', 'clear', 'smooth', 'sophisticated', 'clean' and 'neat'.
Also, a result of finding commonality among the types, we have found that there
are types preferred not only for color but also for material and appearance. In
future studies, it is expected that various methods can be derived from emotional
ergonomic design which is stimulated by various generations' emotions by
studying patterns, brand preference trends and colors.

Keywords: Aesthetic sensibility · Emotional ergonomics · User satisfaction
User evaluation

1 Introduction

In the 4th industry revolution, the mobile phone which is a computer in the hand is
evolving and changing very rapidly in its function, appearance and ideal fashion trend.
The satisfaction of the visual sensibility of the case of the mobile phone has a great
influence on the user convenience, purchase intention and product development.
Mobile phone companies with large sales volumes such as i-phone (Apple®) and
smart-phone (Samsung®, etc.) have been made especially considering the type of case
and the emotional characteristics of customers when designing the phone case.

Recently, in terms of sensibility and ergonomics, trends in color, patterns, and size
changes in the material of mobile phone cases have been investigated in consumer
mobile phone case use case studies and existing mobile phone case design satisfaction
surveys.

© Springer International Publishing AG, part of Springer Nature 2019
T. Z. Ahram and C. Falcão (Eds.): AHFE 2018, AISC 794, pp. 404–411, 2019.
https://doi.org/10.1007/978-3-319-94947-5_41

In this paper, we assume that there will be a connection between the type of mobile phone case with high sales volume and the type of people's pursuit in the design of mobile phone case. And with the change of color case consumption of mobile phone cases over the past few years, we have conducted consumer cell phone case use cases investigation and satisfaction survey of existing mobile phone case design.

2 Literature Review

2.1 Concepts of Aesthetic Sensibility in Emotional Ergonomics

Emotional design is also influenced by the four pleasures, identified in Designing Pleasurable Products by Patrick W. Jordan. In this book Patrick W. Jordan builds on the work of Lionel Tiger to identify the four kinds of pleasures. Jordan describes these as "modes of motivation that enhance a product or a service [1].

Life is unenjoyable without appreciating what we do, and it is human intuition to seek pleasure." The idea of incorporating pleasure into products is to provide the buyer with an added experience. Patrick W. Jordan points out in his book that a product should be more than something functional and/or aesthetic pleasing and it should evoke an emotion through the use of pleasures. Although it is hard to achieve all four pleasures into one product, by simply focusing on one, it might be what can bring a product from being chosen over another [2].

The four pleasures that could be implemented into products or a service are as follows. First, Physio-pleasure deals with the body and pleasure derived from the sensory organs. This includes taste, touch, and smell, as well as sexual and sensual pleasure. In the context of products, these pleasures can be associated with tactile properties (the way interaction with the product feels) or olfactory properties (the leather smell in a new car, for example). Second, Socio-pleasure is the enjoyment derived from the company of others. Products can facilitate social interaction in a number of ways, either through providing a service that brings people together (a coffee-maker enabling a host to provide their guests with fresh coffee) or by being a talking point in and of itself. Third, Psycho-pleasure is defined as pleasure which is gained from the accomplishment of a task. In a product context, psycho-pleasure relates to the extent in which a product can help in task completion and make the accomplishment a satisfying experience. This pleasure may also take into account the efficiency with which a task can be completed (a word processor with built-in formatting decreasing the amount of time spent on creating a document, for example). Fourth, Ideo-pleasure refers to pleasure derived from theoretical entities such as books, music, and art. It may relate to the aesthetics of a product and the values it embodies. A product made of bio-degradable material, for example, can be seen as holding value in the environment which, in turn, may appeal to someone who wishes to be environmentally responsible [2, 3].

Also, Norman's approach is based on classical ABC model of attitudes. However, he changed the concept to be suitable for application in design. The three dimensions have new names (visceral, behavioral and reflective level) and partially new content. In the book, Norman shows that design of most objects are perceived on all three levels (dimensions). Therefore a good design should address all three levels: Visceral design-product appearance, behavioral design-usability, and reflective design-self-image (examples; Google: playful or anti-corporate, Apple's iPod: stylish or avant-garde) [3, 4] (Fig. 1).

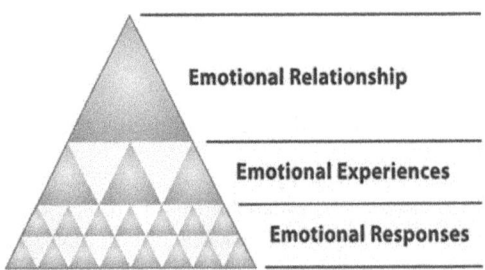

Fig. 1. Emotional design: Norman's approach. Source: http://t.co/VPKhoR249d, http://t.co/FDjkBxhfPx in

By the way, a new word which is "cell-cessories" has emerged that combines accessories and cell phones. People tend to view cell phone cases among various phone sets as the best means of expressing their individuality, and are increasingly inclined to express themselves in various cases.

Among these best practitioners, design is viewed as the art and science of putting all the pieces together: technical, financial, operational and emotional. Currently, so many companies already lavish quite a bit of expertise on the technical, financial and operational aspects of what they do, there is the equal focus on the emotional connection with customers that stands out as novel. Further, among such design-focused companies, this newly coequal dimension influences and informs the others, producing new and unexpected results.

2.2   Human-Centered Industrial Design

Also, emotional Design has been called hedonic design, affective design, affective human factors design, human-centered design, empathetic design, and focused on the influence of emotions on the way we interact with objects. Industrial design focuses on the target user group in the market, which often excludes users who do not belong to the mainstream. Such users include people who cannot use the mainstream products due to physical, cognitive, cultural, educational, financial obstacles, and other challenges. They have never attracted major industries as the market segment is extremely small for mass production and the marketing is not cost effective [6, 8].

Contests classic approaches that treat human behavior as 'stimulus-response' and consider emotions as noise.

Vyas and van der Veer (2005) suggested APEC (Aesthetic, Practical, Emotional, Cognitive) Aspect Framework that 'Aesthetic Aspect' is visceral appreciations based on sensory information only, naturally determined and skin-deep beauty, 'Practical Aspect' is physical activities a user is capable of with respect to the system and exploits usability & functionality, 'Emotional Aspect' is related to emotions such as joy, anger, disgust, etc., and helps in the decision making, and 'Cognitive Aspect' is involved interpretation, information processing, problem solving, use of memory, etc., and beauty within [2, 4, 7] (Fig. 2).

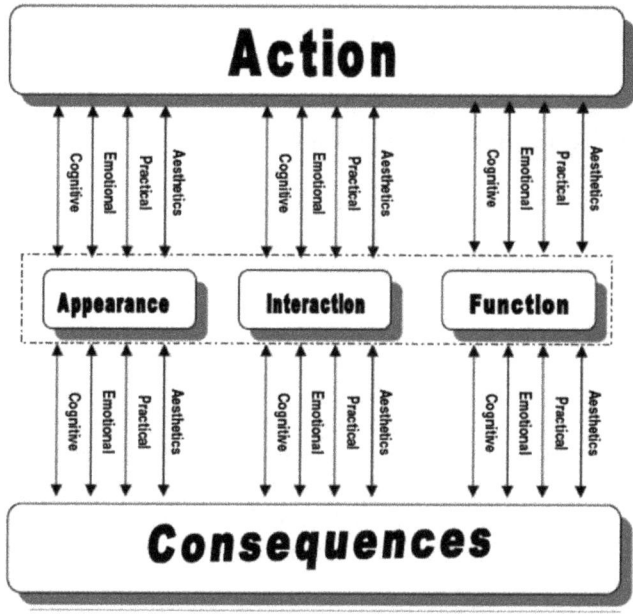

Fig. 2. APEC (Aesthetic, Practical, Emotional, Cognitive) Framework By Vyas and van der Veer (2005) Source: http://t.co/VPKhoR249d, http://t.co/FDjkBxhfPx in

In the usability aspects, 'Emotional design' is as an extension to standard usability practices, 'Standard practice' eliminate sources of frustration by addressing them in the design phase, and 'Additional practice' make application deal with unavoidable user frustration by addressing the user's emotions [7–9].

Recently, it has been difficult to find a big difference in the performance of all commodities in order to compare their excellence. As a result, when consumers purchase something, the design has a big influence on the desire to purchase the product as well as the performance of the item. The demand for change in fashion has risen dramatically according to the rapidly changing fashion. Fashion plays a major role in expressing the personality of young people.

According to these times, the smartphone, which has become a necessity for modern people, could not be free from design. Younger people are using smartphones as a means of revealing themselves beyond simple cell phones. However, due to the high price and the limitation of their own design, it was very insufficient to satisfy their individuality expressing desires.

Cell phone cases are made of many colors, shapes and materials, and are loved by younger generations who pursue their own style. The mobile phone case is not a simple protection function for the mobile phone, but design of 'Cell phone cases' which can satisfy the users' sensibility by focusing on the functions as the accessories for expressing the personality and the style has been developed so far.

Therefore, this study investigated the correlation between user's emotions, preference of color, and color of mobile phone case with high sales volume. It was aimed to investigate whether the attributes of the mobile phone case, such as the preferred color, appearance, and material of the mobile phone case that user's desire, cause users' desire to purchase.

3 Research Method and Procedure

The focus of this research is to understand how consumers are currently reflecting the color, material, and external design of a mobile phone case, and how to improve the future mobile phone case design accordingly.

3.1 Survey Method

Through 'Google Drive' survey method, we obtained 185 questionnaires and conducted questionnaires directly at the school to meet the age and gender ratios. A total of 185 subjects were sampled from 95 male and 90 female, 14–19' aged 59 male, 20–24' aged 60 male, 25–29' aged male and 66 female. They were 138 students, and 41 employees, and the proportions are 74.6% and 22.2%, respectively. The number of students is from middle school to college students, which is roughly equivalent to the young age of adolescents aged.

3.2 Contents of Questionnaires

The questionnaire consisted of six preferred colors for the subjects, the four most preferred designs for the currently used cell phone case, and the 7-scale SD differential method for the emotional vocabulary of the currently used cell phone case. To be marked.

Using SD (semantic differential) method, the other questionnaire consisted of the following adjective pairs of the following 15 items were checked on a seven-point scale: dark-bright, mellow-gorgeous, hazy-clear, moderate-intense, chunky-chic, dizzy-neat, drab-neat, cool-warm, classical-modern, deep-pale, tremendous-fascinating, heavy-light, complex-monotonous, masculine-feminine (Fig. 3).

4 Results and Discussion

4.1 Evaluation of User' Preferences of Colors

Figure 4(A) shows the preferred cell phone case color according to sex among questionnaire statistics. Of the total 95 men, Y was 5, Y/O was 2, O was 1, O/R was 3, R was 4, R/V was none, V was 1, B/V 4 were B, 4 were B, 7 were G, 5 were G, and 59 were achromatic. In the case of women, there were 7, 0, 2, 8, 3, 3, 1, 0, 6, 7, 4 (subjects) per each color and 49 subjects out of 90 subjects, respectively. More than half of all men and women prefer achromatic colors.

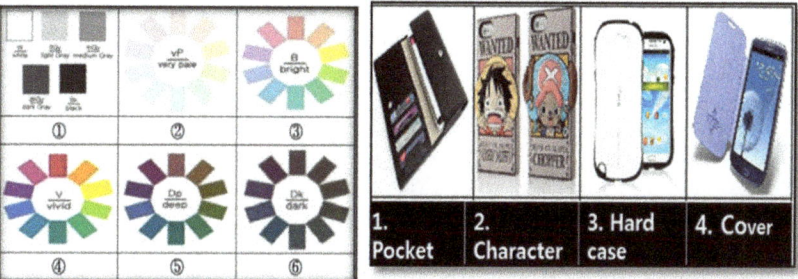

(A) Six preferred colors   (B) Four most preferred appearance designs

Fig. 3. Patterns of two questionnaire areas; (A) Six preferred colors, (B) Four most preferred appearance designs

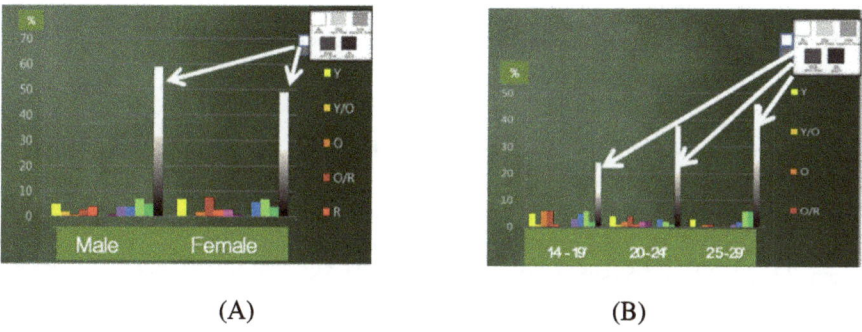

(A)   (B)

Fig. 4. Evaluation of user' preferences of colors according to sex (A) and the aged (B)

Figure 4(B) shows the preferred cell phone case colors according to age in the questionnaire statistics. O/R was 6 subjects, R was 6 subjects, R/V was 1 subjects, V was none, and V was none, respectively. B/V for 3 subjects, B for 5 subjects, B/G for 6 subjects, G for 2 subjects, and achromatic for 24 subjects.

As for color preference of mobile phone case according to sex, more than half of people who prefer achromatic color are most preferred.

4.2   Evaluation of User' Preferences of Appearance Designs of Mobile Phone Case

The respondents' responses to the four design patterns of survey statistics were 32 subjects for the retention type, 35 subjects for the character type, 80 subjects for the hard case type, and 38 subjects for the cover type. The proportions are 17.3%, 18.9%, 43.2% and 20.5%, respectively. Hard case type was the most preferred design with 43.2% (Fig. 5).

Of the 95 men, 14 (subjects) were in the retention type, 18 subjects were in the character type, 43 subjects in the hard case type and 20 subjects in the cover type. Among the total of 90 women, 18, 17, 37, and 18 (subjects) respectively. Both males

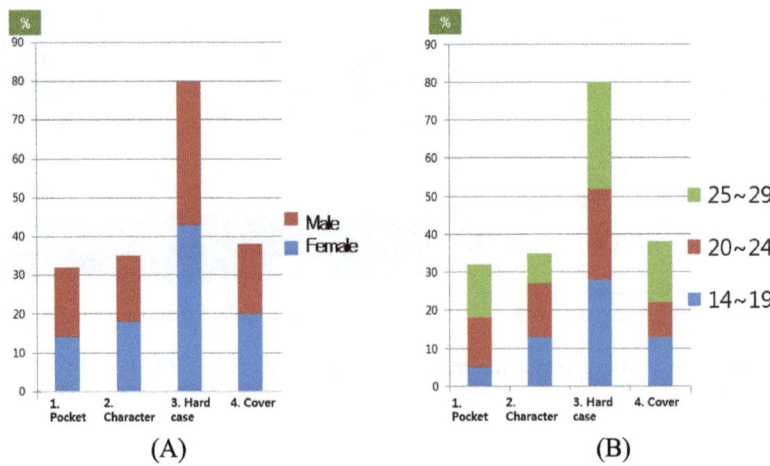

Fig. 5. Evaluation of user' preferences of appearance designs of mobile phone case according to sex (A) and the aged (B)

and females preferred the hard case type to be ranked first, and the ratio was 45.26% for males and 41.11% for females.

In case of men, 29 cases of jelly type, 29 cases of leather type, 30 cases of plastic type, and 5 cases of other cases were found in 95 cases of male cases. Of the total of 90 women, 33, 22, 30, and 5 (subjects) were women. 31.58% chose plastic type as the first rank of male and 36.67% chose jelly type as the first rank of female.

4.3   Analysis of Aesthetic Sensibility

The result of SD (semantic differential) method analysis, emotional vocabulary(dark-bright, mellow-gorgeous, hazy-clear, moderate-intense, chunky-chic, dizzy-neat, drab-neat, cool-warm, classical-modern, deep-pale, tremendous-fascinating, heavy-light, complex-monotonous, masculine-feminine) required especially in mobile phone cases appeared 'clear', 'smooth', 'clean' and 'neat'.

If you change your cell phone case in the future, you can see the emotional feelings of 'passivity and intensity' about the desired cell phone case color according to the Riccart scale. 153 cases (82.7%) of the total 185 cases responded that they prefer a cell phone case with a smooth feeling, but only six responded to the feeling that it was very smooth. 16 of the 185 respondents (8%) preferred to feel intense, with relatively no preference. The user' of 'dizziness and tidiness' of the total 185 cases, 185 (100%) preferred to use a cell phone case with a clean appearance. Of those, 59 (32%) answered 'a little clean' and 109 (59% Responded most to 'clean'. Seventeen (9%) responded the least with 'very clean'.

5 Conclusion

Accordingly, the above results could surmise to be used as a basic data, and is able to use in the design of mobile phone cases that can stimulate emotions in the teenagers and adult ages from the future. These findings seemed to have an extreme preference over other emotional vocabulary, and producers should focus on making these feelings when making cell phone cases.

In addition, further research is conducted along with the study of pattern, brand preference tendency as well as color, more various methods can be derived in the emotional ergonomics design production that stimulates sensitivity of various generations respectively.

References

1. Emotional Design: People and Things - jnd.org. www.jnd.org. Accessed 03 Apr 2017
2. Jordan, P.: Designing Pleasurable Products: An Intro to the New Human Factors. Taylor & Francis, London (2010). ISBN 978-0415298872
3. Norman, D.A.: Emotional Design. Basic Books, New York (2005). ISBN 0-465-05136-7
4. Jordan, P.: Pleasure With Products; Beyond Usability. Taylor & Francis, London (2004). ISBN 9780203302279
5. Cho, N.H., Kim, M.T., Kim, H.S.: The effect of pleasant product scent on consumer's product evaluation. Korean J. Sci. Emot. Sensib. 34(1), 1–25 (2005)
6. Kim, R.H.: Development and emotional evaluation of scented clothing using microcapsules. Procedia Manuf. 3, 558–565 (2015)
7. van Gorp, T.: Emotional Design with A.C.T. – Part 2. http://t.co/VPKhoR249d, http://t.co/FDjkBxhfPx
8. ISO 13207: Human-Centred Design Processes for Interactive Systems. ISO, Geneva (1999)
9. Gabbard, J.L., Swan, J.E.: Usability engineering for augmented reality: employing user-based studios to inform design. IEEE Trans. Vis. Comput. Graph. 14(3), 513–525 (2008)

Research on the Optimization Method of Website Based on User Experience

Chun-Fu Li and Ya-Qi Jiang[⊠]

Huazhong University of Science and Technology,
Wuhan 430074, People's Republic of China
309722939@qq.com

Abstract. With the popularity of Web 2.0 applications, there is a new trend in the way the Internet is used: creating a more user-centered approach to content management, information sharing, communications, teamwork, and more. Interaction and experience in web design become even more important. Positive user experience allows users to easily and efficiently complete the task, increase intimacy, comfort and sense of success, thereby enhancing customer satisfaction and enhance brand reputation. Website user experience design has become the respectable business occupation of the market, the key to winning customers. At present, most of the design of the Web interface in our country still stays in the imitation stage. Research on the Web user experience is only confined to some fragmented design theories, and there is a big gap compared with the foreign industry. In the face of new technologies and new needs of users, academics and industrial designers face enormous challenges. How to provide a systematic and effective method of development for the design and implementation of website user experience has become the focus of this article. This article is a bold attempt at the integration of psychological, behavioral, artistic and technical multidisciplinary approaches in the design of web user interface. Combined with the experimental analysis of a large number of well-known website examples, this article is validated both theoretically and practically this paper propose the scientific and important design principles and methods.

Keywords: Interface design · User experience · Interactive design

1 Research Background and Significance

The economic development has promoted the attention of all walks of life to the user experience, especially the experience in interactive activities. The Internet industry, which is inextricably linked with people's lives, is no exception to be involved in the upsurge of user experience. Those who have embraced the Internet culture have put more demands on the website. Their use of Internet products has not only focused on work efficiency, but also has become more and more concerned with the experience of the interactive process - whether it is pleasant to use.

Research on the optimization of website interface in China still only stays at the theoretical or one-sided level, which has certain limitations. The information structure of the website is not well organized, and the functional and task design lack of careful

© Springer International Publishing AG, part of Springer Nature 2019
T. Z. Ahram and C. Falcão (Eds.): AHFE 2018, AISC 794, pp. 412–421, 2019.
https://doi.org/10.1007/978-3-319-94947-5_42

and in-depth analysis, making it difficult for users to use it. There are many places that are not sufficiently humanized. Therefore, it is particularly important to explore how to design an interactive web interface with high efficiency, ease of use, aesthetics, comfort, technology and art that is highly harmonious and uniform. It is especially important to establish a set of relevant principles and methods to improve the user experience, in order to change the unclear, uninteresting chaotic visual and function of our website.

2   User Experience and Research Methods of Web Interface Design

2.1   User Experience in Web Interface Design

With the continuous development of webpage technology, the complexity of webpage design has also increased. Every stage of development has to meet a huge challenge. With the Web2.0 application in full swing today, more and more researchers and designs have been designed. The public are putting their hands on the issue of how to make websites more attractive and enduring vitality and come up with disparate views. If there is barely one standard measure of web interface design, that is the user experience.

Concept of User Experience. The user experience in the web interface refers to the purely subjective feelings created during the interaction between the user and the web interface, including the user's brand impression of the website and the extent of problems, doubts and bugs that they can tolerate. In the specific design means a lot of things: the emotional framework, the exchange of information, feedback and control, etc. Simply put, a good web interface design not only has its unique style and taste, but also simple, comfortable, Freedom, to give users a sense of control over everything, but do not feel the existence of the interface.

Basics of User Experience. Website usability is the basis of the user experience. The first thing to consider when designing a web page is the rationality of the site, which is the availability of the site. The concept of usability has changed people's understanding of the interactive process, and the user-centered Web design concept has been deeply rooted in people's minds. The most famous usability principle is Steve Krug's three laws of web usability: Do not let me think; It does not matter how many times you click, as long as each click is a choice without thinking and unmistakable; Remove half of the text on each page and remove half of the rest.

Purpose of User Experience. The user experience has far-reaching implications over usability. Any user has certain expectations and goals before interacting with the site, and meeting the goals is a part of the user's expectations. The fulfillment of goals is accomplished by performing the tasks that users need in order to interact with the various features, features, or functionality provided by the site, product, or application in order to accomplish these tasks, resulting in a positive or negative experience. In practice, however, user expectations are derived from prior experience and the credibility of the product or service, and when the expectation is consistent with the user experience, a positive experience is generated; when the conflict occurs between them, the user experiences a negative experience. The user experience requires not only realizing these expectations of users, but also pursuing the surpassing of user expectations, such as the surprise, identities, delights, thrills and the like raised by some scholars.

2.2   Research Methods of Web Interface Design User Experience

Psychology. Responsive design of the interface, we must have a clearer understanding of the user, to understand the way sensory information, how to understand and process information, learning and memory and reasoning process, in order to enhance user-friendly computer interaction. User psychology research is tantamount to solve these problems. The content of user psychology includes motivation psychology, cognitive psychology, user model and so on. Through irrational factors and rational factors in the cognitive psychology of the users, the problems in the human-computer interaction can be well solved, the errors are prevented, the flow is fluent and the interface is more friendly.

Design Sociology. Design sociology studies the essential connection and interaction between design and society. Internet users throughout all countries, ethnic groups and all sectors of society. Due to the different social life forms, occupations, living habits and accepted cultural and educational backgrounds of different groups of people living in different cultural lifestyles, the demand for aesthetic information is also unlike. These factors will directly affect people's way of doing anything on the web.

Physiology. Physiological research and human body-related issues, including human morphological parameters, human perception characteristics, response characteristics, psychological characteristics and living habits. In the design of web user experience, the study of body size can be utilized to design body comfort and operational process when using web pages. The human perception research can be used to design the rationality and ease of web page information and design element layout. The study of psychological characteristics needs to make sure that the user can perform tasks and receive information efficiently during the operation. Physiological aspects of the web interface design are mainly reflected in visual and manual operations.

Art Design. Web art design is a modern interdisciplinary discipline of design and network. In recent years, it is attracted people's attention with the development of the network. It itself takes the network as a carrier and puts all kinds of information in the fastest and most convenient way. To convey to the audience, under the requirements of this standard, gradually produced an aesthetic demand. The design should be based on psychology, refer to the target group's mental model and task, and use the principles of aesthetics in terms of color, spatial layout, visual flow, etc., and combine the site's specific performance requirements to develop the art design of the interface. A concise, beautiful, and full-fledged home page can meet the most basic and strongest physiological need of humans from the aesthetic point of view. This allows users to have an aesthetic pleasure at the beginning of use and psychologically improve user satisfaction. At the same time, it also enhances users' willingness to find out. And the use of psychological hints, to a certain extent, improves the user's usability.

3   Functional Design Principles Based on User Experience

3.1   Page Concise Principle (Principle 1)

Functional Simplicity. In the scope layer stage, to determine the functional scope of the first to meet the most commonly used user, the site's most basic functions to meet the needs of most users. These functions are generally only accounted for "20%" of the list of functions in the preliminary planning, which is the most effective way to show the work effect and should be satisfied first. The remaining 80% of the general functions are mostly enjoyed by expert users and should be "Hide it." This form not only simplifies the interface, but also reduces the learning load of novice and intermediate users, when they inadvertently found these extensions will not help but to experience the surprise, to achieve the user experience of fun claim.

The Simplicity of Information Layout. Under normal circumstances, human visual capacity per second is about four units of information, if the visual information received beyond the optical capacity, people will react within the heart, causing unpleasant. Information is designed to be concise and clear to enable users to efficiently obtain and quickly make decisions. If the information is complicated, the user will bear a large amount of information burden, causing information overload and affecting the user's efficiency in understanding the information. Therefore, we need to ensure that information is as concise and effective as possible (Figs. 1, 2, 3, 4, 5, 6 and 7).

Fig. 1. The "sliding door" layout of Netease website

Reasonable and standardized layout can achieve modest page effect. Organizing information into a good layout is the foundation of a website and should be decided

well before you think about the look and feel. There are many excellent layout resolution techniques available today that can help create a neatly organized content layout.

Simplification of the Task Flow. Simple tasks are more likely to form and accept,
which is universally accepted theory of cognition. Do not appear as far as possible does
not seem, even if the need to appear also use the most concise way to appear. The
interactive design with reasonable flow and simple operation can give users a sense of
pleasure and a sense of accomplishment, and accelerate the speed at which users learn
to use the website. There are six key points in streamlining the task flow: important
functions are placed in the visual center, attention is saved, the operation steps are
properly related, the operation experience is consistent, the current task is highlighted,
and the interactive controls are highlighted.

In addition, the function description of the control should be directly represented on
the control, instead of the auxiliary static text, which can draw the attention of the user
and better understand the function of the control.

3.2   User Customary Support Principles (Principle 2)

User Guides. There are for two types of support for user habits: follow the habits in
real life and follow social habits on the Internet.

Due to the physiological structure of the human wrist and the vast majority of
people using the right hand, and the two regions are the most comfortable and natural
right hand action range, the corresponding can also be projected upwards of the natural
range and down the interval. In addition, according to people's reading habit from left
to right and from top to bottom, the stoppage of important information or visual flow
should be arranged in the upper left and upper part of the page with a higher attention
rate in order to improve the efficiency of information transmission.

Users have accumulated some knowledge and skills from past experience with
using websites, forming habits and expectations. Most of these practices follow people's cognitive structure and thinking style, and because of its effectiveness as a
practice, should be taken seriously and take full advantage of to reduce the cognitive
burden on users.

The Use of Icons. It is an eternal way and way for mankind to grasp the world by
using the straightforward, sentimental, superficial and understanding features of images
to recognize things. Metaphorical icons are streamlined figures that reflect well-known
real-life scenes. They take full advantage of the concepts and knowledge in life, such as
the speaker icon representing "Volume" and the shopping cart icon representing the
purchase. It is more intuitive and more vivid than textual explanations.

3.3   Principle of Interactive Fluency (Principle 3)

Friendly Feedback. Timely feedback helps the user to assess whether the previously
performed operation helps to get closer to the target so that the user can readjust the
operation or proceed to the next operation step. Feedback system information that
enables the user to know the status of the task and anticipate the next possible action,

generally appears on the form and refreshed page. Its design is essential in helping the user to complete the task smoothly and efficiently, and enhancing the user's satisfaction.

Provide Timely Help. When new users learn how to use the website, they may find themselves unfamiliar and overwhelmed with lack of cognition. They provide novice guides to speed up their familiarization with website functions. Extra functions or complex tasks provide operation diagrams and reduce the user's learning burden. On the user error page provide the reasons for the problem and give suggestions, provide links to the help center or other sections that may be of interest to the user, so as to relieve the user's feeling of being complacent; provide up to two exit links according to the psychological needs and operating habits of the user; to avoid too many choices under the user's psychological sensation.

4   User Experience Based Information Construction Principles

4.1   Clear Website Architecture (Principle 4)

Rationality of Website Level. The site hierarchy should be simplified to reach the destination with the fewest clicks (ideally just two or three clicks) to obtain the content you want. Too many clicks will cause the user's discomfort. The actual rate of content access will decrease as the complexity of the web page grows, and the more difficult it is for users to gain access to them. Therefore, we should make every effort to simplify the webpage and try to connect to the specific content with the minimum number of clicks.

Fig. 2. The navigation bar of IBM website

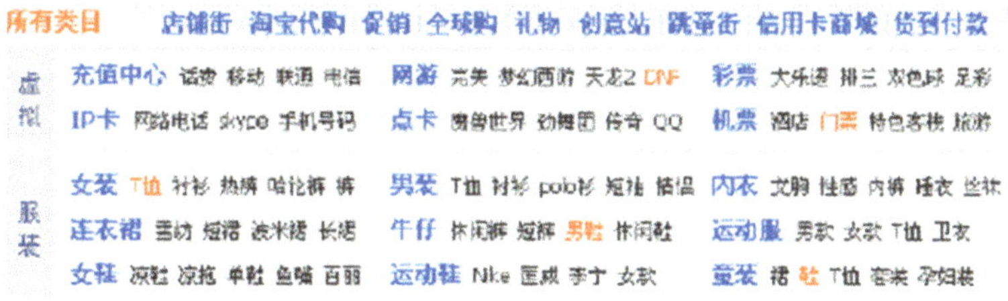

Fig. 3. The Category map of Taobao website

The Friendliness of the Navigation System. The role of the navigation system in the website is equivalent to the role of the road sign in real life. A well-designed navigation system gives the user a sense of belonging in the space while also increasing the efficiency of the user's access to information. The establishment of user loyalty to Important. The following points should be taken into account when designing: Establishing a multi-directional navigation system including global navigation, assisted navigation, contextual navigation, footer navigation, site map, etc., helps users to establish an overall awareness of the distribution of information on the website; Coherence in one website; maintain the comprehensibility of navigation tabs; maintain ease of navigation and prevent disruptions in the use of the process.

Fig. 4. The tag of Yahoo website

4.2   The Principle of Efficient Information Communication (Principle 5)

Provide Key Information on the Home Page. Under normal circumstances, the home page is the entrance to find information. Jacob Nielsen test of the user shows that the user in 40% of cases will first go to the home page of the site. This requires that the information on the front page must provide the user with the most relevant information according to the user's need.

Improve the Efficiency of the First Page Information Scanning. Based on the physiological basis of visual system and the study of eye movement system, vision has the following 9 features: from the center of the space to the periphery of the transition, starting from the center can easily find and grasp the overall situation and character-istics of things; visual perception of light and color is extremely sensitive; from bright to dark, from warm to cold, from solid color to gray, is the visual habits of the reflection of the color attributes; first pay attention to the contrast between the modeling elements of the strong part, after the attention of the weak part of contrast; Recognition, information clear; abstract form is biased toward rationality, serious abstract; from left to right, from top to bottom order; line of sight exercise habits from near to far, from the

Fig. 5. The website with disorder information

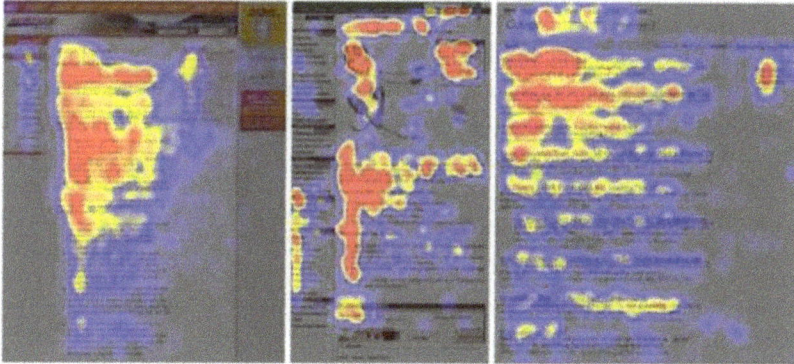

Fig. 6. The "F"-type visual flow

moving to the static, from real to virtual; line of sight clockwise movement; from convex to concave, from positive to negative.

Improve the Efficiency of Text Reading. Text as the main means of information transmission, its own layout design has a significant impact on the user's smooth browsing experience. Should take full account of the user's physical and mental state when reading information, through the establishment of a suitable layout effect, to create a relaxed atmosphere, in order to allow users to read in a pleasant mood. Avoid information overload caused by reading disabilities; as much as possible the use of a list to display information; article title must be prominent; text color set.

5   Visual Art Design Principles Based on User Experience

5.1   Harmonious Beauty of the Overall Style (Principle 6)

The Use of Color. When you feel the space environment, people first notice color, and then they notice the shape of the object and other factors. In other words, the color of the visual stimuli plays the role of the first message. Therefore, in the shaping of the website style, the color will play a more significant role than other modeling elements. Color is associative, divided into specific associations and abstract associations; colors have a positive psychological feeling; colors have cultural differences. Color affects people's sense of spirituality, the website color design must be consistent with the user's lifestyle and aesthetic taste, so as to create a sense of comfort, sense of completeness and beauty.

Layout Artistic. Web pages as a layout, you can include text, symbols, pictures, animation, buttons and other rich elements, if simply listed on a page, will only give people a messy effect. According to the needs of the content, these elements should be arranged and laid out rationally according to a certain order so as to present to the user in an organic whole. Content and form to achieve the state of coordination is the true success of web design. Superb layout design full of musical beauty, without any loss of practicality at the same time, this layout design can make beautiful reverie, beautiful passion, happiness and art to get infected.

5.2   Art Visual Hierarchy (Principle 7)

Web page visual hierarchy can be obtained from the following effects: highlights and reflection effects, gradients, blur effects, three-dimensional border effects, texture effects.

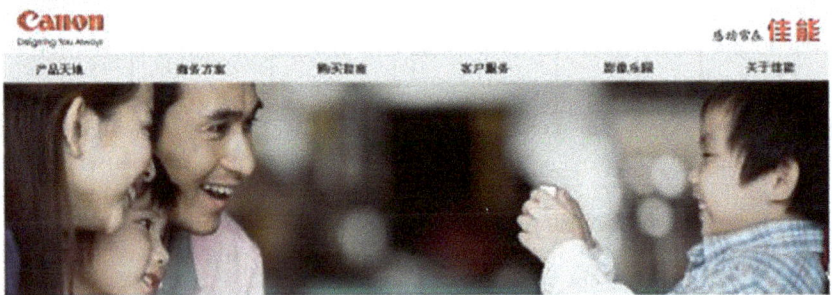

Fig. 7. The fuzzy effects of Canon

6 Conclusion

Based on a comprehensive analysis of a large number of websites, a set of key principles and implementation methods of user experience based on actual web page interface development process are put forward in this paper, using the theoretical knowledge about website design in such disciplines as design psychology, aesthetic study and interactive design., Sub-site functions, interactive processes, information transmission, visual arts effects are described in four aspects, including the user's mental model, the principle of visual flow, the color of the psychological aspects of the principle of a more in-depth analysis and provides the corresponding Technical solutions, and through the analysis of the effect of a large number of website examples, verified the scientifically and effectiveness of the principles proposed in this paper.

References

1. Conference on designing for user experience [EB/OL], 11 June 2006. http://www.dux2005. org/
2. Dam, V.N., Evers, V., Florann, A.: Cultural user experience Issues in e-government designing for a multi-cultural society. LNCS, vol. 3081 (2003)
3. Hassenzahl, M.: The quality of interactive products: hedonic needs, emotions and experience. In: Ghaoui, C. (ed.) Encyclopedia of Human-Computer Interaction, pp. 652–660. Idea Group, Hershey (2005)
4. Scherer, K.R.: Cognitive components of emotion. In: Davidson, R.J., Goldsmith, H., Scherer, K.R. (eds.) Handbook of the Affective Sciences. Oxford University Press, New York (2004)
5. Morville, P., Rosenfeld, L.: Information Architecture for the World Wide Web (2004)
6. Rubinoff, R.: How To Quantify The User Experience[EB/OL] (2004). http://articles. sitepoint.com/article/quantify-user-experience
7. User Experience 2006 Conference[EB/OL], 11 June 2006. http://www.nngroup.com/events/ seattle/agenda.htm
8. Vyas, D., Gerrit, C., Van Der V.: APEC: A framework for designing experience[EB/OL], 11 June 2006. http://www.infosci.cornell.edu/place/15_DVyas2005.pdf
9. Garrett, J.J.: The Elements of User Experience: User-Centered Design for the Web. New Riders Press, Berkeley (2002)
10. Nielsen, J.: F-Shaped Pattern For Reading Web Content[EB/OL], 17 Apr 2006. http://www. useit.com/alertbox/reading_pattern.html

Research on Shared Product Design Based on Service Design Concept-Illustrated by the Case of "Ofo" Design

Qian Ji$^{(\boxtimes)}$, Jiayu Zheng, and Yu Zhang

Industrial Design Department, School of Mechanical Science and Technology,
Huazhong University of Science and Technology, Wuhan 430074
Hubei Province, China
{jiqian, jiayuzheng, zhangyu0815}@hust.edu.cn

Abstract. With the rapid development of sharing economy and mobile Internet technology, a large number of shared products have appeared in many fields in China. Taking shared bicycle for example, which is booming in China nowadays as a typical case to analyze, while for the shared bicycle service in the market are diversiform and immature, which has led to uncertainty, complexity and poor user experience in the process of using shared bicycles for users. This paper demonstrates the current research and development status of shared bicycles and focuses on the user experience of yellow-framed bicycle Ofo, one of the famous shared bicycle brands in China, then analyzes the using process through the user survey, customer journey map and stakeholder map, finally puts forward suggestions and opinions of the improvement of Ofo service.

Keywords: Shared bicycle · Ofo · Service design

1 Introduction

About three decades ago, China was known as the "Bicycle Kingdom". But the two-wheeled mode of transports popularity began to fade, with many bikes soon replaced by their fuel-powered competitors. But recent years have seen a revival of the humble bike across China, with an increasing number of people choosing cycling instead of driving to schools, to workplaces or to do sightseeing. The introduction of bike-sharing schemes, pioneered by start-ups like Ofo and Mobike, has brought the trend to a new level, but there are problems in the process of development, such as unreasonable bicycle distribution, poor maintenance service and uncomfortable riding experience, which cause disappointing and poor user experience. Shared bicycle is a systematic service, which involves relevant elements of people, bicycle, software and infrastructure and so on, while service design is a systematic thinking mode, which can plan and organize the related factors effectively such as people, infrastructures and communications so as to improve user experience and service quality [1].

Therefore, the application of service design in the optimization of shared bicycles is benefit and significant to make it more sustainable in the long run development. In this paper, a representative shared bicycle brand Ofo is selected as an example, as one of the

© Springer International Publishing AG, part of Springer Nature 2019
T. Z. Ahram and C. Falcão (Eds.): AHFE 2018, AISC 794, pp. 422–431, 2019.
https://doi.org/10.1007/978-3-319-94947-5_43

earliest brands to propose the idea of shared bicycle, it is also a leader in the shared bicycles market.

This paper aims to analyze the using process through the user survey, customer journey map and stakeholder map, propose the modified design scheme on the improvement of Ofo service.

2 Research Background

2.1 Brief Introduction to the Development of Shared Bicycles

With the economic development, there are more and more vehicles on the streets, while people enjoy convenience of it, which also aggravates the pressure of urban traffic. In order to alleviate the traffic pressure, the government and the bicycle company launched rented bicycle project in 2007 with high cost but low utilization; With the rapid development of mobile Internet, since 2015 Ofo has launched Internet shared bikes and replaced the rented bicycle for its convenience, low-cost and high-technology [2]. Shared bicycle provides an effective solution to the "last mile" problem, which refers to the final leg of a person's journey, now it is becoming a new way instead of using cars for people to get around in short or medium distance. The Table 1 shows the brief comparison of rented bicycle and shared bicycle.

Table 1. Listing the product, time, operator, brand, usage media of rented bicycle and shared bicycle.

Product	Rented bicycle	Shared bicycle
Time	2007	2015
Operator	Government/Bicycle enterprise	Internet company
Brand	Wuhan communal bicycle	Ofo
Usage media	IC card	Mobile phone

2.2 Overview of Service Design

Service design is a system for visualizing services to be given proper position in the market, which was initially proposed in the paper How to Design a Service [3]. In the design field, service design was defined by the international association for design research, that it's the function and form of the service from the customer's point of view, the goal is to ensure that the product is useful, usable, and desirable to customers, while the service provider feels effective, efficient, and recognizable [4].

Service design has developed in the fields of computer technology, communication and industrial design, and its theories, methods and tools also related to these fields. Nowadays, the concept of service design is widely used in product development, restaurant service, tourism experience and many other aspects.

2.3 Research Status of Bicycle Sharing

In the recent two years, the study of shared bicycles has become prevalent in China, scholars from different fields have discussed the current situation and optimization plan of shared bicycles from various aspects such as law, economy, science, technology and management. In the design related field, a paper proposed a human-based shared bicycle lock design from the product design. [5]. The concept of service design has been applied in the study of service design on changing of sharing and public transport for commuters, it has provided a theoretical basis for the study of the service design for commuters [6]. In some papers, the statistical methods has been used to optimize user experience maps and models on the reliability and rationality of shared bicycles [7].

This paper aims to explore how to apply the concept of service design [8] to the improvement of service quality of shared bicycles based on the example of Ofo and puts forward the optimization ideas and solutions for it.

3   Ofo Preliminary Investigation and Analysis

The brand Ofo implemented shared bicycles in 2015 initially focused on campus market where has a large number of idle bikes, it called for the teachers and students to transfer their own bicycles to Ofo platform, namely to join "Ofo shared bicycle" in exchange for free access to all shared bicycles. The team also launched unified bicycles which are equipped with sensors and smart locks. In 2016, Ofo started entering the city market, nowadays Ofo has already covered many universities across the world and expanded its business to cities and overseas.

The shared bicycle report[1] shows that the age of nearly 90% users is 18 to 45, among them 55% is 18 to 30, 35% is 30 to 45. It indicates that the shared bicycles not only cover the young people, but also the middle-aged people are also widely used. With shared bicycle business model is put forward in China, fast development of Shared bicycle from the second half of 2016, nearly 25 kinds of brands shared bicycles appeared in the market. There are some brands survived in the fierce competition in the shared bicycle market. Taking Mobike, Ofo, Hello Bike, Youon as the example, the 4 brands are developing rapidly in the market. Their product idea, target users, current market share and brand impression have been shown in Table 2.

The table shows clearly that the 4 brands have clear user positioning, make use of high technology to create more easy-to-use bicycles, so that they can quickly occupy the market. Mobike and Ofo enter the shared bicycle market earlier in China, attracted a

---

[1] The First Quarter of 2017 China's Major Urban Cycling Report by Ofo United Transportation Research Institute.

Table 2. The analysis of bike competitive products is introduced in 4 aspects.

Brand	Mobike	Ofo	Hello Bike	Youon
Product idea	Let bike go back to the city	No strange corner in the world	Cycling becomes a life style	Green travel guide
Target users	White-collar workers in the first-tier cities	College students and teachers; first-tier cities	Second and third-tier city users	All age users
Market share (2017)	36.5%	40.8%	<10%	<10%
Brand impression	Good quality healthy high-tech	Vitality youth loveliness	Comfortable easy to ride	Tradition reliability

large number of users successfully. Hello Bike enters the market later, but it mainly focuses on the third and fourth tier cities. Youon has occupied the rented bicycle market long time ago, which has a large number of elder users.

The advantage of Ofo is quickly occupied the campus market, there are many teachers and students using Ofo. Therefore, Ofo needs to create more methods to attract these target users. Ofo and Mobike are competing in the city market, the data shows that about 50% of the users still use both brands, 15.8% of the users only use Ofo, while 18.9% only use Mobike[2]. Mobike pursues technological innovation, Ofo not merely concentrates on the promotion of hardware and software technology but also promotes more interesting activities. It still needs to explore and confirm the needs of target users, make further efforts to optimize services for users to keep existing users and develop potential users.

4   Analysis of Ofo from the Perspective of Service Design

4.1   User Survey

Service design is a user-centered research and design process, so insight into users' needs is one of the important principles for optimizing the service experience. User survey is to collect, analyze and study the respondents' suggestions and opinions through various research methods, the purpose is to understand the users' characteristics and the process of using the products of Ofo shared bicycle.

The core users of Ofo are on campus, their basic use of shared bicycles is on campus and are familiar with the cycling route. They want ofo will be able to meet the basic riding needs, improve the quality of cycling and being more interesting.

The second core users of Ofo are office workers who commute to work, often ride from home to the bus station or subway station, the riding time and route is fixed, there are more than one shared bicycle software in their mobile phones, occasionally ride

---

[2] 2017 Q1 China Shared Bicycle Industry User Monitoring Report released by Trustdata.

outing on weekend. They hope that the bike will improve its quality and make it more comfortable and safer.

In the process of implementing user research, we can closely observe the usage behavior of multiple users in a real environment through field research. In this paper, we profile a typical user without disturbing his riding status, meanwhile record the cycling process, then ask about his feelings after riding and at last draw the user experience chart (Fig. 1).

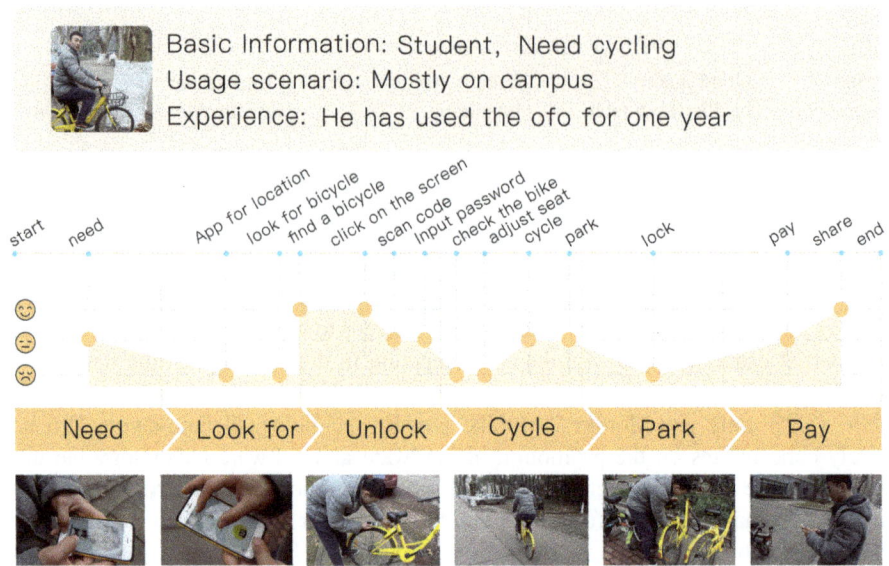

Fig. 1. User experience chart is divided into three parts, the top is the user's basic information, including user background, using scene, using experience, the middle section is the process of riding and feeling, the bottom part is the photo record.

4.2   User Journey Map

User Experience Journey Map is a design tool for carding user scenarios and experience problems. It can disassemble fuzzy requirements into user behaviors, user feelings, usage scenarios and other elements visual expression, user behavior indicates the pain and opportunities points [9].

Through observing the process of using Ofo, user journey map in the Fig. 2 tells the story of the core user's experience: from initial contact, through the process of engagement and into a long-term relationship. It demonstrates a clear picture of where the user has come from and what they are trying to achieve.

There are 6 main stages including generating bicycle demand, finding a car, unlocking, cycling, locking and paying. There are 10 main actives including clicking the App, bicycle positioning, code scanning, password entering, unlocking, riding, parking, locking, paying, sharing, however in actual process, users may encounter

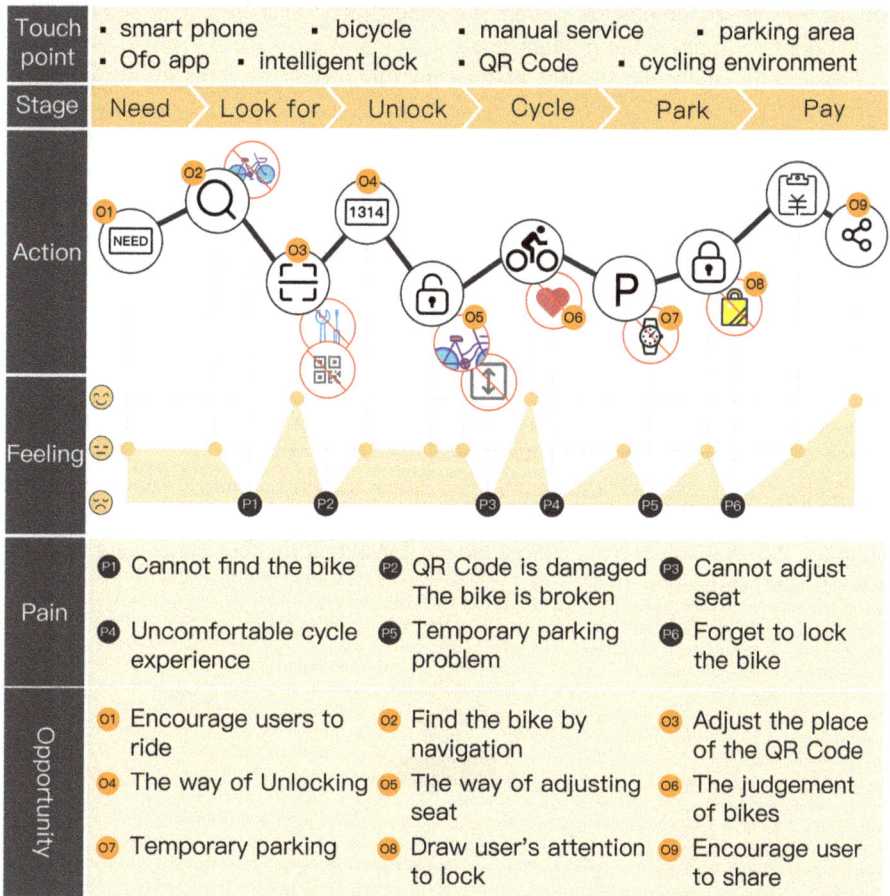

Fig. 2. The user journey map divided into six parts, including touch point, stage, action, feeling, pain and opportunity.

unexpected problems which will result in dissatisfactory or frustrated experience then even terminate the service.

Through the subdivision of these 6 stages of behavior and the analysis of emotional experience, we can summarize the opportunity points in the using process and find the pain points.

The user journey map above figure depicts the user's emotions by describing the whole process of using the bicycle. The six pain points at the lowest point shows the users' depressed. P1: Not be able to find a bicycle quickly when needed due to the inaccurate GPS positioning. P2: The possible failure of scanning the two-dimensional code due to the poor light condition or the damage of the code. P3: Users can't adjust bicycle seats due to the rust of the bicycle seat. P4: Uncomfortable cycling experience due to not considering the application of ergonomics and the poor maintenance management. P5: The needs of temporary short-term parking are unmet. P6: Forgot to lock the bicycle. In the process of improving the need to analyze 6 pain points to find the solutions to improve the product service.

Through the analyzing of user journey map to specify the opportunity points to improve the product. O1: Encourage more users to use the shared bicycles through activities. O2: Provide effective, quick and accurate bicycle location app. O3: The location and protection of scanning code to make the process easy for the users. O4: The bicycle can combine the in-put password and Bluetooth unlock. O5: The way to adjust the height of the seat. O6: The assessment criteria of bicycle condition. O7: Install temporary parking device. O8: Install the reminder to lock the bicycle device. O9: Encourage users to share.

4.3 Stakeholder Map

Service system should not only focus on service users, service providers, managers as well as other stakeholders. We will get a more comprehensive view of the service design plan if various stakeholders involved in can be fully mobilized. Shared bicycle stakeholders mainly include enterprises, bicycle manufacturers, government, operators and users, the stakeholder map explore the relationship of the four parties and the links between them (Fig. 3).

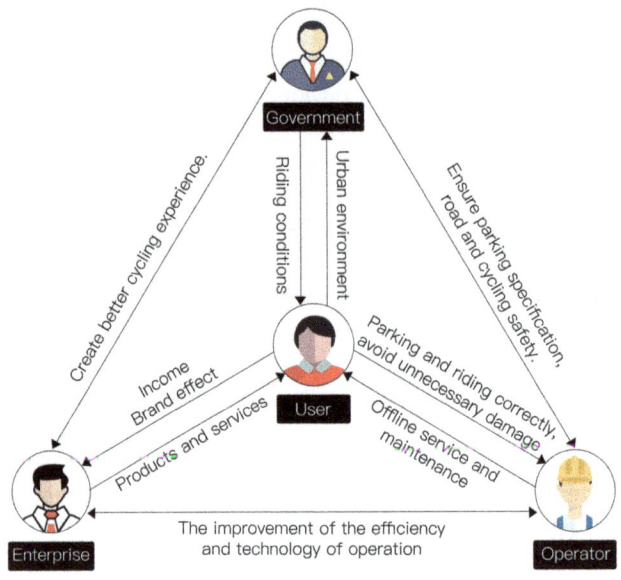

Fig. 3. Stakeholder Map shows the connection between user, government, enterprise and the operator.

The prospect of the user, operator, enterprise and the government:

(1) User: Reach to destination easily, safely and efficiently.
(2) Operator: Reduce the damage of bicycles and provide maintain and repair service efficiently.
(3) Enterprise: Provide bicycle service timely, comfortably, accurately and easily.
(4) Government: Provide multiple paths to make sure security and convenience for the users to relief congestion in cities.

The relationship of the user, operator, enterprise and the government:

(1) Users and Enterprises: Enterprise needs to produce more products that meet the needs of users and meet their emotional needs. While attracting users to use bicycles, enterprises should establish a clear reward and punishment mechanism to guide and regulate users' behaviors.
(2) User and Operator: The user is both the user and the maintainer of the shared bicycle. Ofo should encourage users to be a maintainer of the shared bicycle by using the reward mechanism, and guides the user to connect the operator timely online and offline when the bicycle need repaired.
(3) Business and government: Ofo should work with the government to set up "recommended parking spots" in cities, which will help to regulate the parking order.

5 Ofo Optimization Based on Service Design

Through the analysis of user research, user journeys and stakeholder map analysis about Ofo, several design ideas and optimization development suggestions has been discussed from three aspects as technology, enjoyment and cooperation.

5.1 Upgrade Technology

As Ofo continues to improve its product technology, it has launched "Bluetooth unlock + password unlocking" Smart lock to enhance the competitiveness of their products, Ofo need to continuously improve product technology, grasp the opportunity points and solve pain points.

(1) Available bicycle search: Inaccurate bicycles positioning technology, unreasonable distribution as well as unresponsive operation and maintenance which will lead to the problems for users to locate available bicycles quickly and accurately, Ofo needs to provide more precise positioning according to different cycling features, arrange bicycle distribution through the rational use of user data, and continuously improve the operation and maintenance scientifically and efficiently, provide visual display, cycling scheduling remediation to minimize the idleness and reduce vacancy rate.
(2) Riding process: Through the user survey, we notice that a small group of users need to use the shared bicycle from time to time within 1 h, their parking time is

always short within 5 min, but they still need to pay for an hour. Therefore, a temporary parking device may be developed for the users who need to temporarily park within 5 min and avoid repeating the process of finding, locating and paying for multiple bicycle service.

5.2 Add Design Element

Design can enhance emotional communication and the fun for the users. Products can facilitate social interaction emotional connection with the object enjoyable and interesting of design.

(1) Increasing the interaction of product: add more interactive elements to their products. For example, invite users take part in the design of wheel hub of bicycle, collect ideas from the users, increase the user's interactive needs for shared cycling.
(2) Encouraging users to choose Ofo: In addition to meeting basic riding needs, Ofo should hold various cycling activities to stimulate interest of using Ofo to make Ofo not only travel tool but also a partner in their life.
(3) Sharing and promotion: promotion activities through the line, co-designers to design wheel patterns, highlighting the art of the wheel; encourage users to join Ofo sharing program; increase riding share incentives, cycling and sports data combined, Increase dating mechanism.
(4) Lock design: The locks could be redesigned with different alarming sounds to remind and guide users to lock timely. They can be designed by the patterns of different bicycles. The curiosity will guide users to explore the lock-off sounds and then lock the bicycles.

6 Conclusion

The purpose of this paper is to discuss present situation of shared bicycle from the perspective of service design and put forward improvement ideas. Its sustainable development requires the cooperation of user, enterprise and government. The enterprise needs to improve the technology of shared bike continuously and meets the emotional needs of users. The government needs to improve the urban cycling infrastructure constantly and establish the credibility mechanism with the enterprise. User also needs to contribute their own strength to the development of Shared bicycle, and become a part of the maintenance of Shared bicycle.

Acknowledgments. This research was supported by National Natural Science Foundation of China (Grant No: 51708236) and Foundation of Huazhong University of Science and Technology (Grant No: 2016YXMS273)

References

1. Gao, Y., Xu, X.F.: Service design: a new concept of contemporary design. J. Acad. Lit., 140–147 (2014)
2. Qin, Z., Wang, Q.: Synergy mechanism in the vision of sharing economy: taking shared bikes for example. J. Reform, 124–134 (2017)
3. Shostack, G.L.: How to design a service. Eur. J. Mark. **16**(1), 49–63 (1981)
4. Erlhoff, M., Marshall, T.: Design Dictionary. Birkhäuser, Basel (2008)
5. He, W.X., Wang, J.F., Wang, W.J.: Hybridized bicycle lock design based on humanity. C. Ind. Des., 48–51 (2017)
6. Sun, B., Wang, Y.Y.: A preliminary study on service design of "Sharing + Public Transport" interchange pattern for workers. J. Des. 51–53 (2017)
7. Wu, C.M., Chen, L., Li, P.: User experience map model in sharing product service design. J. Packag. Eng., 62–66 (2017)
8. Liu, Y., Li, K., Ren, H.: Analysis of bike-sharing system from the perspective of service design. J. Packag. Eng. **38**(10), 62–66 (2017)
9. Wang, Y.M., Hu, W.F., Tang, J., Liang, Q.: Travel postcard service design based on user experience trip. Packag. Eng. **37**(22), 158–163 (2016)

Preferred Height and Angle of Touch Screen

Yahui Bai[1], Yinxia Li[1], Huimin Hu[2,3(✉)], Na Lin[2,3], Haimei Wu[2,3], and Pu Hong[1]

[1] School of Mechanical Engineering, Zhengzhou University, Zhengzhou, Henan, China
1298687634@qq.com, liyxmail@126.com, 15090566727@163.com
[2] Ergonomics Laboratory, China National Institute of Standardization, Beijing, China
{huhm, linna, wuhm}@cnis.gov.cn
[3] National Key Laboratory of Human Factors and Ergonomics (LHFE), Beijing, China

Abstract. Compared with the traditional control panel, touch screen provides better user experience for the operator. The heights and angles of touch screen were the main factors in structure design. A device with adjustable heights and angles were designed to provide different heights and angles in this study, and 12 participants with different heights took part in the experiment. The subjective score about heights and the preferred angles of touch screen were recorded during the experiment. Data analyzing results shown the preferred heights were 90 cm–125 cm from ground and the preferred angles at each height can be calculated by the equation y = −0.7345 x + 224.9816. The verification tested the patterns and shown the accuracy. The results can provide reference for the structure design of touch screen on standing posture.

Keywords: Preferred heights · Preferred angles · Touch screen
Subjective · Ergonomics

1 Introduction

Touch screen is widely used in many fields, such as manufacturing industry and services. LG, HTC and Apple Inc all released new touch screen mobile phones in 2007, while Microsoft announced its surface computing initiative. Later many more mobile phone companies followed suit. Today, the touch screen can be seen in each corner, such as the tablet PCs, the ATMs, the ticket vendor and so on. Compared with the traditional control panel, touch screen provides better user experience for the operator. The touch screen can change the interaction context easily and the soft buttons in touch screen are obvious advantages [1].

However, the design of the touch screen is also essential for operating efficiency and accuracy. As one kind the control panel, the design of touch screen should be suitable for human operation and meet the aesthetic and cognitive spiritual needs of people. In related work, Wang [2] announced that the field of vision, operation requirement and placement forms were the main factors that need to be considered in

© Springer International Publishing AG, part of Springer Nature 2019
T. Z. Ahram and C. Falcão (Eds.): AHFE 2018, AISC 794, pp. 432–438, 2019.
https://doi.org/10.1007/978-3-319-94947-5_44

the design of control panels. The design of control panel needs to satisfy the human abilities. Qin [3] mentioned that the design of Multi-Media Console based on ergonomics can provide a safe and comfortable teaching environment for operators. The ergonomic principles provided a guideline for the products design, but the principles were not suitable for all the users and situations. In different conditions, the gestures and tasks of the operators were not the same. The standing posture is demanded when using the ATMS, and the siting posture is demanded when driving a car, where the touch screen was all needed to operate. Moreover, the anthropometry data of individual varied a lot. An important issue in ergonomic design of products is to identify the factors that lead to human comfort and discomfort or preferred or not [4].

When interacting with a touch screen, the heights and angles were the main factors in structure design. The reachable places on standing posture can be found in relevant research, and the heights design of touch screen can reference them. ISO 13406-2 [5] mentioned the comfortable gaze angles range is about 0° to about 45° when viewing, and that is useful in angle design of visual displays. For touch screen, the angles were related much to the human ability, such as the reachable and comfortable places, which need to be researched a lot. The major aim of this study is to find the preferred height and angle of touch screen, which can provide a reference for the ergonomics design of touch screen.

2   Methods

2.1   Experimental Design

This study mainly focused on the standing posture when using the touch screen, just like using an ATMs. In this study, a device with adjustable heights and angles was designed to provide different heights and angles for research. Screen height was defined as the distance from the center of the touch screen to the ground. Screen angle was defined as the angles from horizontal table level to the touch screen, see Fig. 1.

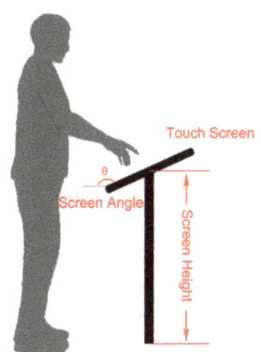

Fig. 1. The experiment arrangement

There were 9 experimental heights adjusted by the experimenter from 70 cm to 150 cm, increased 10 cm per time, and 1 preferred heights adjusted by the subjects themselves. The range of heights 70 cm to 150 cm includes the simulation comfortable range of human in JACK software. The touch screen angles varied from 90° to 270°, which can be regulated by the participants.

2.2   Participants

12 participants with half male and female took part in this experiment. Their age varies from 20 to 32 and heights from 152 cm to 180 cm. All had corrected 0.8 or better visual acuity with normal color vision. In this study, the heights of participants covered almost people heights of Chinese [6].

2.3   Environment

Surveys of actual illumination levels showed that most of VDT-equipped offices are within the range of 300–500 lx [7]. In this experiment, the average ambient light illumination was set as 460 lx.

2.4   Task and Procedure

Each participant was required to proceed with the following steps:

(1) Measure the participants' anthropometry data, record the results and introduce the procedure to the participant.
(2) Adjust the touch screen height to 70 cm. Set the angle of touch screen θ to 180° and set the Baidu Map in homepage.
(3) The participates walked from one meter away from the screen to a comfortable distance to use the Baidu Map app to search for a destination.
(4) The standard posture was natural stance. The participants adjusted the touch screen angles θ to the best touching angles and then evaluated the comfort of touch screen height according to the 7 levels subjective score, see Table 1.
(5) Record the touch screen angles θ that the participants adjusted, and the subjective score of touch screen heights.
(6) Increase the display heights by every 10 cm each time to 150 cm, repeat steps (2) to (5) for the other heights.
(7) When the process of height adjust from 70 cm to 150 cm were finished, each participant need to adjust the touch screen to the preferred heights without limitation.
(8) At the preferred heights, the touch screen angle θ also need to be adjusted to the preferred angle. Record the height and angle adjusted by the participant.

Table 1. 7 levels subjective score

1	2	3	4	5	6	7
Very uncomfortable	Uncomfortable	Less comfortable	General	More comfortable	Comfortable	Very comfortable

The experiment required 1 h to complete, per participant. All participants were paid money for their contributions.

2.5 Data Analysis

The preferred height and angle of touch screen were analyzed in this study. Preferred height was the ones whose subjective score over 5. The preferred touch screen angle was the angles adjusted by the participant at different heights. The regular pattern was concluded fist through the specified heights from 70 cm to 150 cm with the step size 10 cm, and then the last group of data were used to verify the pattern.

3 Results

The data of touch screen height and angle were analyzed in different ways. The results of the relationships between the subjective score and height, the preferred angle and height are as follows.

3.1 Preferred Height

The 12 participants scored the height of touch screen at the preferred angle after the experiment task, so 12 subjective score were collected at each height. The changing trend of each participants were almost the same with firstly increasing to the top then decreasing when the height of touch screen. The average subjective score of 12 participants at each height are shown in Fig. 2.

The heights of touch screen had significant effect on the subjective score, which can be seen in Fig. 2. The subjective score of touch screen height increases from 70 cm to 110 cm, then decreases until the heights of 150 cm. The average subjective score was below 5 (more comfortable) at the height of 70 cm–90 cm. The preferred height whose

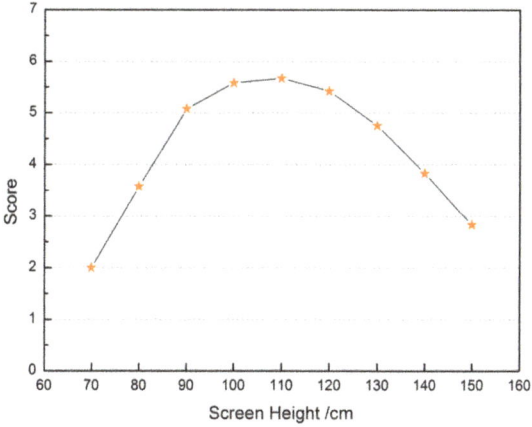

Fig. 2. The subjective score of touch screen heights

average subjective score over 5 is around 90 cm–125 cm. Since the experiment step of height was 10 cm each time, the score of height at around 125 cm was predicted value, 5. With touch screen height between 125 cm and 150 cm, the subjective scores are below 5.

The heights of participants varied from 152 cm to 180 cm, which including most people. The preferred heights were the heights whose average subjective score over 5. The preferred height was a range between 90 cm to 125 cm in this study. The results were obtained based on the standing posture and normal conditions.

3.2   Preferred Touch Screen Angle

The preferred touch screen angle was adjusted by the participants themselves. They are required to adjust the angles according to their own operation preference including the conditions of joint angles. The coefficient of correlation of height and the preferred touch screen angle of 12 participants were obtained by calculating, and the results can be seen in Table 2.

Table 2. Coefficient of correlation of height and the preferred touch screen angle

No.	1	2	3	4	5	6	7	8	9	10	11	12
R	−0.98	−0.98	−0.95	−0.98	−0.94	−0.90	−0.96	−0.95	−0.92	−0.93	−0.96	−0.96

The preferred touch screen angles have strong correlation with the touch screen heights, and all the coefficient of correlation are over −0.9. With the increasing of touch screen heights, the preferred touch screen angles of all participants decreasing. The preferred touch screen angles at each heights of all participants were shown in Fig. 3. At the same heights, the preferred touch screen angles of all participants varied in 20°. However, the outlier was inescapable, like the cube in the height of 120 cm. Most of the preferred touch screen angles were in the ranges.

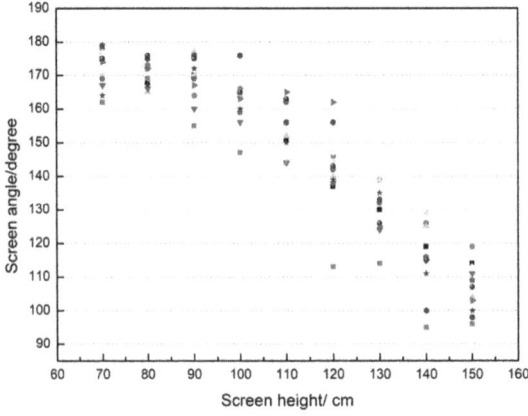

Fig. 3. The preferred touch screen angle

To get the regularities of preferred angle at each height, the average of preferred touch screen angle at each height of 12 participants were calculated and the linear regression equation is shown below in formula (1):

$$y = -0.7345\,x + 224.9816. \tag{1}$$

x is touch screen heights and y is preferred touch screen angles at that height. The R-square of the linear regression equation is 0.9138, which means the regression linear fits the observed value. The distribution of average value at each height is shown in Fig. 4, and the linear regression equation is also plotted in the diagram, the red line in Fig. 4.

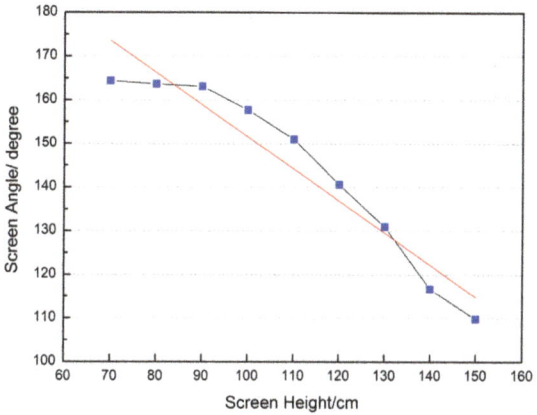

Fig. 4. The preferred touch screen angles

3.3   Verification

The preferred heights and angle were obtained through experiment and data analysis. There was one set of data collected during the experiment to verify the correction of the results in Sects. 3.1 and 3.2. The heights and angles adjusted by the participants are shown in Table 3.

The preferred heights of 12 participants in Table 3 are between 94 cm and 123 cm, which are included in the preferred range of heights obtained in Sect. 3.1. Correspondingly, the preferred touch screen angles calculated by the regression equation in Sect. 3.2 were around the experiment data within 10°. The verification results had shown the correction of the pattern about the preferred height and angle of touch screen and can be used in the design of touch screen in some situations.

Table 3. Heights and angles adjusted by the participants

	1	2	3	4	5	6	7	8	9	10	11	12
Height/cm	98	118	119	112	117	117	123	111	94	117	109	108
Angle/°	155	142	147	147	140	147	137	137	173	140	144	156

4 Conclusions

The preferred height and angle of touch screen were studied in this study, and some patterns were also concluded and verified. The results in this study can provide a reference for the ergonomic design of touch screen. When it comes to the design of touch screen on standing posture, the following results may be helpful:

(1) The preferred touch screen height is 90 cm–125 cm from the ground.
(2) The preferred touch screen angle can be predicted by formula (1) and can varied in 10°.
(3) The preferred touch screen height and angle are all ranges not a certain value.

The research methods in this study mainly used the subjective measurement to obtain data, thus the results may not be repeated in other people due to the differences of individuals. Besides, there were 12 participants took part in the experiment, the results can be influenced some people and the outlier values were not eliminated when analyzing. Cares need to be take when the experiment is related to people and when using of the results.

Acknowledgments. This research is supported by General Administration of Quality Supervision, Inspection and Quarantine of the People's Republic of China (AQSIQ) science and technology planning project (2016QK177), China National Institute of Standardization through the "*special funds for the basic R&D undertakings by welfare research institutions*" (522016Y-4488), and 2017 National Quality Infrastructure (2017NQI) Project (2017YFF0206603).

References

1. Baudisch, P., Hinckley, K.P., Sarin, R., et al.: Operating touch screen interfaces: US, US 7692629 B2 (2010)
2. Wang, K.Q.: Design of control panel for nc machine tools based on ergonomics. J. (2003)
3. Qin, Q.: Design of multimedia console based on ergonomics. J. Packag. Eng. (2007)
4. Bisht, D.S., Khan, M.R.: Ergonomic assessment methods for the evaluation of hand held industrial products: a review. J. Lect. Notes Eng. Comput. Sci. **2204**(1), 559–564 (2013)
5. BS EN ISO 13406-2:2002. Ergonomic requirements for work with visual displays based on flat panels Part 2: Ergonomic requirements for flat panel displays
6. GB/T 10000-1988. Human Dimensions of Chinese Adults
7. Laubli, Th., Hunting, W., Grandjean, E.: Visual impairments in VDU operators related to environmental conditions. In: Grandjean, E., Vigliani, E. (eds.) Ergonomics Aspects of Visual Display Terminals. Taylor & Francis, London (1982)

Design of Proton Therapy Procedure Based on Service Design Theory

Xinxiong Liu and Wanru Wang[✉]

Department of Industrial Design,
Huazhong University of Science and Technology,
No. 1037, Luoyu Road, Wuhan, Hubei Province, China
{xxliu,m201570705}@hust.edu.cn

Abstract. Proton therapy has been developing rapidly in recent years. A better design of the proton therapy center will reduce the cost and improve the treatment effect. In this paper, we propose a new proton therapy service based on the service design theory. By applying service touchpoint optimization, adding new service touchpoints and service procedure optimization, the proposed new proton service has the potential of meeting demands of both patients and the medical stuff. The treatment effect and the treatment experience will be improved in a large extent.

Keywords: Proton therapy · Service design · User centered

1 Introduction

Proton therapy [1–3] has been regarded as an effective way for curing cancer. With the development of technology, more and more specialized proton therapy hospitals have been established. Different to the treatment method of common diseases like taking medicine or having an operation, proton therapy is conducted in a relatively complex way that the cancer cell should be irradiated by the proton ray. Every treatment plan is an outcome of multi-disciplinary team cooperation. Thus, the hospital operates in the way that patients should make an appointment first. However, based on our field survey and deep interview of patients in radio therapy hospital, current procedure of overall service is not satisfying enough and has the potential of improvement.

The procedure of proton therapy mainly contains five parts: appointment, clinic, simulated treatment, getting treatment and recovery. A more reasonable and friendly procedure will help improving the effect of treatment. In this paper, the service design theory was proposed to be utilized in guiding the optimization of the overall procedure. According to the service design theory [4, 5], different parts of the procedure should be regarded as a whole service rather than separate parts. By establishing the service blueprints of proton therapy, service touchpoints of different parts are extracted to be used for optimization.

In this paper, we propose a new proton therapy service based on the service design theory. Main motivations of the proposed service are as follows: (1). Optimization on current proton therapy service touchpoints is conducted, which makes current system more efficient; (2). More service touchpoints are added to the traditional proton therapy

© Springer International Publishing AG, part of Springer Nature 2019
T. Z. Ahram and C. Falcão (Eds.): AHFE 2018, AISC 794, pp. 439–450, 2019.
https://doi.org/10.1007/978-3-319-94947-5_45

service, which meets the demand of patients and medical stuff and improves the effect of the treatment; (3). Service procedure is optimized to reduce the time for patients and technicians staying in the treatment room, which makes more people get treated in a limited time period and protect people's health.

2 Proton Therapy Procedure and Touchpoint Analysis

In this section, we propose the proton therapy service model based on our survey. The standard proton therapy contains four steps: appointment, clinic and simulated treatment, proton therapy treatment and recovery. The proton therapy center provides different services in different steps. The service contains some problems from the view of service design theory. The following field survey is based on the proton therapy service model and the touchpoint analysis in this section.

2.1 Appointment Service Analysis

Proton therapy center mainly applies online system to offer appointment service. The potential patient needs to fill in the appointment registration form through a website. The registration form is commonly composed of information about personal status, intended appointment date, specialized requirement, medical report and medical images. When the registration form is finished, the appointment result will be delivered to the patient by the customer service stuff through telephones.

The online system of proton therapy still has some problems. Firstly, the utilization rate of current website is relatively low. The patient information required on the website could be added. High material gathering efficiency will lead to a better outpatient service. Secondly, the description of the required information is not explicitly illustrated. Detailed expression of the required information needs to be contained. At last, the potential of the online system could be explored. More functions related to the treatment procedure could be added to the system.

2.2 Clinic and Simulated Treatment

When the patient arrives at the clinic, then the doctor will diagnose the patient to justify whether the patient is suitable for proton therapy. Some papers related to knowing the fact need to be signed by the patient. Then, the doctor will make an appointment for simulated treatment. In the simulated treatment period, the purpose is to find the right gesture for the treatment, design the suitable mold, target the cancer tissue position and create the treatment plan.

Before the patient arrives at the hospital, the patient needs to bring the needed material in for the outpatient service. Guidance about the material preparation before the clinic is essential for improving the treatment efficiency. The guidance is based on the bidirectional communication channel for patients and doctors. And this channel is lacked in current proton therapy center.

During the simulated treatment procedure, the patient has to maintain one gesture. The patient feels exhausted and tired. When the cancer tissue is targeted, the commonly used method for marking is drawing a cross marker on the skin. The patient cannot clean the skin before the treatment. Since the treatment mainly lasts for 1 month, more advanced methods like injection marking are applied. But injection marking will make the patient feel painful.

When the simulated treatment is finished, the treatment plan has to been created by doctors, physicists and technicians. The procedure lasts for around one week. During the waiting period, the patient may get anxious and nervous. Some patients may even doubt the professional ethics about the patient.

2.3   Getting Treatment

Patients have to get clothes changed before entering into the treatment room. A specific place needs to be designed for patient changing clothes. The privacy protection method needs to specifically considered. In addition, the privacy of the patient may be seen from the screen of CCTV in the observation room. The physicist has to use CCTV to observe the status of the patient.

After changing clothes, the patient enters the treatment room and lies on the treatment bed. The technician helps with the patient to put on the mold. This process usually lasts for a few minutes, which takes a large part in the whole treatment process. More efficient procedure of proton therapy treatment can be beneficial to increasing the number of treated patients in a constrained period.

Then the patient has to stay stable to during the treatment procedure. Some methods need to be taken to reduce the level of fatigue of the patient. During the treatment, the treatment machine rotates above the patient to find the angle for radiating the beam. The physicist stays in the observation room to supervise the treatment process.

2.4   Recovery

During the recovery treatment period, the medical team will meet the patient every a few weeks. Even not in the appointment date, the patient can advise the nurse. The communication channel for the return visit should be maintained. The recovery plan contains three parts including exercise, nutrition and phycology. The recovery plan is customized.

2.5   Service Touchpoint of Proton Therapy

Touchpoint is the main focus of the service design. It is the connection between the service and the user. The user receives service through touchpoints. Based on the proton therapy service procedure, touchpoints of proton therapy service are proposed.

Service touchpoints of proton therapy mainly contain three types: environment, treatment machine and the medical stuff. By analyzing the experience of the proton therapy patient, the main service touchpoint is proposed in Fig. 1.

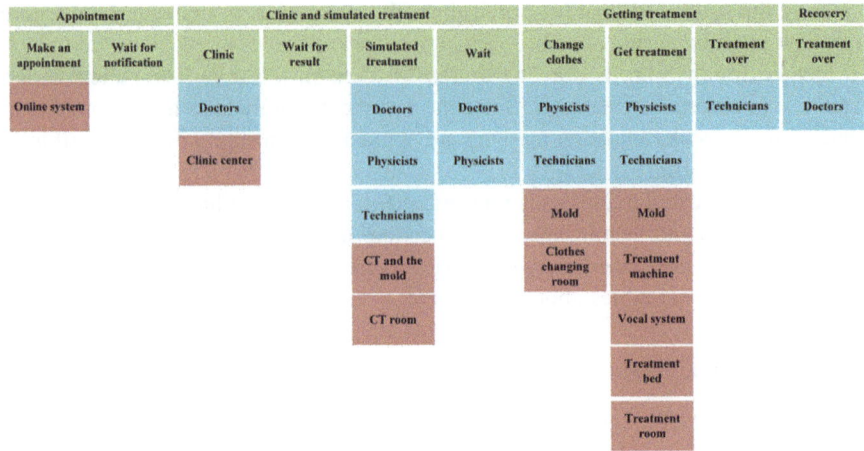

Fig. 1. Service touchpoints of proton therapy service.

In the appointment period, the patient receives appointment service through the online system touchpoint. After waiting for the appointment notification, the patient arrives at the clinic for outpatient system. In the clinic and outpatient service period, the patient will contact with the waiting room, clinic room and doctor. After the clinic, the doctor will make an appointment for simulated treatment. When receiving simulated treatment, the patient will get in touch with the touchpoint of physicists, technicians, doctors, mold, CT facility and the simulated treatment room environment. After finishing the simulated treatment, the doctor and the physicist will combine to make the treatment plan for the patient. This procedure is not obvious for the patient.

In the treatment procedure, the patient firstly will get in touch with the doctor and the technician to change clothes. Then the patient will enter the treatment room, lie on the treatment bed, install the mold and get proton therapy. In this procedure, the patient will get in touch with many types of touchpoints. The medical stuff includes the physicist and the technician. The treatment machine includes the treatment bed, the proton therapy machine and the mold. The environment includes treatment room, light system and the vocal system. After treatment, the patient will leave the treatment room with the help of the technician. In the recovery period, the patient will ask the doctor for recovery advice.

3   Field Research of Radio Therapy Center

Based on the principle of service design, insights about the current service are needed to optimize services. We conducted a field research on the Central Hospital of Wuhan and the First Hospital of Hebei Medical University. We studied their oncology department and radiological department. Although proton therapy and radio therapy have differences in treatment mechanism, the overall procedure and the design of the treatment room are similar. Since the proton therapy center has not been widely used in

China, we at last selected the similar radio therapy center to find insights for designing the proton therapy center.

The field research was conducted to get the image material for the radio therapy room. For the appointment period and the simulated treatment step, the queuing mechanism of the two-surveyed hospital is by doctor calling the patients. Online appointment system is not established. Thus, the main focus of our group is to study the treatment environment and the structure.

In the Central Hospital of Wuhan, the radio therapy center will receive medical records of patients waiting to get treatment. The sequence of these medical records is arranged by the doctor. In the First Hospital of Hebei Medical University, the sequence of the patient is decided by the physicist. In conclusion, the arrangement mechanism of two hospitals is simple. Cutting in the line is inevitable in current mechanism.

The patient has to firstly get the customized mold and then wait for being called by the doctor. The zone for arranging customized molds is shown in Fig. 2 The zone for arranging customized molds usually lies in the entrance of the treatment room at two hospitals. Many molds are placed according to the shape of the mold. From Fig. 2, it can be seen that different molds are overlapped and the place for one mold is not stable. In realistic environment, this will cause the potential for cross infection between different patients and the missing of molds. In addition, the zone for arranging customized molds takes up the limited space for the entrance, which makes the entrance more crowded. This planar structure affects the motion line of technicians and the patient.

Fig. 2. The mold arrangement zone in the Central Hospital of Wuhan (left) and the First Hospital of Hebei Medical University (right).

After getting the mold, the patient will change clothes for the treatment. In the season of winter, the patient has to take off the clothes in order to get the required precision for treatment. Currently, the two hospitals do not offer a specific cloth changing room for the patient, as shown in Fig. 3. Patients have to change their clothes usually beside the treatment bed. When one patient finishes the treatment and the other patient enters the treatment room, the changing cloth behavior causes some privacy leakage between patients. In addition, the technician also gets in touch with the privacy of the patient. In conclusion, the privacy protection needs to be improved for the radio therapy room.

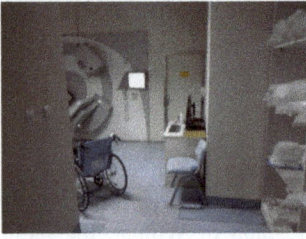

Fig. 3. The cloth changing place in the Central Hospital of Wuhan (left) and the First Hospital of Hebei Medical University (right).

After changing clothes, the patient needs to climb on the treatment, install the mold and lie on the treatment bed with the help of the technician. The patient usually climbs on the treatment bed with the help of an assisting facility since the treatment bed is relatively high for the person to lie directly, as shown in Fig. 4. But the assisting facility is easy to move, which makes the patient control the balance of body difficultly.

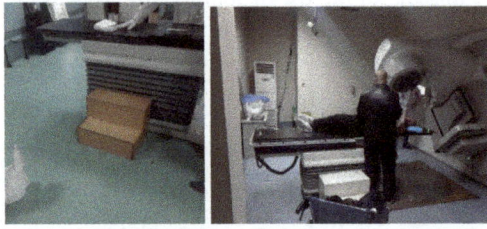

Fig. 4. The assisting facility in the Central Hospital of Wuhan (left) and the First Hospital of Hebei Medical University (right).

Treatment bed is the facility that lasts for the overall treatment procedure. The treatment bed is shown in Fig. 5. The touch feeling of the treatment bed is relatively cold. The shape is planar and lacks the protection mechanism at the two sides of the bed. When the patient is in large size, two arms of the patient lacks the needed space. In conclusion, the treatment bed lacks the consideration for the feeling of the patient and has the potential for cross infection.

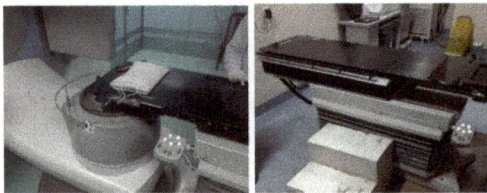

Fig. 5. The treatment bed in the Central Hospital of Wuhan(left) and the First Hospital of Hebei Medical University(right).

The researched hospital adopts the treatment machine which needs to rotates twice to finish one treatment, as shown in Fig. 6. The noise is low enough to affect people's hearing. The treatment machine is equipped with red light positioning system to cooperate with the technician to position the patient. During the treatment period, the machine does not offer vocal indication of current procedure. We found that the patient usually judged the treatment progress by the rotation status of the treatment machine. The treatment machine rotates in a low speed and lasts for 2 to 5 min. The exact time is related to the disease of the patient. In conclusion, the operation time of the treatment machine is short and the lack of indication for treatment progress cause the patient to move their body in advance which reduces the precision level of the treatment.

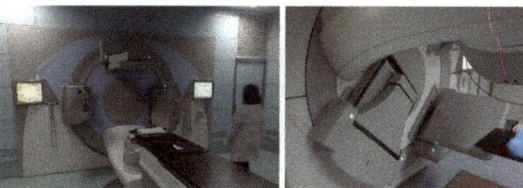

Fig. 6. The treatment machine in the Central Hospital of Wuhan (left) and the First Hospital of Hebei Medical University (right).

For the light system in the treatment room, as shown in Fig. 7, the Central Hospital of Wuhan applies a two-level light system, one for positioning and one for treatment. The First Hospital of Hebei Medical University only applies one level light system. The design of light system can be enhanced to meet the phycology status of the patient. For the room environment, the cultural atmosphere is not enough in current radio treatment room. Even though the First Hospital of Hebei Medical University has conducted some explorations, the cultural atmosphere needs to be improved in the treatment room.

Fig. 7. The light system in the Central Hospital of Wuhan (left), the First Hospital of Hebei Medical University (middle) and the cultural painting in the First Hospital of Hebei Medical University (right).

4   Proposed Proton Therapy Service Blueprint

Based on the analysis above, the proton therapy service blueprint based on service design is proposed in this section. The proposed service blueprint will satisfy the need of the patient, the doctor, the physicist and the technician in a large extent. The resource is optimized in our proposed proton therapy service blueprint.

4.1   Appointment Service Blueprint

The appointment service blueprint is shown in Fig. 8. The patient behavior contains making an appointment online, filling in the information, communicating with the doctor and receiving notification. When making an appointment, the patient will fill in the information form according to the indication through the online system touchpoint. The online system will gather the information and allocate the patient to one doctor. The doctor will diagnose in a preliminary way for judging whether the patient is suitable for proton therapy. The patient will receive the diagnose result and some instructions and guidance of the doctor.

Through establishing a channel for communication between doctors and patients, the online system improves the efficiency of the material preparation and reduces the phycology pressure of the patient.

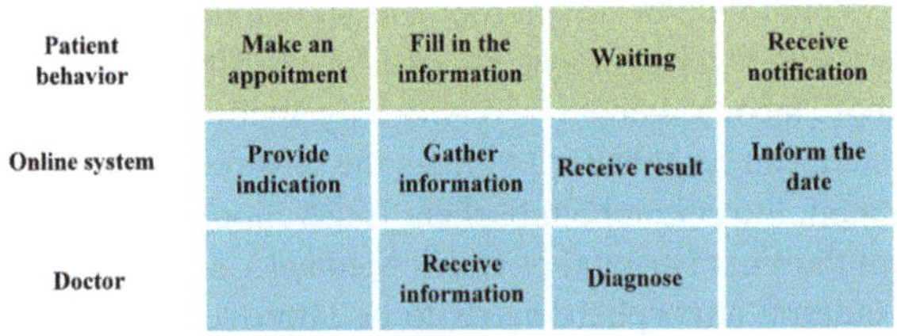

Patient behavior	Make an appoitment	Fill in the information	Waiting	Receive notification
Online system	Provide indication	Gather information	Receive result	Inform the date
Doctor		Receive information	Diagnose	

Fig. 8. The proposed appointment service blueprint.

4.2   Clinic and Simulated Treatment Service Blueprint

In the phase of clinic, the online system touchpoint is added. The service blueprint of the clinic phase is shown in Fig. 9. After the patient is diagnosed by the doctor, the doctor will offer the patient papers to sign. The diagnosed result and the related papers will be uploaded to the online system by the doctor. Meanwhile, the doctor will make an appointment for the simulated treatment. The appointment result can be enquired through the online system. When the appointment is confirmed, both the online system and the doctor will notify the patient. The application of online system touchpoint makes the clinic information more transparent to the patient. The time consumption of every department is visible by the patient. This mechanism provides the patient an expectation about current service and reduces the phycology anxiety.

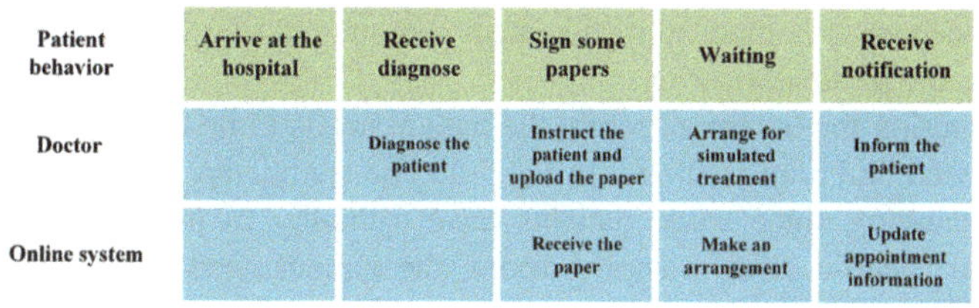

Patient behavior	Arrive at the hospital	Receive diagnose	Sign some papers	Waiting	Receive notification
Doctor		Diagnose the patient	Instruct the patient and upload the paper	Arrange for simulated treatment	Inform the patient
Online system			Receive the paper	Make an arrangement	Update appointment information

Fig. 9. The proposed clinic service blueprint.

In the phase of simulated treatment, the patient will be allocated with physicists and technicians who are responsible for the patient. The allocated physicists and technicians will get notification and the information about the patient. The basic information has been uploaded by the doctor in the clinic phase. The online system touchpoint is proposed to be used in the simulated treatment. Image materials, treatment prescription and the CAD source file of the mold will be contained on the online system. The patient can get access to these information through the online system touchpoint. In addition, current progress of treatment is presented in the online system. In conclusion, the online system provides a more efficient way for information transportation and representation and improves the efficiency of multiple departments.

From the view of the doctor, the work load of doctors needs to be represented online. The doctor needs to evaluate the progress of designing treatment plan and show current progress of the treatment plan. The progress can be quantified in percentage. This mechanism can increase the level of transparence and comfort patients' mood. The service blueprint of the aforementioned mechanism is shown in Fig. 10.

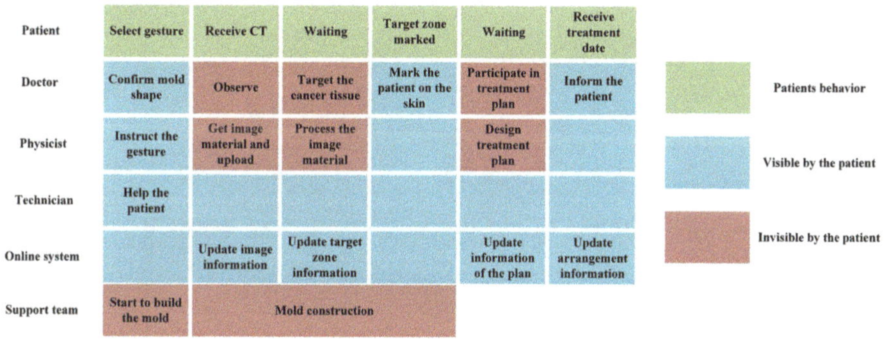

Patient	Select gesture	Receive CT	Waiting	Target zone marked	Waiting	Receive treatment date
Doctor	Confirm mold shape	Observe	Target the cancer tissue	Mark the patient on the skin	Participate in treatment plan	Inform the patient
Physicist	Instruct the gesture	Get image material and upload	Process the image material		Design treatment plan	
Technician	Help the patient					
Online system		Update image information	Update target zone information		Update information of the plan	Update arrangement information
Support team	Start to build the mold	Mold construction				

Legend: Patients behavior — Visible by the patient — Invisible by the patient

Fig. 10. The proposed simulated treatment service blueprint.

4.3 Treatment Room Service Blueprint

Treatment room plays an essential role in the proton therapy service. The cancer actually gets cured in this phase. In this section, the new proton therapy service procedure is proposed based on the service touchpoint optimization. The proton therapy treatment procedure is composed of two steps: before entering the treatment room and after entering the treatment room.

Before entering the treatment room, the patient needs to firstly change clothes. The proposed service blueprint is shown in Fig. 11. The technician should help the patient with cloth changing since some patients have difficulty in motion. After changing clothes in the specific cloth changing room, the patient needs to lie on the mobile treatment bed and gets the mold installed with the help of the technician. Other technicians should adjust the height of the treatment bed to a standard level in preparations for the following treatment in the treatment room. During this period, the patient does not have to maintain a stable gesture and only needs to lie on the mobile treatment bed and wait for the last patient who has already entered the treatment room to finish t. When the last patient finishes the treatment, one technician is responsible for moving the last patient out of the treatment room and helping the patient take off the mold. Two other technicians move the current patient into the treatment room.

The proposed procedure before treatment optimizes the overall service sequence by moving the mold installation from after entering the treatment room to before entering the treatment room. And this optimization is based on the updating from stable treatment bed to mobile treatment bed. By applying this mechanism, more time in the treatment room can be saved since the time-consuming activity which means installing mold and climbing on the bed is parallel with the treatment procedure of the last patient. What's more, the time consumed by last patient's getting off the treatment bed and taking off the mold is also saved. The treatment room only relates to the curing activity. In conclusion, the proposed mechanism reduces the curing time of average patient and improves the average number of patients in one specific time period. Thus, the cost on average patient is reduced and the promotion of proton therapy is enhanced.

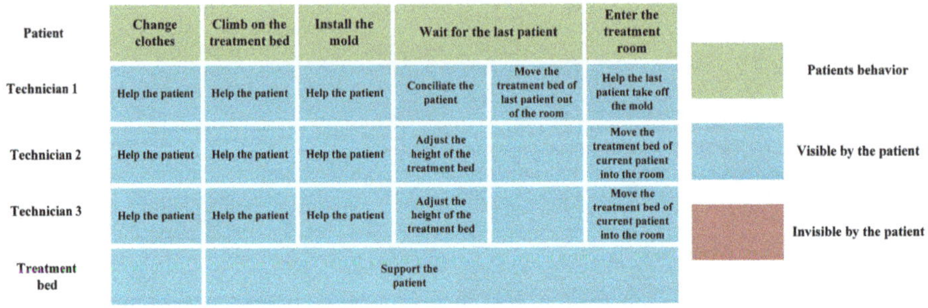

Patient	Change clothes	Climb on the treatment bed	Install the mold	Wait for the last patient		Enter the treatment room		
Technician 1	Help the patient	Help the patient	Help the patient	Conciliate the patient	Move the treatment bed of last patient out of the room	Help the last patient take off the mold		Patients behavior
Technician 2	Help the patient	Help the patient	Help the patient	Adjust the height of the treatment bed		Move the treatment bed of current patient into the room		Visible by the patient
Technician 3	Help the patient	Help the patient	Help the patient	Adjust the height of the treatment bed		Move the treatment bed of current patient into the room		Invisible by the patient
Treatment bed				Support the patient				

Fig. 11. The service blueprint before the patient enters the proton therapy treatment room.

After the patient entering into the treatment room, the patient only needs to lie on the bed. The service blueprint after the patient enters the proton therapy treatment room is shown in Fig. 12. Technicians need to move the mobile treatment bed to the standard position of the treatment room in order to get fixed. The fixing of mobile treatment is essential for the precision of treatment. Thus, the fixing methods need to be carefully designed. After fixing the treatment bed, technicians who help the patient enter the treatment room should leave the treatment room since the treatment room should not contain any people while operating. While the patient is getting treatment, the physicist needs to give short instructions and notifications about current progress in the observation room. After the treatment, the patient is moved out of the treatment room and takes off the mold with the help of the technician.

The proposed mechanism needs three technicians in principle. The average work load of technician is reduced. In addition, this mechanism reduces the time that the technician stays in the treatment room. Thus, the health of both the patient and the technician can be protected. The work load of the physicist is improved a little since the physicist needs to communicate with patient more frequently. But the content of communication with different patients is similar, the content can be decided at first. In another way, common communication content could be delivered using the recording method.

During the treatment period, the vocal system can play the customized list of music for the patient to reduce the level of pressure of the patient. Light system can also apply the customized style to the room. The customized information can be gathered through online system.

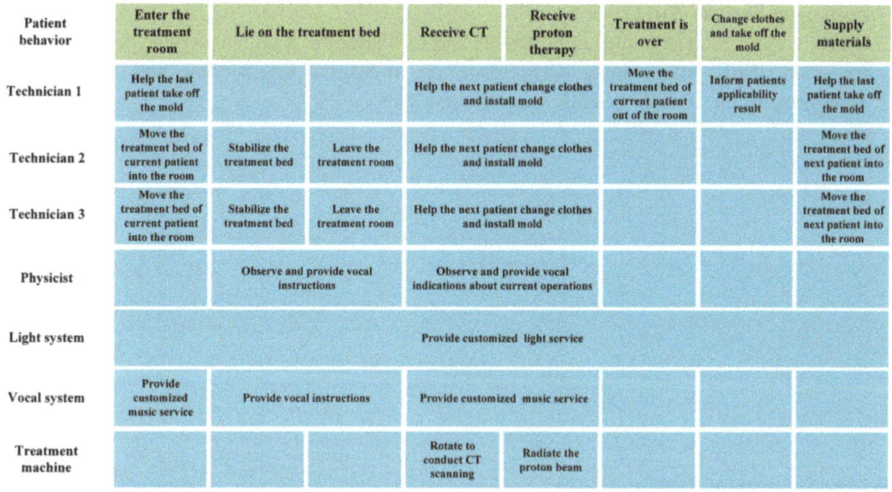

Patient behavior	Enter the treatment room	Lie on the treatment bed		Receive CT	Receive proton therapy	Treatment is over	Change clothes and take off the mold	Supply materials
Technician 1	Help the last patient take off the mold			Help the next patient change clothes and install mold		Move the treatment bed of current patient out of the room	Inform patients applicability result	Help the last patient take off the mold
Technician 2	Move the treatment bed of current patient into the room	Stabilize the treatment bed	Leave the treatment room	Help the next patient change clothes and install mold				Move the treatment bed of next patient into the room
Technician 3	Move the treatment bed of current patient into the room	Stabilize the treatment bed	Leave the treatment room	Help the next patient change clothes and install mold				Move the treatment bed of next patient into the room
Physicist		Observe and provide vocal instructions		Observe and provide vocal indications about current operations				
Light system		Provide customized light service						
Vocal system	Provide customized music service	Provide vocal instructions		Provide customized music service				
Treatment machine				Rotate to conduct CT scanning	Radiate the proton beam			

Fig. 12. The service blueprint after the patient enters the proton therapy treatment room.

4.4   Recovery Service Blueprint

The patient also feels worried and anxious in the recovery period. Regular medical test is essential for reducing the pressure. We proposed to apply the online system in the recovery period. The patient can contact with the responsible team to make an appointment for a return visit. The return visit can also proceed in online or offline ways. What's more, the recovery plan should be designed and uploaded by the doctor. Through online system, the doctor could supervise the execution progress of the recovery plan.

5 Conclusion

In this paper, the new proton therapy service is proposed based on the service design theory. By applying the method like service touchpoint optimization, service procedure optimization and adding new service touchpoint, the proposed proton therapy service has improved the treatment efficiency and treatment effect. Future direction includes designing more advanced treatment machine and the standardization.

References

1. Peterson, S.W., Polf, J., Bues, M., Ciangaru, G., Archambault, L., Beddar, S., Smith, A.: Experimental validation of a monte carlo proton therapy nozzle model incorporating magnetically steered protons. Phys. Med. Biol. **54**, 3217–3229 (2009)
2. Jongen, Y., Laycock, S., Abs, M.: The proton therapy system for the NPTC: equipment description and progress report. Nucl. Instrum. Methods Phys. Res. **113**, 522–525 (1996)
3. Smith, A.R.: Vision 20/20: proton therapy. Med. Phys. **36**, 55–568 (2009). Published for the American Association of Physicists in Medicine by the American Institute of Physics, USA
4. Stickdorn, M., Schneider, J.: This Is Service Design Thinking: Basics, Tools, Cases. Loan/open Shelves (2011)
5. Qi, Z., Kay, C.T.: The application of tools and techniques in a unified service design theory. In: 2008 IEEE International Conference on Industrial Engineering and Engineering Management, pp. 930–934. IEEE, Piscataway (2008)

Usability Recommendations for a Learning Management Systems (LMS) - A Case Study with the LMS of IFPE

Marcelo Penha[✉] and Walter Franklin Marques Correia[✉]

Universidade Federal de Pernambuco (UFPE), Av. Prof. Moraes Rego,
Recife 1235, Brazil
marcelopenha.unibratec@gmail.com, wfmcl0@gmail.com

Abstract. This article presents a case study of ergonomic evaluation of the usability of the Learning Management Systems used at the Instituto Federal de Pernambuco (IFPE), based on the MOODLE platform. In order to identify the usability problems of this system, were performed heuristic usability evaluations with experts and user tests. From the results obtained in the realized evaluations, it was possible to elaborate a list of recommendations based on the Nielsen heuristics to be applied in the interface design of the LMS, aiming to offer a better usability for the users of the system.

Keywords: Usability evaluation · Learning management systems

1 Introduction

E-learning emerged as an alternative modality of teaching with the objective of overcoming the limitations of attendance of regular education. A different alternative to face-to-face teaching that takes place between four walls of a school.

To mediate the interaction between those involved in the process of e-learning, according to [1], may be necessary to use a Learning Management Systems (LMS). A LMS for its features and tools available, is a tool of interaction in the learning process, based on the features available on the Internet.

That one of the problems of the advancement of e-learning, argues [2], is that it is in the hands of engineers and computer science professionals. In general, these professionals do not have major concerns about two fundamental aspects in a process of interaction: perception and usability. This results in faulty Virtual Environments for navigability and interactivity, generating the occurrence of noise in the human-machine interaction, during the learning process performed from an AVA.

In electronic learning, according [3], interaction does not happen by chance. It must be intentionally planned and expressed visually and functionally in the course interface or learning unit. According to [4], the evaluation of an interactive artifact (in this case, an LMS) helps to ensure that it will satisfy user needs based on usability recommendations to be applied for its improvement.

© Springer International Publishing AG, part of Springer Nature 2019
T. Z. Ahram and C. Falcão (Eds.): AHFE 2018, AISC 794, pp. 451–460, 2019.
https://doi.org/10.1007/978-3-319-94947-5_46

2 Theoretical Structure

This topic explains, from the literature, the theoretical basis of the questions relevant to reach the objective of the article, an evaluation and survey of usability recommendations for the virtual environment of IFPE.

2.1 Usability Evaluation Models

An artifact in development may constitute an upgrade/redesign of something already available in the market or something totally new. In the case of new artifacts, it is sought to evaluate the need of the market, understanding the needs of users and their requirements and if they appreciate the artifact.

On the other hand, when the artifact in question fits in the case of an update, changes are more limited and the greater attention is focused on the improvement of the product as a whole. In this case, the evaluations tend to aim to compare the performance and attitudes of users with previous versions.

According to [4], evaluating an artifact helps to ensure that it will meet user needs. [5] states that ergonomics assessment techniques are diagnostic and are based on verifications and inspections of the interfaces seeking problems of interaction between the user and the system.

Four models are related by [5] that can be used in interface evaluations: "fast and dirty" evaluations, usability tests, field studies and predictive evaluation.

"Fast and dirty" evaluation: an informal assessment with users or consultants, at any stage of the project. Through feedback obtained upon request from the designers, it is confirmed that the applied ideas are in accordance with the needs of the users. The data obtained in this technique are generally descriptive and informal. The advantage of this technique is to obtain data in a short time.

Usability tests: this technique seeks to evaluate the performance of the typical users of the interface in the accomplishment of the tasks. It was the dominant approach in the 1980s. The number of errors and the time to complete the task are the issues usually assessed during the tasks, from observations and filming. Virtually all activities are recorded. The tests are performed in an extremely controlled manner, in the laboratory. The evaluator has great responsibility in the tests because it has control of variables such as the location, the module of the system to be evaluated, whether or not the user contacts the other person. These characteristics however have a disadvantage to the technique: the difficulty in performing it, since there is a dependence of participants and a large number of data to be analyzed.

Field studies: evaluation carried out in a real environment, aiming to increase the understanding of what they do naturally and how technology impacts on these activities. Usually this type of evaluation is applied when you want to determine design requirements. Qualitative techniques such as interviews, observations and questionnaires are used to obtain data. In this technique, the evaluator can be a participant or a person immersed in the environment.

Predictive evaluation: guided by tried and tested heuristics, experienced evaluators apply their knowledge about typical users, proposing to predict usability problems. Predictive assessment can also be performed with another approach, involving theoretical-based models. The main feature of this model is that users do not need to be present. This feature makes the process faster and relatively inexpensive. Heuristic evaluation has become popular in recent years. Usability recommendations were designed with the main objective of evaluating screen-based artifacts. Therefore, some of them are still applicable. Others, however, are not appropriate. New sets of heuristics are needed to evaluate different classes of interactive artifacts. These heuristics should be based on a combination of usability and user goals. This technique generates lists of identified problems, usually with solution recommendations.

2.2   Nielsen Heuristics

Based on the analysis of 249 usability problems [6], Nielsen's heuristic analysis is a usability assessment technique based on a set of principles used to consider the elements of a user interface.

According to Nielsen, according to the service of these heuristics, it can be affirmed that the analyzed interface presents a good usability. Otherwise, the interface will display problems. These heuristics resist time because they depend on human behavior, which changes little, and these changes occur slowly.

The Ten fundamental principles [6] that must be fulfilled by an interface are:

H1-Visibility of system status
H2-Compatibility between the system and the real world
H3-Freedom and control to the user
H4 Consistency and standards
H5-Support for the user recognize, diagnose and recover from errors
H6-Error prevention
H7-Recognize instead remind
H8-Flexibility and efficiency of use
H9-Design aesthetic and minimalist
H10-Help and documentation

Nielsen's heuristics are widely used in research related to the human-machine interaction process, and consequently usability. According to [7], in an article about mapping research on interaction design in distance education in the international scenario, it was found that Nielsen is the most used reference in research, being part of 60% of the cases.

3   Case Study

To perform a better evaluation of an artifact, it is necessary to set goals. According to [4], the goals should guide the evaluation. In this way, the evaluation goals determined for this research were:

- Select a virtual environment of Distance Education;
- Establish tasks to be performed to evaluate the selected environment;
- Define the evaluation techniques to be used;

- Evaluate the interface of the selected environment according to the established criteria;
- Identify the usability problems of the evaluated environment.

3.1 Object of Research

The IFPE offers courses in the distance on the technical level, higher and postgraduate. The evaluations performed on this cooperative research were carried out using the environment of the graduate in Environmental Management. The virtual environment has the IFPE MOODLE platform software.

In this environment, students have the following features at your disposal:

- Contact: e-mail address to support the environment.
- News: Local users to post announcements.
- Messages: access tool for communication among the participants of the environment that are part of the student's contact list.
- My Courses: A list of the disciplines in which the student is enrolled, allowing access to specific content thereof.
- Users Online: Access the list and profile of the course participants connected and option for sending messages.
- Administration: Access to the table notes, reports and student profile.
- Participants: List of participants accessed the course.
- Activities: Access to specific types of activities undertaken in the discipline (chats, forums, surveys, questionnaires, resources and tasks).
- Calendar: Indications for major events such as exams and due dates of jobs.

All options described above are common to all students and follow a standard content, except for the agenda of the course that will have its content varies according to discipline.

3.2 Tasks to Evaluate

The tasks to be performed by users should be as follows:

1 - Access the virtual environment IFPE: http://dead.ifpe.edu.br/moodle/
2 - Log into the environment from the user specified;
3 - Edit your profile (change image, insert description, change password);
4 - Access the Course "Sistemas de Informações Geográficas";
5 - Access the Course Calendar and find out the commitments of the month of May;
6 - Access the Material "Aulas 1 e 2";
7 - Perform the deployment of Task "Envio de arquivo 6";
8 - Enter a post on the sixth week, the topic Classroom attendance;
9 - Enter a message in the Chat online tutoring, one week;
10 - Send a Message to Tutor Carlos Viana;
11 - Visit the Notes;
12 - Log off.

3.3   Choice of the Evaluation Paradigm and Techniques

According to [4], the choice of techniques to be used in the evaluation process should consider issues such as material feasibility, technical team and time limitation. At the same time, they argue that different techniques can be used in pursuit of different perspectives and points of view.

The different techniques should be used in a complementary way, thus adding the benefits of each one [8]. Thus, this case study used the techniques of Heuristic Assessment and Cooperative Evaluation.

3.4   Cooperative Evaluation of the LMS (IFPE-PE)

The cooperative evaluation was carried out with eight users. The list of tasks to be performed were previously passed on to users. The average time spent by users to perform the proposed tasks was 25 min. After analyzing the problems identified from the assessments carried out by cooperative users, it was observed that only heuristic H10 was not violated [9]. With respect to frequency, the heuristics H1, H8 and H9 are those with the highest number of occurrences, each in four different.

Figure 1 shows the number of heuristics violated in each of the tasks performed by users. Note that the tasks "Sign", "Edit Profile" and "Send message" showed the largest number of heuristics violated, each with four.

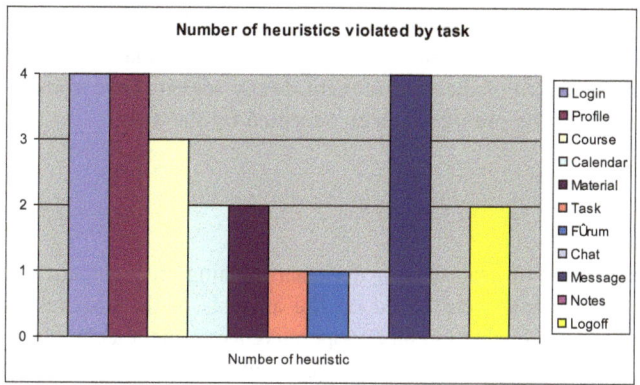

Fig. 1. Number of heuristics violated by task.

3.5   Heurist Evaluation of the LMS (IFPE-PE)

Assessments by evaluators generated a list of 71 problems identified. Some of these problems were common to more than one appraiser. The intersection of all the problems resulted from a list of 54 issues of usability [10].

After analysis of the consolidated list of issues identified from heuristic evaluation performed in the LMS of IFPE, it was observed that all heuristics were violated. Figure 2 shows the percentage for each of the heuristics. Note that the heuristics

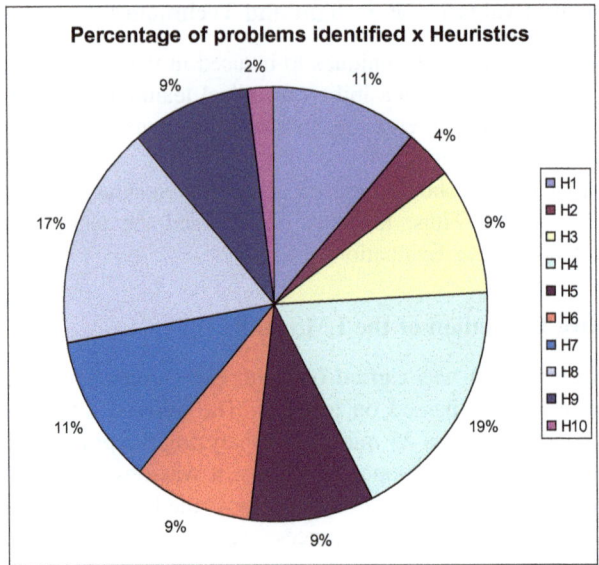

Fig. 2. Percentage of problems identified x Heuristics.

Consistency and Standards (H4) and Flexibility and efficiency of use (H8) showed the largest number of problems detected with 19% and 17%, respectively.

The high percentage of problems identified in H4 is mainly due to the lack of consistency of the layout of the environment. Many screens have completely different layouts in relation to the previous screen accessed by the user.

4   Results

The recommendations of usability for a virtual learning environment obtained in this research and listed below were based on the analysis of the data resulting from the application of heuristic evaluation and cooperative assessment techniques, as well as in the bibliography researched. The recommendations are classified according to the heuristics of Nielsen.

Some of the problems identified in the evaluations can be solved from the same recommendation.

H1 - Visibility of system status

Using feedback in the right way provides the visibility required for good interaction [4].

- Reposition the indication that the user is logged in to the top of the layout;
- Define labels that identify the function of system links;
- Define messages informing the user of the result of each action performed, for example, in the profile update;

- Identify the pages of the environment with titles in the upper left corner of the pages, using a readable source;
- View the progress of the tasks that need to be loaded (access to material, task download).

H2 - Compatibility between the system and the real world

According to [12], the labeling system defines the terminology and visual signs for each information element and to support the navigation of the user, considering the available space and the user's understanding in the creation of the page labeling WEB.

- Avoid using abbreviations for course names;
- Avoid using technical terms, such as "Version without frames and Javascript", available on chat pages;
- Define icons for the display function of one or all weeks of the course and their meaning is easy to identify.

H3 - Freedom and user control

According to [13], Web users, in general, clicking a link expect the new site page to appear instead of the last one. In this situation, if you want to undo this action, use the Back button. The exception with regard to new windows are PDF documents and the like. [10] places this principle with an ethical question.

- Insert a "Back" button to access the previous page visited;
- Allow the student to define how materials will be opened in formats other than HTML (.PDF, .DOC). In the same window, or in another;
- Allow access to all environment functions from any window.

H4 - Consistency and standards

According to [13], if users are accustomed to prevailing design standards and conventions, they will expect to find them on other sites. Among these conventions, the name of the institution in the upper left corner with a direct access link to the homepage. The system must be predictable [11].

- Reposition the environment access option to the top of the environment home page;
- Define a default layout that is maintained on all pages in the environment;
- Define a standard format for course content;
- Reposition the "Exit" link to the upper right corner of the layout, below the institution mark;
- Set a default position for breadcrumb. Preferably in the upper left corner of the layout, below the institution mark;
- Position the search field specific to each function of the environment so that the dependence of this field is clear;
- Insert a link in the institution's trademark to access the homepage of the environment;
- Reposition the calendar function to the left side of the layout, next to the other functions of the environment.

H5 - Support for the user to recognize, diagnose and recover errors

The system should state the reason or nature of the error committed, what should have been done and what to do to reverse the error [5].

- Define error messages in clear language (without codes), informing the problem and suggesting the solution to it;
- Define clear result message not found for pages that can not display the content requested by the student;
- Configure non-environment-related pages to open in a new window;
- Position the "Help" link highlighted.

H6 - Error prevention

Rather than a good error message and avoiding it being committed [11]. Users, according to [13], should receive structured guidance to avoid making mistakes.

- Clearly identify the meaning of the symbols used in the mandatory forms fields;
- Present a clarification message if the student performs any illegal operation.

H7 - Recognize instead of remembering

According to [13], the more detailed the descriptions, the more users discard them. As important as the description of the interface items, it is to show which of these elements are links [11].

- Present short and clear titles and denominations;
- Insert a button titled "Edit profile";
- Use the blue color and underline options only on links;
- Naming tasks, chats and forums clearly, avoiding misinterpretation.

H8 - Flexibility and efficiency of use

A strategy that facilitates good site navigation, according to [11] is the placement of direct links to a small number of high priority tasks placed on the home page. Direct links make browsing easier and shorter. According to [13], the search system is one of the most important elements of a site. It is particularly useful for users who know exactly what they want.

- Add a general search system, which is present in all screens of the environment;
- Insert anchors in pages with large amount of information and extensive scrolling;
- Insert a "Find" function in the environment;
- Insert links in the "Messaging" and "Calendar" field titles for direct access to their functions;
- Provide an alternative means of accessing information in a multimedia format;
- Provide direct access to the chosen chat window;
- Insert the "Change Image" function on the "Edit Profile" page;
- Include in the calendar an option that allows direct access to any month.

H9 - Esthetic and minimalist design

According to [4], removing elements that can be discarded without affecting the overall function of the system can be quite useful. You should eliminate confusing design elements and make the most use of design conventions. [13] argue that animations should serve a purpose for the user. There is no text size that appeals equally to all types of people. It is better to choose a larger source to not discard some users.

- Avoid using banners with rarely needed or confusing information;
- Avoid large amounts of information on the login page of the environment;
- Avoid using many different font sizes;
- Standardize the use of colors in the environment;
- Use a larger font size on the "Exit" button;
- Distribute the functions of the environment in the layout in a standardized way;
- Position important resources, such as the home page access option, highlighted;
- Maintain standardization in graphic elements;
- Avoid using serif fonts.

H10 - Help and documentation
According to [11], the help resource must be available for all tasks that the user will perform, although most of the time it does not need to be consulted.

- Add help resource related functions available on the page accessed.
- Remove the "Moodle" link, located in the footer of the screen. It serves as access to the software documentation home page, and is in English, making it unnecessary for the student.

5   Conclusions

The present case study was driven by Nielsen ergonomic criteria and two evaluation techniques were used: the heuristic evaluation, performed by specialists; and the cooperative assessment carried out with users.

The techniques chosen were efficient in pointing out several usability problems, some common to the two evaluation techniques, others identified particularly by one of them. Were observed violations of basic principles of design and usability, content as very extensive and sometimes unnecessary, lack of standard in layout, messages and inappropriate language inconsistent placement of some features.

The results obtained from the tests applied in the case study showed that the virtual environment of IFPE has enjoyed usability deficiencies, creating an interplay of low quality and, at times, unpleasant for the students. Some of the tasks posed ended up not being made by some participants.

From the problems identified, was elaborated a list of 48 usability recommendations, covering all 10 heuristics used as a usability reference was made. It is hoped that the usability recommendations obtained in this research contribute to the improvement of the virtual environment of any institution that offers the modality of e-learning.

References

1. Barbosa, A.L., Mendes, L.S.: Ambientes Virtuais de Aprendizagem. In: 20% à distância: e agora: orientações práticas para o uso de tecnologia de educação a distância/Alda Carlini, Rita Maria Tarcia. Pearson Education do Brasil, São Paulo, pp. 161–170 (2010)
2. Harasin, L., et al.: Redes de aprendizagem: um guia para ensino e aprendizagem on-line. Editora Senac São Paulo, São Paulo (2005)
3. Filatro, A.: Design instrucional na prática. Pearson Education do Brasil, São Paulo (2008)
4. Preece, J., Rogers, Y., Sharp, H.: Design de interação – Além da interação homem-computador. Bookman, Porto Alegre (2005)
5. Cybis, W., Betiol, A.H., Faust, R.: Ergonomia e Usabilidade, conhecimentos, métodos e aplicações. Novatec Editora LTDA, São Paulo (2007)
6. Nielsen, J.: Ten Usability Heuristics. http://www.useit.com/papers/heuristic/heuristic_list.html
7. Penha, M., Campos, F., Correia, W.F.M.: Mapeamento da pesquisa sobre design de interação na educação a distância no cenário internacional. Revista Científica Tecnologus **5**, 1–24 (2010)
8. Andrade, A.L.L.: Usabilidade de interfaces Web: Avaliação heurística no jornalismo on-line. E-papers, Rio de Janeiro (2007)
9. Penha, M., Correia, W.F.M., Barros, M.L.N., Soares, M.M., Campos, F.: Ergonomics evaluation of usability with users - applications of the technique of cooperative evaluation. In: 15th International Conference on Human-Computer Interaction, Las Vegas, DUXU/HCII 2013, A. Marcus, New York, pp. 379–388 (2013)
10. Penha, M., Correia, W.F.M., Campos, F., Barros, M.L.N.: Heuristic evaluation of usability - a case study with the Learning Management Systems (LMS) of IFPE. Int. J. Hum. Soc. Sci. **4**, 295–303 (2014)
11. Santa Rosa, J.G., Moraes, A.M.: Avaliação e projeto no design de interfaces. Rio de Janeiro 2AB (2008)
12. Agner, L.: Ergodesign e arquitetura de informação: Trabalhando com o usuário. Quartet, Rio de Janeiro (2009)
13. Nielsen, J., Loranger, H.: Usabilidade na Web - Projetando Websites com Qualidade. Campus, Rio de Janeiro (2006)

User Experience Design Manifesto

Paulo Maldonado[1,2,3(✉)]

[1] Centro de Investigação em Território, Arquitetura e Design (CITAD),
Universidades Lusíada, Rua da Junqueira, 188-198, 1349-001 Lisbon, Portugal
paulomaldonado@inspaedia.com
[2] Faculdade de Arquitetura, Centro de Investigação em Arquitetura,
Urbanismo e Design (CIAUD), Universidade de Lisboa, Rua Sá Nogueira,
Polo Universitário, Alto da Ajuda, 1349-055 Lisbon, Portugal
[3] Escola de Artes, Departamento de Artes Visuais e Design, Centro de História
de Arte e Investigação Avançada (CHAIA), Universidade de Évora,
Palácio do Vimioso, Largo Marquês de Marialva 8, 7000-809 Évora, Portugal

Abstract. The rapid change of the elements of context requires an urgent reflection on "Experience" and "Meaning with Value". The aim of this article is to critically review User Experience Design (UXD) principles as a multidisciplinary field of study, and contribute by a Manifesto to an inspiring vision of the future. Some authors will be reviewed in order to list and discuss the different ideas to redefine the desirable skills to an enlightened practice and to influence a new ethos of "Be Innovation" to a better world. The article retrieves and revisits a Design Manifesto made on the basis of a strategic vision for Design based on distinctive design competencies. This Manifesto was in the origin of the platform www.inspaedia.com.

Keywords: Manifesto · User Experience Design · Design · Innovation
Strategic vision · Distinctive design competencies

1 About a Strategic Vision for Design

20 years is the temporal arc that separates us from the reflection that in the scope of the research project of our masters degree in Design with the title «Design: a strategic vision» [1] was at the origin of the platform Inspædia (www.inspaedia.com). In this article we revisit this reflection and the Manifesto that resulted from it in the sense of provoking the discussion about the construction of the research processes, summoning, complementarily, to the reading of other publications that we made on the subject User Experience Design underlying the process of research and development of the Inspædia platform [2–11].

With regard to a strategic vision for Design, we used the form of a manifesto[1] as a matrix to communicate the concept we designate by new «distinctive design competencies»,[2] starting from the «interaction between the organization of creativity, the management of knowledge and the impact of new technologies»[3] and the concept of «utopia as a strategy», which fights the capitalist/consumerist model, replacing it with an «ideology of constructors» (Lester Thurow).[4] We advocated a central ideology, a vision of the future and its alignment with implementation, following the recommendations of James Collins. The Manifesto was not, nor was it meant to be, a final document, because such a position was inconsistent with the ideology that supported it. The collective knowledge resulting from the dynamics of interactivity would be in charge of perfecting the proposed model, introducing to it, continuously, the adjustments that the partners considered necessary. Thus, the idea, the process and the results would be, in themselves, Design.

In 1975, N. Cross, when considering the emergence of a post-industrial society, wondered about the possible need to invent a new Design process [16] in the near future. He referred to Architecture and was far from suspecting the profound changes to occur in the production-consumption process. Simultaneously, other questions arose as to which objects to project in the future and which perspectives to carry out the professional activity. Thus, in 1973, H. Kraus predicted that the industrial designer of the future would stop designing objects of use in close collaboration with other experts to create «jobs, man-machine systems, complete organizations».

---

[1] Charles Handy suggests the use of the words «meta» or «manifesto» when one considers the word «vision» too pompous [12], probably because he mistrusts the connotation of the word «vision» in the wake of the German idealist tradition «conception of the world». We have therefore proposed transposing the term «manifest» to a less usual form (since it is usually applied to a cultural, social or political area): we put it at the service of presenting the proposed concept, in order to incite, summon participation.

[2] Design is a structuring factor of distinctive competence because the design process we reinvented was underpinning the business idea we proposed.

[3] By the way, see the following excerpt from an interview with John Kao (JK), led by Jorge Nascimento Rodrigues (JNR): «JNR [...] if we are moving from the era of competitive advantage to the era of creative advantage is it that Michael Porter and the other competition theorists were wrong? JK - [...] they all talk about creativity ... but they do not say «what» and «how» to do, that is, how to manage creativity. The key issue is no longer to consider creativity as a priority, but to know how to create and manage it systematically. [Jamming] is at the intersection of three axes: the organization of creativity; the management of knowledge; and the impact of new technologies. The last two axes facilitate the first. [...] Design is something established, in architecture, in products, in graphics. It needs to be extended to business. [...] Do not we talk about the design of the organization, the design of the intranets, the design of a culture of collaboration? So the business leader has to look at himself as a designer.» [13] For an identical perspective, see Hammer [14].

[4] It proposes utopia as a strategy, which means instituting a revolutionary process of vision, in order to allow the flourishing of democracy and the reconstruction of a more just and humanized society. He argues that the capitalist/consumerist model will tend towards a «builder ideology», in which governments will play an important but no more important role; the heroes of the future will be able to build the new industries (the «gray-mass industries»). This coordinated action will focus on «human skills, technologies and basic infrastructures» [15]. This author recovers the schumperterian sense of the term «entrepreneur» and enriches it: the new entrepreneur is a «builder» whose skills are based on knowhow and knowledge.

Retracting in time but still in the 70s, T. A. Markus (1972) typified three modes of performance and socio-professional framing of the designer:[5]

«The first role is essentially conservative, centred around the continue dominance of the professional institutions. In such a role the designer remains unconnected with either clients or makers. [...] The opposite to this conservative aproach is actively to seek changes in society which would result in the end of professionalism as we know it. Such a revolutionary aproach would lead the designer to associate himself directly with user groups. Since this kind of designer is also likely to believe in a decentralised society he would be happiest when dealing with the disadvantage, such as the tenents of slum clearence areas, or the revolutionary forsakes his position of independence and power. He no longer sees himself as a leader but as a campaigner and spokesman. [...] The thirth, middle, path lies between these two extremes, and is much more difficult to identify except in vague terms. In this role the designer remains a professionally qualified specialist but tries to envolve the users of his design in his process. These more participatory approaches to design may include a whole range of relatively new techniques, ranging the public inquiry through gaming and simulation to the recent computer-aided design procedures. [...] Designers following this approach are likely to have abandoned the traditional idea that the individual designer is dominant in the process, but they may still believe they have some specialised decision making skills to offer» [16].

Contrary to what was said by T. A. Markus in the 1970s, the second modality appears today (1997) to be extremely timely. Before we realize its potentialities and interest for the strategic vision we have proposed, we have tried to disassemble its argument. Markus considered it very risky for the professional survival of the designers and, for that reason, did not recommend it. He feared for «lose his hand» by opening the creative process «to more voices». Contexts and opportunities had changed radically, but it was not only for this reason that we were convinced that the future of the designer's profession would be in the sharing of decisions, even the project decisions. The digital economy has revolutionized this logic by creating the conditions for consumers to be present in the various moments of the process, from conception and production, to the marketing and distribution of products and services. However, it is not only for this reason that we intend to include them in the process and remunerate them, depending on their participation, it is not only because of this pressure that we will have to go to the trailer for fear of «losing our hand».

Resuming the problem of participation and the socio-professional framework, the strategic vision that we proposed for the design opened up much more voices than the

---

[5] The following text refers to the professional activity of the architect. In the English language, the word design applies to any project activity. While acknowledging specifics in industrial design, also recognized by those who understand design in this broad sense, we propose that the reader extrapolate the thinking of Markus to the industrial design and the pursuit of the professional activity of designers working for the Industry. We refer to Tomás Maldonado and to the reexamination of his 1972 reexamination [17] and to Victor Margolin [18], among other authors and works that could be called for the most consistent attempts to define the term design.

claimed by T. A. Markus.[6] The new technologies allowed it and the brand-new digital economy required it, otherwise major opportunities would be lost.

The profound transformations that took place led to a reflection on the design scenarios (ways of designing and designers relating to customers and users/consumers).

The business idea that we proposed resulted from the observation, interpretation, reinvention and application[7] of a vast field of theories and practices that we tried to expose in the possible detail. It was based on the concept of «global thinking and action»,[8] centered on the new potentialities that were part of the so-called digital age. Bet on creativity, correlative of flexibility and competence.[9] It was not a panacea or a magic word but an imperative that stemmed from the need to change.

---

[6] The underlying motivations have nothing to do with those that animated the experiences called for in the text: contexts (political, social, economic) and possibilities were, at the time, completely different. The proposed way to ensure the participation of allies may face difficulties (which we will try to identify later). However, it is not part of the same type of constraints that guided some of the experiences previously mentioned: the process will be totally open and participated, the access is spontaneous, reason why it seems to be difficult for us to have room for paternalistic or authoritarian attitudes. The contributions of the partners are worth by themselves and not by the importance of their person or personality.

[7] Our research did not coincide with the application of the idea, so only then will it be finished (if it ever will). In the digital state, the investigation does not precede the launch. It is part of the launch, i. and. integrates with the launch-and-learn strategy [19].

[8] For an opinion to the contrary, but that is exactly on the same assumption - the connectivity that the Net allows (see Chuck Martin's positioning, which coincides with John Naisbitt's, «local thinking and global action») in answering the question of Jorge Nascimento Rodrigues to Chuck Martin: In the 80's, we learned to «think global and act locally». Now he tells us that we must «think local and global action» an idea also advocated by John Naisbitt. Why this inversion? Answer: «because the Net makes this possible. This is now a mere evidence.» [20].

[9] «Today all companies are looking for a new advantage, delicate and dangerous, but vital: that of creativity. Global competition increasingly depends on the ability of nations to mobilize their ideas and talents to create creative organizations. [...] Creativity begins with the generation of ideas … It is crucial to import new data from the outside. is to diffuse and impregnate by the company an aspiration of creativity … The desire for creativity must become the norm and not the exception …. Non-creative companies have three possibilities: to buy innovative companies; ally with them; or buy stars. Appealing to outside actors to participate in a specific project has become commonplace… It is up to the leaders to establish a new «temporary culture», indispensable for refreshing new ideas and ideas. [...] The new information technologies favor the generation of ideas by developing access to knowledge and informal exchanges. [...] As in jazz, cyberspace is a meritocracy of talents that are both competitive and cooperative. Power no longer comes from knowledge - to which everyone has access - but from the creativity extended by the network. [...] Companies of the future can design their structure and culture in a flexible way, in function of their creative projects. It will only be the sum of your ideas, knowledge and abilities. The organization of the future will be like a factory of ideas, whose starting point is the unknown. You do not know the products you are going to build, nor the competitors you are going to face: it recreates itself permanently. The leader must manage contrary tendencies like freedom versus discipline, fixed goals versus flexibility, collective responsibilities, safety versus risk, acquisition versus innovation, experience versus novelty, and normalization. experimentation.» [13].

It was intended to find/define a context[10] for the design process (actors and environment)[11,12] and the motivations and capacities for an interaction strategy (culture, information and knowledge). That is to say, based on a global strategic vision, we intend to carefully select the (dynamic, flexible and creative)[13] principles capable of determining the most appropriate competences for a design process («distinctive competencies»),[14] with a well defined objective: without innovation necessarily having to go through the invention of new technologies. However, there is no innovation without creativity. The epochal context seemed to propitiate,[15] at the outset, the success of proposals such as ours.

---

[10] According to Chuck Martin, in the digital state is the context that determines value, not content: «context is the combination of all the surrounding factors: time; the place; the relevance of the content; the technology used. It is fluid, malleable, reactive, and interactive, just like the Net itself.» [20].

[11] Or «procedural paradigm», advocated by Tom Peters and also by Kees Van Der Heijden: «Success is more closely related to a good process than to the discovery of an «optimal strategy.» [21].

[12] Knut Holt thus synthesizes the process of visionary design: «when you invent something realy new, you create a need. The big trick is not to invent to satisfy a need - anybody can do that. The trick is to recognize that need that people do not realize is a need.» [22]. This optimism is qualified that we refrain from adjectivizing out of respect for all those who suffer the dissatisfaction of vital needs and for those who continue to find in them their reason for being.

[13] «Soozoo (criativity) is the new battle cry, companies strive to gain or retain their share of the global market.» [22].

[14] Gary Hamel prefers the concept of «distinctive competencies» rather than «strategic business units» [23]. Its scope fits better with our idea. «Three business strategies have to interface with designing: develop market/ user understanding thoughly before design development; focus on commercially viable translation of ideas, i.e. innovation for successful commercial products; design a well-integrated organization to support product development in the milieu of ever changing conditions, with product systems or cascading development from one product to the next.» [22]. Still on «distinctive competences».

[15] In ostinato, John Kao explains why our era is that of creativity. In his opinion, there are eight great arguments that justify it: «this is the age of creativity because that's where information technology wants to go next. [...] because it's the age of knowlege. And in an era of prizes knowledge, creativity adds value to knowledge and makes it progressively more useful. [...] because companies are increasingly forced to rapidly reinvent themselves to achieve growth. [...] because many workers today feel creative jobs, and talented people are mobile as never before. [...] because of the new primacy of design. [...] because there has been a change of regime in the marketplace. The customer is the boss now - discerning, demanding, and no more loyal than he or she has to be. The new boss has only one question: So what are you going to do for me tomorrow? Only creativity can give the answer. [...] because the subtext of global competition is increasingly about a nation's ability to mobilize its ideas, talents, and creative organizations. A company that ignores the global creativity map is spurning an important set of strategic considerations. [...] This is the age of creativity because management is transforming its role from controller to emancipator - of creativity. This is the new managerial mind – set.» [24].

2 Manifesto

1. The process is open (free).

It is based on the connectivity that the Internet allows and empower.

2. There are no business or business areas defined at the outset.

Herein lies part of the innovation component that we intend to introduce. Distinctive competence[16] will be what it has to be, what it wants to be. What will determine the process of definition will be a network of people that interact for this purpose/common goal. As a consequence, preferred business or business areas are not initially defined.

3. The distinctive competence is locus free.

There are no physical or virtual barriers between research, development, production, distribution and consumption. Anyone can be anywhere and practically at the same time. There is no rigid sequence of actions between the moment the idea arises and the consumption of the «functionalities» created.[17] On the other hand, it is hoped to promote an interaction of events, generating other events.

4. It is not initiated through a pre-established or existing organization.

This principle stems from a non a priori definition of the distinctive competencies to be developed. The alliance of individuals is an initial datum but not an end in itself, only a mean. The definition and configuration of the group/partners will be made later: the proposed model approximates the (amplified) concept of «virtual organization» (Francis Fukuiama), which allows introducing a great dynamics and flexibility in the process.

5. Intelligence acts on information, transforming it into knowledge.

The partners involved in the process (knowledge workers), interacting in a network, contribute to transform the information available in the network into knowledge. The grouping of different knowledge can be organized into thematic clusters to enhance creativity, using the virtualities inherent in the (digital) concept of «pulling».

---

[16] From now on, the concept of «distinctive competence» often arises as a substitute for «business areas» or «strategic business units» because it conveys the necessary breadth to the idea's development. Since design is the process underlying the definition of distinctive competence (s), it is the very design process (reinvented) that interests us.

[17] The term «functionality» replaces «products and services». Flexibility and recurrence underlie the proposed model. «The ultimately successful U.S. organizations abandoned the step-by-step model of R&D, created dedicated integration teams, and shrank the role of their research and manufacturing organizations in choosing technologies.» [25].

6. **Innovation is based on the constant and elaborate exploitation of intelligence and creativity.**[18]

The aim is to achieve innovation through the development of new ideas, stimulation of intelligence and individual and collective expression, using collaboration as «creative advantage» (John Kao).

7. **The exploration of intelligence and creativity will be extended to a great diversity of competences, knowledge and experiences connected in network, with a horizontal hierarchical structure and with a sponsor or coordinator of competences.**

The wealth of the intellectual potential of the network will be proportional to the quality, diversity and predisposition to share individual intellectual capital. It presupposes the definition of collective norms of conduct and responsibility in order to make the process operative and minimize «noise».

8. **The areas of opportunity serve the development of functionalities,[19] not the reverse.**

By promoting diversity of competencies, the process can «pull» organizations with a productive vocation that can be associated with the business. In doing so, they may have a purpose different from that which presides over their reason for existing.[20]

9. **There are no constraints at the outset, as distinctive competencies are not defined in advance.**

The constraints caused by the constraints associated with traditional processes (market, technology, distribution) no longer make sense in this new, open and interactive process. The full opening of the field of hypotheses and possibilities frees creativity in the pursuit of innovation.

10. **The technologies associated with production and distribution will be a consequence of the factors underlying the distinctive competencies to be defined. Technologies that meet these requirements will be selected.**

The definition of the technologies necessary for the commercial viability of the «functionalities» depends on the process, that is, depends on the dynamics found for a given functionality solution. Once «distinctive competence» and hence «functionalities» are defined, it is possible to characterize, define or invent the technologies which incorporate or give it a body. This factor will determine which technologies to

---

[18] On the new challenges to making information operational, Peter Drucker adds: «The development of rigorous methods for gathering and analyzing outsider information will increasingly become a major challenge for business and for information experts.» [26].

[19] Gary Hamel prefers the concept of «areas of opportunity» rather than «industries» and «features» rather than «products and services» [23]. The adoption of its proposal is linked to the semantic enlargement that new designations always presuppose.

[20] «Core ideology» should not change, whereas «core competence» can and should.

choose and not the inverse. However, there is nothing to prevent technology being invented, as nothing prevents industry from reinventing itself.

11. **The fragmentation of competencies and functionalities in components is privileged and shares are shared by alliance partners.**

Irrespective of the stage at which the process is taking place, it is essential that all actors have an overall view of «distinctive competencies». When extending its scope to more actors, it is essential that the new specific knowledge stimulate shared knowledge, contributing to the enrichment of the collective (continuous learning). The fragmentation of competences does not mean that each one contributes to his or her knowledge in order to respond to part of the process: it means that each one contributes with all his knowledge to validate a shared concept for which he is especially well prepared.

12. **The partners in the «areas of opportunity» (who may be coopetitors) may be geographically located anywhere on a global scale and selected primarily by their competencies.**

This principle is a consequence of the locus-free business concept. Partners can be co-opetitors because they can collaborate and compete simultaneously within the alliance: they collaborate because they are involved in it, they are part of it; compete because they can eventually offer competing products or services (on their own initiative - remote hypothesis - because they are involved in other similar processes or because in the business partners with similar competencies are involved). Competence is a decisive factor for a given individual, organization or institution to be admitted as partners.

13. **The end-users of the new «functionalities» are directly involved in detecting the opportunities for «distinctive competencies».**

More than that, they are called from the outset to jointly build the strategic vision that gives shape to a certain «distinctive competence»: they are the soul of the strategic design vision.

14. **Temporary alliances will be established around a certain distinctive competence.**

As distinctive competencies will be defined, temporary alliances will be formed, depending on the individual knowledge of each partner and their suitability for the work to be carried out jointly.

15. **The alliances will be constituted with well defined purposes, with individuals, groups of individuals, framed or not by organizations.**

Innovation is the central purpose that should bring together the group of individuals around the alliance.[21] The individual objectives, although they may be different from

---

[21] The same objectives identified by Chuck Martin apply, as they are also intended to create: «great brands on the Net [...] great products [...] useful service [...] what is [not] expected [...] what is [not] needed [yet]» [19].

individual to individual or from organization to organization, should be governed by the principles of interaction between the organization of creativity, the management of knowledge and the impact of new technologies (John Kao).

16. **The capital required for the creation of «distinctive competencies» will be primarily the «intellectual capital» of each partner, but it may be necessary to ensure the participation of financial capital.**

Intellectual capital underlies the process by being an integral part of the allies. The connectivity of knowledge workers in networks increases the intellectual capital invested by each ally. The financial capital that may be necessary for the development of the idea can be drawn from risk societies, institutional investors or individual investors. Investors who consider profit as a central purpose will be excluded from the outset.[22] Those who take the investment as a natural consequence of a purpose of existence centered on audacious goals, but properly framed in the central purpose (humanization of the framework of life, work, leisure and rest) will be welcome.

17. **«Intellectual capital» is measured in units of credit corresponding to units in «distinctive competencies». Efficient mechanisms will be found to measure the individual intellectual capital invested and to allow the conversion of this intellectual capital into credit units (financial capital).**

This is a sensitive issue, which requires wide and participative discussion, although it is always possible to change what is established throughout the process. By involving the sharing of intellectual property, the right place to discuss and seek the solution is the network itself.[23]

18. **It is not necessary for partners to enter all at the beginning of the business, or to stay in it to the end.**

Just as the workers of knowledge are dynamic, so is the alliance. The continuous entry and exit of partners increases the knowledge which, in turn, is put at the service of alliance partners and, more comprehensively, of the consumers.

19. **The interaction resulting from the application of the concept f «think and act globally» allows the overlap of several stages, in time and space, with the consequent reduction of costs and optimization of results.**

In this context, «thinking and acting globally» means working in multidisciplinary teams, locus-free, high performance, enhanced by the capacity for interaction that the new media and modes of collaboration enable. Thus, it is possible to «pull» the consumer into the process, developing with him and for him, one-to-one, innovative concepts; in the same way, what is created reaches the consumer, avoiding intermediaries. In short, the process allows shortening the inputs and outputs, which is essential in face of global competition.

---

[22] «Profits are the lifeblood of any business, but life consists of more than keeping the blood flowing; otherwise, it would not be worth living.» [26].

[23] Also in this case the concept of learning organization applies.

20. **The connection system also serves to continuously check the excellence of the functionalities.**

Connectivity and simulation technologies enable the sharing and verification of real-time information to define the distinctive competencies, test the effectiveness and sustainability of the concepts that characterize the functionalities, characterize and identify their specifications, and gauge the applicable production technologies.[24]

21. **There is a need to create a database to save ideas to use later.**

The amount of knowledge produced (at high speed) will feed the «ideas factory» (John Kao), stored in a database of free access for partners, which can be reused in the future.

22. **The aim is to stimulate creativity aimed at innovation and lifelong learning.**

Innovation, creativity and quality have no limits. They go through the way they integrate into the process, regardless of the results.

3   Conclusions

We have tried to establish the principles and purposes that underpin the proposed idea. We leave open the need (or not) to code another design method to apply it to the design processes triggered by the first process (the object of our research). On a provisional basis, we admit that the methodologies available to develop new «functionalities»[25] could maintain their operationality, provided that adjustments were made that were necessary, albeit intuitively.[26] To develop them, one would have to experiment with the proposed model and later theorize. We can not predict whether it would be possible (or even necessary) to develop this rationalization effort to apply it to such an open and ever changing process. There might be conclusive answers from their experimentation and the sharing of these concerns with some of the allies: we believed in the

---

[24] «In the next decade, the most important new sense-making tools will be those that help people visualize and simulate. Visualization techniques reduce vast and obscure pools of data into easily comprehended images. And simulation systems will become intellectual training wheels for executives, allowing them to experiment with strategies in the forgiving world of cyberspace, in much the same way that pilots in the Gulf War ran practice missions before flying the real thing.» [26].

[25] We have already mentioned the scarcity of contributions, within the framework of methodologies that consider the profound changes in the production-consumption processes. Many methodological models produced in the 1960s and 1970s, long before the democratization of computer use (in design, production and sale) and the emergence of the medium that revolutionized this revolution - the Internet, are available. The most recent studies, especially those by German authors, point to the need to reflect on the methodological question but do not codify new processes. Thus, it would be useless to describe those available (Alexander, Jones, Bonsieppe, Löbach, Bürdek, Baxter, among others) to support the elaboration of a new theory of projection.

[26] It is not possible to determine whether it will involve grouping, skipping, flipping, or transforming the various steps. It is referred to the small article «L'intuizione del metodo» by Sanmorì [27], in which the author questions the possibility of practicing a project without a codified methodology.

potentialities arising from the experimentation of the design process we proposed and the subsequent enrichment in terms of learning.

Enzo Paci's commentary on the occasion of the 3rd International Congress on Aesthetics (Venice, 1956) is particularly illuminating and prefiguring in this regard:

«One of the most serious dangers of the contemporary cultural situation is the intellectual projection of predetermined discourses, and of methodological plans, on the level of concrete experience. This is not to deny the value of methodology and technique. It only means that the technical and methodological tools should not be badly materialized and placed outside their field of application as if nature and the living experience of man were the matrix of every technique and method, but, at the on the contrary, a world extracted from the method is considered real instead of instrumental.» [28].

Acknowledgements. CITAD – Centro de Investigação em Território, Arquitetura e Design, Universidades de Lusíada, Portugal; CIAUD – Centro de Investigação em Arquitetura, Urbanismo e Design, Faculdade de Arquitetura, Universidade de Lisboa, Portugal. This research is financed by national funds from the FCT – Fundação para a Ciência e a Tecnologia, Portugal, within the scope of the project UID/AUR/04026/2013.

References

1. Maldonado, P.: Design: uma visão estratégica (design: a strategic vision). Msc Dissertation, Universidade do Porto (1997)
2. Maldonado, P.: Inspædia: design, inovação et cetera (Inspædia: Innovation, Design and So on). Universidade Lusíada Editora, Lisboa (2017)
3. Maldonado, P., Ferrão, L., Ermida, P.: Inspædia: changing the landscape of cultural reflection and influence through user experience design. In: Rebelo, F., Soares, M. (eds.) Advances in Ergonomics in Design. AHFE 2017. Advances in Intelligent Systems and Computing, vol. 588, pp. 462–468. Springer, Cham (2018). https://doi.org/10.1007/978-3-319-60582-1_46
4. Maldonado, P., Teixeira, F., Duarte, J.P., Câmara, A., Correia, N., Ferrão, L., Ermida, P., Passos, M.: Inspædia report: an inspired research itinerary. In: Rebelo, F., Soares, M., (eds.) Advances in Ergonomics in Design. AHFE 2017. Advances in Intelligent Systems and Computing, vol. 588, pp. 432–442. Springer, Cham (2017). https://doi.org/10.1007/978-3-319-60582-1_43
5. Maldonado, P., et al.: Inspaedia [internet] (2016). https://www.inspaedia.com
6. Maldonado, P., et al.: Inspædia: [almost] everything about simplicity, playfulness and inspiration. In: Soares, M. et al. (eds.) Advances in Ergonomics Modeling, Usability and Special Populations. Advances in Intelligent Systems and Computing, vol. 486, pp. 231–243. Springer, Cham (2016). https://doi.org/10.1007/978-3-319-41685-4_21
7. Maldonado, P., Teixeira, F., Silva, F.M., Ferrão, L., Ermida, P., Passos, M.: Inspædia user experience design (UXD). In: Procedia Manufactoring, 6th International Conference on Applied Human Factors and Ergonomics (AHFE 2015) and the Affiliated Conferences, vol. 3, pp. 6044–6051 (2015). https://doi.org/10.1016/j.promfg.2015.07.727
8. Maldonado, P., Silva, F.M., Gonçalves, F.: Inspædia, inspiring a collaborative intelligence network: designing the user experience. In: Ahram, T., Karwowski, W., Marek, T. (eds.) Proceedings of the 5th International Conference on Applied Human Factors and Ergonomics AHFE 2014, Kraków, Poland, 19–23 July 2014, pp. 463–472 (2014)

9. Maldonado, P., Ferrão, L.: Inspædia: uma Rede de Inteligência Colaborativa Inspiradora (Inspaedia: a collaborative intelligence network). In: Actas de Diseño – III Congreso Latinoamericano de Enseñanza del Diseño, Año VIII, vol. 15, pp. 193–197. Universidad de Palermo, Buenos Aires (2013)

10. Maldonado, P.: Inovação, Design et cetera (Innovation, Design and so on). Ph.D. Dissertation, Universidade Técnica de Lisboa (2012)

11. Maldonado, P.: Strategic design: an innovation and design process flowchart. In: CIPED VI Congresso Internacional de Pesquisa em Design Livro de Resumos. CIPED VI Congresso Internacional de Pesquisa em Design, pp. 292–293. CIAUD, Lisbon (2011)

12. Handy, C.: A Era da Irracionalidade ou a Gestão do Futuro (The Age of Irrationality or the Management of the Future). Edições CETOP, Mem Martins (1992)

13. Rodrigues, J.N.: Jamming com o Professor Kao. Executive Digest, Ano II, Nº 23, Lisboa, September 1996

14. Hammer, M.: E depois da Reengenharia?. Executive Digest, Ano II, Nº 22, Lisboa, August 1996

15. Thurow, L.: O Futuro do Capitalismo. Executive Digest, Ano II, Nº 18, Lisboa, April 1996

16. Lawson, B.: How Designers Think - The Process Demystified, 2nd edn. Butterworth Architecture, Oxford (1995)

17. Maldonado, T.: Disegno industriale: un riesame, 2nd edn. Feltrinelli Editore, Milão (1992)

18. Margolin, V. (ed.): Design Discourse – History, Theory and Criticism. The University Press, Chicago (1989)

19. Martin, C.: The Digital Estate - Strategies for Competing Surviving, and Thriving in an Internetworked World. Mcgraw-Hill, New York (1997)

20. Rodrigues, J.N.: A Internet tornou-se o Quinto Poder. Executive Digest, Ano III, Nº 35, Lisboa, September 1997

21. Rodrigues, J.N.: O Regresso da Gestão por Cenários. Executive Digest, Ano III, Nº 27, Lisboa, January 1997

22. Oakley, M. (ed.): Design Management - A Handbook of Issues and Methods. Blackwell Reference, Oxford/Cambridge (1990)

23. Gibson, R. (ed.): Rethinking the Future - Rethinking Business, Principles, Competition, Control & Complexity, Leadership, Markets and the World. Nicholas Brealy Publishing, London (1997)

24. Kao, J.: Jamming - The Art and Discipline of Business Creativity, col. "Harper Business". Harper Collins Publishers, New York (1997)

25. Iansiti, M., West, J.: Technology Integration: Turning Great Research into Great Products. Harvard Business Review, Boston, May–June 1997

26. Drucker, P. et al.: Looking Ahead: Implications of the Present. Harvard Business Review, Boston, September–October 1997

27. Sanmorì, M.: L' intuizione del metodo. Stile Industria, Ano II, Nº 7, Milão, September 1996

28. Morello, A.: Alla ricerca di Enzo Paci - l' estetica, la tecnica e il loro rapporto con il mondo. Stile Industria, Ano I, Nº 3, Milão, May 1995

The Development of a Hybrid Approach to Usability Assessment: Leveraging a Heuristic Guidance Framework for End User Feedback

Beth F. Wheeler Atkinson[1(✉)], Mitchell J. Tindall[1],
and Emily C. Anania[2]

[1] Naval Air Warfare Center Training Systems Division, Orlando, USA
{beth.atkinson,mitch.tindall}@navy.mil
[2] DSE, Inc., Tampa, USA
emily.c.anania.ctr@navy.mil

Abstract. The implementation of complex user interfaces within Navy operational and training domains is often challenging. As with other domains, development teams seek a balance between meeting the functional needs of communities while striving for good usability for end users. To increase the usability of interfaces, our team is developing a hybrid survey-based approach that yields benefits from expert led heuristic assessments and domain relevant feedback from end users. The resulting system is a survey containing 200 items assessing the usability of nine heuristic categories. This paper will provide background on the existing methods considered and the resulting survey framework. Further, the preliminary findings on the benefits of this hybrid approach from initial validation studies will be discussed, focusing on the ability of the system to provide developers with valuable quantitative information as well as specific qualitative information meant for fixing, adjusting and enhancing system functions.

Keywords: Usability · Heuristic evaluation · Psychometric validation
Human-Systems integration · Systems engineering

1 Introduction

Usability evaluations have become commonplace in a world transitioning to graphical user interfaces for a variety of work functions (e.g., performance assessment, payment tracking, conference submissions, data sharing). Ideally, usability assessments are performed by domain experts with the guidance of a human factors or usability specialist. However, resource limitations or tight schedules may mean this is not feasible. Off-the-shelf options for assessing the usability of a new software system (e.g., survey-based assessments) are typically limited in their detail. This results in a lack of adaptability and comprehensiveness for system function analysis and difficulty interpreting results. As an alternative to survey-based assessments, researchers in the field of usability often implement heuristic-based approaches. While these assessments improve the flexibility and comprehensiveness of usability feedback, they require a small team of human factors specialist to provide input, and often lack end user's perspectives. To address these limitations and address a need to become more fiscally efficient while increasing capability, the U.S. Navy funded a research and development project meant to strike the appropriate balance between the various goals of any usability assessment (e.g., comprehensive, adaptable, inexpensive). As a result, researchers at the Naval Air Warfare Center Training Systems Division (NAWCTSD) conducted a quantitative review of various types of usability assessments with the intention of developing a novel approach that leveraged the strengths and addressed the short-comings of past approaches. The resulting system (i.e., Experienced-based Questionnaire for Usability Assessments Targeting Elaborations - EQUATE) can be described as a hybrid approach combining survey-based approaches and heuristic-based evaluations. In the next sections we will review the development, strengths and weaknesses of both survey-based and heuristic approaches to usability assessment. Additionally, we will describe how that review informed the initial development of the EQUATE system. Lastly, we will describe the steps needed to properly validate the resulting EQUATE system.

2 Motivation

The initial development of the EQUATE system started with an extensive literature review. Our research team wanted to identify what has been done in the past to pull from strengths of past approaches to develop a novel, hybrid approach. In the following sections, we will review the research and practice that ended up informing the EQUATE and specifically what aspects of those systems were leveraged in developing the new system.

Survey-Based Usability Assessment. Survey-based usability assessments- short, self-report, multiple-choice, Likert scale surveys meant to quickly assess the usability of any type of software interface [4] (e.g., the System Usability Scale (SUS) (Brooke 1996) and the Post Study System Usability Questionnaire (PSSUQ) [10] are common tools to usability evaluation likely because of their reduced cost, efficiency of administration and ease of interpretation. While such benefits are difficult to ignore, these summative assessments tend to offer little valuable feedback to program developers, especially early on in the development cycle due to the lack of prescriptive data [7]. Hornbaek [8] conducted a review of the current practices in measuring usability where he addressed several limitations with extant approaches. He contended that the construct validity of usability surveys may be questionable, as they appear to be general measures of satisfaction as opposed to measuring the actual usability of a system or program. Related to this notion, Hornbaek [8] noted that these surveys cover usability too broadly to be useful for the development and enhancement of a system. In addition to these issues, Hornbaek [8] lists the lack of focus on effectiveness and efficiency and the lack of approaches addressing the *learnability* dimension of usability as problematic issues with existing tools.

Our review of the literature regarding survey-based usability succinctly points out the strengths and weaknesses of this approach to usability evaluation. Concerning the development of the EQUATE, we hoped to leverage the benefits of a survey-based approach in terms of cost, efficiency and ease of interpretation, while addressing the aforementioned limitations. To do this, the team reviewed the literature and practice of heuristic approaches to assessment. In the next section, we will briefly review the practice and literature behind heuristic approaches and we will talk specifically about how features of heuristic approaches can be combined with survey-based approaches to form a hybrid method that leverages the strengths of each.

Heuristic-Based Usability Assessment. Heuristic evaluation [12, 14] is an approach to usability assessment where experts use previously validated rules to guide system evaluations (e.g., visibility of system status, match between system and real world, consistency and standards; [12]). Unlike many survey-based approaches, heuristic evaluations can offer detail that is useful to program developers at the early stages of the development process and when making subsequent enhancements because the output relies on the description of the experts conducting the analysis. However, with this flexibility in the level of detail included in the output comes inconsistency. Heuristic evaluations conducted by domain experts without usability education or experience yield the best results (i.e., most usability issues found) when they are conducted by at least three evaluators [14], while usability specialists or usability specialists with domain expertise only require a single evaluator, with a survey-based approach, for the best results [12]. As result, heuristic-based usability assessment can be both time consuming and expensive relative to survey-based approaches.

The review of the strengths and limitations of the two usability assessment approaches offer contrast that is useful in informing a hybrid method. It is important to note that there are other approaches to usability assessment (e.g., story-boarding, cognitive walkthrough) not considered when developing the EQUATE. These approaches offer benefits when implemented throughout the development lifecycle [2], but were beyond the goals of this research effort.

In general, survey-based approaches are inexpensive, and easy to administer and provide a general understanding of a system's usability in a timely manner. On the other hand, they are not detailed enough to provide useful feedback to program developers at every stage in the development process. Heuristic evaluations offer incredibly detailed outputs but come at a cost in both time and money. The contrasts between these two approaches inspired our research into a hybrid approach. Specifically, we set out to develop a system that can: be administered to end users or novice populations at any point in the development process; provide both summative and detailed feedback to program developers; and allow a single human factors/usability specialist to easily interpret data provided by a group of users. Such a system would be of great value to the Navy, as it would allow acquisition programs to gain *expert* usability data with domain user expertise accounted for in a single data collection, thereby reducing schedule and resource implications.

3 Survey Development Process

The starting place for the team was to revisit the Multiple Heuristic Evaluation Table (MHET) (Atkinson et al. [1]). The MHET integrated four seminal approaches to heuristic-based evaluation for graphical user interfaces (GUIs) (i.e., [11, 16, 18]; principles based on Edward Tufte's visual display work from UW Computing & Communications [19]) into a single table. The table provided the research team with a guided framework for heuristic evaluation in the form of *do's* and *don'ts*. The conceptual analysis that resulted in the MHET also provided a baseline of 12 heuristic categories (i.e., Software-User Interaction, Learnability, Cognition Facilitation, User Control & Software Flexibility, System-Real World Match, Graphic Design, Navigation & Exiting, Consistency, Defaults, System-Software Interaction, Help & Documentation, and Error Management) that would be considered when formulating EQUATE dimensions. Limitations of the MHET included a focus only on four existing heuristic approaches for input, and a lack of empirical validation of the content. The result of expanding this approach led to the development of the User Interface - Table for Evaluating & Analyzing Composite Heuristics (UI-TEACH).

Step 1: Developing the UI-TEACH. The development of the UI-TEACH started by revisiting the literature to expand the number of approaches considered in the development of a heuristic guidance table. Our research team reviewed heuristic-based evaluation approaches including, but not limited to, "The Evaluation Checklist" [15], "Research-Based Web Design & Usability Guidelines" [9], "Audience Centered Heuristic: Older Adults" [5], "Hedonomics: The Power of Positive and Pleasurable Ergonomics" [6], and "Designing the User Interface: Fourth Edition: Strategies for Effective Human-Computer Interaction" [17]. The goal of the review was to generate a comprehensive list of heuristic guidelines and survey items that would inform a card sort analysis conducted with Human Factors subject matter experts (SMEs). Twenty usability experts provided data during an open card sort to inform the closed card sort data collection completed by nine SMEs. The card sort data was analyzed using both hierarchical cluster analysis and confirmatory factor analysis [3], resulting in nine heuristic categories (see Table 1) and over 200 usability guidance items in the form of *do's* and *don'ts* that would later inform survey items. The nine categories were *graphic design and aesthetics, error handling and feedback, user interaction control, learnability, effectiveness of developmental characteristics, memorability and cognitive facilitation, user efficiency, consistency* and *help*. The individual items (*do's* and *don'ts*) were categorized in one or none of the preceding heuristic dimensions based on conceptual linkages identified during data analysis. It is these items that are used in the next phase of EQUATE development.

Step 2: Developing the Initial EQUATE System. The *do's* and *don'ts* that are meant to guide designers and evaluators during system development were rewritten to read as items on a usability survey. The directions for completing the survey read as follows: "Based on your experience, please indicate your level of agreement with each statement." *Do's* were positively coded (e.g., The design provided a pleasant experience) while *don'ts* were negatively coded (e.g., There was too much clutter on the display).

Table 1. User Interface - Table for Evaluating & Analyzing Composite Heuristics (UI-TEACH) Dimensions & Definitions.

Heuristic	Definition
Graphic Design & Aesthetics	Interface display elements (e.g., color, text, graphics) and layout support a positive user experience
Error Handling & Feedback	System feedback on status and errors supports user's understanding of how to interact with the system
User Interaction Control	Mechanisms that allow the user to feel in control of actions and system preferences
Learnability	System design and aids support users learning how to use the system
Effectiveness of Developmental Characteristics	Characteristics of the hardware/software compatibility that affects the ability of the system to deliver the intended functionality and detect errors
Memorability & Cognitive Facilitation	System design helps ease learning and memory load (short-term and long-term memory)
User Efficiency	System design and functionality that supports completion of tasks with minimal time and effort
Consistency	System information and actions are consistently located and formatted throughout the interface
Help	Readily accessible instructions or clarifying information that are easy to use and support task completion

The resulting survey contains 200 items that are currently being evaluated for reliability (i.e., internal consistency and divergent validity) and construct validity (i.e., overlap with extant systems purporting to measure similar dimensions of usability). While the items contain more detail than traditional usability surveys, the research team did not want to rely solely on survey items alone to deliver important feedback to designers at every stage in development. As a result, the survey was broken into multiple dimensions consistent with the UI-TEACH dimensions; the only exception was that the *Effectiveness of Developmental Characteristics* was omitted from the survey after lengthy discussions with our Human Factors experts due to the limited insight that users would have about these items (e.g., the software code causes the system to slow down).

End users completing the EQUATE are first required to answer survey-based questions pertaining to each dimension of UI-TEACH. They are then prompted to provide free-response feedback in space provided to elaborate on any issues they uncovered, to include a prompt to provide an assessment of the criticality of the issue and to offer suggestions for fixing it. While this free-response portion of the EQUATE may seem subtle and unimportant. The research team felt strongly that the addition of this to an existing survey addressed several issues that afflict traditional survey-based approaches. Before end users provide free-response feedback, the items themselves act as heuristic prompts. While the items may provide some information, the opportunity for free-response alone lends an opportunity for users to provide the detail that is often missing in survey approaches. In addition, prompting end users to detail the severity of an issue and ideas for fixing it makes the enhancement and development process far more efficient. Identifying an issue alone does little to help a program designer prioritize and fix the issue. In our experience, after issues are identified, a back and forth ensues between program evaluators and designers that can be lengthy, and without end user input may not come to the *right* solution. For this reason, while prompting participants to provide this information seems simple, it may go a long way in making the feedback process more efficient and successful (Fig. 1).

	Strongly Disagree			Strongly Agree	N/A
5. Overall, I found the system provided informative warnings/alerts when errors were identified.	1	2	3	4	5	0
Error messages were prominently displayed.	1	2	3	4	5	0
Messages did not clearly explain what error occurred to cause the warning.	1	2	3	4	5	0
The system provided enough information to determine why errors occurred.	1	2	3	4	5	0
Symbols used to indicate errors were not familiar.	1	2	3	4	5	0
Error messages contained information about how to correct the error.	1	2	3	4	5	0
Errors messages were delayed causing multiple errors.	1	2	3	4	5	0
Brief error messages were informative.	1	2	3	4	5	0
Error messages were unclear.	1	2	3	4	5	0

In the area below, please 1) describe any specific issues you observed (e.g., location, type of problem), 2) indicate how critical of an issue, and 3) make suggestions for how to fix the issue.

Fig. 1. Example of EQUATE layout.

4 Future Directions and Practical Uses

The EQUATE is undergoing a full scale psychometric evaluation to ensure the internal consistency, discriminant validity and overall construct, and criterion-related validity of the measure. Preliminary analyses (i.e., Confirmatory Factor Analyses) are encouraging. That is, individual items on the survey are strongly correlated with their respective heuristic dimensions and heuristic dimensions are not so highly correlated with one another that they would need to be collapsed into fewer than eight dimensions. As part of this evaluation we are also empirically evaluating the utility of the EQUATE relative to other heuristic and survey-based approaches. We purport that the EQUATE will produce a greater quantity of issues and a greater breadth of issues (i.e., issues that span more than a single heuristic dimension) than other approaches all while reducing the time and cost to implement usability testing when compared to a full-scale heuristic-based evaluation. Such a tool should prove useful in applied arenas, such as the military, where the engineering of systems often outpaces the concern for human factors.

As a part of these analyses, we have used the EQUATE to conduct usability evaluations on two very different system functions. This was done to show the utility of the approach across system types/functions. The first system is meant for the entry and tracking of performance related and non-descript data for Navy aircrews. The second is a simulation-based strategy game for Marine Corp mission planning and execution. The resulting usability evaluations yielded actionable results for both the initial interface design and enhancements to overall system functionality. In both cases, the usability evaluations conducted using the EQUATE delivered large quantities of issues, descriptive details and suggested fix data for developers addressing the issues, and supplemental information that assists the design team in prioritizing issues. This allowed the development teams to implement changes and enhancement to the systems that positively affected efficiency and effectiveness of users employing the systems to complete tasks, without significant impacts to budget or schedule. While these initial results are anecdotal, they provide evidence that the EQUATE system is functioning as intended.

References

1. Atkinson, B.F.W., Bennett, T.O., Bahr, G.S., Nelson, M.M.W.: Development of a multiple heuristics evaluation table (MHET) to support software development and usability analysis. In: Universal Acess in Human Computer Interaction. Coping with Diversity, pp. 563–572. Springer, Heidelberg
2. Wheeler Atkinson, B.F., Kaste, K.P.: The importance of usability analysis in functional design: State-of-the-practice vs. state-of-the-possible. Accepted for publication in the NAVAIR Journal (2014)
3. Wheeler Atkinson, B.F., Tindall, M.J., Kaste, K.P.: Data analysis for survey development: a comparison of hierarchical cluster analysis and confirmatory factor analysis. Poster extended abstract submitted to the Human Factors & Ergonomics Society Annual Conference (2015)
4. Bangor, A., Kortum, P.T., Miller, J.T.: An empirical evaluation of the system usability scale. Int. J. Hum. Comput. Interact. 24(6), 574–594 (2008)
5. Chisnell, D., Redish, G.: AARP Audience-Centered Heuristics: Older Adults (2004)
6. Hancock, P.A., Pepe, A.A., Murphy, L.L.: Hedonomics: the power of positive and pleasurable ergonomics. Ergon. Des. 13(1), 8–14 (2005)
7. Hix, D., Hartson, H.R.: Developing User Interfaces: Ensuring Usability Through Product & Process. John Wiley & Sons, Inc., New Jersey (1993)
8. Hornbæk, K.: Current practice in measuring usability: challenges to usability studies and research. Int. J. Hum. Comput. Stud. 64(2), 79–102 (2006)
9. Leavitt, M.O., Shneiderman, B.: Research-based web design & usability guidelines. US Department of Health and Human Services (2006)
10. Lewis, J.R.: Psychometric evaluation of the PSSUQ using data from five years of usability studies. Int. J. Hum. Comput. Interact. 14(3–4), 463–488 (2002)
11. Nielson, J.: Usability Engineering. Academic Press, San Diego (1993)
12. Nielsen, J.: Usability Engineering. Elsevier, New York (1994)
13. Nielsen, J.: Finding usability problems through heuristic evaluation. In: Proceedings of the SIGCHI Conference on Human Factors in Computing Systems, pp. 373–380. ACM, June 1992
14. Nielsen, J., Molich, R.: Heuristic evaluation of user interfaces. In: Proceedings of the SIGCHI Conference on Human Factors in Computing Systems, pp. 249–256. ACM, March 1990

15. Ravden, S., Johnson, G.: Evaluating Usability of Human-Computer Interfaces: A Practical Method. Halsted Press, New York (1989)
16. Shneiderman, B.: Designing the User Interface: Strategies for Effective Human-Computer Interaction. Addison Wesley Longman, Reading (1998)
17. Shneiderman, S.B., Plaisant, C.: Designing the user Interface, 4th edn. Pearson Addison Wesley, USA (2005)
18. Tognazzini, B.: First Principles of Interaction Design. Interaction design Solutions for the Real World, AskTog (2003)
19. UW Computing & Communications: Graphics and web design based on Edward Tufte's principles. Retrieved October 4, 2006, from the University of Washington, UW Computer (1999). Training Web site: http://www.washington.edu/computing/training/560/zz-tufte.html
20. Zabed Ahmed, S.M.: A comparison of usability techniques for evaluating information retrieval system interfaces. Perform. Meas. Metrics 9(1), 48–58 (2008)

A Pilot Naturalistic Study of PC Mouse Usability

Denis A. Coelho[1,2(✉)] and Miguel L. Lourenço[2,3]

[1] Human Technology Group, Department of Electromechanical Engineering,
Universidade da Beira Interior, 6201-001 Covilhã, Portugal
dac@ubi.pt
[2] C-MAST: Centre for Mechanical and Aerospace Science and Technologies,
Universidade da Beira Interior, 6201-001 Covilhã, Portugal
denis.a.coelho@gmail.com, mlopes@ipg.pt
[3] Engineering and Technology Technical Scientific Unit and Research Unit
for Inland Development, Technology and Management School,
Guarda Polytechnic Institute, 6300-559 Guarda, Portugal

Abstract. An uncontrolled study took place in course settings with 30 3rd year undergraduates. Standard graphical tasks were generated from dedicated software. Subjects used their own and other subjects' pointing devices. Pointing efficiency was calculated within each 2 to 4 subjects group. Hypothetically, subjects would not experience improved efficiency in pointing and dragging tasks when switching to other devices from their own device. The other devices might share the archetype of the owned device, but differ in dimensions and shape details, or activation thresholds. Reasons are suggested for cases where the hypothesis is not verified in the results. Literature review and prior experimental results tentatively explains the subjects' improvement in efficiency when changing to unfamiliar pointing devices. Familiarity with the tasks and improved fitness with borrowed device are potential motivators for efficiency gains observed that do not comply with the hypothesis. A snapshot of PC mice used by students is provided.

Keywords: Human factors · Ergonomics · Human-systems integration
User experience · Computer pointing device · Efficiency

1 Introduction

A naturalistic study is a type of study in which the researchers observe and record very carefully some behavior or phenomenon, sometimes over a prolonged period, in its natural setting while interfering as little as possible with the subjects or phenomena [1]. Pilot studies are small-scale, preliminary studies which aim to investigate whether crucial components of a main study - usually a randomized controlled trial (RCT) – will be feasible and they may be used as an attempt to predict an appropriate sample size for a full-scale project or to improve upon various aspects of the study design [2].

Hence, the current paper, reporting on a pilot naturalistic study of PC mouse usability, does not make use of special controls, except for the graphical tasks used, which are standardized and software generated. It springs from a wealth of controlled

© Springer International Publishing AG, part of Springer Nature 2019
T. Z. Ahram and C. Falcão (Eds.): AHFE 2018, AISC 794, pp. 481–493, 2019.
https://doi.org/10.1007/978-3-319-94947-5_49

studies carried out in the scope of previous research by the authors [3–12], and holding both a validation character, regarding those previous studies, as well as an opportunity to study the phenomenon of computer handheld pointing device interaction with the users in terms of efficiency in their natural settings. The analysis of this pilot naturalistic study may hence present new questions and guide the establishment of future studies to advance the state of the art and bring the frontier of knowledge in the domain further ahead.

Efficiency is taken as one of the fundamental components of usability; usability evaluation involves measuring user performance (effectiveness and efficiency) and registering the user's perception of satisfaction related aspects [13, 14]. In this paper, for the sake of conciseness, focus will be given to efficiency, but effectiveness is a component of efficiency, hence two of the three components of usability are considered and included. This notwithstanding, results are only given for efficiency in percentages.

The hypothesis behind the study reported in this paper is that subjects would not experience improved efficiency in pointing and dragging tasks when switching to other devices from their own device. The other devices might have the same archetype, but have different dimensions and shape details, as well as activation thresholds. The study tests this hypothesis making use of a naturalistic approach. The remainder of the paper is comprised of a methods section, presentation and analysis of results, discussion and conclusion sections. The results are presented on a group by group basis first and then systematized to enable richer insights to be drawn from the data collected in a naturalistic, uncontrolled and pilot approach.

2   Method and Limitations

An uncontrolled study was set up in a course setting. Thirty third year undergraduate industrial design students participated for course credit (15 male and 15 female). Standard graphical tasks were generated through a dedicated software application [3]. Subjects used their own pointing device, plus the pointing device of one, two or three other subjects (in two cases of males, their regular pointing device was a touchpad, which was removed from the hypothesis testing analysis, but their use of other devices was included to compute overall values of efficiency). Efficiency in pointing device use was calculated within each small group of 2 to 4 subjects setting according to the equations detailed in previous studies [3, 12] and shown in Eqs. 1 and 2.

$$efa = 1 - \frac{No.\ FailedTargets}{No.\ TotalTargets} \tag{1}$$

$$efi = efa \times \frac{minimum\ mean\ completion\ TIME}{mean\ completion\ TIME(subject)} \tag{2}$$

efa – effectiveness of pointing and effectiveness of dragging; efi – efficiency
   No. FailedTargets – number of failed targets by the subject for the particular task
   No. TotalTargets – total number of targets to be hit for the particular task

minimum mean completion TIME – lowest mean completion time across the whole set of replications of participant-device combinations (for the particular task and group)

mean completion TIME (subject) – mean time to complete the particular task for the participant-device combination.

Participants, 3rd year industrial design undergraduate students, had experience of Computer Aided Design software, where pointing and dragging tasks are used consistently and abundantly. All participants had normal or corrected to normal vision and none reported complaints of a musculoskeletal natures in the upper extremities at the time the study took place (in late May, early June of 2017). They were invited to form small groups amongst them (group size was limited to 4 maximum, and 2 minimum, with no indication of gender distribution provided). Instructions included making use of the standardized graphical software provided to perform the tasks generated that would enable computation of efficiency, using their own handheld pointing device as well as that of the other members of their group. The software automatically collected data on errors and on time to complete tasks. This data was used to compute the effectiveness and the efficiency of the combinations of subject-device-task. For the sake of simplicity in the analysis, only efficiency results are reported in this paper. Pointing tasks tested included pointing at large, medium and small targets. In the analysis, the efficiency scores are averaged for one encompassing pointing task. A similar approach was pursued for the tasks of dragging with the left, middle and right mouse buttons, which were averaged for the sake of obtaining a condensed dragging efficiency score.

Due to screen size differences, the dragging tasks were not performed across all groups, as these required higher graphical resolutions, and subjects used their own computers, which in most cases were laptops. Some of the devices included in the study were cordless, creating room for two distortions in the pointing speed, one concerned with the speed and traffic of the cordless transmission between the device and the computer, and the other concerned with the level of charge of the battery in the pointing device. These are some of the limitations springing from the uncontrolled setup in which the study took place.

3 Results and Analysis

In the presentation of efficiency results, devices are identified both as "own", when in use by the owner and as "othersXX" (e.g. "others01", "others02", etc.) when used by other participants in the group. All computer handheld devices were optical. Due to screen size inconsistencies, some of the results reported for the dragging tasks were not considered and are absent from the results. A total of 30 participants, and 28 pointing devices were involved in the experiments, with group size ranging from 2 to 4). There were 3 groups with 2 participants, 4 groups with 3 participants and 3 groups with 4 participants. Overall, the simple average including all 88 combinations (from a total of 30 subjects, 15 female and 15 male) of participant-device pointing efficiency was 76.8% (sd = 11.8%). For dragging efficiency, there were 34 combinations of participant-device (from a total of 11 subjects, 7 male and 4 female) resulting in an average dragging efficiency of 78.5% (sd = 17.3%).

3.1   Results

Tables 1, 2, 3, 4, 5, 6, 7, 8, 9 and 10 present results regarding hypothesis compliance as well as efficiency values per subject-device combination considering the three pointing tasks (pointing at large, medium and small targets) aggregated in one (pointing) through simple averaging. The Tables also include the aggregated dragging efficiency (where applicable), resulting from the simple average of the dragging with the left, middle and right button tasks, where data was available. Interspersed with the presentation of the results in the tabular format are Figs. 1, 2, 3, 4, 5, 6 and 7, showing the image of most of the devices used (some of the images were not made available for this report) with identification of their respective owner participant in Tables 1, 2, 3, 4, 5, 6, 7, 8, 9 and 10.

Table 1.   Efficiency results for group A (m-male; f-female).

Subject-device	Pointing efficiency	Hypothesis compliant	Dragging efficiency	Hypothesis compliant
01m-own	94%	no	100%	yes
01m-others02	89%		83%	
01m-others03	97%		100%	
02m-others01	89%		83%	
02m-own	92%	no	100%	yes
02m-others03	97%		100%	
03m-others01	86%		92%	
03m-others02	89%		100%	
03m-own	89%	yes	100%	yes

 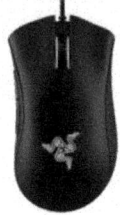 

Fig. 1. Pointing devices 01, 02, and 03 (from left to right) owned and used by group A participants in the naturalistic experiments, with group average pointing efficiency values of 89.7%, 90.0% and 94.3% and dragging efficiency values of 91.7%, 94.3% and 100% respectively.

Group by group (those with PC mouse images available), we can see which devices have better performance. In groups with sufficient device information available and where at least some members were not hypothesis compliant, suggestions on what have caused the non-compliance are made. In group A (Table 1, Fig. 1), an all-male group, it is the smallest device (03) that excels in both pointing and dragging.

This corroborates the findings brought forward by previous studies [3, 4] indicating that smaller devices, relative to the user's hand size enable higher efficiency and hence improved performance. Group B (Table 2, Fig. 2) shows much higher efficiency for the male participant than for the female participant in both devices (04 and 05), but it is the female participant's device (05) that excels in terms of performance. Group C (Table 3), an all-female group shows levels of pointing efficiency at lower value than the ones obtained for all male groups, such as A.

Table 2. Pointing efficiency results for group B (m-male; f-female).

Subject-device	Pointing efficiency	Hypothesis compliant
04m-own	100%	yes
04m-others05	93%	–
05f-others04	51%	–
05f-own	70%	yes

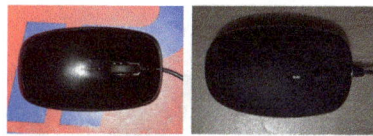

Fig. 2. Pointing devices 04 and 05 (from left to right) owned and used by group B participants in the naturalistic experiments, with group average pointing efficiency values of 75.5% and 81.5% respectively.

Table 3. Efficiency results for group C (m-male; f-female), group average pointing efficiency values of 77.3%, 80.3% and 66.7%, respectively for devices 06, 07 and 08.

Subject-device	Pointing efficiency	Hypothesis compliant
06f-own	79%	no
06f-others07	86%	
06f-others08	68%	
07f-others06	90%	
07f-own	98%	yes
07f-others08	73%	
08f-others06	63%	
08f-others07	57%	
08f-own	59%	no

Group D (Table 4), like group A, show a smaller range of variation across subject-device combinations when compared to mixed sex groups, which is suggested to be due to it being a same sex group (all male). The simple, no frills design of device 13, used by mixed sexes group E (Fig. 3, Table 5), attains the highest pointing efficiency in the group, while a somewhat similar device, albeit with a non-slip surface (device 15) excels in dragging efficiency within the group.

Table 4. Efficiency results for group D (m-male; f-female), group average pointing efficiency values of 76.0%, 76.3% and 80.0%, respectively for devices 09, 10 and 11.

Subject-device	Pointing efficiency	Hypothesis compliant
09m-own	88%	yes
09m-others10	68%	
09m-others11	72%	
10m-others09	74%	
10m-own	64%	no
10m-others11	91%	
11m-others09	88%	
11m-others10	72%	
11m-own	80%	no
12m-others09	88%	
12m-others10	72%	
12m-others11	80%	

Fig. 3. Pointing devices 13, 14, 15 and 16 (from left to right) owned and used by group E participants in the naturalistic experiments, with group average pointing efficiency values of 82.8%, 68.8%, 78.8% and 72.0% and dragging efficiency values of 68.8%, 72.5%, 79.5% and 62.8% respectively.

In group F (Table 6, Fig. 4), an all-male group, no frills device 18 scores the highest average pointing efficiency, while device 17 (bigger buttons with concave finger grooves) excels for dragging performance. Group G (Table 7), a mixed group, elects device 21 as the best performing in pointing tasks; this is a shiny but contoured at the edges simple box design. In group H (Table 8, Fig. 6) device 24 (a simple device but contoured for thumb support) attains the highest average group pointing efficiency.

Table 5. Efficiency results for group E (m-male; f-female).

Subject-device	Pointing effic.	Hyp. compliant	Dragging effic.	Hyp. compliant
13f-own	90%	yes	93%	yes
13f-others14	60%		78%	
13f-others15	72%		75%	
13f-others16	80%		87%	
14 m-others13	83%		64%	
14 m-own	77%	no	83%	no

(continued)

Table 5. (*continued*)

Subject-device	Pointing effic.	Hyp. compliant	Dragging effic.	Hyp. compliant
14 m-others15	90%		85%	
14 m-others16	68%		80%	
15f-others13	80%		57%	
15f-others14	77%		72%	
15f-own	74%	no	69%	no
15f-others16	65%		58%	
16f-others13	78%		61%	
16f-others14	61%		57%	
16f-others15	79%		89%	
16f-own	75%	no	26%	no

Table 6. Efficiency results for group F (m-male; f-female).

Subject-device	Pointing efficiency	Hypothesis compliant	Dragging efficiency	Hypothesis compliant
17m-own	77%	no	100%	yes
17m-others18	81%		65%	
17m-others19	70%		91%	
18m-others17	77%		87%	
18m-own	86%	yes	65%	yes
18m-others19	77%		65%	
19m-others17	64%		71%	
19m-others18	65%		60%	
19m-own	73%	yes	74%	yes

Fig. 4. Pointing devices 17, 18 and 19 (from left to right) owned and used by group F participants in the naturalistic experiments, with group average pointing efficiency values of 72.7%, 77.3% and 73.3% and dragging efficiency values of 86.0%, 63.3% and 76.7% respectively.

Table 7. Efficiency results for group G (m-male; f-female).

Subject-device	Pointing efficiency	Hypothesis compliant
20f-own	60%	no
20f-others21	55%	
20f-others22	80%	
21m-others20	41%	
21m-own	80%	yes
21m-others22	61%	
22f-others20	67%	
22f-others21	93%	
22f-own	75%	no
23m-others20	78%	
23m-others21	86%	
23m-others22	80%	

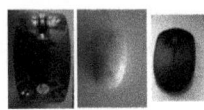

Fig. 5. Pointing devices 20, 21 and 22 (from left to right) owned and used by group G participants in the naturalistic experiments, with group average pointing efficiency values of 61.5%, 78.5% and 72.0% respectively.

Table 8. Pointing efficiency results for group H (m-male; f-female).

Subject-device	Pointing efficiency	Hypothesis compliant
24f-own	83%	yes
24f-others25	58%	
25f-others24	71%	
25f-own	72%	yes

Fig. 6. Pointing devices 24 and 25 (from left to right) owned and used by group H participants in the naturalistic experiments, with group average pointing efficiency values of 77.0% and 65.0% respectively.

Table 9. Pointing efficiency results for group I (m-male; f-female).

Subject-device	Pointing efficiency	Hypothesis compliant
26f-own	82%	yes
26f-others27	72%	
27m-others26	63%	
27m-own	80%	yes

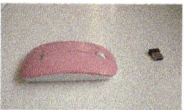

Fig. 7. Pointing devices 26 and 27 (from left to right) owned and used by group I participants in the naturalistic experiments, with group average pointing efficiency values of 72.5% and 76.0% respectively.

Table 10. Efficiency results for group J (m-male; f-female), group average pointing efficiency values of 80.7%, 68.0% and 83.7%, respectively for devices 28, 29 and 30.

Subject-device	Pointing efficiency	Hypothesis compliant
28f-own	88%	yes
28f-others29	68%	
28f-others30	72%	
29f-others28	74%	
29f-own	64%	no
29f-others30	91%	
30f-others28	80%	
30f-others29	72%	
30f-own	88%	Yes

3.2   Statistical Analysis

The statistical analysis of results is supported by IBM SPSS software. Average efficiency of pointing tasks for males (44 combinations of participant-device) was 79.8% (sd = 11.8%), while it was only 73.9% (sd = 11.1%) for females. For dragging efficiency, males (22 participant device-combinations) averaged 84.0% (sd = 14.3%), with females attaining on average only 68.5% (sd = 18.45%). Differences across sexes are statistically significant for both pointing (p = 0.012) and dragging (p = 0.017) efficiency (Independent samples Mann-Whitney U-test). The Chi square test was deployed, considering compliance with the hypothesis in pointing and dragging and looking at the entire sample, as well as looking within each sex (Table 11). A total of 28 subjects are involved in the analysis. The only statistically significant effect found was for the male cohort, in what concerns dragging (this task is more dependent on experience and careful sensory-motor coordination between eye and hand, as well as positioning and size of device and buttons).

Table 11. Chi square test results relating to the test of the study's hypothesis (*- p < 0.05).

Categories	Pointing hypot. compliance (n)	Chi square statistic, significance	Dragging hypot. compliance (n)	Chi square statistic, significance
Overall yes	15	test st. 0.143	8	test st. 2.273
Overall no	13	p = 0.705	3	p = 0.132
Female yes	8	test st. 0.067	1	test st. 0.333
Female no	7	p = 0.796	2	p = 0.564
Male yes	7	test st. 0.077	7	test st. 4.5
Male no	6	p = 0.782	1	p = 0.034*

The related samples McNemar test was also applied to the results of testing of the underlying hypothesis for the study, by means of a cross contingency Table (Table 12). The result was not statistically significant (p = 0.125) with a test statistic of 2.25. The small number of subjects that could be involved springs from the aforementioned limitations of the dragging graphical task vis-à-vis screen resolution of the participants' PCs. No discordant cases were found in the Southwest corner of the 2 by 2 contingency Table, but there is about one third of cases in the Northeast corner. This notwithstanding, the test shows no statistically significant dependency between compliance with the hypothesis in dragging and compliance with the hypothesis in pointing. Hence, pointing and dragging efficiency are not necessarily linked for the same device. Pearson correlation analysis for the 34 available combinations shows a correlation factor of 0.56 (p = 0.001), indicating that the cohort might be too small, somewhat precluding statistically significant results.

Table 12. Cross contingency Table regarding pointing efficiency and dragging efficiency hypothesis compliance (n = 11).

		Dragging efficiency compliance with hypothesis		
Point. effic. compl. w. hypot.		Yes	No	Total
	No	3	4	7
	Yes	0	4	4
	Total	3	8	11

4   Discussion

Differences between overall average efficiencies obtained for pointing and dragging, with the latter unexpectedly attaining higher values, may spring from the longer times it takes in performing a dragging operation, which compensates for more error proneness of the task of pointing. However, this may also be an undesirable consequence of the study not having been controlled. Despite being mostly statistically non-significant, the results of analysis show that there is statistical significance in the acceptance of the hypothesis formulated for the study, but only for the male cohort, who performed the dragging tasks (8 subjects). The significant differences in overall average efficiency across both sexes found, corroborate previous findings from controlled studies performed by the authors [3, 4]. These are explained by the smaller hand size on average of females compared to males, requiring relatively more effort to operate the devices and thus resulting in lower efficiency values. These results stand in line with the concurrent information that is provided by observational studies combining ergonomic and psychosocial factors in the computerized workplace, supporting the notion of a differentiated etiology of musculoskeletal disorders according to sex [15]. This notwithstanding, given that all pointing devices in this study are variations from the traditional archetype, but with different sizes, as this was an uncontrolled study, it was not possible to study the relation between hand size and mouse parameters.

The results of correlation analysis between pointing and dragging efficiencies for the available combinations (a subset of the data involving 34 cases of the total of 88 cases) showed a statistically significant association between these performance factors valued at 0.56. Although this a good correlation, the space left for a perfect correlation perhaps indicates that there may be specific device design features that come into play in the interaction with the user during pointing tasks, which are not necessarily the same that are critical for performance during dragging tasks.

5 Conclusion

The hypothesis underlying this study was that subjects would not experience improved efficiency in pointing and dragging tasks when switching from their own device to other devices with a similar archetype but different size and details in design. The results statistically prove the hypothesis only in what concerns dragging efficiency and only for males. Based on literature review, and prior experimental results, authors suggest tentative explanations to support the understanding of cases of subjects' improvement in efficiency when changing to unfamiliar pointing devices. Variables such as the contour (including grooves for the fingers at the left and right button and contoured body side for thumb support) and the surface finish of the device (polished slippery or mate non-slippery) are suggested as relevant, together with relative size of the device in relation to the size of the hand, corroborating previous studies.

The study showed a snapshot of actual pointing device models used by industrial design students. It shows a very similar set of devices, following the conventional archetype, with variations in size and some shape details as well as surface finish. corded and cordless differences as well as differences in DPI resolution are also present but were not controlled in the study, which had a linking aspect across groups focused on the standardized graphical tasks and the process to obtain efficiency values only. Statistically significant differences were found between male and female participants, who on average have smaller hands, for both pointing and dragging efficiency, supporting previous studies reporting on a positive hand size effect on efficiency in computer handheld pointing device use. Future studies, in particular randomized control trials (RCT) should involve larger numbers of subjects to avoid the pitfalls incurred from small numbers of comprehensive cases in the study.

Acknowledgments. The current study was funded in part by Fundação para a Ciência e a Tecnologia project UID/EMS/00151/2013 C-MAST. The authors express their thanks to the subjects participating in the study. The graphical software used for task generation was developed by Noel Lopes and Miguel Lourenço and is available from https://sourceforge.net/projects/mouse-test

References

1. Sussman, R.: Observational methods. Res. Meth. Environ. Psychol. **13**, 9–27 (2015)
2. Swallow, V.M., Knafl, K., Santacroce, S., Campbell, M., Hall, A.G., Smith, T., Carolan, I.: An interactive health communication application for supporting parents managing childhood long-term conditions: outcomes of a randomized controlled feasibility trial. JMIR Res. Protoc. **3**(4), e69 (2014)
3. Lourenço, L.M.L.: Desenvolvimento e Análise Ergonómica de Dispositivos Manuais Apontadores para Computador [Development and Ergonomic Analysis of Computer Handheld Pointing Devices]. Doctoral Dissertation. Industrial Engineering and Management. School of Engineering. Universidade da Beira Interior, Covilhã, Portugal (2016)
4. Lourenço, M.L., Pitarma, R.A., Coelho, D.A.: Association of hand size with usability assessment parameters of a standard handheld computer pointing device. Occup. Saf. Hyg. IV. **30**, 339–343 (2016)
5. Lourenço, M.L., Pitarma, R.A., Coelho, D.A.: Horizontal and vertical handheld pointing devices comparison for increasing human systems integration at the design stage. In: Advances in Human Factors and System Interactions. Advances in Intelligent Systems and Computing, vol. 497, pp. 15–24. Springer, Cham (2017)
6. Coelho, D.A., Lourenço, M.L., Nunes, I.L.: Psychometric analysis of scales for usability evaluation of pointing devices. In: International Conference on Applied Human Factors and Ergonomics. Advances in Intelligent Systems and Computing, vol. 592, pp. 419–426. Springer, Cham (2018)
7. Lourenço, M.L., Coelho, D.A.: Association of objective and subjective measures in usability evaluation of a standard PC mouse. In: Europe Chapter of Human Factors & Ergonomics Society (2015). http://www.hfes-europe.org/wp-content/uploads/2015/10/PosterLourenco2015.pdf
8. Lourenço, M.L., Pitarma, R.A., Coelho, D.A.: Development of a new ergonomic computer mouse. In: International Conference on Applied Human Factors and Ergonomics. Advances in Intelligent Systems and Computing, vol. 592, pp. 457–468. Springer, Cham (2018)
9. Lourenço, M.L., Coelho, D.A.: Research methodologies in the ergonomic development and evaluation of PC mice. In: International Conference on Applied Human Factors and Ergonomics. Advances in Intelligent Systems and Computing, vol. 592, pp. 448–456. Springer, Cham (2018)
10. Lourenço, M.L., Coelho, D.A.: Performance evaluation of PC mice. In: Occupational Safety and Hygiene V: Selected papers from the International Symposium on Occupational Safety and Hygiene, pp. 157–161. CRC Press (2017)

11. Lourenço, M.L., Coelho, D.A.: S-EMG of forearm muscles activity in conventional pc mouse use. In: International Conference on Applied Human Factors and Ergonomics. Advances in Intelligent Systems and Computing, vol. 592, pp. 439–447. Springer, Cham (2018)
12. Coelho, D.A., Lourenço, M.L.: A Tentative Efficiency Index for Pointing Device Use in Computer Aided Design – A pilot study. Work: A Journal of Prevention, Assessment and Rehabilitation (2018, in press)
13. ISO 9241-11: Ergonomic Requirements for Office Work with Visual Display Terminals (VDTs) – Part 11 Guidance on Usability. International Organization for Standardization (ISO), Geneva, Switzerland (1996)
14. ISO 9241-9: Ergonomic Requirements for Office Work with Visual Display Terminals (VDTs) – Part 11: Non-keyboard Input Device Requirements International Organization for Standardization (ISO), Geneva, Switzerland (2000)
15. Lima, T.M., Coelho, D.A.: Ergonomic and psychosocial factors and musculoskeletal complaints in public sector administration–a joint monitoring approach with analysis of association. Int. J. Industr. Ergon. **31**(66), 85–94 (2018)

Appropriate Operating Force of Knob in Certain Conditions

Huimin Hu[1,2(✉)], Pu Hong[3], Aiping Yang[4], Hong Luo[1,2], Yahui Bai[3], and Yinxia Li[3]

[1] Ergonomics Laboratory, China National Institute of Standardization, Beijing, China
{huhm, luohong}@cnis.gov.cn
[2] AQSIQ Key Laboratory of Human Factors and Ergonomics, Beijing, China
[3] School of Mechanical Engineering, Zhengzhou University, Zhengzhou, Henan, China
15090566727@163.com, 1298687634@qq.com,
liyxmail@126.com
[4] College of Robotics, Beijing Union University, Beijing, China
jdtaiping@buu.edu.cn

Abstract. In the process of driving a car, people often use rotary knob to adjust the temperature of air-conditioning or music volume, etc. This operation is very frequent and necessary. As we all know, when we make changes on the knob, the height, angle, size and operating force of the knob will affect the comfort of the operation knob. Therefore, in order to improve use experience, it is necessary to study the comfortable operating force of knobs. This paper uses MARK-10 force measuring equipment (torque meter, measuring rotating moment), sets the test bench and sets up different knob heights (500 mm, 700 mm, 900 mm), angles (20°, 30°, 40°), sizes (16 mm, 27 mm, 36 mm), operating forces (1N, 3N, 7N) and other parameters range. Then invite 10 subjects to participate in the experiment, recording the subjective evaluation on the different operating force, finding that the subjective score of the 1N and 3N is higher. Based on this research result the appropriate knob operating force which is between 1N and 3N is concluded.

Keywords: Knob · Comfort operating force · Different conditions
Appropriate

1 Introduction

Knob is widely used in all aspects of life. There are many knobs used in the automotive control interface, such as air-conditioning temperature adjustment, which offers convenience for the car driving. However, the operating force of the knob affects the comfort, high efficiency and safety when people are driving, it is clear that the determination of the comfortable operating force is particularly important in the design process of knob.

Gurari [1] analyzed the relationship between the angle of attack, the motor gain and the knob size and finds that the change of each parameter will cause the remaining two

© Springer International Publishing AG, part of Springer Nature 2019
T. Z. Ahram and C. Falcão (Eds.): AHFE 2018, AISC 794, pp. 494–503, 2019.
https://doi.org/10.1007/978-3-319-94947-5_50

parameters to change with each other. Feng [2] introduced a method to test the hand feeling of buttons and knobs by 6-axis machine hand and 6-axis force and torque sensor. According to the input signal of force and torque, the method can dynamically adjust the position direction and displacement of the hand, eliminate the distortion of the measurement data caused by lateral force unbalance and accurately measure the parameter characteristics of the button and knob. Jituo Zheng presented the implementation scheme of several common key force and knob torque parameters test and points out their advantages, disadvantages and relevant technical points. It can be found that the torque of the knob plays an important role in the design of the knob, and it is necessary to study further and determine an appropriate operating force.

Li [3] put forward that the peak torque of the multi-angle knob is recommended as: 0.007–0.03 Nm and the peak torque of positioning knob is recommended as: 0.1–0.2 Nm. Furthermore, the peak torque is recommended being designed as three times as the process torque to form a clear sense when switching. 《JB/T 3907-2008》 [4] stipulates that the torque of the knob should be less than 0.3 Nm when the knob size is less than 15 mm. The torque should be less than 0.5 Nm when the knob size is 15–25 mm. The torque should be less than 0.6 Nm when the knob size is more than 25 mm. Ng [5] studied the size and shape of the knob, and the results show that workers mostly preferred using the small 5-lobes knob.

Therefore, in the design process of the knob, the main factors that need to be considered are the height, angle, dimension and operating force of the knob. The main objective of this study is to find the appropriate knob operation force by carrying out small sample test under the above four conditions and to provide reference for the design of car knobs.

2 Method

2.1 Experimental Design

To realize the change of knob height, angle, dimension and operating force, in this study, an test bench with adjustable height and angle is designed to provide different height and angles, and a knob with adjustable size and operating force is designed to provide different size and operating force. The knob can be adjusted to three heights of 500 mm, 700 mm and 900 mm. The angle between the plane of the knob and the vertical plane can be adjusted to 20°, 30° and 40°. The knob size can be adjusted to 16 mm, 27 mm and 36 mm. The knob damping range is about 0–15N, and 1N, 3N and 7N are selected. 10 subjects took part in the experiment, and each subject seated in a 25 mm high car seat with a seat belt fastened, experiencing the operation force of the knob by rotating knob (see Fig. 1). And each subject is required to give 81 scores to evaluate the operating force of the knob (−2—too small, −1—slightly smaller, 0—suitable, 1—slightly larger, 2—too large). A total of 810 subjective scores in the 81 groups were got in this experiment.

Fig. 1. The experiment arrangement

2.2 Subject

Ten subjects with half male and half female took part in this experiment. The average age of the subjects was 23, and the average male total arm length was 170.4 cm, and the average female total arm length was 155.4 cm. All subjects were right-handed, so the right hands are selected to do the test.

Under the guidance of the experimenter, the subjects were well acquainted with the contents of the experiment. In order to carry out the experiment better, each subject was required to be without medical history of skeletal muscle, and in the last 6 months, they should have no physical injury and discomfort. In terms of dress, the subjects were required to wear baggy clothes or roll up their sleeves to bare wrist and ankle joints. In terms of posture, subjects were required to sit in the car seat with a relaxed posture to ensure that the hips fit the seat face and back fit chair back. Subjects' information can be seen in Table 1.

Table 1. Subjects' information

Gender	Height (cm)	Weight (kg)	BMI	Total arm length (cm)	Age	Dominant hand
Male	167	49	17.6	163	23	Right
Female	160	50	19.5	148	25	Right
Male	178	80	25.2	175	25	Right
Male	172	87	29.4	171	23	Right
Male	170	60	20.8	167	23	Right
Male	178	90	28.4	176	25	Right
Female	165	70	25.7	162	22	Right
Female	153	50	21.4	148	23	Right
Female	170	60	20.8	160	23	Right
Female	162	58	22.1	159	22	Right

2.3 Apparatus

The test bench as shown in Fig. 2 was set up in order to achieve the following experimental conditions: the height from the ground to the knob are respectively 500 mm, 700 mm and 900 mm, the angle between knob and vertical plane are respectively 20°, 30° and 40°, the knob size are respectively 16 mm, 27 mm and 36 mm, and the knob operation force are respectively 1N, 3N and 7N.

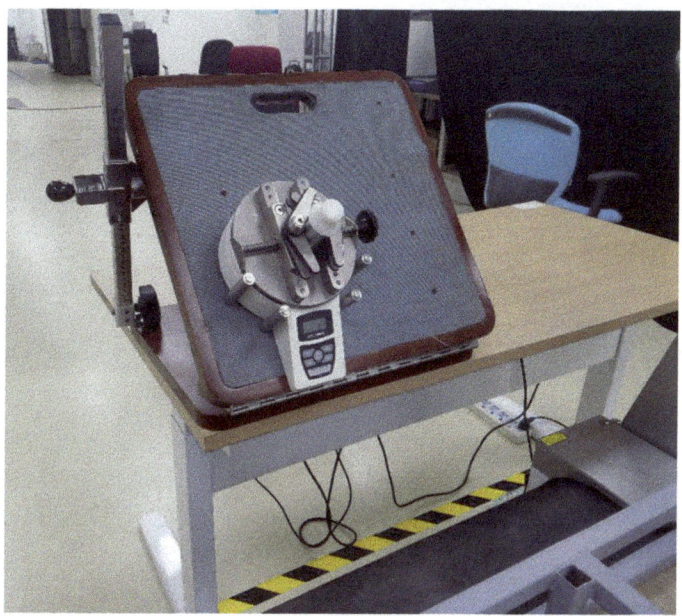

Fig. 2. Test bench

Height. Use the lift table to realize the height adjustment of the knob control plane to provide three heights of the knob (500 mm, 700 mm and 900 mm).

Angle. Use a specially developed rotary panel to achieve the angle adjustment of the knob control plane (as shown in Fig. 3), then change the angle of the panel by changing the height of the support lever to change the angle between the knob control plane and the vertical plane. The angle is 40° when the height of the lever is 14 cm. The angle is 30° when the height is 19 cm and the angle is 20° when the height is 23 cm.

Size. Make three knobs with three size (including knob cover and main, as shown in Fig. 4), and knob cover size is 16 mm, 27 mm, 36 mm. Three knob size can be achieved by changing the knob cover.

Operating Force. The torque of the knob was measured by the Mark-10 torque meter, as shown in Fig. 5. Its measuring range is 0–11.5 Nm, the accuracy is 0.3%, and the sampling frequency is 7000 Hz. Turn the knob when its clamp tightens the knob, and its display screen will show the torque value.

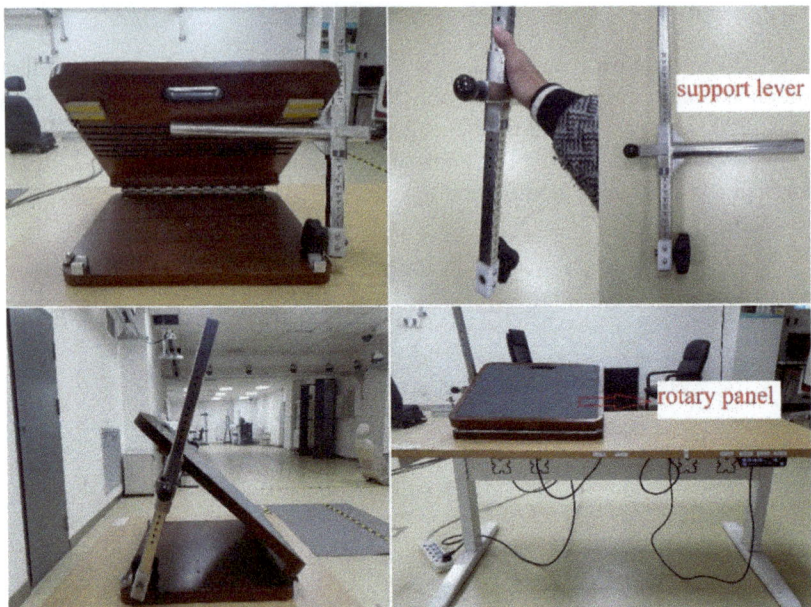

Fig. 3. Rotary panel

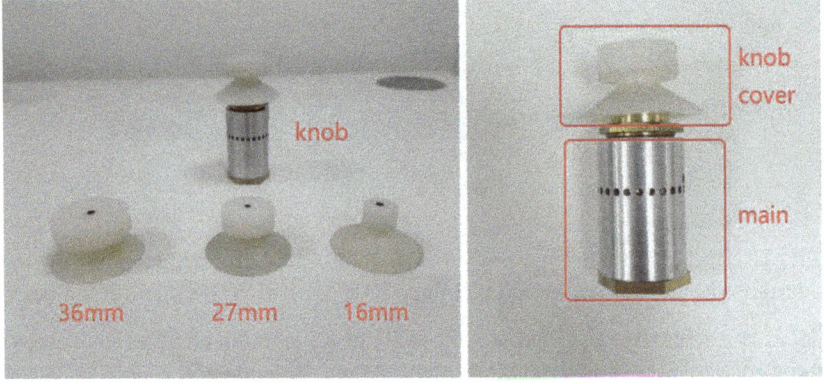

Fig. 4. Knob

The Seat. A car seat with safety belt was set first. The angle of seat back was 25°, and the height of seat surface was 25 cm from the ground (see Fig. 6). When the subject was sitting on the seat in driving posture, the height of hips was around 30 cm and varied depending on different people. The arrangement of the car seat properly offered a comfortable condition.

Fig. 5. Mark-10 torque meter

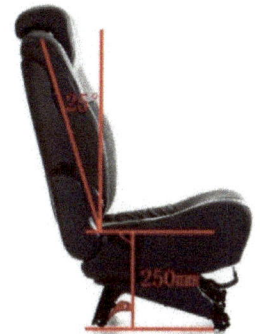

Fig. 6. Arrangement of the seat

2.4   Task and Procedure

Instruct the subjects to complete the following experiment steps:

1. Record the sex, height, weight, age, total arm length and dominant hand of subjects and introduce the experimental contents to the subjects.
2. Adjust test bench to height 500 mm, angle 20°, size 16 mm and operation force 1N.
3. Arrange subjects to sit in the car seat, then fasten the seat belts, adjust the distance between the seat and the test bench to be comfortable status and require subjects to be relax posture in the car seat, ensuring that the hips fit seat surface and back fit chair back.
4. Guide the subjects to experience the knob with a height of 500 mm, angle 20°, size 16 mm and operation of 1N, evaluate the operation force of the knob according to the 5 levels subjective score(see Table 2), and make records.
5. Change the test conditions to allow the subjects to experience the knob operating force of all the conditions, a total of 81 times.

Table 2. 5 levels of subjective score

-2	-1	0	1	2
Too small	Slightly small	Suitable	Slightly large	Too large

2.5   Data Analysis

There are three heights, three angles, three sizes and three operating forces for each subject to evaluate. Accordingly, 81 subjective scores were gotten for turning knob from each subject. So, there are 810 subjective scores from 10 subjects.

In this study, a total of two different methods which are defined as "average method" and "proportional method" were used to analyze the experimental results. Firstly, in each group score, calculate the average value score with their absolute value and it can be found that the corresponding knob operating force is the most appropriate when this score is close to 0. In addition, "proportional method" was used to analyze the score by counting the proportion of frequency of 0 in each group score, and the corresponding knob operating force is the most appropriate when this proportion is the highest. Combining two methods, we can determine the most appropriate knob operating force under a certain condition.

3   Results

3.1   Average Method

The score for each subject was divided into three groups, which were 1N, 3N and 7N, because all subjective scores were evaluated for three kinds of knob operating force. Then, calculate the average value score of the absolute value of 27 subjective scores in each group. The results were obtained, as shown in Table 3.

Table 3. Average value of subjective scores

Subjects	1N score	3N score	7N score
1	1.85	0.81	1.30
2	0.26	0.81	1.85
3	1.00	0.81	1.89
4	0.00	0.04	0.37
5	1.26	0.22	1.74
6	1.44	0.81	1.89
7	0.85	0.93	1.74
8	0.81	0.89	1.96
9	0.37	0.93	1.96
10	0.89	0.26	1.74
Average	0.87	0.65	1.64

It was found from Table 3 that the 3N score was the closest to 0, 7N score was the farthest to 0. Therefore, it can be preliminarily concluded that the knob is the most appropriate when knob operating force is 3N. In addition, the score gap of 1N and 3N is not large, so it can be concluded that the knob is also appropriate when knob operating force is 1N, but not the most.

3.2   Proportion Method

The 810 subjective scores of all subjects were divided into three groups: 1N, 3N and 7N, with 270 subjective scores in each group. In each group, the frequency of the five types of scores was counted, as shown in Fig. 7, 8 and 9. In the subjective scores distribution of 1N, 3N and 7N, the number (such as 72,81 and so on) respectively represents the times of occurrence of the corresponding subjective score.

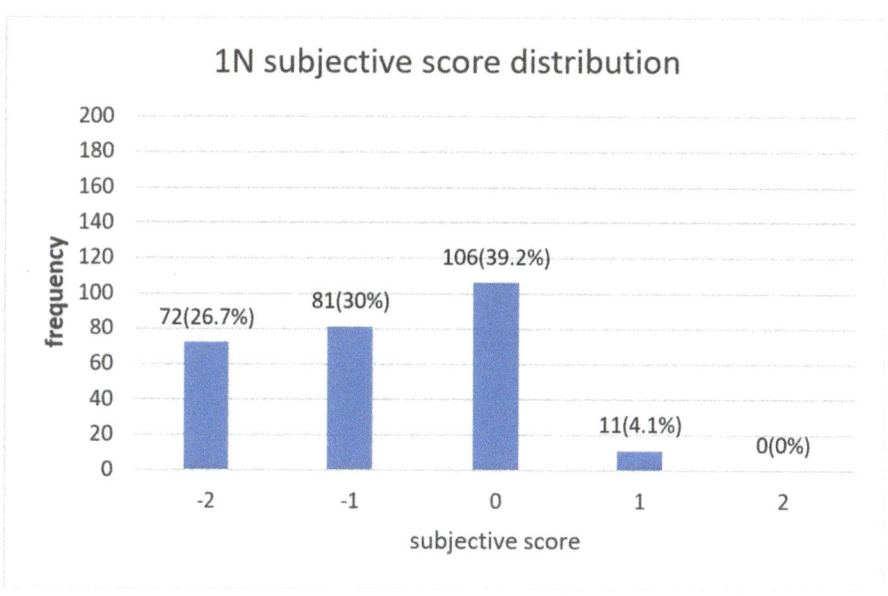

Fig. 7. 1N subjective score distribution

Figure 7 shows that the subjective score for 1N operating force is mainly distributed in 2 (26.7%), −1 (30%) and 0 (39.2%), total 95.9%, and the proportion of 0 is the highest, which is 39.2%. Figure 8 shows that the subjective score for 3N operations is mainly distributed in 0 (39.6%) and 1 (47.4%), totally 87%, and the proportion of 1is the highest, which is 47.4%. Figure 9 shows that the subjective score for 7N operations is mainly distributed in 2 (72.6%). From the above, 10 subjects generally think that the 7N operating force is too large and 1N and 3N operating force is more appropriate, however, the 1N operating force is slightly smaller and 3N operating force is slightly larger.

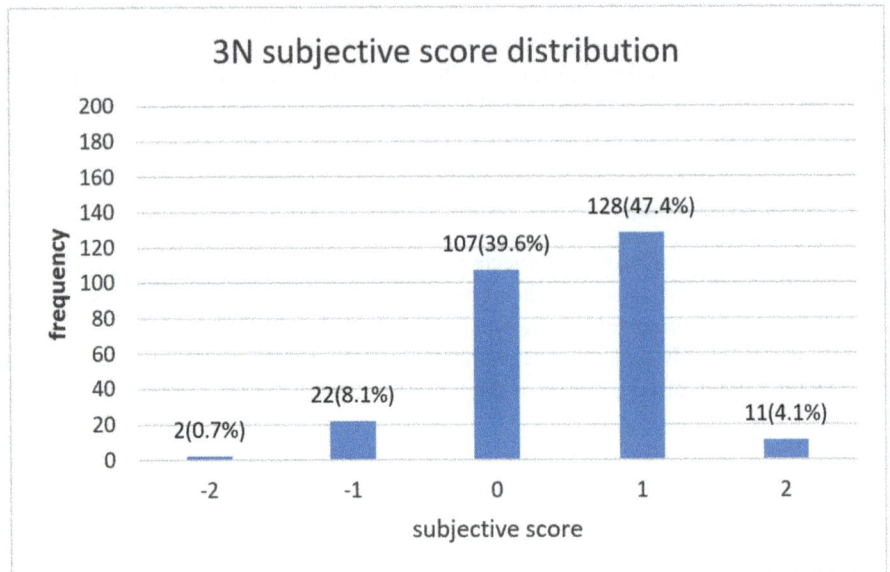

Fig. 8. 3N subjective score distribution

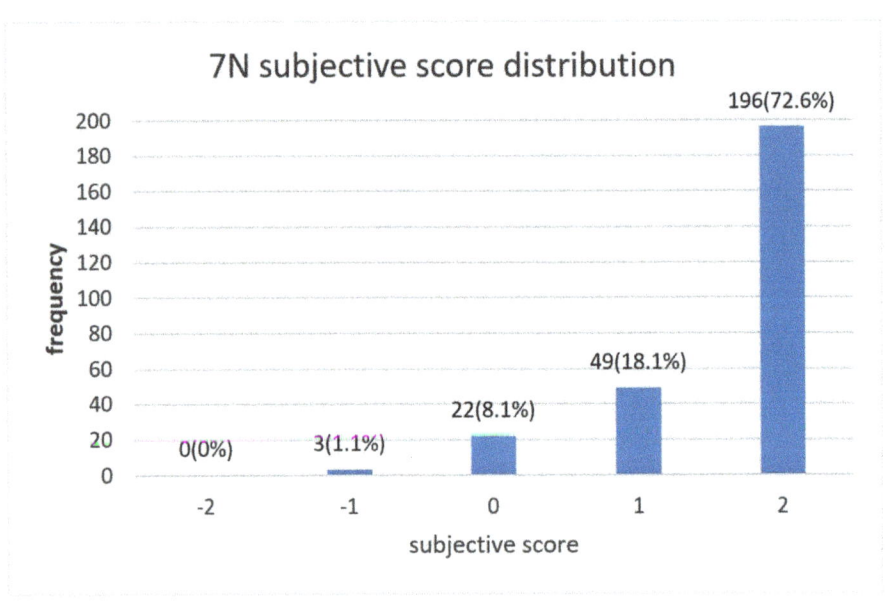

Fig. 9. 7N subjective score distribution

4   Discussion

Through the above results, it can be preliminarily concluded that the appropriate knob operating force is 1N and 3N, and the results are not concentrated in 1N or 3N. It can be conjectured that the most appropriate knob operating force will be between 1N and 3N. After analysis, there are 2 reasons for these results. First, the experimental results can't converge to an exact value because of the small number of the subjects, second, preliminary choice of knob operating force is not reasonable, and step is too large. With the increase of the number of subjects, when the operating force is selected more reasonable and the step is smaller, it will be possible to obtain an exact value of the operation force.

Acknowledgments. This research is supported by 2017 National Quality Infrastructure (2017NQI) project (2017YFF0206603 and 2017YFF0206506) and China National Institute of Standardization through the "special funds for the basic R&D undertakings by welfare research institutions" (522017Y-5278 and 522016Y-4488).

References

1. Gurari, N., Okamura, A.M., et al.: Human performance in a knob-turning task. In: EuroHaptics Conference, 2007 and Symposium on Haptic Interfaces for Virtual Environment and Teleoperator Systems. World Haptics 2007. Second Joint, pp. 96–101 (2007)
2. Feng, S.: A button and knob hand-feeling test method based on 6 axis' smechanical arm. Autom. Instrum. **72**(11), 67–69 (2016)
3. Li, G.-L., Liu, K.-J., Fan, X.: Research on vehicle knob torque characteristic curve. Auto Electr. Parts **12**(2), 5–7 (2013)
4. JBT 3907-2008 [S]
5. Ng, P.K., Boon, Q.H., Chai, K.X., et al.: The roles of shape and size in the pinch effort of screw knobs. Appl. Mech. Mater. **2851**(465), 1202–1206 (2014)

Study on the Handle Test Sample of Furniture for Ergonomics Experiment

Zhiyu Xu[1], Na Yu[1], Huimin Hu[2,3(✉)], Yahui Bai[4], and Pu Hong[4]

[1] College of Furnishings and Industrial Design, Nanjing Forestry University,
Nanjing, China
546282512@qq.com, 48048907@qq.com
[2] Ergonomics Laboratory, China National Institute of Standardization,
Beijing, China
huhm@cnis.gov.cn
[3] AQSIQ Key Laboratory of Human Factors and Ergonomics, Beijing, China
[4] School of Mechanical Engineering, Zhengzhou University,
Zhengzhou, Henan, China
1298687634@qq.com, 912886665@qq.com

Abstract. Handle is used in cabinet furniture frequently, and its comfort plays an important role when use cabinet furniture. At present, ergonomic design of handle lack of relevant standards and normative guidance, so it is necessary to conduct related ergonomic experimental researches. Before executing the ergonomic experiment, it is necessary to design the test sample for ergonomic experiments, including the size of the characteristic dimensions of the experimental sample and the number of samples. The main objective of this study is to obtain the recommend value of experimental sample parameters of the typical handles based on the users' physiological characteristics and experience evaluation.

Keywords: Human factors · Handle · Furniture · Test sample

1 Introduction

1.1 Research Background

At present, the researches on furniture handles at home and abroad are mostly about the form of handles and the appearance design of the handles. However, researches on the ergonomics of furniture handles is seriously lacking. There is no data basis and theoretical guidance for the size and installation position of cabinet furniture handles.

1.2 Domestic Research Status

Chen analyzed furniture design elements based on semantics, including product feasibility, environmental harmony, product culture characteristics, information accuracy, product care and so on [1]. Li and Jin combined the practical analysis process of Kansei Engineering with the shape design of the home handle, and used this method to guide the design [2, 3]. Through experimental research, Huimin Hu and others determine ergonomic design requirements and evaluation methods of manual grasp structure for

© Springer International Publishing AG, part of Springer Nature 2019
T. Z. Ahram and C. Falcão (Eds.): AHFE 2018, AISC 794, pp. 504–515, 2019.
https://doi.org/10.1007/978-3-319-94947-5_51

manual handling of heavy objects, so as to provide reference for ergonomic design and evaluation of manual grasp structure [4]. By analyzing the hand structure and motion characteristics, Guoqiang Tao and others established the grip criterion in the handle grip, and established the hand geometry model for the handle grip. Using ABAQUS finite element analysis software, meshing and setting process of grasping and boundary conditions, the finite element model is established. Finally, they get the pressure distribution of hand grip under virtual simulation, and verify that it is consistent with the actual similar grasp situation. A new virtual test approach, is sought for the measurement of hand surface pressure [5].

1.3 Research Status in Foreign Countries

Paschoarelli and others have studied the differences in human effort perception to quantify the usability and generality of door handles under different handle types [6]. Chang and others put forward a kind of door classification scheme and archetypal task analysis of door use based on man/door interaction, and put forward suggestions for restoring torque of door and door handle location [7]. Guo and others studied the influence of operation type and handle shape of high speed train controller on the driver's comfort [8].

The ergonomics research on furniture handle is badly missing at home and abroad, therefore, the ergonomics index and the size and quantity of the experimental sample in this research play an important role in ergonomics research of cabinet furniture handle.

1.4 Research Methods and Purposes

In this paper, the common types of cabinet furniture handle are classified by means of literature investigation and user investigation. This study use the literature, human body measurement, simulation prototype, experience and other methods to determine the indexes, the sample size and number of commonly used cabinet furniture handle and provide research support for the further study of ergonomics dimensions and installation position of cabinet furniture handle.

2 Research on Typical Handle Type

2.1 Handle Classification

Through market and network research, a large number of handles are investigated and collected, and the handles are divided into 7 categories according to the hand-handle interaction (Tables 1, 3 and 4).

Table 1. Handle type and hand-handle interaction

Serial Number	Handle Types	Hand-handle interaction
1		The two fingers of the index finger and middle finger bend to the inside of the handle, and the thumb is pressed on the outside of the handle to push and pull.
2		The index finger and the middle finger are close to one side of the handle, and the thumb finger is pressed on the other side of the handle to push and pull.
3		The two fingers of the index finger and middle finger are bent into the grooves, and the thumb is pressed on the outer surface to push and pull.
4		The three fingers of the thumb, the index finger and the middle finger hold the handle for pushing and pulling.
5		The forefinger holds the ring for pulling, the action of pushing generally acts directly on the panel.
6		The side of the index finger is close to the inside of the hanging piece, the thumb finger is pressed on the outer side of the hanging piece for pulling,and the action of pushing is usually directly acted on the panel.
7		The two fingers of the index finger and middle finger bend to the inside of the handle, and the thumb is pressed on the outside of the handle to pull,and the action of pushing is usually directly acted on the panel.

2.2 Market Research

The market research sites including IKEA, Red Star Macalline, Easyhome. Research objects are including handles of bookcases, cabinets, file cabinets, wardrobe and cabinets. A total of 24 brands were surveyed. They were divided into three categories based on different styles. The research contents include the main handle types and their usage frequency in the market. In a piece of furniture, usage frequency of the same type of handle is one. If 4 No.1 handles are used in a wardrobe, the usage frequency of No.1 handle is one.

Table 2. The system of handle ergonomics indexes

Handle Type	Diagram of ergonomics indexes	ergonomics indexes
No.1 handle (circular)		Diameter (D_{1c}) Distance From the Panel (d_{1c}) Length (L_{1c})
No.1 handle (square)		Thickness (T_{1S}) Distance From the Panel (d_{1s}) Width (W_{1s}) Length (L_{1s})
No.2 handle		Depth (d_2) Length (L_2) Width (W_2) Thickness (T_2)
No.3 handle		Length (L_3) Width (W_3) Internal Height (H_3) Internal Angle (A_3) Depth (d_3)
No.4 handle		Diameter (D_4) Distance From the Panel (d_4)
No.5 handle		Distance From the Panel (d_5) Diameter (D_5) Thickness (T_5)

Table 3. Hand dimension and related handle ergonomics indexes

Hand dimension	Reference value	Related handle ergonomics indexes
Finger abdominal thickness of distal finger joint of Middle finger	Average of measurements from 10 subjects 15 mm	$(d_{1c})(d_{1s})(d_3)$ (W_3)
Two finger distal joint width of the index and middle fingers	According to the coefficient, the quotient of two finger distal joint width of the index and middle fingers divided by distal joint width of the index, p95 is calculated to be 34 mm.	$(L_1)(L_2)(L_3)$
1/2 distal finger joint length of thumb	According to the coefficient, the quotient of distal finger joint length of thumb divided by length of index, p95 is calculated to be 18 mm.	$(d_2)(d_4)$
1/2 distal finger joint length of the middle finger	According to the coefficient, the quotient of distal finger joint length of the middle finger divided by length of index, p95 is calculated to be 15 mm	(H_3)
Distal finger joint width of the index	In the GB10000-1988, p95 of male is 21 mm	(D_5)
1/2 distal finger joint length of the index	According to the coefficient, the quotient of distal finger joint length of the index divided by length of index, p50 is calculated to be 12 mm	(d_5)

(1) The Modern Concise Style Furniture Handle

The use frequency of the modern concise style cabinet furniture handle showed that the use frequency of No.1 handle is 45%, No.3 handle and No.4 handle are both 20%, No.2 handle accounts for 9%, and the use frequency of other types of handles is under 5%.

(2) The Chinese Style Furniture Handle

The use frequency of the Chinese style cabinet furniture handle showed that the highest use frequency is 43% by No.3 handle, followed by No.1 handle 29% and No.4 handle 11%, No.2 handle and No.5 handle are both 5%, and other categories are all less than 5%.

(3) The European Style Furniture Handle

The use frequency from high to low are No.4 handle 39%, No.1 handle 20%, No.7 handle 17%, No.3 handle 10% and No.6 handle 7%. The use frequency of other types is less than 5%.

Table 4. Handle ergonomics indexes and related pre experiment

Handle Ergonomics Indexes	Results of Pre Experiment and Analysis of Pre experimental Results
Thickness (T_{1s})	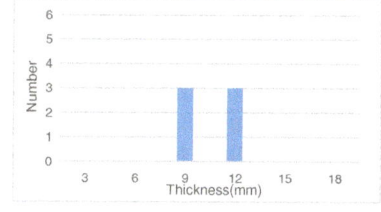 From the pre test results, it can be seen that on the most comfortable thickness selection, 3 people chose 9mm and 3 chose 12mm, so set the optimum value of thickness index to 10mm. The step is initially set based on the literature research results in 4.1, and confirmed that subjects can distinguished the 3mm step in the course of the experiment, the results can be considered in the preparation of No.1 handle test sample that we can choose 3mm or above size as step size of thickness.
Width (W_{1s})	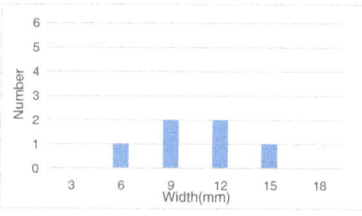 From the pre test results, it can be seen that the most comfortable width is distributed near 10mm, so set the optimum value of width index to 10mm. In the process of experiment, the step size of 3mm can be distinguished, therefore, it can be considered in the preparation of No.1 handle test sample that we can choose 3mm or above size as step size of width.
Diameter (D_{1c})	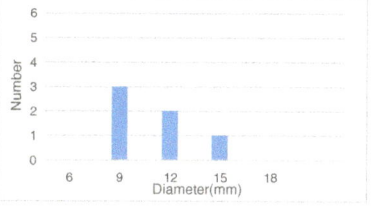 From the pre test results, it can be seen that the most comfortable diameter is distributed around 12mm, so set the optimum value of diameter index to 12mm. In the process of experiment, the step size of 3mm can be distinguished, therefore, it can be considered in the preparation of No.1 handle test sample that we can choose 3mm or above size as step size of diameter.
Thickness (T_2)	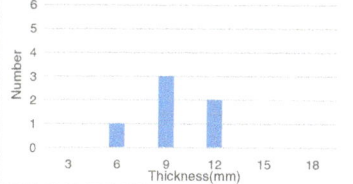 From the pre test results of thickness comfort of No. 2 handle, it can be seen that the most comfortable thickness is distributed around 9mm, so set the optimum value of thickness index to 9mm.

Width
(W$_2$)

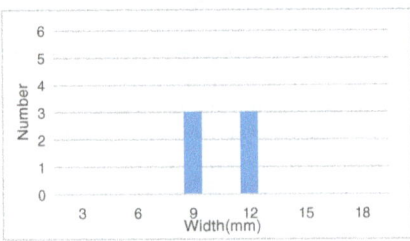

From the pre test results, it can be seen that on the most comfortable width selection, 3 people chose 9mm and 3 chose 12mm. If there is a height difference between the thickness and the width, the operation will be more comfortable, so set the op-timum value of width index to 12mm. In the process of experiment, the step size of 3mm can be distinguished, therefore, it can be considered in the preparation of No.2 handle test sample that we can choose 3mm or above size as step size of width.

Depth
(d$_2$)

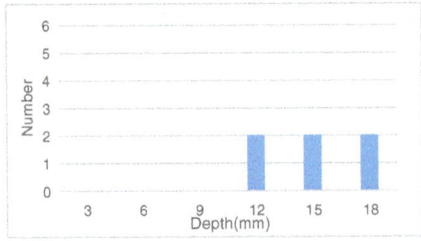

From the pre test results of depth comfort of No. 2 handle, it can be seen that the choice of the most comfortable depth is greater than 12mm. The minimum value of the depth of the experimental sample can be set to 9mm. In order to ensure that the depth index does not affect the evaluation of other indicators, the optimum value of depth index is set as the maximum value in the sample.In the process of experiment, the step size of 3mm can be distinguished, that is, 3mm is greater than the discriminating threshold. Therefore, it can be considered in the preparation of No.2 handle test sample that we can choose 3mm or above size as step size of depth.

Internal
Angle
(A$_3$)

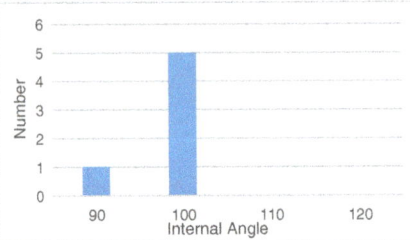

From the pre test results , it can be seen that the most comfortable internal angle is distributed on 100°, so set the optimum value of internal angle index to 100°. In the process of experiment, the step size of 3mm can be distinguished. It can be seen from the results that the step size can be reduced properly to make the result more accurate.

Internal
Height
（H$_3$）

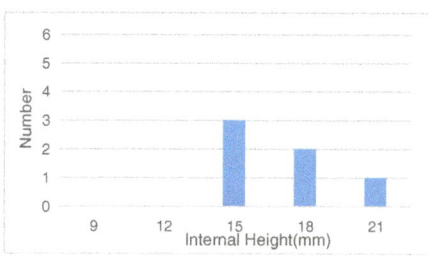

From the pre test results, it can be seen that the choice of the most comfortable internal height is greater than 15mm. In order to ensure that the internal height index does not affect the evaluation of other indicators, the optimum value of internal height index is set as the maximum value in the sample.In the process of experiment, the step size of 3mm can be distinguished, therefore, it can be considered in the preparation of No.3 handle test sample that we can choose 3mm or above size as step size of internal height.

Diameter
（D$_4$）

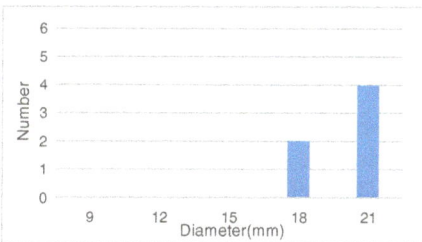

From the pre test results, it can be seen that the most comfortable diameter is distributed on 21mm, so set the optimum value of diameter index to21mm. According to a set of test sample size of index should cover the optimal value or unilateral limit value mentioned above, so set the minimum value of diameter index to 15mm. In the process of experiment, the step size of 3mm can be distinguished, therefore, it can be considered in the preparation of No.4 handle test sample that we can choose 3mm or above size as step size of diameter.

Thickness
（T$_5$）

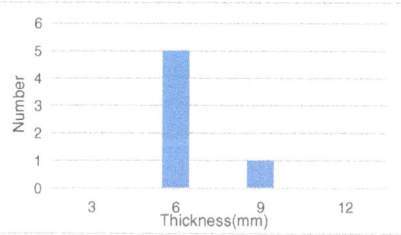

From the pre test results, it can be seen that the most comfortable thickness is distributed on 6mm, so set the optimum value of thickness index to 6mm. In the process of experiment, the step size of 3mm can be distinguished. According to the experimental results, the step size design can be appropriately reduced to get more accurate experimental results. Therefore, the step size of the experimental sample is set to 2mm.

2.3   Investigation on the Use of Household Cabinet Furniture Handle

Based on the survey results of the type of handle in Sect. 2.1, the survey of the use of different types of handles in the family was carried out by questionnaire. A total of 35 target users were surveyed. In the 35 questionnaires, 74% families were decorated with modern concise style. From the statistical results, no matter which style of the cabinet furniture, No.1 handle, No.2 handle, No.3 handle and No.4 handle have a higher rate of use.

2.4 Summary of Typical Handle Types

Market Research and user questionnaire survey showed that the handles of No.1, No.2, No.3 and No.4 are more widely used. In the market research of Chinese style cabinet furniture, the No.5 handle has a proportion of 5%. In the market research of European style cabinet furniture, the No.7 handle has a 17% proportion, because the No.7 handle is only used on the drawer, and the operation way is similar to the No.1 handle, the ergonomics index can refer to the No.1 handle. To sum up, the handle types of this study are finally defined as No.1 handle, No.2 handle, No.3 handle, No.4 handle, and No.5 handle.

3 Ergonomics Indexes of Handles

Refining typical man-machine interaction tasks based on the user's hand-handle interaction in the the use of handle. The tasks are summarized and decomposed to determine the ergonomic factors and indicators that affect the man-machine interaction. The system of handle ergonomics indexes was initially established, as shown in Table 2.

4 Quantity and Step Size of Handle Test Sample for Ergonomics Experiment

Before carrying out the ergonomics experiment of handles based on the user experience, we need to make experiment samples of different sizes, such as the length, width, thickness, and so on, based on the ergonomics indexes of the handle. Based on ergonomics theory analysis and related research results at home and abroad, it is preliminarily determined that when designing the step size of handle test sample, it is necessary to take into account the discrimination threshold under the specific operation. The design of the step size of the related ergonomics indexes should be greater than the discrimination threshold under the corresponding operation of the hand, so that the design of the step size is effective. At the same time, the minimum and maximum of the ergonomics indexes should cover the optimal value or the unilateral limit value of the target value. The number of testing samples needs to consider the feasibility and efficiency of the experiment.

4.1 Hand Discrimination Threshold

Panday [9] and others Studied the discrimination threshold of hand under four conditions. It can be seen in Fig. 1, the discrimination threshold of the hand is between 3 mm–4 mm in the B operation. The operative parts of the handle in this study are basically the same as those in the B operation. Therefore, this paper preliminarily determines the comfort test step size of the handle sample according to the results of this research.

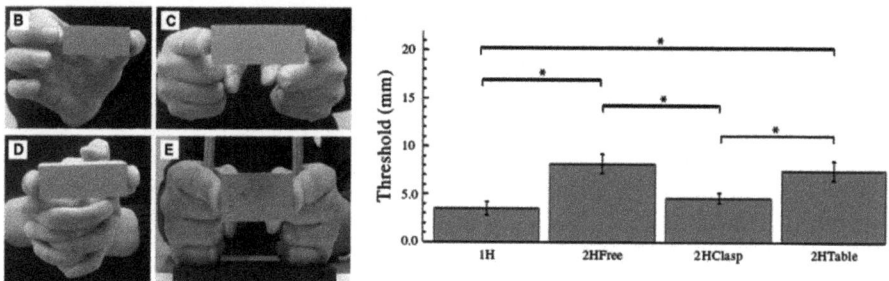

Fig. 1. Hand discrimination threshold

4.2   Research on the Size of Handle Design

There is a close correlation between some size of the handle and the size of the hand. In this study, the part size of the hand which may affect the ergonomics size of the handle is measured. When designing the minimum size of the handle size, we need to consider the size of the hand that can satisfy most of the population, therefore, 10 adult men were selected as the test subjects. On the basis of the hand size measurement results, calculate the relevant relationship between hand size, calculate the corresponding correlation coefficients, combined with the size of Chinese adults in standard GB/T 10000-1988, we calculated the size of hand that was not listed in GB/T 10000-1988 when designing the handle size.

4.3   Study on the Step Size and Optimum Value of Experimental Sample

In the specific work on some indexes of the handle during a test, a set of test sample size of index should cover the optimal value or unilateral limit value, and in order to reduce the influence of other indicators to the experimental results, need to ensure that other indicators in the reasonable range. Therefore, prior to the processing of an experimental sample, a preliminary test is needed to determine the preferred value of the index.

   In this pre - experiment, a pre - experimental simulation test sample was made with a lightweight stone plastic clay. A total of 6 subjects, including 3 men and 3 women, were tested for comfort assessment of use for the part ergonomics indexes of the 5 types handles identified by this study.

4.4   Recommended Size of Handle Test Sample of Furniture
           for Ergonomics Experiment

According to the summary of the relationship between hand sizes and related handle ergonomics indexes in 4.2 (in order to meet the comfort, the optimal value of ergo-nomics indexes related to hand sizes are set as maximum test sample) and analysis of related pre experimental results in 4.3, recommended size of various ergonomic indexes of test sample for handle ergonomics experiment are shown in Table 5.

Table 5. Recommended size of test sample for handle ergonomics index experiment

Handle types	Ergonomics indexes	Test sample size of handle ergonomics index experiment (mm)					Optimum value of ergonomics index (mm)
No.1 handle (square)	Thickness (T_{1s})	6	9	12	15	18	10
	Width (W_{1s})	6	9	12	15	18	10
	Distance from the pane (d_{1s})	9	12	15	18	21	21
	Length (L_{1s})	28	31	34	37	40	40
No.1 handle (circular)	Diameter (D_{1c})	6	9	12	15	18	12
	Distance from the pane (d_{1c})	9	12	15	18	21	21
No.2 handle	Thickness (T_2)	9					9
	Width (W_2)	6	9	12	15	18	12
	Depth (d_2)	9	12	15	18	21	21
	Length (L_2)	28	31	34	37	40	40
No.3 handle	Internal angle (A_3)	90	95	100	105	110	100
	Internal height (H_3)	9	12	15	18	21	21
	Depth (d_3)	12	15	18	21	24	24
	Width (W_3)	12	15	18	21	24	24
	Length (L_3)	28	31	34	37	40	40
No.4 handle	Diameter (D_4)	15	18	21	24	27	21
	Distance from the panel (d_4)	9	12	15	18	21	21
No.5 handle	Thickness (T_5)	4	6	8	10	12	6
	Diameter (D_5)	15	18	21	24	27	24
	Distance from the panel (d_5)	6	9	12	15	18	18

5 Conclusion

Based on literature research and user investigation, this research carried out the study on the handle test sample of furniture for ergonomics experiment. Combined with the research results, the recommended values of design, such as step size, number and feature size of the test samples, were determined. The results of this study can provide the basis and reference for furniture handle and similar handle test.

Acknowledgments. This research is supported by 2017 National Quality Infrastructure (2017NQI) project (2017YFF0206603 and 2017YFF0206506) and China National Institute of Standardization through the "special funds for the basic R&D undertakings by welfare research institutions" (522017Y-5278).

References

1. Chen, W., Shulan, Y.U., Boming, X.U.: Study on the furniture handle design based on product semantics. J. Art Des. (2013)
2. Li, D.: A Study of Household Handles Design from the Perspective of Kansei Engineering. Master Thesis, Kunming University of Science and Technology, Kunming, Yunnan, China (2011)
3. Jin, X.: A Study of Utensils Handles Design based on the Perspective of Kansei Engineering —The Example of Furniture Handles. Master Thesis, Shenyang Aerospace University, Shenyang, Liaoning, China (2012)
4. Hui-Min, H.U., et al.: Experimental research on man-machine adaptation of manual grip structure. Chin. J. Ergon. (2016)
5. Tao, G.Q., Li, J.-Y., Jiang, X.F.: Research on the finite element model of grasp-hand for tool handle. Mod. Manuf. Eng. **12**(12), 31–34 (2011)
6. Paschoarelli, L.C., Santos, R., Bruno, P.: Influence of door handles design in effort perception: accessibility and usability. Work **41**(Suppl 1), 4825 (2012)
7. Chang, S.K., Drury, C.G.: Task demands and human capabilities in door use. Appl. Ergon. **38**(3), 325 (2007)
8. Guo, B., Tian, L., Fang, W.: Effects of operation type and handle shape of the driver controllers of high-speed train on the drivers' comfort. Int. J. Ind. Ergon. **58**, 1–11 (2017)
9. Panday, V., Tiest, W.M.B., Kappers, A.M.: Bimanual and unimanual length perception. Exp. Brain Res. **232**(9), 2827–2833 (2014)

Research on the Standard of Refrigerator Noise Quality Evaluation

Xin Zhang[1($\boxtimes$)], Shuiyuan Yu[2], Linghua Ran[1], and Huimin Hu[1]

[1] AQSIQ Key Laboratory of Human Factors and Ergonomics, China National Institute of Standardization, Beijing 100191, China
{zhangx, ranlh, huhm}@cnis.gov.cn
[2] Communication University of China, Beijing 100024, China
yusy@cuc.edu.cn

Abstract. Noise is one of the most important quality indicators for a refrigerator, which will seriously affect user's experience and is one of the hot spots of consumer complaints. In this paper, we develop a refrigerator noise evaluation method more effective than A-weighted sound pressure level. Its results are more closely related to the subjective feelings of users. Through the subjective evaluation of the noise quality of different types of refrigerators, we found that, compared with A-weighted sound pressure level, annoyance and loudness of the refrigerator noise are more related with each other. On this basis, we conducted a noise loudness subjective evaluation experiment to determine the function relationship between refrigerator noise annoyance and loudness and to measure the noise loudness threshold user can accept of different types of refrigerators. The evaluation result according the method is better consistent with users' subjective feeling and the after-sales of the enterprise. The research results of this paper may provide reference for the development of new standard for refrigerator noise evaluation.

Keywords: Refrigerator noise · Sound quality · Loudness threshold
Evaluation standard

1 Introduction

Noise is one of the most important quality indicators for household appliances. With the improvement of the living material level and the progress of manufacturing technology, consumers have increasing requirements on the control of refrigerator noise. The refrigerator noise evaluation is the basis of noise control. Over the years, the method of noise control has been focused on reducing acoustic radiation of refrigerator, and the noise evaluation is also performed based on A-weighted sound pressure level [1, 2]. A-weighted sound pressure level is based on the reaction of the person to the pure sound while the actual noise is often complex sound with complex frequency components, and mutual masking effects exist between the frequency components. Therefore, A-weighted sound pressure level as the evaluation indicator cannot well explain many noise phenomena and cannot describe the user's experience of different noise well. In some cases, noise with the same A-weighted sound pressure level may

© Springer International Publishing AG, part of Springer Nature 2019
T. Z. Ahram and C. Falcão (Eds.): AHFE 2018, AISC 794, pp. 516–524, 2019.
https://doi.org/10.1007/978-3-319-94947-5_52

give us completely different subjective feelings. Some refrigerators with low A-weighted sound pressure level in pre-delivery inspection, sound very noisy with high degree of annoyance when the users actually use them at home. This brings great troubles to the refrigerator manufacturers in noise control, and there is the urgent need to find a more suitable noise evaluation indicator.

Sound quality refers to the human auditory perception of sound signals, and it lays more emphasis on the subjectivity of human's judgment of the sound characteristics. There are many physical factors affecting the sound quality of noise, and loudness is one of the important physical evaluation indicators. It has been found in this thesis through the subjective evaluation tests of noise quality of many refrigerators in different types that refrigerator noise annoyance and loudness have a better correlation when compared with the A-weighted sound level. On this basis, this thesis further determines the function relationship between refrigerator noise annoyance and loudness and the noise loudness threshold of refrigerators in different types through the subjective evaluation experiment of noise loudness. The research results of this thesis can provide reference and basis for the development of the new standard for refrigerator noise assessment.

2 Experimental Method

2.1 Refrigerator Sample Selection

In order to determine the functional relationship between refrigerator noise annoyance and loudness better, the types of experimental refrigerator sample noise signal shall be as many as possible, and the differences shall be as large as possible. In other word, the brands and models of refrigerator samples shall be as many as possible and the structural and component differences between them shall be as large as possible, so that all types of refrigerator noise can be covered. In this research, the operating noise of a total of 49 refrigerators was collected. With regard to the cooling mode, the air cooling refrigerators and direct cooling refrigerators were included. In respect of style and volume, the single-door refrigerators, double-door refrigerators, three-door refrigerators, multi-door refrigerators and side-by-side combination refrigerators were included. The details are shown in Table 1.

Table 1. Experimental refrigerator samples

Types	Direct cooling refrigerator				Air cooling refrigerator			
Structure	Single-door	Double-door	Three-door	Multi-door	Double-door	Three-door	Multi-door	Side-by-side combination
Quantity	7	5	7	1	4	4	9	12

2.2   Original Noise Signal Recording

The original noise signal was recorded in a semi-anechoic chamber, and the recording equipment mainly consisted of the sound quality head & torso simulator (B&K-4100-D) and microphone (B&K-4189). A total of 3 channels' signals were recorded at the sampling rate of 32 KHz with the quantization bit of 16 bit. Two channels were the for the microphone in the left and right ears of the sound quality head & torso simulator, and the other channel was the external microphone placed in the middle at the top of head of the head & torso simulator. The signal recorded by the head & torso simulator was used for making the listening signal in the subjective evaluation of noise sound quality, and the signal recorded by the external microphone was used for analyzing the signal spectrum. The head & torso simulator placement was set mainly based on the actual location of refrigerator for the user's normal use. The head & torso simulator was placed in the middle of the front of refrigerator with its tragion 50 cm away from the refrigerator door and 150 cm from the ground. 150 cm is the average ear height of Chinese adults (Fig. 1).

Fig. 1. Refrigerator noise signal recording

The operating condition of refrigerator during noise recording was kept in compliance with the provisions of IEC 60704-1: 2010 [1].

2.3   Subjects Selection

In this research, a total of 45 people aged 25–45 years with normal hearing were randomly selected as experimental subjects, and the ratio of male to female was 1 : 1.

2.4   Subjective Evaluation Experiment of Noise Sound Quality

The experiment used the rating method for the subjective evaluation of sound quality, the subjects carried out rating in accordance with the established rating criteria according to their own subjective feelings of the signal sound quality:

- Very good (1 point)
- Good (2 points)
- General (3 points)
- Barely acceptable (4 points)
- Unacceptable (5 points)

The test stimulation signal duration was 2 s, and the beginning and end of each stimulus signal were multiplied by an exponential function time window for time-domain smoothing. Stimulus signals were played back using a Sennheiser HD650 monitor headphone.

3 Relationship Between Noise Annoyance and Loudness and A-Weighted Sound Level

3.1 Relationship Between Refrigerator Noise Annoyance and A-Weighted Sound Level

The test results of relationship between refrigerator noise annoyance and A-weighted sound level are shown in Fig. 2.

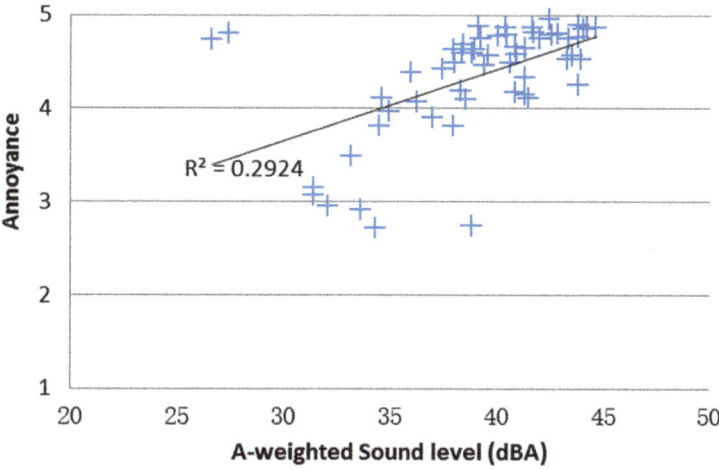

Fig. 2. Relationship between refrigerator noise annoyance and A-weighted sound level

As can be seen from Fig. 2, there was linear relationship with A-weighted sound level and annoyance of most refrigerators, but there were still the noise of some refrigerators. Although the A-weighted sound level was low, there was a high degree of annoyance. Statistics and analysis of after-sales data also showed that the user complaints rate of these refrigerators were higher. It shows that the noise quality evaluation indicator and noise control strategies based on A-weighted sound level have some defects and they are not suitable for some refrigerators.

3.2   Relationship Between Refrigerator Noise Annoyance and Loudness

In consideration of the defects in the noise quality evaluation indicator based on A-weighted sound level, we need new noise parameters to evaluate and control the noise quality of refrigerator. This parameter should have stronger corresponding relation with annoyance, and has more concise and fully recognized calculation method so as to facilitate the subsequent standard implementation. The experimental results showed that loudness was a good choice, especially when considering that there is not so many difference in the manufacturing process and material quality of various household refrigerators in the market, and thus the composition of spectral structure of the refrigerators' noise is similar and the loudness will have better applicability. Figure 3 is the test results of refrigerator noise loudness and annoyance.

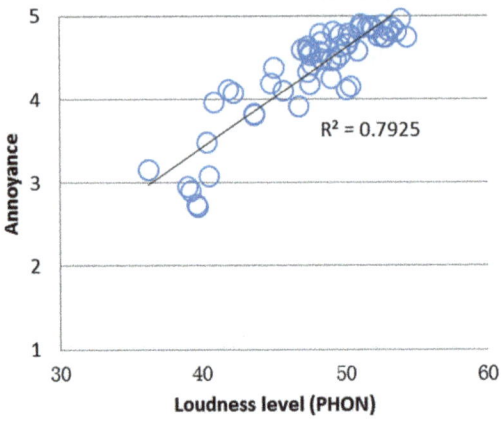

Fig. 3. Relationship between loudness level and annoyance of refrigerator noise

3.3   Sum-Up

It can be seen from the above that the relationship between loudness and annoyance has a better linear correlation than the relationship between A-weighted sound level and annoyance. The fitting results showed that the linear fitting goodness of loudness and annoyance was 0.7925, while the goodness of A-weighted sound level and annoyance was only 0.2924, so the loudness is more suitable to be used as an indicator for assessment of the sound quality of refrigerator noise than the A-weighted sound level.

4   Refrigerator Noise Loudness Threshold

In this study, the loudness threshold of refrigerator noise was defined as follows: under daytime quiet conditions (ambient noise level <40 dB), if the refrigerator noise exceeded the upper limit, it would be unacceptable to a large number of users. The loudness and annoyance function curve was in s shape, i.e. there was a section in which the loudness and annoyance had a good linear relationship, and outside of which the loudness and annoyance had little effect. We sought this section through psychoacoustic experiments and calculated the linear function relationship. The experimental results are shown in Fig. 4. In the experiment, we systematically modified the power of each noise so that its loudness reached 30 Phon, 35 Phon, 40 Phon, 45 Phon and 50 Phon, respectively, and then measured their annoyances with the use of subjective evaluation experiment of noise quality. Then, the mean of annoyance of all the noise measured at each loudness value was linearly fitted to the loudness.

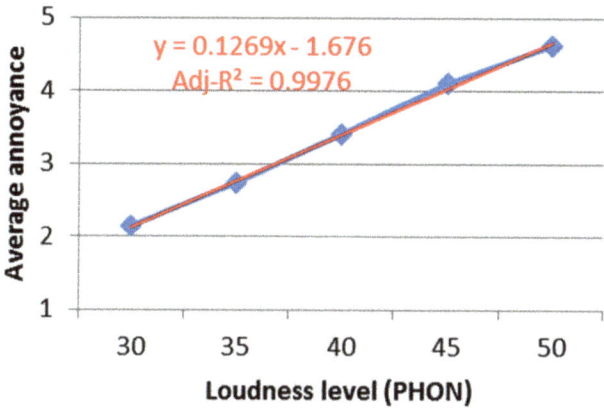

Fig. 4. Linear fitting relationship of average loudness and annoyance

It can be seen from Fig. 4 that the good linear relationship (goodness of fit 0.9976) is reflected between them in an area with moderate loudness (over annoyance 3), and this relationship curve can be used as a basis for formulating the noise quality standard based on loudness.

The determined loudness threshold should ensure the high detection rate and the low false detection rate of the refrigerator with non-conforming noise. If the annoyance average score is more than 4, it means a large number of users gave 5 points and thought the refrigerator noise level was unacceptable. Therefore, according to the definition of the maximum loudness limit of refrigerator, the cumulative lines of 3 and 4-point annoyance can be used as the detection limit and false detection limit of non-conforming refrigerator:

- If the average annoyance score above 3, the refrigerator will be deemed as exceeding the loudness limit;
- If the measured loudness level of refrigerator exceeds the established loudness threshold but its average annoyance score is less than 4, then the refrigerator will be deemed as being misjudged as exceeding the limit.

It is found in our experiments that there was heterogeneity (systematic difference) in the relationship between noise loudness and annoyance of different refrigerators, and that the distribution of loudness limits of all refrigerators was relatively discrete, as shown in Figs. 5 and 6. This is because the annoyance is not only closely related to the loudness, but also related to the parameters such as spectral components and time domain changes, etc. The differences in compressor, cabinet structure and cooling method will affect the relationship between refrigerator noise loudness and annoyance. If all the refrigerators use the same loudness threshold, the noise assessment results will be greatly deviated, therefore it is necessary to classify the signals according to the signal characteristics of different refrigerators, and then provide corresponding loudness thresholds for each type of refrigerator, respectively. In each type of refrigerator, the determined loudness value should ensure the high detection rate and the low false detection rate, otherwise the type should be subdivided or reclassified.

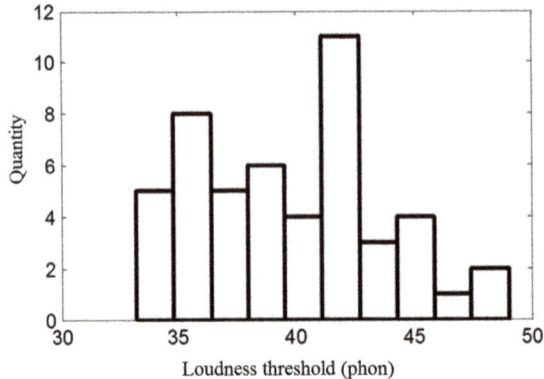

Fig. 5. Distribution histogram of refrigerator loudness threshold

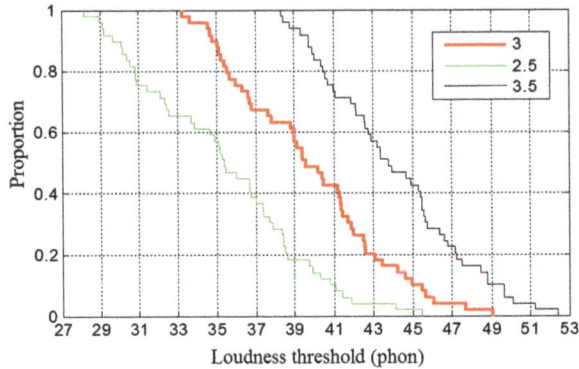

Fig. 6. Threshold cumulative curve of refrigerator loudness

The experimental results show that there is a significant difference in the relationship between noise loudness and annoyance of the direct cooling and air cooling refrigerators, and that there is also difference in the relationship between noise loudness and annoyance in case of different refrigerator volumes. Table 2 gives the final refrigerator noise loudness threshold on this basis.

Table 2. Refrigerator noise loudness threshold

Types		Loudness threshold
Direct cooling	Volume <200 L	45 Phon
	Volume >200 L	43 Phon
Air cooling		42 Phon

5 Summary and Outlook

Based on the research and test in this thesis, it is found that the loudness is more suitable than the A-weighted sound level to be used as the assessment indicator of refrigerator noise quality, it has a better correlation with the subjective annoyance of refrigerator noise, and that its detection results are closer to the subjective feelings of people. With the help of refrigerator enterprises, we compared the results of tests based on loudness with the results of tests based on A-weighted sound level, and the results of tests based on loudness showed a high consistence with the statistical results of user complaints. This also verifies from another perspective that the loudness is more suitable to be used as an evaluation indicator of the sound quality of refrigerator noise. In addition, the loudness calculation method is also relatively simple and has been widely recognized. Furthermore, using it as an assessment indicator can also ensure the consistency of the test results.

Subject to the limitation of research funding and time period, the refrigerator noise signals collected in this research was of small amount, and more refrigerators should be added in the follow-up experiments to evaluate the sound quality so as to determine more accurate standard of loudness threshold of refrigerators. Taking into account the greater differences between the user's home environment and the laboratory environment, the test methods should also be further improved and amended in this regard.

Acknowledgment. This work is supported by Presidential Foundation of China National Institute of Standardization (project number 522017Y-5279-2017).

References

1. IEC 60704-1:2010 Household and similar electrical appliances: Test code for the determination of airborne acoustical noise - Part 1: General requirements
2. IEC 62552:2015 Household refrigerating appliances: Characteristics and test methods
3. Zwicker, E., Fastl, H.: Psychoacoustics. Facts and Models, 2nd edn. Springer, Heidelberg (1999)
4. ISO 226:2003: Acoustics-normal equal loudness level contours
5. ANSI S3.4:2007 American National Standard Procedure for the Computation of Loudness of Steady Sounds Accredited Standards
6. ISO R/532 B: Method for calculating loudness level (1990)

7. DIN 45631/A1-2010: Calculation of loudness level and loudness from the sound spectrum - Zwicker method - Amendment 1: Calculation of the loudness of time-variant sound
8. Mao, D.: Progress in sound quality research and application. Tech. Acoust. **26**(1), 2 (2007)
9. Mao, D.: Recent progress in hearing perception of loudness. Tech. Acoust. **28**(6), 12 (2009)

The Interface Design of Mobile Library: A Case Study

Siyu Yang[1], Peng An[1], and Zhe Chen[2(✉)]

[1] China University of Mining and Technology, Beijing 100083, China
[2] School of Economics and Management, Beihang University,
Beijing 100191, China
zhechen@buaa.edu.cn

Abstract. This study aims to improve user experience by seeking users' needs. We want to design a new attractive mobile library interface with more usability and characters of China University of Mining and Technology (Beijing) and give design criteria of the mobile library. We design the first interface by reviewing current mobile libraries and literature. We interviewed 30 students for three design stages. Finally, we designed a mobile library prototype using Axure RP and give our design guidelines. This study aims to develop suitable interface and functions for the mobile library, using China University of Mining and Technology (Beijing) as a case. We give our design guideline for the mobile library, which is useful to the development of university mobile libraries.

Keywords: Mobile library · User experience · Interface design
Prototype

1 Introduction

The mobile library becomes popular due to the development of mobile technologies. Yan [2], believe that mobile library is the inevitable trend of library development and the design of mobile library should focus on practicality and creativity besides the basic query capabilities. Wang [1] introduced principles of user interface design include user-centered, good usability, information feedback, and certain artistic design. We evaluated the current mobile library from 9 colleges and universities in China from Apple Store. The vast majority provide query function and interface design does not have a considerable user experience (Table 1).

Based on the characteristics of China University of Mining and Technology (Beijing), we design the mobile library interface of it. After getting our students try some mobile libraries of other colleges, we figured out the shortcomings of interface design and human-computer interaction of other universities' mobile libraries through interviews.

By reading numerous references and interviewing 30 students, we have made many iterations of the interface design. We get the feedback of users and improve our interface design before we design the whole app instead of collecting users' feedback after using it. Thus, it can save software development costs and design the interface which can meet users' needs. This can be used as a reference for other colleges and

© Springer International Publishing AG, part of Springer Nature 2019
T. Z. Ahram and C. Falcão (Eds.): AHFE 2018, AISC 794, pp. 525–534, 2019.
https://doi.org/10.1007/978-3-319-94947-5_53

Table 1. Features of mobile library in 9 universities.

Universities	Features
Tsinghua University	The user can change functions of the homepage The interface is hard to recognize
Jilin University	Learning Progress An account can bind other apps
Nanjing University	Recommendation of the library Clearly classification of the resource
Ningbo University	Forum
Inner Mongolia University	Audio books Introduction of the mobile library Cloud disk
Shanxi University	Similar to Jilin University
Shanghai International Business and Economics University	1. ISBN Retrieval 2. Recommendation from readers
Southeast University	Keywords based search Recommendation from readers
Nanjing University of Aeronautics and Astronautics	The ranking list of books

universities to develop mobile libraries with school characteristics and to promote the development of mobile libraries in Chinese universities.

2 Methodology

We carried out a literature review and investigated the basic requirements of users. On this basis, 15 students, majored in different subjects from Jilin University and Nanjing University, were pre-interviewed to gather the user experience of the mobile library for their university. Unclear to use or even don't know the existence of the mobile library reflects that the development of mobile library is greatly hindered. The bad user feedback shows the importance of great user experience with good interface design. We invited four students in China University of Mining and Technology (Beijing) to try the mobile libraries of Jilin University and Nanjing University, and we interviewed with them. The convenience and concision can be the most attractive request for them.

Interviews with 30 individuals were used as the source for our paper prototype. Respondents are undergraduates (15) and postgraduates (15) in our school. According to the proportion of the number of each college, we choose different professional students. The interview was divided into three stages and data from a group of ten respondents was used as a data basis for each generation of paper prototypes. We counted the top three functions most requested by interviewers and used them as a basis for analyzing students' rigid needs on a statistical table. In the first stage, we designed the first interface and got their suggestion. In the second stage, based on their suggestion, we designed the second interface and let them contrast them. In the third stage, we asked more specific questions like their preference for colors and designed the final interface. Contrast with predecessors' design criteria; we analyze the similarity and difference.

3   Results

3.1   Function

Functions Chose by Students

After interviewing 30 students from different majors and different grades, we found that there are three functions very popular among those students. We found "Bibliographic Retrieval," "Download Books" and "Borrow books" are the necessary functions for the mobile library in Table 2.

Table 2. Functions chose by 30 students.

Function	Undergraduates	Postgraduates	Total
Bibliographic Retrieval	12	13	25
Borrow books	12	9	21
Reservation/renewal	10	10	20
Notice of new books	3	4	7
Download Books	10	12	22
Database	7	6	13
Journal subscription	6	5	11
Announcement	5	6	11
Timetable of the library	8	5	13
Audio books	4	7	11
Recommendation from readers	2	5	7
Book new book	2	5	7
Personalized Recommendation	5	6	11
Search records of borrowed books	3	5	8
Self-copy	5	7	12
Personal comment on books	1	2	3
Forum	7	5	12
Public courses	7	5	12
Activities	1	1	2
Exhibition	1	1	2
Feedback box	2	5	7
Countdown of book lending time	6	9	15
Stories of the school	5	0	5
Learning process	3	2	5
Punch in reading	0	0	0
punch for the accolade	4	3	7
Share source with other universities nearby	6	5	11
Look for vacant seats	9	4	13
Recommendation from celebrities	4	3	7

Meanwhile, we found there are six functions which were chosen less than 20% students. These functions are "Stories of the school" (16.7%), "Learning process" (16.7%), "Personal comment of books"(10%), "Activities"(6.7%), "Exhibition "(6.7%) and "Punch in reading"(0%).

Importance of Functions

By interview 30 students and let them rank functions they chose to see the most important functions in their eyes. Then we marked the top 3 of the rank and made the table. We can see from Table 3 that "Bibliographic Retrieval" ,"Borrow books" ," Reservation/renewal" and "Download Books" are the most important functions in their eyes no matter they are undergraduates or postgraduates.

Table 3. Importance of functions chose by 30 students.

Function	Undergraduates	Postgraduates	Total
Bibliographic Retrieval	**8**	**9**	**17**
Borrow books	**10**	**9**	**19**
Reservation/renewal	**5**	**7**	**12**
Notice of new books	0	0	0
Download Books	4	5	9
Database	3	2	5
Journal subscription	0	1	1
Announcement	0	0	0
Timetable of the library	2	1	3
Audio books	0	2	2
Recommendation from readers	0	0	0
Book new book	0	0	0
Personalized Recommendation	2	1	3
Search records of borrowed books	0	1	1
Self-copy	1	1	2
Personal comment on books	0	1	1
Forum	0	1	1
Public courses	2	1	3
Activities	0	0	0
Exhibition	0	0	0
Feedback box	0	1	1
Countdown of book lending time	1	1	2
Stories of the school	0	0	0
Learning process	0	0	0
Punch in reading	0	0	0
punch for the accolade	0	0	0
Share source with other universities nearby	1	0	1
Look for vacant seats	3	1	4
Recommendation from celebrities	3	0	3

Adjustment Based on the Analysis

We can cancel "Punch in reading" because no one chooses it. However, if users can get an award from punching, they would like do it. That's why there is a part of students chose "Punch for the accolade."

We can change some functions into a small program so that users can use them when they need. Such as "Recommendation from readers," "Book new book," "punch for accolade" and "Stories of the school."

Because "Activities" and "Exhibition" is similar to each other, so we can merge them into one function and set it in a small program.

Although "Personal comment of books" has low rate, "Forum" (a place for users to communicate with book review) has a higher rate, so this function is still needed.

"Learning process" has low rate among postgraduates and "Feedback box" has low rate among undergraduates, so we put the two functions at the bottom of the interface. If they want to look through it, they can easily find.

3.2   Interface Design

The First Interface Design

In the first stage, based on other interface design of the mobile library and reading app, we designed our first mobile library's interface through the paper prototype.

This prototype has three main interfaces. The first interface is "Homepage." Main functions are put on the homepage, and they are divided into different modules so users can choose functions easily. Users can add more functions if they click the "Add." The second interface is "Study." On the top of "Study," there are two areas (subscription area and download area). The third interface is "Person Center." On the top right corner, the user can change sets of the mobile library. Besides the set is an icon look like a bell. When it close to the deadline for book lending time, this bell will remind

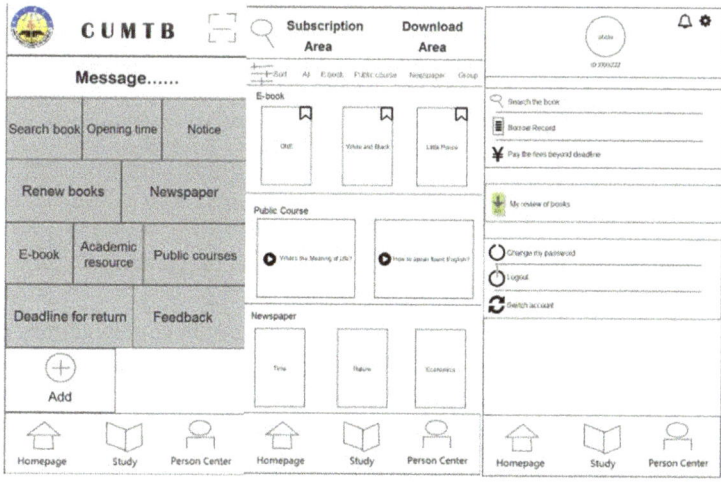

Fig. 1. The first interface.

users to return the book. In "Person Center," user can search books, look through his borrow record and pay the fees beyond the deadline. The user can also write his review of books (Fig. 1).

The Second Interface Design

In the second stage, after interviewing ten persons about functions and interface, we designed an interface with keywords style and small programs, and we add more functions to complete our mobile library.

The second design has four main interfaces. Users can search through keywords, and they can see all materials by vague-search. They can also input words by voice. Under the box, there are four icons which can link to corresponding content. At the bottom of the homepage is recommendation formed by the data that users have searched. We add "Learning Process" at the bottom of "Study" so that users can see how long they have learned from mobile library and read their book review. The third interface is "Information." Users can look through renew books recommendation and notice for an exhibition. They can communicate with other students in "Book review" and give suggestions on "Feedback box." The fourth interface is "Person Center." Based on the first design, we add "Small program" and "New book reservation." Users can use functions directly through small programs. If there are some books that users want to read, but the library hasn't, users can reserve it (Fig. 2).

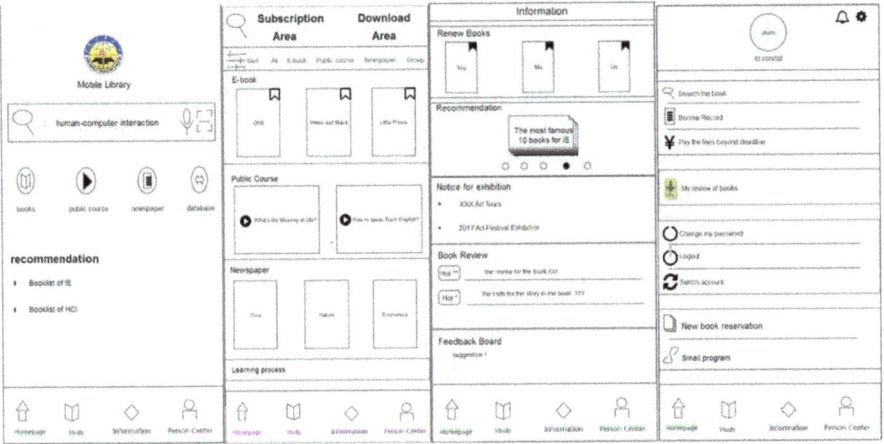

Fig. 2. The second interface.

The Third Interface Design

In the third stage, we summarize interviewees' suggestion about the interface and their preference of functions. We find the second prototype can meet most students' demand. So we optimize the second prototype to get the final interface of the mobile library. Every icon at the bottom will become black when users choose it.

The first interface adds small programs on the top. When users pull-down the interface, they can see small programs. If not, these small programs would be hidden and this will let the interface more concise. Under the search box, we add some labels about exams that are hot among students. These labels can change with time. We add color to the icons to make them colorful. Because of different colors can make the different vision, users can choose it easily. The second interface is similar to last one. We only add reading process on bookmark so users can know the rate of the process. The third interface is the same as the second design. Our data reveals that "E-book" is popular than "Public course," and "Public course" is popular than "Journals" among students. We don't need to change this interface. The fourth interface has changed functions order and added one function (Renewal) and one icon (Message). We put "Reservation" and "Renewal" together. Let "Small program" as a special module. We put functions of logging at the bottom. On the top left, a user can receive messages of a book review and important information (Fig. 3).

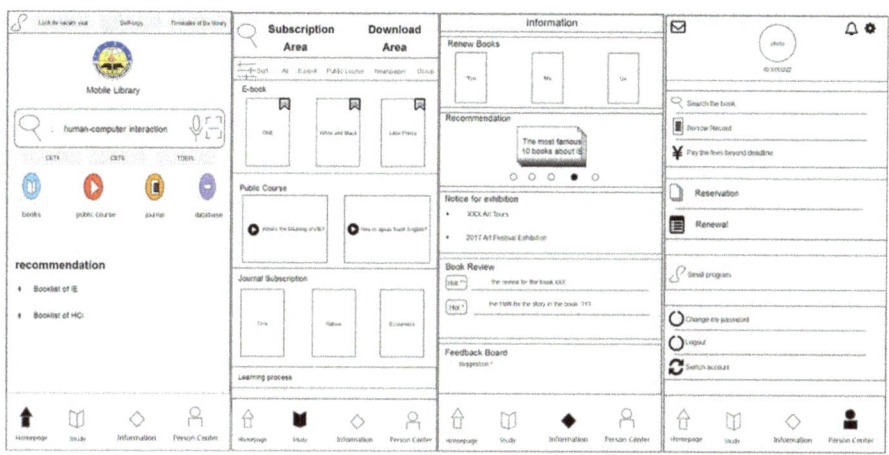

Fig. 3. The final interface.

4 Results

4.1 Design Criteria from Previous Studies [3–10]

- **Design guidelines centered on user experience**

  Interface design is based on user needs, not functions.

- **Consistency of interface design**

  Consistent use of color, consistent appearance of elements, and consistent interaction.

- **Good information feedback**

  Technical level: The system response time directly affects the speed and fluency of information feedback and design level.

- **Design art**
- **Design availability**
- **Moderate amount of information principle**
- **The principle of interactive continuity**

Interaction refers to the interaction between the user, and the design work or system should ensure that the interactive conduct was smooth, effective, continuous.

- **Targeted**

Provide a variety of push services for user habits.

- **Practicality and creativity**

Special services that meet the needs of users such as digital collection resource push, online exhibition and seminar promotion, interactive user community, popular science platform, virtual technology experience, information literacy training tools.

4.2   Design Criteria for the Study

- **Easy to use**

The mobile library should be easy to search. The way to search should follow advanced way. Now, keywords based search and voice input are easy to use and accord with users' operating habits.

- **Creative and simple way for user-defined**

Nowadays, small programs are hot in China. So we learn from its way and design our own small programs. All of our functions can be small programs so that user can choose any function they want. The functions which are frequently used can be found on the top of the homepage when user pull-down the interface. Because of this function, no matter the user is undergraduate or postgraduate, this mobile library can be suitable for him.

- **In line with users' operating habits**

The interface of a database can imitate the interface of academic resource on a computer. The way we use e-book can imitate Kindle, and the way we log in the mobile library can be similar to the way we log into our campus internet.

- **Effective way for users**

We can add maps which match to each room. When the user looks for one book on the mobile library, he can see the serial number and corresponding map. A map would save much time.

- **Guidance for beginner**

If there are some new and special functions on the mobile library, we should give some guidance for new users when they first go into our interface.

- **Easy to understand**

The icons on the interface should reflect the meaning of corresponding functions. If the icon has vague meaning, we should remind the user at first and add words to the icon if possible.

- **Hot labels**

The most popular demand should set the label on the homepage, and these labels can be changed with students' demand so that they can find the useful information directly.

- **Personalized recommends**

We can let users choose some interesting field so we can recommend books match to them. And we can analyze the context they search most to predict what they want and show it on the homepage.

- **Reflect the school's features**

For example, we can put school badge on the homepage and design the unique LOGO of this school. And when there are some important events in school, we can design a picture of this event and show it when user log in the mobile library.

- **Connect with students' training plan**

We can invite excellent teachers from different majors to recommend books so that students can learn more about their majors. Students can have public courses on mobile library and have exams, at last, if they pass the exam, they can get a credit for this course. Students can also apply for some lectures to get some credit.

4.3 Similarity of Criteria

1. Centered on user experience.
2. Easy to use
3. Personalized Service
4. Design Art: Color, layout, school's features
5. Good information feedback: icons, introductions
6. Good interaction.

4.4 Difference of Criteria

1. The mobile library can link to school's training plan for undergraduates and postgraduates. Credits are vital to every student. If the school can connect it with a mobile library, there will be many students use it with little publicity.
2. Add redundant information in information retrieval. There are lots of books on different subjects, but some reading room hasn't mapped to guide students. Finding a book is difficult. If we can add maps after a user searches a book, it will save time.
3. Consider most user' demands about exams. There are some exams important for most students. If undergraduates can't pass CET4, they cannot get their diploma.

4. Small program. Different students have a different attitude toward the mobile library. Some students only want to search books. But other students want to search more information on mobile library. A small program is an idea from WeChat. We can make special functions as a small program. In this way, the mobile library can meet most students' needs.

4.5 Problems of the Study

1. We add a database in a mobile library. But it's hard to read the whole article on our phone, so the mobile library should provide abstract and summary. Users can find articles they want and then look for this article easily on their computer later, and database won't take up too much memory space of the phone.
2. There are many students want to look for vacant seats on mobile library. It's difficult to achieve as buying cinema tickets online. So, we plan to add QR code on each desk in the library. When students use it, they scan the code. We get the feedback and deal with the information. Then students can look for vacant seats online.
3. We didn't design the mobile library's LOGO. We think it's a sign of the school. If we let more students join in the design, the LOGO will be more typical.

Acknowledgments. This study is supported by the National Nature Foundation of China grant 71401018, the Social Science Foundation Beijing grant 16YYC04, and China Scholarship Council.

References

1. Wang, C.: The Research on Multimedia-interaction Design Based on Graphical Interface. Shanghai Jiao Tong University, Shanghai (2010)
2. Yan, L.: APP mobile services of libraries at home and abroad: comparative analysis and enlightenment. Inf. Doc. Serv. **23**(6), 85–88 (2013)
3. Yan, L., Peng, W.: Talking about the organization and retrieval of multimedia information in digital libraries. Sci-Tech Inf. Dev. Econ. **27**, 14–16 (2007)
4. Chen, H.: Summary of user behavior research in mobile library at home and abroad. Libr. Inf. Serv. **22**, 135–144 (2016)
5. Song, W.: The Empirical Study on Influence Factors of User Behavior on Mobile Library. Nanjing University, Nanjing (2016)
6. Li, A.: Mobile library service in key Chinese academic libraries. J. Acad. Librariansh. **39**, 223–226 (2013)
7. Kubat, G.: The mobile future of university libraries and an analysis of the Turkish case. Inf. Learn. Sci. **118**, 120–140 (2017)
8. Paterson, L., Low, B.: Student attitudes towards mobile library services for smartphones. J. Libr. Hi Tech **3**, 412–423 (2011)
9. Yoon, H.-Y.: User acceptance of mobile library applications in academic libraries: an application of the technology acceptance model. J. Acad. Librariansh. **42**, 687–693 (2016)
10. Singh Negi, D.: Using mobile technologies in libraries and information centers. Libr. Hi Tech News **5**, 14–16 (2014)

Design of Location-Based Audio Guide System for City Tourism

Shiori Furuta$^{(\boxtimes)}$ and Katsuhiko Ogawa

Keio University, 5322 Endo, Fujisawa-shi, Kanagawa 252-0882, Japan
{shioshio, ogw}@sfc.keio.ac.jp

Abstract. Using a smartphone while walking is dangerous. If smartphones could be used in this environment as "tools for listening," this would enable people to locate exciting opportunities in the city. This study aims to design a location-based audio guide system that combines music playlists that are tailored depending on the city and recorded commentary depending on the location. We call this system "Street Ongaku and Spot Oshaberi (SOSO)", which allows the users to select a favorite program according to the situation from multiple SOSO programs and enjoys walking while listening to it. In this paper, the system design and its experimental results are presented.

Keywords: Smartphone · Audio guide · Music · Recorded commentary
Media

1 Introduction

How do you source information while sightseeing? We believe tourists acquire information through several means such as guidebooks, the Internet, and social networking services (SNSs) before sightseeing. Nevertheless, tourists, especially first-time visitors, prefer to go sightseeing along well-known routes. Locals, however, are familiar with some noteworthy locations and shops unknown to tourists. For example, there may be a small bar that serves delicious sake or a unique location for clicking beautiful photos unbeknownst to tourists. We want to make such local knowledge about interesting places easily available to tourists.

Audio guides are a contemporary solution to providing information used in many museums. While visitors view the exhibits, the audio guide provides an explanation. Such a service is not commonly used in the field of tourism.

This paper proposes a location-based audio guide system. We called this "Street Ongaku and Spot Oshaberi (SOSO)". In Japanese, "Ongaku" means music and "Oshaberi" means chatting. In this system, the recorded audio commentary is mapped to various places on the map. The user walks around the city while listening to music playlists tailored to the area in which they are walking. When the location where the audio commentary is mapped coincides with the positional information of the tourist's GPS, the volume of the music decreases and audio commentary for that position flows.

A conventional study on "Podwalk" services [1] considers services providing audio to accompany walks around city streets. Podwalk services are based on the premise that podcast audio is downloaded to the playback equipment in advance, and one can walk

© Springer International Publishing AG, part of Springer Nature 2019
T. Z. Ahram and C. Falcão (Eds.): AHFE 2018, AISC 794, pp. 535–545, 2019.
https://doi.org/10.1007/978-3-319-94947-5_54

while watching the printed map. As the length of the audio is fixed, the walking speed and audio often do not match.

A conventional study on "SkyDesk Media Trek" services [2] considers a location-based audio guide system. This service is a smartphone application. The user downloads the guide program in the application and listens to this program in town. There are two differences between this service and our research. First, our service is a web application, which eliminates the need for prior downloads. Second, our service incorporates music that can be enjoyed while walking in the city.

2   System Concept and Design

2.1   System Concept

SOSO is a system that allows the user to walk around the city listening to their chosen music playlist and provides location-appropriate audio of local people's talk in certain places. With SOSO, the user can learn what local people enjoy from the people who enjoy it. Furthermore, it can be expected that this will lead to the discovery of new places and new experiences in urban walking. The process is depicted in Fig. 1.

Fig. 1. Using the "Street Ongaku and Spot Oshaberi" (SOSO) guide system. (1) User walks while listening to music. (2) Corresponding locations lead to local commentary from popular places being played to the user. (3) The user listens and learns from local knowledge. (4) The user can act on their new knowledge to access interesting new experiences.

Our earlier research [3] investigated a service that enabled users were to walk around the streets learning about places using only location-based audio commentary. However, we found that if only audio commentary was used, there was no connection between each audio commentary item, and a sense of unity of service was lacking. Therefore, we decided to put music between audio commentary items, providing a pleasant continuity.

2.2   System Operation

SOSO is a web service that can map texts, sounds, and photos to maps. The program improved on a prior service, "Whispath" [4], developed by our laboratory. When the user sees SOSO on the browser of the smartphone, it automatically acquires the user's position information and matches this to available locations for which recorded audio commentary exist. The appropriate audio content plays automatically when the user comes to the place where the content exists. As shown in Fig. 2, by mapping the audio commentary locations, it is possible to listen to the audio regardless of the walk speed of the user. The green icons indicate locations to which audio commentary is mapped. Audio commentary flows automatically when the user arrives at these locations.

Fig. 2. Example of the SOSO browser screen mapping audio content locations.

The flow from production to the user is shown in Fig. 3. Registration information specifying the place the caller wants the user to hear their audio commentary is registered on the server, and the user's role is simply reading the information and listening to the audio content.

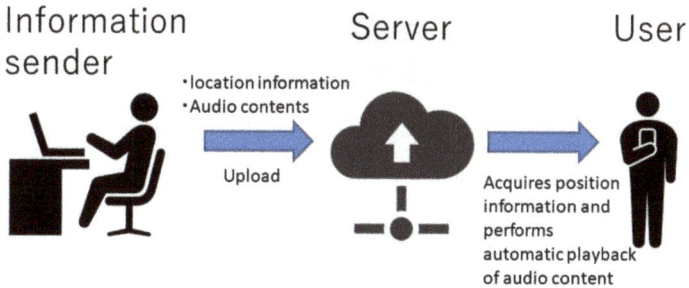

Fig. 3. Flow from SOSO program production to end-user system operation.

2.3   System Design

In this paper, SOSO uses music playlists and audio commentary created by locals in tourist spots, and tourists listen to these with SOSO as they walk around the city. Two improvements were made to Whispath to create the SOSO system, as previously explained. Whispath's functionality corresponded to the "Spot" function in SOSO. A playlist playback function for music (the "Street" part of SOSO) was added. Furthermore, while the audio commentary is being played back, the volume of music is lowered.

SOSO's music playback function loads music onto the server, making it available for end-user playback. Users can select music playlists of locally composed music when using the service. As both the music playlist and the audio commentary program are prepared, the user can freely combine them while they walk. Figure 4 shows SOSO's program selection page.

Fig. 4. Example of a program selection page. Users can select a program page by clicking on an image of a face.

The volume of the music is automatically lowered when the recorded audio commentary is played back in order to allow the user to hear it. Although it may be possible to temporarily stop music playback, there is a feeling of incompatibility when music is interrupted. Therefore, when playing back audio commentary, we merely reduced the volume of the music so that it flows in the background.

We designed SOSO to play audio commentary for approximately 5–15 s, based on the study in [5]. There are two reasons for this. First, if the playback time of the audio commentary is too long, music that flows between different audio commentary elements cannot be heard. As music is an important element of SOSO, we set the playing time to facilitate musical continuity. Second, when the playback time of the audio commentary is long, the user's location and the commentary being played back, may no longer correspond appropriately. For example, it is assumed that you can walk past a small shop in 10 s. If audio content is played for 30 s, the user passes by the store and can become confused because the audio no longer matches their location. To make time to play music, we spaced content items such that there is at least 20 s of walking time between each item.

3   Experimental Methods

3.1   Experiment Place

This experiment was conducted in Enoshima island which is a representative tourist place near the university with which the authors are affiliated. The topography of Enoshima is mountainous, with many stairs. Places often frequented by tourists visiting Enoshima include assorted seafood rice bowl restaurants, and the Enoshima shrine. The audio commentary program used this time was produced by one of the authors as a local university student. The page the participants was viewing on the smartphone is depicted in Fig. 5 (with recorded audio commentary denoted by the speaker icons). The single speech bubble icon denotes speech-synthesized audio commentary, prepared for comparison with human-voiced audio commentary. We asked the subject to walk from the location of the speech bubble, guided by the audio icons, toward the lighthouse on the top of Enoshima (denoted by the star in Fig. 5). The purpose of the experiment was to determine the improvement of the SOSO system and its content and to investigate whether SOSO provided a pleasant urban walking experience for the user.

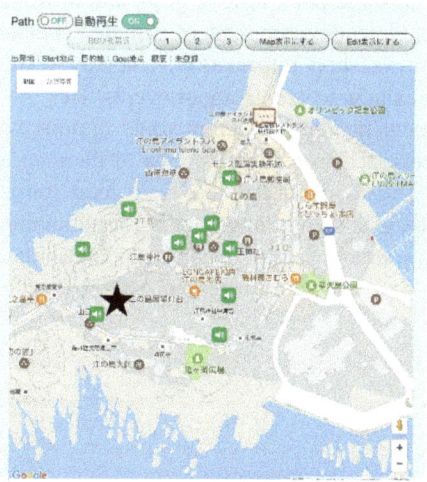

Fig. 5. User's SOSO smartphone screen.

3.2   Participants

Five participants took part in the experiment, two males and three females. They were all either visiting Enoshima for the first time or returning there for the first time and so had little implicit knowledge about Enoshima.

Each participant opened a web page with their smartphone, placed an ear-phone in one ear, selected a music playlist (buttons 1 to 3 in Fig. 7), and then departed. We followed the participants, as shown in Fig. 6, in order to respond to problems and observe. Participants were instructed to behave as sight-seers, wandering to a shop or taking pictures of the scenery. We asked oral questions about what was worrisome about the behavior of the participants during the experiment. After the experiment, we administered a participant questionnaire.

Fig. 6. Participants walking in Enoshima during the experiment.

3.3   Contents

Three types of music playlists were prepared for the experiment using local music: stage animation songs, Japanese pop music (J-pop), and instrumental befitting the local atmosphere. In addition, the audio commentary included content from several different genres including directions, shop information, empathy with the user, commentary about tourist attractions, announcements about events, etc.

Also, using audio content, three different route options were made available. In the audio commentary mapped to the circled area of the map in Fig. 7, the following directions are given. "The road on the left is a good tourist route, including the popular Enoshima shrine as a sightseeing spot before climbing the stairs that follow. To the right, on a narrow street used by residents of Enoshima Enter, you can see the ocean."

Fig. 7. Three routes that the users can select: from the circled area facing the bottom of the map, the leftmost voice guidance path is numbered 1, the middle path is 2, and the rightmost road is 3.

4   Experiment Results

The music playlists and routes selected by the five participants are listed in Table 1.

Table 1. Contents selected by participants

Participant number	Music	Route
No. 1	Animation song	2
No. 2	Animation song and J-pop	2
No. 3	Animation song	1
No. 4	Animation song and J-pop	1
No. 5	J-pop and instrumental	1

We orally asked participant's behavior during the experiment we are interested in. After completing the experiment, we conducted a questionnaire. In the questionnaire, we asked the participants about three main topics: the music content, about the audio content, and the pleasure they experienced while walking around the streets. The results are detailed in the next section.

4.1   Music Contents

In the question "It's fun to choose a playlist made by others," four people answered "fun," and one person answered "some fun." They enjoyed choosing their music playlists, as this allowed them to hear new music they were unfamiliar with.

The question "About the discomfort to sound being heard during the reproduction of music" had four people who answered "There was no discomfort," and only one who felt "There was almost no discomfort". The reason was that it sounded natural that the sound overlapped the music. For example, car navigation is the same whether you are playing music or not.

The question "Content I want to listen when walking around a street" allowed multiple answers. Three people selected, "Only Music," and five selected "Music and Audio Content"; none selected "Audio Content Only."

4.2   Audio Contents

Participants indicated the information that they wanted to hear in the audio content as follows: five people selected "Information on what kind of store is where", four people selected "Directions," three people selected the options "Photogenic place information" and "Local people's story," two people selected "Information on the whole tourist spot" and "Information suitable for time like information on bargain sale of goods," one person selected "City event information," and none selected "Information irrelevant to the city."

Regarding the question, "How did you choose the way when you came to the branch point?", four people answered with the option "We used route guidance for audio content." The last participant answered "unconsciously chosen the way"; people who chose unconsciously almost never came to Enoshima; hence, they saw the street and told they chose a road that seemed to have many passersby.

When asked "Were you interested in surroundings by listening to the content?", there were four people "interested" and only one who "had a little interest".

4.3   Pleasure of Walking Around the Street

Regarding the question "Was there something interesting while walking in the town," two people answered "Something interesting," two people answered, "Nothing," and one person answered "Somewhat interesting."

A specific item of interest included "a shrine being demolished and carrying out construction".

Regarding the question "Were you interested in taking pictures?", three people seletcted "There were times when you wanted to take pictures," and two seletcted "You sometimes wanted to take a little picture." When asked what they had photographed or wanted to photograph during the experiments, participants mentioned several things including "sea," "track running along narrow paths," "tiles," "landscapes of their feet," and "cats".

In the question "How to feel the time at the time of the experiment," two people said that "the experimental time felt short," two people said that "the experimental time felt somewhat short," and one person said "It was a little longer."

5   Discussion

5.1   Usability

Usability considerations are described here. During our experiments GPS readings could not be accurately determined on participants' smartphones on several occasions, impacting the timeous flow of the recorded audio commentary in some cases. The topography of Enoshima, a mountainous shape, could be responsible for this. When looking down at Enoshima from directly above, if location information about altitude has not been acquired, there is almost no difference in coordinates for the spoken audio to be played, even if a temporal gap exists between content items. This problem cannot be solved by expanding the range audio commentary is reproduced, as this leads to audio from neighboring audio locations being reproduced in some cases. Audio content need to be mapped to topographically homogeneous areas with as little undulation as much as possible. Alternatively, the installation of a beacon to increase positional accuracy would be necessary.

The users found they were able to use it almost without feeling uncomfortable about hearing audio content during music playback. In a daily life, most people use the navigation application while listening to music on the smartphone; thus, we think that they could use the system without any discomfort.

5.2   Contents

First, about music content. Experimental results showed that all participants selected a music playlist (animation song or J-pop) that was relevant to their location. When listening to reasons for choosing a playlist in an oral interview, there were many responses that "I am interested in music that fits the place." Also, when listening to a playlist made by someone other than themselves, participants indicated pleasure that the user "may be able to meet unknown songs." It might be fun for users to select music playlists.

Next, regarding audio content. When participants changed the sightseeing route on the way, many people referred to the audio content. From that, it was appropriate to arrange the audio content on the sidewalk. One of the participants chose a person who knew little about Enoshima, and thus Route 3 was not chosen. There is a sense of security in feeling that if Route 1 or Route 2 is chosen, the participant will avoid to get lost. A participant said, "Route 3 was interested in. However, as the weather was likely to rain, I thought it would be better to have the experiment finished earlier. And I had to go to the goal. Hence, I decided to go the way I'm not going to get lost." If we want to guide people to secluded sightseeing spots, it may be necessary to eliminate the goal or to provide carefully guiding directions so that the user does not become anxious.

Also, when the participants listened to the information about the Enoshima shrine, which is a sightseeing spot, and the information of the shop, it was often to see the surroundings. This suggests that audio commentary that ties into locational information is useful. Meanwhile, as for empathic content such as "The stairs here is long," we noted that participants walked without stopping while listening.

Contemporary popularity of apps like Instagram [6], a photographic SNS, motivates many people to want to take beautiful pictures to share online. In terms of whether SOSO's assistance in finding photogenic places was significant, we noted that some participants took pictures they wanted to post in Instagram during the experiment, in places such as spots where the sea could be seen, spots where a red bridge grew green, etc. It is confirmed that the participants who took these pictures posted at least one picture of Enoshima to Instagram after the experiment, leading us to conclude that photogenic place information was considered also useful content by the participants.

5.3   Pleasures of Walking Around the Street

The experimental results indicate enjoyment of urban walking. A participant said that the shrine in construction in Fig. 8, which they found while walking and searching for

Fig. 8. Shrine where the construction was taking place.

audio content, was interesting. For example, encounters with animals such as stray cats were answered to be enjoyed, even there is no audio commentary.

From the answer about content, it was clear that the audio commentary raised the participants' interest in their surroundings. Also, many participants felt the passage of time was short. Before walking, the participants are instructed to go to the lighthouse of the mountain, they felt like a distant place. However, by the end of the experiment some people made remarks like "Has the experiment has finished already?" SOSO would be able to make users more enjoyable in the tourist area.

6 Conclusion

In this paper, we proposed "Street Ongaku and Spot Oshaberi (SOSO)," which was a combination of music playlists tailored to the city and recorded audio commentary regarding particular spots. The design concept of SOS is that the contents creator and the user are collaborated through information and music.

Although it was a small-scale experiment, these preliminary results suggested that SOSO did evoke pleasure for users who were walking around the streets. We observed that users looked forward to place-dependent content. Therefore, future content shoud be created with an emphasis on place characteristics.

In future development, we intend to increase the experiment scale, seeking opinions from a larger sample of general users rather than a small group. Next, we will implement the SOSO program at another location (Otaru, Hokkaido's sightseeing spot) and create multiple programs for users in order to choose what they want to hear.

References

1. Kato, F.: Capture, share, and experience: "Podwalk" as a medium for flaneurs. In: PICS (Pervasive Image Capture and Sharing) Workshop, UbiComp2006, Orange Coungy, CA (2006)
2. Sawayama, A., Shimada, Y., Takaya, K., Ueda, K.: An Audio Guide Service That Can Be Created Together with Local Communities. No. 25 Technical report, Fuji Xerox (2016)

3. Furuta, S., Ogawa, K.: Proposal of internet radio walking around street while listening to voice of virtual idols. In: 19th International Conference on Human-Computer Interaction, vol. 714, pp. 366–373. Posters' Extended Abstracts, Vancouver (2017)

4. Utsumi, S., Ogawa, K.: The experiment of vicarious experience by using place's second sound channel media, "Whispath". Inst. Electron. Inf. Commun. Eng. 181–185 (2016). LOIS2015-94

5. Miyasaka, K., Ogawa, K.: Design and evaluation of "social networking radio" to provide voice and sound of the location based information. In: The Tenth International Conference on Digital Society and eGovernments, ICDS 2016, pp. 72–77, Venice (2016)

6. Instagram. https://www.instagram.com/. Accessed 23 Feb 2018

#MeToo: An App to Enhancing Women Safety

Javed Anjum Sheikh[1(✉)] and Zonia Fayyaz[2]

[1] Capital University of Science and Technology, Islamabad, Pakistan
Javed.Anjum@cust.edu.pk
[2] The University of Lahore, Gujrat Campus, Gujrat, Pakistan
zoniafayyaz@gmail.com

Abstract. Few years back, women harassment was not an issue. Nowadays, women harassment is recognize as a legitimate human rights issue and threat to women's well-being. The #MeToo campaign launched by actor Alyssa Milano to drew the world's attention to women harassment, with a hashtag that went viral. Harassment is unwanted actions and comments by strangers. Due to the fear of harassment, women are paying to stay safe as women don't walk independently as men. Pakistani women were among of them. This research will illustrate the harassment problems faced by working women and will examine the available solutions with the context of Pakistani women. There is no evidence that these apps have the power to decrease incidents, despite what the apps developers' claim. This research will also examine these apps to find out why women still do not feel safer as the apps claims.

Keywords: #MeToo · Safety apps · Women centered design · Women safety
Usability · Design · User experience

1 Introduction

This is From Stone Age until yet women are enthusiastically and providing support to men in their difficulties. In order to continue breath in a sensible manner it's harder for middle class families to earn bread and butter [1]. In early ages, women confined to their kitchen. Very few numbers of women had the access to higher education and they forced to be the compassion of their fathers or husbands' attitudes towards women and work. Most of the women do not enter in job market due to social and cultural aspects [2]. Now the world's perception regarding work and women has changed. Women contribute their countries' economy the same way as men can do. From past few years, there has been an enormous invasion of the concept work and women in every corner of the world. In public services, a larger number of women are taking part in it. Rapidly different countries are encouraging their women to join the labor market and implementing the equality rules in all over in order to get outstanding results [3].

Most of the world's population consists of women and only few percent are working outside the house. The reason behind this is women faced many of the problems while working outside the house. These problems make them resist not to working outside [3]. Some of the major problems present in this paper.

© Springer International Publishing AG, part of Springer Nature 2019
T. Z. Ahram and C. Falcão (Eds.): AHFE 2018, AISC 794, pp. 546–553, 2019.
https://doi.org/10.1007/978-3-319-94947-5_55

Work-life balance is one of the issues for working women, as they have to work for both their career and for their family, which demands long working hours and this would leads towards a norm [4]. One of the most important and key problem-working people faces is their security issues. Here in this paper we will describe security issues in detail with three major security problems; Physical harassment, mental harassment and bullying [5]. To overcome these problems we have proposed some solutions. These solutions will lessen the security problems women have to face at workplace [6].

This paper organized the follows: Sect. 2 illustrates the problems faced by working women and comparison among existing security applications. Section 3 describes the methods referred to women security, the technologies etc. Section 4 concludes the paper.

2 Background

Women are doing well in every field including space discovery and rocket science. They are also playing a vital role in the economic development of the country and their input is not less than their male counterparts [4]. Still there are several issues, which are facing by women even today.

- Most of the time women are not consider equal as men at workplace and treated inferior. There are salary inequalities even both graduate from same college with same grades [5].
- Another issue, which working-women have to face, is balancing the work life and their personal life [5]. Working-women of today accomplish responsibilities of families and try to remain fully involved in their careers coping up with the competing demands of their multiple roles. Women are in terrific problems in order to accomplish their different role simultaneously, which lead them towards mental pressure. As a result family becomes organizational stakeholder [6].
- Working-women suffer from other issue that is lack of family support. At times, the family doesn't support women to leave the domestic work and go to office. They also resist for women working until late in office which affects their performance and promotion [6].
- Different problems such as attitude of the society members, prejudice and non-recognition that the working-women encounter with regard to their status and role in the economic life adversely affect the utilization of their talents and work capabilities. These problems may reduce the efficiency of working women and act as hindrance for entering the females in different jobs.

2.1 Issues

Security issues are the major concern for working-women. In the current situation, it is very much important to save women from harassment and violence [6]. Few years back, women harassment was not an issue. Nowadays, women harassment is recognize as a legitimate human rights issue and threat to women's well-being. The #MeToo campaign launched by actress Alyssa Milano to drew the world's attention to women harassment, with a hashtag that went viral. Harassment is unwanted actions and comments by strangers. Due to the fear of harassment, women are paying to stay safe as women don't walk independently as men. Pakistani women were among of them. This research will illustrate the harassment problems faced by working women and will examine the available solutions with the context of Pakistani women. It is also to be noted that, there is no evidence that these apps have the power to decrease incidents, despite what the apps developers' claim. This research will also examine these apps to find out why women still do not feel safer as the apps claims. Although government but still safety provides rules and regulations is the major concern [13]. Some of the major safety issues are:

Harassment. Any conduct from boss, coworker, vendor or customer whose action or communication puts down an employee is always unwelcome [10, 15]. Here in this section we will describe different types of harassment, which women face at workplace. Snooping with employee's talent to do her work is also considered harassment [7, 11].

- **Mental harassment:** Mental and verbal harassment in the workplace refers to sexually degrading comments, such as whistling or intimidating a person, including giving a pejorative name to a person [12]. It also includes certain political statements, dirty jokes and even some types of art that people take as unpleasant [8].
- **Physical harassment:** It is less common than verbal harassment, but it can often be more severe [14]. Physical harassment associated with sexual harassment, like touching, grouping, hitting, pushing etc. most importantly victim is touched in an inappropriate way against her will [9].

Absence of role model at workplace. Usually we learn things by following other's footsteps. Either it is job or business, junior ones follow their seniors. In this male dominated culture, women do not easily find a female role model for themselves to progress in their jobs, rather they have to seek help from male colleagues. Therefore, the variations in the communication style of men to women hinder their growth at one place.

Transport Issues. In Pakistan, we do not have any women friendly transport service so they can travel alone to their job. Not only working-women but young students also face this problem due to which they suffer in their professions.

Acceptance as Working Professionals. Most men in our society are yet to come to terms that women are capable of working with them, shoulder to shoulder, in any professional sphere independently. They still visualize women as individuals who are bound to kitchen domains and other domestic affairs.

Balancing Work-Family Life. Women are still considered as family manager back home no matter how high managerial position they attain in their profession. They expect to take care of home and look after family matters timely.

Travelling Unacceptability for Work. Most of the women cannot travel or go on hours without having to answer uncomfortable questions by family and peers. Married women, who have a flourishing career and are dominant in their field, suffer a lot due to this. So, their job obligations entirely depend upon the causal support from their family members. But in case of married men, they stay outside their home city on long official tours for days without raising eyebrows, but they show disapproval for equally-successful wife. As an outcome, women have to opt out of their jobs which offer travelling and settling in other cities.

Safety of Working Women. There is still concern for safety of women who go on official business tours. Women travelling alone consider vulnerable and become a chapter of talk for their male chauvinist colleagues. Even if the trip is official, checking into hotel alone is a problem itself for them. Many hotels do not allot room to single women due to their safety concerns.

Unequal Pay. One of the raging topics among the context of problems is that of equal pay, which most women suffer from. Officially, same work assigned, woman and men are equally paid. However, gender discrimination is rampant as many companies still do not adhere to these guidelines and pay women less than their male colleagues pay.

Discrimination at Workplace. Working-women face discrimination at their place blatantly. Sexual harassment at workplace is a major issue they face. Women employees, working in night shift face serious problems. There are no proper steps taken to disregard this issue as a result harassment tends to increase at work place.

Apparently, there are many solutions, which could provide the better security app for working-women. However, these apps (Table 1) have not enough functionality that could provide security to working-women [24–27]. Our pilot study shows that women still do not feel safe by using apps, as these are not according to their culture and fear to

Table 1. Comparison of Safety Apps

App	Tracking	Audi recording	Online complaint	Scream alarm	Notification	Fake call	Cost	Real time alerts
SheSafe	Yes	No	No	No	Yes	Yes	Free	Yes
Guardly	Yes	Yes	No	No	Yes	No	Free	No
Life 360	Yes	No	No	No	Yes	No	Free	No
Hollaback	Yes	No	No	No	Yes/Images	No	Paid	No
VithU	Yes	No	No	No	Yes	No	Free	Yes
Fight back	No	No	No	No	Yes/SMS	No	Paid	No
Raksha	Yes	No	No	No	Yes/SMS	Buzzer	Paid	Yes
Be safe	Yes	No	No	No	Yes	Yes	Free	No
Street safe	Yes	Yes	No	No	Yes	Yes	Paid	No
Scream alarm	No	Yes	No	No	Yes	No	Paid	No
Safety pin	Yes	No	No	No	Yes	No	Free	No
Vanitha Alert	Yes	No	No	No	Yes/SMS	No	Free	No
Women's security	Yes	Yes	No	No	Yes	Yes	Paid	No

lose their jobs. Therefore, it is a harassment app with privacy and guide according to the situation features.

Apart from available apps there are significant barriers exist for women and girls accessing digital tools [30].

3 Solution

The problems of working women Security issues in both developed and developing countries consider the major concern for working women. In the current situation, it is very much important to save women from harassment and violence [8]. Rules and regulations are although provided by Government but still safety is the major concern [13]. Some of the major safety issues are physical harassment and, mental harassment. At the workplace, comments refer to sexually degrading comments, such as whistling or intimidating a person, including giving a pejorative name to a person [12] is objectionable. It also includes certain political statements, dirty jokes and even some types of art that people cover the three main stages of harassment: when a woman notices, being harassed and in the take as unpleasant [8]. Physical harassment: Physical harassment is severe than verbal/mental [17]. Physical harassment means touching in an inappropriate way against the will [9].

To ensure the women security we have proposed some of the methods, which could be beneficial for the safety reasons of working women. Here in this section we will discuss the methods and technologies for women security.

- **Online complaint**: To save time and to eradicate corruption, online complaint management system provides a way to solve the problem online [20]. The core purpose of online complaint management system is to make complaint easier to coordinate, monitor, track and resolve and to provide company with effective tools to identify and target the problem [16].
- **Position detecting**: Tracking is important, to notify friends by, Global Positioning System (GPS) [21].
- **Scream alarm:** It is not always possible to make a call in emergency there are certain situations where you can just trigger a button to find help. Scream alarm is designed to create a loud noise or alarm when it's switched on [18]. This button is in most of the cases is referred as panic button. The alarm is a high-decibel noise that's meant to warn other people that someone needs help [20].
- **Database** will have contact numbers and voice keyword. User registers a contact list of people to whom user wants to ask for help and keyword or voice is saved for recognition purpose. Contacts and keywords are saved in database. Database is stored in mobile memory. Database used is SQLite database. There should be at least two contact numbers in database [28].
- **Voice Recognition:** Voice recognition module is use to recognize keyword spoken by the user. Keyword spoken by the victim will compare with the registered keyword. This keyword will match with converted text. If keyword matched then message will send [29].

- **Location Tracking and Address Finding Module:** It requires GPS enabled mobile. Location will tracked using GPS. Using longitude and latitude location is searched and an actual address is given via message. Internet is essential in others mobile. GPS in disable system will not find the exact location of user. It will just send the longitude and latitude of the location [22, 23].
- **Message Sending Module:** The GPS Application Program Interface (API) fetches the longitude and latitude coordinates [17]. Pre-stored emergency message is send to registered contact numbers along with the longitude and latitude and an exact address of user. If network is not available on user's mobile then message goes in queue and message is send when available. When message is send then notification send [19].

Women are doing well in every field including space discovery and rocket science. They are also playing a vital role in the economic development of the country and their input is not less than their male counterparts [4].

4 Conclusion

Women's security is very important for the sake of country's progress and also for their own self confidence, to ensure their security their should be an app which make them enable to walk freely without any fear. In this research we will compare different available apps for women security purpose and still there is a need to develope a new app which could overcome the existing problem and also by adding some more features which are helpful. Here we have described some of the important features which are missing in existing apps. By producing with such app we can make a remarkable step in women security and they can feel secure while working or walking outside.

References

1. Rizwani, S., Sabir, M.: Attitudes towards employment of women in an urban perspective. In: Proceedings 10th Sociological Annual Conf. Held at Lyallpur (Faisalabad) (1976)
2. Pearl, J., Linda, S.: Professional women: the continuing struggle for acceptance and equality. J. Acad. Bus. Ethics **1**, 98–111 (2009). http://www.aabri.com/manuscripts/08056.pdf
3. Arnove, R.F., Torres, C.A., Franz, S. (eds.): Comparative Education: The Dialectic of the Global and the Local. Rowman & Littlefield Publishers (2012)
4. Nabi, B.N., Abdullah, M.A., Gopang, N.: The study of problems of working women in Hyderabad city. Women Annu. Res. J. (2011). University of Sindh, Jamshoro
5. Bailyn, L., Drago, R., Kochan, T.A.: Integrating work and family life – a holistic approach. A Report of the Sloan Work-Family Policy Network, pp. 1–10, 14 September 2001
6. Friedman, S.D., Greenhaus, J.H.: Work and Family—Allies or Enemies? What Happens When Business Professionals Confront Life Choices. Oxford University Press, New York (2000)
7. Zimmerman, T.: Marital equality and satisfaction in stay-at-home mothers and stay-at-home father families. Contemp. Fam. Ther. **22**, 337–354 (2000)

8. Nussbaum, M.: Capabilities as fundamental entitlements. Feminist Econ. **9**(2–3), 33–59 (2003)
9. International Labour Organization (ILO): Sexual Harassment in the Workplace in Nepal, International Labour Organization, Kathmandu (2004). www.ilo.org/kathmandu/whatwedo/publications/WCMS_113780/lang–en/index.htm
10. The Advocates for Human Rights. Law and policy on street harassment (2013). http://www.stopvaw.org/law_policy_street_harassment
11. Ayres, M., Friedman, C., Leaper, C.: Individual and situational factors related to young women's likelihood of confronting sexism in their everyday lives. Sex Roles **61**(7–8), 449–460 (2009)
12. Crouch, M.: Sexual harassment in public spaces. Soc. Philos. Today **25**, 137–148 (2009)
13. Davis, D.: The harm that has no name: street harassment, embodiment, and African American women. UCLA Women's Law J. **4**, 133–427 (1994)
14. DeKeseredy, W.S.: Feminist contributions to understanding woman abuse: myths, controversies, and realities. Aggressive Violent Behav. **16**, 297–302 (2011)
15. Livingston, B.A., Wagner, K.C., Diaz, S.T.: When street harassment comes indoors: a sample of New York City service agency and union responses to street harassment, Cornell IRL, Ithaca, NY (2012)
16. Nasr, O., Alkhider, E.: Online complaint management system. **2**(6) (2015)
17. Kim, H., Howland, P., Park, H.: Dimension reduction in text classification with support vector machines. J. Machine Learn. Res. **6**, 37–53 (2005)
18. Combarro, E.F., Montanes, E., Díaz, I., Ranilla, J., Mones, R.: Introducing a family of linear measures for feature selection in text categorization. IEEE Trans. Knowl. Data Eng. **17**(9), 1223–1232 (2005)
19. Sebastiani, F.: Machine learning in automated text categorization. ACM Comput. Surv. **34**(1), 1–47 (2002)
20. Nasr, O., Alkhider, E.: Online complaint management system. IJISET Int. J. Innovative Sci. Eng. Technol. **2**(6) (2015)
21. Li, N., Chen, G.: Sharing location in online social networks. In: IEEE Conference on Network, pp. 20–25, September/October 2010. ISBN:0890-8044/10
22. Tekawade, A., Tutake, A., Shinde, R., Dhole, P., Hirve, S.: Mobile tracking application for locating friends using LBS. Int. J. Innovative Res. Comput. Commun. Eng. **1**(2), 303–308 (2013). ISSN: 2320 -9798
23. Siriteanu, A., Iftene. A.: Meetyou-social networking on android. In: 11th RoEduNet International Conference, Sinaia, 17–19 January 2013, pp 1–6 (2013). ISBN: 978-1-4673-6114-9. https://doi.org/10.1109/roedunet.2013.6511763
24. "Fightback" Android App Developed ByCanvasM Technologies, 26 June 2013. http://www.fightbackmobile.com/welcome
25. "Guardly" Android App Developed By Guardly Corp., 28 January 2014. https://www.guardly.com/
26. "OnWatch" Android App Developed By OnWatch, 10 November 2012. https://play.google.com/store/apps/details?id=com.onwatch
27. "Life 360 – Family Locator" Android App Developed ByLife360, 20 February 2014. https://www.life360.com/family-locator/
28. Baccam, T.: Making database security an IT security priority, November 2009. https://www.sans.org/reading-room/whitepapers/analyst/making-database-security-security-priority-34835

29. Akash, S.A., Al-Zihad, M., Adhikary, T., Razzaque, M.A., Sharmin, A.: HearMe: a smart mobile application for mitigating women harassment. In: 2016 IEEE International WIE Conference on Electrical and Computer Engineering (WIECON-ECE), pp. 87–90 (2016)
30. Avis, W.: Digital tools and improving women's safety and access to support services (GSDRC Helpdesk Research Report 1,415) Birmingham, GSDRC, University of Birmingham, UK (2017)

Seeing Patterns for Guiding Users and Avoiding Pitfalls Trough Design

Tingyi S. Lin[1,2(✉)]

[1] Design Department, National Taiwan University of Science and Technology,
43, Keelung Rd., Taipei City 106, Taiwan
tingyi.desk@gmail.com
[2] Division of the Humanities, University of Chicago, 1010 East 59th Street,
Chicago, IL 60637, USA

Abstract. New ideas and services are evolving with ever-changing technological development. The relationship between human and machine is constantly
shifting the way we think and act. With ultimate goal of building a service
model within a shared economy, this paper focuses on finding patterns and
defining pitfalls through visual analysis and information design processes so as
to implement creative brainstorming at a later stage to solve the problem. This
pilot stage investigates targeted users' interactions and behaviors through onsite
observation and survey interviews. Mobility service and line management are
raised in order to reshape quality service flow and to design adequate information. With those efforts, we understand the relationship between the community and its users, discuss the measurement and the attributes of information,
and synthesize the ways communication with visual elements and forms can
serve us better.

Keywords: Visual analysis · Information design · Service model
Communication

1 Introduction

With the rapid change in technology and the popularity of mobile devices and wear-
ables, the applications of Internet of Things allow various objects, things and services
to be able to connect to each other anytime and anywhere. With the heated discussion
of Internet of Things (IoT), Internet of Everything (IoE), and Internet of Anything
(IoX), the waves of new technology come into our lives one after another. Many
scenarios we thought in arts and science fiction are now happening in or around our
lives. While Google Home's and Amazon Alexa's are joining many households, now
people are expecting domestic robots, self-driving cars, flying taxi, and more in the
near future. Partnered with transportation departments, 3M's development and testing
of the Smart Street concept for the coming of autonomous vehicle (self-driving technology) is leading to innovation engineering evolution [1]. Fictitious science fiction
based products no longer seem novel, but already in our lives. Falkan [2] argues that
"technology is not formed in isolation from society— technology and society are
formed and transformed simultaneously and in correlation." New ideas and services are

© Springer International Publishing AG, part of Springer Nature 2019
T. Z. Ahram and C. Falcão (Eds.): AHFE 2018, AISC 794, pp. 554–563, 2019.
https://doi.org/10.1007/978-3-319-94947-5_56

evolving with this ever-changing technological development, with which the human-machine collaborative relationship is also sequentially altering and transforming. The fluctuations between human and machine relationship is constantly shifting the way we think and act. How users react and interact in an environment with manmade object, and designated visuals are growing issues for securing a high quality of life.

The new inter-connectivity and information exchange creates not only excitement but curiosity and anxiety. Considering the social and behavioral changes mentioned above, we then start to think, "what kind of lifestyle we will have?" "what kind of life do I want to have?" and "what services do I need in my life?" Design plays an integral role assisting us to skyrocket technology and to make critical decisions wisely.

The efficiency and the effectiveness of information enhances both production and management. Users are more highly motivated to take action when they receive and understand information. Users identify the quality of high or low service with their experiences when they retrieve and respond to information. This paper focuses on finding patterns and defining pitfalls through visual analysis and information design processes. The ultimate goal is to build a service model in view of a shared-economy-so as to implement creative brainstorming at latter stage to solve problems. This pilot stage therefore investigates targeted users' interactions and behaviors through onsite observation and survey interviews. Although not included in this paper, role-play with a focus group will be implemented later during the prototyping stage to gain a deeper understanding.

2 Thinking and Analysis with Visuals

The dynamic relationship of advance and retreat between artistic and scientific views on visual representation never ends. As Arnheim suggests to consider "vision as a creative activity of the human mind [3] " from an embodied and experiential viewpoint on perception and cognitive metaphor theory, we therefore can perceive general patterns by weaving context and modify information to develop the overall structure.

2.1 Visual Thinking

We constantly represent, describe, and perceive our thriving three-dimensional world through the use of two dimensional images, screens and scripts. New technology is taking us into an era of 'seeing' and 'feeling' a three dimensional environment allowing us to perceive the fullness of the real world. Through the rising trends of VR and AI, the techniques of flatness have been being an important vehicle for us to undertake the perceptual investigation for knowing the world [4]. The skill of visual thinking allows us to discover tangible and intangible objects, reasonable and unreasonable situations, as well as interconnected and relationships. Understanding operative iconicity, meaningful storytelling and functional visual language, and the connection of existing iconicity, visuality, spatiality and graphism provides us with the clues and signals that bring us answers and solutions. The "inscribed surfaces" of those two-dimensional visual representations, as Krämer claims, remain "major principles of modern technology" [5].

2.2   Visual Analysis

In order to implement the analytical and design thinking processes, visual tools can fulfill the needs of Hyerle's nine human cognitive qualities including: (1) metacognition, (2) constructing abstractions, (3) storing information outside the body, (4) systems thinking, (5) problem finding, (6) reciprocal learning, (7) inventing, (8) deriving meaning from experiences, and (9) altering response patterns [6]. In doing so, graphically displayed thinking processes are the forms of metacognition. Secondly, visual thinking skills can assist users to organize, find patterns and to make sense of large amounts of information. Third, visual tools are ways to store so that the information to be recalled and rearranged in a more functional way to communicate. Forth, good use of visual tools enhances thinking and analytical processes by providing us to simultaneously see various angles of the system. Better decision making usually comes from holistic considerations in this capacity of parts-whole relationships. Fifth, visual mapping is the technique to layout the sequences, showing clusters of diversity and representing the alternatives. Sixth, visual thinking processes own the features of multimodality, which often encourages interactivity and productivity. Seventh, visual tools are used to elaborate the visual thinking processes, in order to help retrieve, manage, and expand information. Eighth, the skill of visual thinking enhances users to reflect and to accumulate experiences. Ninth, visual thinking processes encourage empathy and flexibility, which allow alternative responses and stimulate creative growth.

2.3   User Behavior

Users require a series of information depending on the roles they play and the tasks they need to accomplish. The service model that we are targeting to build is a spatial-shared eco-system in a small community setting. Managers, retailers, employers, and customers are potential users within the system. Industry has been sensitive and eager to know how to refine their business strategy through analyzing and understanding their customers. In doing so, many decision makers hold the teamwork between market-focused consideration and customer-driven action in high regard [7]. The comprehensive consideration of the situation often involves customer satisfaction, market-orientation planning, horizontal competition strategy, buyers' behavior analysis...etc. This pilot stage therefore focuses on knowing the customers' behavior among potential users in order to manage the sales force and co-create a friendly environment for consuming and working. According to Engel, Blackwell, and Kollat, connecting influencing consumer behaviors should include: (1) making contact, (2) shaping consumer opinions, and (3) helping customers to remember [8]. Good communication with users can stimulate motivation, facilitate attitudes, (re)shape beliefs, and embody feelings. In order to build the well bonded connection, understanding the users is the key to build special rapport with them. On-site examination helps researchers and designers to know and to update both group and individual behaviors patterns through the flow of technological, cultural, ethnical, and social influences.

3 The Experiment

The revolution of digitalization and online services allows us to connect human and machine, to innovate new ways of operation, and to facilitate advanced lifestyles onward. With various widespread online applications, savvy shops and restaurants boost their market venues by connecting on and off-line services. With those successful individual cases in mind, it will be beneficial to consider a bigger picture to benefit larger amounts of a variety of people. The food court is a public setting which serves food and beverages at multiple stands. Food courts often provide convenient dining experiences with economic, affordable prices in the community. They often exist in university campuses and in larger commercial spaces such as malls and department stores. Diners share the open space with a relaxing atmosphere. Implementing a new service with updated technology can enhance the quality of the dinning and working environment in one of the food courts at Taiwan Tech. This study investigates on-site targeted users' interactions and behaviors in a campus food court setting.

Methods. In order to understand users' interactions and their behavior patterns, onsite observations and survey interviews were conducted. Visual analysis and information design processes were employed to understand users' behavioral patterns, to define the breaches and to develop the strategies for further stages, including prototyping, testing, revision, and implementation. Data from onsite observation and survey interviews were examined, cross-referenced and analyzed with the blueprints, then the concepts of information design thinking processes including inventory, strategizing, planning, testing, filtering and guiding were applied. This process of digging for a deeper understanding provides views and strategies and allows for a new design concept to be formed.

4 Results and Analysis

Non-participatory observation was conducted three peak time-flames (8:30–9:00, 12:00–12:30, 17:30–18:00) and a random off-peak hour for five regular academic days. The major observation features include crowd flow, queue line, food ordering procedure, preparing & waiting, pick-up, seating, operation condition, and the time spent in the stages above. The results were analyzed and visually translated to depict how people proceed their buying, how the queue forms in the environment, and how they interact with each other. For example, Fig. 1 shows the crowd flow and the density of queue line in the morning (8:30–9:00). There is an average of 120 people within the space (12,749.31 ft^2, aka., 1184.45 m^2) in that half hour, averaging 61 s/p for waiting and 6:01 min/p from food ordering to pick-up. Figure 2 shows the crowd flow and the density of queue line at noon (12:00–12:30). There is an average of 811 people within the space in that half hour, averaging 82 s/p for waiting and 3:33 min/p from food ordering to pick-up. Figure 3 shows the crowd flow and the density of queue line in the evening (17:30–18:00). There is an average of 342 people within the space in that half hour, averaging 37 s/p for waiting and 3:48 min/p from food ordering to pick-up.

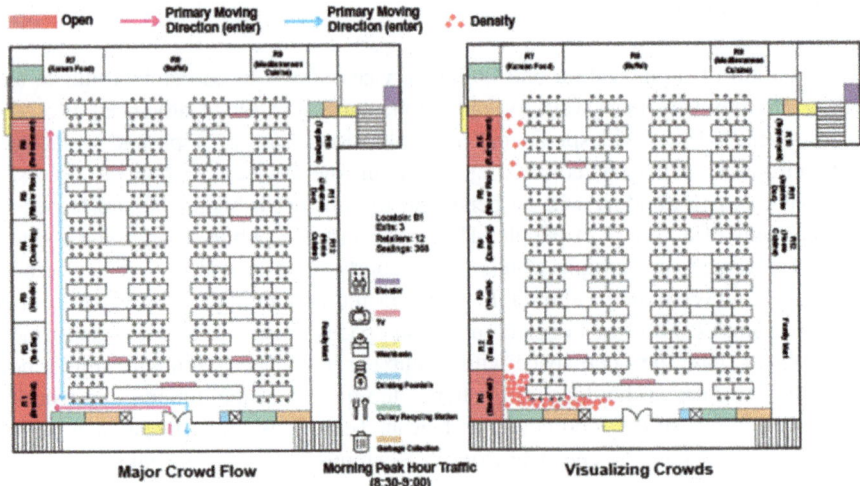

Fig. 1. The crowd flow and the density of queue line in the morning. An average of 120 people within the space during 8:30–9:00, averaging 61 s/p for waiting and 6:01 min/p from food ordering to pick-up.

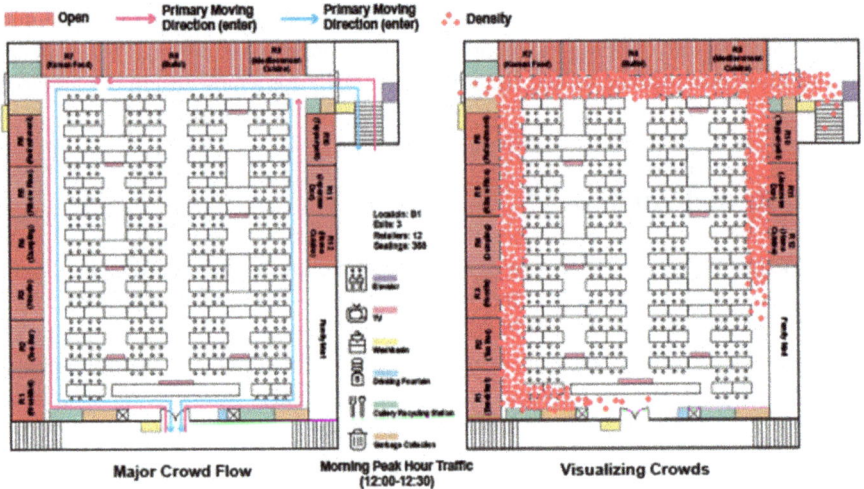

Fig. 2. The crowd flow and the density of queue line at noon. An average of 811 people within the space during 12:00–12:30, averaging 82 s/p for waiting and 3:33 min/p from food ordering to pick-up.

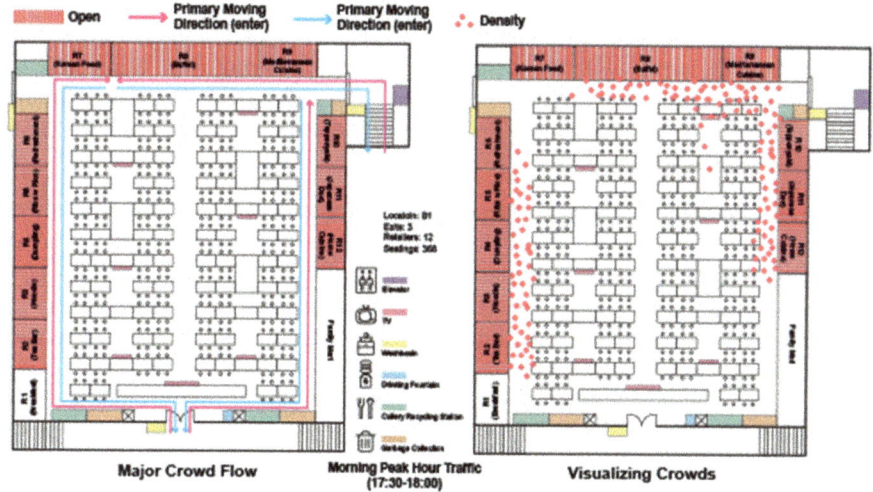

Fig. 3. The crowd flow and the density of queue line in the evening. An average of 342 people within the space during 17:30–18:00, averaging 37 s/p for waiting and 3:48 min/p from food ordering to pick-up.

Thirty subjects (15 males and 15 females) were invited to a semi-structured interview after the simulation of a purchasing task. Results show the most desired services are: (1) *adequate information* between retailers and buyers: need to make the information accurate, instant, transparent, and understandable; (2) *quality service flow*: to make operating process easier and purchasing experience smoother; and (3) *enjoyable environment*: to provide a hassle-free space. Juxtaposing the results with onsite observation and survey interview, a journey map is created from synthesizing the interactions described above. The map itself as well as the inventory and thinking process will be a great fundamental resource for further analytical processes such as planning, testing, filtering and guiding throughout the information design act.

Problems, Solution and Propositions. Obstacles are easier to seen when visualized. The purpose of identifying the problems is to look for the right solutions. Since this experiment falls on a semi-commercial/community setting, studies from consumer and user behaviors are helpful. Nicosia's model of buyer behavior claims the importance of four elements: (1) transfer of information, (2) search for and evaluation of information, (3) purchase decision, and (4) consumption and feedback [9]. We can design information in a way that makes it easier for the users to understand allowing for an enhanced the flow of service, assisting buyers to make informed decisions while boosting sales. We therefore focus on how to reshape the *quality of service flow* and design the *adequate information* to provide patrons with a user-friendly information environment. Although we cannot change the physical space in this project, it is possible to create a more pleasant experience and enjoyable environment for the users.

It is horrible to enter a chaotic, crowded environment to begin your dining experience. The worse thing is that you have to wait there with no idea where to go or what to do. One of the biggest obstacles is the crowd control in the peak times. The information design would work on mobility service and help create more efficient line management. The former will allow the ordering and preparing jobs to happen anywhere and at anytime without a long wait in the physical space; the latter will arrange the on-site circumstance in an orderly manner, reduce frustratingly long waiting times, and minimal grouchy attendees.

Mobility Service. The mobility service can (1) sell/order food anywhere; (2) process payment online/offline; (3) advanced food prepared before arrival; (4) digital advertisement, promotion, and reminder; and (5) social data tools to connect larger scope of activities and management. From the standpoint of the service provider, employees can accomplish their routine duties and daily activities (e.g., food and beverage orders, transmission of tasks to the kitchen, payment transaction process, printing bill…etc.) with a sense of self-accomplish and team collaboration. Buyers and visitors will save time and enjoy their purchasing journey (e.g., search for and evaluation of information, ordering action, payment and receipt, waiting and pick-up, and seating for enjoy…etc.) with adequate, transparent, and accessible information. Three components of consumer decision making (CDM) model by Milner and Rosenstreich are inputs, processes and outcomes [10]. Inputs include purchase situation (contextual and environmental variables), consumer characteristics (psychological and social influences) and information source (marketing mix and interpersonal). Processes include need arousal, information utility, criteria development and evaluation of alternatives. Outcomes includes the decision (to buy or not to buy), the purchase itself and post-decision evaluation. Milkman, Chugh, and Bazerman claim their new model is a more appropriate to decision-making aids that allow consumers to make decisions [11]. Information utility plays an important role in interacting with other processes, with which the quality of information does have a strong influence on their evaluation of the service.

In order to establish the mobility service, the points of sale (POS) are defined for a standalone machine or a network of input and output devices and service plans for before (inputs), during (processes), and after (outcome). The experience of consumption includes from the point of sale (POS) step I to IV (Table 1). The feedback during and after those consumption acts influences most users' attitude and their motivation. The online platform and application nowadays provide more power for direct judgments, praises or complaints, with which makes the information even more transparenct. In Table 1, POS I (select) and POS (pay) are the actions during the decision making processes. Communication plays the key role to reinforce the action.

Table 1. The points of sale in mobility service.

Processes & Service	Before	During	After
Promotion	Deliver/Receive	Deliver/Receive	Deliver/Receive
POS I (select)	Sell/Order	Sell/Order	–
POS II (pay)	Earning/Purchase	Earning/Purchase	–
POS III (proceed)	–	Preparation/Wait	–
POS IV (complete)	–	Deliver/Pick-up	Deliver/Pick-up
Feedback	–	–	Receive/Deliver

However, POS III (proceed) and POS IV (complete) are the stages for the service reaching out and to be 'felt'. The moments of contact make significant influences for the receivers to judge. Those POSs therefore should be carefully considered when depending on word-of mouth and developing both return and new customers.

Line Management. One of the techniques for a big event to lessen the big crowd is to have multiple lanes and keep the movement flowing. The more lanes you have, the faster you will be able to get attendees enter and pass through your event. Several vendors in food courts setup multiple check-in lanes similar to large events. However, food court is not the entrance. Keeping lines to a manageable length and to wait time low are the key points.

When utilizing mobility service, prepare for variations in crowd fluctuations by keeping online and offline services in chronological order. Make several changes in the fixed layout of the venue. It would be very helpful to direct patrons where to enter and exit. By establishing this, efficient online information will help those with reserved (online purchase) tickets options by having a separate fast track lane. It is important to set expectations, so users are informed ahead of time of what time they need to arrive at the venue. Clear on-site information will keep those buying on-site from slowing down the line of pre-purchased ticketholders. Preparing a pleasant and manageable waiting area can ease hassles as well. That is to say, providing an inviting way to enter is as important as providing a good way to communicate the exit. In order to keep the patrons flowing through the space, we need to pay attention to bottle-neck areas and to make sure they do not create traffic jams. Technology cannot make up for a poorly organized space. However, well designed signage can appoint a traffic controller, with which to assess the entrances and exits, and assist users to figure out the best way to keep the line and/or lines defined and moving. We want all users stay comfortable and safe. When it works well, the users often know what is going on without having to ask.

5 Conclusion

Through visual analysis and information design thinking process, this article investigates targeted users' interactions and behaviors through onsite observation and survey interviews. It establishes a source for discussions and creative brainstorming at later stage to solve current problems. Besides crowd flow and lines, the three most desirable improvements are adequate information between retailers and buyers, quality service flow to make operating process easier and purchasing experience smoother, and enjoyable environment for working and dining. The objective is to reshape the quality service flow and to design the adequate information to provide users with an environment filled with easy to understand information. Mobility service and line management are raised to facilitate an enjoyable market place and dinning space in a community/campus setting, inspire motivation in a working environment and to use technology better.

The existing systems are either on paper or a semi-manual system. Most business information, such as shifting and communication, is generated by employers. Utilizing the new concept of mobility service, retailers can save money and time, eliminate arithmetic errors, easily track single money being spent, and make the interaction between clients and staff better. From the consumer point of view, the new system promotes simplified ordering and interaction between retailers and their clients.

In recent years, online platforms have been established in many restaurants to lessen operating costs, to grow profits, and most importantly, to gain a better reputation. New service for a community setting will benefit the larger population. It will require a multi-tier architecture equipped with standalone software with different modules, with which to handle staff activities. A friendly web service as well as mobile application will take care of clients. Delivery service like Uber Eats, Seamless, and GrubHub are populating in the delivery market. They are good examples that force us to think about expanding and developing collaborative services, such as senior friendly program, for a friendly and more sophisticated community.

This experiment establishes a new service model for sharing, to find a way to organize and to delineate invisible information such as service and experiences, and to present the information visually. Good service needs a friendly interface to deliver. The amiability is constructed by design to be (1) systematically inter-connected and functionable, (2) easy for touch screen displays, (3) easy to learn and to use, and (4) multi-modeled for various users with multiple views. In order to make life easier and users' experience more enjoyable, design should not only make things beautiful, but functional and inclusive. Applying those efforts, we understand the relationship between the community and its users, discuss the measurement and the attributes of information, and synthesize the ways communication with visual elements and visual forms provide better services. It is a milestone for further study that will verify the designed model and the tool kit for better communication.

Acknowledgments. Thankful for the opportunity in participating the Sustainable and Innovative Community Project from Ministry of the Interior, invited by Taiwan Tech. With the inspiration and motivation, the research received supports from the Ministry of Science & Technology, ROC, Taiwan, project MOST 106-2410-H-011-025. The appreciation also extends to VIDlab project assistant, Ching-chun Shen, who helped up in data collection and visualization. A special thanks goes to my dear colleague Sharon Placko-Steckel, who like to read my articles, listen to my stories, and support me without any hesitation.

References

1. WAYMO/Google. https://waymo.com/ (2018). https://www.google.com/selfdrivingcar/. Accessed 23 Feb 2018
2. Falkan, K.: Design History: Understanding Theory and Method, p. 55. Berg, New York (2010)
3. Arnheim, R.: Art and Visual Perception: A Psychology of the Creative Eye. University of California Press, Berkeley, pp. 44, 46 (1974)
4. Summers, D.: Real Spaces: World Art History and the Rise of Western Modernism. Phaidon Press, London (2003)
5. Krämer, S.: Trace, writing, digram: reflections on spatiality, intuition, graphical practices and thinking. In: Benedek, A., Nyíri, K. (eds.) The Power of the Image: Emotion, Expression, Explanation, p. 6. Peter Lang, New York, Frankfurt am Main (2014)

6. Hyerle, D.: Visual Tools for Constructing Knowledge. Association for Supervision and Curriculum Development, Alexandria (1996)
7. Kotler, P.: Marketing Management: Millennium Edition, 10th edn. Prentice-Hall, Upper Saddle River, (2000)
8. Engel, J.F., Blackwell, R.D., Kollat, D.T.: Consumer Behavior, 3rd edn. The Dryden Press, New York (1978)
9. Nicosia, F.M.: Consumer Decision Processes: Marketing and Advertising Implications. Prentice-Hall, Englewood Cliffs (1966)
10. Milner, T., Rosenstreich, D.: A review of consumer decision-making models and development of a new model for financial services. J. Finan. Serv. Mark. **18**(2), 106–120 (2013)
11. Milkman, K.L., Chugh, D., Bazerman, M.H.: How can decision making be improved? Perspect. Psychol. Sci. **4**(4), 379–383 (2009)

Usability and Interaction Evaluation on Breakfast Delivery Mobile App: Users' Experience Expectations

Marcelo A. Guimarães[✉] and Adriano B. Renzi

SENAC – Serviço Nacional de Aprendizagem Comercial,
Interaction Design Program, Rua Santa Luzia 735, Rio de Janeiro, Brazil
maguimaraes@gmail.com, adrianorenzi@gmail.com

Abstract. The experimental project Fornada, a system for breakfast delivery, is part of the post-graduate interaction design program at Senac, which encourages students to work with project development, and its several research stages, considering market's opportunities, society's needs and users' expectations. This paper describes, based on users' experience journey planning, the usability evaluation phase, using the Think-aloud Protocol method. The method execution is detailed, following Villanueva (2009) and Renzi and Freitas' (2014) directives. Results are presented following the sequential interaction with a mobile within the experience journey.

Keywords: Human-computer interaction · Usability · User experience
Think-aloud Protocol

1 Breakfast Rituals and Bread Consumption

According to Mortimer [1], breakfast, if compared to other eating rituals such as dinner and banquets, has no literature or chronicles about. Historians surface that, on the final period of the medieval era, most people did not have breakfast. Housekeepers from noble families would specify who was authorized, and who was not, to have breakfast. The author shows some examples: in 1412-13, in the house of Lady Alice de Bryenne, only 6 of the 20 residents were allowed to have breakfast; The Dutch of York, of the Cicely house, extended the privilege of breakfast only to the principal officials, such as ladies, the dean, the chaplain, the admiral and the marshal, when present; In the black book of Edward IV, is given special attention to who had permission to have breakfast.

From historic document findings, it is noted that throughout the centuries, the breakfast was considered a privilege, until the industrial revolution turned it into habit. According to eLondon, the traditional English breakfast became a fast fuel for the working class, who consumed an average of 3000 calories each morning, which would be spent throughout the forenoon until lunch break. The gradual change of the morning eating habits of the migrating workforce, from country to the city, reverberated to health effects, as the use of boiling water to make coffee or tea, diminished the incidence of disease.

© Springer International Publishing AG, part of Springer Nature 2019
T. Z. Ahram and C. Falcão (Eds.): AHFE 2018, AISC 794, pp. 564–573, 2019.
https://doi.org/10.1007/978-3-319-94947-5_57

Eventually, the breakfast became part of most social rituals and presents in its preparation and its nutritional composition, diverse meanings to different cultures. Drouard [2] points out the cultural French breakfast resistance against the more nutritious Anglo-Saxon choice, as the traditional café latte, croissant and butter marks their cultural identity.

The breakfast rituals in Brazil came with the Portuguese colonization and throughout the years became a mix of many cultures, encompassing its European heritage with natives, Africans and immigrants from many parts of the world. Bakeries, mainly related to Portuguese and Italian immigrants, are the main provider of nourishments related to Brazilian breakfast and are present in almost every corner of any city. The bakeries' structure, products and social part in the neighborhood can be related to the Portuguese *padarias*, the Italian cafes and French *boulangerie patisseries*. As a tradition, Brazilians tend to walk to the nearest bakery (*padaria*) and buy the minimum amount of fresh products related to breakfast for immediate consumption in their residence.

According to the Brazilian Bakery and Boulangeries Association (ABIP) [3], during the first 14 years of the 21st century, bakeries throughout the country evolved radically in their business management, bringing innovation to the business and new products to attend consumers' new expectations. It soon became the second largest food distribution network. ABIP surfaces the importance of the sector in Brazilian economy:

- Bakery business represents 36% of all food industry
- 60,000 bakeries, of a total 63,200 in the country, are small business with an annual total income of 84,7 billions of Reais
- The sector generates an average of 700,000 employments, in which 245,000 are directly involved with production.
- 76% of Brazilian consumes bread on their breakfast and 98% buy baked goods for general consumption
- The French bread represents 52% of all bread and its consumption per capita is 22,61 kg per year
- The diversity of products offered in bakeries is smaller only to super markets
- Bakeries represent 79% of food produced in the sector, industrial bakeries represents 14% and super markets, only 7%
- With the increase of wheat costs in Brazil, owners are investing in alternative sources, like corn and manioc

The association (ABIP) conducted a quantitative research regarding bread consumption and bakeries, surfacing the ten most common reasons Brazilians prefer to purchase bread and related products directly from bakeries. Home proximity and work proximity are the top reasons, corroborating the social ritual of going to the nearby bakery to buy fresh bread. Hygiene and cleaning conditions, good service, product quality and variety of bread are also well regarded by the interviewed, although in numbers far bellow the first two reasons (Fig. 1).

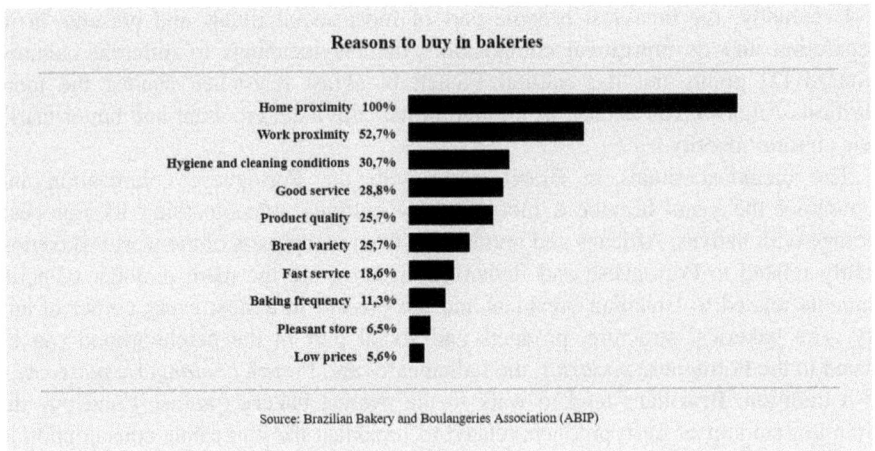

Fig. 1. The top 10 reasons Brazilians buy breakfast products at *padarias*. Source: ABIP.

2   Digital Food Delivery

Although telephones are still majorly used to order food, the delivery service has been increasingly adopting digital systems as basis for expanding business. For instance, the largest on the Brazilian market, IFood, gathers over 1,5 thousand restaurants from 15 states of Brazil. The app connects the two sides of the food business: the producer and the consumer.

Data released from the Brazilian Bars and Restaurants Association (ABRASEL), shows growth of revenues from the food delivery apps reaching 1 billion per year, with a 12% annual increase and expectations of attaining over 10 billion of Reais this year. The growth is so attractive that draws attention and investment from companies well established in other sectors, such as Uber Eats. For restaurants, the association with delivery apps represents a marketing tool and the possibility of increasing distribution and sales.

Even with the noticeable growth of the bakery business and its cultural importance to Brazilian breakfast ritual, food delivery apps in Brazil concentrate their efforts mostly on other options: pizzas, sandwiches, Japanese cuisine, finger food, lunch break and dinner.

From the disassociation of food delivery from breakfast as a business opportunity, a quantitative research showed users are interested in having a more indulged breakfast into their routine. The collected data on users preferences and expectations was the first step for further investigation and basis for the project proposition of Breakfast Delivery: Fornada.

The project is part of the post-graduate interaction design program at Senac, which encourages students to work with project development on several research stages, considering market's opportunities, society's needs and users' expectations. The whole project experiment program intends to take students through different phases of developing interactive projects, allying research with industry practices. It can be resumed in 4 major phases: (1) understanding market and society, (2) mapping users' mental model and needs, (3) structuring and developing project prototype, (4) UX testing and adjustments. This paper presents the development and results of the prototype UX testing, based on research and method directives from Villanueva [4], Renzi and Freitas [5], Renzi [6] and Vermeeren *et al.* [7].

Parallel to identifying project opportunities and companies with similar services (BedandBreakfast–London, Brekkie—Bangalore, The Breakfast Company–United Kingdom, Breakfastbay–Gurgaon), a quantitative survey with 416 respondents and affinities diagram helped identify users' mental model and categorize them in 3 user types (personas):

- A business executive, with little spare time in his/her mornings, likes the idea of scheduling breakfast products to optimize his/her time management. He/she will have breakfast in his/her office from time to time.
- A fitness person, who prefers fresh products in the morning, selected by their nutritional value. He/she is concerned with his/her daily amount of calories, sugars, gluten, fat and nutritional information.
- A Bon Vivant young person, who doesn't follow a strict schedule for breakfast and likes to have a many options to choose and gastronomic visual appeal.

The project proposal (Fornada) is based on the idea of an ecosystem [6, 8, 9], where consumers interact using a mobile app from one side and bakery managers interact using a PC or tablet on the other side. The Fornada system aims to bring the food delivery business to the breakfast rituals, linking users, interested in a broader diversity of products delivered to their home or office, to bakeries, interested in expanding the reach of their business. According to the applied survey, Bed & Breakfast and 4–5 star Brazilian hotels' breakfast are a reference for users' expectancies regarding breakfast delivery service.

The planning of the project needs to go beyond the direct human-computer interaction and understand the whole user experience journey, as it is pervasive to artifacts, environments, systems and actors [10]. Although user experience is subjective, and subtle to individual variations and external factors, the mapped users' preferences founded decisions on information architecture, interaction flow and prototype, to propose a main UX journey (Fig. 2): (1) the user wakes up; (2) if with breakfast deliver in mind, opens Fornada app, checks delivery time, bakery choices and orders breakfast; (2b) bakery manager receives and check information on PC or tablet; (3) user prepares his/herself for work on week days, or extends bed time on weekends; (3b) bakery staff prepare the breakfast and send it to user's address; (4) user checks the status of his/her purchase; (5) user receives breakfast at door; (6) user eats breakfast; (7) user rates service in the app.

Touchpoints are the users' interactions with the ecosystem [6, 8, 9] to build their experience. From mapping possibilities of the pervasive experience, the main touchpoints are expected to occur: when users use the app to select and order breakfast, when users go back to the app and check the order status, when receives the order at home and, if chosen, when users go back to the app to rate de service. Although not a direct touchpoint, the reception of the breakfast order by the bakery manager influences the experience indirectly.

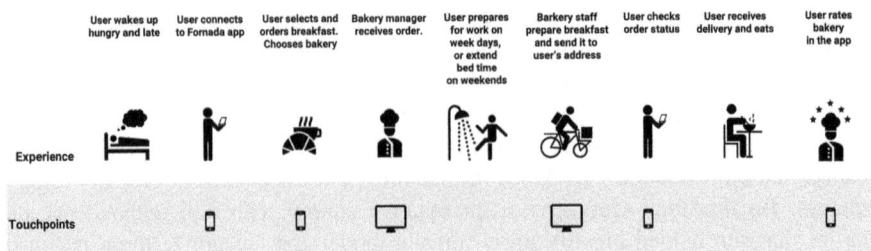

Fig. 2. Planned user experience journey and the touchpoints within the ecosystem.

Understanding the full pervasive experience and possible touchpoints, helped determine directions for the structure of the Fornada app and develop a prototype. For the prototype usability testing, the Think-aloud Protocol method was executed with real users, in order to better point interaction problems with the project.

3   Think-Aloud Protocol

There are different ways of testing the usability of systems. Veermeer et al. mapped and categorized 96 methods, that could be applied during different phases of a project. Even though the project was planned with theories of cross-channel pervasive experience in mind, the usability testing was focused on the use of the system, from consumer's side, with a smartphone.

From all possibilities of prototype testing, the Think-aloud Protocol is the most suitable choice to help observe usability and interaction problems from users' point of view. According to Villanueva [4], the technique consists of a researcher observing 1–4 users doing specific tasks within a controlled environment. The user's actions and thoughts are to be described verbally aloud by him/herself on real time. The researcher records the user actions by written notifications, video or voice recorder.

Filming and voice recording have the advantage of capturing the exact steps and descriptions of users, while written notifications depends on the researcher experience with observing reactions and quickness in writing down relevant actions of the experiment [5]. For this research, the chosen direction was to use voice recording and written notifications, in order to maintain users in an informal environment.

When noticing some reluctancy from users in verbalizing actions and thoughts during the Think-aloud Protocol, questions related to the users' actions were placed to keep the flow of verbalization of their thoughts [5, 11]. It is important to prepare a set of tasks for users to accomplish, in order to simulate the use of the system and check specific features, specific interactions and action recognition.

Four users were selected, based on the 3 persona profiles, and invited to participate in this research phase. The participants were 3 women and 1 man with age range between 28 and 35 years old. All of them consider breakfast an important meal in their daily routine and declare not to miss one.

The participants were to perform a simulation of selecting and purchasing breakfast using the Fornada app. The Think-aloud Protocol was executed individually and the duration of each session was between 10 to 16 min. The steps of the task to be closely observed were: do users understand what is the app? Do users cognitively understand icons and actions to take? What is the visual perception of users regarding colors, types, size, icons and visual impact? Is the process of selecting and ordering a choice of breakfast an easy and smooth experience? How do users explore the options and follow the order status?

Before each session, participants were given the basic information about app's purpose of breakfast delivery and the task to perform, but nothing further. Each user should start at the entrance page for login (simulating a first time interaction) and questioned about the visual impact of the product's brand and first impressions. From there, users should browse through breakfast options and select one for ordering and editing, if necessary. After selecting, users chose the bakery to deliver. There are 2 available filtering options to help users select the source of breakfast: geographic location and rating points. From this point on, users can check information about the selected breakfast. After choosing the bakery, users can select to have the breakfast delivered immediately or schedule it. Sequentially, users should confirm the order, pay and finish.

4   Results

As expected, with the use of Think-aloud Protocol, the 4 participants showed similarities in interaction, affordance perception and expectancies using the app. This paper presents the most relevant comments, in sequential order of the interaction process of the proposed task.

4.1   Entrance

All participants expressed positive reactions to the visual impact on the first contact with the app (Fig. 3). The choices of color and the logo were considered well representative of the breakfast culture. Three of them reported frequent use of tutorials in form of slides when using a system for the first time and gladly would search for the possibility in the app, if available. Concerning the options of login to the system, half of participants prefer to use facebook and the other half prefer to use their gmail.

4.2   Breakfast Menu

Entering the breakfast menu to browse options was natural and without interaction doubts. There are 4 breakfast options in the prototype and ¾ of participants explored all of them using the arrows for sliding the options (Fig. 3). One of the users suggested there should be a description of breakfast items under each photo option and reached using scroll down. Although visually representative of the each breakfast option, it was pointed out that is not easily perceive the quantitative differences between breakfast option just by looking at the photos: "the option *larica* certainly has more items than

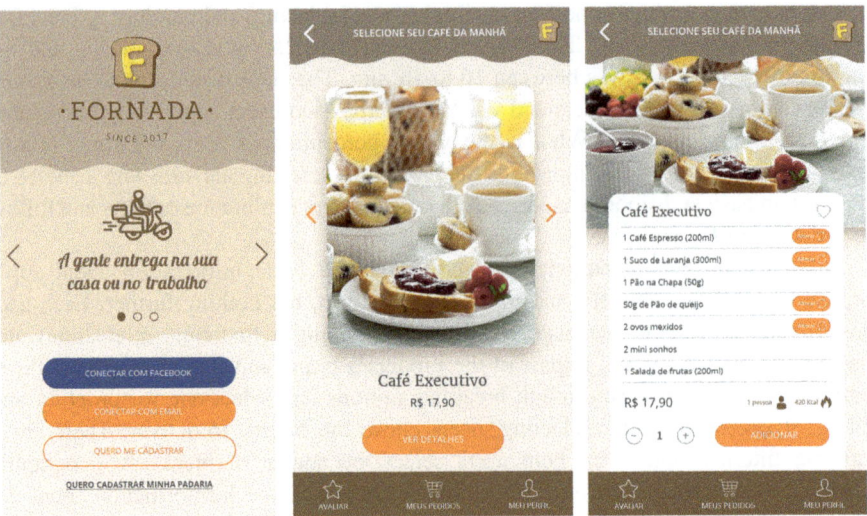

Fig. 3. Entrance page, breakfast menu page and breakfast details and editing page

the option *excutiva*, but the photo doesn't communicate the quantity, nor volume of items". Also at this point, a participant questions about how to choose the bakery.

4.3   Breakfast Details and Editing

When clicking for details on a breakfast option, users access its description and detailed items with the possibility of changing some of the items. The description includes the ideal number of people for the chosen option and its calories information. After editing any necessary items, users can select the option using the continue button and place the order (Fig. 3).

Three users did not understand that the flaming icon referred to calories and perceived it as an indication of how spicy the breakfast was, as a natural correlation to this kind of representation on restaurants' menus. Half of users thought the size of fonts were too small and had to pinch zoom to read the information.

The editing of items was perceived immediately by half of participants and easily done by clicking a small button on the side of each item and selecting the available options for replacement. But one participant was not sure how changing an item would affect the final price. The same one brings back his doubts of when would be the time to choose the bakery, referring to his experience with IFood: "when thinking of food delivery, I relate to my experience with IFood, where a list of restaurants comes before the meal options"

4.4   Bakery Selection

After selecting the breakfast, users are directed to the bakery selection page, where they can select by home proximity, using a location map, or by rating points, shown in the map. The map is centered on the users' location and shows all registered bakery options in a 3-kilometer radius, that can be zoomed in or out. Each bakery option shows its rating points and can be clicked for more details. In this page, users can still change the breakfast order or add a new one (Fig. 4).

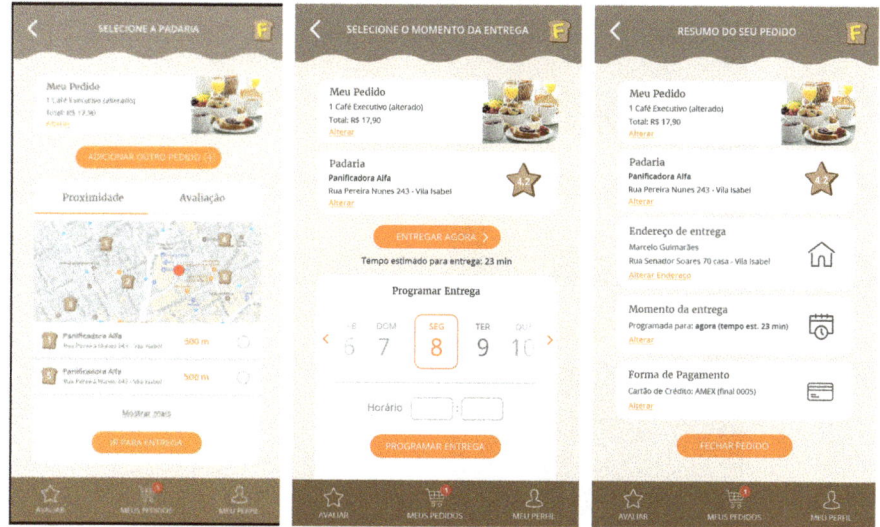

Fig. 4. Bakery selection page, Immediate or scheduled delivery and breakfast order final confirmation.

All participants were surprised by the interaction sequence and the project's business model of presenting the same 4 breakfast options, independently of chosen bakery: "I though it was weird not to choose the bakery first, but it is much more practical to choose a breakfast first, and then pick available bakeries"; "The app functions inverted if compared to other food delivery apps, where each supplier presents its own menu. Here is the opposite: the breakfast menu is already set and each bakery has to suit to the 4 options". Three of the participants declared that this business model is more practical and helps to save time.

After selecting the bakery, users can select either "deliver now" or schedule a delivery date and time (Fig. 4). All participants thought that the possibility of scheduling the delivery was really relevant. Two of them added the possibility of using the app to send breakfast to a different address and surprise a friend or a relative. Clicking on either option brings up a modal screen with details of the order for confirmation, or final editing (Fig. 4). One user asked if the payment would occur on this step.

4.5 Status of Delivery

When confirming the order info, the purchased breakfast info goes to the chosen bakery system. Users can follow the status of their order in real time (Fig. 5). All users understood and approved the feature.

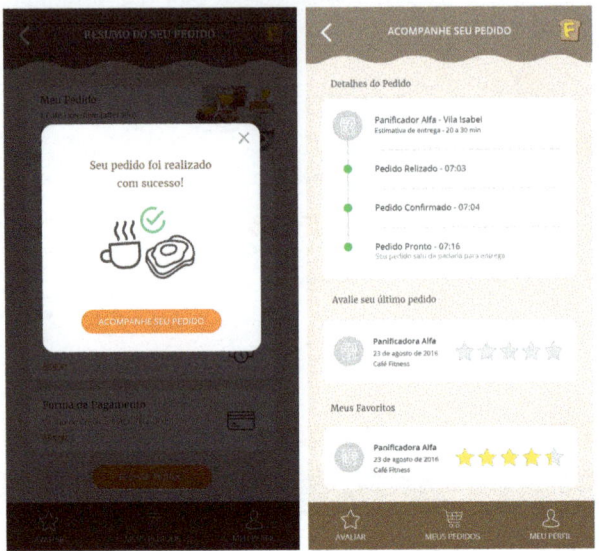

Fig. 5. Following the breakfast status

5   Conclusion

The usability testing using Think-aloud Protocol with real users helped validate the project proposal, most of its interaction flow and features in the app. Choosing participants related to representing profiles of users, based on previous quantitative research and their mental models, was crucial to observe real interactions and comprehend users' flow in the app.

The projection of a pervasive experience journey for UX projects is essential to map all possible touchpoints with the ecosystem. Determining the touchpoints throughout the experience journey can help clarify the premises of connecting with the system, the devices of connection, possible interoperability, context of use, sense of use within each context, the parts of the journey that develops without the influence of users and the actors and spaces integrated to the journey. Understanding the whole, helps determine the possible role of the parts and how each short story can help build a blended experience.

In this project, mapping the whole journey influenced choices of structure and task flow for the app. And, as a consequence, the collected testing results using Think-aloud Protocol, validated decisions with just minor adjustments for the system. The participants expressed the interaction with the app feels natural and cognitively obvious. Even when coming across an interactive flow that differed from their cultural conventions. Users made sense of it and understood the change as an improvement to their needs.

The few proposed adjustments for the app, based on the Think-aloud Protocol results, are quick to manage and does not affect the UX journey nor the planned structure:

- the possibility of a slide tutorial in the entrance page, specially for first time users;
- make sure it is clear to users that selecting a breakfast option comes before the selection of bakery in the task flow;
- substitute the calories icon;
- increase the font size.

References

1. Mortimer, I.: How the Tudors invented breakfast. BBC History Magazine (2003). http://www.historyextra.com/period/tudor/how-the-tudors-invented-breakfast/
2. Drouard, A.: Naissance et evolution du petit déjeuner en France. Cahier de Nutrition at diétetic **35**(3), 167–171 (1999)
3. Associação Brasileira da Indústria da Panificação. http://www.abip.org.br/
4. Villanueva, R.: Think-aloud protocol and heuristic evaluation of non-immersive, desktop photo-realistic virtual environments. Dissertation (Master of Science) - University of Otago, Dunedin - New Zealand (2004)
5. Renzi, Adriano Bernardo, Freitas, Sydney: Affordances and gestural interaction on multi-touch interface systems: building new mental models. In: Marcus, Aaron (ed.) DUXU 2014. LNCS, vol. 8518, pp. 615–623. Springer, Cham (2014). https://doi.org/10.1007/978-3-319-07626-3_58
6. Renzi, A.B.: Experiência do usuário: a jornada de Designers nos processos de gestão de suas empresas de pequeno porte utilizando sistema fantasiado em ecossistema de interação cross-channel. DSc. thesis. 239 p. Escola Superior de Desenho Industrial. Rio de Janeiro, RJ (2016)
7. Vermeeren, A.P.O.S., Law, E.L.C., Roto, V., Obrist, M.H., Vananen-Vainio-Mattila, K.: User experience evaluation methods: current state and development needs. In: Proceedings: NordiCHI 2010, October 16–20 (2010)
8. Resmini, A., Rosatti, L.: Pervasive information architecture – Designing cross-channel user experiences. In: Morgan Kaufmann – inprint of Elsevier, Burlington, MA (2011)
9. Renzi, A.B.: Experiência do usuário: construção da jornada pervasiva em um ecossitema. In: Proceedings SPGD 2017, vol. 1. Rio de Janeiro (2017). https://www.even3.com.br/anais/spgd_2017/59572-experiencia-do-usuario-construcao-da-jornada-pervasiva-em-um-ecossistema
10. Benyon, D., Resmini, A.: User Experience in Cross-channel Ecosystems, that paper was presented at the 31st British Human Computer Interaction Conference, Sunderland, UK, 3–6 July (2017)
11. Xiao, D.Y.: Experiencing the library in a panorama virtual reality environment. Library Hi Tech **18**(2), 177–184 (2000)

Automated Smartphone Keyboard Error Corrections

Vered Aharonson[1,2(✉)], Rotem Rousseau[3], and Eran Aharonson[3]

[1] Department of Electrical Engineering,
Afeka Tel Aviv Academic College of Engineering, Tel Aviv, Israel
[2] School of Electrical and Information Engineering,
University of the Witwatersrand, Johannesburg, South Africa
vered.aharonson@wits.ac.za
[3] Department of Software Engineering,
Afeka Tel Aviv Academic College of Engineering, Tel Aviv, Israel
erana@afeka.ac.il

Abstract. The usability of smartphone's touch keyboard is often hampered by typing mistakes resulting from the small size of the virtual keys relatively to the user's finger size. Although this problem has been addressed in various methods, an optimal solution in terms of both accuracy and user experience, however, has not been achieved. We developed an algorithm that predicts users typing intentions based on a statistical geometrical modeling of the touch points area. The algorithm builds a user-adaptive virtual location of the key based on deviations probability computation. An uncertainly measure activates a language statistics engine to enhance the prediction. The algorithm was integrated into the default Android® keyboard and was tested on users. Typing error rate using the implemented algorithms was reduced by 23.1% on average. The proposed method can enhance typing accuracy and user experience and may facilitate and improve the design of smaller and cheaper touch based smartphones.

Keywords: Smart keyboard · Enhanced typing experiences
Typing error corrections · Location statistics

1 Introduction

A vast majority of smartphones applications use text inputs and entail typing. The user experience when typing on a smartphones' virtual keyboard is often hampered by typing errors. One reason for these errors is the small size of the keys relatively to the user's finger [1, 2]. Another may be shifts in the user's finger placement due to eye – finger coordination [3–5].

A prevalent solution to this problem, which is implemented in many smartphones is a language-based auto-correction of the errors [6], which predicts the letter typed according to its highest probability in the word or sentence (Fig. 1).

Many users, however, find this solution irritating, since the predicted word often does not match the user's intention, especially when it comes to non-vocabulary words like names and abbreviations, or to uncommon languages [7]. Although this technique

© Springer International Publishing AG, part of Springer Nature 2019
T. Z. Ahram and C. Falcão (Eds.): AHFE 2018, AISC 794, pp. 574–580, 2019.
https://doi.org/10.1007/978-3-319-94947-5_58

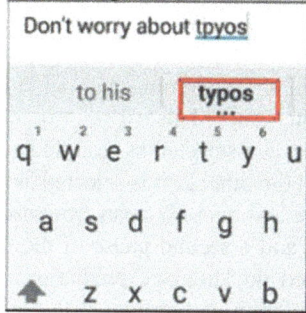

Fig. 1. Illustration of a keyboard language based auto-correction

improved considerably over the years, especially where a user's language adaptation was implemented, (i.e. SwiftKey®, Google® keyboard, Fleksy®, Swype®, Minuum®) these limitations have not yet been fully resolved.

Recent studies explored the association between the geometrical properties of a user's touch points and the intended key. Azenkot et al. [4] investigated the average deviations between the positions of each key to the users' touch points. Their study illustrated various patterns in the offsets between the different touchpoints and their intended keys. The authors proposed a method called 'Remulation' for real time corrections based on their users' studies [8]. The method is based, however, on prior user's typing patterns database and is not personalized to a specific user.

The goal of our study is to develop and implement an algorithm that can provide typing accuracy while maintaining the user experience. The algorithm is user adaptive and does not necessitate training.

2   Methods

Our design entails an adaptive keyboard that adjusts for key offsets in the individual user typing. The keyboard learns the association between the user's touch locations and the intended key. A hybrid configuration design combines this geometrical analysis of the user's touch point and a language analysis of the typed keys. An unsupervised learning design does not require prior knowledge or data base.

2.1   Algorithm

An The algorithm produces an "intended key" estimation for each location of the user's touch, denoted Touchpoint. The computation is based on a first order Markov chain model, updating for each additional touchpoint. The initial state, or TouchPoint, of the chain is the center of the keyboard key. A geometrical location estimation, denoted TouchMap, is computed as the weighted average of the current and last TouchPoints for each key (Eq. 1). The weights are heuristically chosen, based on the observation that the current touch points have more strength than past ones.

$$TouchMap = 0.9(TouchPoint_{current}) + 0.1(TouchPoint_{last}) \qquad (1)$$

The association of every new TouchPoint, with an "intended key" is based on the TouchPoint's distance from all TouchMaps:

The three smallest distances are selected as 'candidates'. If one of these 3 values is smaller by more than 5% from the other 2, it is selected as the "intended key". If two of the three or all three values are less than 5% apart from each other in terms of distance, the decision is of uncertainty and a second phase of the algorithm is activated

The second phase is based on language prediction: each of the "intended key" candidates is tested using its bigram and trigram with the previous letters in the typed word and the probability of these n-grams is retrieved from a database dictionary. The highest probability ranks the chosen candidate. If two candidate letters have identical probability, one of them is randomly chosen.

A flow diagram of the algorithm is presented in Fig. 2.

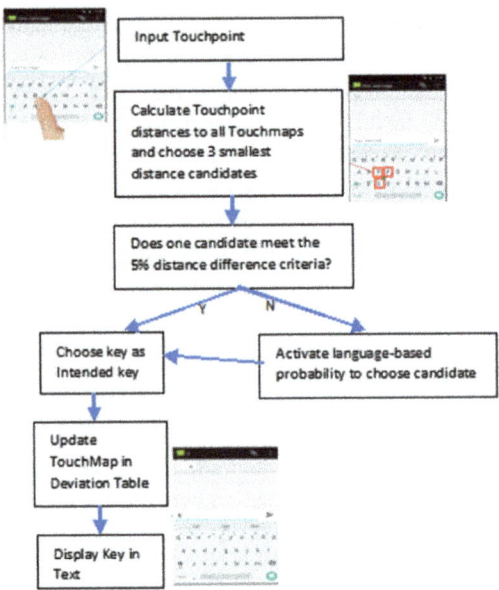

Fig. 2. Algorithm flow chart

2.2   Implementation

Signal The system includes three main components: a user interface, which is the original user interface of the smartphone's default keyboard, the algorithm's implementation and a database for later analysis and testing.

The implementation is designed to fit Android smartphones from version 4.1 or higher, which can accommodate the majority of Android devices available today.

The algorithm was implemented using Android Studio® and Google®'s open source default keyboard software.

The user interface is identical to the standard Smartphone keyboard except of an optional additional key which activate a new debug visual tool called TouchpointView. The tool allows a qualitative examination of the TouchPoints. As illustrated in Fig. 3, the additional key displays a circle and an "x" which lays over the keyboard a display of the actual touch points of the user's typing and clears them, respectively.

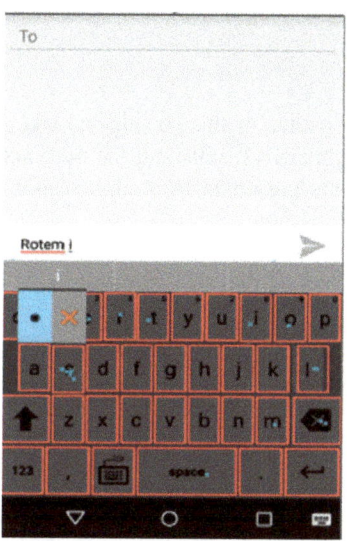

Fig. 3. TouchpointView tool display of the touch points (small blue dots) upon pressing the additional button (circle and "x").

The TouchpointView tool is also implemented as an option in the user interface, for users who wish to observe their own key touch distributions.

2.3   Performance Evaluation

Ten users participated in a performance evaluation experiment. The users were asked to type a set of 21 short sentences on an Android smartphone, using the default smartphone keyboard. The sentences contained 86 words and 405 letters which were uniformly distributed among the 26 English letters. The users did not see the result of their typing, except from an asterisk symbol indication that a letter was typed. The same experiment was performed twice, on two different smartphones, with different screen sizes. The Smartphones used were Galaxy S5 – as a large screen example (5.1 in.) and Posh Micro X SN40 – the smallest Android smartphone (2.4 in.). The keyboard key sizes are 182×108 pixels and 52×24 pixels. For the large and small screens, respectively.

The performance, in terms of error rate, was compared between the default keyboard and when running the algorithm. In order to evaluate the contribution of the language based key prediction to the basic geometrical model, the performance of the two versions of the algorithm: with and without the language layer were compared, for the two screen sizes. The performance measure was error rate: ratio of key errors to the number of TouchPoints, in percentage.

Error rates were compared between the two keyboard implementations: standard and enhanced, and between the two screen sizes.

3   Results

The Fig. 4 presents the error rates of the 10 subjects when typing on the two screen sizes. Both the standard keyboard's results and the ones acquired when implementing our algorithm ("enhanced" are presented for both screen's experiments.

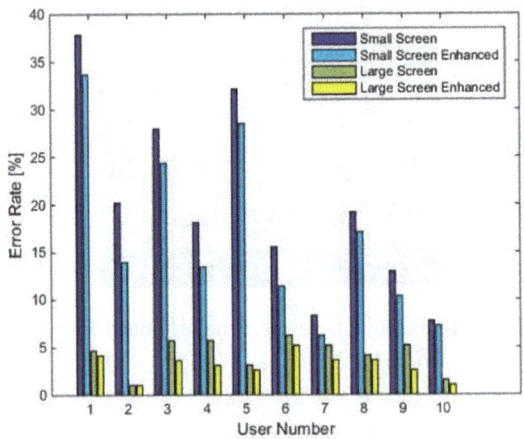

Fig. 4. Bar graph of the users' error rate for the two screen sizes and for the standard and enhanced keyboards implementations.

The error rates' means and standard deviations (SD) for the two keyboard implementations, standard and enhanced, and for the two screen sizes are presented in Table 1.

Table 1. Average users error rates.

Error rate [%] mean ± SD	Standard keyboard	Enhanced keyboard
Small screen	20.0 ± 9.9	16.63 ± 9.2
Large screen	4.2 ± 1.6	3.0 ± 1.1

It is obvious that the smaller screen induces significantly more errors. The results indicate that error rate was reduced when the enhanced keyboard was implemented, for both screen sizes. The reductions in error rate were 18.1% and 22.9% for the small and large keyboard, respectively.

A further analyzes examined which of the correction layers: geometric or language based contributed to the error correction. The results are presented in Table 2.

Table 2. Relative contribution to error rate corrections.

Error rate correction improvement [%]	Geometric correction algorithm	Language-based correction algorithm
Small screen	29.38	70.62
Large screen	77.50	22.50

The results indicate that the geometric correction is able to attend to more than 75% of the typing errors for the large screen, whereas in the small screen case, most errors necessitated an activation of the language-based correction.

4 Discussion and Conclusions

We presented an algorithm that could correct typing errors on touchscreen, primarily by examining the geometrical distribution of a user's TouchPoints and determining the user's intended key according to distances minimization. The algorithm is implemented in a hybrid structure, such that in case of uncertainty in the geometrical decision, a language-based decision is activated, which takes into account the letters bi-grams probabilities. The algorithm is continuously updating its decision with each TouchPoint employing an unsupervised decision logic, and therefore does not require any training by the user.

The experiment was performed on two screen sizes and indicated that the algorithm was able to substantially reduce the error rates in both screens.

When the contribution of each part of the hybrid system; geometric and linguistic on the performance was examined, a linguistic error correction was needed for 70% of the error corrections in the small screen as compared to 22.5% of the cases in the large screen experiments. As could be predicted, the uncertainly in the geometric decision is larger for smaller keys and hence a language based decision is required in most cases for the smaller screens.

The results imply that this enhanced keyboard is more accurate and more user-adaptive than the standard keyboard of Android smartphones available in the market today. Its implementation is transparent to the user and does not necessitate any user interface changes.

Healthy subjects' typing errors are not large in size and their rate, especially on large keyboards, is relatively small. A further study will examine if our system could be augmented to accommodate people with both vision and motor disorders or disabilities [9].

References

1. Sears, A., Revis, D., Swatski, J., Crittenden, R., Shneiderman, B.: Investigating touchscreen typing: the effect of keyboard size on typing speed. Behaviour & Information Technology **12**(1), 17–22 (1993)
2. Findlater, L., Wobbrock, J. O., Wigdor, D.: Typing on flat glass: examining ten-finger expert typing patterns on touch surfaces, pp. 2453–2462
3. Kwon, S., Lee, D., Chung, M.K.: Effect of key size and activation area on the performance of a regional error correction method in a touch-screen QWERTY keyboard. Int. J. Ind. Ergon. **39**(5), 888–893 (2009)
4. Azenkot, S., Zhai, S.: Touch behavior with different postures on soft smartphone keyboards, pp. 251–260
5. Nicolau, H., Jorge, J.: Touch typing using thumbs: understanding the effect of mobility and hand posture, pp. 2683–2686
6. Kukich, K.: Techniques for automatically correcting words in text. ACM Computing Surveys (CSUR) **24**(4), 377–439 (1992)
7. Tetariy, E., Bar-Yosef, Y., Silber-Varod, V., Gishri, M., Alon-Lavi, R., Aharonson, V., Opher, I., Moyal, A.: Cross-language phoneme mapping for phonetic search keyword spotting in continuous speech of under-resourced languages. Artificial Intelligence Research **4**(2), p72 (2015)
8. Bi, X., Azenkot, S., Partridge, K., Zhai, S.: Octopus: evaluating touchscreen keyboard correction and recognition algorithms via, pp. 543–552
9. Trewin, S., Pain, H.: Keyboard and mouse errors due to motor disabilities. Int. J. Hum Comput Stud. **50**(2), 109–144 (1999)

Virtual Reality and Interaction Design

In the Journey of User Center Design
for the Virtual Environment

Norma Antunano[(⊠)]

School of Sciences and Humanities and School of Business, University of
Phoenix, Phoenix, USA
normaat.ieee@gmail.com

Abstract. Research of interactions in virtual systems, including the value for
customers considering engagements, usability testing experiences, and the work
processed in virtual environments (including in academic environments and
transactional workflows) build up the motivation for this research. With digitization rising, our dependency on virtual or on-line interactions continues
expanding. According to Internet World State (2017), there are over 3.4 billion
worldwide web users, and English is the most common language (26%), followed by Chinese (21%), and then by Spanish (7.7%). More and more parts of
our daily life experiences are becoming digital, and promising integration of
virtual environments are motivating further growth. Such context is driving
significant changes in our human life ecosystems. The International Ergonomics
Association (2017) and ISO 6385:2016 (2016) characterize Ergonomics as a
discipline committed to understand the human interactions in a system. With
application of principles, theory, design methods, and data analyses a vision
with primary focus on the human experiences and well-being should be further
embraced, considering alternatives to optimize performance of the hosting
digital system as well. The insights from software developers on practices and
challenges addressing customer experience (including scenarios when customer
may not explicitly communicate his/her expectations), coupled with experience
using on line education systems and other virtual services are analyzed. In
support of this study, over 500 web sites representing variety of organizations
are explored with the goal of finding opportunities to improve the overall user's
experience. Through the study we identify factors driving web site performance
from the users' perspective; web designers and architectures should consider
these factors to drive excellence in the experience of the virtual journey.

Keywords: Virtual environment · Human factors · Context · Smart grid
Software developers · Systems engineering · User · Customer experience
Accessibility · Virtual ergonomics · Web site design · Web site performance

1 Introduction

The growing advantages of digitalization is prompting consumers, enterprises (of all
sizes, and practically in all the verticals), community focused organizations, and
government entities to gradually tap into digital capabilities. The customer's digital
experience impacts outcomes as the overall end to end value chains continue increasing

© Springer International Publishing AG, part of Springer Nature 2019
T. Z. Ahram and C. Falcão (Eds.): AHFE 2018, AISC 794, pp. 583–592, 2019.
https://doi.org/10.1007/978-3-319-94947-5_59

their dependency on virtual applications. Reliability and availability of digital systems are more explicitly impacting e-services, e-commerce or client's bottom line outcomes. In this study, performances of web sites from samples representing various types of enterprises are explored to identify web design accessibility design characteristics potentially leading to customer experience. As digitalization continues disrupting in the world, design and infrastructure performance, the accessibility of digital content in the web are becoming further relevant in our daily lives. The definition of accessibility is to make available of the web content to everyone, regardless of their ability conditions. Accessibility should not be associated with disable conditions. Even if only disable population may be accounted, ongoing studies find about 20% of the population has some kind of disability [35]. Disability may be cognitive, motion, visual or hearing based.

Web sites occasionally perform well in some dimension (for example responsiveness speed during the initial engagements), but the web site(s) may drop performance once users are in their journey. Some examples where web site performance drops, impacting customers' experience include: (a) slow processing of a transaction (the user has to wait for the processing of the system, (b) when the customers are directed to unresponsive locations or digital links that may not be available (sometimes these are broken), (c) when the system expels the user (or presents a banner with message like "reload", or it shows an error code number interpretable by the web developer or technical support services, (d) when the user has to go through different screens or views to find the information he or she needs to complete an action (designed as part of the work flow or driven by the particular process flows, terms of use or policies, or a procedures) to complete the journey or achieve the objective(s) that led her/him to the web site, and there could be other type of circumstances. The organizations marketing and selling the requested products or web services through on line transactions (such as on line retailing, on line service subscription, pay for consumption) tend to closely monitor the performance of the web site as their short and long term success have direct connection/relationship to the user's experience in their sponsored web site(s). However, when the on line system is not prepared to maintain its service availability (meaning readiness to act in a timely basis, seamlessly for the user with the vision of maintaining performance reliability) in a virtual interaction, the irregularities caused by design or performance of the service hosting platform and/or system surface up to the user through some unfavorable experience. This factor may be more vulnerable in sectors where high and dynamic volume of records and transactions are processed; in these cases the organization may miss the opportunity of addressing or correcting in timely basis no satisfactory user experiences. For example in an on line educational setting where students have to commit to a course at the time of registering, the on line support system of the course may present vulnerabilities after the student registered. Student may experience vulnerabilities such as frozen screen, slow responsiveness, time out, broken links and possible other circumstances through the course, surfacing up marginal or no adequate performance. These vulnerabilities can impact students' experience. Other areas where web performance vulnerabilities impact relevant aspects of life in user's experience include the growing digitalization of banking or health care management services. From the users' experience perspective, parts of the digital infrastructure or work flows have not been properly tested, and/or these may not be designed accounting for the human journey in the virtual web system. The systems supporting such important services in the life of the user, appear being driven by bottom line goals focusing on capitalization through digital transformations without carefully reviewing the impact of customer's experience. There are many other examples like the growing trend of implementing on line self-services, including payment for a variety of basic living services such as for housing, transportation, insurance, and more. Harper [10] has brought earlier the considerations of actual usability and accessibility into consideration when designing interactive web design on line services. This study explores how performance and accessibility of web based platforms may impact customers' experience and outcome results.

2 Background and Related Work

As far as usability, the International Standards Organization, ISO 9241-11/1998 defines usability as "the extent to which a product can be used by specified users to achieve specified goals, with effectiveness, efficiency and satisfaction in a specified context of use". With the significant growth of mobility the viewing experience of web pages in desktop computer screen can discourage the viewer (the user) because views appear to be designed for significantly smaller screen (for mobile devices). Other instance is when web sites load heavy images slowing down responsiveness while processing the requested service through the intended user virtual engagement. In other instances the digital or virtual location works well when certain browsers are used, but performs poorly using other browser(s). Many virtual applications have been designed to gather information in a very structured way (influenced at large by storage, architecture design, and traceability convenience). However, from the users' and overall value chain perspectives, this architecture can cause efficiency challenges, and potential unsatisfactory performance from the user's perspective because users need to feed information in the way the system expects (sometimes users have to manually enter multiple times the same information, and in top of this, occasionally a technical issue such as loss of server availability prevents saving of information already entered by the user on line, like it is the case of the information entered to process a transaction). Also, the information may be accessible in a language no well-known by significant part of the population. According to United States' 2003 National Center of Education Statistics (the most recent study performed by U.S. Department of education), there were 11 million adults non-literate in English when currently most of the on line information available in United States is in English.

Human readers tend to scan (browse) for information, browsing is different than reading printed materials. According to Weinreich et al. [39], in an average visit the user typically reads at the most 28% of the wording contained at the location being accessed. Nielsen [20] concluded that Web readers tend to follow predominantly an "F" reading pattern, per an eye tracking study following views of 323users on thousands of Web pages.

More than two decades ago Nielsen found that web site responsiveness was one of the key user's experiences; up to today this attribute remains as a leading customer experience performance indicator. Nielsen [20] and Krug [15] concur that the web page should be intuitive for the user, it should nurture an environment free to navigate, encouraged when it is easy to understand. In 2008, the Pew Internet and American Life Project estimated that 75 to 80% of users in the world have looked for health related information online, and 75% of online users with a chronic disease or a disability reported they have recently searched for health information. Moreover, such finding may influence how an illness or other condition may be treated as the patient may consider findings on line to describe his/her symptoms. Pew Research [25] estimates that two thirds of US population 65 years old or older use the internet, and more than 50% of the senior population use the internet. Pew's study shows upward usage trends of tablets, smart phones, and social media among this segment of the population (Pew Research 2016 study). Romano et al. [26] have studied the effect of age on Web usability, and on performance when tracking eyes' dynamics while using web content. They conclude that aging population tend to have more fixations at the center of the screen than younger people. In the retail space, a McKinsey's study conducted in 25 countries involving 25,000 respondents (February 2017) in France and in United Kingdom found higher groceries acquisitions on line than in United States. According to interview of Christian Wanner, pioneer of online groceries in Europe (Garcia Lopez and others' interview, McKinsey 2017): "Website ergonomics and transactional behavior are similar across geographic and cultures, so you can leverage similar system in different countries". Benda et al. [4] with focus on the use of portals by the general public have found that purpose and focus on the journey of the users are the key initial steps that will drive ergonomic quality experience for the average user. With respect Web responsiveness, since early nineties Nielsen [19, 20] has identified that a friendly web environment where the user can intuitively navigate without having to figure out what to do next (or how to do it) should be the vision of user experience designers; humans continue being better than machines for color image segmentation, including background and foreground separation. A study including texture, contours, and color segmentation found that the highest accuracy for color boundary detection (especially for gray images) comes from humans [27]. The studies from Nielsen, inputs from software developers (including micro service designers and web designers) continue finding that data processing driven by more complex tasks and/or widgets can influence web site responsiveness.

3   Analysis and Study

We used primarily Wave Aim™ and Web Accessibility Checker™ open tools to help assess accessibility of web sites, considering the web design accessibility principles in alignment to WCAG 2.0b (Authoring Tool for Accessibility Guidelines). This means that the content displayed by the Web site is expected to be perceivable, operable, understandable and robust. Data was gathered on over 500 public domain web sites using Google™ Chrome browser in laptop and desktop computers. The data gathered by the accessibility tools were standardized for the descriptive statistics and for the analysis. The web sites reviewed were classified in the retail, media, and non-profit organization segments. Over 90% of these web sites present the content in English, although in some cases the parent organization is based in a non-English based country like China, Germany, Sweden, Australia and other countries. As Spanish is the largest minority population in United States (US Census Bureau, [29]), our study also included a sub-set of web sites in Spanish; the difference on language did present significant performance differences. The performance metric was the number of errors identified by the tool(s) on each web site, and the corresponding type of the errors. The spread of the total number of errors reported for all the samples (without segmentation) is shown by Fig. 1. When a web site was not available for the assessment other web site was identified (either searching on line alternative web site from the same sector). To confirm repeatability and reproducibility of the initial performance results reported by the tool(s), performance was re-assessed later (in a month) for a small sub-set (50 web sites); performance results were stable and the same. The total number of errors was calculated adding the errors under the six categories: errors, alert errors, structural errors, HTM5 and ARIA errors, Contrast errors.

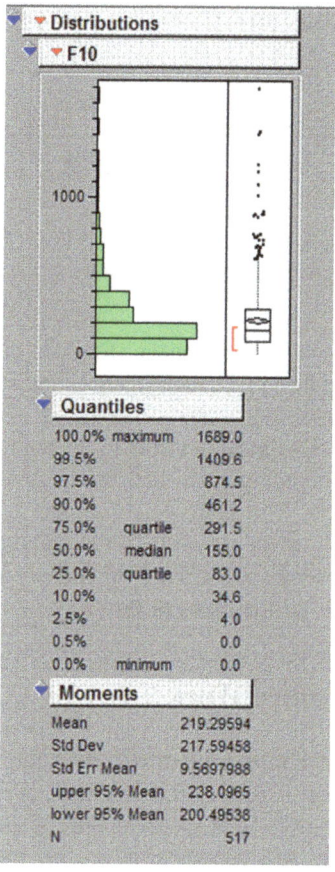

Fig. 1. Distribution of the total accessibility errors from 517 sampled web sites

ARIA stands for Accessibility Rich Internet Applications; it houses a set of attributes a web site is expected to have to be accessible (WCAG 2.0, Government wide accessibility program 508); it facilitates access when native HTMLs cannot handle it (it enhances semantics of non-semantic content, enabling description of the web page like heading, tables, zones), it provides status of widgets (like if there is a popup), and also facilitates drag and drops.

Contrast in this context refers to the contrast the foreground text is expected to have with respect the background colors (WCAG 2.0). Structural errors trigger structure of heading tabs in HTML, when errors are present these can cause accessibility with assisted technologies (like screen readers, head pointers or touchscreens). The spread of the total number of errors (name "Total") was driven primarily by Structural errors, followed by Alerts errors as shown by Fig. 2.

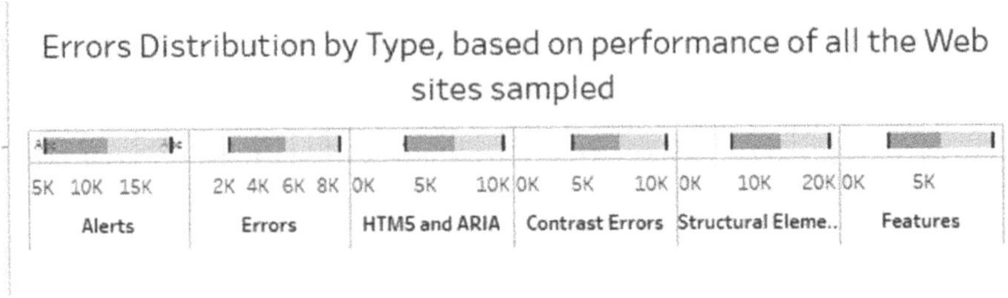

Fig. 2. Distribution of accessibility errors by type

Errors may be reported when the language (English, other) is not identified in the web page. Alerts highlighted design features (like headings used for formatting writing style) that may present accessibility issues.

Figures 3, 4 and 5 show the spread of errors by segment for the web sites of Media, Non-Profit organizations (including web sites from credited university educational institutions), and Retail sectors. The retail sector samples represent 68% of the 517 web sites samples, the Non- Profit organizations 27.46%, and the Media web sites represent 4.5%. The larger spread of errors was driven by Alert type of errors. Retail segment shows the highest spread for all the types of errors.

4   Conclusions and Future Work

Although the study was limited in number of web site sampled for evaluation and types of browser and device used for accessing the web sites, overall in the context of virtual experience these insights bring to light the dynamics of the type of errors influencing the users' experience. It improves our perspective over the contributing web accessibility factors adding up to the overall user's experience. Considering the limitations of the study, we could not find significant statistical difference on how each accessibility factor performs on the sampled web sites. Overall, this small empirical study supports what Nielsen, McKinsey and others have been identifying; geographical locations do

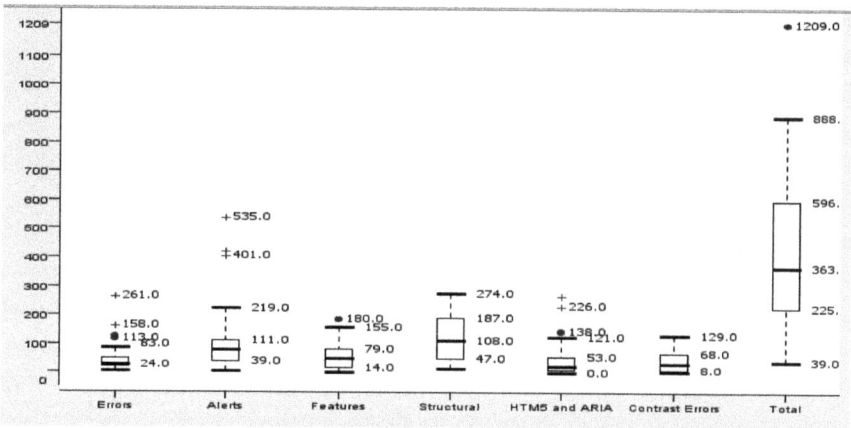

Fig. 3. Box plots by error type for media segment web sites sampled

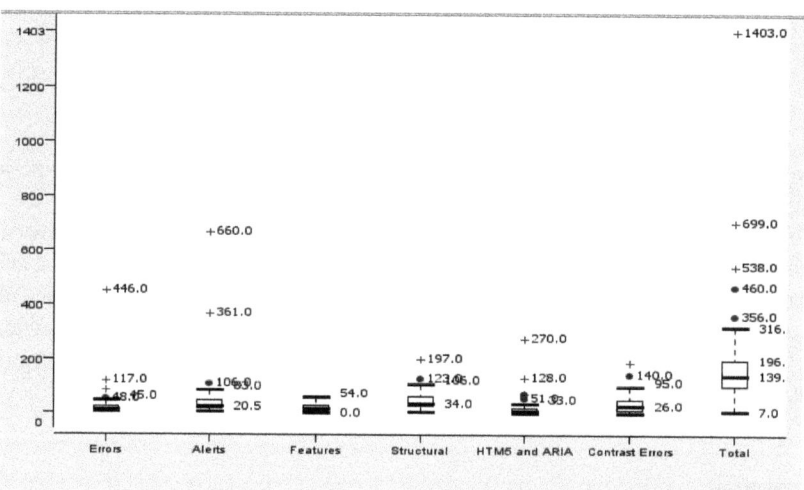

Fig. 4. Box plots by error type for nonprofit organization web site samples.

not matter. Recent Forrester's study [6] found strong relationship between the quality of customers' experience to profitability (over 5X better than the laggards based on information gathered from over 4000 professionals across various organizations and functional disciplines). Now, the focus on understanding users' journeys in the virtual environment appears to be priority (usability was priority in earlier years). As use of digitization continues rising, and overall demographics across the world continue adapting and/or embracing digital capabilities, relevance of how web sites are designed is becoming a stronger factor at personal and organizational levels. This implies the need of further elevate the focus on human factors with ethical responsibility of designers, product owners, web designers, software developers, marketing, sales, artists

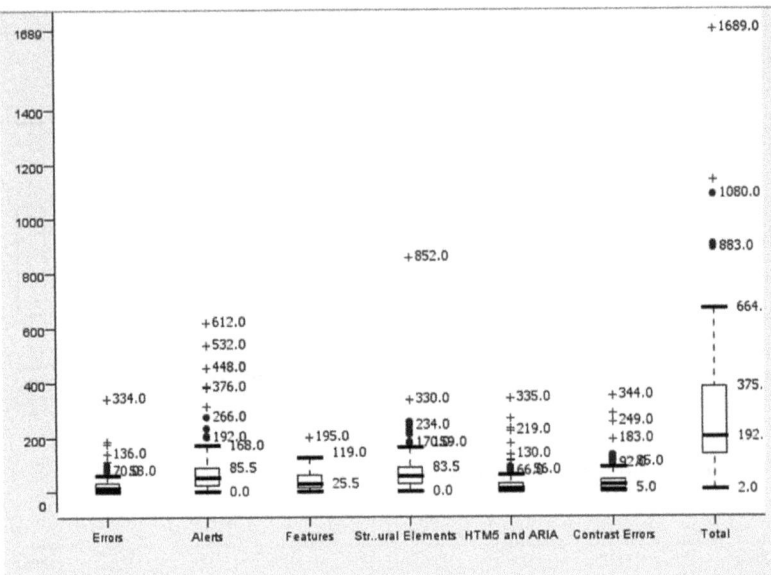

Fig. 5. Box plots by error type based on retail segment web site samples

and other dedicated professionals directly or indirectly impacting the experience through the journey in virtual environments. Operational optimization for the best possible outcomes has been a key priority for factories, facilities, cities, businesses and for other locations hosting human engaging settings, the virtual experience must also be optimized. Performance assessment of other 500 web sites (including a subset of the same web sites selected in this study) is part of our future work. We expect to add accessibility evaluation in small screens (smart phones and tablet devices) on a sub-set sample as well.

References

1. Accessibility Checker. https://achecker.ca/checker/
2. Allison, J.: Interaction in communication technologies and virtual learning environments: human factors. Q. Rev. Distance Educ. **13**(2), 125–128 (2012)
3. Antunano, N.: Ergonomic factors in virtual intensive interactions. In: Advances in Intelligent Systems & Computing, vol. 489. Springer, Cham (2016)
4. Benda, P., Šimek, P., Masner, J., Vaněk, J.: Analysis of eAGRI web portal ergonomics and presentation of information in terms of the general public. AGRIS On-line Pap. in Econ. Inform. **9**(4), 3–13 (2017). http://dx.doi.org.contentproxy.phoenix.edu/10.7160/aol.2017.090401
5. Fogg, B.J., et al.: Mobile Persuasion: 20 Perspectives on the Future of Behavior Change (2007)

6. Forrester: The US customer Experience Index, 2017, How Brand builds Loyalty with the Quality of Their Experience, 1 August 2017. https://www.forrester.com/report/The+US +Customer+Experience+Index+2017/-/E-RES136424?objectid=RES136424

7. French, T., Taylor, A., et al.: Developing a design brief for a virtual hospice using design tools and methods: a preliminary exploration, Copyright University of Cincinnati in behalf of Visible Language, pp. 97–111, April 2012

8. Garcia, L.E., Said, K., Westphely, K.: How to win in online grocery: advice from a pioneer, interview from December 2014 (2014). Published by McKinsey in 2017

9. Government Wide Section 508 Accessibility Program. https://www.section508.gov/

10. Harper, S.: Is there design-for-all. Univ. Access Inf. Soc. **6**(1), 111–113 (2007)

11. Harrati, N., Ladjailia, A., et al.: Exploring user satisfaction for e-Learning systems via usage-based metrics and system usability scale analysis. Comput. Hum. Behav. **61**, 463–471 (2016)

12. Internet Live Stats (2017). www.InternetLiveStats.com

13. ISO 9241-11: Ergonomics for Human Interactions (1998). www.iso.org

14. Jeng, T.: Interactive architecture: spaces that sense, think and respond to change. In: Gu, N., Wang, X. (eds.) Computational Design Methods, Technology Applications in CAM, CAE Education, pp. 257–273. IGI Global, Hershey (2012)

15. Krug, S.: Don't Make Me Think!: A Common Sense Approach to Web Usability, 2nd edn. New Riders Publishing, Berkeley (2006). ISBN 03-213-4475-8

16. Maher, M., Gu, N.: Designing Adaptive Virtual Worlds. De Gruyter, 4 June 2014. Available Pro Quest

17. Maher, M.L., Pauline, M., et al.: Scaling up from individual design to collaborative design to collective design. In: Proceedings of DCC 2010, Stuttgart, Germany, pp. 581–600 (2010)

18. Mauve©. http://giove.isti.cnr.it:8080/MauveWeb/

19. Nielsen, J.: Usability metrics and methodologies. ACM SIGCHI Bull. **23**(2), 53–69 (1991). ISSN 0736-6906

20. Nielsen, J.: Usability Engineering, 1st edn. AP Professional, Boston (1993). ISBN 01-251-8406-9

21. Nielsen, J.: F-shaped Pattern for Reading Web Content, 17 April 2008

22. Nielsen, J., Pernice, K.: Eyetracking Web Usability. New Riders, Berkeley (2009)

23. Nielsen, J.: Website Response Times, 21 June 2010

24. Pew American Life Project: The engaged e-Patient population (2008). http://www.pewinternet.org/topcis/Health.aspx

25. Pew Research Center, Internet and Technology: Tech Adoption Climbs Among Older Adults, 17 May 2017. http://www.pewinternet.org/2017/05/17/tech-adoption-climbs-among-older-adults/

26. Romano, J.C., Jans, M.E., et al.: Age-related differences in eye tracking and usability performance: webistie usability for older adults. Int. J. Hum. Comput. Interact. **29**, 541–548 (2013)

27. Saber, E., Vantaram, S.R.: An adaptive Bayesian clustering and multivariate region merging-based technique for efficient segmentation of color images. In: IEEE International Conference on Acoustics, Speech and Signal Processing (ICASSP), Prague, Czech Republic, pp. 1077–1089 (2011)

28. Taylor, H.A., et al.: Implementation of a user-centered framework in the development of a web-based health information database and call center. J. Biomed. Inform. **44**, 897–908 (2011)

29. United States Census Bureau 2015. https://www.census.gov/newsroom/facts-for-features/2015/cb15-ff18.html

30. US Department of Education: The health literacy of America's adults: results from the 2003 National Assessment of Adult Literacy. Publication No. 2006-483 (2006)
31. Usable Web (2017)
32. Usability (2017). www.usability.gov
33. User Testing: 2017 UX, User Research Industry Survey. www.usertesting.com
34. Vantaram, S.R., Saber, E.: Survey of contemporary trends in color image segmentation. J. Electron. Imaging **21**(4), 2012
35. Wave© (2017). www.waveaim.org
36. WebAccessibilityChecker© (2017). http://www.stanford.edu/group/accessibility/cgi-bin/accessibilitychecker/checker/index.php
37. Web Accessibility Initiative. https://www.w3.org/WAI/AU/
38. WCAG (Web Content Accessibility Guidelines). https://www.w3.org/WAI/intro/wcag
39. Weinreich, H., et al.: Not quite the average: an empirical study of web use. ACM Trans. Web **2**(1) (2008). Article 5
40. Younghwa, L., Chen, A.K.: Usability design and psychological ownership of a virtual world. J. Manag. Inf. Syst. **28**(3), 269–308 (2011)

Analysis of the Relationship Between Content and Interaction in the Usability Design of 360° Videos

Nicholas Caporusso[1(✉)], Meng Ding[1], Matthew Clarke[1],
Gordon Carlson[1], Vitoantonio Bevilacqua[2],
and Gianpaolo Francesco Trotta[2]

[1] Fort Hays State University, 600 Park Street, Hays, USA
{n_caporusso,mkclarke2,gscarlson}@fhsu.edu,
m_ding6@mail.fhsu.edu
[2] Polytechnic University of Bari, Via Edoardo Orabona, 4, Bari, Italy
{vitoantonio.bevilacqua,
gianpaolofrancesco.trotta}@poliba.it

Abstract. In the recent years, 360° images and video became a popular format for designing, producing, and consuming information. Immersive videos enable users to control the viewing angle at playback, and experience content in a unique fashion. Therefore, understanding how viewers interact with 360° video is crucial for improving their design, production, distribution, and consumption. In this paper, we introduce a model for categorizing 360° video based on the type of points of interest in the scene, and we present a study in which we analyzed heatmaps and changes in the viewing angle to identify the key features of regions of interest, the common interaction patterns within video categories, and the relationship between content design and engagement.

Keywords: Information visualization · Heatmap · Viewing angle
Immersive video

1 Introduction

Thanks to recent technology advances in video cameras, full 360° images and videos rapidly gained popularity as the best tools for experiencing locations (e.g., museums), environments (e.g., the rainforest), and activities (e.g., skiing) in an immersive fashion. Nowadays, affordable equipment supports shooting, producing, and playing high-quality 360° content with minimal technical effort. Moreover, most digital media platforms support uploading, playing, streaming, and even interacting with 360° videos using standard protocols. In the recent years, the production of 360° content and its consumption through popular platforms has significantly increased, and the integration of immersive video in online communities, mobile applications, and head-mounted devices is contributing to rendering 360 video a standard for storytelling. According to a recent report [1], 360° videos already are in a dominant position (99.37%) compared to VR content. In addition, according to recent reports, as the market of 360° cameras

© Springer International Publishing AG, part of Springer Nature 2019
T. Z. Ahram and C. Falcão (Eds.): AHFE 2018, AISC 794, pp. 593–602, 2019.
https://doi.org/10.1007/978-3-319-94947-5_60

will grow over 1500% over the next six years, this opens new opportunities and scenarios for content production and utilization [2].

Several studies focused on technological aspects of 360° videos, such as, development of hardware rigs and mounts for optimal multi-camera alignment, research on software for stitching images from multiple sources, and implementation of algorithms for optimizing video streams. Conversely, less attention has been dedicated to human factors and, specifically, to the design and fruition of 360° videos from a user interaction standpoint, though enhancing the quality of the experience is a key competitive factor [3]. As a result, despite the amount of material being produced both by amateurs and by professionals, there are no guidelines about how to create storyboards to fully unlock the potential of 360° content. Moreover, there are no best practices about how to organize Points of Interests (POIs) and visual cues in the scene so that users can change their viewing angle (VA) and navigate the story according to the intent of the director. Consequently, besides some experimental attempts, most of the experiential aspects of 360° content are still unclear and usability of 360° videos is yet to be investigated.

In this paper, we focus on the human aspects of 360° video, and we analyze the relationship between the content of images and user interaction (i.e., viewing angle). Specifically, the objective of our study is two-fold: by considering type, number, positioning, and movement patterns of POIs in the scene, we associate the video with a category that describes the syntactic complexity of the scene, that is, the amount and type of POIs in the video (i.e., no POIs, single fixed POI, single moving POI, multiple fixed POIs, or multiple moving POIs). This, in turn, together with the semantics of the video, might help predict user interaction dynamics and, thus, design both the story and the scene to include visual and auditory cues that guide users' viewing angle to help them experience the content.

Furthermore, we detail the results of a study in which we investigated the correlation between video design (i.e., syntactic complexity and content type) and interaction dynamics (i.e., how users navigate the scene). In our experiment, we tracked changes in viewing angles to generate heatmaps that reflect users' orientation, and we analyzed heatmap patterns within each category to validate the initial classification in terms of syntactic complexity. Also, we discuss qualitative information collected from participants using questionnaires aimed at understanding the relationship between syntactic complexity and overall performance of the video experience, evaluated on a two-dimensional emotion space. The consistency in our findings might suggest ways to approach the design of 360° videos based on a specific desired narrative outcome.

2   Related Work

Indeed, the main feature introduced by 360° video is providing viewers with opportunities to control and change the viewing angle at playback. Therefore, as discussed by several studies, at any given time viewers focus on a specific portion of the video: a common property for several types of interactive videos is that there is a region of interest (ROI) currently viewed by the user [4] whereas the others are available, though not visible. This, in turn, has several implications in terms of pipeline of: (1) narrative design, (2) video recording, (3) data storage and transfer, (4) content visualization,

(5) user experience, and (6) information perception and processing. Consequently, 360° content require a completely different approach to their design, development, and distribution, with respect to traditional video. For instance, the availability of content over a full 360° angle, which offers a completely immersive viewing experience, requires data transmission of parts of the video that will not be displayed to the user, resulting in bandwidth consumption, increased loading time [5, 6], and potential quality reduction in the ROI. Also, the design and production of content should accurately consider the positioning of POI in the video, to avoid placing items that are crucial to the narrative in areas which potentially will be outside of the current ROI, at playback. To this end, the authors of [4] found that movement, sound and lighting cues from the fixation regions are the basic and effective methods for directing user's attention. Moreover, as immersive videos place the viewer at the center of the scene, the spatial organization of content becomes crucial to ensure a comfortable experience: [7, 8] reported that a distance of three meters between the user viewing point and items offered a good balance of being close enough to see clearly and creating a sense of immersion, whereas objects that are at a shorter distance may be perceived as unnaturally close and cause discomfort as they invade viewer's virtual personal space. Furthermore, being introduced in a realistic immersive environment may bias the audience and create the feeling that the scene is happening at that moment. As a result, in watching videos featuring people, viewers' engagement increases if they are acknowledged by the characters in the scene or if there is eye contact with them [7, 8].

Analyzing viewing patterns is crucial to understanding interaction with 360° video and, consequently, to improving the quality of content design, production and consumption. Recent studies focused on common ROIs, and they demonstrated that users' viewing directions are closely correlated for most of the videos [8]. In [9], the authors analyzed users' behavior when watching 360° videos in 5 categories based on content, that is, exploration, static focus, moving focus, rides, and miscellaneous. They utilized orientation and velocity parameters to compare angle distribution, Point of View (POV), and exploration phase, and they found common patterns in user interaction, that is, (1) angles distribution was highly dependent on video content, (2) viewpoints had minimal changes if there were clear characters in the scene, (3) at the beginning of a new video, viewers explored the scene to understand where the focus should be, and (4) the viewer tended to make large rotations towards the front than towards the back of a video.

Furthermore, other studies utilized heatmaps, which are a data visualization method that provides an intuitive way to identify regions of high and low concentration of a parameter, such as, ROI: areas of an image are color-coded based on the weighted number of ensemble members in that specific region [10], to show the point density interpolation within the area. Heatmaps, also known as intensity maps [11] are extensively utilized in several applications, such as, geographic data visualization for highlighting density of houses, crime reports, or roads or utility lines influencing a town or wildlife habitat [12, 13]. Heatmaps have been utilized in Human-Computer Interaction to study attention [14] and to produce maps of observation patterns, in combination with eye-tracking and gaze acquisition devices. Among their benefits, heatmaps render extremely easy to interpret the distribution over an area, because they use color coding to produce maps that quickly elicit relations and stimulate visually comparisons, facilitating differentiation between ROIs that received more attention and areas with less or no fixations. However, the main drawback of heatmaps is that they are not suitable for dynamic stimuli, such as, videos or systems whose interface changes as users interact with them [14]. Nevertheless, current video distribution platforms integrated heatmaps as a convenient visualization technique for analyzing the consumption of videos: as they aggregate information but conserve the essence of the forecast [10], heatmaps are offered to video publishers to visually support the human decision maker in understanding patterns of visualization and ways to improve the design of content for better audience engagement.

In this paper, we describe a study in which we analyzed the ROIs of 360° videos using static visualization methods (i.e., heatmaps) and the dynamic components of user interaction (i.e., time and rotation of the viewing angle) to identify the key features for content categorization and to achieve a classification of immersive content based on common user interaction patterns.

3   Study

We designed an observational study to evaluate the appropriateness and applicability of our model, and to compare it with categorizations presented in the most recent literature [9], which utilizes 5 classes, that is, exploration, static focus, moving focus, rides, and miscellaneous. Conversely, our model has four categories:

- contemplation, in which there are no explicit POIs, there is a fixed POV, and the video results in limited frame changes; as a result, the users would contemplate the scene in relation to implicit visual cues; gestalt
- observation: one fixed POI dominates the scene, which is watched from a fixed POV;
- tracking, in which the viewer has a fixed POV and the movement of one POI, which could consist in a group of elements, is expected to determine changes in the ROI;
- exploration, in which viewers have a fixed POV and are presented with a scene consisting on multiple moving POIs, or they are on a moving POV (e.g., in a car).

In our model, adjacent categories can overlap, so that an exploration video can contain contemplative elements, depending on the design of the content.

A custom web-based 360° player was developed using Three.js [X], an open source library based on WebGL, which supports rendering video as texture material. Therefore, by mapping equirectangular frames on the surface of an empty sphere, the 3d engine enables the user to view the spherical projection from the inside, and to control the viewing angle using standard input peripherals or acceleration sensors (i.e., on a mobile device). We implemented a client-side script that continuously tracks the viewing angle and converts it into a pixel location in the equirectangular image. Width and height coordinates are stored in a flat file with their timestamp relative to the beginning of the video. For this experiment, we utilized a sampling rate of 10 Hz.

A set of 16 videos were selected among material available on the Internet, with the intent of maximizing differences between them in terms of type (e.g., contemplative, experiential), content (e.g., nature, music, story), number of points of interest (i.e., no POIs, one fixed POI, one moving POI, multiple fixed POIs, and multiple moving POIs), and amount and speed of action. Also, the selection considered the categories identified by other studies. Fragments from each video were extracted to standardize their length to one minute and to make sure that the scene is captured from one shooting angle, only. They were compressed to 1080p (1920 × 1080 pixels) variable-bitrate progressive MP4 formats at a data rate of 30 fps.

In this study, we did not acquire gaze using an eye tracking device (ETD), to replicate the same information that video players can record when users watch 360 videos, and to render our solution applicable and comparable to currently available distribution and analytics platforms. Also, we did not use a head-mounted display (HMD), because the majority of users interact with immersive videos using standard screens, due both to market penetration of this technology and to the limited availability of HMDs in the interaction.

A total of 29 subjects participated to the experiment in person. However, the study had a mortality of 6 who could not continue their involvement. Therefore, data from 23 participants were recorded and analyzed. The majority of subjects (65.22%) aged 25–35, 26.08% were 18–24, and 8.70% were 35–44. The group included 43.48% females and 56.52% males. Participants had a high education level (which we associated to a factor of technology use) 56.52% had master's degree, 39.13% had a bachelor's degree, whereas others accounted for 4.35%.

The study was realized in a dedicated room with no distractions and optimal light conditions, where participants were comfortably seated on a swivel chair, in front of a 20" display playing videos at full screen. Subjects were asked to watch the videos, which were played in a random order. After each task, they were presented with a short questionnaire asking them to evaluate the video in terms of (1) perceived duration (length in seconds), (2) perceived engagement (on a Likert scale), (3) clarity of the content (on a Likert scale), (4) quantity of content (on a Likert scale). Also, they were asked to categorize the video according to the model described in Fig. 1.

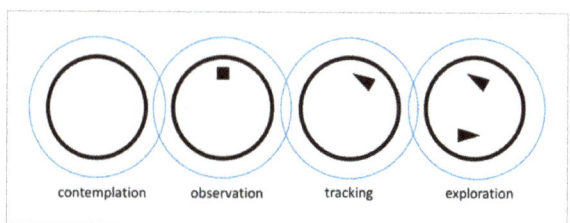

Fig. 1. Categorization model for 360° content depending on the type of elements in the scene. The square indicates a fixed primary POI, whereas triangles represent moving POIs.

4 Results and Discussion

A total of 296 videos were displayed to participants. Data from ROIs were acquired and answers to the questionnaires after individual videos were collected. Table 1 summarizes responses to the survey. Interestingly, they show strong correlation with the type of category and with the heatmaps (see Fig. 2). Moreover, results have a strong internal consistency. Although the duration of each video was 60 s, viewers perceived them differently. In accordance with studies about perception of time in relation to the engagement of an experience, content perceived as less engaging was rated as lasting longer than one minute, whereas users associated a shorter duration with videos that they considered more interesting. Video 4 (contemplation), which featured a lecture, rated as the longest, compared to videos 13 (exploration, featuring a roller-coaster ride), and two of the tracking videos, 10 and 9 (consisting of a hip-hop dance and a film-like computer graphics-generated scene, respectively), which were perceived as the shortest. In addition to content and engagement, there could be other reasons for this difference. For instance, as the audio track was removed, viewers might have perceived videos in which sound was is a prominent component (e.g., concert and ballet, such as videos 6 and 7, respectively) as lasting longer, because of the different experience compared to an actual situation. Also, information elicited by cues in the videos and associated with memories, beliefs, and personal history, might influence the perceived duration: as an example, video 1 (contemplation), containing a still image performed better than video 4 (contemplation). In addition to quantifying duration in seconds, users were asked to evaluate whether the video was too long or too short, using a Likert scale. The results were consistent with the data for perceived duration. In addition,

Table 1. Viewers' responses for 360° videos

Video	Perceived duration	Engagement	Content
1	59.25 ± 11.04	1.95 ± 1.05	4.65 ± 0.67
2	59.77 ± 10.74	2.09 ± 0.97	4.41 ± 0.67
3	59.35 ± 11.21	2.09 ± 0.95	4.39 ± 0.78
4	61.00 ± 10.46	2.35 ± 1.23	3.6 ± 0.99
5	59.52 ± 9.86	2.86 ± 0.96	3.14 ± 0.91
6	57.38 ± 9.30	2.86 ± 0.96	3.43 ± 0.98
7	57.61 ± 8.38	3.74 ± 0.96	2.61 ± 0.84
8	58.86 ± 8.99	2.82 ± 1.22	3.55 ± 0.86
9	52.39 ± 8.38	4.09 ± 1.16	2.13 ± 0.92
10	54.35 ± 7.43	4.43 ± 0.66	1.91 ± 0.79
11	56.90 ± 9.55	3.43 ± 1.21	2.57 ± 1.08
12	55.87 ± 8.87	2.96 ± 1.02	3.00 ± 1.00
13	50.00 ± 6.22	4.26 ± 0.81	2.61 ± 1.03
14	57.50 ± 8.69	3.82 ± 0.91	3.09 ± 0.97
15	58.10 ± 9.93	3.86 ± 0.79	2.81 ± 1.17
16	57.50 ± 8.69	3.91 ± 0.87	2.68 ± 1.17

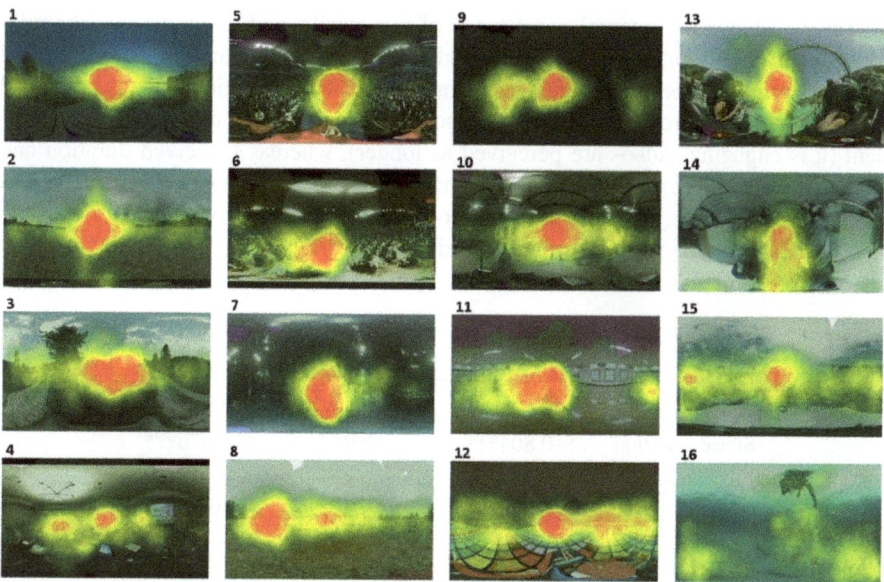

Fig. 2. Heatmap visualization of the videos utilized in the experiment: tracks of VAs from multiple users have been utilized to generate attention maps. The four categories are represented in the columns: contemplation (1–4), observation (5–8), tracking (9–12), and exploration (13–16). Features that include movement (quantity of changes in pixels) result in changes in ROIs and in different concentration (red indicates high persistence of the VA). Videos are organized in order of their dispersion coefficient (d), described in Fig. 3.

users' responses on engagement have a strong negative correlation with the duration, as the most engaging videos were perceived to last less than the least interesting ones.

In addition, we asked participants to rate video clarity and to evaluate whether it was easy to interpret the scene or content was confusing. Videos 13 (exploration of a roller coaster ride), 9 (tracking of a movie scene), and 6 (observation of concert) were the ranked first, whereas video 12 (a ride on a Super Mario Bros cart) was regarded as the most confusing video. This is because the POV was moving on a car, but the viewing angle was not aligned with the direction of the cart. As a result, participants were changing their VA to match the direction faced by the car, otherwise they faced the sides or the back of the car. Video 12 is an example of overlapping between the tracking and exploration categories, because it involves a moving POV, though viewers have to change VA to track the direction of the car. However, most of the videos featured simple content or stories.

Furthermore, participants were asked to describe the amount of content, to evaluate whether there was too much (values closer to 1) or too little (values closer to 5) happening, or if the quantity of information in the video was appropriate (values close to 3). Although there are some outliers, responses from participants show correlation with the categories of the videos. Data from user interaction patterns confirm the information explicitly stated by subjects, though groups account for the sign of correlation.

Table 2 shows the internal consistency of our results, and it summarizes the relationship between duration, engagement, and content. From our findings, we identified negative correlation between perceived duration and stated duration (videos perceived as lasting less are perceived as shorter), and between perceived duration and engagement (less engaging videos are perceived as longer), whereas perceived duration and content show positive correlation (if content is obvious, then the video is perceived as boring). The negative sign of the correlation between the perceived duration and the stated duration is due to the sorting utilized in the scale.

Table 2. Spearman's R statistics analysis for 360° videos

Factors	Perceived duration	Engagement	Content
Perceived duration	——	−0.781***	0.797***
Stated duration	−0.804***	0.917***	−0.876***
Engagement	−0.781***	——	−0.91***
Clarity	−0.31	0.391	−0.415
Content	0.797***	−0.91***	——

Note: *** means $p < 0.001$

Heatmaps obtained from multiple VA tracks of different users (see Fig. 2) revealed insightful information about user experience with the immersive video and their interaction patterns with the content of 360° scenes. Heatmaps showed interesting similarities due to natural overlapping occurring between adjacent categories.

From our findings, we can conclude that both perceived and stated duration have a positive correlation with data points describing the VAs and, specifically, with the dispersion (d) in ROIs shown in the heatmaps (Fig. 2). Moreover, depending on the category, the dispersion coefficient represents specific features of the video, which, in turn, can be utilized to modulate the experience. Specifically, in contemplation videos, when there are no specific events, dispersion reflects the presence of visual cues in the scene (e.g., roads, forests, and trails) that attract the user, because they might reveal some action. Conversely, in observation videos, dispersion is moderate and it is triggered by events, such as, an applause, or by elements that instantaneously become POIs. Changes in VAs are elicited by movement of items in the scene that do not belong to the main POI, and their effect is stronger when they are humans. In tracking videos, dispersion is elicited by the interaction of elements in the scene: patterns in VAs are less noisy when the story involves movement of a single element, or multiple elements move within one ROI. Finally, for exploration videos, if the POV moves, dispersion in VAs is negatively correlated with the speed of the POV.

Based on our findings, we argue that the categorization achieved by [9] can be further improved by incorporating our model, which has the benefit of being more abstract and, potentially, content agnostic. This, in turn, might enable automatic identification of ROIs based on correlation between a limited training set of user interaction patterns and the corresponding features extracted from frame-by-frame pixel changes. Consequently, this would enable implementing compression codecs and streaming protocols that could deliver ROI-based variable-bitrate quality.

Moreover, as shown in Fig. 3, users have different patterns in exploring the scene, and visual cues have a different role depending on the amount of information and on the speed of movement and elements change in the scene. For instance, in exploration videos, the relevance of visual cues and even moving elements decreases with the speed of the POV, whereas the importance of implicit and distal visual cues increases in contemplation videos. In observation and in tracking videos, changes in VAs indicate lower levels of engagement. Especially in the latter category, dispersion in viewers' ROIs might reveal issues in the spatial organization of content, or in the design of the scene.

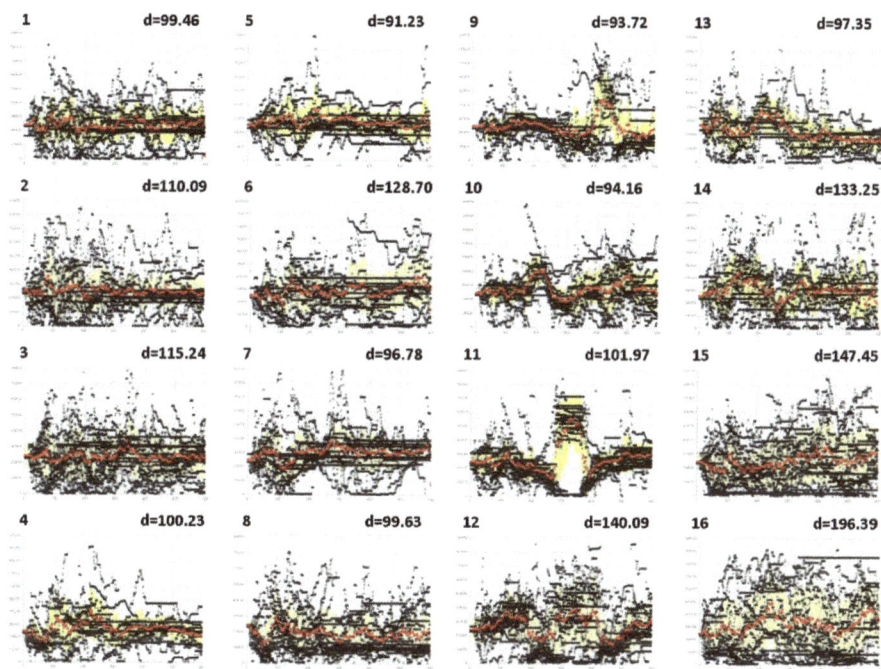

Fig. 3. Scatterplot visualization of ROIs over time, represented as the pixel in the middle of the region of interest at any given time. Videos are arranged as in Fig. 2.

5   Conclusion

In the recent years, great emphasis has been given to technological features of 360° video (e.g., resolution, format, and devices), whereas less attention has been dedicated to content. Nevertheless, the increasing community of producers, directors, and video makers - both professional and amateur - are starting to investigate how to fully understand and unlock the potential of 360° video.

In this paper, we focused on the usability of 360 video by evaluating how users interact with immersive scenes. To this end, we proposed a model for categorizing immersive videos based on the number and type of points of interest, and we detail the results of a study focusing on how users change their viewing angles and interact with regions of interest, to analyze interaction with different categories of video and different types of content. From our findings, we can confirm the robustness of our categorization system, which could be utilized, together with rotation patterns in viewing angles and with dispersion in ROIs to predict viewers' engagement.

Our work aimed at establishing a framework that can be utilized in future studies to investigate aspects involved in the production, recording, and consumption of 360° video in the areas such as, journalism, narrative storytelling, entertainment, and interaction.

References

1. HUAWEI iLab: "VR data report," HUAWEI Report (2016). https://mp.weixin.qq.com/s/tcsm9NIECa7d1L7gZekrrQ
2. The MPEG Virtual Reality Ad-hoc Group: "Summary of survey on virtual reality," in ISO/IEC JTC 1/SC 29/WG 11 N16542 (2016)
3. Kaasinen, E., Roto, V., Hakulinen, J., Heimonen, T., Jokinen, J.P., Karvonen, H., Keskinen, T., Koskinen, H., Lu, Y., Saariluoma, P.: Defining user experience goals to guide the design of industrial systems. Behav. Inf. Technol. 34, 976–991 (2015)
4. Sheikh, A., Brown, A., Watson, Z., Evans, M.: Directing attention in 360-degree video (2016)
5. Bao, Y., Wu, H., Zhang, T., Ramli, A.A., Liu, X.: Shooting a moving target: motion-prediction-based transmission for 360-degree videos. In: 2016 IEEE International Conference on Big Data (Big Data), pp. 1161–1170. IEEE (2016)
6. Bao, Y., Zhang, T., Pande, A., Wu, H., Liu, X.: Motion-prediction-based multicast for 360-degree video transmissions. In: 2017 14th Annual IEEE International Conference on Sensing, Communication, and Networking (SECON), pp. 1–9. IEEE (2017)
7. Bailenson, J.N., Blascovich, J., Beall, A.C., Loomis, J.M.: Interpersonal distance in immersive virtual environments. Pers. Soc. Psychol. Bull. 29(7), 819–833 (2003)
8. Wilcox, L.M., Allison, R.S., Elfassy, S., Grelik, C.: Personal space in virtual reality. ACM Trans. Appl. Percept. (TAP) 3(4), 412–428 (2006)
9. Almquist, M., Almquist, V.: Analysis of 360° Video Viewing Behaviours (2018)
10. Köpp, C., von Mettenheim, H.J., Breitner, M.H.: Decision analytics with heatmap visualization for multi-step ensemble data. Bus. Inf. Syst. Eng. 6(3), 131–140 (2014)
11. Yeap, E., Uy, I.: "Marker Clustering and Heatmaps: New Features in the Google Maps Android API Utility Library." Google Geo Developers (2014). Accessed Apr 2014
12. ArcGIS, E.S.R.I.: 10.1. Redlands. ESRI, California (2012)
13. DeBoer, M.: Understanding the heat map. Cartogr. Perspect. 80, 39–43 (2015)
14. Tula, A.D., Kurauchi, A., Coutinho, F., Morimoto, C.: Heatmap explorer: an interactive gaze data visualization tool for the evaluation of computer interfaces. In: Proceedings of the 15th Brazilian Symposium on Human Factors in Computer Systems, p. 24. ACM (2016)

Evaluation of Usability and Workload Associated with Paper Strips as Compared to Virtual Flight Strips Used for Ramp Operations

Victoria Dulchinos[✉]

SJSURF/NASA Ames Research Center, Moffett Field, CA, USA
victoria.l.dulchinos@nasa.gov

Abstract. This paper describes a study comparing the use of paper strips with virtual flight strips depicted on a new user interface, the Ramp Traffic Console (RTC), designed for use by ramp controllers to be used in place of paper strips. A Human-In-the-Loop (HITL) experiment was performed as the fifth in a series of six HITL simulation studies designed to evaluate a pushback Decision Support Tool (DST) concept for Charlotte Douglas International Airport (CLT). Workload and usability were assessed in post-run and post-study questionnaires. In the RTC virtual flight strip condition, post-run questionnaire results show lower workload ratings across all aspects of workload; additionally, a trend is found toward increased usability ratings. Post-study questionnaire results indicate a preference for RTC over paper strips. Additional research is suggested with more training runs and a greater number of participants to increase statistical power. It is also suggested that this new technology be re-evaluated as a part of the ATD-2 field testing activities.

Keywords: Human factors · Human-systems integration
Decision support tool · Usability · Workload

1 Introduction

New technologies developed for use by Air Traffic Controllers (ATC) and airline ramp operators are studied in a Human in the Loop (HITL) simulation study. The Ramp Traffic Console (RTC), shown in Fig. 1 below, was designed along with the Spot and Runway Departure Advisor (SARDA) Decision Support Tool (DST) proposed to aid ramp controllers in reducing taxi delay. SARDA was first evaluated as a decision support tool for air traffic controllers to meter flights from the spot to the runway (Hayashi et al. 2013).

Air Traffic Control Towers (ATCT) are equipped with multiple electronic systems that have been developed over time to facilitate controllers in the management of air traffic. Advanced Electronic Flight Strips (AEFS) is one such technology that is likely to be subsumed into Terminal Flight Data Management (TFDM) as a part of a larger effort to integrate multiple existing electronic systems. In a 2012 study of a prototype ATCT TFDM system, Controller-Pilot communications were used to measure cognitive workload (Lockande 2016). This study found that controllers utilizing the prototype TFDM system reported lower workload than the control group. While RTC is designed for use by airline operators, like AEFS and TFDM, RTC is intended to replace paper strips with a digitally integrated information source to present integrated flight data. In the current study, SARDA advisories are presented to the ramp controller as a tactical surface scheduler (DST) designed to meter flights from the gate. The RTC has a novel user interface displayed on a 27" multi-touch screen monitor, used by ramp controllers in place of paper strips and paper maps, and includes the SARDA pushback advisories.

During simulated operations, ramp controllers gave instructions to pilots via radio communications to manage traffic and ensure airplanes were safely separated while efficiently taxiing to their destination. This task required the controllers to engage in a variety of high-level cognitive functions, including planning, managing, monitoring, problem solving, and coordinating with other ramp controllers, pilots, and air traffic controllers. The CLT ramp is divided into four sectors, North, East, South and West, with most airplanes needing to taxi through multiple sectors. Ramp controllers hand off airplanes to each other at the sector boundaries. Handoffs are also made to air traffic controllers at various points, called spots, intersecting with the Federal Aviation Administration controlled active movement area on their way to and from the arrival or departure runway. Outbound departure flights are handed off to the Air Traffic Controller (ATC) at the spots and inbound arrival flights are received from the ATC at the spots and directed to their gate. Consequently, the ramp controllers were required to communicate with other sector controllers as well as air traffic controllers and multiple pilots to efficiently manage all the departure and arrival flights to and from their gates on the ramp. The RTC and SARDA concept were developed initially for use at the Charlotte Douglas International Airport (CLT).

The simulation study reported in this paper is one in a series of studies to evaluate SARDA and RTC from the ramp controller's point of view. Human-in-the-Loop (HITL) simulations are used as a safe and controlled environment to evaluate new concepts and decision support tools. The goal of the present study was to evaluate virtual flight strips on RTC as compared to the use of paper strips in ramp traffic management.

The research questions explored here are regarding the effect of using virtual flight strips on RTC as compared to using paper strips shown in Fig. 2 below, on the workload and usability ratings of the ramp controller participants.

Fig. 1. RTC with virtual strips

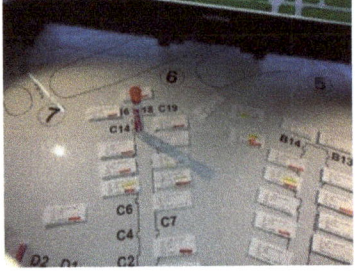

Fig. 2. Paper strips and paper map

2 Methods

The virtual flight strips as presented on RTC were tested in a HITL simulation study in Future Flight Central (FFC), a high-fidelity tower simulator at NASA Ames Research Center. This study included eight 90-minute data collection runs over three days. There were two RTC training sessions for a total of 3 h and 20 min of controller training using RTC. There were four ramp controller participants. In four of the data collection runs, the ramp controllers used paper strips and paper maps while controlling ramp traffic, and in the other four runs, the ramp controllers used the virtual flight strips on RTC. There were two traffic scenarios used in the simulation and each was repeated twice in the paper condition and in the RTC condition. Two participants were active ramp controllers from CLT, a third was a retired FAA controller, and the fourth participant was an active ramp controller from another airport. The four ramp controller participants used the RTC in the simulated ramp operations environment while usability and workload data was collected from the users under the two different conditions. In one condition the participants used paper strips and paper map, while in the second condition participants used the virtual flight strips and movable map on RTC. The two ramp controllers who were current CLT controllers were rotated through sector assignments such that each worked both scenarios in the paper and RTC conditions. The other two ramp controllers who were not active CLT controllers remained in one of the "less busy" sectors that were deemed to have less impact on the operation. Post-run and post-study workload and usability questionnaires were administered to all four of the sector controllers.

User workload is commonly assessed with subjective measures, which require the participants to report on their subjective psychological experience. These measures include self-reported subjective ratings on certain scales, such as the NASA Task Load Index (TLX) (Hart and Staveland 1988). Workload for the purposes of the present study is defined by four components of the NASA-TLX (Task Load Index). The four components include Mental Demand (Thinking, deciding, calculating, searching, etc.), Physical Demand (Hands and arm movement, force), Temporal Demand (Time pressure), and Frustration (Stress, annoyance, irritation). Controllers were asked to rate each of the four components of their workload after every run on a scale of 1–10. For example, see Fig. 3 for the "mental demand" question response format. A performance sub-scale was not included.

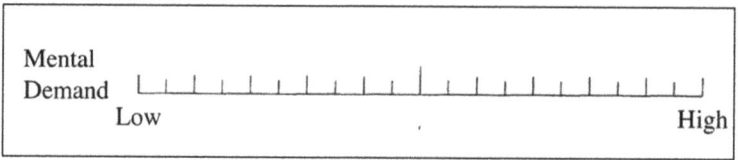

Fig. 3. Post run questionnaire workload question format

Along with workload, usability of the RTC was also assessed. There are several definitions of usability (Jeng 2005, provides a good review of various definitions). In this paper, the definition used by the International Organization for Standardization (ISO 1998) will be followed. It defines usability as the extent to which the users of a product are able to work *effectively*, *efficiently*, and with *satisfaction.* Following the definition used by the International Organization for Standardization (ISO 1998), usability for the purposes of this paper is defined by three aspects of usability, effectiveness, efficiency, and satisfaction. Traffic management performance questions were included in the post run questionnaire with the aim of determining the "effectiveness" aspect of usability. Resources and efficiency questions were included in the post-run questionnaire with the aim of determining the "efficiency" aspect of usability. The post-study survey questions were designed to assess the "satisfaction" aspect of usability. After each run, the controllers were asked questions regarding their traffic management performance and resources and efficiency using a response format with a scale of 1 "Referred to Always" to 7 "Referred to Never." For example, one Traffic Management and Performance aspect of Usability is assessed by the controller's response to the question shown in Fig. 4 below:

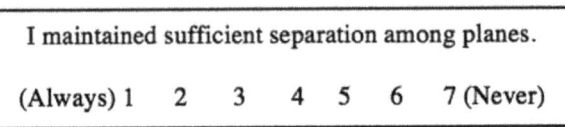

Fig. 4. Post-Run questionnaire usability question format

Post Run and Post Study questionnaire responses were gathered and the results were analyzed to assess controller workload and usability ratings under both conditions, virtual flight strips on RTC and paper strips. To determine the effect of condition (Paper or RTC) on controller workload and usability ratings, mean post run responses on the workload and usability related questions were collected from all four sector controllers and a 2 X 2 Analysis of Variance (ANOVA) was conducted with sector as a between-subject variable to determine if there was a main effect of condition.

3   Results

The mean post run workload ratings and ANOVA results shown in Table 1 and are graphed with standard error bars at a 95% confidence level in Fig. 5 below. These results show that the mean workload ratings for the RTC condition are lower than the mean ratings for the Paper condition across all four components of workload. With respect to the Mental Demand aspect of workload, the participants reported a higher mean workload rating of 5.7 for the Paper condition as compared to a mean workload rating of 3.9 in the RTC condition however, as can be seen in Table 1, this was not a statistically significant main effect. There was a statistically significant main effect across the other three aspects of workload. With respect to the Time Pressure aspect of workload, the participants reported a higher mean workload rating of 4.9 in the Paper condition as compared to a mean rating of 2.4 in the RTC condition. With respect to the Physical Demand aspect of workload, the participants reported a higher mean workload rating of 4.6 in the Paper condition, and 2.8 in the RTC condition. Finally, looking at the Frustration aspect of workload, the participants reported a higher mean workload rating of 3.6 in the Paper condition, and 1.3 in the RTC condition.

Table 1. Mean workload response all sectors

| Mean Participant Workload Ratings Across Four Aspects of Workload | | | | | |
Aspect of workload	Mean response paper	SE paper	Mean response RTC	SE RTC	$F(1,3)=$
Mental demand	5.7	0.82	3.9	1.67	3.59, p = .155
Time pressure	4.9	0.57	2.4	0.50	48.46, *p = .006
Physical demand	4.6	1.32	2.8	1.43	84.26, *p = .003
Frustration	3.6	0.31	1.3	0.34	29.73, *p = .012

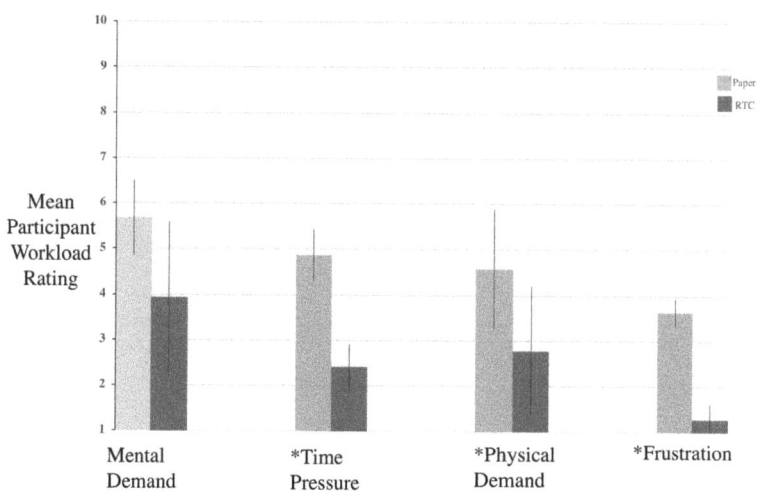

Fig. 5. Mean participant workload rating

Because the response scale for the post run usability questions was presented in reverse order such that "Always" is the lower anchor (1) on the scale, and "Never" is the upper anchor (7) on the scale, for ease of discussion, an inverse scale of the means is reported in this paper to account for the opposite phrasing of the questions.

The mean usability ratings of the post run traffic management and performance questions, meant to assess the "effectiveness" aspect of usability, were higher in the RTC condition as compared to the Paper condition for all of the seven questions. The means and standard errors are shown in Table 2 and graphed in Fig. 6 below. The

Table 2. Participant ratings of traffic management and performance

Traffic Management Performance Questions Mean Response with Standard Error and F values					
Question	Mean paper	S.E.	Mean RTC	S.E.	F (1,3)=
1. Maintained separation	6.7	0.157	6.9	0.125	9, p = .058
2. Maintained flow	6.1	0.373	6.6	0.295	12, *p = .04
3. Minimized delay	5.9	0.329	6.5	0.25	22.09, *p = .018
4. Avoided grid-lock	6.6	0.161	6.9	0.063	6.82, p = .088
5. Maintained pressure on runway	5.9	0.258	6.7	0.237	54, *p = .005
6. Metered departures	6.2	0.493	6.6	0.12	.73, p = .456
7. Responded promptly	6.8	0.188	6.9	0.063	.33, p = .604

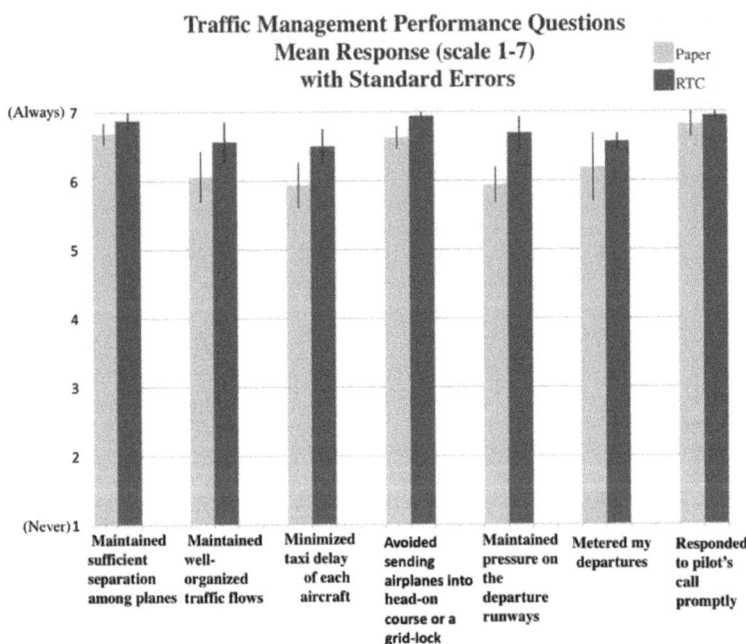

Fig. 6. Mean participant ratings of traffic management performance

results of the analysis showed a statistically significant main effect of condition for questions 2, 3, and 5 that asked about "maintaining organized traffic flow," "minimizing taxi delay," and "maintaining pressure on the runways" respectively (see Table 2). Looking at question 2 which asked if the participant "maintained well organized traffic flows," the participants reported a higher rating of 6.6 for RTC as compared to a mean rating of 6.1 in the Paper condition. Looking at question number 3 which asked if the participant "minimized taxi delay of each aircraft," the participants reported a higher mean rating of 6.5 in the RTC condition than the mean rating of 5.9 in the paper condition. For question number 5 which asked if the participant "maintained pressure on the departure runways," the participants reported a higher mean rating of 6.7 in the RTC condition than the mean rating 5.9 in the paper condition.

All of the other traffic management questions had higher mean usability ratings in the RTC condition as compared to the paper condition, although this difference was not statistically significant (see Table 2). For question number 1 which asked if the participants "maintained sufficient separation among planes," the participants reported a higher mean rating of 6.9 for the RTC condition than the mean rating of 6.7 for the Paper condition. For question number 4 which asked if the participant "avoided sending airplanes into head on course or gridlock", the participants reported a higher mean response of 6.9 in the RTC condition than the mean rating of 6.6 in the Paper condition. For question number 6 which asked if the participant "metered their departures", the participants reported a higher mean response of 6.6 in the RTC condition than the mean response of 6.2 in the paper condition. Finally, for question number 7 which asked if the participant "responded to the pilots call promptly", the participants reported a higher mean response of 6.9 in the RTC condition than the mean response of 6.8 in the Paper condition. Looking at the results overall for the Traffic Management questions, there is a trend toward increased mean usability ratings in the RTC condition as compared to the paper condition for the traffic management and performance questions which were meant to assess the "effectiveness" aspect of usability, with the mean participants rating being higher in the RTC than the paper condition for all of these questions.

The mean participant response values for the post run usability resources and efficiency questions meant to assess the "efficiency" aspect of usability are shown Table 3 and graphed in Fig. 7 below. The mean rating was higher in the Paper condition for questions 3 and 4, and the mean was the same for RTC and Paper conditions for question 6. However, none of these results demonstrated a statistically significant main effect of condition on participant usability ratings (See Table 3).

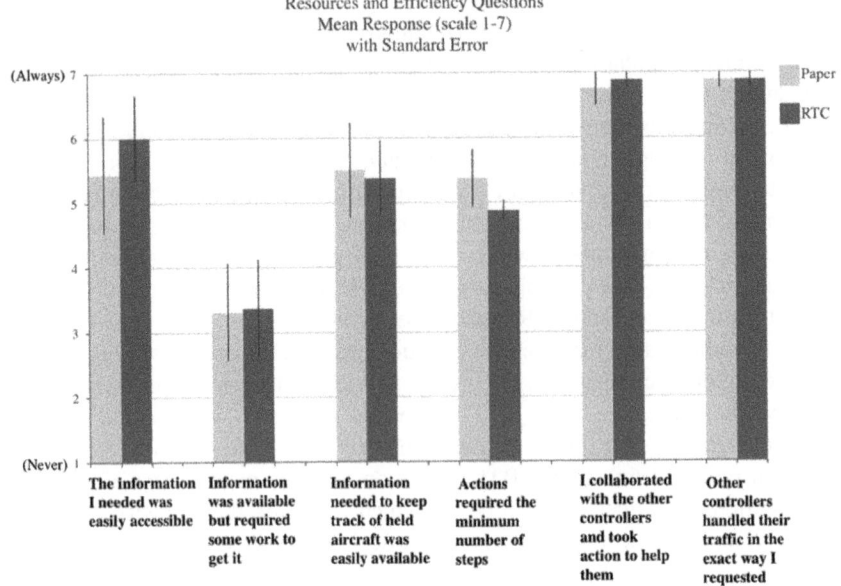

Fig. 7. Resources and efficiency participant mean response

Table 3. Resources and efficiency mean participant response

Resources and efficiency questions mean response with standard error and F values

Question	Mean paper	S.E.	Mean RTC	S.E.	F (1,3)=
1. Information accessible	5.4	0.904	6.0	0.654	1.86, p = .266
2. Information available, but required work	3.3	0.753	3.4	0.74	.03, p = .878
3. Held Aircraft information available	5.5	0.729	5.4	0.582	.16, p = .718
4. Actions required minimum steps	5.4	0.439	4.9	0.161	1.85, p = .267
5. Collaborated	6.8	0.25	6.9	0.125	.27, p = .638
6. Others handled traffic as expected	6.9	0.125	6.9	0.125	0, p = 1.0

Questions 1, 2, and 5 of the resources and efficiency questions resulted in a higher mean rating in the RTC virtual strip condition as compared to the paper strip condition. Looking at the Resources and Efficiency question 1 which asked if "the information needed was easily accessible," the participants reported a higher rating of 6.0 for RTC virtual strips as compared to a mean rating of 5.4 in the paper strip condition. Similarly, looking at question 2 which asked if "the information was available but required some work to get to it," the participants reported a mean rating of 3.4 for RTC and 3.3 for Paper. Question number 5 asked the participants if "they collaborated with other controllers and took action to help them," the participants reported a higher rating of 6.9 in the RTC virtual strip condition as compared to a rating of 6.8 in the Paper condition. Questions 3 and 4 of the Resources and Efficiency questions the results show a higher mean rating in the Paper condition as compared to the RTC condition.

Looking at question 3 which asked "if information need to keep track of held aircraft was available," the participants reported a higher mean rating of 5.5 in the Paper condition as compared to the mean rating of 5.4 in the RTC condition. The Resources and Efficiency question 4 asked "if the actions required the minimum number of steps," with a higher mean participant rating of 5.4 in the Paper condition as compared to the mean RTC rating of 4.9. Finally, for question 6 which asked "if other controllers handled traffic in the way it was requested," the mean participant rating was the same in both Paper and RTC conditions with a mean rating of 6.88 for both RTC and Paper.

To assess the satisfaction aspect of usability, a set of 18 specific preference questions were included in the post study questionnaire. The responses were collected from all four controller participants with responses on a scale of 1 (Prefer Paper) to 7 (Prefer RTC). The results shown in Fig. 8 above indicate that very high level of satisfaction ratings were achieved for all the questions ranging from tracking aircraft status, and being aware of the direction of the flight, to managing sector handoff to ease of reading of information.

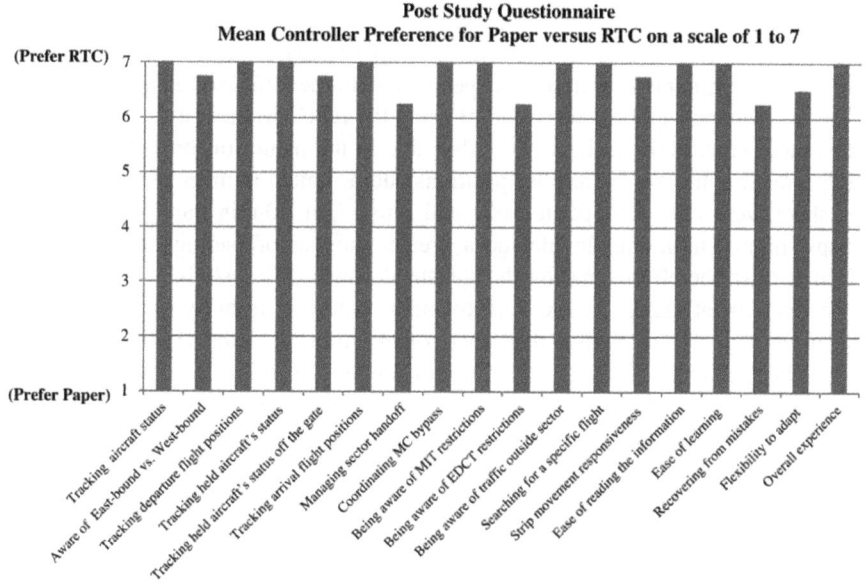

Fig. 8. Post study questionnaire mean participant satisfaction ratings

In sum, results from the Post Run questionnaire indicate lower workload ratings for RTC condition, with only one of the workload elements not statistically significantly lower. Usability ratings for Traffic management performance questions are lower in the RTC condition than in the paper condition showing a preference for RTC over Paper, with not all of the questions showing a statistically significant difference. Usability ratings for Resources and efficiency questions showed mixed results. Post Study Usability responses and satisfaction ratings indicated a clear preference for RTC.

4 Discussion

The mean participant ratings for workload were lower in the RTC virtual strips condition as compared to the Paper condition for all four aspects of workload. There was a statistically significant main effect of condition for all aspects of workload measured except for the mental demand aspect of workload, which was similar for paper and virtual strips. It is possible that this mental workload result would decrease with increased training and increased familiarity. The participants had a minimal amount of training with the RTC virtual strips prior to the data collection. The total amount of time spent training with RTC was 3 h and 20 min; it is possible that with more time training the participants might have reported lower mean mental demand workload rating for RTC condition as compared to the Paper condition resulting in a statistically significant main effect. The participants in this study had been using only traditional paper strips to manage traffic in their experience as professional ramp and air traffic controllers, and RTC was a new tool. The participant ratings for mental demand aspect of workload were lower in the RTC condition than in the paper condition, however this was not a statistically significant difference, perhaps more time training in preparation for the data collection runs, or a greater number of data collection runs might have allowed the participants to gain more experience with the tool resulting in a decrease in the mental demand aspect of workload of using the RTC virtual strips to perform their role as ramp controllers in the HITL. Also, due to the nature of the simulation study with a limited number of controller positions and a limited number of data collection runs, there were only four participants and only eight 90-min data collection runs. Perhaps, future studies might include a greater number of participants and or data collection runs, thereby increasing the statistical power of the study.

The participant ratings for the "effectiveness" aspect of usability were higher in the RTC virtual strip condition than the Paper condition for all of the Traffic Management Performance questions, with statistically significant results for some of these questions. The trend shows that RTC was more efficient than paper on all questions except for two. The lower RTC rating regarding managing the strips was possibly due to lack of familiarity and usage; potentially the participants did not perceive a difference in the efficiency between the two conditions (RTC virtual strips and Paper strips) or the lack of sufficient data in this study.

Looking at the results of the Resources and Efficiency questions in relation to the results of the Traffic Management questions, the Traffic Management questions received a more consistently favorable and statistically significant positive rating for RTC than the Resources and Efficiency questions, perhaps the participants found using the RTC virtual strips to be more effective than using the paper strips. At the same time, these results might be interpreted to indicate that for some aspects of efficiency, the results were not a clear indication of a preference for RTC. Again, perhaps this is a function of the participants being new to the RTC virtual strips and given more time and experience using the RTC virtual strips, the participants rating of the efficiency aspect of usability might improve. Participants' ratings from the post study question-naire for the "satisfaction" aspect of usability indicate a definite preference for the RTC over the Paper condition. Overall these results indicate a trend towards increased mean participant Usability ratings when using the RTC virtual strips as compared to using the paper strips across the three aspects of Usability assessed: effectiveness, efficiency and satisfaction.

As in the TFDM prototype system study by Lockande (2012), the workload results from the current study indicate reduced workload in the RTC virtual strip condition as compared to corresponding baseline or paper strip condition. Similar to the Lockande (2012) study, one possibility is that a reduction in workload is a function of the RTC displaying data on the virtual flight strips that is digitally updated. Like the TBFM prototype used by Lockande, the RTC also integrates other operational data and pre-sents it to the ramp controller in real time such that the ramp controller is not seeking out and verifying information regarding, for instance, Traffic Management Initiatives, or airport configuration, thereby reducing overall workload. The workload results indicating reduced Workload when using RTC along with the Usability results indi-cating a trend toward increased Usability when using RTC seem to indicate that the participants favored the RTC virtual strips as compared to the Paper condition. Future studies of the RTC may benefit from more training runs, as well as having either a greater number of participants or a greater number of data collection runs to increase the statistical power of the analyses.

Recently, RTC has undergone a design refactoring, removing the touch capability, and going to a mouse only design. This refactoring was prompted by a couple of reasons. During the HITL testing of RTC, feedback from some of the controllers indicated that they prefer using the mouse over touch screen functionality. Also, it was decided to a larger 32' screen size for screen sharing with another technology in the field. Going to a larger screen meant possible degradation of touch screen precision along with possible increased fatigue while using the larger display. The controller feedback information along with deciding to go to a larger screen size resulted in the decision to go to a mouse only design. The SARDA tactical surface scheduler has also undergone some development and maturation as it has been integrated along with the RTC with a set of other Air Traffic Management Technologies as a part of NASA's ATD-2 effort (Malik et al. 2016). The ATD-2 Phase One field testing began in September of 2017 where RTC is currently in use by ramp controllers at CLT. Given that additional development and maturation has been completed on the RTC and the tactical scheduler tool, it will be important to follow up on this study to determine the impact of this refactoring on ramp controller user workload and usability ratings.

Acknowledgments. The author acknowledges the work of the team of people who made this research possible. I express my special thanks to Miwa Hayashi, Yoon Jung, Savita Verma, Katherine Lee and Victoriana Delosantos.

References

Hart, S.G., Staveland, L.E.: Development of a NASA-TLX (task load index): results of empirical and theoretical research. In: Hancock, P.S., Meshkati, N. (eds.) Human Mental Workload, pp. 139–183. Elsevier Science Publishers B.V., Amsterdam (1988)

International Organization for Standardization: Ergonomic requirements for office work with visual display terminals (VDTs)–part 11: guidance on usability. ISO 9241-11, Geneva, Switzerland (1998)

Jeng, J.: Usability assessment of academic digital libraries: effectiveness, efficiency, satisfaction, and learnability. Int. J. Libr. Inf. Serv. **55**, 96–121 (2005)

Lokhande, K., Reynolds, H.J.: Cognitive workload and visual attention analyses of the air traffic control tower flight data manager (TFDM) prototype demonstration. In: Proceedings of the Human Factors and Ergonomics Society 56th Annual Meeting (2012)

Hayashi, M., et al.: Usability Evaluation of the Spot and Runway Departure Advisor (SARDA) Concept in a Dallas/Fort Worth Airport Tower Simulation. ATM Seminar (2013)

Malik, W.A., Lee, H., Jung, Y.C.: Runway Scheduling for Charlotte Douglas International Airport. AIAA-2016-4073, 2016 AIAA Aviation and Aeronautics Forum and Exposition, Washington D.C., 13–17 June 2016

The Development of an Online Questionnaire for End Users About the Visual Perception of Informational Ergonomics and Its Attributes in Graphic Brands

João Carlos Riccó Plácido da Silva$^{(\boxtimes)}$, Luis Carlos Paschoarelli,
Valéria Ramos Friso, and José Carlos Plácido da Silva

UNESP - Univ. Estadual Paulista, Av. Eng. Luiz E. C. Coube, 14-01,
17033-360 Bauru, Sao Paulo, Brazil
joaocarlos_placido@hotmail.com

Abstract. The ultimate goal of graphic development is a more effective interface and better end-user understanding. For this, several methods and theories were developed to give support to the graphic developers in this task, but these are not measured in scientific evaluative form, being supported by conventions. A scientific method is needed to understand how the user assimilates the message developed in a graphic object by the designer. This study focuses on graphic objects developed with a method based only on informational ergonomics. The resulting graphic marks were presented to end users in an online questionnaire that raised their impressions, allowing them to cross-reference the data with the attributes considered in the development. The results showed a correct presentation of the items requested in the briefing by the experimental group, which used the method directed to the development of these graphic brands.

Keywords: End user · Information ergonomics · Visual perception
Graphic marks · Questionnaire

1 Introduction

Reaching the end-user is the first reason for a graphic development. The image is developed in such a way as to create a message so that the reader can read, feel and understand it, and understanding how it is processed is interesting for Design. Several methods and theories have arisen to provide graphic developers with this task, but these are not measured in an evaluative way, supported by historical conventions and few scientific tests with concrete methods.

Studies, such as Forrattini [1] and Gentil [2], affirm that the damages of information overload are not restricted only to spaces and products, but can also affect the mental health of the user, of unnecessary information. They also say that the biggest problem is not the existence of information, but its lack of control.

The use of the design method characterizes the work done by designers, whether this graphic or product, and this aspect allows a functional product. Free or intuitive methods on the other hand generate products with illustrative image, contemplative objects or a product that is not sure of its functionality, ignoring the functional value of the design. Some professionals have ventured into using project development methods that are not based on a scientific study, so their results, the final products, will be arbitrarily submitted to users.

Facing this gap, it was observed the need to evaluate a specific scientific method to understand how the user assimilates the message developed in a graphic object by the designer through a semantic differential questionnaire for end users.

2 Graphic Brands and Informational Ergonomics

The graphic mark is the symbolic representation of an institution or product, something that can be identified immediately, as a symbol, an icon or a word. It consists of a sensitive sign, that is, a junction of verbal information with visual information, a linguistic sign used to designate, verbalize, write and internalize an institution, so that it is close and identifiable by the user [3].

Informational ergonomics becomes responsible for the visibility, readability, comprehension and quantification, prioritization and coordination, standardization, compatibility and consistency of the symbolic components, such as alphanumeric characters and iconographic symbols, which are widely used in the signaling, security and orientation system [4].

These models of cognition use some paradigms for their functionality as attention, perception, understanding and memorization. Attention is related to the level of alertness of the organism. Perception is the relation to the cultural context of the individual. The understanding is related to the correspondence of the senses of the message attributed by the source and the memorization by the selective retention of the message, as can be visualized in Fig. 1, information processing model proposed by Alves and adapted by the author.

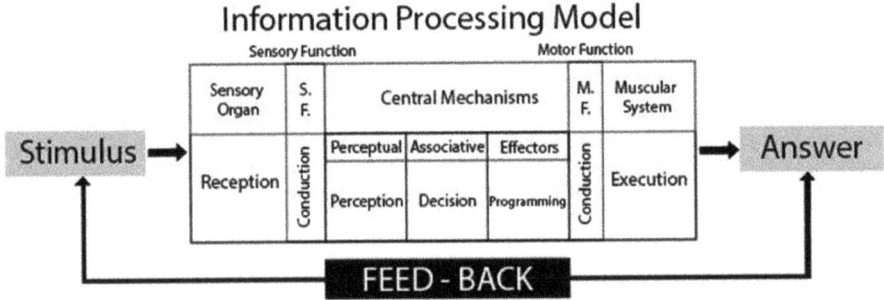

Fig. 1. Information Processing Model (Source: Alves, 1985, adapted by the author)

This same model can be used in the understanding of visual identities, which usually present information about the institution to be identified, so that the user can understand and associate with something already seen and processed by his cognitive system. The use of pictograms and icons in visual identities can facilitate this understanding.

Informational ergonomics involves a series of aspects and principles, which deal, in particular, with the whole relationship at the interface between man and technology, where the visual and auditory environment in information processing are preponderant to an action or activity [5].

3   Development Workshops

For the study it was necessary to develop a graphic design workshop with two groups of students of design courses from universities in the city of Bauru, SP. The first development of graphic marks was done in a free way, without methodological guidance and the second with a directed methodology, as seen in Fig. 2. Students were

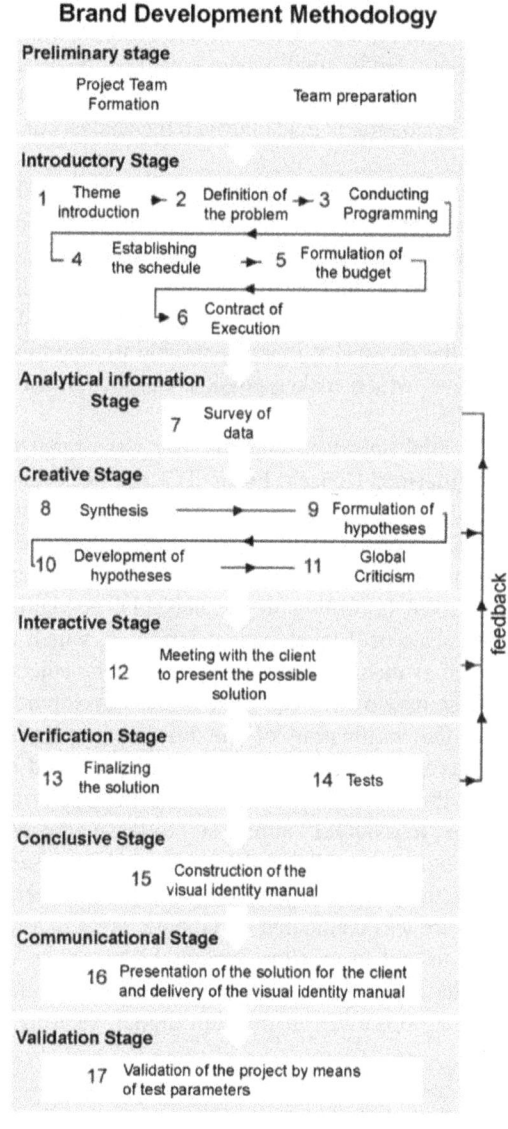

Fig. 2. Brands Methodology development

instructed to develop the second group of graphic marks to consider the attributes referring to information ergonomics and the adjectives present in the image created. These results were evaluated according to a specific chart mark evaluation developed by Silva [5], the same author of the methodology, in which he selected 4 marks for evaluation of end users, of which 2 of the control group and 2 of the experimental group being of the same sector, for later comparisons.

4   End User Questionnaire

The objective of the questionnaire was to evaluate the perception of end users about each graphic brand, in order to understand if the attributes requested in the briefing were actually perceived by this individual. The test selected for this evaluation was the scale that has been diffused in some studies of perception of symbolic values related to products that is the Semantic Differential, where pairs of bipolar adjectives are placed by a Likert scale, which according to Tullis and Albert [6] can be composed of 5 or 7 points. And generally its evaluation is made through the analysis of the values of means and of factorial analysis.

For the design area, this technique is used to evaluate users' feelings regarding both products and interfaces and brands [7]. It has been used by other researchers to verify specific aspects of the shape of the products, including style, color and other important attributes for the area. The great challenge of using the semantic differential technique is in the selection of the correct adjectives to be used, in this case Tullis and Albert [6] consider that the use of the dictionary is indispensable to ascertain the possible antonyms for a given adjective, which makes possible a selection according to what one wishes to investigate [8].

The Semantic Differential (DS) test with end user was organized in 6 phases. The first one consists of the Informed Consent Form (TCLE), the second the personal data, which includes the gender, age group, region, training and training area. The third, fourth, fifth and sixth parts consist of the graphic marks being evaluated. There are 6 bipolar descriptor pairs, arranged at the extremities and among which there were seven anchors to be marked according to the subjects' perception. The use of the bipolar scale on a likert scale of 7 levels was determined, which allowed the organization of the adjectives of evaluation, power and action. The adjectives inserted in the table were the same ones previously requested in the development briefing, besides three more that aimed the verification of the information ergonomics, adding 18 adjectives, but 12 adjectives per brand to be compared presented in Fig. 3.

In this DS protocol the order of presentation of the marks and terms was randomized, through the site "RANDOM" aiming not to leave adjectives with positive or negative characteristics only one side of the scale, and also seeking to avoid any comparison bias between the pairs of adjectives, who also had their order randomized.

The entire questionnaire was conducted on Google forms in a way to make it more practical and accessible to end users from various locations, as well as providing a general control of how the research is being conducted without the need for multiple researchers, and thus being disclosed in medium digital, such as social networks, e-mails and messengers.

Adjectives of the Semantic Differential Questionnaire		
	Positive	**Negative**
Common	Easy to view	Difficult to view
	The name is legible	Name is not legible
	Trustworthy	Unreliable
"Samurai"	Elegant	Sloppy
	Clean	Dirty
	Oriental Restaurant	Martial Arts Academy
"Diginfo"	Technological	Classic
	Harmonic	Messy
	Computer shop	Computer Brand

Fig. 3. Adjectives selected for the questionnaire

5   Results

The questionnaire was applied in a time frame of 30 days with a large number of followers, which showed the efficiency of the online questionnaire and generated less biased results, since there is no participation of an interlocutor. The result showed a correct presentation of the items requested in the briefing by the experimental group, which used the method directed to the development of these graphic brands. The method of analysis of the questionnaire was followed by specialists who confirmed the result, validating the construction and application of the questionnaire developed for analysis of visual perception.

The results were expressed in infographics which demonstrate the means of responses in a ruler where the negative aspects were allocated to the left whereas the positives to the right differed from how the user was asked to evaluate the mark where the standardization was randomized. To facilitate the understanding of the individual results, when the mean of the point of the rule goes to the positive adjective was represented by the blue color, while it goes towards the negative part it turns red, as seen in Fig. 4.

The result of the semantic differential in its total places the experimental brand as the best in 11 of the 12 analyzes performed, or 91.7% of the research with the end users, reaffirming that the directed development method allows a project that achieves

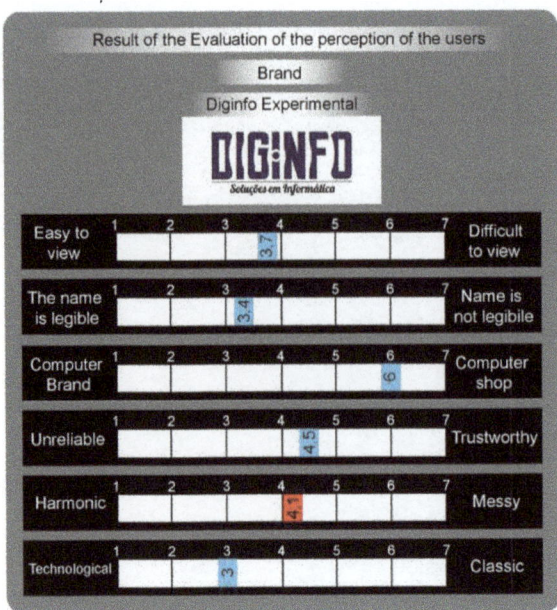

Fig. 4. Result expressed in infographic of the questionnaire for the end user of the Diginfo brand

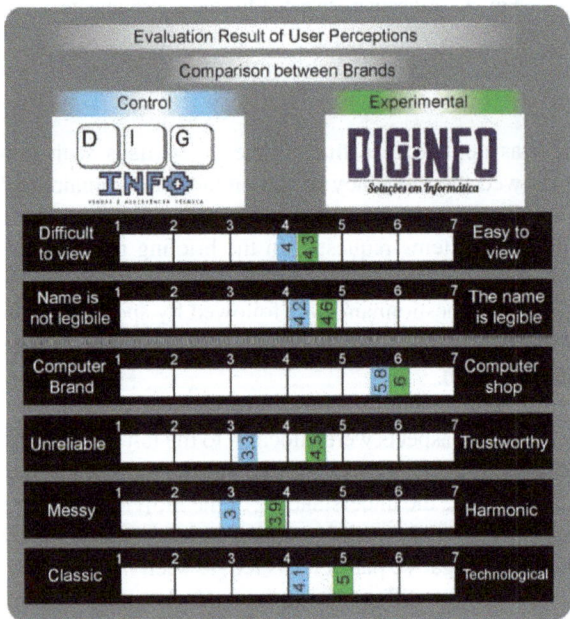

Fig. 5. Comparison of the results obtained by the marks in the semantic differential evaluation.

better levels of acceptance and perception of the qualities that the brand had to represent, corroborating with the results of the specialists, obtained by the DELPHI method. The mark of the experimental group in both cases is more faithful to the needs requested in the briefing, this can be evaluated in Fig. 5, which presents the comparative between the marks and the impressions of the end user.

6   Conclusive Notes

The results showed a more positive user perception for the brands of the experimental group in the two cases studied, which allows to affirm that the experimental development method reaches the points requested by the briefing, allowing the user to better read and understand an understanding of the correct attributes that the company seeks to achieve.

The attributes were rearranged in a way to be more understandable and clear, maintaining the direction of the requested briefing, facilitating the understanding of the end users, objects of study at this stage. These attributes were organized in an online questionnaire and divulged so that they could express their perception in relation to the graphic brands developed. There was concern that the message informed by the brand would be the same as requested in the briefing. The results indicate responses from all over Brazil, with an audience of different schooling and ages, concluding that the preference for the GE brand was 100%, validating in the second method the directed experimental methodology.

It is concluded from the methods used to evaluate projects and attributes that a directed design method allows a more objective development that meets the characteristics that the institution (client) intends to pass, achieving a level of comprehensibility of the general public and, consequently, a better usability of the information.

References

1. Forattini, O.P.: Qualidade de vida e meio urbano. A cidade de São Paulo, Brasil. Rev. Saúde Pública, vol. 25, no. 2, São Paulo, abr (1991)
2. Gentil, P.A.B.: Poluição Visual é Crime. Portal Clubjus, Brasília–DF (2008)
3. Costa, J.: A imagem da marca: Um fenômeno social. Trad.: Osvaldo Antonio Rosiano. Edições Rosari, São Paulo (2008)
4. Santos, N., Fialho, F.: Manual de Análise Ergonômica no Trabalho. Curitiba: Gênesis Editora, 2a edição (1997)
5. Silva, J.C.R.P.: Diretrizes para análise e desenvolvimento de identidade visual – contribuições para o design ergonômico. Dissertação de mestrado do Programa de Pós-Graduação em Design - FAAC - UNESP, Bauru (2012)
6. Tullis, T., Albert, W.: Measuring the User Experience: Collecting, Analyzing, and Presenting Usability Metrics. Morgan Kaufman, Burlington (2008)

7. Santa Rosa, J.G., Moraes, A.D.: Design participativo, técnicas para inclusão de usuários no processo de ergodesign de interfaces. 1a. ed. Rio Books, Rio de Janeiro (2012)
8. Lanutti, J.N.L., Campos, L.F.A., Pereira, D.D., Paschoarelli, L.C.: Análise de usabilidade do juicy salif a partir de teste de diferencial semântico emdiferentes níveis de interação. In: 12o Congresso Internacional Ergodesign / USIHC - Ergonomia, Design, Usabilidade e Interação Humano- Computador. Natal. 12o Congresso Internacional Ergodesign/USIHC (2012)

Accessibility in Chatbots: The State of the Art in Favor of Users with Visual Impairment

Cecília Torres$^{(\boxtimes)}$, Walter Franklin, and Laura Martins

Universidade Federal de Pernambuco, Recife, PE, Brazil
cvtb@cin.ufpe.br, ergonomia@terra.com,
bmartins.laura@gmail.com

Abstract. Society has been experiencing a great technological advance in the most diverse areas and, clearly, the development of accessibility for software and applications does not seem to follow this speed. In fact, systems sometimes do not embrace people with some kind of disability, and this is a problem that should be on the agenda of every designer and system designer when thinking about user experience. Chatbots are conversational interfaces on which users communicate with a robotic entity through text, either designed with artificial intelligence or not. However, how does a blind user interact with Chatbots? How should this interaction be carried on? What to expect when users' needs are challenged by physical barriers worse than what affects the common user? This article aims to present a systematic review of the existing literature on Chatbots, conversational interfaces and the inclusion of accessibility in these interfaces.

Keywords: Accessibility · Chatbots · Conversational interface
Visual impairment · Smartphones · User centered design
Rapid systematic review

1 Introduction

1.1 Accessibility and Smartphone Uses

In the widely diversified world where we live today, where each person is in a very specific context, we should not assume that everybody interacts with digital products in the same way. When we think about inclusion and diversity, accessibility should be the keyword. A product considered affordable is a product that can be used by all kinds of users. However, in most cases, there is little effort being made to design products that work well for everybody: from the common user to a range of disabled persons.

It is estimated that there are around 40 to 45 million blind people in the world today, according to the World Health Organization. By 2020, it is expected that those numbers will rise to 75 million of blind people and 225 million with low-vision.

"According to data from the 2010 Demographic Census, there were in Brazil 45,606,048 people with at least one of the investigated deficiencies, which makes up 23.9% of the Brazilian population. The visual impairment was the one that most affected the population, where 35,774,392 people reported having difficulty seeing, even with the use of glasses or contact lenses, which is equivalent to 18.8% of the

© Springer International Publishing AG, part of Springer Nature 2019
T. Z. Ahram and C. Falcão (Eds.): AHFE 2018, AISC 794, pp. 623–635, 2019.
https://doi.org/10.1007/978-3-319-94947-5_63

Brazilian population. Of this amount, 6,562,910 people had severe visual impairment, 506,337 of whom were blind (0.3% of the population), and 6,056,533 had great difficulty in seeing (3.2%)" [7].

The visually impaired are a significant part of the population, who also are users of all the available technologies, especially mobile devices and smartphones. Thinking of ways to make the use of these devices easier is essential in an inclusive world. Designers play a role of crucial importance in the development process of inclusive and accessible interfaces. To understand those users' difficulties, to sympathize with them and, especially, to know how they interact with those devices are the first steps to succeed in the making of an accessible interface.

The increase in the quantity and variety of mobile devices available on the market comes with a huge range of applications designed to make life easier for people. These include users who are blind, deaf or otherwise disabled. Although there are no statistics on the adoption of mobile devices by people with disabilities, it is known that more than 100,000 people who are either blind or suffer from low-vision have used iPhone since the introduction of VoiceOver (iOS screen reader) and zoom options in 2008 (U. B. of Engraving and Printing, 2011). Despite this huge growth in the number of devices, applications and users, accessibility has not yet become a priority in the technology development.

"Being accessible means making your system, with all data and resources, available for anyone, notwithstanding the way they use it or the difficulties they may face." [3]. Cunningham [3] also adds that, once accessibility becomes a goal within development, the project can be developed without data overload and the resulting system can become even better for the common users. To include accessibility in design development should be a goal for any designer who projects solutions for current problems.

1.2   Design Process

Design process is a human activity that goes back to primordial times and has evolved throughout the centuries. Whereas our ancestors built stone tools, we are in the present able to project intangible artifacts such as software and applications for mobile phones. The development of artifacts projected by human beings, from the stone tools to the complex interfaces that we project these days, has walked in tandem with the evolution of interaction design.

Preece et al. [16, p. 8] define interaction design as "to project interactive products to support the way people communicate and interact on a daily basis, either at home or at work". Interaction design also allows "to create experiences to better and widen the way people work, interact and communicate to each other". Winogard [19], in turn, describes interaction design as "projecting spaces for human communication and interaction."

Design process, according to Löbach [11], is both a creative and troubleshooting process. There is a well-defined problem; information about the problem is gathered together, analyzed and its parts are creatively related to each other; alternatives for the solution are brought about and, at last, the alternative judged the most adequate is developed.

This cycle is iterative, as it is possible to reach back or forward to any stage during the project. To put the user at the center of the project is also to include them in the process. The user-centered design is "a philosophy based on their needs and interests, which gives special attention to the question of making products that are within their grasp and are easy to use" [14]. There are several theories regarding design principles. A sizable part of these studies deals with determining what designers should take into consideration when creating an interactive system.

Norman [14] defined design principles as "to make it sure that (1) the user is able to find out what to do and (2) that they have the conditions to know "what is happening", and proceeds to describe them as such:

- **Visibility:** the more visible the functions are, the more able users will be to proceed;
- **Feedback:** related to the visibility concept, it refers to making information feed back to the user as to what action was made and what was achieved;
- **Restrictions:** it is about determining the ways to delimit the kind of interaction that may happen at any time;
- **Charting:** it is about the relation between controls and their effects;
- **Consistence:** it is about interfaces that have similar operations, with similar elements for the performance of similar tasks;
- **Affordance:** term applied to refer to an object's attribute that allows for people to understand how to use it. It is about elements that are self-explanatory as to how the user is supposed to interact with them.

The aim of methods and techniques directed towards evaluation is to verify the experience of using a system, product or service and their interaction with people. Thus, according to Nielsen [13], "usability is a quality attribute that gauges how easy to use is a given interface" and "the measure of a user's experience quality when interacting with a product or system." That is to say that usability is associated with the employment of methods that help facilitate the use during the process of conception of a system, product or service.

As per Nielsen [13], usability is linked to the following factors:

- **Efficacy:** being able to accomplish what is expected from the product.
- **Efficiency:** the way by which the system eases users into completing their tasks by using as few steps as needed in order to reach their goal.
- **Security:** to protect the user from dangerous conditions and undesired situations.
- **Utility:** it has to do with how the system provides the right kind of functionality, in such a way that the users may be able to do what is needed or wanted.
- **Learning capability:** how easy is to learn to use the system.
- **Memorization capability:** being easy to remember how to use the system once one already has learnt it.

The use of design principles applied to accessibility influences directly the development of interfaces both easier to interact and more efficient to use, which ensues more benefits for the user.

1.3 Conversational Interfaces

A conversational interface is any one that works in the manner of a conversation between a human and a machine. These interfaces allow for the user to interact with intelligent devices (which may range from chatting with a robot to even objects that answer questions asked by users) through spoken language. Instead of communication by means of non-human terms with syntax controlled by specific command lines, it flows like a real conversation between two persons.

For a long time, conversational interfaces were only a sight into the future thought up by researchers in fields such as speech technology and artificial intelligence, but until not long ago those intellectual forays were rather based on science fiction's books and movies.

"Since the mid-1950s, artificial intelligence (AI) researchers have struggled to conquer the challenge to build computers capable of intelligent behavior. AI has gone through cycles of euphory and rejection, having had some initial accomplishments followed by some dramatic failures" [12, p. 16].

Many technological advances have contributed towards the increase in number of conversational interfaces, besides the users' increasing approval to make use of such interfaces. There are currently two such interfaces: Voice Assistants, by which the user speaks and the interface gives answers (like, for instance, Siri and Ok Google, which operate, respectively, on the iOS and Android platforms) and Chatbots, interfaces where the user interacts through texting.

Chatbots ("bot" standing for a shortened form of "robot"), also known as chatterbots, simulate real talk, wherein the user inputs some text and the Chatbot outputs an answer. Even though some Chatbots are developed with the aim of deceiving the user into thinking there is an actual chat between two persons taking place, this is not considered sound practice within a Chatbot project.

Most Chatbots interact through texting with users, though it has been common practice the implementation of buttons and menus that anticipate the users' decisions and make the conversation run faster. It is also possible to include avatars and empathetic answers so that the Chatbot appears to have something of a human personality.

According to Mc Teal et al. [12], Chatbots had their start with a system developed by Weizenbaum (1966) called ELIZA, which simulates convincingly the type of conversation a therapist would carry on. ELIZA inspired a whole generation of Chatbots developers since then.

"Chatbots have been increasingly used in such areas as education, data recovery, business and e-commerce, for instance, as automatized online assistants to complement or even replace call centers' human-based services" [12 p. 16].

Chatbots are conversational interfaces whose functioning is dictated by rules or generated through artificial intelligence. The difference between both kinds of interface is as follows:

Rule-based:

- Narrower output, responses are given only to specific commands;
- Follow well-defined navigation charts;
- If the user makes a mistake, the system won't be able to interpret the input;
- Their intelligence goes only so far as the code allows for.

Artificial Intelligence:

- Those systems have an "artificial mind", that is, there is no need for the user to be the most precise in what they say because the bot learns and understands natural language, not only command lines;
- Chatbots learn and become more intelligent the longer they have conversations with users.

However, how can blind users interact with those interfaces? Do designers and developers have accessibility in mind in their projects? The importance of this kind of research resides in investigating how accessible those conversational interfaces are. Besides, this line of inquiry focused on the research's social impact into the future not only improves the systems' usability and consequent accessibility, but also makes them better for the general public.

2   Justification and Relevance

Previous bibliographic research shows that guidelines for accessible projects are still insufficiently explored and little known; interfaces projected specially for a public with specific needs are almost non-existent. For blind and low-vision users, the most used solutions are the TTS (Text to Speech), a system that converts written text into voice, and the screen readers, which describe through sound one interface's content and its interactions.

ISO 9241 defines usability and deals with requisites and recommendations for user-centered design principles and activities related to the cycle of interactive systems, such as: definition of use context, creation of requisites and solutions, tests and software evaluations.

The benefits of a system following those recommendations include increased productivity, a rise in the well-being of users, less stress, better accessibility and decreased margin for mistakes.

ISO 9241 describes six key principles that make it sure that a given project is user-centered:

- Project based on an explicit understanding about the users, tasks and use context;
- Users involved in all the development process;
- The project is conducted and improved by means of a user-centered evaluation;
- The project is iterative;
- The design tackles the whole of the user experience;
- The team is multidisciplinary, composed of people with different abilities and perspectives in relation to the project.

Usability and a good user experience are much more than the simplification of an interface. The ISO 9241 defines user experience as the perception and the responses of the user that come about from the use of a product, system or service, which include their beliefs, emotions, preferences, behavior, physical and psychological responses that happen during and after the use.

Preece et al. [16] list as desirable aspects in a user's experience systems that are: satisfying, pleasant, attractive, comfortable, exciting/thrilling, interesting, helpful, funny, provoking, surprising, rewarding, stimulating, challenging, that promote sociability, reward creativity, be emotionally fulfilling and cognitively challenging. On the other hand, the authoresses empathize that systems cannot be: boring, frustrating, irritating, infantile, unpleasant, patronizing, that make the user feel stupid or that be too glossy and artificial.

The main goal in developing products and systems with those desirable features lies in the experience the user will have while interacting with the system. Usability is the key factor in improving efficacy, efficiency and satisfaction during the use of a given interface. Accessible interfaces improve the user experience because they make the product easy to use, better their efficiency and efficacy and, lastly, keep the user satisfied with their use. In short, they widen the totality of users who benefit from their use experience.

"Good design is good citizenship" [5]. Only by being conscious of their duties, rights and their role in society, is the designer able to contribute with their work in an actual improvement in people's lives, most of all those who are handicapped somehow. If a fourth of the world population have some kind of disability, this number is too high to be simply put aside.

With such context within sight, it is paramount for designers to be active citizens, interested and engaged in society. That means they should be able to change the world around them. To have accessibility as a goal is not charity, it is an investment. The cost to include accessibility in a project after it has been finished is much bigger than it would be if it had been thought of since its inception. It is crucial for the designer to take part in this context as an agent of change, that they become able to perceive their role in society and their work's strength to make people's lives better.

3   Goals

The general goal of this research is to analyze the accessibility of Chatbots in the context of smartphones by means of criteria of accessibility and heuristics of usability with a focus on blind users, understanding what are the main aspects of use that make the best of experiences and what aspects can be re-studied and improved.

The specific goals are:

- To identify and analyze the studies and guidelines concerning accessibility in the context of mobile devices, with focus on the Android and iOS systems;
- To identify and analyze interactions and interfaces of Chatbots with focus on accessibility, by means of guidelines and heuristics found during a systematic review;

- To investigate what are the challenges that blind users meet when using Chatbots;
- To check aspects that can be utilized in a universal way in accessible interfaces;
- To suggest a guide of good practices involving guidelines and patterns for interactions and interfaces to be used during the development of Chatbots that have accessibility as a goal.

4   Rapid Systematic Review of Literature

The systematic review is a type of investigation focused on a well-defined theme, which aims to identify, select, evaluate and synthetize the relevant evidences available for approaching the question or a specific problem. It is "the application of scientific strategies that allow to delineate the frame of reference for selection of articles, to evaluate them from a critical viewpoint and summarize all relevant studies about a specific topic" [2, p. 126].

A systematic review is a kind of research that takes as its data source the existent literature on a given theme. Contrarily to the non-systematic process, systematic review is done in a meticulous and formal way, through application of explicit and systematized methods for search, critical evaluation and synthesis of the selected information. In order to do so, it is necessary to stablish a method for the reviewing protocol and to follow it rigorously.

In line with Sackett et al. [17], a research based on evidences leads to an unbiased evaluation and a synthesis of empirical outcomes relevant for a given research question by means of a process of systematic literature, reviewing and integration of the evidence into professional practice. Besides, once access to summaries of all studies on a given theme is achieved, the systematic reviews widen the range of relevant outcomes, consequently preventing the research to become restricted to only a part of the literature.

"Rapid reviews are a form of evidence synthesis that may provide more timely information for decision making compared with standard systematic reviews." (AHRQ, 2013). This method of review varies in terms of the time needed to complete it and is typically done in less than 5 weeks. When there is not enough time to undertake a systematic review or when it is not practical to synthesize evidence, a rapid review speeds up the process by omitting some steps that are mandatory in the systematic review.

4.1   Differences Between A Rapid Review and A Systematic Review

The basic difference between a rapid review and a systematic review is relative to the execution time and the rigor of the methodology. A rapid review takes 5 weeks maximum. The amount of time needed "depends on many factors such as but not limited to: resources available, the quantity and quality of the literature, and the expertise or experience of reviewers" [4]. Sources are limited due to search time constraints, though transparent and reproducible search methods are still used. As it is done in the systematic review, the rapid review is based on inclusion and exclusion criteria, critical and rigorous appraisal but limited time.

4.2   Methodology

Are Chatbots accessible to blind users? How do these users interact with those interfaces? In the search of the state of the art to answer these research questions, a charting of the literature was initially made.

This stage aimed at exploring as freely as possible the available literature, so as to identify possible relevant work within the studied theme.

During the research through relevant articles and books, it became a necessity to define what searching strategy to adopt. In the exploring research, books and articles possibly relevant for the research's progress were found, but nothing specifically related to Chatbot accessibility came up.

The searching strategy was shaped by choosing search engines, by formulating search terms and by intersecting keywords and the whole of the retrieved results. The used terms had keywords such as "accessibility", "Chatbot", "chatterbot" and "conversational interface", generating the following search strings:

- Accessibility Chatbot
- Accessibility Chatterbot
- Accessibility "Conversational Interface"

Besides the English keywords, the same corresponding Portuguese keywords were used: "acessibilidade" and "interface conversacional". The words "Chatbot" e "chatterbot" do not have equivalents in Portuguese:

- Acessibilidade Chatbot
- Acessibilidade Chatterbot
- Acessibilidade "Interface Conversacional"

The used strings were the same along all search engines due to the way each one indexes their results.

4.3   Search for Primary Studies on the Search Engines

The first step was to do the research through the search engines considered the most important and relevant to the technology, design and computer science fields. The choice of research bases was decided having the extent of the scope taken in consideration. Therefore, Periódico Capes and Scopus were chosen for the charting's exploratory stage. This was followed by research done on the ACM Digital Library and IEEE Xplore Digital Library, since both are libraries equipped with specific material in the fields of computer science and system development.

Searching was done by previously defined keywords and the formulating of search strings, which consists of combining two or more keywords. These strings also must be made in a manner specific to each engine, as sidestepping this factor can lead to very different results. The used filters also narrow considerably the number of found articles.

Besides keywords, some searches used the release date as a filtering criterion, taking in consideration articles and books ranging from 2007 to 2017.

4.4   Selection and Evaluation of the Publications

After the research steps, came the article selection stage. With the intention of further refining the researches, some criteria were stablished to help delimit the most relevant results.

The studies regarded as most important for reviewing the state of the art were chosen and then their titles and summaries were read to eliminate the irrelevant ones. In the case where the summary was insufficiently informative but there was still the perception it could be useful, both introduction and conclusion were read. After this first selection of relevant studies, each and every one must be read to determine whether they comply with both the exclusion and inclusion criteria in order to decide which ones will actually comprise the review's primary bases.

For it to be included, a publication should conform to all stablished criteria. After a publication fits the inclusion criteria, it is then compared to the exclusion criteria, whereby it would be excluded from the former selection in the case of having a positive response to at least one of the latter criteria. Thus, a collection of publications able to pass into the analyze and extraction stage would be comprised of all those that complied with the inclusion criteria while being entirely devoid of any positive response to the exclusion ones.

Inclusion criteria were:

- Publication period being the one from 2007 to 2017;
- Used language being either English or Portuguese;
- Researches dealing with interaction with conversational interfaces;
- Researches about accessibility in conversational interfaces.

Exclusion criteria were:

- Publications that are just extended summaries;
- Publications that are posters;
- Publications unavailable for free.

To be able to evaluate studies through criteria of exclusion and inclusion, it is important to verify each publication's quality as well as the quality and relevance of the analyzed text.

4.5   Extraction of Information from Publications

This stage intended to identify and choose the relevant information from the selected material to analyze it. For this purpose, one can try to answer the research questions with information already present in the studies or to hold information that carries importance for the research's continuation.

4.6   Results

Through keywords and search strings, 95 articles were chosen. After analyzing titles and summaries, a number of publications were identified as capable of adherence and relevance to the research following inclusion and exclusion criteria. So, after extracting information from every publication, a total of 25 publications were arrived at, which will be used in the research's development:

- 06 Articles related to the use of conversational interfaces in the health area;
- 05 Articles related to the use of conversational interfaces as an aiding tool in the educational area;
- 04 Articles related exclusively to accessibility or assistance to persons with disabilities.

The remaining articles refer to frameworks, patterns and innovations in the area of artificial intelligence applied to conversational interfaces. These may be useful for future research.

No publications about Chatbot accessibility were found. The theme is so much in its beginning that academic studies on it are still non-existent.

Based on the strings, series of searches were made complying with the criteria already mentioned in the methodology part. Some bases have different search operations, which demands applying some filters so that satisfying results are reached (Table 1).

Table 1. Results on search engines such as obtained by applying inclusion and exclusion criteria.

Base	Keywords/Strings	Filters	Partial result	Post-selection results
Periódicos Capes	Accessibility AND Chatbot	2007–2017	33	5
Periódicos Capes	Accessibility AND Chatterbot	2007–2017	19	2
Periódicos Capes	Accessibility AND "Conversational Interface"	2007–2017	28	6
Periódicos Capes	Acessibilidade AND Chatbot	2007–2017	0	0
Periódicos Capes	Acessibilidade AND Chatterbot	2007–2017	0	0
Periódicos Capes	Acessibilidade AND Chatterbot	2007–2017	0	0
Periódicos Capes	Acessibilidade AND "Interface Conversacional"	2007–2017	0	0
Scopus	Accessibility AND Chatbot	2007–2017 Title, Keywords, Abstract	3	2
Scopus	Accessibility AND Chatterbot	–	0	0
Scopus	Accessibility AND "Conversational Interface"	2007–2017 Title, Keywords, Abstract	4	2
Scopus	Acessibilidade AND Chatbot	2007–2017	0	0

(*continued*)

Table 1. (*continued*)

Base	Keywords/Strings	Filters	Partial result	Post-selection results
Scopus	Acessibilidade AND Chatterbot	2007–2017	0	0
Scopus	Acessibilidade AND "Interface Conversacional"	2007–2017	0	0
ACM	Accessibility Chatbot	2007–2017 Full Text	143	–
ACM	Accessibility Chatbot	2007–2017 Abstract	5	5
ACM	Accessibility Chatterbot	2007–2017 Full Text	41	–
ACM	Accessibility Chatterbot	2007–2017 Abstract	1	1
ACM	Accessibility "Conversational Interface"	2007–2017 Full Text	127	–
ACM	Accessibility "Conversational Interface"	2007–2017 Abstract	2	2
IEEE	Accessibility Chatbot		0	0
IEEE	Accessibility Chatterbot		0	0
IEEE	Accessibility "Conversational Interface"		0	0
TOTAL			95	25

5   Conclusions

Smartphones are increasingly present in people's daily life, which include those with disabilities, who comprise about one fourth of the world population. Of those, about 45 million are blind, which makes them an important public to focus on during development of products and services. To make applications and smartphones accessible has become a prerequisite for most companies these days, since the cost to include accessibility in the development of applications as soon as the project's inception is quite low. It is also already known that including accessibility in digital products improves usability even for those not disabled.

In this context, companies are progressively including virtual assistants to help their clients solve problems. They are called Chatbots, conversational interfaces with which users interact by texting. The question, though, is whether those interfaces are being projected with accessibility in sight. How do users interact with them? What are the biggest difficulties? These are questions to be answered as the research progresses.

In the review of the state of the art, no specific article or book were found on Chatbot accessibility. As it is an extremely new and almost unchartered theme, there is no academic studies on it, which reinforces the necessity of starting a general research on accessibility in conversational interfaces. What was found of literature on Chatbots and conversational interfaces was included in the review by the force of its relevance to future studies.

In order to fill the void in researches on accessibility and as a means to foster future research routes, there will be deeper assessments of the maturity level in the users' interaction with those interfaces and how much accessible they are, through the utilization of accessibility analyzing and tests with blind users.

References

1. Ahrq.gov: EPC Evidence-Based Reports | Agency for Healthcare Research & Quality (2018). https://www.ahrq.gov/research/findings/evidence-based-reports/index.html. Accessed 16 Dec 2017
2. Botelho, L.L.R., Cunha, C.C.A., Macedo, M.: O método da revisão integrativa nos estudos organizacionais. Gestão e Sociedade. Belo Horizonte, Gestão e Sociedade (2011)
3. Cunningham, K.: Accessibility Handbook. O'Reilly, California (2012)
4. Grant, M.J., Booth, A.: A typology of reviews: an analysis of 14 review types and associated methodologies. Health Inf. Libr. J. **26**(2), 91–108 (2009)
5. Heller, S.: VIENNE, Veronique. Citizen Designer. Allworth Press, Nova York (2003)
6. Holone, H.: Inclusion by accessible social media. In: International Conference on Computers for Handicapped Persons, pp. 554–556. Springer, Heidelberg (2012)
7. Instituto brasileiro de geografia e estatística – IBGE.: Censo Demográfico 2010. http://biblioteca.ibge.gov.br/visualizacao/periodicos/99/cd_2010_resultados_gerais_amostra.pdf
8. Who.int: WHO | Blindness: Vision 2020 - The Global Initiative for the Elimination of Avoidable Blindness (2018). http://www.who.int/mediacentre/factsheets/fs213/en/. Accessed 16 Dec 2017
9. ISO. ISO 9241-210:2010: Ergonomics of human-system interaction—Part 210: Human-centred design for interactive systems (2010)
10. Khangura, S., Konnyu, K., Cushman, R., Grimshaw, J., Moher, D.: Evidence summaries: the evolution of a rapid review approach. Syst. Rev. **1**(1) (2012)
11. Löbach, B.: Design Industrial: bases para a configuração dos produtos industriais. Tradução de Freddy Van Camp. Edgard Blücher, São Paulo (2001)
12. McTear, M., Callejas, Z., Griol, D.: The Conversational Interface: Talking to Smart Devices. Springer International Publishing Switzerland, Switzerland (2016)
13. Nielsen, J.: Heuristic evaluation. In: Mack, R., Nielsen, J. (eds.) Usability Inspection Methods, pp. 25–62. John Wiley & Sons, New York (1994)
14. Norman, D.A.: O Design do Dia a Dia. Rocco, Rio de Janeiro (2006)
15. Prates, R.O., de Souza, C.S., Barbosa, S.D.J.: A Method for Evaluating the Communicability of User Interfaces. Interactions 7, 1, pp. 31–38. ACM Press (2000)
16. Preece, J., Rogers, Y., Sharp, H.: Design de Interação: além da interação humano-computador. Bookman, Porto Alegre (2013)
17. Sackett, D.L., Strauss, S.E., Richardson, W.S., Rosernberg, W., Haynes, R.B.: Evidence-Based Medicine How to Practice and Teach EBM, 2nd edn. Churchill Livingstone, Edinburgh (2000)
18. Glaucoma Research Foundation: iPhone App Helps Blind and Visually Impaired to Identify US Currency (2018). https://www.glaucoma.org/news/iphone-app-helps-blind-and-visually-impaired-to-identify-us-currency.php. Accessed 16 Dec 2017
19. Winogard, T.: From Computing machinery to interaction design. Em: Denning, P., Metcalfe, R. (eds.) Beyond Calculation: the Next Fifty Years of Computing. Springer-Verlag, Amsterdã (1997)

20. Patch, K., Spellman, J., Wahlbin, K.: Web Content Accessibility Guidelines (WCAG) 2.0 (2018). https://www.w3.org/TR/WCAG20/. Accessed 16 Dec 2017
21. W3.org: Mobile Accessibility: How WCAG 2.0 and Other W3C/WAI Guidelines Apply to Mobile (2018). http://www.w3.org/TR/mobile-accessibility-mapping/. Accessed 16 Dec 2017

User Experience in Healthcare and Learning

Usability Assessment of a Portable Corneal Topography Device

Carlos Aceves-González[1(✉)], Zuli T. Galindo-Estupiñan[2],
Irma C. Landa-Avila[3], Citlali Díaz-Gutiérrez[4],
and Stephanie Daphne Prado-Jiménez[2]

[1] Ergonomics Research Centre, University of Guadalajara, Guadalajara, Mexico
c.aceves@academicos.udg.mx
[2] Master in Ergonomics, University of Guadalajara, Guadalajara, Mexico
Zulig89@gmail.com, daphne_90@hotmail.com
[3] Facultad de Ingenieria, Universidad Panamericana, Guadalajara, Mexico
ilanda@up.edu.mx
[4] Bleps Vision, Mexico City, Mexico
citlali@blepsvision.com

Abstract. This study aimed to identify design and usability issues of a portable corneal topography device, its software and manual. A usability test with a think aloud protocol was carried out by sixteen experienced optometrists. After using the device, the PSSUQ and SEQ questionnaires were applied. The results show a positive mean score for all the usability dimensions, being the top three elements of the higher evaluation the interface quality, usefulness and overall satisfaction. Further, based on the results it is possible to improve the device design by reducing the size of the handgrip and changing the location of control for image capturing. Results also allow enhancing the feedback provided by software when using the device and organise better the information in its manual. Overall, this study strengthens the idea that implementing usability tests as a key element on the design process to recognizing the user' needs, and thus, improving medical device systems.

Keywords: Usability · Portable medical devices · Optometry
Human factors

1 Introduction

Nowadays, 217 millions of people live globally with moderate to high visual impairments [1], mainly caused by imperfections on the form and refraction of the cornea [2]. The most frequent diagnoses are astigmatism, myopia, hyperopia and keratoconus. In Mexico, these diagnoses are the second disability affecting about 1.561.000 people [3].

Corneal topography device is the most common apparatus used for the diagnosis of the diseases in ophthalmology. This device allows to measure the form, curvature and refractive power of the cornea of patients. However, conventional instruments for corneal topography should remain static to correctly performance critical functions such as calibration and size and weight. These actions increase the complexity of use

© Springer International Publishing AG, part of Springer Nature 2019
T. Z. Ahram and C. Falcão (Eds.): AHFE 2018, AISC 794, pp. 639–650, 2019.
https://doi.org/10.1007/978-3-319-94947-5_64

due to a patient should remain in the same position and adjust their posture accord to the machine. In addition, traditional corneal topography equipment is voluminous and its transportation capacity is extremely limited.

A new portable corneal topography named TOCO (initials of corneal topography in Spanish) was developed in Mexico trying to solve the limitations of static topographies [4]. The first proposal was a device for measuring the cornea of newly born and children due to the absence of instruments for this population [5]. Afterwards, the research team noticed the feasibility of using the device, not only with children, but also with adults [6]. The device works with a null screen technology that connects to the computer software through USB power output. The weight of this product is less than a kilogram, and it doesn't require batteries or an extra element.

According to the ergonomics discipline, design should take advantage of the technology to develop ease of use, safety and comfortable medical devices with the purpose to attend more patients without compromises their safety and experience. As part of an ergonomic design methodology, products should face usability test to minimize the risk to potential troubles due to ergonomic adjustments before the product arrives to the market [7–10]. As a result, a usability test will offer design solutions that respond to the user needs. Normally, medical devices include interaction with hardware, control panel and software [11]; therefore, a holistic usability test of medical devices should be done with tasks and within the environments that are closest to reality and it should include both patients and medical providers. Consequently, the aim of this study was to identify design and usability issues of a portable corneal topography and to suggest design recommendations to solve them.

2 Method and Materials

2.1 Participants

Sixteen specialists in optometry (10 men and 6 women) performed the usability test. An inclusion criterion was to have a minimum of 1 year of experience in the use of traditional corneal topographers. All the participants agreed to participate voluntarily in the study and signed the informed consent where it was established that the test to be performed did not imply any risk.

2.2 Protocol

A think aloud protocol was carried out to determine the strengths and opportunities identified to improve the product based on the experience resulting from the interaction with it [11, 12]. Both questionnaires use a scale of 7 points with affirmations in the extremes "Strongly agree" (1) and "Strongly disagree" (7).

The test allowed identifying and recording the number of times the participants made an error in each task. Additionally, the Post-Study System Usability Questionnaire (PSSUQ) was applied to evaluate the satisfaction of the users with respect to the usability of the system [13] and the Single Ease Question (SEQ) to assess the difficult level of a task for users [14]. Table 1 shows the stages of the usability test and the resources used in each one.

Table 1. Test protocol

Stage	Description	Resources
Introduction to the test	The purpose and stages of the test were explained. The participants read and signed the informed consent	-Checklist with the stages of the test -Informed consent
Think aloud- device use	The participants were instructed to comment on each of the actions, also to express doubts and problems during the use of the device. The researcher provided some prompts when necessary. A Hierarchical Task Analysis (HTA) was developed and used by the researchers as a tool to guide the protocol	-Checklist with the stages of the test -Maqueta of the Software -Prototype of the device (See Fig. 1) - HTA (see Fig. 2) -User manual -Video camera
PSSUQ	The participant answered the 19 questions of the PSSUQ format on a scale of 1 to 7	- Checklist with the stages of the test - PSSUQ
SEQ	The participant performed a specific task: taking the image of the eye of three volunteer patients. Then he answered the SEQ question	-Prototype of the device -Volunteer participants

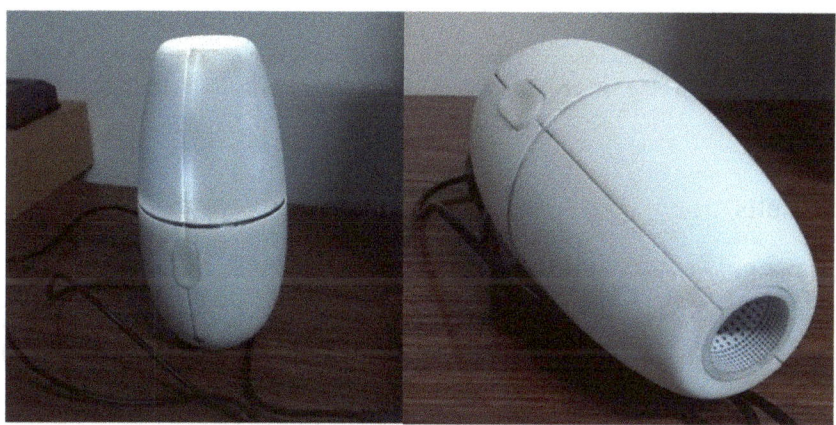

Fig. 1. TOCO device prototype

2.3   Data Analysis

The results from this study comprised different data sets which required different methods of analysis. Quantitative data from questionnaires were descriptively analysed using SPSS V19. The process to identify and quantify mistakes was undertaken using the HTA and observing the actions that users have done.

A thematic analysis [15] was undertaken to the exploration of the qualitative data. Based on the elements of the system four categories were generated to classify the

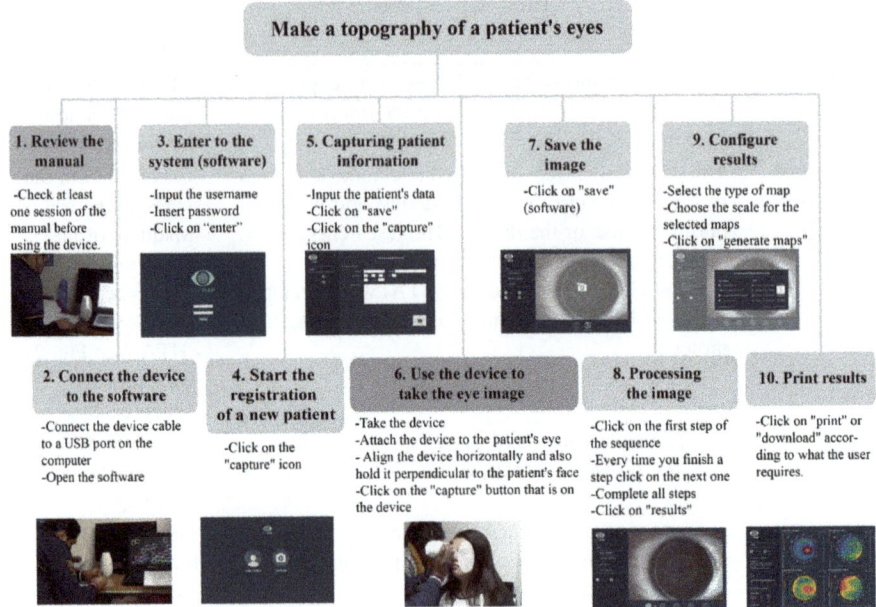

Fig. 2. HTA of TOCO device

comments of the optometrists. Three were based on the characteristics of software, device and manual, the other was related to the interaction of at least two of those elements.

3   Results

3.1   The Effectiveness of the Tasks

Table 2 shows the number of participants who made the mistake for each task, the reason expressed by the participants about their action and an image that represents that moment.

3.2   User Satisfaction Score

The results of the scores of each item of the Post-Study System Usability Questionnaire are grouped into three factors as shown in Table 3.

The task of using the device for taking the eye image (task 9) was evaluated with the Single Easy Question. This task was chosen since it is the moment in which an interaction occurs between the optometrist, the device and the patient. The optometrists were asked how easy or difficult it was to perform this task on a scale of 1 (very easy) to 7 (very difficult). The result obtained was an average of 2.94 (SD 1.48).

Table 2. Errors while using TOCO's system

Task	E*	Error description	Reasons expressed by users
1. Review the manual before using the product	15	Do not review or use the manual when using the product	They consider that it is not necessary because it must be very similar to traditional topographers
2. Connect the device to the software	0		
3. Enter to the system (software)	0		
4. Start the registration of a new patient	5	The label is not recognized to enter information of a new patient	The label "capture" is only for take the eye image in comparison and no for enter the information of a new patient
5. Capturing patient information	9	The user did not press the "save" icon before pressing the "capture" icon	By pressing "capture" patient information should be saved automatically
6. Use the device to take the eye image	7	The device was not properly aligned	There is no indicator that the device should be positioned horizontally
7. Save the image	5	The "save" icon was not pressed	The "save" icon is not visible
8. Processing the image	13	Do not follow the sequence of steps and press the "results" icon	It was not clear that the menu was a sequence of steps to process the captured image
9. Configure results	1	Do not press the "Generate maps" icon	It is not clear that the "generate maps" icon is to obtain the results
10. Print results	0		

* Errors: Number of participants that made an error

3.3   Results of the Thematic Analysis

Table 4 contains the results of the thematic analysis of the comments made by the participants when they were using TOCO. The comments are organised according to the characteristics of software, device and manual, and the last category is related to the interaction of at least two of those elements. It is also shown the number of times that a comment was mentioned.

Table 3. PSSUQ Scores

Statement	Average	Factors
1. Overall, I am satisfied with how easy it is to use this system (device, software, manual)	1.81	System Usefulness SYSUSE (Items 1–8) Score: 1.62
2. It was simple to use this system	1.44	
3. I could effectively complete the tasks and scenarios using this system	2.06	
4. I was able to efficiently complete the tasks and scenarios using this system	2.0	
5. I was able to complete the tasks and scenarios quickly using this system	1,5	
6. I felt comfortable using this system	1.81	
7. It was easy to learn to use this system	1.0	
8. I believe I could become productive quickly using this system	1.31	
9. The system gave error messages that clearly told me how to fix problems	4.56	Information Quality INFOQUAL (Items 9–15) Score: 2.23
10. Whenever I made a mistake using the system, I could recover easily and quickly	2.88	
11. The information (such as on-line help, on-screen messages, and other documentation/manual) provided with this system was clear	2.13	
12. It was easy to find the information I needed	1.81	
13. The information provided for the system was easy to understand	1.31	
14. The information was effective in helping me complete the tasks and scenarios	1.44	
15. The organization of information on the system screens was clear	1.50	
16. The interface of this software was pleasant	1.56	Interface Quality INTERQUAL (Items 16–18) Score: 1.58
17. I liked using the interface of this software	1.25	
18. This software has all the functions and capabilities I expect it to have	1.94	
19. Overall, I am satisfied with this system	1.63	
Overall items 1–19	**1.84**	

Table 4. Comments raised by participants while using the system

Category	Raised comments
Software characteristics	- It is not clear what the ROI center is, what it means or how it should be located (n = 10)
	- After making the right eye shot, the system should send to capture the image of the left eye (n = 10)
	- The "save" icon should be next to the "capture" icon (n = 5)
	- In the "image processing" screen, the buttons and the sequence of steps are confused with the background since they have similar colors (n = 3)
	- Age should appear automatically once the date of birth is entered (n = 2)
	- The date on which the data was recorded should be shown (n = 2)
	- There should be a slot to indicate whether the taking of two eyes or a specific one will be done (n = 2)
	- The configuration of the maps required for each patient must be adjustable to the needs of each specialist (n = 2)
	- Add a box to write the diagnosis (n = 1)
	- Add a box to write the name of the person doing the study (n = 1)
	- The patient's last name should appear before the name (n = 1)
	- Add a box to describe if the pupils are dilated or not (n = 1)
	- Two captures of the patient's eye should be required to ensure that a good shot is taken (n = 1)
	- The possibility of removing the noise from the eyelashes (n = 1)
	- Change the name of "image processing" to "image analysis" (n = 1)
	- Once the maps are generated, they should be saved automatically (n = 1)
	- Add the patient's age on the results sheet that is printed (n = 1)
	- Before saving/printing/downloading the results, the system must ask if it wants to take another shot, in case the results are not as expected (n = 1)
	- it must be clear which eye is the one that is being captured (n = 1)
Device characteristic	- There must be an element that allows to recharge the device in the face of the patient to have better stability (n = 12)
	- Reduce the size of the front of the device to have more space to place the fingers and thus be able to open the patient's eyelids (n = 13)
	- Improve the focus of the device, regardless of the distance a good shot should be made (n = 10)
	- The capture button should be within reach of the fingers (n = 9)
	- Add a handle that allows to better support the device, which would also help to make the capture faster n = 7
	- Add an indicator or a guide inside the null screen to ask the patient to look there (n = 7)
	- Consider the tremors of the hand and the sensitivity of the equipment as elements that make it difficult to have a good capture of the eye image (n = 5)
	- The device is very thick, this hinders a good grip and control (n = 4)

(*continued*)

Table 4. (*continued*)

Category	Raised comments
	- The image that is captured should cover the largest area of the eye, which depends on the opening of the eyelids (n = 4) - It would be ideal not to have to interact so much with the patient's eye and eyelids. Be less invasive (n = 3) - Add a sensor that indicates the appropriate distance to where the device should be placed to the patient's eye (n = 3) - The dimension of the back side of the device must be smaller to improve handling (n = 2) - The cable should be reinforced at the junction with the device and be longer (n = 2) - Do not use cable, use Bluetooth connection instead (n = 1) - Add something more comfortable or less rigid to the front of the device (the one that has contact with the patient's eye) (n = 1) - Change the colour of the device so that it does not get so dirty (n = 1) - Add a handkerchief to be able to clean the device after using it (n = 1)
Manual characteristics	- The quick guide should be outside the manual (n = 6) - Add information about the care and cleaning of the device (n = 5) - The quick guide must be more specific and with images (n = 4) - Add more images on how the device should be used (n = 4) - Add the details of the calibration of the device (n = 3) - Add and explain how external light affects the results of the capture of the image (n = 3) - Add a section of frequently asked questions, information about maintenance and repair of the device, resistance to impacts and restrictions of use (n = 1) - The parts of the device should come at the beginning of the manual (n = 1) - The nomenclature to refer to the eyes is commonly OD and OS by its acronym in Latin (n = 1)
Interaction between elements	- It is important the feedback of the connection or disconnection between the device and the software (n = 12) - Indicator on the device or software to know how to align the device (n = 10) - Add more information about the alignment of the device and the patient (n = 9) - The software should alert on which is the best image to capture (n = 6) - The manual should be included in the software (n = 3) - Generate an application so that both: doctor and patient can have the results (n = 1)

4   Discussion

The aim of this study was to identify design and usability issues of TOCO device its software and manual and to suggest design recommendations to improve its use. The results allowed to recognise the elements that affect the effectiveness and satisfaction during the use of the device.

4.1   Software

Five of the eight tasks that were performed with the software represented difficulty for the participants. "Processing the image" was the task in which the more participants made an error. In this task, a sequence of steps have to be performed, which was not understood by some participants. They omitted some steps and did click directly on the "results" icon. This situation relates with what Jordan [16] points out about the importance of prioritizing the elements according to the differences in functions. In this case, it is important to show clearly to the users that there is a series of steps that must be followed to proceed with the subsequent task.

Some recommendations to improve this aspect of the interface are (1) to list the sequence of actions [17]; (2) use color as a differentiating element between one action and another [16] and, (3) produce a change in the size of the buttons as the sequence of steps progresses.

Likewise, users did not identify the "save" button in two of the tasks. This can be explained by the fact that they pressed the "capture" button, which was closer to the last action they had taken. This is related to the principle of proximity that must exist between elements that are used in the same task [18]. On the other hand, some participants did not use the "save" icon since they assumed that the data should be saved automatically. Therefore, a mental model must be proposed according to the user's experience and knowledge with this type of software [19, 20].

Some icons generated confusion among the participants. Therefore, it is necessary to consider the use of familiar language for users [18]. Most of the comments related to the software were about the need of having more information about the patients, which should be considered by the designers to improve it. This is also related to the result of question 18 of the PSSUQ, which reports that the software interface does not have everything that users expect (1.94, the highest score in the INTERQUAL factor).

4.2   Device

Regarding the only task that involved the use of the product, it was found that 7 of 16 participants made an error in its use, mainly due to inadequate alignment of the same. The feedback of the interaction established between the device and the software was practically null. This is also reflected in the score of the questions of the INFOQUAL factor reflect the lack of error messages that help the users in correcting the problems (4.56 points), and/or the difficulty to correct the errors (2.88 points). Therefore, it is suggested to implement feedback mechanisms to correctly point out to users what they are doing [16, 18, 21]. This can be through the use of indicators, messages on the screen or auditory feedback as appropriate to the action.

On the other hand, the need to modify the dimensions of the product (the diameter of the front, middle and distal part) was recurrent, due to the difficulty that the participants had in grasping the device. This converges with the recommendations for a precise grip [22], which is essential to bring the device closer to the patient's eye. Additionally, the participants suggested the inclusion of auxiliary elements to the device that allows them to recharge the device on the patient's face to provide better stability or the design of a gun-type handle to better support the device. On the other hand, the need to modify the dimensions of the product (the diameter of the front, middle and distal part) was recurrent, due to the difficulty that the participants had in grasping the device. This converges with the recommendations for a precise grip [22], which is essential to bring the device closer to the patient's eye. Additionally, the participants suggested the inclusion of auxiliary elements to the device that allows them to recharge the device on the patient's face to provide better stability or the design of a gun-type handle to better support the device. These requests may arise because the demand of the task exceeds the resources of the users [16], therefore they propose additional elements that reinforce the ability to stabilize and be accurate during the manipulation of the device.

4.3   Manual

The manual was used by only one of the participants, which is in line with the literature that holds that health professionals are little interested in reading instructions for the use of a product (laypeople) [11]. Likewise, it is common for users to be willing to learn by trial and error, looking only for the minimum elements necessary to complete the task [20]. The optometrists pointed out that the quick guide should be outside the manual with images and specific information that would allow them to have a clear idea of how to use the device. The assessment regarding to how clear was the information obtain one of the highest values in the INFOQUAL factor (2.13 points). In particular, it has been pointed out that an interface with graphics has greater usability than one based on only text (characters) [23] and help reducing the learning curve [18].

The participants suggested including more information to the manual. However, the suggested information is already there, but they could not identify it. This can be explained by the absence of a clear hierarchy in the titles [24, 25], and the lack of content selection according to user's needs [17, 24]. According to Winklund, Kendel and Strochlic [11] the design of the instructions should influence the interaction that the user will have with the device. Therefore, the inclusion of instructional videos and training for the user as a complementary element can be considered [26] and consider the design attributes of the documents and guidelines for instructions [17].

Although the participants made mistakes in 7 of the 10 tasks, the results of the PSSUQ showed that the participants' perception is very close to the positive extreme "totally satisfied". This can be explained by the bias generated by the questionnaires because users tend to give general answers to general questions [27]. However, when the participants were asked to specifically perform the task of taking the picture with volunteers, the scores moved to the point of totally disagreeing (2.94 average points). This may be due to the fact that when interacting with patients, specialists noticed more impediments to use due to the characteristics of the product, which were not noticed during the first interaction with the device.

5   Conclusions

The usability assessment of the three elements that comprises the TOCO system made it possible to demonstrate some areas of opportunity to improve the use of this device. The recommendations to increase the usability of this portable corneal topography are (1) use of language according to the knowledge of the users (software and manual); (2) software design according to the mental model of the users; (3) generate a quick guide with illustrative graphics; (4) improve the dimensions of the device; and (5) implementing elements (physical and cognitive) that provide precision and stability when approaching the patient's eye.

Acknowledgments. This work was supported by CONACYT through the *Fondo de Innovación Tecnológica* FIT [grant number ECO-2016-C01-274898]

References

1. Bourne, R.R., Flaxman, S.R., Braithwaite, T., Cicinelli, M.V., Das, A., Jonas, J.B., Naidoo, K.: Magnitude, temporal trends, and projections of the global prevalence of blindness and distance and near vision impairment: a systematic review and meta-analysis. Lancet Global Health **5**(9), e888–e897 (2017)
2. de la Salud, A.M.: Salud ocular universal: un plan de acción mundial 2014–2019 (2013)
3. Inegi: La discapacidad en México (2014)
4. Estrada, A.J.: Amilcar Estrada Molina, Diseño y Construcción de un Videoqueratómetro portátil para uso con lactantes, Tesis de Maestría en Ciencias Físicas, UNAM, México (2010)
5. Estrada, A.J., Díaz, R., Ramírez, M.: Neonatal videokeratometer: basic design of a portable device and evaluation method. In: Brandan, M.E., Herrera-Martínez, F., Ramírez, V., Rodríguez-Villafuerte, M. (eds.) Eleventh Mexican Symposium on Medical Physics, AIP Conference Proceedings, vol. 1310, pp. 56–59 (2010)
6. Estrada, A.J.: Queratometría Infantil con Pantallas Nulas. Doctorado en Ingeniería (Instrumentación), Posgrado en Ingeniería, UNAM, México (2014)
7. Cushman, W.H., Rosenberg, D.J.: Human factors in product design. In: Advances in Human Factors/Ergonomics, vol. 14 (1991)
8. Brooke, R.E., Isherwood, S., Herbert, N.C., Raynor, D.K., Knapp, P.: Hearing aid instruction booklets: employing usability testing to determine effectiveness. Am. J. Audiol. **21**(2), 206–214 (2012)
9. Fries, R.C.: Reliable Design of Medical Devices. CRC Press, Boca Raton (2012)
10. Carayon, P., Bass, E.J., Bellandi, T., Gurses, A.P., Hallbeck, M.S., Mollo, V.: Sociotechnical systems analysis in health care: a research agenda. IIE Trans. Healthc. Syst. Eng. **1**(3), 145–160 (2011)
11. Wiklund, M.E., Kendler, J., Strochlic, A.Y.: Usability Testing of Medical Devices. CRC Press, Boca Raton (2015)
12. Rubin, J., Chisnell, D.: Handbook of Usability Testing: How to Plan, Design, and Conduct Effective Tests. Wiley, Indianapolis (2008)
13. Lewis, J.R.: IBM computer usability satisfaction questionnaires: psychometric evaluation and instructions for use. Int. J. Hum.-Comput. Interact. **7**(1), 57–78 (1995)

14. Sauro, J., Dumas, J.S.: Comparison of three one-question, post-task usability questionnaires. In: Proceedings of the SIGCHI Conference on Human Factors in Computing Systems, pp. 1599–1608. ACM (2009)

15. Braun, V., Clarke, V.: Using thematic analysis in psychology. Qual. Res. Psychol. **3**(2), 77–101 (2006)

16. Jordan, P.W.: An Introduction to Usability. CRC Press, Boca Raton (1998)

17. Robinson, P.A.: Writing and Designing Manuals and Warnings 4e. CRC Press, Boca Raton (2009)

18. Nielsen, J.: Usability Engineering. Elsevier, New York (1994)

19. Vu, K.P.L., Proctor, R.W.: Handbook of Human Factors in Web Design. CRC Press, Boca Raton (2011)

20. Rieman, J., Young, R.M., Howes, A.: A dual-space model of iteratively deepening exploratory learning. Int. J. Hum Comput Stud. **44**(6), 743–775 (1996)

21. Saffer, D.: Microinteractions: Designing with Details. O'Reilly Media, Inc. (2013)

22. Fernandez, J.E., Marley, R.J.: Applied occupational ergonomics: a textbook. International Journal of Industrial Engineering Press (2007)

23. Kools, M., van de Wiel, M.W., Ruiter, R.A., Kok, G.: Pictures and text in instructions for medical devices: effects on recall and actual performance. Patient Educ. Couns. **64**(1), 104–111 (2006)

24. Nielsen, J., Loranger, H.: Prioritizing Web Usability. Pearson Education, London (2006)

25. Ganier, F.: Factors affecting the processing of procedural instructions: implications for document design. IEEE Trans. Prof. Commun. **47**(1), 15–26 (2004)

26. Palmiter, S., Elkerton, J., Baggett, P.: Animated demonstrations vs written instructions for learning procedural tasks: a preliminary investigation. Int. J. Man Mach. Stud. **34**(5), 687–701 (1991)

27. Wright. P.: Printed instructions: can research make the difference? En Zwaga, H., Boersema, T., Hoonhout, H. (eds.) Visual Information for Everyday Use: Design and Research Perspectives, pp. 45–66. CRC Press (2003)

Designing and Developing a Prototype of Parents and Teachers Communication Application for Early Childhood

Manutchanok Jongprasithporn[1]([⊠]), Nantakrit Yodpijit[2],
Kristin Halligan[3], and Teppakorn Sittiwanchai[2]

[1] Department of Industrial Engineering, Faculty of Engineering,
King Mongkut's Institute of Technology Ladkrabang, Bangkok, Thailand
mjongpra@gmail.com
[2] Center for Innovation in Human Factors Engineering and Ergonomics,
Department of Industrial Engineering, Faculty of Engineering,
King Mongkut's University of Technology North Bangkok, Bangkok, Thailand
[3] Well International School, Bangna Campus, Bangkok, Thailand

Abstract. The purpose of this project is to develop Parents and Teachers Communication (PTC) application for early childhood. An application development process has three phases. The first phase is application design phase: the questionnaire had been used to investigate the main requirements from the users (teachers and parents). The application's function comparison was another method which had been applied in design phase. Five school communication applications had been compared to find for strong features in each application. Human factors engineering was used for designing PTC application's user interfaces in order to understand user capabilities and limitations. The second phase was application coding phase: PTC application was developed under IOS operation system by using the X-CODE program. The third phase was application usability study phase: in this phase, the usability study test was measured in two steps. First, the System Usability Scale (SUS) was distributed to 10 volunteers during application development. Second, another usability test was done by an adopted SUS questionnaire. It was distributed to 6 volunteers after they had experience with PTC application about two week period. Results showed that PTC application had high usability scale during application development (SUS, 85) and after four weeks experience (adopted SUS, 80). SUS during application development showed that PTC application was in excellent usability scale. The adopted SUS questionnaire can be interpreted that users have a great experience with PTC application. Parent-Teacher communication application for early childhood could assist to decrease cap of communication between home and school efficiently.

Keywords: Application · Communication · Early childhood · Usability

© Springer International Publishing AG, part of Springer Nature 2019
T. Z. Ahram and C. Falcão (Eds.): AHFE 2018, AISC 794, pp. 651–663, 2019.
https://doi.org/10.1007/978-3-319-94947-5_65

1 Introduction

Effective communication is essential for building school-family partnerships and children's development. There are number of ways that parents and teachers can communicate such as phone calls, school contact books, e-mails, and meeting with teachers. These methods are good ways to cooperate with teachers and keep informed about your child's progress. Written communication technique such as report and school notebook is an easy and traditional method to transfer student's information, however, it is an one way communication [1]. Two-way communication such as phone calls and parent-teacher conference is one-on-one communication technique that develops trust between parents and teachers [2]. Although, the chance to discuss students progress one-on-one is an opportunity to understand more about our children and share information, it requires a lot of time and effort on both parent and teacher sides.

Effective parent and teacher communication benefits parents, teachers, and children. The school's communication and interactions with parents affects the quality of parent involvement with their children's learning. Parents also benefit from clear communication to know how to help and support their children learning [3]. Teachers could learn more about children's' needs which is information they can apply to those children at schools. Several studies [3–12] showed that parent involvement benefits children including raising their academic achievement, increased motivation for learning, improved behavior, and positive attitude about learning in general. Studies show that students do better in school when parents communicate often with teachers. The applies especially to early years students aged between 2–5 years old. These students require full understanding from their parents and teachers to enhance their learning, physical, social-emotional and cognitive development.

Communicating through technology is an innovative and time-efficient way to transfer children's information and gain better understanding about students [13]. Technology can create unlimited opportunities to communicate between parents and teachers [14]. Nowadays, the emerging use of smartphone technology makes communication between parents and teachers more efficient, less costly and time consuming. Smartphones are one of the most important technologies for daily life. It was reported that the smartphones users increased from 36% in 2013 to 70% by 2020 [15]. Smartphone has been used for accessing to internet and applications. The traditional communication such as Voice, SMS, MMS had been replaced by e-Communication such as WhatsApp, LINE and Skype. These e-Communication provide better user experience such as multimedia objects (images, audio, video and text), information management (contact and chatroom), and social network sharing.

Effective communication between parents and teachers will enhance early child development. Therefore, this paper presents the Parents-Teachers Communication (PTC) application for early childhood development based on human factors and usability engineering approach. PTC application will be implemented at Well International School, Bangna Campus. The proposed application aims to improve communication between parents and teachers by sending e-report instead of traditional communications. Optimal understanding between parents and teachers will results in better long-term outcomes for children's social-emotional, behavioral and academic development.

2   Methods

Parents and Teachers Communication (PTC) application development process, shown in Fig. 1, has three phases including application design phase, application coding phase, and application usability study phase.

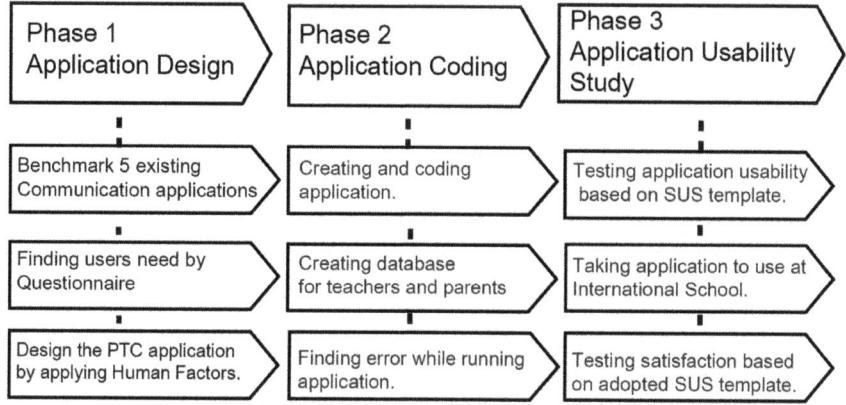

Fig. 1. PTC application development process

2.1   PTC Application Design Phase

In phase 1, an application design phase had three steps. In the first step, it was benchmarking, helping to identify the most important features from the existing applications. Five existing school communication applications were benchmarked in this study. They included Buzzmob, Remind safe-classroom communication, Google education, Edmodo, and Class messenger. Table 1 shows top three most important features from benchmarking among the five existing applications. The three most important features including reminder, real time post, and discussion, were created in PTC application for early child learning.

Table 1. Top three most important features from benchmarking among five existing school communication applications

Application features	Buzz mob	Remind safe classroom communication	Google education	Edmodo	Class messenger	Total
Reminder	✓	✓	✓	✓	✓	5
Real time post	✓	✓			✓	✓	4
Discussion		✓	✓	✓	✓	4

In the second step, the questionnaires were constructed using a five point Likert scale to allow individual to express how much one agrees or disagrees with each statement. Then the questionnaires were distributed to eight teachers and twenty-seven parents to ask about their needs related PTC application. The parents and teachers questionnaires along with their feedbacks were shown in Tables 2 and 3, respectively.

Table 2. Questionnaire for teachers to identify their needs about PTC application

No.	PTC application features	Sastisfaction level				
		5	4	3	2	1
1	Included "A medication box" will help the teacher about when and how much their child need medicine when they sick from parents	5 62.50%	2 25%	- 0%	1 12.50%	- 0%
2	Included "Story book during circle time" that you can share with parents to enhance kids understanding	2 25%	2 25%	2 25%	2 25%	- 0%
3	Included "Section seperator" that teacher can separate class or section in 1 application	1 12.50%	5 62.20%	1 12.50%	1 12.50%	- 0%
4	Included "Calendar" that teacher setting activities of kids around 1 year with student information in it	3 37.50%	3 37.50%	1 12.50%	1 12.50%	- 0%
5	Included "Timeline and share" that teacher can share picture and video of the student to parents	5 62.50%	3 37.50%	- 0%	- 0%	- 0%
6	Included "Time check-in, Check-out" that teacher can take photo of kids when they arrived at school to make parents acknowledge	2 25%	2 25%	3 37.50%	1 12.50%	- 0%
7	Included "Discussion and voting" that teacher can make pole for voting the activities of kids at school	4 50%	- 0%	2 25%	- 0%	2 25%
8	Included "Communication channel" that teacher can contact to parent directly and group of parents	4 50%	3 37.50%	1 12.50%	- 0%	- 0%
9	Included "Reminder" that teacher can remind parents about homework, activities, etc. of students and remind parents conferencing at school	6 75%	1 12.50%	1 12.50%	- 0%	- 0%
10	Included "Language translators" that teacher type by your own language to decrease the time of translating	5 62.50%	1 12.50%	- 0%	- 0%	2 25%

5 is the most requirement option that you want from application to 1 is the least respectively.

The survey showed that 75% of teachers needed the "Reminder" feature. About 62.5% of teachers though that "A medication box", "Timeline and share", and "Language translators" should be constructed in PTC application. 50% of teachers strongly agreed that "Discussion and Voting" and "Communication Channel" are the most important features.

The parent questionnaire showed 77.78% of parents though "A medication box" and "Communication Channel" will be very helpful features in PTC application. 59.26% of parents strongly agreed that "Reminder" and Kids Information" should be constructed in PTC application. 55.56 and 51.85% of parents needed "Calendar" and "Story Book" features in PTC application, respectively. Therefore, PTC features were designed and constructed based on benchmarking and questionnaires results.

Table 3. Questionnaire for parents to identify their needs about PTC application

No.	PTC application features	Sasticfaction level				
		5	4	3	2	1
1	Included "A medication box" that parents can inform a teacher about when and how much your child need medicine	21 77.78%	4 14.81%	2 7.40%	- 0%	- 0%
2	Included "Story Book" that parents can help your kids to improved their language skill, creativity and inspiration	14 51.85%	9 33.33%	4 14.81%	- 0%	- 0%
3	Included "Kids information" that parents would like to share with teachers. Ex: Behavior, progress in the subject, etc.	16 59.26%	6 22.22%	5 18.52%	- 0%	- 0%
4	Included "Calendar" that parents can acknowledge activities of kids around 1 year with student information in it	15 55.56%	4 14.81%	7 25.93%	- 0%	1 3.07%
5	Included "Timeline and share" that parents can share capture and save the picture or video of your kids	9 33.33%	11 40.74%	5 18.52%	1 3.07%	1 3.07%
6	Included "Time check-in and check-out " that parents can check arrival of the kids to school or the time from school to home	12 44.44%	7 25.93%	7 25.93%	1 3.07%	- 0%
7	Included "Discussion and voting" that parents can vote activities of kids at school and comment your ideas	10 37.04%	11 40.74%	5 18.52%	- 0%	1 3.07%
8	Included "Communication channel" that parents can contact to teacher or others parent directly	21 77.78%	4 14.81%	1 3.07%	1 3.07%	- 0%
9	Included "Reminder" that help parents remind homework, activities, etc. of students and remind parents conferencing at school	16 59.26%	8 29.63%	3 11.11%	- 0%	- 0%
10	Included "Language translators" that can open application from computer and understanding the other language	10 37.04%	8 29.63%	4 14.81%	4 14.81%	1 3.07%

5 is the most requirement option that you want from application to 1 is the least respectively.

In the last step of design phase, human factors were applied into conceptual and visual design to develop PTC application to match users' conceptual model and user interface. In conceptual design, developers have designed various operating systems within the application by taken the results from benchmarking and questionnaires as the important parts to meet the needs of users. Moreover, PTC application was designed based on the actual group of users, parents and teachers for early children school as shown in Fig. 2. When the design of application was not matched with users' conceptual model, the application will be too complex for the user group. In visual design, the display formats such as main screen, screen color, icon location, icon size, screen contrast sensitivity, and flow of events are vital to efficient of PTC application. Moreover, developers must take into account the coordination between the application and the user interface, so the communication between teachers and parents are simple and responsive.

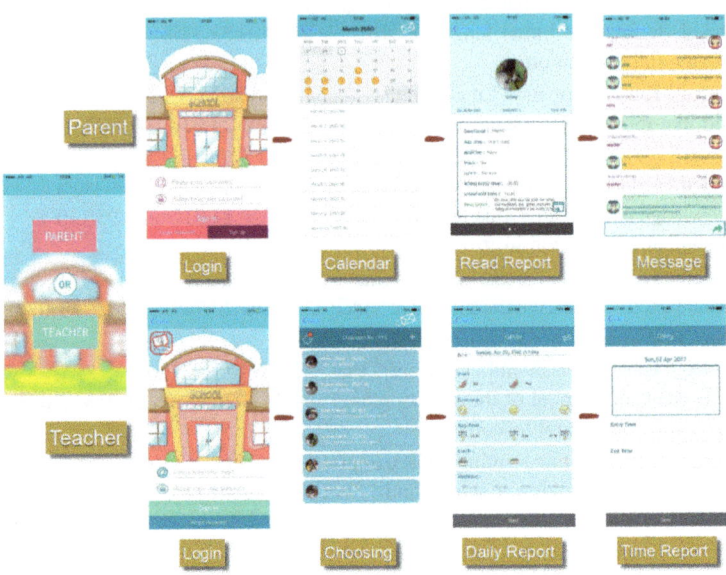

Fig. 2. Conceptual and visual design in Parents-Teachers Communication application

2.2 PTC Application Coding Phase

In the first step of coding phase, Swift 4 and Xcode program was used to develop PTC application. PTC application runs under iOS system. Swift is easy code which use for writing software that is fast and safe by design. Swift can make programming simple things easy, and difficult things possible. It was developed by Swift.org, with source code, a bug tracker, mailing lists, and regular development builds [16]. Xcode is an integrated development environment (IDE), containing software development tools developed by Apple for developing software for iOS system [17].

Creating PTC application database is the next step in coding phase. The Firebase real-time database was utilized to store student, parent, teacher, school, and communication data. Firebase is a cloud-hosted database which stores data as JSON and synchronize in real-time to every connected user [18]. Table 4 shows an example of database structure for student's information.

Table 4. An example of database structure for student's information in PTC application.

Student			Student report							
Full name	Nickname	Date of birth	Date	Snack	Lunch	Emotional	Nap	Medication	Note	Picture
Virginia Yodpijit	Ginny	11-Mar-13	10-Jan-15	Yes	Eat a lot	Happy	1 h	No	:)	:)
.	.	.	.	.	.	.	.	.	.	.
.	.	.	.	.	.	.	.	.	.	.
.	.	.	.	.	.	.	.	.	.	.

2.3 PTC Application Usability Study Phase

The System Usability Scale (SUS) was developed by John Brooke in 1986 [19]. The SUS has 10 items, with odd-numbered questions (1, 3, 5, 7 and 9) worded positively and even-numbered questions (2, 4, 6, 8 and 10) worded negatively. The 5-point scales numbered from 1 as Strongly disagree to 5 as Strongly agree were used in SUS. In case that subject fails to respond to any question, assign it a 3, the center of the rating scale. In each question, score contribution will range from 0 to 4. The score contribution in positively-worded questions is the scale position minus 1. For negatively-worded questions, it is 5 minus the scale position. The overall SUS score will compute by multiply the sum of the question score contributions by 2.5. Therefore, SUS scores range from 0 to 100 in 2.5-point increments. The recommend ranges for SUS scores are separated into three ranges including Not acceptable (0–64), Acceptable (64–84), and Excellent (85–100). The System Usability Scale (SUS) was used to test usability of PTC application. The following is the list of 10 SUS questions that were distributed to 10 users (7 parents and 3 teachers) during PTC application development.

1. I think that I would like to use this system frequently.
2. I found the system unnecessarily complex.
3. I thought the system was easy to use.
4. I think that I would need the support of a technical person to be able to use this system.
5. I found the various functions in this system were well integrated.
6. I thought there was too much inconsistency in this system.
7. I would imagine that most people would learn to use this system very quickly.
8. I found the system very cumbersome to use.
9. I felt very confident using the system.
10. I needed to learn a lot of things before I could get going with this system.

When PTC application completed, users had experience with PTC about two weeks. The adopted SUS, created for this study, were distributed to the same group of user for 5 parents and 1 teacher. The score method was computed as describe in previous paragraph. The list of adopted SUS questions is following.

1. I thought login function was easy to use.
2. I found that it was not complicate to add students in this application.
3. I thought that it was easy to approve parents (*for teachers*) in the system and get approve to use the system.
4. I need a support to write (*for teachers*) and read (*for parents*) the report.
5. I found that easy to record (*for teachers*) and read (*for parents*) the entry-time and exit-time.
6. I needed to learn a lot to use image report.
7. I found that I felt very confident to send messages.
8. It was hard to read the messages from parents and teachers.
9. I felt very positive to learn and use PTC application.
10. I thought it was much inconsistency in PTC application.

3   Results

3.1   Parents and Teachers Communication (PTC) Application

The users of PTC application are parents and teachers. The application was designed and created in two main parts including parent and teacher parts. The login page for parent was applied color coding with pink color. Teacher login was applied color coding with blue color as shown in Fig. 3.

Fig. 3. Parents-Teachers Communication application (a) PTC starting page, (b) Teacher login page, and (c) Parent login page

The teacher and parent parts in PTC application illustrate in Figs. 4 and 5, respectively. Each teacher have to be registered by a school administration before starting using PTC application. Username could be teacher's e-mail and password was generated by random. Password could be changed after the first successful login. To start daily report, teachers have to select student names to go to the report and comment pages. Student's information in report page is including snack, emotional, nap time, lunch, medication, comments, picture, and enter/exit time.

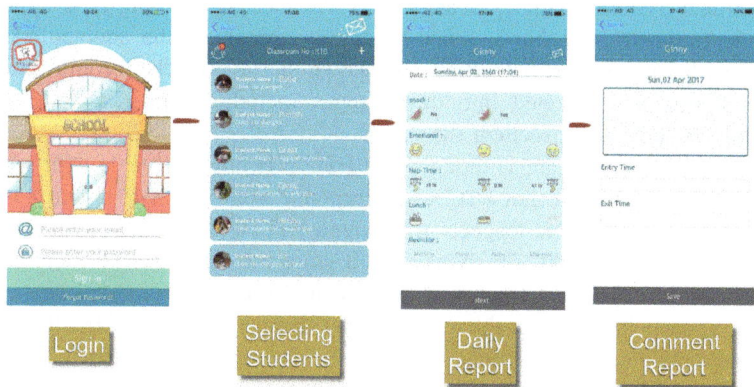

Fig. 4. PTC in the teacher part (login, selecting students, daily report, and comment)

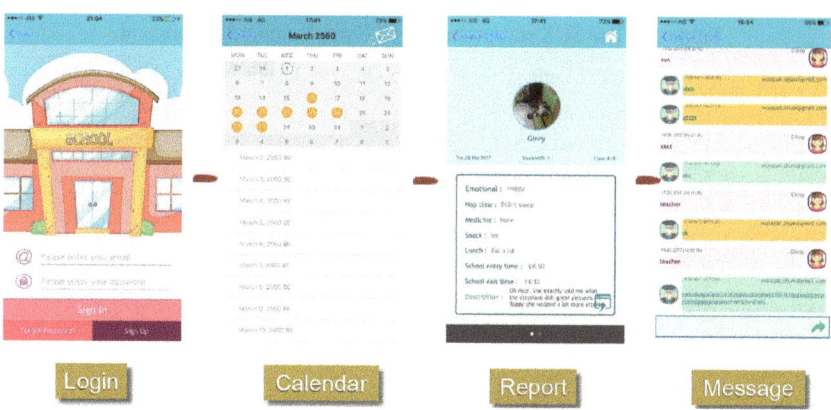

Fig. 5. PTC in the parent part (login, calendar, daily report, and message)

In parent part, to start using PTC, parents have to register into the system and wait for an approved notification along with a randomly-generated password, sent by email. Username could be parent's e-mail. Password could be changed after the first successful login. After successful login, parents can choose the date that they would like to see report. The orange circle in calendar shows the day that teachers written report. In the report page, student's information was display in one page. For message, teachers can send message to all parents as an announcement (color coding as Green) or to any parent as a private message (color coding as Yellow). Parents can only send message to teacher (color coding as Pink) however parent cannot send message to other parents.

3.2 PTC Application Usability Study

The traditional System Usability Scale (SUS) was measured during PTC application development, shown in Table 5. After PTC was completed, the adopted SUS was measured with 5 parents and 1 teachers and shown in Table 6. The recommend ranges for SUS scores are separated into three ranges including Not acceptable (0–64), Acceptable (64–84), and Excellent (85–100).

One teacher and one parent) thought that PTC was not acceptable with scores 60 and 62.5. Two parents felt that PTC was acceptable with scores 67.5 and 85. There were three parents and three teachers who found PTC was excellent with scores 90, 92.5, 100, 95, 97.5, and 100. The average SUS score was 85. It can be interpreted that PTC application is excellent communication method for parents and teachers.

After experiencing PTC for three weeks, four parents and one teacher thought that PTC was acceptable with scores 72.5, 75, 80, 80 and 82.5. One parent felt that PTC was excellent with scores of 90. The average adopted SUS score was 80. It can be interpreted that PTC application is acceptable to use as communication tool for parents and teachers.

Table 5. The SUS scores during PTC application development

SUS	User1	User2	User3	User4	User5	User6	User7	User8	User9	User10
Q1	5	4	3	5	5	5	5	5	4	5
Q2	1	1	4	1	1	1	5	1	1	1
Q3	5	4	4	1	5	5	5	5	5	5
Q4	2	3	2	2	4	1	5	1	1	1
Q5	4	5	3	4	5	5	4	4	5	5
Q6	1	1	2	1	1	1	3	1	1	1
Q7	5	5	4	1	5	5	5	4	5	5
Q8	1	1	2	1	1	1	1	1	1	1
Q9	3	5	2	2	5	5	5	5	5	5
Q10	1	3	2	1	1	1	5	1	1	1
Scores:	90	85	60	67.5	92.5	100	62.5	95	97.5	100
Average:	85									

Table 6. The adopted SUS scores after users experienced completed PTC application

Adopted SUS	User1	User2	User3	User4	User5	User6
Q1	4	4	5	4	4	4
Q2	3	4	4	4	4	4
Q3	3	4	4	5	4	4
Q4	5	4	4	5	3	4
Q5	5	4	5	4	4	4
Q6	4	5	4	4	4	5
Q7	4	4	5	5	4	4
Q8	4	4	4	5	4	4
Q9	4	5	4	5	4	5
Q10	4	4	4	5	4	4
Scores:	75	80	82.5	90	72.5	80
Average:	80					

4   Discussion

The benefits of using a PTC application are vast for early year's students. Due to the fact that the PTC application is user friendly and enables two way communications, it will allow students to develop more holistically in regards to their social-emotional, physical and cognitive development. Parents will be able to report on how their children are doing at home in these areas and teachers can respond to any questions parents have about what is happening at school. Studies show that children thrive when there is a home school connection.

Social-emotional development is one of the main focuses at school for early learners. If parents and teachers communicate daily about the topics (i.e. sharing, caring, being principled learners, etc.) the class is engaging in at school then parents can collaborate with the teachers to enhance the child's social- emotional development. If the child is struggling in a particular area or needs additional support from home, the PTC application will allow for consistent and clear communication without delays that often occur in more traditional means such as communication books and emails. Physical development, especially in the areas of fine and gross motor skill development are also a focus in early year's centers. If a student needs to work on physical development, the PTC application allows teachers and parents to identify and work on the need together. Daily updates are easy using the application. Progress can be reported in two way communication instantaneously.

The topics, themes and key concepts that are covered in class can be announced to parents daily using the PTC application and this will allow parents to re-teach the topics at home when needed. Parents will be able to read books to their children that are related to the daily lesson when teachers report on what they covered in class. This is important at our school where we utilize an inquiry based approach in our lessons and the lesson are adjusted according to the interest of the students. Basically, lessons are subject to change as students drive their own learning. This application is effective in communicating daily lessons, topics, vocabulary that arose, key concepts and learner's attitudes.

The feedback from the K1 teacher who used the application for three weeks was positive. She believed that the application would allow for clear and consistent communication with her students' parents. She said that is was less time consuming and one benefit was that she could update communication even after the children went home for the day, whereas communication books have to be completed while the kids are still at school. An additional benefit in the context of our school is that some of parents are not fluent in English and the application allows parents to use a translation application easily by pasting the comments in a translation application. The parents and the teacher provided positive feedback about the PTC application.

5   Conclusion and Future Works

In conclusion, parents and teachers communication application is a model and sustainability way to transfer information between school and home. When parents and teachers understand about their children very well, the most benefit will reflect to children in terms of social-emotional, physical, behavioral and academic development. There are three limitations in this study. Firstly, the users had only three weeks to experience PTC application due to school schedule. Secondly, PTC application was developed to run on iOS system. It cannot run on Android system. Finally, the less number of users take part in this study. The Android users who are willing to experience PTC application cannot use it.

In the future, users should have longer period of time to experience PTC application. They will have more time to learn and feel more comfortable with the system. This application should be developed for both iOS and Android systems to suit more users. Finally, PTC application should add more features which help enhance active learning and share student's live activities in classrooms to parents.

Acknowledgments. This work is funded by Smart City 4.0, Thailand. Also thank you for support by Budsakorn Chukaew and Worapat Thummakriengkrai.

References

1. Williams, V.I., Cartledge, G.: Passing notes–to parents. Teach. Except. Child. **30**, 30–34 (1997)
2. Lawrence-Lightfoot, S.: Building bridges from school to home. Instr. -Prim. **114**, 24–28 (2014)
3. Graham-Clay, S.: Communicating with parents: strategies for teachers. Sch. Community J. **15**, 117 (2005)
4. Sirvani, H.: The effect of teachers communication with parents on students' mathematics achievement. Am. Second. Educ. **36**, 31–46 (2007)
5. Machen, S.M., Wilson, J.D., Notar, C.E.: Parental involvement in the classroom. J. Instr. Psychol. **32**, 13–16 (2005)

6. Hara, S.R., Burke, D.J.: Parent involvement: the key to improved student achievement. Sch. Community J. **8**, 9–19 (1998)
7. Coots, J.J.: Family resources and parent participation in schooling activities for their children with developmental delays. J. Spec. Educ. **31**, 498–520 (1998)
8. Epstein, J.L.: Improving family and community involvement in secondary schools. Educ. Dig. **73**, 9–13 (2008)
9. Finn, J.D.: Parental engagement that makes a difference. Educ. Leadersh. **55**, 20 (1998)
10. Herman, J.L., Yeh, J.P.: Some effects of parent involvement in schools. Urban Rev. **15**, 11–17 (1983)
11. Michael, S., Dittus, P., Epstein, J.: Family and community involvement in schools: results from the school health policies and programs study 2006. J. Sch. Health **77**, 567–579 (2007)
12. Epstein, J.L., Dauber, S.L.: School programs and teacher practices of parent involvement in inner-city elementary and middle schools. Elem. Sch. J. **91**, 289 (1991)
13. Ramirez, F.: Technology and parent involvement. Clear. House **75**, 30–31 (2001)
14. Brewer, W.R., Kallick, B.: Technology's promise for reporting student learning. Yearb. Assoc. Superv. Curric. Dev. **1996**, 178–187 (1996)
15. Statista: Number of smartphone users worldwide 2014–2020. https://www.statista.com/statistics/330695/number-of-smartphone-users-worldwide/
16. Swift.org: About Swift. https://swift.org/about/
17. Apple: Xcode - Apple Developer. https://developer.apple.com/xcode/
18. Google: Firebase. https://firebase.google.com/
19. Brooke, J.: System usability scale (SUS): a quick-and-dirty method of system evaluation user information. In: Usability Evaluation in Industry, pp. 4–7 (1996)

Research on User Comfort of Intelligent Toilet Based on Ergonomics

Zhongting Wang[1(✉)], Haimei Wu[1], Wang Wei[2], Ling Luo[1],
Yunhong Zhang[1], and Chaoyi Zhao[1]

[1] AQSIQ Key Laboratory of Human Factors and Ergonomics (CNIS),
Beijing, China
{wangzht, wuhm, luoling, zhangyh, zhaochy}@cnis.gov.cn
[2] China Standard Certification Co. Ltd., Beijing, China
wangwei@csc.org.cn

Abstract. An intelligent toilet was an upgrade product on the ordinary toilet, with all the functions of a common toilet combined with automatic flushing, seat ring heating, hip cleaning, female cleaning and drying with warm air and so on, which attracts the researchers' interests recently. In the past, there are rarely few user comfort researches about intelligent toilet based on ergonomics. This study focuses on the research method of ergonomics to explore the comfort design of the intelligent toilet from user experience.

Keywords: User comfort · Intelligent toilet · Ergonomics · Body pressure

1 Introduction

Nowadays, people have increasing demanding about products, while whose basic function can no longer satisfy their variety requirements. People begin to pursue user friendly and pleasure experience while using products. For the toilet, people have put forward higher requirements. Modern toilet has turned into an intelligent one, an upgrade product on the ordinary toilet, with all the functions of a common toilet combined with automatic flushing, seat ring heating, hip cleaning, female cleaning and drying with warm air and so on [1].

Intelligent toilet originated in the United States and developed widely in Japan. Recently, with the rapid development of the national economy, the intelligent toilet has also begun to receive the attention of the domestic, which also attracts the researchers' interests. Liu (2009) [2] Zhang (2011) [3] and S. Zhu et al. (2016) [4] studied the control system of the intelligent toilet. Wang (2005) [5] proposed the scheme of intelligent closestool system applied in detecting and defending the inferior health state. Cao (2015) [6] proposed targeted intelligent home-based care in the aging toilet gasket design. Xu (2007) [7] discussed the characteristic of health bathroom, including the toilet product. Zhu (2010) [8] proposed to focus on human emotional needs in product design. Bai (2010) [9] and Liu (2017) [10] studied the operating interface of intelligent toilet. In the past, there are rarely few user comfort researches about intelligent toilet based on ergonomics. This study focuses on the research method of ergonomics to explore the comfort design of the intelligent toilet from user experience.

© Springer International Publishing AG, part of Springer Nature 2019
T. Z. Ahram and C. Falcão (Eds.): AHFE 2018, AISC 794, pp. 664–672, 2019.
https://doi.org/10.1007/978-3-319-94947-5_66

The word 'ergonomics' comes from the Greek: ergos, work; nomos, natural law, which concerned with the design of artifacts and environments for human use in general. The ergonomic approach to design may be summarized in: the principle of user-centered design (see Fig. 1) [11]. In this situation, ergonomic design should focus on the users and their interaction with products and environments used in work and everyday living.

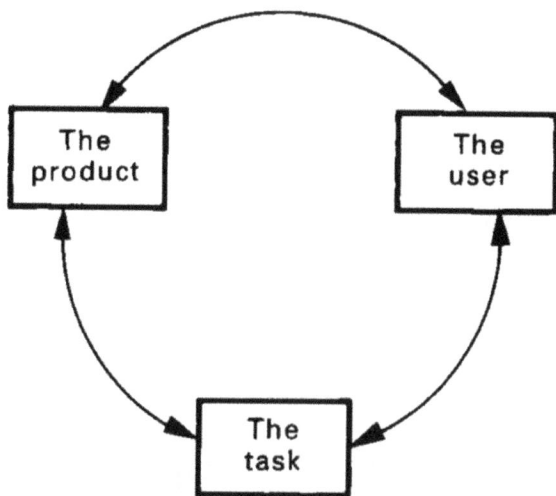

Fig. 1. User-centered design: the product, the user, the task

User Experience refers to the internal and subjective feeling a person has when using a product or encountering a service [12]. It is a complete process and varies from person to person. Traditional questionnaires are the most often used in the context of design and design research, both for obtaining quantitative and qualitative data within the wide variety of approaches to assess user experience [13].

This paper adopted a new method to use body pressure distribution measurement system to assess the comfort of the toilet seat, combined with a subjective questionnaire. Body pressure distribution measurement has been wildly used to evaluate the comfort of products, such as chairs, sofa, and mattress, which has been testified as a reliable objective method for products comfort evaluation [14–18]. Besides, for operating interface and functional experience, subjective questionnaires were applied, and with a post - Test interview.

2 Experiment

To explore the factors that affect the comfort of the intelligent toilet and the potential problems in the design of the current intelligent toilet products, three types of intelligent toilet seat, which were installed on the common toilet with ceramic base at a uniform height. The parameters of the three products (represented by the letter A-C) are shown in the following table (see Table 1).

Table 1. The parameters of the three products

Parameters	A	B	C
Seat ring size (length × width × height*)	410 × 370 × 30	370 × 379 × 40	378 × 360 × 32
Operation mode	Operation interface		
Basic function	seat ring heating, hip cleaning, female cleaning, drying with warm air		

*the unit of the seat ring size is mm.

2.1 Experimental Environment

The experiment was carried out in an office building toilet environment. The three types of intelligent toilet seat were installed on the common toilet with ceramic base at a uniform height. To avoid the experimental bias, the brand identity of the products was covered (see Fig. 2).

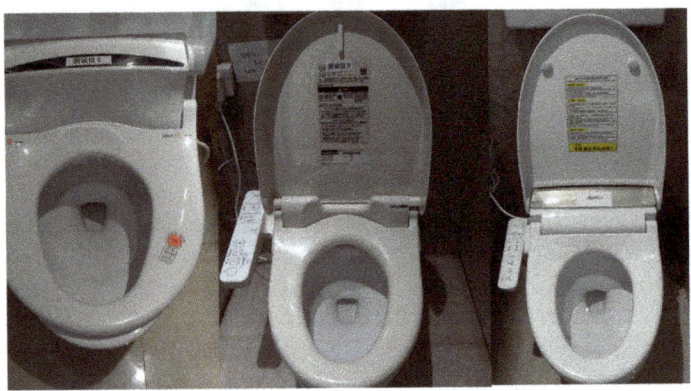

Fig. 2. The environment of the experiment

2.2 Participants

To reduce gender differences, all participants were females and with mean age of 34 years ranging from 21 to 47. A total of 30 participants involved in the experiment. To cover the broader population characteristics, the participants were of various body shapes, with BMI distribution as Fig. 3. Average height was 162.7 cm, weight 56.7 kg.

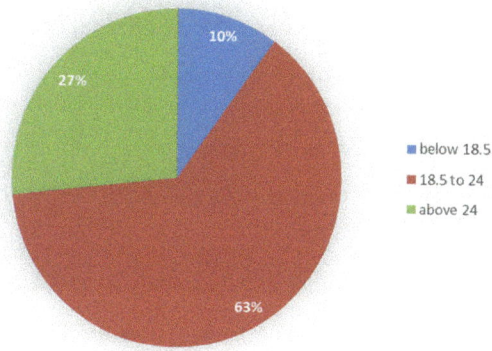

Fig. 3. The BMI distribution of the participants

2.3   Procedures

Before the experiment, participants were told about the whole procedure of the experiment. Then, several body dimensions were measured, including Popliteal Height, Hip Breadth, Buttock to Back of Knee (sitting), the values were as Table 2. The Body Pressure Test was carried out, where participants were asked to sit on the toilet seat with natural sitting posture. Body pressure distribution data was collected with the use of German Novel seat cushion pressure distribution measurement system. The data collection was carried out after the pressure distribution was relatively stable (see Fig. 4). Lastly, participants were asked to use each intelligent toilet and filled out the questionnaire, respectively. Each of the participants received A post - Test interview.

Table 2. The dimension of the participants

Dimension	Average value	Range value
Weight (kg)	57.4	47.2–77.6
Height (cm)	161.4	150.7–172.7
Popliteal Height (cm)	403	374–436
Hip Breadth (cm)	368	330–414
Buttock to Back of Knee (cm)	432	373–477

2.4   Data Analysis

In this research, the combination of user's objective evaluation methods and subjective experience evaluation were adopted to analyze the relationship between seat ring pressure and the severity of comfort under different intelligent toilet seat ring pressures. Subjective experience evaluation is to assess the degree of comfort under different seat ring conditions based on the participant' subjective feelings. A 5-point scale of the subjective questionnaire was asked and recorded to evaluate their subjective feelings on Seat Ring comfort. For the part of Operating Interface and Functional Experience of

Fig. 4. Scenario of the Body Pressure Test

different toilet, Subjective evaluation was adopted as the same as in the Seat Ring comfort evaluation. A repeated ANOVA were used for statistical analysis. SPSS (16.0 J, SPSS Inc.) was used for calculation.

3   Results

The following is the result of subjective and objective measurement we obtained in the experiment.

3.1   The Result of the Comfort for the Seat Ring of the Three Intelligent Toilets

3.1.1   Pressure Distribution Results of Seat Ring Surface

The average pressure on the seat ring surface of all the participants in different toilet type was obtained by processing the collected pressure distribution data on seat ring surface. The results are shown in Fig. 5.

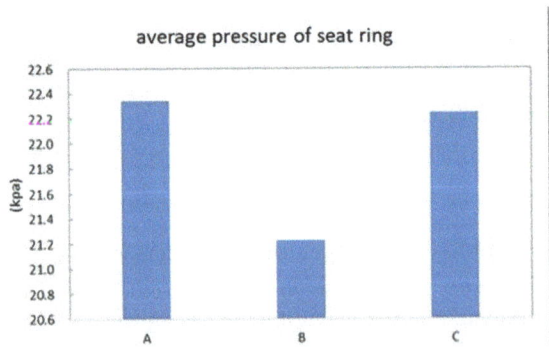

Fig. 5. The average pressure of the three seat rings of the toilet

It can be inferred from the figure that the average pressure was various on the seat ring surface of different toilet type. The repeated measure ANOVA shows significant main effects of different toilet type on seat ring comfort (F $(2,58)$ = 3.804, P = 0.028 < 0.05). Further, the LSD post hoc test results (see Table 3) showed that the seat ring of toilet A and C had more comfort than B, but there was no significant difference in toilet A and C.

Table 3. Results of the LSD post hoc test on the average pressure for the seat ring surface

Toilet type	LSD post hoc test result	
	Std. error	Sig.
A vs. B	.388	.008*
C vs. B	.478	.043*
A vs. C	.459	.852

3.1.2 The Subjective Results of the Comfort for the Seat Ring

The result of the subjective evaluations of the comfort for the Seat Ring was shown in Fig. 6. Figure 6 shows the average subjective scores of all the participants in different toilet types of the seat ring.

It can be inferred from the figure that the subjective evaluations of the comfort for the Seat Ring was various. The repeated measure ANOVA shows significant main effects of different toilet type on seat ring comfort (F $(2,58)$ = 4.275, P = 0.019 < 0.05). Further, the LSD post hoc test results (see Table 4) showed that Participants felt that the seat ring of toilet C had more comfort than B, but there was no significant difference in toilet A to B or C. During the post interview, participants reported that the comfort of the Seat Ring for toilet B was not good, which was mainly related to the shape of the seat ring, with an upward angle in the front of the seat ring.

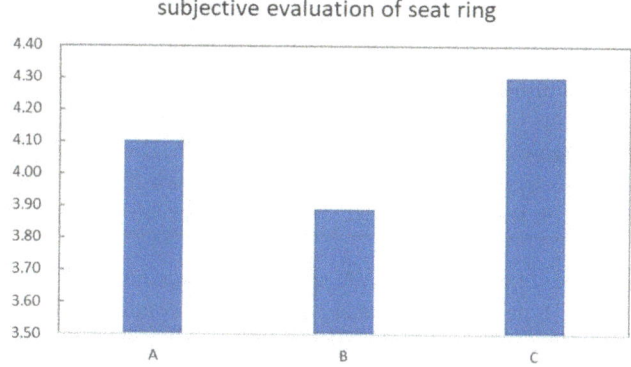

Fig. 6. The subjective evaluation of the comfort for the Seat Ring

Table 4. Results of the LSD post hoc test on the subjective evaluation for the seat ring comfort

Toilet type	LSD post hoc test result	
	Std. error	Sig.
A vs. B	.154	.170
C vs. B	.143	.007**
A vs. C	.126	.129

3.2 The Result of the Comfort for the Operating Interface and Functional Experience of the Three Intelligent Toilets

The result of the subjective evaluations of the comfort for the Operating Interface and Functional Experience was shown in Fig. 7. Figure 7 shows the average subjective scores of all the participants in different toilet types of the Operating Interface and Functional Experience.

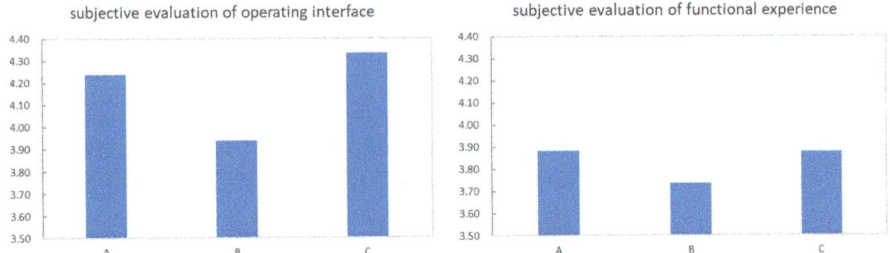

Fig. 7. The subjective evaluation of the comfort for the Operating Interface and Functional Experience

It can be inferred from the figure that the subjective evaluations of the comfort for the Operating Interface was various. The repeated measure ANOVA shows significant main effects of different toilet type on Operating Interface comfort (F $(2,58)$ = 7.315, P = 0.002 < 0.01). Further, the LSD post hoc test results (see Table 5) showed that Participants felt that the Operating Interface of toilet A and C had more comfort than B, but there was no significant difference in toilet A to C. During the post interview, almost all participants mentioned the dark background color of Operating Interface for C was too deep, which leaded to the text and symbols on the panel could not be seen clearly.

While, the repeated measure ANOVA shows there was no significant main effects of different toilet type on functional experience comfort (F $(2,58)$ = 0.941, P = 0.396 > 0.05). In the post interview, participants pointed out that little difference existed among the three products of the hip washing function, they complained that even if you adjusted the flushing position, the experience of female washing function was still poor. The function of drying with warm air was terrible. It deduced that the air temperature was the main cause.

Table 5. Results of the LSD post hoc test on the subjective evaluation of the comfort for the Operating Interface

Toilet type	LSD post hoc test result	
	Std. error	Sig.
A vs. B	.110	.008**
C vs. B	.125	.004**
A vs. C	.091	.370

4 Discussion and Conclusion

The present study was concerned with user comfort of intelligent toilet based on Ergonomics. The intelligent toilet was divided into three parts -the Seat Ring, Operating Interface and Functional Experience, to explore the comfort of intelligent toilet. According to the standard GB-10000 of Human Dimension of Chinese Adults, 30 female volunteers were recruited, and they could represent the wildly female population of the physical characters in domestic. This study has employed an objective method called Body Pressure Test to evaluate the comfort of toilet Seat Ring. The results from the experiment showed that the Body Pressure Test could be used in the comfort of toilet Seat Ring evaluation. The objective indicator could efficiently identify the severe comfort induced by different toilets. More importantly, it was shown that the objective measurement had reliable performance in evaluating the comfort of the toilet Seat Ring and coincided quite well with the subjective evaluation. It was worthwhile to be recommended as a research approach in this field.

In conclusion, the results of this study demonstrated that the comfort of different toilets was various. The statistical results showed that there had significant main effects in the comfort of the Seat Ring and the usability of the Operating Interface for the three products. However, there was no statistical difference among the three products in Functional Experience. Unfortunately, it was a pity to found that the function design of intelligent toilet products was too poor to meet the users' needs. We proposed that researchers and the designers should focus on user-centered principles to improve the function design.

In future, it might be convenient to develop a suitable seat cushion pressure distribution measurement system (like Seat-Ring-Shaped seat cushion) for objective evaluation of seat ring comfort to keep abreast of progress in technology. In addition, further study may focus on the design of the operating panel, such as the relationship between the installation position and the accessibility threshold of the arm in the sitting position, the scope of human vision and the design of keys on the panel and so on.

Acknowledgments. We would like to thank the participants who took part in the experiment for their many valuable comments. We gratefully acknowledge the financial support from National Key R&D Program of China (2017YFF0206603) and China National Institute of Standardization President Funds Project (242016Y-4700).

References

1. Bai, J.J., et al.: Design of intelligent toilet for wheelchair-bed. J. Mach. Des. Res. **26**(1), 121–124 (2010)
2. Liu, Zh.W.: Research on the anti-interference techniques of embedded system based on ATmega64. Zhejiang University of Technology (2009)
3. Zhang, Y.Z.H.: The design and implementation of smart toilet based on ATmega128. Zhejiang University of Technology (2011)
4. Zhu, S., Chen, Y., Wang, H., Xie, F.: Design of control system for an intelligent closestool. In: Huang, B., Yao, Y. (eds.) Proceedings of the 5th International Conference on Electrical Engineering and Automatic Control. LNEE, vol. 367, pp. 743–749. Springer, Heidelberg (2016). https://doi.org/10.1007/978-3-662-48768-6_83
5. Wang, H.Y.: Research of the Intelligent network Closestool system. Guangdong University of Technology (2005)
6. Cao, W.D.: Research and design of the internet based on the aging of the smart toilet. Hubei University of Technology (2015)
7. Xu, H.: The study on the constructing health life in modern bathroom. Jiangnan University (2007)
8. Zhu, F.J.: The research of skin experience design in bathroom products. Jiangnan University (2010)
9. Liu, Y.J., Zhang, X.D.: Human-computer interaction design study on the operating interface of health intelligent toilet for the elderly. J. Art Des. **2**, 102–104 (2017)
10. Chen, Y., et al.: The engineering design of control panel on the Intelligent toilet based on visual angle. J. China Ventur. Cap. (29) (2017)
11. Pheasant, S.T.: Bodyspace: Anthropometry, Ergonomics, and the Design of Work. Taylor & Francis, London (1998)
12. Chen, Y., You, F., Wang, J., Schroeter, R.: Measuring user experience in situ: use emotion data to assess user experience. J. Adv. Usability User Exp. (2018)
13. Bargas-Avila, J.A., Hornbaek, K.: Old wine in new bottles or novel challenges: a critical analysis of empirical studies of user experience. In: Proceedings of the SIGCHI Conference on Human Factors in Computing Systems, pp. 2689–2698. ACM Press, New York (2011)
14. Park, S.J., Kim, C.B.: The evaluation of seating comfort by the objective measures. J. Sae Tech. Pap. Ser. (1997)
15. Na, S., Lim, S., Choi, H.S., Min, K.C.: Evaluation of driver's discomfort and postural change using dynamic body pressure distribution. Int. J. Ind. Ergon. **35**(12), 1085–1096 (2005)
16. Qin, T., Zhang, J.G., Dai, Y.X.: Research on body pressure distribution at different angles of wheelchair seat surface. J. Appl. Mech. Mater. **477–478**, 345–348 (2013). ISSN 1662-7482
17. Hu, H., Luo, L., Yao, Y., Zhao, C., Wu, H., Zhang, X., Ran, L., Wang, R.: Research on multi-factor sofa inclination comfort based on user experience. In: Ahram, T., Falcão, C. (eds.) AHFE 2017. AISC, vol. 607, pp. 698–708. Springer, Cham (2018). https://doi.org/10.1007/978-3-319-60492-3_66
18. Hu, H., Yao, Y., Luo, L., Ran, L., Zhao, C., Zhang, X., Wang, R.: Research on pressure comfort of sofa based on body pressure distribution and subjective experience. In: Duffy, Vincent G. (ed.) DHM 2017. LNCS, vol. 10286, pp. 26–38. Springer, Cham (2017). https://doi.org/10.1007/978-3-319-58463-8_3

Satisfaction Analysis for Using Educational Serious Games for Teaching Wound Treatment

Gabriel Candido da Silva[1], Lúcia Paloma Freitas da Silva[1],
Nicolau Calado Jofilsan[1], Walter Franklin M. Correia[3],
Alex Sandro Gomes[2], and Amadeu S. Campos Filho[1,4(✉)]

[1] Faculdade São Miguel, Recife, PE, Brazil
gabcandidods@gmail.com, lpgameart@gmail.com,
ncj@cin.ufpe.br
[2] Centro de Informática, UFPE, Recife, PE, Brazil
asg@cin.ufpe.br
[3] Centro de Artes e Comunicação, UFPE, Recife, PE, Brazil
ergonomia@terra.com.br
[4] Núcleo de Telessaúde, UFPE, Recife, PE, Brazil
amadeu.campos@nutes.ufpe.br

Abstract. This article aims to describe the production process of the educational game "treat well!", idealized for learning in higher education institutions. Research was done with students of the health courses, which helped to prove the direct effect in the improvement of cognitive functions such as memory, attention, perception, among others. The acquired advantages of technology, when well used are unimaginable, especially when used for education. The barriers encountered by this tool to realize its real application were perceptible, but these were worked on and perfected to find a balance between education and fun. We can infer that this tool, when properly applied, is able to attract and perpetuate information in students in any educational field. In this context, an educational game was developed that serves as a support to the learning process of the students of nursing courses. The goal of this research was to analyze the usability and satisfaction of the educational game "Treat Well!" which teaches the treatment of a simple wound with nursing students. For the development of the project, the methodology used was based on software engineering practices, User Centered Design and Usability and Satisfaction Analysis. The study was also based on a qualitative and quantitative approach with exploratory character and also statistical. The qualitative variable used to capture the perception of users in the study was made in an empirical way of observing the search for relevant and convenient data obtained through experience observed. The quantitative variables used to analyze usability were the effectiveness, time of use and user perception through the Attrakdiff questionnaire. The usability test was performed with high fidelity game prototype with 10 volunteers in a college in Recife Brazil. From the results generated we can understand that improvements can be made to a greater identification and interaction of the user with the proposed game.

Keywords: Human factors · Usability · User experience

© Springer International Publishing AG, part of Springer Nature 2019
T. Z. Ahram and C. Falcão (Eds.): AHFE 2018, AISC 794, pp. 673–682, 2019.
https://doi.org/10.1007/978-3-319-94947-5_67

1 Introduction

Currently, new information and communication technologies have been used in education with the aim of facilitating the teaching-learning process of the human. Knowledge can be considered infinite, starting from the premise that the human will never learn everything. But the search for knowledge can be made easy and even fun. The days when the massive and repetitive lessons were mandatory are gone. Today, the student has the right to choose his future after high school, but the way the teacher prepares him for this future influences his choice.

The National Curricular Guidelines of the Undergraduate Nursing Course in Brazil were consolidated with a breakthrough in the field of education in 2001, with the Resolution of the National Education Council No. 3 of November 7, [1]. With the technological advances, the training of nurses has been adapted and its application in a multiprofessional way has been of enormous weight for the advancement of our history. In this scenario, it can be observed that the teaching model and pedagogical principles of the training of the qualified nursing professional in conjunction with the technology is of great relevance so that it can act in a multiprofessional way according to the needs of the Unified Health System. With this, tools like applications and games have been developed and designed to further support the permanence of knowledge and encourage the student to look for it.

This scenario was only applied in Brazil a few years later, the main barriers that made the arrival of technology in classrooms were the costs and difficulty of adapting technology in didactic and educational scope. The lack of motivation also had influence in the little use of technology, teachers usually close to the new tool, refusing to abandon the primary means of teaching. This fear of the new reflects mainly on how the student will arrive at the knowledge, making arduous and discouraging the search for more.

The current reality of Brazil allows these barriers to be minimized both in the private and in the public educational sphere. In 2005, the UCA project - one computer per student - began to be planned for the dissemination of low-cost computers in order to intensify information technology in classrooms. This facilitated that public schools could work in the new teaching model, and even optimize the previous model. Public high school students could then have their first real access outside of laboratories [2].

With the application of an educational game in nursing monitoring, Cavalcante [3] carried out qualitative and quantitative researches with the proposal of facilitating the exchange of knowledge and stimulating critical thinking through the ludic. The game consisted of 46 multicolored letters with wounds, clinical cases, questions, etc., and were acquired 3 points if the answer given is correct, 2 points if it is partially correct and 1 point if it equals one third of the correct answer. Thanks to the application of this game, it was perceived by the author that when monitoring occurs associated with the use of educational strategies, a positive result in learning occurs, evidenced by the gradual increase of correct answers shown in her study.

According to research by Cogo [4] and Campos Filho [5], nursing classes are normally done in a very expositive way where the student ends up being forced into a routine of observing and repeating. The proposal of incorporating games into teaching activities contributes to reduce the time of expositive and mass classes, in addition to which digital tools with educational purposes support and reinforce in the learning process. Thus, students can access the content of the class with a more detailed presentation and with visual and auditory resources that can be a challenge and generate a curiosity in learning, simulating the problem digitally before performing it in the classroom.

With the use of the technology being used in an educational context, a game project was elaborated as part of the Human-Computer Interaction discipline that aims to assist in the educational method of nursing technical education students. This article aims to evaluate the usability and satisfaction of the educational game "treat well", where we intend to describe the facts and phenomena observed and analyzed, using a questionnaire as a form of direct response, to collect assertive data and argumentation.

2 Methodology

The methodology used in the research was of the descriptive type where it intends to describe the facts and phenomena observed and analyzed. To perform this study we were first used the software engineering method that aimed to develop the game "Treat well!". The development was also based on Human-Centered Design [6] practices which is a process consisting of a set of techniques divided into three phases (Listen, Create and Implement) used to create new solutions and modes of interaction. In the listening phase, the techniques of bibliographic analysis, survey of the competitors and interviews with users were carried out to define the necessary requirements for the creation of the game. In the creation phase we used the techniques of context analysis, personas and scenarios with specialists for the generation of interactive prototypes using the technique of paper prototyping, wireframe and high fidelity prototyping. In the implementation phase, the game was built Construct 2 while the images for 2D games were built as the Photoshop CS6 tool. The creation and implementation phases were carried out in partnership with students of the course of development of digital games.

The study was also based on a qualitative and quantitative approach where it was exploratory and also statistical through the method of usability analysis and user perception. The qualitative variable used to capture the perception of users in the study was done in an empirical way of observing the search for relevant and convenient data obtained through experience. The quantitative variables used to analyze usability were the effectiveness, time of use and user perception through the Attrakdiff questionnaire (Gerhardt and Silveira 2009).

The AttrakDiff has the ability to measure the user's perception about the intended object and according to Valentim [7, p. 4] "allows to evaluate the attractiveness through the different aspects of an application". The questionnaire is divided into three dimensions: i. the Pragmatics Quality (QP) referring to the quality of the application and the desired objectives reached by the user; ii. the Hedonic Quality that is divided into Stimulus (QH-S) that points to where the object meets the needs of the user and promote an interest, motivation, etc. and Hedonic Quality of Identity (QE-I) that indicates the extent to which an identification occurred of user need as application;

iii. Attractiveness (ATT) that indicates the general value of the application, based on the perception of quality. In addition, AttrakDiff is composed of twenty-eight pairs of words grouped in the dimensions where each pair of words were placed at the ends of the scale with a semantic differential of seven points (−3 to 3, with 0 being the - point).

2.1 Scenario

The evaluation of usability and game satisfaction was performed with the voluntary participation of 10 nursing students from a higher education institution, all of the metropolitan area of Recife, in November 2017. The only exclusion criterion was not to have been approved in the discipline of human anatomy. The test happened several times during the period described and was divided into 3 stages and it was also proposed to the student to perform a task: successfully complete the treatment of a wound in the game, using the necessary equipment in the correct order.

In the first stage the student played the game for free time, without the interference of the test applicators. This is done so that we can observe the actual experience of the user, identifying their difficulties and facilities in the use of the game. We used the Think Aloud technique in this first step, which consists in letting the user speak everything he thinks while testing the product. In the second step, students completed Attrakdiff's standardized satisfaction test aimed at identifying usability issues, analyzing the user experience, and determining the participant's satisfaction with the product. In the third and last stage, the student was asked to report orally or write his opinion about the game, his criticisms, modifications and suggestions that he would make and everything else he judges to be relevant, where the notes were made and analyzed.

3   Results

3.1   The Game "Treat Well"

It was proposed for the students of the fourth period of the Digital Games course of the Faculty of São Miguel the challenge of creating an educational game to aid in the educational method of the students of the courses of health of the same faculty in question. At first, we used the IHC (Human-Computer Interaction) concepts to build a user profile, and rely on it to make our application compatible with customer interest.

From the information of the user profile, the group of students of the course of Digital Games elaborated an educational game for the students of the course of health of the Faculty São Miguel. The game itself has the role of facilitating some lessons and adding knowledge to these students.

After defining the user profile, a brainstorming was conducted to understand the ideas of how to build an application. Together with the choice of application, a Game Design Document (GDD) was developed, containing all valid game information for better organization and functionality. With the documents and ideas ready, the initialization of the actual project could be initialized. The following sprites were produced for the game:

These elements (sprites) were made in a drawing and illustration tool, and then applied in an interface in the structural part of the game. The main idea of the game is to use visual perception to understand the subject matter. The objective is to allow the student to solve the problem proposed, as shown in Fig. 1, in the treatment of a wound, using the equipment and utensils necessary for it. To play only the mouse whose cursor is replaced in the game by a procedure glove is used. By clicking the left mouse button the player can select one of the equipments at a time and he must drag them to the wound or other interaction item that is present on the screen. By pressing the 'M' key at any point in the gameplay, the player can return to the main menu (Figs. 2 and 3).

Fig. 1. Main menu of the game treat well!

Fig. 2. Elements of the game that serve to treat wounds

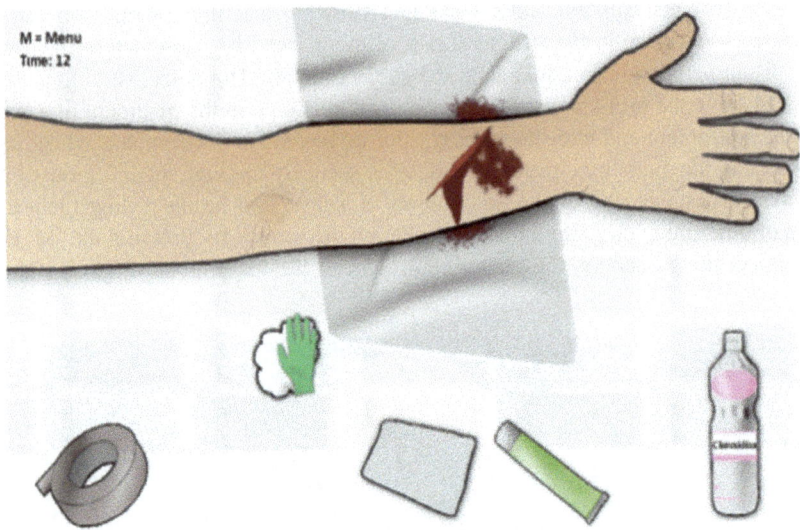

Fig. 3. Gameplay screen of the game treat well!

There are rules for the student to complete a phase, such as using the right materials in the correct order, absorbing the information, and responding coherently. In case of error the equipment returns to its place of origin and the player can continue trying. The procedure time is taken into account as competitiveness, the ideal is to complete this process in the shortest time possible.

3.2 Usability Assessment

The evaluation of usability and user satisfaction was performed through an experiment with ten volunteer students who had the same profile as the predefined target audience of the nursing course of different faculties. After students use the "treat well!" game, they were invited to fill in AttrackDiff, a protocol that allows us to evaluate user usability and satisfaction. All participants had knowledge of some kind of current technology that helps in the use of the developed game. All of them were residents of Recife.

The questionnaire is made based on a scale of numbers, where they are grouped horizontally from 1 to 7 and correspond to a characteristic of both the right side, the left. In addition, the study was done in a way that did not influence the decision of the volunteer, letting him respond according to his own impression created when testing the product, in order to give us a valid and accurate result on the developed.

The attrakdiff questionnaire was answered by 10 students from different schools who participated in the usability test of the "treat well!" and the results can be seen in the figures below.

According to AttrakDiff (2018) and the portfolio-representation (Fig. 4), it can be considered that the game has a balance in the two dimensions and that is quite close to what would be desired. The confidence interval of pragmatic quality (0.58) is higher

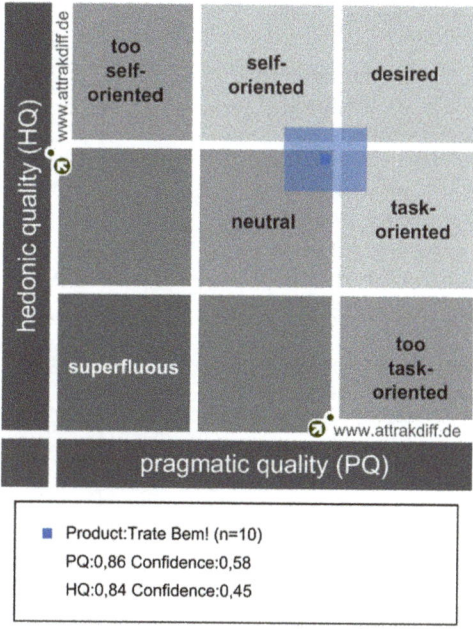

Fig. 4. Portfolio-representation

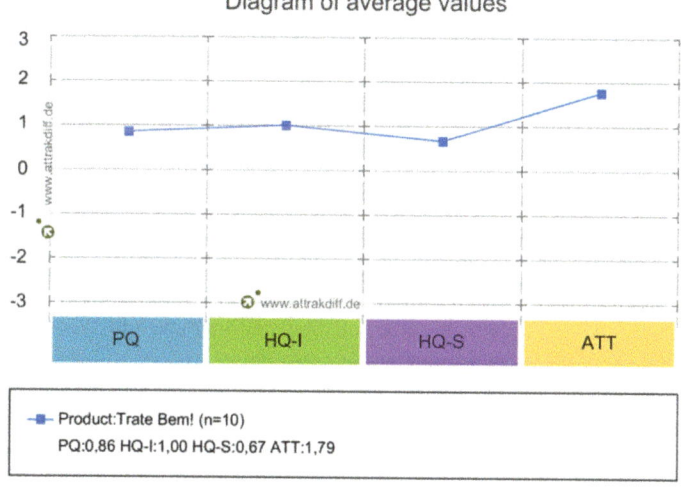

Fig. 5. Diagram of average values.

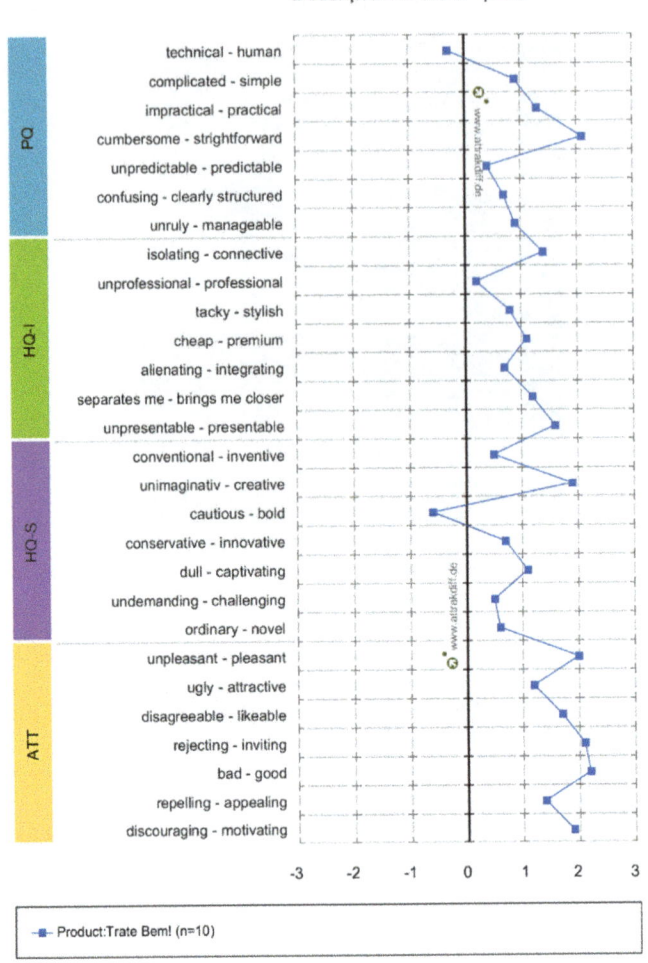

Fig. 6. Description of word – pairs

than that of hedonic quality (0.45), but both have low values. This can be attributed to the different classifications given by the users, indicating that even though there are spaces for product improvement in terms of their hedonic and pragmatic quality, they are well balanced and satisfactory. On the other hand, the confidence interval with respect to experience is great. This could be attributed to divergences of opinions among users or the fact of having a limited sample.

According to the Values Diagram (Fig. 5), we realized that the Pragmatic Quality (PQ), which indicates the satisfaction of the user in achieving his goals with the application, indicated a score of 0.86, where the one that stood out the most it was his practicality, and especially his directness should be done. However it was noticeable some points to be worked as the user's difficulty in predicting what should be done.

Quality Identity Hedonic (HQ-I) indicating the user ID level with the application had a score of 1.00, with the highlights of their connectivity and a good presentation, which was enough to satisfy students who tested yet, the ratings indicate a lack of a more professional look for the game.

In Hedonic Stimulus Quality (HQ-S), which evaluates if the application is original, interesting and stimulating, it obtained a score of 0.67. The main items that stood out were the creativity and the captivating feeling when playing, but the game was too cautious, demonstrating that the game even innovating and fulfilling, in a way, the proposal of being a captivating Serious Game has a weak visual identity that may be being improved.

The Attractiveness (ATT), which says how much the game was attractive to the user had a score of 1.79, with basically all the features of that area at a desired and satisfactory level.

According to the Description of Peer Words (Fig. 6), most of the items had a positive score with the exception of "Technician – Human" and "Cautious – Bold" items that gave values below zero (negative). The fact that the game was characterized as technical and cautious occurred because the development of the game was focused enough on the task itself. On the other hand, the items "farfetched – direct", "pleasant – nasty", "rejectable – inviting" and "bad – good" had a positive positive score because the game facilitated the understanding of the content for the students, become less dense. With these data, in order to improve the player's interaction with the product, we understand that we need to make an improvement: making the game less technical and more fun.

4 Conclusion

After completing the proposed activities, we come to some conclusions about how important is the use of technology in the educational process, be it children or adults, for better education and stimulation. The advantages of educational games applied in the pedagogical process, seeking to develop the individual's cognitive abilities. The types of categorizations of games in childhood are essential for the correct application of an activity.

The research was really a big challenge, although we needed an initial amount of time to understand the content and develop the proposed game. The students worked together with the teachers who helped in a pleasurable way in the development of the application.

At the end of the prototype presented here, we find that our development options are much broader and more diverse than we had previously thought. The possible reach to the most diverse oudience brings a new perspective of work and planning. Through empathy, we can overcome horizons and empower more and more what saves lives: education.

References

1. BRASIL. Conselho Nacional de Educação. Câmara de Educação Superior. Resolução n. 3, de Institui Diretrizes Curriculares Nacionais do Curso de Graduação em Enfermagem. Brasília. (2001). http://portal.mec.gov.br/cne/arquivos/pdf/CES03.pdf. Accessed 7 Nov 2001
2. Lavinas, L., Veiga, A.: Desafios do modelo brasileiro de inclusão digital pela escola. Cadernos de Pesquisa vol. 43(149), 542–569 (2013). http://publicacoes.fcc.org.br/ojs/index.php/cp/article/view/2665
3. Cavalcante, S.N., Morais, H.C.C.: Tecnologia em saúde sobre o tratamento de feridas: estratégia educativa na monitoria de enfermagem. Encontro de Extensão, Docência e Iniciação Científica (EEDIC), 12 (2016). http://publicacoesacademicas.fcrs.edu.br/index.php/eedic/article/view/968
4. Silveira, M.S., Cogo, A.L.P.: Contribuições das tecnologias educacionais digitais no ensino de habilidades de enfermagem: revisão integrativa. Rev. Gaúcha Enferm. vol. 38(2), Porto Alegre (2016)
5. Lemos, W.B., Farias Junior, I.H., Campos Filho, A.S.: Uma Proposta de um Serious Game no Auxílio do Aprendizado da Anatomia Humana. In: VI Congresso Brasileiro de Informática na Educação (CBIE 2017). Anais do XXVIII Simpósio Brasileiro de Informática na Educação (SBIE 2017). Recife, Brasil (2017)
6. HCD Methods. http://www.hcdconnect.org/methods. Accessed 23 Jan 2018
7. Valentim; N.M.C., Silva, W.A.F., Conte, T.: Avaliando a Experiência do Usuário e a Usabilidade de um Aplicativo Web Móvel: Um Relato de Experiência. XVIII Congresso Ibero-Americano em Engenharia de Software, vol. 1 (2015)
8. AttrackDiff. http://attrakdiff.de/sience-en.html. Accessed 22 Feb 2018

Development and Experiential Analysis of a Chinese Customer Satisfaction Model for Medical Service Industry

Ruifeng Yu[1], Jacky Y. K. Ng[2(✉)], Alan H. S. Chan[2], and Yifan Tian[1]

[1] Department of Industrial Engineering, Tsinghua University, Beijing, China
[2] Department of Systems Engineering and Engineering Management,
City University of Hong Kong, Hong Kong, China
jacky.ngyk@my.cityu.edu.hk

Abstract. The purpose of this research was to construct a Chinese customer satisfaction model for medical services and to identify the key elements that influence customer satisfaction. Eighty seven valid responses from inpatients were collected using a survey questionnaire. Structural equation modeling was employed to estimate and adjust the proposed model. The results showed that perceived quality, expected quality, and information were three important antecedents influencing customer satisfaction, and that perceived quality was the strongest antecedent of customer satisfaction and that loyalty was the strongest consequence. Information had relatively low direct effect on customer satisfaction while it indirectly affected customer satisfaction considerably through the other antecedents of perceived quality and expected quality. This research provides hospital managers with information on the aspects to be emphasized with regarding to customer satisfaction and, also, highlights the importance of increasing customer satisfaction when seeking customer loyalty.

Keywords: Customer satisfaction · Structural equation modeling
Medical service

1 Introduction

As the customer-oriented perspective is accepted widely by service providers, the concept of customer satisfaction has become an important research topic in service industries. The key motivation for the growing emphasis on the importance of customer retention and building long-term relationships with existing customers is that higher customer satisfaction results in higher market share and profit, as well as customer loyalty, in a highly competitive market [1, 2]. Reichheld and Sasser [3] reported that service industries could increase their profits up to 85% by reducing the customer defection rate by 5%. Numerous studies have shown that satisfying customer requirements is closely related to the long-term success of a firm [4, 5].

The quality of health care service is the foundation and core of the management of medical institutions. Patient satisfaction has been highlighted as an important objective of healthcare and as a useful indicator of service quality [6, 7]. Satisfaction with quality of health care provision can be conceptualized as the degree of congruence between a

© Springer International Publishing AG, part of Springer Nature 2019
T. Z. Ahram and C. Falcão (Eds.): AHFE 2018, AISC 794, pp. 683–696, 2019.
https://doi.org/10.1007/978-3-319-94947-5_68

patient's expectation of services and care and his perception of the services and care received [8]. There have been many studies on this topic and some factors that influence satisfaction evaluation have been identified and analyzed [9–12].

From the customer perspective, the Customer Satisfaction Index (CSI) is an important indicator of medical institution management, and provides an evaluation of treatment results and medical service quality. The survival and development of a hospital largely depends on customers' post-purchase evaluation of the service. Satisfied customers return and tell other people about their experiences [13]. Customer satisfaction surveys make it easier to understand customer requirements in terms of what customers need, prefer, expect, or demand with respect to the service they receive and their interactions with the service delivery process [14].

Different customer satisfaction index models define different constructs. The first national CSI was the Swedish Customer Satisfaction Barometer (SCSB) which contained two antecedents and two consequences of satisfaction; however, the American Customer Satisfaction Index (ACSI) included three antecedents and two consequences of satisfaction [15]. The European Customer Satisfaction Index (ECSI) is similar to ACSI except that it incorporates company image as an antecedent and ignores the complaint behavior variable in consequences [16]. The Chinese Customer Satisfaction Index (CCSI) was developed in 2003 by the China Enterprise Research Center. The CCSI survey was administered to 36 industries and 271 companies [17]. The fundamental difference between the CCSI and ECSI is that the company image construct of ECSI is replaced by the brand image construct in the CCSI model. Since the CCSI is a measure of the nation's total consumption experience, the effect of different industry structures when measuring overall customer satisfaction in a specific industry is taken into account. Thus, in order to measure the customer satisfaction of Chinese medical industry, a specific new model needs developing.

In this research, a model of customer satisfaction index for Chinese medical services (CSI-CMS) was developed by simultaneously exploring the effects of antecedents (perceived quality, expected quality, information, and perceived value) and consequences (customer loyalty and customer complaints) of customer satisfaction to explain the results and implications of customer satisfaction. Customer complaints were incorporated into the model based on the consideration that it is of great importance to user experience. The proposed model was evaluated using the structural equation modeling, and the CSI score was calculated to determine customer satisfaction level with hospital service quality.

2   Research Model and Hypotheses

The Chinese Customer Satisfaction Index constitutes the framework for the CSI-CMS model. The theoretical and empirical methods used in this research are primarily based on the studies of Fornell et al. [13], Robert and Jami [18], and Türkyilmaz and Özkan [19] who investigated the antecedents and consequences of customer satisfaction. Figure 1 shows the seven constructs and the 13 path hypotheses in the CSI-CMS model. The antecedent constructs of customer satisfaction are perceived quality (PQ), expected quality (EQ), information (IN) and perceived value (PV). The consequences

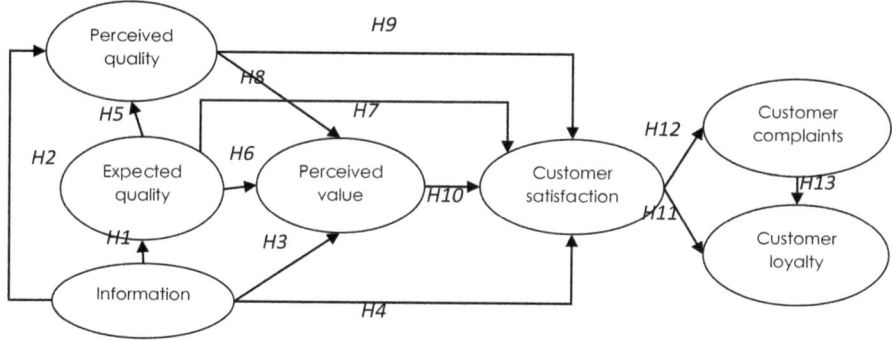

Fig. 1. CSI-CMS model.

are customer complaints (CC) and customer loyalty (CL). Customer satisfaction is the center of a chain with each path representing a hypothesized causal relationship.

2.1   Information

The construct of information contains the concept of image, which is defined as the brand name and an individual's mental representation of knowledge, feelings and overall perception of a particular company [20]. It also includes the availability of information to customers from the hospital and the authenticity of the information.

The theory of consumer behavior points out that, consumers will actively and comprehensively collect all different kinds of information if they have motivation to purchase [21]. It has been widely acknowledged that the amount of information collected has effect on perceived quality [22]. If patients get enough positive content about a medical service, the perceived quality of that service is expected to improve, and vice versa. The authenticity of information can also influence customer satisfaction. In general, false information will lead to negative psychological reactions in consumer, and result in lower customer satisfaction. In addition, to some degree, information represents the true capabilities of a service provider, and is a driving factor of expected quality. Based on these perspectives, the following hypotheses are proposed for content:

H1. Information has a positive effect on expected quality.
H2. Information has a positive effect on perceived quality.
H3. Information has a positive effect on perceived value.
H4. Information has a positive effect on customer satisfaction.

2.2   Expected Quality

Expected quality is the result of the prior experience of patients and/or the brand name of the hospital. Prior experience includes nonexperiential information available through sources such as websites, word of mouth, advertisements and forecasts of the service provider's ability [13]. Williams et al. [23] reported that the relationship between patient expectations and patient satisfaction was not a simple direct link. In SCSB and

ACSI, customer expectations are used to model the antecedents of satisfaction, with the assumption that expected quality has a direct effect on perceived quality and value, and customer satisfaction is positively affected by expected quality [24, 25]. Thus, the following hypotheses are proposed:

H5. Expected quality significantly affects perceived quality.

H6. Expected quality significantly affects perceived value.

H7. Expected quality significantly affects customer satisfaction.

2.3   Perceived Quality

Perceived quality is the post-purchase evaluation of recent experiences. It is expected that perceived quality has a positive association with perceived value and customer satisfaction [13, 26, 27]. Perceived quality reflects patients' practical experience and cognition of quality of medical services against the background of the overall process of the provision of medical services. We addressed the following hypotheses:

H8. Perceived quality has a favorable influence on perceived value.

H9. Perceived quality has a favorable influence on customer satisfaction.

2.4   Perceived Value

Perceived value is described as the perceived level of quality relative to the price paid. In medical services, perceived value is the patient's assessment of hospital service based on perception of what is paid. Equity theory [28] points out that a higher customer satisfaction would be obtained if consumers received more value than they spent. Generally, perceived value is measured in two aspects. One is the rating of the price paid for the quality perceived, and the other is a rating of the quality perceived for the price paid [13]. In previous studies, perceived value is expected to have a positive effect on satisfaction [29]. Thus, we propose:

H10. Perceived value significantly affects customer satisfaction.

2.5   Customer Satisfaction and Loyalty

Customer satisfaction is the generally pleasurable emotional state resulting from consumption-related feelings of adequate fulfillment [30]; it is the degree of overall pleasure or contentment felt by the customer resulting from the ability of the product to fulfill the consumer's desires, expectations and needs [31]. Here, it measures the quality of the services as experienced by the patients. It consists of three evaluation indexes, i.e., overall satisfaction, fulfillment of expectations, and actual perception versus the ideal. Loyalty describes the types of potential behavioral responses that satisfied patients may take. Customer loyalty evaluates intention to repeat using the service and intention to recommend hospital to others. There is a general consensus that higher customer satisfaction exercises a positive influence on customer loyalty [32–34]. Therefore, we propose:

H11. A positive relationship exists between customer satisfaction and customer loyalty.

2.6   Customer Complaints

Consumer complaint is described as an action that involves negative communication about a product or service consumption or experience [35]. Many researchers have pointed out the negative relationship between satisfaction and complaints [13]. However, to some degree, the relationship between loyalty and complaints depends on the effectiveness of the way in which complaints are handled. Some research has indicated that if customers were satisfied with the handling of consumer complaints, dissatisfied customers can be converted into satisfied customers and the providers' reputation improved [36]. Furthermore, effective handling of consumer complaints can avoid the spread of negative word-of-mouth opinions and enhance economic profitability. The opposite effects are to be expected if complaints are handled ineffectively. Therefore, we propose the following hypotheses:

H12. A negative relationship exists between customer satisfaction and customer complaints.

H13. Customer complaints have a negative effect on customer loyalty.

The seven constructs in the conceptual CSI-CMS model are latent variables that cannot be observed directly and they are described indirectly by observable variables. In order to increase reliability, each construct is estimated by using multiple questions rather than with a single question. The latent variables and their related observable variables used in the conceptual model are given in Table 1.

Table 1. Latent variables, observable variables and their relations.

Latent variables and structural model equations	Observable variables	Measurement model
Information (ξ_1)	x_1: sufficiency	$x_{1i} = \lambda_{1i}\xi_1 + \delta_{1i}$
	x_2: decision-making	
	x_3: brand	
EQ (η_1) $\eta_1 = \gamma_{11}\xi_1 + \zeta_1$	y_{11}: for fulfillment of personal need	$y_{1i} = \lambda_{1i}\eta_1 + \varepsilon_{1i}$
	y_{12}: for service quality	
	y_{13}: for technical quality	
PQ (η_2) $\eta_2 = \beta_{21}\eta_1 + \gamma_{21}\xi_1 + \zeta_2$	y_{21}: reliability	$y_{2i} = \lambda_{2i}\eta_2 + \varepsilon_{2i}$
	y_{22}: responsiveness	
	y_{23}: tangibility	
	y_{24}: assurance	
	y_{25}: empathy	
PV (η_3) $\eta_3 = \beta_{31}\eta_1 + \beta_{32}\eta_2 + \gamma_{31}\xi_1 + \zeta_3$	y_{31}: rating of the price paid	$y_{3i} = \lambda_{3i}\eta_3 + \varepsilon_{3i}$
	y_{32}: rating of the quality perceived	
CSI (η_4) $\eta_4 = \beta_{41}\eta_1 + \beta_{42}\eta_2 + \beta_{43}\eta_3 + \gamma_{41}\xi_1 + \zeta_4$	y_{41}: overall satisfaction	$y_{4i} = \lambda_{4i}\eta_4 + \varepsilon_{4i}$
	y_{42}: fulfillment of expectations	
	y_{43}: actual perception versus ideal	
CC (η_5) $\eta_5 = \beta_{54}\eta_4 + \zeta_5$	y_{51}: Frequency of complaints	$y_{5i} = \lambda_{5i}\eta_5 + \varepsilon_{5i}$
	y_{52}: complaints handling	
CL (η_6) $\eta_6 = \beta_{64}\eta_4 + \beta_{65}\eta_5 + \zeta_6$	y_{61}: recommendation	$y_{6i} = \lambda_{6i}\eta_6 + \varepsilon_{6i}$
	y_{62}: repeat intention	

3   Method

3.1   Data Collection

The questions used in this study to describe indirectly the seven latent variables were adapted from existing literature to improve content validity [37]. The questionnaire included two sections. Items were measured on a five-point Likert scale. Section 1 consisted of seven blocks of questions to measure the constructs while Sect. 2 included some demographic questions.

3.2   Participants

Inpatients were selected as the sample based on the following considerations. Compared to outpatients, inpatients stay longer in hospital, receive multiple services, and are likely to know medical services better. Therefore, it is to be expected that their assessments of hospital services are relatively comprehensive and accurate. Data were gathered in a middle-level hospital located in northern China. One hundred and ten questionnaires were distributed using well established sampling method [38]. All returned questionnaires were scrutinized and those that had missing values were removed. As a result, 87 valid responses were used for data analysis. Of the respondents, 53.9% were male and 46.1% were female. The mean age of respondents was 44.6. With respect to education, 90.2% were secondary level. Most of respondents regarded medical expense and waiting time as important. Familiarity was the most important reason for choosing a hospital.

3.3   Measures

In order to test the reliability of the research instrument, we used the Cronbach's alpha coefficient which determines the internal consistency of the scales. The obtained alpha values varied from 0.712 to 0.948, which were considered as acceptable [39] and supported the reliability of relationships among measurement variables and the latent variables.

4   Results

4.1   Model Analysis

The software AMOS was used to conduct data analysis. First, the reliability of the measurement model was examined using confirmatory factor analysis (CFA), and then we examined the structural model to validate the hypotheses. CFA was conducted to assess model fitness. In the testing process, several common model fit indices were used as criteria to judge measurement reliability. Table 2 shows the goodness-of-fit statistics for the measurement model and the recommended values. As indicated, all but one of the fit indices (χ^2/df) did not satisfy its criterion for a good fit, which revealed that the structural model needs to be further improved.

Table 2. Goodness-of-fit indices.

Fit indices	χ^2/df	GFI	AGFI	CFI	NFI	IFI	RMSEA
Recommended values	<3	>0.90	>0.90	>0.90	>0.90	>0.90	<0.08
Actual values	3.06	0.910	0.921	0.944	0.919	0.944	0.072

Notes: χ^2/df is the ratio between the Chi-square and degrees of freedom, GFI is Goodness of Fit Index, AGFI is the Adjusted Goodness of Fit Index, CFI is the Comparative Fit Index, NFI is the Normed Fit Index, RMSEA is the Root Mean Square Error of Approximation.

Raw data showed evidence of normal distribution and Structural Equation Modeling (SEM) analysis was conducted using the maximum likelihood method to estimate parameters and examine relationships. Table 3 shows the estimates of regression weights between parameters, standard error, critical ratios, and p value of the SEM analysis. The five insignificant paths were perceived quality to perceived value, information to perceived value, perceived value to customer satisfaction, expected quality to customer satisfaction, customer complaints to customer loyalty.

Table 3. Regression weights between parameters of the SEM.

Parameter	Estimate	S.E.	C.R.	P
Information → Expected quality	.148	.034	4.371	***
Expected quality → Perceived quality	.191	.051	3.755	***
Information → Perceived quality	.651	.045	14.581	***
Perceived quality → Perceived value	−.138	.178	−.775	.438
Information → Perceived value	.044	.137	.322	.748
Expected quality → Perceived value	.426	.114	3.750	***
Perceived value → Customer satisfaction	−.032	.023	−1.415	.157
Information → Customer satisfaction	.349	.061	5.730	***
Perceived quality → Customer satisfaction	.843	.091	9.296	***
Expected quality → Customer satisfaction	−.058	.047	−1.227	.220
Customer satisfaction → Customer Complaints	−.710	.065	−10.884	***
Customer satisfaction → Customer Loyalty	.879	.110	7.975	***
Customer Complaints → Customer Loyalty	.119	.137	.869	.385

Note: ***indicates significance at 0.01 levels.

The three paths relating to perceived values were not significant ($p > 0.05$), which indicated perceived value was not a significant attribute of the model. Thus, the variable of perceived value was removed from the structure. Similarly, the other two non-significant paths were also deleted. After revising the model by deleting the non-significant paths, the result of regression weights revealed that all path coefficients were significant at $p < 0.05$. In order to improve goodness-of-fit statistics of the revised model, the modification indices were adopted for revise model. Based on the principle of modification indices (MI), adding residual error path of the greatest modification indices value can reduce χ^2 value substantially [40].

The greatest MI value was for the path between reliability aspect of perceived quality and responsiveness aspects of perceived quality, which suggested that the χ^2 value could be reduced substantially if the residual error path was added. In practice, poor quality facilities for medical services in hospital usually lead to the longer response times, the result being customer dissatisfaction with long waiting times. In view of the connection between the two residual errors, a path between reliability and responsiveness was added. Using a similar approach, two paths were added, one between the assurance aspect of perceived quality and empathy aspect of perceived quality, and another between overall satisfaction aspect of CSI and the ideal aspect of CSI. Finally, we arrived at the optimal model as shown in Fig. 2. In addition, Table 4 shows results contrasting the initial and optimal model fit indices. As shown in Table 4, all indices showed improvement and satisfied the criteria for their acceptance levels.

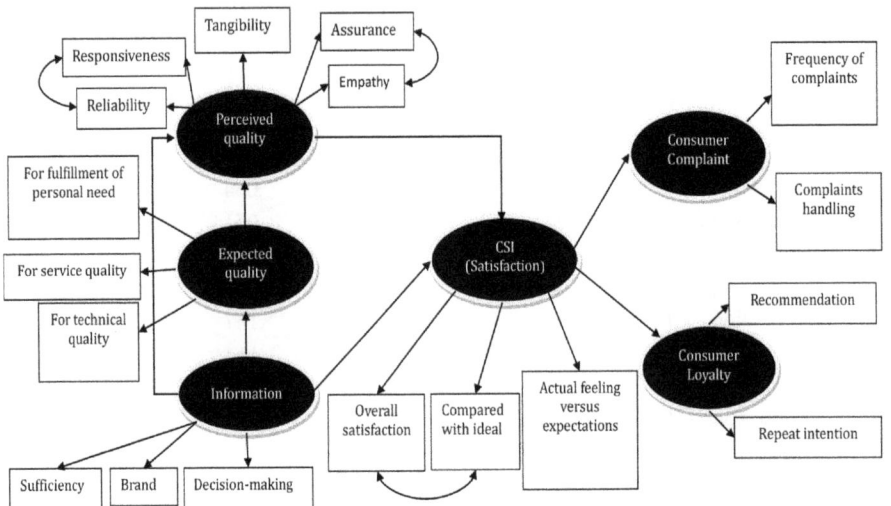

Fig. 2. Latent variables and their related observable variables in the optimal CSI-CMS model.

Table 4. Goodness of the fit indices for the initial and optimal model

Fit indices	χ^2/df	GFI	AGFI	CFI	NFI	IFI	RMSEA
Recommended values	<3	>0.90	>0.90	>0.90	>0.90	>0.90	<0.08
Initial model	3.06	0.910	0.921	0.944	0.919	0.944	0.072
Optimal model	2.79	0.90	0.932	0.954	0.931	0.955	0.069

4.2   Analysis of Direct and Indirect Effects

The hypothesized relationships among the constructs in the optimal model are shown in Fig. 3. Compared with Fig. 1, several hypotheses are absent in the optimal model. During the model modification stage, the construct "Perceived value" was deleted and the corresponding four hypotheses (H3, H6, H7 and H10) were eliminated. In addition, the paths from customer complaints to customer loyalty (H13) and from expected quality to customer satisfaction (H8) were also taken out because of their non-significant coefficients.

Note: *indicates significance at 0.05 levels

Fig. 3.   Test results of the optimal CSI-CMS model.

As expected, perceived quality had a significant positive impact on customer satisfaction ($\beta = 0.84$); thus, supporting H9. In addition, the model indicated that information played an important role in increasing perceived quality (H2, $\beta = 0.64$). Customer satisfaction was found to be a significant predictor in determining customer loyalty with $\beta = 0.83$, supporting hypothesis H11. Hypothesis H12 indicated the linkages between customer satisfaction and customer complaints. This path had estimate of −0.72 which demonstrated that customer satisfaction negatively influenced complaints. Compared to the other constructs, information had a relatively small effect on customer satisfaction (H4, $\beta = 0.35$). This may have been because a patient may tend to assess a hospital more on the basis of perceptions of what is received and what is actually given, rather than on the brand of a hospital. Also, the path coefficient between information and expected quality was relatively small, with $\beta = 0.15$. The probable reason for this is that people are used to going to a hospital familiar to them and, therefore, have already had plenty of information about the hospital. A noticeably weak relationship (0.18) existed between expected quality and perceived quality, indicating that expected quality had a low effect on perceived quality. This relationship was consistent with the findings of previous CSI studies [19, 41]. Direct and indirect effects between constructs are displayed in Table 5. Direct effects were association of one factor with another specified factor. Indirect effects were association of one factor with another mediated factor through other variables. Total effect was represented by the sum of direct and indirect effects.

As indicated in Table 5, perceived quality had the highest value for direct influence on customer satisfaction. Also, although information had a relatively small value as a direct effect on the customer satisfaction, it had a considerable value as an indirect effect. In fact, it was the top ranked total effect, indicating that service providers need to pay close attention to their image and to ensuring that information released is reliable and sufficient.

Table 5. Causal relationships among the constructs.

Path	Direct effect	Indirect effect	Total effect
Information → Perceived quality	0.643	0.026	0.669
Information → Customer satisfaction	0.346	0.565	0.912
Expected quality → Perceived quality	0.179	–	0.179
Expected quality → Customer satisfaction	–	0.151	0.151
Expected quality → Customer Complaints	–	–0.109	–0.109
Perceived quality → Customer satisfaction	0.844	–	0.844
Perceived quality → Customer Complaints	–	–0.609	–0.609
Customer Complaints → Information	–	–0.658	–0.658
Customer satisfaction → Customer Complaints	–0.721	–	–0.721
Customer satisfaction → Customer Loyalty	0.827	–	0.827
Customer Loyalty → Information	–	0.754	0.754
Customer Loyalty → Perceived quality	–	0.698	0.698
Customer Loyalty → Expected quality	–	0.125	0.125

4.3   Customer Satisfaction Index

In the structural equation modeling, the CSI was related to three measurement variables, i.e., overall satisfaction, fulfillment of expectations, and actual perception versus ideal. The measurement equations regarding the relationship between observable variables and latent CSI were extracted to calculate CSI. CSI(yi) was denoted as the degree of satisfaction for a specified evaluation index yi. The CSI(yi) = 1 when customers were extremely satisfied with the service; the CSI(yi) = 0 when customers were extremely dissatisfied with the service; in between CSI(y_i) $\in$ (0, 1) [1].

Adopting linear interpolation, standardizing the degree of satisfaction, and using the five-point Likert scale, the CSI(yi) was calculated as follows:

$$\text{CSI(yi)} = \frac{E[y_i] - \text{Min}[y_i]}{\text{Max}[y_i] - \text{Min}[y_i]} = \frac{E[y_i] - 1}{4}$$

In order to find the relationship between overall satisfaction index and the related three observed variables, the following formula was used:

$$\text{CSI} = \frac{\sum_1^3 w_i y_i - 1}{4} \times 100\%$$

Where $w_i = \frac{v_i}{\sum_1^3 v_i}$, $v_i = \frac{\lambda_{ji}}{\lambda_{11}^2 + \lambda_{12}^2 + \lambda_{13}^2}$, λ was the regression coefficient between latent variable and observed variable, yi represented the relevant particular measurement variables of customer satisfaction.

As a result of measuring the CSI of medical services in the middle-level hospital used for this study, the CSI turned out to be 74.20. This result showed that overall customer satisfaction was high.

5   General Discussion

This research constructed a Chinese customer satisfaction model for medical services (CSI-CMS) and identified the key variables that drive customer satisfaction as well as their relationships to each other. The results from Structure Equation Modeling (SEM) suggested that customer satisfaction for inpatients in the medical service industry was based on many factors such as perceived quality, expected quality, and information. It was also closely related to post-purchase behaviors such as customer complaint and customer loyalty.

The path coefficients provided reliable evidence that perceived quality, expected quality, and information were three important antecedents influencing customer satisfaction. Perceived quality had the strongest direct effect on customer satisfaction among the three constructs. This result was consistent with that of Gallarza and Saura [42]. Thus, managers should pay more attention to the control of quality. Since the quality of medical services of a hospital includes many departments; good coordination and cooperation among different departments in a hospital should be emphasized, in addition to enhancing the quality of service of a single department. In addition to the relatively low direct effect of information (0.15) it indirectly affected customer satisfaction through the other antecedents of perceived quality and expected quality. The path coefficient between information and customer satisfaction was highly significant with the total effect, being 0.912. This indicated that information played an important role in influencing customer satisfaction. Thus, managers should focus on publicizing information. Information should not be merely understood as publicizing a brand, but other aspects of information such as the amount of information and authenticity should also be viewed as important.

Customer satisfaction was found to have a negative influence on customer complaints and a positive effect on customer loyalty, confirming the findings of Fornell et al. [13] and Hirschman [43]. A noticeably strong relationship (−0.72) existed between customer satisfaction and customer complaints, indicating managers should always take appropriate measures to resolve complaints and pay enough attention to customer needs. Although service failures are inevitable with associated negative consequences, effective handling of customer complaints can transform dissatisfied customers into satisfied customers. Thus managers should focus on effective complaints management to maintain long-term relationship with customers and improve service performance. Customer satisfaction had a strong positive relationship (0.83) with loyalty. Since loyalty involves repeated use of a service provider and recommending that provider to others, managers should be aware of the impact of word-of-mouth recommendations and devote greater effort to maintaining customer loyalty.

While the Chinese customer satisfaction model for medical services was built on theory, that is well established in the literature concerning numerous parallel fields and the results here indicated that the data fitted the model well, this research has limits. This study was conducted in a specific setting, a middle-level hospital located in northern China. The conclusions of this study may be limited to hospitals at this level in China. Future research, with regard to China, should provide more studies to assess and validate the robustness of the proposed model in hospitals at other levels.

References

1. Fornell, C.A.: National satisfaction barometer: the Swedish experience. J. Mark. **56**, 6–21 (1992)
2. Yang, C.C., Yang, K.J., Cheng, L.Y.: Holistically integrated model and strategic objectives for service business. TQM J. **22**, 72–88 (2010)
3. Reichheld, F.F., Sasser, W.E.: Zero defections: quality comes to services. Harv. Bus. Rev. **68**, 105–111 (1990)
4. Kanoe, N.: The customer satisfaction. Thai Cosmet. Manuf. Assoc. J. **3**, 33 (2003)
5. Cheng, L.Y., Yang, C.C., Teng, H.M.: An integrated model for customer relationship management: an analysis and empirical study. Hum. Factors Ergon. Manuf. Serv. Ind. **23**, 462–481 (2013)
6. Anastasios, M., Elizabeth, D.E., Papathanassogloub, C.L.: Evaluation of patient satisfaction with nursing care: quantitative or qualitative approach. Int. J. Nurs. Stud. **41**, 355–367 (2004)
7. Heidegger, T., Saal, D., Nuebling, M.: Patient satisfaction with anaesthesia care: what is patient satisfaction, how should it be measured, and what is the evidence for assuring high patient satisfaction? Best Pract. Res. Clin. Anaesthesiol. **20**, 331–346 (2006)
8. Abiodun, A.J.: Patients' satisfaction with quality attributes of primary health care services in Nigeria. J. Health Manag. **12**, 39–54 (2010)
9. Fisher, A.W.: Patients' evaluation of outpatient medical care. J. Med. Educ. **46**, 238–244 (1971)
10. Fletcher, R.H., O'Malley, M.S., Earp, J.A., Littleton, T.A.: Patients' priorities for medical care. Med. Care **21**, 234–242 (1983)
11. Linn, L.S., DiMatteo, M.R., Chang, B.L., Cope, D.W.: Consumer values and subsequent satisfaction ratings of physician behavior. Med. Care **22**, 804–812 (1984)
12. Lee, W.I., Shih, B.Y., Chung, Y.S.: The exploration of consumers' behavior in choosing hospital by the application of neural network. Expert Syst. Appl. **34**, 806–816 (2008)
13. Fornell, C., Johnson, M.D., Anderson, E.W., Cha, J., Bryant, B.E.: The American customer satisfaction index: nature, purpose and findings. J. Mark. **60**, 7–18 (1996)
14. Kessler, S.: Measuring and Managing Customer Satisfaction: Going for the Gold. ASQ Quality Press, Milwaukee (1996)
15. Hsu, S.H.: Developing an index for online customer satisfaction: adaptation of American Customer Satisfaction Index. Expert Syst. Appl. **34**, 3033–3042 (2008)
16. Eklöf, J., Westlund, A.H.: The pan European customer satisfaction index program. Total Qual. Manag. **13**, 1099–1106 (2002)

17. Zhao, P., Li, C.Q., Yu, X.Z.: Guide to China customer satisfaction index. China Standard Press, Beijing (2003)
18. Robert, C., Jami, L.D.: An examination of hospital satisfaction with blood suppliers. Transfusion **44**, 1648–1655 (2004)
19. Türkyilmaz, A., Özkan, C.: Development of a customer satisfaction index model: an application to the Turkish mobile phone sector. Ind. Manag. Data Syst. **107**, 672–687 (2007)
20. Fakeye, P.C., Crompton, J.L.: Images differences between prospective, first-time and repeat visitors to the lower Rio Grande Valley. J. Travel Res. **30**, 10–16 (1991)
21. Hoyer, W.D., Macinnis, D.J.: Consumer Behavior. Houghton Mifflin Company, Boston (2003)
22. Bosque, I.A.R., Martín, H.S., Collado, J.: The role of expectations in the consumer satisfaction formation process: empirical evidence in the travel agency sector. Tour. Manag. **27**, 410–419 (2006)
23. Williams, B., Coyle, J., Healy, D.: The meaning of patient satisfaction: an explanation of high reported levels. Soc. Sci. Med. **47**, 1351–1359 (1998)
24. Anderson, E.W., Fornell, C., Lehmann, D.R.: Customer satisfaction, market share, and profitability: findings from Sweden. J. Mark. **58**, 53–66 (1994)
25. Sanchez-Hernandez, R.M., Martinez-Tur, V., Peiro, J.M., Moliner, C.: Linking functional and relational service quality to customer satisfaction and loyalty: differences between men and women. Psychol. Rep. **106**, 598–610 (2010)
26. Tsai, M.C., Chen, L.F., Chan, Y.H., Lin, S.P.: Looking for potential service quality gaps to improve customer satisfaction by using a new GA approach. Total Qual. Manag. Bus. Excell. **22**, 941–956 (2011)
27. Chen, R., Hsiao, J., Hwang, H.: Measuring customer satisfaction of internet banking in Taiwan: scale development and validation. Total Qual. Manag. Bus. Excell. **23**, 749–767 (2012)
28. Oliver, R.L., Swan, J.E.: Consumer perceptions of interpersonal equity and satisfaction in transactions: a field survey approach. J. Mark. **53**, 21–35 (1989)
29. Lee, C.K., Yoon, Y.S., Lee, S.K.: Investigating the relationships among perceived value, satisfaction, and recommendations: the case of the Korean DMZ. Tour. Manag. **28**, 204–214 (2007)
30. Oliver, R.L.: Satisfaction: A Behavioral Perspective on the Consumer. McGraw-Hill Press, New York (1997)
31. Hellier, P.K., Geursen, G.M., Carr, R.A., Rickard, J.A.: Customer repurchase intention, a general structural equation model. Eur. J. Mark. **37**, 1762–1800 (2003)
32. Anderson, E.W., Fornell, C.: Foundations of the American customer satisfaction index. J. Total Qual. Meas. **11**, 869–882 (2000)
33. Chen, C.F., Cheng, L.T.: A study on mobile phone service loyalty in Taiwan. Total Qual. Manag. Bus. Excell. **23**, 807–819 (2012)
34. Setó-Pamies, D.: Customer loyalty to service providers: examining the role of service quality, customer satisfaction and trust. Total Qual. Manag. Bus. Excell. **23**, 1257–1271 (2012)
35. Landon, E.L.: The direction of consumer complaint research. In: Olsen, J.C. (ed.) Advances in Consumer Research, vol. 7, pp. 335–338 (1980)
36. Fornell, C.R., Bookstein, F.L.: Two structural equation model: LISREL and PLS applied to consumer exit-voice theory. J. Mark. Res. **19**, 440–452 (1982)
37. Taner, T., Antony, J.: Comparing public and private hospital care service quality in Turkey. Leadersh. Health Serv. **19**, 1–10 (2006)
38. Parasuraman, A., Grewal, D., Krishnan, R.: Marketing Research. South-Western College Publishing, Boston (2006)

39. Bayraktar, E., Tatoglu, E., Turkyilmaz, A., Delen, D., Zaim, S.: Measuring the efficiency of customer satisfaction and loyalty for mobile phone brands with DEA. Expert Syst. Appl. **39**, 99–106 (2012)
40. Hox, J.J., Bechger, T.M.: An introduction to Structural Equation Modeling. Fam. Sci. Rev. **11**, 354–373 (1998)
41. Johnson, M.D., Gustafsson, A., Andressen, T.W., Lervik, L., Cha, J.: The evolution and future of national customer satisfaction index models. J. Econ. Psychol. **22**, 217–245 (2001)
42. Gallarza, M.G., Saura, I.G.: Value dimensions, perceived value, satisfaction and loyalty: an investigation of university students' travel behavior. Tour. Manag. **27**, 437–452 (2006)
43. Hirschman, A.O.: Exit, Voice and Loyalty: Responses to Decline in Firms, Organizations and States. Harvard University Press, Cambridge (1970)

User Experience and Visualization in Automotive Industry

Drivers Quickly Trust Autonomous Cars

Robert Broström[✉], Annie Rydström, and Christoffer Kopp

User Experience Development Center, Volvo Car Corporation,
SE-405 31 Gothenburg, Sweden
{robert.brostrom, annie.rydstrom,
christoffer.kopp}@volvocars.com

Abstract. Successful introduction of autonomous cars require autonomous technology that users experience as trustful and useful. The aim of the study reported in this paper was to explore if drivers trust a fully autonomous car and if they experience that in-vehicle tasks can be conveniently carried out when in full autonomous mode. The test was conducted on a test track and an autonomous research car was used. The car was capable of handling the test track driving environment with full autonomy. When in full autonomous mode the participants got to engage in individually selected tasks, such as use media display, read, eat, drink and carry out work tasks with their own portable devices. The results show that participant trust the autonomous car and they find it convenient to conduct in-vehicle tasks while in full autonomous mode.

Keywords: Autonomous Driving · Trust · User experience · Glance behaviour

1 Introduction

Significant advances in technology have made autonomous cars a practical reality and the premise is that autonomous driving (AD) can improve safety, enhance driving efficiency and reduce pollution [1, 2]. In AD research, the focus has been on technical development and operational safety. However, the potential benefits of AD are all dependent on the willingness to use the autonomous systems, stressing the importance of user centred research. Recent user centered research studies have predominantly been focusing on transitions of control between the driver and the automation [e.g. 3–5]. Most studies have been conducted in a simulator setting and have been exploring human-machine interaction (HMI) designs for handover of control. In addition, these studies have had a performance perspective, e.g. focused on the time it takes to resume control of the car after a transition between different levels of autonomy. Other studies have given valuable input on users' attitudes and expectations towards AD and their intention of use [6, 7]. However, these studies have not included real experience of such systems. Previous research has shown that an autonomous car, on which the autonomous mode is frequently activated, needs be experienced as trustful and convenient, in accordance with Muir [8]. To our knowledge, no reported studies have investigated people's experiences of driving a fully autonomous car in a real setting. Hence, the study described in this paper is novel in the sense that it is conducted in a "real" AD context on a test track with

© Springer International Publishing AG, part of Springer Nature 2019
T. Z. Ahram and C. Falcão (Eds.): AHFE 2018, AISC 794, pp. 699–705, 2019.
https://doi.org/10.1007/978-3-319-94947-5_69

the aim to explore if drivers trust a fully autonomous car and if they experience that it is convenient to do in-vehicle tasks when in level.

Trust is a central concept in the design of new technologies and is considered to greatly influence the adoption of AD systems. From other domains, it has been revealed that trust plays a prominent role when it comes to using automation. In this paper, we measure trust in accordance with measures used in previous research, such as self-reported, glance behaviour and seating postures [9–11].

2 Method

2.1 Participants

Six Volvo Cars employees, not working with research and development, were recruited to take part in the study. The test group comprised three women and three men ranging in age from 28 to 59 years (M = 48.2, SD = 11.4). The participants were all regular commuters with previous experience of diving assistance functionality, but no experience of autonomous cars.

2.2 Equipment

The study was performed at a rural test track at AstaZero (http://www.astazero.com/), a test environment placed in the western part of Sweden. The rural tarmac road at AstaZero is 5.7 km long and has few changes in the surroundings (Fig. 1).

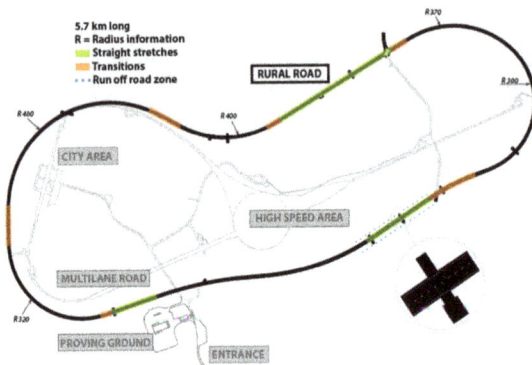

Fig. 1. The rural test track at AstaZero.

The study was conducted using an autonomous research Volvo XC90 car, which is capable at handling the test track driving environment with full autonomy. The car behave as a level 4 car according to the SAE definition [12]. Such a car shall be able to handle all situations without any expectation on intervention from the driver. In this particular test setting, a safety driver positioned in the rear seat continuously monitors the car and is ready to intervene if needed. A lead car, an XC90, was also included in the study (Fig. 2).

Fig. 2. The research car and the lead car.

Two GoPro Hero 3+ cameras were used to capture the behaviour and reactions of the participants. One camera was mounted on the dashboard facing the participant and one was mounted at the right front seat capturing the participant and the road ahead (Fig. 3).

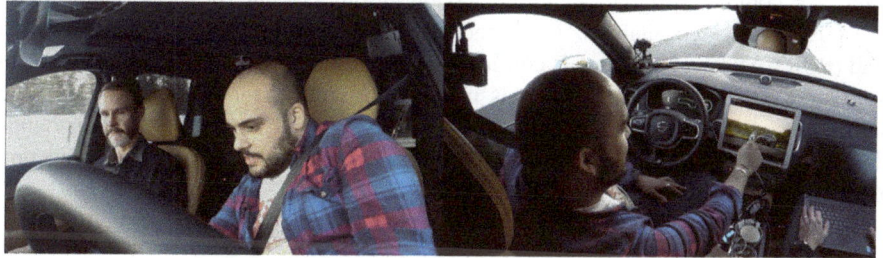

Fig. 3. The camera views in the research car.

2.3 Procedure

Before the test occasion, a pre-study one-to-one telephone interview was conducted with each participant. The interviews focused on the participants expectations on what they imagine they would like to do during full autonomous driving in their daily commute. The participants envisioned themselves using media, such as listening to music, watching movies, reading, surf the web, eat, drink and work using their own devices such as smartphone and laptop. To be able to meet the participant's expectations, an in-car media center based on a touch screen was mounted in the car, coffee and snacks were brought into the car and the participants were asked to bring their personal devices.

The test leader met up participants by the AstaZero reception, where he gave them a brief overview of autonomous driving in general. The test leader then drove the participant to the rural road track and at the same time he gave a description of the study procedure. The participants were also given the opportunity to ask questions. By the research car the participants were asked to be seated in the driver's seat, the test leader was seated in the passenger seat, and the test leader started out with a dialog about what items they had brought into the car. A brief introduction was given to the interaction with the research car, how to switch between three levels of support: no support, assisted driving and full autonomous driving. They were informed to follow the lead car when driving with no support and with assistance. The lead car had a speed of 60–70 km/h.

In total, the participants drove on the rural road for 90 min. The first part of 45 min consisted of a mixed test switching between the three levels. The second 45-min part consisted solely of a full autonomous driving test. During this time, the participants got to engage in their individual selected tasks. A short debriefing interview was held at the end of the test.

2.4   Measures and Analysis

The analysis in the present paper focused on the second part of the study, the fully autonomous driving test. After the test the participants were asked to fill in their level of comfort in conducting in-vehicle tasks on a seven point Likert scale (I feel comfortable conducting in-vehicle tasks while in autonomous mode, 1 = completely disagree, 7 = completely agree). Glance behaviour during task interaction was carried out in a post-drive video analysis. The glance measure used was percent glance time on road (POR), where a low value is considered a good indication of high level of trust. In addition, the post-drive video analysis included an observation of the posture of the participants. It was noted to which extent the participants hovered with their hands close to the steering wheel and foot over the brake pedal, since this gives an indication of whether the participants trust the car or not.

3   Results and Discussion

Previous research [e.g. 3–5] have focused on the transition between different levels of automation, and other studies have given valuable input on users' attitudes and expectations towards AD [e.g. 6, 7]. This study also explored transitions between different levels of automation; however, the main aim was to investigate if drivers trust a fully autonomous car in a "real" AD setting. In addition, the purpose of the present study was to investigate if participants felt comfortable doing in-vehicle tasks when the car drove fully by itself, i.e. level 4 automation [12].

The result from the subjective Likert scale clearly shows that all participants are comfortable doing in-vehicle tasks while in full autonomous mode (Median = 7). In addition, the video analysis of the posture of the participants shows that most of the participants relatively quickly got into a relaxed seating position and that they did not hover with their hands close to the steering wheel or the foot over the brake pedal. This indicate that the participants felt comfortable and trusted the car behaviour [10].

In addition, the interviews strengthen the result that participants quickly trusted the car. As an example, one participant stated:

"I completely forgot that I am driving and that I am sitting in a car."

In terms of glance behaviour, the results show a clear difference between non-visual tasks (NVT) and visually engaging tasks (VET) (Table 1).

Table 1. Driver glance behaviour while in autonomous mode in terms of percent glance time on road. NVT: non-visual task, VET: visually engaging task, POR: percent eye time on road.

Participant	NVT POR	Description	VET POR	Description
1	n/a	No task	1.8	Watch TV news in in-car media center
2	86.8	Drink coffee	0.8	Watch TV series in in-car media center
3	88.3	Drink coffee	0.3	Read magazine
4	90.0	Drink coffee	0.7	Watch TV series in in-car media center
5	100.0	Drink coffee	16.7	Put on makeup
6	96.0	Phone Call	0.8	Email and messaging on phone
Mean	92.2	n/a	3.5	n/a

When occupied with the NVT, such as drinking coffee and engaging in a phone call, all participants looked more than 85% of the time towards the forward roadway (POR > 85). Hence, for the NVTs drivers spend long periods looking at the road even if they do not have to, probably since this is a natural position. Even if the glance data for NVTs indicate lack of trust to the autonomous car, the Likert and the posture data show that participants indeed trusted it. One can assume that trust can be high even if the driver looks out of the front window and onto the forward roadway.

For the VETs, such as watch TV series in the in-car media center or read a magazine, glances away from the task towards the forward roadway could indicate under-trust since the driver moves focus away from the VET towards the forward roadway [11]. Quite surprisingly, a majority of the participants locked less than two percent of the time towards the forward roadway (POR < 2, Table 1). This result was confirmed both by the subjective Likert scale (Median = 7) and the objective posture data which showed that participants really trust the car.

A bit unexpected, the POR value for participant 5 differed significantly from the other VETs (POR = 16.7). All tasks where selected by the participants as tasks that they were willing to do while the car was in fully autonomous mode. Putting on makeup is obviously a very different task from reading a magazine. When putting on makeup, the makeup mirror is positioned in the sun visor and the eyes are directed towards the road. This may bring particular interest in looking at road compared to when reading a magazine or watch TV.

In terms of exposure to an autonomous car, participants seem to adopt to and trust the autonomous car in a much shorter time than expected. One explanation of this could be the monotonous driving environment, with few traffic events and little surrounding traffic. On the other hand, the test track is a very realistic setting since it is a real road, and involves real risks if the car would go off road. Also the fact that the car repeatedly showed that it could handle all scenarios on the road, could have contribute to the rapid build-up of trust.

Based on the results in this paper it seems like a task carried out in an autonomous car can be much more immersive if it is motivating and has relevance for the participant. The following quote from a participant reflects these findings quite well:

> "It really feels like I am on a train, I feel safe. The first round it was very new of course, but now there are no surprises. The magazine that I am interested in makes me relaxed. The feeling of trust came so fast, it was as if I did not even feel the urge to check if the vehicle was doing right. I am surprised."

4   Conclusion and Future Work

As a conclusion, the results from this study indicate that to gain trust in the autonomous car and to experience the autonomy as convenient, (1) autonomy have to be experienced for a certain amount of time, (2) the car should be able to repeatedly handle scenarios, and (3) the time in autonomy should be experienced as valuable.

In future research it is important to go beyond testing the experiences of autonomous driving at test tracks and enter the real world. It is also of importance to address long-term effects of using an autonomous car in everyday life to get evidence and input for future designs. Other research topics to consider are the kinematics of the automated vehicle (e.g. how aggressive one should be when passing another vehicle), adjustment possibilities (e.g. operational, tactical and strategic), and what information is needed from the car to enable the creation of proper trust between the driver and the vehicle. In addition, trust needs to be studied in a wider context, taking into account the multi-dimensionality of the construct.

Acknowledgments. This study was conducted within the project Human Expectations and Experiences of Autonomous Driving (HEAD). The project is financially supported by the Swedish strategic vehicle research and innovation programme (FFI). FFI is a partnership programme run jointly by the Swedish state and the Swedish automotive industry that funds research, innovation and development with an emphasis on climate, the environment and safety.

References

1. Davila, A., Nombela, M.: Platooning - safe and eco-friendly mobility. In: SAE World Congress & Exhibition, Detroit (2012)
2. Rupp, J.D., King, A.G.: Autonomous driving – a practical roadmap. In: SAE Convergence, Detroit (2010)
3. Gold, C., Körber, M., Lechner, D., Bengler, K.: Taking over control from highly automated vehicles in complex traffic situations: the role of traffic density. Hum. Factors **4**, 642–652 (2016)

4. Merat, N., Lai, A., Daly, F., Carsten, O.: Transition to manual: driver behaviour when resuming control from a highly automated vehicle. Transp. Res. Part F Traffic Psychol. Behav. **27**, 274–282 (2014)
5. Walch, M., Lange, K., Baumann, M., Weber, M.: Autonomous driving: investigating the feasibility of car-driver handover assistance. In: the 7th International Conference on Automotive User Interfaces and Interactive Vehicular Applications – AutomotiveUI 2015, pp. 11–18. ACM, New York (2015)
6. Pettersson, I.: Travelling from fascination to new meanings: understanding user expectations through a case study of autonomous cars. Int. J. Des. **11**, 1–11 (2017)
7. Wallgren, P., Rexfelt, O.: A qualitative study on car drivers' attitudes towards and expectations of automated cars (2018, in preparation)
8. Muir, B.M.: Trust between humans and machines, and the design of decision aids. Int. J. Man Mach. Stud. **27**, 527–539 (1987)
9. Gold, C., Körber, M., Hohenberger, C., Lechner, D., Bengler, K.: Trust in automation - before and after the experience of take-over. In: the 6th International Conference on Applied Human Factors and Ergonomics - AHFE 2015, pp. 3025–3032. Elsevier (2015)
10. Helldin, T., Falkman, G., Riveiro, M., Davidsson, S.: Presenting system uncertainty in automotive UIs for supporting trust calibration in autonomous driving. In: the 5th International Conference on Automotive User Interfaces and Interactive Vehicular Applications – AutomotiveUI 2013, pp. 210–217. ACM, New York (2013)
11. Morando, A., Victor, T., Dozza, M.: A reference model for driver attention in automation: Glance behavior changes during lateral and longitudinal assistance (2018, in preparation)
12. SAE International (http://standards.sae.org): Taxonomy and Definitions for Terms Related to Driving Automation Systems for On-Road Motor Vehicles, SAE J3016_201609 (2016)

Scenario-Based User Experience Research in Automobile Interior Lighting Innovation

Bo Ouyang[✉] and Yun He

School of Design, South China University of Technology, Guangzhou, China
{ouyangbo, hebin}@scut.edu.cn

Abstract. To implement advanced technology into appropriate design with WOW user experience, an approach of scenario-based User Experience Design (UED) method was proposed for further exploring the application of user experiences in automobile interior lighting innovation. The related literatures of User Experience Design (UED), scenario-based design and scenario-based prototype methods were analyzed. Then, the Scenario-based user experience research was divided into three dimensions including target user, behavior pattern, and user scenario. Through quantitative research and qualitative research, including user observation, interview, and questionnaires as well as data collection and processing, the typical UED scenario model was constructed, and the scenario-based automobile interior lighting system was proposed. Finally, some innovation design concepts and a scale size prototype were developed to verify the effectiveness and practicality of the scenario-based UED research of automobile interior lighting innovation.

Keywords: Scenario · User Experience Design · User behavior
Automobile interior lighting

1 Introduction

With the developing of social economy, we are in the time of experience economy. Except consuming goods and products, consumers are more focusing on individual emotional experience beyond the products [1]. As a typical mass production product, and a carrier of economy, technology and humanity, car industry is also paying more and more attention to user experience. Traditionally, user experience of car is about itself: styling, performance, configuration etc. However, with the developing of artificial intelligent, IOT (internet of things), user experience of car is more than product itself, is about service, activity and environment which connect product with users.

The experience economy brings a number of changes for design and innovation. User experience design (UED) has received extensive attention in design study and practice among many fields in the recent years. It is a comprehensive design approach including product design, service, behavior and environment related. Every design elements is based on individual or group's demands, willing, believe, knowledge, technical, experience and awareness. User experience design emphasizes user oriental, and brings new perspectives and methods for product design and related services design [2].

© Springer International Publishing AG, part of Springer Nature 2019
T. Z. Ahram and C. Falcão (Eds.): AHFE 2018, AISC 794, pp. 706–714, 2019.
https://doi.org/10.1007/978-3-319-94947-5_70

However, UED should have accomplished in the certain social scenario. It is an activity based on social circumstance, and it relies on UED users to go through board "symbols connect" in particular social culture scenarios. Without connection of scenario, user experience is empty and meaningless. In different scenarios, user emotion and behavior are different, and it directly affects the results of UED.

This paper bases on a design research project supported by Audi, a luxury car brand. In order to get innovative interior lighting concepts, a scenario-based user experience research was conducted. A typical UED scenario model was built, and the scenario-based automobile interior lighting system was proposed. Finally, some innovation design concepts and a scale size prototype were developed to verify the effectiveness and practicality of the scenario-based UED research of automobile interior lighting innovation.

2 User Experience Design

User Experience (UX) refers to a person's emotions and attitudes about using a particular product, system or service in the whole product life cycle. It includes the practical, experiential, affective, meaningful and valuable aspects of human–product interaction and ownership. Additionally, it includes a person's perceptions of system aspects such as utility, ease of use and efficiency. User experience is dynamic as it is constantly modified over time due to different circumstances and changes to individual systems as well as the wider usage context in which they can be found. In the end user experience is about how the user interacts with and experiences the product.

The UX concept begin with human-computer interaction (HCI)。In The international standard on ergonomics of human system interaction, ISO 9241-210 [3], defines user experience as "a person's perceptions and responses that result from the use or anticipated use of a product, system or service". According to the ISO definition, user experience includes all the users' emotions, beliefs, preferences, perceptions, physical and psychological responses, behaviors and accomplishments that occur before, during and after use. The ISO also list three factors that influence user experience: system, user and the context of use.

Although UX started in the computer science field from 40th 20 century, it has expanded to boarder areas. Based on UX concept, User experience design (UED) was proposed to get better user experience in solving complex problems. In fact, UED is the process of enhancing user satisfaction by improving the usability, accessibility, and pleasure provided in the interaction with the product or the system [4]. User experience design encompasses traditional human–computer interaction (HCI) design, and extends it by addressing all aspects of a product or service as perceived by users [5].

In academic, Garrett [6] describe UX has several experiences such as brand identities, information usability, functionality, context, etc. Normal [7] has a theory that UX has three different levels: instinct level, behavior level and emotional level. In industry, high-tech IT companies such as Apple, Google, IBM are recognized step ahead leaders in UED. Such UED practices in those companies have engaged for many years. With the rising of internet and information technology, UED is widely accepted not only in traditional product companies, but also in organizations, such as banks, government administrations, NGO organizations.

2.1 Scenario-Based Design

UED is a design approach based on user center, emphasis on perceiving product or system from user's perspective. By this method to get the customer's true needs, and to better even creative a completely new user experience. Traditionally, we use marketing research to get massive data, and get user's need through process the data. This is a typical objective way to look at use experience. In this way, designers like an irrelevant third party. The disadvantage of this method is it it's difficult to build a sympathy connection between designers and users. The final design usually become a reflection of designer's own knowledge system and personal feeling to products or system.

Scenario building can lead the participants to the effect of turning their feelings. Scenario is about people and people's activity, their story [8]. Scenario based design is a systematic approach centered on scenario. User's perception of product and service system can be show off by complete story. Scenario-based design provides an effective communication and design method to help designer fully understand user's mind and realize design target correctly.

The most important part of scenario-based design is how to construct scenario and analyze the user experience model in this scenario. Right scenario can help designer to clearer design goal, and easier to lead user to emerge in scenario. In such scenario, designer and user will communicate more effectively. Furthermore, user will be more actively to participate in design process. In this paper, according the trend of automobile interior lighting technology, we try to build some typical user scenario to get more customer insight with users and based on scenario to develop innovation concept prototype.

2.2 Scenario-Based Prototype

A prototype is an early sample, model, or release of a product built to test a concept or process or to act as a thing to be replicated or learned from [9]. It is a term used in a variety of contexts, including semantics, design, electronics, and software programming. A prototype is generally used to evaluate a new design to enhance precision by system analysts and users [10]. Prototyping serves to provide specifications for a real, working system rather than a theoretical one. In some design workflow models, creating a prototype (a process sometimes called materialization) is the step between the formalization and the evaluation of an idea [1].

Scenario-based prototyping is not only a traditional prototype. In this method, scenario model become a kind of prototype itself. In any design stages, such as analyze, design, testing and evaluation, scenario model always is used as a foundation. It is not only just a testing tool used by designer, but also a platform that users and designers can participate together, to describe, communicate in the same circumstance. Through this way, the design iteration will be more effective and save time and budget for product develop. In this paper, from the early beginning of user questionnaires and concept develop, scenario-based prototype is used to design, test and improve design concept.

3   Scenario-Based User Experience Design Model

Scenario-based UX design method mainly focuses on the interaction between user and system. This interaction is affected each other to users and system in typical scenario. Scenario-based UX design model can be explained in three dimensions. (Fig. 1.)

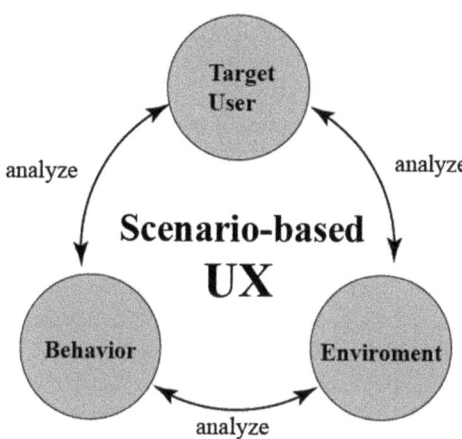

Fig. 1. Scenario-based UX model

Target user: user is a changeable factor and represents diversity of human being. Different user has different feature, including physiological characteristic (sex, age, size) and psychological characteristic (personality). It is nearly impossible to design a product or system to satisfy everyone. However, a group of users could have common feature that is social characteristic, people's activity, emotion, status, understanding and expecting about a task, and lifestyle. Find out who are target users is the foundation of exploring meaningful user experience.

Behavior pattern: people's behaviors are usually not change too much in a relevant period. It is close to human social characteristic. A group of people if they have similar education and culture background, similar growing background, the same social status, they will have similar and steady behavior or lifestyle. On the contrary, product or system is growing faster. Technology develops rapidly that there is a mismatch between human behavior and product. That is why behavior pattern study is an another key of get better user experience.

Environment: Environments Include physical environment and virtual environment. Such as natural condition, man-made infrastructures, social organization, fashion trend, etc.

There three combinations becomes so call user scenario. It is an interaction pattern that typical users use product or system in a certain circumstance. Through scenario-based user experience design, we can find innovative solutions to support product even generate a new business-mode. For example, the Chinese bike-sharing mode, it is an innovative product-service system. The essential user scenario of it is: unlike other sharing schemes around the world, China's bicycles can be picked up and left anywhere, making them convenient for users.

4 Case Study

Brand identity is also kind of user experience that built base on whole images which company tries to transfer and communicate to their customers. However, sometimes there is a mismatch between brand and customers. Especially when brand meets culture barrier and regional differences, the mismatch will be amplify. Since Audi entered the Chinese market, Audi has always represented the Chinese government official vehicle for a long time. In addition, because of this situation, the cars of Audi always give people the sense of dignity and steady. However, with the world famous brand of automobile entering the Chinese market, competition is becoming increasingly fierce, the single market strategy cannot let a company survive in the future. Audi is also the case, trying to break their traditional image, establishing a new image in front of consumers, especially for young people, interpreting "Vorsprung durch Technik" in the context of a new era. We try to understand the users in south China, digging out their concerns about automotive interior lighting. Based on these researches, we also use the latest lighting technology to create a new interior of car in order to enhance the driving experience.

4.1 Scenario Emerged Questionnaire and Interview

Questionnaire is a typical marketing research method to get information from users. However, if you want get useful result from traditional questionnaire, you need have massive samples from interview. In order to get customer insight from small samples, scenario emerged questionnaires, interviews are used, and we iterate questionnaires for five times. Each time we adjust a little bit according to the last time feedback and analyze.

The reason we do the iteration because we find people has difficulty to connect the advanced lighting technology with their life experience. Especially in China, car market develops so fast. Even though car ownership is growing so rapidly, many Chinese people do not understand car. How to connect car life to their own life is the key to help them to tell us their true needs about car interior lighting.

The first version is a questionnaire of lighting tendency. We try to use this questionnaire to infer the consumption trends in Guangzhou, such as whether they are interested in the lighting concept from Audi (show pictures and videos). It shows that the consumers in Guangzhou are concerned more about practicability and feasibility of products. Based on this, we should be focus on whether it is a practical design concept that can achieve in the next five to ten years when concerning the users in Guangzhou.

In the second version of questionnaire, we take a similar approach as the first time we do not directly asked about the lights inside the car. Instead, we start from the lights that we are familiar with in daily life (such as bedroom lights). That is because many owners usually are not aware of the lights inside the car because compared with other objects; the light seems to be not so important. But with pictures and videos of these familiar lights, we can arouse their memories of lights in cars and they will realize light seems to be an indispensable part of the car, then we gradually change the questions to the car lights and the result of questioning will be much better (Fig. 2).

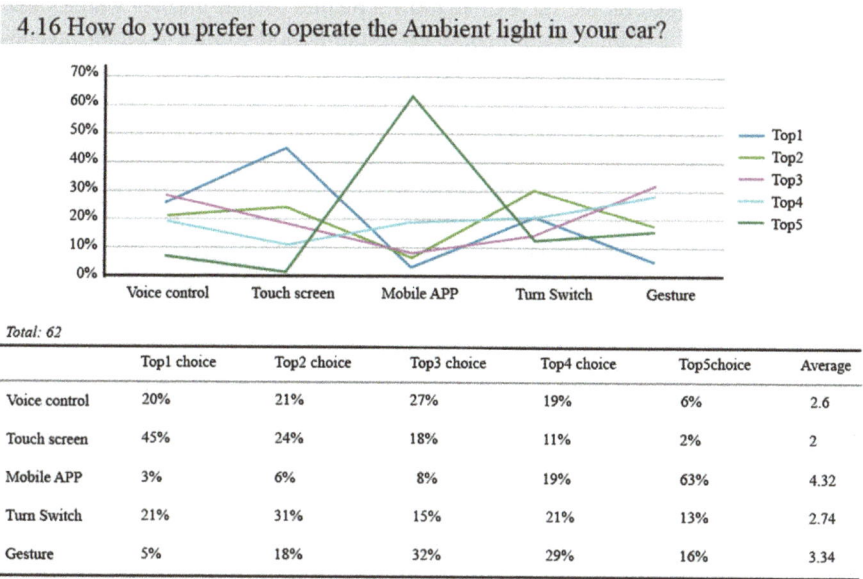

4.16 How do you prefer to operate the Ambient light in your car?

Total: 62

	Top1 choice	Top2 choice	Top3 choice	Top4 choice	Top5choice	Average
Voice control	20%	21%	27%	19%	6%	2.6
Touch screen	45%	24%	18%	11%	2%	2
Mobile APP	3%	6%	8%	19%	63%	4.32
Turn Switch	21%	31%	15%	21%	13%	2.74
Gesture	5%	18%	32%	29%	16%	3.34

Fig. 2. The datas of the final questionnaire (part of the result)

4.2   Scenario-Based UX Analysis

According to the feedback of early questionnaires, we make the final questionnaires. We collect all the datas of every question and organize the data.

After the questionnaire survey, we know the attitudes of consumers towards each question and we start to make a diagram of target groups & scenario combinations. (Fig. 3). We have total 22 scenario-based user experience mode toward three different target user groups. These 22 scenarios could be inspirations for innovative automobile interior lighting concept.

Target Group	Scenarios Waiting				Scenarios Stress & Safety				Emotional*			Scenarios Activity				
					Physical Tiredness *											
	Make up*	Pick-up *	Playing Kids **	selfie *	Darkness*	Jet lag*	Marathon Day*	Long Distance *	Extreme Weather *	Traffic Jam *	Pressure **	Destination ***	Work in Car ***	Wechat Mobile *	Dating Fun *	Dating*
	A	B	C	D	E1	E2	E3	E4	F1	F2	F3	G	H	I	J	K
1. Lifestyle	√	√		√		√				√		√		√	√	√
2. Young Family		√	√			√				√		√		√	√	
3. Business	√	√				√				√			√	√		

Explanation: * priority 1   ** priority 2   *** priority 3

Fig. 3. Scenario-based UX analysis

4.3   Scenario-Based UX Concept Design

In this case, each scenario combination is use as a prototype, to develop 22 different automobile interior lighting concepts. Here is one of examples of these 22 designs. The target group is business people and scenario is B (waiting, pick-up).

User journey: Based on the survey, we draw a picture of user journey which shows the typical situation in China when people go to the airport to pick-up their business client. It is difficult to find each other (Fig. 4).

03:00 pm	03:40 pm	04:10 pm
The phone call comes, I have to pick up a client.	I arrive at the airport, but there are so many cars and people, I can't find my client!	Finally I find my client, but it takes a long time, he seems not very happy, I feel so sorry.

Fig. 4. Pick-up user journey

Design concept: the lighting system is design in two parts, finding and welcome. Using holographic project and other advanced lighting technology that Audi provides to us. The new concept provide a easy way to let business client to identify their welcome partner, and satisfy the very important convention for business in China, give your client VIP service. (Fig. 5)

With the new concept design, we summarize a new optimized User journey (Fig. 6).

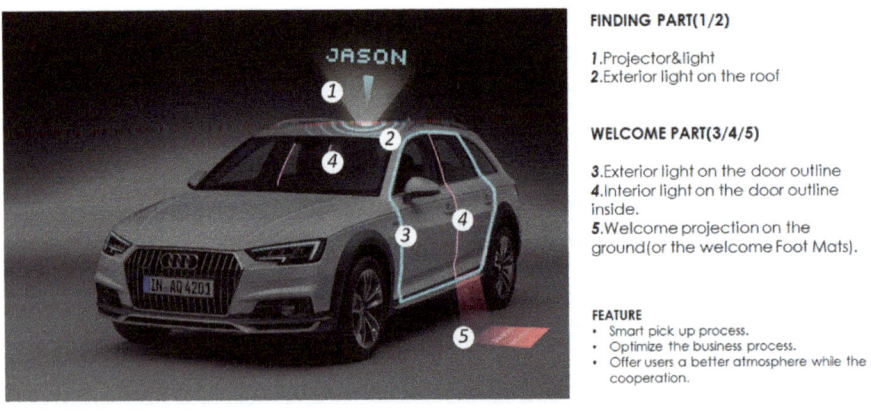

Fig. 5. Business_pick-up lighting concept

Fig. 6. The optimized user journey

5 Conclusion

Through theoretical analysis and case study, Scenario-based user experience research can help designer better understand users to get creative ideas and new experience, especially when we develop a relative small system in very effective way. In the case study, by using scenario-based approach, we also found some interesting thing about Chinese consumers towards automobile. We found out that the consumers in Guangzhou are very concerned about practicability and feasibility of products. We divided those scenarios into three parts including waiting, stress or safety, activity and divided target groups into three parts including lifestyle, young family, and business. Based on our scenarios and target groups, we designed some concepts to solve these problems and the concepts are based on Audi's latest lighting technology and other new technologies. At the same time, we are aware that digging the concerns (problems) and expectations of consumers in the future is a very difficult and complex process because consumers always have strong stereotypes and do not know what they really want. So the work also showed that scenario-based user experience research is not only to meet the customer's needs, but also to create new experience to surpass user's expectation.

References

1. Pine II, B.J., Gilmore, J.: The experience economy: work is theatre & every business as stage. Harvard Business School Press, Boston (1999)
2. Kuniavsky, M.: Observing the user experience: a practitioner's guide to user research. Morgan Kaufmann, San Francisco (2003)
3. International Organization for Standardization: Ergonomics of human system interaction - Part 210: Human-centered design for interactive systems (2009)
4. Kujala, S., Roto, V., Väänänen-Vainio-Mattila, K., Karapanos, E., Sinneläa, A.: UX Curve: A method for evaluating long-term user experience. J. Interact. Comput. **23**(5), 473–483 (2014)
5. Nardi, B.A.: The use of scenarios in design. J. ACMSIG-CHI Bull. **24**(4), 13–14 (1992)
6. Garrett, J.: The elements of user experience: user-centered design for the Web. New Riders, Berkeley (2002)
7. Norman, D.A.: Emotional design: why we love or hate everyday things. Basic Books, New York (2004)
8. Carroll, J.M.: Scenario-based design: envisioning work and technology in system development. Wiley, New York (1995)

9. Blackwell, A.H., Manar, E., (eds.) "Prototype". UXL Encyclopedia of Science (3rd edn.). Accessed 13 July 2015

10. Gero, J.S.: Design Prototypes: A Knowledge Representation Schema for Design. AI Mag. **11**(4), 26 (1990)

11. Soares, M.M., Rebelo, F.: Advances in Usability Evaluation, p. 482. CRC Press, Boca Raton (2012)

Usability Research of In-vehicle 3D Interactive Gestures

Hao Tan[1(✉)] and Qin Zhang[2]

[1] State Key Laboratory of Advanced Designand Manufacturing for Vehicle
Body, Hunan University, Changsha, China
Htan@hnu.edu.cn
[2] School of Design, Hunan University, Changsha, China
zhangqin@hnu.edu.cn

Abstract. With the infrared system, computer vision, ultrasonic array and other kinds of technology rapidly developed for gesture recognition, gesture control as a kind of natural interaction has been a well-respected research topic of in-vehicle interaction. In this paper, we will propose evaluation indices based on the previous works about 3D gesture control usability, and conduct usability experiments in an indoor environment via driving simulator where three kinds of data will be collected during the test, namely distraction while driving, performance of risk and subjective assessment. Then an information entropy method is used to process data. Through this method, 9 usability indicators are obtained to measure the usability of in-vehicle 3D gesture interaction, providing a reference for usability evaluation of in-vehicle gesture control. A set of 3D gesture applications that accord with the habits of Chinese drivers and can be applied to in-vehicle gesture control are also proposed.

Keywords: In-vehicle · 3D gesture control · Usability · Interactive

1 Introduction

Finger-tip control and touch screen have been two main ways of in-vehicle interaction during the past decades. However, with the rapid development of the Internet and automobile industry, in-vehicle interaction has become more and more complex and multi-functional. People need to deal with a lot of tasks during driving, but buttons and touch screens, which need accurate operation interaction, are more likely to cause traffic accidents and affect the user experience [1]. Therefore, carmakers are rapidly integrating 3D gesture recognition technology in their infotainment systems. Gesture control, as a natural method of human-computer interaction, are used in cars where users' gestures are identified through car's camera or sensor. Differed from haptics communication methods including button and touch screens, gesture control has a prominent advantage of being natural and intuitive and not requiring precise operation.

Google released car control technology of gesture recognition in 2013. Then Ford and Intel worked together to develop gesture recognition technology named Mobii on-board system, so do BMW 7-series. The gesture interaction has become a most cutting-edge technology as well as a most popular trend among all the other in-vehicle

© Springer International Publishing AG, part of Springer Nature 2019
T. Z. Ahram and C. Falcão (Eds.): AHFE 2018, AISC 794, pp. 715–726, 2019.
https://doi.org/10.1007/978-3-319-94947-5_71

interactive technologies. But at present, gesture interaction has been designed in vastly different ways and it should also be noticed that some gestures we usually use in our daily life may not be suitable to be applied to cars. Therefore, it's significant and urgent to develop a set of usability evaluation guidelines regarding in-vehicle 3D gesture interaction as well as to propose a set of gestures that can be well-adapted to a particular group of people.

This paper includes 7 sections. Section 2 mainly presents the background of 3D gesture control. Section 3 outlines the usability indices based on previous works and how these indices can be refined. Section 4 covers usability experiments and how information entropy method can be used to analyze the differences among evaluation indices, while Sect. 5 analyzes results of experiments. The conclusion and future work are included in Sects. 6 and 7 respectively.

2 Related Research

The international organization for standardization (ISO 9241-11) established the most widely used usability evaluation index for human-computer interaction. The standards are described as the extent to which a product can be used by specified users to achieve specified goals with effectiveness (task completion by users), efficiency (task in time) and satisfaction (response of user in term of experience) in a specified context of use (users, tasks, equipment & environments) [2]. In addition, Nielsen, the founder of the usability engineering, also proposed four design indicators for 3D gesture usability, including "learnability", "efficiency", "error rate" and "safety", which can be verified via experiment [3]. Helman I Stern proposed balancing machine factors, claiming that using gestures while driving will bring driver operation load (as hand remains suspended), and those less comfortable gestures will lead to even more operation load. He also introduced four availability indices to evaluate gestures, including intuitiveness, comfort, learnability and recognition rate [4]. Jun Ma et al. taking driver's attention, task execution and subjective evaluation into consideration, put forward that car 3D gesture interaction should be easy to learn and easy to remember, and should be efficient and comfortable [5].

However, current gesture control evaluation criteria proposed by ISO and Nielsen are initially designed for a general gesture-control context, as opposed to be specifically applied to in-vehicle applications [2, 3]. This drawback will severely compromise the reliability of these evaluation indices for in-vehicle gesture control, as there exist a great many differences between gesture control in vehicle and that in other environments. On the other hand, those vehicle-related criteria tend to focus on specific aspects and still cannot be viewed as a universal evaluation system for in-vehicle gesture control [4, 5]. Considering a great variety of gestures adopted on different car platforms, it is undoubtedly significant and valuable to establish a series of universal usability evaluation indices for in-vehicle gesture control.

3   Usability Indices Establishment

Given that in-vehicle interaction should be done in an effective and accurate way so as not to compromise safety of driving, effectiveness and efficiency should be considered as two significant evaluation indices of in-vehicle gesture control usability. Particularly, error rate can be used to measure the effectiveness mentioned above. Safety is always top priority for driving, so it should also be added as another important evaluation index. Moreover, given that the feedback of the 3D gesture interaction is usually not obvious and that there exist few ways to provide gesture instructions for drivers while driving, it is equally important that the gesture should be easy to remember. Finally, intuitiveness, comfort, learnability and memorability which are mentioned above should also be assessed subjectively as a measurement of the level of driver's satisfaction regarding gesture control. Based on this, evaluation guidelines of in-vehicle 3D gesture usability and corresponding definitions are established in Table 1. It should be pointed out that technological limitations of gesture recognition are not considered so that gesture recognition rate is viewed as a usability index.

Table 1. Usability guidelines of in-vehicle gesture control

Usability index	Definition
Effectiveness	The rate of driver successfully and correctly completing the task during driving
Efficiency	The time to complete a task correctly
Safety	The degree of driver's concentration while completing a task
Subjective Satisfaction	The subjective evaluation given by the user regarding intuitiveness, comfort, learnability and memorability

To quantify the above usability indices of in-vehicle gesture control, three aspects of data are collected: how concentrated driver is during gesture control, how well the gesture control operation is handled as well as driver's subjective evaluation.

Particularly, regarding driving safety, a survey by the U.S. highway safety administration (NHTSA) found that nearly 80 percent of all driving accidents and 60 percent of near-driving accidents are related to driver distraction [6]. The literature [7] indicates that we can distinguish whether the driver is distracted by analyzing the steering wheel angle standard deviation, eye position deviation, and the duration of driver's sight point towards one certain direction (defined as "dwell time"). Therefore, during the driving process, the gesture operation can also be evaluated through task completion rate, number of errors and operation time.

All the indicators mentioned above are summarized in Table 2:

Table 2. Usability indices of in-vehicle gesture control

Topic	Usability index		Definition
Distraction while driving	Safety	1. Steering wheel angle standard deviation	The standard deviation of steering wheel angle while the task is being performed
		2. Eye position deviation	The deviation of eye position while the task is being performed
		3. Dwell time	The duration of driver's sight point towards one certain direction
Performance of task	Effectiveness	4. Completion rate	The ratio of completing a task correctly
		5. Errors	The number of errors before successfully completing a task
	Efficiency	6. Operation time	The time required to perform a task correctly
Subjective assessment	Subjective assessment	7. Memorability	How easy to remember a specific gesture
		8. Comfort	How comfortable to use a specific gesture
		9. Learnability	How easy to learn a specific gesture
		10. Intuitiveness	How well a specific gesture matches user's intuition

4 Experiment

4.1 A Preliminary List of Selected Gestures

In this study, we first summarize the applications of the gesture interaction involved in different brands of cars, and 16 most common tasks which can be controlled via gesture during driving are listed as below. Then 50 drivers are asked to use a random but most intuitive gesture for the selected 16 tasks. Finally, we count three most popular gestures for each task as the candidates for the gesture usability evaluation, as shown in Fig. 1:

4.2 Test Method

There are mainly two ways to conduct gesture-control task experiments. The first way is real road test: drivers do real tasks on real road and their performance data will be collected while driving. The second way is simulated driving test: divers need to do gesture-control tasks in a driving simulator. The second way has been widely adopted in research as it provides a safe environment for subjects and researchers without compromising authenticity. Therefore, in this experiment we implement simulated driving test to find out a set of gesture usability indices.

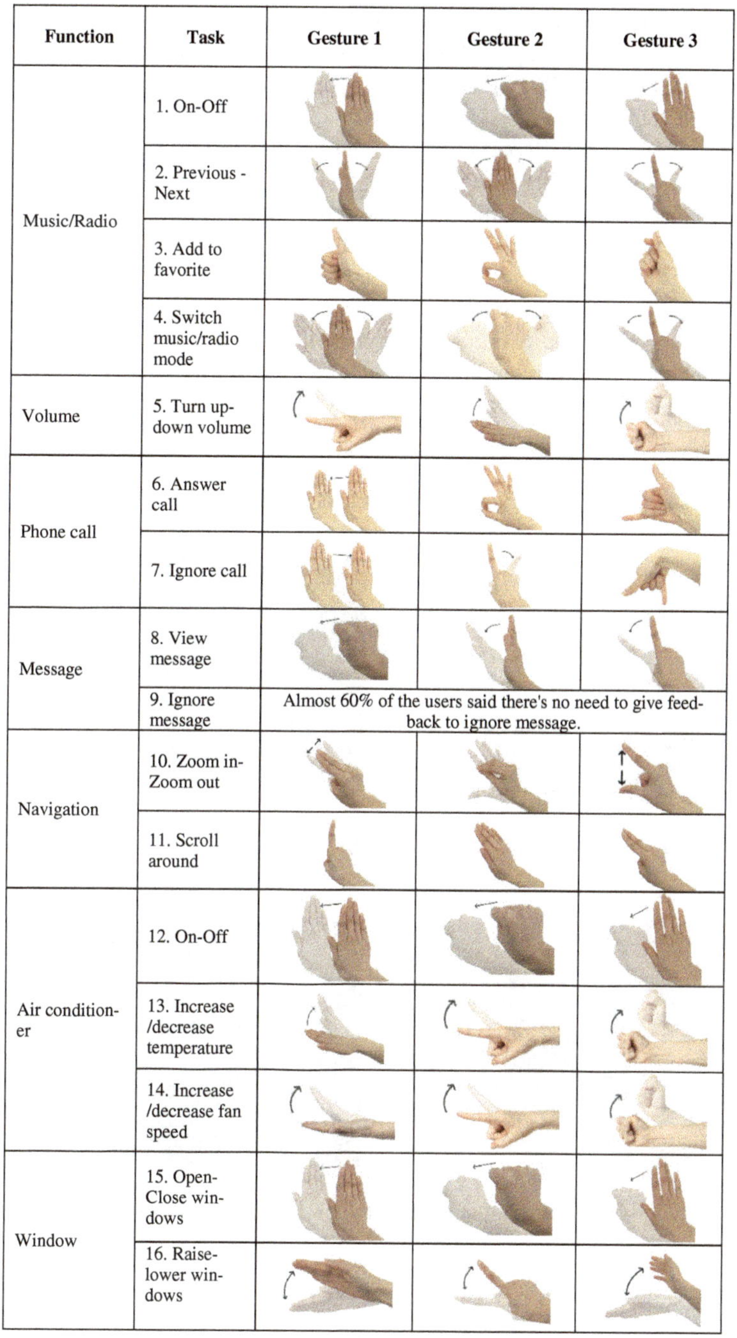

Function	Task	Gesture 1	Gesture 2	Gesture 3
Music/Radio	1. On-Off			
	2. Previous - Next			
	3. Add to favorite			
	4. Switch music/radio mode			
Volume	5. Turn up-down volume			
Phone call	6. Answer call			
	7. Ignore call			
Message	8. View message			
	9. Ignore message	Almost 60% of the users said there's no need to give feedback to ignore message.		
Navigation	10. Zoom in-Zoom out			
	11. Scroll around			
Air conditioner	12. On-Off			
	13. Increase /decrease temperature			
	14. Increase /decrease fan speed			
Window	15. Open-Close windows			
	16. Raise-lower windows			

Fig. 1. Example of gesture database

The experiment is conducted in a quiet indoor environment, with one computer and one Leap Motion sensor. Leap Motion sensor is connected to computer as a peripheral hardware whereas the computer provides software support for 3D gesture recognition. In the experiment, the simulated road is a straight road section of highway, and driving data is collected by the simulator. 18 participants were selected aged from 20 to 45 years old, all having at least 2 years' driving experience, including 12 male drivers and 6 female drivers. This 2:1 ratio is a rough approximation of the actual ration of men vs. women driving in China. They all obey traffic rules and are used to right-handed operation.

Before a gesture test is started, participants are required to go through pre-experiment training. During the training, participants are asked to practice gesture-control operations for 30 min using one of the gestures from our preliminary list described in Sect. 4.1 for each task and are free to drive in the simulator so as to get familiar with it. Besides, to ensure that the test results are not affected in the case where participants find it hard to memorize too many gestures at a time, participants are required to memorize gestures corresponding to only one task, and once finishing that task, they pause for one minute and memorize gestures corresponding to the next task (Figs. 2 and 3).

Fig. 2. (a) Driving simulator and (b) Simulated driving on highway

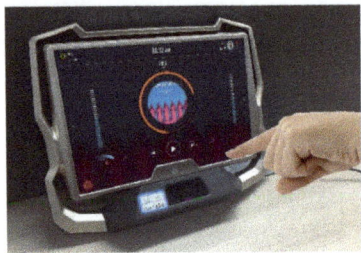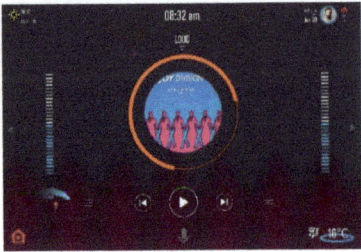

Fig. 3. (a) Infotainment system prototype and (b) Music player interface

4.3   Information Entropy Method

In this study, the information entropy method is used to analyze differences among evaluation indices. The entropy theory first appeared in the field of thermodynamics. Shannon applied this theory to information theory, stating that information entropy can be used to measure the degree of chaos of a system [8]. In other words, information entropy can be used to characterize the differences among various systems regarding a particular index.

Generally speaking, if a certain index has a better ability to differentiate multiple systems, its information entropy value is small, and the weight value of entropy is relatively high. On the contrary, if a certain index cannot well differentiate multiple systems, its information entropy is large and the weight of entropy is relatively low [9, 10].

We utilize the information entropy method to evaluate the previously mentioned 10 usability indices. These 10 usability indices are demonstrated in Fig. 4.

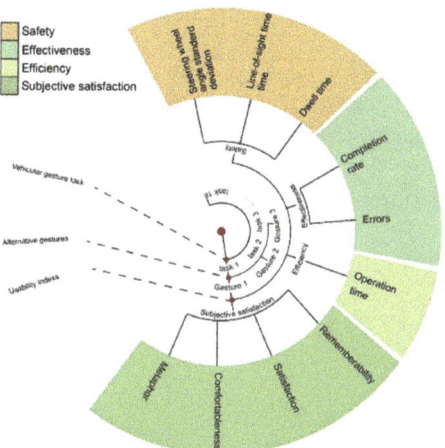

Fig. 4. Hierarchical structure of usability indices

We first construct an evaluation matrix to describe the scores of 3 gestures regarding 10 evaluation indices in each task. We denote this matrix as $X = (x_{ij})_{10 \times 3}$:

$$
X = \begin{vmatrix}
x_{11} & x_{12} & x_{13} \\
x_{21} & x_{22} & x_{23} \\
\cdots & \cdots & \cdots \\
x_{10\,1} & x_{10\,2} & x_{10\,3}
\end{vmatrix}
\tag{1}
$$

where $x_{ij}(i = 1, 2, \ldots, 10; j = 1, 2, 3)$ denotes the score of gesture j regarding index i.

We can continue to define the entropy of index i as follows:

$$
H_i = -k \sum_{j=1}^{3} f_{ij} \ln f_{ij}, i = 1, 2, \ldots, 10
\tag{2}
$$

where $f_{ij} = x_{ij} / \sum_{j=1}^{3} x_{ij}$, $k = 1/\ln 3$. Specially, if $f_{ij} = 0$, then we define:

$$
f_{ij} \ln f_{ij} := 0
$$

Finally, we can compute the weight of entropy of index i using the following equation:

$$w_i = \frac{1 - H_i}{10 - \sum_{i=1}^{10} H_i},\tag{3}$$

where $0 \leq w_i \leq 1$, $\sum_{i=1}^{10} w_i = 1$ 。

5 Results

After gathering scores for 3 gestures regarding 10 indices in all 16 tasks, average values of each score are calculated over 50 participants. The information entropy method is then applied to calculating the weight of entropy of each evaluation index to differentiate three different gestures for each task. Results are shown in Fig. 5.

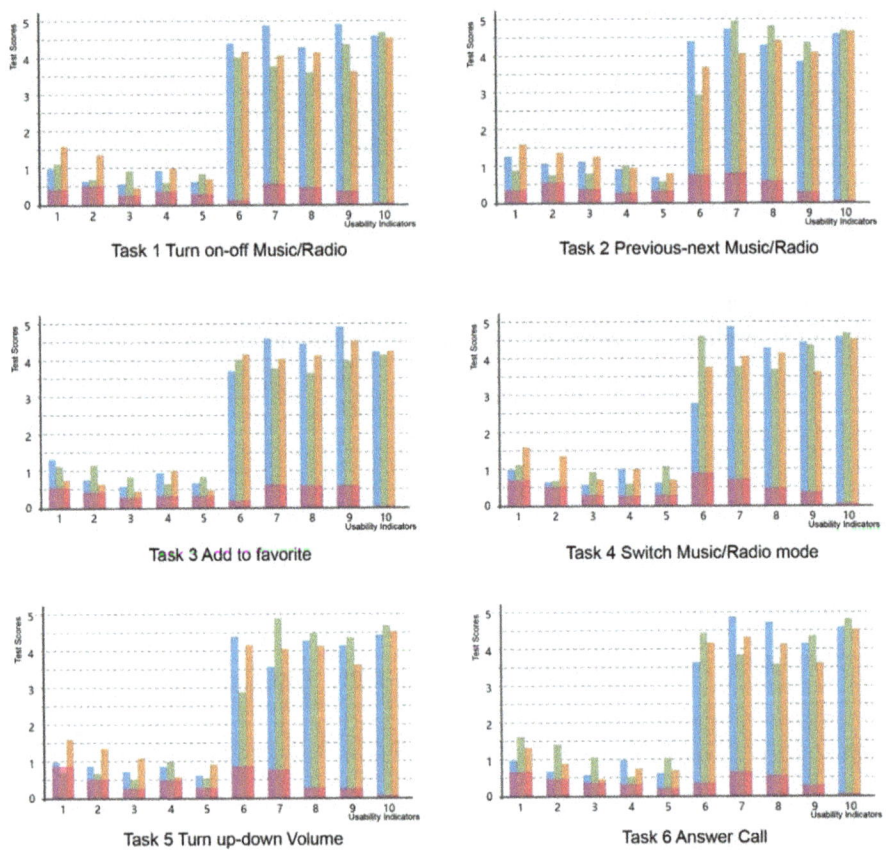

Fig. 5. Usability indices screening results

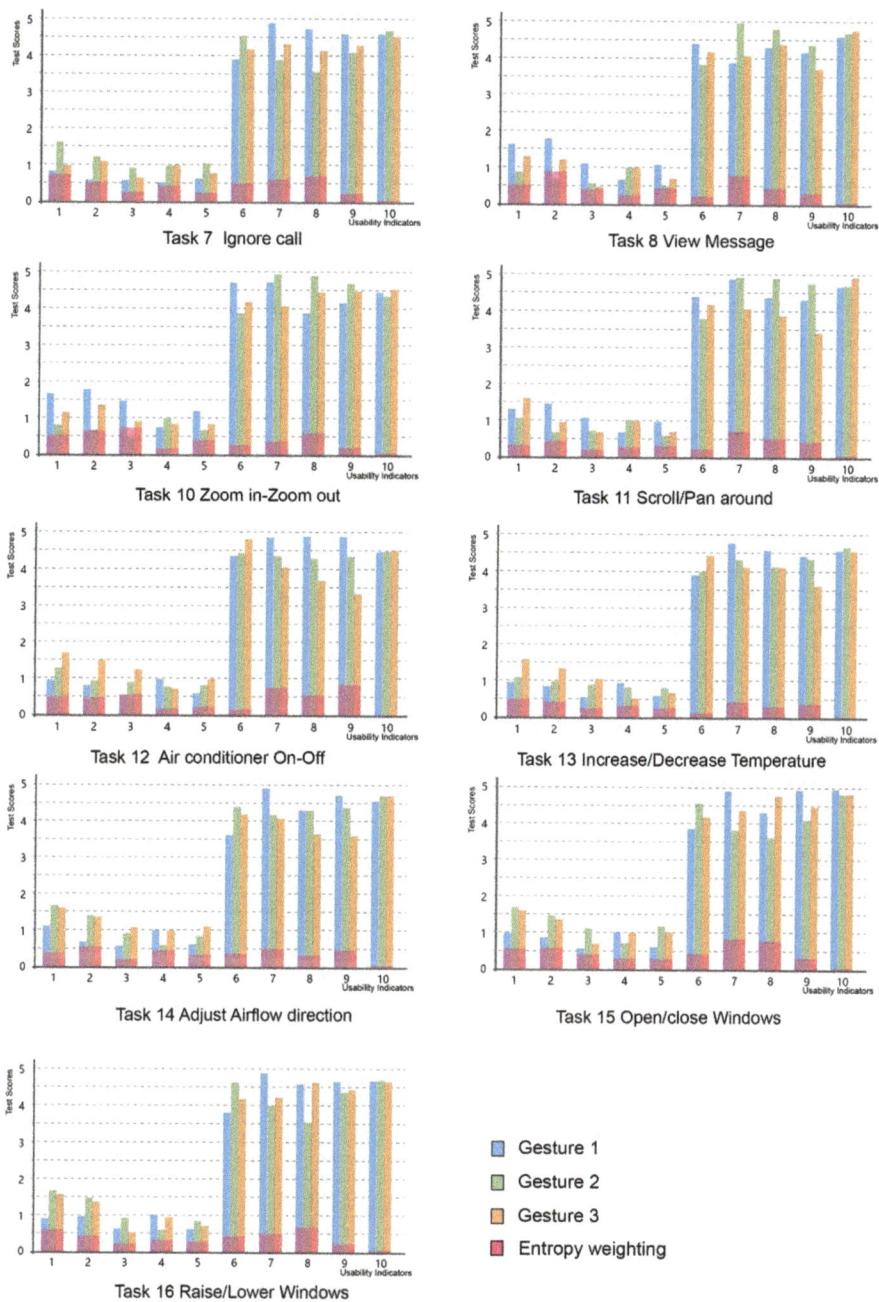

Fig. 6. The x-axis represents the usability index number and the y-axis represents scores of three gestures and their corresponding entropy weighting. (Index #1: Steering wheel angle standard deviation; Index #2: Eye position deviation; Index #3: Dwell time; Index #4: Completion rate. Index #5: Errors; Index #6: Operation time; Index #7: Memorability; Index #8: Comfort; Index #9: Learnability; Index #10: Intuitiveness.)

Based on the analysis of data, the following conclusions can be drawn.

Firstly, it is shown that intuitiveness (evaluation index #10) has a relatively small weight of entropy in each task, indicating that this index cannot effectively distinguish different gestures in each task. Therefore, this index should be deleted from the series of in-vehicle gesture control evaluation indices. The remaining indicators have relatively high entropy weights, meaning that they can effectively differentiate among gestures regarding various tasks. Therefore, these 9 indicators which include steering wheel angle standard deviation, eye position deviation, dwell time, completion rate, errors, operation time, memorability, comfort and learnability are appropriate to be taken as 3D in-vehicle gesture control usability indices (Fig. 6).

Secondly, in order to find people's preference towards gestures, we categorize 16 tasks into 4 classes as described in Table 3. For example, form Fig. 7, we can discover that tasks number 1, 6, 7, 8, 15 are talking about starting or ending a task and thus can be categorized as a single class: On-off. For each class, we examine various gestures by calculating their total scores of nine evaluation indices in each task that is involved in this class. The gesture with the highest score is viewed as the most preferred one in its class of tasks, which should lead to least distraction, help driver maintain the best driving performance, and receive the most positive subjective assessment from driver as well. The most preferred gestures in each of the 4 classes are also described in Table 3.

Table 3. Task classification

Task classification	Definition	Task number	Gesture preference
On-off	Start or end a task	1,6,7,8,15	
Switch	Switch from one state to another	2,4,14	
Adjustment	minor adjustments in the same state	5,12,13,16	
Other		3,10,11	(3) (10) (11)

It should also be noted that for task 9 "Ignore Message", more than 60 percent of the participants view it as unnecessary to respond via gestures when a message notification appears as that notification will automatically go away itself after a short period. Based on this piece of statistics, it is determined that gesture for "Ignore Message" task could be omitted.

Thirdly, results we've calculated above (Fig. 7) also show that the subjective satisfaction scores are over 3.5 out of 5 for all the 16 tasks. This demonstrates that in-vehicle gesture control technology has reached an acceptable level of usability from the standpoint of Chinese consumers.

6   Conclusion

In this paper, we propose a series of indices to evaluate usability of in-vehicle gesture interaction and implemented usability experiments in an indoor driving simulator where three kinds of data are collected during the test, namely distraction while driving, performance of task and subjective assessment. The information entropy method is applied for data processing and leads to a final list of 9 in-vehicle gesture control usability indices, including steering wheel angle standard deviation, eye position deviation, dwell time, completion rate, errors, operation time, memorability, comfort and learnability. Based on these 9 indices, a set of 3D gestures that accord with the habits of Chinese drivers and can be applied to in-vehicle environments are proposed, providing a valuable and reliable reference for in-vehicle gesture control design in future.

7   Future Work

Research into usability of gesture interaction is still at an early stage and there exist limited studies towards this topic that can be taken as reference. Although this study fills in the gap of in-vehicle gesture control research, there still remain some limitations that we need to address in the future. As this study is conducted in a driving simulation lab, there still exists a certain degree of deviation from the real-world testing scenario. In order to achieve results that can best describe real-world scenarios, road tests would be implemented. This would be done as the future work.

Acknowledgments. The research was supported by National Key Technologies R&D Program of China (2015BAH22F01), the State Key Laboratory of Advanced Design and Manufacturing for Vehicle Body Funded Projects.

References

1. May, K.R., Gable, T.M., Walker, B.N.: A multimodal air gesture interface for in vehicle menu navigation. In: The International Conference (2014)
2. Bevan, N.: International standards for HCI and usability. Int. J. Hum. - Comput. Stud. **55**(4), 533–552 (2001)

3. Nielsen, M., Störring, M., Moeslund, T.B., et al.: A procedure for developing intuitive and ergonomic gesture interfaces for HCI. Lecture Notes in Computer Science, 17(17), 1445–1453 (2004)
4. Stern, H.I., Wachs, J.P., Edan, Y.: Optimal hand gesture vocabulary design using psychophysiological and technical factors. In: International Conference on Automatic Face and Gesture Recognition, pp. 257–262. IEEE Computer Society (2006)
5. Ma, J., Du, Y.: Study on the evaluation method of in-vehicle gesture control. In: IEEE International Conference on Control Science and Systems Engineering, pp. 145–148. IEEE (2017)
6. Klauer, S.G., Dingus, T.A., Neale, T.V., et al.: The Impact of Driver Inattention on Near-Crash/Crash Risk: An Analysis Using the 100-Car Naturalistic Driving Study Data. U.S. Department of Transportation Washington D.C. (2006)
7. Ma, Y., Fu, R., et al.: Rating driving distraction tasks based on distraction impact on drivers. J. Transp. Inf. Saf. 32(5), 47–51 (2014)
8. Shannon, C.E.A.: A mathematical theory of communication. AT&T Tech. J. ACM Sigmob. Mob. Comput. Commun. Rev. 5(1), 3–55 (2001)
9. Qiu, W.H.: Management Decision and Applied Entropy, pp. 193–196. China Machine Press, Beijing (2002)
10. Shannon, C.: A Mathematical Theory of Communications. University of Illinois P. (1948)

Eye Tracking and Visualization

Application of Eye Tracking Technology in Naturalistic Usability Assessment of an Academic Library Website

Nima Ahmadi[✉], Matthew R. Romoser, Lindsay M. Guarnieri,
Theresa G. Kry, and Emily I. Porter-Fyke

Western New England University, Springfield, MA, USA
{nima.ahmadi,matthew.romoser,lindsay.roberts,
theresa.kry,emily.porter-fyke}@wne.edu

Abstract. The current research assesses the usability of an academic library website. This approach is based on visualization of visual attention and learnability of the website. Ten freshmen students, as novice users, and three librarians, as expert users, participated in this research. Seventeen tasks were developed; these tasks were informational searches, some completed up to two or three times in order to assess the learning curves of participants. The librarians' task times were treated as the benchmark "golden times" that a novice user would approach with experience. An eye tracker was used to assess usability. It was of particular interest to (1) identify the areas of the library website that are not clear to novice users and (2) to measure the learning curve of various informational searches. The results of this research indicated how quickly novice users learned tasks and which areas of the library webpages attract more attention.

Keywords: Academic library · Usability study · Eye tracking technology
Learnability · Visual search

1 Introduction

In fall 2017, a team consisting of researchers from the Center for Advanced Training Research and Naturalistic Studies (ATRANS) and librarians at Western New England University (WNEU) carried out a usability study on the D'Amour Library website. D'Amour Library is the main library at Western New England University. It serves undergraduate and graduate students, as well as staff and faculty in the Colleges of Arts and Sciences, Business, Engineering, and Pharmacy and Health Sciences.

The D'Amour Library website (https://www1.wne.edu/library/) is intended to be the primary point of access to databases and other electronic resources, the catalog, research guides and tutorials, and information about library services and policies. Freshmen students are introduced to the library website as part of a required multi-session information literacy program. This is not the case for transfer students and graduate students, who often learn to navigate the website with limited assistance from

© Springer International Publishing AG, part of Springer Nature 2019
T. Z. Ahram and C. Falcão (Eds.): AHFE 2018, AISC 794, pp. 729–740, 2019.
https://doi.org/10.1007/978-3-319-94947-5_72

faculty or library staff. It is necessary, therefore, that the website be well organized and easy to navigate.

Academic library websites act as a gateway to the educational and scholarly information users need to succeed in their academic endeavors, be they students, faculty, or staff members. The library website is multi-tiered, providing first access to the tools (databases, catalogs, etc.) needed to access information, then allowing users to use these tools to find the information (electronic journal articles, call numbers for books, etc.) they need. Rosenfield and Morville found that information architecture (defined as the way information and material are grouped, arranged, labeled, and presented to users) has a major impact on the users' ability to accomplish a task on the website [1]. In academia, meanwhile, the internet is the main means through which to obtain information, and scholars and researchers regularly use library websites to access scholarly and education information [2]. Therefore, the architecture and design of a library website is essential to the success of its users.

A usability study was performed on the library website using eye tracking technology. The purpose of the study was to determine whether freshmen students with limited exposure to the library website would be able to complete routine library-related tasks and eventually narrow the time gap between themselves and librarians. The study focuses on evaluating the learnability of the website as measured by the time required to complete repeated tasks and the mapping of eye-fixations while completing the tasks.

2 Literature Review

User-centered design is a well-accepted concept in library website designs [3]. The main characteristic of a user-centered interface is usability. It indicates an attribute of a system, which is quality. A usability study evaluates a system or product in terms of ease-of-use. Several models and frameworks were presented for usability assessment. Nielsen presented one of the most cited models in 1998. This model has five components: learnability, memorability, users' satisfaction, the rate of errors, and efficiency. The International Organization for Standardization (ISO) presented another model as a formal document. The usability model of ISO 9241-11 has three attributes: effectiveness, efficiency and users' satisfaction [4]. Effectiveness is the user's ability to accomplish a task. Nielsen recognized this ability as the most significant metric in usability assessment [5]. Efficiency is the user's effort to accomplish a task. It commonly measures the time that a user requires performing a task [6]. Generally, models of usability share common components and attributes. While, Oulanov and Pajarillo [7] adapted usefulness, supportiveness, and intrusiveness as attributes of usability in addition to effectiveness and users' satisfaction, Brinck's [8] postulated adaptability, efficiency, and helpfulness as usability criteria.

An academic library is a usable website, provided: (1) users can accomplish tasks quickly and easily with minimum efforts, workload, and errors, (2) users feel pleased and satisfied while using the website and after finishing a task, (3) library's contents, services, and guides meet expectations of users, and (4) the website is easy to learn (learnability) [9]. In the domain of Library and Information Science (LIS), usability studies have been the subject of much research. Some research presented models and frameworks for usability assessments of the academic library websites. Others investigated usability approaches as well as usability metrics and discussed their applications in assessment of information science, namely academic library websites, databases, and search engines.

One of the first usability studies in the field of Library and Information Science dates back to 1997. Eliasen et al. (1997) studied the usability problem of the University of Washington's library by using questionnaires. The authors were of the opinion that most navigation problems were caused by inappropriate terminology and layout. It was of particular interest to observe how terminology and layout of a website assist students to properly select databases. The students had better performance, namely with rudimentary library instruction, when descriptive labels were used. Furthermore, a content grouping of online sources was found more effective than subject grouping [10]. In a 2001 study, McMullen carried out a usability study on the Roger Williams University library website and reported how it was conducted. The researcher first did a usability study to find out the problems of the website by methods such as observation, think-aloud technique, survey and interviewing students. Then, a prototype was prepared and cognitive walkthrough and heuristic evaluation methods were used to evaluate the designed website. It was reported that students got frustrated when a task took more time than they expected. Furthermore, the reported weakness of observation and think-aloud was that students thought they were monitored, and it affected their search behaviors [11].

Susan Augustine and Courtney Greene were two reference librarians of the University of Illinois at Chicago and they did a usability study in 2002. Verbal data, as well as performance data (time on task and number of mouse clicks), were collected and analyzed. The objective of the study was to understand users' search behaviors. The striking result of the study was that instead of searching through the hierarchical order of the library website and through pages, students preferred to use the internal engine of library's website. The result is outstanding whereas it might change the priority of making a website more usable or having a more powerful search engine [12]. In 2002, two librarians carried out a usability study at Western Michigan University. It was of particular interest to understand article search behaviors of participants. They used formal usability testing and verbal protocol. The observation of participants' search behaviors while navigating library databases indicated a tendency to apply web search behavior. Furthermore, in most cases, participants did not navigate the bottom of the library webpage by scrolling down [13]. In 2005, three librarians of Hunter College conducted two usability studies on the library website. The first usability study was performed to get feedback from users. Based on received feedback, the website was redesigned. Then, they carried out the second study to measure whether the website's usability had improved or not. The objective of the study was to evaluate features of the library website in terms of ease-of-use and clarity of functions. They expected the number of false starts from the library homepage to reduce after implementing modifications. Participants were asked to apply verbal protocol. Furthermore, they were interviewed at the end of the study. In addition, quantitative data such as number of mouse click and time on tasks was collected. As result of the studies, the layout of the library website was modified and a virtual tour of the library was added to the main page to familiarize users who visit the library website [14].

In a 2006 study, Tsakonas and Papatheodorou proposed a usability framework for Electronic Information Service (EIS). In the presented framework, the relationship between a system, the contents and the users were discussed and defined in terms of usability, usefulness, and performance. The authors explained that attributes of a system and its content are two parameters that affect users' behaviors. To validate the framework, researchers surveyed and asked participants to rank each attribute based on importance. The results showed that ease-of-use, easy to navigate, and learnability of the system were the most important usability attributes of EIS, in order of gratitude [15]. In 2008, librarians of Oregon State University conducted an experiment and compared participants' search behaviors on a metasearch system, website of library and Google Scholar. It was of particular interest to understand reasons and factors that affect undergraduate students to choose a search system for class assignments. The outcome of the study indicated that familiarity with a commercial web-based search engine is an important factor to select a search system over other options. This factor shapes performance expectations of novice users. Speed is a factor that makes a search engine desirable to students. In addition, they expected to be able to check available contents and material in a system. Furthermore, it was reported that the complexity and disorganization of the library website dissatisfied users [16]. Another usability study was conducted on Main Library of the St. Augustine Campus of the University of the West Indies (UWI) in 2009. The objective of the study was to identify the weaknesses of the website to redesign it. The usability assessment was performed based on a heuristics approach, key task and card sort. The authors reported the effectiveness of this approach on the academic library website [17].

In a 2010 study on a digital library, the relationship between three attributes of usability (efficiency, effectiveness, and satisfaction) was investigated. It figured out that there was a strong relationship between satisfaction and efficiency as well as satisfaction and effectiveness. It was noticed that the correlation between satisfaction and effectiveness was stronger than the correlation between satisfaction and efficiency [18]. In 2011, Joo et al. presented a survey-based model for academic libraries' usability assessment. This model measures three attributes of usability: effectiveness, efficiency, and learnability. Six questions were developed per attribute to measure their extent. These questions were developed to quantify the weight of each attribute in a usability study for a given system [19]. In 2016, another study using questionnaire survey and checklist was carried out on the website of University of Delhi's library to find out usability problems. In addition to three common attributes of usability (efficiency, effectiveness, and satisfaction) accessibility, learnability and usefulness were measured [20]. In 2017, three librarians of the University of Houston-Clear Lake applied eye tracking technology and verbal protocol to understand how students accomplish complex tasks by using LibGuide. The authors interviewed students at the end of the study sessions in order to get more information on their search behaviors. The researchers used heat maps to visualize the attention distribution of students while using the library website [21].

Most of the reviewed studies on the usability assessments of academic libraries applied subjective approaches to get feedback from users on the library website.

In other words, users reported and explained their actions and thoughts or filled surveys out. As alluded above, these methods have some drawbacks.

In the domain of academic libraries, librarians require applying an approach that enables them to perform a naturalistic assessment of searching behaviors of actual users. Application of eye tracking technology as a non-invasive approach allows librarians to investigate library webpages in qualitative and quantitative methods. Collecting eye movement data and performance data assists librarians in identifying features of the website that need modification. In this research, the usability of an academic website was investigated in terms of learnability. It was of particular interest to figure out how quickly a novice learns a task on a website and how quickly a novice user would approach the golden performance of librarians in terms of time. The outcome of this study will be used to address usability issues of the library homepages.

3   Method

3.1   Participants

Novice and expert users of the website took part in the experiment. Novice users were freshmen students and expert users were librarians of D'Amour Library. A total number of 10 freshmen students, aged 18 to 19 years old, were recruited during the fall semester of 2017–2018, prior to beginning library instruction sessions. The group of novice students was gender-balanced. As compensation for participating in the study, students' names were entered into a raffle to win one of four gift cards. On the other spectrum, three librarians participated in the study. At the time of the experiment, they were working as Access Services and Electronic Resources Librarian, Archives and Emerging Technology Librarian, and Information Literacy Librarian.

3.2   Experimental Procedure

After obtaining Institutional Review Board (IRB) approval, D'Amour Library began to invite students to participate in the study. Due to having a limited time, the library tried to inform students through University Posts, the library blog, Instagram, Twitter, and a table in the library's lobby. In these posts, students who were willing to participate in the study were invited to send an e-mail to a librarian. Then, the librarian contacted them and shared a link to Google Form to get their demographic information. Then, based on received information, librarians scheduled an appointment for qualified students at the Center for Advanced Training Research and Naturalistic Studies (ATRANS) at Western New England University.

To develop experimental tasks, a committee of three librarians was formed. These librarians prepared a pool of potential tasks, based on their experience, received feedback, questions from students, and objectives of the study. The committee selected 17 tasks. These tasks were informational searches, some done up to two or three times to assess learning curves of participants. These 17 tasks were randomized in a way that was impossible for a given user to get similar tasks back-to-back. A chronological order was used for presenting the tasks. First, users received nine tasks at a time.

Next, they received five tasks, which were similar to five of the original nine tasks, and accomplished them for the second time. Finally, participants performed three more tasks for the third time. The list of tasks is presented in Table 1.

On the day of the experiment, the study administrator met students. Once the consent form was signed and the subject student ID was checked, then the study administrator briefed participants on experimental tasks. Then, participants wore eye tracking. Once the eye tracker was calibrated, the experiment was initiated. Participants received one task at a time. There was a 30-s time lapse between two tasks.

Table 1. The list of experimental tasks

The experimental tasks
Performing Tasks for 1st trial
1- Find the call number for the book "The Grapes of Wrath"
2- Find the Library's hours
3- Find the Library Director's email address
4- Find a way to contact the Reference Librarians to ask a question or get help
5- Find the Library blog
6- Find information about printing at D'Amour Library
7- Find the course reserve list for Professor Beagle
8- Find the Library's "Reserve a study room" form
9- Find resources about "global warming"
Performing Tasks for 2nd trial
10- Find the call number for the book "To Kill a Mockingbird"
11- Find the Staff Assistant's email address
12- Find the course reserve list for Professor Clark
13- Find the Library's "Make a Suggestion" form
14- Find resources about "The Civil War"
Performing Tasks for 3rd trial
15- Find the call number for the book "The Hobbit"
16- Find the course reserve list for Professor Doe
17- Find resources about "artificial intelligence"

4   Apparatus

The eye tracking device was Tobii Pro Glasses 2. This apparatus is a 40-g head-mounted eye tracker. The sampling rate of the eye tracker was 50 Hz. The eye tracker's horizontal and vertical visual field of view was more than 160 and 70°, respectively.

5   Library Homepage

The library homepage provides information about the physical library and helps users navigate online sources. The library homepage has nine areas: (1) "Find It@ D'Amour Library" (a discovery service that searches across many Library databases and the catalog); (2) "Left Hand Navigation" (provides access to five subpages); (3) "Ask a Librarian" (includes a chat widget and contact information for the reference librarians);

(4) "Research" (has three links to electronic resources including databases, the catalog, and research guides); (5) "I Want to Find …" (provides information on course reserves, journals, and answered common questions); (6) "Quick Links"; (to Google Scholar, library forms, and to a few digital services); (7) "D'Amour Library Hours"; (8) "Library Archives"; and (9) "D'Amour Library Blog".

6   Usability Metrics

To investigate the visual search of participants (novice and expert users) as well as the learnability of the website, two groups of metrics were defined: (1) eye movements' metrics, and (2) performance metrics. The eye movements' metrics were fixation count and fixation duration. The areas on the library homepage were defined as Areas of Interest (AOIs) (Fig. 1, Table 2), and eye movements' metrics were measured on each of them. Furthermore, performance data of participants, such as time on task and number of mouse clicks, were measured for each task. The library homepage is shown in Fig. 1.

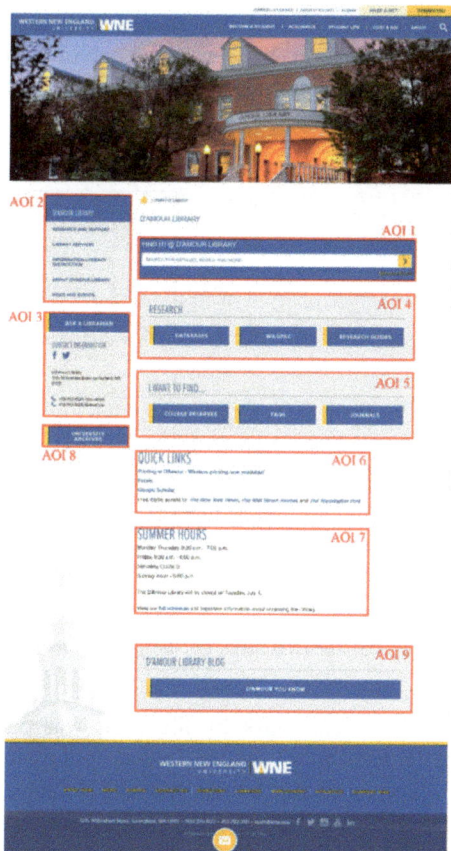

Fig. 1. D'Amour Library homepage

Table 2. List of AOIs on D'Amour Library Website

AOI 1	AOI 2	AOI 3
Find It@ D'Amour Library	Left Hand Navigation	Ask a Librarian
AOI 4	AOI 5	AOI 6
Research	I want to Find …	Quick Links
AOI 7	AOI 8	AOI 9
D'Amour Library Hours	Library Archives	D'Amour Library Blog

7   Results

The results are presented in two sections. The first section visualizes times spent on tasks by plotting the learning curve. The plotted figure indicates the learning effect on the times on tasks. Then, in the second part, visual search and visual attention are graphed and the effects of learnability on the scanning pattern is plotted.

7.1   Learning Curve

The outcome of the study was promising in terms of learnability. Three tasks were repeated three times in a 30-min study. The results for two of these tasks are presented: finding a resource for a given topic and finding a course reserve of a given professor.

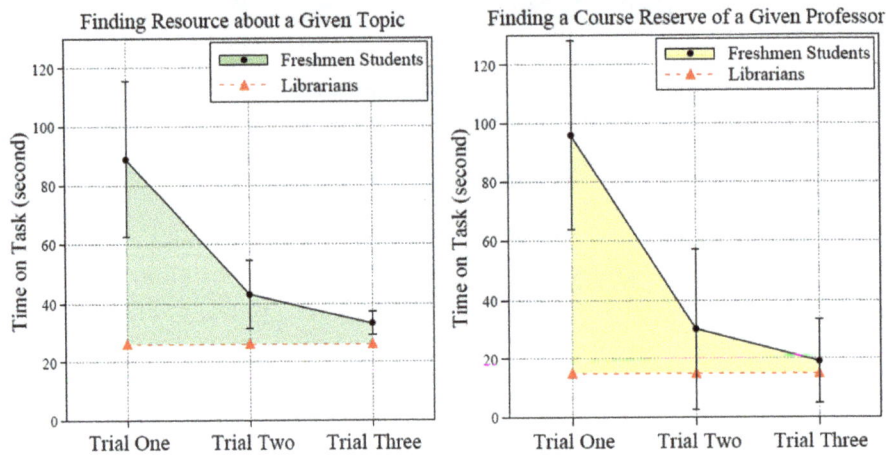

Fig. 2. Learning curves for two tasks on the D'Amour Library webpage

The average time spent on each task by librarians was considered the "golden time". This time was fairly constant in each trial. In case of finding a resource about a given topic, the average time for a novice user, who did not have any prior experience of working with the library website, stood at 96 s. These times are shown in Fig. 2.

The librarians' time was only four seconds less than novice users. In other words, after only three trials, novice users could narrow the gap between their times and the librarians' times. A similar declining trend was noticed when participants looked for the course reserves of a given professor. In the first trial, novice users spend 96 s to find a professor's course reserves. This time continued to decline, but less steeply, to 43 s in the second trial. By the third trial, this reduced to 33 s, which was only 7 s higher than the librarians' golden time on these tasks.

7.2 Visual Search

In this part, the effects of learnability of the website were investigated on participants' visual search. The eye movements of novice and expert users were compared when they were trying to find a resource. This task was repeated three times. The detailed result is presented in Table 3. In this table, novice eye movements were compared with expert ones in terms of fixation count and fixation duration.

Table 3. Fixation count (number) and dwell time (millisecond) when finding a resource

	AOI 1	AOI 2	AOI 3	AOI 4	AOI 5	AOI 6	AOI 7	AOI 8	AOI 9
Trail 1									
Fixation count	93.1	134.6	18.3	86.3	48.3	21.0	2.5	4.2	1.9
Expert	64.0	–	–	–	–	–	–	–	–
Novice	96.0	148.1	20.1	94.9	53.1	23.1	2.7	4.6	2.1
Dwell time	835.4	1,668.7	558.9	424.6	255.2	166.8	34.9	92.0	42.0
Expert	846.2	–	–	–	–	–	–	–	–
Novice	834.0	1,854.1	621.0	477.7	297.7	194.6	38.8	102.2	46.7
Trail 2									
Fixation count	87.7	83.8	14.1	41.5	14.1	–	–	–	–
Expert	61.0	–	–	91.0	–	–	–	–	–
Novice	90.4	92.2	15.5	36.6	15.5	–	–	–	–
Dwell time	1,010.4	1,627.1	681.3	560.8	165.5	–	–	–	–
Expert	526.9	–	–	1,301.4	–	–	–	–	–
Novice	1,064.1	1,859.5	749.4	455.0	189.2	–	–	–	–
Trial 3									
Fixation count	157.7	22.4	3.0	57.0	22.4	–	–	–	–
Expert	64.0	–	–	–	–	–	–	–	–
Novice	169.4	25.3	3.4	64.1	25.3	–	–	–	–
Dwell time	1,013.9	474.4	360.0	482.1	267.2	–	–	–	–
Expert	755.0	–	–	–	–	–	–	–	–
Novice	1,042.7	569.3	400.0	535.6	300.6	–	–	–	–

In all three trials, the librarians had similar scanning patterns. They limited their gaze points to critical regions. For the first time, the librarians only scanned one area of the D'Amour webpage: "Find It@ D'Amour Library". While novices scanned all of the nine regions of the library webpage. The highest proportion of fixation counts were on "Left Hand Navigation" (33%), "Find It@ D'Amour Library" (23%), and "Research" (21%). "Left Hand Navigation" was the most fixated region of the webpage. On this area, dwell time was longer than other areas. It indicates that participants needed more time to understand this area. The second rank in terms of dwell time is for "Find It@ D'Amour Library". In the second trial, the librarians' scanning patterns were similar to the first trial; they only scanned two areas: "Find It@ D'Amour Library", and "Research". Novices, this time, limited visual search to only five areas: "Find It@ D'Amour Library", "Left Hand Navigation", "Ask a Librarian", "Research", and "I Want to Find...". This time, the most visited area was "Find It@ D'Amour Library" and "Left Hand Navigation", respectively, with 36% and 35%. This indicates that novice users remember the path on the webpage they used before. The novice users spent more time on "Left Hand Navigation" and "Find It@ D'Amour Library", with respect to order. In the third trial, there is clear indication that the novices' scanning patterns did not occur due to chance. This time, the most visited area was "Find It@ D'Amour Library". It means that novice users could recognize and remember the optimal path only after three trials. The proportion of fixation count on "Find It@ D'Amour Library" was 65%, then 24% on "Research". In other words, 89% of all fixation was in these areas of the library webpage. The spent time on the library webpage to find the right path declined 29% in the second trial compared to the first trial. For the third trial, participants needed half of the time they spent in the first trial (Fig. 3).

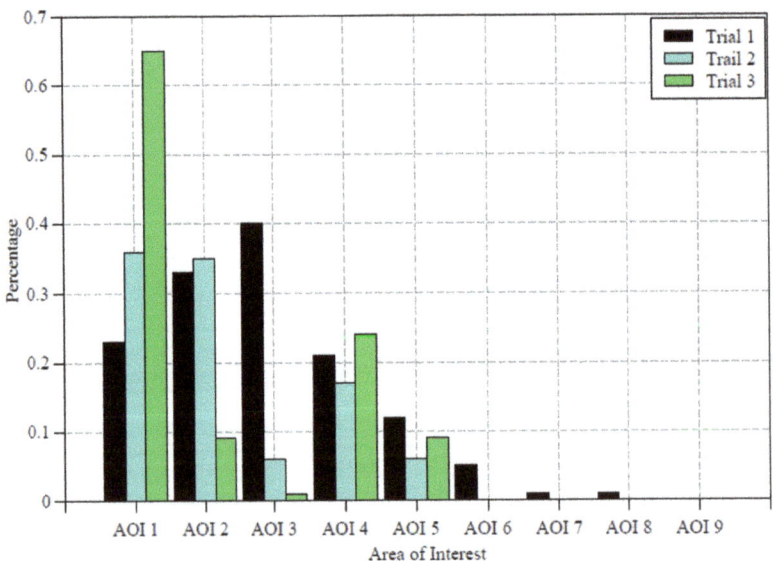

Fig. 3. Proportion of novice users' fixation count on areas of D'Amour Library

8 Discussion and Conclusion

There are several methods to evaluate the usability of academic libraries. Some usability approaches focus on survey and questionnaires. The issue with the subjective approach is that the results do not provide direct feedback on the performance of participants. Besides, the result of the said usability approach is not that much clear and useful to librarians to modify the webpages of libraries. For instance, one measure of ease-of-use of an academic library is learnability. One way to visualize this attribute of usability is measuring learnability. If a novice user's time on a given task approaches a librarian's times as an expert user of the system, then it means the website has a good design. This measure is clear and useful to the librarians as they can monitor the improvements of the novice users' performance of tasks by visualizing their time. Furthermore, in the proposed approached for measuring usability, there is a baseline. The baseline is a golden time for librarians.

In the previous usability studies, ease-of-use of the system only measured through surveys. The question was, how much should a system improve, or which areas of the website need modification? In other words, how should an acceptable time be defined for each task? In this research, we used a multi-approach in usability assessment. This approach is a combination of eye tracking technology with an emphasis on the measurement of the learnability of academic libraries websites. In this multi-approach, we repeated a task three times and then compared how fast a user could approach the golden time of that task. In this approach, the capability of novice users to narrow the gap between their times on tasks with librarians is the main measure. If novices require several trials to narrow a time gap, then it indicates that there could be a potential to improve the usability of library webpage. In addition to visualization of learnability of website, visualization of visual attention assist librarian to understand what areas of the website could be misleading or what areas are more demanding areas than other areas of the website. A usability assessment based on visualization of visual attention and learnability of the library website could improve the quality of the academic website.

References

1. Rosenfeld, L., Morville, P.: Information Architecture for the World Wide Web. O'Reilly Media Inc, Sebastopol (2002)
2. Lee, J.-Y., Han, S.-H., Joo, S.-H.: The analysis of the information users' needs and information seeking behavior in he field of science and technology. J. Korean Soc. Inf. Manag. **25**, 127–141 (2008)
3. George, C.: User-Centred Library Websites: Usability Evaluation Methods. Elsevier, New York (2008)
4. Usabilitynet.org: ISO 9241-11: Guidance on Usability. http://www.usabilitynet.org/tools/r_international.htm#20282
5. Nielsen, J.: Success rate: the simplest usability metric. Jakob Nielsen's Alertbox, 18 (2001)
6. Kortum, P.: Usability Assessment: How to Measure the Usability of Products, Services, and Systems. Human Factors and Ergonomics Society (2016)

7. Oulanov, A., Pajarillo, E.J.Y.: CUNY+ Web: usability study of the web-based GUI version of the bibliographic database of the City University of New York (CUNY). Electron. Libr. **20**, 481–487 (2002)

8. Brinck, T., Gergle, D., Wood, S.D.: Usability for the Web: Designing Web Sites that Work. Elsevier, New York (2001)

9. Pant, A.: Usability evaluation of an academic library website: experience with the Central Science Library, University of Delhi. Electron. Libr. **33**, 896–915 (2015)

10. Eliasen, K., McKinstry, J., Fraser, B.M., Babbitt, E.P.: Navigating online menus: a quantitative experiment. Coll. Res. Libr. **58**, 509–516 (1997)

11. McMullen, S.: Usability testing in a library web site redesign project. Ref. Serv. Rev. **29**, 7–22 (2001)

12. Augustine, S., Greene, C.: Discovering how students search a library web site: a usability case study. Coll. Res. Libr. **63**, 354–365 (2002)

13. Cockrell, B.J., Jayne, E.A.: How do I find an article? Insights from a web usability study. J. Acad. Librariansh. **28**, 122–132 (2002)

14. Cobus, L., Dent, V.F., Ondrusek, A.: How twenty-eight users helped redesign an academic library web site: a usability study. Ref. User Serv. Q. 232–246 (2005)

15. Tsakonas, G., Papatheodorou, C.: Analysing and evaluating usefulness and usability in electronic information services. J. Inf. Sci. **32**, 400–419 (2006)

16. Jung, S., Herlocker, J.L., Webster, J., Mellinger, M., Frumkin, J.: LibraryFind: system design and usability testing of academic metasearch system. J. Assoc. Inf. Sci. Technol. **59**, 375–389 (2008)

17. Buchanan, S., Salako, A.: Evaluating the usability and usefulness of a digital library. Libr. Rev. **58**, 638–651 (2009)

18. Joo, S.: How are usability elements-efficiency, effectiveness, and satisfaction-correlated with each other in the context of digital libraries? Proc. Assoc. Inf. Sci. Technol. **47**, 1–2 (2010)

19. Joo, S., Lin, S., Lu, K.: A usability evaluation model for academic library websites: efficiency, effectiveness and learnability. J. Libr. Inf. Stud. **9**, 11–26 (2011)

20. Iqbal, M., Warraich, N.F.: Usability evaluation of an academic library website: a case of the University of the Punjab. Pak. J. Inf. Manag. Libr. PJIML. **13** (2016)

21. Ford, L., Holland, J., Iakovakis, C.: "I don't know what i'm looking at": understanding student libguide use with eye-tracking software. In: Association of College and Research Libraries (ACRL) (2017)

Log-Based Process Visualization

Johannes Schwank[✉], Sebastian Schöffel, and Achim Ebert

Computer Graphics and HCI, Technische Universität Kaiserslautern,
Kaiserslautern, Germany
{schwank, schoeffel, ebert}@cs.uni-kl.de

Abstract. Understanding processes is very important since todays processes are getting more and more complex. Therefore, a visual representation is often helpful. In many cases, the processes are similar and rely on the same components in the same system structure. Current tools lack in particular with respect to the presentation of multiple processes that are based on the same involved structure but slightly differ in the sequence of events. In this paper we introduce a tool for visualizing script-based animated processes. The underlying structure can be reused by only modifying the script containing the process' events. The highly adaptable visualizations allow a direct and clear view on a process and intuitive functionalities to traverse through its steps. Finally, we show its usage in two case studies from different domains.

Keywords: Process visualization · Log visualization · Visualization tool

1 Introduction

Processes can be found in diverse systems, contexts, and environments, including data transfer in networks and between companies, or production processes in industry. In general, the term *process* has many different definitions. In [18] process is defined as "A series of actions, changes, or functions bringing about a result" [19] defines a process as "a series of actions that you take in order to achieve a result". These definitions include, like many others, the result as the central goal of a process.

We define a process as a sequence of events performed by different connected components communicating with each other. A process can be of arbitrary nature. It can, for example, be physical such as a package delivery process with physical components (e.g., company, customer) and connections (e.g., company sends package to customer). However, a process can also be virtual, e.g., in software systems where data is transferred or mixed, material flow control, or home automatization, where both, virtual data and physical objects are transferred.

Thus, processes are often illustrated by flow diagrams [17]. Nowadays, several applications such as Microsoft PowerPoint, LibreOffice Impress, or Apple Keynote provide various features for visualizing processes using extensive animation possibilities. In all those applications, it is obviously possible to position the needed components, customize them, or show and hide connections using animations. Users can choose out of many styles for the components and animations in a large catalogue. The resulting process illustration often ends up in a complex structure containing many

© Springer International Publishing AG, part of Springer Nature 2019
T. Z. Ahram and C. Falcão (Eds.): AHFE 2018, AISC 794, pp. 741–751, 2019.
https://doi.org/10.1007/978-3-319-94947-5_73

components and connections with many steps. Furthermore, for visualizing time sequences, animations are needed that show and hide elements or move elements along a path. Often, the list of animations can get confusing very fast. Once an existing process presentation needs to be edited or duplicated to present a slightly different scenario, time-consuming manual editing is needed in which it is often very hard to find the right points to modify.

To overcome this issue, this paper presents a novel tool to present processes based on a set of components and their connections. It allows easy modifications to build different versions of the process visualizations by reusing the structure and only modifying the events.

2   Related Work

The visualization of processes has a long history in diverse domains such as business processes [3, 4] or processes in software [2]. In [5], Matkovic et al. present an approach to monitor processes by combining multiple virtual instruments showing process data on a dashboard, e.g., temperature data over time. They extend several existing virtual instruments with, e.g., history encoding (presenting values of the near past in parallel) and apply a Focus + Context approach combined with different levels of detail to their dashboard. However, our tool does not visualize process data such as temperature, instead it visualizes the process itself embedded in the overall system structure.

Bobrik et al. highlight that customizable appearance is an important aspect in process visualization [6, 7]. In contrast to our approach, they visualize a single process as an individual node in a node-link diagram. The approach in the paper at hand handles a single process step as a connection between two components.

As defined in the introduction, we assume that a process consists of a sequence of events. Automatically generated and recorded sequences of events or actions are commonly called a log. Usually, it includes timestamps and descriptions of events, behaviors, etc. in the system and is stored in human-readable files.

The visualization of log files has been investigated in different ways. In many cases, log file visualization and analysis are used for anomaly detection [8, 9, 11], to uncover patterns [10, 12] or get impressions of the behavior of the recorded system [1, 10]. One example is the *histogram matrix* by Frei and Rennhard [8], used to visualize the content of textual log files. Therefore, they use characteristics such as number of words in the log entry or the number of characters. Similarly, most log-based visualizations focus on the visualization of the contents of the log files.

Siirtola et al. present a visualization tool to analyze textual log files of an e-learning platform [10]. They propose a node-link diagram representation to visualize the tasks of the users. The transitions between tasks are visualized as simple arrows labelled with the numbers of the corresponding transitions which is encoded in the width of the arrow as well. However, the underlying structure of the platform is not reflected in the resulting visualization.

In the domain of software engineering, the term sequence is strongly related to UML sequence diagrams, a visualization to show interactions focusing on the exchange of messages between a number of lifelines [13]. The lifelines represent the involved components of the system. Often, these diagrams show a user and the system.

Yue et al. [14] propose an approach to automatically derive such sequence diagrams from textual use cases. The use cases to be transformed require to follow a specific use case modeling approach containing defined key words, attributes, etc. Without such a predefined structure, an automatic transformation is not possible. Elallaoui et al. come to a similar conclusion when presenting their approach to automatically generate sequence diagrams based on user stories [15].

Similar to the visualizations generated by our proposed tool are data flow diagrams (DFDs). A DFD is built from elementary blocks that can be connected and interpreted as a directed graph [17]. Its nodes represent entities or processes, the edges represent data flows.

There are ways to automatically generate DFDs in Microsoft Visio [16]. However, this requires exactly defined and structured Excel files. Labels, shapes, etc. have to be specified for each node and each link has to be stored and specified in an additional list as well.

3   Concept and Implementation

The tool presented in this paper allows users to create visualizations to show a sequence of process events. The processes are represented by components and connections. Components form the structure of the environment, while connections link the components and are generated based on events. The specified sequence of events forms a script. Many scripts can be applied to the same structure.

The generated visualization provides an overview of the structure. The level of detail of the structure depends on the detail of the events in the script and how many components are defined in the configuration file. The tool provides different interaction possibilities which supports the users in presenting the visualization, e.g., tracing through the script (Fig. 1).

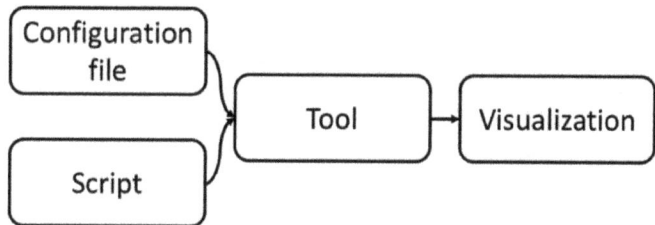

Fig. 1. Workflow overview. A configuration file and a script serve as input for the tool which results in an interactive visualization.

In addition to the overview area, there are two detail areas: one area for details about the script (Fig. 2b) and the other detail area where information about certain visualization elements are provided (Fig. 2c).

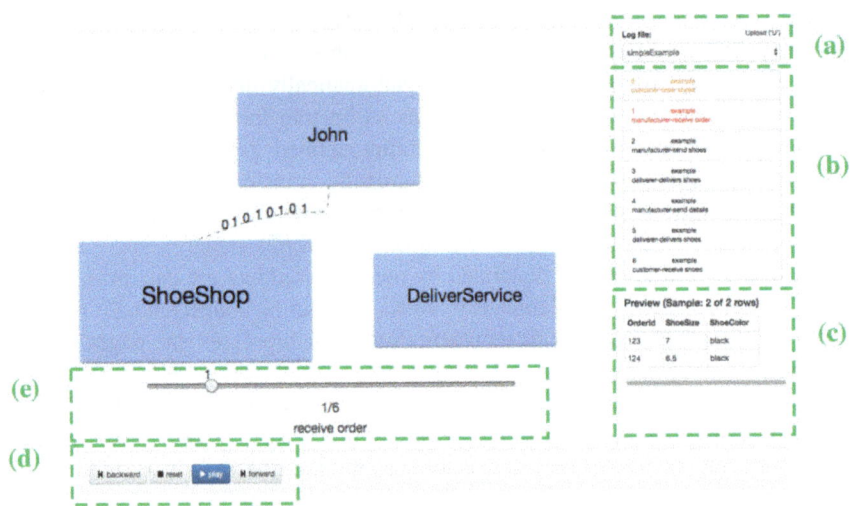

Fig. 2. Screenshot of the tool. The visualization shows the structure of a very simple example consisting of a customer John ordering shoes in an online shoe shop which are delivered by a delivery service. (a) – (e) mark the different areas: (a) list of available scripts, (b) event list of the selected script, (c) detail information about a selected connection, (d), (e) navigation control to trace through the script.

The tool is designed as a web-based solution to be platform-independent. The visualization is composed in the background and can be divided into two steps. The first step creates the structure, i.e., the components (see Sect. 3.1), while in the second step, the connections are generated visualizing the events of a script (see Sect. 3.2).

The tool needs a configuration file and at least one script file to generate the visualization. These files are described in the following two sections. Afterwards, the functionality of the tool is presented in detail.

3.1   Step 1: Structure Setup

The configuration file is loaded at the beginning of the tool's usage – the first step of the visualization generation process. It contains all necessary specifications to build the structure for the visualization, i.e., the underlying information for each individual component. Each component description includes all necessary information to visualize the component, such as the component's position, its width and height, and colors. Optionally, images and texts (as "label" attributes) can be specified to be shown in the visualization. The configuration file is based on the human-readable JSON format.

Components can have sub-components. These are collected into an array and added to the "children" attribute of the component. Positions in sub-components are given relative to the position of the containing ("parent") component.

Example of a specification for a component representing "John". The "id" attribute is used as identifier; the "label" is visualized on the component.

```
{
    "id": "customer",
    "label": "John",
    "color": "green",
    "x": 400,
    "y": 0,
    "w": 200,
    "h": 100,
    "children": []
}
```

3.2   Step 2: Connection Setup and Usage

As soon as a user selects a script from the list, the second step starts. The selected script is loaded and the containing events are parsed. Each of the events has an "id" attribute which relates to the "id" attribute of one of the components.

The visualization shows the connections one after the other in the order as they appear in the script and not in parallel (as in, e.g., DFDs). The visualization determines whether to show a connection. Once, the script reaches an event, the tool searches for the match of the event's "id" attribute with the corresponding component's "id" attribute. This component is then marked as active. An active component has a connection if the following two conditions are fulfilled. First, there is a previous event. Second, the connection of the previous event's component to the current active component is allowed. A list of possible and forbidden connections between certain components can be created by the user.

A visualized connection is then styled following the attributes provided in the event and the specifications provided in the configuration file, e.g., the event contains a "data" attribute and the configuration defines small black circles as symbols whenever there is a "data" attribute.

The user is able to trace and navigate through the list of entries in the script. Since the list of entries can be very long depending on the use case, the tool provides different options for navigation.

Navigation. Similar to a simple music player navigation, the tool offers Play/Pause, Forward, and Back buttons to trace step-by-step through the script entries, either automatically or manually (see Fig. 2d). Furthermore, a range slider enables users to move to a certain event in the process (see Fig. 2e). Additionally, a list is given in which the script entries are formatted in an easy-to-read table (see Fig. 2c). In this list, the current visualized entry is colored. By selecting one item, the visualization jumps to the corresponding step (as like selecting an event using the slider).

The navigation through the script and the visualization of the corresponding connections in the structure already support the user in presenting the process.

Animation. To support the user in drawing the viewer's attention to the currently active component and connections, animations are automatically created. In our tool, we realize animations for active connections by small symbols that are moving along the actual visualized paths of the connections from one component to another. By this, the direction of the connection is pointed out as well. The symbols can be defined by the user. They can be either simple forms such as circles, rectangles, letters, numbers, or even small images (icons). The specific symbol is not described in the script entry itself. Instead, the symbols can be assigned to certain attributes in the configuration file.

Each active component is highlighted by increasing its size keeping the width-height ratio. The user has the possibility to define additionally a highlight color by adding an "highlight" attribute in the configuration file.

Script file. The script file contains a list of entries, i.e., the events of the desired process. Each event contains specifications that determine its visualization. The only required attribute is an attribute to identify the component that has caused the event. This is needed to assign the event to the component in the visualization. There can be unlimited additional attributes (such as data or description). These consist of the detail information which can be shown on demand.

Example for the first script entry with the shoe order of John.

```
{
  'component': 'customer',
  'description': 'order shoes',
  'data': 'shoe size 7, black, US$ 45.00'
}
```

The entries in the script file are separated by a new line. As mentioned above, a process is a sequence of events. As the example demonstrates, each event is recorded and readable as a JSON object itself.

Example of how the script with the shoe order of John could look like.

```
{'component': 'customer', 'description': 'order shoes', ...
{'component': 'manufacturer', 'description': 'send shoes...
{'component': 'deliverer', 'description': 'delivers shoe...
{'component': 'customer', 'description': 'receives shoes...
```

Overall, the visualization created by the tool strongly depends on the inputs. The elements of the visualization are defined by the user and, thus, are very flexible and customizable. In terms of interaction and animation, the tool provides different options to support users and viewers.

4   Case Studies

4.1   Case Study I: Home Automatization

Home automatization is getting more and more interesting and affordable for everyone. Sensors for temperature, air humidity, as well as light switches, radiators, television, etc., are combined with a smart home server to measure and manage home devices via smartphones or automatically by defined rules.

The components of the home automatization are placed in different rooms. In the visualization, the smart home server is shown outside of any room, assume it is placed in another room which is not necessary to visualize. Furthermore, there is a balcony with an outdoor temperature sensor and a bedroom containing two windows, a heating unit, a room thermostat, and a light (see Fig. 3). The windows are attached with a sensor to know if the window is open or closed, the room thermostat is controlling the heating devices in the corresponding room, in this case only one radiator.

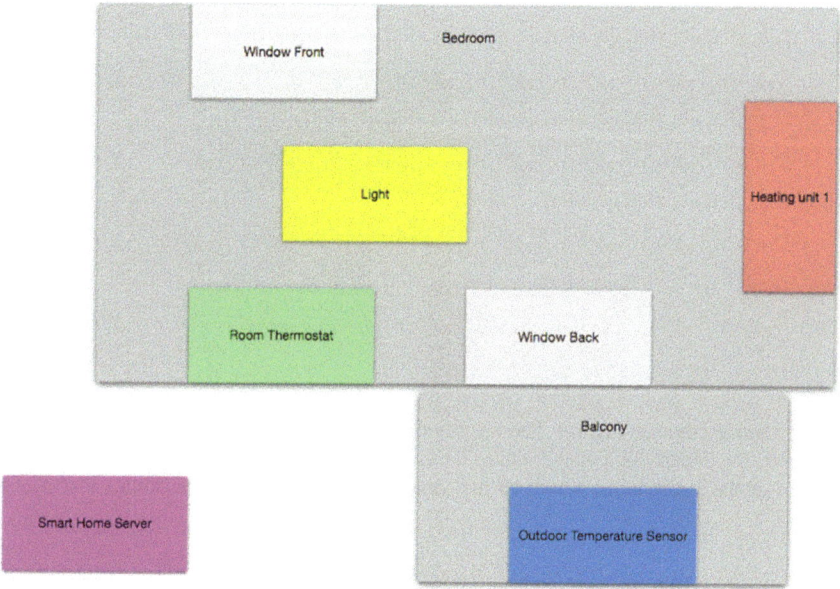

Fig. 3. Structure of the components of the home automatization case study.

A usual scenario might be the following: A salesman wants to demonstrate a potential customer how the smart home system works with the help of the open-window example. This means, in case the window is opened, the radiator should be adapted to save energy. In detail, the window sensor of the front window recognizes that the window has been opened and sends a message to the smart home server. The user set up a rule for that case. The server then queries the outside temperature from the sensor on the balcony. After the server received the current temperature, it sends a new target temperature value to the room thermostat. This passes the newly received value to the radiator (Fig. 4).

The visualization of this scenario demonstrates the home automatization system and which devices are involved for a specific task. Regarding the bedroom, this scenario includes the window, the room thermostat, and the radiator. However, the same structure could be reused to visualize a different process, e.g., manage that the light is switched off as soon as the back window opens. This is possible with only small effort by simply changing the script while the structure stays untouched.

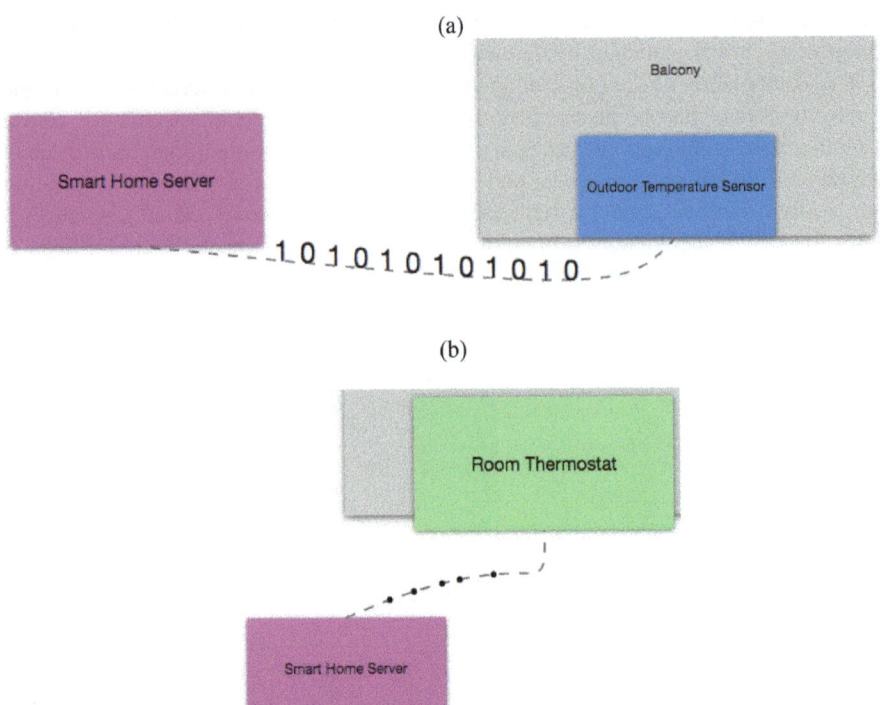

Fig. 4. Example connections. (a) The answer of the outdoor temperature sensor to the smart home server, the underlying event contains data and is therefore visualized by numbers. (b) The instruction of the smart home server to the room thermostat, small black circles visualize the instruction.

4.2   Case Study II: Production Optimization

The second case study is set in the domain of production optimization. The tool has been successfully applied in the research project PRO-OPT which has been funded by the German Federal Ministry for Economic Affairs and Technology. In this context, the tool and the visualization has been mainly used for presentation purposes, i.e., to show the functionality of the developed system. It consists of different companies which communicate via a shared platform (see Fig. 5). Users of the companies can send queries via the platform to all connected companies and get results.

Usually in industry, companies are not interested in sharing or publishing their data even if it might help to solve a common problem together. Therefore, it is important to be able to show which data leave the company and are visible "outside". For this, the possibility to show detailed information of a log entry is used. A sample of the data that is returned for a query is added to the log entry and can be demanded by manually selecting the appropriate connections, resulting in showing the data in the detail area as described in Sect. 3. Furthermore, the symbols moving along the connection paths are adapted. If the connection represents a query, it is visualized by grey circles (see Fig. 6).

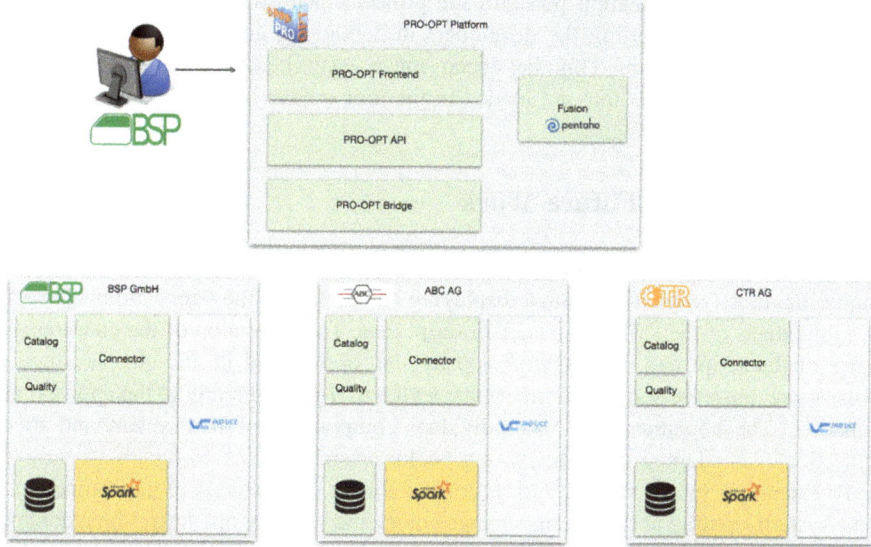

Fig. 5. Example screenshot of the PRO-OPT project. The tool is used to visualized three companies and their shared communication platform (above the companies). Each company is modeled as one component containing several sub-components.

If there is instead a "data" attribute in the corresponding log entry and thus sample data, 1 and 0 in black color are used (see Fig. 7).

Using the tool, project presenters were able to show how the system handles queries and their results, i.e., that there are algorithms applied to pseudonymize, anonymize, and aggregate the company's data and protect it. Additionally, the visualization helps to understand which components of the system are involved when handling queries.

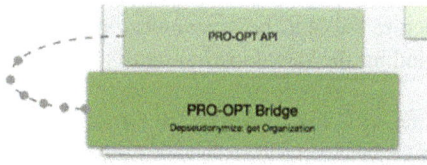

Fig. 6. Query visualization: grey circles move along the connection path.

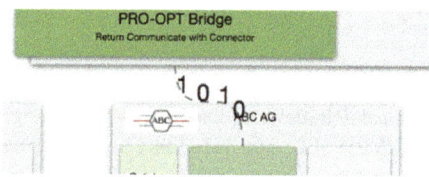

Fig. 7. Result data visualization: numbers move along the connection path to notify that data is transferred on this connection.

During the implementation phase of the project's platform, the tool and its visualization has been supporting the developers in debugging and controlling the functionality of the platform. This has been made easy because the log entries – summarized as the scripts that are loaded in the tool – have been generated automatically by the software system.

5   Conclusion and Future Work

In the paper at hand, we presented a tool that enables process visualizations representing the underlying structure and displaying automatically the connections between the components of the structure based on script files. The animation of the consecutive entries of the script can be triggered step by step or watched in an auto-play mode. Furthermore, the selection of certain events, let the user jump directly to the appropriate animation. The appearance of the individual components of the system and their connections can adapted individually, e.g., color or size.

To extend the tool, a visual editor is conceivable to set positions of the components of the system via drag and drop and change colors and sizes by opening simple context menus. The number of customizable elements can be increased as well, e.g., the exact path of possible connections.

Concerning the visualization, more interaction possibilities can be implemented. A search functionality can ease to find the correct "entry of interest" in a script. The search for the number of a specific entry is as imaginable as the search for certain key words in the specification of an entry. Furthermore, filtering for different aspects such as components can be added.

By applying the concept of semantic zooming, the visualization could potentially get more depth. It could be possible to zoom in one component to get more detailed information about the flow situation between sub-components.

Currently, the script files are manually generated to define the desired process. In future work, the software systems could be extended to create the scripts themselves like they write events into log files. These automatically created scripts – following the correct format and structure – could be presented directly and additionally used for debugging and understanding the developed system.

Acknowledgments. This research was funded by the German Federal Ministry for Economic Affairs and Technology in the context of the technology program "Smart Data - Innovations in Data", grant no. 01MD15004E.

References

1. Bao, T., Gardner, W.B.: Log visualization tool for message-passing programming in pilot. In: 2017 IEEE International Parallel and Distributed Processing Symposium Workshops (IPDPSW), Lake Buena Vista, FL, pp. 331–338 (2017). https://doi.org/10.1109/ipdpsw.2017.52
2. Moher, T.G.: PROVIDE: a process visualization and debugging environment. IEEE Trans. Software Eng. **14**(6), 849–857 (1988). https://doi.org/10.1109/32.6163

3. Barrett, J.L.: PROCESS VISUALIZATION getting the vision right is key. Inf. Syst. Manage. **11**(2), 14–23 (1994)
4. Foley, J.D., McMath, C.F.: Dynamic process visualization. IEEE Comput. Graphics Appl. **6** (3), 16–25 (1986)
5. Matkovic, K., Hauser, H., Sainitzer, R., Groller, M.E.: Process visualization with levels of detail. In: IEEE Symposium on Information Visualization, INFOVIS 2002, pp. 67–70 (2002)
6. Bobrik, R., Reichert, M., Bauer, T.: View-based process visualization. In: Alonso, G., Dadam, P., Rosemann, M. (eds.) BPM 2007. LNCS, vol. 4714, pp. 88–95. Springer, Heidelberg (2007)
7. Bobrik, R., Bauer, T., Reichert, M.: Proviado – Personalized and configurable visualizations of business processes. In: Bauknecht K., Pröll B., Werthner H. (eds.) EC-Web 2006. LNCS, vol. 4082, pp. 61–71. Springer, Heidelberg (2006)
8. Frei, A., Rennhard, M.: Histogram matrix: log file visualization for anomaly detection. In: 2008 Third International Conference on Availability, Reliability and Security (2008). https://doi.org/10.1109/ares.2008.148
9. Takada, T., Koike, H.: Tudumi: information visualization system for monitoring and auditing computer logs. In: Proceedings of Sixth International Conference on Information Visualisation, pp. 570–576 (2002). https://doi.org/10.1109/iv.2002.1028831
10. Siirtola, H., Räihä, K.J., Surakka, V., Vanhala, T.: Flexible method for producing static visualizations of log data. In: IEEE 2008 12th International Conference Information Visualisation, pp. 127–132 (2008). https://doi.org/10.1109/iv.2008.42
11. Hanniel, J.J., Widagdo, T.E., Asnar, Y.D.W.: Information system log visualization to monitor anomalous user activity based on time. In: Proceedings of 2014 International Conference on Data and Software Engineering, ICODSE (2014). https://doi.org/10.1109/icodse.2014.7062673
12. Lee, J., Jeon, J., Lee, C., Lee, J., Cho, J., Lee, K.: A study on efficient log visualization using D3 component against APT: how to visualize security logs efficiently? In: Proceedings of 2016 International Conference on Platform Technology and Service, PlatCon (2016). https://doi.org/10.1109/platcon.2016.7456778
13. UML Sequence Diagrams. https://www.uml-diagrams.org/sequence-diagrams.html
14. Yue, T., Briand, L.C., Labiche, Y.: Automatically deriving UML sequence diagrams from use cases. Quality Engineering, pp. 1–17, December 2014
15. Elallaoui, M., Nafil, K., Touahni, R.: Automatic generation of UML sequence diagrams from user stories in scrum process. In: 2015 10th International Conference on Intelligent Systems: Theories and Applications, SITA (2015)
16. Automating Diagrams with Visio. http://boxesandarrows.com/automating-diagrams-with-visio/
17. Bruza, P.D., van der Weide, Th.P.: The semantics of data flow diagrams. In: Prakash, N. (ed.) Proceedings of the International Conference on Management of Data, Hyderabad, India (1989)
18. The free dictionary. https://www.thefreedictionary.com/process
19. Cambridge Dictionary. https://dictionary.cambridge.org/dictionary/english/process

Bullet Graph Versus Gauges Graph: Evaluation Human Information Processing of Industrial Visualization Based on Eye-Tracking Methods

Lei Wu[1(✉)], Lingli Guo[2], Hao Fang[2], and Lijun Mou[1]

[1] School of Mechanical Science and Engineering,
Huazhong University of Science and Technology, Wuhan 430074,
People's Republic of China
Lei.wu@hust.edu.cn
[2] School of Arts and Communication, China University of Geosciences,
Wuhan 430074, People's Republic of China

Abstract. This paper reports on an experimental study on industrial information visualization interface to measure the mechanism of information style, information complexity and task complexity on human information processing. Based on eye-tracking method, we conducted an experimental research study. The independent variables were information style, information complexity and task complexity. The dependent variables included time to first fixation and subjective feelings. A total of 40 subjects participated in the experiment. The main findings of this study were as follows: (1) information style, information complexity and task complexity significantly influenced the time to first fixation ($P < 0.05$); (2) there is significant interaction between information style*information complexity, information style*task complexity, information complexity*task complexity ($P < 0.05$); (3) the bullet graph provides more efficient reading than gauges graphs. Furthermore, the research results could provide an approach of using eye-tracking method for evaluation information visualization in relevant industrial areas.

Keywords: Industrial visualization · Evaluation method
Human information processing · Cognitive psychology · Eye-tracking

1 Introduction

The collecting and understanding of data is one of the basic ways of human perception and cognition in the world. The unprecedented and complex information challenges the human's ability of information processing [1–4]. Data visualization is the integration of computer graphics, human-computer interaction, cognitive psychology, semiotics and graphic design [5, 6]. According to the definition by Thomas and Cook, data visualization is the technology that assists users in analyzing and reasoning complex data through visualization interface [7]. "The top 10 challenges in visual analytics" points out that the core themes of visual analysis will be focused on the depth fusion of human cognition and visualization in the future [8].

© Springer International Publishing AG, part of Springer Nature 2019
T. Z. Ahram and C. Falcão (Eds.): AHFE 2018, AISC 794, pp. 752–762, 2019.
https://doi.org/10.1007/978-3-319-94947-5_74

China ministry of industry issued that the key tasks of "China made 2025", which proposed to the development of industrial data visualization in manufacturing service platform [9]. Industrial visualization interface is a critical factor in intelligent manufacture information systems. Industrial data visualization focus on the manufacturing data collection, extraction and feature analysis, present into the form of graphical, which help users to monitor, judge and make decisions. Furthermore, the industrial data visualization interface is the important way of factory operators to obtain production information and real-time monitoring. In summary, the key factors of industrial data visualization mainly have those of characteristics: (1) the environmental parameters; (2) the maintenance records; (3) the operating conditions data; (4) the equipment performance data; (5) other equipment operating data. However, the difficulty of the visualization of industrial data is how to transform the complex information along with the human's mental mode, transform data into the visual information elements that are easy to perception and cognition.

In this research, we conducted an eye-tracking experiment in industrial visualization interface. Using the eye-tracking method, we were able to analyze the human information processing and task solution strategies of participants during real task procedure. The theoretical framework of this study is shown in Fig. 1.

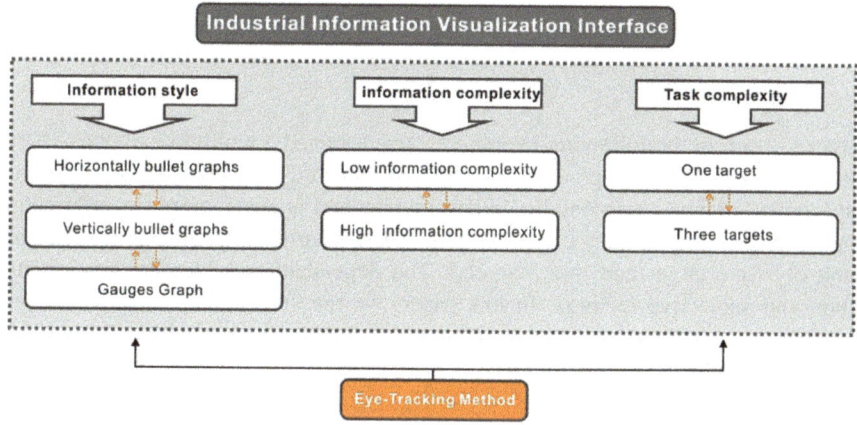

Fig. 1. The theoretical framework of this study

2   Related Research

In the field of evaluation methods for information visualization systems, Ware [10] introduced a methodology for evaluating the cognitive cost of graph aesthetics. Ghoniem [11] described taxonomy of generic graph related tasks along with a computer-based evaluation designed to assess the readability of two representations of graphs. Heer [12] investigated techniques for visualizing time series data. Goldberg [13] used eye-tracking method to compare radial and linear graphs visualization. Burch et al. [14] used eye-tracking methods to evaluate participants' reactions to different tree diagrams. Giacomin [15] evaluated the visual representation of five thermodynamic visualizations and the impact of different scenarios on the user's emotional perception. Cölln [16] focused on the comparison of traditional engineering drawings with CAD visualization in terms of user performance and eye movements in an applied context. Raschke [17] discussed three methodologies for analyzing the vast amount of eye tracking data from the visual analytics perspective. Steichen [18] provided an analysis of user eye gaze data aimed at identifying behavioral patterns. Netzel [19] presented the results of an eye tracking study that compares different visualization methods.

Information visualization helping people explore and explain data through visual interface that exploits the capabilities of the human perceptual system. Furthermore, Kurzhals [20] investigated recent scientific publications from the main visual analytics conferences and journals. Urribarri [21] presented metrics that evaluates data representation visibility considering glyph visibility in scatterplots. Quispel [22] investigated the relationship between familiarity, perceived ease of use and attractiveness of graph designs in two target groups. Huang [23] proposed to derive aesthetics based on graph internal structural features.

Based on the eye-tracking research reviewed above, we chose to use the eye-tracking metrics of time to first fixation to evaluate operators' visual behavior.

3 Experiment Method

3.1 Definitions

In order to confirm the following hypotheses, we designed a multiple-variable (3*2*2) experiment study. The independent variables in this study are information style (vertically bullet graphs, horizontally bullet graphs and gauges graphs), information complexity (low information complexity and high information complexity) and task complexity (one target and three targets). The dependent variables are time to first fixation and subjective feelings. In this paper, we measure human information processing using eye-tracking data (objective metric) combined with a subjective feelings questionnaire (subjective metric).

3.2 Hypotheses

The main hypotheses of this study are as follows:
H1:   information style significantly affects the time to first fixation.
H2:   information complexity significantly affects the time to first fixation.
H3:   task complexity significantly affects the time to first fixation.

3.3 Participants

A total of 40 students at china university of geosciences were randomly selected to participate in this experiment. There were 19 male and 21 female students, ages 18–29 (mean age = 23.58, SD = 1.97). Male subjects accounted for 47.5% of the study, female subjects accounted for the remaining 52.5%. All participants had normal or corrected-to-normal color vision, 8 participants wore glasses and 2 of them wore contact lenses. None of the participants had prior eye surgeries or eye problems such as "droopy" eyelids.

3.4 Equipment and Environment

Eye movements were sampled at 60 Hz using the Tobii X60 eye tracker (Sweden, 14.4-inch screen size, 16:9 screen resolution, 1600*900pixels). Data gather by the eye tracker can be analyzed using Tobii Studio version 3.3.1. The test environment was a quiet psychology laboratory without noise and interference. Participants sat in front of the Tobii X60 monitor in chairs set at a distance of about 50–60 cm from the monitor. Desk and monitor position and height were fixed. The chair was adjustment to fit participant's natural angles of elbow and knee. The room was artificially illuminated and only a minimum of objects was contained inside. Participants were instructed to switch off their mobile phones to reduce possible distractions.

3.5 Stimuli and Task

As the experimental material in this research, we chose the information visualization interface of monitoring systems in the CNC machine tools. Twelve prototypes of the information visualization interface were generated based on different independent variables that was undergoing compare evaluation the visual searching mode and task solution strategies. Temperature monitoring represents one of the most important statuses in the production process of the CNC machine tools. Therefore, we chose "the temperature alarm" as the target visual searching graphs. For example, A of gauges graphs is in statuses of temperature alarm, B and C are within the statues of tolerable temperature range, as shown in Fig. 2. We designed the eye-tracking task of finding the temperature alarm graphs in the information visualization interface. Determine the relative speed and effectiveness of interpreting temperature alarm data when presenting in different information type, information complexity and task complexity.

We designed a within-subjects study design with 40 participants. All participants were involved in all the levels of eye-tracking task. We counterbalanced the twelve experiment stimuli using a random method to compensate for learning effects. Participants were asked to complete twelve visual search tasks in random order. Each task was conducted one time for each participant. During each task, there are 5 s of break time (gray screen). In each task, the temperature alarm graphs are defined as the AOI (area of interest). Time to first fixation in the experiment is defined as the time between the interface showing up and the participant visually identifying the temperature alarm graphs. There was one or three fixed AOI in each of the interface screen based on the different level of task complexity. In addition, the location of the AOI was adjusted in each complexity of the interface to eliminate learning effects.

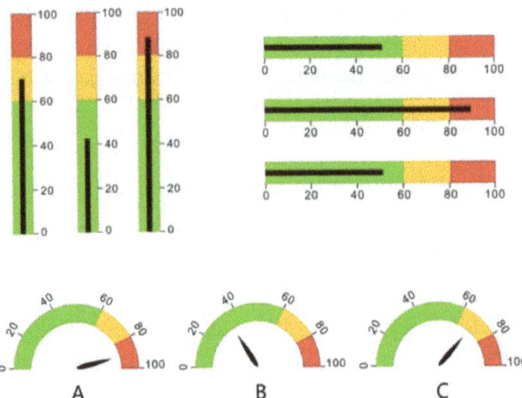

A   B   C

Fig. 2. Types of temperature alarm graphs in the information visualization

3.6 Procedure

Before the experiment began, participants were asked to read an introduction of the experiment requirements and then sign the "experimental consent". Next, they read a short manual about the experiment stimuli to insure they were able to understand and solve the given task. When the participant was ready, we started the eye-tracking experiment. The participants were asked to find the temperature alarm graphs in each task as soon as possible. We recorded the time and the gaze plot it took them to find the AOI (temperature alarm graphs).

Until the participants confirmed they had found the temperature alarm graphs, and then they needed to tap "Space Bar" by their forefinger of the left hand to the next task, see Fig. 3. After taping the "Space Bar" of keyboard, participants entered the next task. After all the twelve tasks were finished, participants were immediately asked to complete the subjective feelings questionnaire. The questionnaire answers were recorded by the Tobii Studio software 3.3.1. Finally, 40 participants completed all the 480 tasks (40 participants *12 tasks).

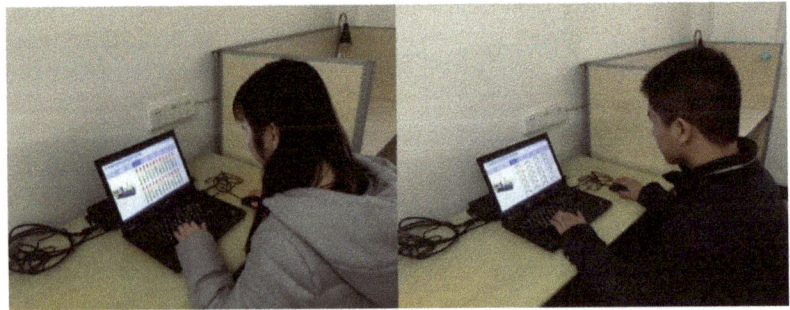

Fig. 3. Participant in the experiment and the environment

4   Experiment Results

4.1   Gaze Plot Analysis

The gaze plot showed the sequence and position of fixations on the information visualization interface. The size of the dots indicates the length of fixation duration. Meanwhile, the numbers in the dots represent the order of the fixations. Different colors represent different participates. The frequency of visual gaze manipulation was related to the length of the fixation duration and difficulty of understanding information visualization interface.

Firstly, we compared the gaze plot from the three information type. The visual search trajectory of the vertically bullet graphs is distributed along the horizontal direction. Then, the visual search of the horizontally bullet graphs is distributed along the longitudinal direction. However, the visual search trajectory of the gauges graphs is distributed horizontally and longitudinally, which is much more complicated.

Secondly, we compared the gaze plot from each of the information complexity. The gaze plot visualization showed that as the information complexity increased, the gaze plot gradually become more complicated and disorderly. The results indicate that as information complexity increased, there is an increase in the user's cognitive workload and the visual search tracks become longer and more dispersed over the interface area.

Thirdly, we compared the gaze plot from each of the task complexity. It is found that there is significant difference between the one target task and the three targets. Since the user's cognitive capacity is limited, increasing the task complexity decreases the user's effective visual focusing, see Fig. 4.

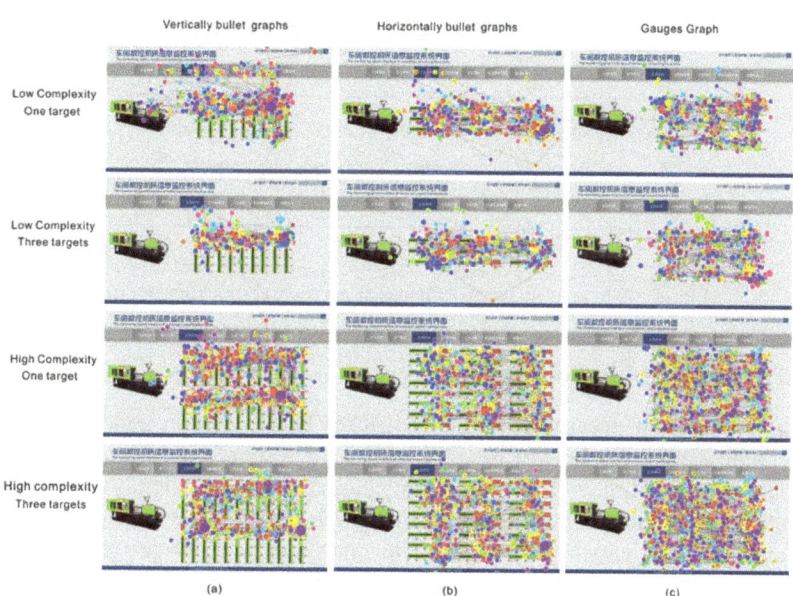

Fig. 4. The gaze plot data based on the eye-tracking device

4.2   Heat Map Analysis

The heat map used different colors to show the number of fixations participants made in certain areas of the information visualization interface or how long they fixated within that area. Red indicates the highest number of fixations and the longest time, green indicates the least fixations and the shortest time, with varying color levels in between.

In these representations, hot zones with higher density indicate where users focused their gaze with a higher frequency. From the analysis below, we can clearly see the heat map of the gauges graphs contains more red hot zones. Also, it appeared that the number of hot zones increased as the information complexity increased. This is most likely caused by the increased difficulty in finding the AOI (temperature alarm graphs). The operators were paying more attention resource on different way-finding areas. These search strategies become significantly more complex during the increase of task complexity, see Fig. 5.

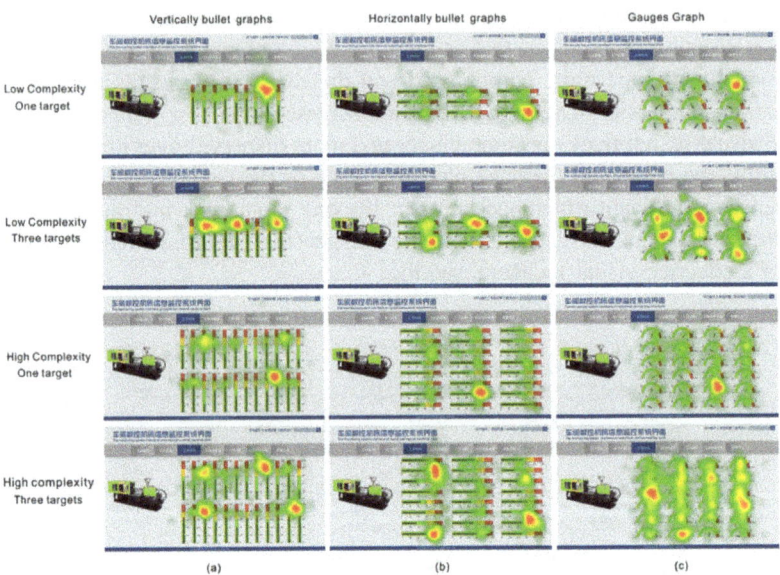

Fig. 5. The heat maps of AOI based on the eye-tracking device

4.3   Time to First Fixation

Time to first fixation measures how long it takes before a participant notice on the "temperature alarm" for the first time. Firstly, at the low information complexity and one target level, the time to first fixation of gauges graphs was slightly lower than the vertically bullet graphs and horizontally bullet graph. Secondly, at the high information complexity and one target level, the time to first fixation of gauges graphs was significantly higher than the vertically bullet graphs and horizontally bullet graph. Thirdly, at the low information complexity and three targets level, the time to first fixation was showing a relationship of slow linear growth.

Furthermore, at the high information complexity and three targets level, the visual search strategies of gauges graphs showed these differences, the time to first fixation being significantly longer compared to two types of bullet graphs and three targets. At last, the time to first fixation of the horizontally bullet graphs was the shortest in high information complexity and three targets level.

To further validate the research hypotheses, repeated measures were used to examine the association between information style, information complexity and task complexity. The results suggest that information style produces a significant effect, F $(2, 78) = 31.83$, P value <0.05, $\eta^2 = 0.45$; information complexity also produces a significant effect, F $(1, 39) = 183.72$, P value <0.05, $\eta^2 = 0.83$; At last, task complexity produces a significant effect, F $(1, 39) = 40.40$, P value <0.05, $\eta^2 = 0.51$. Furthermore, there is significant interaction between information style and information complexity, F $(2, 78) = 17.31$, P value <0.05. There is also significant interaction between information style and task complexity, F $(2, 78) = 8.82$, P value <0.05. Finally, there is significant interaction between information complexity and task complexity, F $(1, 39) = 16.58$, P value <0.05, see Fig. 6.

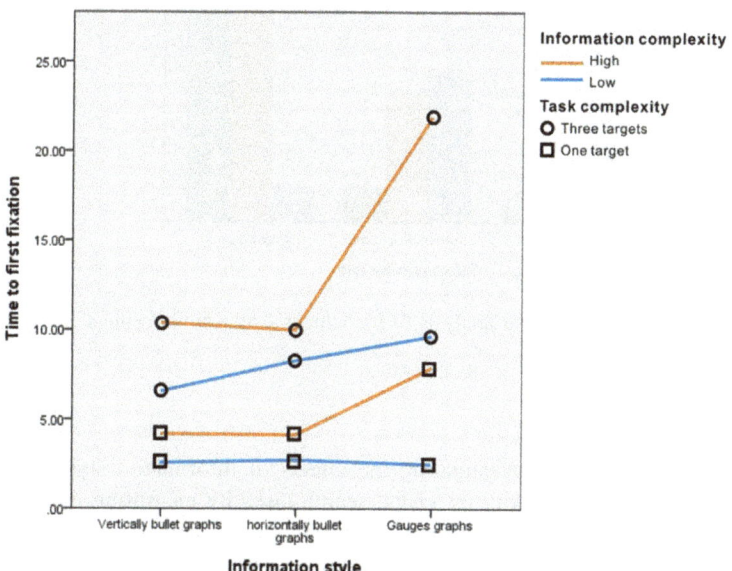

Fig. 6. The line chart of the time to first fixation based on independent variables

4.4   Subjective Feelings Analysis

Satisfaction refers to the observation and evaluation of a user's multidimensional emotional experience of the information visualization interface. Cognitive workload refers to the psychological pressure on a user during a particular operation or completion of a complex task. Aesthetic factors include the user's aesthetic attitude and feelings.

Through the chi-square test results, we can see Pearson Chi-Square = 54.397, p value <0.05, indicating that there are significant differences in the three information styles. The result showed that there is a significant impact on subjective evaluation. The participant has significantly felt higher levels of cognitive workload in gauges graphs. In addition, participant felt significantly higher levels of satisfaction in vertically bullet graphs and horizontally bullet graphs. It should be noticed that many participant felt that the aesthetic effect of gauges graphs is the best, see Fig. 7.

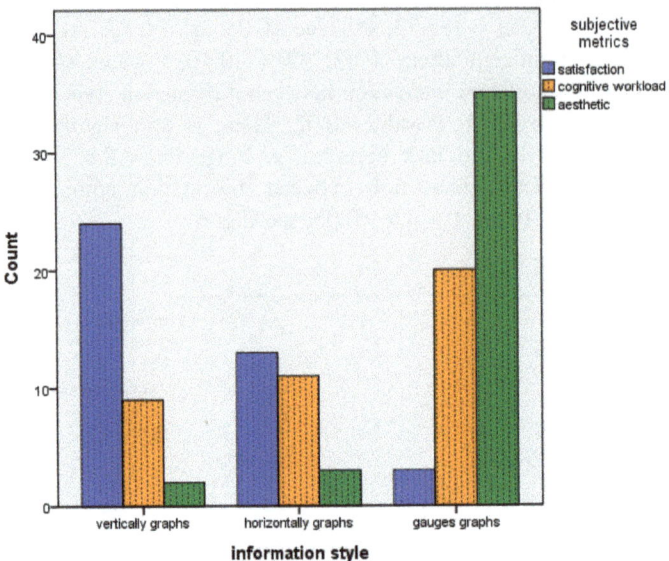

Fig. 7. The bar analysis of participant's subjective feelings

5 Conclusion

This paper presents study investigating the effects of information style, information complexity and task complexity on visual search tasks by measuring eye movements and subjective feelings. The aim of this research is to assess the impact of both independent variables on human information processing in industrial information visualization interface. In this study, eye-tracking technology revealed how operators visually processed different independent variables levels of industrial visual analytics systems.

The main task of industrial information visualization is to help users extract useful information from the complex manufacturing data. The information visualization could potentially increase the understanding of the digital production status and ultimately enhance decision making for the use of industrial visual interfaces. A statistical analysis of the data showed that information style, information complexity and task complexity significantly affected the human information processing.

The main findings of this study were as follows:

(1) Information style, information complexity and task complexity significantly influenced the time to first fixation, P value <0.05. The results indicated that information style, information complexity and task complexity significantly affects the operator's visual searching strategy.

(2) There is significant interaction effect between information style*information complexity, P value <0.05; information style*task complexity, P value <0.05; information complexity*task complexity, P value <0.05.

(3) The participant felt higher levels of satisfaction in vertically bullet graphs and horizontally bullet graphs. However, participant has felt higher levels of cognitive workload in gauges graphs. Finally, many participants felt that the aesthetic effect of gauges graphs is the best.

(4) The bullet graph (both vertically bullet graphs and horizontally bullet graphs) provides an effective display of data in a small space. Its linear design not only gives it a small footprint, but also supports more efficient reading than gauges graphs.

Based on the human factors theory and eye tracking method, this research presents the eye-tracking evaluation method in information visualization for improving clarity, utility and effectiveness. Furthermore, the study provides an approach of using eye-tracking evaluation method for information visualization interface in relevant industrial areas. Future research could include additional biofeedback measures, such as ERP and FMRI research method. Although this study has its limitation, we hope that it can serve as a basis for future studies.

Acknowledgments. The research financial supports from the Natural Science Youth Foundation of Hubei Province (2017CFB276) and CES-Kingfar Excellent Young Scholar Joint Research Funding (CES-KF-2016-2018).

References

1. Shen, H., Zhang, X., Chen, W., Yuan, X., Wang, W.: Preface to the visualization and visual analysis. J. Softw. **27**(5), 1059–1060 (2016)

2. Ren, L., Du, Y., Ma, S., Zhang, X., Dai, G.: Visual analytics towards big data. J. Softw. **25**(9), 1909–1936 (2014)

3. Dai, G., Chen, W., Hong, W., Liu, S., Qu, H., Yuan, X., Zhang, J., Zhang, K.: Information visualization and visual analytics: challenges and opportunities. Chin. Sci. Inf. Sci. **43**(1), 178–184 (2013)

4. Xiaoru, Y., Xin, Z., He, X., et al.: Visualization research frontier and prospects. E-Sci. Technol. Appl. **2**(4), 3–13 (2011)

5. Ziemkiewicz, C., Kosara, R.: The shaping of information by visual metaphors. IEEE Trans. Vis. Comput. Graph. **14**(6), 1269–1276 (2008)

6. Keim, D., Andrienko, G., Fekete, J.D., et al.: Visual analytics: definition, process, and challenges. In: Information Visualization, pp. 154–175. Springer, Heidelberg (2008)

7. Cook, K.A., Thomas, J.J.: Illuminating the path: The research and development agenda for visual analytics. Pacific Northwest National Laboratory (PNNL), Richland, WA, US (2005)

8. Wong, P.C., Shen, H.W., Johnson, C.R., et al.: The top 10 challenges in extreme-scale visual analytics. IEEE Comput. Graph. Appl. **32**(4), 63 (2012)

9. Ministry of Industry and Information Technology on the release of industrial restructuring and upgrading in 2016 (China Manufacturing 2025) key project guide, the Ministry of Industry and Communications [2016] 433 (2016)

10. Ware, C., Purchase, H., Colpoys, L., et al.: Cognitive measurements of graph aesthetics. Inf. Vis. **1**(2), 103–110 (2002)

11. Ghoniem, M., Fekete, J.D., Castagliola, P.: On the readability of graphs using node-link and matrix-based representations: a controlled experiment and statistical analysis. Inf. Vis. **4**(2), 114–135 (2005)

12. Heer, J., Kong, N., Agrawala, M.: Sizing the horizon: the effects of chart size and layering on the graphical perception of time series visualizations. In: Proceedings of the SIGCHI Conference on Human Factors in Computing Systems, pp. 1303–1312. ACM (2009)

13. Goldberg, J., Helfman, J.: Eye tracking for visualization evaluation: reading values on linear versus radial graphs. Inf. Vis. **10**(3), 182–195 (2011)

14. Burch, M., Konevtsova, N., Heinrich, J., et al.: Evaluation of traditional, orthogonal, and radial tree diagrams by an eye tracking study. IEEE Trans. Vis. Comput. Graph. **17**(12), 2440–2448 (2011)

15. Giacomin, J., Bertola, D.: Human emotional response to energy visualisations. Int. J. Ind. Ergon. **42**(6), 542–552 (2012)

16. Cölln, M.C., Kusch, K., Helmert, J.R., et al.: Comparing two types of engineering visualizations: task-related manipulations matter. Appl. Ergon. **43**(1), 48–56 (2012)

17. Raschke, M., Blascheck, T., Burch, M.: Visual analysis of eye tracking data. In: Handbook of Human Centric Visualization, pp. 391–409. Springer, New York (2014)

18. Steichen, B., Wu, M.M.A., Toker, D., et al.: Te, Te, Hi, Hi: eye gaze sequence analysis for informing user-adaptive information visualizations. In: International Conference on User Modeling, Adaptation, and Personalization, pp. 183–194. Springer International Publishing (2014)

19. Netzel, R., Burch, M., Weiskopf, D.: Comparative eye tracking study on node-link visualizations of trajectories. IEEE Trans. Vis. Comput. Graph. **20**(12), 2221–2230 (2014)

20. Kurzhals, K., Fisher, B., Burch, M., et al.: Eye tracking evaluation of visual analytics. Inf. Vis. **15**(4), 340–358 (2016)

21. Urribarri, D.K., Castro, S.M.: Prediction of data visibility in two-dimensional scatterplots. Inf. Vis. (2016). https://doi.org/10.1177/1473871616638892

22. Quispel, A., Maes, A., Schilperoord, J.: Graph and chart aesthetics for experts and laymen in design: the role of familiarity and perceived ease of use. Inf. Vis. **15**(3), 238–252 (2016)

23. Huang, W., Huang, M.L., Lin, C.C.: Evaluating overall quality of graph visualizations based on aesthetics aggregation. Inf. Sci. **330**, 444–454 (2016)

Effectiveness of Eye-Gaze Input Method: Comparison of Speed and Accuracy Among Three Eye-Gaze Input Method

Atsuo Murata and Makoto Moriwaka[✉]

Department of Intelligent Mechanical Systems, School of Engineering,
Okayama University, 3-1-1, Tsushimanaka, Kita-ward, Okayama, Japan
{murata,moriwaka}@iims.sys.okayama-u.ac.jp

Abstract. Effectiveness of eye-gaze input methods was examined in click, drag, and menu selection tasks. In a click task, three eye-gaze methods were (c)-(i) eye-gaze input with fixation, (c)-(ii) eye-gaze input with pressing BS key, and (c)-(iii) eye-gaze input with voice (voice1). Method (d)-(i) eye-gaze input with pressing BS key and (d)-(ii) eye-gaze input with voice (voice1) were compared for the drag task. In the menus selection task, the performance was compared between Method (m)-(i) eye-gaze input with voice (voice1) and (m)-(ii) eye-gaze input with voice (voice2: uttering one of the following menu items: "save", "print", "cut", "copy", and "paste"). The pointing time in the click task increased according to the following order: (c)-(i) eye-gaze input with fixation, (c)-(ii) eye-gaze input with pressing BS key, and (c)-(iii) eye-gaze input with voice (voice1). The pointing accuracy of (c)-(i) was nearly equal to 100% and by far better than that of Method (c)-(ii) and (c)-(iii). Concerning the drag, Method (d)-(i) tended to be faster than Method (d)-(ii). However, the pointing accuracy of both methods was not satisfactory and ranged from 70% to 80%. This indicated that Method (d)-(i) and (d)-(ii) must be further improved when used for the drag task. The pointing time in the menu selection task did not differ significantly between Method (m)-(i) and (m)-(ii). The pointing accuracy of Method (m)-(ii) was by far higher than that of Method (m)-(i) when the target size was small. The larger target size tended to lead to faster and accurate pointing for all three tasks. It seems that the better pointing method differs according to the eye-gaze method. Other than the click task, the pointing accuracy was at most 90%. Therefore, future research should propose an effective method to increase the prediction accuracy for both drag and menu selection tasks.

Keywords: HCI · Click · Drag · Menu selection · Pointing time
Prediction accuracy · Subjective rating

1 Introduction

We can naturally direct our eyes toward the location of a target to be pointed and eye movements are faster than hand movements. When pointing an object in human-computer interactions, an eye-gaze input system is more intuitive and faster than that a mouse [1–8]. Sibert and Jacob [2] and Murata [3] found that the pointing time was

© Springer International Publishing AG, part of Springer Nature 2019
T. Z. Ahram and C. Falcão (Eds.): AHFE 2018, AISC 794, pp. 763–772, 2019.
https://doi.org/10.1007/978-3-319-94947-5_75

faster using gaze with short dwell times less than 150 ms than that using a mouse. Agustin et al. [9] evaluated the potential of gaze input for game interaction. They suggested that there is a potential for gaze input in game interaction, given a sufficiently accurate and responsive eye tracker and a well-designed interface.

There are still a lot of problems we must overcome so that such an input system can be put into practical use in actual HCI tasks. When dragging or menu selection tasks are executed, unlike click tasks, it is very difficult to accomplish these tasks with only eye-gaze. The promising candidate of inputs in drag or menu selection task is key press or voice input. However, the effective additional input method to be used in drag or menu selection tasks has not been fully explored until now.

This study attempted to compare the pointing performance (speed and accuracy) of a variety of eye-gaze input methods in three HCI tasks (click, drag, and menu selection). For the click task, the eye-gaze input methods were (c)-(i) eye-gaze input with fixation, (c)-(ii) eye-gaze input with pressing BS key, and (c)-(iii) eye-gaze input with voice (voice1: uttering only a command "left" corresponding left click of a mouse). The performance measures in the drag task were compared between Method (d)-(i) eye-gaze input with pressing BS key and (d)-(ii) eye-gaze input with voice (voice1). In the menus selection task, the performance was compared between Method (m)-(i) eye-gaze input with voice (voice1) and (m)-(ii) eye-gaze input with voice (voice2: uttering one of the following menu items: "save", "print", "cut", "copy", and "paste"). The input method and the target size were within-subject experimental variables. It was examined how these variables affected pointing performance (speed and accuracy) and subjective rating on satisfaction and workload. Some limitations of the present eye-gaze input technology were pointed out, and an effective method was discussed to resolve these limitations. Implications were also made for the HCI design of eye-gaze input system especially when the system is used together with other input modalities such as voice input or key press.

2   Method

2.1   Participants

Three undergraduates took part in the experiment. All of them are using PC more than four hours a day. They had no orthopedic or neurological diseases.

2.2   Apparatus

Using EMR-AT VOXER (Nac Image Technology), an eye-gaze input interface was developed. Visual C# (Microsift) was used as a programming language. This apparatus enables us to determine eye movements and fixation by measuring the reflection of low-level infrared light (800 nm), and also admits the head movements within a pre-determined range. The eye-tracker was connected with a personal computer (HP, DX5150MT) with a 15-inch (303 mm × 231 mm) CRT. The resolution was 1024 x 768pixel. Another personal computer was also connected to the eye-tracker via a RS232C port to develop an eye-gaze input system. The line of gaze, via a RS232C port, is output to this computer with a sampling frequency of 60 Hz. The illumination on the keyboard of a personal was about 200 lx, and the mean brightness on CRT was about 100 cd/m². The viewing distance was set to about 70 cm.

2.3   Design and Procedure

The calibration of eye tracker (camera) was conducted so that eye movements (visual line of gaze) could be measured accurately for each condition of three tasks (click, drag, and menu selection). After the practice session was over according to the judgment of experimenters, the participants began the experimental session. The eye gaze data sampled with a sampling frequency of 60 Hz were filtered using a 5-point moving average. The evaluation measures were the pointing time for the correct trials and common to three tasks. The error trials were excluded from further analysis.

The experimental factors were the input method and the target size for all of three tasks. All were within-subject factors. The click, drag, and menu selection tasks were conducted in this order. In addition to the eye-gaze input, the click task was also conducted with a mouse. After each task was completed, the participants were required to evaluate subjective rating of satisfaction and workload for each eye-gaze input and a mouse using a 5-point scale. The satisfaction scores 1 and 5 represent very dissatisfactory and very satisfactory, respectively. The workload scores 1 and 5 represent very high workload and very low workload, respectively.

In the click task, the order of performance of six eye-gaze input conditions (target size and input method) and two mouse conditions (target size) was randomized across the participants. In the drag and menu selection tasks, the order of performance of four eye-gaze input conditions (target size and input method) and two mouse conditions (target size) was randomized across the participants.

In the click task, for each of six eye-gaze input conditions and two mouse conditions, the participant was required to carry out the click task 10 times for each of eight directions (upper, lower, left, right, upper left, upper right, lower left, and lower right). The eight kinds of directions were randomly presented to the participant. We followed the procedure below for both drag and menu selection tasks. For each of eight eye-gaze input conditions and two mouse conditions, the participant was required to carry out the click task 10 times for each of eight directions (upper, lower, left, right, upper left, upper right, lower left, and lower right). The eight kinds of directions were randomly presented to the participant.

In the click task, the following three eye-gaze methods were used: (c)-(i) eye-gaze input with fixation, (c)-(ii) eye-gaze input with pressing BS key, and (c)-(iii) eye-gaze input with voice (voice1). The 100 ms fixation of a target terminated one trial of pointing in Method (c)-(i). The pressing of BS key and the voice input (utterance of "left") terminated one trial in Method (c)-(ii) and (c)-(iii), respectively. Method (d)-(i) eye-gaze input with pressing BS key and (d)-(ii) eye-gaze input with voice (voice1) were used in the drag task. In the menus selection task, the performance was compared between Method (m)-(i) eye-gaze input with voice (voice1) and (m)-(ii) eye-gaze input with voice (voice2: uttering one of the following menu items: "save", "print", "cut", "copy", and "paste"). Three methods (c)-(i)-(c)-(iii) were compared for the click task. Method (d)-(i) and (d)-(ii) were compared for the drag task. In the menus selection task, the performance was compared between Method (m)-(i) and (m)-(ii).

3 Results

As similar results were obtained for three participants, the results of Participant A are shown below. The reaction time of click task (Participant A) is compared between small and large target sizes and among input methods in Fig. 1. The percentage correct (Participant A) is plotted as a function of input method and target size in Fig. 2. In Fig. 3, the satisfaction and workload score (Participant A) are depicted as a function of input method and target size.

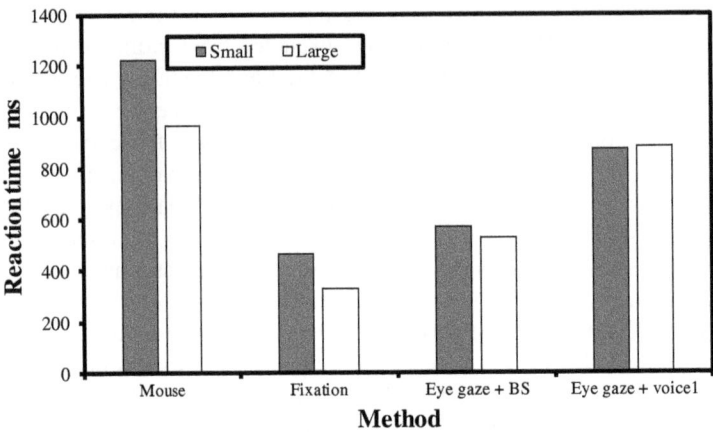

Fig. 1. Reaction time as a function of input method and target size (Participant A: click).

The reaction time of drag task (Participant A) is shown as a function of target size and input method in Fig. 4. Figure 5 compares the percentage correct (Participant A) among input method and between target sizes. In Fig. 6, the satisfaction and workload score (Participant A) are depicted as a function of input method and target size.

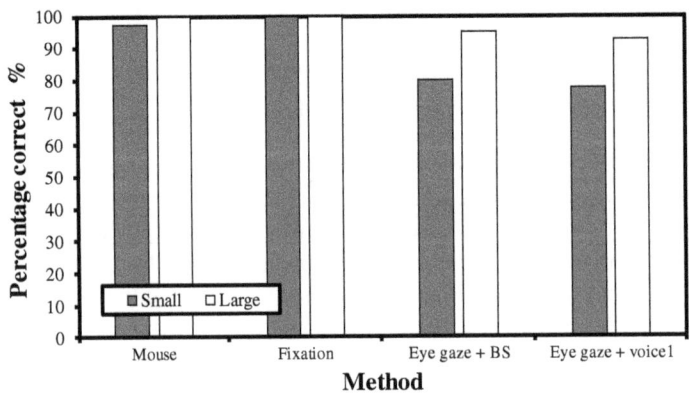

Fig. 2. Percentage correct as a function of input method and target size (Participant A: click).

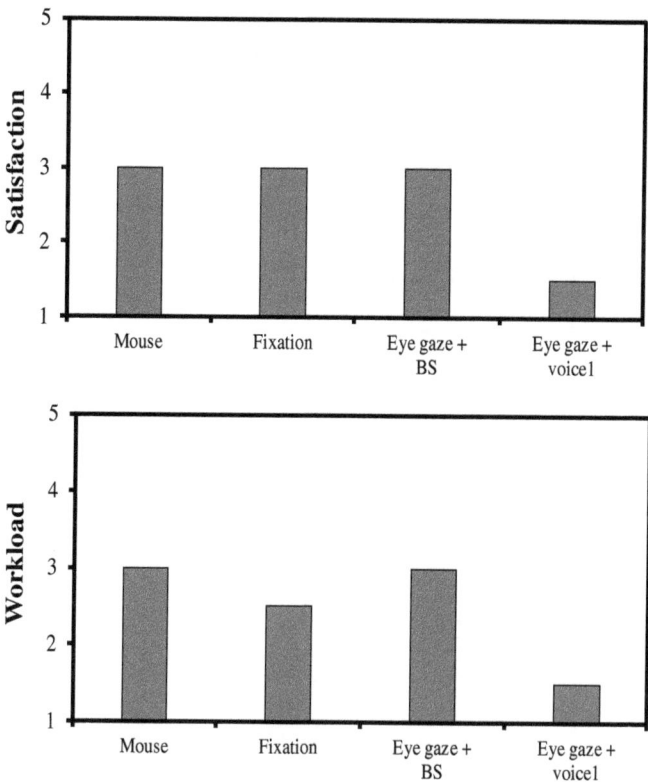

Fig. 3. Subjective rating of satisfaction and workload as a function of input method (Participant A: click). Upper: Satisfaction score, Lower: Workload score.

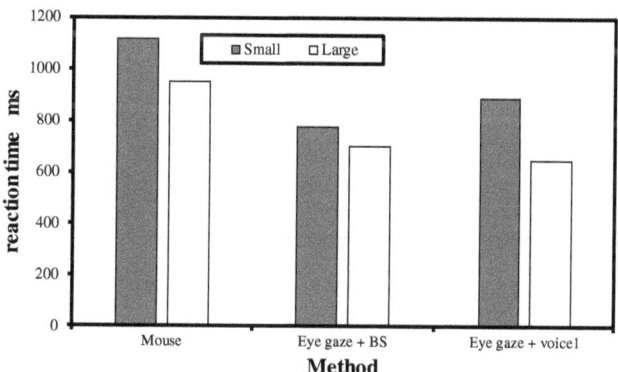

Fig. 4. Reaction time as a function of input method and target size (Participant A: drag).

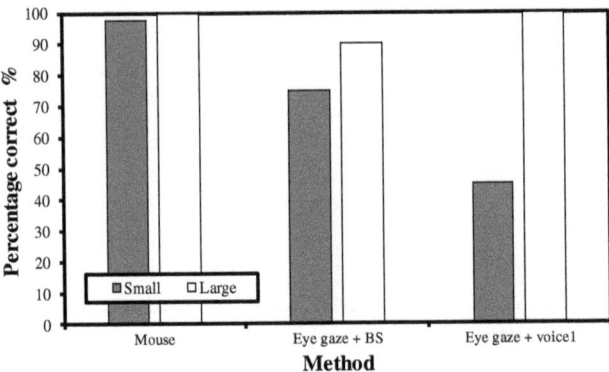

Fig. 5. Percentage correct as a function of input method and target size (Participant A: drag).

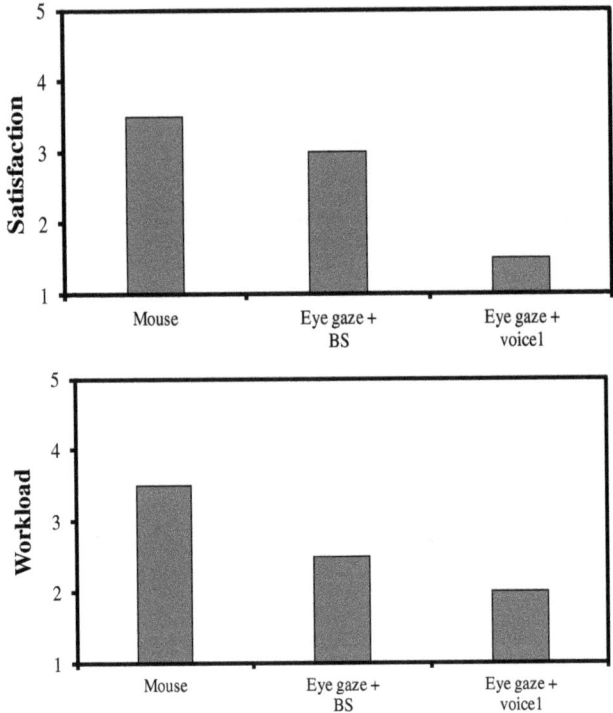

Fig. 6. (a) Subjective rating of satisfaction and workload as a function of input method (Participant A: drag, Satisfaction score). (b) Subjective rating of satisfaction and workload as a function of input method and target size (Participant A: drag, Workload score).

The reaction time of drag task (Participant A) is shown as a function of target size and input method in Fig. 7. Figure 8 compares the percentage correct (Participant A) among input method and between target sizes. The satisfaction and workload score (Participant A) are depicted as a function of input method and target size (see Fig. 9).

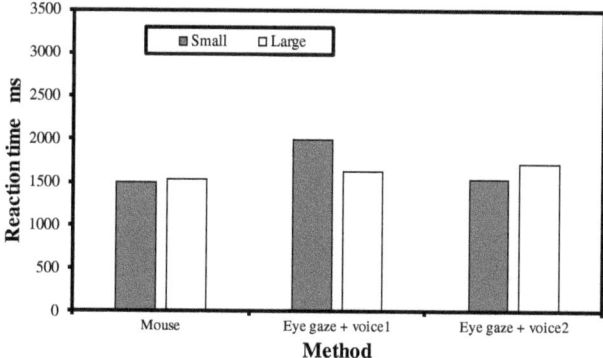

Fig. 7. Reaction time as a function of input method and target size (Participant A: menu selection).

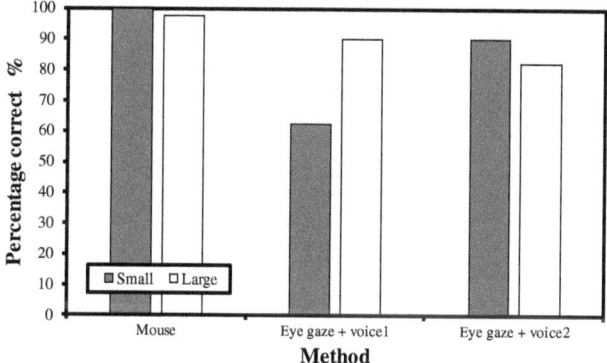

Fig. 8. Percentage correct as a function of input method and target size (Participant A: menu selection).

4 Discussion

The pointing time in the click task increased according to the following order: (c)-(i) eye-gaze input with fixation, (c)-(ii) eye-gaze input with pressing BS key, and (c)-(iii) eye-gaze input with voice (voice1). The pointing time was shorter than that of the mouse. The pointing accuracy of mouse and (c)-(i) eye-gaze input with fixation was nearly equal to 100% and by far higher than that of Method (c)-(ii) and (c)-(iii). As far as this experiment is concerned, the target size did not have a major effect on the reaction time. The percentage correct of Method (c)-(ii) and (c)-(iii) was affected by the target size. The subjective rating of satisfaction and workload was especially poor for

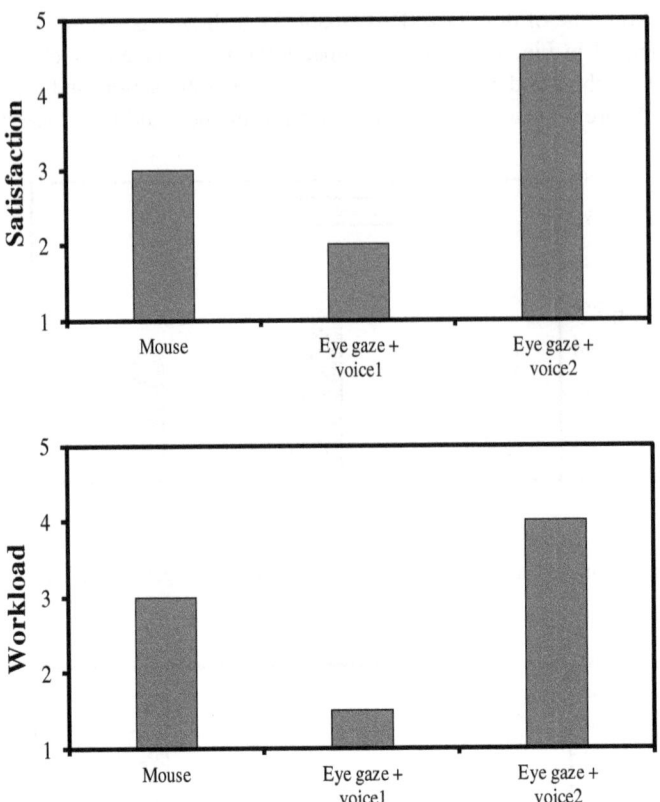

Fig. 9. Subjective rating of satisfaction and workload as a function of input method (Participant A: menu selection). Upper: Satisfaction score, Lower: Workload score.

Method (c)-(iii). The results mean that the performance of eye-gaze input is degraded when eye gaze was combined with voice input.

Concerning the drag task, Method (d)-(i) tended to be faster than Method (d)-(ii). This is also suggestive of the disadvantage of voice input combined with the eye-gaze input. The pointing accuracy of both methods was not satisfactory and ranged from 70% to 80%. This indicated that Method (d)-(i) and (d)-(ii) must be further improved when used for the drag task. Even in this task, the subjective rating of satisfaction and workload was especially poor for Method (c)-(ii) that combined voice with eye gaze.

As for the menu selection task, the reaction time did not differ significantly between Method (m)-(i) and (m)-(ii). Moreover, unlike click and drag tasks, the reaction time was longer than that of the mouse. It must also be noted that the pointing accuracy of Method (m)-(ii) was by far lower than that of Method (m)-(i) especially when the target size was small. The larger target size tended to lead to faster and accurate pointing. However, the results indicates that the eye-gaze input system in such a menu selection task is not sufficient as it is and further improvement is necessary to put the eye-gaze input into practical use.

It seems that the better pointing method differs according to the type of tasks. Especially when eye-gaze input must be combined with voice, this input method should be further elaborated. Other than the click task, the pointing accuracy was at most 90%. Therefore, the results suggest that future research should be conducted to propose an effective method to increase the prediction accuracy for both drag and menu selection tasks. The low prediction accuracy when eye gaze must be combined with voice for all of three tasks must be attributed to the coupling of subtle movement (jittering) of cursor and other inputs such as key press or uttering when a cursor was within a target. A promising candidate of methods to increase the prediction accuracy and shorten the reaction time (pointing time) is discussed below.

Although using only the eye gaze is more natural and desirable as suggested by Bader and Beyerer [10], it is practically difficult to use only the eye gaze especially when the interactive tasks are complicated as in drag and menu selection tasks. When executing such complicated tasks, the combination of eye-gaze input with voice input or key pressing is one solution. However, such a technique has disadvantages and could not attain satisfactory performances as demonstrated in this study. When involuntary eye movements during fixation occur simultaneous with the utterance for voice input or key press, it is highly possible that the fixation point is deviated from the target, and thus the accuracy and speed of pointing is degraded. This must be caused by the interference of the visual system and the muscular or auditory system. This must lead to inaccurate and slow pointing operation. Although Partala et al. [11] studied the benefit of combining gaze pointing and facial-muscle EMG clicking compared to mouse input in target acquisition tasks, a very high error rate (34%) was observed. Surakka et al. [12] extended the previous study by Partala et al. [11] with a more detailed Fitts' law analysis, and showed that gaze-EMG input combination was more effective for long-distance movement than the mouse, but for short distances the mouse was more effective. This study also used an eye-gaze combined with muscular response. This unnatural setting might lead to low prediction accuracy.

It is speculated that such an event (interference of eye gaze with utterance or key pressing) further makes it irritating for users of eye-gaze input system to execute complicated tasks such as menu selection tasks. It is, therefore, expected that the prevention of involuntary eye movement during a fixation within a target will further improve the usability or performance of eye-gaze input system from both pointing speed and accuracy. Although Kumar et al. [13] attempted to improve the pointing accuracy of eye-gaze input using a smoothing technique of eye gazes, the pointing accuracy remained at most 92%. This needs to be further improved. Therefore, future research should explore the effectiveness of prevention method of subtle cursor movements within a target (jittering) such as an automatic lock of cursor movement within a target area.

References

1. Zhai, S., Morimoto, C., Ihde, S.: Manual and gaze input cascaded (MAGIC) pointing. In: Proceedings of SIGCHI Conference on Human Factors in Computing Systems, CHI 1999, pp. 246–253. ACM Press, New York (1999)
2. Sibert, L.E., Jacob, R.J.K.: Evaluation of eye gaze interaction. In: Proceedings of CHI 2000, pp. 281–288 (2000)
3. Murata, A.: Eye-gaze input versus mouse: cursor control as a function of age. Int. J. Hum. Comput. Interact. **21**(1), 1–14 (2006)
4. Murata, A., Miyake, T.: Effectiveness of eye-gaze input system-identification of conditions that assures high pointing accuracy and movement directional effect. In: Proceedings of 4th International Workshop on Computational Intelligence & Applications, pp. 127–132 (2008)
5. Murata, A., Moriwaka, M.: Basic study for development of web browser suitable for eye-gaze input system-identification of optimal click method. In: Proceedings of 5th International Workshop on Computational Intelligence & Applications, pp. 302–305 (2009)
6. Jacob, R.J.K.: What you look at is what you get: eye movement- based interaction technique. In: Proceedings of ACM CHI 1990, pp. 11–18 (1990)
7. Jacob, R.J.K.: The use of eye movements in human-computer interaction techniques: what you look at is what you get. ACM Trans. Inf. Syst. **9**, 152–169 (1991)
8. Jacob, R.J.K., Sibert, L.E., Mcfarlanes, D.C., Mullen, M.P.: Integrality and reparability of input devices. ACM Trans. Comput. Hum. Interact. **1**(1), 2–26 (1994)
9. Agustin, S.J., Mateo, C.J., Hansen, J.P., Villanueva, A.: Evaluation of the potential of gaze input for game interaction. Psychnol. J. **7**(2), 213–236 (2009)
10. Bader, T., Beyerer, J.: Natural gaze behavior as input modality for human-computer interaction. In: Nakano, Y.I., Contai, C., Bader, T. (eds.) Eye Gaze in Intelligent User Interfaces-Gaze-Based Analysis, Models and Applications, pp. 161–183, Springer, New York (2013). https://doi.org/10.1007/978-1-4471-4784-8_9
11. Partala, T., Aula, A., Surakka, V.: Combined voluntary gaze direction and facial muscle activity as a new pointing technique. In: Hirose, M. (ed.) Interact 2001, pp. 100–107. IOS Press, Amsterdam (2001)
12. Surakka, V., Illi, M., PIsokoski, P.: Gazing and frowning as a new human-computer interaction technique. ACM Trans. Appl. Percept. **1**, 40–56 (2004)
13. Kumar, M., Klingner, Puranik, J.R., Winograd, T., Paepcke, A.: Improving the accuracy of gaze input for interactio. In: Proceedings of the 2008 symposium on Eye tracking research & applications, pp. 65–68. Savannah, GA (2008)

Assistive Technology and Design Solutions

A Digital Assistance System Providing Step-by-Step Support for People with Disabilities in Production Tasks

Volkan Aksu[1(✉)], Sascha Jenderny[1], Björn Kroll[1], and Carsten Röcker[1,2]

[1] Fraunhofer IOSB-INA, Lemgo, Germany
{volkan.aksu,sascha.jenderny,bjoern.kroll,
carsten.roecker}@iosb-ina.fraunhofer.de
[2] inIT - Institute Industrial IT, OWL University
of Applied Sciences, Lemgo, Germany

Abstract. The use of Assistive Technology (AT) plays a significant role in the advancement of greater independence for individuals with disabilities in their work life. In particular, digital step-by-step support can enable people to perform production tasks that were formerly difficult to accomplish. In this paper, we focused on finding a solution for a specific production process. To this end, we set up a prototype assistive system for performing a cutting task which provides step-by-step support for people with disabilities. In an evaluation study with impaired people, we investigated how our assistive system affects the task efficiency as well as participants' subjective evaluation of perceived mental effort and system usability. Results show advantages for step-by-step support with regard to users' task efficiency and subjective evaluation.

Keywords: Assistive Technology · People with disabilities · User-centered Design · Human-computer-interaction · Production · Step-by-step instructions

1 Introduction

Approximately 15% of the world population are suffering from at least one disability. Unfortunately, this percentage is expected to grow rapidly in the coming years due to population ageing, chronic health conditions such as cancer and mental disorders [1]. In the industrialized countries, around 80 million people suffer from a disability (ranging from mild to severe) and face significant challenges in almost every area of society such as in employment [2, 3]. People with disabilities are usually economically disadvantaged and experience higher rates of unemployment (17.4%) than people without disabilities (10.2%) [4, 5]. On the other hand, technological advancement in recent years offers great potential in enhancing job opportunities for disabled people based on their abilities and resources [6]. In this context, assistive devices and technologies can play a key role in enabling the inclusion of people with disabilities as active and independent participants in the labour market.

© Springer International Publishing AG, part of Springer Nature 2019
T. Z. Ahram and C. Falcão (Eds.): AHFE 2018, AISC 794, pp. 775–785, 2019.
https://doi.org/10.1007/978-3-319-94947-5_76

In this paper, we introduce how such an assistive system can promote both quality of work and independence of people with disabilities by carrying out complex industrial tasks with great care and precision. In cooperation with an organization for handicapped people, we developed a hardware-based assistant system with a graphical user interface that supports people with disabilities producing high-quality jewelry boxes using step-by-step video instructions. Without an assistive system, disabled workers need stringent monitoring by attendants. Nevertheless, a significant proportion of jewelry boxes has to be sorted out because they are faulty, inaccurate or daubed with traces of glue. With the introduction of a digital assistance system we aimed at supporting workers in producing the jewelry boxes with absolute precision (see Fig. 1).

Fig. 1. Producing a jewelry box: raw material (*left*), cut material (*middle*), glued and folded material to a jewelry box (*right*).

The paper is organized as follows: In Sect. 2 we give an overview of related work. Section 3 describes the prototype system and the methodology of our evaluation study. Results are presented in Sect. 4. Finally, we discuss the results and draw conclusions in Sect. 5.

2 Related Work

In recent years, a large number of assistive systems for people with special needs have been developed for supporting them in everyday life activities that rely on a variety of different technologies such as mobile devices [7, 8], speech recognition [9–11], gesture recognition [12, 13], augmented reality (AR) [14, 15], virtual reality (VR) [16, 17] and autonomous robot systems [18, 19]. A general review about assistive technology systems for people with disabilities is provided by Sauer et al. [20]. However, assistive systems for inclusion of people with disabilities into the regular labour market have not yet been investigated to this extend. In the following, we present the related work regarding available assistance systems for impaired people in industrial environments.

One line of research addresses in-situ projection for workplaces [21]. For instance, Korn et al. [22] investigated the potential of thereof in a sheltered work organization. They used a toolkit for measuring the performance of impaired persons and built a prototype system projecting work instructions directly into the workplace [23]. Subsequently, they analyzed the effect of in-situ projection on participant's work quality and acceptance of the system. Results with regard to work quality were heterogeneous:

some participants could reduce their assembly time and error rates through the system, while others were overwhelmed by using the prototype and performed worse. With respect to system acceptance, however, all participants indicated that they would like to retry the system. Furthermore, Baechler et al. [24] evaluated different pictogram visualizations for order picking tasks with cognitive disabled employees. In a comparative wizard-of-oz study, 24 employees tested four picking visualizations: pick-by-projection, pick-by-paper, pick-by-light and pick-by-display [25]. Dependent variables such as picking time, error rate and participants' perceived mental effort in using the system were measured. In contrast to other methods, participants made almost no mistakes with the pick-by-projection.

With regard to picking time and subjective mental effort, pick-by-light was the first, pick-by-projection the second-best method. Funk et al. [26] investigated the impact of in-situ-projection instructions on workers with disabilities in an assembly scenario. In a user study with 64 participants, they compared a contour-, a video- and a pictorial-visualization to a control group using no visual feedback. They found that participants made fewer errors and were faster using the contour-visualization in an assembling task.

These "conventional" assistive systems in production environments focus mostly on technical aspects of the assembly. To make work more attractive and to increase motivation of impaired workers, different design approaches of motion-controlled gamification[1] have been recently introduced for disabled workers in production (e.g. the tetris design [27], the circle design [28] and the pyramid design [29]). The results reveal that there is a common tendency towards higher work speed and motivation of workers with disabilities, when gamification components are integrated for future implementations into the production process.

In spite of the fact that there are various research projects about augmented-based assistive systems for impaired people, there is no previous work combining a computer-based system with manual support for specific tasks like cutting, folding and gluing. In this paper, we explored how to find a reliable solution that can be also transferred to similar production tasks in a modified form. Subsequently, we tested the prototype in an evaluation study with disabled workers from a sheltered work organization.

3 Materials and Methods

3.1 Prototype System

The prototype system was developed with user-centered design methodology. First, thinking-aloud tests were conducted with three disabled users and two attendants producing jewelry boxes using conventional methods such as scalpel, wood glue and manual folding. On this basis, user requirements, needs and problems were analyzed as a starting point for the novel prototype. Based on this input we created a hardware-based prototype which supports users in three different stages: cutting, glueing and

---

[1] "[...] the use of video game elements in non-gaming systems to improve user experience (UX) and user engagement" [32]

folding. The following paragraphs describes the general approach and implementation concept in more detail.

In a first step, the cutting process was analyzed and adapted to the needs of people with special needs. The requirement analysis led to the creation of a 3D mould (see Fig. 2a) with a hard plastic template that workers can place on the fiberboard to cut it precisely along the line (see Fig. 2b). As a cutting tool, we printed two 3D handles which allows for interchangeable blades (see Fig. 2c). The user can choose between a 45° and −45° angle blade depending on which line they are cutting. The mould is equipped with remote controlled LEDs using Arduino and Bluetooth technology in order to guide the worker through the cutting steps by presenting video instructions on a mobile device. The video instructions were recorded from a bird's-eye view in HD resolution. A brief interruption of one second was added to the end of each step to enable users getting a better temporal orientation while cutting the fiberboard. To facilitate the user playing the instructions in a simple manner, we implemented a software with a user-friendly graphical interface using C++ and Qt (see Fig. 2d). With the click of a button integrated in the mould, the user can skip to the next step after performing the current subtask. Blinking LEDs before each step support the user finding the right position to cut quickly. Furthermore, the software allows attendants to edit or create new work instructions.

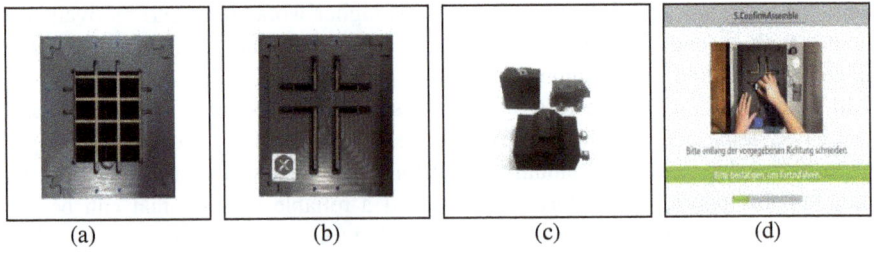

| (a) | (b) | (c) | (d) |

Fig. 2. Prototype system: (a) 3D mould, (b) hard plastic template, (c) cutting tool, (d) graphical user interface.

For the glueing task, we used an automatic glue dispenser (Drifton 2000-D)[2] with timer control and foot pedal. It regulates the dosage of the adhesive with the air pressure and thus enables accurate application of the glue (see Fig. 3). To facilitate the last stage of folding, we printed a 3D folding aid that allows the user to form the glued fiberboard into a box shape. The aid consists of two identical moulds, only differing in height, that are mounted directly one above the other: Firstly, the glued fiberboard is placed the enclosed square on the one mould and is pressed carefully down with one finger until the desired box shape is achieved. Subsequently, the other mould is fitted to the top side of the box. A soft cloth is used to avoid scratching and to achieve a more stable fixation of the box (see Fig. 3).

---

[2] http://www.drifton.eu/.

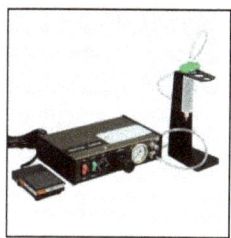

Fig. 3. Automatic glue dispenser (*left*), folding aid with a soft cloth (*middle*), folding process (*right*)

In the present study we focus on the cutting process which turned out to be the most demanding and complex subtask of the production process. First solution approaches of glueing and folding tasks are described and discussed, however they are not included in this study.

3.2   Study Design

Next, we conducted an evaluation study applying a within-subject design with two experimental conditions. The hardware-based assistance system was used either with or without the step-by-step support as described in the previous section. As dependent variables we measured the following:

- *Mental Effort:* Participants' perceived mental effort in conducting the specific tasks, was assessed with the SEA scale ("Subjectively Perceived Effort" [30]). The one-dimensional scale ranges between 0 ("no cognitive effort") and 220 ("maximum cognitive effort").
- *Usability of the system*: To assess systems' usability, we employed an adapted version of the standardized ten-item SUS questionnaire ("System Usability Scale" [31]) for people with impairments. Ratings on ten items are given on 5-point Likert scales ranging from 1 ("strongly disagree") to 5 ("strongly agree"). Bangor and colleagues suggested the following interpretation of SUS scores [9]: <50: Not acceptable; 50–70: Marginal; >70: Acceptable

Finally, we also investigated participants' efficiency in performing tasks. To this end, we measured how long it took them to complete the tasks (time on task), whether they succeeded or failed at a task (task success) and whether they solved the task without help (task accuracy).

3.3   Procedure

The study was conducted as part of a workshop for handicapped people. First, participants were welcomed by the experimenter. Prior to participation, all participants were given a brief description about the aim and procedure of the study. Then, the experimenter demonstrated how to use the system with and without step-by-step support and clarified all outstanding issues until the participants felt confident in their understanding and handling of the prototype. The demonstration phase was carried out to ensure that all participants have an equal foundation of experience in using the prototype. The order of experimental conditions was randomized. Subsequently, participants were asked to conduct the same cutting procedure with and without step-by-step support (see Fig. 4). The condition with step-by-step support contained 17 subtasks (see Table 1).

Table 1. Overview of the 17 subtasks in the condition with step-by-step support.

Step-Nr.	Description
01	Please check your work material
02	Please put the blank into the mould
03	Please place the template on the mould
04	Please take the scalpel with the digit 1. (blade with an inclination of 45°)
05–08	Please cut the blank in the specified direction (the corresponding LEDs are blinking)
09	Please take the scalpel with the digit 2. (blade with an inclination of −5°)
10–14	Please cut the blank in the specified direction. (the corresponding LEDs are blinking)
15	Please put the scalpel back
16	Please remove the template from the mould
17	Please remove any residues from the mould

Fig. 4. The evaluation study: Demonstration of the cutting process (*left*), participant performing the cutting task (*right*).

After finishing the task, participants gave a post-task rating of their perceived cognitive workload on the SEA scale and were then asked to fill out the SUS questionnaire. Finally, participants were debriefed and thanked for their time.

3.4   Participants

Five German speaking male participants with different levels of cognitive disability took part in this study. They ranged in age from 20 to 21. The participants mean age was 20.6 years ($SD = 0.55$). One of the participants had a physical disability and used a manual wheelchair. The study was conducted at the Werkstätte of Lebenshilfe Detmold e.V., a German sheltered work organization supervising about 890 workers with cognitive and motoric limitations. Neither of the participants had previous experience with our supporting system.

4 Results

In the following we report results regarding the effect of experimental conditions on (1) measures of task efficiency, (2) subjective evaluations (measured with standardized inventories). Due to the small sample size, we could only use descriptive statistics to assess general trends in both conditions. Therefore, the statistical findings can not be used to infer significance.

Task Effeciency
Here, we investigated participants' efficiency in performing the cutting task with regard to *time on task*, *task accuracy*, and *task completeness*. All three variables were noted by the experimenter who observed the participants' activities on cutting fiberboards. Results are visualized in Fig. 5.

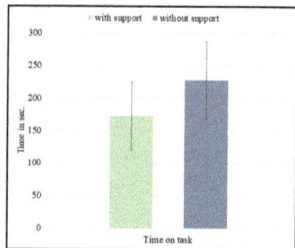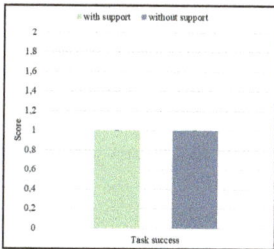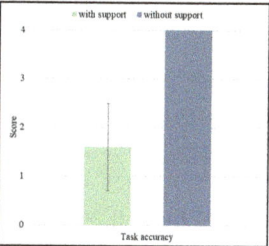

Fig. 5. Comparison of mean values and standard deviation for the variables time on task (*left*), task success (*middle*) and task accuracy (*right*).

Time on task: We compared the sum of participants' completion time for finishing the cutting task across experimental conditions. The mean task completion time in the "with support"-condition was 161 s while tasks performed without support took about 218 s on average.

Task success: Task success measures whether participants succeeded or failed at a task. A score of 1 was given for "full success" on a task, 2 for "partial success" and 3 for "no success" (see Fig. 2). In both experimental conditions, participants' success rate was maximal ($M = 1$, $SD = .00$).

Task accuracy: Participants' accuracy in task performance was measured as follows: 1 for "participant solved the task without help", 2 for "participant solved the task with trial & error", 3 for "participant solved the task with a single hint of the lab member" and 4 for "participant solved the task with constant support of the lab member". On average, participants' accuracy in the "with-support"-condition ($M = 1.6$, $SD = .894$) was higher as in the "without-support"-condition ($M = 4$, $SD = .00$).

Subjective Evaluation

Participants' subjective evaluation of the interaction was measured with standardized inventories in the dimensions of perceived mental effort and usability of the prototype (see Fig. 6).

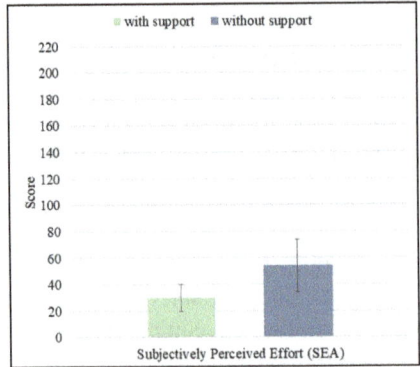 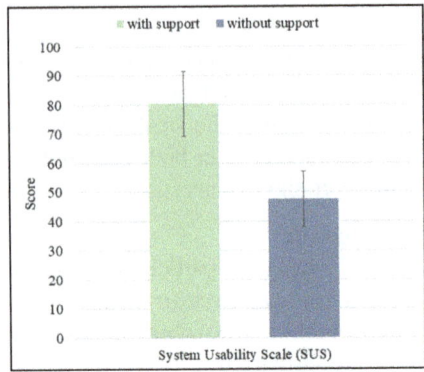

Fig. 6. Comparison of Subjective Perceived Effort (SEA, *left*) and of System Usability Scale (SUS, *right*) in "with-support" and "without-support"-condition.

Mental effort: We measured participants' perceived mental effort in task performance after the cutting task with the SEA scale ranging from 0 ("no effort") to 220 ("extremely high effort"). On average, participants judged their cognitive effort in the "with support"-condition lower ($M = 30.0$, $SD = 10.00$) than in the "without support" condition ($M = 54.0$, $SD = 19.49$).

Usability of the system: We employed the SUS questionnaire to measure the usability of our prototype on a 5-point Likert scale from 1 ("fully disagree") to 5 ("fully agree"). Mean values and standard deviations are visualized in Fig. 6. On average, participants rated the "with support"-condition ($M = 80.5$, $SD = 11.096$) better than the "without support"-condition ($M = 47.5$, $SD = 9.519$).

5   Discussion

In this paper, we investigated the potential of step-by-step support for people with disability using an assistive system in production. We applied a user-centered methodology and implemented a prototype which enables disabled workers to perform a specific subtask of cutting a square fiberboard for producing jewelry boxes. In an evaluation study, we compared our assistive system being employed either with or without step-by-step instruction. Our results can be summarized in two major points.

First, participants' efficiency on using the prototype was assessed with regard to *time on task, task accuracy* and *task completeness*. Along the dimension *time on task* and *time accuracy*, the "with-support" condition was rated more efficiently than "without-support" condition. However, we could not observe any difference between both conditions across the *task completeness*.

Second, participants' subjective evaluation of the interaction was assessed with regard to *mental effort* and *system usability*. Again, our results showed a clear advantage for the interaction "with-support". Participants' perceived cognitive effort using the assistive system with support was lower as compared to the condition without support. The advantage of the "with-support" condition was also found in terms of system usability: Here, the "with-support"- condition was rated higher than "without-support" condition.

Overall, our assistive system *with* step-by-step support showed to have several benefits over the assistive system *without* support. That is, the supportive technology we developed seems to be a helpful aid for disabled workers. Nevertheless, our pilot study has some limitations we plan to overcome in future work.

First, due to the small amount of five participants, only trends could be seen regarding differences in both conditions. Therefore, we conclude that the same study should be repeated with a larger amount of participants. However, there are major discrepancies between individuals with cognitive disabilities. Thus, it will significantly complicate to obtain accurate and reliable evaluation results.

Second, based on the fact that the regular SUS questionnaire could be too complex for participants with disabilities, we used a modified version of the questionnaire in our study. Nonetheless, the participants didn't have the cognitive ability to read, completely understand and fill in the questionnaire. Therefore, the experimenter read out each question from the questionnaire and noted participants' answers. In some cases, participants gave uninterpretable responses to the experimenter which may affect the results. The same applies to placing a check mark on the SEA scale to measure participants' mental effort. Future research includes a study with special questionnaires for people living with cognitive impairments.

Even if there are still a few obstacles to overcome, we could show in this paper that an assistance system can enhance the quality of work and thus improve job opportunities for people with disability to employment. Our assistance system can be seen as a successful reference project for other companies and organizations planning to employ handicapped people and thereby offers new possibilities for inclusion.

Acknowledgments. We thank the sheltered work organization "Werkstätte of Lebenshilfe Detmold e.V." for participating in the evaluation and proving pedagogical support during the project. Special thanks go to Mario Heinz for supporting the hardware development.

References

1. World Health Organization and World Bank: World Report on Usability (2011)
2. European Commission: European Disability Strategy 2010-2020: A Renewed Commitment to a Barrier-Free Europe. Europa.Eu, pp. 1–9 (2010)
3. Lazar, J., Jaeger, P.: Reducing barriers to online access for people with disabilities. Issues Sci. Technol. **27**(2), 68–82 (2011)

4. Fritz, D., Miller, U., Gude, A., Pruisken, A., Rischewski, D.: Making poverty reduction inclusive: experiences from Cambodia, Tanzania and Vietnam. J. Int. Dev. **21**(5), 673–684 (2009)
5. Roulstone, A., Prideaux, S., Priestley, M.: Employment situation of disabled people in european countries (2009)
6. Cavalier, A.: The application of technology in the classroom and workplace: unvoiced premises and ethical issues. In: Images of the Disabled: Disabling Images, pp. 129–141 (1987)
7. Aksu, V., Jenderny, S., Martinetz, S., Röcker, C.: Providing context-sensitive mobile assistance for people with disabilities in the workplace (2018)
8. Lewis, C., Sullivan, J., Hoehl, J.: Mobile technology for people with cognitive disabilities and their caregivers - HCI issues. In: Lecture Notes in Computer Science (Including Subseries Lecture Notes in Artificial Intelligence and Lecture Notes in Bioinformatics). LNCS, no. PART 1, vol. 5614, pp. 385–394 (2009)
9. Lopez, A., Rodriguez, I., Ferrero, F.J., Valledor, M., Campo, J.C.: Low-cost system based on electro-oculography for communication of disabled people. In: 2014 IEEE 11th International Multi-Conference on Systems, Signals and Devices, SSD 2014 (2014)
10. Noyes, J.M., Haigh, R., Starr, A.F.: automatic speech recognition for disabled people. Appl. Ergon. **20**(4), 293–298 (1989)
11. Hawley, M.S., Green, P.D., Enderby, P., Cunningham, S., Moore, R.K.: Speech technology for e-inclusion of people with physical disabilities and disordered speech. Int. Speech Commun. Assoc. **5**, 445–448 (2005)
12. Chattoraj, S., Vishwakarma, K., Paul, T.: Assistive system for physically disabled people using gesture recognition. In: 2017 IEEE 2nd International Conference on Signal and Image Processing (ICSIP), pp. 60–65 (2017)
13. Bien, Z., Do, J.-H., Kim, J.-B., Stefanov, D.: User-friendly interaction/interface control of intelligent home for movement-disabled people (2009)
14. Arvanitis, T.N., Petrou, A., Knight, J.F., Savas, S., Sotiriou, S., Gargalakos, M., Gialouri, E.: Human factors and qualitative pedagogical evaluation of a mobile augmented reality system for science education used by learners with physical disabilities. Pers. Ubiquit. Comput. **13**(3), 243–250 (2009)
15. Tang, L.Z.W., Ang, K.S., Amirul, M., Yusoff, M.B.M., Tng, C.K., Alyas, M.D.B.M., Lim, J.G., Kyaw, P.K., Folianto, F.: Augmented reality control home (ARCH) for disabled and elderlies. In: 2015 IEEE 10th International Conference on Intelligent Sensors, Sensor Networks and Information Processing, ISSNIP 2015 (2015)
16. Covaci, A., Kramer, D., Augusto, J.C., Rus, S., Braun, A.: Assessing real world imagery in virtual environments for people with cognitive disabilities. In: Proceedings of 2015 International Conference on Intelligent Environments, IE 2015, pp. 41–48 (2015)
17. Loomis, J.M., Golledge, R.G., Klatzky, R.L.: Personal guidance system for the visually impaired using GPS, GIS, and VR technologies. In: Proceedings of the First Annual International Conference Virtual Reality and Persons with Disabilities, pp. 17–18 (1993)
18. Chang, P.H., Park, H.S.: Development of a robotic arm for handicapped people: a task-oriented design approach. Auton. Robots **15**(1), 81–92 (2003)
19. Yanco, H.A.: Wheelesley: a robotic wheelchair system: indoor navigation and user interface. In: Assistive Technology and Artificial Intelligence, pp. 256–268. Springer, Heidelberg (1998)
20. Sauer, A.L., Parks, A., Heyn, P.C.: Assistive technology effects on the employment outcomes for people with cognitive disabilities: a systematic review. Disabil. Rehabil. Assist. Technol. **5**(6), 377–391 (2010)

21. Fellmann, M., Robert, S., Büttner, S., Mucha, H., Röcker, C.: Towards a framework for assistance systems to support work processes in smart factories. In: LNCS, vol. 10410 (2017)
22. Korn, O., Schmidt, A., Hörz, T.: The potentials of in-situ-projection for augmented workplaces in production. In: CHI 2013 Extended Abstracts on Human Factors in Computing Systems on - CHI EA 2013, p. 979 (2013)
23. Korn, O., Schmidt, A., Hörz, T., Kaupp, D.: Assistive system experiment designer ASED: a toolkit for the quantitative evaluation of enhanced assistive systems for impaired persons in production. In: Proceedings of the 14th International ACM SIGACCESS Conference on Computers and Accessibility, pp. 259–260 (2012)
24. Baechler, A., Baechler, L., Kurtz, P., Kruell, G., Heidenreich, T., Hoerz, T.: A study about the comprehensibility of pictograms for order picking processes with disabled people and people with altered performance, pp. 69–80. Springer, Cham (2015)
25. Baechler, A., Baechler, L., Autenrieth, S., Kurtz, P., Kruell, G., Hoerz, T., Heidenreich, T.: The development and evaluation of an assistance system for manual order picking - called pick-by-projection - with employees with cognitive disabilities, pp. 321–328. Springer, Cham (2016)
26. Funk, M., Bächler, A., Bächler, L., Korn, O., Krieger, C., Heidenreich, T., Schmidt, A: Comparing projected in-situ feedback at the manual assembly workplace with impaired workers. In: Proceedings of the 8th ACM International Conference on PErvasive Technologies Related to Assistive Environments - PETRA 2015, pp. 1–8 (2015)
27. Korn, O.: Industrial playgrounds: how gamification helps to enrich work for elderly or impaired persons in production. In: Proceedings of the 4th ACM SIGCHI Symposium on Engineering Interactive Computing Systems - EICS 2012, p. 4 (2012)
28. Korn, O., Funk, M., Schmidt, A: Towards a gamification of industrial production. A comparative study in sheltered work environments. In: Proceedings of the 7th ACM SIGCHI Symposium on Engineering Interactive Computing Systems - EICS 2015, June, pp. 84–93 (2015)
29. Korn, O., Funk, M., Schmidt, A.: Design approaches for the gamification of production environments: a study focusing on acceptance. In: Proceedings of the 8th ACM International Conference on PErvasive Technologies Related to Assistive Environments, pp. 6:1–6:7 (2015)
30. Eilers, K., Nachreiner, F., Hänecke, K.: Entwicklung und Überprüfung einer Skala zur Erfassung subjektiv erlebter Anstrengung. Z. Arbeitswiss. **Jg. 40**(H. 4), 215–224 (1986)
31. Lewis, J.R., Sauro, J.: The factor structure of the system usability scale. In: Lecture Notes in Computer Science (Including Subseries Lecture Notes in Artificial Intelligence and Lecture Notes in Bioinformatics). LNCS, vol. 5619, pp. 94–103 (2009)
32. Deterding, S., Sicart, M., Nacke, L., O'Hara, K., Dixon, D.: Gamification. Using game-design elements in non-gaming contexts. In: Proceedings of the 2011 Annual Conference Extended Abstracts on Human Factors in Computing Systems - CHI EA 2011, p. 2425 (2011)

My Intelligent Home (MiiHome) Project

Ipek Caliskanelli[1(✉)], Samia Nefti-Meziani[1], Jonathan Drake[2],
and Anthony Hodgson[3]

[1] System and Advanced Robotics Research Centre, University of Salford,
Manchester, UK
{i.caliskanelli, s.nefti-meziani}@salford.ac.uk
[2] Business Development, Salix Homes, Salford, Manchester, UK
jonathan.drake@salixhomes.org
[3] Salford Royal NHS Trust, Salford, Manchester, UK
anthony.hodgson@srft.nhs.uk

Abstract. The cost and complexity of healthcare delivery to a globally ageing
population is driving developments in the field of assistive living technologies.
This paper focuses on the concept and the system infrastructure of the MiiHome
project, which assists person-centered integrated care in the home. Working
closely with a care organization delivering integrated health and social care and
the housing sector, the MiiHome project aims to prevent problems before they
escalate to the detriment of the person by using digital technologies.

Keywords: Assistive technology · Assistive technology for daily living
Physical wellbeing · Emotional wellbeing · Activity monitoring
End-user experience · Behavioral pattern detection · Sensor fusion
Data aggregation

1 Introduction

Our autonomous remote monitoring and decision support system deploys low-cost, off-the-shelf embedded technologies in the home to provide affordable solutions to allow
older people to live at home for longer. We have three objectives: (1) to evaluate the
effectiveness of using MS Kinect and other sensors for remote monitoring of the elderly
in the home; (2) to assist clinical teams with the assessment of performance of routine
activities of daily living in the periods between clinic visits [1], (3) to maintain or
improve elderly quality of life, and promote independent living in their own home. In
this study, we demonstrate our findings of the effectiveness of the remote monitoring.

To deliver these objectives we have worked closely with all stakeholders including
technologists, health and social care delivery teams, those who supply social and
affordable housing and most importantly the older people who are potential benefi-
ciaries of the technology.

The participant centric approach is at the heart of the health and social care mandate
of Salford Together.

© Springer International Publishing AG, part of Springer Nature 2019
T. Z. Ahram and C. Falcão (Eds.): AHFE 2018, AISC 794, pp. 786–797, 2019.
https://doi.org/10.1007/978-3-319-94947-5_77

1.1   Setting

Salford Royal NHS Foundation Trust delivers health and social care services in the City of Salford, Greater Manchester, UK. It operates under the umbrella of the Northern Care Alliance NHS Group comprising the Care Organisations of Salford, Bury & Rochdale, Oldham and North Manchester. In Salford, around 70% of the population live in areas classified as highly deprived and disadvantaged. The population of Salford therefore experiences health and well-being that is worse than the national average in the UK. People in the city are likely to die earlier and live longer with disability [2].

The Salford care organisation not only provides hospital care, it plays a much broader role in the locality and is supporting the establishment of a new integrated model of care. It is working closely with the local Council to develop an Integrated Care Organisation (ICO) called "Salford Together" [3]. Salford Together is a partnership between Salford City Council, NHS Salford Clinical Commissioning Group, Salford Royal NHS Foundation Trust, Salford Primary Care Together and Greater Manchester Mental Health NHS Foundation Trust.

The partnership is working to transform the health and social care system in Salford by integrating health and social care, bringing the services of GPs, nursing, social care, mental health, community based services and voluntary organisations into a more joined up system that focuses on a person's individual needs and provides them with the support to manage their own care.

Salford Together conducted a community engagement exercise in 2017 called the "The Big Health and Social Care Conversation" [4]. The aim was to provide early opportunities for active, open, dialogue on developing health and social care plans in Salford and to allow service users, carers and other stakeholders to input to and be involved in the transformation process. In total 4200 people were directly engaged with face-to-face and this was backed up by more than 19,000 interactions online (website and social media). From this some 1671 Salford people filled in a written questionnaire providing some valuable data, which has been analysed and which Salford Together partners are planning to consider as part of the developing transformation plans. We have adopted some of the output from this exercise to inform the direction of our MiiHome project in Salford.

The exercise told us that almost a half of respondents (48%/805) attend hospital repeatedly due to an ongoing condition either for themselves, a relative or someone they care for. Of these half (51.4%) felt their experience could have been better if they had more support at home. Therefore, we have explored how digital technologies may help the people of Salford mindful that the solution should be applicable more widely across the UK National Health Service (NHS). We have also incorporated the vision for the local health and social care ecosystem expressed in its locality plan [2]. Relevant objectives from this plan are quoted as follows:

- Put outcomes for people at the heart of the way we work and the care we provide.
- Maximise the use of effective digital technology.
- Ensure Salford learns and develops, using data and intelligence sourced from across the public, private and voluntary sectors.
- Enable care and support to be accessed as close to home as possible.

The result of this approach will be that more care will be delivered in a community setting, largely in people's homes, with a corresponding reduction in unplanned demand for hospital care and expensive packages of social care; improved quality of life for users and carers and significant increase in the ability of people to manage their own condition.

1.2   Approach

Our autonomous remote monitoring and decision support system deploys low-cost, off-the-shelf embedded technologies in the home to provide affordable solutions to allow older people to live at home for longer.

The MiiHome project assists person-centred integrated care in the home. The project depends on co-creation of solutions by close working between participants, clinical teams, managers from the social housing sector and technology experts in sensing, autonomous systems and machine learning. Through this close working with a care organization delivering integrated health and social care and the housing sector, the MiiHome project aims to prevent healthcare problems before they escalate to the detriment of the person by using at-market sensors and machine learning technologies.

The project focuses on a preventative population health philosophy. This necessitates both individual behaviour change as well as changes to care pathways. Because environmental and infrastructural issues are important, engagement with a wide range of stakeholders is being sought. Working closely with the housing sector is important because the project will rely on implementation of structural changes (albeit modest ones) to homes. However, engagement is also important because it expands the range of agency. Through mobilization of the tenants of our project partner, Salix Homes, a registered social housing provider based in Salford that owns 8,500 properties, the project has greatly added to the richness of the intervention and driven forward recruitment of participants. Amplification of the agency of such tenant/participants is pivotal to the broader aims in developing a proactive approach to maintenance of health and well-being as well as sustainability of the NHS.

This paper focuses on one aspect of the development of system infrastructure as part of the MiiHome project that assists person-centered integrated care in the home. The MiiHome project aims to prevent problems before they escalate to the detriment of the person by using digital technologies viewed eventually as an integral component of the home, not as infrastructure fitted into a home. By adopting this approach the aim is to encourage social house builders to adopt digital technology in the home as standard in future builds. It applies a philosophy of co-creation with participants [5] and with all stakeholders to develop a product that not only meets clinical requirements but is acceptable to the participants living in the home.

2   MiiHome Ecosystem

The MiiHome project is a joint venture between University of Salford, Salford Royal NHS Foundation Trust and Salix Homes. Infrastructure consists of The Living Laboratory a living space within the university campus laboratories that is fully furnished and with a fully functional kitchen supplied by Salix Homes. The suite is heavily sensorised with monitoring at every level including motion sensors, door sensors, appliance monitors, water flow meters, Internet of Things appliances such as fridges and dishwashers etc. This is being used as a test bed to develop software, networks and machine learning. Older people including those living with dementia and their carers visited the laboratory to begin co-creation of the programme of work in August 2015.

There is also a heavily sensorised one bedroomed home (Smart Home 1, Fig. 1) situated within the community directly adjacent to the hospital campus that has been provided by Salix Homes. Here participants can live in the short-term to engage in experiments or evaluate systems in terms of functionality.

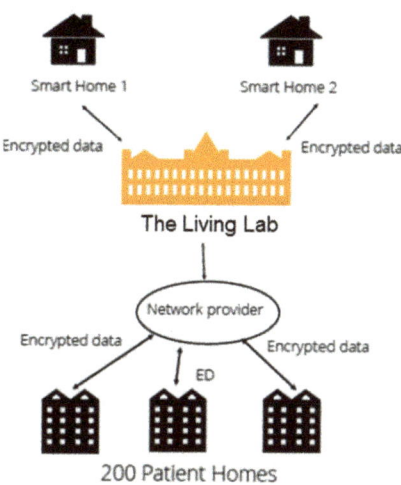

Fig. 1. MiiHome configuration.

Furthermore, we have a pool of Kinect sensors, mini-computers that power the sensors and internet connectivity that is being installed in up to 200 homes, see Fig. 2. The system is turn-key and is installed by Salix Homes working with our technicians. The participant living in the home does not need to have any knowledge of computers, the internet or how to use a mobile computing device of any sort. In this way, we can offer the system to any participant regardless of educational attainment or socioeco-nomic standing.

Figure 3 illustrates the information flow of the MiiHome project. We have created two smart homes for pilot testing, one adjacent to the hospital campus and used for temporary living (testing or clinical evaluation). The two smart homes deploy Microsoft Kinect, and a number of sensors (motion, magnetic, temperature, tilt, air quality, touch, sound, moisture, water flow). MiiHome is a large-scale project up to 200 participant/patient's homes throughout Salford deploying a more limited collection of sensory arrays. Some homes deploy care-on-call (a health service call-centre that

Fig. 2. Participant's homes.

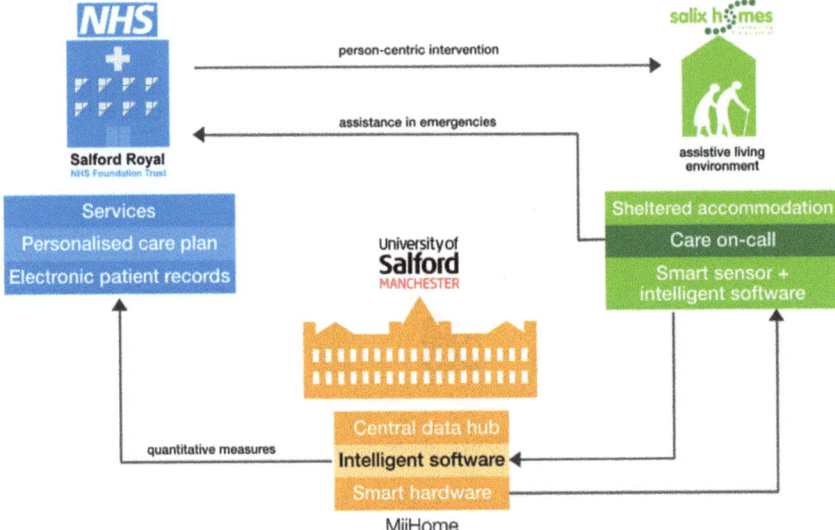

Fig. 3. MiiHome information flow.

provides emergency assistance, currently based on pendant alarms - eventually planned to be triggered by MiiHome).

Given the complexity of the MiiHome ecosystem, the NHS and the Salix Homes contributions are significant and will be described in other, more appropriate venues. In this paper, we focus on the technological aspects and what is more relevant from a computer scientist perspective.

2.1 Platform Model

MiiHome algorithms perform a range of clinical and wellbeing assessments. Figure 4 shows the three-tier system model. Patient/occupier data are collected using an array of sensor devices and processed using software modules. Mostly passive sensing devices

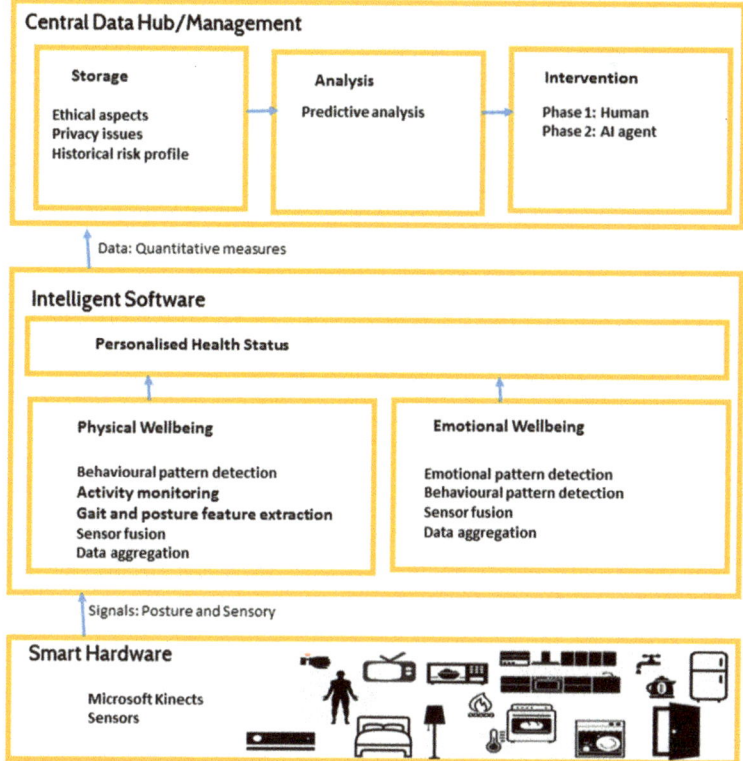

Fig. 4. Three-tier conceptual MiiHome platform.

such as temperature, tilt, air, touch, sound, moisture, water flow sensors as well as the Microsoft Kinects are preferred within the MiiHome project, as opposed to wearable sensors. Features include posture and balance, gait features and behavioral pattern analysis. Quantitative measures assess the progression of specific disorders, assist clinical decisions as well as offering personalised predictive analysis.

Raw and processed patient data are used to model the trajectory of health status. Data is hosted on UoS servers but will be moved to NHS systems as a plugin to the electronic patient record (EPR). This patient data will be fused and aggregated with EPR. Clinician input and clinical decision support systems then update the individual's care plan. We will examine the feasibility and benefit of the MiiHome system in maintaining quality of elderly life.

Figure 5 illustrates a portion of the system developed in Java [6]. It consists of analysis modules (Speed, Fall Detection, Furniture Crawling, Gait-Speed, and Posture Activity), a database (MongoDB), separate reporters for each module (pulling the outputs of the analysis modules from the database, hourly, daily and weekly), log files (used by the reporters and analysis modules to log the system state including errors). Windows task schedulers ensure the proposed system runs 24/7 and restarts if it shuts down for any reason.

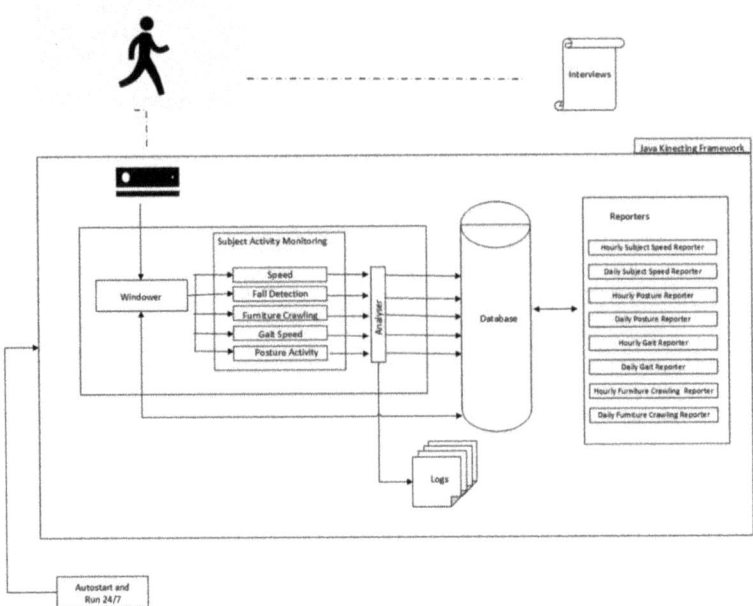

Fig. 5. System architecture of the Kinecting Frailty Framework [6] of MiiHome.

2.1.1 Features Detected

The features detected by the system range from composite signals drawn from the electricity and water supply to the dwelling that can be interrogated to give information about specific behaviours (for example see [7]); For example filling a kettle with water and switching it on to boil the water in it.

We opted to also detect broad indicators of health that were potentially accessible to measurement rather than try to de-convolute information into activities of daily living (ADL) as has been done by many smart home installations and systems [8, 9].

It is well established that several simple tests of physical performance are strongly associated with the onset of long-term functional decline and disability [10]. Ample evidence supports the potential use of physical performance measures in risk assessment strategies that can identify subgroups of older persons, initially independent in all ADLs, who are at increased risk for decline into disability or even death [10]. We sought to investigate whether real-time measurement of such physical performance was also reflective of acute decline as a potential signal triggering proactive interventions to prevent or mitigate such decline. The following behaviours and attributes were identified for monitoring using the Kinect. Items were selected because they fulfilled the following criteria: simple to implement, acceptable load on computing power, clinically relevant to a wide range of situations, and potential for giving acceptable predictive values in assessing risk of adverse outcomes.

2.1.2   Gait Speed and Gait Disorders

Gait or walking speed is a common clinical measure and has been described as the sixth vital sign [11]. Timed walking tests are an important measure in comprehensive geriatric assessment [12]. Changes in walking speed mark a critical point in personal performance and the assessment of gait speed has the potential to serve as a key indicator in mapping the trajectory of health and function in ageing and disease [13]. Systematic reviews have shown that it reliably predicts disability, cognitive impairment, institutionalisation, falls, hospital admission and mortality [14, 15]. Furthermore, a declining trajectory of gait speed is also associated with adverse events such as mortality [16]. Furniture crawling (cruising in North America) is a classic adaptive response to more severe levels of gait disorder [17]. In response to severe postural or gait instability patients hold on to furniture, walls and door handles in order to stabilize themselves while moving around the home. Acutely it is a marker of severe loss of stability and chronically an adaptive response to neurological disorders such as ataxia [17]. We considered using this approach after advice from clinical colleagues because the arms are used for support in furniture crawling and this may be easy to detect using Kinect in a real home setting.

2.1.3   Sit-to-Stand (SiSt)

Rising to a stand from a sitting position is one of the most common activities one performs in a home every day. Sit-to-stand and stand-to-sit are two of the most mechanically demanding activities undertaken in daily life. The ability to perform a sit-to-stand (SiSt) is therefore an important skill. In elderly people, without disability the inability to perform this basic skill can lead to institutionalization, impaired functioning in activities of daily living (ADL), and impaired mobility. Objective measures of lower-extremity function such as the SiSt are highly predictive of subsequent disability [18, 19]. Aging causes a loss of muscle strength or of muscle power [20]. The SiSt depends on quadriceps femoris and trunk musculature. Muscle strength, which is the ability to generate force and muscle power, which is the ability to generate this force rapidly are associated with falling [21, 22]. There is also evidence for the contribution of both muscle strength and power to the ability of older people to maintain balance and posture [20]. Balance and posture are also key components of SiSt [23] alongside muscle strength and power. We evaluated SiSt using Kinect taking into account biomechanical considerations [24] but also being mindful that we performed assessments in a real home where variables such as the height of the chair could not be controlled. Our approach was to use the results as part of a composite marker for general health and not as a marker for falls risk as has been done by others [25, 26].

3   Results

Figures 6a, 6b and 7 analyses one of the participants skeletal data (acquired by Kinect V2) between 17 Nov. 17 and 5 Jan 2018. For all three figures, the median is shown. Figure 6a illustrates the histogram for hourly subject speed (m/s) when the participant walks in his/her lounge where the Kinect camera is fitted. The number of low speed occurrences (frequency ~ 500) is large. It is due to the participant standing stationary or the acceleration or deceleration phases of the participant's walk.

Figure 6b enlarges the same data presented in Fig. 6a in order to show the distribution of the speed in detail. We observe two sets of similar patterns in this set of data: there are two peaks (~ 0.2 and ~ 0.5 m/s) and there is gradual declination of frequency between ~ 0.2 to ~ 0.4 and ~ 0.5 to ~ 0.8. We are examining whether the peaks represent changes in the participants' general health. One of the peaks when speed is ~ 0.5 m/s for example might occur at times when the participant is walking faster (when the participants muscles are flexible enough and in a good day when s/he is not

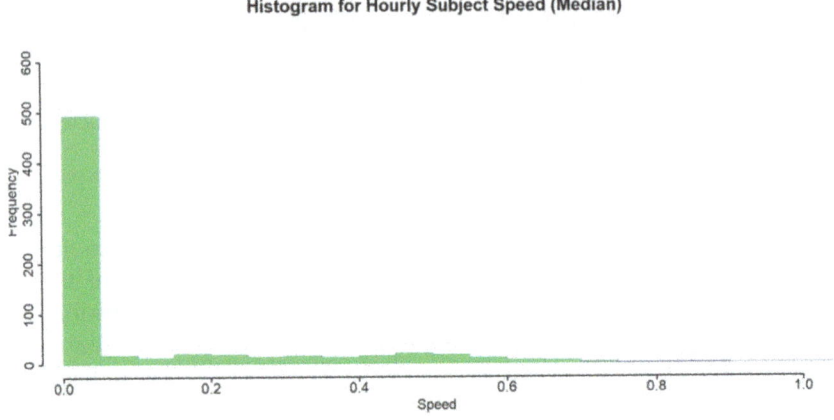

Fig. 6a. Histogram for hourly subject speed over 17 Nov 2017–5 Jan 2018.

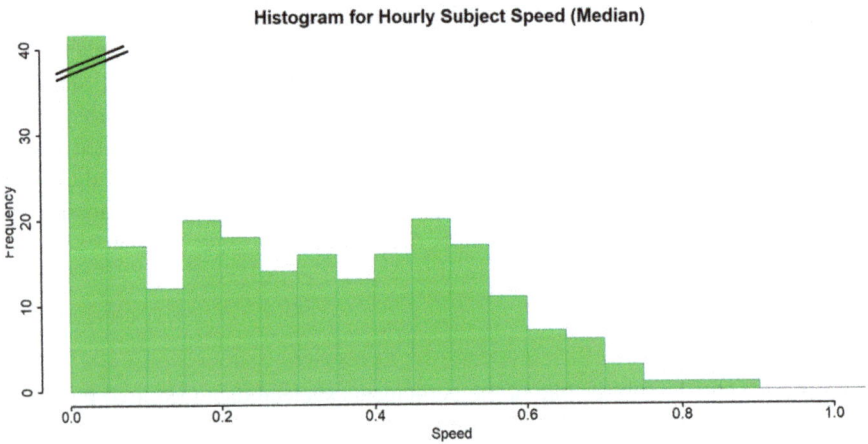

Fig. 6b. Expanded view of the histogram for hourly subject speed over 17 Nov 2017–5 Jan 2018.

feeling any stiffness). Whereas the lower peak might occurs when the participant's health is not as good (when s/he is suffering from a back pain or during mornings when the muscles are not flexible yet).

Figure 7 shows the daily walking speed over the given period and it is calculated based on the hourly subject speed outcomes. Daily subject speed is calculated at 23:59 pm every night and the median of the hours where there is an activity detected by the Kinect is calculated. For example, if the participant is sleeping between 23 pm–7 am in a room where there is no Kinect camera fitted, then is no activity data, hence the first 7 h of the morning will contribute no data. For the same day, there will be maximum of 17 hourly subject speed data points collected (when the participant appears in front of the Kinect at any point during the rest of the day) from which we then calculate the median of the daily subject speed. There are several days where the median daily speed is (close to or equal to) zero and there are days where there daily subject speed is over 0.4 m/s. We believe the major reason for this is that the participant is not spending same amount of time at home, which would mean that the data we have been collecting varies in terms of quantity. We have noted that collecting data in real world situation results in artefacts such as biologically implausible in-home walking speeds (>5 m/s). It is for this reason that we chose to report median values. Our interpretation of the daily outcomes is that if the participant is spending less time in front of the Kinect, it is less likely that we will have good samples because we are likely to have noise and lower accuracy.

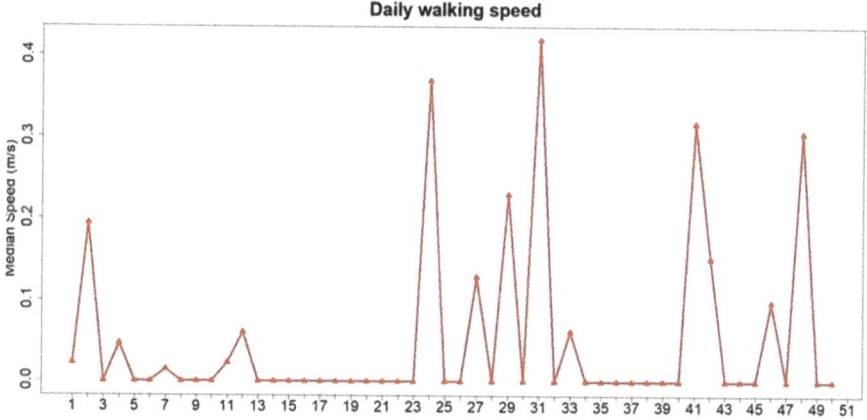

Fig. 7. Daily subject speed over 17 Nov 2017–5 Jan 2018.

In conclusion we have shown proof of concept that the MiiHome system can collect long-term data about selected participant behaviour that can now be analysed for clinical relevance over a larger cohort of participants.

Acknowledgements. The authors would like to thank the group of Salix Homes tenants (represented by George) who were participants in this project. The study was supported by a University of Salford (ethics approval no: STR1617-114), Higher Education Funding Council for England (HEFCE) grant. We would like to thank Salix Homes for their generous contributions including by members of the IT department and IMS department who helped set up the smarthome and install technology into the participants homes, as well as providing the property to create the dedicated smarthome. Partial funding for this study was provided by an award from NHS Salford Clinical Commissioning Group.

References

1. Gulrez, T., et al.: Can autonomous sensor systems improve the well-being of people living at home with neurodegenerative disorders? In: International Conference on Cross-Cultural Design, pp. 649–658. Springer, Cham (2016)
2. Greater Manchester Health and Social Care Devolution Locality Plan for Salford August 2017 March 2018. http://www.salfordccg.nhs.uk/salford-locality-plan
3. Salford Together, March 2018. http://www.salfordtogether.com/
4. Salford's Big Health and Social Care, Conversation, March 2018. http://www.salfordtogether.com/survey-big-health-and-care-conversation/
5. Giebel, C., et al.: Effective public involvement in the HoST-D Programme for dementia home care support: From proposal and design to methods of data collection (innovative practice). Dementia (2017). https://doi.org/10.1177/1471301216687698
6. Caliskanelli, I., Meziani, S.-N., Hodgson, A.: Kinecting Frailty: A Pilot Study on Frailty, in Accepted for the HCI 2018 (2018)
7. Hunt, T.D., et al.: A minimally intrusive monitoring system that utilizes electricity consumption as a proxy for wellbeing. J. Appl. Comput. Inf. Technol. **18**(2) (2014)
8. Chan, M., et al.: A review of smart homes—present state and future challenges. Comput. Methods Progr. Biomed. **91**(1), 55–81 (2008)
9. Sanchez, V., Pfeiffer, C., Skeie, N.-O.: A review of smart house analysis methods for assisting older people living alone. J. Sens. Actuator Netw. **6**(3), 11 (2017)
10. Gill, T., Williams, C., Tinetti, M.E.: Assessing risk for the onset of functional dependence among older adults: the role of physical performance. J. Am. Geriatr. Soc. **43**, 603–609 (1995)
11. Fritz, S.P.L., Lusardi, M.: White paper: "Walking Speed: the Sixth Vital Sign". J. Geriatr. Phys. Ther. **32**(2), 2–5 (2009)
12. Peel, N.M., Kuys, S.S., Klein, K.: Gait speed as a measure in geriatric assessment in clinical settings: a systematic review. J. Gerontol. A Biol. Sci. Med. Sci. **68**(1), 39–46 (2013)
13. Studenski, S.: Bradypedia: is gait speed ready for clinical use? J. Nutrition Health Aging **13**(10), 878–880 (2009)
14. Studenski, S., et al.: Gait speed and survival in older adults. JAMA **305**(1), 50–58 (2011)
15. Van Kan, G.A., et al.: Gait speed at usual pace as a predictor of adverse outcomes in community-dwelling older people an International Academy on Nutrition and Aging (IANA) Task Force. J. Nutrition Health Aging **13**(10), 881–889 (2009)
16. White, D.K., et al.: Trajectories of gait speed predict mortality in well-functioning older adults: the health, aging and body composition study. J. Gerontol. A Biol. Sci. Med. Sci. **68**(4), 456–464 (2013)
17. Briggs, R., O'Neill, D.: Vascular gait dyspraxia. Clin. Med. **14**(2), 200–202 (2014)

18. Guralnik, J.M., et al.: Lower-extremity function in persons over the age of 70 years as a predictor of subsequent disability. N. Engl. J. Med. **332**(9), 556–562 (1995)
19. Guralnik, J.M., et al.: A short physical performance battery assessing lower extremity function: Association with self-reported disability and prediction of mortality and nursing home admission. J. Gerontol. **49**(2), M85–M94 (1994)
20. Orr, R.: Contribution of muscle weakness to postural instability in the elderly. A systematic review. Eur. J. Phys. Rehabil. Med. **46**(2), 183–220 (2010)
21. Moreland, J.D., et al.: Muscle weakness and falls in older adults: a systematic review and meta-analysis. J. Am. Geriatr. Soc. **52**(7), 1121–1129 (2004)
22. Skelton, D.A., Kennedy, J., Rutherford, O.M.: Explosive power and asymmetry in leg muscle function in frequent fallers and non-fallers aged over 65. Age Ageing **31**(2), 119–125 (2002)
23. Lord, S.R., et al.: Sit-to-stand performance depends on sensation, speed, balance, and psychological status in addition to strength in older people. J. Gerontol. A Biol. Sci. Med. Sci. **57**(8), M539–M543 (2002)
24. Janssen, W.G.M., Bussmann, H.B.J., Stam, H.J.: Determinants of the sit-to-stand movement: a review. Phys. Ther. **82**(9), 866–879 (2002)
25. Ejupi, A., et al.: Kinect-based five-times-sit-to-stand test for clinical and in-home assessment of fall risk in older people. Gerontology **62**(1), 118–124 (2016)
26. Ejupi, A., et al.: A Kinect and inertial sensor-based system for the self-assessment of fall risk: a home-based study in older people. Hum. Comput. Interact. **31**(3–4), 261–293 (2016)

Co-designing: Working with Braille Users in the Design of a Device to Teach Braille

Rhianne M. Lopez[1(✉)], Shane D. Pinder[1,2], and T. Claire Davies[1]

[1] BDAT Lab, Mechanical and Materials Engineering,
Queen's University, Kingston, ON K7L 3N6, Canada
{rll4,shane.pinder,claire.davies}@queensu.ca
[2] Department of Physics, Engineering Physics, and Astronomy,
Queen's University, Kingston, ON K7L 3N6, Canada

Abstract. The objective of this research was to develop a "paper-based" prototype design of a device that will eventually be developed to teach braille in a co-design process that involved end-user input throughout the process. Questionnaires aimed to explore the use of assistive technologies to help learn or teach braille. Features of existing assistive technologies were identified by the participants. Taking these features into consideration, seven conceptual design solutions were developed by six designers. A weighted evaluation matrix (WEM) ranked potential designs. A weighting for each design feature was calculated using the frequency of that feature. The responses from each participant group were weighted equally. Two semi-structured interviews were conducted with braille teachers. The design preferred by both the teachers was ranked fifth in the weighted evaluation matrix. Designs that were ranked poorly according to the WEM, were actually ranked highly by the end-users. The co-design process was essential in identifying these differences.

Keywords: Co-design · Braille device · Visual impairment

1 Introduction

According to the World Health Organization (WHO), there are an estimated 285 million people in the world living with visual impairments [1]. In Canada it is estimated that half a million people live with significant vision loss that affects their quality of life [2]. Many people with visual impairments (low vision or blindness) have difficulty gathering information from their surroundings or communicating with others [3–5]. Education, job opportunities and social participation are limited as a result [2–4]. In September of 2016, Digital Learning for Development and All Children Reading released a "Grand Challenge for Development" to explore the barriers that children with sensory disabilities in the Philippines face that impede their ability to learn and read. This "Grand Challenge" identified the need for improved assistive technologies to help children with sensory disabilities learn to read. In the Philippines, close to half a million people are blind and many more have low vision [6]. Often, children in the Philippines with visual impairments (VI) do not have access to the necessary resources needed to assist with learning how to read [7, 8].

© Springer International Publishing AG, part of Springer Nature 2019
T. Z. Ahram and C. Falcão (Eds.): AHFE 2018, AISC 794, pp. 798–807, 2019.
https://doi.org/10.1007/978-3-319-94947-5_78

It is estimated that as much as 30% of assistive devices are abandoned [9]. Abandonment could result if the condition of the individual worsens, but also for more social aspects including a lack of acceptance of the device, a mismatch between the user and the technology and a lack of technical support and training [9–11]. To reduce device abandonment, multiple studies have shown that incorporating end-users in the evaluation, prescription and/or creation of their own devices can lead to lower abandonment rates [10, 12]. Co-design is "an umbrella term covering both 'community design' and 'participatory design'. As such, [co-design is defined as] the effort to combine the views, input and skills of people with many different perspectives to address a specific problem" [13].

Assistive technology is an umbrella term that encompasses both assistive products (or devices) and related services [14]. The International Classification of Functioning, Disability and Health (ICF) defines assistive products or technology as "any product, instrument, equipment or technology adapted or specially designed for improving the functioning of a disabled person" [15]. To help improve the quality of life of persons living with VI, there has been an increase in the development of assistive technologies [16, 17]. Some areas of assistive technology development for the visually impaired or blind include mobility, navigation, object recognition, social interaction and printed information access [8, 16].

Braille is a system, invented by 1809 by Louis Braille, that uses raised dots to allow persons who are visually impaired to read [18, 19]. Letters or symbols are formed within a space called a cell [18, 19]. A full cell consists of six dots, arranged in a formation of two columns and three rows [18, 19]. Depending on which dots are raised, symbols representing either an alphabet letter, number, punctuation mark, or whole word can be formed [18, 19]. To read braille, readers gently glide their fingers over a surface, such as paper, embossed with braille to feel the patterns of raised dots [18, 19].

In 2010, it was reported that in Canada, only 10% of persons with visual impairments could read braille [20]. The lack of braille education is partially attributed to several social factors. In 2012, Rogers created a survey for pupils who were learning braille and print simultaneously. Rogers' findings revealed that it was unlikely for some pupils to use braille because it would make them appear different from their peers. Another factor leading to the rejection of braille was family influences. Some parents of children learning braille thought of braille as an "outward sign of [their children's] difficulties and wanted [their children] to be 'normal'." For children with parents who did not support braille learning, there was slow progress when learning braille which sometimes led to the full rejection of braille [21].

From the perspective of persons with vision impairment in the Philippines and the call for support through All Children Reading, "Literacy unlocks human potential and is the cornerstone of development. It leads to better health, broadens employment opportunities, and creates safer and more stable societies" [22]. The lack of reading skills development of children during primary education often results in poor educational progress and eventually, limited economic opportunities [22]. A study by Ryles suggests that if children learn at a young age to use braille as a primary means of communication, they develop lifelong reading skills which can lead to better employment opportunities [23].

Russomanno *et al.* stated that for blind computer users, the preferred method of "reading" electronic text is through speech because "it is relatively inexpensive (or even free) when compared to braille and requires no additional hardware" [24]. However, this method of "reading" may not be optimal due to mind wandering. Varao-Sousa *et al.* described mind wandering as the process where a reader finds his or her mind wandering to thoughts unrelated to the text they are reading [25]. In another study by the same authors, findings showed that participants experienced less mind wandering when engaged in active modes of reading such as reading aloud in comparison to listening [26]. Thus, Russomanno suggested that active reading of braille could be better for information transmission for a blind reader and could lead to better comprehension of text compared to plain listening [24].

Unfortunately for blind or visually impaired persons in the Philippines hoping to learn to read, there is a lack of accessibility to specialized materials necessary for this process [8]. This is a problem because "the inability for a blind person to read is one of their greatest disappointments" [8]. ACR GCD believes that many of the barriers that children with sensory disabilities face in regards to learning how to read could be addressed through technology-enabled innovations [27]. However, what are these barriers that children with visual impairments in the Philippines and Canada are facing in regards to using existing devices that help them learn to read? Also, how do the barriers of the people of the Philippines compare to the barriers that persons with visual impairments in Canada face? The overall aim of this research project is to develop a device that can be used by children in the Philippines to learn to read in braille. The first research questions in this process include:

(i) what are the needs of the individuals when learning braille?
(ii) what devices are currently used in the Philippines?
(iii) can the co-design process enable us to develop a design to meet the require needs?

2 Methods

This section showcases the process conducted by the study investigator to develop and choose a design to prototype. The steps taken include creating and distributing a preliminary questionnaire, using QFD matrices and the Kano method, engaging in the parallel design process, using a weighted evaluation matrix and finally conducting semi-structured interviews.

2.1 Questionnaires

Questionnaires that aimed to identify the barriers that children with visual impairment in the Philippines and Canada face regarding using devices that help them learn to read were distributed to both countries through project partners.

Ethics approval for this questionnaire was granted from Queen's University's General Research Ethics Board under the Project Title: GMECH-040-16 Technology-enabled innovations to assist children with visual impairments or blindness in the Philippines.

Three different questionnaires were created to determine the barriers that people in the Philippines and Canada face in regards to using existing devices that help them learn to read. The three questionnaires were similar to each other, but phrased differently to target each participant group. All questionnaires comprised of four main sections: General Information, Specialized Schools or Programs, Reading Environment and Reading Devices. For the questionnaire for teachers, there was an additional section: Teaching Reading to Persons with visual impairment.

2.2   Participants

Participants were recruited by partners in the Philippines and Canada who worked with persons with visual impairment. Letters of Information for the study were given to these partners, and if these partners believed an individual qualified for the study, the partners would show them the Letter of Information and allow the individual to decide whether to participate.

Questionnaires were distributed in the Philippines as the need for improved assistive technologies was identified by ACR GCD. However, as the study investigator was located in Canada, questionnaires were sent to Canadians as well.

Individuals were invited to complete the questionnaires if they were either:

- below the age of 18, living with visual impairment and learning how to read;
- any age, living with visual impairment and had already learned how to read;
- or teaching reading to persons with visual impairment (e.g. teachers, parents of children with VI, etc.)

3   Results and Analysis

3.1   Questionnaires

The questionnaires were collated and needs were drawn from the results. Table 1 shows the frequency with which each identified need was mentioned.

Table 1. Frequency of questionnaire responses.

Device needs	Total
Accessibility	9
• Affordable	22
• Available in the country	7
	38
Portability	10
• Not bulky	8
• Lightweight	5
• Good battery life	5
• Does not need to be plugged in	1

(*continued*)

Table 1. (*continued*)

Device needs	Total
	29
Durability	7
• Hardware	0
– Not fragile	3
– Does not rip easily	1
• Software	0
– No technical errors/malfunctions	3
– No viruses	3
– No upgrades required	1
	18
Usability (Easy to use)	7
• Adjustable letter size/zoom feature	11
• Adjustable font	7
• Convenient to use/handy	5
• Adjustable colour contrast	4
• Can get work done easily	3
• Easy to navigate/good layout	1
• Anti-glare	1
	39
Function	0
• Feedback	3
– Can read Tagalog/any language	2
– Good pronunciation	2
– Talks back (text to Braille, or Text to Speech)	3
– Descriptive/Emotive speech	1
• Effective learning tool	3
– Tactile learning feature/pictorial	2
• Multifunctional	7
– Screen Reader/Reads content	12
– Dictionary/Thesaurus	4
– Info accessible through multiple devices	2
– Internet	1
– Helps locate research on computer	1
• Reliable	1
	48

3.2 Quality Function Deployment

The QFD method identified device needs from the questionnaires and prioritized them among and between participant groups. Three QFD matrices were developed, one for each participant group. For each QFD matrix, the identified device (customer requirements) were mapped to the technical specifications and identified with either strong, moderate or weak relationships. A high technical specification score (relative to other technical specification scores) indicated targets to increase customer satisfaction.

3.3 Kano Analysis

The Kano model was used to obtain a visual representation of how each identified device need would contribute to customer satisfaction. Device needs were distributed in one of four groups relating to varying customer satisfaction. These groups were: Attractive, One-Dimensional, Indifferent and Must Haves. Attractive needs add satisfaction to the customer, but their absence does not detract from customer satisfaction [28]. One-dimensional needs can either satisfy or dissatisfy the customer based on whether or not they are present [28]. Indifferent features neither add nor detract from the satisfaction of the customer [29]. Must Haves are needs that on their own cannot satisfy a customer, but will lead to customer dissatisfaction if they are missing [28]. The Kano model is shown in Fig. 1.

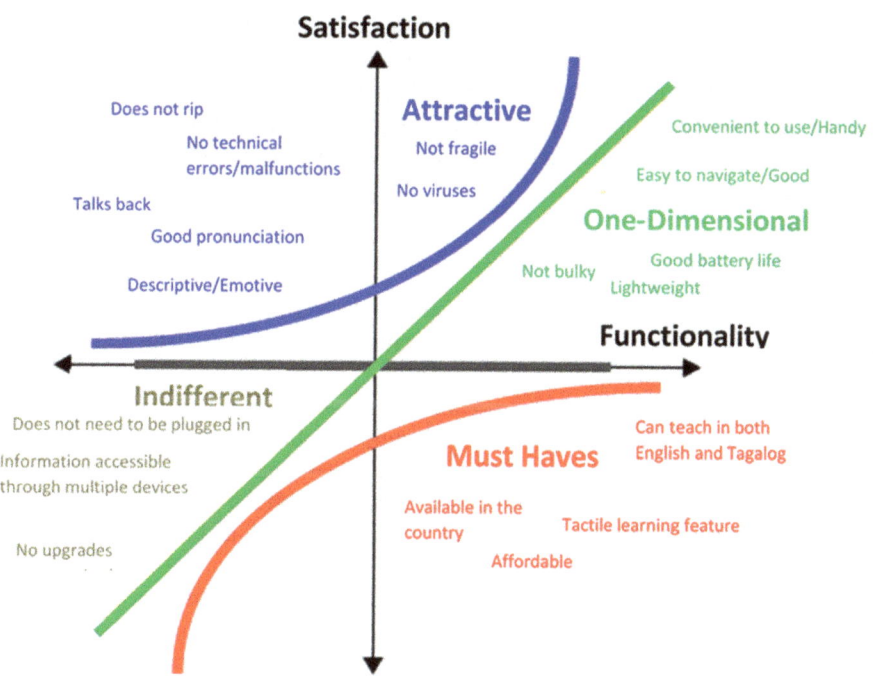

Fig. 1. Kano model of device needs identified by participants

4 Parallel Design Process

Both the QFD and the Kano were presented to a design team and informed the parallel design process. When presenting the results of the Kano model to her lab group, the research investigator explained how all device needs in the group 'Must haves' were necessary for any of the ideas to be brainstormed. All other device needs could increase user satisfaction, but were not necessary for a final device. Seven individuals participated in the brainstorming session. From this session, 17 unique device ideas were identified.

After the preliminary device ideas had been presented, the study investigator, and the members of the BDAT Laboratory brainstormed new device ideas using the best qualities from the preliminary devices. Seven "merged devices" were established.

4.1 Weighted Evaluation Matrix

A weighted evaluation matrix was used to determine which conceptual design would best satisfy the needs identified. Weightings of each need were calculated based on how often they were mentioned in the questionnaires. Scores of 0-2 were given to each device for each need depending on how well the device would be able to satisfy each need. A score of 0 meant that theoretically the device would not be able to meet the need, a score of 1 meant that it would somewhat meet the need and a score of 2 meant that it would be able to meet the need. Following the assignment of scores, each score was multiplied by the weighting of each need, and total scores for each design were calculated.

4.2 Semi-structured Interviews

Concurrently, semi-structured interviews were conducted with two braille teachers, one in the Philippines (Maria) and one in Canada (Susan) to identify how they teach braille and to aid in the identification of the best design.

When talking with Maria, she explained that students first has to develop tactile skills. They feel different fabric pieces to identify similarities and differences. After tactile skills are sufficiently developed, the students learn the braille alphabet using toy eggs and egg cartons. Once the students were familiar with the different braille letters, they learned how to use a slate and stylus to write in braille and a brailler to type in braille.

The semi-structured interview with Susan revealed similarities regarding how children with visual impairment develop braille skills when learning to read between the Philippines and Canada. Children taught by Susan were first trained to recognize shapes and textures. Once her students had sufficiently developed their tactile skills, they learned the braille alphabet. Susan used a paint pallet with six holes (three rows, two columns) and small balls to teach braille. The balls, when placed in the pallet in specific arrangements, would represent the dots of a braille cell for the different letters. This activity was the same as the egg carton activity explained by Maria.

4.3   Design Selection

Based on the weighted evaluation matrix, the investigator determined that the design in Fig. 2a would be most beneficial and acceptable to the end users. However, this was in conflict with the suggestions made during the interview process. Both interviewees preferred device b as shown in Fig. 2. This disparity shows the importance of obtaining feedback on a regular basis throughout the design process.

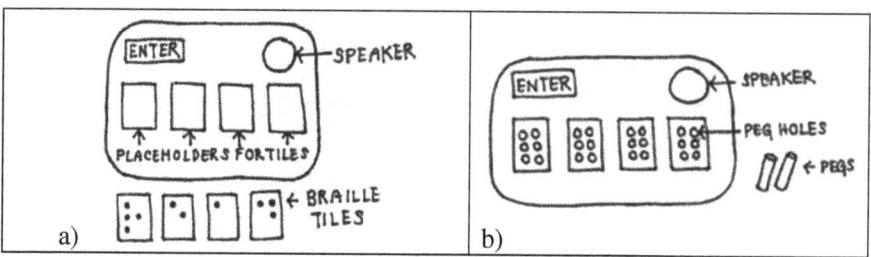

Fig. 2. The designs chosen through the weighted evaluation matrix and the design preferred by teachers of braille.

5   Discussion

The study investigator, end-users and other stakeholders were involved in each step of the co-design process. Results from the questionnaires and semi-structured interviews all contributed to the creation and selection of the designs. Of the 42 participants from the Philippines who completed the questionnaire, 17 were blind. It was determined that to best meet the needs of these participants, the device should be able to teach braille. Forty-three device needs were identified by the participants in the Philippines. These needs were sorted into five categories: accessibility, portability, durability, usability and function.

To inform the parallel design process, QFD and Kano was used. To satisfy the greatest number of and/or most important device needs identified by the participants, the technical specifications that needed to be prioritized included 'cost', 'engagement level' and 'material strength'. Although there were many device needs identified from the questionnaires, using the Kano model allowed the researcher to determine which ones to prioritize.

When comparing the results from the WEM and the semi-structured interview, it was observed that the design rankings were different. This suggests the importance of including the stakeholders in the design process and strengthens the importance of co-design. There are several advantages to involving users in the design process. For example, findings by Kujala found that the co-design process could reduce the number of iterations needed; improve the levels of acceptance of the design; and reduce the time and cost of development through the identification of design flaws early in the process [30]. Additional advantages included better concept generation and an overall increase in user satisfaction [11, 31].

Ideally, end-users would be engaged in all parts of the design process including the creation of the QFD matrices, WEM and prototyping stages. However, additional time and monetary costs necessary for these engagements to occur reduce the ability to include end-users in all stages of the process. Thus, there is a trade-off between how much input can be sought and integrated.

6 Conclusions

Overall, findings from this study revealed that engaging in the co-design process can help reduce designer biases regarding the designer's interpretation of what the user wants. It is recommended that designers, end-users and other stakeholders be involved in the device development process to ensure usable, fit-for-use devices that stakeholders actually want, are created.

Acknowledgments. This project was funded by the Natural Sciences and Engineering Research Council of Canada, NSERC RGPIN-2016-04669.

References

1. WHO | World Health Organization: WHO (2017)
2. CNIB: Fast Facts about Vision Loss (2015). http://www.cnib.ca/en/about/media/vision-loss/pages/default.aspx. Accessed 13 Sep 2017
3. World Health Organization: Socio economic aspects of blindness and visual impairment. WHO (2017)
4. Naraine, M.D., Lindsay, P.H., Cdso, F.: Social inclusion of employees who are blind or low vision. Disabil. Soc. **26**(4), 389–403 (2011)
5. World Health Organization: Global Initiative for the Elimination of Avoidable Blindness (2007)
6. Resources for the Blind: Blindness in the Philippines (2017). http://blind.org.ph/blind_phil.html. Accessed 06 Sep 2017
7. All Children Reading: A Grand Challenge for Development. https://docs.google.com/forms/d/e/1FAIpQLSfSzC9Q-N4AFWb9NSIBAVSK0Vvh4FeprFHmVNSQqcbBP4397g/viewform
8. Resources for the Blind: Rehabilitation | Training and Rehabilitation (2017). http://blind.org.ph/projprog/projprog_tr_r.html. Accessed 06 Sep 2017
9. Phillips, B., Zhao, H.: Predictors of assistive technology abandonment. Appl. Res. Assist. Technol. **5**, 36–45 (1993)
10. Verza, R., Carvalho, M.L., Battaglia, M., Uccelli, M.M.: An interdisciplinary approach to evaluating the need for assistive technology reduces equipment abandonment (2006)
11. Choi, Y.M.: Utilizing end user input in early product development. Procedia Manuf. **3**, 2244–2250 (2015)
12. da Costa, C.R., Ferreira, F.M.R.M., Bortolus, M.V., Carvalho, M.G.R.: Dispositivos de tecnologia assistiva: fatores relacionados ao abandono. Cad. Ter. Ocup. da UFSCar **23**(3), 611–624 (2015)

13. Bradwell, P., Marr, S.: Making the most of collaboration an international survey of public service co-design making the most of collaboration: an international survey of public service co-design (2008)
14. Borg, J., Berman-Bieler, R., Khasnabis, C., Mitra, G., Myhill, W.N., Raja, D.S., Burlyaeva-Norman, A., Chadha, S., Cormency, C., Cote, A., Farkas, A., Mariotti, S., Mirza, Z., Mitra, S., Moller, H., Alarcos Cieza Moreno, M., Mukherjee, A.K., Mullally, S., Olusanya, B.O., Pupulin, A., Sabbe, L., Shankar, A., Steenbeek, M., Tucker, M.: Assistive Technology for Children with Disabilities Assistive Technology for Children with Disabilities: Creating Opportunities for Education, Inclusion and Participation A discussion paper
15. ICF: Chapter 1 Products and Technology. Available: http://apps.who.int/classifications/icfbrowser/. Accessed 07 Sep 2017
16. Bhowmick, A., Hazarika, S.M., Bhowmick, B.A.: An insight into assistive technology for the visually impaired and blind people: state-of-the-art and future trends. J. Multimodal User Interfaces **11**, 149–172 (2017)
17. Chakraborti, P., Toprakci, H.A.K., Yang, P., Di Spigna, N., Franzon, P., Ghosh, T.: A compact dielectric elastomer tubular actuator for refreshable Braille displays. Sens. Actuators A Phys. **179**, 151–157 (2012)
18. What Is Braille? - American Foundation for the Blind (2017). http://www.afb.org/info/living-with-vision-loss/braille/what-is-braille/123. Accessed 30 Aug 2017
19. CNIB - About the Braille System (2017). http://www.cnib.ca/en/living/braille/braille-system/Pages/default.aspx. Accessed 30 Aug 2017
20. Mulholland, A.: Few blind Canadians reading braille | CTV News (2010). http://www.ctvnews.ca/with-new-technology-few-blind-canadians-read-braille-1.503149. Accessed 15 Sep 2017
21. Rogers, S.: Learning Braille and print together - the mainstream issues. Br. J. Vis. Impair. **25**, 120–132 (2012)
22. All Children Reading, "The Problem of Literacy" (2018). https://allchildrenreading.org/about-us/problem/. Accessed 07 Sep 2017
23. Ryles, R.: The impact of braille reading skills on employment, income, education, and habuts. J. Vis. Impair. Blind. **90**(3), 219 (1996)
24. Russomanno, A., O'modhrain, S., Gillespie, R.B., Rodger, M.W.M.: Refreshing Refreshable Braille Displays (2015)
25. Varao-Sousa, G.J., Kingston, T.L., Solman, A.: Re-reading after mind wandering. Can. J. Exp. Psychol. **71**(3), 203 (2017)
26. Varao Sousa, T.L., Carriere, J.S.A., Smilek, D.: The way we encounter reading material influences how frequently we mind wander. Front. Psychol. **4**, 892 (2013)
27. Resources for the Blind: All Children Reading: A Grand Challenge for Development (2018). http://allchildrenreading.org/innovators/resources-for-the-blind-inc/
28. Kano Model Tutorial – ASQ. http://asq.org/learn-about-quality/qfd-quality-function-deployment/overview/kano-model.html. Accessed 02 Aug 2017
29. The Complete Guide to the Kano Model - Folding Burritos. https://foldingburritos.com/kano-model/. Accessed 02 Aug 2017
30. Kujala, S.: User involvement: a review of the benefits and challenges (2003)
31. Lowdermilk, T.: User-Centered Design: A Developer's Guide to Building User-Friendly Applications. O'Reilly Media Inc., Sebastopol (2013)

Reducing Scanning Keyboard Input Errors with Extended Start Dwell-Time

Frode Eika Sandnes[1,2（✉）], Evelyn Eika[1], and Fausto Orsi Medola[3]

[1] Faculty of Technology, Art and Design, Department of Computer Science,
OsloMet–Oslo Metropolitan University, Oslo, Norway
{frodes,Evelyn.Eika}@hioa.no
[2] Westerdals Oslo School of Art, Communication and Technology,
Oslo, Norway
[3] UNESP-São Paulo State University, Bauru, Brazil
fausto.medola@faac.unesp.br

Abstract. Some individuals with reduced motor function rely on scanning keyboards to operate computers. A problem observed with scanning keyboards is that errors typically occur during the first group or first cell of a group. This paper proposes to reduce such errors by introducing longer dwell-times for the first element in scan sequences. The paper theoretically explores several designs and evaluates their effect on overall text entry performance.

Keywords: Scanning keyboards · Motor disability · Dwell-time

1 Introduction

Conventional usage of computers usually comprises users' actions of typing and clicking. To perform such task in a satisfactory and efficient way, a certain level of motor control provided by sensory-motor integration is needed. Individuals with high levels of movement impairment have limited access to the independent and efficient use of a computer. In a study with individuals with quadriplegia due to spinal cord injury (spine levels of C5 to C7), it was found that high levels of motor impairment limit the independence and efficiency in computer usage not only in typing tasks, but also in the operation of mouse, cables, and accessories [1]. The current study focuses on text entry.

Individuals with motor disabilities may be unable to enter text using conventional keyboards. One common aid is scanning input [2], which allows the user to input text and control computer with just a single switch. The principle of scanning keyboards is that the characters are displayed in a regular grid. These elements are then highlighted in turn for a time interval known as the dwell-time; when an element is highlighted, it can be selected by the user by using the switch. One practical scanning pattern is to first scan through the row, followed by scanning through the cells of a selected row. With such a configuration, the user selects a character in two steps.

One problem that has been observed in the literature is that users typically make more mistakes with the first element of a scan, being it the first row or the first cell of a row [3].

© Springer International Publishing AG, part of Springer Nature 2019
T. Z. Ahram and C. Falcão (Eds.): AHFE 2018, AISC 794, pp. 808–817, 2019.
https://doi.org/10.1007/978-3-319-94947-5_79

One possible explanation for this could be that the elements usually are scanned with the same dwell-time for each element; since the dwell-time is the bottleneck in scanning keyboards, one usually tries to keep the dwell time as small as possible. However, research into text input rhythmic patterns has found that text is often inputted according to some rhythmic pattern [4], and that there typically is a longer delay between characters than for the steps within characters [5]. One possible cause is that users may need a small mental break- or pause between consecutive characters. However, the scan keyboard paradigm is different to non-scanning methods in that the system drives the text entry, not the user. As soon as a character is input, the new character input starts immediately without pause.

This theoretical study proposes to introduce a longer dwell-time for the first element of a scan sequence. Based on empirical results from the literature, this study models the effect of introducing such a delay on both error rates and text entry performance.

2 Background

The design of universally accessible technologies [6] such as self-service kiosks in public spaces [7–9] and smart home technologies [10] focus on general characteristics of colour contrast [11–13] and text readability [14–16]. Such efforts also include the design of assistive technologies for specific groups. Broadly speaking, assistive technologies can be divided into those that address sensory disabilities including low vision [17–19], blindness [20–22], mental disabilities [23] such as dyslexia [24–26], and motor disabilities [27, 28]. One iconic assistive technology for mobility impairments is the wheelchair. Wheelchair mobility research typically focuses on manual wheelchair design [29, 30], handrim-activated power assisted wheelchair design [31, 32], wheelchair configuration [33], handrim design [34, 35], hand pressure [36], and manoeuvrability [37].

There are several approaches for assisting individuals with reduced motor function to input text on computers such as chording [38–40], tapping [41, 42], menu driven text entry [43], ambiguous keyboards [44], text prediction [45], adjusting the keyboard layout [46], and scanning keyboards [47–49]. One advantage of scanning keyboards is that they support recognition rather than recall [50, 51].

Scanning keyboards are slow compared to other types of text entry techniques and most of the research attention has thus focused on optimizing the virtual keyboard layout [52–54]. Errors then become even more costly and critical. Several studies have addressed issues involving errors [50, 55]. It has been found that specific keyboard based error-correcting mechanisms such as reverse-scanning-direction buttons and abort buttons are not effective [55].

One challenge with scanning input research is to obtain disabled participants and having too short experimental sessions [55, 56]. Consequently, researchers have proposed theoretic models [3, 56, 57]. This study exploits the scan steps per character measure introduced by MacKenzie [57] and a model for errors based on empirical data proposed by Francis [3].

3 Method

Francis [3] proposed a model for the probability of making a correct decision for a given location on a virtual keyboard. This model was based on empirical data from a between-subjects experiment with 3×60 students which had to select numbers on a 8×8 grid of numbers with dwell-times of 200, 350, and 500 ms. Slightly longer dwell-times were used in a study of a four-key ambiguous keyboard [44] ranging from 700 to 1100 ms. The results were analysed using a genetic programming algorithm, which yielded the following expression:

$$p(correct) = a + b \left[\sqrt{R} + \sqrt{C} + \frac{9(10)^{1/4}}{\sqrt{D + (RC/3)^{1/4}}} \right]^{1/4} \tag{1}$$

where R and C are the row and column numbers, D is the dwell time and a and b are constants with the values of 23.43 and -11.40, respectively. In simple terms, Francis observed that error rates were higher for all the elements along the first row and the first column compared to the other cells in the matrix. One explanation for this is that users get too little time to make the selection when the cursor is first displayed on the left side.

Since the first row is missed in a similar manner, all the cells in this row are missed. This study refers to Eq. (1) as $P(C, R, D)$ since the probability of success can be considered a function of C, R, and D.

Clearly, the scanning input bottleneck is the dwell-time since the total time to input a character is the sum of all the dwell-times for all the individual scan steps. One useful measure is the scan-steps-per-character (SPC) [22]

$$SPC = \sum_i f_i s_i \tag{2}$$

where f is the frequency of character i, and s is the number of scanning steps to reach character i. This measure assumes uniform dwell-times. This phenomenon is illustrated in Fig. 1, which shows how the text entry speed in terms of word per minute changes as function of dwell-time. A rough estimate of a lower bound for the words-per-minute was estimated using

$$WPM = \frac{60}{5D \times SPC} \tag{3}$$

Clearly, shorter dwell-times give faster text entry speeds, while the decrease in text entry speed becomes less prominent with larger delays.

The analysis in this study is based on a QWERTY scanning keyboard due to users' familiarity with QWERTY [58]. The scanning keyboard comprises six groups, namely QWERT, YUIOP, ASDFG, HJKL, ZXCVB, and NM. The two-step left-to-right scanning procedure [59] is illustrated in Fig. 2. Although the analysis is based on this QWERTY design, the general patterns will apply similarly to other layouts.

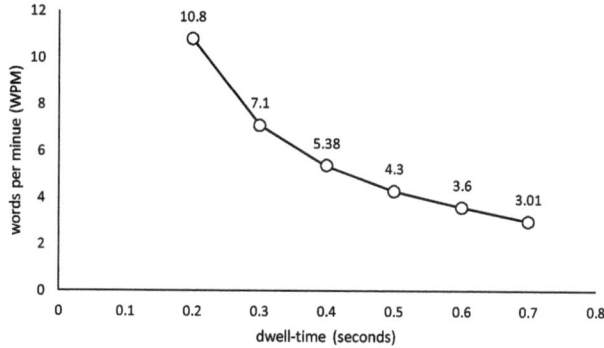

Fig. 1. Productivity in terms of words per minute as a function of uniform dwell-time with a QWERTY scanning keyboard.

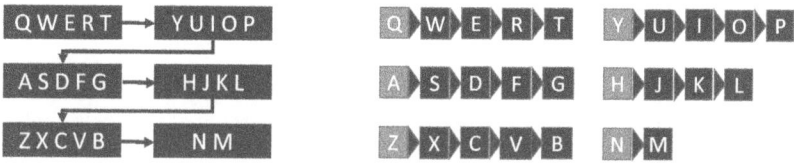

Fig. 2. Scanning keyboard: group sequence (left), cell sequence (right).

This study assumes that the probability of success can be increased by making the dwell-times of the first elements longer. Using Francis' model [3] in Eq. (1), the probabilities of success were plotted as a function of dwell-times for the first element in the first group (QWERT) or $P(1, R, D)$, second group (YUIOP) or $P(2, R, D)$ and third group (ASDFG) or $P(3, R, D)$, respectively (see Fig. 3). All the plots show that the probability of success associated with the first cell in the group (black) is lower than the success probabilities for the remaining cells (grey). Moreover, Fig. 3 shows that the success probabilities for the first groups are lower than the probabilities for subsequent groups.

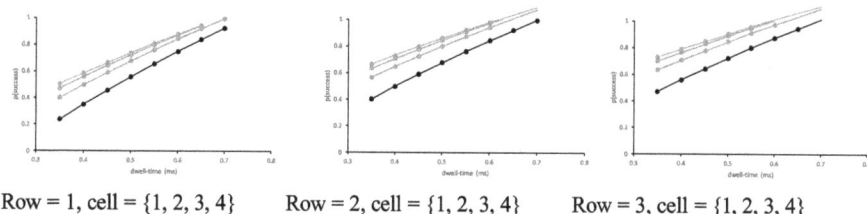

Row = 1, cell = {1, 2, 3, 4}   Row = 2, cell = {1, 2, 3, 4}   Row = 3, cell = {1, 2, 3, 4}

Fig. 3. Probability of success as a function of dwell-time for the first four elements at column 1, 2 and 3 respectively.

By using Francis' model [3] in Eq. (1) the dwell-times needed to achieve probabilities of success at levels of 0.8 and 0.9 are estimated and shown in Fig. 4, namely

$$D = Q(P, R) \tag{4}$$

where D is the dwell-time given by the function Q of the probability level P and cell number R. To achieve a success probability of 0.9, the first element needs a dwell time of 0.7 s, the second element needs a dwell time of about 0.65 s and the remaining elements need dwell times of just over 0.6 s.

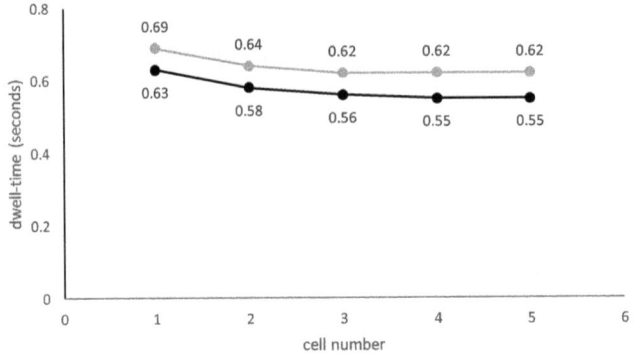

Fig. 4. Dwell-time as a function of cell number with success probabilities of 0.8 (black) and 0.9 (grey), respectively.

Note that these predictions are based on a mathematical model that is a best-fit of the empirical data. One may make a simplification where only the first element is associated with a longer dwell-time and the remaining elements are assigned a shorter dwell-time, based on the theory of mental break between consecutive characters. With such simplifications, at what levels should the first and the subsequent dwell-times be set to achieve a text entry speed? Using Eqs. (2) and (3) we can redefine SPC in terms of two dwell times D_1 and D_2:

$$SPC = \sum_{i \in C_1} f_i s_i + \sum_{j \in C_2} f_j s_j \tag{5}$$

where C_1 and C_2 are the set of characters with dwell times of D_1 and D_2, respectively. The two sums can be considered constant since they are independent of the dwell time.

$$SPC = K_1 + K_2 \tag{6}$$

If we substitute this into the equation for *WPM*, we get

$$WPM = \frac{60}{5(D_1 K_1 + D_2 K_2)} \tag{7}$$

Solving for D_1 we get

$$D_1 = \frac{60}{5 \times WPM \times K_1} - \frac{K_2}{K_1} D_2 \tag{8}$$

By simplifying the expression in terms of constants, it is apparent that there is a linear relationship

$$D_1 = \frac{1}{WPM} c_1 - c_2 D_2 \tag{9}$$

Clearly, the relationship between D_1 and D_2 is independent of the WPM and is linear with a negative slope. The slope is the ratio of the SPCs for the two-character groups. Moreover, the range is inversely proportional to the WPM.

Figure 5 shows the function of dwell-time of the first cell plotted against the dwell-times for the remaining cells along a scan for 3.01, 4.3, and 5.3 words per minute, respectively. As seen, the higher text entry speeds, the shorter the dwell times. Moreover, the shorter the dwell times, the shorter the range of variability.

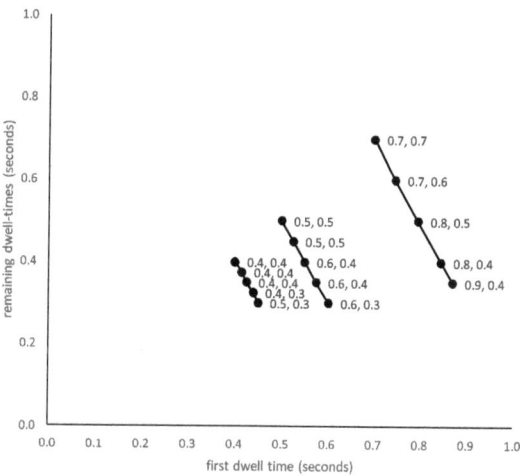

Fig. 5. The trade-off between the length of the dwell-time for the first cell and remaining cells for different speeds (3.01, 4.3, and 5.3 words per minute).

Figure 4 shows that for the QWERTY layout, there is an approximate 2:1 relationship between the dwell-time of the first cell versus remaining cells. That is, to increase the dwell time of the first cell, the dwell time of each of the remaining cells are halved similarly. Sincethere must be a lower bound on the dwell-time for the remaining cells, say 0.3 s, there is also an upper bound on the maximum dwell-time for the first cell.

4   Conclusions

This study explored the effects of extending the dwell-time for the first elements in a sequence during scanning keyboard text entry. By extending the dwell-time on the first element, it is assumed that the probability of successful decision for these first elements increases, as the user is given more time to get ready for inputting the new character. The model shows that it is possible to increase the dwell-time of the first element without sacrificing performance if the remaining dwell-times are reduced. For the QWERTY design explored, the increase in dwell-time of the first element must therefore be compensated by a half reduction in dwell-time for each of the remaining elements. As with most theoretical studies [60–63], the theoretical results herein must be confirmed with observations of actual users.

References

1. Medola, F.O., Lanutti, J., Bentim, C.G., Sardella, A., Franchinni, A.E., Paschoarelli, L.C.: Experiences, problems and solutions in computer usage by subjects with tetraplegia. In: Marcus, A. (ed.) Design, User Experience, and Usability: Users and Interactions. DUXU 2015. LNCS, vol. 9187, pp 131–137. Springer, Cham (2015)
2. Polacek, O., Sporka, A.J., Slavik, P.: Text input for motor-impaired people. Univ. Access Inf. Soc. 16, 51–72 (2017)
3. Francis, G., Johnson, E.: Speed–accuracy tradeoffs in specialized keyboards. Int. J. Hum. Comput. Stud. 69, 526–538 (2011)
4. Sandnes, F.E., Jian, H.L.: Pair-wise variability index: evaluating the cognitive difficulty of using mobile text entry systems. In: International Conference on MobileHCI 2004. LNCS, vol. 3160, pp. 347–350. Springer, Heidelberg (2004)
5. Sandnes, F.E.: Human performance characteristics of three-finger chord sequences. Procedia Manuf. 3, 4228–4235 (2015)
6. Whitney, G., Keith, S., Bühler, C., Hewer, S., Lhotska, L., Miesenberger, K., Sandnes, F.E., Stephanidis, C., Velasco, C.A.: Twenty five years of training and education in ICT design for all and assistive technology. Technol. Disabil. 3, 163–170 (2011)
7. Sandnes, F.E., Jian, H.L., Huang, Y.P., Huang, Y.M.: User interface design for public kiosks: an evaluation of the Taiwan high speed rail ticket vending machine. J. Inf. Sci. Eng. 26, 307–321 (2010)
8. Hagen, S., Sandnes, F.E.: Toward accessible self-service kiosks through intelligent user interfaces. Pers. Ubiquit. Comput. 14, 715–721 (2010)
9. Hagen, S., Sandnes, F.E.: Visual scoping and personal space on shared tabletop surfaces. J. Ambient Intell. Humaniz. Comput. 3, 95–102 (2012)
10. Sandnes, F.E., Herstad, J., Stangeland, A.M., Orsi Medola, F.: UbiWheel: a simple context-aware universal control concept for smart home appliances that encourages active living. In: Proceedings of Smartworld 2017, pp. 446–451. IEEE (2017)
11. Sandnes, F.E.: Understanding WCAG2.0 color contrast requirements through 3D color space visualization. Stud. Health Technol. Inform. 229, 366–375 (2016)
12. Sandnes, F.E., Zhao, A.: A contrast colour selection scheme for WCAG2. 0-compliant web designs based on HSV-half-planes. In: Proceedings of SMC 2015, pp. 1233–1237. IEEE (2015)

13. Sandnes, F.E., Zhao, A.: An interactive color picker that ensures WCAG2.0 compliant color contrast levels. Procedia-Comput. Sci. **67**, 87–94 (2015)
14. Eika, E.: Universally designed text on the web: towards readability criteria based on antipatterns. Stud. Health Technol. Inform. **229**, 461–470 (2016)
15. Eika, E., Sandnes, F.E.: Assessing the reading level of web texts for WCAG2.0 compliance-can it be done automatically? In: Di Bucchianico, G., Kercher, P. (eds.) Advances in Design for Inclusion. Advances in Intelligent Systems and Computing, vol. 500, pp. 361–371. Springer, Cham (2016)
16. Eika, E., Sandnes, F.E.: Authoring WCAG2.0-compliant texts for the web through text readability visualization. In: Antona, M., Stephanidis, C. (eds.) UAHCI 2016. LNCS, vol. 9737, pp. 49–58. Springer, Cham (2016)
17. Sandnes, F.E.: Designing GUIs for low vision by simulating reduced visual acuity: reduced resolution versus shrinking. Stud. Health Technol. Inform. **217**, 274–281 (2015)
18. Sandnes, F.E.: What do low-vision users really want from smart glasses? faces, text and perhaps no glasses at all. In: Miesenberger, K., Bühler, C., Penaz, P. (eds.) ICCHP 2016. LNCS, vol. 9758, pp. 187–194. Springer, Cham (2016)
19. Sandnes F.E., Eika, E.: Head-mounted augmented reality displays on the cheap: a DIY approach to sketching and prototyping low-vision assistive technologies. In: Antona, M., Stephanidis, C. (eds.) UAHCI 2017, LNCS, vol. 10278, pp. 168–186, Springer, Cham (2017)
20. Gomez, J.V., Sandnes, F.E.: RoboGuideDog: guiding blind users through physical environments with laser range scanners. Procedia Comput. Sci. **14**, 218–225 (2012)
21. Sandnes, F.E., Tan, T.B., Johansen, A., Sulic, E., Vesterhus, E., Iversen, E.R.: Making touch-based kiosks accessible to blind users through simple gestures. Univ. Access Inf. Soc. **11**, 421–431 (2012)
22. Lin, M.W., Cheng, Y.M., Yu, W., Sandnes, F.E.: Investigation into the feasibility of using tactons to provide navigation cues in pedestrian situations. In: Proceedings of the 20th Australasian Conference on Computer-Human Interaction: Designing for Habitus and Habitat, pp. 299–302. ACM (2008)
23. Sandnes, F.E., Lundh, M.V.: Calendars for individuals with cognitive disabilities: a comparison of table view and list view. In: Proceedings of the 17th International ACM SIGACCESS Conference on Computers & Accessibility, pp. 329–330. ACM (2015)
24. Berget, G., Mulvey, F., Sandnes, F.E.: Is visual content in textual search interfaces beneficial to dyslexic users? Int. J. Hum.-Comput. Stud. **92–93**, 17–29 (2016)
25. Berget, G., Sandnes, F.E.: Do autocomplete functions reduce the impact of dyslexia on information searching behaviour? A case of Google. J. Am. Soc. Inf. Sci. Technol. **67**, 2320–2328 (2016)
26. Berget, G., Sandnes, F.E.: Searching databases without query-building aids: implications for dyslexic users. Inf. Res. 20 (2015)
27. Bertolaccini, G., Sandnes, F., Paschoarelli, L., Medola, F.: A descriptive study on the influence of wheelchair design and movement trajectory on the upper limbs' joint angles. In: International Conference on Applied Human Factors and Ergonomics, pp. 645–651. Springer, Cham (2017)
28. Medola, F.O., Busto, R.M., Marçal, Â.F., Achour Junior, A., Dourado, A.C.: The sport on quality of life of individuals with spinal cord injury: a case series. Revista Brasileira de Medicina do Esporte **17**, 254–256 (2011)
29. Medola, F.O., Elui, V.M.C., da Silva Santana, C., Fortulan, C.A.: Aspects of manual wheelchair configuration affecting mobility: a review. J. Phys. Ther. Sci. **26**, 313–318 (2014)

30. Lanutti, J.N., Medola, F.O., Gonçalves, D.D., da Silva, L.M., Nicholl, A.R., Paschoarelli, L.C.: The significance of manual wheelchairs: a comparative study on male and female users. Procedia Manuf. **3**, 6079–6085 (2015)

31. Medola, F.O., Purquerio, B.M., Elui, V.M., Fortulan, C.A.: Conceptual project of a servo-controlled power-assisted wheelchair. In: IEEE RAS & EMBS International Conference on Biomedical Robotics and Biomechatronics, pp. 450–454. IEEE (2014)

32. Lahr, G.J.G., Medola, F.O., Sandnes, F.E., Elui, V.M.C., Fortulan, C.A.: Servomotor assistance in the improvement of manual wheelchair mobility. Stud. Health Technol. Inf. **242**, 786–792 (2017)

33. da Silva Bertolaccini, G., Nakajima, R.K., de Carvalho Filho, I.F.P., Paschoarelli, L.C., Medola, F.O.: The influence of seat height, trunk inclination and hip posture on the activity of the superior trapezius and longissimus. J. Phys. Ther. Sci. **28**, 1602–1606 (2016)

34. Medola, F.O., Silva, D.C., Fortulan, C.A., Elui, V.M.C., Paschoarelli, L.C.: The influence of handrim design on the contact forces on hands' surface: a preliminary study. Int. J. Ind. Ergon. **44**, 851–856 (2014)

35. Medola, F.O., Fortulan, C.A., Purquerio, B.D.M., Elui, V.M.C.: A new design for an old concept of wheelchair pushrim. Disabil. Rehabil. Assist. Technol. **7**, 234–241 (2012)

36. Medola, F.O., Paschoarelli, L.C., Silv, D.C., Elui, V.M.C., Fortulan, A.: Pressure on hands during manual wheelchair propulsion: a comparative study with two types of handrim. In: European Seating Symposium, pp. 63–65 (2011)

37. Medola, F.O., Dao, P.V., Caspall, J.J., Sprigle, S.: Partitioning kinetic energy during freewheeling wheelchair maneuvers. IEEE Trans. Neural Syst. Rehabil. Eng. **22**, 326–333 (2014)

38. Sandnes, F.E.: Can spatial mnemonics accelerate the learning of text input chords? In: Proceedings of the Working Conference on Advanced Visual Interfaces, pp. 245–249. ACM (2006)

39. Sandnes, F.E., Huang, Y.P.: Chording with spatial mnemonics: automatic error correction for eyes-free text entry. J. Inf. Sci. Eng. **22**, 1015–1031 (2006)

40. Sandnes, F.E., Huang, Y.P.: Chord level error correction for portable Braille devices. Electron. Lett. **42**, 82–83 (2006)

41. Sandnes, F.E., Medola, F.O.: Exploring russian tap-code text entry adaptions for users with reduced target hitting accuracy. In: Proceedings of the 7th International Conference on Software Development and Technologies for Enhancing Accessibility and Fighting Info-exclusion, pp. 33–38. ACM (2016)

42. Levine, S., Gauger, J., Bowers, L., Khan, K.: A comparison of Mouthstick and Morse code text inputs. Augment. Altern. Commun. **2**, 51–55 (1986)

43. Sandnes, F.E., Thorkildssen, H.W., Arvei, A., Buverad, J.O.: Techniques for fast and easy mobile text-entry with three-keys. In: Proceedings of the 37th Annual Hawaii International Conference on System Sciences. IEEE (2004)

44. Mackenzie, I.S., Felzer, T.: SAK: Scanning ambiguous keyboard for efficient one-key text entry. ACM Trans. Comput.-Hum. Interact. (TOCHI) **17**(3), 11 (2010)

45. Sandnes, F.E.: Reflective text entry: a simple low effort predictive input method based on flexible abbreviations. Procedia Comput. Sci. **67**, 105–112 (2015)

46. Sandnes, F.E.: Effects of common keyboard layouts on physical effort: Implications for kiosks and Internet banking. In: Sandnes, F.E., Lunde, M. Tollefsen, M., Hauge, A.M., Øverby, E., Brynn, R. (eds.) The proceedings of Unitech2010: International Conference on Universal Technologies, pp. 91–100. Tapir Academic Publishers, Norway (2010)

47. Chiapparino, C., Stasolla, F., de Pace, C., Lancioni, G.E.: A touch pad and a scanning keyboard emulator to facilitate writing by a woman with extensive motor disability. Life Span Disabil. **14**, 45–54 (2011)

48. Felzer, T., Rinderknecht, S.: 3DScan: an environment control system supporting persons with severe motor impairments. In: Proceedings of the 11th international ACM SIGACCESS conference on Computers and accessibility, pp. 213–214. ACM (2009)
49. Zhang, X., Fang, K., Francis, G.: How to optimize switch virtual keyboards to trade off speed and accuracy. Cognit. Res. Princ. Implic. **1**, 6 (2016)
50. Jones, P.E.: Virtual keyboard with scanning and augmented by prediction. In: Proceedings of the 2nd European Conference on Disability, Virtual Reality and Associated Technologies, pp. 45–51 (1998)
51. Sandnes, F.E., Medola, F.O.: Effects of Optimizing the scan-path on scanning key-boards with QWERTY-Layout for English text. Stud. Health Technol. Inf. **242**, 930–938 (2017)
52. Higger, M., Moghadamfalahi, M., Quivira, F., Erdogmus, D.: Fast switch scanning keyboards: minimal expected query decision trees (2016). arXiv preprint arXiv:1606.02552
53. Hamidi, F., Baljko, M.: Reverse-engineering scanning keyboards. In: International Conference on Computers for Handicapped Persons, pp. 315–322. Springer, Heidelberg (2012)
54. Baljko, M., Tam, A.: Indirect text entry using one or two keys. In Proceedings of the 8th international ACM SIGACCESS conference on Computers and accessibility, pp. 18–25. ACM (2006)
55. Simpson, R.C., Mankowski, R., Koester, H.H.: Modeling one-switch row-column scanning with errors and error correction methods. Open Rehabil. J. **4**, 1–12 (2011)
56. Bhattacharya, S., Samanta, D., Basu, A.: Performance models for automatic evaluation of virtual scanning keyboards. IEEE Trans. Neural Syst. Rehabil. Eng. **16**, 510–519 (2008)
57. MacKenzie, I.S.: Modeling text input for single-switch scanning. In: International Conference on Computers for Handicapped Persons, pp. 423–430. Springer, Heidelberg (2012)
58. Sandnes, F.E., Aubert, A.: Bimanual text entry using game controllers: relying on users' spatial familiarity with QWERTY. Interact. Comput. **19**, 140–150 (2007)
59. Sandnes, F.E.: Directional bias in scrolling tasks: a study of users' scrolling behaviour using a mobile text-entry strategy. Behav. Inf. Technol. **27**, 387–393 (2008)
60. Sandnes, F.E.: Evaluating mobile text entry strategies with finite state automata. In: Proceedings of the 7th international conference on MobileHCI 2005, pp. 115–121. ACM (2005)
61. Sandnes, F.E., Sinnen, O.: A new strategy for multiprocessor scheduling of cyclic task graphs. Int. J. High Perform. Comput. Netw. **3**, 62–71 (2005)
62. Sandnes, F. E.: Scheduling Partially Ordered Events in a Randomised Framework: Empirical Results and Implications for Automatic Configuration Management. In: Proceedings of LISA, pp. 47–62. USENIX (2001)
63. Rebreyend, P., Sandnes, F.E., Megson, G.M.: Static multiprocessor task graph scheduling in the genetic paradigm: a comparison of genotype representations. Laboratoire de l'Informatique du Parallelisme, Research report no. 98–25. Ecole Normale Superieure de Lyon (1998)

Variations in Vital Signs Associated with the Postural Changes When Using a Stand-up Wheelchair in Patients with Spinal Cord Injury

Thalía San Antonio[1]([⊠]), Fernando Urrutia[2], Anita Larrea[2],
Víctor Espín[1], and María Augusta Latta[3]

[1] Faculty of Civil and Mechanical Engineering, Universidad Técnica de Ambato,
Ambato, Ecuador
{t.sanantonio,victorrespin}@uta.edu.ec
[2] Faculty of Computing, Electronic, and Industrial Engineering, Universidad
Técnica de Ambato, Ambato, Ecuador
{fernandourrutia,anitallarea}@uta.edu.ec
[3] Faculty of Health Sciences, Universidad Técnica de Ambato, Ambato, Ecuador
mariaalatta@uta.edu.ec

Abstract. Postural changes are important for the health of patients who are unable to walk, since they avoid complications in the acute stage. Transitions (standing/sitting/lying down) help vital functions remain in normal ranges. In Ecuador, rehabilitation in people with spinal cord injury is not mandatory, which leads to drawbacks, such as the appearance of bedsores and a decrease in muscle strength; both of them are associated with osteotendinous injuries that reduce the individual's independence. The aim of this study was to quantify changes in vital signs in patients with spinal cord injury in chronic stage when changing from a sitting to a standing position. The sample consisted of 10 patients with dorso-lumbar injury. Results show a significant improvement in blood pressure when standing up. Patients reported great satisfaction of being in a standing position and remarked the potential usefulness of this position to perform daily tasks or work because of the increase in the reach range of their upper limbs.

Keywords: Postural changes · Vital signs · Spinal cord injury

1 Introduction

Postural changes (standing/sitting/lying down) are frequent in the daily routine, as they positively influence some physiological functions and, therefore, the overall health of people [1]. Vital signs, such as blood pressure, respiratory rate, body temperature, oxygen saturation, and heart rate are health condition indicators and can be obtained in a minimally invasive way for the patient. Among them, the heart rate and blood pressure are reliable tools to evaluate hemodynamic changes in the autonomic nervous system, and tend to be sensitive to postural changes because of their relation with the stimulation of gravity [2].

© Springer International Publishing AG, part of Springer Nature 2019
T. Z. Ahram and C. Falcão (Eds.): AHFE 2018, AISC 794, pp. 818–823, 2019.
https://doi.org/10.1007/978-3-319-94947-5_80

With the purpose of improving the health of people with disabilities (PwD) who have limitations in the mobility of their lower limbs which causes them to be generally in sitting or lying down position, several types of technical aids have been developed as facilitators of body mobility; these devises are aimed to expand the freedom of their movements, thus improving their physical and mental health and, therefore, their general living conditions [3]. One of these technical aids are the standing up devices, which have been shown to improve muscular movement stimulation because loads are forced to go through the leg bones slowing down the inevitable osteoporosis associated with the immobility of the lower limbs [4]. Other positive effects of staying frequently in standing position are that the cardiovascular system strengthen by activating the circulation of the cardiac and peripheral components, avoiding cardiopulmonary complications [5]; and, reduce the inflammation of the lower limbs and the occurrence of injuries derived from the sitting position preventing pressure ulcers [6] and improving renal function, among others [3, 7].

In wheelchairs users, the prolonged sitting position is an important factor for the development and worsening of back pain [8], deterioration of the osseous system associated with fractures, and weight gain due to low mobility. The main causes of mortality in people with spinal cord injuries are urogenital infections [9], the occurrence and development of skin infections (pressure sores), and respiratory and cardiovascular diseases. By the regular used of an standing up devices, it is possible to place the PwD in a vertical position, so they will have a stable balance by resting on their feet forcing the functional work of the lower limbs and increasing the reach of the upper limbs together with the positive psychological effects of being as the same height as its interlocutors [10–12].

The purpose of this study is to quantify changes in physiological parameters: blood pressure, oxygen saturation, and heart rate when the person with spinal cord injury changes from a sitting to a standing position. The equipment used to place the patient in a bipedal position was designed and built in a previous phase, on the basis of the anthropometric measurements of the sample [13, 14].

2 Methodology

Sample Description

The sample consisted of 10 patients, 8 males and 2 females with complete dorso-lumbar spinal cord injury, who have neither adopted a standing position since they had the injury, nor received physiotherapy assistance. They live in Tungurahua province, Ecuador; their age ranges between 34 and 66 years of age. Among them, 60% reports to have a medium to high physical activity level, and the remaining 40% reports to have a low physical activity level.

The evaluation of ethical procedures was carried out with the Bioethics Committee of Universidad Técnica de Ambato. The interviews were used to gather general information about their personal data, weight, height, mobility ease, date of the injury; then, the tests were performed. All the patients were informed about the procedure that would be followed for collecting data required for the study, and all of them were subjected to the same tests in similar conditions.

Data Collection
The data collection (blood pressure, heart rate, and oxygen saturation) is done with a portable equipment Connex Spot Monitor Welch Allyn 901058 vital monitor core. The routines performed with each patient are described below:

- Sitting on the wheelchair, an exercise routine of upper limbs consisting in flexion, extension, adduction, abduction and rotations of shoulder, elbow, wrist and hand with a 4 lb weight was carried out; 10 repetitions were made per exercise. After 2 min, the vital signs of blood pressure, oxygen saturation, and pulse were taken, and a spirometry was performed.
- The equipment developed in a previous research project [13, 14] is presented to the patient, it consist of a wheelchair with an electrical system to put the user in a standing position. Emphasis is placed on security systems and how they are used.
- If the patient requires it, he/she is assisted in the transfer to the stand up wheelchair. Then the security systems, which vary according to the height of the injury, are placed and given indications of the standing system so that the patient stands up at the speed that feels comfortable. There is a waiting time of between 10 and 15 min for the person to feel comfortable, safe, and calm in this bipedal position; then, the exercises and data collection, previously described, are repeated in the standing position.

3   Results

Table 1 summarizes the results of the physiological parameters measured in the sitting and standing positions, for all the evaluated patients with spinal cord injury; since all of them are adults between 34 and 66 years of age, normal values can be set for each parameter, which appears in the title in parentheses.

The results show that, in the case of pulse, blood pressure, and oxygen saturation, the values are close to normal when the patient is standing, but they are not the normal values. The variations obtained did not show sensitivity to patient characteristics, such as age, place of injury, time since the injury occurred or level of physical activity.

Figures 1 and 2 shows that pulse and oxygen saturation values are sensitive to postural changes, registering a maximum variation of 8.75% and 10.5%, respectively. Blood pressure is more sensitive to postural changes, registering maximum variation values of 15% for systolic BP and 16.25% for diastolic BP.

In the case of systolic BP, the difference between mean values and the normal value is 12.43% in the sitting position and 5% in the standing position. When it comes to diastolic BP, the differences are less than 2.63% in the sitting position and 0.73% in the standing position. This indicates that in both cases the values improve when standing and that the systolic blood pressure is much more sensitive to postural changes, which is expected due to the fact that in the standing position limbs are farther away from the heart.

Table 1. Values of vital signs in patients, both sitting and standing up

Age	Sex	Level of injury	Time elapsed since injury	Blood pressure (120/80 mmHg)		O₂ saturation (95 to 100%)		Heart rate (80 l/min)	
				Seated	Standing	Seated	Standing	Seated	Standing
55	F	T12	10	124/78	110/99	92	91	84	91
52	M	T6- T7	31	125/75	123/78	85	86	64	65
43	M	T9- T10	18	159/85	131/61	94	94	69	75
34	M	T5- T6	8	122/83	121/78	93	93	83	88
39	M	T5- T7	17	133/84	129/82	89	91	80	85
58	M	T5	13	125/74	130/85	89	90	101	108
39	M	T11	11	123/83	125/85	94	95	66	67
66	M	L1	7	163/96	161/79	95	92	73	76
36	M	L3	9	132/79	104/69	92	92	73	80
50	F	L5- L6	15	143/84	126/87	93	95	64	65
Mean values			13.9	134.9/82.1	126/80.3	91.6	91.9	75.7	80

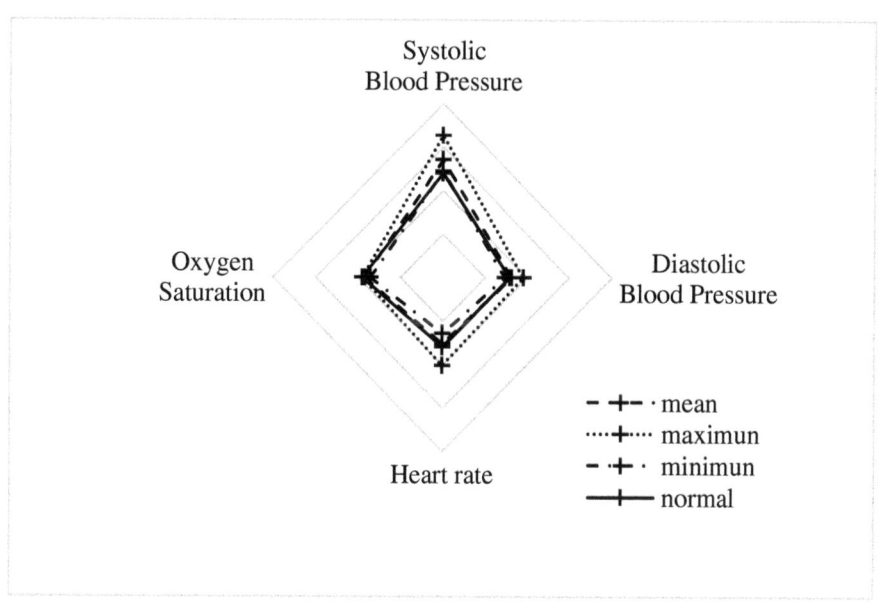

Fig. 1. Comparison of vital signs values in the sitting position

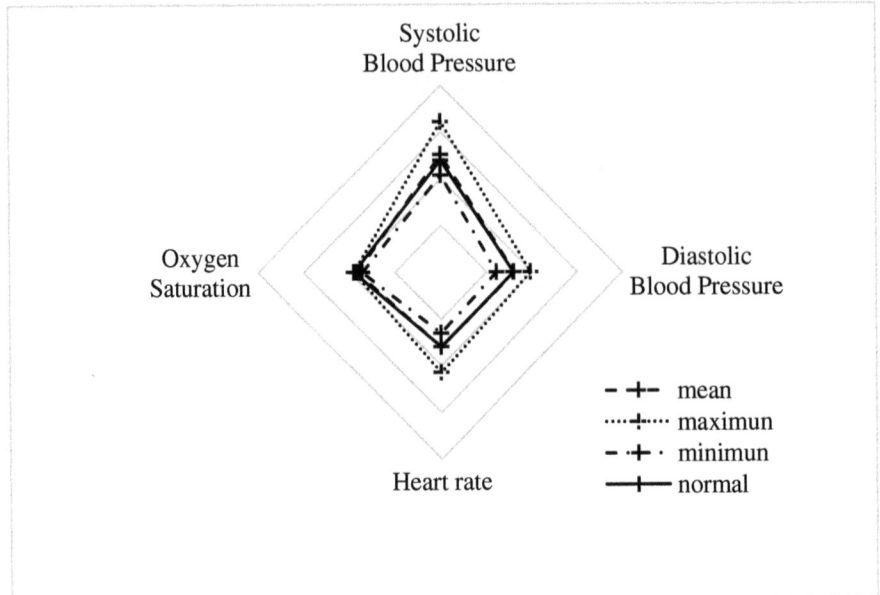

Fig. 2. Comparison of the vital signs values in the standing position

4   Discussion

From a physiological point of view, a therapy with standing devices is necessary, especially for heart functioning, as it presents more variations that are closer to normal values. In addition, a great psychological effect is produced when people can adopt the standing position because it provides a wider reach range to their upper limbs, and they can be at the same level as their conversation partners; these two characteristics can be helpful in their reintegration into the labor market.

Given that almost all users had not stood up since they acquired the injury, 13.9 years in average, it took an extended amount of time (more than 15 min) for users to calm down; however, the psychological state of the users was altered, so this could had influenced the measurements.

The detected variations in blood pressure values when changing from a sitting to a standing position are similar to the ones reported by [16] who carried out this analysis when going from a lying down to a sitting position. In both cases, values tend to normalize when leaving the sitting position.

Acknowledgments. This work is being funded by Dirección de Investigación y Desarrollo (DIDE), of Universidad Técnica de Ambato under the project: *"Implementation of ergonomic criteria in the design and marketing of equipment to assist people with limited mobility: case standup wheelchair".*

References

1. Boccardi, S., Ferrarin, M.: Recuperación de la bipedestación y de la marcha en el paciente parapléjico. EMC-Kinesiterapia-Medicina Física **26**(3), 1–10 (2005)
2. Alfonso, I., del Carmen, J., Aldana Vilas, L., García Gutiérrez, E., Vázquez Díaz Granados, G.: Frecuencia cardiaca en la bipedestación activa inmediata en jóvenes sanos. Revista Cubana de Medicina Militar **39**(2), 104–115 (2010)
3. Pabón, L.D.M., Ángel, V.A.H.: representaciones sociales de la silla de ruedas para la persona con lesion medular. Revista Colombiana de Rehabilitación **12**(1), 96–102 (2017)
4. Arroyo Riano, M.O., Martín Fraile, E., Alcaraz Rouseelet, M.A., Pascual Gómez, F.: Ortesis de bipedestación y marcha en la lesión medular. Rehabilitación **32**(6), 437–451 (1998)
5. McKee, K.E., Hackney, M.E.: The effects of adapted tango on spatial cognition and disease severity in Parkinson's disease. J. Mot. Behav. **45**(6), 519–529 (2013)
6. Russo, J.L., Cezeaux, J.L.: Design of a wheelchair pressure monitoring system. In: Bioengineering Conference, Proceedings of the 2010 IEEE 36th Annual Northeast, pp. 1–2. IEEE, March 2010
7. Oñate, N.A., Beltran, M.A.: Equipamiento para la discapacidad: propuesta de un bipedestador eléctrico infantil. XIV International Congress on Project Engineering, Madrid (2010)
8. Medola, F.O., Pisconti, F., Elui, V.M.C., Santana, C.D.S.: Dolor en individuos con lesión de la médula espinal: un estudio descriptivo. Revista Iberoamericana de Fisioterapia y Kinesiología **13**(2), 58–62 (2010)
9. Gros Herguido, N., Pereira Cunill, J.L., Barranco Moreno, A., Socas Macias, M., Morales-Conde, S., García-Luna, P.P.: Paciente con paraplejia y obesidad mórbida: nuevo reto en la cirugía bariátrica. Nutrición Hospitalaria **29**(6), 1447–1449 (2014)
10. Argote, K.Q., Torres, A.R., Torres, A.G.: Diseño de un prototipo de bipedestador para pacientes pediátricos con espina bífida. Revista de la Facultad de Medicina **61**(4), 423–429 (2013)
11. Contreras, L.A.L., Casallas, E.C., Prieto, G.S.: Estudio de los rangos articulares en la bipedestación estática en personas normales vs. Amputados transtibiales. Tecnura **17**, 60–68 (2013)
12. Morales, R., Becedas, J., Feliu, V.: automatización de un prototipo de silla con accionamiento electrico y capacidad de transferencia a la cama. XXV Jornadas de Automática (2004)
13. Urrutia, F., San Antonio, T., Latta, M. A., Ortiz, P., López, J., Urrutia, P.: User centered design of a wheelchair based in an anthropometric study. In: 2015 CHILEAN Conference on Electrical, Electronics Engineering, Information and Communication Technologies (CHILECON), pp. 235–243. IEEE, October 2015
14. San Antonio, T., Urrutia, F., Larrea, A.: Ergonomic analysis for people with physical disabilities when the wheelchair is considered as their workstation. In: Ecuador Technical Chapters Meeting (ETCM), IEEE, pp. 1–6. IEEE, October 2016
15. Vicén, J.A., Luesma, R.B., Lasaosa, P.L.: Análisis de las coherencias entre la variabilidad del ritmo cardiaco, la variabilidad de la presión sanguínea y la respiración. In: XXIV Congreso Anual de la Sociedad Española de Ingeniería Biomédica, pp. 37–40 (2006)
16. Castiglioni, P., Di Rienzo, M., Veicsteinas, A., Parati, G., Merati, G.: Mechanisms of blood pressure and heart rate variability: an insight from low-level paraplegia. Am. J. Physiol. Regul. Integr. Comp. Physiol. **292**(4), R1502–R1509 (2007)

We Have Built It, But They Have Not Come: Examining the Adoption and Use of Assistive Technologies for Informal Family Caregivers

Pamela Wisniewski[(⊠)], Celia Linton, Aditi Chokshi,
Brielle Perlingieri, Varadraj Gurupur, and Meghan Gabriel

University of Central Florida, 4000 Central Florida Blvd.,
Orlando, FL 32816, USA
{pamwis,varadraj.gurupur,meghan.gabriel}@ucf.edu,
{celinton,aditi.chokshi,briellep}@knights.ucf.edu

Abstract. We conducted interviews with 14 informal family caregivers of elderly Alzheimer's and dementia patients in the U.S. to understand opportunities to increase the adoption and use of assistive technologies (ATs) in the home. We identified three key themes: (1) Most of the caregivers were interested in adopting assistive technologies, but they did not know where to begin; healthcare providers gave little to no guidance. (2) Caregivers demonstrated a need for assistive technologies that enabled or enhanced remote caregiving, as many were adult children who worked full-time and had to leave their elderly parent at home, unattended during the day. (3) While caregivers rarely adopted assistive technologies designed specifically for caregiving, they often repurposed everyday technologies (e.g., home security systems, calendar applications) to aid in care. These findings provide insights for how we can better support the use of assistive technologies by informal family caregivers.

Keywords: Assistive technology · Informal caregiving · Alzheimer's disease
Dementia

1 Introduction

As the number of patients diagnosed with a neurodegenerative disease rises, so does the number of family caregivers. Many Alzheimer's disease or dementia patients reside at a home and rely only on the care of a loved one. There are approximately 43.5 million informal caregivers in the United States; 15.7 million of whom provide unpaid care for someone who has Alzheimer's disease or dementia [1]. Informal caregivers often take care of loved ones as in-home patients, which creates significant caregiver burden, especially when care requires constant supervision [2]. Caring for a patient with a neurodegenerative disease is quite difficult and the majority of family caregivers have little to no formal training. While numerous assistive technologies (ATs) have been developed to ease this burden, the adoption and use of these technologies has been markedly low, for reasons ranging from caregivers' lack of technical expertise to the prohibitive cost of assistive devices and monitoring services [3, 4]. This study explored in-home and remote technology adoptions and the use by the caregivers to support the patients. We go beyond studying the adoption and use of ATs in the home to examining the unmet needs of informal caregivers and opportunities to improve assistive technologies, so they can better meet these needs.

2   Related Literature

We provide a brief summary of the literature that gives an overview of the ATs available for informal family caregiving to contextualize our study. Then, we conduct an in-depth review of the existing literature most closely related to our own.

2.1   Assistive Technologies for Informal Family Caregiving

Clark et al. [5] outlined the challenges in the long term and post-acute caregiving processes to motivate an imperative need for further analysis of current ATs to help with informal caregiving of patients with Alzheimer's and dementia. One of the areas where caregivers need support for caring for in-home patients with Alzheimer's and dementia is the need to constantly monitor patients to ensure their safety. Slip and falls, burns upon household equipment, and night wandering are realistic occurrences that cause distress to both patients and their caregivers. Past research has summarized various ATs for managing these concerns, including motion sensors with remote alarms, wireless cameras with handheld LCD and night vision, proximity range alarms, wireless home security sensors, and Global Positioning Systems (GPS) [6]. In addition, Robinson et al. [7] indicated that the role of assistive technology and Information and Communication Technology (ICT), including memory aids and home improvement modifications are seen as promising possibilities for helping people with dementia age in place at home by compensating for the patient's functional losses [8–10]. A number of studies (e.g., [11–13]) have examined ATs for in-home caregiving of Alzheimer's and dementia patients, focusing on the affordances of these technologies, and at times, introducing them into in-home caregiving situations to assess their effectiveness. However, fewer studies have studied actual use (or non-use) of these ATs when an intervention is not present. We summarize these studies, which are most closely related to our own work, in the next section.

2.2   User-Focused Studies with Informal Family Caregivers

Research that has taken a more user-centered approach to examining ATs for informal family caregivers of elderly Alzheimer's and dementia patients demonstrates an interesting dichotomy between users' perceptions versus their actual adoption and use of these technologies. In 2015, Mao et al. [14] studied the perceived usefulness of 82 assistive technology devices and, based on a survey of 72 caregivers of dementia patients, found that they were "generally amenable" to using these devices. Similarly, in 2017, Mulvenna et al. [15], examined the ethical and privacy implications of using video surveillance in the home to monitor people with dementia. They found that caregivers were generally in favor and willing to use such monitoring technologies. However, a limitation of these works is that they solicited feedback on hypothetical or perceived use, as opposed to actual use, of such ATs. In contrast, researchers who studied actual adoption and use of in-home ATs found that it was relatively low and have identified a number of barriers to adoption. For instance, Edlund and Björklund [3] conducted a study in Sweden and found that family caregivers often do not use assistive technology because they have a lack of technical knowledge. In addition, Olsson et al. [4] found that caregivers were not able to use ATs because they were expensive, and they could not afford them. Gibson et al. [16] found that caregivers had negative perceptions about costs, were unsure the best time to adopt ATs, and were unsure how ATs could best be used. We also adopt a user-centric, qualitative approach to understanding the use of ATs by informal caregivers in the home. We extend beyond the existing literature by examining non-use [17] and remote use of ATs, which lends insight as to missed opportunities for caregiver support.

3 Methods

3.1 Study Overview

We conducted a semi-structured interview study with informal caregivers of Alzheimer and dementia patients. The inclusion criteria for this study was that the participant needed to be the primary in-home caregiver of a patient with Alzheimer's or dementia in the past year and over the age of eighteen. Adoption and use of ATs was not a requirement for participation, as we were interested in understanding the perspectives of both users and non-users [17]. The interview began by asking caregivers questions about the nature of the informal caregiving relationship, such as how long they had been a caregiver and the progression of the patient's condition. Next, we asked participants about their level of comfort with technology and whether they used ATs for in-home care or for providing care remotely. We followed up with questions regarding the benefits and challenges associated with using these ATs. We concluded the interview with questions related to how participants learned about or were introduced to ATs. For those who were adopters, we asked how these technologies impacted their caregiving duties. For those who reported low usage, we asked why they chose not to adopt. The study was approved by our university's Institutional Review Board (IRB), and participants received a $20 Target gift card for their time.

3.2 Data Collection and Analysis

Participants were recruited from Central Florida via Alzheimer and dementia organizations, as well as through word-of-mouth and social media postings. Interviews took place during 2015–2016 and were conducted in-person or over the phone. We performed a qualitative, thematic content analysis [18] to extract emergent theme from the interviews that extended beyond the findings from prior literature, noting many of the themes that emerged from our interviews mirrored some of the findings from past work (e.g., [3, 4]). For instance, our results confirmed the low adoption rates and many of the barriers to adoption (e.g., privacy concerns, technical expertise, and cost) from the prior literature. These themes were noted but de-emphasized over the themes that had not been highlighted in prior literature. The principal investigators (Drs. Wisniewski, Gurupur, and Gabriel) trained and supervised five undergraduates and one graduate research assistant to recruit participants, conduct semi-structured interviews, transcribe audio recorded interviews, and qualitatively analyze the data based on the salient themes identified from their initial analyses of the interview data. The research assistants extracted illustrative quotes from the interviews to include in the presentation of our paper. All quotes are anonymized and any personally identiable information has been removed to protect the identities of our participants.

4   Results

Below, we describe our participants and the types of ATs used to support informal caregiving, then provide an in-depth analysis related to our three emergent themes.

4.1   Participant Profiles

Our sample consisted of fourteen informal caregivers of elderly Alzheimer's and dementia patients. The majority of participants were female with three who identified as male. This distribution is consistent with past research that shows that women are more likely to become informal caregivers than men [19]. The relationships of caregivers to Alzheimer's and dementia patients included spouses, children, and siblings. Most participants self-identified as white or Caucasian with two participants (P8 and P12) who were of Hispanic descent. Additional descriptive statistics, such as gender, age, employment status, and income level are summarized in Table 1.

Table 1. Descriptive characteristics of study participants.

ID	Gender	Age range	Relation to patient	Employment status	Income level	Patient's condition
1	Female	60–69	Daughter	Full-Time	$50 K–$79 K	Moderate/Advanced
2	Male	60–69	Son-in-law	Full-Time	$50 K–$79 K	Advanced (Deceased)
3	Male	50–59	Son	Retired	$100 K+	Advanced
4	Male	60–69	Husband	Retired	$80 K–$99 K	Mild
5	Female	70+	Wife	Retired	$100 K+	Mild
6	Female	70+	Wife	Retired	$50 K–$79 K	Advanced
7	Female	70+	Sister	Retired	Not reported	Mild
8	Female	60+	Daughter	Retired	Not reported	Moderate
9*	Female	50+	Daughter	Full-Time	$80 K–$99 K	Mild
10	Female	50+	Daughter	Part-Time	Not reported	Advanced
11	Female	50+	Daughter	Full-Time	$30 K–$49 K	Advanced
12	Female	40–49	Daughter	Full-Time	$100 K+	Mild
13	Female	70+	Wife	Retired	$35 K	Advanced
14	Female	70+	Wife	Retired	$80 K–$99 K	Mild/Moderate

* Both parents of P9 exhibited mild symptoms of dementia under her care.

4.2 Caregiver Narratives of Assistive Technology Use and Non-use

Many of our participants were adult children who took on the caregiving responsibilities for their parents in the same home. For instance, P1 moved in to care take her mother six years ago after her father's death. She works full-time as a specialist in emerging technologies, but she still struggles to find the right in-home and remote technologies to help her care for her mother. Currently, she only uses an *"ADT like Life Alert system."* During her interview she said her immediate need is for remote ATs to monitor her mother while she is at work. In Sect. 4.2, we describe some of the barriers as to why she has yet to adopt many in-home or remote ATs, even though she is well-versed with technology. P11 is similar to P1 in that she works full-time and has been the primary caregiver for her mother for three years. Unlike P1, P11 uses an in-home camera system with remote monitoring from her phone, so that she can get out of the house while making sure her mother is safe at home. She also considers herself *"very comfortable"* with new technologies, and uses the internet to take care of her mother's needs, such as understanding her insurance benefits. Both P1 and P11 will be used as examples for the need for ATs for remote caregiving (Sect. 4.3).

Other participants cared for parents with milder symptoms, so did not feel they needed ATs at this time. P12's mother recently moved in with her after exhibiting early signs of dementia. She is quite comfortable using technology, but currently only uses a cell phone and landline to manage communication with mother. She sets reminders on her mother's cell phone to take her pills, but she often calls the landline just in case her mother did not pay attention to the reminder. She plans to install a home security system with cameras and has begun researching different options that fit her budget. She also helps her mother manage simple tasks, such as writing checks and using an ATM card. Both of P9's parents have mild symptoms of dementia. While she does not live with them, she helps manage their day-to-day affairs, such as using a calendar application to manage their doctor's appointments. Her husband helps her manage her parents' finances. Otherwise, she says she is quite comfortable with technology but does not use ATs because they are not yet necessary.

When caregivers (adult children or spouses) had loved ones in the advanced stages of their disease, many felt like constant supervision and managed care were more of a necessity than using ATs. P10 helped her father care take her late mother in the advanced stages of dementia. In the last five years of her mother's life, her mother was confined to a wheelchair and unable to walk. Therefore, she said, *"I didn't have a concern of her getting out of the home and getting away."* The only technology she mentioned using to support her mother's caregiving was a calendar app on her phone to help keep track of her mother's doctor's appointments. Otherwise, since her mother was *"never left alone,"* human care (primarily provided by her father) replaced the need for ATs. P8 has lived with her mother as her primary caregiver for seven years. Her mother stays in bed most of the time and has a Certified Nursing Assistant that comes to administer care periodically. P8 admits that she is "not very good" with technology and prefers to keep an eye on her mother. She has an emergency security alert keychain for herself and a home security system. However, she only uses the home security system when her mother is not home, in fear that she would get scared or confused if she set it off (see Sect. 4.3).

Similarly, P2's mother-in-law recently passed away, but he was her primary caregiver for approximately a year and a half prior to her death. Due to his mother-in-law's need for constant supervision, P2 did not express the need for remote ATs (Sect. 4.3). P3's mother lives in a managed care facility, which also replaced the need for most ATs. His mother has a Life Alert and he uses internet-enabled technologies to handle all of his mother's affairs, such as *"her insurance, Medicaid, drug plan, and bank statements."*

The remainder of our participants were spouses or siblings (P7) of the loved ones they cared for. Some of these participants made use good use of ATs in the home. For example, P4's wife was diagnosed with dementia in 2014, and ATs they use to support her care include, a Fitbit to track her daily activity, a cell phone *"as a simple means for communication,"* and calendar to keep track of important events. While he uses these technologies and considers himself *"generally very good"* with technology, he gets frustrated that his wife will only use a landline and will not use the calendar. P6's husband has become completely dependent on her the past six years. To manage his care, she embraces a number of advanced ATs; she personally installed a camera system throughout the house, as well a home security system that she can monitor and send her alerts via her mobile phone, and has her husband carry a GPS tracking device on his keychain.

In contrast, other caregivers were less likely to use ATs, often due to lack of knowledge. P5's husband was diagnosed with mild cognitive impairment and symptoms of Alzheimer's about ten years ago. He is still fairly self-sufficient, able to bathe himself and take walks for up to three and a half miles a day; however, he has *"given up"* trying to perform daily functions, like managing finances, because *"he gets frustrated and angry."* She was proficient with desktop computing technologies that allowed her to manage their daily affairs, but she was uncomfortable with more modern technologies, such as smartphones. She says that her husband *"carries a flip phone and he knows how to use it still."* In terms of ATs, P5 was unaware of technologies that might be able to help them (see Sect. 4.2). She noted that if he got to the point where he needed a home monitoring system to ensure his safety, she would rather give up their home and move into a retirement home. P7 is the twin sister of her brother, who was diagnosed with early stages of Alzheimer's; she lived with him for three years. She emphasized that her brother's wealth was a key factor in the use/non-use of assistive technology in their shared home. Most of his needs were taken care of by hired professionals within the home, but he had a paging system in the house in the case that someone was not immediately available, so that they could come attend to his needs. P7 took only a supervisory role in her brother's care, but she also admitted that she did not use ATs *"because I am not capable."*

Many participants who admitted not being tech savvy were also the ones who felt like they had no need for ATs. P13 is the caregiver for her husband, who is in the advanced stages of his disease. She generally does not use technology much in the home and did not report using any ATs. This was partly due to her lack of knowledge, as well as her limited income. We discuss her perspective about in more depth in Sect. 4.2. She is in constant supervision of her husband and only leaves him when the home health aid is present to supervise him. Similarly, P14 is also a caregiver for her husband, who was diagnosed with mild cognitive impairment. Similar to other participants, she says she is comfortable with technology, but has not embraced newer technologies, such as smart phones or ATs. She explains, that at her age and her husband's mild condition, she simply does not have a need for it. Table 2 summarizes the relatively low usage of assistive technologies by our participants for informal family caregiving.

Table 2. The use and non-use of assistive technologies for informal caregiving

Assistive technologies used (or not used)	Participants
No usage of assistive technologies reported	13, 14
Productivity software (e.g., personal finance, calendars, reminders etc.)	3, 4, 5, 9, 10, 11
Cell phones and landlines	1, 4, 12
Emergency alert system (e.g., Life alert)	1, 3, 8
Camera-based home security system (with remote monitoring)	6, 11
Home-based security system (without remote monitoring)	2
Wearable health trackers (e.g., Fitbit)	4
GPS tracking device	6
Other (e.g., home paging system, remote control A/C unit)	7

4.3   A Willingness to Adopt but Real Barriers Outweighed Perceived Benefits

Overall, most of our participants recounted numerous barriers to adoption that prevented them from adopting ATs. For instance, P2 was also concerned about costs; however, he was even more skeptical of the benefits that ATs would provide given his mother-in-laws advanced stages of her disease. For cost and functionality, it made more sense to invest the funds in a professional in-home caregiver:

"At the expense of that we could make sure she had a caregiver around, that was one of the prohibitive factors with installing something like that." –P2

P2 and P3 both highlighted the trade-off between the cost of managed care versus the need for ATs in the advanced stages of their loved one's condition, suggesting that managed care was preferable to relying on ATs for long-term care. P6, however, was an exception to this case, likely due to her proficiency with technology and desire to keep her husband out of an assisted living facility as long as possible. Meanwhile, P7 noted that her brother had enough money to have people take care of him before he was sick, so had no need for ATs now:

"He was in a position that he could afford to have help and wives. Once he got sick, he's spoiled really." –P7

P2 also noted that, *"there is not a lot of resources,"* in terms of guidance from medical professionals or doctors. Therefore, the burden falls of the caregiver to explore technologies that might fit their needs. P5 expressed similar frustrations when we asked her about the use of ATs for in-home care:

"Well sure, if I knew what they were. What technologies are you suggesting that I should be using?... No nobody has told me anything. I have no clue what." –P5

Similarly, P8 said, *"No one has suggested anything,"* and P12 concluded that she would rather get information from her doctor than having to search the internet:

"I try not to depend on information that I see online, I would rather get it from a doctor because there is so much out there that sometimes there is conflicting information so I would rather get the information from a doctor first." –P12

In contrast, P6 leveraged multiple Alzheimer's and dementia organizations to find ATs to help her care for her husband:

"I check the Alzheimer's reading room, Alzheimer's and dementia associations. I go to a lot of different places looking for information for him." –P6

P1 was interested and highly capable of using ATs for caregiving her mother, but a number of critical factors prevented meaningful adoption. First, she simply did not have the time or energy to research the available options. Working full-time and caregiving her mother left little time for anything else. She admitted that she knew of community Alzheimer's and dementia resources that provided this type of information but never had a chance to go. Another issue was having to make *"decisions by consensus"* with her brothers and sisters. Even though she had power of attorney and was her mother's trustee, she felt like she had to be diplomatic on the decisions that were made about her mother's care. She also expressed concerns related to violating her mother's privacy and personal finances, which both contributed to her non-use of ATs. Being a specialist in emerging technologies, she was somewhat frustrated by these barriers, concluding, *"I need to just get something implemented."*

Multiple participants (P1, P3, P4, P7) noted that a barrier to adoption was their loved ones' inability or unwillingness to use ATs that would benefit them. Therefore, any technologies that would be a viable solution would need to be targeted solely to the caregiver, not requiring any interaction on the part of the patient. P13 summarized all of the above themes to show how many caregivers simply felt the benefits did not warrant the costs of adopting ATs in the home:

"I don't know, because I don't know what's available. And if it was anything that he had to use at the other end, he wouldn't know how... Well, I don't see how it would help me. I have to be with him all the time anyway." –P13

She did not know what technologies were available, did not see the need for them, and since she acted as her husband's constant companion, could not justify the cost.

4.4 The Need for Assistive Technologies for Remote Caregiving

A number of participants (e.g., P1, P4, P5, P6, P11) were interested in ATs that helped them get out of the home, rather than technologies for enhanced in-home care. P1 explained why she made so many sacrifices to keep her mother at home, which came with conflicted emotions of whether continuing her job was sustainable:

"I've curbed a lot of things to take care of her, but I promised my dad I would. [She] deserves to be taken care of. Her main desire is to stay at home. I enjoy my work, but there is part of me

saying maybe I should retire early and help my mom, but then there is part of me that just loves what I'm doing." –P1

Being a technology specialist, she offered a number of ideas that she thought would be beneficial to her. For instance, she suggested a wearable blood pressure monitor for her mother, so that she could automatically monitor her mother's blood pressure when she was not home. She emphasized the importance of ATs that she could use but did not have to be used by her mother. A number of other participants reiterated the idea that any ATs they used would have to not require interaction from the patient.

Similar to P1, P11 also worked while caretaking her mother. A remote monitoring camera system gave her a peace of mind about her mother's safety when she was not home, but she desired a better solution that did not require her constant attention:

"I wish I had something where my physical eyes don't have to be on her, so I can focus on trying to work because I still have to make a living and try to be a caregiver." –P11

P6 was also able to leverage ATs, combined with hired part-time caregivers, to gain some freedom from her constant caregiving duties for her husband:

"I can leave depending on the time of day and what's going on. Generally, I do not leave him alone. I have the security system on my phone, so I know where he is at all times and I also know what the caregivers are doing at all times." –P6

In contrast, even though P5's husband was only in the early stages of Alzheimer's, she felt that his condition greatly altered her life, and she was unaware of ATs that might be able to give her more independence:

"My retirement is totally different than I thought it would be. I like to go out and do things, I like to play tennis, golf, but I can't do that anymore... Here you make your plans your whole life for when you're going to retire, your career, and it changes, this diagnosis. It changed my whole life." –P5

P4 is also retired, but there are times when he would like to make sure his wife is safe and can reach him in case of an emergency when he is away from her. Therefore, he also expressed the need for communication mechanisms that allowed him to know how she was doing when they were apart. With similar backgrounds (fairly affluent, 70 + year-old retired women caretaking their husbands) P5 and P6 illustrate a stark contrast in how ATs have the potential to improve the quality of a caregiver's life.

4.5   Repurposing Every Day Technologies to Aid in Caregiving

A number of participants opted to repurpose every day and fairly low-tech technologies to aid in caregiving. The most common was the use of home security systems to prevent their loved ones from wandering:

"We have an alarm system that we don't use except for (noise) an alarm system we don't typically use or haven't used except for when mom started to get into the wandering stage...it's like a wander guard for personal home." –P2

Some participants (e.g., P7, P8), however, expressed concern about home security systems because their loved one could inadvertently set them off:

"I could put on the alarm, it says 'stay,' but I'm afraid that my mother would get up and open the door, and it would set off the alarm. You know that was the toughest thing about the alarm, if she happened to be walking around." –P8

Many caregivers (e.g., P3, P4, P9, P10, P11, P12) also relied on online and technologies and client-based applications that helped them manage their loved one's affairs since the patient could no longer do so on their own. This was true for patients who ranged from early stages of Alzheimer's and dementia to advanced stages. For instance, P4's wife only exhibited mild symptoms of dementia, so could handle most self-care tasks, such as bathing and cooking, but he refused to maintain higher level executive functioning tasks, such as finances, scheduling, taking pills, and shopping. Therefore, he had to manage daily schedules for the two of them:

"Calendaring is very important to me. We go over a weekly calendar and have her write down what are the important appointments from now. Being able to track appointments. Being able to make her aware that we do have things to do on certain days and time. This way is not shocking to her. Even though she won't necessarily remember it. That removes the disturbance factor." –P4

In contrast, P3's mother lived in managed care due to her advanced condition, so he did not need ATs to keep her safe in the home, but he did use technology to remotely manage all of her daily affairs:

"I handle her insurance, Medicaid, drug plan, and I'm using technology to do all that. Lots of paperwork. I can get her bank statements online." –P3

Participants consistently highlighted the need for tools to help them manage their parents' daily affairs more so than the need for traditional ATs that directly aided with patient care. Therefore, we discuss the implications of this finding and our other themes in the next section.

5   Discussion

All of the caregivers in our study were unique in terms of their needs, the needs of their loved one, their level of proficiency with technology, and their perspectives about the benefits and drawbacks of ATs. Overall, we found what appeared to be a curvilinear relationship between the need for ATs and the patient's condition. Generally, ATs were not perceived as needed in the early stages of the patient's disease (e.g., P4), but were also less useful in advanced stages (e.g., P2, P3) when constant supervision and managed care was necessary. We saw the greatest need and benefit of ATs for caregiving when patients were in the moderate stages of their condition and their caregivers needed to be away from the home for extended periods of time. For instance, in the cases of P1 and P11, who were both adult children who had to work full-time while caregiving their mothers.

Our three themes highlighted in this paper illustrate: (1) the need for reduced barriers to adoption for ATs, especially in the sense of doctors and other organizations providing caregivers the necessary resources to assess whether various ATs meet with unique needs, (2) the need for ATs that aid in remote caregiving, which can give caregivers more freedom to leave the home and simultaneously maintain their caregiving duties, and (3) the need to reconceptualize ATs based on existing systems to enhance their capabilities for caregiving. To address these needs, researchers, practitioners, and clinicians should form partnerships with non-profit and community-based organizations to possibly create an easy-to-use website built on top of a recommender system [20] that takes into account the unique needs of each family (e.g., income, stage of disease, technology expertise, task-orientation, etc.) to suggest the best ATs based on these given parameters. Doctors or case managers could potentially walk patients and their caregivers through this process to help them find the ATs best suited to their needs. Additionally, ATs that aid in remote care need further development because this was a pain point identified by many informal caregivers. Many of our interviewees had to balance their day-to-day obligations, including full-time jobs, with providing care. As such, more solutions need to be designed that take into consideration the unique context of care in a way that is respectful of all families, especially those who may not have significant financial resources.

Finally, most of our participants expressed the need for technologies that are generally not considered "assistive" but are necessary for managing the day-to-day affairs of their loved ones. Therefore, we recommend future research consider designing collaborative software that meets the unique needs of caregivers and their loved ones, such as calendaring applications that allow caregivers to keep track of important doctor's appointments electronically, but are connected to a physical in-home scheduling display for the elderly loved one. A digital-physical hybrid display would better meet the needs of caregivers who are more comfortable with technology and their loved ones who are unable to interact with technology. Further, financial account management and other important application-based or online platforms should consider having a way to transition ownership from the patient to the caregiver. Similar to how Facebook now provides an option for legacy account management [21] (so that others can manage the account upon the death of a loved one), these products should give caregivers the ability to manage the day-to-day lives of Alzheimer's and dementia patients, who are no longer able to take care of their own affairs while they are living.

6   Conclusion

As a society, we should work to make assistive technologies more accessible to caregivers who could truly benefit from them. This means reducing barriers to adoption, which include knowing where to get started, lower costs, and making ATs more intuitive and usable for novice technology users. One way to do this may be to integrate AT capabilities into existing technologies, such as home security systems, so that as users age, systems evolve to meet their needs. Another missed opportunity is patient and caregiver education for ATs by physicians and social workers caring for these patients. At the same time, it is equally important that we continue to develop programs and policies to support caregivers who opt to not adopt assistive technologies to care for their loved ones.

References

1. Family Caregiver Alliance: Caregiver Statistics: Demographics. https://www.caregiver.org/caregiver-statistics-demographics
2. Williams, A., Sethi, B., Duggleby, W., Ploeg, J., Markle-Reid, M., Peacock, S., Ghosh, S.: A canadian qualitative study exploring the diversity of the experience of family caregivers of older adults with multiple chronic conditions using a social location perspective. Int. J. Equity Health **15**, 40 (2016)
3. Edlund, C., Björklund, A.: Family caregivers' conceptions of usage of and information on products, technology and web-based services. Technol. Disabil. **23**, 205–214 (2011)
4. Olsson, A., Engström, M., Skovdahl, K., Lampic, C.: My, your and our needs for safety and security: relatives' reflections on using information and communication technology in dementia care. Scand. J. Caring Sci. **26**, 104–112 (2012)
5. Clark, S., Elswick, S., Gabriel, M., Gurupur, V., Wisniewski, P.: Transitions of care: a patient-centered perspective of health information systems that support post-acute care. J. Integr. Des. Process Sci. **20**, 95–110 (2016)
6. McKenzie, B., Bowen, M.E., Keys, K., Bulat, T.: Safe home program: a suite of technologies to support extended home care of persons with dementia. Am. J. Alzheimers Dis. Demen. **28**, 348–354 (2013)
7. Robinson, L., Brittain, K., Lindsay, S., Jackson, D., Olivier, P.: Keeping In Touch Everyday (KITE) project: developing assistive technologies with people with dementia and their carers to promote independence. Int. Psychogeriatr. **21**, 494–502 (2009)
8. Cahill, S.: "It gives me a sense of independence" Findings from Ireland on the use and usefulness of assistive technology for people with dementia (2007)
9. Cahill, S.: Technology in dementia care (2007)
10. Gitlin, L.N., Chee, Y.K.: Use of adaptive equipment in caring for persons with dementia at home. Alzheimer's Care Q. **7**, 32–40 (2006)
11. Novitzky, P., Smeaton, A.F., Chen, C., Irving, K., Jacquemard, T., O'Brolcháin, F., O'Mathúna, D., Gordijn, B.: A review of contemporary work on the ethics of ambient assisted living technologies for people with dementia. Sci. Eng. Ethics **21**, 707–765 (2015)
12. Horvath, K.J., Trudeau, S.A., Rudolph, J.L., Trudeau, P.A., Duffy, M.E., Berlowitz, D.: Clinical Trial of a Home Safety Toolkit for Alzheimer's Disease. https://www.hindawi.com/journals/ijad/2013/913606/
13. Bossen, A.L., Kim, H., Williams, K.N., Steinhoff, A.E., Strieker, M.: Emerging roles for telemedicine and smart technologies in dementia care. Smart Homecare Technol. Telehealth **3**, 49–57 (2015)
14. Mao, H.-F., Chang, L.-H., Yao, G., Chen, W.-Y., Huang, W.-N.W.: Indicators of perceived useful dementia care assistive technology: caregivers' perspectives. Geriatr. Gerontol. Int. **15**, 1049–1057 (2015)
15. Mulvenna, M., Hutton, A., Coates, V., Martin, S., Todd, S., Bond, R., Moorhead, A.: Views of caregivers on the ethics of assistive technology used for home surveillance of people living with dementia. Neuroethics **10**, 255–266 (2017)
16. Gibson, G., Dickinson, C., Brittain, K., Robinson, L.: The everyday use of assistive technology by people with dementia and their family carers: a qualitative study. BMC Geriatr. **15**, 89 (2015)
17. Baumer, E.P.S., Burrell, J., Ames, M.G., Brubaker, J.R., Dourish, P.: On the importance and implications of studying technology non-use. Interact. New Vis. Hum.-Comput. Interact. **22**, 52 (2015)

18. Braun, V., Clarke, V.: Using thematic analysis in psychology. Q. Res. Psychol. **3**, 77–101 (2006)
19. Sharma, N., Chakrabarti, S., Grover, S.: Gender differences in caregiving among family - caregivers of people with mental illnesses. World J. Psychiatry **6**, 7–17 (2016)
20. Duan, L., Street, W.N., Xu, E.: Healthcare information systems: data mining methods in the creation of a clinical recommender system. Enterp. Inf. Syst. **5**, 169–181 (2011)
21. Brubaker, J.R., Hayes, G.R., Dourish, P.: Beyond the grave: facebook as a site for the expansion of death and mourning. Inf. Soc. **29**, 152–163 (2013)

Touchscreen-Based Haptic Information Access for Assisting Blind and Visually-Impaired Users: Perceptual Parameters and Design Guidelines

Hari Prasath Palani[1,2(✉)], Jennifer L. Tennison[3], G. Bernard Giudice[2], and Nicholas A. Giudice[1,2]

[1] Spatial Informatics Program: School of Computing and Information Science, The University of Maine, Orono, ME 04469, USA
{hariprasath.palani,nicholas.giudice}@maine.edu
[2] VEMI Lab, The University of Maine, Orono, ME 04469, USA
bernie.giudice@gmail.com
[3] Department of Aerospace and Mechanical Engineering, Saint-Louis University, St. Louis, MO, USA
jen.tennison@slu.edu

Abstract. Touchscreen-based smart devices, such as smartphones and tablets, offer great promise for providing blind and visually-impaired (BVI) users with a means for accessing graphics non-visually. However, they also offer novel challenges as they were primarily developed for use as a visual interface. This paper studies key usability parameters governing accurate rendering of haptically-perceivable graphical materials. Three psychophysically-motivated usability studies, incorporating 46 BVI participants, were conducted that identified three key parameters for accurate rendering of vibrotactile lines. Results suggested that the best performance and greatest perceptual salience is obtained with vibrotactile feedback based on: (1) a minimum width of 1 mm for detecting lines, (2) a minimum gap of 4 mm for discriminating lines rendered parallel to each other, and (3) a minimum angular separation (i.e., cord length) of 4 mm for discriminating oriented lines. Findings provide foundational guidelines for converting/rendering visual graphical materials on touchscreen-based interfaces for supporting haptic/vibrotactile information access.

Keywords: Assistive technology · Haptic information access
Haptic interaction · Multimodal interface · Design guidelines

1 Introduction

Accessing graphical information is a major challenge for blind and visually-impaired (BVI) individuals. Text-to-speech programs such as JAWS for Windows (www.freedomscientific.com), VoiceOver for iOS-based devices (www.apple.com/accessibility/voiceover), and TalkBack for Android devices (www.google.com/accessibility/), have largely resolved the issue of non-visual access to textual and verbal information. However, there are currently no analogous technologies for providing

© Springer International Publishing AG, part of Springer Nature 2019
T. Z. Ahram and C. Falcão (Eds.): AHFE 2018, AISC 794, pp. 837–847, 2019.
https://doi.org/10.1007/978-3-319-94947-5_82

similar nonvisual access to graphical information. For decades, many researchers, developers, and companies have attempted to resolve this issue, but the solutions that have been developed have made little progress in reaching BVI end-users [1–3]. This is problematic, as a substantial amount of informational content used in educational settings, the workplace, or in myriad everyday activities is presented in graphical formats. Thus, unless new graphical access solutions are developed, more than 12 million BVI people in the U.S. (and 285 million worldwide) will continue to experience negative consequences of this accessibility gap on their educational, vocational, and navigational needs/success [4–6]. A considerable amount of work has been carried out on techniques for converting 2D graphical information, such as graphs and maps, into tangible versions that are developed and rendered using thermoform machines, tactile embossers, force-feedback devices, refreshable haptic displays, surface-haptic displays, tactile-shape displays and vibrotactile displays [7–12]. While these approaches have demonstrated utility in supporting access to graphical content, they also suffer from various shortcomings, such as significant expense, non-portability, and lack of ability to render graphics in a dynamic, real-time context [3, 13, 14].

Advancements in touchscreen-based smart computing devices (such as smartphones and tablets) have transformed the way the BVI demographic interacts with digital information. Most of these touchscreen devices are capable of providing multimodal feedback (i.e., through visual, auditory, and haptic cuing). In supporting universal design principles, most of these devices are also designed with accessibility features as part of the out-of-the box native interface, such as Apple's VoiceOver and Google's TalkBack. Owing to these advantages, and the ability to leverage many of these built-in multimodal features to support other tasks, there has been growing interest among researchers and developers in utilizing touchscreen-based smart devices as the core computational platform for providing non-visual graphical access to BVI users. Solutions have been developed based on auditory cues [15–17], vibratory cues [18–20], or combinations of the two [21–23]. Several recent approaches have also utilized electro-static screen overlays that were coupled with touchscreen devices to generate frictional forces between the contact finger and the screen [24, 25].

These solutions offer great promise as the touchscreen devices underlying this new wave of information-access are widely available at a reasonable cost and are capable of providing portable, refreshable graphical information. While promising, they also come with some inherent non-visual accessibility challenges, as the underlying device was primarily developed for use as an interactive visual display. It is argued here that the haptic feedback capabilities of touchscreen devices are vastly under-utilized. If implemented based on principled knowledge of human perceptual characteristics, as is being studied in the current research, the haptic modality could be a highly effective primary interaction style with these devices for BVI users, as well as for supporting sighted users in eyes-free applications [26]. One of the major challenges with touchscreen-based, non-visual graphical access is that the display is a flat, featureless surface, which does not provide the meaningful cutaneous stimulation that one would receive from apprehending traditional raised tangible graphics. To overcome this limitation, touchscreen-based interactions must rely on extrinsic feedback (e.g., vibration, audio, or electrostatic frictional cues) to indicate contact with an on-screen graphical element. This extrinsic feedback means that it is more difficult to distinguish fine detail and precise spatial information on a touchscreen that would otherwise be easily discernible from physical access using tangible graphics or from visual perception on the same touchscreen display.

For accurate non-visual haptic interpretation of the on-screen rendered graphical elements, users must follow a three-step process: (1) employ proprioception (i.e., force, position and motion sensors) to keep track of their finger position within some frame of reference, defined by the body or external landmarks such as the display frame, (2) extract the spatial information by synchronously interpreting the vibrotactile cues that innervate pacinian corpuscles in the fingertip, and (3) interpret the on-screen stimuli by associating the perceived sensory information with the on-screen graphical element [21, 30]. Because of these differences, graphical materials rendered on touchscreen-based interfaces should be schematized and rendered differently from techniques used for creating traditional tangible graphics. Although several studies have shown initial efficacy of utilizing touchscreen-based devices to address the non-visual graphical accessibility issue [18–20], they have all utilized different parameters for their evaluations. For instance, a ~ 0.35 in. (which is 8 times the size of traditional embossed graphical lines) was utilized as the optimal line width for rendering and accessing shapes, graphs and maps using a *Vibro-Audio Interface* (VAI) on a 7.0 in. android galaxy tablet [27, 28]. Similarly, a target size of ~ 0.17 in. (48pixel) was used in the *Timbremap* project for map exploration using an iPhone [16] and a rendering width of ~ 0.20 in. was used for shape identification in the *GraVVITAS* project, which was based on a Dell Latitude XT touchscreen Tablet [18]. For these alternative non-visual access solutions to succeed, it is crucial that the underlying graphical material is schematized and rendered based on perceptual parameters that are empirically identified to support accurate haptic perception on touchscreen-based displays. Towards this end, this paper conducted three psychophysically-motivated experiments to investigate three key perceptual parameters for detecting on-screen vibrotactile lines (Exp 1), discriminating straight vibrotactile lines (Exp 2) and discriminating oriented vibrotactile lines (Exp 3). The Institutional Review Board (IRB) of the University of Maine approved all three studies and all 46 participants gave informed consent and were paid for their participation.

2   Experiment 1: Line Detection

Lines are a foundational element and a crucial spatial construct for rendering graphical materials such as graphs and maps (see Fig. 1. for a sample transit map). The ability to detect distinct lines using vibrotactile feedback is a key process for supporting haptic information extraction on touchscreen-based non-visual interfaces. To support accurate haptic perception and apprehension of the overall graphical information, each vibrotactile line must be rendered at a minimum width that not only supports detection but also preserves the spatial structure and topology of the original visual graphical rendering. Accordingly, experiment 1 was designed to identify the minimum threshold for rendering graphical lines that best supports detection via vibrotactile cuing.

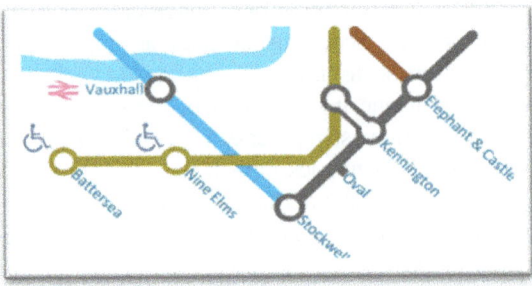

Fig. 1. Sample transit map

2.1 Method

Twenty blind and visually-impaired participants (nine males and eleven females, ages 27–74) were recruited for the study. Seven different line widths (0.125, 0.25, 0.5, 1, 2, 4, and 8 mm) were compared. The seven line widths were chosen to reflect a meaningful range, e.g. the smallest width of 0.125 mm is approximately equivalent to the size of a single pixel on most touchscreen displays. From this base, the stimuli increased linearly by a factor of 2 up to 8 mm, which is known from empirical studies to be sufficient to perform the three-step process described earlier [27, 28, 31]. The vibrotactile lines were all rendered using an experimental prototype, called a vibro-audio interface (see [27] for technical details and implementation of the VAI), which was implemented on a 5.6 in. Galaxy Note4 Edge Android phablet (with a screen resolution of 524 ppi). The vibratory feedback was triggered using Immersion Corp's (www.immersion.com) universal haptic layer (UHL). On-screen contact with the vibrotactile lines triggered constant vibratory feedback based on the UHL effect "Engine1_100" which uses a repeating loop at 250 Hz with 100% power. The study followed a within-subjects design, where each participant performed 84 line counting trials (resulting in 360 observations for each tested line width). In each trial, the randomly generated lines were rendered on the device screen. Participants were asked to move their finger across the screen from left to right at a constant speed, to count the number of vibrotactile lines perceived during this scan and to verbally indicate this number to the experimenter. Participants performed 5 practice trials before performing the 84 experimental trials. Each participant took between 15 and 30 min to complete the entire experiment. Based on this design, line detection accuracy was compared between the 7 line widths.

2.2 Results and Discussion

A one-way repeated measures ANOVA comparing detection accuracy across the seven tested line widths revealed a statistically significant difference, $F(6, 1526) = 89.913$, $p < 0.001$, $\eta2 = 0.261$. Subsequent post-hoc t-tests with Bonferroni correction indicated that line widths 0.125, 0.25 and 0.5 mm were statistically different in detection accuracy from each other and exhibited significantly lower detection accuracy than the remaining four line widths (all $ps < 0.05$). However, there were no statistically

Table 1. Mean detection accuracy and standard deviation across tested line widths

Length (mm)	Mean	SD
0.0125	0.39	0.489
0.25	0.54	0.499
0.5	0.75	0.432
1	0.94	0.237
2	0.93	0.261
4	0.94	0.237
8	0.96	0.188

significant differences (all $ps > 0.05$) observed in detection accuracy between line widths of 1, 2, 4, and 8 mm (see Table 1 for means and SDs).

These results indicate that rendering graphical (vibrotactile) lines at a width of 1 mm is sufficient for tasks requiring simple line detection. While adopting a line width wider than 1 mm may improve saliency, it will also consume more screen space than necessary. Since touchscreen devices have limited screen real-estate, we argue that adopting wider than a 1 mm line width is a poor design decision.

3 Experiment 2: Discrimination of Vibrotactile Lines Rendered Parallel to Each Other

Graphical materials often have multiple lines rendered in close proximity to each other. For instance, consider the transit map depicted in Fig. 1, where there are three different transit lines that make up the actual map. To be recognized as a distinct transit line, each of the lines must be separated from its adjacent line by a gap wider than the minimum perceivable vibrotactile gap width. If the transit lines of this example were to be rendered too close to each other on the touchscreen display, they will be haptically perceived as one line, owing to the sparse spatial resolution of touch. On the other hand, rendering them further apart, using too large of an inter-line gap, is a poor design decision, as it consumes unnecessary screen space on the limited information density displays available on touchscreen-based devices. In addition to the actual gap width, the width of the bounding vibrotactile lines might also influence the perception of the gap. This is because the vibrotactile feedback on touchscreen devices is generated via actuation of an embedded vibratory motor, which has a temporal lag in turning the motor on or off. Any lag due to turning the motor on or off, could in principle, create a spurious perception of a line being narrower or wider than its actual size. Depending on the width of the bounding vibrotactile lines, this spurious haptic perception could mask the gap between them, resulting in the two lines being incorrectly perceived as one. Accordingly, the second experiment was designed to identify the minimal gap width that supports discrimination of two or more vibrotactile lines rendered parallel to each other while also evaluating whether the width of adjacent lines causes spurious haptic perception due to the lag in vibrotactile feedback.

3.1   Method

Eighteen blind and visually-impaired participants (seven males and eleven females, ages 27–74) were recruited for the study. Five gap widths (i.e., 0.25, 0.5, 1, 2, and 4 mm) were compared. The gap widths were chosen such that 1 mm (as was found in experiment (1) was kept as the median value and increased (or decreased) by a factor of two. The apparatus, implementation, and procedure was similar to that of experiment 1. To assess the effect of temporal lag in triggering vibrotactile feedback and to better characterize and understand the relation of line width on gap detection accuracy, the five gap separations were tested across three different line widths (i.e., 1, 2, and 4 mm). A gap trial could have 1, 2, or 3 pairs of lines. The line widths and gap widths were held constant within each trial. To prevent learning effects, 9 dummy trials (i.e., trials where the rendered stimuli did not have gaps) were added to the 45 gap detection trials (5 gaps by 3 line widths by 3 line pairs). In each trial, randomly generated lines were rendered on the screen. Participants were asked to move their finger across the screen from left to right at a constant speed, to count the vibrotactile lines perceived during the scan, and to verbally indicate this count to the experimenter. Participants performed 5 practice trials before performing the 54 experimental trials, which resulted in 324 observations for each tested gap width (i.e., 6 instances for each of the 3 line widths by 18 participants). Each participant took between 20 and 30 min to perform the task. Based on this design, the accuracy in gap detection was compared as a function of: (1) gap width (i.e., the space between a pair of parallel vibrotactile lines), and (2) the vibrotactile line width.

3.2   Results and Discussion

A one-way repeated measures ANOVA comparing the detection accuracy across the five tested gap widths revealed a statistically significant difference between the gap widths $F(4, 805) = 16.859, p < 0.001, \eta2 = 0.077$. Subsequent post-hoc paired sample t-tests with Bonferroni correction indicated that gap widths 0.25 and 0.5 mm were statistically different in detection accuracy from each other and exhibited reliably lower detection accuracy than the remaining two gap widths (all $ps < 0.05$). Of the tested gap widths, only the 2 mm and 4 mm gap widths exhibited detection accuracy greater than is required by traditional psychophysical procedures (i.e., 75% detection accuracy [32]). While the trend of these data suggests that further increasing the gap width would likely lead to a corresponding increase in detection accuracy, it will also consume excessive screen real estate and will eventually reduce the efficiency (and practicality) of using touchscreen-based nonvisual graphical access solutions. The results also clearly demonstrate that gap detection accuracy was significantly influenced by the width of the bounding vibrotactile lines, with wider lines exhibiting higher detection rates. On comparing the detection accuracy across the three line widths, a repeated measures ANOVA revealed a statistically significant difference between line widths $F(2, 807) = 31.323, p < 0.001, \eta2 = 0.072$. Data here suggest that the detection accuracy increased with an increase in line widths. These findings suggest that gap detection is not only dependent on the width of the gap but also on the width of the bounding lines. We interpret these results as demonstrating that the line and gap width parameters should not be treated separately when creating/authoring vibrotactile graphical information. While exp1 suggested that a 1 mm width is sufficient for detection of individual line, the data here suggests that a line width of at least 2 mm, in conjunction with an inter-line gap of at least 4 mm, should be maintained for distinguishing distinct parallel lines.

4 Experiment 3: Discrimination of Oriented Vibrotactile Lines

As stated earlier, with the extrinsic cuing mechanism employed on touchscreen devices, users can only detect whether the touched location is on or off of an on-screen graphical element but they cannot directly perceive any other meaningful information such as stimulus width/length/orientation/angle. To extract this type of information from touchscreen devices, users must perform exploratory procedures (Eps), which are a stereotyped pattern of manual exploration observed when people are asked to learn about a particular object property during voluntary manual exploration without vision [33]. While experiments 1 and 2 indicated the minimum line and gap widths for detection of parallel vibrotactile lines, it is not clear whether these parameters are generalizable to oriented vibrotactile lines and angular graphical elements (For example, see the green and yellow transit lines on Fig. 1.). For identifying such oriented lines and judging the angle subtended between them, users typically employ a 'circling' strategy, where they move their finger in a circular pattern around the intersection as their exploratory procedure [21, 22, 31]. Based on this exploration strategy, we posit here that the arc of the circle formed between two oriented vibrotactile lines will be perceived by the user as the angular magnitude subtended between the two lines. The cord length (and by extension the angular separation between two oriented lines) is a variable that is dependent on both the angle (θ) subtended between oriented lines and the radius (r) of the circle formed by the user while performing their exploratory procedure to apprehend the vertex/intersection of the lines. From a geometric standpoint, the straight-line distance between two angled lines is the cord length (cord length = $2r \sin (\theta/2)$), a variable that depends on: (1) θ - angle subtended between the lines, (2) r – the radius of the traced circle, or (3) both 1 and 2. In theory, the 4 mm gap width identified in exp-2 should be translated into a 4 mm cord length for accurate detection of distinct oriented lines. However, the cord length can vary depending on the angle, the radius, or both. For instance, an angle of 5° will lead to a 4 mm cord length with a 1-in. radius circle, and an angle of 2° will lead to a 4 mm cord length with a 2-in. radius circle. Accordingly, experiment 3 was designed to assess the influence of the angle, radius, and cord length on users' ability to discriminate oriented vibrotactile lines.

4.1 Method

Eight blind and visually-impaired participants (three males and five females, ages 25–74) were recruited for the study. The stimulus set was designed as a simple network map where multiple vibrotactile lines were converging to/diverging from an intersection point at the center. The number of lines in each stimuli ranged from 5 to 9 based on

Miller's "The Magical Number Seven, Plus or Minus Two" [34]. As stated earlier, the radius was set as a constant value of 1-in. and 2-in. for conditions 1 and 2 respectively. At a radius of 1-in. from the intersection, the minimum gap width of 4 mm (i.e., cord length in this context) was translated to an angular magnitude of $\sim 9°$. Similarly, at a 2-in. radius, the gap width of 4 mm width was translated to a $\sim 5°$ angular magnitude. To evaluate the influence of cord length (i.e., gap) on the perception of oriented lines, two additional angles (2° and 22°) were also added to the stimulus set that approximately translated to the 4 mm gap width at a radius of 0.5-in. and 4-in. (i.e., the radius of the two primary conditions increased and decreased by a factor of 2).

The stimuli were all rendered using the vibro-audio interface implemented on a 10.1 in. Galaxy Tab 3 Android Tablet (with a screen resolution of 264 ppi). For controlling the circle radius in each condition and for assisting users with the circling strategy, two circular paper stickers of 4 mm width (one at 1-in. from the center and the other at 2-in. from the center) were affixed on the screen (see Fig. 2). In addition, the intersection point (center of the screen) was also demarcated with a paper sticker of 10 mm radius. To assist participants with orienting themselves on the screen, each circle had a start point (indicated by a tactile marker at the 5 o'clock position). A trial rendered 5, 6, 7, 8, or 9 lines on the screen. In each trial, the angular magnitude between adjacent lines was kept constant irrespective of line number. The order of the conditions (1-in. versus 2-in. radius) was balanced across the participants and the order of stimuli presentation in each condition was randomized. Each participant performed 4 practice trials before performing 28 oriented line counting trials in each condition (resulting in 180 observations for each tested angular magnitude). Each participant took between 20 and 40 min to complete the entire experiment. Based on this design, oriented line detection accuracy was compared as a function of 4 angular magnitudes and across 2 circling conditions.

Fig. 2. Experimental device setup with sample experimental stimuli

4.2   Results and Discussion

A one-way repeated measures ANOVA comparing the line counting accuracy across the four tested angles revealed a statistically significant difference for both circling conditions. The f and p values are as follows,

$$For\ the\ 1 - inch\ circular\ path,\ F(3, 220) = 10.057, p < 0.001,$$
$$For\ the\ 2 - inch\ circular\ path,\ F(3, 220) = 8.574, p < 0.001,$$

Post-hoc t-tests with Bonferroni correction revealed that the difference in line counting accuracy between observations with a 2° angle compared to the other three angles was significant ($p < 0.001$). However, there were no significant differences between the other three angles (5°, 9°, and 22°). Overall, findings indicate that a 4 mm cord length must be maintained to detect/discriminate oriented vibrotactile lines using a circling strategy.

5 Conclusion

Smartphone usage among the BVI demographic has sharply increased in recent years, going from 12% in 2009 to 82% in 2014 [35]. Given the magnitude of this touchscreen device adoption/usage trend among BVI users, it is of utmost importance to investigate and identify the key usability parameters and cognitive abilities pertinent to maximizing accurate use of these interfaces. This paper described three experiments that assessed three key usability parameters for non-visually detecting and discriminating graphical elements using vibrotactile cues on commercial touchscreen interfaces. Overall, results showed that a width of 1 mm is sufficient for detecting on-screen graphical elements using vibratory feedback (Exp 1), but a line width of 2 mm along with a 4 mm inter-line gap must be maintained for accurate detection and discrimination of distinct vibrotactile lines that are parallel to each other (Exp 2). Similarly, experiment 3 suggested that a 4 mm cord length (similar to the 4 mm gap width) must be maintained for accurate detection and discrimination of oriented vibrotactile lines. It is important to consider that these parameters are not just based on cutaneous sensation but represent the value at which a user can effectively perform the three-step process of employing proprioception to keep track of their finger position within some frame of reference, extracting the spatial information by synchronously interpreting the vibrotactile cues, and associating the perceived sensory information with on-screen graphical elements.

This work adds to the growing corpus of research demonstrating the efficacy of these interfaces as the latest category of information-access technology. To best utilize this technological trend, findings from this work provide much-needed foundational guidelines for converting/rendering visual graphical elements on touchscreen-based interfaces for supporting haptic information access. These findings, along with other work in this domain, are the first step towards development of a set of robust design guidelines to provide improved haptic access on touchscreen devices supporting a wide range of non-visual applications.

Acknowledgments. We acknowledge support from NSF grants CHS-1425337 and ECR DCL Level 2 1644471 on this project.

References

1. Perkins: Perkins Museum. http://www.perkins.org/
2. Giudice, N.A., Legge, G.E.: Blind navigation and the role of technology. In: Helal, A., Mokhtari, M., Abdulrazak, B. (eds.) Engineering Handbook of Smart Technology for Aging, Disability, and Independence, pp. 479–500. Wiley (2008)
3. O'Modhrain, S., Giudice, N.A., Gardner, J.A., Legge, G.E.: Designing media for visually-impaired users of refreshable touch displays: possibilities and pitfalls. Trans. Haptics. **8**, 248–257 (2015)
4. Kaye, H.S., Kang, T., LaPlante, M.P.: Mobility device use in the United States. Disability Statistics Report (14), Washington, D.C., USA (2000)
5. Clark-Carter, D.D., Heyes, A.D., Howarth, C.I.: The efficiency and walking speed of visually impaired people. Ergonomics **29**, 779–789 (1986)
6. World Health Organization: Visual impairment and blindness Fact Sheet (2011). http://www.who.int/mediacentre/factsheets/fs282/en/
7. Rowell, J., Ungar, S.: The world of touch: an international survey of tactile maps. Part 1: production. Br. J. Vis. Impair. **21**, 98–104 (2003)
8. Rowell, J., Ungar, S.: The world of touch: an international survey of tactile maps. Part 2: design. Br. J. Vis. Impair. **21**, 105–110 (2003)
9. Braille Authority of North America: Guidelines and Standards for Tactile Graphics (2010). www.brailleauthority.org/tg
10. Bach-Y-Rita, P., Collins, C.C., Saunders, F.A., White, B., Scadden, L.: Vision substitution by tactile image projection (1969)
11. Hasser, C.: HAPTAC: A Haptic Tactile Display for the Presentation of Two-Dimensional Virtual or Remote Environments (1995)
12. Phantom: Phantom Omni. http://geomagic.com/en/products-landing-pages/sensable
13. Zeng, L., Weber, G.: Audio-haptic browser for a geographical information system. In: Computers Helping People with Special Needs, pp. 466–473 (2010)
14. Rastogi, R., Pawluk, D.T.V.: Toward an improved haptic zooming algorithm for graphical information accessed by individuals who are blind and visually impaired. Assist. Technol. **25**, 9–15 (2013)
15. Williamson, J.R., Crossan, A., Brewster, S.: Multimodal mobile interactions: usability studies in real world settings. In: Proceedings 13th International Conference Multimodal Interfaces, ICMI 2011, pp. 361–368 (2011)
16. Su, J., Rosenzweig, A., Goel, A., Lara, E.D., Truong, K.N.: Timbremap: enabling the visually-impaired to use maps on touch-enabled devices. In: Proceedings of the 12th International Conference on Human Computer Interaction with Mobile Devices and Services, pp. 17–26. ACM (2010)
17. Hoggan, E., Brewster, S.: Designing audio and tactile crossmodal icons for mobile devices. In: Proceedings 9th International Conference Multimodal Interfaces, ICMI 2007, p. 162 (2007)
18. Goncu, C., Marriott, K.: GraVVITAS: generic multi-touch presentation of accessible graphics. In: Lecture Notes in Computer Science, vol. 6946, pp. 30–48 (2011)
19. Tennison, J.L., Gorlewicz, J.L.: Toward non-visual graphics representations on vibratory touchscreens: shape exploration and identification. In: Bello, F., Kajimoto, H., Visell, Y. (eds.) Haptics: Perception, Devices, Control, and Applications: 10th International Conference, EuroHaptics 2016, London, UK, July 4–7 2016, Proceedings, Part II, pp. 384–395. Springer International Publishing, Cham (2016)

20. Gershon, P., Klatzky, R.L., Palani, H.P., Giudice, N.A.: Visual, tangible, and touch-screen: comparison of platforms for displaying simple graphics. Assist. Technol. **28**, 1–6 (2016)
21. Palani, H.P., Giudice, N.A.: Principles for designing large-format refreshable haptic graphics using touchscreen devices. ACM Trans. Access. Comput. **9**, 1–25 (2017)
22. Palani, H.P., Giudice, N.A.: Evaluation of non-visual panning operations using touch-screen devices. In: Proceedings 16th International ACM SIGACCESS Conference on Computers & Accessibility. ACM (2014)
23. Palani, H.P., Giudice, U., Giudice, N.A.: Evaluation of non-visual zooming operations on touchscreen devices. In: Universal Access in Human-Computer Interaction. Interaction Techniques and Environments: 10th International Conference, UAHCI 2016, Held as Part of HCI International 2016, Toronto, ON, Canada, 17–22 July 2016, Proceedings, Part II, pp. 162–174. Springer International Publishing (2016)
24. Mullenbach, J., Shultz, C., Colgate, J.E., Piper, A.M.: Exploring affective communication through variable - friction surface haptics. In: Proceedings SIGCHI Conference on Human Factors in Computing Systems, pp. 3963–3972 (2014)
25. Xu, C., Israr, A., Poupyrev, I., Bau, O., Harrison, C.: Tactile display for the visually impaired using TeslaTouch. In: Proceedings CHI EA 2011, pp. 317–322 (2011)
26. Challis, B.: Tactile interaction. In: Soegaard, M., et al. (eds.) Encyclopedia of Human-Computer Interaction, 2nd edn. (2012)
27. Giudice, N.A., Palani, H.P., Brenner, E., Kramer, K.M.: Learning non-visual graphical information using a touch-based vibro-audio interface. In: Proceedings 14th International ACM SIGACCESS Conference on Computers and Accessibility, pp. 103–110. ACM Press, New York (2012)
28. Palani, H.P.: Making Graphical Information Accessible without Vision using Touch-Based devices, Unpublished Masters Thesis (2013)
29. Loomis, J.M., Klatzky, R.L., Giudice, N.A.: Sensory substitution of vision: importance of perceptual and cognitive processing. In: Manduchi, R., Kurniawan, S. (eds.) Assistive Technology for Blindness and Low Vision, pp. 162–191. CRC, Boca Raton (2012)
30. Klatzky, R.L., Giudice, N.A., Bennett, C.R., Loomis, J.M.: Touch-screen technology for the dynamic display of 2D spatial information without vision: promise and progress. Multisens. Res. **27**, 359–378 (2014)
31. Raja, M.K.: The development and validation of a new smartphone based non-visual spatial interface for learning indoor layouts. Unpublished Masters Thesis (2011)
32. Gescheider, G.A.: Psychophysics: The Fundamentals. Lawrence Erlbaum Associates Publishers, Mahwah (1997)
33. Lederman, S.J., Klatzky, R.L.: Hand movements: a window into haptic object recognition. Cogn. Psychol. **19**, 342–368 (1987)
34. Miller, G.A.: The magical number seven, plus or minus two: some limits on our capacity for processing information. Psychol. Rev. **63** (1956)
35. WebAim: WebAim: Screen Reader User Survey #5 Results. http://webaim.org/projects/screenreadersurvey5/

Towards a Macroscopic View of Using an Assistive Technology for Mobility for Its Development: Assessing Users' and Co-users' Experience

Anne-Marie Hébert[1(✉)], Philippe Archambault[1], and Dahlia Kairy[2]

[1] School of Physical and Occupational Therapy,
McGill University, Montreal, Canada
anne-marie.hebert@mail.mcgill.ca,
philippe.archambault@mcgill.ca
[2] Rehabilitation School, University of Montreal,
Montreal, Canada
dahlia.kairy@umontreal.ca

Abstract. This paper aims to propose an approach to include a holistic user experience (UX) perspective in the development of an assistive technology for mobility, more precisely an intelligent powered wheelchair (IPW). First, the UX related to the anticipated experience (powered wheelchair users, caregivers, clinicians) was obtained in a previous study in order to define technological and functional specifications. Second, the UX related to the episodic and momentary experience were highlighted within a Living lab approach. In that framework, we tested with end-users an IPW in an ecological setting, a mall. For our next step, we propose a methodology to apprehend the cumulative experience. This study will take into account co-users, i.e. co-workers, other pedestrians, etc., that most studies do not take into account and that, in diverse ecological settings.

Keywords: Assistive technology · Intelligent powered wheelchair
System assessment · User experience · Technological specifications

1 Introduction

This paper proposes a methodological approach to include the feedback of the user experience (UX) in the development of an assistive technology (AT) for mobility, an intelligent power wheelchair (IPW). We aim to widen the concept of the UX and of other stakeholders sharing space with the IPW (co-users) in order to obtain insights for the development of the IPW. This will allow the implementation of functions and the development of interaction modalities that directly satisfy the real needs of the IPW (co-)users. The user-centered design approach focuses on the necessity to take into account the specific needs of certain populations, e.g. users with sensorimotor, cognitive disabilities, etc. [1]. Indeed, studies points out that some populations are denied the use of powered wheelchairs (PW) due to safety problems [2, 3].

© Springer International Publishing AG, part of Springer Nature 2019
T. Z. Ahram and C. Falcão (Eds.): AHFE 2018, AISC 794, pp. 848–859, 2019.
https://doi.org/10.1007/978-3-319-94947-5_83

The new challenge is to consider a holistic perspective of UX, i.e. anticipated, momentary, episodic and cumulative experiences; it involves going beyond the instrumental characteristics of the technology, i.e. the realization of a task, in order to access the non-instrumental needs of users. We believe that those non-instrumental needs, translated into specifications, could enhance the overall quality of the AT. For example, the potential contributions of new technologies, particularly the intelligent powered wheelchair (IPW), could range from improved self-confidence to increased social participation [2]. In this study, these benefits were highlighted for IPW use by diverse users with cognitive, physical or visual limitations.

In order to contribute to the development of IPWs, a multidisciplinary team from the fields of Computer Science, Engineering and Rehabilitation is developing an IPW prototype (Fig. 1). This prototype has intelligent navigation features, which combines technologies related to artificial intelligence (AI) and robotics. Built from a PW available on the market, it has several interfaces, i.e. a joystick and a tactile tablet. It is equipped with several navigation sensors, i.e. laser rangefinder and sonar. The prototype makes it possible to determine trajectories, to follow a planned path, to avoid static and dynamic obstacles, to cross doors and to follow a specific object such as a wall, a person or a group of people.

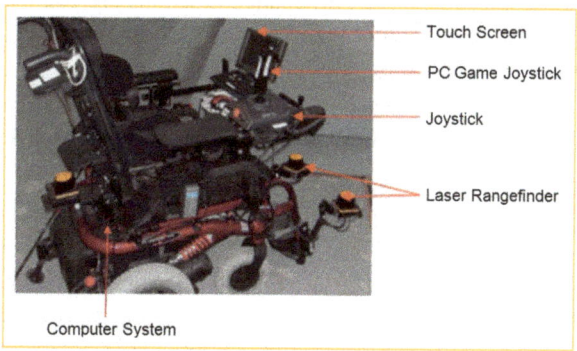

Fig. 1. The IPW prototype and its components.

Hitherto, our team's studies have examined several perspectives including technological aspects such as the human-robot interaction (HRI) and intelligent modules of the IPW and a more social perspective of users, caregivers and clinicians to validate the functions to be implemented [4–9]. The purpose of this paper is twofold. It aims to describe a process of an assistive technology development from an ergonomic perspective, and more specifically, user experience (UX). In addition, it proposes a methodology for our next steps. The overall goal is to broaden the concept of UX for the development of mobility aids to match the real needs of users and co-users, i.e. non-users who are in some way impacted by the system being used such as family members, caregivers, pedestrians sharing public space, etc., from a developmental perspective. At first, the concepts related to the UX will be discussed and then studies and/or methodology regarding these concepts will be presented based on the development of the IPW prototype.

2 The UX of Assistive Technology for Mobility

The methods used to study the UX offer a first-person perspective, that of the users. But what about co-experience, the experience of non-users who are in some way also impacted by the system being used? Can the same methods be used to describe the experience of the co-user, i.e. family, caregivers, pedestrians sharing of public space, etc.? Can users' needs be re-appropriated and/or reinterpreted for other groups, e.g. co-users? In this vein, the first-person perspective could be enriched by a third-person perspective, i.e. family, caregiver, pedestrians, etc. However, how can the UX be enriched by this third-person perspective and therefore impact the design of a product or service?

Inclusive design requires not only to converge on the perception and judgments of the quality of the design of a product or service, but also cooperation with users in the production of the object or service designed [10]. With this in mind, how can we attempt to include the different parties involved in a wide range of ecological environments where IPW is being used? These latter situations are complex; they include various people sharing the spatial environment where an IPW user can navigate. A potential answer lies in the analysis of the holistic UX.

2.1 Beyond Traditional UX to Consider the Non-instrumental Aspects

Since the beginning of research focusing on human-machine interaction (HMI) and ergonomics of technologies for productivity, the object has generally been to perform tasks efficiently, effectively and with satisfaction, i.e. the instrumental characteristics of a product [1]. From this perspective, tasks are the focal point of user-centered analyzes and usability assessments, and thus of the instrumental quality of products [11].

In the context of assistive technology for mobility, it was emphasized that providing appropriate, well-designed and well-equipped assistive devices increases mobility, and brings new opportunities for education, work, learning and social life for wheelchair users [3]. Thus, the emergence of perspectives beyond the instrumental quality of technology is of particular importance.

The new challenge of this 'holistic' UX perspective therefore involves going beyond the instrumental characteristics of technologies in order to identify, define, prioritize and operationalize the non-instrumental needs of users to better translate them into product attributes and specifications [11]. From this perspective, technological and functional specifications could have a real impact on the overall quality that surrounds the instrumental quality.

2.2 Social Span of the UX: Considering the Co-experience

The use of technologies does not only involve direct uses, but also indirect use; it has an impact on the social environment beyond the immediate user. It is considered that the UX includes several protagonists involved in the use of a system. This UX not only covers the active user (e.g. IPW user), but also passive users (e.g. caregiver, clinician, pedestrians) involved in the social context of a device's use [12]. This characteristic is described by the term "co-experience", which refers to situations in which experiences are interpreted as situated and socially constructed [12].

2.3   Temporal Span of the UX

The UX spans over a longer timeframe than simply the use of a device. There is in fact a temporal spectrum where the UX can take several facets ranging from anticipated, momentary, episodic experience to cumulative experience [12]:

- The term "anticipated experience" refers to the possible use of a specific brand's product, using the presentation or demonstration of a new product. In this vein, it is therefore indirect experiences that can extend beyond the use of a product by including reflections on its potential use. In this context, users can have indirect experiences with a device even before having contact with it.
- The "momentary experience" is the experience when using a product or service.
- "Episodic experience" refers to the evaluation of a specific use episode of a system.
- The "cumulative experience" represents an overall view of the system built from multiple uses.

These different UXs shed a different light to the development of a product. Indeed, we can access imagined experiences, experiencing, reflection on a specific experience and a memory of a collection of various uses respectively to identify different users' needs in products' use cycle.

2.4   UX of an IPW: Studies of the Holistic UX

IPWs differ from conventional PWs because they have features that allow users to navigate autonomously or semi-autonomously, for example by avoiding obstacles, performing parking maneuvers, and so on. Thus, individuals who cannot control their wheelchair autonomously, either by a lack of dexterity, motor control of the arm, or due to cognitive or orthopedic limitations, could benefit from an IPW in order to increase their autonomy. So far, no IPW model is available on the market.

IPW prototypes have been the subject of recent studies to assess navigation, AI and different human-robot interactions (HRI) involved in their use, i.e. brain-machine interface, voice, joystick, tactile (e.g. see [2, 7, 8]). Until now, few of these investigations have involved future users in the evaluation [2, 4, 7, 9], and little research has been done in an ecological approach, i.e. in a context that reflects the real environments of the future users [2, 13]. Indeed, studies related to the development of IPWs have examined few real environments, but without the use of an IPW by the participants [4–6], a restricted spectrum of clinical situations with the use of the IPW by the end-users, e.g. rehabilitation center [7], long-term care center durations [13], and a number have used healthy subjects [14].

3   The Four Temporal Spans of UX

3.1   UX Anticipated: User Needs, Caregivers and Clinicians

Studies have highlighted the anticipated needs of users and co-users of IPW in relation to instrumental quality, i.e. transport, but also the non-instrumental quality, i.e. the psychosocial dimension of the use of IPW. In this line, the anticipated UX and co-experience were apprehended in our studies, in order to identify (co-)users' needs.

Methodology: Viewing a Video of the Product Followed by a Semi-structured Interview. Studies carried out by members of our research team [4–6] aligned with the psychosocial dimension of the IPW illustrated the main functions of the IPW prototype [7] through viewing of a 4-min video (see [5] for more information on the methodology). The video was presented to active users of IPW and co-users, i.e. clinicians and caregivers. Following the video, facilitators and barriers to using an IPW perceived by different (co-) users were discussed during semi-structured interviews guided by the Consortium for Assistive Technology Outcomes Measures Research (CATOR) model [15].

Anticipated Needs Identified. Using this approach, four categories of needs were identified, as describe below.

Mobility and Autonomy with an IPW. Mobility and independence needs are reported in many studies [2]. Along these lines, the studies [4–6] highlighted several themes raised by potential future users of IPW. These themes cover different navigation situations that are considered as problematic. These situations include restricted space management, obstacle avoidance and crowds, as well as navigation by following an object, e.g. wall, person or group.

Correspondence and Concordance of the IPW with the Context of the User. The need for correspondence and concordance can be taken into account from two perspectives. They can be related to the real activity of the user, i.e. the activity carried out at a specific moment, but also to the state of the user, i.e. his/her cognitive state as the level of fatigue to perform a maneuver, his/her uncertainty in the execution of a maneuver. The study [5] highlights the need for users to find features that correspond to their actual activities. Moreover, the study [2] emphasizes that the behavior of an IPW must be consistent with the purpose of its use, i.e. as an aid to mobility, training or as an evaluation tool. This requires a flexibility of the system to take into account several populations, objectives and contexts. For example, the feature of following an object was considered relevant only in the context where a caregiver or other assistant accompanies the IPW user.

Social Participation. Mobility difficulties are seen as a predictor of activities of daily living and can lead to decreased opportunities for socialization and social isolation [2]. The mobility provided by the use of intelligent powered devices can increase participation in daily activities and quality of life [2, 6]. Mainly, it was pointed out that an IPW may lead to an increase in the opportunities for social participation, a change in the experience of the participation and a reduction of the risks of accidents during participations [6]. Indeed, social contexts can be better managed with such functions as obstacle avoidance and navigational assistance implemented in an IPW.

Illustration of the Reliability of the System. A need derived directly from HMI is part of the reliability of the system. This refers, amongst other things, to the feedback given to the users about the state of the system. Findings suggest that several users would like to be reassured of the reliability of the system before their appropriation [4, 5]. They underline the need for a training period and an alert mechanism which would inform the users of any IPW'S malfunctions. This is intended to increase the users' feeling of security [16].

Technological Specifications to Implement the Identified Needs. From the identified needs, we highlighted several technological and functional specifications that should be taken into account in the further development of the IPW. These specifications come from more or less complex AI components and human-robot interaction (HRI).

It should be noted that studies on anticipated needs have involved end-users and people related to the users of the IPW. However, the studies carried out have not yet taken into account the potential needs of people unrelated to IPW users such as the pedestrians and passersby of the IPW entourage.

Different Functions to Support the Autonomy of the User. A common problem reported by PW users is navigation in the presence of obstacles or restricted spaces [4–6]. In order to provide users with increased autonomy, the IPW prototype has been equipped with functions such as obstacle avoidance and a tracking module. The former function allows the user to avoid damaging either themselves, the PW/IPW or the adjacent/near environment. The obstacle avoidance function allows the user to navigate in tight spaces without colliding with objects in the environment and manage navigation in busy areas. As for the tracking module, it allows the IPW to navigate by following a wall [7].

Flexibility of the System to Match the User's Context. In order to address the need for matching the system characteristics to the user's context, implementing the possibility to enabling/disabling the 'smart features' could be considered. In this sense, 'smart features' might be suitable for a larger population than reported in the study [5]. Indeed, it is reported that the presence of all functions would be non-essential for certain populations [2]. In contrast, the physical, perceptual and/or cognitive limitations associated with, for example, advancing age may hinder an individual from acquiring or maintaining the skills necessary to maneuver a PW.

In the case of obstacle avoidance and tracking functions, disabling these functions could allow the user to preserve and/or increase his/her navigational abilities in constrained environments and contexts. This flexibility of the system could also be appropriate in cases where the user experiences a temporary fatigue or more permanently sees his/her state of health deteriorate.

With respect to the activity of the user, the users could activate a function implemented in the prototype of 'group tracking' if he/she is in the situation where they must or can actually follow a group. In this case, the system would automatically regulate the navigation speed of the IPW.

In addition, a component of the system provides human-robot cooperation. The latter would allow the system to modulate the functions activated by the user to preserve the security of the user if compromised.

Control and Feedback to Emphasize System Reliability and Security. The reliability of the system can be enhanced through several possibilities depending on what type of system behavior should be reliable. Indeed, we can increase the reliability of the system via a remote control or via feedback for the user.

The use of a remote-control device would allow a clinician, caregiver or another person to take control if the need arises to secure the user's safety. In this case, the training period for example could be monitored and thus the user's safety would be supervised and/or maintained by a protagonist of the nearby environment.

Regarding the feedback, the system must provide users with information related to its status at all times [16]. In fact, the user should preferably have information related to the state of the system and to all the components of the system, e.g. the battery level, the activated functions and the engaged path. Furthermore, an onboard camera device is being developed to provide the visual information needed by the user on the space available behind the PW/IPW. This information device will allow, among other things, users to maneuver in reverse with less difficulty if they have mobility restrictions for moving their head. All this information should be included in the user interface (UI).

3.2   Momentary and Episodic UX: End Users and Ecological Environment

In the continuity of studies regarding IPW development, we tested the IPW prototype with real end-users. In addition, we proceeded with scenarios of use in an ecological environment, a shopping center. In this framework, our goal was to capture the momentary and episodic UX. This enables us to confirm and expand on the previous results obtained in laboratory and clinical settings. Furthermore, it will provide the opportunity to validate or not the HRI and functions implemented in the IPW prototype developed by our team.

The fit between the functions, the interaction with the IPW and the users' needs was investigated, amongst others. The goal was to collect insights on the use of the IPW from real users in an ecological setting for further development of the IPW.

Methodology: An in Vivo Approach and Self-confrontation Interviews. Twelve participants were recruited. They had all been using a PW in the community for at least one year, were 18 years of age or older, were able to express themselves in French or English, and had a long-term severe mobility limitation. Ethics approval was obtained from the research ethics board for the Centre for Interdisciplinary Research in Rehabilitation of Greater Montreal. All the participants provided written informed consent.

We resorted to a validated outcome measurement instrument to guide our scenarios in the ecological setting where users drove the IPW. The Wheelchair Skills Test (WST) [17] has been developed to assess manual and PW performance using a standardized course. It proposes a set of 30 tasks that are considered as parts of daily activities, e.g. rolls forwards, turns while rolls backward, reaches high object, etc. It also offers evaluation forms for clinicians to rate manual wheelchair or PW users.

Each WST task is scored on 3-point scale. Fourteen tasks were selected amongst these as relevant for the IPW and the ecological setting. The WST tasks were administered and scored by clinicians. The tasks performed were video recorded by several cameras; we video-recorded a subjective perspective integrating the use of the interface and an objective perspective on the spatial context of navigation. A researcher could take control of the IPW in cases where the participant's safety was compromised. The users had to maneuver the IPW using a HRI. The interface is illustrated in Fig. 2.

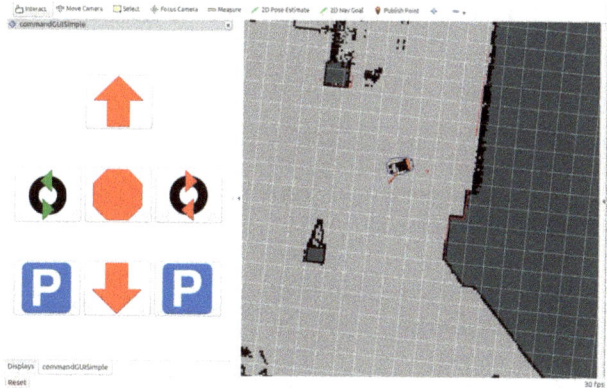

Fig. 2. The IPW's tested HRI.

After performing the WST tasks, researchers administered the System Usability Scale questionnaire (SUS) [18, 19] in order to gain insights on the usability of the system. In addition, researchers conducted self-confrontation interviews with a sub-set of participants (n = 5). These interviews consisted of watching video-recordings of WST tasks they performed with the IPW. The videos were used as a support to ask the participant to comment on his/her use of the IPW in order to (1) reveal the underlying cognitive processes and (2) provide an understanding of activity [20, 21]. With the SUS and the self-confrontation interview, we were able to collect information on the episodic and momentary experience respectively.

Overall Usability of the System. The SUS score ranges from 0 to 100 [19]. In order to scrutinize the usability score, McLellan, Muddimer and Peres [22] defined the range of scores as follows: not usable (score < 65), usable (65 ≤ score < 85) and excellent (score ≥ 85). In regards to the overall usability of the IPW for daily tasks from the WST, our sample highlighted that the tested system is usable (M = 71.60; SD = ± 9.73). It was considered as not usable for 2 participants (M = 56.25; SD = ± 5.30), usable for 9 participants (M = 74.72; SD = ± 6.31) and excellent for 1 participant (M = 90).

Testing Functions and HRI Implemented from the Identified Anticipated Needs.
The ecological approach with the WST and questionnaire enable us to verify the correspondences of some highlighted needs regarding mobility.

The WST allowed users to perform common tasks in the shopping mall. All the participants were able to perform the fourteen WST tasks with the IPW. We can highlight that participants scored lower in three tasks than in the other tasks (Table 1): the turn while moving backwards, rolling over a long distance and parking. We could suggest that the complexity of the task (for the two former ones) or of the automation of the task (the function of parking) could have triggered some difficulties for the users. Nevertheless, all participants were able to complete the tasks.

Table 1. Means scores on the WST tasks.

WST tasks	N	Means	Standard deviation
Roll forward	10	1.80	.42
Roll backward	10	2.00	.00
Turn in place	10	1.90	.32
Turn while forward	10	1.40	.52
Turn while backward	10	1.10	.57
Parking	10	0.90	.32
Reach high object	10	1.40	.52
Roll long distance	10	1.20	.42
Avoid obstacle	9	1.67	.71
Ascend side slope	10	1.80	.42
Descend side slope	9	1.78	.44
Across side slope	10	1.80	.42
Roll on soft surface	10	1.90	.32
Roll on threshold	10	2.00	.00

The next objective of our team will be to focus on the self-confrontation interviews. This will allow us to identify the unanticipated needs of real users and to translate them into technological and functional specifications for the development of the IPW. In addition, it will define and validate the various interfaces and user features that are in line with previously identified needs.

3.3   Cumulative UX: A Diversity of Ecological Settings

The cumulative experience can be approached with end users in a variety of ecological frameworks relevant to them, e.g. home, work, leisure. The diversity of ecological frameworks, but also co-users will make a considerable contribution to the current data. The same methods used for momentary and episodic UX will be adopted in order to highlight the cumulative experience. However, the scenario with the tasks performed by end-users will be chosen depending on the ecological settings.

We elaborated a list from the self-confrontation interviews in order to prepare scenarios of evaluation for the IPW. The list includes public library, transport, shops such as pharmacy, supermarket or home retail shop, museum, cinema and schools. However, several elements should be taken into account:

The Cost of in Vivo Studies. It is important to note a certain heaviness in terms of time required when setting up in vivo studies. Indeed, this approach requires the monitoring of participants, but also a considerable set of data to be processed later. In addition, the physical limitations of the participants (e.g. transferring to another wheelchair), adapted transport of participant and the prototype to the site, the team ensuring participant safety and the proper functioning of the equipment in ecological environments are significant factors to take into account. Moreover, the environmental constraints and diversity of ecological frameworks and financial cost of deploying such a setup must also be taken into consideration.

Aiming for Complementarity of Data: First-person and Third-person Perspectives. As for the co-experience, it will be taken into account to highlight other users' needs and translate them into specifications for the development of the IPW. Indeed, this type of population involved in the co-experience, e.g. caregiver or pedestrians, may have knowledge on end-users, on the use of motorized wheelchairs and/or on the space shared with wheelchair users. In addition to the methods put in place to study the needs of end-users of IPW, the study of co-experience will be apprehended with an additional method, that is the allo-confrontation interview, i.e. viewing the activities of others and commenting on them [20]. The contribution of the allo-confrontation interview will bring a new third-person perspective on the needs of users and thus enrich the pool of needs to be implemented in the next IPW prototype.

4 Conclusion

The development of new mobility assistance technologies can benefit from collaborative work conducted by multidisciplinary research teams. Indeed, each discipline can enrich the evolution of technology by contributing different perspectives related to each discipline. In addition, the involvement of different (co-)users in the development and inclusive design process is highlighted in the considerable contributions to the IPW development. By considering the needs that evolve with the use of new technologies, the 'holistic' UX may allow for the feedback of the various (co-)users to be taken into account throughout the process of designing a product such as the IPW.

Acknowledgment. We would like to thank the CRIR-Living Lab Vivant team who have supported our project from the beginning. We also thank the management of Place Alexis-Nihon for welcoming us into their space. These projects and the first author were funded by FQRNT (INTER Strategic Network), NSERC Canadian Network on Field Robotics (NCFRN), and NCE AGE-WELL. Without the team of students and clinicians, the ever-changing project would not have been possible.

References

1. Barcenilla, J., Bastien, J.-M.-C.: L'acceptabilité des nouvelles technologies: quelles relations avec l'ergonomie, l'utilisabilité et l'expérience utilisateur ? Le travail Humain **4**(72), 311–331 (2009)
2. Simpson, R.C.: Smart wheelchairs: a literature review. J. Rehabil. Res. Dev. **42**(4), 423–436 (2005)
3. World Health Organization: Guideline on the Provision of Manual Wheelchairs in Less-Ressourced Settings. WHO Press, Geneva (2008)
4. Kairy, D., et al.: Users' perspectives of intelligent power wheelchair use for social participation. In: Conference Proceeding RESNA 36th International Conference on Technology and Disability, Bellevue (2013)
5. Kairy, D., et al.: Exploring powered wheelchair users and their caregivers' perspectives on potential intelligent power wheelchair use: a qualitative study. Int. J. Environ. Res. Public Health **11**(2), 2244–2261 (2014)
6. Rushton, P.W., et al.: The potential impact of intelligent power wheelchair use on social participation: perspective of users, caregivers and clinicians. Disabil. Rehabil. Assist. Technol. **10**(3), 191–197 (2014)
7. Boucher, P., et al.: Design and validation of an intelligent wheelchair towards a clinically-functional outcome. J. NeuroEng. Rehabil. **10**, 58 (2013)
8. Pineau, J., Atrash, A., Kaplow, R., Villemure, J.: On the design and validation of an intelligent powered wheelchair: lessons from the SmartWheeler project. In: Brain, Body and Machine: Proceedings of an International Symposium on the Occasion of the 25th Anniversary of McGill University Centre for Intelligent Machines Series: Advances in Intelligent and Soft Computing, vol. 83, pp 259–270 (2010)
9. Tsang, E., Ong, S.C.W., Pineau, J.: Design and evaluation of a flexible interface for spatial navigation. In: Conference on Computer Robot Vision, Toronto (2012)
10. Heylighen, A., Bianchin, M.: How does inclusive design relate to good design? Des. Stud. **34**, 93–110 (2013)
11. Hassenzahl, M., Tractinsky, N.: User experience - a research agenda. Behav. Inf. Technol. **25**(2), 91–97 (2006)
12. Roto, V., Law, E., Vermeeren, A., Hoonhout, J.: User experience white paper bringing clarity to the concept of user experience. In: Result from Dagstuhl Seminar on Demarcating User Experience, Dagstuhl, pp. 1–12 (2011)
13. Wang, R.H., Mihailidis, A., Dutta, T., Fernie, G.R.: Usability testing of multimodal feedback interface and simulated collision-avoidance power wheelchair for long-term-care home residents with cognitive impairments. J. Rehabil. Res. Dev. **48**(7), 801–822 (2011)
14. Guerdira, Y., Farcy, R., Desailly, E., Bellik, Y.: Évaluation cinématique d'une interface tactile pour le pilotage d'un fauteuil roulant électrique: une étude pilote. In: AFIHM 29th Conference francophone sur l'Interaction Homme-Machine, Poitiers, pp. 281–290 (2017)
15. Jutai, J.W., Fuhrer, M.J., Demers, L., Scherer, M.J., DeRuyter, F.: Toward a taxonomy of assistive technology device outcomes. Am. J. Phys. Med. Rehabil. **84**(4), 294–302 (2005)
16. Bastien, J.-M.-C., Scapin, D.L.: Ergonomic criteria for the evaluation of human-computer interfaces. Technical report Institut National de Recherche en Information et Automatique RT-0156 (1993)
17. Programme d'habiltés en fauteuil roulant. http://www.wheelchairskillsprogram.ca/
18. Finstad, K.: The system usability scale and non-native english speakers. J. Usability Stud. **1**(4), 185–188 (2006)
19. Brooke, J.: SUS: a quick and dirty usability scale. Usability Eval. Ind. **189**(194), 4–7 (1996)

20. Mollo, V., Falzon, P.: Auto- and allo-confrontation as tools for reflective activities. Appl. Ergon. **35**(6), 531–540 (2004)
21. Rix, G., Biache, M.-J.: Enregistrement en perspective subjective située et entretien en re-situ subjectif: une methodologie de la constitution de l'expérience. Intellectica **1**(38), 363–396 (2004)
22. McLellan, S., Muddimer, A., Peres, S.C.: The effect of experience on system usability scale ratings. J. Usability Stud. **7**(2), 56–67 (2012)

Digitalize Limits for Increased Capability: Technology to Overcome Human Mechanisms

Mila Stepanovic[✉] and Venere Ferraro

Department of Design at Politecnico di Milano, Milano, Italy
mila.stepanovic@mail.polimi.it

Abstract. Senses help us go through everyday life. Thanks to our natural senses we can orientate in the space, recognize danger or pleasurable experience. In this text we will treat problematics of loss or damage of the Sense of Smell (Olfactory System) and investigate how does this event change everyday life and perception of the world around us. The objective of this text is to show importance of empathy related to the complexity of the pathology and the importance of the design for Assistive Technological (AT) devices; in this particular case, investigate possibilities of technological devices in improving quality of life by increasing one's capabilities related to Human Mechanisms, where we refer to smell experience. In regard, we will describe the Case Study based on workers' perception of the risk in Coating Plants, that largely rely on their sense of smell, which as a consequence impacts their health condition.

Keywords: Human capabilities · Olfactory system · Assistive technology
Olfactory loss · Perception · Occupational disease · Prevention

1 Introduction

Since antique Greece, philosophers as Plato and Aristotle discussed about the phenomena of senses and sensation, and its importance in knowing the world and ourselves. Plato claimed that the consciousness stands exactly for *the sense* [1]. He considered that the truth is related to what is perceived [1]. On contrary, Aristotle believed that the smell is a secondary sense for humans because it will never be as accurate as animals' one [2].

It is important to emphasize the fact that the Sensory perception is a biological phenomena [3]. All forms of the life have sensory perception [3, 4]. Plants are sensing the Sun; animals are capable of sensing danger, pray or partner for coupling; even single-celled amoeboid is capable of sensing [3, 4]. Also humans use their senses to focus on outside world, orient in the space and communicate with each other and environment; but the difference comparing to animals is that humans have the capability of experiencing on higher cognitive level [3–5]. It is thanks to our senses that we have a perception of ourselves, of "being in the world" [4].

If senses are so important for humans, what happens when one or more senses are loss or damaged? - Certainly this event changes our everyday life, the way in which we are interacting with each other, with objects and environment [6].

© Springer International Publishing AG, part of Springer Nature 2019
T. Z. Ahram and C. Falcão (Eds.): AHFE 2018, AISC 794, pp. 860–870, 2019.
https://doi.org/10.1007/978-3-319-94947-5_84

Technological evolution helped development of devices that are able to replace some human mechanisms or organs, and assist in completing activities. For instance, some fields of design and technology application are focusing on individuals with specific problem or impairment. Thus, we can consider that technology has a potential to be a digital extension of human natural capabilities or mechanisms [7, 8].

Assistive Technology (AT) aims at improving users with impairment but also support the life and daily interaction of individuals. There are different types of Assistive technology (AT) and mode of uses. From automatic digital to analog, from commercial to personal, from replacing the body part to technological wearable device ecc.

In this text we will investigate problematics that occur when the sense of smell is lost or damaged, its causes, and finally propose a technology as a solution for improving the life of individuals with olfaction impairment.

This research will be supported with the Case Study that is referring to the workers in Coating Plants, which after long exposure to polluted air have reduced sensitivity to odors and as a consequence also the low perception of the risk.

2   Sense of Smell

In all moments of our lives we are perceiving different kinds of smell. Sense of smell is considered to be a secondary sense compared to audition and vision [9]. Still it is the very important factor in experiencing the world and people around us. We are realizing ourselves trough our smell but also we are relating to others by smell; like smell of our mother, people that we frequent regularly [10].

Animals are using olfactory system not for abstract experiences but for survival, reproduction, and discovering sources of nutrition [3]. Differently, humans are living experiences trough their senses, evoking memories and emotions [4, 5].

We perceive odors principally as pleasant and unpleasant [4, 5]. Human sense of smell can indicate whether there is a danger in the environment and generally the quality of the things that we consuming or being in contact with. Even tough our ability to sense is very important it is not always reliable, and not only, it is certainly less developed than animals' one. This means that even healthy olfaction system is not able of detecting all odors and certainly not all with a same intensity.

Sense of smell is in a tight relation with a sense of taste. Eating information is geathered also from sense of smell (olfaction receptors) and this is contributing the eating experience by emphasizing the taste of food [4]. This means that there is more possibility that if we do not like the smell of something, we also won't like the taste of it [4, 5].

In the following text we will observe all these biological and cultural facts of odors and human sense of smell, but also pathologies and possible difficulties that occur due loss of capability of sensing.

2.1   Importance of the Sense of Smell: Individual and Cultural Factors

Human are capable of recognizing over thousand different odors; but we can also loose completely the sense of smell [5]. The main role of our senses is to unconscious monitor what is around us; senses help us to focus our attention on specific thing and describe feelings and space [4, 5]. Each thing that is sensed is translated in perception - perception is the image or experience of what is sensed.

Our nose is able of recognizing both near, direct odors (e.g. spoiled food) as those sparse in the air (e.g. smoke, pollution) [4]. This function of our nose is enabling the olfactory epithelium with receptors that is detecting molecules and register scents [4].

It is considered that the odor perception - the experience and memory evoked trough sense of smell - is not as accurate as auditory, visual or tactile input [4].

Scholars are arguing about if there is a Mental Imaging for the sense of smell [11]. Mental imaging is creation of mental representations that are strongly based on imager's will [12]. Some researchers are claiming that there is odor imaging while some others retain that it is not possible to generate odorlike mental images, or that this sort of imagination is rather poor [11, 12]. Beside imaging as a mediator in recognizing odors, it is also difficult to find a correct linguistic designation for describing odors [10]. It might be that these two phenomena, of odor imaging and linguistic characterization of odors, are related [6]. The emotional vividness may be exchanged with the imaging [4]; there so, it is believed that lack of semantic information it is very difficult to classify precisely odors. Odor imaging would be a semantic mediator that help in recognizing quality of what we are smelling, or provide higher accuracy of human sensing [12]. Anyway, encoding smells in our memory is not only possible but normal, we are all encoding different smells within possibilities of our nose and sensitivity [13]. Sensory perception is a strongly subjective experience of individuals [4, 5]. Fortunately the most of us are able of recognizing odors and use the sense of smell in efficaciously [10]. Odors that we meet more frequently during the life are also more easily recognizable (i.e. fragrances based on lemon or roses); on contrary, too frequent experiencing of same odor, or exposure to it, may also decrease sensitivity of nasal epithelium [5]. The fact that there is a memory and experience related to sense of smell is confirming that there is a relation between sensory system, lifestyle and environment [3–5]. Human sensory system is detecting the quality of environment and the individual is adopting to it [3].

Some levels of pollution, fire or spoiled food a healthy olfaction system is recognizing, actually these are the most important functions of our olfaction system [13]. Human brain has a role of interpreter of what is sensed [4, 9]. Perception about something is created in our brain and it is an active process depended on a lot of factors, both biological and cultural [3, 4, 9].

The quality of life certainly change in anosmic patients (patients with damaged olfactory system) comparing to normosmic (normal functioning olfaction system) [13]. Especially segments of life related to food and social relations are showed to be poorer in anosmic patients [13]. Other very important segments of life influenced negatively by olfaction impairment are safety, eating and personal hygiene [13].

Inhalation as a process of breathing was very important for Christian and Jewish tradition and religion [14]. The Bible is describing The Creation of Man with a scene in which the God breaths the soul into Adam (Gen. 1:7). Whereas in the Old testament fragrances were related to dark realm of sexuality [9].

In Jewish tradition the smell is strongly related to the Sabbath, where this sacred day was different from other weekdays with introduction of fragrant herbs [14].

In the Jewish tradition was also believed that the smell is bringing the seal of Lord, or as later was taught, the nose is a seal of God set on our faces [9].

Historically, people became more interested in smell of themselves and their clothes from the end of eighteenth and beginning of nineteenth century [15]. This due the reduced economical and social differences and the first commercialization of soaps, deodorants and generally products for hygiene [15]. This was the historical period in which we started the war against smells; not only with personal hygiene but also with smell of environment. Smell of our habitat changed as cause of pollution and due usage of fragrances to reduce bad odors.

Related to historical evolution of smell it is evident that the smell had its social, significant and semiotic function and that these were changing in time [15]. Like from more basic musk odors to elaborated floral fragrances for personal use ecc [15].

This phenomena is related also with a fact of how close we are to the nature today. Scholars are claiming that now a days we have less developed sensitivity to odors that our ancestors because we are distant from nature [15].

Problems in perceiving smell can occur in few ways: when there are few odors in the same environment and they are getting neutralized, when there is some physical problem or ill-ness that damaged Olfaction system or when we are exposed to certain odors for a long period so it cannot be perceived anymore or not with the same intensity as before. But beside physiological characteristics and diseases that cause damage of olfaction system, smelling as a human cognitive action is almost distinct in developed societies. We are not capable of bringing accurate conclusions about the quality of what we are smelling, as it was once where the perception about odor was strongly focused on survival, reproduction, orientation and in general nature and its characteristics.

2.2   Causes of Loss or Damage of Sense of Smell (Olfactory Disorders)

Olfaction disorder is quite common as pathology, while its consequences are significantly less known [13]. Loss or damage of sense of taste and smell is not considered for disability. Sensory impairments are related to following: deafness and hearing impairment, blindness and visual impairment and deafblindness.

Even if the sense of smell is not playing that important role in medical sense, it is very important for experiencing the world. The system that is enabling smell sense is called Olfaction System. For functioning of Olfac-tion System it is very important the role of the brain (neurons). The role of brain is to interpret smells – create perception [4, 5].

The Olfaction system is divided in two parts: Central and Peripheral. The receptors in the nose and olfaction nerve are considered as Peripheral part, the part that is receiving and sending information to the brain. The Central part is the Olfaction Bulb where is the brain part associated with memory and feelings [4]. Olfactory performance is determined by interactions occurring in peripheral part (early processing) and central part (higher order processing) [16]. There are different etiologies for olfactory dysfunction that can be caused by damaging one of two extremities of information processing (peripheral or central), or it can be caused by some other pathology (physical or cognitive). Typical diseases that are causing olfaction dysfunction are sinonasal disease, upper respiratory tract infections, exposure to hazards, neurodegenerative diseases but also it can be a result of congenital nature [16]. These are very direct causes of olfaction damaging, but there are also some other diseases that may lead to loss of sense of smell, as Alzheimer and Parkinson, especially in elderly population.

What was of particularly interest in this research is olfactory loss due long exposure to hazards. Exposures to pollution and hazards is very common way that brings to olfactory damage. During a long exposures to particular odor human develops habit to it. Developed habit is reducing a sensitivity to specific odor and perception about the risk (hazard substances present in the air) decreases as a consequence. Being less aware of pollution is certainly increasing the possibility of higher exposure and there so more severe disease progress may occur.

What we propose in this research is that when there is a limit of human natural senses there is a necessity to introduce technology able of replacing or increasing human capability, and for introduction of technological solution it is necessary to observe pathology and causes. Value of Technology implementation is in providing the objective data instead of human sensing that is based on subjective experience.

3 General Observation of Assistive Technology (AT) and Its Potential

Assistive Technology (AT) is defined as any technology, or technological device, service or equipment, commercial or not, able to assist the user with some sort of disability (physical or cognitive) by maintaining, increasing or replacing damaged human function or mechanism [7, 17, 18]. Basically the role of Assistive Technology is to make actions easier or possible to do [13].

Assistive Technologies certainly had its growth with a technological and economical development that brought to changes in assessment and costs, giving a possibility to everyone being a part of it [19, 20]. Particularly development of microelectronics contributed in creating new devices of small size with high performances but also reduction of costs of production and electronics [20]. In the past years we were witnesses of pervasive diffusion of technology and development of IoT (Internet of things). These events make believe that the technology may have advantage or utility for all kind of purposes, even if most of the time it is not like this. It is needed to create right condition for having effective results from technological products, as creating technologies that are used only when indeed it is needed and that can guarantee good quality products that correspond to technical standards and market, and user requirements [17]. Successful technological devices have to adopt to the capacities of the end users, his level of knowledge of informatics and physical and cognitive capabilities [17].

Beside technical and technological characteristics and innovation, very important factor for creating good Assistive Technological products is Design and Engineering as disciplines, with its processes and methods.

When designing assistive devices or applications it is important to take in consideration empathy toward pathology and human condition and engagement of the user in design process. Understanding if the level of innovation and generally application of technology is comprehensive, feasible and accurate is possible by introduction of prototypes and engagement of the user in design process.

Assistive Technologies have both social and individual importance. The formal factor is not less important than technical and functional one. Assistive Technology is not invisible, neither neutral [21]. It is the part of the user and often it is very visible and perceived by the users as stigma, there so it becomes a barrier between the user and environment and the main reason of abandonment [21]. Successful assistive device should fall in a background of daily life, not to give the impact.

Assistive Technologies even if dedicated to personal use often involve also the family of the user. In order to avoid collective frustration these devices have to be easy and comprehensive for use [14].

Beside home use Assistive Technologies can be part of the healthcare system or have a form of a service provided to the user [7]. These applications are often used for rehabilitation purposes [13, 17].

There are numerous digital assistive devices that support different needs. From those more common as hearing device to advanced body prosthesis. Other fields of application of Assistive Technologies are related to the vision, orthopedic and among all the mobility sector. Another field of interest is cognitive and motor disabilities, both in case of elderly (Alzheimer or Dementia) and congenial and pathological problematics, where the most common assistance is applied in sense of education, social participation and communication [14].

4   Case Study: Plurisensorial Device to Prevent Occupational Disease

In the previous text we investigated the field of Assistive Technologies and in particular sense of smell, characteristics of olfactory system and experience related to it. Observing the causes of loss of sense of smell we noticed that typical cases are related to long exposure to hazards which cause serial infections that may result as damaged olfactory system (1), often it happens as a process of ageing (2), as a result of other pathologies (3), and due injuries that damage peripheral or central part of olfactory system (4). Here we will focus on the risk of losing or damaging sense of smell as a cause of long exposure to hazard substances in specific working environment.

Our research was developed on findings evolved from the Transnational Research Project (SAFERA Joint Call 2014) called "POD-Plurisensorial Device to Prevent Occupational Disease" with a focus on specific environment and group of users - workers in Coating Plant. These workers are highly exposed to Volatile Organic Compounds (VOC) that are present in the environment. In Coating Plant environment is highly recommended wearing of Personal Protection in order to prevent numerous health problems that can occur after long time exposure. These pathologies are always related to respiratory tract, and occur initially in form of irritation, allergy, chronical bronchitis, asthma, chronic obstructive pulmonary disease (COPD) and in worst case lung cancer.

In the following text we will show the Case Study based on the context of Coating Plants, where is common decrease of sensitivity to smell as a cause of long exposure to hazards, which can lead to develop more severe diseases. We will suggest as a solution to this problem an assistive device based on sensing technology and design process based on empathy and user engagement.

4.1 Problem Identifying: Empathy Phase

A report of the Scientific Committee on Occupational Exposure Limits (SCOEL) highlights that workrelated exposures are estimated to account for about 15% of all adult respiratory diseases. The last INAIL (Istituto Nazionale Assicurazione Infortuni sul Lavoro) report on occupational diseases showed that the working places with the highest percentage of respiratory diseases are agriculture, manufacturing and transportation sectors [22].

Our research started from these facts and went forward to observe workers and the environment in which they are working. Objective of the User Session was to understand why still today there are so many cases of occupational disease within this sector, despite provided protective equipment and general evolution in health prevention methods. In order to understand general problematics and characteristics of Coating Plants we proceeded with observation of workers during the working performance that was followed up with the interview formulated in the way to understand the perception of workers toward health, risk and prevention and on the other part understand their level of technological devices manipulation and comprehension.

Geathered results showed that generally risk and health perception of workers is on low level because it is not easily perceivable. Situations in which they are perceiving more exposure to risk are those in which there is a overspray. The visual perception is overwhelming the smell perception. Visible representation of paint is a sort of semantic information about the exposure risk to workers.

Even tough wearing of Personal Protective Equipment (PPE), is highly recommended within Coating Plants, workers are rather avoiding to wear it because they do not see the realistic utility of it.

Interesting fact that came out of our discussion with workers is that they were not perceiving the smell with the same intensity as us, that we were there for the first time. After a long time exposure they lost sensitivity on odors in Coating Plant. This condition is preventing them from responsible health behavior and increase possibility of developing respiratory diseases in future.

Here we found as a suitable solution application of technology with a purpose to inform and alert worker about environmental pollution, make more visible risks that they are exposed to.

4.2 Development of Device

Results from the Empathy phase brought us to the complex solution based on sensor technology. Here we will briefly describe the system and than we will focus on the main part of it which is the Electronic Nose Device able to detect Volatile Organic Compounds (VOC) and provide the real-time feedback to the user. Aim of this kind of system is to communicate the risk trough objective data to user and during the time change user's perception and raise consciousness about health risks.

The system that we will introduce is the Plurisensorial Device to prevent Occupational Disease which is an interactive and protective wearable system that consists of four parts: Protective Mask (1), Electronic Nose Device (2), Respiratory Chest Band (3), Mobile Application (4). All these parts of the system communicate one with another and provide a feedback trough the Mobile Application to the user, about the environment and personal vital parameters, in the linear comparable way so that the worker can have a complete image about how environment impacts his health compared to when the protective mask is worn and not. This sort of system is based on sensor technology and Bluetooth for information transfer.

The most important part of this system is Electronic Nose Device which is the "brain" of the system. Electronic Nose Device (Fig. 1) is developed as a support device for worker that offer real-time feedback about the air quality.

Fig. 1. Electronic Nose Device prototype

This device is a mediator to inform the user about the level of hazard substances in the air and remind him to put the protection. Device's sensor is calibrated to detect three levels of air quality: good (1), the air is getting polluted (2) and the high exposure to hazards (3). The device is giving visual and tactile feedback in form of colored LED and vibrations, set respect to three levels in the mood of three colors – green, yellow, red – and three vibration level – low, medium, high (Fig. 2). Vibrations of the device are designed to be continued in order to force the user to look at it, because the only way to stop vibration is to press a button on it.

Here we tried to activate other senses (vision, touch) because the sense of smell is not validating in accurate way the condition. As workers claimed, only when the risk is visible or the clothes that they are wearing is dirty of colors they remember that they are exposed to toxic substances. Giving a real - time visual feedback tries to operate on this level, emphasize the visual perception, while with vibration is activated the tactile one, and in this way provide the objective data related what is not perceivable with a nose.

Detecting the polluted air as a human physiological function is not enough developed to diverse all kind of hazards, and not all with a same accuracy. In the case of our workers this is even less valid because their developed habit to this kind of

Fig. 2. Use cases Electronic Nose Device

odors, or as many of them have already irritated or damaged epithelium their sensitivity on odors is on very low level. Purpose of technology in this case is to arrive where human mechanisms are not able and prevent development of more sever diseases. Electronic Nose Device is a sort of assistance to our nose and perception.

5   General Discussion

In this text we wanted to show an alternative application of assistive technology and emphasize the problematics of loss or damage of sense of smell that is often taken for grant.

The sense of smell has certainly great impact on our lives and experiencing. Problematics of loss or damage of sense of smell are not related only on hedonistic characteristic of smell experience, but also to a numerous risks that one is exposed to when not being able to perceive odors as smoke, hazard substances, gas, spoiled food ecc. Not being able of perceiving polluted environment may lead to a continuous exposure and to development of various diseases or intoxication.

We showed in our Case Study the context in which workers often suffer a disease after long exposure to hazards because they smell perception is not accurate anymore.

For this particular case we developed Device able to detect Volatile Organic Compounds (VOC). Potential of this kind of device is in its flexibility and versatility - possibility of integration in different sectors. Value is in its capacity to go beyond human limits and give precise, objective data and provide immediate feedback. Having a possibility of real - time feedback is in reducing the time gap and provide continuous assistance.

There are many sectors in which this problematics is present and that it can be resolved by introducing sensor technology and devices designed to increase capabilities, that are strongly based on understanding the complexity of the pathology.

6   Conclusions

In this paper we interpolated principally two arguments: importance of the empathy toward pathology as in this particular case Olfactory Disorders (1), and design of Assistive Technologies (2).

We were interested in understanding the daily problematics of the people living with damage or loss of sense of smell, how they experience the world and potential risks that they are exposed to.

Our findings are that, even if it is taken for grant, sense of smell is very important human factor and that the damage of it may bring a numerous complications, from those related to pleasurable experiences as eating, interpersonal relations, habitat recognition; to those related to danger perceiving. With a loss of sense of smell all these characteristics and activities of our life deny.

Assistive Technology (AT) has a potential to become a protagonist from where human limits begin. Introduction of Assistive Technology in lives of people with reduced sensitivity or complete dysfunction to odors may effectively increase capabilities and independence. It could prevent from eating spoiled food, to recognizing fire smoke and different hazard situations.

When one of the senses are lost there is a necessity to emphasize function of other sense, as we saw from the example of workers in Coating Plants. They are perceiving the risk related to hazards only when there is an visual input as overspray, or dirt of paint on themselves. There so we offered VOC (Volatile Organic Compound) sensor equipped device that gives a real - time visual and tactile feedback to workers about the level of hazards.

Technological potential, or the potential of technological devices (i.e. Assistive Technology) is in the possibility to give objective data to the user and an immediate feedback, instead to rely on subjectivity of our own senses. Another added value of technology implementation lies in the fact that in this way there is a possibility to arrive to perform both on individual and social lever, and on social participation of individuals.

References

1. Plato: Teeteto, o sulla scenza, Natoli, S. Feltrinlli (2000). (Theaetetus)
2. Kemp, S.: A medieval controversy about odor. J. Behav. Sci. **33**, 211–219 (1997)
3. Friedrich, G.B.: Sensory perception: adaption to lifestyle and habitat. In: Barth, F.G., Giampieri-Deutsch, P., Klein, H.D. (eds.) Sensory Perception, Mind and Matter, pp. 89–91. Springer, Vienna (2012). https://doi.org/10.1007/978-3-211-99751-2_6
4. Cardello, V.A., Wise, M.P.: Taste, smell and chemesthesis, In: Schifferstein, N.J., Hekkert, P. (eds.) Product Experience, pp. 91–122. Elsevier (2008)
5. Rot, N.: Opšta psihologija, Zavod za udžbenike Beograd, (2000). (General psychology)
6. Coesby, J., Johnston, S.S., Dunn, L.M.: Sensory processing disorders and social partecipation. Am. J. Occup. Ther. **64**(3), 462–473 (2010)

7. Alper, S., Raharinirina, S.: Assistive technology for individuals with disabilities: a review and synthesis of the literature. J. Spec. Educ. Technol. **21**, 47–64 (2006). https://doi.org/10.1177/016264340602100204

8. Schon, D.: Technology Change - New Heraclitus. Delacorte Press, New York (1967)

9. Jutte, R.: The sense of smell in historical perspective. In: Barth, F.G., Giampieri-Deutsch, P., Klein, H.D. (eds.) Sensory Perception, Mind and Matter, pp. 313–329. Springer, Vienna (2012). https://doi.org/10.1007/978-3-211-99751-2_18

10. Rosenblum. D.L.: Lo straordinario potere dei nostri sensi. Bollati Boringhieri (2011). (See what I'm saying: The Extraordinary Power of our Five Senses)

11. Flohr, E.L.R., Arshamian, A., Wieser, M.J., Hummel, C., Larsson, M., Muan, A., Muhlberger, A., Hummel, T.: The fate of inner nose: odor imagery in patients with olfactory loss. Neuroscience **268**, 118–127 (2014). https://doi.org/10.1016/j.neuroscience.2014.03.018

12. Djordjevic, J., Zatorre, R.J., Petrides, M., Jones–Gotman, M.: The mind's nose, effects of odor and visual imagery on odor detection. Psychol. Sci. **15**(3), 143–148 (2004). https://doi.org/10.1111/j.0956-7976.2004.01503001.x

13. Hummel, T., Nordin, S.: Olfactory disorders and their consequences for quality of life, in Acta Oto/Laryngologica, pp. 116–121 (2005). https://doi.org/10.1093/chemse/bjt072

14. Trends and differential Use of Assistive Technology Devices: United States (1994)

15. Jenner, M.S.R.: Follow your nose? smelling, and their histories. Am. Hist. Rev. **116**, 335–351 (2011). https://doi.org/10.1086/ahr.116.2.335

16. Yao, L., Pinto, J.M., Yi, X., Peng, P., Wei, Y.: Gray matter volume reduction of olfactory cortices in patients with idiopathic olfactory loss. Chem Senses **39**(9), 755–760 (2104). https://doi.org/10.1093/bju047

17. Bitelli, C., Guerreschi, M., Rossi, A.: Manuale degli ausili elettronici ed informatici, Tecnologie assistive a support della qualità della vita, GLIC (2016). (Handbook of assistive electronics and informatics)

18. Williams, J.M.: Past: present and future of assistive technology, In: Eizmendi, G., et al. (eds.) Challenges for Assistive Technology, pp. 20–25. IOS Press (2007)

19. Gelderblom, G.J., de Witte, P.: The assesment of assistive technology outcomes, effects and costs. Technol. Disabil. **14**, 91–94 (2002)

20. Schmeikal, B., Haart, H.H., Richter, W.: Impact of Technology on Society. Pergamon Press, Oxford (1983)

21. Polgar, M.J.: The myth of neutral technology. In: Oishi, M.M.K., Mitchell, M.I., Van der Loos, H.F.M (eds.) Design and use of Assistive Technology. Social, Technical, Ethical and Economic Challenges, pp. 17–23. Springer, New York (2010). https://doi.org/10.1007/978-1-4419-7031-2_2

22. Authors: Persuasive Technology as a key to increase Working Health Condition. The Case study of a Wearable System to prevent Respiratory Disease. Des. J. **20**, 439–2450 (2017). https://doi.org/10.1080/14606925.2017.1352757

Modeling Augmentative Communication with Amazon Lex and Polly

Ahmad Abualsamid$^{(\boxtimes)}$ and Charles E. Hughes

Modeling and Simulation Department, University of Central Florida,
Orlando, FL, USA
ahmad@abualsamid.com, ceh@cs.ucf.edu

Abstract. We present LEXY, a framework for developing multi-actor, natural language, augmentative communication chatbots. The LEXY framework is unique in that it allows for the building of conversational chatbots whose model of the world is constructed conversationally by a caregiver. The resulting chatbot can then conversationally interact with persons with special needs, such as persons on the autism spectrum or with cognitive disabilities. The framework utilizes Amazon's Lex and Poly allowing for spoken as well as typed conversations.

Keywords: Conversational chatbots · Special needs · Autism
Cognitive disability · Amazon Alexa

1 Introduction

Augmentative and alternative communication devices, such as storyboards, communication boards, and picture exchange communication systems are often used with persons on the autism spectrum, as well as people with cognitive disabilities not diagnosed with autism, to aid in dealing with communication and language challenges such as repetitive questions, echolalia, anxiety, short-term memory challenges and cognitive challenges [1–3]. We introduce LEXY, a framework for modeling augmentative communication multi-actor conversational chatbots, using the nascent technologies of Amazon Lex and Amazon Polly. The framework can be used to build purpose-specific augmentative communication chatbot systems. Amazon Lex is a new service from Amazon Web Services (AWS) for building conversational interfaces through parsing utterances and inferring intents. Amazon Polly is another recent service from AWS that turns text into lifelike speech, allowing for the development of applications that speak in a natural, lifelike speech. AWS Lex and Polly are built on the same technologies that power the Amazon Echo, more commonly known as Alexa, an artificial intelligence powered device that sits at homes or offices and responds to spoken commands, such as "what's the weather like today", and also allows for home automation [4]. The nascent nature of all those technologies means there is no published research yet on how persons with autism interact with those devices. The primary author's anecdotal experience, including with his own two children on the autism spectrum, shows that persons with autism interact with Alexa as a person, asking it questions, greeting it, and requesting home automation commands from it. Google

© Springer International Publishing AG, part of Springer Nature 2019
T. Z. Ahram and C. Falcão (Eds.): AHFE 2018, AISC 794, pp. 871–879, 2019.
https://doi.org/10.1007/978-3-319-94947-5_85

recently released a similar device, called the Google Assistant. Apple announced they will be releasing their own device in early 2018. The proliferation [5], affordability, easy setup, and natural, conversational interface of those devices, makes them a natural candidate for developing conversational assistive technologies for persons with autism and related disorders. We developed LEXY, a framework for modeling augmentative communication chatbots using Lex and Polly to support persons with autism with their needs in repetitive questions, echolalia, I want-communications, first/then-communications, calming-communication, communicating feelings, and communicating actions. In addition to being the first conversational framework we are aware of that caters to the special needs population, LEXY has a distinctive feature not found in current chatbots. While current chatbots respond to requests from users via built-in business logic, machine learning, artificial intelligence, and integration with other systems [6], LEXY uniquely allows caregivers to dynamically configure the model of the world that is used in turn to fulfil the requests of the individuals with special needs interacting with the chatbots.

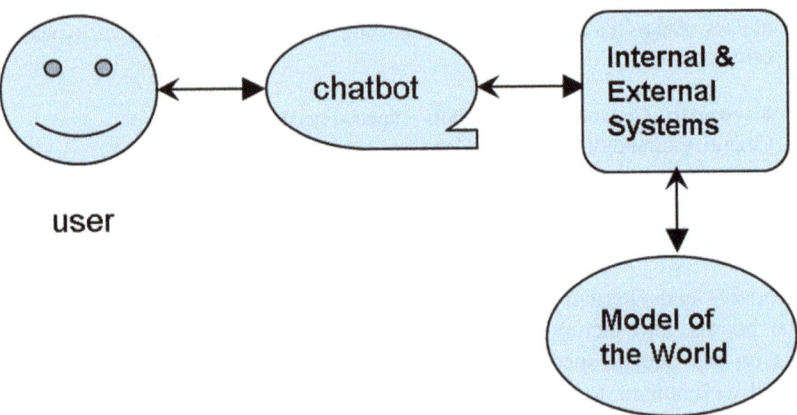

Fig. 1. A typical chatbot flow.

Figure 1 shows an illustrative, very simplified, flow of a typical chatbot where one user interacts with the chatbot. The chatbot's logic would interact with internal business logic and external systems, build a model of the world based on business logic and input from external systems, and interact with the end user accordingly. Figure 2 highlights the distinction of the LEXY framework which adds interactions with other actors, in this case the caregiver, to augment the model of the world.

The current state of the art in conversational interfaces, whether at home such as Amazon Alexa, or in a business setting, such as customer service chatbots, is centered around the fulfillment of a single intent. An intent is the basic unit of work in conversational interfaces. An intent is modeled by a collection of utterances. Each utterance is composed of natural language words and slots. An example of an intent would be "order ice cream". Sample utterances to model the intent could be "I want ice cream", "I would like to order ice cream", "I want 3 scoops of ice cream", "I would like

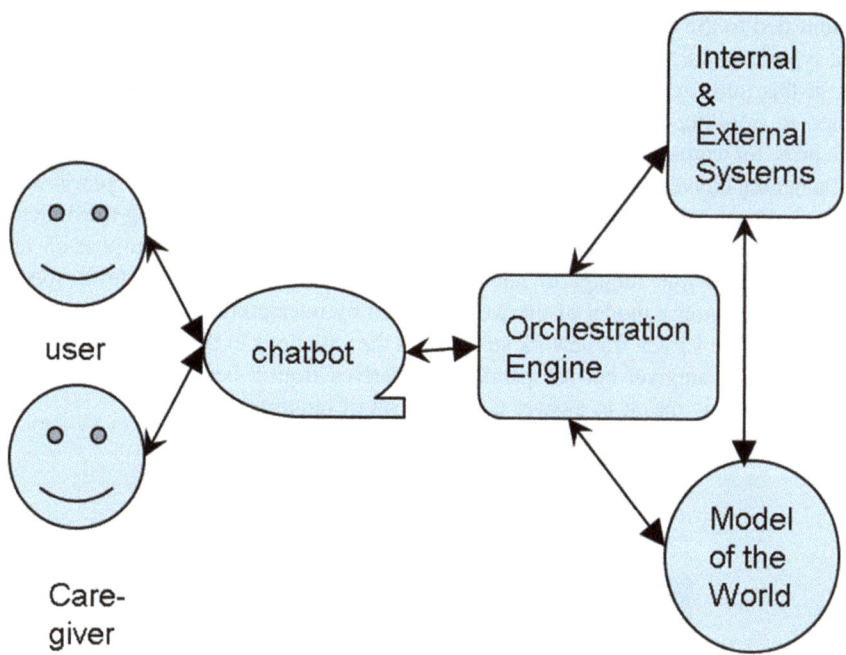

Fig. 2. A chatbot flow in the LEXY framework.

to order 2 scoops of chocolate ice cream". In order to allow for dynamic natural conversions the last two utterances can be modeled with slots, or placeholders, representing the number of scoops and flavors as follows: "I want {count} scoops of ice cream" and "I would like to order {count} scoops of {flavor} ice cream". In order for the chatbot to fulfill the intent it needs to resolve the two slots, {count} and {flavor}. If those slots are present in the conversational request, for example: "I would like to order 2 scoops of chocolate ice cream", then the chatbot can fulfill the intent without further interaction. If on the other hand, the utterance was missing some of the information, for example, "I want ice cream" then the chatbot would prompt for the number of scoops and the desired flavor. For simplicity, we assume the intent is fulfilled once the bot infers the intent from the utterance. In real life scenarios, a further action, such as a person procuring the ice cream, would be the final step in the process.

2 The LEXY Framework

In order to provide an interaction that goes beyond fulfilling a simple intent, a chatbot must maintain context, session, a state machine and a model of the world. Systems such as Amazon Alexa have large teams of developers that build sophisticated business logic and leverage machine learning and external sources to build context and a model of the world allowing it to respond to requests. For example, if a user asks Alexa to "Alexa, add detergent to my shopping cart", Alexa would already know the user from the account tied to the device, it would know their shopping history, allowing it to know what type and size of detergent to add, and would have access to their shopping cart. Yet, at this time, Alexa cannot have a multi-turn conversation out of the box, and is limited to fulfilling a single intent. The global Amazon Alexa competition offered a large prize for teams from universities around the globe to develop applications that can hold a natural conversation with Alexa. The winning team in 2017, after a year's worth of development, won by managing a 10 min conversation [7] highlighting the difficulty of developing a natural conversation system.. Our framework, being a purpose-specific framework does not attempt to handle generic requests but only to fulfil specific requests leveraging a model of the world created by interacting with a caregiver. The generality of the LEXY framework stems from the variations in the model of the world created by the caregiver but its specificity is derived from it being a purpose-specific framework that is meant to answer a specific set of requests.

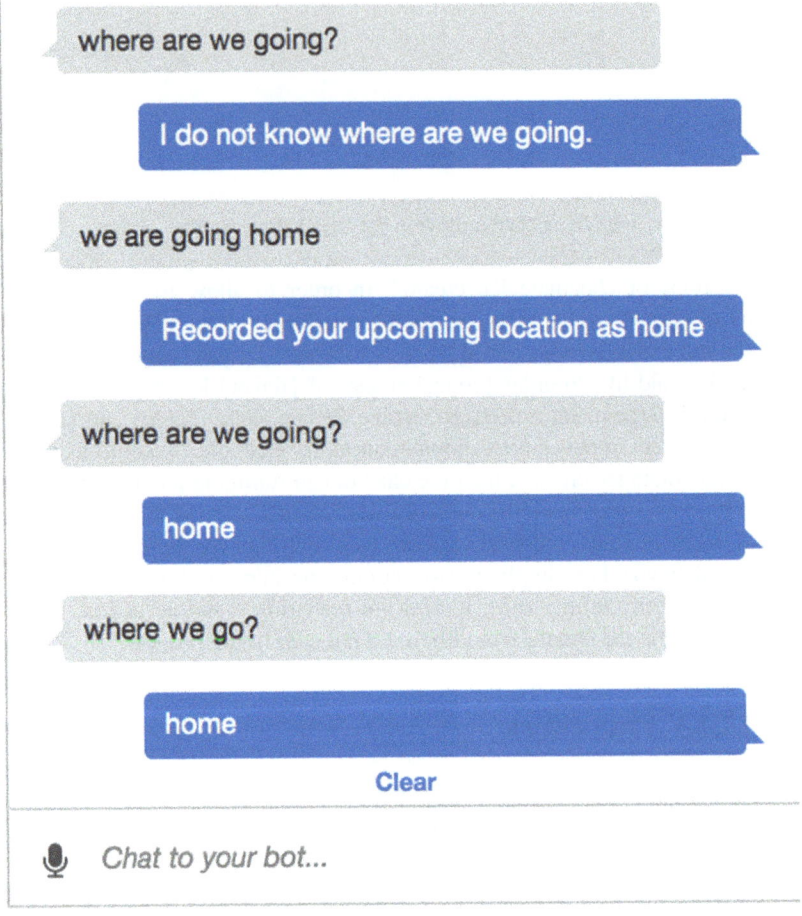

Fig. 3. A sample conversation between a prototype chatbot and two actors, the child with autism and their caregiver.

Figure 3 shows a sample conversation between a prototype chatbot that implements the LEXY framework and two actors, a child with autism and a caregiver. The conversation can be a spoken voice conversation, a typed text conversation, or a mixture of both. The illustrative conversation starts with the child asking "where are we going?". The model of the world has not been seeded by the caregiver yet, so the chatbot does not know and it responds with "I do not know where we are going". The caregiver then tells the chatbot that "we are going home". Subsequent questions by the child would then receive the correct answer from the chatbot. The example in the figure shows the chatbot, recognizing different variations of the question, answering the question correctly and repeatedly. In this specific example, the second variation of the question, "where we go?" is not part of the chatbot configuration but was resolved correctly by Amazon Lex to match the intent "where are we going?". Resolving unspecified utterances is an important component of the Amazon Lex service. Without this component, a person developing the chatbot in this example would have to specify all acceptable utterances, with the chatbot failing to respond to any utterances not explicitly specified by the developers. In this particular instance the prototype chatbot was built with the following three utterances: "where are we going", "going where", "where". From those utterances, Lex resolved the utterance "where we go?" to the intent attached to those three utterances. It is important to note that there are design trade-offs when deciding on what utterances are used to model an intent. The more flexible, less specific an utterance, the more likely Lex will match against the wrong intent in a chatbot that has multiple intents.

The LEXY framework is composed of: (1) pre-configured intents, utterances and slots, (2) an orchestration engine, (3) a business logic engine, and (4) a set of purpose-specific finite state machines. The orchestration engine and the business logic engine are developed using AWS Lambda, a service for developing serverless cloud functions. The finite state machines are implemented using Lambda functions and DynamoDB, a NoSQL cloud database. The purpose-specific activities handled by the LEXY Framework are based on popular activity boards, based on the picture exchange communication systems, often used by occupational therapists and caregivers to aid children with autism [3]. The set of activities supported by the LEXY framework at this time are "First/Then", "Schedule", and "Coping".

Figure 4 illustrates the general architecture of the Lexy framework. The same user interface is used to interact with end users, such as children with autism, and caregivers. The user interface can be a smartphone app, an Alexa device, a web application or other custom developed applications. All conversational interactions flow the AWX Lex service along with LEXY's pre-configured utterances and slot types to resolve a user input utterance into an intent. Once an intent is resolved, the orchestration engine takes over to decide how to handle the intent by invoking the appropriate chatbot. Each chatbot would then apply its business logic to update the applicable state machine and the model of the world. The model of the world and state machines are available to all chatbots which is what allows LEXY to interact with multiple actors successfully. The system's state, composed of the model of the world, the state machines, and ancillary data, is saved to DynamoDB for persistence. The Orchestration engine, user chatbots, caregiver chatbots, model of the world, and state machines, are developed using stateless cloud functions. This is why they need to persist their state to a database.

The state is hydrated from the database upon cold invokes of those cloud functions. Under the hood, AWS implements lambda functions as Docker containers that are disposed of after a period of inactivity, currently 15 min. If the system's load increases AWS would instantiate multiple copies of the Docker containers, each needing to hydrate its state from the NoSQL DynamoDB. For simplicity we did not include the session management and user identification components in the diagram but those components are also part of the framework.

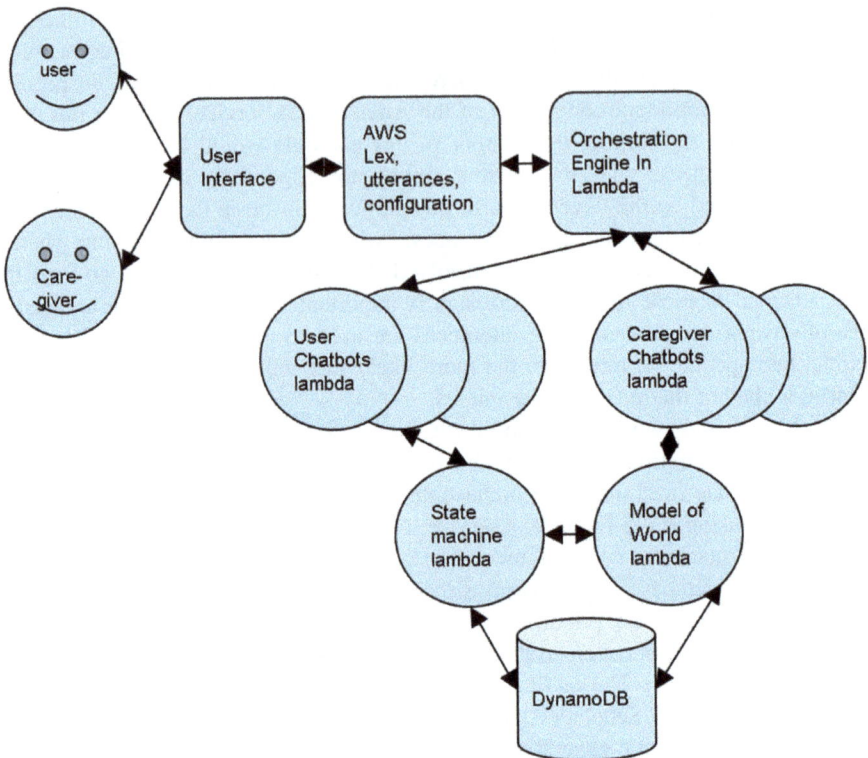

Fig. 4. General Architecture of the LEXY Framework.

2.1   LEXY Supported Activities

At this time the LEXY framework supports the following activities: "First/Then", "Schedule", and "Coping". First/Then activity boards are typically structured around two activities, the sought activity and the required activity. For example, to watch t.v. first you have to do homework. In a typical environment, the activity board would contain pairs of pictures of activities. For example, a picture of a t.v. next to a picture of a child studying. When the child requests to watch t.v. the caregiver would point to the activity board and reiterate that to watch t.v. the child must complete the required activity of doing homework. As these boards are typically physical boards with physical pictures on them, they are relatively easy to create at home by the caregiver but are static by nature and practicality limits them to a small set of activities. In LEXY, the "First/Then" activity is dynamic and is configured by the caregiver. Figures 5 through Fig. 8 show a typical interaction with the prototype chatbot for a First/Then activity. In Fig. 5, the child asks the chatbot to watch t.v. The chatbot, having not yet built a model of the world, is not aware of the requirements for watching t.v. and thus informs the child they need to ask their parents. In Fig. 6, the parent builds the model of the world by telling the chatbot that study is a requirement for watching t.v. In Fig. 7, the child repeats the request. Since the model of the world has been built, the chatbot responds differently this time. Instead of asking the child to "Ask your parents", this time the chatbot informs the child that the child needs to study first. The conversation concludes in Fig. 8, when the child has completed studying and the chatbot responses appropriately informing the child that she can watch tv.

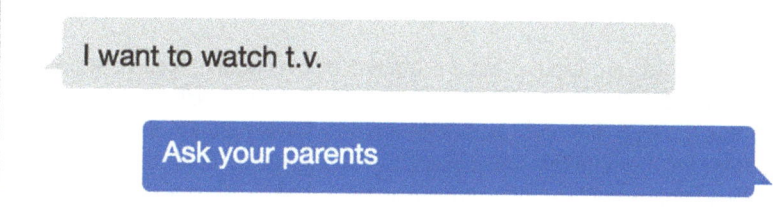

Fig. 5. The child informing the chatbot that the child wants to watch tv.

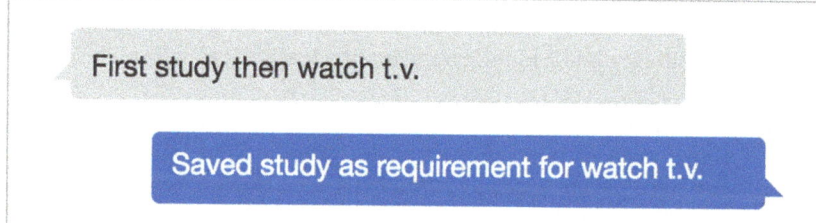

Fig. 6. The parent is building the model of the world conversationally.

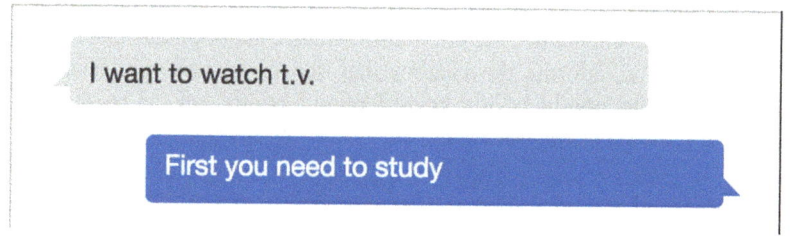

Fig. 7. The child repeating the request, after the model of the world has been built.

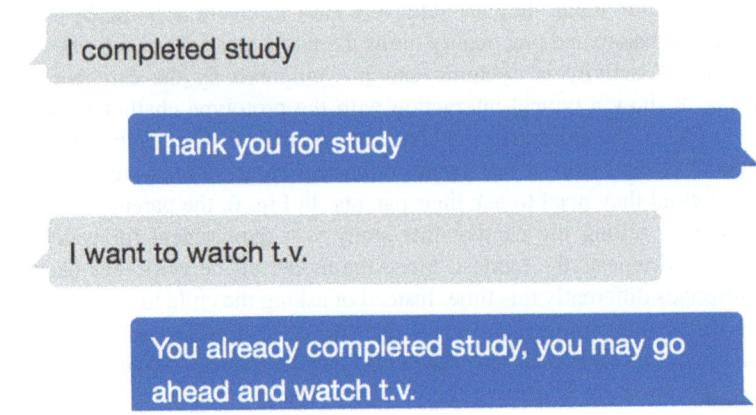

Fig. 8. The child interacting with the chatbot after completing the required chore.

Schedule boards are used to list a sequence of events, they can be used on a daily basis, to describe a daily schedule, for a specific complex task, or for a sequence of events within the day. A daily schedule board may have a picture of breakfast, a school bus, a classroom, a school bus and a home to indicate to the child they need to eat, go to school, spend the day at school and come back home. At school, another schedule board may contain a picture of books, lunchbox, playground and books again to indicate that the child will have some classroom time, then lunch time, then recess, and finally back to the classroom. In the LEXY framework, the schedule boards are an extension of the First/Then activity. The caregiver would conversationally specify to the chatbot what are the events, and their sequence, that comprise the schedule, the chatbot would correspondingly interact with the child informing her of the next activity when she asks. Both the child and the caregiver, or a teacher, can inform the chatbot of completed activities.

Coping boards are used to identify emotional state and present options for the child depending on their state of mind. For example, when a child is feeling anxious, a coping board can have picture options for "taking a deep breath", "counting till 10", or "go to a quiet place". Coping boards are implemented in LEXY using the more general If/Then activity. The If/Then activity implements the inverse workflow of the First/Then activity. While in the First/Then activity the child seeks a desired outcome, in the If/Then activity, the triggered event, for example feeling anxious, would trigger the outcome. Additionally an If/Then activity can have multiple options, all of which are enumerated to the child when the child informs the chatbot of her state of mind. For example, if the child informs the chatbot that she is anxious, the chatbot would responding with something like: "If you are feeling anxious you can try taking a deep breath, counting till 10 or finding a quiet place for few minutes."

3   Conclusions and Future Thoughts

We introduced LEXY, a Framework for modeling multi-actor conversational chatbots using Amazon Lex and Amazon Poly. The novelty of Lexy is that it allows human actors to model the world for the benefit of other humans, specifically individuals with special needs, interacting with the chatbots. Using the LEXY framework we modeled typical augmentative communication typically built with picture exchange systems and storyboards. The proflication of conversational devices, such as the Amazon Echo, makes it a natural evolution for Alternative and Augmentative Communication (AAC) community to build tools using those conversational devices and LEXY and is one of the first steps in that direction. While we successfully modeled several popular activities using LEXY, there are several challenges and opportunities ahead. Picture exchange systems are popular because many individuals with special needs are visual learners, and/or nonverbal individuals, this fact presents a challenge for a system that is by its nature conversational and not visual. Additionally, the picture exchange systems are easily accessible in terms of cost, availability, simplicity and requirements. Systems built on the LEXY framework will by nature be more costly, and less available than a simple paper based picture exchange system. The current norm of chatbot interactions dictates that chatbots do not initiate conversations with users. While this is not necessarily a shortcoming when we are modeling simple storyboards, if the conversational systems become successful one can envision use cases where it would be helpful for the chatbot to initiate conversation with the child. Finally while the LEXY framework presents an opportunity to build applications that can be utilized at home or school, we are in the early stages of that process and there is currently no applications, yet, that are available for the public to use based on the LEXY framework.

References

1. Flippin, M., et al.: Effectiveness of the picture exchange communication system (PECS) on communication and speech for children with autism spectrum disorders: a meta-analysis. Am. J. Speech Lang. Pathol. 19(2), 178 (2010)
2. Light, J.C., et al.: Augmentative and alternative communication to support receptive and expressive communication for people with autism. J. Commun. Disord. 31(2), 153–180 (1998)
3. Dooley, P., Wilczenski, F.L., Torem, C.: Using an activity schedule to smooth school transitions. J. Posit. Behav. Interv. 3(1), 57–61 (2001)
4. Dempsey, P.: The teardown Amazon echo digital personal assistant [Teardown Consumer Electronics]. Eng. Technol. 10(2), 88–89 (2015). https://doi.org/10.1049/et.2015.0231
5. Perez, S.: Amazon sold 'millions' of Alexa devices over the holiday shopping weekend. TechCrunch (2018). https://techcrunch.com/2017/11/28/amazon-sold-millions-of-alexa-devices-over-the-holiday-shopping-weekend/. Accessed 4 Feb 2018
6. Yan, M., Castro, P., Cheng, P., Ishakian, V.: Building a chatbot with serverless computing. In: Proceedings of the 1st International Workshop on Mashups of Things and APIs, p. 5. ACM (2016)
7. Developer.amazon.com (2018). 2017 Alexa Prize. http://developer.amazon.com/alexaprize/2017-alexa-prize. Accessed 4 Feb 2018

An Age Adapting Electrolarynx – A Feasibility Study

Pieter Coetzee[1(✉)], Joice Lamfel[1], David M. Rubin[1],
and Vered Aharonson[1,2]

[1] Biomedical Engineering Research Group, School of Electrical and Information
Engineering, University of the Witwatersrand, Johannesburg, South Africa
coetzpeter@gmail.com, joicelamfel@gmail.com,
{david.rubin, vered.aharonson}@wits.ac.za
[2] Department of Electrical Engineering,
Afeka Tel Aviv Academic College of Engineering, Tel Aviv, Israel

Abstract. We propose a mathematical model for voice aging that could be used in the design of an age-adapting Electrolarynx. Voice data from public figures, at the ages of 30, 40, 50 and 60 years old, were acquired from a YouTube corpus. The voice processing consisted of an extraction of 70 Mel-Frequency Cepstral Coefficients (MFCCs) and a computation of their statistical features. ANOVA F-tests were used to determine which of these features change with age. Significant differences between age groups were found only for the first 40 MFCCs. The aging model was then constructed using non-linear regression and an averaged quadratic polynomial fit on these coefficients. Model age-adapted voices were reconstructed from the young dataset speakers' voices and compared to their voices at older ages. The model was validated by the correlation between speakers' MFCCs at older ages and the model-aged MFCCs. The average correlation results were in the range of 0.62 to 0.93. The results imply that the first 40 MFCCs are more susceptible to age related changes and that the proposed model has the potential to enhance the Electrolarynx by providing age adaptation as the speaker grows older.

Keywords: Human factors · Electrolarynx · Speech processing

1   Introduction

A laryngectomy on laryngeal cancer patients is often followed by use of an Electrolarynx - a mechanism that produces a synthesized voice. The produced voice is, however, very artificial and does not alter as the patient gets older, which may cause psycho-social issues, especially in young patients [1, 2]. In this research we studied a mathematical voice aging model that could be used in the design of an age adapting Electrolarynx to reduce some unwanted social consequences.

Speech features extracted for age discrimination purposes are classified as prosodic, spectral or glottal [3]. Fundamental frequency (F_0), pitch (reciprocal of F_0) and loudness are prosodic features [4, 5]. A "U" shape can be used to describe the change of fundamental frequency with age in males. A young boy's F_0 starts off high, declines

© Springer International Publishing AG, part of Springer Nature 2019
T. Z. Ahram and C. Falcão (Eds.): AHFE 2018, AISC 794, pp. 880–886, 2019.
https://doi.org/10.1007/978-3-319-94947-5_86

from young adulthood into middle age, then arises again due to the loss of flexibility of the vocal cords. Females between the ages of 20–50 have a moderately stable F_0, which declines thereafter due to hormonal changes (mass of vocal cords increase during menopause). The loudness of the voice decreases with age due to the reduced activity of the vocalis muscle in the larynx [6].

Spectral features of speech are commonly studied using either Mel-Frequency Cepstral Coefficients (MFCCs) or Linear Predictive Cepstral Coefficients (LPCCs) [3]. The extraction of these features reveal vocal tract information such as its shape, spectral bandwidths, etc. which can be used for speaker age estimations [3]. A speaker's age can also be determined through the extraction of glottal features. Evident effects of an aging voice include: increased hoarseness, vocal tremors or shakiness and breathy voices. Dehydration of the vocal cords cause irregular vibrations, resulting in increased hoarseness of the voice. Variations in the blood supply to the larynx with age lessens vocal cord control, which could result in shakiness of the voice. The articular surfaces become thinner with age allowing air to escape during speech. Hence, the increased breathing in the speech of the elderly [6].

Several age recognition systems, aiming to determine the most efficient age-discriminating features, have been proposed in the past 2 decades. A comparative analysis of age group recognition [7] using various features such as MFCCs, LPCCs and Support Vector Machines (SVM) for classification yielded age recognition accuracies of 91.39% for the MFCCs as feature vectors and 84.69% for the LPCCs. Two other studies that employed MFCC features for age recognition [3, 8] respectively improved recognition performance accuracies to 92% and 94.6%. A performance comparison of MFCC features and Shifted Delta Cepstral (SDC) features [4] showed that under noisy speech conditions SDC features yield better performance whereas MFCC features were more appropriate for clean speech data. The trade-off between accuracy and complexity is also an important consideration in cepstral feature selection: LPCCs are more cost effective computationally whereas MFCCs yield more accurate age discrimination results.

An attempt to determine the most relevant parameters that can be used for voice age recognition, extracted from the glottal signal is presented by Mendoza et al. [9]: time domain, frequency domain and parameters representing perturbations in the fundamental frequency were compared to determine which parameters were the best age discriminators. For females, the best age discriminating parameters were shimmer and time domain parameters (amplitude and opening quotient). The greatest differences between the male age groups were time domain parameters (closing phase, opening and speed quotient). Age recognition in the combined female and male database yielded a 91.66% accuracy.

Although various age recognition systems have been explored and designed, no voice aging models currently exist nor have any been studied for usage in the framework of an Electrolarynx. In this research we explored the changes in MFCC statistics as speakers get older to create voice aging models, for both males and females.

2 Voice Aging Model Design

Obtaining the voice aging model relies on the ubiquitous MFCC extraction method, which consists of an initial conditioning and pre-processing phase [10] followed by regression analysis. The MFCC features are selected due to their high age discrimination accuracy and the extraction of 70 MFCCs provides a good balance between computational cost and reconstruction quality [11, 12].

2.1 Regression Analysis

An ANOVA F-test, significance value of $p = 0.05$ was used on each MFCC for the 4 different ages of study to assess whether their means were changing. MFCCs that failed the Null Hypothesis were flagged for regression analysis.

The mean and 3 standard deviations were calculated, 7 points in total, for each of the flagged MFCCs. These points represent 99.7% of the distributed data and this compressed representation was found to accelerate the computation for nonlinear regression. A polynomial fit was chosen, with a quadratic order determined by the lowest Root-Mean-Square Error (RMSE) criteria. Higher order polynomials were not considered as they produce estimated results which are highly sensitive, have less attractive weights compared to local linear regressions and can result in misleading confidence levels, thus increasing the possibility of error [13].

Each of the quadratic polynomials produced for the various MFCCs of the training dataset, were then averaged for male and female speakers separately. This produces a curve which is representative of the mean aging effects on any subject for that particular word. The y-intercept term needed to be adjusted, as this value was found to be speaker dependent and is less pertinent to the speaker age and the change in the means [14].

The adjustment is made by comparing the modelled result, of the mean MFCC of the speakers' voices at the younger age, with the means at older ages. The curve is then shifted accordingly. Finally, using the averaged standard deviation of each MFCC, frame magnitudes for each MFCC are created by randomly choosing ones that are normally distributed around the model curve.

2.2 Speech Waveform Reconstruction from MFCCs

Signal reconstruction (from MFCCs to time domain waveforms) was performed to allow for subjective listening tests. The reconstruction procedure is adapted from [12]. To recover the magnitude spectrum the Moore-Penrose pseudo inverse and $l2$ norm minimization least-squares estimate has been used. Using white Gaussian noise as an initial estimate, the Least Squares Estimate Inverse Short-Time Fourier Transform Magnitude (LSE-ISTFTM) algorithm is adapted for phase reconstruction purposes. Integration of the recovered magnitude and phase spectrum results in speech frame estimates. Thereafter, an estimate of the entire speech waveform in the time domain is obtained by the application of an overlap-add procedure of the estimated frames [12, 15].

2.3   Objective and Subjective Testing

An objective testing metric included a cross correlation measurement between the modified-aged and original voice (MFCC and time domain signals) to determine similarity.

Subjective testing involved the use of 3 listeners. The listeners were presented with the original young dataset speakers voice, their voice from older ages and the reconstructed model-aged voice. They were asked to give their subjective impression on the similarity between the 3 voices as well as their estimate for the age of the speaker in each recording.

This study received an ethics waiver from the Human Research Ethics Committee (Medical) at the University of the Witwatersrand, Johannesburg.

3   Results

The ANOVA F-test results for the MFCC features extracted from the training set of recordings, of both male and female speakers, showed that the first 40 MFCCs were significantly different across different ages, in 80% of the utterances. The features of MFCCs 41 to 70 could not reject the null hypothesis, at the chosen level of significance.

The first 40 MFCCs were therefore used to reconstruct an aged speech from the young speaker's speech utterances. The averaged correlations between the model-aged utterances and the utterances recorded at the speakers' older age (60 years old), calculated for the test set of utterances, are presented in Table 1. The correlations are in the range of 0.62 to 0.93 which indicates moderate to high correlation.

Table 1. Average correlations of MFCCs between the model-produced and recorded (at 60-year-old) speech utterances.

Test Subject	Correlation
Female 1	0.64
Female 2	0.93
Male 1	0.62
Male 2	0.79

Figure 1 illustrates the MFCCs of an utterance of a female speaker from the test database (Female 2). The MFCCs of the utterance recorded at the age of 60 years old is presented in the upper graph and the reconstructed utterance in the lower graph. A visual comparison supports the high correlation between the two signals in terms of MFCCs (0.93).

The correlations between the recorded speech utterances of the test set and the model-aged MFCC reconstructed speech, using Gang's algorithm [16], yielded poor correlation values (0.03 to 0.12). The listening tests confirmed that these reconstructed speech utterances were barely audible and contained a great amount of noise that, in

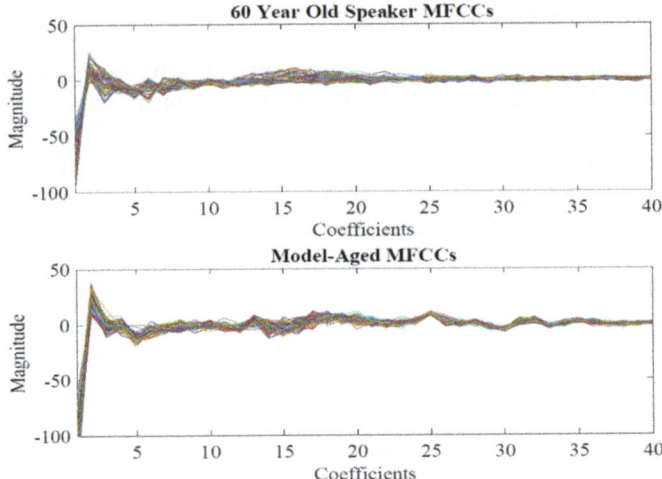

Fig. 1. MFCC magnitudes of all utterances of speaker "Female 2" at the age of 60-years-old (upper graph) and the model-produced aged voice (lower graph).

some cases, prohibited the understanding of the words spoken. The reconstruction algorithm was tested on the unaltered recorded speech (70 MFCCs) and further confirmed the inadequacy of this algorithm to produce faithful speech.

Figure 2 portrays this finding: The recorded speech at 60 years old is shown in the top graph and the reconstructed speech, from its 70 unaltered MFCCs, is shown in the lower graph.

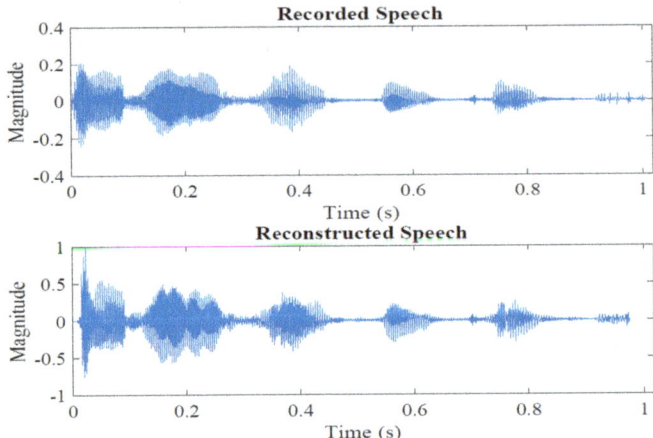

Fig. 2. Utterance recorded at 60-years-old (upper graph) and unaltered MFCC reconstructed speech using Gang's Reconstruction Algorithm (lower graph) for Female 2.

Although similarity can be observed between the two signals, the correlation between them is poor (0.47). The listening tests confirmed this as all listeners commented that they were able to recognize the words spoken but that the speech had a poor quality and a distinct echo which prevented them from reliably commenting upon the speaker's age.

4   Discussion and Conclusions

A mathematical model for speech aging based on changing the speech MFCC features was proposed. The model was trained and then validated on a test dataset using correlation between model-aged MFCCs and MFCCs of speech recorded at older age, for the same utterances. The results indicated moderate to high correlations for both male and female speakers' MFCCs. This finding implies the model's potential to successfully age a young Electrolarynx voice into an older one. The model may improve by increasing the training data as well as an inclusion of more age groups. The quality of the recordings may also contribute to the accuracy of the model: using stereophonic recordings rather than monophonic recordings will ensure that acoustic information is not lost in the conversion from the former to the latter and using recordings sampled at a rate of 32 kHz will better match the human auditory capabilities (majority of adults can only hear up to 12–14 kHz) [17] and this may contain more age-related information.

The results indicate that the first 40 MFCCs, out of 70, contain the greatest amount of age-related acoustic information. The reconstruction of speech from the aging model-MFCCs yielded only poor correlation to the speakers' recordings at older ages. This may be due to both the method of transformation from the cepstral to the time domain used in this study [12] as well as the construction of each frame magnitude of a modeled MFCC, from randomly chosen magnitudes that conform to a normal distribution. These limitations could be further explored using different reconstruction algorithms as well as a formulation of a method that can be used to better place the distributed points. A possible method could be to compare each generated point of an MFCC with its nearest neighbors from the non-aged, corresponding MFCC points). The current preliminary feasibility study thus illustrates potential for an automated voice aging capability which could be implemented in an Electrolarynx.

References

1. Goode, R.L.: Artificial laryngeal devices in post-laryngectomy rehabilitation. Laryngoscope **85**(4), 677–689 (1975)
2. Kaye, R., Tang, C.G., Sinclair, C.F.: The electrolarynx: voice restoration after total laryngectomy. Med. Dev. (Auck) **10**, 133–140 (2017)
3. Mittal, T., Barthwal, A., Koolagudi, S.G.: Age approximation from speech using Gaussian mixture models. In: Proceedings of the 2nd International Conference on Advanced Computing, Networking and Security, pp. 74–78, December 2013

4. Erokyar, H.: Age and gender recognition for speech applications based on support vector machines. Graduate Theses and Dissertations (2014). http://scholarcommons.usf.edu/etd/5356. Accessed 05 Dec 2017

5. Bocklet, Z.V., Stemmer, T., Aeissler, V., Noeth, E.: Age and gender recognition based on multiple systems-early vs. late fusion. In: Interspeech 2010 (2010)

6. Eadie, T.: Characteristics of the aging female voice. J. Speech Lang. Pathol. Audiol. **24**(4), 162–179 (2000)

7. Lee, M.-W., Kwak, K.-C.: Performance comparison of gender and age group recognition for human-robot interaction. Int. J. Adv. Comput. Sci. Appl. **3**, 12 (2012)

8. Kim, H.J., Bae, K., Yoon, H.S.: Age and gender classification for a home-robot service. In: 16th IEEE International Conference on Robot & Human Interactive Communication, pp. 122–126, August 2007

9. Mendoza, L.A.F., Cataldo, E., Vellasco, M., Silva, M.A., Cañón, A.D.O., de Seixas, J.M.: Classification of voice aging using ANN and glottal signal parameters. In: ANDESCON 2010, pp. 1–5. IEEE (2010)

10. Fairhurst, M., Erbilek, M., Da Costa-Abreu, M.: Selective review and analysis of aging effects in biometric system implementation. IEEE Trans. Hum. Mach. Syst. **45**, 294–303 (2015)

11. Boucheron, L.E., De Leon, P.L., Sandoval, S.: Low bit-rate speech coding through quantization of mel-frequency cepstral coefficients. IEEE Trans. Audio Speech Lang. Process. **20**(2), 610–619 (2012)

12. Gang, M., Zhang, X., Yang, J., Zou, X.: Speech reconstruction from mel-frequency cepstral coefficients via 1-norm minimization. In: IEEE 17th International Workshop on Multimedia Signal Processing (MMSP), pp. 1–5, October 2015

13. Gelman, A., Imbens, G.: Why high-order polynomials should not be used in regression discontinuity designs. J. Bus. Econ. Stat. **2017**. https://doi.org/10.1080/07350015.2017.1366909. Accessed 27 Oct 2017

14. Ganchev, T., Fakotakis, N., Kokkinakis, G.: Comparative evaluation of various MFCC implementations on the speaker verification task. In: Proceedings of SPECOM, pp. 191–194, October 2005

15. Griffin, D., Lim, J.: Signal estimation from modified short-time Fourier transform. IEEE Trans. Acoust. Speech Signal Process. **32**(2), 236–243 (1984)

16. Gang, M.: Reconstruct speech from MFCCs (v.3), October 2016. https://www.Mathworks.com/matlabcentral/fileexchange/53186-invmfccs?requestedDomain=www.mathworks.com. Accessed 28 Oct 2017

17. Gordon-Salant, G.S., Frisina, R.D., Fay, R.R., Popper, A.: The Aging Auditory System, p. 115. Springer (2010)

Designing and Creating a Prototype of Robotic Skeleton Systems for Computerized Lower Limb Prosthesis

Yatip Auarmorn[1], Nantakrit Yodpijit[1(✉)],
and Manutchanok Jongprasithporn[2]

[1] Center for Innovation in Human Factors Engineering and Ergonomics,
Department of Industrial Engineering, Faculty of Engineering, King Mongkut's
University of Technology North Bangkok, Bangkok, Thailand
nantakrit.y@eng.kmutnb.ac.th, nantakrit@gmail.com
[2] Department of Industrial Engineering, Faculty of Engineering,
King Mongkut's Institute of Technology Ladkrabang, Bangkok, Thailand

Abstract. Recent statistics reveal that more than 40 million people of all ages from all over the world suffer amputations, and only 5% of them in need have access to assistive products. Common causes of amputation include congenital deformities, vascular diseases, diabetes, and accidents. A prosthesis is an assistive device that is used to substitute and restore the normal functions of the missing part of body. A prosthesis can be either functional or cosmetic, and can be either attached to the body externally or implanted surgically. Without prosthetic innovations, lower-limb amputees are required to have ordinary ambulatory skills sufficient to perform the basic movement functions and the ability to stand, walk, sit, reach with hands and arms, and manipulate (lift, carry, move) light to medium weights. Unfortunately, some amputees are unable even to stand and use their prosthesis to meet their demand. It has been found that traditional prostheses are mostly passive products with limitations and drawbacks. Recent research suggests that a trend in the utilization of active actuators/prostheses can overcome limitations and drawbacks of passive prostheses. Major functions of active prostheses are real-time intent recognitions, control strategies, torque requirements reduction, and energy saving. This paper presents the design of robotic skeletal system of the lower limb for computerized prosthetic leg development. The major structure of robotic skeletal systems consists of lower-limb joints modules, kinematics sensors and kinetics sensors. By taking advantage of the modern microprocessor-based controllers and low-power transmitters, the new robotic skeleton systems are created with better control performance in nearly real-time basis, lighter weight, higher flexibility, greater range of movement around a join, and better safety. The preliminary experiments are performed to evaluate the new design of robotic skeletal system by measuring sensors data while walking on ground, and climbing up-down stairs. Findings indicate the new design of a robotic skeletal system can provide the useful biomechanics parameters, which can be used for the development of computerized lower limb prosthesis.

Keywords: Robotic skeleton systems · Computerized lower limb prosthesis

© Springer International Publishing AG, part of Springer Nature 2019
T. Z. Ahram and C. Falcão (Eds.): AHFE 2018, AISC 794, pp. 887–898, 2019.
https://doi.org/10.1007/978-3-319-94947-5_87

1   Introduction

The exoskeleton is a mechanical orthotics system that is attached to the exterior of a human body to improve the muscular functions of the wearer. Nowadays, exoskeleton have been studied in many countries, including US, Japan, and Europe, and currently being researched in various type of applications, including industries [1], military [2, 3], medicine [4, 5], and rehabilitation [6, 7]. Lower extremity exoskeleton can be classified by muscular functions support as power augmentation systems and power assistance.

Recent research reveals that a trend in the utilization of active actuators prostheses. Can overcome limitations and drawbacks of passive prostheses. However, in order to create active prostheses, precise and sufficient kinematics and kinetics data is required for being used in development of key functions, including real-time intent recognitions, control strategies, torque requirements reduction, and energy saving.

The study aims to design the lower limb exoskeletal system for computerized prosthetic leg development. This exoskeletal is power augmentation type. The spring systems are installed at each lower-limb joints modules. All joints modules consist of motion sensors and force sensors which can provide kinematics and kinetics data. The microprocessor-based controller system with wireless transmitter are developed to collect all data (Fig. 1).

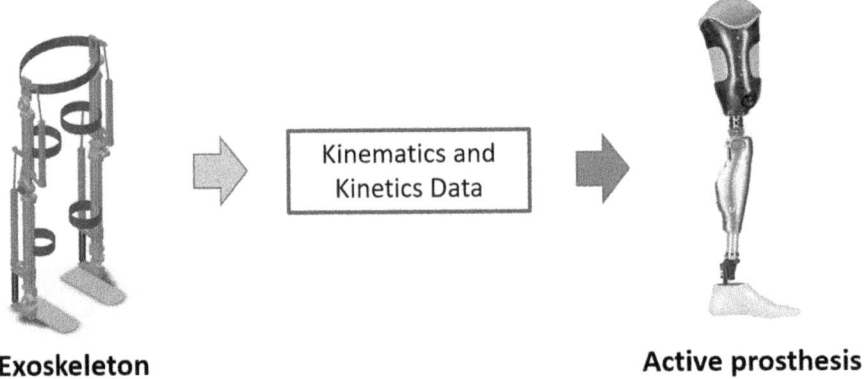

Exoskeleton   **Active prosthesis**

Fig. 1. The kinetic and kinetic data obtained from study this exoskeleton

2   Methods

2.1   Human Walking

Gait Cycle [8]. Walking is a natural evolution. Gait Cycle is divided into 2 stages. Stance phase takes 60% of walking distance. Starting from the interval Heel strike 0% Toe off 60%. Swing phase is the average duration is 40% of the walking cycle. From the Toe off to 60% to 100% heel strike (Fig. 2).

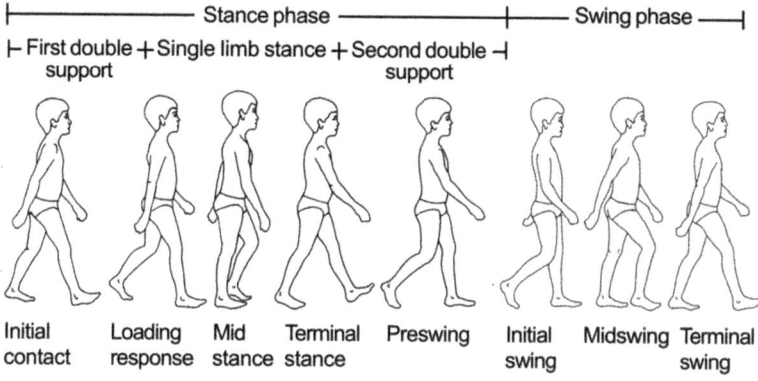

Fig. 2. Gait cycle [9]

The Cycle Gait Data (CGD). The presented human walking medium on flat ground. Subject is 20 healthy adult. Age from 20–72 years, body mass approximately 68.5 kg and height approximately 171 cm [10].

Ankle. Figure 3 (Left) shows moment of ankle data in gait cycle. The peak toque of ankle approximate 90 Nm that is the highest torque in three joint (ankle, knee and hip). Range of peak torque start at heel off (At this point of heel off the floor.) until toe off (At this point of toe off the floor.). The ankle torque is almost entirely planter extension. Therefore, get idea is to put the direction of the spring. Figure 3 (Right) Shown degree of ankle data at each point in gait cycle. Range of peak degree at toe off and initial swing.

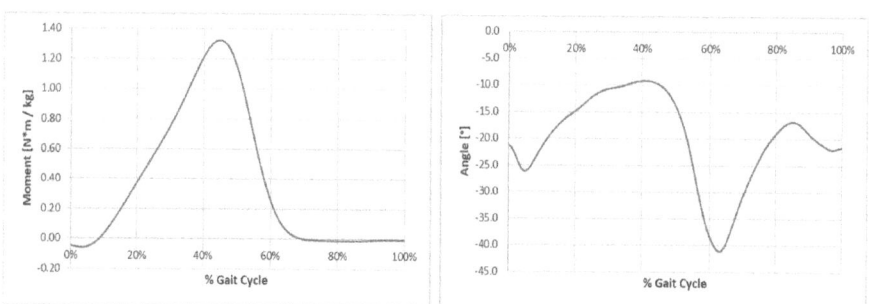

Fig. 3. Moment of ankle during walking (Left). Angle of ankle during walking (Right).

Knee. Figure 4 (Left) shows moment of knee help in control extension for protect knee fall between walk range stance in gait cycle. The peak toque of ankle approximate 34 Nm at stance phase. All most torque phase at positive graph phase (At extension). Figure 4 (Right) shows degree of knee data at each point in gait cycle. The peak degree of knee approximate 63° that most degree in three joint (ankle, knee and hip). Because the most degree help in swing phase for clear ground.

Hip. Figure 5 (Left) shows moment of hip has both extension and flexion in gait cycle. The peak torque of extension at early stance has approximate 35 Nm because hip get load direct from the upper part of the body and tried to exert stepped forward. The peak torque of flexion at the end stance phase has approximate 55 Nm. Figure 5 (Right) shows degree of hip data at each point in gait cycle. Mostly movement at positive graph phase (At flexion) and peak degree of flexion has approximate 34° before swing phase. Extension is negative graph phase and peck degree of the extension at early heel off until initial swing.

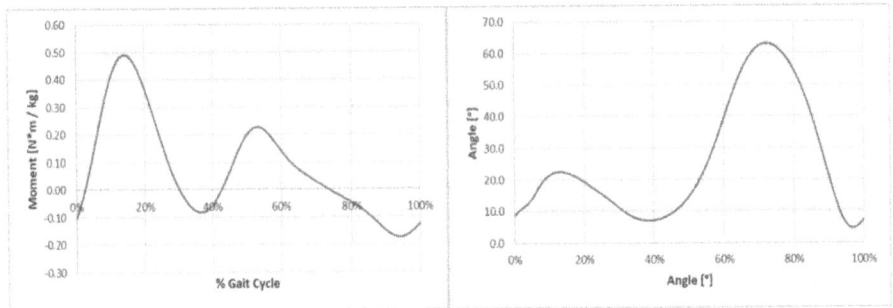

Fig. 4. Moment of knee during walking (Left). Angle of knee during walking (Right).

2.2   Range of Motion

The Exoskeleton joint ranges of motion are determined by examining human joint ranges of motion. At least the Exoskeleton joint range of motion should be equal to the human range of motion during walking [10]. Safety dictates that the Exoskeleton range of motion should not be more than the human range of motion [11] (Table 1).

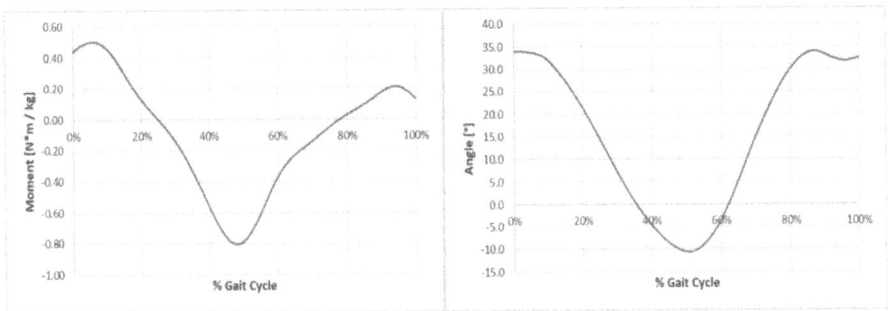

Fig. 5. Moment of hip during walking (Left). Angle of hip during walking (Right).

Table 1. Exoskeleton joint range of motion

	Human walking maximum [10]	The Exoskeleton maximum	Human males maximum [11]
Ankle dorsi flexion	9.1°	15°	15°
Ankle planter extension	41.1°	45°	45°
Knee flexion	63°	100°	118°
Hip flexion	34°	100°	116.5°
Hip extension	10.7°	20°	Not available°

2.3   Exoskeletal Design

Structure. The exoskeleton has three degrees of freedom include 1 DOF at hip, 1 DOF at knee, and 1 DOF at ankle (Fig. 6).

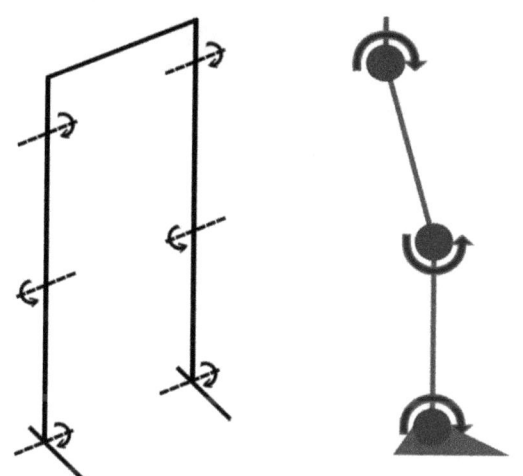

Fig. 6. Degree of freedom and direction of the spring load

Each joint has movement and torque are not the same. So that consider add spring each point dissimilar. Spring direction show at Fig. 6. Nevertheless, spring direction of knee joint was different from another joint. Knee joint required both direct force but choose put one sided at extension. Because torque of extension greater than flexion and did not want to load the wearer.

Figures 7 and 8 are an overall model of Exoskeleton. The following sections discuss the critical features of the major components.

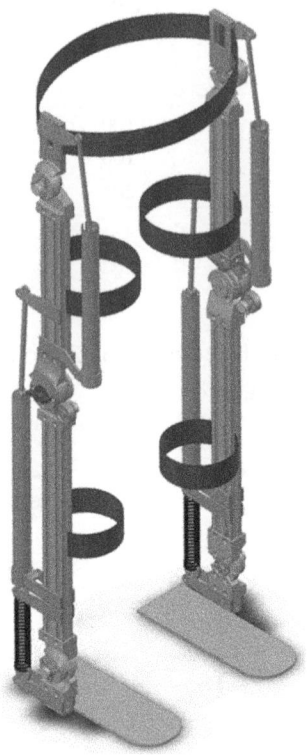

Fig. 7. The Exoskeleton overall model (isometric view)

Sensors. Potentiometers and strain gauges are attached to hip, knee, and ankle joints for kinematics and kinetics data collections.

Ankle. The ankle joint was single degree of freedom (DOF). The key component is the spring. Select tension spring constant k approximately 15 N/m connect with arm of 70 mm for transform the linear stiffness to effective rotary stiffness at the ankle joint. (Figure 9) During the walking cycle, the plantar flexion spring stored energy during controlled dorsiflexion, and that energy was subsequently released to assist the human foot during powered plantar flexion. Material used aluminium alloy high strength type (AL7075) in structure. Exoskeleton had ROM stopper for protect more than the human range of motion.

Shank and Thigh used standard part aluminium profile convenient purchase, inexpensive and easy adjustment of length for shank and thigh according to the anthropometry data of each user [11–14].

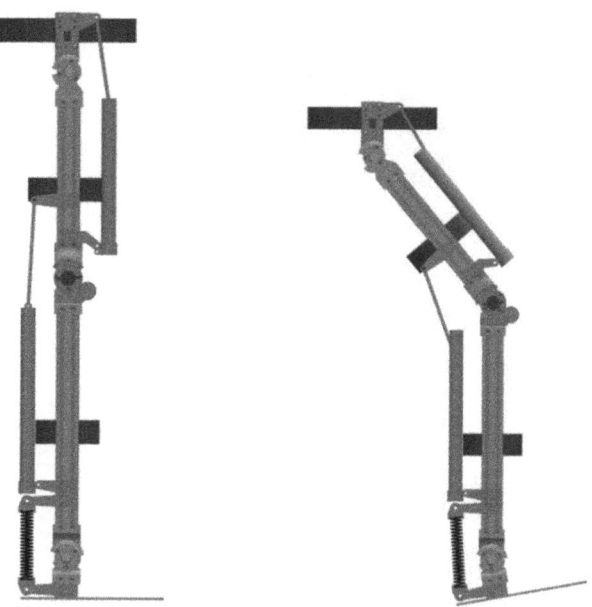

Fig. 8. The exoskeleton overall model (side view)

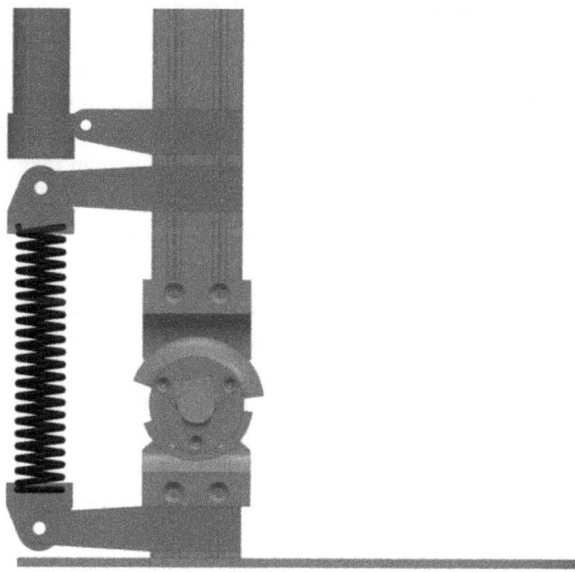

Fig. 9. Ankle joint (side view)

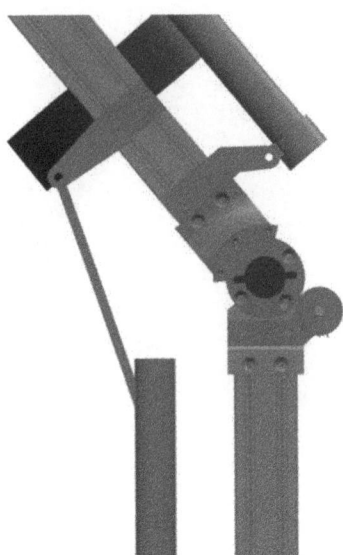

Fig. 10. Knee joint (side view)

Knee. The Knee joint was single degree of freedom (DOF). Select type compression spring and spring constant $k_{kneespring}$ approximately 2.5 N/m connect with arm of 60 mm for transform the linear stiffness to effective rotary stiffnesses at the knee joint and spring had movement in cylinder. (Figure 10) Used $k_{kneespring}$ value low because of knee. The required knee torque has both extension and flexion. The highest peak torque is extension. Consider adding a spring force that reduces peak torque extension at early stance. Material selected aluminium alloy high strength type (AL7075) in structure. Exoskeleton had ROM stopper for protect more than the human range of motion.

Figure 11 shows mechanism for lock position angle every 20°. The mechanism for this design has the advantage of able quick release the lock. Using springs to help release.

Hip. The hip joint was single degree of freedom (DOF). Select type tension spring and spring constant $k_{hipspring}$ approximately 15 N/m connect with arm of 50 mm for transform the linear stiffness to effective rotary stiffnesses at the hip joint and spring had movement in cylinder. (Figure 12) The hip flexion spring stored energy during hip extension and released that energy to assist hip flexion movements in walking. Material selected aluminium alloy high strength type (AL7075) in structure. Exoskeleton had ROM stopper for protect more than the human range of motion.

Fig. 11. Knee position locking mechanism

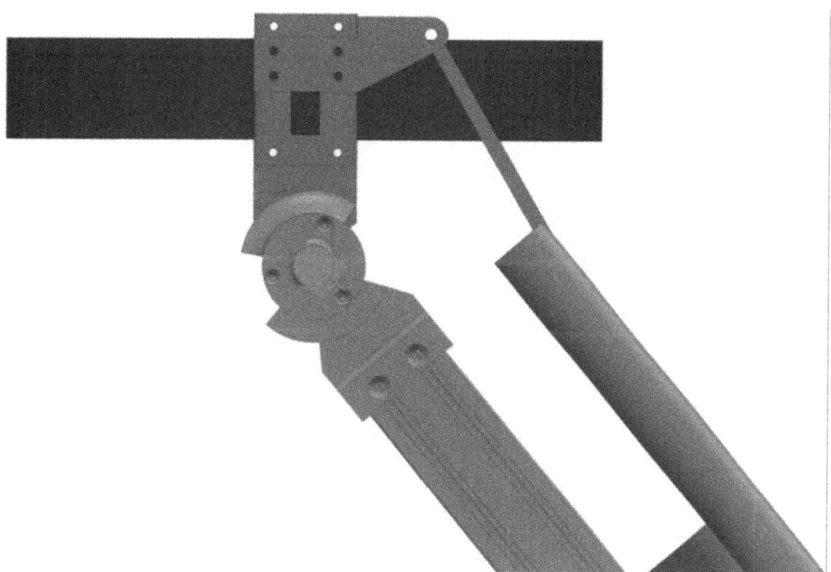

Fig. 12. Hip joint (side view)

2.4   Biomechanical Model Analysis

In calculate we use the torque and range of motion from the gait cycle data. Moreover, the length of the arm from CAD design. Used Solidworks in the design. We consider only sagittal plane.

Calculating Springs at ankle knee and hip by $F = k (S_o - S)$ is given below.

$$F_{spring} = k_{kneespring} (S_o - S)(\cos \theta) \tag{1}$$

Calculate the torque from the springs at ankle knee and hip by $\Sigma M = F\,r$ is given below.

$$\Sigma M_{spring} = F_{spring}(r_x \cos \theta + r_y \sin \theta) \tag{2}$$

Total torque calculation at ankle knee and hip by $\Sigma M_{totel} = 0$ is given below.

$$\Sigma M_{totel} = M_{gait\,cycle\,data} - M_{spring} \tag{3}$$

Use excel to create formulas for calculating torque occurring at the ankle and knee to calculate all walking ranges and then plot the values obtained.

3   Biomechanics Results

Ankle. When wearing exoskeleton, the peak torque of planter extension were reduce about 20 Nm or about 23% (Fig. 13).

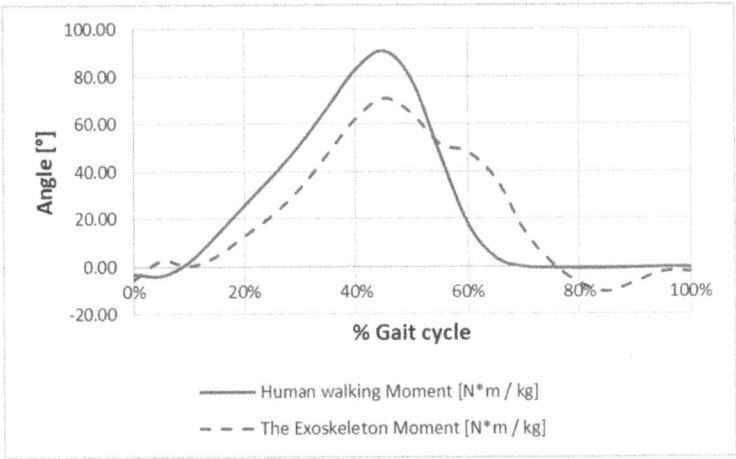

Fig. 13. Moment of ankle during walking comparison between human and exoskeleton

Knee. When wearing exoskeleton, the peak torque of extension were reduce about 7 Nm or about 20% but torque of flexion was increased. Because required knee torque had both extension and flexion (Fig. 14).

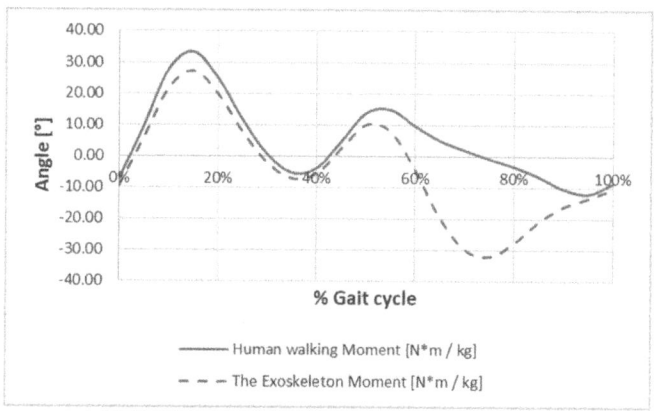

Fig. 14. Moment of knee during walking comparison between human and exoskeleton

Hip. When wearing exoskeleton, the peak torque of extension were reduce about 30 Nm or about 88% and the peak torque of flexion were reduce about 23 Nm or about 42% (Fig. 15).

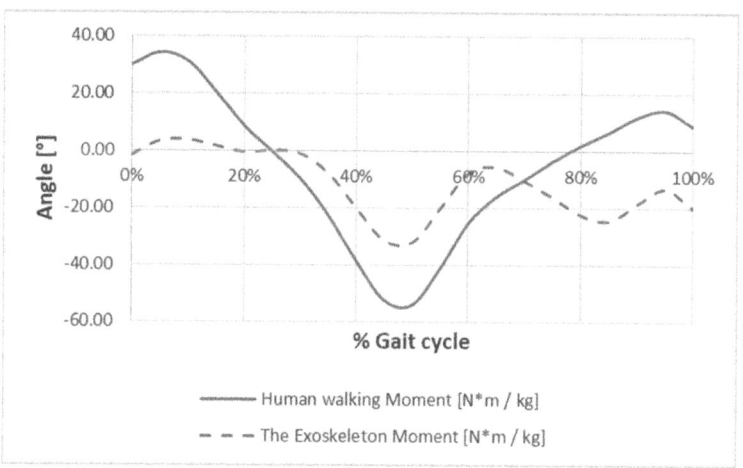

Fig. 15. Moment of hip during walking comparison between human and exoskeleton

Acknowledgments. This research was funded by King Mongkut's University of Technology North Bangkok. Contract no. KMUTNB-61-GOV-C2-50.

References

1. Ekso Bionics: An exoskeleton bionic suit or a wearable robot that helps people walk again. https://eksobionics.com/
2. Zoss, A.B., Kazerooni, H., Chu, A.: Biomechanical design of the berkeley lower extremity exoskeletong (BLEEX). IEEE/ASME Trans. Mechatron. **11**, 128–138 (2006)
3. Walsh, C.J., Endo, K., Herr, H.: A quasi-passive leg exoskeleton for load-carrying augmentation. Int. J. Humanoid Robot. **4**, 487–506 (2007)
4. Jezernik, S., Colombo, G., Keller, T., Frueh, H., Morari, M.: Robotic orthosis lokomat: a rehabilitation and research tool. Neuromodul. Technol. Neural Interface **6**, 108–115 (2003)
5. Hocoma: Lokomat®. https://www.hocoma.com/solutions/lokomat/
6. Kim, W., Lee, H., Kim, D., Han, J., Han, C.: Mechanical design of the Hanyang exoskeleton assistive robot (HEXAR). In: 2014 14th International Conference on Control, Automation and Systems (ICCAS 2014), pp. 479–484. IEEE (2014)
7. Sankai, Y.: HAL: hybrid assistive limb based on cybernics. Presented at the (2010)
8. Inman, V.T.: Human Walking (1981)
9. Zhang, L.Q., Wang, G.: Dynamic and static control of the human knee joint in abduction-adduction. J. Biomech. **34**, 1107–1115 (2001)
10. Bovi, G., Rabuffetti, M., Mazzoleni, P., Ferrarin, M.: A multiple-task gait analysis approach: kinematic, kinetic and EMG reference data for healthy young and adult subjects. Gait Posture **33**, 6–13 (2011)
11. Wagner, D., Birt, J.A., Duncanson, J.P.: Developmental Systems System Resources Corporation (SRC) National Technical Information Service. Security, vol. 1007 (1996)
12. Intaranont, K.: Ergonomics (2005)
13. Yodpijit, N., Bunterngchit, Y., Lockhart, T.E.: Anthropometry of Thai technical university students. In: IIE Annual Conference and Exhibition 2004, vol. 9, pp. 899–904 (2004)
14. Winter, D.A.: Anthropometry. Biomech. Mot. Control Hum. Mov. 82–106 (2009)

Interdisciplinary-Based Development of User-Friendly Customized 3D Printed Upper Limb Prosthesis

Letícia Alcará da Silva[1(✉)], Fausto Orsi Medola[1],
Osmar Vicente Rodrigues[1], Ana Cláudia Tavares Rodrigues[2],
and Frode Eika Sandnes[3]

[1] Department of Design, São Paulo State University (UNESP),
Av. Eng. Luiz Edmundo C. Coube 14-01, Bauru 17033-360, Brazil
lalcar4@gmail.com, {fausto.medola,
osmar}@faac.unesp.br
[2] Specialized Center for Rehabilitation SORRI-Bauru,
Av. Nações Unidas, 53-40, Bauru 17033-130, Brazil
cacaautavares@gmail.com
[3] Oslo Metropolitan University (OsloMet), Pilestredet 46, 0167 Oslo, Norway
Frode-Eika.Sandnes@hioa.no

Abstract. This study reports on the development of a customized transradial mechanical prosthesis for a patient with bilateral transradial amputation, with a design process involving interdisciplinary project combining the areas of product design, rehabilitation, rapid prototyping, and ergonomics. The process started with clinical assessment of the patient's characteristics, functional abilities and needs, as well as interviews exploring patient's preferences. Next, a first version of the prosthesis was prototyped and tested, and a second phase started aiming to solve the problems and improve the prosthesis design. A second test was then executed, revealing that the second prototype utilized a more appropriate and comfortable coupling with the limb, easy activation of the hand prehension by elbow flexion and met the child's parameters and expectations. It was possible to manufacture personalized parts for the user faster and more precisely, making them more suitable for development of discrete prostheses, which must be custom fabricated.

Keywords: Design · Prosthesis · Assistive technologies · 3D Printing

1 Introduction

Global data show that more than one billion people in the world experience deficiencies of different types and degrees [1]. In low-income and developing countries, many people do not have access to assistive devices and, therefore, have limited independence and social participation. Assistive Technologies (AT) have the potential to promote social inclusion and dignified life.

While lower limb prostheses work satisfactorily for most of the users, substituting the complexity of hand function remains a difficult challenge. Most upper limb

© Springer International Publishing AG, part of Springer Nature 2019
T. Z. Ahram and C. Falcão (Eds.): AHFE 2018, AISC 794, pp. 899–908, 2019.
https://doi.org/10.1007/978-3-319-94947-5_88

prostheses are heavy, difficult to operate and do not allow for efficient and satisfactory functioning in daily activities. As a result, many people with unilateral upper limb amputation tend to adapt to perform daily tasks with one hand. The upper limb prosthesis, in many cases, are not used and, finally, abandoned.

Usability problems related to upper limb prosthesis have been reported [2]. Also, studies have addressed the stigma associated to the design of assistive technologies, highlighting the need of customizable design not limited to practical concerns, but also focusing on best meet the user's preferences, thus benefiting acceptance and satisfaction [3].

3D Printing and Rapid Prototyping (RP) are technologies that started to be applied to the field of prosthetics and can enhance the fabrication process by enabling tailored (unique and customized) parts to be produced faster, more accurately, and more sustainably compared to the conventional modeling and prototyping process, making them more suitable for the development of prostheses.

This paper describes the interdisciplinary collaborative development of a customized transradial mechanical prosthesis for an infant patient with bilateral upper limb amputation (forearm level), combining knowledge and approaches of areas of product design, rehabilitation, rapid prototyping, and ergonomics.

2   Development

This paper reports on the development of a customized low cost upper limb prosthesis based on interdisciplinarity approach between Product Design and Rehabilitation. Specific knowledge and practices of both areas were applied to a linear development process, as shown by Fig. 1.

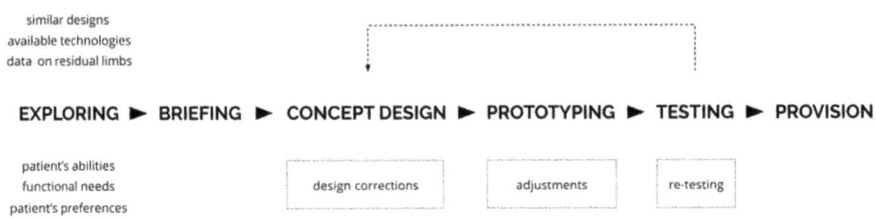

Fig. 1. Design process of the upper limb prosthesis development.

2.1   Exploring

This case involved a patient who, with a year and a month, had her upper and lower limbs amputated, due to the occurrence of meningococcemia, an infection caused by the bacterium Neisseria meningitidis (meningococcus), the same type of bacteria that can cause meningitis. The disease has caused sequelae in the patient, bilateral transtibial amputation in the upper third of the leg (calf area, tibia and fibula are cut) and bilateral transradial amputation (bony section between the elbow and wrist joint. It may be proximal, middle, or distal). SORRI Bauru began follow-up treatment in 2009, the year that the amputations occurred. The patient had prosthesis on both of her lower limbs but still did not have the upper limbs protected.

First, an interview was conducted to learn about the patient's needs and expectations. The patient had some complaints about her inability to perform daily activities alone such as conducting her personal hygiene, using the toilet, changing clothes, using zippers, closing buttons, brushing her hair, to mention a few. In addition, she complained about perspiring too much when she writes, since she uses both stumps to hold the pencil, which is distracting and taxes her energy. The patient takes athletic classes, loves to play with her dolls, watch cartoons, and is extremely active despite her condition.

The patient had an unsuccessful previous experience with a standard mechanical prosthesis similar to the one presented in Fig. 2.

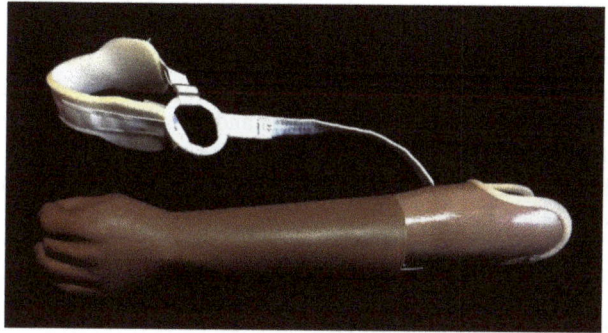

Fig. 2. Standard mechanical prosthesis provided by the Public Health System.

Mechanical prostheses are the most suitable for children because of their adaptive characteristics [4].

2.2   Briefing

Due to the needs of both the patient and the available technologies, we focused on the design of an upper limb prosthesis that could activate hand grasp by the movement of elbow flexion and extension (mechanical prosthesis). In this case, it is better to work with a mechanical prosthesis because the patient needed a functional prosthesis that went beyond simply having a stump (aesthetic/passive prosthesis), since the patient still has the natural movement of the elbow.

To support proper decisions during the design process, the main aspects of the prosthesis design were separated into three categories: technical, ergonomic, and aesthetic aspects (Table 1). From the technical perspective, it was important that the prosthesis met the following requirements: easy maintenance; resistant material; low cost; simple manufacturing elements - printed parts should be as simple as possible.

From an ergonomic point of view, the prosthesis should be as light and easy to activate as possible [5]. Also, it should comfortably fit the left limb, as the stump is longer than the right side, thus facilitating the elbow flexion (which is the movement

Table 1. Briefing results.

Technical	Ergonomic	Aesthetics
Easy maintenance	Right stump	Playful ("toy")
Resistant material	Lightweight	Customizable according to user
Low cost	Simple hygiene	
Simple manufacturing elements	Comfortable	
Durable Open Source, reproducible Efficient	Easy attachment Finger tips coated for better grip	

that activates the hand grasping). Simple hygiene, easy attachment/placement and increased friction on fingers and palm to facilitate the user to hold objects are other important ergonomic features to be presented in the prosthesis design. Finally, the aesthetic aspects were crucial for the acceptance and satisfaction of the user, especially for a child. In this context, the prosthesis should look playful, so that the child could see it as a "toy" that could help her in her daily activities. The prosthesis design should therefore consider the preferences of the child, such as colors [6] and cartoon characters.

2.3   Concept Design and Prototyping

A test was performed to better analyze the Unlimbited Arm model, which after research was the best option among the transradial open-source prostheses available. For the construction of the prosthesis, the instructions available on the prosthesis page on Thingiverse was followed, a website dedicated to the sharing of digital files created by its users.

Some user measures were required for the creation of the prosthesis, such as biceps circumference, forearm length, and hand length. This data was applied in Customizer, a Thingiverse application that allows files to be modified. First, it must be chosen to which side (left or right) the prosthesis will be made, then place the measurements (all in mm).

The prosthesis consists of a total of 34 pieces, divided into: fingers (5), phalanges (5), palm (1), forearm (1), long pins (13), circular pins (4), template, clamp (1) and tensioning pins (3).

The printing was done at CADEP – the Center for Advanced Product Development at UNESP Bauru using the CubeX Duo 3D Printer and at Fab Lab Livre SP, Cidade Tiradentes unit, using a Sethi3D AiP 3D Printer. Both machines used the same material, namely white and blue PLA (Polyactic Acid) plastic filament.

To perform the assembly, it was followed the instructions presented online (Fig. 3).

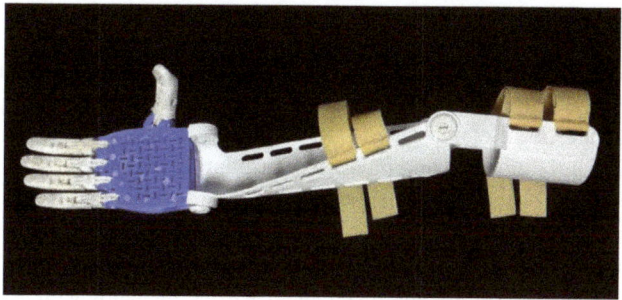

Fig. 3. Unlimbited Arm assembled.

2.4 Testing

For user testing, some objects that are easy to pick up was chosen so as not to generate frustration, such as Lego blocks, a glass, a jar, and a textured cylinder.

The prosthesis was worn by the patient, who soon began to open and close her hand, even before explaining how to perform the movement.

It was noticed that the prosthesis could be better fixed by simply applying one more Velcro to the proximal part of the forearm, so that the user did not have to apply so much force to generate the movement of the fingers. It was also sanded the proximal part of the forearm because it was limiting the movement of the user. These procedures facilitated biceps flexion.

The test was also done in the left stump, where we had a better result, being able to close the fingers more easily (Fig. 4).

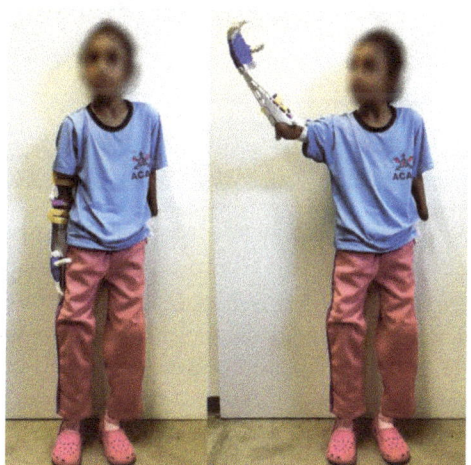

Fig. 4. User test.

2.5   Final Concept Design and Prototyping

After testing with the user, we concluded on how the final prosthesis should be. It must be made for the left stump; the proximal forearm should be reduced approximately 4 cm in length; the distal clamp should be curved for added comfort during use; the palm of the hand and fingers should have more grip/larger contact surface. For this final prosthesis, the holes for the Velcro inserts can be removed because rivets can be used; the thumb should be rotated approximately 10° upwards, so that it becomes more natural and improves the grip; edges must be smoothed; modification of the fingers, by adding another phalange (joint); the mechanism will be the same as it works perfectly well. To improve the forearm and armband, a cast of the user's stump was made. We would scan three-dimensionally and apply these parameters obtained to adapt the geometry of the forearm to perfectly fit the patient.

For the prosthesis to meet the aspects of the briefing, it must have few parts; simple form; construction, mechanisms, assembly, and simple accessories; have less material (reducing thickness and dimensions); the material used should be plastic (easy to clean) and be lightweight; to be comfortable. In addition, it needs to be aesthetically attractive according to the user, allowing customization for user needs and specifications.

The aim was to redesign the prosthesis, with greater playfulness, to refer more to a toy than to a prosthesis, consequently becoming more attractive to children.

Stump Mold Making. For the dimensions of the stump of the user in a more precise way, the user stump mold was made in SORRI - Bauru. By obtaining the positive mold, it is possible to mold the forearm and the armband in it and consequently the fixation on your stump and arm will be perfected. Thus, the user will be able to open and close the prosthesis more easily (Fig. 5).

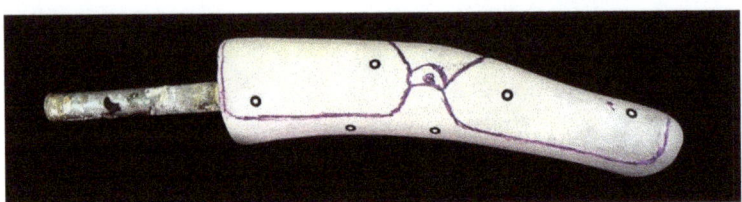

Fig. 5.   Positive mold with reference points for scanning.

Mold Scanning. The mold was digitalized at CADEP, to obtain a digital polygon mesh. For this process, the 3D mobile optical scanning system, GOM, model ATOS I 2M was used, which obtains three-dimensional data of an object quickly and accurately due to the high resolution. An outlet points of this equipment can get up to two million points, and each individual outlet is added to the set of previous measurements, resulting in a dense cloud of points. The parameters obtained will be used to adapt the geometry of the prosthesis to fit perfectly to the stump of the patient (Fig. 6).

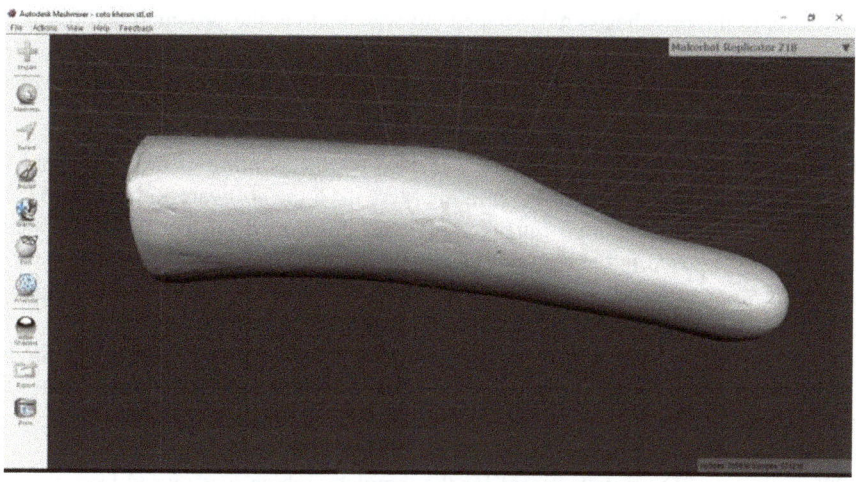

Fig. 6. Virtual mold open in the software Meshmixer.

Redesign of the Prosthesis. For the editing of the pieces, the software Rhinoceros and Meshmixer were used. The only files that did not require any modification were the pins and the template. In all the pieces the edges and the mesh of triangles were softened.

On the fingers the cat paws were applied to increase the grip and the contact surface. In the phalanges the paws in the center of the lower part, however, a cavity was made where the paws should stay, to print them in a material different from the phalanges. In the palm, the vector of a drawing of the "Marie kitten" was applied and the general geometry was modified, as the inner canals, which were reduced for passage of the nylon threads, besides removing the mesh from the palm, and creating an internal mesh, giving flexibility to the piece. In the forearm, it was used as the basis for the modification of geometry the scanning of the forearm mold. It was reduced 3 cm in length, and a mesh of kitten paws was applied to reduce material and transmit

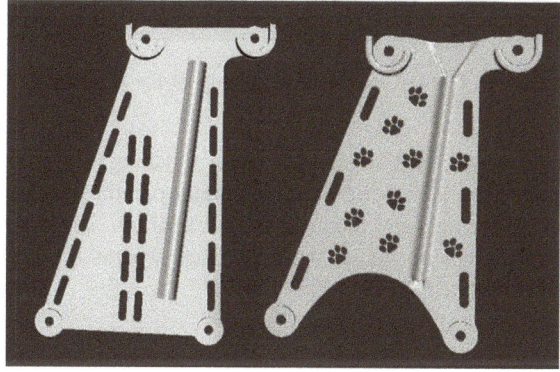

Fig. 7. Forearm. On the left, before the redesign; on the right, after.

playfulness to the prosthesis. A U-shaped cut was made at the proximal and distal but softer end. The channel of the nylon wires was reduced. Only three cavities were retained for the passage of Velcro. In the clamp, a "U" cut was made from the distal end of the piece and applied the user's name and project logo (Fig. 7).

3D Printing. The Sethi S3 3D Printer was used for the printing of the pieces and for the preparation of the pieces for the printing were used the software Repetier and Cura 2.6.2. The files were imported into the software, positioned on the 3D printer table. Support was only used in the clamp because it does not have self-support. The settings for printing were adjusted, such as layer thickness, print speed, among others, and then the pieces were sliced by the software in 2D layers, finally ready for printing.

The use of PLA was preferred as it is biodegradable and non-toxic, as well as having a lower molding temperature compared to ABS (Acrylonitrile Butadiene Styrene) filament. Also, the TPU (Thermoplastic Polyurethane) filament was used to have flexibility in some of the pieces.

All parts were 3D printed, except for the template, which was reused from the first prototype (Fig. 8).

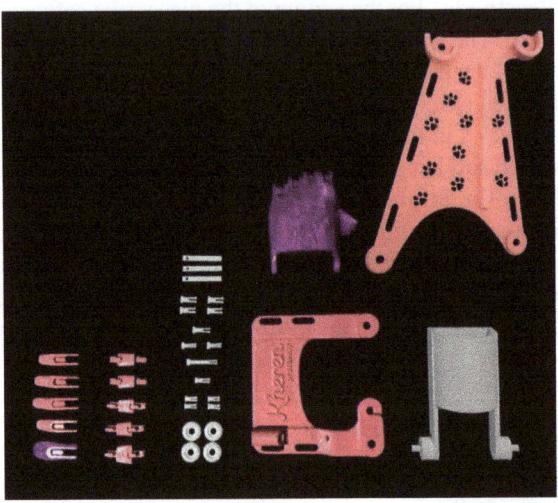

Fig. 8. Parts ready for the assembly of the prosthesis.

Assembly. First the pieces were sanded for better finishing. Only the molding of the forearm was different from the assembly of the similar prosthesis, since the plaster cast was used to shape the forearm and the armband. In addition, since the paws meant for application to the phalanges needed to be 3D printed separately, they had to be fixed together with Cyanoacrylate (Fig. 9).

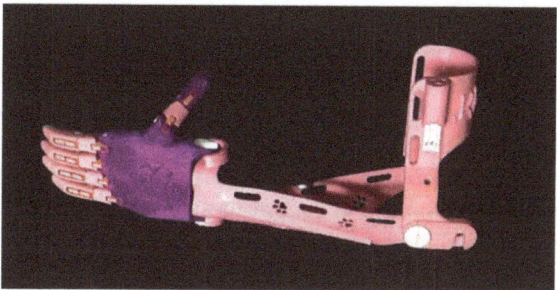

Fig. 9. Final prosthesis assembled.

2.6   Provision

The final test was conducted at SORRI in the presence of a prosthetist and occupational therapist. The prosthesis was delivered to the patient, who was delighted with the result.

The user was asked to perform some tasks, such as picking up simple objects such as an eraser. Another task we asked the user to perform was writing. One of the main complaints was that when she wrote, she perspired a lot because she held the pencil with the two stumps. An adapter was placed in the pencil and the user was able to write without difficulty. Also using an adapter, the user was able to close zippers and buttons, another complaint from her (Fig. 10).

Fig. 10. User writing with the prosthesis.

3 Conclusion

The manufacture of the prosthesis was simple, all executed by 3D printers, while only the assembly was done by hand. The durability is greater because the Velcro can be regulated for the fixation of the prosthesis on the user. The prosthesis is extremely efficient as it does not require much effort from the user to perform the movement of opening and closing the fingers (hold) compared to the prosthesis made by SORRI. Another aspect elaborated upon was the lightness, compared to the conventional prosthesis.

The cat paw was applied on the fingertips and phalanges to create a texture that facilitated the grip of objects, because during the test prosthesis the fingers were smooth, causing objects to slip out of the hand. The handle of the prosthesis was much improved compared to the Unlimbited Arm, due to the increased contact surface using the flexible plastic filament (TPU) in the palm of the hand. It was used the character and the colors requested by the wearer, making it playful, making it feel like more like a toy than a prosthesis. Another great differential between the final prosthesis and the Unlimbited Arm was the use of the user's stump mold to mold the forearm, making the part the exact shape for the wearer's stump. According to the wearer, the prosthesis is comfortable, and the placement of the prosthesis is as simple as just adjusting the Velcro. A great result was obtained, the user was satisfied with the prosthesis in every aspect.

Simple maintenance of the prosthesis can be easily performed by the wearer's parents, replacing the elastics and Velcro when necessary. Also, the prosthesis can be sanitized with water at room temperature and soap.

We conclude that 3D Printing and Rapid Prototyping can contribute greatly to the manufacturing process of assistive technologies, mainly prostheses, as we demonstrated in this work, streamlining development processes, and reducing product costs.

In addition, 3D Printing and RP technologies have contributed to the execution of tailor-made and user-tailored parts which is much faster and more accurate compared to the conventional manual prosthetic process.

Acknowledgments. The would like to thank the Norwegian Centre for International Cooperation in Education (UTF-2016-long-term/10053) for the financial support.

References

1. World Health Organization. http://www.who.int/disabilities/world_report/2011/report.pdf
2. Biddiss, E., Chau, T.: Upper-limb prosthetics: critical factors in device abandonment. Am. J. Phys. Med. Rehabil. **86**, 977–987 (2007)
3. Lanutti, N.L., Medola, F.O., Gonçalves, D.D., Silva, M.L., Nicholl, A.R.J., Paschoarelli, L. C.: The significance of manual wheelchairs: a comparative study on male and female users. Procedia Manufact. **3**, 6079–6085 (2015)
4. Andrade, F.L.: Amputações adquiridas na infância. In: Teixeira, et al. TO na Reabilitação Física. Rocca, São Paulo (2003)
5. Biddiss, E., Beaton, D., Chau, T.: Consumer design priorities for upper limb prosthetics. Disabil. Rehabil. Assist. Technol. **2**, 346–357 (2006)
6. Biddiss, E., Chau, T.: Upper extremity prosthesis use and abandonment: a survey of the last 25 years. Prosthet. Orthot. Int. **31**, 236–257 (2007)

A Software Based on Eye Gaze to Evaluate Mathematics in Children with Cerebral Palsy in Inclusive Education

Omar Alvarado-Cando[⊠], G. Belén Jara, Paúl Barzallo, and Hugo Torres-Salamea

Escuela de Ingeniería Electrónica, Universidad del Azuay, Cuenca, Ecuador
oalvarado@uazuay.edu.ec

Abstract. The government of Ecuador has promoted inclusive education in public and private education centers, ensuring that all children can access education and not be discriminated by their disability. Teachers are responsible for planning, support and reorganized the curriculum according to the needs of each child included. The inclusion process for children with cerebral palsy (CP) and speech difficulties have been a little difficult because they cannot communicate properly with the teachers and the evaluation process is not clear and objective. In this paper, through the use of eye tracking technology provide by Irisbond, we present an educational software to evaluate mathematics in children with CP. The software performs the questions in written and audible way, and it allows the students to select and/or match the correct answer; children with CP from 5 to 7 years tested the program and they got a better rating than not using it.

Keywords: Eye gazed · Inclusive education · Cerebral palsy
Assistive technology

1 Introduction

The International Classification of Functioning, Disability and Health (ICF) defines disability as an umbrella term for impairments, activity limitations and participation restrictions [1]. The World Health Organization (WHO) define the Cerebral Palsy (CP) as a disability and it's considered a neurological disorder cause by a non-progressive and non-contagious brain injury or malformation in the development of the child's brain and described by loss or impairment of motor function.

The motor disabilities of CP are usually accompanied by difficulties in communication, cognitive impairment and seizure disorders. In addition, CP is the most frequent serious disorder that affects the family, education and social life of the child [2]. The spastic cerebral palsy, people find difficult to control some or all of their muscle, hinders communication almost entirely, which limited the personal development of the patient and specially the academic performance [3, 4].

Inclusion of all person in educational process is a right of all human beings; the government of Ecuador has promoted the inclusive education in all levels. In the

© Springer International Publishing AG, part of Springer Nature 2019
T. Z. Ahram and C. Falcão (Eds.): AHFE 2018, AISC 794, pp. 909–915, 2019.
https://doi.org/10.1007/978-3-319-94947-5_89

province of Azuay, there are eight educational institutions for people with disability, called special institutions [5]. Some of these centers are not inclusive because they are focus on a single type of disability, for example, the Institute of Cerebral Palsy of Azuay (IPCA) where all students suffer cerebral palsy and this is not an inclusive education because they do not socialize and share in the classroom with students without needs.

The Ministry of Education of Ecuador has guidelines for the educational inclusion process of people with educational needs with or without disabilities [6]. This guide explains the administrative process, promotion and evaluation of students in a general way, but it does not mention about the support for assistive technology or Human-Computer Interface (HCI) that allows a more natural communication with computer and people for students with speech difficulties and cerebral palsy.

It is hard to access assistive technology for a person suffering a disability because the basic salary is around $350 dollars monthly and it is too low to acquire any kind of technology. For example, the IPCA has a population of 86 people with multiple disabilities in the institution and they are in poverty quintiles 1 and 2. In Azuay province, there are 20634 people with physical, intellectual and language disabilities. All these people need an assistive technology to communicate their desires, feelings and emotions (Fig. 1).

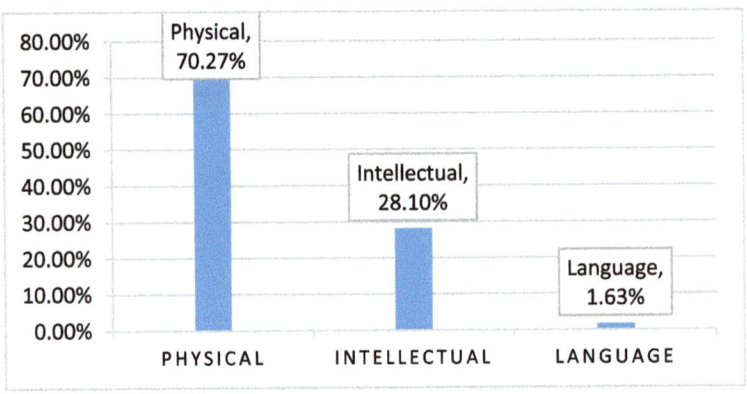

Fig. 1. Disability in Azuay province of Ecuador

Augmentative and Alternative Communication (AAC) are tools and methods that improve the intercommunication of patients with disabilities who are nonverbal or whose speech is not understandable like person with CP. The lack of communication does not help the educational process and teachers always have a subjective feedback of the knowledge acquired from the students, because they don't know for sure what answer the students want to say.

In this paper, we present a new software to evaluate mathematic in children between 5 and 7 years old. It is organized as follows: the assistive technology, IRIS-BOND, is described in detail in Sect. 2. The software designs is presented in Sect. 3, followed by the results and discussion in Sect. 4. Finally, conclusion are in Sect. 5.

2   Irisbond: Eye Gazed System

Irisbond is a system that enables the control of the computer through the movement of the eyes, that is, a "hands-free" system that works in a similar way to a normal mouse, which instead of being handled with the hands is handled with the eyes, manipulating the movement of the pointer. This system benefits people who have severe mobility limitations and can not use a conventional mouse to management a computer. This tool can be used with children with CP and their can interact with a software in the computer. This system consists of a camera and synchronized LEDs that emit an innocuous infrared light; it is connect by the USB port to a computer or tablet (Fig. 2).

Fig. 2. Irisbond hardware system

The Irisbond system has an application, Irisbond Primma, that performs the image processing and control algorithms, so that the movement of the eyes are transformed into Cartesian coordinates of the mouse position on the screen and this allows determining where the user is viewing and position the mouse in that place. In addition, the software can determine the event that the user wants to perform like simple click, double click, click for fixation, etc.

Initial calibration of the Irisbond is necessary to adapt the control algorithm to the conditions of the user, to recognize the movement of the user's eyes and to specify the distance between the patient and the computer. The Primma software makes the calibration in two steps: positioning and calibrate. In addition, it can has as many users as it wants, so various persons of the same educational institution can use the system, Fig. 3.

The first step configures the distance of the student from the camera; this can be recalibrated at any time, which allows having a better effectiveness when controlling the pointer of the computer, Fig. 4. The second step is to follow and object that moves along the margins of the screen, which determinates the limit of the coordinates; this can be done with different calibrations points: 16 points, 9 points and 5 points. After this, the IRISBOND can be used as mouser in any software or App.

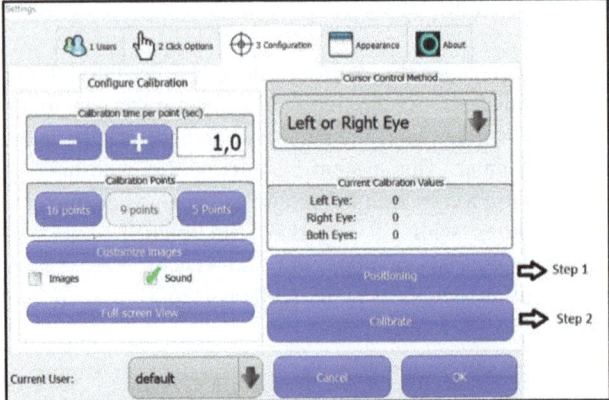

Fig. 3. Primma settings for initial calibration.

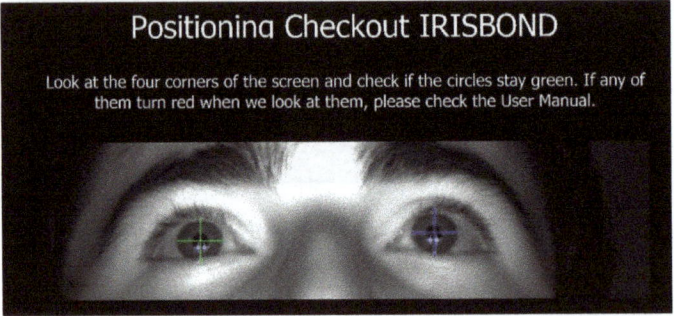

Fig. 4. Positioning checkout IRISBOND.

3   Software Design

The development of an educational software must have three important aspects to be considered: pedagogical, subject and age of the students. Students with PC needs a tutor in the classroom, so the education and content is personalized but they need to interact and socialize with their classmate. The subject that we want to evaluate is mathematics because it is very difficult for the teacher to know the knowledge, specialty in student between 5 and 7 years old. The teachers request is that the software has some exercise and quiz to evaluate the student.

The software has a home page where the teacher can choose between quiz and exercise, each option has different types of question and content: addition, color, number recognitions, sequence. If the records of the student were need, a log would be necessary; the software will evaluate according to the teacher criteria, which have to be load before the quiz or exercise (Fig. 5).

The evaluation of students with CP can not be as rigid as that of a student without needs. They need more attempts or more time to solve a problem, for that reason if the

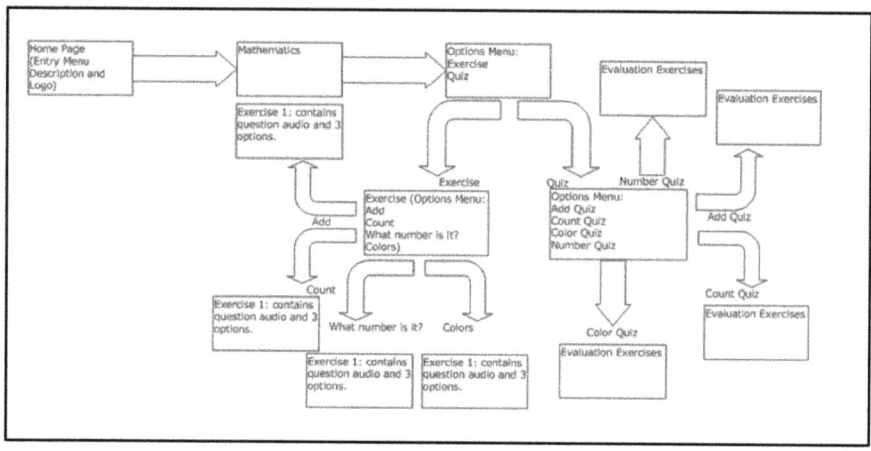

Fig. 5. Diagram Block of the Home page.

student get an incorrect answer the program will be in a loop until it succeeds. Of course, the software makes a count so the teacher can know what content has to be reinforced because the purpose of evaluation is to know the debilities of the students in order to strengthen their knowledge (Fig. 6).

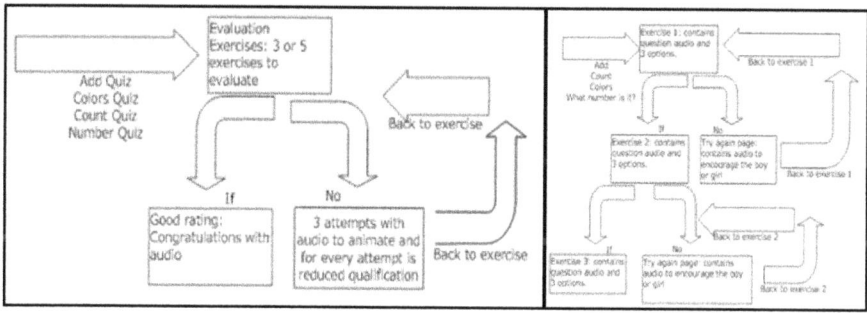

Fig. 6. Diagram Block of quiz and exercise.

4   Evaluation and Results

The proposed system has been put up for trial with 5 children with CP. They have been provided an introduction for the exercises and they work on this for two month in 1 h sessions 3 times a week. This practice allows the students to know the methodology of the exercises and to feel confident about the use of the eye-gazed system. At the beginning, the question were presented in plain text, which seemed to be bored to the students; the images were changed to gifs format.

In Fig. 7, the graphical model of the questions for exercises and quizzes are presented. The section **"a"** in the following figure is for the question, which can be read and the software reproduce it audibly. The **"b"** box is the graph type gif that allows a better understanding of the people with CP and the **"c"** box are the possible answers to the questions.

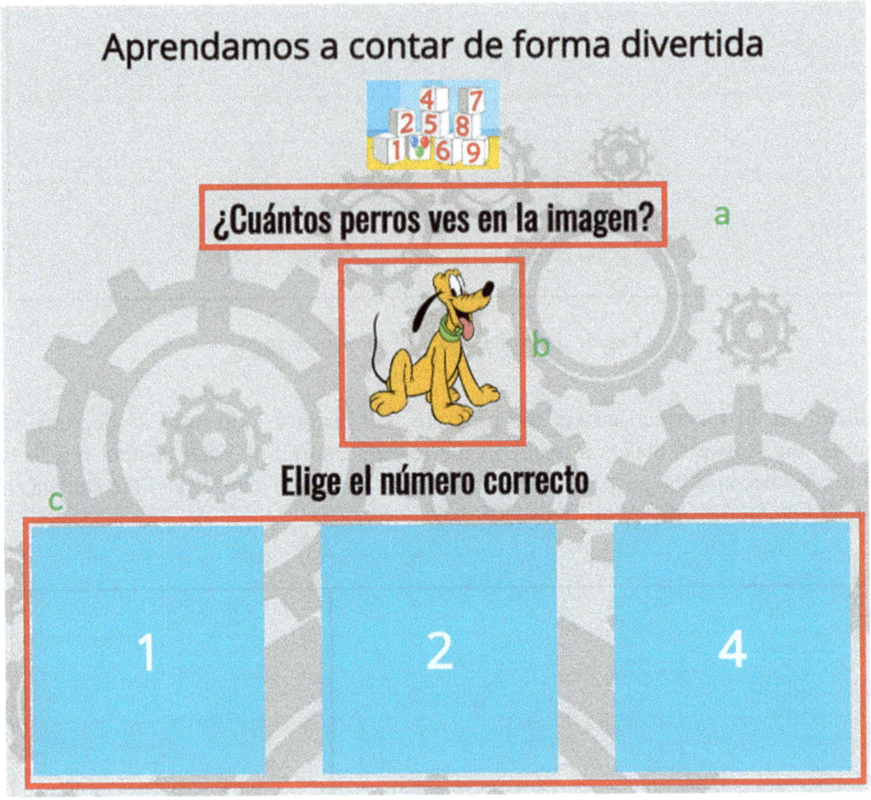

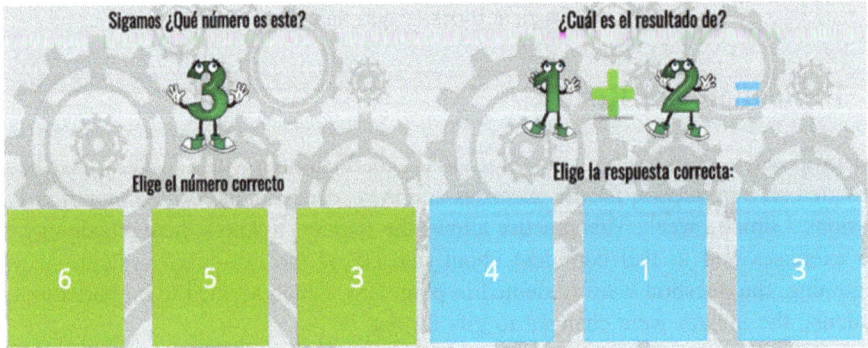

Fig. 7. Format of the questions

Fig. 8. Format question of the quiz and exercise

All topics were developed through the same methodology: question, gif image and possible answer, Fig. 8. The students were evaluated through software and in the traditional way without the help of TICs, the result are presented in Table 1.

Table 1. Traditional evaluation and eye tracking evaluation system.

Student	Traditional evaluation	Eye-tracking evaluation
Student 1	60%	76%
Student 2	62%	80%
Student 3	75%	72%
Student 4	76%	85%
Student 5	85%	90%

5   Conclusion

Children with cerebral palsy are given away from the educational process, thinking that they do not have the necessary knowledge to pass the course. In this study, we can demonstrate that four of the five students get a higher grade using the eye-gazed system than using the traditional evaluations methods. Students using an augmentative and alternative communication system based on eye tracking can improve their school performance; student 3 improved his grade by 18%.

In the other hand, student 3 has a lower grade using the AAC system because she don't feel confident with technology and gifs does not like her. It is very important to emphasize that two of the students evaluated in this paper who failed the quiz through the traditional evaluation methods (basic pictograms); they pass the quiz using the help of TICs.

References

1. Borblik, J., Shabalina, O., Kultsova, M.: Assistive technology software for people with intellectual or development disabilities: design of user interfaces for mobile applications. In: 2015 6th International Conference on Information, Intelligence, Systems and Applications (IISA), Corfu (n.d.)
2. Ministerio de Educación del Ecuador: Instituciones de Educación Especial, 05 January 2018. https://educacion.gob.ec/instituciones-de-educacion-especial/
3. Rosenbaum, P., Paneth, N., Leviton, A., Goldstein, M., Bax, M.: A report: the definition and classification of cerebral palsy April 2016. Dev. Med. Child Neurol. Suppl. **49**(109), 8–14 (2007)
4. Sakib Khan, S., Haque Sunny, M., Shifat Hossain, M., Hossain, E., Ahmad, M.: Nose tracking cursor control for the people with disabilities: an improved HCI. In: 3rd International Conference on Electrical Information and Communication Technology (EICT), Khulna (2017)
5. Subsecretaría de Educación Especializada e Inclusiva: Instructivo para la evaluación y promoción de estudiantes con necesidades educativas especiales. Ministerio de Educación del Ecuador, Quito (2016)
6. World Health Organization: Disability and health, January 2018. http://www.who.int/mediacentre/factsheets/fs352/en/. Accessed 03 January 2018

Research on the Environmental Thermal Comfort Based on Manikin

Rui Wang[1(✉)], Chaoyi Zhao[1], Huimin Hu[1], Yifen Qiu[2],
and Xueliang Cheng[3]

[1] Ergonomics Laboratory, China National Institute of Standardization,
Beijing, China
{wangrui,zhaochy,huhm}@cnis.gov.cn
[2] Laboratory for Man-Machine-Environment Engineering, School of Aeronautic
Science and Engineering, Beihang University, Beijing, China
qiuyifen@buaa.edu.cn
[3] School of Mechanical Engineering, Beijing Institute of Technology,
Beijing 100081, China
392462026@qq.com

Abstract. The subjective evaluation and objective test are the main methods to study the thermal comfort sensation of the environment. The Predicted mean votes (PMV) are calculated by thermal environment parameters obtained from the test points in general objective tests. However, the unstable factors of indoor environment make the description and prediction comprehensively and accurately impossible by the traditional methods. Through the physical and physiological characteristics of Chinese population in the Chinese adult human database, 50 percentile male adults' three dimensional physical data was selected to establish the manikin model. Controlling strategy was formulated according to the heat balance equation and thermal manikin test system was designed to test the exchanged heat between human body and environment in an unsteady and heterogeneous environment. In the specific heat and humidity environment created by the laboratory, the predicted mean vote (PMV) of the current environment was −0.8 by the thermal manikin test system. At the same time, the subjective evaluation tests were conducted in the environment and the thermal sensation vote (TSV) was −0.65. The relative error was only 2.1% between the subjective evaluations and the objective test results of the manikin.

Keywords: Thermal comfort · Thermal manikin · Predicted mean vote
Thermal sensation vote

1 Introduction

According to figures, we spend more than 80% time of our life indoors, in which the environmental qualities, like sound, light, thermal environment and air quality, all have a significant impact on our physical and mental health, comfort and work efficiency. Physiological research shows that in a comfortable thermal environment, it's good for thinking, observation and operation. The subjective and objective evaluation methods are always used to evaluate the thermal environment in the research. The results of the

© Springer International Publishing AG, part of Springer Nature 2019
T. Z. Ahram and C. Falcão (Eds.): AHFE 2018, AISC 794, pp. 916–924, 2019.
https://doi.org/10.1007/978-3-319-94947-5_90

subjective evaluation are usually discrete, which are obtained through the process and synthesis of the subjective evaluation scales. The subjective evaluation method can assess the comfort of the current environment intuitively and clearly. The objective evaluation is completed by the experiments, and the results are in good stability and consistency, which guarantees the results comparability and reproducibility even in different laboratories. Consequently, it is the main method to study the thermal comfort of the environment by the combination of subjective evaluation and objective test.

When it comes to the objective evaluation method, currently, the majority of researchers use the stable thermal environment created by an artificial climate chamber to get the evaluation. Indoor thermal environment parameters (air temperature, humidity, air velocity and average radiant temperature) are measured by the arranged measuring points, and the indoor thermal environment evaluation indexes, such as the predicted mean vote, the air vertical temperature difference value, temperature fluctuation and temperature uniformity, are leaded to give the thermal environment evaluation more comprehensively and objectively. But the actual indoor thermal environment is often unstable because of the numerous unstable factors, like the asymmetric radiation, the partial airflow and the partial cooling factor. Accordingly, it is difficult for the general thermal environment parameters to describe and predict the thermal comfort extent accurately in such environment. However, in unsteady conditions, the heat exchange between the human and the environment can be tested directly by the thermal manikin which is similar to human in the shape and it can predict the thermal comfort degree of the human. Through the manikin, the heat exchange process between the human body and the environment can be simulated and the thermal comfort can be evaluated scientifically. What is more, the effect of individual difference is avoided and the accuracy is ensured. So, it has been generally accepted that the thermal manikin is an indispensable method in the human body efficacy research.

Many foreign institutions have developed thermal manikin. In Finland, Ehab Foda et al. used thermal manikin that are based on skin temperature control method to test the thermal comfort and energy efficiency of the partial floor heating system, and the results were used to conduct the design of the system. In Hungarian, Edit Barna et al. found the warm floors and cold walls had combined effects on human thermal comfort during their research by thermal manikin. The results showed that the cold wall can reduce the partial average equivalent temperature of the hand and the face by 2°, and the vertical cold wall has much greater influence than the warm floor on the thermal comfort sense. In Swiss, Bogerd et al. used thermal manikin to study how the speed, head elevation, and hair condition affect the forced convection heat loss of a motorcycle helmet.

At Coimbra University in Portugal, Oliveira et al. made a comparative study on the heat loss between the static and dynamic posture of the thermal manikin in the natural convection and the forced convection condition. The result showed that in natural convection condition, the average convective heat transfer coefficient of the static body was about 3.5 W/m2·K, while the dynamic body was about 4.5 W/m2·K. In Singapore, Cheong et al. used thermal manikin with independent heating control method in the 26 sections to study the thermal comfort of the human in the displacement ventilation room with different temperature distributions. The result showed that the partial thermal discomfort in the cold was more obvious than the warm. In France, Elabbassi et al.

studied the effect of electric blanket on the dry heat loss of the newborn body by thermal manikin. Matsunaga et al. evaluated the thermal comfort in the vehicle by combining the thermal manikin test with numerical simulation and subjective experience.

Nowadays, most thermal manikins in research institutions were based on European and American people's physiological characteristics instead of Chinese. So, it has been imperative to develop and build the thermal manikin system based on Chinese people's physiological characteristics to evaluate the thermal comfort of the environment objectively. In this thesis, the manikin model was established through the 50 percentile male adults' three dimensional physical data from the Chinese adult human database. Controlling strategy was formulated according to the heat balance equation and thermal manikin test system was designed to test the exchanged heat between human body and environment in an unsteady and heterogeneous environment. And the feasibility of the test system to evaluate the environment was verified through subjective evaluations and objective experimental tests.

2 The Physical Model of Thermal Manikin

The thermal manikin was designed according to the 50 percentile China's male adults' physical data, including 16 independent control sections, such as the head, the left upper arm, the right upper arm, the left forearm, the right forearm, the left hand, the right hand, the chest, the back, the hip, the left thigh, the right thigh, the left leg, the right leg, the left foot and the right foot. There were active joints, like the hip, the knee, the elbow and other joints. It included standing and sitting posture. The heating of each area in the manikin was controlled individually by low voltage power and the temperature sensors were arranged to measure the surface temperature. There were various sensors in the regions with different heat transfer conditions, like the legs, the torso, the hips, etc. The heat flow and surface temperature of each region were measured and controlled by a computer. The manikin's structures and partitions are shown in Fig. 1. Its main joints can be adjusted, and the posture can be adjusted according to the measured demand.

3 The Control Principle of the Thermal Manikin

The manikin is controlled according to the heat balance equation in the comfortable condition of human body, and its surface temperature depends on the heat exchange between the body and the surrounding environment. Without considering the external work of the human body, the heat balance equation in the comfortable condition is as follows:

$$M = Q_t = Q_{res} + E_s + Q \tag{1}$$

$$Q_{res} = 1.7 \times 10^{-5}M(5867 - P_a) + 0.0014M(34 - t_a) \tag{2}$$

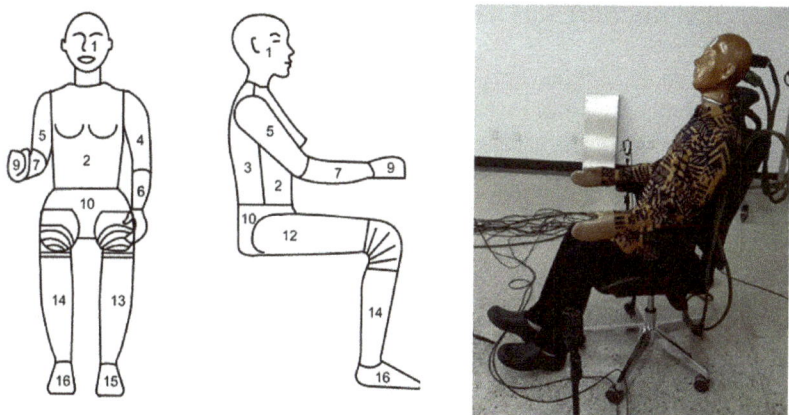

Fig. 1. The sections of the thermal manikin partition and the entire structure

$$E_s = 3.05 \times 10^{-3}(5733 - 6.99M - P_a) + 0.42(M - 58.15) \qquad (3)$$

In the formula: M is the metabolic heat production of body (W/m2); Qt is the total heat transfer between the body and the environment (W/m2); Qres is the respiratory heat exchange of the body, (W/m2); Es is the evaporative heat exchange of the skin (W/m2); Q is the convection and radiant heat transfer between the body and the environment (W/m2); Pa is the partial pressure of the vapour in the environment, taken as 1500 Pa; ta is the air temperature, taken as 20°C.

From the above three formulas, the relationship of the total heat dissipation (Qt) and the convection radiation heat transfer (Q) between the body and the environment is as follows:

$$Q_t = 1.96Q - 21.56 \qquad (4)$$

In comfortable conditions, the relationship between the average superficial skin temperature (tsk) and total heat dissipation (Qt) is as follows:

$$t_{sk} = 35.77 - 0.028Q_t \qquad (5)$$

From the above two formulas, the following formula can be got:

$$t_{sk} = 36.4 - 0.054Q \qquad (6)$$

This formula is the controlling equation to adjust the manikin's surface temperature. The surface temperature of the manikin depends on the heat dissipation to the environment and the heating power of the manikin. According to the heating power of each region (Qn) and the surface temperature (tsk,n) of the its different regions, the equivalent space temperature $(t_{eq,whole})$ of the manikin can be calculated.

4 The Experimental Environmental Conditions

Because the indoor environment test is to simulate the home environment, the experimental room were built, including a bed.

The test procedures and matters are as follows:

(1) Before the testing, the thermal manikin and its clothing thermal resistance should be calibrated in the laboratory environment. After calibration, the position and dressing cannot be changed.
(2) In order to avoid the influence of wall heat accumulation in the environmental comfort test, the wall should be preheated for a period to ensure its temperature reaches the specified value.
(3) In the test, the thermal manikin and environment parameters are used to evaluate the thermal comfort at first, and then the subjective thermal comfort evaluation of the test participants is carried out in the same environmental conditions.

In the outdoor environmental simulation:The Table 1 shows the outdoor environmental conditions of the air conditioner thermal comfort evaluation laboratory. After the operation of the tested air conditioner, the ambient temperature and humidity of the outside room shall conform to the specified requirements in Table 1. When the thermal comfort technology of the manikin is under experiment, the outdoor environment should be in the summer condition.

Table 1. Laboratory operating conditions.

Environmental parameters	Refrigeration condition (in summer condition)
Outdoor dry-bulb temperature (°C)	35 ± 0.5
Outdoor wet-bulb temperature (°C)	24 ± 0.5

The average indoor temperature of the experiment was 24.1 °C. In the test, the clothes of the manikin included a single vest, spring or autumn shirts, underwear, spring trousers, thin socks and single layer shoes. Calibration test of clothing thermal resistance was conducted before test and the result was 0.58 col. In the verification test of the thermal comfort test, the thermal manikin was adopted sitting posture, and the participants sat quietly or did some mild activities, like reading. The intensity was 1.2 Met.

5 The Experimental Process

The manikin was located in the center of the room. After the environmental parameters being stable, thermal manikin was used to evaluate the thermal comfort of the air conditioning environment. After the test output of the test system steadily, the thermal comfort evaluation results of the current environment can be obtained.

Subjective experimental process:

(1) The participants wore the lab clothes in a room and then enter the experiment room.
(2) The experimenter explained the test procedures and matters to the participants, and more attentions were paid to the method of judging the thermal sensation for the participants' suitable and accurate judgment.
(3) Let the participants adapt the environmental conditions. Observed the indoor air temperature and the black globe temperature by the temperature testing system. After the stability of the relative data more than 5 min, the participants were informed to make the first subjective sensation evaluations. After 10 min, 15 min, 20 min, 25 min, 30 min, 45 min, 60 min, 75 min, 90 min, subjective sensation evaluations also needed to be taken.

Attention: Participants must be in the same position as the manikin.

24 participants were selected, and they are adult male from 20 to 50 years old, which the specific age distribution are in the Table 2. Because of the relationship

Table 2. The age distribution of the participants

20–29 years'old	30–39 years'old	40–49 years'old
8	12	4

between the thermal comfort sensation and body mass index (BMI), the participants' BMI are counted in Table 3.

Before the test, participants changed the summer clothes and adjusted themselves in the setting outdoor environment more than 30 min. During the time, referring to the thermal comfort subjective test table, the test assistant who was in charge of the test, introduced the test process, and informed the participants what they need to record during the test. Results were taken by 7 levels scoring system to describe the thermal sensation, which is in Table 4.

6   The Experiment Results

The results of the 24 participants' test data were processed and the thermal sensation of the human body in the summer environment varied for the time they spent indoors, as shown in Fig. 2:

Because the outside temperature was 35°, participants felt comfortable and cool when they entered the room. As time went on, most participants still felt comfortable within 30 min, and the average thermal sensation vote of the 12 participants fluctuated within ±0.2. With the increase of the indoor time, the overall heat sensation of the participants came into being cool after 30 min, and the average TSV began to decrease gradually. That is because of the experimental requirements. The participants were engaged in meditation, reading or other mild activities in the a bit cooler environment, so

Table 3. The BMI distribution of the participants

Slightly thinner <18.5	Normal 18.5–23.9	Slightly heavier 24–27	Obesity 27.1–32	Extremely obesity >32
0	12	8	4	0

Table 4. 7 levels of the thermal sensation

Cold	Cool	Slightly cool but comfortable	Comfortable	Slightly hot but comfortable	Slightly hot	Blazing
−3	−2	−1	0	1	2	3

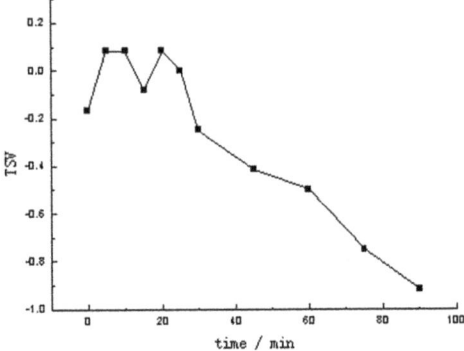

Fig. 2. The change of whole-body thermal sensation with time in the test environment

the longer they spent, the cooler sensation they would have. Over 90 min, the average TSV of the whole body had dropped to −0.92. According to the ISO14505 which regulates the ergonomics of the thermal environment, the subjective test results of thermal comfort experience should be stemmed from the participants coming into the environment 30 min to 90 min. According to this regulation, the environmental thermal comfort of the test was calculated and the average thermal sensation vote was −0.65.

The thermal manikin was used to evaluate the thermal comfort of current environment and the results can be seen in the Fig. 3.

In the Fig. 3, about one hour after the beginning, the equivalent space temperature and the predicted mean vote (PMV) of the manikin tend to be stabilized. When the test system is stable, the average equivalent space temperature of the body is about 22°, and the PMV is about −0.8. Comparing the test results about the predicted mean vote with the subjective evaluation test, the relative error is only 2.1% between the subjective evaluations and the objective test results of the manikin, which demonstrates that thermal manikin can be used to test the thermal comfort of the environment accurately.

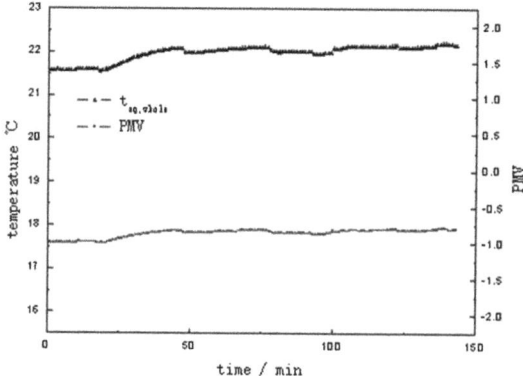

Fig. 3. The test results of the environmental thermal comfort by the manikin' evaluation.

7   Conclusions

Through Chinese adult human body database, 50 percentile male adults' three dimensional physical data was selected to establish the thermal manikin physical model. The control strategy was formulated according to the heat balance equation of the human body under the comfortable condition and a thermal manikin test system was designed to test the exchanged heat between human body and environment in unsteady and heterogeneous thermal environment. Creating a specific heat and humidity environment in an artificial environment laboratory, test the predicted mean vote (PMV) of the current environment by using the thermal manikin test system, which was −0.8. At the same time, the subjective evaluation tests were conducted in the environment and the thermal sensation vote was −0.65. The relative error was only 2.1% between the subjective evaluations and the objective test results of the manikin, which demonstrates that thermal manikin can be used to test the thermal comfort of the environment with various merits, like convenient operation and high accuracy.

Acknowledgments. This research is supported by "Special funds for the basic R&D undertakings by welfare research institutions" (522017Y-5276, 522016Y-4488 and 712016Y-4940) and General Administration of Quality Supervision, Inspection and Quarantine of the People's Republic of China (AQSIQ) science and technology planning project for 2017 (2017QK157 and 2016QK177).

References

1. Xiaolin, X., Baizhan, L.: Influence of indoor thermal environment on thermal comfort of human body. J. Chongqing Univ. **4**(28), 102–105 (2005)
2. Li, S., Lian, Z.: Discussions on the application of Fanger's thermal comfortable theory. In: Shanghai Refrigeration Institute Academic Annual Conference (2007)
3. Holmér, I., Nilsson, H., Bohm, M., et al.: Thermal aspects of vehicle comfort. Appl. Hum. Sci. J. Physiol. Anthropol. **14**(4), 159–165 (1995)

4. Hai, Y., Runbai, W.: Evaluation indices of thermal environment based on thermal manikin. Chin. J. Ergon. **11**(2), 26–28 (2005)
5. Tanabe, S., Zhang, H., Arens, E.A., et al.: Evaluating thermal environments by using a thermal manikin with controlled skin surface temperature. ASHRAE Trans. **100**, 39–48 (1994)
6. Zhu, Y.: Building Environment. China Building Industry Press (2010)
7. Nilsson, H.O., Holmér, I.: Comfort climate evaluation with thermal manikin methods and computer simulation models. Indoor Air **13**(1), 28–37 (2003)
8. Zhaohua, Z.: Thermal manikin application in the thermal comfort evaluation. China Pers. Prot. Equip. **1**, 23–25 (2008)
9. Foda, E., Kai, S.: Design strategy for maximizing the energy-efficiency of a localized floor-heating system using a thermal manikin with human thermoregulatory control. Energy Build. **51**(8), 111–121 (2012)
10. Barna, E.: Combined effect of two local discomfort parameters studied with a thermal manikin and human subjects. Energy Build. **51**(4), 234–241 (2012)
11. Bogerd, C.P., Brühwiler, P.A.: The role of head tilt, hair and wind speed on forced convective heat loss through full-face motorcycle helmets: a thermal manikin study. Int. J. Ind. Ergon. **38**(3), 346–353 (2008)
12. Oliveira, A.V.M., Gaspar, A.R., Francisco, S.C., et al.: Analysis of natural and forced convection heat losses from a thermal manikin: comparative assessment of the static and dynamic postures. J. Wind Eng. Ind. Aerodyn. **132**, 66–76 (2014)
13. Cheong, K.W.D., Yu, W.J., Kosonen, R., et al.: Assessment of thermal environment using a thermal manikin in a field environment chamber served by displacement ventilation system. Build. Environ. **41**(12), 1661–1670 (2006)
14. Elabbassi, E.B., Delanaud, S., Chardon, K., et al.: Electrically heated blanket in neonatal care: assessment of the reduction of dry heat loss from a thermal manikin. Elsevier Ergon. Book **05**, 431–435 (2005)
15. Matsunaga, K., Sudo, F., Yoshizumi, S., et al.: Evaluating thermal comfort in vehicles by subjective experiment, thermal manikin, and numerical manikin. JSAE Revi. **17**(4), 455 (1996)
16. ISO 14505: Ergonomics of the thermal environment—Evaluation of thermal environments in vehicles (2007)

Specific Dyslexia Exploratory Test (TEDE): Two Tasks Using Augmented Reality

Maritzol Tenemaza[(⊠)], Rosa Navarrete, Erika Jaramillo,
and Andrés Rodriguez

Departamento de Informática y Ciencias de la Computacion,
Escuela Politécnica Nacional, Quito, Ecuador
{maritzol.tenemaza, rosa.navarrete, erika.jaramillo01,
andres.rodriguez02}@epn.edu.ec

Abstract. Dyslexia is a specific learning disability characterized by the difficulty to accurately and fluently recognize words due to a deficiency in the phonological component of language. This disability has a global impact and affects around 5% of the population. This study proposes to test if applying an Augmented Reality (AR) interface it may improve the identification of reading-writing disorders in school-age children to anticipate dyslexia diagnosis. Based on this premise, a mobile application that tests two errors of the Specific Dyslexia Exploratory Test (TEDE) has been developed. This software was tested on children from a school located in Quito- Ecuador. The results of this test prove that this software gets a better detection of the reading and writing disorders than using a manual inspection. The main contribution of this work is the validation of the helpfulness of software with an AR interface for the early detection of this learning disability.

Keywords: Augmented reality · AR · Dyslexia
Specific Dyslexia Exploratory Test · TEDE

1 Introduction

Dyslexia is a specific learning disability of neurobiological origin characterized by reading and writing disorder due to difficulties on an accurate and fluent word recognition [1]. The difficulties associated with dyslexia decrease language sufficiency for spelling, comprehension and impact negatively on the academic performance of people of all ages [2, 3]. The causes of the problems are centered in cerebral dysfunction or emotional behavioral disturbance [2].

In particular, developmental dyslexia refers to a reading disability that occurs in about 5% [5] to 15% of the world population [6]. In this case, reading and writing skills are usually below what is expected of an individual with an average intelligence quotient [4].

Although there are software applications to conduct an early recognition of dyslexia, the detection is still conducted by applying manual tests. To encourage the use of software for this purpose, this work pursues to prove the efficiency of a mobile application with augmented reality (AR) interface to conduct a dyslexia test.

© Springer International Publishing AG, part of Springer Nature 2019
T. Z. Ahram and C. Falcão (Eds.): AHFE 2018, AISC 794, pp. 925–933, 2019.
https://doi.org/10.1007/978-3-319-94947-5_91

AR provides a novel interface approach used in a wide range of fields, such as medicine, entertainment, robot planning, education, and more [7–9].

The scope of this mobile application covers only two tests of the Dyslexia Exploratory Test Specifies (TEDE); therefore, it does not pursue a conclusive diagnosis of dyslexia.

To validate the results of its application in comparison with manual exploration, this mobile application was applied in a school in Quito, Ecuador to a group of children who had reading problems. The results obtained show the relevance of this technology to help in detection of dyslexia.

The following research questions (RQ) are proposed:

- RQ1: Are there differences in results when the child performs the manual test versus the test using the mobile application?
- RQ2: Is the AR interface decisive when checking the usability (especially ease of use and usefulness) of the mobile application?

This article is organized as follows: Sect. 2 presents concepts involved; Sect. 3 describes the materials and method employed in this study; Sect. 4 shows the results; and, finally, Sect. 5 includes the conclusions.

2   Concepts Involved

2.1   Dyslexia

Dyslexia refers to a retardation disorder of reading, writing, spelling, but it is not the result of mental retardation [2, 3, 10]. The symptoms related to a dyslexia diagnosis are incoordination, left-right confusions and poor sequencing. It is described as a neuronal syndrome [4], although at this time, it is reported that genetic and environmental factors are involved with dyslexia development [11].

Reading requires the acquisition of good orthographic skills for recognizing the visual form of words, which allows one to access their meaning directly. It also requires the development of good phonological skills for sounding out unfamiliar words using knowledge of letter-sound conversion rules [4]. People lack these competencies because of dyslexia.

2.2   Augmented Reality

Augmented reality (AR) allows the user to see the real world, with virtual objects superimposed upon or composited with the real world. An AR interface has the following three characteristics: (1) it combines real and virtual, (2) it is interactive in real time, and (3) it is registered in three dimensions [7].

AR systems generally comprises a pattern disposed on an object surface. A computer or a mobile device receives an input and presents an AR output to a user, and a tracker for detecting the pattern [12]. In this work, we use patterns for presenting the words or letters.

AR enhances a user's perception of an interaction with the real world. By using a computer as a tool, the information conveyed by the virtual objects helps a user perform real world tasks [7] in an easier interface [7, 13]. The educational values of AR are not only based on the use of technology but also closely related to how it is designed, implemented and integrated into formal and informal learning settings [9]. For this reason, AR is considered as a helpful tool when presenting information to a child with dyslexia.

3   Materials and Method

3.1   Mobile Application

This paper presents a mobile application with AR interface to help detect dyslexia in school education students. Figure 1 displays the model view controller architecture. The model is associated with a set of scripts for data management. The view is developed with an AR interface system, and the controller listens to the events triggered by the view (or another external origin) and executes the appropriate reaction to these events.

The mobile application was developed under the understanding of dyslexia and how this affects learning as well as the phonological processing and the ability of the child to identify and manipulate the sound structure of words.

This preliminary prototype was developed in Spanish to test it with students who are native Spanish speakers.

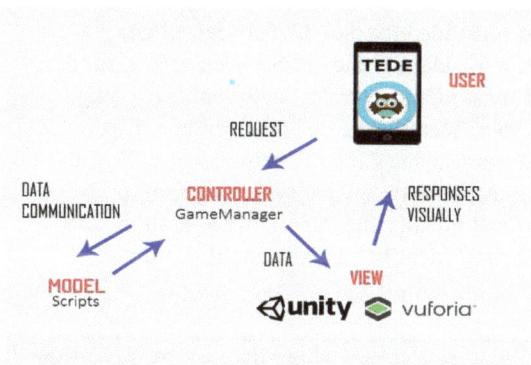

Fig. 1. View-controller model

3.2   Participants

This study took place in an elementary school in Quito, Ecuador. Participants were eleven children with developmental dyslexia (dyslexic readers). The children were between eight and eleven years old with an average intelligence coefficient graded as normal. The participants and their parents were informed and consent participation. In addition, an authority, two teachers and a psychologist of the institution observed the tests.

3.3   Context of Study

The methodological design of this research considered the diagnosed criteria for specific disorders of dyslexia. This mobile application runs two tasks included in the error section of Dyslexia Exploratory Test Specifies TEDE [14].

The goal of the test is to determine the reading level of the child based on increasing complexity. The results serve as a guide to design the corrective treatment plan on dyslexia signs in oral reading. The test consists of 171 items divided into two parts: reader level with 100 items and specific errors with 71. The test also has two parts: in the first part, the degree of difficulty in reading that the child can solve is determined. In the second, the errors made by the child are detected.

The test was applied to eleven students in two ways: (a) Receives a printed page of the test to resolve it with paper and pencil; and, (b) Uses the mobile application. Teachers, authorities, and psychologists observe the test performance.

The items implemented were:

3.3.1. Confusing letters by sound at the beginning of the word: The examiner names a word without meaning and the child must point the letter with which the word begins by choosing among the possibilities that are displayed in the interface. The items that allow to detect this aspect are:

- Chado: the child must identify the "ch" between y, j, s, ll, ch.
- Deco: the child must identify the "d" between f, d, t, l, n.
- Fido: the child must identify the "f" between f, j, v, b, s.
- Llotio: the child must identify the "ll" between ll, ch, j, g.
- Tarpo: the child must identify the "t" between c, k, t, m, d.
- Gupa: the child must identify the "g" between y, r, j, m,g.
- Boso: the child must identify the "b" between b, t, f, p.
- Jallón: the child must identify the "j" between g, y, ll, j, f.
- Pola: the child must identify "p" between s, t, b, m, p
- Mite: the child must identify the "m" between s, m, n, l, b.
- Ñuma: the child must identify "ñ" between ll, j, m, ch.

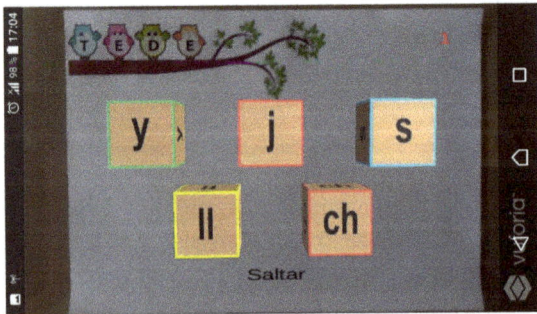

Fig. 2. Confusing letter test

The student must listen and choose the letter he hears. If the answer is correct, a counter of correct letters is registered, otherwise, an incorrect letter is incremented to the counter. This is observed in Fig. 2 and Fig. 3, respectively.

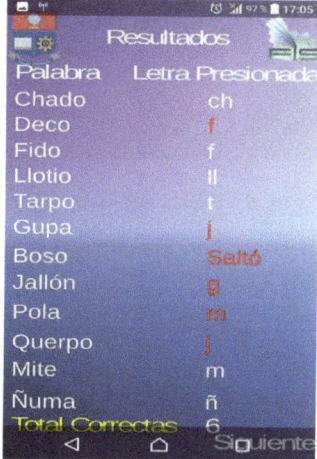

Fig. 3. Confusing letter test results

3.3.2. Reversion of complete words: The examiner can detect if the child reverses the words by using these items: la, sol, se, as, nos, los, al, es, son, le, sal. The words were presented with their letters in disorder, it is observed in Fig. 4. The student must place the letters of the words in the correct order. If he did it correctly, the number of accurate formed words is counted, and if there was an error, it is also registered. It is shown in Figs. 5 and 6.

Fig. 4. Complete word test

Fig. 5. Exchanged letter test

Fig. 6. Complete word result test

3.4   Protocol

The examiner was present all time. Testing took place in a quiet, well-lit room to avoid disturbing elements from the external environment. The examiner provided the clarifications and instructions that were needed. The raw score on the tests is the number of words read correctly. If the child does not indicate or say they do not know, they must register. For each time interval, the children received the indication to go ahead and the result was registered. If the child lost the line, this inconvenience was recorded, and the child received the indication "Let's continue here".

The examiner should be aware of the movement of the child's eyes, since many errors may be due to the loss of line and not to difficulties in recognizing the stimulus. The mobile prototype records the letter selected by the child, but the examiner must also record it.

4   Results

In the following, each research questions proposed in this work is analyzed

Are there differences in results when the child performing the manual test versus the test using the mobile application?

The participating children have dyslexia detected by teachers and psychologists of the institution. Figure 7 shows, by using the mobile application, in the first test 45% of children improve their rate, 27% get the same rate and 27% reduce their rate, with respect to paper and pencil handling. Considering that this test requires listening comprehension. These results show that some child with dyslexia usually has better listening comprehension than reading comprehension; without ignoring that, they have difficulty decoding and reading words with precision or fluency. In this test the child should have the ability to decode words into smaller parts and choose the beginning letter.

Our first significant observation was that all students showed their preference with the prototype in a mobile device rather than on a paper. The mobile application helps children with difficulties in reading and writing, concentrate in developing the test.

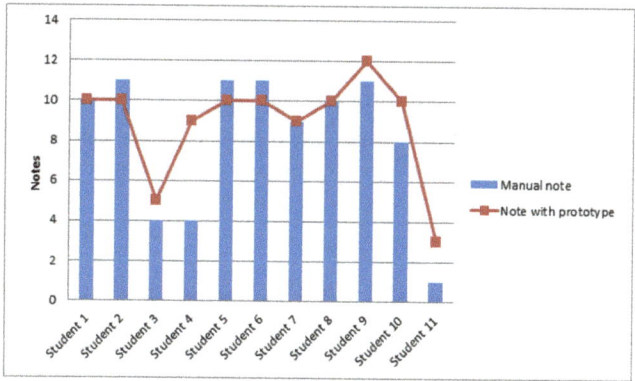

Fig. 7. Results of Confusing letters by sound at the beginning of the word test

The results on *Reversion of complete words test* are presented in Fig. 8. Word recognition is the ability to read individual printed words. It is also called word reading or word identification. Word recognition tests require students to read individual words printed on a list. In the test, 45% of the children obtained a better grade and 65% the same grade as in the paper and pencil test. This shows that, the children are more precise, but they are still slow in reading words. The point of interest is the time it takes the child to form the word. The test does not present clues, to help student decipher the word.

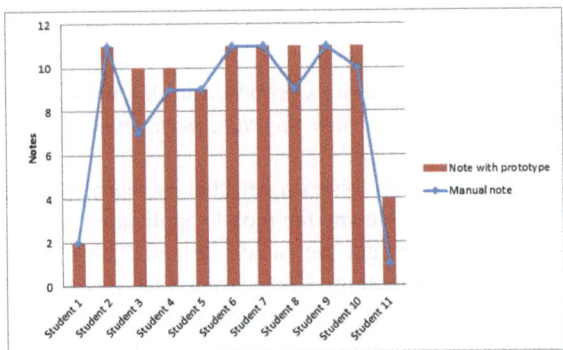

Fig. 8. Results reversal of complete words test

Is the AR interface decisive when checking the usability (especially ease of use and usefulness) the mobile application?

The results of the surveys determine that the mobile application is easier and more useful for both groups of students and observers: parents, teachers, manager, and psychologist. In general, students are motivated to use technology. The interface with AR improves this type of applications.

We tested a relatively small sample and did not show the fact those children with reading difficulties often have additional attention problems that can interfere with the test results. Therefore, it is imperative to use the mobile application with other groups of children in different schools so that more psychologists can participate in the adjustment of the mobile application. It is planned to include more tests and complete the current ones.

5 Conclusions

In this work, we present a mobile application to develop testing activities for children with reading difficulties in order to conduct dyslexia signs detection. This study embraces two TEDE tests to apply to the group of children with reading and writing difficulties: detection for reversion of complete words and confusing letters by sound.

The students who participated in the test felt more motivated interacting with the mobile application, mainly because of the novelty of AR technology. The AR interface becomes attractive for children because it combines the real and virtual in 3D and the deploy of interface with colors and sounds. This fact confirms the helpfulness of mobile application with an AR interface to improve the attention and performance of a child during the tests. Moreover, the availability of this software for mobile devices makes it affordable and easy to use for psychologists and teachers.

Although this is a work in progress, the results obtained from the participants motivate us to continue with the research. The results showed progress in students, not only in the recognition of words and reading but also in phonological decoding. The students also gained experience and got familiar with the mobile application.

In the near future, more tests for dyslexia detection will be incorporated to improve the mobile application.

Acknowledgments. The authors gratefully acknowledge the financial support provided by the Escuela Politécnica Nacional, for the development of project PII-16-04 "Aceptabilidad de la A2R".

Thanks to the School "Republic of Argentina" of Amaguaña, Pichincha-Ecuador Thanks to its authorities, teachers, parents and children participants. Thanks to MSc María Torres for the support given in English language.

References

1. Lyon, G.R., Shaywitz, S.E., Shaywitz, B.A.: A definition of dyslexia. Ann. Dyslexia **53**(1), 1–14 (2003)
2. Kirk, S., et al.: Educating exceptional children. Cengage Learning (2011)
3. Hammill, D.D.: On defining learning disabilities: an emerging consensus. J. Learn. Disabil. **23**(2), 74–84 (1990)
4. Stein, J.: The magnocellular theory of developmental dyslexia. Dyslexia **7**(1), 12–36 (2001)
5. Van Ingelghem, M., et al.: Psychophysical evidence for a general temporal processing deficit in children with dyslexia. NeuroReport **12**(16), 3603–3607 (2001)
6. Skiada, R., et al.: EasyLexia: a mobile application for children with learning difficulties. Procedia Comput. Sci. **27**, 218–228 (2014)
7. Azuma, R.T.: A survey of augmented reality. Presence Teleoperators Virtual Environ. **6**(4), 355–385 (1997)
8. Lin, C.-Y., et al.: Augmented reality in educational activities for children with disabilities. Displays **42**, 51–54 (2016)
9. Wu, H.-K., et al.: Current status, opportunities and challenges of augmented reality in education. Comput. Educ. **62**, 41–49 (2013)
10. Bateman, B.: Three approaches to diagnosis and educational planning for children with learning disabilities. Acad. Ther. Q. **2**(4), 215–222 (1967)
11. Dilnot, J., et al.: Child and environmental risk factors predicting readiness for learning in children at high risk of dyslexia. Dev. Psychopathol. **29**(1), 235–244 (2017)
12. Meisner, J., Donnelly, W.P., Roosen, R.: Augmented reality technology. Google Patents (2003)
13. Brooks Jr., F.P.: The computer scientist as toolsmith II. Commun. ACM **39**(3), 61–68 (1996)
14. Condemarín, M., Blomquist, M.: La dislexia. In: Condemarín, M., Chadwick, M., Milicic, N. (eds.) Manual deLectura correctiva. Editorial Universitaria, Santiago de Chile 1970 (1978)

Mobile Application for Crowdmapping Accessibility Places and Generation of Accessible Routes

Nigel da Silva Lima[1](✉), João Pedro Caldas Leite[2],
Anselmo Cardoso de Paiva[1], Ivana Marcia Oliveira Maia[2],
Aristófanes Corrêa Silva[1], Geraldo Braz Junior[1],
and Cláudio de Souza Baptista[3]

[1] Applied Computer Center (NCA), Federal University of Maranhão (UFMA),
1966, Bacanga, São Luís, MA, Brazil
nigelnaiguel.comp@gmail.com
[2] Furniture Design and Visual Communication, Federal Institute of Maranhão
(IFMA), 4, Monte Castelo, São Luís, MA, Brazil
[3] Department of Systems and Computing, Federal Institute of Campina Grande
(UFCG), 882, Bodocongó 5809970, Campina Grande, PB, Brazil

Abstract. Accessibility, as an essential attribute in a society, guarantees an improvement in people's quality of life. Therefore, it must be present in the most diverse means of social interaction, such as transportation, communication and information. Ensuring that these assemblies are accessible to wheelchairs or people with low mobility generates positive social outcomes and contributes to inclusive and sustainable development in a social setting. This work aims the development of a crowdmapping application based on mobile apps, which enables the user photograph and rate locations in the city that present obstacles for reduced mobility people (accessibility traps) making hard or even impossible for disabled people to bypass them, especially wheelchairs users. Based on the collected data provided by the crowdmapping initiative, the application permits the user visualize heat maps that demonstrate the most accessible places, combine the geographic locations and their rates, and then be able to elaborate the best accessible routes. The proposed mobile app is a powerful tool to help disabled people avoid obstacles in the available paths. Also, this collaborative system is important to easily indicate the current access problems around the city and serve as support to monitoring and maintenance of these locations. An interesting future work might involve expanding the system to other accessibility aspects apart from mobility, such as blind or deaf people.

Keywords: Geographic · Information systems · Accessibility
Mobile application

© Springer International Publishing AG, part of Springer Nature 2019
T. Z. Ahram and C. Falcão (Eds.): AHFE 2018, AISC 794, pp. 934–942, 2019.
https://doi.org/10.1007/978-3-319-94947-5_92

1 Introduction

Urban development has been growing fast in the last decades. It is estimated that 66% of world's population will be living in cities by 2050 [1]. The way of how infrastructure and facilities are planned in cities directly affects social well being, since basic activities as going to shops or more essential ones as establishing accessible routes to go to hospitals.

The growth of the number of cars and population has resulted in a dispute of space in urban environment, which has raised the awareness of the importance of sustainable development, as it may guarantee the quality of social well being in terms of accessibility [2].

The concept of accessibility plays an important role to generate social equality, where all population, undiscriminated, must have equal access to public spaces as more acceptable as possible. However, around 15% of world's population live with disability, this is an amount of over 1 billion of people around the world, living with disability, face a lack of accessibility in urban environment [1]. This raises the need for the development of projects that focus on promoting equality for disabled people, which can certificate positive outcomes and inclusive results in social settings.

Efforts to improve accessibility for disabled people may involve a vast set of fields of work, from architectural projects to the use of information and communication technologies (ICT) to assist in better decision making. The use of information technologies has served as a great benefit to promote social equality in many terms, Its access is able to accelerate social and economic progress of all individuals [3]. The evolution of ICT has expanded the ground base for ideas that, along with effective technology tools, can achieve prospective results for disabled people.

One of these effective tools is the use of Geographic Information Systems (GIS), which are systems that integrate a set of tools to manipulate, analyze, manage and present geographic information [4]. Once this geographic data is analyzed, it enables the possibility of prescribing actions related to spaces. To plan and execute projects that ensure accessibility for low mobility people, GIS technology can serve as an alternative to understanding urban spaces and point to affordable solutions to improve accessibility in specific locations.

Capturing geographic information in terms of identifying places with low accessibility may demand a long time and high human resource, which can become a harder task in big urban centers. On the other hand, these places have constant daily human traffic and it is in constantly increase the number of mobile phone users. It is forecast that in 2017 the number of users of cellphones will reach 4.77 billion [5]. Due to these facts, emerging technologies based on crowd inputs over the Internet have become potential tools in projects that use information technologies to promote social digital innovation and social impacts. One of them is Crowdmapping, which is a mapping technology that combines crowd inputs and geographic data to create real time digital maps in a specific context that serve as tools for different social groups map their own communities and consider management problems and urban planning, then improve the process of decision making and problems solutions [6].

The merge of these technologies is capable of resulting in a tool able to support the initiative of a scenario built on mobility access by the collaboration of mobile phone users indicating locations with low accessibility. The objective of this paper is to describe the development of a mobile application that enables users to photograph urban locations that present accessibility traps to people with low mobility, especially wheelchair users, assign a rate to that place, and classify by indicating the type of trap that place presents. Based on the crowd inputs submitted by the users, the mobile application permits the users visualize heatmaps of these obstacles around the city, and establish the most accessible routes. The proposed mobile app is a powerful tool to help disabled people avoid obstacles in the available paths. Also, this collaborative system is important to easily indicate the current access problems around the city and serve as support to monitoring and maintenance of these locations.

2   Related Work

Crowdsourcing is a process where a large group of people volunteer in an online activity by submitting their inputs related to a specific context [6]. Thus, crowdmapping, as a category of crowdsourcing, is the process where the information from the group of volunteers is combined with geographic data, which results in a collaborative digital map with information relevant to that community. In the last few years, an expressive amount of projects have emerged and empowered communities in events related to crimes, wars, elections or natural disasters [7].

WikiCrimes is a project that offers an online common space where people submit information about occurrences of criminal activities in locations tracked by them. The main goal of the project is to serve as an empowerment tool for people to collaboratively participate in a platform to generate useful information to everyone and create a bridge between citizen participation and open government [8].

Flooding Points is crowdmapping system that aims to provide a Web map showing the flooding points in the city of São Paulo at the moment of flooding event. These locations information are provided by Web users in near real time. The system has demonstrated itself as a solution tool to exchange information about the problems of inundation and flooding in near real time in the city of São Paulo [9].

Another example of crowdmapping usage was the accountability in a Nigerian election. It was tested the effects of crowdsourcing election monitoring in the form of population inputs related to failures, abuses, and success through the Ushahidi open-source geographical information systems (GIS) platform in regard to the 2011 election. It was observed that the number of crowdmap reports generated by the population is correlated with the increased number of election participation in the Nigerian presidential election of 2011 [10].

All the projects discussed in the literature aim at empowering communities through the use of geographic information systems (GIS) as a tool to bridge the population's participation to open governments. The use of crowdmapping opens a huge opportunity to new platforms that can actually identify problems regard to social settings of urban planning. The application described in this paper is able to assist in the cause of accessibility problems in urban centers by encouraging citizens to identify possible trap

access throughout the city, provides the feature of photographing the place for visualization and input for comments, and offers the users an alternative by showing the
most accessible routes, avoiding the most problematic places.

3   The Mobile System for Accessibility Rating

This section introduces the proposed mobile application, a crowdmapping system that
indicates, classifies and rates accessibility traps, and suggests the most accessible routes
in urban centers. The system has been developed to be a hybrid mobile application, i.e.
it is been built with the combination of Web technologies to be run on any operating
system. The referred system also has some features provided by Leaflet library for
mapping visualization.

The idea of this application is to permit users - handing their cellphones - to
photograph a location that presents accessibility trap for low mobility people, to mark a
map marker on that location, to rate it by its level of difficulty, and to classify it by the
type of the trap (e.g., holes, missing ramps).

In the next subsections, the architecture of the system and some feature details are
described.

3.1   System Architecture

The proposed mobile application has been developed as a client-server system, which
provides conditions to easily integrate new technologies into the system and it facilitates the management of the features. By that, the system is divided into 3 layers:
Application Layer, Logical Layer, and Persistence Layer. Figure 1 shows the system's
architecture.

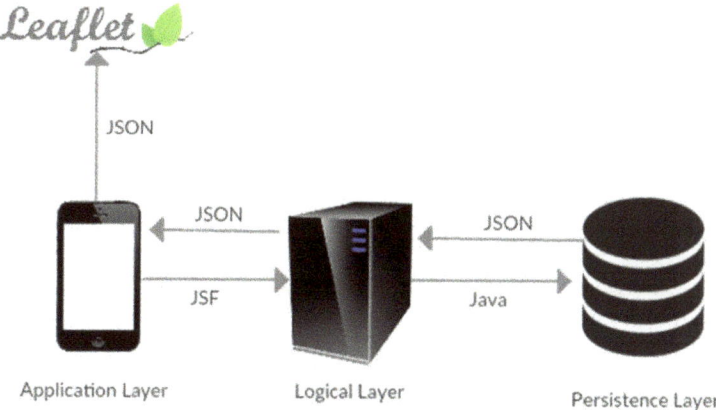

Fig. 1. System architecture

The Application Layer is responsible for the user interface, manipulation and visualization of all information involved. The visualization of spatial data (markers and routes) is done using Leaflet JavaScript library. On the home screen of the system, it automatically detects the current location of the user. This localization is done through the combination of cellphone's GPS and a plugin included in Leaflet. In the case the GPS is not turned on, the user can tap on the map and add a new marker anyway. The Application Layer requests all the information to the Logical Layer, manipulates the response in order to display it to the user as more suitable as possible.

The Logical Layer plays the role of receiving and processing the requests done by the Application Layer, all through HTTP requests. Once it receives a request, the Logical Layer sends the data request to the Persistence Layer, which interprets the request and generates a response to the Application Layer. This workflow requires internet access to work properly.

Finally, the Persistence Layer includes a relational database with support to spatial data (PostGIS/PostgreSQL). The job of this layer is to provide access, management and storing of spatial data.

3.2 Conceptual Project for Spatial Database

The database was projected with 4 entities, as shown in Fig. 2. The entity USER represents the users in the system, with information related to login access and personal info. The entity MARKER represents the location where the accessibility trap takes place, and stores data such as the url for the picture, rate, comment, and a foreign key for the type of trap.

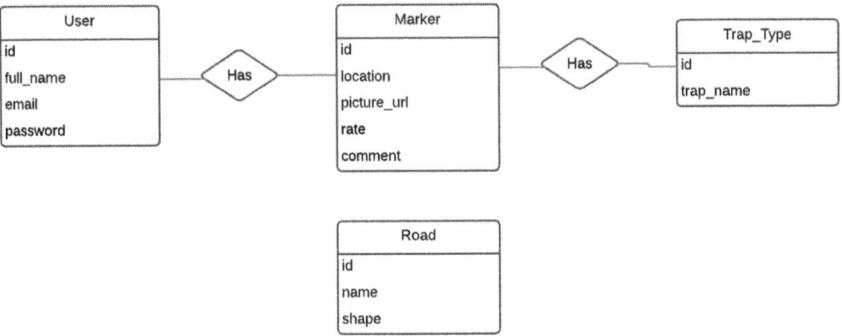

Fig. 2. Conceptual model

The entity TRAP_TYPE stores all possible types of traps within this context, such as holes, missing ramps, missing sign, etc. Finally, the entity ROAD represents the roads imported from Open Street Map.

3.3 Feature of Adding Marker

The main feature of this mobile application is adding a marker on the map, as it is the feature that sustains the crowdsourcing nature of the system. For this purpose, the system provides the users an interface to add a location marker on the map and its attributes.

On the home screen, shown in Fig. 3, the map is plotted based on current user's location. At the top, a menu bar is placed and includes the button 'Add Rating Marker'. Once this button is tapped, a dialog box is opened and shows a form with all input fields for that location, such as: upload picture, rate scale, type of trap, and optional comments. After inputting the required information, the marker is added.

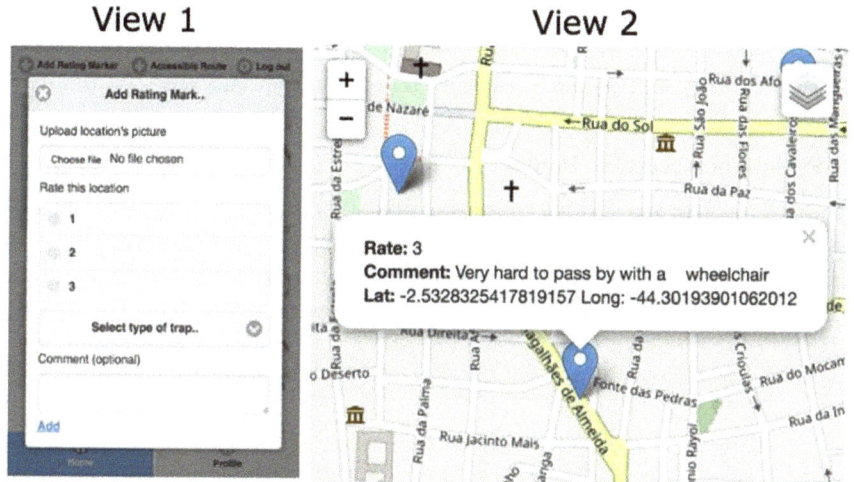

Fig. 3. View 1 shows the adding rating marker form. View 2 shows a marker's popup and its info.

All the markers available on the map can be tapped. This action opens a popup with some info related to that location: rate, comment, and latitude and longitude.

3.4 Layer Visualization: Heatmap and Markers Layer

Depending on the urban centers, some may have high problematic mobility access or not. This can result in a large number of markers on the map, which can be noisy and uncomfortable for visualization in high scale. Therefore, the mobile application provides two map layers for visualization: heatmap and markers only. The heatmap is shown in Fig. 4.

The heatmap visualization presents the concentration of problematic places. Depending on the location, the layer may show some places in red, which indicates high concentration of low mobility access. Whereas, the markers visualization shows the actual markers submitted by the application users.

Fig. 4. Heatmap based on the accessibility trap's markers

3.5 Feature of Accessible Routes

This feature provides the user the suggestion of a route that has low chances of having accessibility traps based on the crowd inputs provided by the system users. The user inputs the departure and destination locations and the system responds with a route, shown on the map, that is most suitable for people with low mobility, as shown in Fig. 5.

Fig. 5. Most accessible route between two locations

4   Conclusion and Further Work

This paper presents the development of a mobile application that aims at empowering disabled people and people with low mobility in urban centers through the use of crowdmapping technology. The objective is to provide essential means to communities of users with disability or low mobility problems to contribute on building a map of accessibility traps, showing problems that these people may find in their daily-based activities. Moreover, the proposed application may be used by disabled people to find the best route between two locations, avoiding the mapped traps.

People with low mobility face a lack of accessibility in urban centers in a daily-base, and this application can serve as bridge between these people and open government. This collaboration is able to come up with best solutions suitable for communities around the world.

The described mobile application still has some limitations, e.g., it is narrowed to mobility problems and can be improved to assist deaf and blind people as well.

Further works include the improvement of the generation of the most accessible routes feature, expanding it to more specific types of urban mobility places, like sidewalks and crosswalks. Also, include more sophisticated queries that suit to each kind of mobility problem, e.g., generation of different heatmaps based on user's choice. Other aspects to be worked on include to expand the feature of classification to mobility places that present good quality of access, which can help on the generation of accessible routes, and finally launch the mobile application to be tested and validated by the interested communities.

References

1. Disability, Accessibility and Sustainable Urban Development | United Nations Enable, https://www.un.org/development/desa/disabilities/resources/disability-accessibility-and-sustainable-urban-development.html
2. Almeida, E.P., Giacomini, L.B., Bortoluzzi, M.G.: Mobilidade e Acessibilidade Urbana
3. Alvarez, L.: Developing the network for growth and equality of opportunity, https://reports.weforum.org/global-information-technology-report-2015/1-6-developing-the-network-for-growth-and-equality-of-opportunity/
4. Goodchild, M.F.: Geographic information systems and science. Today Tomorrow. Ann GIS **15**, 3–9 (2009)
5. Number of mobile phone users worldwide 2013–2019 | Statista, https://www.statista.com/statistics/274774/forecast-of-mobile-phone-users-worldwide/
6. Junior, C.P., Holanda, G., Spitz, R.: Crowdmapping e mapeamento colaborativo em iniciativas de inovação social no Brasil. In: XX Congreso de la Sociedad Iberoamericana de Gráfica Digital, pp. 969–974 (2016)
7. Shahid, A.R., Elbanna, A.: The Impact of Crowdsourcing on Organisational Practices: The Case of Crowdmapping. AIS Electronic Library (2015)
8. Furtado, V., Caminha, C., Ayres, L.: Open Government and Citizen Participation in Law Enforcement via Crowd Mapping - IEEE Xplore Document, http://ieeexplore.ieee.org/abstract/document/6285930/?reload=true

9. Hirata, E., Giannotti, M.A., Larocca, A., Quintanilha, J.A.: Flooding and inundation collaborative mapping - use of the Crowdmap/Ushahidi platform in the city of Sao Paulo, Brazil. J. Flood Risk Manag. **26**, n/a-n/a (2015)

10. Bailard, C.S., Livingston, S.: Crowdsourcing accountability in a Nigerian election. J. Inf. Technol. Polit. **11**, 349–367 (2014)

Development of a Tracking Sound Game System for Exercise Support of the Visually Impaired Using Kinect

Kazuki Miyamoto[✉], Kodai Ito, and Michiko Ohkura

Shibaura Institute of Technology,
3-7-5, Toyosu, Koto-Ku, Tokyo 135-8548, Japan
{mal7113, nbl5501}@shibaura-it.ac.jp,
ohkura@sic.shibaura-it.ac.jp

Abstract. Even though many visually impaired people have a desire to exercise, they often cannot because they don't have enough time or facilities. Based on this background, we developed an exercise support system that the visually impaired can use by themselves at home. However, previous systems require a dedicated stationary bicycle that we can't obtain anymore. Therefore, we developed a new system using Kinect v2, which renders dedicated stationary bicycles obsolete, and any stationary bicycle can be used. We experimentally evaluated our system and described our conclusions.

Keywords: Visually impaired · Support system · Sports

1 Introduction

Many visually impaired people have a desire to exercise. However, they often don't have enough time or facilities. Effective exercise must be done regularly. Based on this background, we developed an exercise support system that the visually impaired can use alone at home [1]. We developed an exercise support game that encourages continuous exercise through fun activities with a dedicated stationary bicycle device for use at home because it doesn't need much space. Our system is controlled by a PC, which receives such data as the degree of the handle bar lean and the number of pedal rotations and sends sound data to headphones.

Our evaluation experiment results suggest that our system provided enjoyable support for continued exercise for the visually impaired. However, it requires a dedicated stationary bicycle that we can't obtain anymore. Therefore, we developed a new system using Kinect v2. We attached sensors to a dedicated stationary bicycle for detecting pedal and handlebar movements. In the new system, Kinect v2 sends to a PC such data as the degree of the user's body lean and the number of pedal rotations. Therefore, a dedicated stationary bicycle becomes unnecessary; any stationary bicycle can be used. The system also needs no sensors. Moreover, users can get a very thorough workout because they can operate the system by gestures with their entire body.

© Springer International Publishing AG, part of Springer Nature 2019
T. Z. Ahram and C. Falcão (Eds.): AHFE 2018, AISC 794, pp. 943–948, 2019.
https://doi.org/10.1007/978-3-319-94947-5_93

2 System Development

2.1 Outline

Figure 1 shows a system overview. The following is our system flow:

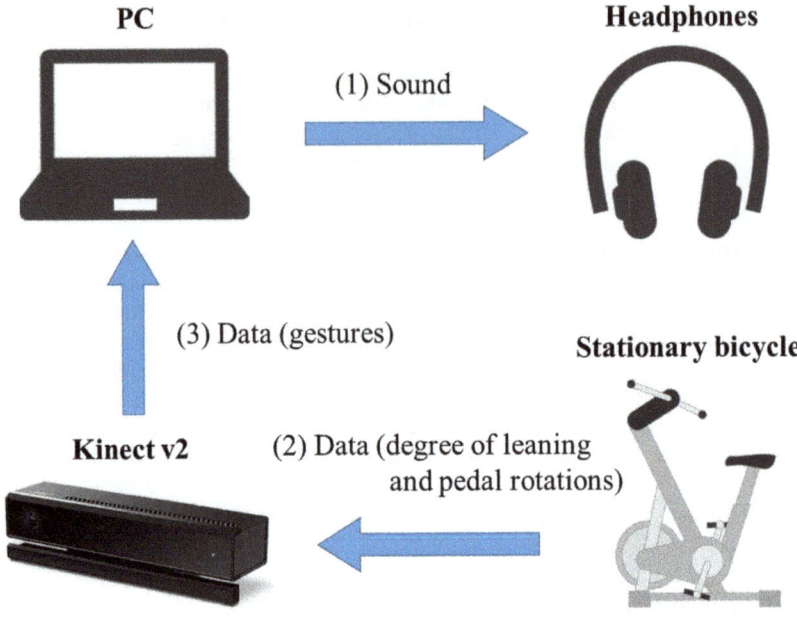

Fig. 1. Overview

(1) The PC outputs a target sound and background music (BGM) through headphones.
(2) Kinect v2 obtains such data as the degree of leaning of the user's body and the number of pedal rotations.
(3) Kinect v2 sends the acquired data to the PC as gesture data.

2.2 System Operations

In this system, the users operate the game through the following gestures:

    I. Users step forward by pedaling while leaning their upper body forward.
    II. They step backward by pedaling while leaning their upper body backward.
    III. They turn right or left by pedaling while leaning their upper body right or left.
    IV. They select or decide the game mode by raising their left or right hands.

2.3 Outline of Game

The game in our previously developed system for the visually impaired employed 3D sound localization. Since it got favorable comments, we also employed 3D sound localization for our new system's game [2, 3]. In this game, users pedal a bicycle to select the target sounds they are approaching by leaning their body and pedaling. The target sounds emerge from eight directions in increments of 45 degrees from the player's median plane. The game consists of three levels. The degree of difficulty differs for each level: easy, normal, and hard.

The following is the game flow:

1. Users wear headphones and ride a bicycle-type device.
2. We calibrate the Kinect v2.
3. The PC outputs a target sound that is assumed to be located around the users.
4. Users pedal the bicycle to select a approaching target by leaning their body and pedaling.
5. When users capture the target, it disappears, the PC outputs a sound that denotes success and gives a score. If a certain period of time elapses during which the users fail to capture the target, it disappears automatically, and then the PC outputs a sound that denotes failure.
6. A new target appears.
7. After a certain number of targets have appeared, one stage is finished, and the next begins.
8. When the three stages are finished, the users receive their scores by voice.

3 Evaluation

3.1 Experimental Method

Our experimental method is as follows:

i. The experimenter explains how to play the game to the participants.
ii. They wore headphones, rode a stationary bicycle, and practiced the game. They can stop practicing at any time.
iii. After they finish practicing, the game starts.
iv. After the game is finished, the headphones are removed, and participants get off the bicycle and answer a questionnaire.

3.2 Experimental Results

We experimentally evaluated our new system with ten participants: four were completely blind, one had amblyopia, and five had normal sight (age 21–33 years, mean = 25.0, SD = 3.8). We recorded the system's log data and performed five-point Likert scale questionnaires. Figure 2 shows the evaluation results about the game's enjoyment. It was evaluated as enjoyable by both the visually impaired and the sighted. Figure 3 shows the evaluation results about the game's ease of operation. It was

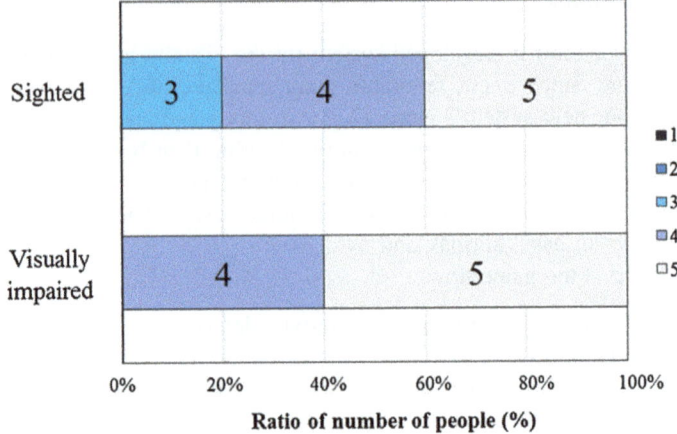

Fig. 2. Questionnaire results about game's fun (1 is very boring, 5 is very enjoyable)

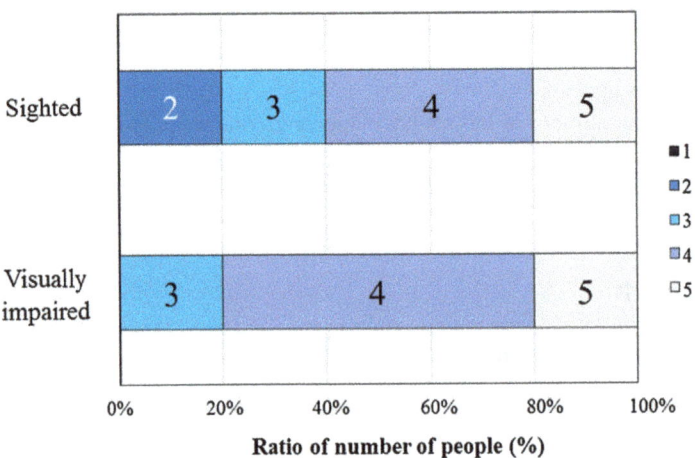

Fig. 3. Questionnaire results about game's ease of operation (1 is very hard, 5 is very easy)

evaluated as easy to operate by both the visually impaired and the sighted. Figure 4 shows the evaluation results about difficulty. Many thought that it was difficult. In addition, the log data showed that the visually impaired had shorter average required times to identify the targets than the sighted.

3.3 System Improvement

Based on the questionnaire results, we made the following improvements:

- Added a function to control pedaling speed. The faster the users pedal, the faster they move through the game.
- Changed the appearance position of the targets. They no longer appear behind the users.

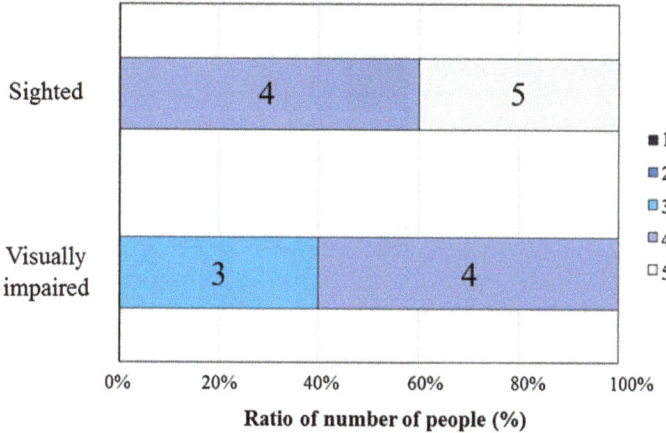

Fig. 4. Questionnaire results about game's difficulty (1 is very easy, 5 is very hard)

- Increased the time before a target automatically disappears.
- Added a function that raised the degree of difficulty if users played the game well. If they haven't performed well, the difficulty level remains easy.
- Changed the number of levels from three to two to shorten the game time.

3.4 Evaluation Experiment of Improved System

We re-evaluated our system after the above improvements with five participants: four were completely blind and one had amblyopia (age 19–32 years, mean = 25.6, SD = 5.6). We recorded the system's log data and performed five-point Likert scale questionnaires. Figure 5 shows the questionnaire results. No participants felt that the game was boring or hard to operate. However, one person described the game's progress as monotonous and too easy. In addition, we presented our system at a popular exhibition for visually impaired adults. 15 participants experienced our system and answered questionnaires. Figure 6

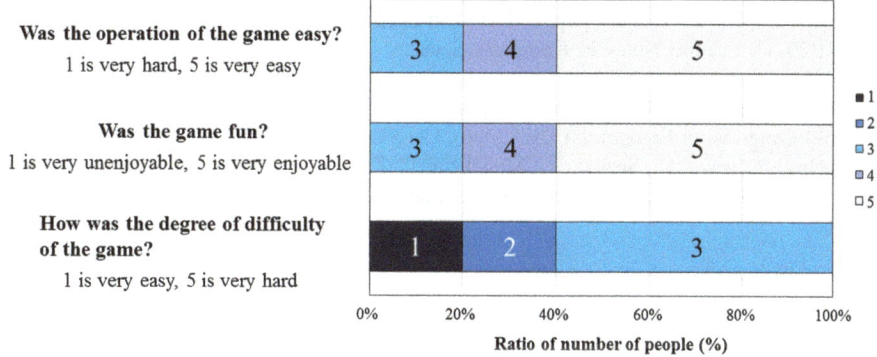

Fig. 5. Questionnaire results (re-experiment)

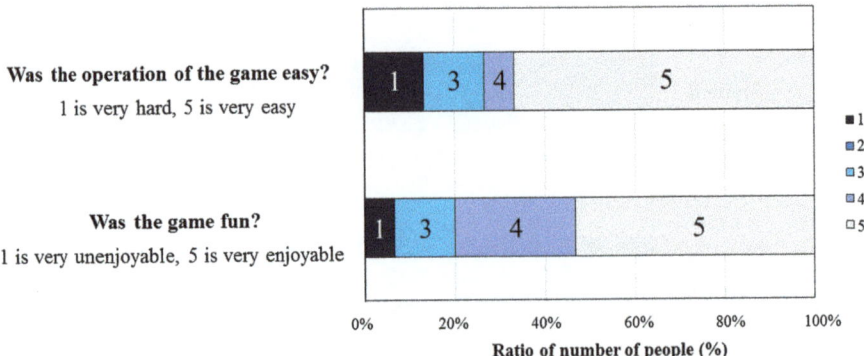

Fig. 6. Questionnaire results (exhibition for visually impaired)

shows the questionnaire results. For the visually impaired older adults, the operation of the game was easy and enjoyable.

4   Conclusions

We developed a system that encourages the visually impaired to exercise at home. In this new system, a dedicated stationary bicycle is unnecessary, and any stationary bicycle can be used. The system also needs no sensors. Our evaluation experiment results suggest that it provided enjoyable support for continued exercise for the visually impaired. We will continue to improve our system for future marketability.

Acknowledgements. We thank the students at Special Needs Education School for the Visually Impaired, University of Tsukuba who participated in our experiment.

References

1. Ikegami, Y., Ito, K., Ishii, H., Ohkura, M.: Development of a Tracking Sound Game for Exercise Support of Visually Impaired, Human Interface and the Management of Information (HCII2011), Lecture Notes in Computer Science, vol. 6772. Springer, Berlin, Heidelberg (2011)
2. Ishii, H., Inde, M., Ohkura, M.: Development of a game for the visually impaired. In: 16th World Congress on Ergonomics (IEA2006), CD-ROM, Maastricht (July, 2006)
3. Ohuchi, M., Iwaya, Y., Suzuki, Y., Munekata, T.: Training effect of a virtual auditory game on sound localization ability of the visually impaired. In: Proc. of ICAD2005, pp. 283–286 (2005)

Assistive Design Solutions and Prosthetic Environments

Empowering Design Solutions for Orbital Epitheses, Avoiding the Uncanny Valley

Julie Snykers[1(✉)], Yvonne Motzkus[2], Marieke Van Camp[1],
and Kristof Vaes[1]

[1] Faculty Design Sciences, University of Antwerp,
Ambtmanstr. 1, 2000 Antwerp, Belgium
`julie.snykers@student.uantwerpen.be`,
`{marieke.vancamp,kristof.vaes}@uantwerpen.be`
[2] Berliner Zentrum für künstliche Gesichtsteile, Epithetiklabor
in der MKG-Chirurgie, Augustenburger Platz 1, 13 353 Berlin, Germany
`yvonne.motzkus@googlemail.com`

Abstract. Current humanlike orbital epitheses are unnoticeable from afar. When bystanders come closer and discover the epithesis, they often react distressed. This can be explained by the uncanny valley effect. Humanlike epitheses fall in the depths of the uncanny valley and potentially evoke negative responses from bystanders. Instead of hiding the epithesis, this study suggests a shift towards empowering designer epitheses located on the left side of the uncanny valley. A literature study was conducted to determine the state-of-the-art on facial epitheses and stigma-free design. In addition, interviews were organized with six anaplastologists and five orbital epithesis wearers. Design workshops with stakeholders allowed to derive three promising design directions for orbital design epitheses. Additionally, lead users identified three specific use cases: professional, personal/casual, and social/festive use. These explorations resulted in a better understanding of future challenges and values.

Keywords: Uncanny valley effect · Facial prosthesis · Orbital epithesis
Empowerment · Stigma · Inclusive design

1 Introduction

There are numerous ways to acquire a facial defect, but they are mostly the result of cancer, trauma or congenital malformation [1]. Researches have proven that people with facial defects experience problems during social interactions. Facial expressions and appearance play an important role in social interactions. As such, people with facial defects have a higher risk of developing psychological issues such as a lowered self-esteem and depression [2–4]. Facial epitheses are used to replace lost or absent facial tissue and restore the patient's facial appearance as much as possible. Continuous advances to increase the degree of human realism for facial epitheses, results in the development of highly realistic-appearing facial epitheses. Hiding patient's physical defects and disabilities is common in the medical world [3, 5]. Despite that such

© Springer International Publishing AG, part of Springer Nature 2019
T. Z. Ahram and C. Falcão (Eds.): AHFE 2018, AISC 794, pp. 951–959, 2019.
https://doi.org/10.1007/978-3-319-94947-5_94

camouflaging techniques are widely used, alternative approaches for restoring physical/facial defects should be investigated.

Facial epitheses are unnoticeable from afar, but at a closer distance they risk being discovered. When bystanders notice the epithesis, they often react distressed or shocked. This reaction can be explained by the uncanny valley effect, which states that slightly imperfect humanlike objects evoke negative emotional responses. So-called humanlike epitheses, often fall in the depths of the uncanny valley and evoke negative responses from bystanders [6] (Fig. 1).

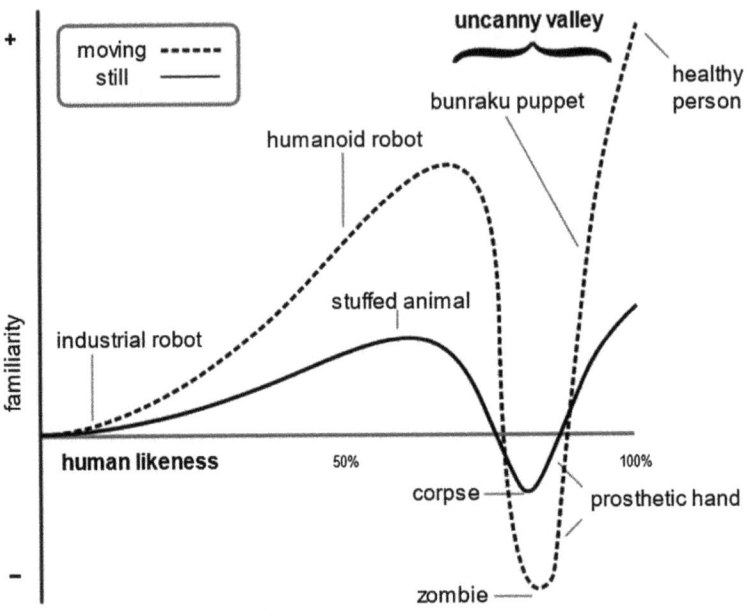

Fig. 1. Visualization of the uncanny valley effect.

The concept of the uncanny valley was presented in 1970 by Masahiro Mori. It recently became a popular subject due to the emergence of humanlike simulations, VR applications and robots. Multiple investigations have been executed to evaluate and reduce the effect of the uncanny valley for these applications [7]. The adoption of this theory for facial epithesis applications was investigated in a pilot study by the Berlin Center for Artificial Facial Parts. Following the web-based study of the Japanese scientists Seyama and Nagayama in 2007, subjects were presented with patient images of differently aestheticized epitheses and the reactions were investigated. Seyama and Nagayama found in their studies that the hypothesis posed by Mori is only verifiable under certain conditions. Their results suggested that to have an almost perfectly realistic human appearance is a necessary but not a sufficient condition for the uncanny valley. As they found out the uncanny valley emerges only when there is an abnormal feature [8].

As expected, an epithesis in the Berlin pilot study that does not achieve a certain degree of similarity to the natural environment, in aesthetic quality, caused irritated reactions that can be explained with the Uncanny Valley. Regardless of this static, web-based study, it is to be presumed that, in a moving comparison or a live experiment, any deviation of the epithesis in facial expressions, e.g. the lack of blink of the eye causes the reaction of the Uncanny Valley, as the viewer finds a different feature. These evoked reactions carry the risk of stigmatizing the wearer [9].

In his work, Mori proposes to stay away from the uncanny valley and create a safe level of affinity by developing non-human like designs. As an example, Mori mentions that in the future prosthetic hands might become fashionable assets instead of pitiful realistic copies [6]. This theory has already been successfully applied to leg prostheses by companies as UNYQ and Alleles [10, 11]. They offer personalized design prostheses in various materials like metal and wood. This approach received positive responses from users. Furthermore, special designs started to appear in fashion shows and famous people like Victoria Modesta and Aimee Mullin use their remarkable prostheses to their advantage. Slowly, non-human like arm prostheses are also appearing on the market. However, this new trend isn't yet adopted in the discipline of facial epitheses.

There are multiple types of facial epitheses, the most common ones are orbital, nasal and ear epithesis. In the case of facial epitheses, there is a risk that they will be discovered as described above, either by aesthetic deviant features or by the inability to adapt to facial expressions. A responsible factor for this are conspicuous transitions of the epithesis to the natural skin area. These transitions are usually made very thin in order to integrate the epithesis as inconspicuously as possible into the natural environment. The mentioned mimic movements make these edges very stressed and lose their resilience after a certain time, so that a gap between epithesis and skin area can arise. Another common problem is the color difference between the patient's skin and the epithesis. The color of the silicon material is not stable and is discolored by liquids such as sweat or the influence of the sun. To minimize these problems, facial epitheses are usually remade every two years [12]. This is a costly undertaking and all too often problems arise much faster. The silicon might already lose its color after 6 months [13]. Furthermore, orbital epitheses have the additional complexity of the static eyeball and eyelid. Compared to other body parts, designing epitheses for the facial area is more challenging, due to the sensitivity and complexity of the human face. Therefore, further research needs to be conducted in this area. This study aims to design empowering orbital epitheses that remain on the left side of the uncanny valley.

2 Challenges Orbital Epithesis

Multiple elements can be responsible for the emergence of stigma. The first step in creating a stigma free orbital epithesis design, is to analyze the current problems and wishes of orbital epithesis wearers. Insights were gathered through literature research and interviews with: two anaplastologists from Belgium, four anaplastologists from Germany, one anaplastologist from the Netherlands, two users from Germany, and three users from the Netherlands. Before executing the interviews, the PAMS (Product

Appraisal Model for Stigma) - a tool that 'unveils' stigma pitfalls and social conflicts between users of stigma-sensitive products and their surroundings - was used to ensure that all the use aspects were considered. The tool allowed to assess three components: product perception, product in use, and reflecting on product use; and how these components are appraised by three stakeholders: product users, bystanders, and society [14]. The most important findings are illustrated in Table 1.

Table 1. Results from the interviews.

Statement	Epithesis wearers (5)	Anaplastologists (6)	Total (11)
Stigmatizing situations occur in the presence of other people			
With unknown people	9	1	10[a]
With acquaintances	1	1	2[a]
With children	3		3[a]
The epithesis should be a communication tool towards bystanders	3	2	5[b]
The design should empower its user	2	2	4[b]
Adoption of the new concept			
Rejection		2	2[b]
Hesitation	2	2	4[b]
Acceptation	3	2	5[b]

[a]Amount of situations mentioned
[b]Amount of people

All the stigmatizing situations that were mentioned by the participants occurred in the presence of other people. Remarkable was how it almost always involved unknown people like random passerby's or first meetings. Their stares and uncomfortable behavior affect the epithesis wearers, even if the wearer feels mentally at ease with her or his epithesis. Three out of five epithesis wearers stated that people stare because they are unfamiliar with the situation. Facial epitheses are uncommon and have less public exposure compared to leg prostheses. Once orbital epithesis wearers get the chance to explain their condition, the other party is often more at ease. Unfortunately, and depending on the situation, it might be difficult or inappropriate to clarify it, for example in public places, during work, or in the presence of children. This explains the wish of epithesis wearers for the designer epithesis to be a communication tool towards bystanders. Instead of hiding the epithesis, they want to show it and explain the situation in a positive way to answer the questioning stares of strangers.

Aside from informing bystanders, the epithesis should also empower the wearer by projecting a strong, confident appearance. Now epithesis wearers are often pitied, but this is not the reaction they want to elicit. Multiple reasons were given for wanting to avoid pity. Mostly because it didn't fit with their own self-image and because they didn't want to be reminded of their defect the entire time. They agreed that people should look at the positive side of the story. For example, many people with orbital epitheses survived cancer and are happy to be alive.

When discussing the new orbital epithesis design concepts, different opinions were voiced between participants. Both anaplastologists from Belgium stated that this approach was not suitable for the target group. According to them this approach was suitable for leg prostheses but not for facial epitheses. This reaction could be related to cultural differences or the controversial approach. Anaplastologists strive to create a perfect copy for their customer, this concept opposes their philosophy. The sample is however too small to draw a conclusion on this matter. The different reactions indicate that the concept is rather innovative and might experience resistance during the adoption process.

3 Design for User Empowerment

After discovering the main problems and wishes of the users, the next challenge is to translate their input into an empowering design. The PIMS (Product Interventions Model for Stigma), a set of 17 stigma-alleviating empowering design interventions was applied [14]. Eight promising design interventions were selected to inspire our design solutions. These interventions were selected based on how well the solution space they offer matches with the product and the design challenges. To increase the impact, a promising design solution will combine several of these interventions (Table 2).

Table 2. Promising design interventions from the PIMS tool.

Design interventions
Strengthen the products individual identity
Strengthen the products brand identity
Integrate additional benefits and experiences
Reshape the product meaning through advances in material technology
Boost the user's social skills
Endow the product user with extra abilities
Increase positive social visibility

In order to turn the collected information into possible designs, co-creation workshops were executed with four anaplastologists, five epithesis users and two family members. Having orbital epithesis users participate during the entire design process empowers them and ensures a better result [15, 16]. During the workshop, participants first created their own designer epithesis by composing a mood board with the aid of card-sorting techniques. Afterwards they gave their opinion on the selected design interventions.

Participants were asked to compose their own design by completing a template containing four fields: situations, styles, materials and colors. First, they were asked to select the situations in which they would like to have a designer epithesis. During the interviews, it was clear that users felt stigmatized in different kinds of situations, during work, during their daily routine, etc. The products that you wear and use influence your identity. They can contribute to your identity in a positive or in a negative way [17]. Depending on the social context, people wear different outfits. Just like Aimee Mullins has a collection of different leg prostheses for different occasions, orbital epithesis wearers would prefer a different design depending on when or where they plan to use it [18]. Afterwards participants selected the style they would like to wear in those situations along with a set of associated materials and colors (Fig. 2 and Table 3).

Fig. 2. The composition of mood boards during the co-creation workshop.

Table 3. Results from the mood boards.

Topic		Epithesis wearers (5)	Family (2)	Anaplastologists (4)	Total (11)
Situation	Work	3			3
	Personal time/casual	2	1	1	4
	Festive event/party	4	1	3	8
	On vacation	1			1
Materials	Leather	1	1		2
	Jeans	3	1		4
	Silk	3	2	1	6
	Silver	2	2	2	6
	Plastics	2		2	4
	Ceramics	1		2	3
	Metal	1	1	2	4
	Rubber	2	1		3
	Wood	3			3
	Glass	1			1
	Light	1			1
	Gold	2			2
	Cupper	1			1

(continued)

Table 3. (*continued*)

Topic		Epithesis wearers (5)	Family (2)	Anaplastologists (4)	Total (11)
Colors	Bleu tint	2	2	3	7
	Red tint	4	2	1	7
	Black	1	2	2	6
	Grey tint	1	2	2	5
	Brow tint	1		1	2
	Green tint			1	1
	Yellow tint	1	1		2
	Purple tint			2	2

The participants indicated three main design directions: professional, personal/ casual, and social/festive use. Eight out of twelve participants would wear a designer epithesis to a festive event. Four participants would wear it during their personal time, with people they know. These two options were evaluated as safe situations, where they would feel comfortable to try out the designer epithesis. This indicates that they are still hesitant towards the concept. Many participants were however surprised by the given examples and felt inspired afterwards. All the orbital epithesis wearers with active working careers composed an epithesis to go to work.

The chosen materials did not confirm our expectations. For designer leg prostheses, wood and metal are popular materials, whereas fabrics were clearly preferred for the orbital epitheses. Participants didn't simply choose an aesthetically pleasing material, they immediately mentioned how the used material should be comfortable to wear.

Even though colors are a personal preference, red, blue and grey tints were very often selected. Two orbital epithesis wearers explained that the color should not divert too much from their skin color.

Contrary to their choice for safe situations and camouflaging colors, half of the orbital epithesis wearers stated that the designer epithesis should catch the eye of bystanders and project a solid image (Table 4).

Table 4. Design interventions selected by the participants.

Design interventions	Epithesis wearers	Family	Anaplastologists	Total
Strengthen the products individual identity	6	2	4	12
Reshape the product meaning through the product's meaningful interaction with other products	5	2	2	9
Boost the user's social skills	3	1	3	7

The design interventions were discussed using simple visualizations. Three interventions were clearly favored by the participants.

Strengthen the products individual identity. As mentioned earlier, epithesis wearers felt pitied. They didn't feel that the realistic epithesis reflected their personality. This matches with the results from a previous study, where approximately half of the participants evaluated their facial epithesis as a negative influence on their identity and distanced themselves from the product [19].

Reshape the product meaning through the product's meaningful interaction with other products. By making references to other wearable products like jewelry, glasses or clothing accents, the epithesis becomes an accessory instead of a medical tool. This approach elicited a lot of positive responses from participants.

Boost the user's social skills. As an example to visualize the concept, an image was provided showing an orbital epithesis with a message on it, explaining how the wearer acquired the defect. Participants reacted very positive to this possibility, it fulfilled their wish to communicate with bystanders. However, they remained unsure if others would share their opinion.

4   Conclusion

Compared to highly realistic orbital epitheses, nonhuman like designs, that remain on the left side of the uncanny valley, have the potential to reduce negative responses from bystanders and avoid stigmatizing the wearer. To examine the feasibility and desirability of this novel approach, professional anaplastologists and patients were interviewed, and several workshops were conducted.

The workshops allowed to derive three promising design directions for orbital design epitheses for three specific use cases that were identified by the participating users/patients, namely: professional, personal/casual, and social/festive use. In addition, the workshops led to a better understanding of future challenges and values that need to be taken into account. Two important design values were exposed. The designs should strengthen the user's individual identity, subsequently increasing user-product attachment and reshaping the product's meaning. Furthermore, the design epithesis should boost the user's confidence and social skills, supporting social empowerment. A well-designed orbital epithesis should have the ability to change bystanders' perception, improving the wearer's self-perception as well.

Further research will be executed to empirically evaluate the effect of non-human like orbital epitheses both on the behaviour of bystanders, and subsequently on the self-perception of the wearer. High-end prototypes of design orbital epitheses will be developed to conduct user tests and observational research.

References

1. Goiato, M., Pesqueira, A., da Silva, C.R., Filho, H., dos Santos, D.M.: Patient satisfaction with maxillofacial prosthesis. Literature review. J. Plast. Reconstr. Aesthet. Surg. **62**, 175–180 (2009)
2. De Sousa, A.: Psychological issues in acquired facial trauma. Indian J. Plast. Surg. **43**, 200 (2010)
3. Lim, S., Lee, D., Oh, K., Nam, B., Bang, S., Mun, G., Pyon, J., Kim, J., Chang Yoon, S., Song, H., Jeon, H.: Concealment, depression and poor quality of life in patients with congenital facial anomalies. J. Plast. Reconstr. Aesthet. Surg. **63**, 1982–1989 (2010)
4. Levine, E., Degutis, L., Pruzinsky, T., Shin, J., Persing, J.: Quality of life and facial trauma. Ann. Plast. Surg. **54**, 502–510 (2005)
5. Pullin, G.: Design meets disability. MIT Press, Cambridge (2011)
6. IEEE Spectrum: Technology, Engineering, and Science News, The Uncanny Valley: The Original Essay by Masahiro Mori. https://spectrum.ieee.org/automaton/robotics/humanoids/the-uncanny-valley
7. Cheetham, M.: Editorial: the uncanny valley hypothesis and beyond. Front. Psychol. (2017)
8. Seyama, J., Nagayama, R.: The uncanny valley: effect of realism on the impression of artificial human faces. Presence: Teleoperator Virtual Environ. **16**, 337–351 (2007)
9. Motzkus, Y., Menzel, K., Voigt, A., Toso, S., Menneking, H., Herzog, M., Adolphs, A., Heiland, M., Raguse, D.: Ausreichend, zweckmäßig, wirtschaftlich - das paradoxe Phänomen des Uncanny Valley als Qualitätsmaßstab in der Epithetik. Epithetik-Kompendium dvbe **2** (in press 2018)
10. UNYQ Armor™: Personalized Prosthetic Covers for Lower Limb Amputees. http://unyq.com/prosthetic-covers/
11. The ALLELES Design Studio - Prosthetic Covers Fashion. https://www.alleles.ca/
12. Veerareddy, C., Nair, K., Reddy, G.: Simplified technique for orbital prosthesis fabrication: a clinical report. J. Prosthodont. **21**, 561–568 (2012)
13. Thaworanunta, S., Shrestha, B.: Orbital prosthesis fabrication: current challenges and future aspects. Open Access Surg. 21 (2016)
14. Vaes, K.: Product Stigmaticity - Understanding, Measuring and Managing Product-Related Stigma. Delft University of Technology - Antwerp University (2014)
15. Ladner, R.: Design for user empowerment. Interactions **22**(2), 24–29 (2015)
16. O'Hern, M.S., Rindfleisch, A.: Customer co-creation: a typology and research agenda. Rev. Market. Res. **6**, 84–106 (2010)
17. Jacobson, S.: Personalized Assisitive Products – Managing Stigma and Expressing the Self. Unigrafia, Helsinki (2014)
18. Aimee Mullins |The Aesthetics of Prosthetics| EWC. https://everwideningcircles.com/2015/10/18/aimee-mullins-12-pairs-of-legs/
19. Wondergema, M., Lieben, G., Bouman, S., van den Brekel, M.W., Lohuis, P.J.: Patients' satisfaction with facial prostheses. Br. J. Oral Maxillofac. Surg. **54**, 394–399 (2015)

Examining Visually Impaired People's Embossed Dots Graphics with a 3D Printer: Physical Measurements and Tactile Observation Assessments

Kazunori Minatani[✉]

National Center for University Entrance Examinations,
Komaba 2-19-23, Meguro-ku, Tokyo 153-8501, Japan
minatani@rd.dnc.ac.jp

Abstract. Objective and significance: This study examines embossed dots graphics produced with a 3D printer. Tactile graphics are materials created for the purpose of providing graphical information to visually impaired people. They are usually made with a braille embosser as embossed dots graphics. Using 3D printing to produce tactile graphics has two benefits. First, tactile graphics made with a 3D printer retain the advantages that literally solid models are hard to achieve. Namely, they are (1) limited in volume and can thus be easily stored (2) can be compiled into books and communicate content that is closely tied to written information. The second is the ability to make use of our cumulative assets in tactile graphics production: techniques and production tools.

Keywords: Visually impaired people · 3D printer · Tactile graphics
Embossed dots graphics

1 Using Digital Fabrication for 3D Tactile Observation: Current Issues

The interest in providing visually impaired people with tactile observation materials has led to more active interest in attempting to make use of digital fabrication technologies. Research and development in this area began in the late 2000's, around the time that 3D printers using fused filament fabrication (FDM) technology started dropping in price-so considerable progress has already been made. Along with these developments have been attempts to produce models of things which cannot be tactilely observed directly, maps, famous pictures and so on [1]. In all cases, there is a shared goal of using the solid models made possible through three-dimensional molding technology for the purpose of tactile observation.

The tactile graphics used in conventional tactile observation materials are no more than raised versions of typical graphic images. They came about with the idea that this conversion would make the graphics intelligible to visually impaired people as well. What this means is that traditional tactile graphics are just two-dimensional objects extended into a third dimension. The research and development that uses digital fabrication, however, is completely different in that it is based on the idea of having

© Springer International Publishing AG, part of Springer Nature 2019
T. Z. Ahram and C. Falcão (Eds.): AHFE 2018, AISC 794, pp. 960–969, 2019.
https://doi.org/10.1007/978-3-319-94947-5_95

visually impaired people touch an actual three-dimensional object that represents something in the world. The expressive potential of this approach thus far exceeds that of existing tactile graphics.

Unfortunately, the products of this research and development have yet to be widely used in real-world situations, either to provide information to visually impaired people or in educational settings for visually impaired children. For the most part, graphical information is provided to visually impaired people in these practical settings via the raised paper methods collectively referred to as tactile graphics—a researcher named as the "2.1D graphic" [2].

There are two factors driving the failure of digital fabrication to catch on as a means of providing information to visually impaired people. The first is the current limitations of 3D printing in terms of printer functionality and usage conditions. The second is limitations stemming from the characteristics of solid models themselves.

The author starts by addressing the current limitations of 3D printing in terms of printer functionality and usage conditions. While FDM 3D printers aimed at general consumers have come down in price, we are unlikely to see any major improvements in performance. More specifically, the lamination pitch is stuck at around 0.1 mm, and printing speed is not likely to improve significantly either.

As the author will see later, these machines take around ten hours to produce tactile observation materials for visually impaired people-essentially a full business day. The current FDM 3D printers also require a tremendous amount of maintenance. The person operating the machine has to perform minute adjustments of the platform or heat bed on which the molded object being created sits, and molding materials that adhere to the extruder and other mechanical components have to be removed. Furthermore, these maintenance tasks take some skill to perform. The typical staff members involved in the education of visually impaired children cannot be expected to have these skills, making the operation of these printers on-site a rather impractical choice.

Now let's look at the second factor, which is limitations stemming from the characteristics of solid models themselves. Compared to tactile graphics, solid models have some serious restrictions in terms of their accessibility for practical use.

Tactile graphics, while raised, are still paper materials, and can be stored in a limited amount of space. This characteristic also makes it easy to manage them in a standardized way, so that tactile graphics are frequently considered to be a kind of book. Information can be provided in a way that is closely tied to written text, including the addition of textual information to the figure itself. For these reasons, tactile graphics are an important means of effectively communicating information to those who use them.

Solid models, on the other hand, are not as easy to store and manage. It is also difficult to create the same close linkages to textual information. This second limitation is a feature of solid models as a method of expression. That said, there are some things that can't be expressed well using tactile graphics. Solid models created with a 3D printer should be a direct and practical solution. As mentioned earlier, research into the use of digital fabrication as a way to provide information to visually impaired people has targeted objects that are rich in three-dimensional detail and difficult to convey using tactile graphics. In short, researchers see them as a way to complement the use of tactile graphics rather than replace them.

2 Aims and Challenges

In this paper, the author will use the term "3D embossed dots graphics" to refer to the embossed dots tactile graphics that we produced with a 3D printer. These are the main focus of our study. Conventional embossed dots graphics made using a braille embosser will be referred to as "paper dots graphics."

While this study aims to assess the effectiveness of using digital fabrication to provide information to visually impaired people, the author also recognizes the value of conventional tactile graphics. This approach led him to see the usefulness of 3D printing in the creation of tactile graphics.

Using 3D printing to produce tactile graphics has three benefits. First, tactile graphics made with a 3D printer retain the advantages that literally solid models are hard to achieve. Namely, they are (1) limited in volume and can thus be easily stored and managed, and (2) can be compiled into books and communicate content that is closely tied to written information. That said, the molding material used in 3D printers is essentially plastic resin, so whether or not the 3D-printed tactile graphics can be compiled into books will depend on the characteristics of the actual molded object's shape itself. This issue is one of the targets of our study.

The second benefit of using 3D printing to produce tactile graphics is the ability to make use of our cumulative assets in tactile graphics production. People have been making tactile graphics for a long time, always with the aim of using expressive methods that are the easiest for visually impaired people to understand. This accumulated experience has resulted in codified know-how that is widely available in the form of manuals. Operational procedures have also been substantially standardized [3]. But when it comes to the production of solid models, not enough progress has been made in terms of standardizing know-how for the purpose of creating products that are easy for visually impaired people to understand. Even if we make tactile graphics using a 3D printer, however, we should be able to take advantage of the accumulated expertise in tactile graphic creation and produce educational materials that are clear and accessible for the visually impaired.

When related utilities are properly arranged, production tools specifically designed to create tactile graphics may be able to be an advantage of production of tactile graphics with a 3D printer [4]. The use of computer software is standard practice when it comes to creating and printing embossed dots graphics with a braille embosser or tactile graphics using swell paper. There are three key benefits to tactile graphics produced with computer software: (1) it is easy to edit the designs, (2) the results can theoretically be replicated endlessly, and (3) the data can be modified and reused.

Of course, 3D CAD software is used in 3D molding in an attempt to get the same benefits as computer software gives tactile printing. The problem is that it takes a good deal of specialized knowledge and skill to operate 3D CAD software. If we used the usual CAD software to produce solid models for visually impaired people, we'd need a professional 3D designer to do the work. In this study, the author had no intention of using general-purpose 3D CAD software to produce the tactile graphics he had in mind, as it would greatly increase workload and difficulty. Instead, the author has focused on developing a way to produce 3D embossed dots graphics with the same simplicity that one can produce tactile graphics on a computer.

One of the advantages that 3D embossed dots graphics have over paper embossed dots graphics is the solidity of the material and the streamlined production method. These features open up new usage possibilities as well.

For example, tactile graphics are sometimes used as site maps for visually impaired people at train stations or other public facilities (see Fig. 1). Paper tactile graphics would not be sufficiently durable for this application, so they're made out of metal instead. But because of the labor and cost involved in producing metal tactile graphics, it's not a practical solution for widespread use. Another problem is that new maps cannot be produced when there are temporary layout changes, such as during reno-vation work.

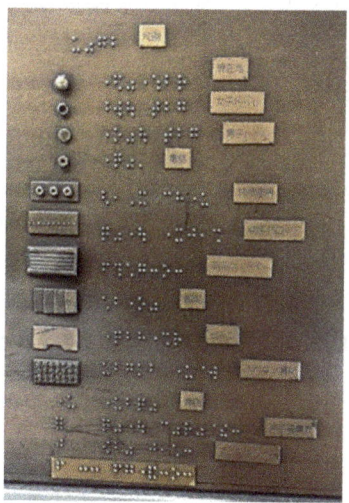

Fig. 1. An example of site maps for visually impaired people at train stations

With 3D embossed dots graphics made from plastic material, however, it would be possible to meet the need for temporary site maps while maintaining the durability needed for that application. In short, they could offer nearly the same durability as metal tactile graphics at a low cost comparable to paper tactile graphics. This gives them the potential to open up new tactile graphic applications as well.

The critical advantage that 3D embossed dots graphics have over paper embossed dots graphics is the ability to freely customize their arrangement. The braille embosser market has two types of machines. The first are those made to create braille text and used to create embossed dots graphics as-is [5]. The second are braille embossers specifically made to have the expressive versatility needed for embossed dots graphics. This type can create all kinds of dot varieties, with machines that can adjust dot height to eight different levels [6] or make dots of different diameters and heights in small, medium, and large [7]. Braille embossers that have these functions frequently use special drawing software to create embossed dots graphics [8, 9].

Braille embossers made specifically to create embossed dots graphics are capable of producing a large variety of dots, but they are still extremely limited compared to past tactile graphics production methods in terms of their expressive ability. In other words, the amount of creativity you have with a braille embosser is limited to the design options available with your particular machine (eight dot levels and three sizes, for example).

Meanwhile, the thermoform tactile maps and zinc plate presses widely used in the past used manual designs, giving them a level of expressive power limited only by the imagination of the person creating the tactile graphics. Printers are a widespread means of producing tactile graphics today, primarily because of how efficiently they are able to turn out graphics and how convenient they are to use-particularly in terms of the ease of edits and modifications. In terms of expressive power, one still cannot overlook the usefulness of longstanding technologies like thermoform.

Meanwhile, the thermoform tactile maps and zinc plate presses widely used in the past used manual designs, giving them a level of expressive power limited only by the imagination of the person creating the tactile graphics. Printers are a widespread means of producing tactile graphics today, primarily because of how efficiently they are able to turn out graphics and how convenient they are to use-particularly in terms of the ease of edits and modifications. In terms of expressive power, one still cannot overlook the usefulness of longstanding technologies like thermoform.

3   Developed Software and Production Process

With above understanding, the author developed a system that would convert data created using embossed dots graphics drawing software for a braille embosser into data that could be used with 3D CAD software [4].

3.1   Environment

Equipment
This study used the following equipment. For the host computer running the software, the author used a machine with an Intel XEON processor and 32 GB of memory running Linux (Debian 9.0). The data was output to the Ultimaker 3 Extended 3D printer [10].

Software
The author developed software that would take data generated by the embossed dots graphics drawing program Edel and convert it into 3D CAD data: Edl2scad. Edel is a piece of drawing software for designing embossed dots graphics for the ESA 721 embosser [9]. Edel allows users to draw embossed dots graphics using ESA 721's three sizes of dots (small, middle and large). An overwhelming percentage of embossed dots graphics in Japan are created using a combination of the Edel program and the ESA 721 machine, so the author felt that if we could convert this Edel data and use it to

create objects on a 3D printer, the majority of people who create embossed dots graphics in Japan would be able to continue doing so using the new technology and without having to learn any new skills.

The author put the 3D CAD data converted from the Edel data into SCAD format. SCAD is the format used by the OpenSCAD CAD software. OpenSCAD is designed to have a programmable CAD environment [11]. Users define 3D data by configuring numerical coordinates that determine the size and arrangement of basic geometrical solids like boxes, cylinders, and spheres. Figure 2 shows a screenshot of some in-progress 3D data on OpenSCAD.

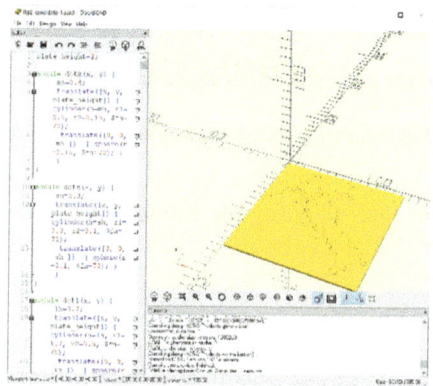

Fig. 2. A screenshot of 3D data on OpenSCAD.

In addition to SCAD format, OpenSCAD data can also be saved in STL format, which is the standard for 3D CAD/CG data. The author sent STL data to the 3D printer for printing.

Figure 3 shows an enlarged photo of the dots from an embossed dots graphics design that were produced using an ESA 721 embosser. When trying to approximate the shape of each of the three dot sizes as basic geometrical forms, the best choice is a hemisphere on the surface of a circular truncated cone. We'll refer to this combination as a "dot modeling solid."

The size and shape of the solids that correspond with each of the three dot sizes can be modified freely once by changing the parameters set in Edl2scad. It's not always practical, but you can also use the original Edel drawing to reverse the small, medium, and large dot sizes. Finally, though the dot modeling solid is the default, it is possible to change the shape to a quadrangular prism or triangular pyramid, for example.

Though these design adjustments are beyond the scope of this study, they should make it possible to change the tactile sensations produced by the finished product.

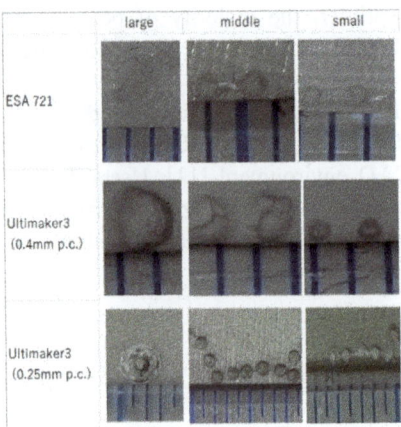

Fig. 3. An enlarged dots from an embossed dots graphics by ESA 721 and 3D printing

3.2   Evaluation of Dot Size on a Sample Embossed Dot Graphic

About the Sample

For the purpose to use as a sample the author found a map showing the power rela-tionships on the Italian peninsula during the 15th century on Wikipedia [12], and used small, medium, and large dots to recreate it as an embossed dots graphic. Figure 4 is given as an example of how the map he made was displayed on Edel. The author chose the Italian peninsula as his subject since he felt that its well-known "boot shape" was a good clue for people to focus on when performing tactile observations and identification.

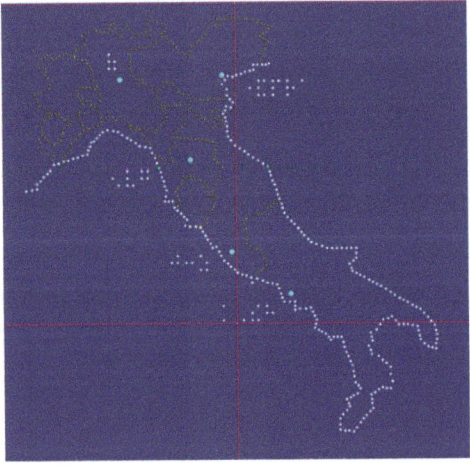

Fig. 4. A map showing the power relationships on the Italian peninsula recreated as an embossed dots graphic

The author used the three different dot sizes as follows. The coastline of the Italian peninsula was depicted using medium dots. The overland borders indicating various spheres of influence were drawn using small dots. Finally, the five major cities (Naples, Rome, Florence, Venice, and Milan) were indicated with large dots. Abbreviations representing the names of each of these cities were written in Japanese braille next to their respective large dots. The text itself was produced using medium dots. He intentionally added text to the map because we wanted to investigate whether it was possible to produce content closely linked to written test using our method.

The then used this data to create a 3D embossed dots graphic. Figure 5 is a picture of that a 3D embossed dots graphic. For comparison, he used the same data to create a conventional embossed dots graphic with the ESA 721 embosser.

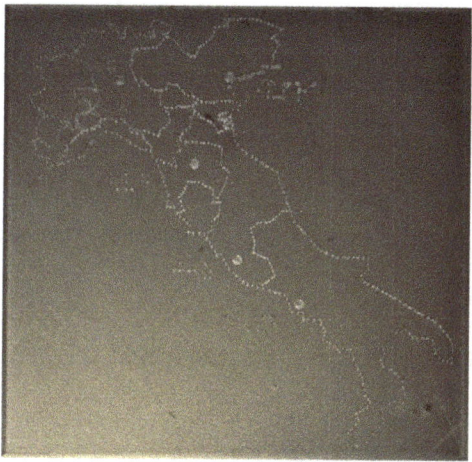

Fig. 5. The 3D embossed dots graphic generated from Fig. 4's map data

Production Process

The conversion from Edel data to STL data was done using the two-stage process described above. Edl2scad was first used to convert Edel data to OpenSCAD data, and then the OpenSCAD program was used to convert the data to STL format.

A model was then created in OpenSCAD using regular polygons to approximate cylinders and cones. By increasing the number of corners on these regular polygons, we were able to create a more precise replica of the original shapes. The OpenSCAD program allows you to define the number of corners on the polygons used to create these solids. Even during this study, which used the dot modeling solids to form the embossed dots, maximizing the number of corners on the regular polygons was ideal.

That said, increasing the number of corners when creating the models also requires more computer resources when converting the data into STL format. The machine the author used for this evaluation had 32 GB of memory, and the map he made had 80 corners. This amount of memory was insufficient to complete the modeling task. We then reduced the number of corners to 70 and tried to process the data again.

The conversion to STL succeeded under these conditions, but took the machine around 250 min to complete.

The author used ABS resin as our production material. Each map took ten hours to produce with an infill of 20% and laminate pitch of 0.1 mm. Extraneous pieces of ABS resin less than a millimeter square also adhered to the surface of the model. The three dot sizes were replicated fairly well, but we had to remove the debris after identifying it by comparing the model to the original drawing.

Meanwhile, on the poorly formed models, some of the dots would peel off or there would be stuck-on debris that could not be removed. These models were scrapped and remade.

Evaluations

On measuring dot size and tactile observation evaluations, the author described those details in another paper [4]. Two significant points on the quality of 3D embossed dots graphics are: (1) they were not inferior to the paper dots graphics and (2) the dots did not sufficiently protrude when the author used the exact size value for the paper dots graphics.

In addition, it is become clear that the quality of 3D embossed dots graphics are dominated by preciseness of a printer's extruder. The Ultimaker 3 supports detachable extruders so-called the print core system [13]. When printing with 0.25 mm print core, results of quality of 3D embossed dots graphics are more precise than printing with 0.4 mm (Fig. 5).

4   Conclusions and Issues for Further Study

This study examined a system for using a 3D printer to create solid models corresponding to the embossed dots graphics produced on a braille embosser. It is possible for current users of embossed dots graphics drawing software to create 3D embossed dots graphics. The combined height of the plate and the dot protruding above it on the 3D embossed dots graphics is just under two millimeters, which makes the 3D models just as easy to store as their paper counterparts. Because it is possible to adjust dot size on the 3D embossed dots graphics, we were able to create a model that allowed users to easily distinguish dots that were difficult to tell apart on the paper graphic. In short, the author's initial results were promising on the whole.

During the course of the study, the author repeated the cycle of (1) adjusting the dot size values, (2) generating and executing 3D model data, and (3) measuring/making tactile observations of the models eleven times. Each cycle was an exploratory test driven by trial and error, and took a minimum of two days to complete. For this reason, we were unable to conduct evaluations with numerous participants this time. Future studies should use the models created by the values the author obtained as a standard and compare them to models created from values that have been modified as deemed appropriate.

Other noteworthy problems with the study include the long amount of time it took to produce the 3D models, the lack of smoothness on the surface of the material, and the debris generated by the printer. These issues are specific to 3D printers that use

FDM technology, and are even more pronounced with devices aimed at the general consumer. Ideally, future studies would use commercial FDM printers or the up-and-coming inkjet 3D printers to create solid models.

Acknowledgments. This work was supported by KAKENHI (17H02005).

References

1. Teshima Y.: Three-Dimensional Tactile Models for Blind People and Recognition of 3D Objects by Touch: Introduction to the Special Thematic Session. ICCHP 2010, LNCS 6180, 517–518 (2010)
2. Reichinger, A., et al.: Computer-Aided Design of Tactile Models - Taxonomy and Case-Studies. ICCHP 2012, LNCS vol. 7382, 497–504 (2012)
3. Eriksson, Y., Strucel, M.: A Guidebook for production of tactile graphics on swellpaper. Talboks-och punktskriftsbiblioteket, Enskede (1994)
4. Minatani, K.: A proposed method for producing embossed dots graphics with a 3D printer. ICCHP 2018 (forthcoming)
5. Basic-D V5: https://www.indexbraille.com/en-us/braille-embossers/basic-d-v5
6. VP Max: https://viewplus.com/?p=23
7. ESA721 Ver'95: http://www.jtr-tenji.co.jp/products/ESA721_Ver95/ (in Japanese)
8. Tiger Software Suite: http://www.viewplus.com/products/software/braille-translator/
9. Edel and its related software http://www7a.biglobe.ne.jp/~EDEL-plus/EdelDownLoad.html (in Japanese, English version of these software are also hosted)
10. Ultimaker 3: https://ultimaker.com/en/products/ultimaker-3
11. OpenSCAD: The Programmers Solid 3D CAD Modeller. http://www.openscad.org/
12. File:Map of Italy (1494): https://commons.wikimedia.org/wiki/File:Map_of_Italy_(1494)-it.svg
13. Ultimaker 3 print cores: https://ultimaker.com/en/products/ultimaker-3-print-cores print-cores

Workplace Accommodation for Autistics: Autistic Autobiography and Technology-Enabled Prosthetic Environments

G. R. Scott[✉]

Naval Postgraduate School, 1 University Circle, Monterey, CA, USA
grscott@nps.edu

Abstract. Even though employers are increasingly recruiting autistic (The author recognizes the sensitivities around the choice of terms. The use of the term autistic in this paper is based on guidance from the Autistic Self-Advocacy Network (ASAN). ASAN holds that autism, unlike many other conditions, is a central feature of an autistic person's identity, not something that they "have." Therefore person-centric terms like "person with autism" are inappropriate [64]) employees, autistic adults have one of the highest unemployment rates in the United States. This paper presents ongoing research by the author: (a) providing a brief overview of current scientific and societal perspectives on autism; (b) describing an on-going qualitative study of autistic autobiographical writings to gain insight into the autistic experience, challenges faced in society, and barriers to employment; and (c) proposing Technology-Enabled Prosthetic Environments (TEPE) as a design concept for the integration of assistive technology for workplace accommodation.

Keywords: Autism · Neurodiversity · Prosthetic environments
Assistive technology · Workplace accommodation

1 Introduction

We are convinced, then, that autistic people have their place in the organism of the social community. They fulfill their role well, perhaps better than anyone else could, and we are talking of people who as children had the greatest difficulties and caused untold worries to their care-givers [1].

Employers increasingly seek the unique talents of people on the autism spectrum. Cognitive traits often associated with autism, such as pattern and error recognition, intense focus, and attention to detail can be particularly applicable in information-centric fields such as data analytics [2–4]. Even with this growing interest, autistic adults have one of the highest unemployment rates in the country, with estimates as high as 90% [5, 6]. The problem stems from the difficulties in accommodating the unique needs of autistic workers. While U.S. law mandates "reasonable" accommodation of differences and disability in the workplace, leaders of organizations can view accommodating the needs of autistic workers as unreasonable. Accommodation often requires continual attention from management and may impose on co-workers.

© Springer International Publishing AG, part of Springer Nature 2019
T. Z. Ahram and C. Falcão (Eds.): AHFE 2018, AISC 794, pp. 970–981, 2019.
https://doi.org/10.1007/978-3-319-94947-5_96

The perceived cost of accommodating autistic workers can over shadow the benefits of hiring them. Autistic workers also report long periods of unemployment and underemployment, difficulties in the hiring process, negative work experiences, and an inability to keep jobs [7].

Work environments are commonly designed to be inclusive of workers with various disabilities. Accommodating environments typically incorporate standardized prosthetic design features (e.g., wheelchair ramps, braille signage) and assistive technology (e.g., voice transcription devices for the deaf). These accommodations increase productivity and help people with disabilities maintain their autonomy and independence. The variety of challenges experienced by adults with autism makes the development of similar, standardized accommodation practices much more difficult. Difficulties in communication and social skills, which are definitive characteristics of autism, also impede efforts by autistic workers and their employers to determine and understand what specific accommodations might be useful [7]. Recent "autism as diversity" perspectives suggest that an approach built upon autistics' descriptions and interpretations of their experiences might prove fruitful.

This paper explores one area of inquiry that is emerging from an on-going research project. Viewing the problem of workplace accommodation as a design challenge (e.g. "How might we" create workplaces that better meet the needs of autistic workers [8]). This paper reflects the part of the design process often termed "divergent thinking" [9]. First, following the design adage of "observe first, design second," [10] I seek to develop a better understanding of the autistic worker, and then from that understanding, to propose a model from which to prototype new assistive technologies.

2   Autism: Diagnosis, Definition, Disorder, and Diversity

Any human kind explained in terms of deviation from the normal is partly descriptive: the kind differs from the usual. However, it is also partly evaluative: the kind differs from what is right; it is worse, or in the case of Galton's deviation from mediocrity, possibly better [11].

Diagnosis of autism is based on the 2013 edition of the American Psychological Association's Diagnostic and Statistical Manual of Mental Disorders (DSM-5). The DSM-5 groups previously separate diagnoses of Pervasive Development Disorder (PPD-NOS), Asperger's Syndrome, and Autism under the diagnosis of Autism Spectrum Disorder (ASD); listing specific diagnostic criteria, and "Severity Levels" from 1 to 3. In the DSM-5 ASD is presented as a spectrum of disorder characterized by "deficits in social communication and social interaction," and "restricted and repetitive patterns of behavior, interests, or activities;" that "impair social or occupational function [12]."

These diagnostic criteria also serve as the principal definition of autism, but many people who themselves have the diagnosis challenge this. Groups of autistic self-advocates challenge the inclusion of autism as a disorder, claiming instead that autism is simply natural variation, which should be accepted and accommodated as society does other types of diversity. Steve Silberman discusses this debate in detail in his recent book, Neurotribes [13]. This neurodiversity view has gained some acceptance within the "neurotypical[1]" research community, although the preponderance of research, and the most widely accepted theories remain based on the principle of autism as disorder.

2.1 Disorder

Spanning over fifty years of research, studies of the causes, traits, and treatments of autism most often focus on individual details; such as differences in genetics [14, 15], neurobiology [16–18], communication [19–21], and social interaction [22–24]. This has led to an ironically "autistic" understanding of autism – that is, an understanding of autism that reflects an often-observed preference for detail over holism in autistics [25, 26]. Researchers acknowledge the multi-dimensional spectrum of autistic traits, but have not reached consensus regarding a comprehensive account of autism.

Three somewhat more general theories of autism are widely cited: (a) Theory of Mind (ToM), developed in 1985 [27] proposes that autistics lack an ability to view situations from the perspectives of others, to "put themselves in another's shoes;" (b) Weak Central Coherence (WCC) [28] describes differences in how perceptions are processed (e.g. local information before global information) and how current context and prior knowledge are integrated; and (c) Executive Dysfunction Theory [29, 30], which does not specifically refute ToM or WCC, highlights these and other deficits in cognitive functions such as working memory and self-regulation in terms of the broadly accepted Executive Function Theory [31]. While these theories are useful to group and explore the relationships between the various manifestations of autism, none attempts to provide a holistic explanation of what autism is, what causes it, or how it works.

A newer approach, Intense World Theory (IWT) proposes that autism is rooted in hyper-sensitivity to stimuli and hyper-functionality of some cognitive processes that results in overload, and manifests in common autistic traits, such as repetitive behaviors, as coping mechanisms to, in effect, "turn-down the volume" [32, 33]. Many autism advocates wanting to de-stigmatize autism have embraced IWT because, while still presenting autism as abnormal, IWT focuses on enhancements rather than deficits.

2.2 Diversity

Perhaps the most dramatic recent trend in autism research is a notable recognition of the "autism as diversity" view espoused by autistic self-advocates. This is a politically charged scientific issue, not unlike the mid-1900's shift regarding homosexuality, which was not removed as a disorder until the 1973 DSM-II [34]. Science does not remain aloof from societal change, and so significant debate is emerging regarding both the scientific utility and ethical implications of how science views autism [35–37].

Neurodiversity activists cite IWT along with research into genetic and environmental causes of autism to support their view that autism is merely normal variation

---

[1] "Neurodiversity" and "Neurotypical" are terms originating from within the autistic self-advocacy community; the first as an alternative to the term ASD, and the second to define those without autism. Neurotypical has since started to gain a broader meaning to describe those without other cognitive differences as well. In this paper, it is used in its original sense of non-autistic, except where noted.

that should not be stigmatized as a disorder. While the genetic contribution or pre-disposition towards autism is significant, unlike some conditions such as Down syndrome, there is no single, clear genetic marker. The genetic risk for autism is itself a spectrum, based on multiple, heterogeneous groupings of alleles that are otherwise part of normal variation [38]. Further, there are no consistent environmental factors that lead to a single etiological model [39] of what we term autism.

In parallel with this activism, research into what is often termed the "cognitive style" of autism departs from the autism as disorder models. Some research extends from, but doesn't rely on disorder models to investigate aspects of better than normal function such as reduced change blindness or attention to detail [25, 26]. Other research, notably that stemming from embodied and enactive cognition models [40, 41], more directly challenges disorder models and claims to have broader explanatory power. The embodied cognition models of autism emphasize how different manifestations of autistic traits, such as echolalia or "stimming" are better explained by differences in cognitive style (e.g. information salience, sense-making, etc.) than by deficits in isolated cognitive processes.

However, researchers have noted some shortfalls in the diversity model. Jaarsma and Welin questioned if the diversity model can be applied equally across the autism spectrum, and proposed a narrow model of neurodiversity, applicable only to the mildly autistic [35]. Baron-Cohen notes this as well, and proposes how to appropriately use the terms "difference," "disability," and "disorder," as they apply to autism. Baron-Cohen proposed reserving the term "disorder" for when there is nothing positive about the situation, using the term "disability" when there are functional deficits that need support, and using the term "difference" when a person is simply atypical [37]. Baron-Cohen also noted a second shortfall: other disorders, such as obsessive-Compulsive Disorder (OCD), are frequently co-morbid with autism. Though these disorders are not autism, there is a risk that they get swept up as autistic diversity and go untreated [37].

Regardless of these critiques, autism advocates (in particular, self-advocates) argue that neurodiversity is a fundamental concept that employers should understand. Baron-Cohen also acknowledged the positive impact of this view. He proposed concepts and directions for autism research directly attributed to the neurodiversity movement, noting, "There is no single way for a brain to be normal." He called for non-stigmatizing models, and more research into autistic capabilities [37]. Similarly, effective workplace accommodation for autism should address both "difference" and "disability." Additionally it should recognize the diversity *within* autism; the myriad manifestations of autism hinder the development of "standard" accommodations such as the design parameters developed for wheelchair accessibility. It is also worth noting that mobility-impaired self-advocates created and fought for wheelchair accessibility standards. Developing workplace accommodation for autism similarly should include the voice of the neurologically diverse.

3 Current Study: The Autistic Perspective

If you begin with the user and set out on a path to look at the broader context of their lives and activities you will suddenly see a whole new set of opportunities to be tapped [42].

The preponderance of both the academic theorizing and public understanding of autism and autistics derives from the body of formal research conducted by neurotypicals. The recent diversity perspective suggests though, that for workplace accommodation to be effective, it is imperative to understand how autism is experienced by autistics. Congruent with this conclusion, a design thinking approach assumes that the users of solutions should have a voice in their design.

This study into workplace accommodation draws on the diversity perspective on autism, the design thinking perspective and a rapidly growing body of autistic self-expression and introspection that can provide the voice of the user. Formal study of autistic self-expression is scant, but Oliver Sacks' An Anthropologist on Mars [43], and Ian Hacking's "Autistic Autobiography," provide examples that indicate the utility of this line of inquiry. As Hacking summarizes, this genre is "creating the language in which to describe the experience of autism. Hacking focused exclusively on four published books, but acknowledged that "the richest habitat" for autistic autobiographical expression is online, it is "awash with autistics chatting about autism [44]." Online autobiography is, by its nature, less polished than published writing, undergoes less editing, and is more raw, more the voice of the author than the editor. Therefore, it more closely reflects the cognitive style of the author at the time of writing. This section presents a very brief overview of an ongoing analysis of publicly available, online posts written by autistics in order to add context, understanding, and meaning to the important, but often-disembodied findings of quantitative research.

3.1 Ethical Considerations

The Naval Postgraduate School Institutional Review Board reviewed this study and determined it to be in an exempt from human subjects research protocols. Additionally, following the guidelines established by the Association of Internet Researchers [45], I determined that the community being studied is inherently vulnerable and therefore implemented additional safeguards:

- This study will not disclose any personally identifiable information (PII) of the bloggers included in the study. This includes authors' names or other contact information, blog names or urls, and organizational affiliations of the authors. Exceptions to this are bloggers who are officially and publically associated with an activist or advocacy group.
- This study will not disclose certain demographic characteristics (e.g. ethnicity, gender identity, or sexual orientation) if combining this information with the small sample size and focus of sample criteria might readily lead to identification of the author.
- This study will not disclose certain details found in the blogs (e.g. admission of questionable activities) if that information would put a blogger at risk should their identity be otherwise discovered.

3.2   Methods

I am using purposeful, snowball sampling [46] to develop the dataset for this analysis. I started by reviewing the Autistic Self-Advocacy Network's (ASAN) social media presence (i.e. the blog on the ASAN website, and the ASAN Facebook and Twitter feeds) identifying nine authors who had multiple recent entries in one or more of the sources specifically addressing their experience with autism. Following an initial review of these posts, I established the following criteria for the inclusion of authors in our dataset: (a) They have an established, current feed (i.e. having more than ten posts, over a period of at least three months, with the last post less than one year old); (b) they frequently write from a first-person perspective about their experience with autism; and (c) they seem (given the information available) to genuinely be autistic[2]. I then identified other potential authors by following links and references provided by the initial set of authors. At the time of this writing, I have included 28 authors in the dataset, providing over 300 individual posts for consideration, but continually seek and identify additional authors who meet the criteria

From this collection of posts, I selected individual posts for coding if they met the following criteria; (a) the post is written (at least primarily) from a first-person perspective; and (b) the post includes description or discussion of the writer's experiences related to autism. At the time of this writing, I have included 74 posts in the coding dataset. Additionally, following the same criteria, I have begun to build a collection of excerpts from published autistic autobiographies for comparison and contrast. These autobiographies include the four reviewed by Hacking [44], and one newer, similar book, Ido in Autismland [47].

I am currently engaged in an iterative cycle of coding and theorizing [48, 49] of this growing dataset. Based on the body of research discussed in Sect. 2, substantial prior reading of autistic self-expression, and personal relationships with autistics[3], I began by creating a draft coding structure of autistic experience. This served primarily as an acknowledgement and formalization of my biases and pre-conceptions; helping me to bracket what I *expect* to find from what the texts actually say [50]. In the first round of coding (N = 12), I followed open coding practices [48], using a priori codes when appropriate, and generating additional codes in-vivo. I then reviewed the results, paying particular attention to the use of a priori codes to ensure they were applied appropriately. Additionally, colleagues less familiar with the subject separately coded several posts for validation. From this I began to consider second-order codes and categories.

---

[2] That is they don't seem to be "self-diagnosed," or "self-identify" as autistic. Some bloggers write specifically about their experience of the diagnosis process, or about experiences with therapists, teachers, and doctors as children. Others either explicitly state they have self-diagnosed, or use phrases like "I'm probably autistic." While these blogs are excluded from this study, exclusion should not be construed as disbelief of, or minimizing the experiences of, the authors. As Hacking discusses in [44, 65], they too contribute to the language we use about, and therefore our understanding of, autism.

[3] The author is the father of a severely autistic 20-year-old woman.

3.3   Early Observations

Currently, I am conducting a second round of coding (N = 74), using a coding structure which has changed significantly from the initial draft (lending confidence to the bracketing approach). The initial coding structure consisted of ten codes grouped into three categories: (a) experience, (b) actions, and (c) meaning. Currently there are 78 first-order codes, more than 60 of which are tentatively nested in one of five categories:

a) Physiological phenomena (e.g. "looking in someone's eyes is painful");
b) Communication with neurotypicals (e.g. "implied communication may be why I misunderstood");
c) Differences (and similarities) between autistics and neurotypicals. (e.g. "I know that at times I am perceived to be being aggressive or unkind);
d) "Inner lives" and external actions (e.g. "I committed myself to becoming a chameleon and "masking" my autistic tendencies"); and
e) Close relationships with a neurotypicals (e.g. "She was my sounding board, and my coach in a number of situations.")

Woven throughout these categories are expressions of stress and anxiety. Stress and anxiety (both included in our a priori coding structure) are pervasively, and bi-directionally cross-coded. Over 80% of excerpts coded with either stress or anxiety are also coded in one of the categories listed above, while between 10–30% of excerpts in those categories are also coded with either stress or anxiety.

Therefore, while stress and anxiety remain a separate category, they also serve them as modifying attributes of other categories. For example, one author wrote about her sensory experience with background noise. First saying, "no one in this situation is doing anything out of the ordinary," (*recognition of difference*), then "all the different streams seem to individually assault my sensory system and cut to the heart of my ability to function," (*physiological phenomena*) and "these situations are more potent in the times when things have been or are extra stressful" (*stress*). The addition of stress changes her ability to function with background noise.

When the authors write about communication difficulties with neurotypicals, they often write about the stress it can cause; or when they write about a close, "sounding board" relationship, they also write about how it reduces anxiety. This is not unique to autism of course, mountains of research explore the linkages between stress and anxiety with other aspects of life for all humans [51]. But, researchers have found that autistics report substantially higher levels of anxiety, and stronger stress response related to their anxieties than do neurotypicals [52]; and so, may indicate that anxiety and stress are an important area of consideration for workplace accommodation.

Relating this back to the design challenge of workplace accommodation, this interconnectedness of stress and anxiety with other aspects of autistic experience may be useful as an indicator of design quality. For example, suppose we design an assistive technology to mitigate an autistic worker's sensitivity to office lighting. The purpose of that technology is to reduce the distraction and discomfort caused by the lighting. The ultimate impact of the accommodation is improved performance and longer retention, but a likely immediate impact is a reduction in stress. Several studies indicate a reduced self-awareness of emotion and stress state suggesting that autistic workers may not be able to express or even realize the impact of things like workplace lighting on their ability to work [53–56]. However, advances in wearable sensors such as fitness bands are making the real-time detection of stress response practical; so physiological measures of stress response may be a useful proxy for evaluating work environments and the effectiveness of modifications to accommodate the differences and disabilities of autistic workers[4].

4   Towards Technology-Enabled Prosthetic Environments

Autistic workers can be very focused on their jobs, and an employer who creates the right environment often get superior performance from them [57].

The diversity of experience within autism, and the interconnectedness of stress with environmental factors such as noise suggest that one approach to workplace accommodation is to create reactive and adaptive environments. For the autistic worker environmental factors can become barriers to employment as daunting as a flight of stairs to someone in a wheelchair. Our society has recognized these environmental barriers to the wheelchair mobility, and in response has modified the environment by building ramps, installing automatic doors, and designing accessible spaces. Designers create *prosthetic environments* to accommodate mobility challenges. Such environments are *prosthetic* in that they are long-term interventions used to improve function, they are not therapies or treatments that seek to heal or cure. They are *environments* in that they surround the client, are embedded in, and facilitate interaction with, otherwise inaccessible features of the ambient environment [58].

Lindsley [59] and Holmes [60] describe the design and use of prosthetic environments for cognitive deficits in clinical and residential care facilities. Hart uses the term to describe environments found in families with autistic children and characterizes such environments as an emergent phenomenon. Hart observes that a parent/child team develops a type of "joint embodiment" in which the parent understands and anticipates the child, allowing them to adapt their surroundings and serve as intermediaries between the autistic child and the neurotypical world [61]. These are descriptions of what we might term Human-Enabled Prosthetic Environments, in that, one or more neurotypical humans (e.g. parents, clinicians, caregivers, etc.) must actively and continually participate to create and maintain the prosthetic environment for the autistic. While these human-enabled prosthetic environments may provide one of the best possible environments in which autistic children can develop, these environments are less practical for adults. Just as environmental prosthetics for mobility support autonomy and independence and intentionally help the mobility-impaired become *less* reliant on others, to be widely adopted and successful, prosthetic environments for autistics will need to support autonomy and independence.

---

[4] This of course has ethical implications regarding employer surveillance of workers and individual privacy.

Therefore, I propose that designers, autistic workers and employers can jointly design, create, and use Technology-Enabled Prosthetic Environments (TEPE) to overcome the barriers to autism in the workplace. Based on the functions of human-enabled prosthetic environments, combined with insights gained from studying autistic autobiography, and advances in assistive technologies, TEPE is a design concept for the integration of assistive technologies to support independence and autonomy; *and* to help facilitate better communication between autistics and neurotypicals[5].

TEPE are learning systems of integrated wearable, mobile, and embedded technologies, designed for long-term, continual use to overcome cognitive, physical, and social challenges resulting from autism or other neurologic diversity. TEPE integrate technologies that: (a) continually monitor the client and ambient environment; (b) adapt the ambient environment; (c) provide or modulate sensory inputs for the client; and (d) help the client understand context, meaning, and intent in social situations. Integration of these functions relies on machine-learning algorithms that compare sensor input with historical readings and outcomes, adjusting the local and individual environment or providing suggestions to the user.

For example, many office workers find that wearing headphones helps to block out distracting noises and improve focus. In a TEPE, the system might monitor the soundscape of the workplace along with the stress state of the user, suggesting headphones and providing and individually generated soundscape that reduces physiological indicators of stress. The TEPE could also control local lighting, monitor and buffer phone and messaging notifications that can be come overwhelming and smoothly break the user's focus when something else needs their attention. Extending into the social, the TEPE could incorporate social cue recognition [62, 63] with work-related information about co-workers to improve office communication.

This is not a single, standardized solution to workplace accommodation; rather TEPE is a design concept. TEPE respond and adapt the environment to the diverse and changing needs of users to improve performance, job satisfaction, and quality of life. The next steps, along with the continuation of the autistic autobiography study, are to develop design considerations and principles for TEPE and to begin to build and test prototypes.

References

1. Asperger, H.: Autistic psychopathy. Childhood. Arch. Psychiatr. Nervenkr. **117**, 76–136 (1944)
2. Noguchi, Y.: Autism can be an asset in the workplace, employers and workers find. (2016) https://www.npr.org/sections/health-shots/2016/05/18/478387452/autism-can-be-an-asset-in-the-workplace-employers-and-workers-find
3. Lam, B.: Companies hiring workers with autism - The Atlantic. https://www.theatlantic.com/business/archive/2016/12/autism-workplace/510959/

---

[5] We do not propose that TEPE can (or should) replace supportive relationships between autistics and neurotypicals with technology.

4. Hedley, D., Wilmot, M., Spoor, J., Dissanayake, C.: Benefits of employing people with autism: the dandelion employment program. La Trobe Univ. Sch. Psychol. Public Heal. Olga Tennison Autism Res. Centre. (2017)

5. Carley, M.J.: Unemployed on the Autism Spectrum: How to Cope Productively with the Effects of Unemployment and Jobhunt with Confidence. Jessica Kingsley Publishers, London (2016)

6. Employment - advancing futures for adults with autism. http://www.afaa-us.org/core-issues/employment

7. Müller, E., Schuler, A., Burton, B.A., Yates, G.B.: Meeting the vocational support needs of individuals with asperger syndrome and other autism spectrum disabilities. J. Vocat. Rehabil. 18, 163–175 (2003)

8. Brown, T.: Design thinking. 86, 84–92 + 141 (2008)

9. Brown, T.: Change by Design. HarperCollins, New York, NY, USA (2009)

10. May, M.: Observe first, deisgn second. Rotman Mag. 39–46 (2012)

11. Hacking, I.: Hacking (1995) - The Looping Effects of Human Kinds.pdf

12. American Psychiatric Association: Diagnostic and Statistical Manual of Mental Disorders: DSM-5. Association, American Psychiatric (2013)

13. Silberman, S.: NeuroTribes: The Legacy of Autism and the Future of Neurodiversity by Steve Silberman| Key Takeaways. Eureka Books, Analysis & Review (2015)

14. Richards, C., Jones, C., Groves, L., Moss, J., Oliver, C.: Prevalence of autism spectrum disorder phenomenology in genetic disorders: a systematic review and meta-analysis. Lancet Psychiatry. 2, 909–916 (2015)

15. Hens, K., Peeters, H., Dierickx, K.: The ethics of complexity. Genetics and autism, a literature review. Am. J. Med. Genet. Part B Neuropsychiatr. Genet. 171, 305–316 (2016)

16. Maski, K.P., Jeste, S.S., Spence, S.J.: Common neurological co-morbidities in autism spectrum disorders. Curr. Opin. Pediatr. 23, 609–15 (2011)

17. Abrahams, B.S., Geschwind, D.H.: Advances in autism genetics: on the threshold of a new neurobiology. Nat. Rev. Genet. 9, 341–55 (2008)

18. Lam, K.S.L., Aman, M.G., Arnold, L.E.: Neurochemical correlates of autistic disorder: a review of the literature. Res. Dev. Disabil. 27, 254–289 (2006)

19. Westerveld, M.F., Trembath, D., Shellshear, L., Paynter, J.: A systematic review of the literature on emergent literacy skills of preschool children with autism spectrum disorder. J. Spec. Educ. 50, 37–48 (2016)

20. Goldstein, H.: Communication intervention for children with autism: a review of treatment efficacy. J. Autism Dev. Disord. 32, 373–396 (2002)

21. van der Meer, L.A.J., Rispoli, M.: Communication interventions involving speech-generating devices for children with autism: a review of the literature. Dev. Neurorehabil. 13, 294–306 (2010)

22. Walton, K.M., Ingersoll, B.R.: Improving social skills in adolescents and adults with autism and severe to profound intellectual disability: a review of the literature. J. Autism Dev. Disord. 43, 594–615 (2013)

23. Meadan, H., Stoner, J.B., Angell, M.E.: Review of literature related to the social, emotional, and behavioral adjustment of siblings of individuals with autism spectrum disorder. J. Dev. Phys. Disabil. 22, 83–100 (2010)

24. Flynn, L., Healy, O.: A review of treatments for deficits in social skills and self-help skills in autism spectrum disorder. Res. Autism Spectr. Disord. 6, 431–441 (2012)

25. Smith, H., Milne, E.: Reduced change blindness suggests enhanced attention to detail in individuals with autism. J. Child Psychol. Psychiatry 50, 300–306 (2009)

26. Baron-Cohen, S., Ashwin, E., Ashwin, C., Tavassoli, T., Chakrabarti, B.: Talent in autism: hyper-systemizing, hyper-attention to detail and sensory hypersensitivity. Philos. Trans. R. Soc. Lond. B Biol. Sci. **364**, 1377–83 (2009)

27. Baron-Cohen, S., Leslie, A.M., Leslie, U.: Does the autistic child have a "theory of mind"? Cognition **21**, 37–46 (1985)

28. Frith, U.: Autism: explaining the enigma (1989)

29. Ozonoff, S., Pennington, B.F., Rogers, S.J.: Executive function deficits in high-functioning autistic individuals: relationship to theory of mind. J. Child Psychol. Psychiatry **32**, 1081–1105 (1991)

30. Pennington, B.F., Ozonoff, S.: Executive functions and developmental psychopathology. J. Child Psychol. Psychiatry **37**, 51–87 (1996)

31. Denckla, M.B.: A Theory and model of executive function: A neuropsychological perspective. (1996)

32. Markram, H., Rinaldi, T., Markram, K.: The intense world syndrome–an alternative hypothesis for Autism. Front. Neurosci. **1**, 77–96 (2007)

33. Markram, K., Markram, H.: The intense world theory – a unifying theory of the neurobiology of Autism. Front. Hum. Neurosci. **4**, 224 (2010)

34. Drescher, J.: Out of DSM: Depathologizing homosexuality. Behav. Sci. (Basel, Switzerland) **5**, 565–75 (2015)

35. Jaarsma, P., Welin, S.: Autism as a natural human variation: reflections on the claims of the neurodiversity movement. Heal. Care Anal. **20**, 20–30 (2012)

36. Solomon, O.: Sense and the senses: anthropology and the study of Autism. Annu. Rev. Anthropol. **39**, 241–259 (2010)

37. Baron-Cohen, S.: Editorial Perspective: Neurodiversity – a revolutionary concept for autism and psychiatry. J. Child Psychol. Psychiatry Allied Discip. **58**, 744–747 (2017)

38. Gaugler, T., Klei, L., Sanders, S.J., Bodea, C.A., Goldberg, A.P., Lee, A.B., Mahajan, M., Manaa, D., Pawitan, Y., Reichert, J., Ripke, S., Sandin, S., Sklar, P., Svantesson, O., Reichenberg, A., Hultman, C.M., Devlin, B., Roeder, K., Buxbaum, J.D.: Most genetic risk for Autism resides with common variation. Nat. Genet. **46**, 881–885 (2014)

39. Kinney, D.K., Barch, D.H., Chayka, B., Napoleon, S., Munir, K.M.: Environmental risk factors for autism: do they help cause de novo genetic mutations that contribute to the disorder? Med. Hypotheses **74**, 102–6 (2010)

40. Eigsti, I.: A review of embodiment in autism spectrum disorders. Front. Psychol. **4**, 1–10 (2013)

41. De Jaegher, H., Jaegher, H.De, Leary, M.R., Practice, P.: Embodiment and sense-making in autism. Front. Integr. Neurosci. **7**, 15 (2013)

42. Fraser, H.: Turning design thinking into design doing. In: Martin, R., Christensen, K. (eds.) Rotman on Design, pp. 116–121. University of Toronto Press, Toronto (2013)

43. Sacks, O.: An anthropologist on Mars : seven paradoxical tales. Knopf (1995)

44. Hacking, I.: Autistic autobiography. Philos. Trans. R. Soc. B Biol. Sci. **364**, 1467–1473 (2009)

45. Markham, A., Buchanan, E.: AOIR guidelines: ethical decision making and internet research. Assoc. Internet Res. (2012)

46. Biernacki, P., Waldorf, D.: Snowball sampling: problems and techniques of chain referral sampling. Sociol. Methods Res. **10**, 141–163 (1981)

47. Kedar, I.: Ido in Autismland: climbing out of autism's silent prison (2012)

48. Strauss, A., Corbin, J.: Basics of qualitative research (1990)

49. Gioia, D.A., Corley, K.G., Hamilton, A.L.: Seeking qualitative rigor in inductive research. Organ. Res. Methods. **16**, 15–31 (2013)

50. Tufford, L., Newman, P.: Bracketing in qualitative research. Qual. Soc. Work Res. Pract. **11**, 80–96 (2012)

51. Hofmann, S.G.: The SAGE Encyclopedia of Theory in Psychology Anxiety Disorders (2017)

52. Gillott, A., Standen, P.: Levels of anxiety and sources of stress in adults with autism. J. Intellect. Disabil. **11**, 359–370 (2007)

53. Lopata, C., Volker, M., Putnam, S., Thomeer, M.L., Nida, R.E., J. A.: Undefined: Effect of social familiarity on salivary cortisol and self-reports of social anxiety and stress in children with high functioning autism spectrum disorders. Springer (2008)

54. Hill, E., Berthoz, S., Frith, U.: Undefined: Brief report: Cognitive processing of own emotions in individuals with autistic spectrum disorder and in their relatives. J. Autism Dev Disord. Springer (2004)

55. Moriguchi, Y., Ohnishi, T., Lane, R., Maeda, M., Mori, T.M.: Undefined: impaired self-awareness and theory of mind: an fMRI study of mentalizing in alexithymia. Neuroimage. Elsevier (2006)

56. Lewis, M., Ramsay, D.S.: Stress reactivity and self-recognition. Child Dev. **68**, 621–629 (1997)

57. Grandin, T.: Thinking in Pictures: And Other Reports From My Life With Autism. Doubleday, New York, New York, USA (1995)

58. Lawton, M.: Planning environments for older people. J. Am. Inst, Plann (1970)

59. Lindsley, O.R.: Direct measurement and prosthesis of retarded behavior. J. Educ. **147**, 62–81 (1964)

60. Holmes, D.L.: Community-based services for children and adults with autism: The Eden family of programs. J. Autism Dev. Disord. **20**, 339–351 (1990)

61. Hart, B.: Autism parents & neurodiversity: radical translation, joint embodiment and the prosthetic environment. Biosocieties **9**, 284–303 (2014)

62. Voss, C., Washington, P., Haber, N., Kline, A., Daniels, J., Fazel, A., De, T., McCarthy, B., Feinstein, C., Winograd, T., Wall, D.: Superpower glass: delivering unobtrusive real-time social cues in wearable systems. In: Ubicomp. pp. 1218–1226 (2016)

63. Washington, P., Voss, C., Haber, N., Tanaka, S., Daniels, J., Feinstein, C., Winograd, T., Wall, D.: A wearable social interaction aid for children with Autism. In: Proceedings of the 2016 CHI Conference Extended Abstracts on Human Factors in Computing Systems - CHI EA'16. pp. 2348–2354 (2016)

64. ASAN Position Statemetns. http://autisticadvocacy.org/home/policy-center/position-statements/

65. Hacking, I.: How we have been learning to talk about Autism: a role for stories. Metaphilosophy **40**, 499–516 (2009)

Biomechanical Models of Computerized Prosthetic Leg

Nantakrit Yodpijit[1]([⊠]), Manutchanok Jongprasithporn[2],
Penpetch Maneewong[1], Nanthanit Faksang[1],
and Teppakorn Sittiwanchai[1]

[1] Center for Innovation in Human Factors Engineering and Ergonomics,
Department of Industrial Engineering, Faculty of Engineering,
King Mongkut's University of Technology North Bangkok, Bangkok, Thailand
nantakrit.y@eng.kmutnb.ac.th, nantakrit@gmail.com
[2] Department of Industrial Engineering, Faculty of Engineering,
King Mongkut's Institute of Technology Ladkrabang, Bangkok, Thailand

Abstract. Computer-controlled prosthetic devices are becoming more popular among people of all ages from all over the world suffer amputations. Only 5% of them in need have access to assistive products. This paper presents an analysis and an assessment of biomechanics of physiological gait and the construction of prosthetic leg. The normal gait patterns are performed to develop dynamic biomechanical models of the computerized prosthetic leg for productive simulation of a transfemoral amputee's gait under different conditions. The purposes of the current research project are to provide and modify biomechanical models for a better understanding of amputee gait and make changes on prosthetic leg components for improving the ambulatory performance. Three-dimensional finite element analysis and Solidworks Simulation are used to investigate kinetic and kinematic characteristics of gait in patients with limb loss. Findings from the current study reveal that most people with lower limb amputations prefer computerized prosthetic leg to passive prosthetic leg.

Keywords: Biomechanics · Biomechanical models · Prosthetic leg
Above knee prostheses

1 Introduction

Walking is the move from one place to another to carry out activities. Leg is a vital organ in walking. The survey of disability of Thai people throughout National Statistical Office in 2012 found that Thailand had a total of 25,000 disabled people, representing 2.3% of the total disabled population [1]. And there is a growing trend. Mostly caused by traffic accidents. Stomp on the bomb Chronic wounds from diabetes And congenital abnormalities Thailand has a tendency to use higher prosthetics as well. The prosthetics is a type of prosthetic device that is made to replace lost organs. People with disabilities are required to use their prostheses to carry out daily activities and to provide them with a higher level of self-help to improve their quality of life and quality of life. As a result, the domestic economy is well-propelled. Therefore, the quality of

© Springer International Publishing AG, part of Springer Nature 2019
T. Z. Ahram and C. Falcão (Eds.): AHFE 2018, AISC 794, pp. 982–993, 2019.
https://doi.org/10.1007/978-3-319-94947-5_97

prosthetics can help solve the problems mentioned above, and so that users can safely and comfortably wear them at a lower price than abroad. It will be a way to improve the quality of life for disabled people who lack opportunities in Thailand.

2   Methods

During this research, biomechanical models of the human body will be necessary. They will be used for theoretical evaluation and compare the unimpaired and impaired ankle and knee joint, for simulations of the leg dynamics and for develop the implementation of an adaptive control for computerized prosthetic leg. The biomechanical model analysis and inverse dynamics of leg and foot segments model during gait cycle. The segments are connected by two hinge joints representing ankle and knee. Mass, inertia, Center of Mass (CoM) and length of each segment are defined using anthropometric data [3, 6, 7]. All segments direction completely described in two dimensions (sagittal plane).

2.1   Walking Theories

Walking has two basic components [12]
   1. The movement of the foot. From one position to another. There is another foot on the ground.
   2. The reaction force from the foot area must be sufficient. To support the body weight
   Walking is a natural evolution. Gait Cycle is divided into 2 stages. As follows

   1. Stance phase: 60% of walking distance. Starting from the interval Heel strike 0% Toe off 60% Include
   - Heel strike is short-lived. From the foot touch the ground. And the first phase of the foot touches the ground.
   - Foot flat to transfer weight to the foot.
   - Mid-stance is the area where only one feet are touching the ground. And weight will be taken to one leg.
   - Heel off is the start when the heel is raised from the ground.
   - Toe off is the toes are raised from the floor.
   2. Swing phase: the average duration is 40% of the walking cycle. From the Toe off to 60% to 100% heel strike. Include.
   - Initial swing (Initial swing) is the initial stage of the foot does not touch the ground.
   - The mid-range of the swing (Mid swing) is the point where the foot is drawn to the bottom of the body
   - The end of the foot swing (Terminal swing) is the moment the forward foot swing forward. Muscles are slowed down. Enter the foot phase (Stance phase) to start the cycle again (Fig. 1).

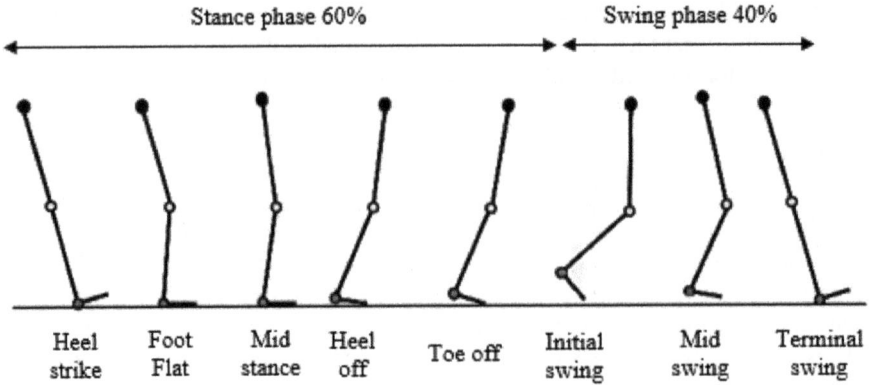

Fig. 1. The normal gait cycle.

2.2   Reference Coordinate System

In gait analysis. A reference coordinate system is required to obtain the desired parameter. The coordinate system used for gait analysis has two types. Global coordinate system is the main coordinate system used to refer to the data of all environments. And Local coordinate system is the coordinate system of each segment in the body. The two types of coordinate systems can be used to calculate different values, such as power moment and various walking parameters (Fig. 2).

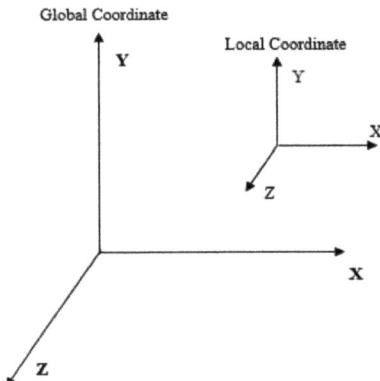

Fig. 2. Coordinate systems

2.3   Data Used to Calculated

Sample Weight. Size Thai is a standard size for the shape of the Thai people, especially from the body survey with 3D Body Scanning technology with the sample of both men and women aged 16 years [4]. In this research, the weight of Thai male in technical university students used the average weight of males at aged 16–25 years. Calculate the mass of each leg part to represent the value of the biomechanical calculation. The mass of each 7 part of the leg use data from the determination of body segment masses and centers of mass [6].

Anthropometry Data. The distance of the center of mass of the body used data from the research Determination of body segment masses and centers of mass. Kitti [5], the length of each segment is based on Anthropometry research of Thai Technical University students. It is a research on measurement two hundred male and two hundred female students of King Mongkut's Institute of Technology North Bangkok. A set of 36 body dimensions based on Pheasant (1990) [7]. The mass and length of each segment and center of mass of prosthetic leg can be found from Solidworks by using the command Mass Properties we will know the mass and position of center of mass of prosthetic and measure distance from proximal and distal point to the center of mass. It is really hard to know the mass and center of mass at thigh segment of the above knee amputation. In this study the hip mass and center of mass values were the same for both human and prosthetics models (Table 1).

Table 1. Anthropometry data of thai male and prosthetic leg

Segment	Length (cm)	Thai male University student			Prosthetic leg model		
		Mass (kg)	Center of mass (cm)		Mass(kg)	Center of mass (cm)	
			Proximal	Distal		Proximal	Distal
Upper leg	48.14	6.57	21.21	26.92	6.57	21.21	26.92
Lower leg	43.94	3.25	18.79	25.15	3.99	15.72	28.22
Foot	25.63	1.00	10.39	15.24	0.341	11.23	16.4

Kinematics Data. Kinematics is that area of mechanics which describes the motion of a body without considering the forces causing the motion [3]. Kinematics variables include linear and angular displacement, velocity, and acceleration. Velocity is the time rate of change in displacement. Acceleration is time rate of change in velocity. Kinematics profiles of joint centers as well as centers of segment masses are common in biomachanical analyses. Kinematics data in this research use acceleration and angular acceleration from The Biomechanics and Motor Control of Human Gait [2] (Fig. 3).

Kinetics Data. Information about the forces. The data used in this study consists of ground reaction force. This is a force that occurs between the foot of the Force platform from the book Biomechanics and Motor Control of Human Gait [2] and the moment of inertia data can be found in Dempster [3]. Data are presented as radius of gyration data as a function of segment length (Fig. 4).

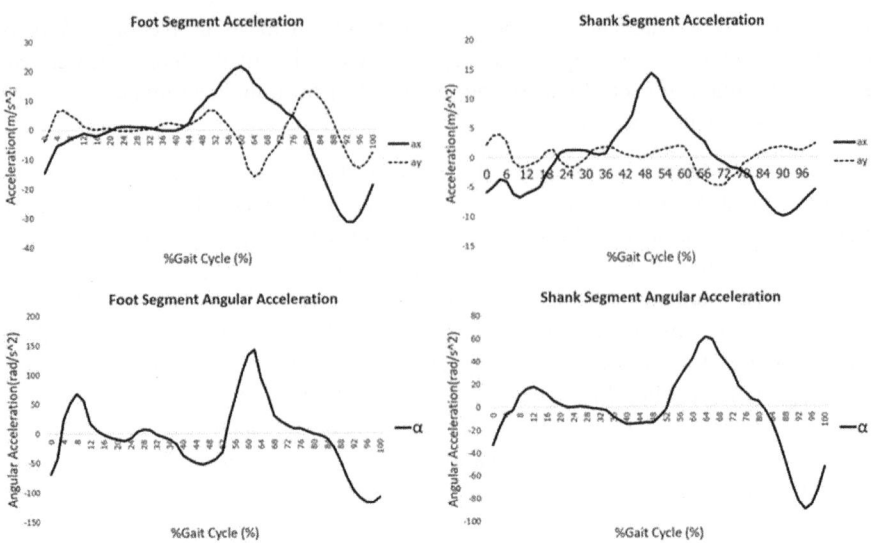

Fig. 3. Kinematics data

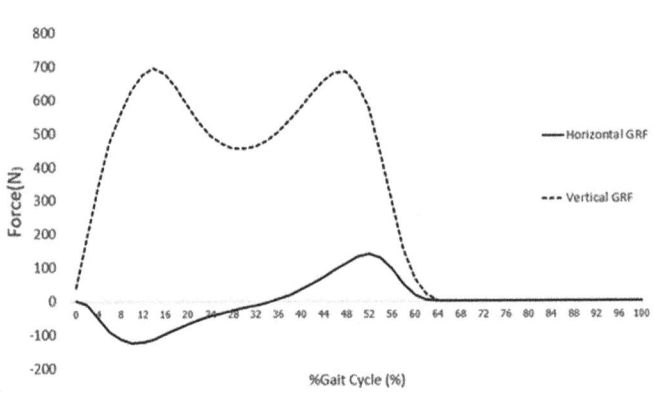

Fig. 4. Ground reaction force

2.4 Calculation of Knee and Ankle Forces and Moment of Joint Using Biomechanical Analysis Models

To calculate we use the human model as shown in Fig. 4. Under the assumption that the human gait is approximated by the motion on the sagittal plane, we consider only sagittal plane. The human model consists of three links, that is, Foot Link, Shank Link and Thigh Link (Fig. 5).

The reaction force on joint in the horizontal axis is calculated by $\sum F_x = ma_x$ is given below.

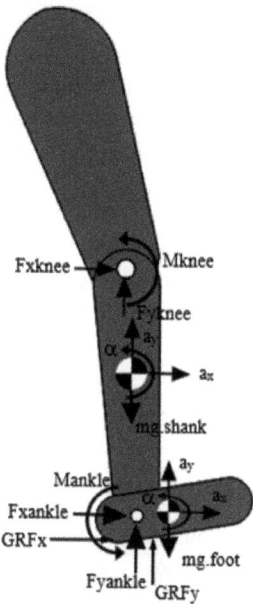

Fig. 5. Biomechanic model

$$F_{xknee} = F_{xankle} + m_{knee}a_{xknee} \tag{1}$$

The reaction force on joint in the vertical axis is calculated by $\sum F_y = ma_y$ is given below.

$$F_{yknee} = F_{yankle} + m_{knee}g + m_{knee}a_{yknee} \tag{2}$$

The moment of joint is calculated by $\sum M_{cm} = I_{cm}\alpha$ is given below.

$$
\begin{aligned}
M_{knee} = {} & M_{ankle} + I_{CM}\alpha - R_{yankle}d_1(\sin\theta) + R_{xankle}d_1(\cos\theta) \\
& - R_{yknee}d_2(\sin\theta) + R_{xknee}d_2(\cos\theta)
\end{aligned} \tag{3}
$$

Assumptions

- Centers of mass remain constant and can be represented by single points.
- Centers of pressure occur at 1/3 of the foot length from the proximal point during heel contact to heel off and occur at 1/3 of the foot length from the distal point during heel off to toe off.

Use excel to create formulas for calculating the force and moment occurring at the ankle and knee to calculate all walking ranges and then plot the values obtained.

3   Results

This project we simulate and calculate the forces and moments that occur while walking in between the knee and ankle prosthetic leg above the knee compared to the human body, using the information of Thailand. The samples used in this study were Thai Technical University students. To study the differences between human legs and prosthetics and to improve the prosthetic prototype. Comparison table for ankle and knee moment between biomechanics model of human leg and prosthetics leg to see the difference of the moment at each stage of walking consist of heel contact, foot flat, mid-stance, heel off, toe off and mid-swing as shown in Table 2 (Figs. 6, 7, 8, 9, 10 and 11).

Table 2. Compare moment between human leg and prosthetics leg.

Segment	Model	Moment (Nm.)					
		Stance phase					Swing phase
		Heel contact	Foot flat	Mid-stance	Heel off	Toe off	Mid swing
Ankle	Human	−3.65	−49.68	−31.95	−83.29	−0.02	1.59
	Prosthetic	−3.26	−52.99	−38.40	−118.65	−3.29	0.69
Knee	Human	−18.07	16.11	14.46	−33.29	13.09	−8.96
	Prosthetic	−14.01	13.472	8.32	−67.36	5.09	−6.13
Segment	Model	Moment (Nm.)					
		Stance phase					Swing phase
		Heel contact	Foot flat	Mid-stance	Heel off	Toe off	Mid swing
Ankle	Human	−3.65	−49.68	−31.95	−83.29	−0.02	1.59
	Prosthetic	−3.26	−52.99	−38.40	−118.65	−3.29	0.69
Knee	Human	−18.07	16.11	14.46	−33.29	13.09	−8.96
	Prosthetic	−14.01	13.472	8.32	−67.36	5.09	−6.13
Segment	Model	Moment (Nm.)					
		Stance phase					Swing phase
		Heel contact	Foot flat	Mid-stance	Heel off	Toe off	Mid swing
Ankle	Human	−3.65	−49.68	−31.95	−83.29	−0.02	1.59
	Prosthetic	−3.26	−52.99	−38.40	−118.65	−3.29	0.69
Knee	Human	−18.07	16.11	14.46	−33.29	13.09	−8.96
	Prosthetic	−14.01	13.472	8.32	−67.36	5.09	−6.13

The stress contour plot and analysis results of the above knee prosthetic are presented in Fig. 12. use peak force that occur in gait cycle that is 656.32 N compressive from top and 691.58 N from bottom of model. Maximum stresses in the parts for the

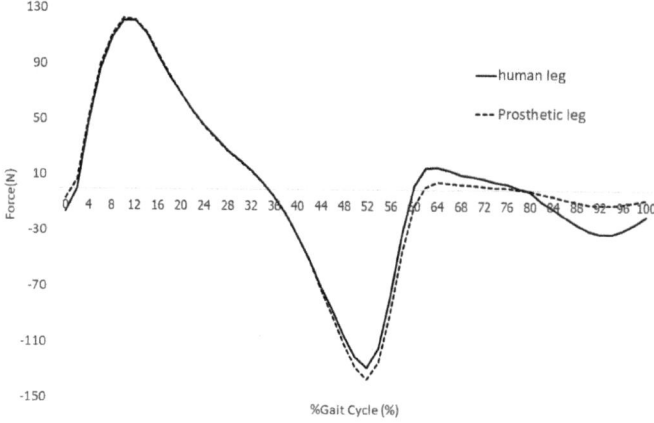

Fig. 6. Horizontal ankle force

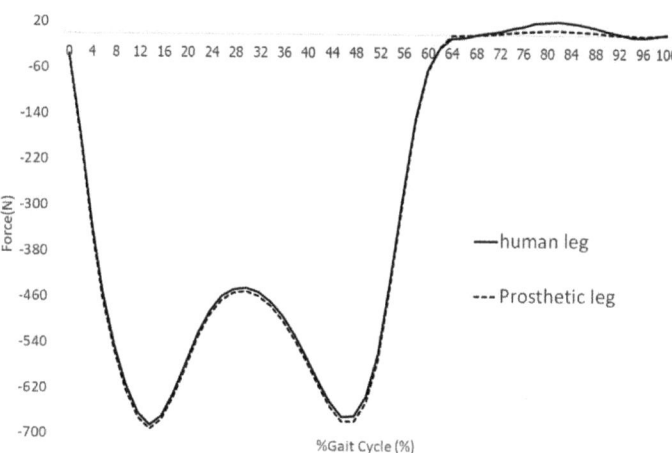

Fig. 7. Vertical ankle force

conditions analyzed are 818.9 MPa as show in Fig. 12c. Based on the stress results, the material selected for the structural parts was the Titanium Ti-6Al-4 V (hardness HB334, 880 MPa yielding limit, 950 MPa ultimate stress limit).

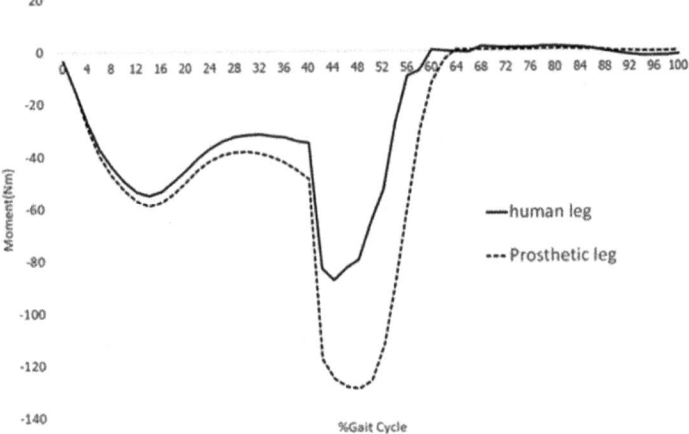

Fig. 8. Ankle Joint Moment of Force

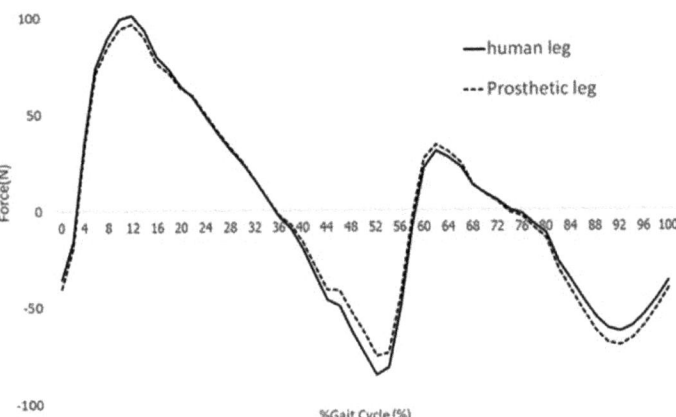

Fig. 9. Horizontal knee force

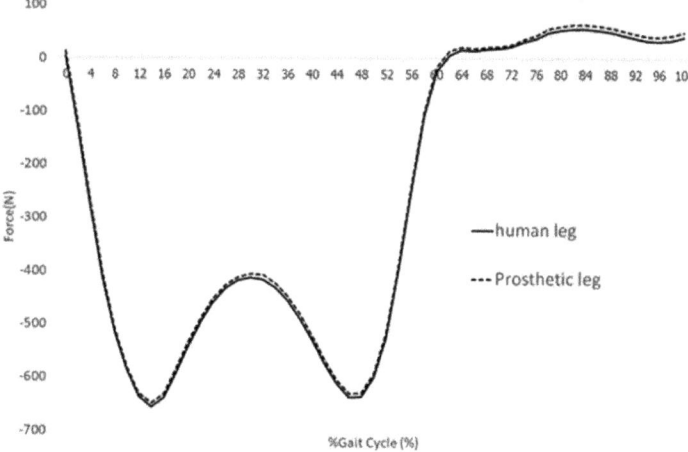

Fig. 10. Vertical knee force

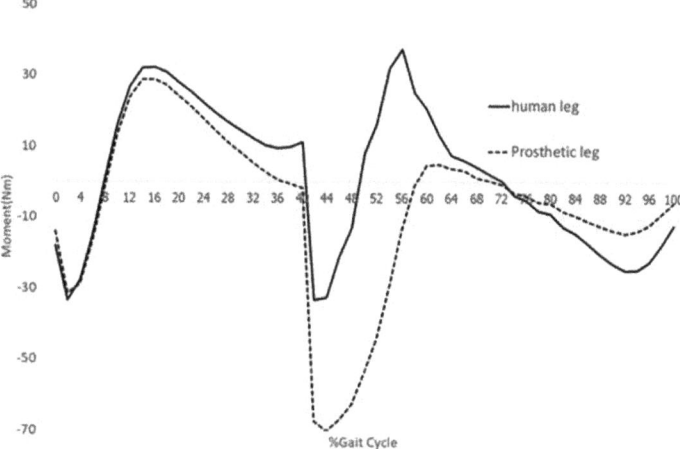

Fig. 11. Knee joint moment of force

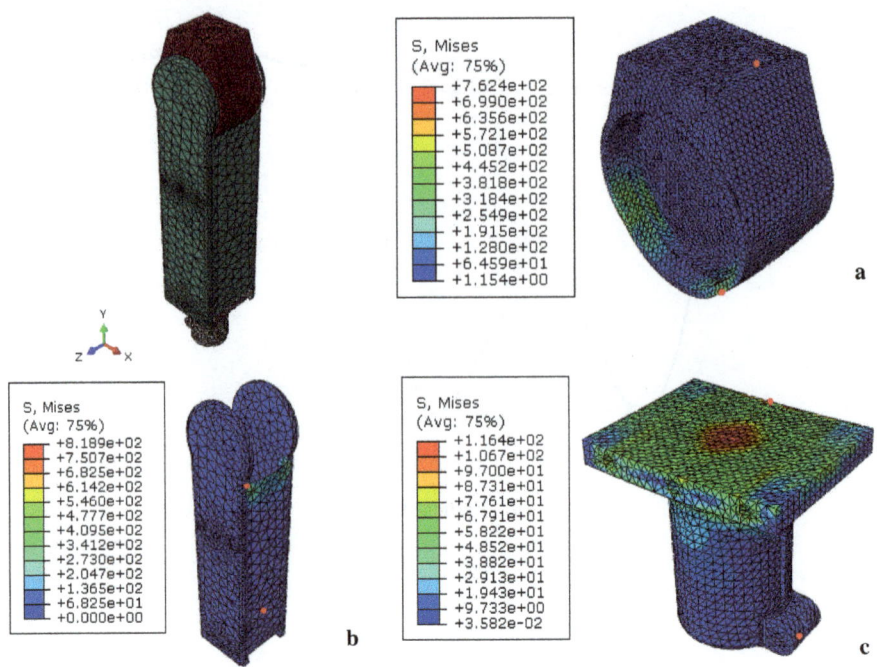

Fig. 12. Stress contour plot maximum and minimum stress at the (a) upper body, (b) body frame, (c) lower body

4 Conclusions and Discussions

The force and moment of the prosthetic leg compared to the human leg obtained by this research is based on the biomechanical model and uses the human body data of Thai people to calculate in. Know the difference of force in the prosthetic leg and human leg. In the horizontal and vertical forces, the appearance and size of the forces are not significantly different because relevant factors are only mass. But in terms of the moment that occurs in each phase the difference is obvious. The factors involved are as follows different angles, moment of inertia and the position of the center of mass. By the first factor, the angles are most noticeable in the moment of the foot. Artificial feet can not bend up and down like human legs, resulting in varying degrees of difference in feet and in the model of the prosthesis, we looked at the foot flat to the floor (degrees between feet to the floor is 0) but in the human foot the angle between foot and floor is not flat enough the plane of foot and floor is about 30degrees, next factors of different size of moment inertia because the different shape of the legs and the prosthetics. Finally the position of the center of mass it make moment arm changes during human model and prosthetics model.

Limitations and Future Work
The data used on this research is old data and the values used in the calculation are the data from other researches, does not include actual data, it may result in discrepancies

in the results. However, the quality of life of prosthetics users in Thailand can be improved and the prostheses are being designed and developed closer to or better than the legs. And will reduce imports of expensive prosthetics from foreign countries. This is to improve the quality of life for disabled people in Thailand. In the future when the prototype of the prosthesis is created. The data used in the calculation can be found in the Kinematic study to determine the true pattern of prosthetic movements. The results of this research can be disseminated and transferred to interested persons and research methods for those who want to study this field.

Acknowledgments. This research is funded by The Faculty of Engineering of KMUTNB and The National Research Council of Thailand (NRCT). Authors would like to thank the CIHFE2 staffs at KMUTNB for helping us in running experiments and providing invaluable technical support on data collection and analysis in this research project.

References

1. National Statistical Office: Number of persons with disabilities having impairments by type of impairments, age group, sex and area, Whole Kingdom (2012)
2. Winter, D.A.: The Biomechanics and Motor Control of Human Gait, pp. 17–41 (1991)
3. Tayyari, F., Smith, J.L.: Occupational Ergonomics Principles and Applications, pp. 40–91 (1997)
4. National Electronics and Computer Technology Center: http://www.sizethailand.org/region_all.html
5. Kitti, I.: Ergonomics. **2548**, 64–89
6. Kitti, I.: Determination of body segment masses and centers of mass (2000)
7. Yodijit, N., Bunterngchit, Y.: Anthropometry of Thai technical university students
8. Bunterngchit, Y., Pimsakul, K.: Anthropometric study of Thai male industrial workers (2006)
9. Hirata, Y., Iwano, T., Tajika, M., Kosuge, K.: Motion control of wearable walking support system with accelerometer based on human model (2008)
10. Chaffin, D.B., Andersson, G.B.J., Martin, B.J.: Occupational Biomechanics. 109–153
11. Vaughan, C.L., Davis, B.L., O'Connor, J.C.: Dynamics of Human Gait. Champaign, Ill: Human Kinetics Publishers
12. Rose, J., Gamble, J.G.: Human Walking (3rd ed.)
13. Obreg, K.: Knee mechnisms for through-knee prostheses
14. Lee, W.C.C., Zhang, M.: Design of monolimb using finite element modeling and statistics-based Taguchi method. Clin. Biomech
15. Harrington, I.J.: A bioengineering analysis of force actions at the knee in normal and pathological gait
16. Mich P. Greene, B.S., M.E., C.P.O., Four Ba

Human-Centered Design of Computerized Prosthetic Leg: A Questionnaire Survey for User Needs Assessment

Nantakrit Yodpijit[1(✉)], Manutchanok Jongprasithporn[2],
Uttapon Khawnuan[1], Teppakorn Sittiwanchai[1],
and Juthamas Siriwatsopon[3]

[1] Department of Industrial Engineering, Faculty of Engineering, Center for Innovation in Human Factors Engineering and Ergonomics, King Mongkut's University of Technology North Bangkok, Bangkok, Thailand
nantakrit.y@eng.kmutnb.ac.th, nantakrit@gmail.com
[2] Department of Industrial Engineering, King Mongkut's Institute of Technology Ladkrabang, Bangkok, Thailand
[3] Sirindhorn School of Prosthetics and Orthotics, Faculty of Medicine Siriraj Hospital, Mahidol University, Bangkok, Thailand

Abstract. In Thailand, there are approximately 2 million people with disabilities and nearly 50,000 people living with lower limb loss. Previous research investigates the quality of life and factors affecting quality of life of transfemoral and transtibial amputees after receiving prosthesis using a method of WHOQOL – BREF – THAI. However, the WHOQOL – BREF – THAI is a questionnaire that is used to investigate quality of life for normal people not for people living with lower limb loss. Recent records show a lack of satisfaction with the traditional passive prosthetic leg and needs of new active prosthesis with different motor functions. The objective of this research is to design and develop a questionnaire, using human-centered design principles for investigating and improving the quality of life of transfemoral and transtibial amputees after receiving prostheses. The new questionnaire is created based upon both prosthesis evaluation questionnaire (PEQ) and trinity amputation and prosthesis experience scales (TAPES). This questionnaire survey explores user's satisfaction, usability study, product appearance, comfort and pain, and cleansing and handling the prosthesis. A total of 24 subjects are randomly selected. Data collection and analysis are made from a list of patients in Sirindhorn School of Prosthetics and Orthotics and Veterans General Hospital in Thailand. From the amputees' point of view, defining their needs is one of the most critical factors in the first step for prosthesis design. Veterans has asked for modern prostheses and they believe that modern prostheses can help them feel more comfortable and natural while performing activities of daily living and improve their quality of life. As a result, this research project provides a list of needs of people living with lower limb loss in Thailand. Findings have indicated that user needs assessment is necessary and critical to make a better computerized lower limb prosthesis design and improve user's satisfaction.

Keywords: Human-centered design · Questionnaire
Computerized prosthetic leg · User needs assessment · Prosthesis design

© Springer International Publishing AG, part of Springer Nature 2019
T. Z. Ahram and C. Falcão (Eds.): AHFE 2018, AISC 794, pp. 994–1005, 2019.
https://doi.org/10.1007/978-3-319-94947-5_98

1 Introduction

Statistics reveal that around 40 million people in the world are disabled and only 5% of them in need have access to assistive products [1]. The most common causes of amputation are congenital deformities, vascular diseases, diabetes, and accidents [2]. In Thailand, there are about 2 million people with disabilities. Of these, 50,000 people living with lower extremity lost [3]. The lower limb amputees are less capable of performing activities of daily living (ADL) in comparison with healthy people and has a tremendous psychological impact [4].

A Prosthesis is an assistive product that developed to substitute the missing part of body which has lost its functional and cosmetic for the amputee. Despite advances in prosthetic innovations, lower-limb amputees have to re-practice elementary ambulatory skills to archive function within the community, and some amputees still unable to use their prosthesis to meet their demands [5]. The traditional prostheses are mostly passive products with some of limitations and drawbacks. It has been found that a trend in the utilization of active actuators/prostheses can overcome some of limitations and drawbacks of traditional prostheses [6–10].

Most of powered prosthetic are designed to meet the biomechanical functions by focus on the technical approach, such as real-time intent recognitions [11–13], control strategies [14–19], reduction of the torque requirements [20], and minimize energy consumption [21]. Recent research paper reveal the human–machine-centered design method approach for powered prosthetic development, which consider both technical factors and human factors [22], result in a distinct change of technical requirement priorities that lead to completely different prosthetic designs. Studies show the number of non-technical factors involved in designing prosthetics that bringing about enormous psychological impacts [4].

In Thailand, Previous research investigates the quality of life and factors affecting quality of life of transfemoral and transtibial amputees after receiving prosthesis using a method of WHOQOL – BREF – THAI [23]. However, the WHOQOL – BREF – THAI is a questionnaire that is used to investigate quality of life for normal people not for people living with lower limb loss. Recent records show a lack of satisfaction with the traditional passive prosthetic leg and needs of new active prosthesis with different motor functions.

The objective of this research is to design and develop a questionnaire, using human-centered design principles for investigating and improving the quality of life of transfemoral and transtibial amputees after receiving prostheses. The new questionnaire is created based upon both prosthesis evaluation questionnaire (PEQ) [24, 25] and trinity amputation and prosthesis experience scales (TAPES) [26]. The custom-built questionnaire focus on transfemoral and transtibial amputees' satisfaction, usability study, product appearance, comfort and pain, and cleansing and handling the prosthesis. This research project provides a list of needs of people living with lower limb loss in Thailand.

2 Methods

The methods of this research is an approach for exploring and understanding the meaning transfemoral and transtibial amputees ascribe to quality of life or prosthesis problem. The process of this research involves design and development a questionnaire, using human-centered design principles for investigating and improving the quality of life of transfemoral and transtibial amputees after receiving prostheses. Data were corrected by this questionnaire. Data collection were analyzed by using statistical that provides a more complete understanding of user needs. Then the researcher interpret survey responses to better meet the needs that is thus important for both transtibial and transfemoral [27].

2.1 Participants

A total of 30 subjects are randomly selected. Participants were recruited from a list of patients in Veterans General Hospital in Thailand. The 24 subjects were included in this study (mean age 56.17 ± 14.87 years). There were two groups of participants in this study. The first group was 11 transfemoral participants, 11 male and female. Mean age was 48.27 years. Participants in the second group were 13 transtibial participants, 11 male and 2 female. Mean age was 63.41 years. Table 1 summarizes show the activity classes of all participants (* have one missing value).

Table 1. Frequencies of activity classes of the participants

	Activity class				
	AK0	AK1	AK2	AK3	AK4
Frequencies	1	0	2	15	5

*Have one missing value

2.2 Instruments

The new questionnaire (Table 2) is created based upon both prosthesis evaluation questionnaire (PEQ) [25] and trinity amputation and prosthesis experience scales (TAPES) [26]. This questionnaire survey explores user's satisfaction, usability study, product appearance, comfort and pain, and cleansing and handling the prosthesis. This questionnaire is composed of 72 items with seven factors: Factor 1, the appearance of the prosthesis (items 1–8). Factor 2, Safety in wearing the prosthesis (items 9–19). Factor 3, Safety in activities of daily living (items 20–35). Factor 4, Proficiency in activities of daily living (items 36–46). Factor 5, Integration between body and prosthesis (items 47–54). Factor 6, Social acceptance (items 55–62). Finally, Factor 7, Desirable prosthetics (items 63–68) which adapt from [22, 28]. Questionnaire using a Likert-type response format with 5 response options (1 = strongly disagree; 5 = strongly agree) [29].

Table 2. Questionnaire

Item	Strongly agree	Agree	Neither agree nor disagree	Disagree	Strongly disagree
Factor 1: The appearance of the prosthesis (items 1–8)					
1. How often your prosthesis broken?	[5]	[4]	[3]	[2]	[1]
2. How often is your prosthesis making noises (squeaking, clicking, etc.)?	[5]	[4]	[3]	[2]	[1]
3. Do you feel that your prosthesis not strong?	[5]	[4]	[3]	[2]	[1]
4. Do you have barrier for the activities because the weight of the prosthesis?	[5]	[4]	[3]	[2]	[1]
5. Do you have barrier for the activities because the size of the prosthesis?	[5]	[4]	[3]	[2]	[1]
6. Are you satisfied with the look of your prosthesis (e.g., colors, shape, softness, hardness, beauty etc.)?	[5]	[4]	[3]	[2]	[1]
7. Do you feel that your prosthesis overpriced?	[5]	[4]	[3]	[2]	[1]
8. Do you feel comfortable with using a prosthesis?	[5]	[4]	[3]	[2]	[1]
Factor 2: Safety in wearing the prosthesis (items 9–19)					
While you wearing your prosthesis…					
9. You can easy wear socket	[5]	[4]	[3]	[2]	[1]
10. Your socket is often not fit	[5]	[4]	[3]	[2]	[1]
11. Your socket fit to your stump	[5]	[4]	[3]	[2]	[1]
12. Your stump sweat	[5]	[4]	[3]	[2]	[1]
13. Your stump have blisters	[5]	[4]	[3]	[2]	[1]
14. Your stump have pressure marks	[5]	[4]	[3]	[2]	[1]
15. Your stump have swellings	[5]	[4]	[3]	[2]	[1]
16. You have pain at the hip on the side of the amputation	[5]	[4]	[3]	[2]	[1]
17. You have pain at the hip of the sound side, because of the prosthesis	[5]	[4]	[3]	[2]	[1]

(*continued*)

Table 2. (*continued*)

Item	Strongly agree	Agree	Neither agree nor disagree	Disagree	Strongly disagree
18. You have pain at the leg of the sound side	[5]	[4]	[3]	[2]	[1]
19. You have pain at the body, because of the prosthesis	[5]	[4]	[3]	[2]	[1]
Factor 3: Safety in activities of daily living (items 20–35)					
While you wearing your prosthesis...					
20. You can stand confidently without feeling falling	[5]	[4]	[3]	[2]	[1]
21. You feel that both legs are carrying the weight equally	[5]	[4]	[3]	[2]	[1]
22. You can walk on flat ground without stumble	[5]	[4]	[3]	[2]	[1]
23. You walk on the rough ground with safely	[5]	[4]	[3]	[2]	[1]
24. You are concerned about walking on slippery ground	[5]	[4]	[3]	[2]	[1]
25. You are sure to walk up the ramp	[5]	[4]	[3]	[2]	[1]
26. You can walk down the ramp with fluently	[5]	[4]	[3]	[2]	[1]
27. You walk up the stairs with dexterous	[5]	[4]	[3]	[2]	[1]
28. You worry about walking down the stairs	[5]	[4]	[3]	[2]	[1]
29. You can sit comfortably without worrying about falling	[5]	[4]	[3]	[2]	[1]
30. You often stumble from walking with a prosthetic	[5]	[4]	[3]	[2]	[1]
31. You are embarrassed to wear your prosthetic to doing activities	[5]	[4]	[3]	[2]	[1]
32. You are sure to live safely	[5]	[4]	[3]	[2]	[1]
33. You need to exert more at your stump while walking	[5]	[4]	[3]	[2]	[1]
34. You need to exert more at your hips while walking with your prosthesis	[5]	[4]	[3]	[2]	[1]
35. You feel tired while walking with your prosthesis	[5]	[4]	[3]	[2]	[1]

(*continued*)

Table 2. (*continued*)

Item	Strongly agree	Agree	Neither agree nor disagree	Disagree	Strongly disagree
Factor 4: Proficiency in activities of daily living (items 36–46)					
While you wearing your prosthesis, You can change the motion and specify the transition level….					
36. Transition from sitting to standing	[5]	[4]	[3]	[2]	[1]
37. Transition from standing to sitting	[5]	[4]	[3]	[2]	[1]
38. Transition from walking to standing	[5]	[4]	[3]	[2]	[1]
39. Transition from standing to walking	[5]	[4]	[3]	[2]	[1]
40. Change you gait speed	[5]	[4]	[3]	[2]	[1]
41. Walking on slippery ground. (e.g., wet tile and floor)	[5]	[4]	[3]	[2]	[1]
42. You usually use the prosthesis for the first step	[5]	[4]	[3]	[2]	[1]
43. Walk up the ramp alternately with both feet	[5]	[4]	[3]	[2]	[1]
44. Walk down the ramp alternately with both feet	[5]	[4]	[3]	[2]	[1]
45. Walk up the stairs alternately with both feet	[5]	[4]	[3]	[2]	[1]
46. Walk down the stairs alternately with both feet	[5]	[4]	[3]	[2]	[1]
Factor 5: Integration between body and prosthesis (items 47–54)					
47. You can feel the structure of the rough ground surface through your prosthesis	[5]	[4]	[3]	[2]	[1]
48. You can feel the structure of the ramp surface through your prosthesis	[5]	[4]	[3]	[2]	[1]
49. You can relieve itching at the corresponding part of the body by scratching the prosthesis	[5]	[4]	[3]	[2]	[1]
50. You can use prosthesis to rake object	[5]	[4]	[3]	[2]	[1]
51. You can control the prosthesis as needed	[5]	[4]	[3]	[2]	[1]
	[5]	[4]	[3]	[2]	[1]

(*continued*)

Table 2. (*continued*)

Item	Strongly agree	Agree	Neither agree nor disagree	Disagree	Strongly disagree
52. You can feel the spatial position of your prosthesis without looking at it					
53. You have the feeling that the prosthesis is part of your body	[5]	[4]	[3]	[2]	[1]
54. You have the feeling that the prosthesis can replace lost legs	[5]	[4]	[3]	[2]	[1]
Factor 6: Social acceptance (items 55–62)					
While you wearing your prosthesis, specify the level of following items…					
55. You can comfortably do activities of daily living	[5]	[4]	[3]	[2]	[1]
56. You can choose to your choice of clothing	[5]	[4]	[3]	[2]	[1]
57. You can do activities alone	[5]	[4]	[3]	[2]	[1]
58. You can take care of someone else (e.g. your partner, a child, or a friend)	[5]	[4]	[3]	[2]	[1]
59. You can do activities together as someone else	[5]	[4]	[3]	[2]	[1]
60. You can do activities as like the normal people	[5]	[4]	[3]	[2]	[1]
61. You feel alienated when you do activities together as someone else	[5]	[4]	[3]	[2]	[1]
62. You are comfortable with prosthesis	[5]	[4]	[3]	[2]	[1]
Factor 7: Desirable prosthetics (items 63–68)					
63. You need prosthetics that moves more easily	[5]	[4]	[3]	[2]	[1]
64. You need prosthetics that make you less exert in movement	[5]	[4]	[3]	[2]	[1]
65. You need a prosthetic leg that can adjusts the walking speed itself	[5]	[4]	[3]	[2]	[1]
66. You need a prosthetic leg that can adjusts the knee resistance itself	[5]	[4]	[3]	[2]	[1]

(*continued*)

Table 2. (*continued*)

Item	Strongly agree	Agree	Neither agree nor disagree	Disagree	Strongly disagree
67. You need prosthetics that is available all day without charging	[5]	[4]	[3]	[2]	[1]
68. You need a prosthetic leg that you can choose materials and colors	[5]	[4]	[3]	[2]	[1]
69. You need a more effective socket	[5]	[4]	[3]	[2]	[1]
70. You can do activities more easily, if you have a computerized prosthetic leg	[5]	[4]	[3]	[2]	[1]
71. You will be more satisfied, if you have a computerized prosthetic leg	[5]	[4]	[3]	[2]	[1]
72. You will be more confident and social, if you have a computerized prosthetic leg	[5]	[4]	[3]	[2]	[1]

2.3   Statistical Analysis

The data were analyzed with the statistical methods for the behavioral sciences [30]. T-test for equality of means were used to compare the differences between transtibial and transfemoral amputees. T-test for equality of means is used to compares the means of transfemoral and transtibial amputees in order to define that there is statistical evidence that transfemoral and transtibial amputees means are significantly different. Skewness and Kurtosis were used to indices reflect an acceptable degree of normality. Skewness is a measure of a dataset's symmetry, or lack of symmetry. A data set or distribution are symmetric, data set or distribution look the same to the right and left of the center point. Kurtosis is a measure of whether the data are fat tails and a low relative to a normal distribution.

3   Results

3.1   Descriptive Statistics

Table 3 shows the descriptive statistics including means, standard deviations, Skewness and Kurtosis of the 7 factors of the quality of life of transfemoral and transtibial amputees after receiving prostheses. It reveals that participants lack of satisfaction in safety in activities of daily living (M = 3.2588), proficiency in activities of daily living (M = 3.2711), safety in wearing the prosthesis (M = 3.3786) and appearance of the

Table 3. Descriptive statistics of each factors of the quality of life of transfemoral and transtibial amputees

Factors	N	Mean	Std. deviation	Skewness		Kurtosis	
				Statistic	Std. error	Statistic	Std. error
F1	24	3.4764	0.63813	−0.204	0.472	−1.028	0.918
F2	24	3.3786	0.44547	0.019	0.472	−0.717	0.918
F3	24	3.2588	0.55984	0.123	0.472	−0.325	0.918
F4	23	3.2711	0.63832	0.479	0.481	0.067	0.935
F5	24	4.0097	0.53281	−0.174	0.472	1.227	0.918
F6	23	3.5885	0.71059	−0.740	0.481	1.130	0.935
F7	23	4.3219	0.74319	−1.140	0.481	0.391	0.935
F1_6	24	3.4963	0.34530	−0.095	0.472	−0.271	0.918

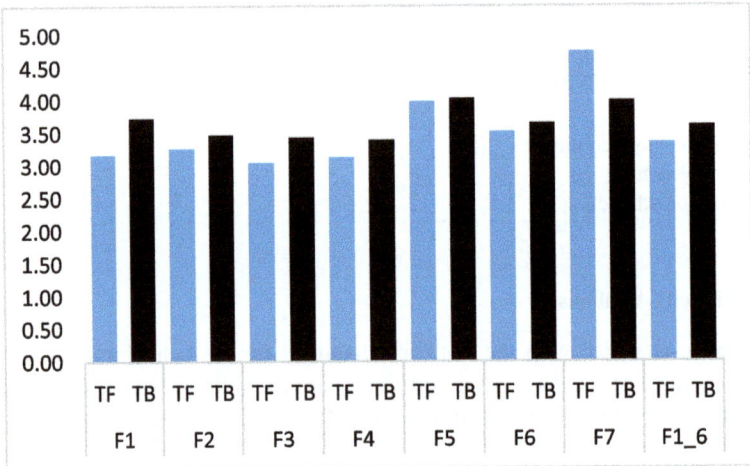

Fig. 1. Satisfaction ratings between two different groups N = 11 transfermoral and N = 13 transtibial of amputees

prosthesis (M = 3.4764). Furthermore, the result demonstrated that the data in this research were normally distributed based on the degrees of Skewness and Kurtosis because both reflect an acceptable degree of normality [31].

Mean of satisfaction ratings (Fig. 1) are demonstrated in participants with different types of amputation (transfemoral and transtibial).

3.2   Difference Between Transfemoral and Transtibial Amputees

In Table 4, No significant differences were found in safety in wearing the prosthesis (F2) (p-value > 0.05), proficiency in activities of daily living (F4) (p-value > 0.05), Integration between body and prosthesis (F5) (p-value > 0.05) and Social acceptance (F6) (p-value > 0.05) between the two groups. Furthermore, significant differences were found in the appearance of the prosthesis (F1) (p-value < 0.05), Safety in activities of

Table 4. Difference between transfemoral and transtibial amputees

Factors	TF		TB		p-value	95% CI difference	
	M	SD	M	SD		Lower	Upper
F1	3.1759	0.68043	3.7308	0.49172	0.030	−1.052	−0.0578
F2	3.2678	0.44892	3.4723	0.43775	0.272	−0.5808	0.17169
F3	3.0511	0.41554	3.4344	0.61958	0.095	−0.839	0.07255
F4	3.1364	0.60405	3.3947	0.66962	0.344	−0.8132	0.29657
F5	3.9870	0.42739	4.0288	0.62532	0.853	−0.5043	0.42065
F6	3.5227	0.54146	3.6488	0.85736	0.681	−0.7548	0.50269
F7	4.7489	0.35305	3.9934	0.80639	0.012	0.18526	1.32577
F1_6	3.3568	0.26898	3.6144	0.36781	0.067	−0.53509	0.01990

daily living (F3) (p-value < 0.05), Desirable prosthetics (F7) (p-value < 0.05) and F 1_6 (p-value < 0.05) between the two groups. Table 4 show t test for equality of means between transfemoral and transtibial amputees.

4 Discussions and Conclusions

In this study, respecting research criteria, the findings have indicated that a lack of satisfaction in the Safety in activities of daily living, Proficiency in activities of daily living, appearance of the prosthesis and Safety in wearing the prosthesis. Transfemoral amputees achieved significantly higher demands on the prosthesis satisfaction than transtibial amputees. These findings differ from the literature, which shows that transfemoral amputees have not higher demands on the prosthesis satisfaction than transtibial amputees [32]. Ratings of the appearance of the prosthesis showed trend in transfemoral amputees. Lower values in satisfaction with the safety in activities of daily living significantly lower than transtibial amputees. Additionally, higher values in the satisfaction with desirable prosthetics significantly higher than transtibial amputees but transtibial amputees have demands desirable prosthetics. As a result, this research project provides a list of needs of people living with lower limb loss in Thailand. Findings have indicated that user needs assessment is necessary and critical to make a better computerized lower limb prosthesis design and improve user's satisfaction.

Acknowledgements. This research was funded by National Research Council of Thailand. The authors thankfully acknowledge the support of the Sirindhorn School of Prosthetics and Orthotics and Veterans General Hospital in Thailand.

References

1. Limbs International: Why Limbs. https://www.limbsinternational.org/why-limbs.html
2. Gailey, R., McFarland, L.V., Cooper, R.A., Czerniecki, J., Gambel, J.M., Hubbard, S., Maynard, C., Smith, D.G., Raya, M., Reiber, G.E.: Unilateral lower-limb loss: Prosthetic device use and functional outcomes in servicemembers from Vietnam war and OIF/OEF conflicts. J. Rehabil. Res. Dev. **47**, 317–332 (2010)

3. Department of Empowerment of Persons with Disabilities: Situation Report on Persons with Disabilities in Thailand
4. Murray, C.D., Fox, J.: Body image and prosthesis satisfaction in the lower limb amputee. Disabil. Rehabil. **24**, 925–931 (2002)
5. Segal, A.D., Orendurff, M.S., Klute, G.K., McDowell, M.L., Pecoraro, J.A., Shofer, J., Czerniecki, J.M.: Kinematic and kinetic comparisons of transfemoral amputee gait using C-Leg® and Mauch SNS® prosthetic knees. J. Rehabil. Res. Dev. **43**, 857–870 (2006)
6. Lawson, B.E., Mitchell, J., Truex, D., Shultz, A., Ledoux, E., Goldfarb, M.: A robotic leg prosthesis: Design, control, and implementation. IEEE Robot. Autom. Mag. **21**, 70–81 (2014)
7. Au, S., Berniker, M., Herr, H.: Powered ankle-foot prosthesis to assist level-ground and stair-descent gaits. Neural Netw. **21**, 654–666 (2008)
8. Au, S.K.S.K., Weber, J., Herr, H.: Powered ankle-foot prosthesis improves walking metabolic economy. IEEE Trans. Robot. **25**, 51–66 (2009)
9. Lawson, B.E., Varol, H.A., Huff, A., Erdemir, E., Goldfarb, M.: Control of stair ascent and descent with a powered transfemoral prosthesis. IEEE Trans. Neural Syst. Rehabil. Eng. **21**, 466–473 (2013)
10. Jiménez-Fabián, R., Verlinden, O.: Review of control algorithms for robotic ankle systems in lower-limb orthoses, prostheses, and exoskeletons. Med. Eng. Phys. **34**, 397–408 (2012)
11. Wentink, E.C., Schut, V.G.H., Prinsen, E.C., Rietman, J.S., Veltink, P.H.: Detection of the onset of gait initiation using kinematic sensors and EMG in transfemoral amputees. Gait Posture **39**, 391–396 (2014)
12. Dindo, H., Lo Presti, L., La Cascia, M., Chella, A., Dedić, R.: Hankelet-based action classification for motor intention recognition. Rob. Auton. Syst. **94**, 120–133 (2017)
13. Varol, H.A.H.A., Sup, F., Goldfarb, M.: Multiclass Real-time intent recognition of a powered lower limb prosthesis. IEEE Trans. Biomed. Eng. **57**, 542–551 (2010)
14. Sup, F., Bohara, A., Goldfarb, M.: Design and control of a powered transfemoral prosthesis. Int. J. Rob. Res. **27**, 263–273 (2008)
15. Tucker, M.R., Olivier, J., Pagel, A., Bleuler, H., Bouri, M., Lambercy, O., del Millán, J.R., Vallery, H., Gassert, R.: Control strategies for active lower extremity prosthetics and orthotics: a review. J. Neuroeng. Rehabil. **12**, 1 (2015)
16. Sup, F., Varol, H.A.H.A., Mitchell, J., Withrow, T.J.T.J., Goldfarb, M.: Preliminary evaluations of a self-contained anthropomorphic transfemoral prosthesis. IEEE/ASME Trans. Mechatronics. **14**, 667–676 (2009)
17. Lawson, B.E., Varol, H.A., Goldfarb, M.: Standing stability enhancement with an intelligent powered transfemoral prosthesis. IEEE Trans. Biomed. Eng. **58**, 2617–2624 (2011)
18. Liu, M., Zhang, F., Datseris, P., Huang, H.: Improving finite state impedance control of active-transfemoral prosthesis using dempster-shafer based state transition rules. J. Intell. Robot. Syst. Theory Appl. **76**, 461–474 (2014)
19. Sun, J., Voglewede, P.A.: Powered transtibial prosthetic device control system design, implementation, and bench testing. J. Med. Device. **8**, 11004 (2013)
20. Jimenez-Fabian, R., Geeroms, J., Flynn, L., Vanderborght, B., Lefeber, D.: Reduction of the torque requirements of an active ankle prosthesis using a parallel spring. Rob. Auton. Syst. **92**, 187–196 (2017)
21. Rouse, E.J., Mooney, L.M., Martinez-Villalpando, E.C., Herr, H.M.: Clutchable series-elastic actuator: Design of a robotic knee prosthesis for minimum energy consumption. In: IEEE International Conference on Rehabilitation Robotics (2013)
22. Beckerle, P., Christ, O., Schürmann, T., Vogt, J., von Stryk, O., Rinderknecht, S.: A human–machine-centered design method for (powered) lower limb prosthetics. Rob. Auton. Syst. **95**, 1–12 (2017)

23. Sirasaporn, P.: Quality of life of trans-femoral and trans-tibial amputees after receiving prosthesis. J. Thai. Rehabil. Med. **20**, 4–9 (2553)

24. Legro, M.W., Reiber, G.D., Smith, D.G., Del Aguila, M., Larsen, J., Boone, D.: Prosthesis evaluation questionnaire for persons with lower limb amputations: assessing prosthesis-related quality of life. Arch. Phys. Med. Rehabil. **79**, 931–938 (1998)

25. Legro, M.W., Reiber, G., del Aguila, M., Ajax, M.J., Boone, D.A., Larsen, J.A., Smith, D. G., Sangeorzan, B., Aguila, M., Megan, J., Boone, D.A., Larsen, J.A., Smith, D.G.: Issues of importance reported by persons with lower limb amputations and prostheses. J. Rehabil. Res. Dev. **36**, 155–163 (1999)

26. Gallagher, P., MacLachlan, M.: Development and psychometric evaluation of the trinity amputation and prosthesis experience scales (TAPES). Rehabil. Psychol. **45**, 130–154 (2000)

27. Creswell, J.: Research Design Qualitative, Quantitative, and Mixed Methods Approaches. Sage, Oaks (2003)

28. Gauthier-Gagnon, C., Grisé, M.C., Potvin, D.: Enabling factors related to prosthetic use by people with transtibial and transfemoral amputation. Arch. Phys. Med. Rehabil. **80**, 706–713 (1999)

29. Harpe, S.E.: How to analyze Likert and other rating scale data. Curr. Pharm. Teach. Learn. **7**, 836–850 (2015)

30. Edwards, A.L.: Statistical Methods for the Behavioral Sciences (1962)

31. Teo, T.: Modelling technology acceptance in education: a study of pre-service teachers. Comput. Educ. **52**, 302–312 (2009)

32. Christ, O., Jokisch, M., Preller, J., Beckerle, P., Wojtusch, J., Rinderknecht, S., Von Stryk, O., Vogt, J.: User-centered prosthetic development: Comprehension of amputees' needs. Biomed. Tech. **57**, 1098–1101 (2012)

Enhancing Voice Quality in Vocal Tract Rehabilitation Device

Bianca Sutcliffe[1], Lindzi Wiggins[1], David M. Rubin[1],
and Vered Aharonson[1,2(✉)]

[1] Biomedical Engineering Research Group, School of Electrical and Information
Engineering, University of the Witwatersrand, Johannesburg, South Africa
{david.rubin, vered.aharonson}@wits.ac.za
[2] Department of Electrical Engineering, Afeka Tel Aviv Academic
College of Engineering, Tel Aviv-Yafo, Israel

Abstract. The assistive devices used for vocal rehabilitation by patients after Laryngectomy produce a distinctly robotic sounding speech. This study aims at introducing human-like qualities into the synthetically generated voices. A simplified source filter model, LPC coefficients and line spectral frequencies were used to characterize the vocal tract and manipulate the acoustic properties of speech. Two different mapping functions were employed: A Gaussian mixture model (GMM) and a linear regression model (LR). Objective and subjective testing showed that both mapping functions produced significant changes in the re-synthesised speech, with the LR mapping producing slightly better results. However, the subjective listening tests indicated that re- synthesized voices improved on the synthetic voice but still lacked human quality. This may imply that the vocal tract model contains only partial information pertaining to the subjective perception of artificiality in speech. Future work is aimed at investigating an elaborate model containing the speech production excitation and radiation signals.

Keywords: Human factors · Laryngectomy · Speech processing

1 Introduction

A laryngectomy involves the complete surgical removal of the larynx. Patients who have undergone a laryngectomy achieve some form of voice functionality after surgery through the use of various voice replacement methodologies such as voice prosthesis, oesophageal speech and the electrolarynx [1, 2].

While all three of these rehabilitation techniques give patients the ability to speak and communicate, they produce robotic sounding voices which can make patients feel self-conscious and adversely affect their social interactions.

In this study we designed and implemented a voice conversion (VC) system that could introduce human-like qualities into a synthetically generated voice. VC is the process whereby a speech signal from a source speaker is transformed into a specified target speaker's sound [3]. VC is based on Fant's source-filter speech model that can capture the acoustic characteristics of human speech [4]. The extracted features are

© Springer International Publishing AG, part of Springer Nature 2019
T. Z. Ahram and C. Falcão (Eds.): AHFE 2018, AISC 794, pp. 1006–1013, 2019.
https://doi.org/10.1007/978-3-319-94947-5_99

optimized using the source (synthetic) voice and the target (human) voice. Re-synthesis of the voice is performed by a manipulation of the extracted features.

2 Methods

Figure 1 presents an overview of the proposed VC.

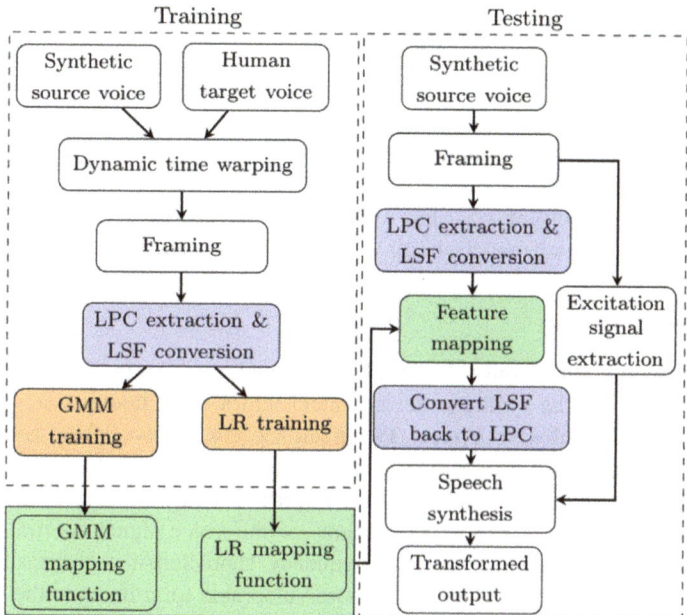

Fig. 1. Flow diagram of the proposed voice conversion algorithm.

2.1 Voice Conversion Model

The source-filter model is used to model the vocal tract and speech production organs as a linear time invariant system whose input x(t) is the glottal air volume and its output is the resultant sound pressure wave y(t) [4, 5]. This model is commonly used in speech processing due to its simplicity and the assumption of independence of source and filter. Equation 1 denotes the transfer function of the model, where $X(\omega)$ is the source of acoustic energy, referred to as the excitation signal or excitation residual, $T(\omega)$ is the vocal tract transfer function, and $R(\omega)$ the radiation characteristic.

$$Y(\omega) = T(\omega)R(\omega)X(\omega) \tag{1}$$

Figure 2 illustrates the anatomical origins of the respective sound sources and filters involved (adapted from [5]) as well as the assumed spectra of the signals and filters involved.

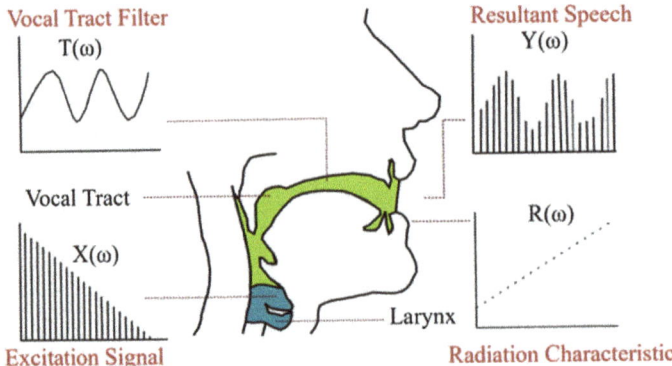

Fig. 2. Diagram illustrating the anatomical origins of the elements in Eq. 1

The harmonic spectrum $X(\omega)$ has a gradient of -12 dB/octave and the radiation characteristic $R(\omega)$, which models the air volume velocity, has a slope of 6 dB/octave [5].

2.2   Voice Conversion

In this preliminary study, only the vocal tract signal was manipulated to explore its capability to convert the synthetic device sound into a more human one. The source filter model was simplified to $\mathbf{Y}(\omega) = \mathbf{T}(\omega)\mathbf{X}(\omega)$ and the effects of $\mathbf{R}(\omega)$ were ignored.

The features extracted to represent $T(\omega)$ were Linear Predictive Coding Coefficients (LPC), which were then converted into Line Spectral Frequencies (LSFs). Both feature sets are considered as reliable representations of the voice signal spectral envelope, which captures the vocal tract and the frequency characteristics of speech [6]. The LPCs, however, have been shown to be more vulnerable to artifacts in the speech re-synthesis stage, whereas LSFs are considered more robust in their interpolation properties, which promoted their usage in speech coding applications [7, 8].

The mapping between the source and target voice feature sets was performed using two methods: Gaussian Mixture Model (GMM) and Linear Regression (LR).

The VC was implemented in Matlab® and was trained and tested on a speech database as described below.

2.3   Data Acquisition

Human speech utterances from the CSTR VCTK speech corpus [9] were used in conjunction with synthetically generated voices created using Microsoft Platform voice *Hazel* via a text to speech generator TextToWav [10]. The training dataset consists of one hundred 1.5 s speech utterances sampled at 48 kHz. 50 from a British woman (target voice) and 50 synthetically generated parallel utterances (source voice). The testing dataset consists of a 4 s utterance from the synthetically generated voice.

This study was performed following ethics clearance from the Human Research Ethics Committee (Medical) at the University of the Witwatersrand, Johannesburg. Protocol reference number M170629.

2.4   Signal Processing

The speaking rate of the TextToWav generator was slower than the human speaker, therefore dynamic time warping (DTW) needed to be performed. The DTW Matlab code [11] was used to warp the synthetically generated voice to the human voice.

The dynamic time warped signals were then segmented into 30 ms frames and LPC coefficients were extracted from each frame. The extracted LPC coefficients were then converted into 50 LSFs. The collected LSF data was then used to train the LR and GMM mapping functions.

In order to train the LR Mapping Function, the LSFs were initially normalised between $[0, \pi]$ to provide a natural linear progression. A mapping function was trained for each of the 50 LSFs. A separate mapping function was developed for and applied to the gain extracted from the signals.

For the GMM Mapping Function the joint distribution of the LSF values for the synthetically generated voice signals and the human recordings was calculated. 10 mixtures were fitted to this distribution since this number provided the best results after trial and error experiments. Mean and covariance were optimized on the training dataset for each mixture in the model using the Expectation Maximisation (EM) algorithm. To prevent singularities, a constant diagonal matrix $\epsilon \cdot I$, where $\epsilon = 0.001$, was added to the covariance matrices after each iteration of the EM algorithm [12].

In the test stage, LSFs of the test dataset speech utterances were calculated and converted using both mapping methods. GMM regression, as described in [12], was used to predict the target human LSF values from the GMM obtained in the training stage. Similarly, the LR mapping function obtained from the training stage was applied to these LSFs.

Once the transformed LSFs were obtained they were converted back to LPC coefficients. In order to re-synthesise the converted synthetically generated voice, the extracted excitation signal is filtered with the vocal tract transfer function.

2.5   Performance Evaluation

The source synthetically generated voice, the target human voice and the two re-synthesised voices produced by the two VC methods were compared, to evaluate their performance. The LSF values as well as the frequency response graphs of these voice signals were compared. Subjective listening tests were conducted by 20 listeners to qualitatively evaluate the "naturalness" of the resynthesized voices and their similarity to the human voice. Each listener was presented with the set of 4 voices 3 times, where the voices played were unlabeled and randomized.

3   Results

The performance of the LR mapping for the LSFs is demonstrated in Fig. 3, for the 8th LSF.

A comparison between the LPC coefficients' frequency response for the four speech signals is presented in Fig. 4. The spectra of the converted voice in both

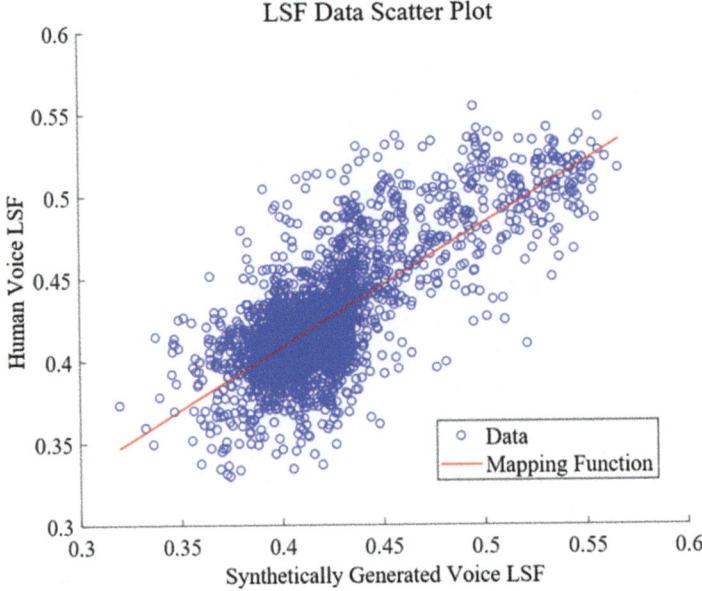

Fig. 3. Synthetically Generated Voices' 8th LSF versus Human Voices' 8th LSF for all the Training Data. The blue dots represent the human and synthetically generated voice LSFs plotted against each other and the red line indicates the mapping function.

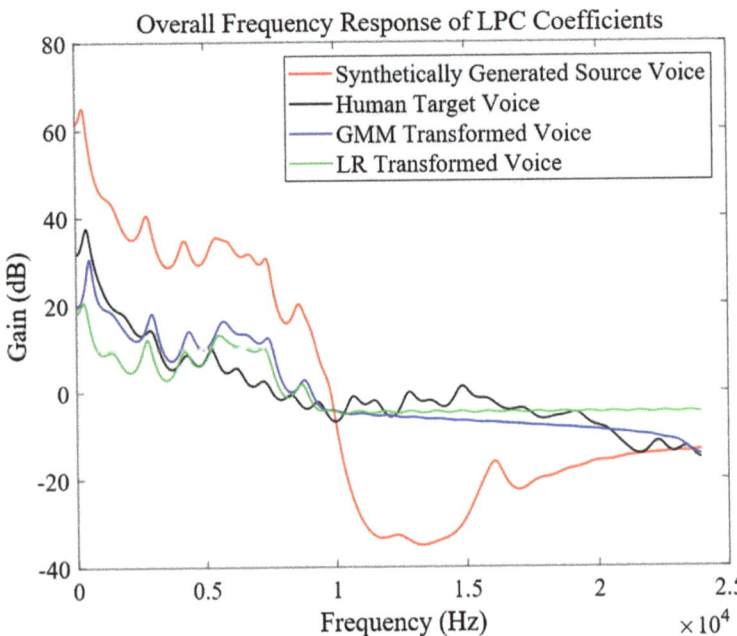

Fig. 4. Spectra of the source, target, LR and GMM converted voice averaged across all utterance frames.

methods resembles the human voice whereas the synthetically generated voice LPC is significantly different.

In a minority of the frames the voice conversion produced poor performance and was more similar to the source synthetic voice than to the target voice. An example is demonstrated in Fig. 5 for two different frames of an utterance from the testing set. Whereas for the frame in Fig. 5a the LR conversion of the synthetic voice LPCs shows distinct similarity to the human voice spectrum, for the other frame (Fig. 5b) this conversion yielded a frequency response which is more similar to the synthetic voice.

The results of the subjective listening tests are presented in Tables 1, 2 and 3.

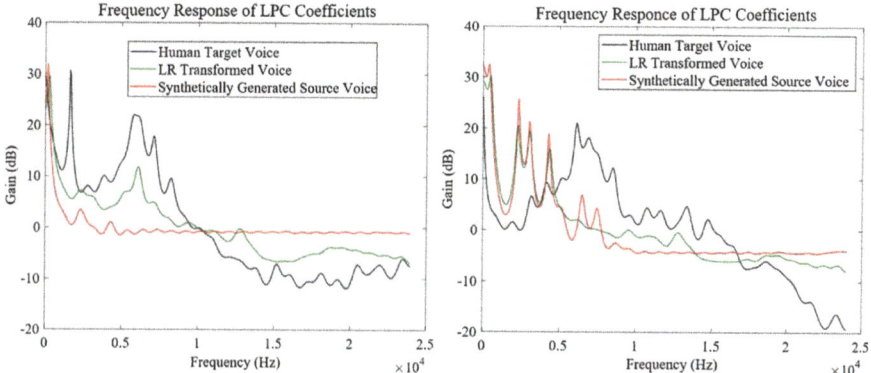

Fig. 5. Spectra of the source, target and LR converted voice for: (**a**) Frame 102, (**b**) Frame 99

Table 1. Results of the subjective listening tests

Voice		Result
Test 1: Which sounds more human?		
Case 1	Synthetically generated voice	70%
	GMM re-synthesised	30%
Case 2	LR re-synthesised	90%
	GMM re-synthesised	10%
Case 3	Synthetically Generated Voice	40%
	LR re-synthesised	60%

Table 2. Results of the subjective listening tests

Voice		Result
Test 2: Which voice has a more natural flow?		
Case 1	Synthetically generated voice	50%
	GMM re-synthesised	50%
Case 2	Synthetically generated voice	20%
	LR re-synthesised	80%

Table 3. Results of the subjective listening tests

Specified voice	Sounds more:	
	Robot	Human
Test 3: Does the specified voice sound closer to a robotic voice or a human voice?		
LR re-synthesised	65%	35%
GMM re-synthesised	70%	30%

4 Discussion

Two voice conversion algorithms were implemented and compared in this study with the aim of introducing human-like qualities into a synthetically generated voice for Laryngotomy patients' rehabilitation.

Several assumptions were made in the study. First, the synthetically generated and human voices are modelled using the source-filter model. Secondly, the model was simplified and only the vocal tract filter $T(\omega)$ of the synthetically generated voice is manipulated assuming that the effects of the radiated signal $R(\omega)$ can be neglected, in the context of a characterization of the "naturalness" in the voice. A third, intrinsic assumption was that the VC employed, which was based on spectral features only, could capture prosodic human voice characteristics such as cadence and emphasis.

These assumptions may be the cause for the results indicating that although the conversion of the LPC coefficients in both methods showed similar spectra to human voice, the voices re-synthesised from these coefficients did not sound human enough to the participants of the listening tests.

The GMM mapping indicated better performance in the spectrum graphs compared to the LR mapping, which may imply that the LSF clusters do not have a linear trend. Moreover, in both methods, a subgroup of the frames did not yield a desirable conversion and their spectra remained similar to the unconverted, synthetic voice.

A comprehensive analysis into the source for these inadequately -converted frames may provide further understanding into the limitations of each method and the comparison between them.

Interestingly, listening tests yielded a subjective perception that contradicted the signals' spectra comparison: 90% of the participants in the listening tests maintained that the re-synthesised voice using LR sounded more human than the re-synthesised voice using GMM.

This may imply that the prosodic properties of speech do not coincide with its spectral properties. A corroboration to this assumption is found in some listeners' observation that the pitch of the resynthesized speech using GMM was higher which promoted their decision of dissimilarity between the voices.

The listening tests indicated that the re-synthesised voice using LR VC has a more human sound than the synthetically generated voice, but is still distinctly different than the human voice. The listeners remarked for both VC mapping re-synthesised voice, that the signals contain noise and clicking sounds, which may have skewed the results.

The noise introduced in the re-synthesis of the voice can be reduced by using an overlap between the frames, in addition to the low pass filtering.

A simplistic model for the excitation signal which serves as an input to the model has been used in our analysis. Some evidence suggested that speech traits such as prosody, cadence and word duration, which are major contributors to the perceptual classification of a voice sounding 'robotic' are carried by the excitation signal. Thus, even where the vocal tract filter model may resemble the human one, the resultant source-filter output sound would still yield a sound with robotic qualities [13]. The inclusion of the radiation signal to the model may further improve the results.

These weaknesses will be addressed in our next studies, to further the goal of converting the robotic sounding voices from the various vocal rehabilitation techniques into more human sounding voices.

The study thus provides a preliminary feasibility of a relatively simple signal processing method to enhance devices used for vocal rehabilitation by patients after Laryngectomy.

References

1. Moore, K., Dalley, A., Agur, A.: Clinically Orientated Anatomy, 7th edn. Lippincott Williams & Wilkins, a Wolters Kluwer, Philadelphia (2014)
2. Hočevar-Boltežar, I., Žargi, M.: Communication after laryngectomy. Radiol. Oncol 35(4), 249–254 (2001)
3. Mohammadi, S., Kain, A.: An overview of voice conversion systems. Speech Commun. **88**, 65–82 (2017)
4. Fant, G.: Acoustic Theory of Speech Production. De Gruyter Mouton, Tubingen (1960)
5. Kent, R.: Vocal tract acoustics. J. Voice 7(2), 97–117 (1993)
6. Ye, H., Young, S.: High quality voice morphing. ICASSP, pp. 9–12 (2004)
7. Tianren, Y., Juanjuan, X. and Wei, L.: The computation of line spectral frequency using the second Chebyshev polynomials. In: 6th International Conference on Signal Processing (2002)
8. Shum, S.: A GMM-STRAIGHT Approach to Voice Conversion. Berkeley University, Berkeley (2009)
9. Veaux, C., Yamagishi, J., MacDonald, K.: CSTR VCTK Corpus. University of Edinburgh . http://homepages.inf.ed.ac.uk/jyamagis/page3/page58/page58.html (2010). Accessed 5 Sept 2017
10. Text to WAV.Dnasoft (2015)
11. Ellis, D.: Dynamic Time Warp (DTW) in Matlab. http://www.ee.columbia.edu/~dpwe/resources/matlab/dtw/ (2003)
12. Kain, A.: High Resolution Voice Transformation, pp. 50–53. Rockford College, Rockford (2001)
13. Percybrooks, W., Moore, E.: Voice conversion with linear prediction residual estimaton. ICASSP, pp. 4673–4676 (2008)

Author Index